Lecture Notes in Electrical Engineering

Volume 673

The book series *Lecture Notes in Electrical Engineering* (LNEE) publishes the latest developments in Electrical Engineering—quickly, informally and in high quality. While original research reported in proceedings and monographs has traditionally formed the core of LNEE, we also encourage authors to submit books devoted to supporting student education and professional training in the various fields and applications areas of electrical engineering. The series cover classical and emerging topics concerning:

- Communication Engineering, Information Theory and Networks
- Electronics Engineering and Microelectronics
- Signal, Image and Speech Processing
- Wireless and Mobile Communication
- Circuits and Systems
- Energy Systems, Power Electronics and Electrical Machines
- Electro-optical Engineering
- Instrumentation Engineering
- Avionics Engineering
- Control Systems
- Internet-of-Things and Cybersecurity
- Biomedical Devices, MEMS and NEMS

For general information about this book series, comments or suggestions, please contact leontina.dicecco@springer.com.

To submit a proposal or request further information, please contact the Publishing Editor in your country:

China

Jasmine Dou, Associate Editor (jasmine.dou@springer.com)

India, Japan, Rest of Asia

Swati Meherishi, Executive Editor (Swati.Meherishi@springer.com)

Southeast Asia, Australia, New Zealand

Ramesh Nath Premnath, Editor (ramesh.premnath@springernature.com)

USA, Canada:

Michael Luby, Senior Editor (michael.luby@springer.com)

All other Countries:

Leontina Di Cecco, Senior Editor (leontina.dicecco@springer.com)

**** Indexing: The books of this series are submitted to ISI Proceedings, EI-Compendex, SCOPUS, MetaPress, Web of Science and Springerlink ****

More information about this series at http://www.springer.com/series/7818

Vijay Nath · J. K. Mandal
Editors

Proceedings of the Fourth International Conference on Microelectronics, Computing and Communication Systems

MCCS 2019

Springer

Editors
Vijay Nath
Department of Electronics
and Communication Engineering
Birla Institute of Technology
Ranchi, Jharkhand, India

J. K. Mandal
Department of Computer Science
and Engineering
Kalyani University
Kolkata, West Bengal, India

ISSN 1876-1100 ISSN 1876-1119 (electronic)
Lecture Notes in Electrical Engineering
ISBN 978-981-15-5545-9 ISBN 978-981-15-5546-6 (eBook)
https://doi.org/10.1007/978-981-15-5546-6

This Springer imprint is published by the registered company Springer Nature Singapore Pte Ltd.
The registered company address is: 152 Beach Road, #21-01/04 Gateway East, Singapore 189721, Singapore

Preface

Our conferences focus on the development of new process and technologies such as Machine learning, Internet of Things, Cloud computing, 5G and beyond which are going to take place in 2020 in worldwide in broad manner. For this purpose our mission is to focus on research to prepare software, compatible hardware, innovative products to meets its feasibility/requirements of coming generations. In which Machine learning is an application of artificial intelligence (AI) that provides systems the ability to automatically learn and improve your experience without being explicitly programmed. Machine learning concentrates on the development of computer programs that can access data and use it learn for themselves. As per need of customer, we focused on solving some of the toughest challenges that hold back machine learning from being it the hand of every developer. Computer vision, languages, recommendations and forecasting can be select for pre training of AI services. It develops the capabilities to build the most comprehensive cloud platform and optimized for machine learning with executing the high performance computer and no compromises on security and anilities.

In this conference good numbers of papers have received. In which peer blind reviewed, registered and presented papers have been accepted for publication in conference proceeding of Springer Scopus book series Lecture Notes in Electrical Engineering (LNEE) and some outstanding papers were selected for SCI Journals like Microsystems Technologies and IETE Technical Review & Technical Research etc. Here, authors are also get the opportunity for extended version of papers for publication in SCI journals listed in conference website. This unique conference is leading by a group of experts that provide the guidance to authors for enhancing or updating their articles of contents. This forum also provides outcome based learning & research strategy.

4th International Conference on **Microelectronics Computing & Communication System (MCCS-2019)** along with **IETE Zonal Seminar & ISF Congress** was organized by Indian Society for VLSI Education (ISVE) & IETE Ranchi Centre with support of BSNL at ARTTC BSNL Ranchi on 11–12th May 2019. Function Chief Guest was the Chief Secretary Government of Jharkhand. Dais was shared by eminent dignitaries Dr. D. K. Tiwari, IAS, Chief Secretary Govt. Of

Jharkhand; Prof. K. T. V. Reddy, President IETE New Delhi; Dr. A. K. S. Chandele, Former President & Rtd. Lt. General, Govt. of India; Prof. J. K. Mandal, Keynote Speaker1, Kalyani University (WB); Prof. Sachidanand Shukla Pro-VC, Dr. RML Avadh University Ayodhya (UP); Dr. Ajay Thakare, Mentor IETE East Zone; Sh. KK Thakur, CGMT BSNL & Chairman IETE Ranchi Centre; Dr. Martin Stava, Brno University of Technology, Keynote Speaker2; Dr. Vijay Nath faculty BIT Mesra and General Chair MCCS-2019 & Honorary Secretary IETE Ranchi, Dr. Anand Kumar Thakur faculty Ranchi University & Honorary Treasurer IETE Ranchi Centre & Organizing Secretary MCCS-2019. Program was started by lighting lamp with all the dignitaries. Sh. K. K. Thakur welcomes all the dignitaries on dais and off the dais and participants came across the country and abroad. He demonstrates the brief activities of IETE Ranchi Centre and Excellence of BSNL. He also focuses the need of research for industry people. Dr. Vijay Nath demonstrated all the technical activities of two days and also focuses the achievement of our conference. He also described the uniqueness of this conference. Dr. J. K. Mandal deliver keynote lecture and demonstrate the research activities, plagiarism related issues and its updates. He also describes the role of advance digital signal processing in ICT. Dr. A. K. S. Chandele explains the need of outstanding research for updates with trend and technology. He appreciates the activities of IETE Ranchi Centre who doing excellent for Society. Dr. K. T. V. Reddy describes the activities of IETE and its future. He takes leading role to establish IETE University in Jharkhand and give quality of technical education in Jharkhand. IETE is well known institutions in the country in the direction of publication, education, and working with Govt. plan. Dr. Ajay Thakare appreciates the activities of Ranchi Centre and also assure for the possible support in IETE Ranchi Centre activities. After the demonstration of individuals souvenir of conference in hardcopy and softcopy were released by dignitaries. We would get the moral support from our dynamic and visionary Chief Minister Govt. of Jharkhand Shri Raghubar Das. We also get support from our Honourable Governor of Jharkhand Smt. Droupadi Murmu for such activities and her message/blessing has published in our souvenir MCCS-2019. IETE, ISVE & BSNL team always support to authors for enhancing the quality of papers, direction of research, capability to write articles to SCI & Scopus levels. This is a great achievement of our team. We also express our thanks to IITs, NITs, BITs and other top technical university of India and abroad and core industries from India and abroad for supporting as reviewers, authors, expert lectures etc. Who have attached once and know the complete procedure of the forum they become fan to serve society for ever. After several review, scrutiny, comments, papers meets the quality of SCI Journals and Scopus book series and at last it will published if authors are ready to incorporate the comments and suggestions. Within five year of journey more than 600 Scopus and more than 150 SCI articles have been published through this platform. We are proud to say that due to our quality of articles editors of journals and books demanding the articles and publishing in their outstanding journals and book series. In this conference several technical session have been conducted by eminent experts and a good number papers have been presented in all technical sessions and each session 6–8 judges was available to judge and guide to authors for their betterment. The technical session was conducted such as Microelectronics & Signal Processing; VLSI Design & Communication System; Artificial Intelligence & Computational System; Embedded System & Internet of Things; Green Energy and Automation by a group 6–8 judges/experts etc.

In this conference authors are invited for the original papers submission and quality of papers has been accepted for presentation. Authors are described their article in above mentioned domain very well. Authors and editors have taken utmost care in presenting the information and acknowledging the original sources whenever necessary. Editors express their gratitude towards to the authors, organizers of IC-MCCS and staff of Springer (India) for publication of this research book/proceeding possible. Readers are requested to provide their valuable feedbacks on the quality of presentation and inadvertent error or omission of information if any. We expect that the book will be welcomed by students as well as practicing engineers/researchers/professors.

Ranchi, India Vijay Nath

Acknowledgements

We extend our thanks to all the authors for contributing to this book/proceeding by sharing their valuable research findings. We specially thank a number of reviewers for promptly reviewing the papers submitted to the conference. We are grateful to the volunteers, invited speakers, session chairs, sponsors, sub-committee members, members of International advisory committee, and members of national advisory committee, members of technical program committee, members of joint secretary, and members of scientific advisory committee for successful conduct of conference. The editor expresses their heartfelt gratitude towards Dynamic & Visionary Chief Minister Shri Raghubar Das of Jharkhand and Honourable Governor of Jharkhand Smt. Droupadi Murmu. The editor also express their heartfelt gratitude towards Dr. K. T. V. Reddy President IETE New Delhi, Dr. A. K. S. Chandel Immediate Past President IETE New Delhi, Smt. Srimati Dagur Former President of IETE New Delhi, Prof. S. Pal Former VC IAT Pune, Dr. D. K. Tiwari, Chief Secretary, Government of Jharkhand, Sri Vinay Kumar Choubey Principal Secretary IT Govt. Of Jharkhand, Sh. Rakesh Sharma Principal Secretary Higher Education Govt. Of Jharkhand, Sh. Sudhir Tripathy Former Chief Secretary Govt. Of Jharkhand, Sh. Sanjay Kumar Jha Former Chairman IETE Ranchi & Executive Engineer Govt. of Jharkhand, Prof. Bernd Michel Micro Materials Centre (MMC) Berlin Germany, Prof. Bharath Bhushan Ohio Eminent Scholar & The Horward D. Winbigler Professor & Director NBLL The Ohio State University Columbus Ohio USA, Prof. P. S. Neelakanta C. Engg. Fellow IEE Florida Atlantic University (FAU) U. S. A., Sh. K. K. Thakur Chairman IETE Ranchi & CGMT BSNL Jharkhand Circle Ranchi, Sh. Prasad Vijay Bhushan Pandey DTO Term Cell1 BSNL Ranchi & Chairman ISVE Ranchi, Prof. A. A. Khan Former VC Ranchi University, Prof. M. K. Mishra VC BIT Mesra, Prof. Gopal Pathak VC Jharkhand Technical University Ranchi, Prof. R. K. Pandey VC Ranchi University, Prof. P. K. Barhai Former VC BIT Mesra, Prof. S. Konar Dean (Faculty Affairs & Sponsored

Research) BIT Mesra, Sh. R. Mishra Former CMD HEC Ranchi, Dr. M. Chakraborty Professor IIT Kharagpur, Dr. Ramgopal Rao Professor IIT Bombay & Director IIT Delhi, Dr. P. Chakraborty Professor IIT BHU & VC BESU Kolkata, Dr. S. Jit Professor IIT BHU, Dr. J. K. Mandal Professor Kalyani University, Dr. Abhijit Biswas Professor Kolkata University, Dr. Subir Kumar Sarkar Professor Jadavpur University, Dr. Gaurav Trivedi Associate Professor IIT Guwahati, Dr. Y. S. Chauhan Associate Professor IIT Kanpur, Dr. B. K. Kaushik Professor IIT Roorkee, Dr. Shree Prakash Tiwari Assistant Professor IIT Jodhpur, Dr. P. Kumar Associate Professor IIT Patna, Dr. M. Bhaskar Professor NIT Trichy, Dr. Adesh Kumar Assistant Professor UPES University Dehradun, Dr. Manish Kumar Associate Professor MMMUT Gorakhpur, Dr. Manish Mishra Professor DDU University Gorakhpur, Dr. Umesh Yadav Professor DDU University Gorakhpur, Prof. D. Acharjee President ISTM Kolkata, Dr. N. Gupta Professor BIT Mesra Ranchi, Dr. S. Pal Professor BIT Mesra, Dr. S. Konar Professor BIT Mesra, Dr. Vibha Rani Gupta Professor BIT Mesra, Dr. B. K. Mishra Principal Jumeritelaya Govt. of Jharkhand, Dr. V. K. Jha BIT Mesra, Sh. Ajay Kumar AGM (admin) ARTTC BSNL Ranchi & Imm past Chairman IETE Ranchi, Dr. P. R. Thakura Executive Member of IETE & ISVE Ranchi & Professor of BIT Mesra Ranchi, Dr. M. Chandra Executive Member of IETE Ranchi & Professor of BIT Mesra Ranchi, Dr. S. K. Ghorai Executive Member of IETE Ranchi & Professor of BIT Mesra Ranchi, Dr. B. Chakraborty Executive Member of IETE Ranchi & Executive Engineer Mecon Ranchi, Dr. S. Chakraborty Executive Member IETE Ranchi & Professor BIT Mesra Ranchi, Dr. Vandana Bhattacharya Professor BIT Mesra Ranchi, Dr. D. Mohanta Professor BIT Mesra, Dr. S. S. Solanki Professor BIT Mesra Ranchi, Dr. S. Pal Professor BIT Mesra Ranchi, Dr. B. Acharya Professor NIT Raipur, Dr. Chandrashekhar NIT Jamshedpur, Dr. S. Kumar Executive Member IETE Ranchi & Associate Professor BIT Mesra Ranchi. Dr. B. K. Bhattacharya Professor NIT Agartala, Dr. Anand Kumar Thakur Treasurer of IETE Ranchi & Assistant Professor & Director FM Ranchi University, Dr. Raj Kumar Singh Executive Member IETE Ranchi & Assistant Professor & Coordinator UGC Refresher Course Ranchi University, Dr. R. K. Lal Associate Professor BIT Mesra Ranchi, Dr. Sudip Sahana, Mesra, Dr. P. Pal, Dr. Amritanjali, Dr. Rishi Sharma, Dr. K. K. Senapati, Dr. S. S. Sahu, Dr. Aminul Islam, Dr. Sudip Kundu, Dr. S. Pushkar, Dr. M. K. Mukul, Dr. K. Bose of BIT Mesra, Dr. Parul Patel Head RESPOND, SAC ISRO Ahmadabad, Dr. H. S. Gupta, Sh. Manish Mehta, Sh. V. Vedam Senior Scientist, and Smt. Ankita Patel SAC ISRO Ahmadabad; Smt. Saroj Treasurer of ISVE Ranchi, Prof. Jyoti Singh Joint Secretary of ISVE Ranchi, Dr. A. K. Pandey Secretary ISVE Ranchi, Sh. Suraj Kumar Saw, Sh. Subro Chakraborty Executive Engineer IBM Bangalore, Sh. Dipayan Ghosh Executive Engineer Aircell, Sh. Ramkrishna Kundu Executive Member of ISVE Ranchi & Design Engineer IBM, Sh. Deepak Prasad SRF CSIR BIT Mesra, Mohd Javed Khan, Ms. Namrata Yadav, Sh. Om Prakash, MentorGraphics Noida, Sh. Rohan Kumar MentorGraphics, Sh. Ankit Kumar Synopsis Bangalore, Sh. Sumit Singh IBM Ahmedabad, Sh. Hiranmoy Kar, Sh. Rajanish Yadav, Sh. Aditya Nath BITT Ranchi, Sh. Shrill Kumar, Sh. Anup Tirkey for their endless support, encouragement, motivation to organize such prestigious event that paved the way for this book on Proceeding of Fourth International Conference on Nanoelectronics, Circuits & Communication Systems (NCCS-2018). At last we express our sincere gratitude towards the staff members of Springer (India) who helped in publishing this book.

Contents

About the Editors

Dr. Vijay Nath was born in Gorakhpur (U.P.) India in 1976. He received his bachelor's degree in Physics and master's degree in Electronics from DDU Gorakhpur University, India, in 1998 and 2001. He also received PGDCN (GM) from MMMUT Gorakhpur in 1999. He received his Ph.D. degree in VLSI Design & Technology from Dr. RML Avadh University Faizabad in association with CEERI, Pilani, in 2008. From 2000 to 2001, he served as project trainee in IC Design Group, CEERI, Pilani, under the guidance of Dr. K. S. Yadav (Senior Scientist). From 2002 to 2006, he severed as Faculty in the Department of Electronics, DDU University Gorakhpur. Presently, he is an Associate Professor in the Department of Electronics and Communication Engineering, Birla Institute of Technology, Mesra, Ranchi (JH), India, and joined this institute in 2006. He has academic and administrative responsibilities such as Professor In-charge of Embedded System Design Lab, Advisor of NAPS (News & Publication Society), member of BIT Mesra Brand Management, Faculty Advisor IEI Student Chapter (ECE) BIT Mesra and Honorary Secretary of IETE Ranchi Centre. He is the recipient of Viveka Nanda Techno Fiesta Award-2002, Young Scientist Award-2004, CCSN Best Paper Award-2013, First Prize in IEI Technical Contest-2013 and Cadence Design Contest-2014, 2016. His research interests include microelectronics, nanoelectronics, analog, digital, mixed CMOS VLSI circuits, low-power VLSI circuits, ultra-low-power CMOS temperature sensor, ADC, DAC, PTAT, CMOS bandgap voltage reference, ASICs, embedded systems designs, smart cardiac pacemaker, smart grids, Internet of things and early stage cancer detection. He has to his credit around 150 publications in reputed Scopus & SCI journals, Scopus books and conferences. He has successfully completed two R&D projects funded by DST New Delhi jointly with DRDL Hyderabad and MHRD New Delhi, and the third project is in ongoing stage funded by RESPOND SAC ISRO Ahmadabad, Government of India. He has developed VLSI Design course under Developing Suitable Pedagogical Methods for Various Classes, Intellectual Calibres & Research in E-learning sponsored by the Ministry of Human Resource Development Government of India. Now the complete course is available in IIT Kharagpur official website and opens for all IITs, NITs, BITs and technical universities of India & abroad. He is the resource person of 16 Winter & Summer School on

Digital VLSI Design. He is the program coordinator and resource person of one day seminar/workshop on Electronic System Design & Manufacturing (ESDM), East Zone of India funded by MHRD New Delhi under the aegis of IETE New Delhi. He has conducted 15 ESDM workshops in IITs NITs, BITs and other university/college of India in FY2015-16. He is the Editor of four books Proceeding of International Conference on Nanoelectronics Circuits & Communication Systems (NCCS-2015) & NCCS-2017 and Proceeding of International Conference on Microelectronics Computing & Communication Systems (MCCS-2015) & MCCS-2018 published in Scopus book series: Lecture Notes of Electrical Engineering, Springer. He is a member of several professional societies and academic bodies including IETE, ISTE, ISVE and IEEE.

Prof. J. K. Mandal received his M.Sc. in Physics from Jadavpur University in 1986 and M.Tech. in Computer Science from the University of Calcutta and was awarded a Ph.D. in Computer Science and Engineering by Jadavpur University in 2000. Currently, he is a Professor of Computer Science and Engineering and was Dean of the Faculty of Engineering, Technology and Management, Kalyani University, West Bengal, for two consecutive terms. He started his career as a Lecturer at NERIST, Arunachal Pradesh, in September 1988 and has 30 years' teaching and research experience. His areas of research include coding theory, data and network security, remote sensing and GIS-based applications, data compression, error correction, visual cryptography, steganography, security in MANET, wireless networks and unify computing. He has been a life member of the Computer Society of India since 1992, CRSI since 2009, ACM since 2012, IEEE since 2013 and a Fellow of IETE since 2012. He has chaired more than 60 sessions at various international conferences and delivered more than 60 experts/invited lectures during the last 5 years. He has acted as a program chair of several international conferences and edited more than 30 proceedings volumes. He is a Reviewer for various international journals and conferences and has published over 360 articles and 6 books. He is one of the Editors for the Springer AISC and CCIS Series. He has published more than 450 articles out of which 160 papers are published in the international journal. He has authored 7 books. 23 scholars were awarded Ph.D. degree under his supervision.

Diode Switch-Based 17-Level Inverter with Lesser Power Electronic Elements

Samrat Paul, Bidyut Mahato, Saikat Majumdar, and K. C. Jana

Abstract In advancement of technology, the multilevel inverter (MLI) has become more popular with respect to conventional two-level inverter due to its unparalleled advantages such as higher efficiency, reduced harmonic distortion, reduced switching stress, lesser electromagnetic interference and ability to high-voltage generation. A large number of semiconductor switches and other circuit components are required to obtain higher stepped output voltage levels that lead the circuit more complicated and costly. Thus, reduced component application in multilevel inverter topology is the greatest challenge towards the researchers. This proposed work introduces a new concept of a 17-level multilevel inverter based on diode switch configuration employing the lesser amount of power electronic components, especially the power switches. To incorporate such new topology, first the simulation is carried out in MATLAB environment and thereafter corresponding experimental prototype with RL-load is also developed and executed in the laboratory. The experimental results are further amalgamated to verify the simulation results.

Keywords Power electronic components · DC/AC power conversion · 17-level inverter · SPWM

1 Introduction

Multilevel inverter is still the research hub in the field of multiple voltage-level generation, and generally it converts DC to AC through different types of architecture [1]. In earlier, two-level inverter was first introduced in the year of 1975 but during operation some serious issues such as higher total harmonic distortion (THD), higher switching frequency, higher dv/dt stress on the switches and higher commutation problem had been noticed in [2, 3]. To overcome the aforesaid limitations, multilevel inverter gained significant prominence along with some additional

S. Paul · B. Mahato (✉) · S. Majumdar · K. C. Jana
Department of Electrical Engineering, Indian School of Mines, Dhanbad, India
e-mail: bidyut1990@gmail.com

V. Nath and J. K. Mandal (eds.), *Proceedings of the Fourth International Conference on Microelectronics, Computing and Communication Systems*, Lecture Notes in Electrical Engineering 673, https://doi.org/10.1007/978-981-15-5546-6_1

advantages like—their capability of high-voltage generation, higher efficiency, lower voltage stress of power switches, lower harmonic current and lower electromagnetic interference (EMI) [4–6]. Few foremost applications are related to the MLIs are flexible AC transmission system (FACTS), renewable energy systems [7], electric drives [8] and active filters [9]. In earlier, three types of classical multilevel inverter termed as neutral-point clamped (NPC) [10], flying capacitor multilevel inverter (FC-MLI) [11] and cascaded H-bridge (CHB) [12] were proposed for architectural benefits. To produce a higher output voltage steps, the neutral-point clamped multilevel inverter (NPC-MLI) was became the first MLI to be introduced but it comprises huge number of clamping diodes that increases cost and circuit complicacy [10]. To improve those problems, a new topology termed as flying capacitor multilevel inverter (FC-MLI) was introduced. The voltage unbalancing problem across the connected capacitors in FC-MLI inherits a major weakness [11]. However, the aforesaid disadvantages are overshadowed by the emergence of cascaded H-bridge MLI due to the fact that any extra power components such as diodes and capacitors are not required in the circuit [12]. Further, this MLI architecture can be categorized into two ways termed as symmetrical cascaded multilevel inverters based on H-bridge (CHB-MLIs) that has the DC sources with equal magnitudes and asymmetrical cascaded H-bridge multilevel inverters (CHB-MLIs) that has the DC sources with unequal magnitudes [13]. Nevertheless, one of the major drawbacks in MLIs is to obtain the higher voltage levels; the greater number of components (number of switches and DC voltage sources) is generally required; thus, the MLIs become more expensive. However, lower-voltage-rated switches are required to design multilevel inverter topology where each semiconductor switch utilizes the corresponding gate drive circuit that causes difficulties in controlling technique. Even, to increase the output voltage level the offline PWM technique also becomes difficult to design of lookup table and calculate time sequences. Thus, to minimize cost and complexity and to achieve better performance along with significant amount of reduced power components count become more challenging in the area of multilevel inverter.

In the proposed work, a novel architecture of MLI topology is introduced which illustrates an diode switch based 17-level MLI in order to generate required voltage levels with significantly less number power component especially the power switches. This new proposed topology is first designed and tested in MATLAB environment, and then an experimental prototype is developed and executed in the laboratory to obtain the practical results. The simulated results in MATLAB are thus verified by the corresponding experimental results with RL-load at the output terminal. To produce appropriate switching pulses and quality output voltage, the level-shifted pulse width modulation (LS-PWM) is used and the pulses are further fed to power semiconductor switches using the dSPACE-1103. In addition, proposed 17-level topology is matched with some other different recent topologies presented in [14–20] and found comparatively better in several aspects.

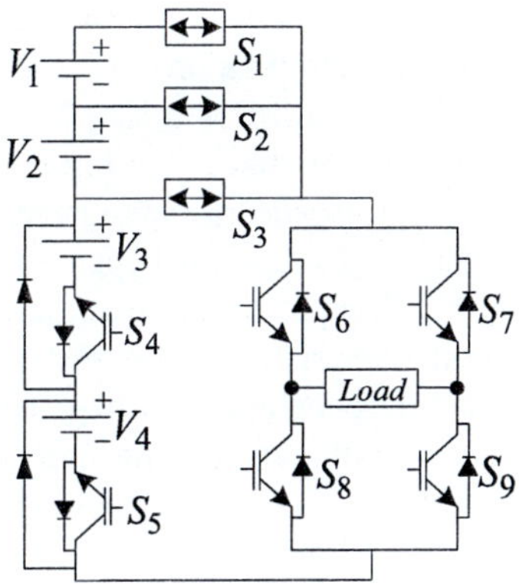

Fig. 1 Structure of proposed 17-level inverter

2 Proposed Inverter Topology

The proposed 17-level inverter based on diode switch topology is depicted in Fig. 1. The basic structure consists of 9 power switches and 4 DC sources. To generate 17 steps in output voltage at the load terminal, two varieties of DC sources ($V_1 = V_2 = 3V_{dc}$ and $V_3 = V_4 = V_{dc}$) are used. Three bidirectional switches like S_1, S_2 and S_3 and six unidirectional switches like S_4, S_5, S_6, S_7, S_8 and S_9 are utilized in the proposed topology.

The proposed scheme employed the same structure of bidirectional switches as described in [14]. The H-bridge circuit is also employed in this proposed topology to obtain the negative voltage levels. To generate 17 voltage levels, proposed topology requires lesser power switches among the compared topologies described in [15–20] except topology studied in [14]. It can be observed that, in [14], only 8 power switches and 4 DC sources are used in the suggested topology to obtain the same voltage levels whereas only nine semiconductor switches along with four input DC supplies are required for the proposed inverter. Even, the number of required bidirectional switches in the proposed topology is also less that associates less conduction losses in the overall inverter system. Furthermore, the other inverters discussed in [15, 17, 18, 20] with reduced power electronic elements require the large number of power components to generate 17 levels across the load terminal. Thus, the proposed diode switch-based topology stands better than the other suggested topologies [14–20].

3 Modes of Operation

The controller method suggested in this paper for the MLI is the technique of pulse width modulation (PWM). In this simple and well-defined technique, triangular carriers are chosen and a reference sine signal is compared to generate the corresponding pulse patterns. To cultivate the pulses in MATLAB, level-shifted PWM

(LS-PWM) is chosen among some of the well-known control strategies, i.e. selective harmonic elimination (SHE) [21], nearest level control (NLC) [22, 23], phase-shifted PWM [24], LS-PWM [25, 26] due to its simpler approach. The control switches are provided with the switching frequency to be 3 kHz. dSPACE-1103 controller is used in the prototype model to generate experimental pulses. For this topology, eighteen triangular carriers are kept upon each other in the fashion of different upper and lower levels. The pulse pattern of the switching devices is obtained from the simulation and is illustrated in Fig. 2a–c. Figure 3 illustrates the various positive modes that generate the different steps of voltage have been well presented and the dark path shows the current path as well as the corresponding voltage generation of the proposed 17-level inverter (symmetrical) topology. The voltage steps are well discussed in the theoretical and pictorial explanations, and the switching pattern of the ON switches is explained too in Table 1.

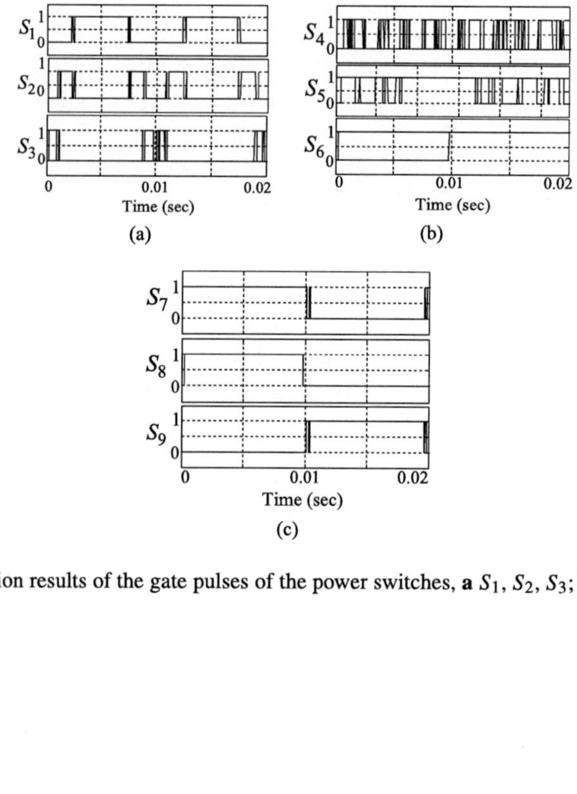

Fig. 2 Simulation results of the gate pulses of the power switches, **a** S_1, S_2, S_3; **b** S_4, S_5, S_6; and **c** S_7, S_8, S_9

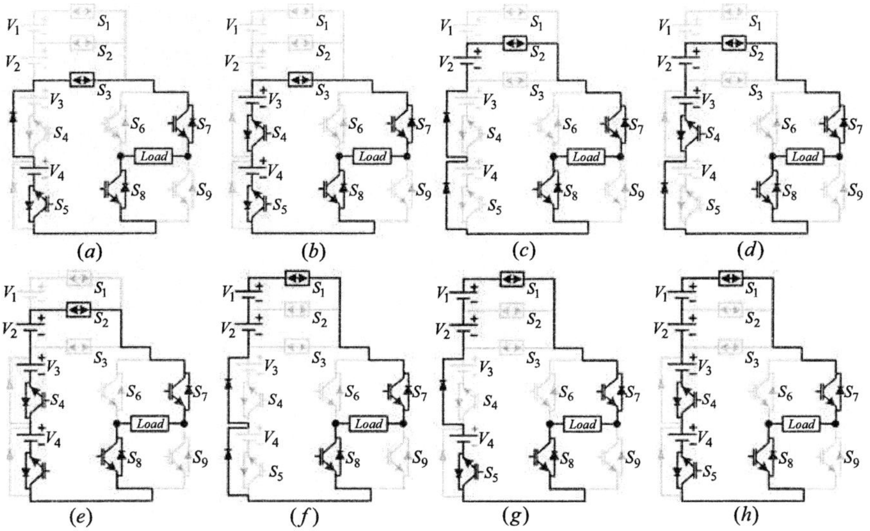

Fig. 3 Different modes of operation (for positive voltage levels)

Table 1 ON/OFF status for the proposed 17-level inverter (asymmetrical; $V_1 = V_2 = 3V_{dc}$; $V_3 = V_4 = V_{dc}$)

ON = 1; OFF = 0									Voltage levels
S_1	S_2	S_3	S_4	S_5	S_6	S_7	S_8	S_9	
1	0	0	1	1	0	1	1	0	$+8V_{dc}$
1	0	0	0	1	0	1	1	0	$+7V_{dc}$
1	0	0	0	0	0	1	1	0	$+6V_{dc}$
0	1	0	1	1	0	1	1	0	$+5V_{dc}$
0	1	0	1	0	0	1	1	0	$+4V_{dc}$
0	1	0	0	0	0	1	1	0	$+3V_{dc}$
0	0	0	1	1	0	1	1	0	$+2V_{dc}$
0	0	1	0	1	0	1	1	0	$+V_{dc}$
0	0	0	0	0	1	1	0	0	0
0	0	1	0	1	1	0	0	1	$-V_{dc}$
0	0	0	1	1	1	0	0	1	$-2V_{dc}$
0	1	0	0	0	1	0	0	1	$-3V_{dc}$
0	1	0	1	0	1	0	0	1	$-4V_{dc}$
0	1	0	1	1	1	0	0	1	$-5V_{dc}$
1	0	0	0	0	1	0	0	1	$-6V_{dc}$
1	0	0	0	1	1	0	0	1	$-7V_{dc}$
1	0	0	1	1	1	0	0	1	$-8V_{dc}$

The voltage levels are well explained and mentioned for the positive-level generation where the switching pulse sequence for the 17-level inverter is specified as ON = 1 and OFF = 0 in Table 1. Various modes of voltage generation required for the voltage levels are clarified beneath.

- To generate $+8V_{dc}$, the power switches S_1, S_4, S_5, S_7 and S_8 are in ON state as depicted in Fig. 3h.
- To generate $+7V_{dc}$, the power switches S_1, S_5, S_7 and S_8 are in ON state as depicted in Fig. 3g.
- To generate $+6V_{dc}$, the power switches S_1, S_7 and S_8 are in ON state as depicted in Fig. 3f.
- To generate $+5V_{dc}$, the power switches S_2, S_4, S_5, S_7 and S_8 are in ON state as portrayed in Fig. 3e.
- To generate $+4V_{dc}$, the power switches S_2, S_4, S_7 and S_8 are in ON state as portrayed in Fig. 3d.
- To generate $+3V_{dc}$, the power switches S_2, S_7 and S_8 are in ON state as portrayed in Fig. 3c.
- To generate $+2V_{dc}$, the power switches S_4, S_5, S_7 and S_8 are in ON state as portrayed in Fig. 3b.
- To generate $+V_{dc}$, the power switches S_3, S_5, S_7 and S_8 are in ON state as portrayed in Fig. 3a.

Similarly, all other levels can be generated according to the switching pulses based on Table 1.

4 Comparison with Other Topologies

The components required for the proposed inverter are compared with few of the well-known inverter topologies that has been recently published in various articles introduced in [14–20]. The generalized formulae of the different power components (total employed input DC supplies, N_{dc}, control switches, N_{sw}, unidirectional switches, N_U, bidirectional switches, N_B) of MLIs reported in [14–20], i.e., are illustrated in Table 2. Figure 4 shows the various structures of the reported topologies for the easier understanding to the readers.

5 Simulation and Experimental Results

It is observed that reduction in power switches reduces the number of driver circuits and makes the circuit less complex, easy to implement and low in cost. A 17-level inverter (asymmetrical input DC supply) is opted to implement in the MATLAB platform having different magnitudes of DC supplies, i.e. $V_3 = V_4 = 3V_{dc} = 116.1$ V and $V_3 = V_4 = V_{dc} = 38.7$ V. Therefore, the peak voltage of the output (maximum

Table 2 Comparison of the various components for 17-level inverter

Types of MLI	Components count			
	(N_{sw})	(N_B)	(N_U)	(N_{DC})
[14]	8	6	2	4
[15]	14	–	14	8
[16]	9	3	6	4
[17]	14	2	12	6
[18]	9	3	6	6
[19]	12	–	12	4
[20]	11	7	4	8
Proposed topology	9	3	6	4

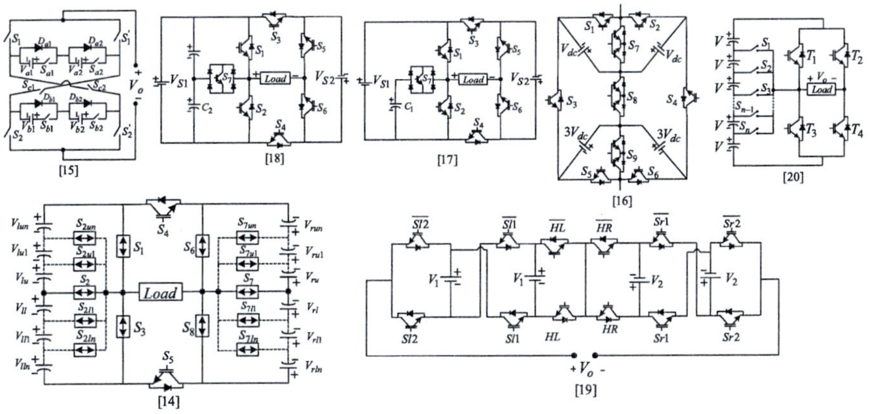

Fig. 4 Compared topologies studied in [14–20]

and minimum) is +310 V and −310 V, respectively. The simulations are performed employing the RL-load, being highly resistive in nature, $R = 100\ \Omega$, $L = 25$ mH. In addition, simulated results of output voltage and load current are incorporated for different modulation indices, M_i= 1 and M_i= 0.6, as illustrated in Fig. 5a–b, respectively.

An experimental set-up for a specimen of 17-level inverter is also developed in laboratory with asymmetrical DC links. The experimental verification of the simulated results is carried out with the same type of load, $R = 100\ \Omega$, $L = 25$ mH. Figure 6a–b illustrates the experimental results of output voltage along with the load current for different modulation indices, M_i= 1 and M_i= 0.6, respectively. It can be noticed that the experimental results are almost similar to the simulation results.

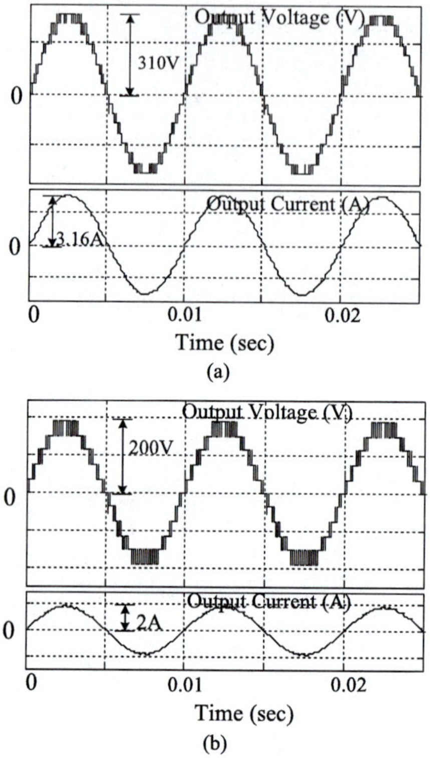

Fig. 5 Simulated results (load voltage and load current) for the proposed 17-level inverter at **a** $M_i = 1$ and **b** $M_i = 0.6$

6 Conclusion

In this paper, the authors suggested a new and generalized structure of inverter that is easily expanded by addition of the aforesaid explained units. This work suggested a new diode-based topology that is able to generate 17 levels, the proposed inverter structure is replicated in the simulation, and the matching experimental outcomes for the same easily verify the effective working of the inverter. This MLI structure shows better outcomes related to the overall required power components. In addition, the generalized equations for each required component are calculated for the proposed inverter as well as for the few recent inverter topologies as reported in [14–20]. Therefore, the comprehensive studies show that the proposed inverter with the reduced power electronic devices is comparatively superior than few newly suggested similar type of multilevel inverters.

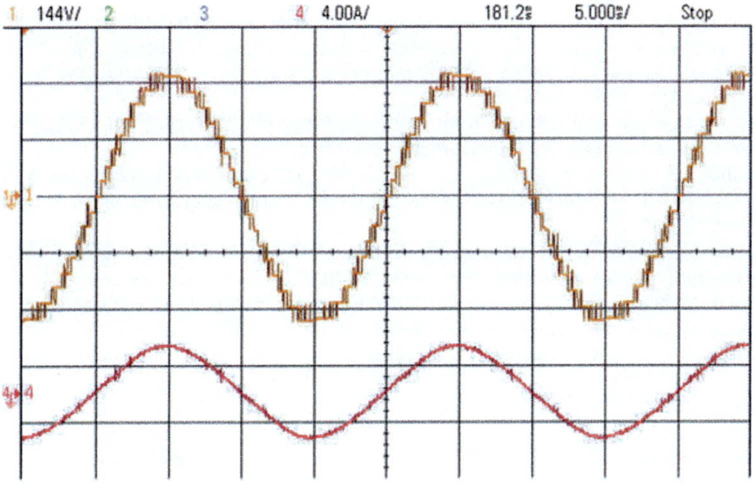

(a)

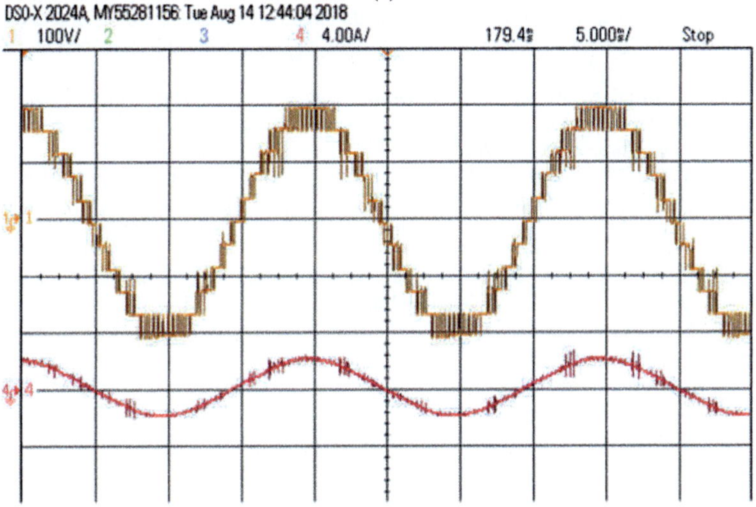

(b)

Fig. 6 Experimental results (load voltage and load current) for the proposed 17-level inverter at **a** $M_{\mathrm{i}} = 1$ and **b** $M_{\mathrm{i}} = 0.6$

References

1. Rashid MH (2004) Power electronics: circuits, devices and application
2. Franquelo LG, Rodriguez J, Leon JI, Kouro S, Portillo R, Prats MA (2008) The age of multilevel converters arrives. IEEE Ind Electron Mag 2(2):28–39
3. Rodríguez J, Lai JS, Peng FZ (2002) Multilevel inverters: a survey of topologies, controls, and applications. IEEE Trans Industr Electron 49(4):724–738

4. Rodríguez J, Bernet S, Wu B, Pontt JO, Kouro S (2007) Multilevel voltage-source-converter topologies for industrial medium-voltage drives. IEEE Trans Industr Electron 54(6):2930–2945
5. Tolbert LM, Peng FZ, Habetler TG (1999) Multilevel converters for large electric drives. IEEE Trans Ind Appl 35(1):36–44
6. Abu-rub H, Member S, Holtz J, Rodriguez J, Baoming G (2010) Medium-voltage multilevel converters—state of the industrial applications, 57(8): 2581–2596
7. MukherjeeA, Mahato B, Sinha A, Jana KC (2017) Comparative performance analysis of PV based grid-tied single phase asymmetrical multilevel inverter using different PWM techniques. J Electr Eng 17(1)
8. Carpita M, Moser D, Marchesoni M, Pellerin M (2008) Multilevel converter for traction applications: small-scale prototype tests results. IEEE Trans Industr Electron 55:2203–2212
9. Jha KK, Mahato B, Prakash P, Jana KC (2016) Hardware implementation of single phase power factor correction system using micro-controller. Int J Power Electron Drive Syst 7(3):790–799
10. Rodriguez J, Bernet S, Steimer PK, Lizama IE (2010) A survey on neutral-point-clamped inverters. IEEE Trans Industr Electron 57(7):2219–2230
11. Malekjamshidi Z, Jafari M, Islam R, Zhu J, Member S (2014) A comparative study on characteristics of major topologies of voltage source multilevel inverters
12. Reza Ahrabi R, Farakhor A, Najafi Ravadanegh S, Ardi H (2015) Symmetric and asymmetric transformer based cascaded multilevel inverter with minimum number of components. IET Power Electron 8(6):1052–1060
13. Mahato B, Raushan R, Jana KC (2017) Modulation and control of multilevel inverter for an open-end winding induction motor with constant voltage levels and harmonics. IET Power Electron 10(1):71–79
14. Mahato B, Mittal S, Majumdar S, Jana KC, Nayak PK (2019) Multilevel inverter with optimal reduction of power semi-conductor switches. In: Renewable energy and its innovative technologies, Springer, Singapore, pp 31–50
15. Arun N, Noel MM (2018) Crisscross switched multilevel inverter using cascaded semi-half-bridge cells. IET Power Electron 11(1):23–32
16. Samadaei E, Sheikholeslami A, Gholamian SA, Adabi J (2018) A square T-type (ST-type) module for asymmetrical multilevel inverters. IEEE Trans Power Electron 33(2):987–996
17. Gautam SP, Kumar L, Gupta S (2018) Single-phase multilevel inverter topologies with self-voltage balancing capabilities. IET Power Electron 11(5):844–855
18. Gautam SP, Gupta S, Sahu LK (2016) Reduction in number of devices for symmetrical and asymmetrical multilevel inverters. IET Power Electronics 9(4):698–709
19. Ajami A, Jannati Oskuee MR, Mokhberdoran A, Van den Bossche A (2014) Developed cascaded multilevel inverter topology to minimise the number of circuit devices and voltage stresses of switches. IET Power Electron 7(2):459–466
20. Alishah RS, Nazarpour D, Hosseini SH, Sabahi M (2015) Reduction of power electronic elements in multilevel converters using a new cascade structure. IEEE Trans Ind Electron 62(1):256–269
21. Dahidah MSA, Konstantinou G, Agelidis VG (2015) A review of multilevel selective harmonic elimination PWM: formulations, solving algorithms, implementation and applications. IEEE Trans Power Electron 30(8):4091–4106
22. Mahato B, Raushan R, Jana KC (2016) Comparative study of asymmetrical configuration of multilevel inverter for different levels. In: 2016 3rd international conference on recent advances in information technology (RAIT), pp 300–303
23. Mahato B, Jana KC, Thakura PR (2018) Constant V/f control and frequency control of isolated winding induction motor using nine-level three-phase inverter. Iran J Sci Technol Trans Electr Eng 6(1): 1–13
24. Jana K, Biswas S, Kar Chowdhury S (2016) Dual reference phase shifted PWM technique for a N-level inverter based grid connected solar photovoltaic system. IET Renew Power Gener 10(7): 928–935
25. Kumar C, Mahato B, Raushan R, Maity T, Jana KC (2016) Comprehensive study of various configurations of three-phase Multilevel inverter for different levels. In: 2016 3rd international conference on recent advances in information technology, RAIT

26. Mahato B, Majumdar S, Jana KC (2018) Carrier-based PWM techniques for multi-level inverters: a comprehensive performance study. J Sci Part A Eng Innov 5(3):101–111

Samrat Paul (Junior Research Fellow)
received his B.Tech and M.E degrees from **SIT, WBUT,** and **Jadavpur University** in the years 2005 and 2008, respectively. He is currently serving as Assistant Professor in **SVIST, Kolkata,** from August 2012. He has been working in **MCKV Institute of Engineering, Howrah,** from August 2007 till July 2012. He is working towards his **Ph.D.** from **IIT (ISM), Dhanbad,** since December 2016 in the areas of multilevel inverter, renewable energy sources and drives. His project interests include multilevel inverters, multilevel rectifier and traction drives.

Bidyut Mahato (Senior Research Fellow)
pursuing Ph.D. from **IIT (ISM), Dhanbad,** received B.Tech and M.E degrees from **WBUT, Kolkata,** and **BIT, Mesra, Ranchi,** in the years 2011 and 2014, respectively. He is **Life Member** of **ISTE, SSI, SESI, IIIE** and **ISCA**; **Member** of **IAENG**; and **Associate Member** of **IEI (India), the IRED** and **IETE.** He is an active **reviewer** of **IEEE Trans. on Industrial Electronics, IEEE Journal of Emerging and Selected Topics in Power Electronics, IET Power Electronics, Electric Power Components and Systems, Electric Power Systems Research,** etc. He has authored and co-authored more than 20 journals/conferences/chapters. His research interests include power electronic converters, multilevel inverters, power factor correction and hybrid electric vehicles.

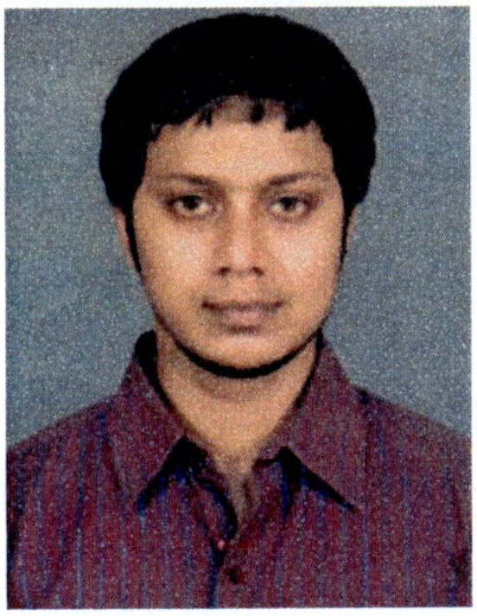

Saikat Majumdar (Senior Research Fellow)
received his B.Tech and M.Tech degrees from **AOT** and **IIT (ISM), Dhanbad,** in the years 2014 and 2016, respectively. He is working towards his **Ph.D.** from **IIT (ISM), Dhanbad,** since 2016 in the areas of multilevel inverter, renewable energy sources and drives.

Kartick Chandra Jana (Asst. Prof.) received his M.Tech and Ph.D. degrees from **NIT Durgapur** and **Jadavpur** University in 2003 and 2013, respectively. He is currently serving as Assistant Professor in **IIT (ISM), Dhanbad,** from June 2012. He has been working in **Birla Institute of Technology, Mesra,** from July 2003 till May 2012. His research is in the areas of modelling and design of multi-level inverter, real-time control of power electronics devices, design and implementation of efficient power converters.

28 GHz 5G Receiver Design Using 65 Nm CMOS and Performance Analysis Through Simulation

S. Pournamy and Navin Kumar

Abstract In this paper, 28 GHz millimeter wave (mmWave) low noise amplifier (LNA) and a down converter for 5G is designed using 65 nm CMOS technology. This design is used with a standard radio receiver chain and each block is tuned to its best performance to obtain the necessary overall specifications. Typical single-stage common source (CS) cascode with degenerative inductor topology is used for LNA to meet high gain requirement. The designed LNA has a peak gain of 16.43 dB, noise figure (NF) of 3.6 dB, good linearity (P_{1dB} of 10 dBm) and −3 dB bandwidth of 2.2 GHz. A modified Gilbert cell topology is used for down converter which gives a conversion gain of −2.2 dB from 27.5 to 29.5 GHz frequency, third-order intercept point IIP3 of 10 dB at 28 GHz from a 1.5 V supply voltage and a 1 Vpp of local oscillator (LO) drive. With the proposed LNA and down converter, a direct conversion receiver is analyzed. Cascaded gain of 39 dB, NF of 4.2 dB, third-order intercept of −23.3 dBm and a low error vector magnitude (EVM) of −37.76 dB is observed for 64 QAM with a received power of −100 dBm for the receiver.

Keywords EVM · LNA · Low noise amplifier · mmWave · 65 nm CMOS · 28 GHz · Down converter

S. Pournamy (✉)
Electronics and Communication Engineering Department, Amrita School of Engineering, Bangalore, India
e-mail: s.pournamy@gmail.com

N. Kumar
Amrita Vishwa Vidyapeetham, Coimbatore, India

V. Nath and J. K. Mandal (eds.), *Proceedings of the Fourth International Conference on Microelectronics, Computing and Communication Systems*, Lecture Notes in Electrical Engineering 673, https://doi.org/10.1007/978-981-15-5546-6_2

1 Introduction

Increased demand for higher data rates, minimum of latency along with constraints of power efficiency are few main drivers of emerging research in the next generation millimeter wave (mmWave) 5G wireless communications. Among several prospective mmWave bands such as 60 GHz, E-band [1]; the 28 GHz (also called as quasi-mmWave band) is perhaps the most attractive [2]. These higher frequency bands offer larger bandwidth to support around 1 Gbps and higher data rate. The architecture of radio frequency (RF) chain for mobile wireless technology, i.e., mmWave 5G radio is yet to be decided in these frequencies.

RF components to be designed to handle this unused or under-utilized high bandwidth to meet the ever-increasing data rate are challenging. One of the important and necessary modules in the radio (transceiver architecture) is low noise amplifier (LNA). The LNA amplifies the small received power signals without adding significant noise by itself and produce adequate signal to noise ratio (SNR) for the demodulator. Additionally, performance parameters of LNA provide an acceptable error vector magnitude (EVM), a measure of how far the received constellation points are from the ideal locations [3]. High gain, wide bandwidth and low noise figure (NF) are some of the key characteristics of an LNA. Designing an LNA using CMOS technology and maintaining the necessary values of gain, bandwidth, NF is challenging. Additionally, LNA precedes a down converter in the RF receiver. The main function of down converter is the frequency translation. Similar to the LNA, the performance parameters of down converter affect the acceptable EVM. The main challenge in the design of these RF components is to reduce EVM by reducing the RF impairments. A stand-alone LNA or down converter design is no longer meaningful as its performance heavily depends on the other circuits in the RF chain.

Many 28 GHz transceiver modules are commercially available as surface mount devices (SMD). In [4], a 28 GHz RF transceiver module is designed and fabricated for 5G beam forming systems. Fabrication is done with commercially available RF SMD components such as LNA, digital analog (DA), and up/down IQ converters. A GaAs HMC519 LNA block which operates in 18–32 GHz offers 25.5 dB gain and NF of 3.5 dB. Similarly, the HMC519 provides 15 dB of gain, 2.8 dB of NF, and has an output IP3 greater than 23dBm with a power supply of 3 V at 65 mA current. In this case, a single-stage CS cascade with shunt-shunt resistive feedback topology is used to obtain wide band (1.6–28 GHz) input impedance matching. Authors used 90 nm CMOS technology to design. HMC1063LP3E from analog devices is a GaAs MMIC IQ mixer for an RF range of 24–28 GHz with following specifications: Local oscillator (LO) frequency can be varied in the range of 21–30 GHz, intermediate frequency (IF) of 3 GHz, 40 dB LO/RF isolation and IIP3 of 17 dBm with a 10 dBm of LO power.

Hong-T K et al. in [5] designed a 28 GHz CMOS direct conversion transceiver with 2×4 patch antenna arrays. Authors used TSMC 28 nm RF CMOS technology to fabricate the design. In this, four-stage LNA (in-CS-CC-CC-CS-out) is designed to obtain high gain and low noise. The input matching is obtained with minimum

NF using a series inductor and a degenerated inductor at first stage CS amplifier. The LNA has gain range 12–24 dB, NF of 5.6 dB at 28 GHz. A Gilbert cell-based IQ down converter with a third active-RC LPF is used for the conversion of RF to baseband. Tow-Thomas biquad approach is used for a wideband characteristic. A low IF of 250 MHz is chosen in this direct conversion receiver to filter out the flicker noise. The measured EVM of this transceiver with an RF power of −70 dBm is −23.1 dB for 64 QAM.

In this work, LNA and a down converter suitable for 28 GHz mmWave 5G communication are designed using 65 nm CMOS technology and simulated. A single-stage common source cascade with a center inductor topology is used in LNA design. The designed is used with standard radio receiver chain and the performance of each block is optimized to obtain the overall necessary specifications of the receiver. The performance parameters of the RF components are optimized to get an acceptable EVM of −37.76 dB for 64 QAM with an RF power of −100 dBm. A gain improvement of 1.6–3.7 dB is obtained in the required frequency band of interest with the modified topology. DC power dissipation and variation of performance due to the bias variation is very minimal. The simulation results show that the design performs better.

The contents of the rest of the paper are as follows. Section 2 presents the design of amplifier, matching circuits, BPF and down converter. A measurement setup to analyze the performance of the designed RF components is presented in Sect. 3. In Sect. 4, the simulation results are presented and compared with some of the existing state of the art. Conclusion is discussed in Sect. 5.

2 Designing of RF Components

In this section, we present the circuit design of LNA, matching block and band pass filter and the down converter.

2.1 Designing of Low Noise Amplifier and Matching Circuits

Single-stage CS cascode amplifier is designed in UMC 65 nm CMOS technology using low leakage, low threshold voltage N_12_LLLVTRF transistors as shown in Fig. 1. The transistors are biased to get a current density of 150 μA/μm. This ensures low NF and high cutoff frequency, f_T [6]. The available gain is made high by selecting the size of NMOS transistors M_0 and M_1 as 30 ×1 μm × 60 nm and 50 × 1 μm × 60 nm, respectively. The number of fingers of transistors M_0 and M_1 is optimized in such a way that the stability factors go beyond 1; a necessary condition for the stability to be assured [7]. The available gain obtained at 28 GHz is 16.78 dB. Also, unity current gain frequency f_T for this work has attained a high value of 110 GHz and guarantees better performance in the operating frequency of 28 GHz [8].

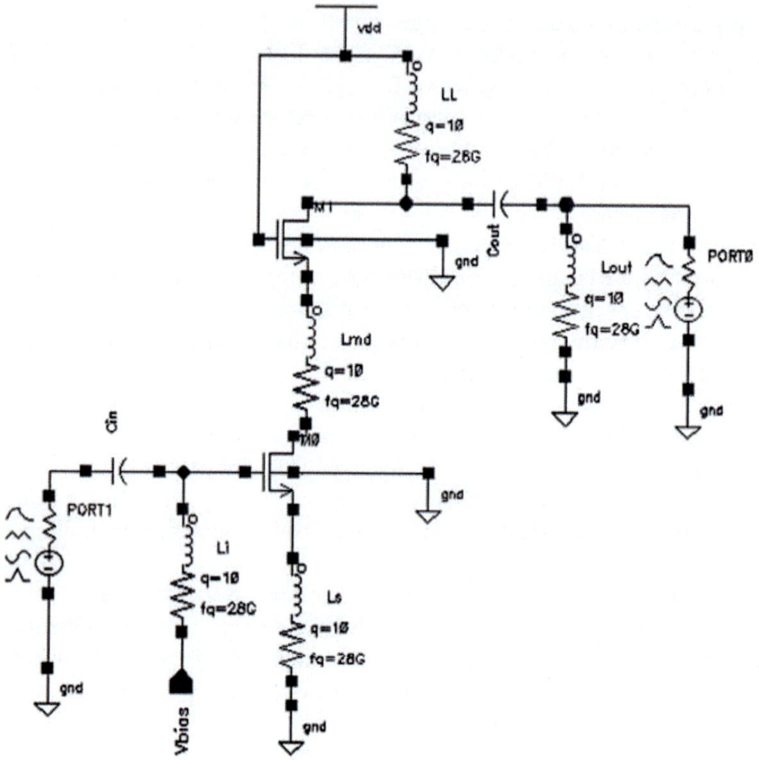

Fig. 1 Simplified schematic of single-stage CS cascade LNA

Small-signal model of the proposed amplifier circuit at the input side is shown in Fig. 2a. The input impedance of the amplifier is given as:

$$Z_{\text{in}} = \frac{\left(1 + S\left(g_m + SC_{\text{gs1}}\right)\right)\left(1 + S^2 L_i C_{\text{in}}\right)}{SC_{in}\left(1 + SL_s g_m + S^2 L_s C_{\text{gs1}} + g_m SL_i + S^2 L_i C_{\text{gs1}}\right)} \tag{1}$$

where g_m is the transconductance of the transistor. The component values of C_{in} of 40 fF, *Li* of 420 pH and *Ls* of 20 pH are chosen to satisfy Eq. (1) at the resonant frequency. The matching circuits are designed to ensure that amplifier delivers the maximum power gain and minimum return loss. LC distributed matching circuits at the output are chosen to resonate at f_0 of 28 GHz along with $L_L = 136$ pH as in Eq. (2). A coupling capacitance of 152 fF as C_{out} and load inductance of 56 pH as L_{out} are selected to satisfy Eq. (2) as given:

$$fo = \frac{1}{2\pi\sqrt{c_{\text{out}} + c_{\text{ds}})L_L}} \tag{2}$$

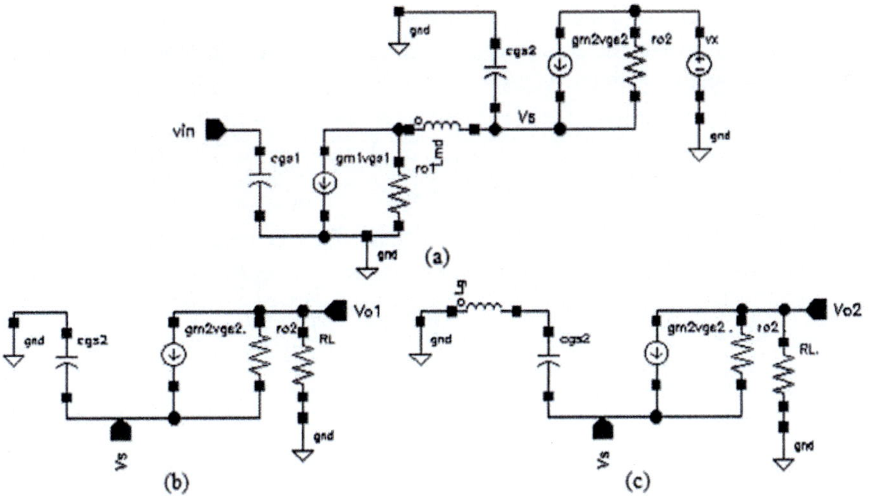

Fig. 2 Small-signal model for **a** calculation of input and output impedance, **b** conventional CG stage, **c** modified CG stage

Parasitic capacitance across drain and source is denoted as c_{ds}. A center inductor L_{md} of 87 pH is included to increase the unity current gain frequency, f_T. All the inductors are chosen to have a Q factor of 10.

2.2 *Modified LNA Circuit for Gain and Bandwidth Improvement*

The designed LNA has been modified to get a reduced EVM. The output resistance of LNA as shown in Fig. 1 is calculated by assuming $\text{v}_{\text{gs1}} = 0$ and connected a voltage v_x to the output terminal to produces a current of i_x as shown in Fig. 2a. The output resistance is given in Eq. (3) and the corresponding upper cutoff frequency f_H is given in Eq. (4).

$$R_{\text{out}} = r_{01} + r_{02} + g_{m2} r_{01} r_{02} \tag{3}$$

$$f_H = \frac{1}{2\pi\sqrt{(C_{\text{out}} + C_{gd2})R_{\text{out}}}} \tag{4}$$

Output resistance of CS and CG sections are denoted as r_{01} and r_{02}. g_{m1} and g_{m2} are the transconductance of CS and CG configurations, respectively. In order to improve the cutoff frequency or to compensate the reduction in gain due to parasitic capacitances at high frequencies, an inductor L_g is added to the gate of CG. The small-signal model of the conventional and the modified CG amplifier of cascode structure are shown in Fig. 2b, c, respectively. The gain of the common gate amplifier in a cascode section is given as [9]:

$$\frac{V_{01}}{V_S} = \left(g_{m2} + \frac{1}{r_{02}}\right)(R_L||r_{02}) \tag{5}$$

The presence of L_g modifies the current equation at the output node in Fig. 2c as:

$$\frac{V_s - V_{02}}{r_0} = g_{m2}v_{gs2} + \frac{V_{02}}{R_L} \tag{6}$$

where v_{gs2}, the voltage across c_{gs2} can be given as:

$$v_{gs2} = \frac{V_s}{1 - \omega^2 L_g c_{gs2}} \tag{7}$$

Thus, the modified gain can be written as:

$$\frac{V_{02}}{V_S} = \left(\frac{g_{m2}}{1 - \omega^2 L_g C_{gs2}} + \frac{1}{r_{02}}\right)(R_L||r_{02}) \tag{8}$$

Comparing Eq. (8) with Eq. (5), a significant improvement in the gain can be observed with the addition of gate inductance. Thus, the modified circuit of LNA for higher gain is shown in Fig. 3. L_g of 250 pH compensates the decreasing impedance of parasitic capacitance and thus extends the gain decrement in high frequency. L_i of 420 pH is used for blocking the RF signal and not allowing to the bias circuit. The degenerating inductor at the source of M_0 helps in gain and stability. The LNA is designed to be operated for a supply voltage, V_{DD} of 1.5 V.

2.3 Designing of Image Rejection Filter

LNA block precedes an image rejection filter and mixer in the receiver. A parallel coupled micro-strip BPF with cutoff frequencies of $f_{c1} = 27.5$ GHz and $f_{c2} = 29.5$ GHz are designed. A pass band gain of −2dB and return loss of <−15dB is observed in the desired band of interest.

2.4 Designing of Down Converter

MOSFET-based Gilbert cell down converter as shown in Fig. 4 is designed and analyzed. The conversion gain A_c, is defined as the ratio of the amplitude of the IF

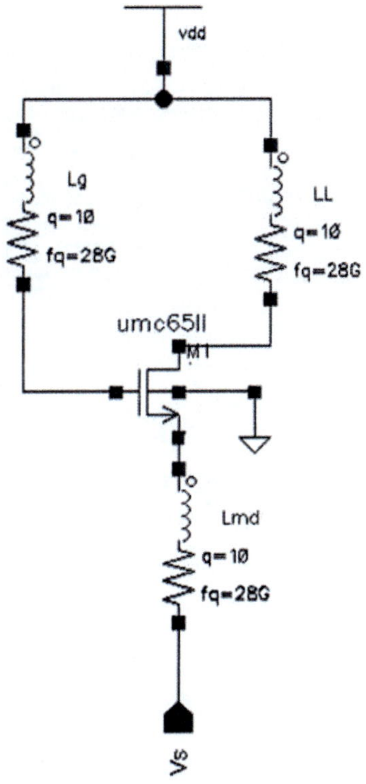

Fig. 3 Modified circuit for gain and bandwidth improvement

signal to the amplitude of RF signal of the Gilbert cell is given as [10]:

$$A_c = \frac{2}{\pi} g_m R_L \tag{9}$$

where g_m is the transconductance of the RF transistors, M_5 and M_6 in Fig. 4. R_L is the output impedance. To get a good conversion gain, an active load is chosen. The linearity component IIP3 is proportional to $2\pi f_{\text{RF}} g_m L_2$ and L_2 is chosen to be 50 pH with a g_m of 17.9 m. Another inductor L_{in} is added in series with the gate of M_5 to tune out the imaginary part of the input impedance to satisfy:

$$L_{in} = \frac{50 C_{\text{gs}}}{g_m} \tag{10}$$

L_{in} is selected as 860 pH.

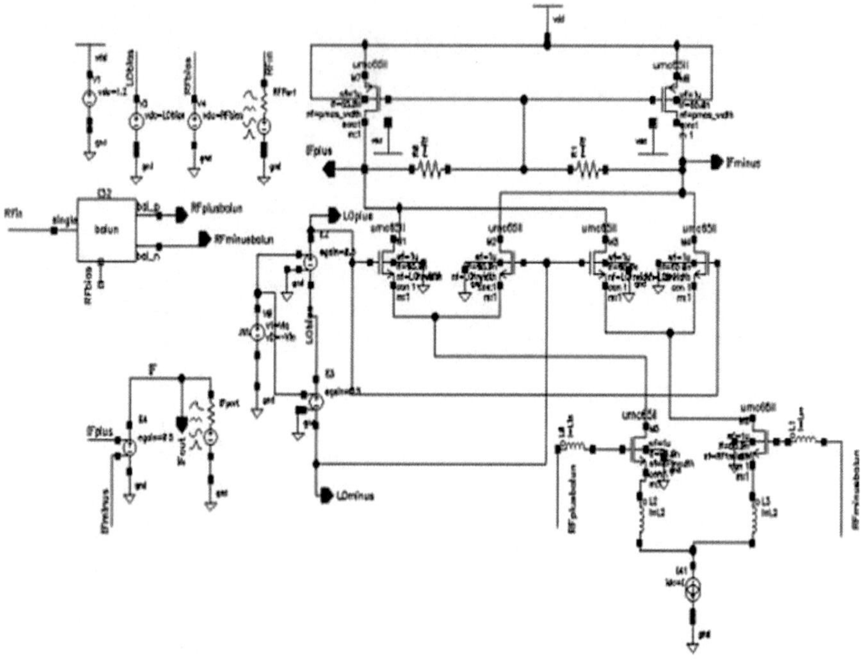

Fig. 4 Schematic of mixer

3 Measurements Setup

A system-level simulation is performed using visual system simulator in AWR design environment. 64 QAM with filtered bank multi-carrier (FBMC) modulated signal is taken as the test signal for the RF chain. FBMC is configured for 2048 subcarriers with a subcarrier spacing of 75 kHz. RF power level is set at −100 dBm. The receiver is configured with the designed LNA followed by the BPF and down converter along with RF attenuators and base band amplifiers. Cascaded gain, NF and receiver linearity are analyzed. The designed RF circuits are further tested for RF imperfections. Since due to imperfections, the location of the received points will change from its ideal on the constellation graph. These imperfections can be measured using EVM. The measurement setup for EVM calculations is shown in Fig. 5.

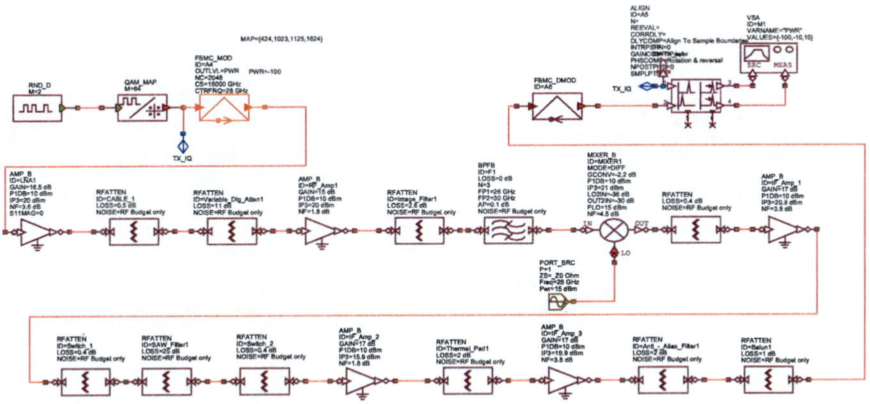

Fig. 5 Receiver chain and measurement setup

4 Results and Discussion

4.1 *Performance of LNA*

Simulation of the designed LNA is carried out using Cadence Virtuoso. Proposed LNA attains a peak gain of 16.43 dB at 28 GHz and more than 15 dB in the band of interest (shown in Fig. 6).

The input return loss of −12dB is maintained to have a better isolation as shown in Fig. 7. The output return loss of −16.7dB is also obtained as shown in Fig. 8. 1dB

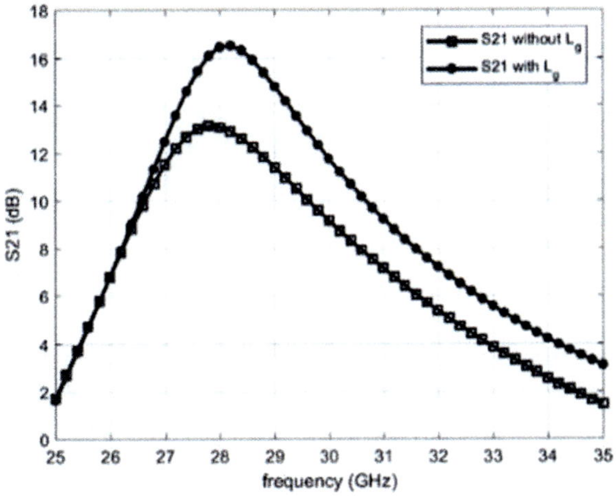

Fig. 6 Gain, S21 of the cascode LNA with and without *Lg*

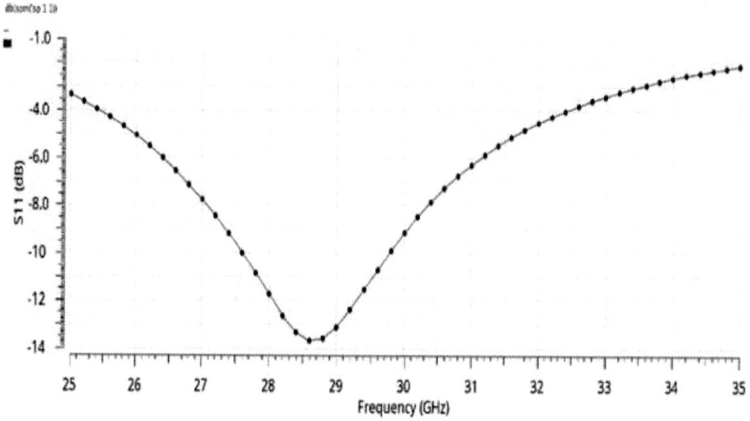

Fig. 7 Input return loss, S11

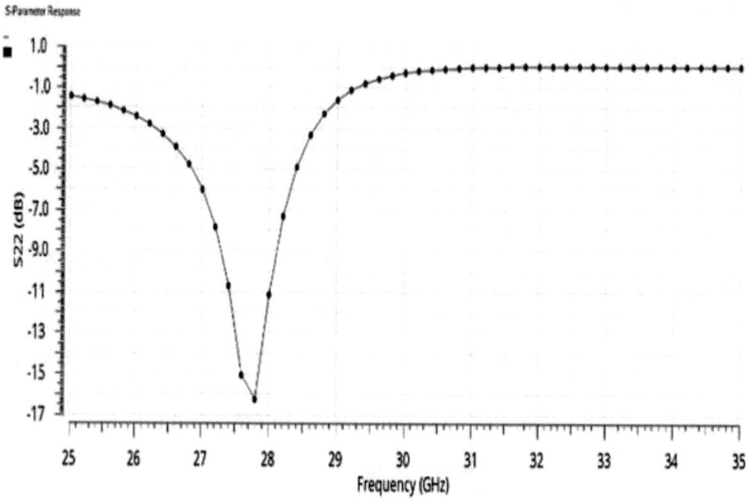

Fig. 8 Output return loss, S22

compression point presents the indication of linearity and is located at 10 dBm and the NF of 3.6 dB is observed as in Figs. 9 and 10.

Stability analysis shows that the designed LNA is unconditionally stable. Both the stability criteria such as stability factor, k_f greater than 1 and alternate stability factor, B_{1f} less than 1 as given in [7] are satisfied for the entire frequency band. This guarantees all the source and load circles to be outside the unit circle in smith chart and is stable for any load variations. The power consumption of the proposed LNA is 7.5 mW from a power supply of 1.5 V.

Robustness to bias voltage variation is illustrated in Fig. 11, where the gain of LNA

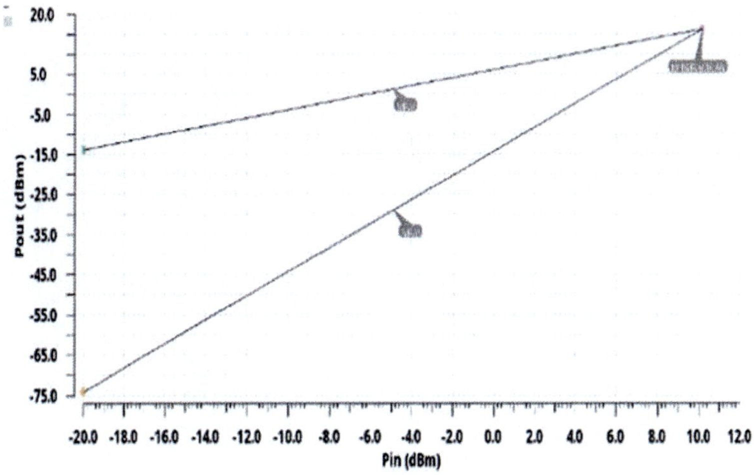

Fig. 9 Plot of P_{out} versus P_{in} to find 1dB compression point

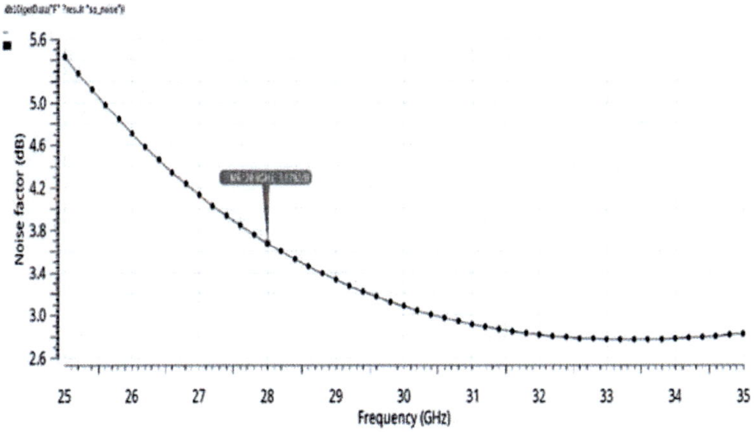

Fig. 10 Noise figure

changes only by 0.4dB when the gate voltage of the input MOSFET, M_0 changes from 0.6 to 0.8 V. The maximum gain is observed at the optimum bias voltage of 0.7 V as v_{bias}.

Table 1 provides the summary of the proposed LNA along with few available designs at 28 GHz. [3, 4] are the products whose LNA specifications are referred, while [5] is fabricated LNA. Our proposed LNA shows better results in simulation including PVT analysis. It is expected that if the designed is fabricated, there might be some drop in the performance. However, most of the performance parameters are of high value. From the literatures, it is noticed that there would not be significant drop in the performance post-layout design and electromagnetic (EM) simulation.

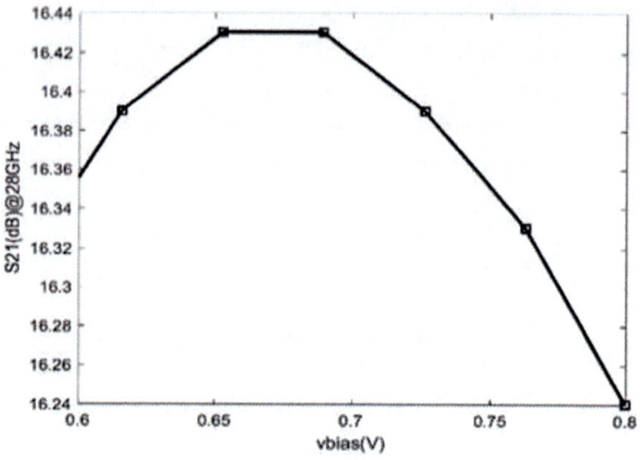

Fig. 11 S_{21} at 28 GHz with variation in bias voltage

Table 1 Comparison of LNA performance with available study

Ref.	Freq (GHz)	Gain (dB)	IIP3 (dBm)	NF (dB)	VDD (V)	Pdc (mW)
[3]	27.5–31	15	–	2.5	3.5	77
[4]	1.6–28	9.6±1.1	−3.4	4.5	3	16.8
[5]	28	21	−1	5.6	1.2	
Modified LNA	28	16.43	10	3.6	1.5	7.5

Table 2 PVT analysis of LNA

PVT analysis (°C)	Peak gain (dB)	NF (dB)
tt, 1.5 V, 27	16.4	3.6
ss, 1.5 V, 27	13.9	3.65
ff, 1.5 V, 27	17.3	3.4
tt, 1.4 V, 27	16.2	3.6
tt, 1.6 V, 27	16.4	3.61
tt, 1.5 V, 26	16.4	3.6
tt, 1.5 V, 28	16.4	3.6

Therefore, we find significant improvements in many performance parameters in the proposed LNA design.

PVT analysis of LNA is performed with different process corners, a variation of 1 V supply and 1 °C of temperature. The result is summarized in Table 2.

4.2 *Performance Parameters of Down Converter*

Down converter output and its spectrum are plotted in Figs. 12 and 13. The spectral

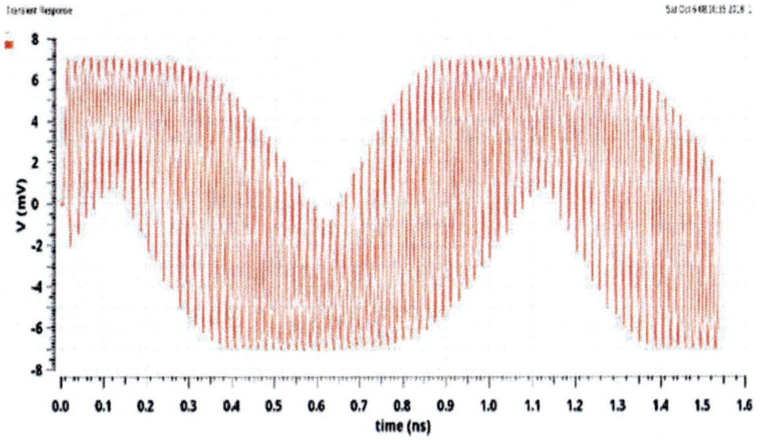

Fig. 12 IF output

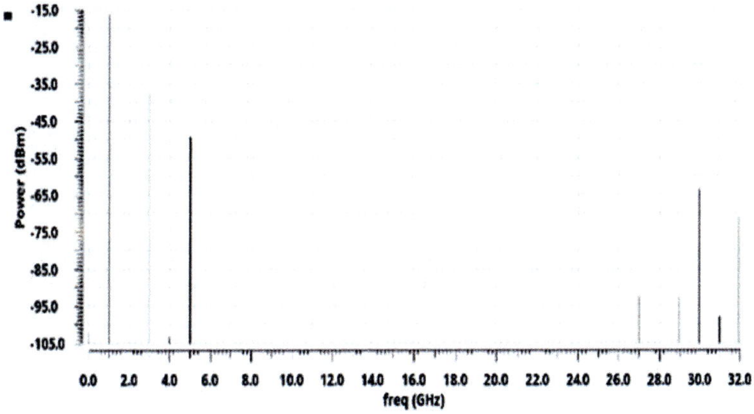

Fig. 13 IF spectrum

purity can be seen in Fig. 13 as the second harmonics are −20 dBm below the IF signal. Since there is matching networks at the RF side, the RF return loss is <−10 dB for the entire band of interest as shown in Fig. 14. Conversion gain of −2.2 dB is obtained for the frequency range of 27.5–29.5 GHz. The NF is around 4.5dB for the band of interest. RF to IF isolation around −36.2 dB is obtained. Linearity of down converter, 1 dB compression point, $P1$dB is found to be at −1.16 dB. Table 3 outlines the comparison of previously reported 28 GHz RF down converters in various technologies and topologies.

PVT analysis of down converter is performed with different process corners, a variation of 0.1 V supply and 1 °C of temperature. Isolation is more or less same in all the corners. Return loss is below −12 dB for all the cases. The process variation on

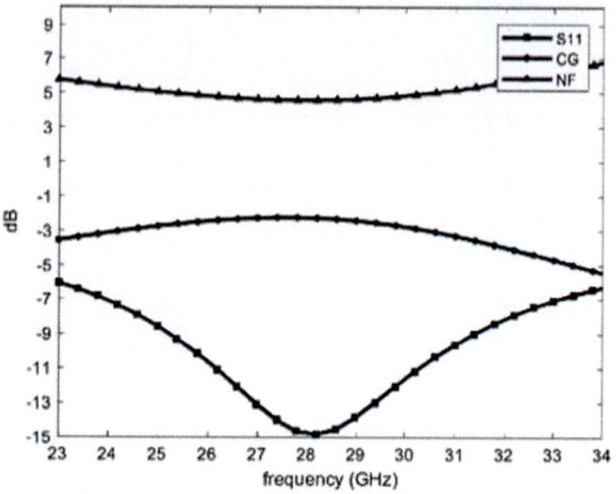

Fig. 14 S-parameters and NF of down converter

Table 3 Comparing the performance of down converter

Ref.	Technology	Freq (GHz)	CG (dB)	P1dB (dBm)	NF (dB)	Isolation (dB)
[5]	28 nm CMOS	28	5 to −2	−12.6	14	–
This work	65 nm CMOS	28	−2.2	−1.16	4.5	−36.2

return loss and CG is shown in Fig. 15a, b. Table 4 summarizes the process variation of down converter.

4.3 Performance Parameters of Receiver

Initial assignment of gain, NF and linearity requirement to individual stages satisfies the required system-level performance parameters. Cumulative gain is found to be 39 dB and NF of the entire chain is only 4.2 dB. Linearity of the receiver is measured at 15.7 dBm of OIP3 and IIP3 of −23.28 dB for a received power level of −100 dBm. Through the constellation and EVM measurement for 64 QAM, we find that the EVM is well below −37.76 dB. These measurements are plotted in Fig. 16. Tables 5 and 6 outline the comparison with some of the available literatures in terms of EVM and cascaded performances, respectively. EVM is comparable or below the existing measurements for a low received power level.

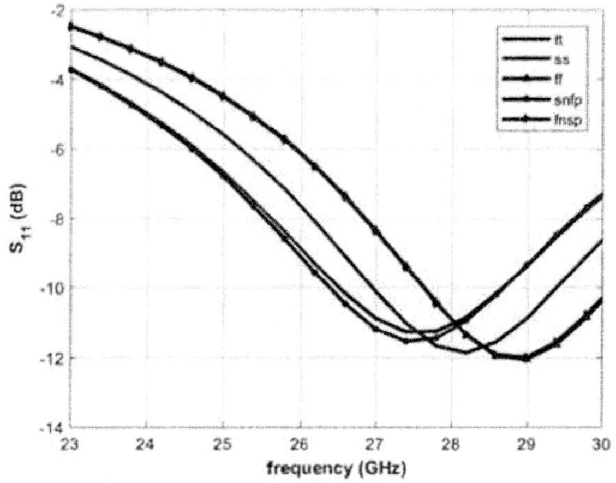

(a)

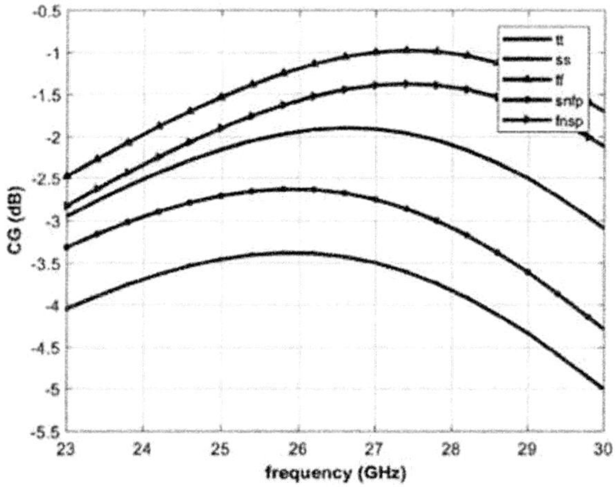

(b)

Fig. 15 **a** Process variation on S_{11}, **b** process variation on CG

Table 4 PVT analysis of down converter

PVT analysis (°C)	Conversion gain (dB)	NF (dB)
tt, 1.5 V, 27	−2.2	4.5
ss, 1.5 V, 27	−3.39	5.67
ff, 1.5 V, 27	−0.99	4.214
spfn, 1.5 V, 27	−1.377	3.92
fpsn, 1.5 V, 27	−2.679	5.137
tt, 1.4 V, 27	−2.45	4.50
tt, 1.6 V, 27	−1.73	4.69
tt, 1.5 V, 26	−1.88	4.55
tt, 1.5 V, 28	−1.9	4.58

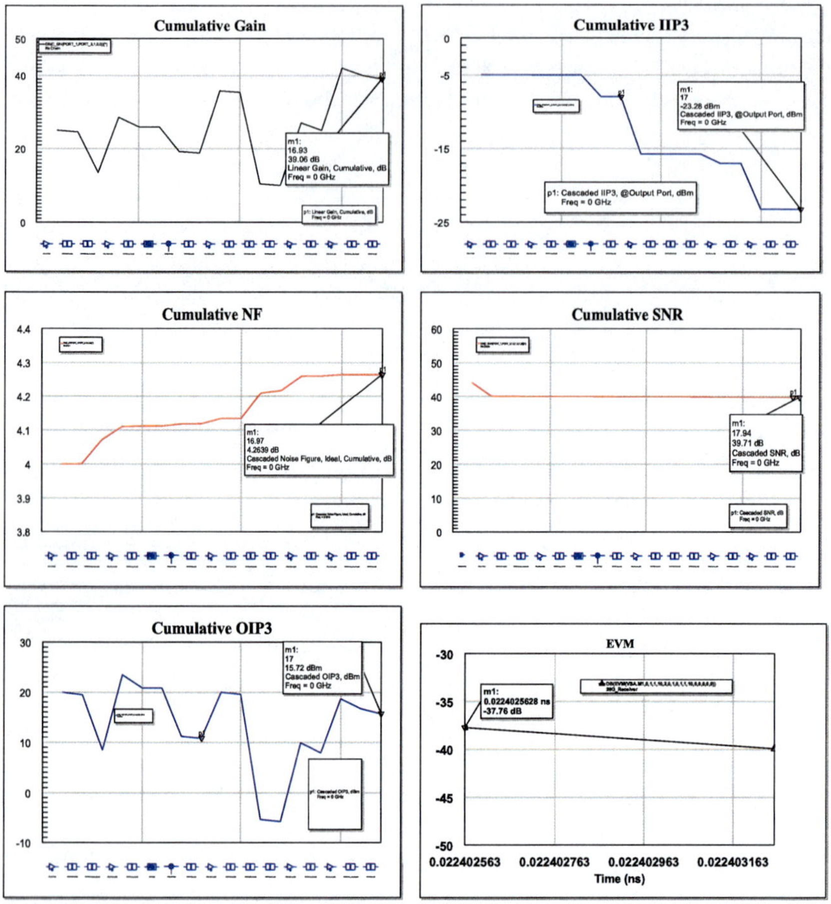

Fig. 16 Receiver performance parameter

Table 5 EVM comparison

Ref.	EVM (dB)	Modulation	Received power level (dBm)
[5] [a]	−33	64 QAM	−70
This work[b]	−37.76	64 QAM	−100

[a]Measurement result, [b]Simulation results

Table 6 Receiver performance

Ref.	Gain (dB)	NF (dB)	IIP3 (dBm)
[11] [a]	28	5	−28
This work[b]	39	4.2	−23.28

[a]Measurement result, [b]Simulation results

5 Conclusion

A 28 GHz low noise amplifier and a down converter suitable for prospective 5G wireless communication have been designed. LNA offers a peak gain of 16.43 dB and a noise figure of 3.6 dB. An improvement of 1.6–3.7 dB in gain is obtained for the proposed circuit as compared to the conventional CS LNA. A −3 dB bandwidth of 2.2 GHz is also obtained with a linearity of IIP3 of 10 dBm. Additionally, the power dissipation of 7.25 mW and a return loss less than −12 dB is also observed around 28 GHz. The output return loss is less than −14 dB. A trade-off between noise figure, gain and bandwidth has been addressed here. Modified Gilbert cell topology is used for down converter which gives a conversion gain of −2.2 dB, P_{1dB} of −1.16 dBm with RF at 28 GHz from a 1.5 V supply voltage. NF is observed at 4.5 dB and isolation is −36 dB. With these RF components, an EVM of −37.7 dB is observed for 64 QAM. Compared with the state of the art, almost same performance is observed using a higher technology and lower RF received power level. Additional work is necessary before the block is ready for tap out. As a future work, post-layout simulation and an EM simulation need to be carried out to find coupling between different structures and the stability issues caused by power networks.

Acknowledgements This research work was supported by Women Scientists Scheme A (WOS-A) SR/WOS-A/ET-82/2016 from Department of Science and Technology (DST), Government of India.

References

1. Rappaport TS, Ben-Dor E, Murdock JN, Qiao Y (2012) 38 GHz and 60 GHz angle-dependent propagation for cellular & peer-to-peer wireless communications. In: IEEE International Conference on Communications (ICC), Ottawa, 2012, pp. 4568–4573
2. Zhao H, Mayzus R, Sun S, Samimi M, Schulz JK, Azar Y, Wang K, Wong GN, Gutierrez F, Jr Rappaport ST (2013) 28 GHz millimeter wave cellular communication measurements for reflection and penetration loss in and around buildings in New York city. In: IEEE international conference on communications (ICC), Budapest, pp 5163–5167
3. Park DW, Gwag GH, Oh HJ, Park IM, Eo YS (2016) 28 GHz RF transceiver module for 5G beam-forming system. In: Asia-Pacific microwave conference (APMC), New Delhi, pp 1–4
4. Chen HK, Lin YS, Lu SS (2010) Analysis and design of a 1.6–28-GHz compact wideband LNA in 90-nm CMOS Using a π-match input network. IEEE Trans Microw Theory Tech 58(8):2092–2104
5. Kim H et al (2018) A 28 GHz CMOS direct conversion transceiver with packaged antenna arrays for 5G cellular system. IEEE J Solid State Circuits 53(5):1245–1259
6. Voinigescu SP, Dickson T, Beerkens R, Khalid I, Westergaard P (2004) A comparison of Si CMOS, SiGe BiCMOS, and InP HBT technologies for high-speed and millimeterwave ICs. In: IEEE SiRF, Sep 2004, pp 111–114
7. Pournamy S, Kumar N (2017) Design of 60 GHz broadband LNA for 5G Cellular using 65 nm CMOS Technology. In: 7th international conference on communication systems and network topologies (CSNT), Nagpur, pp 320–324
8. Yao T, Gordon MQ, Tang KKW, Yau KHK, Yang MT, Schvan P, Voinigescu SP (2007) Algorithmic design of CMOS LNAs and PAs for 60-GHz radio. IEEE J Solid-State Circuits 42(5):1044–1057

9. Shekhar S, Walling JS, Allstot DJ (2006) Bandwidth extension techniques for CMOS amplifiers. IEEE J Solid-State Circuits 41(11):2424–2439
10. http://www.analog.com/media/en/technical-documentation/data-sheets/hmc1063.pdf, Last accessed on 6 Oct 6, 2018
11. http://www.anokiwave.com/products/awmf-0108/index.html, Last accessed on 6 Oct 6, 2018

Study and Implementation of Ladder Logic Conversion to VHDL for Field Programmable Gate Array (FPGA)-Based Programmable Logic Controllers (PLC)

Aditya Kumar, Prashant Kumar, Sompurna Modi, and Vijay Nath

Abstract Programmable Logic Controller (PLC) is commonly used in the modern era of manufacturing and industrial automation. PLC can be programmed in three ways: Ladder logic, function block diagram, and sequential function chart. The most common and popular programming language is ladder logic in PLC. The efficiency of PLC is highly restricted by the speed of microprocessor in real time and the power consumption is quite high. To speed up the PLC performance and flexibility, this paper proposed a new PLC ladder logic design based on FPGA (VHDL). Further with the use of state machine process, the conversion of ladder logic into VHDL has been optimized. There is wide variety and diversity of ladder logic instruction for different PLCs, so a universal converter is required with an extended Boolean expression which will act as a bridge between ladder logic and VHDL. This proposed design will provide improved flexibility, consume less power, increased reliability, faster scanning time, and better performance.

Keywords PLCs · Ladder logic · FPGA · VHDL

A. Kumar (✉) · P. Kumar · V. Nath
Department of ECE, Birla Institute of Technology, Mesra, Ranchi, Jharkhand 835215, India
e-mail: aditya88952@gmail.com

P. Kumar
e-mail: prashant.paccific@gmail.com

V. Nath
e-mail: vijaynath@bitmesra.ac.in

S. Modi
Department of EEE, Birla Institute of Technology, Mesra, Ranchi, Jharkhand 835215, India
e-mail: sompurna27@gmail.com

V. Nath and J. K. Mandal (eds.), *Proceedings of the Fourth International Conference on Microelectronics, Computing and Communication Systems*, Lecture Notes in Electrical Engineering 673, https://doi.org/10.1007/978-981-15-5546-6_3

1 Introduction

PLCs have played a very vital role in the industrial automation as a control element for controlling the logic of the manufacturing system [1]. The most common and popular is ladder logic to program a PLC. In PLC, there is a microprocessor which is embedded in it which decodes the ladder logic and executes the program in a sequential and cyclic manner. The cyclic scan and sequential execution of program limit the PLC performance as it depend on the size of program as well as the processor speed. To speed up the PLCs performance and overcome the drawbacks, this paper proposed a new PLC design which is based on Field Programmable Gate Array (FPGA) in Very High Speed Integrated Circuit Hardware Description Language (VHDL). It can achieve reprogrammable and reconfigurable hardware structure with the advantage of parallel execution. The ladder logic of PLC programming consists of rungs. These rungs are controlled by a FPGA which four side 32 bit four stages RISC pipeline processor which is based on the rung architecture. The proposed design will increase the speed of PLC, increase reliability, increase flexibility, better performance and faster scanning time this is because Field Programmable Gate Array (FPGA) offers reconfigurable and reprogrammable hardware.

2 Related Works

A number of researches are done in this field by various researchers. In 1999, Miyazawa [2] and Ikeshita et al. [3] developed a very inefficient technique to convert PLC ladder logic into VHDL. To build a controller, Chen and Patyra [4] designed a VHDL model for the complete system from the basic and original system. An algorithm was developed by Abdel-Hamid et al. [5] and Kuusilinna et al. for the conversion of FSM into VHDL. Petri net model as alternative of the ladder diagram in manufacturing control was chosen by Adamski et al. All the above researches and studies show that reprogrammable and reconfigurable hardwares have some advantages like size, cost, power consumption, simplicity of program, and parallel execution of FPGA which will improve the PLC scan speed and performance. The current scenario of the researches confined to a very primitive and rough stage due to only introducing the case base conversion from PLC ladder logic into Very High Speed Integrated Circuit Hardware Description Language (VHDL) [6]. Most of technique focuses on obtaining the HDL or RTL architecture of FPGA from the original system requirements. As mostly the PLC programs in the automation and manufacturing sectors are written in the ladder logic format; therefore, it is important to use the ladder logic programs for the existing PLCs system to perform a new PLC design.

3 Methodology

FPGA-Based PLC Design

This paper proposed the advantage of FPGA technology to model new PLC design over the tradition PLC-based software solution. The proposed scheme can improve the flexibility as well as scan timing and reduce the manufacturing cost. For the modelling of new FPGA-based PLC design, FPGA will have to execute parallel with the PLC [7]. For the achievement of this goal, the first priority is to convert the PLC ladder logic into the gate-level digital Boolean expression to get the VHDL code since the architecture of FPGA is reprogrammable and reconfigurable. So it is easy to convert ladder logic programs into RTL architecture and download to the FPGA kit. This implementation will work exactly the same as the original PLC but without the slow traditional method of sequential cyclic scan. For the RTL schematic of FPGA implementation, many softwares are available like Xilinx ISE 8.1, Xilinx ISE 9.1, Xilinx ISE 14.1, Altera, etc., which will extract the RTL schematic into high-level language like Very High Speed Integrated Circuit Hardware Description Language (VHDL) [8]. The architecture of FPGA-based PLC is given in Fig. 1 [9].

Architecture of the Converter

As shown in Fig. 1, the architecture [9] of the converter consist of four main blocks, i.e., the main frame, PLC instruction dictionary, dynamic link libraries (DLL) of Boolean expression to VHDL, and lastly, dynamic link libraries (DLL) of the Boolean

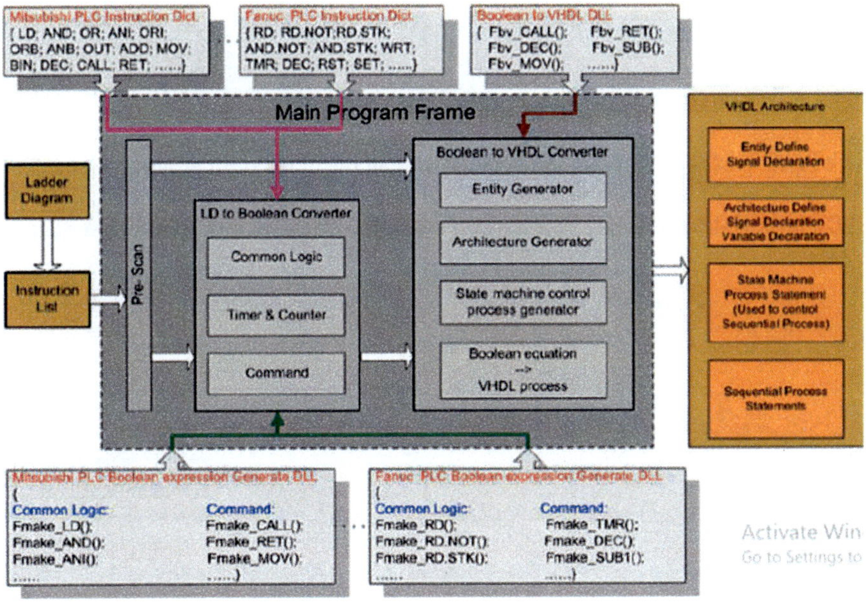

Fig. 1 Architecture of FPGA-based PLC

expression generator. The main frame consists of three blocks, namely the pre-scan block, ladder logic to Boolean converter, and Boolean to VHDL converter. The PLC dictionary is different for different PLCs depending upon their manufacturing companies. When the PLC ladder logic is executed, the main frame calls the PLC instruction dictionary to the ladder logic to Boolean convertor block then this block forwards the information to the Boolean to VHDL convertor for the conversion of PLC ladder logic to VHDL code. This architecture is universal for all the PLCs only we have to change the PLC dictionary for the code conversion of different PLCs to VHDL code.

PLC Ladder Logic Code Conversion into VHDL Code
In the era of industrial automation, timing and speed are two very important aspects. Most of the manufacturing and automation company uses PLC for their work. But the PLC uses are not very efficient in sequential scan time and not flexible as well. To overcome these problems, FPGA-based PLC design is introduced in this paper which will convert the ladder logic code to VHDL code. In the proposed scheme, the FPGA will execute parallel with PLC, and the code conversion will be executed in two steps (1) PLC ladder logic converted to Boolean expression [10]. (2) Boolean expression converted to VHDL code.

The conversion of PLC ladder logic into VHDL code is shown in Fig. 2.

ALGORITHM for the Conversion Process

STEP1 Ladder diagram is introduced.
STEP2 This goes through the 'Instruction list' to the 'Main program frame' for 'Pre scan.'
STEP3 Then, it moves to the block 'LD to Boolean converter.'
STEP4 Here, the LD goes through different sets of instruction dictionary whether it be Mitsubishi or Fanuc, etc., for a common logic in
Between Ladders diagram logics and Boolean logics.
STEP5 This logic is being processed through the defined timer and counter. The count is decremented each clock tick in the case of timers,
while counters have a channel associated to the input that may be counted.

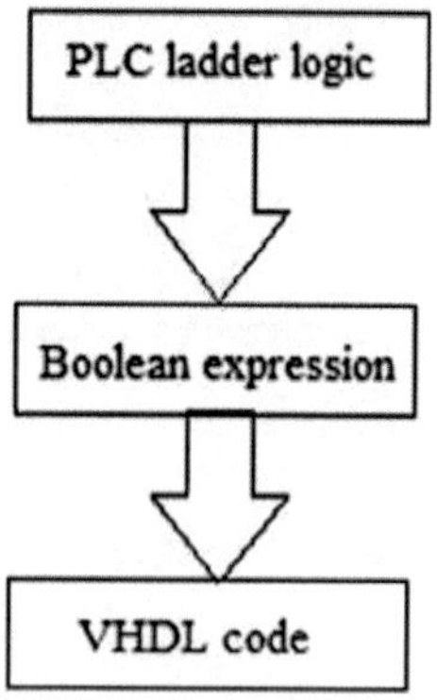

Fig. 2 Steps involving PLC ladder logic conversion into VHDL code

STEP6 This is then processed for the generation of command through the common logic established.

STEP7 The command in Boolean expression is forwarded to the block'Boolean to VHDL Converter.'

STEP8 This command is then processed to 'Entity Generator Block' that would generate entities of discrete event simulation.

STEP9 Then forwarded to 'Architecture Generator Block' to carry out the generated entity for signal or variable declaration.

STEP10 This goes to 'State Machine Control Process Generator' block. It traverses through a sequence of states where the next state is determined upon the present state and input conditions. And provides sequences of output signals to be controlled.

STEP11 This state machine process statement is converted to sequential process statement so that the converted VHDL code could be obtained.

COMPARISION TABLE with other related papers proposed for study

Proposed scheme	Previous related works
1. The process proposed is *universal* in nature. It accepts all the different sets of instruction for a common logic	The proposed papers lack this concept in their process making them bounded for a chosen set of instructions
2. This paper proposed a fast processing conversion process as compared to other study papers till date	The proposed processes by other papers are slow and time taking during their implementation
3. This paper proposes a before check step in addition to other steps for implementation which reduces the chances of error	No before check step was introduced

The conversion of rungs into VHDL code involving certain algorithm and syntaxes that are shown in Table 1.

The symbol description of Table 1 is given in Table 2.

Timer in PLC ladder logic can be synthesized by D-type flip flop as the output is delayed by the clock signal, while the counter is designed by state diagram in FPGA.

4 Results and Simulations

In the simulation, we are using ***LogixPro*** software for PLC ladder logic based on ALLEN BRADLY PLC and ***Xilinx ISE 8.1*** for VHDL code. Toggle switches are used which will avoid the holding logic. If the switches are in ON condition, we will treat as '1' and when the toggle switches are in OFF condition, '0' will be the logic.

From Figs. 3, 4, 5, 6, 7, 8, 9, 10, 11, 12, 13, 14, 15 and 16 show the conversion of PLC ladder logic into Boolean expression, now these expressions will be converted

Table 1 PLC ladder logic, Boolean expression along with VHDL code

PLC ladder logic	Boolean expression	VHDL code syntax
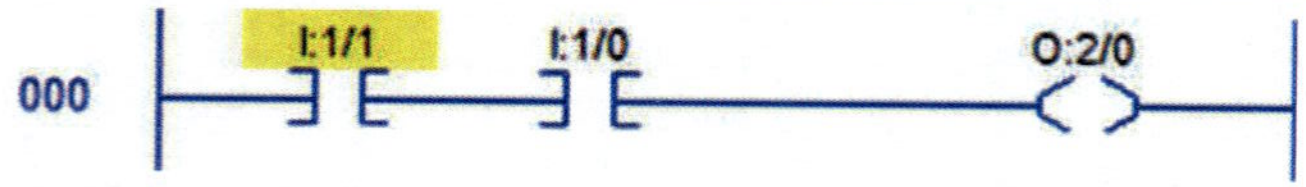	$Y = A.B$	$Y \leq A$ AND B
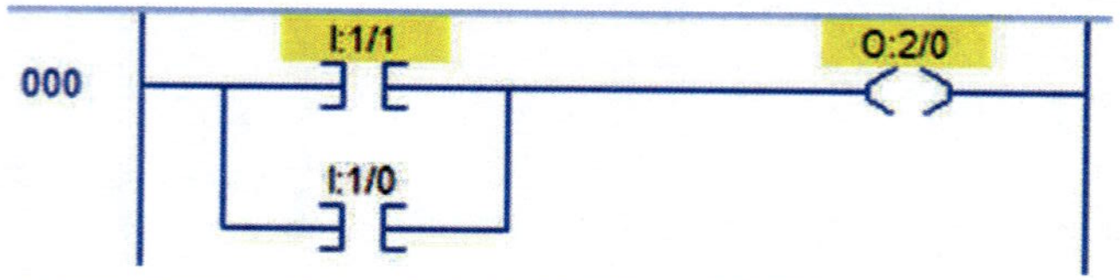	$Y = A + B$	$Y \leq A$ OR B
	$Y = \bar{A}$	$Y \leq$ (NOT A)
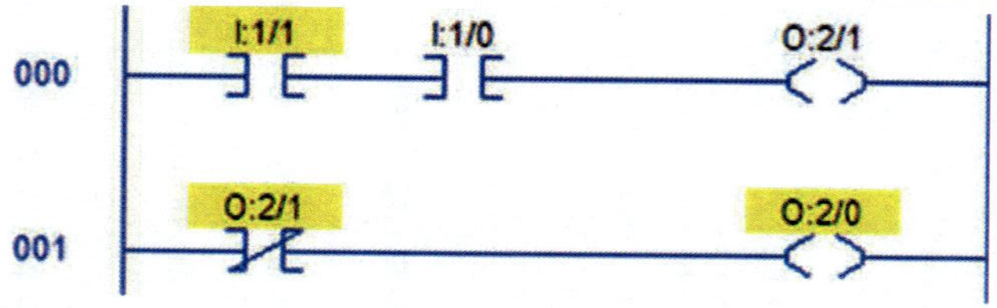	$Y = \sim (A.B)$	$Y \leq A$ NAND B

(continued)

Table 1 (continued)

PLC ladder logic	Boolean expression	VHDL code syntax
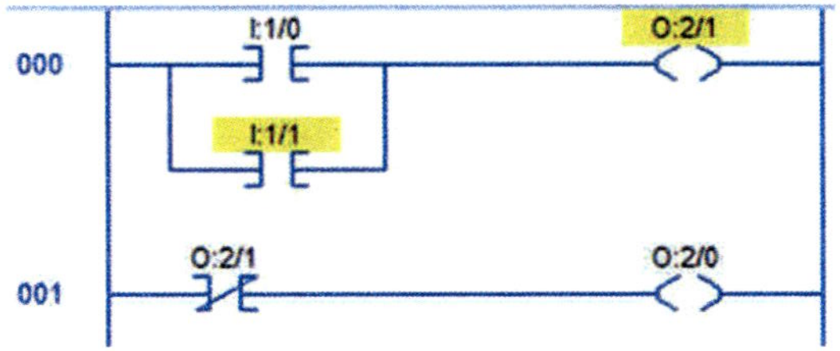	$Y = \sim (A + B)$	$Y \leq A$ NOR B
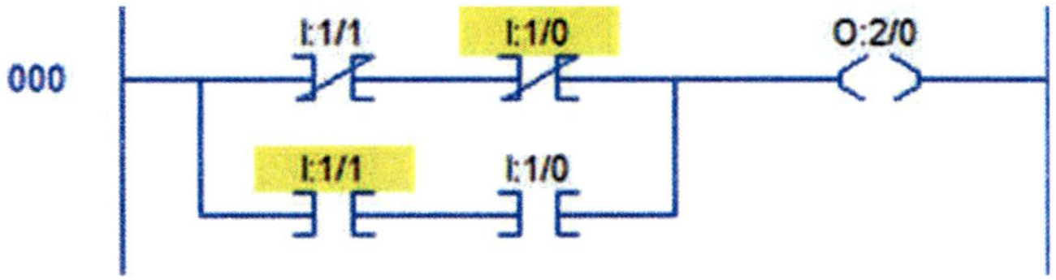	$Y = A \odot B$	$Y \leq A$ XNOR B
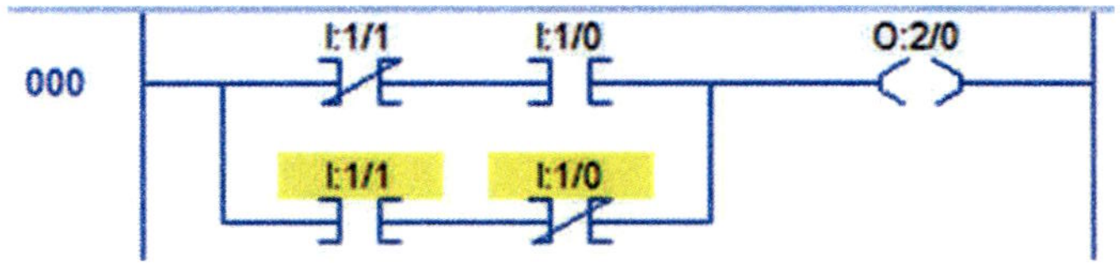	$Y = A \oplus B$	$Y \leq A$ XOR B

Table 2 Detailed description of symbols used in Table 1

Symbol	Description
Y, O:2/0	Output
A, I:1/0	Input 1
B, I:1/1	Input 2
O:2/1	Auxiliary output

Fig. 3 AND-gate implementation with input $A = 1$ and input $B = 1$ with output $Y = 1$

Fig. 4 AND-gate implementation with input $A = 0$ and input $B = 1$ with output $Y = 0$

Fig. 5 OR-gate implementation with input $A = 1$ and input $B = 0$ with output $Y = 1$

Fig. 6 OR-gate implementation with input $A = 0$ and input $B = 0$ with output $Y = 0$

Fig. 7 NOT-gate implementation with input $A = 0$ with output $Y = 1$

Fig. 8 NOT-gate implementation with input $A = 1$ with output $Y = 0$

Fig. 9 NAND-gate implementation with input $A = 0$ and input $B = 1$ with output $Y = 1$

Fig. 10 NAND-gate implementation with input $A = 1$ and input $B = 1$ with auxiliary output $= 1$ and output $Y = 0$

Fig. 11 NOR-gate implementation with input $A = 0$ and input $B = 0$ with output $Y = 1$

Fig. 12 NOR-gate implementation with input $A = 0$ and input $B = 1$ with auxiliary output $= 1$ and output $Y = 0$

Fig. 13 XNOR-gate implementation with input $A = 0$ and input $B = 1$ with output $Y = 0$

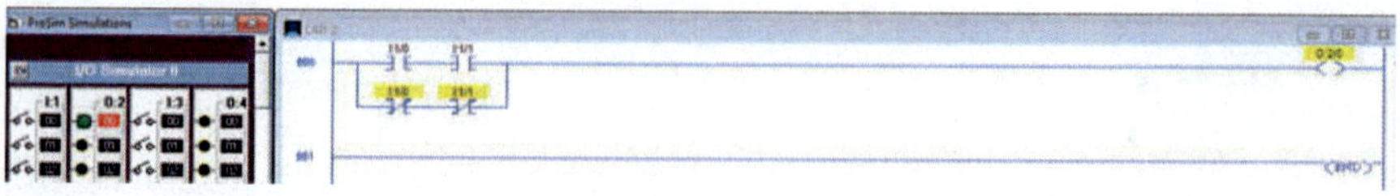

Fig. 14 XNOR-gate implementation with input $A = 0$ and input $B = 0$ with output $Y = 1$

Fig. 15 XOR-gate implementation with input $A = 0$ and input $B = 1$ with output $Y = 1$

Fig. 16 XOR-gate implementation with input $A = 0$ and input $B = 0$ with output $Y = 0$

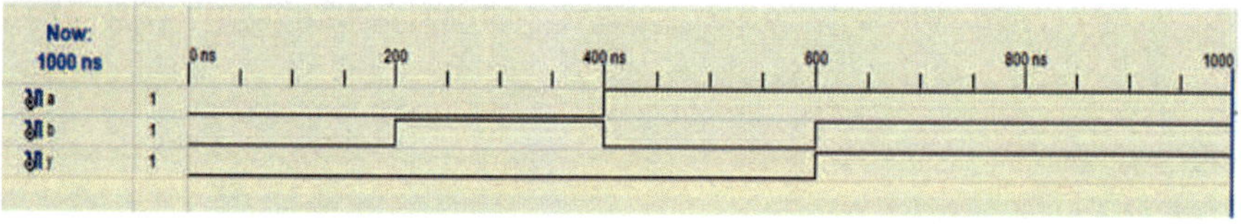

Fig. 17 Waveform of AND gate in Xilinx 8.1i

into VHDL code. All the simulations in the above figures are simulated on ***LogixPro*** simulator software based on ALLEN BRADELY PLCs.

Figures 17, 18, 19, 20, 21, 22 and 23 represent the waveform of combinational circuit after the conversion of Boolean expression into VHDL code. All the simulations in the above figure from Figs. 17, 18, 19, 20, 21, 22 and 23 are simulated on ***Xilinx 8.1i.*** The simulation results are according to our expectation and they are following their respective truth table.

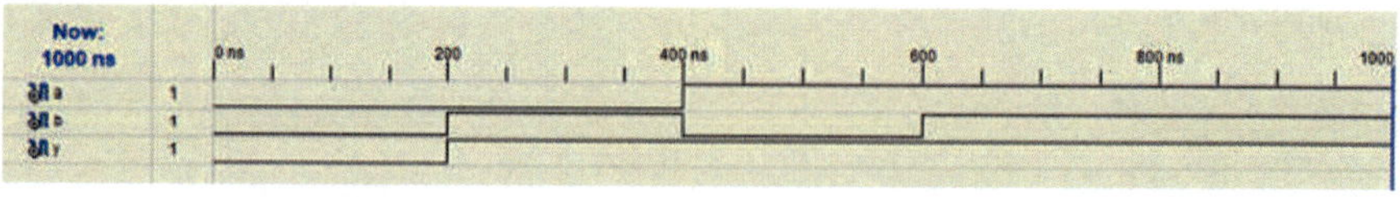

Fig. 18 Waveform of OR gate in Xilinx 8.1i

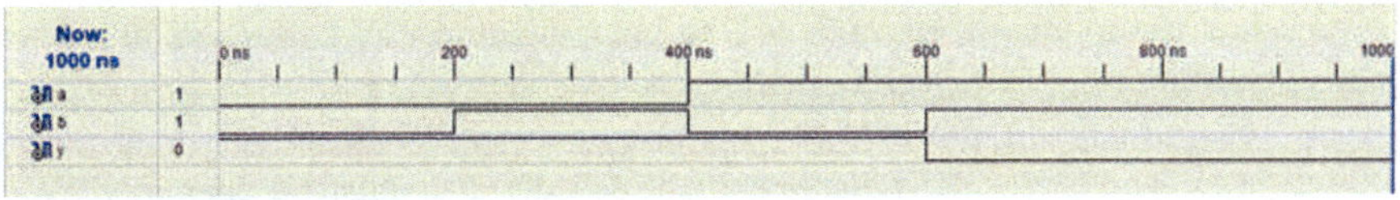

Fig. 19 Waveform of NAND gate in Xilinx 8.1i

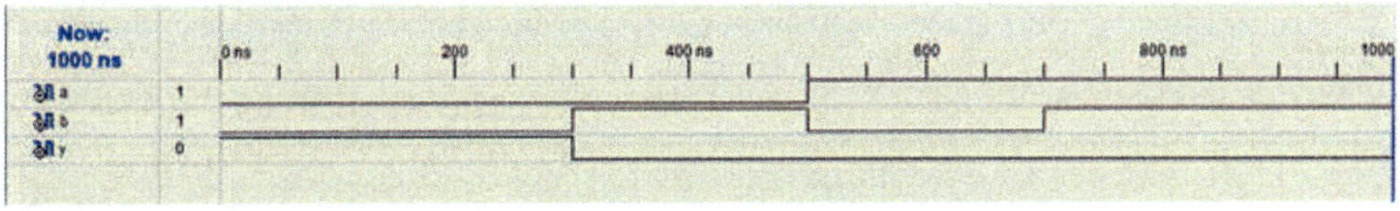

Fig. 20 Waveform of NOR gate in Xilinx 8.1i

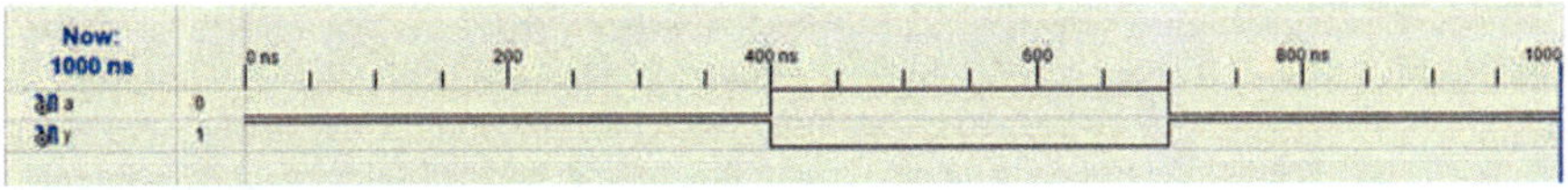

Fig. 21 Waveform of NOT gate in Xilinx 8.1i

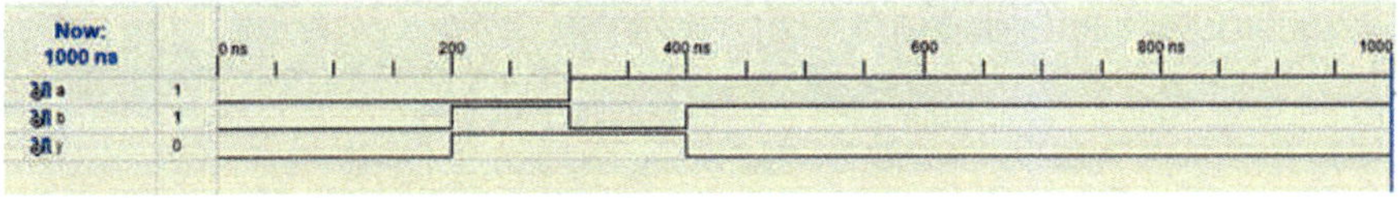

Fig. 22 Waveform of XOR gate in Xilinx 8.1i

5 Conclusion

From the above results and simulation, the performance, flexibility, reliability, and faster sequential scanning are achieved over the traditional microprocessor-based PLC. In this proposed paper, a FPGA-based PLC design is introduced with the conversion of PLC ladder logic into VHDL code. In the process of code conversion, the Boolean expression has played a very important role by acting as a bridge between the PLC ladder logic and VHDL code. For the conversion of PLC ladder logic into a VHDL code, a universal architecture is adopted for this work which will convert the

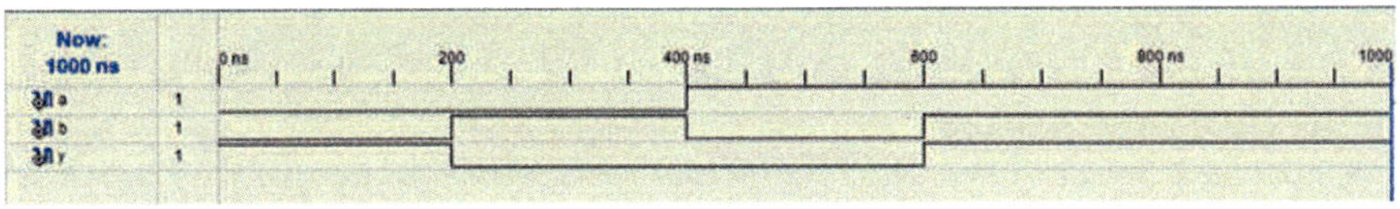

Fig. 23 Waveform of XNOR gate in Xilinx 8.1i

PLC ladder logic into VHDL code with a great advantage of better flexibility, low power consumption, better scan time.

From the practical point of view, this research work is still in very early developing stage. There are many mathematical aspects that must be examined. With the advancement of modern industrial automation era, use of PLC has increased manifolds, so much more research work is needed to improve this work.

Acknowledgements The authors wants to pay their sincere gratitude to the Indian Society OF VLSI Design, BIT MESRA, faculties of ECE dept., our family and friends for their generous and selfless support which makes this work possible. We are also grateful to GOD for enlightening our knowledge in the field of VLSI and Industrial Automation.

References

1. Auslander DM, Pawlowski C, Ridgely J (1998) Reconciling programmable logic controllers with mechatronics control software. In: Proceedings of the 1998 IEEE international conference on control applications, Trieste, Italy, pp 415–420
2. Ikeshita M, Takeda Y, Murakoshi H, Funakubo N, Miyazawa I (1999) Application of FPGA to high-speed programmable controller—development of the conversion program from SFC to Verilog. In: IEEE symposium on emerging technologies and factory automation, Piscataway, NJ, USA, vol 2, pp 1386–1390
3. Miyazawa I, Nagao T, Fukagawa M, Itoh Y, Mizuya T, Sekiguchi T (1999) Implementation of ladder diagram for programmable controller using FPGA. IEEE symposium on emerging technologies and factory automation, Piscataway, NJ, USA, vol 2, pp 1381–1385
4. Chen J, Patyra MJ (1994) VHDL modeling of a multivariable fuzzy logic controller hardware system. In: IEEE international conference on fuzzy systems. Piscataway, NJ, USA, vol 1, pp 129–132
5. Abdel-Hamid AT, Zaki M, Tahar S (2004) A tool converting finite state machine to VHDL. In: IEEE Canadian conference on electrical and computer engineering, Niagra Falls, Canada, vol 4, pp 1907–1910
6. Asbenden PJ (2002) The designer's guide to VHDL. Morgan Kaufmann, San Diego
7. Lee JS, Hsu PL (2005) An improved evaluation of ladder logic diagrams and Petri nets for the sequence controller design in manufacturing systems. Int J AdvManufTechnol 24(3–4):279–287
8. Kuusilinna K, Lahtinen V, Hamalainen T, Saarinen J (2001) Finite state machine encoding for VHDL synthesis. IEE Comput Dig T 148:23–30
9. Shanta S, Dipali S (2005) A new generation of PLC—An FPGA based PLC. In: Proceedings of the SICE annual conference, Okayama, Japan, pp 2367–2370
10. Shanta S, Dipali S (2005) A new generation of PLC—An FPGA based PLC. Proceedings of the SICE annual conference, Okayama, Japan, pp 2367–2370

Automated Monitoring and Regulation of User-Friendly Greenhouse Using Arduino

Priyanshu Kumar, Sonali Saroj, Sanjeev Kumar, and Chandrashekhar Azad

Abstract This paper presents an Arduino-based system which automatically monitors the physical conditions of a greenhouse and controls or regulates those physical conditions accordingly. The system measures the humidity, temperature, soil moisture, etc., of the greenhouse and accordingly controls the various components automatically. Different sensors such as soil moisture, temperature, humidity and LDR are used for measuring the environmental data of greenhouse. Measured data are compared with the standard value favorable for the desired plant/crop, and accordingly components such as heater, exhaust fan and water pump are activated with the help of relays. Greenhouse gardening is spreading widely; hence, using this automatic control system will be very useful for farmers and can contribute to improve productivity and quality of crops.

Keywords Arduino · Automatic control · Desired conditions · Relay · Sensors

1 Introduction

Food is the basic requirement of mankind. The demand for crops is increasing rapidly due to population growth. Global warming and other environmental factors have reported substantial landmass for the production of crops. Hence, crop production is

P. Kumar · S. Saroj
Amity University Jharkhand, Ranchi, India
e-mail: priyanshu.kumar.ece@gmail.com

S. Saroj
e-mail: sonali.saroj@gmail.com

S. Kumar
Sarala Birla University, Ranchi, India
e-mail: sanjeevsingh.ece@gmail.com

C. Azad (✉)
National Institute of Technology Jamshedpur, Jamshedpur, Jharkhand, India
e-mail: csazad.ca@nitjsr.ac.in

V. Nath and J. K. Mandal (eds.), *Proceedings of the Fourth International Conference on Microelectronics, Computing and Communication Systems*, Lecture Notes in Electrical Engineering 673, https://doi.org/10.1007/978-981-15-5546-6_4

one of the utmost problems of twenty-first century. To overcome this problem, now greenhouses are coming more in practice. Greenhouse is a controlled area environment to grow crops. However, operating them manually is quite tedious and time consuming. To address this problem, automated greenhouse is developed.

The work by Taras Teslyuk et al. shows the functioning of an Arduino-based automated greenhouse focused on a GUI-based monitoring system and shows the functioning of an automated monitoring system [1] that is controlled by the system itself in real time.

Automated greenhouse requires automated monitoring and regulation of environmental parameters that directly or indirectly influence the plant's growth and hence its output. Plants need limited range of temperature, soil moisture, light, moisture, air and nutrients for growth. Plants also need some kind of physical support for roots. To autonomously control the climate factors and environment, low-cost electronic components are required. This automated greenhouse system monitors all the micro-climatic conditions inside the greenhouse [2] with the help of some electronic sensors [3] and accordingly activates the components required. The most basic component for building this system is Arduino UNO and sensors. The sensors collect data on the atmosphere. We will be made available to the Arduino board after the data have been filtered [4]. Which then measures and compares the current values of the controlled variables to the specified threshold value, the resulting actuator will be enabled to restore the optimal state if any of the controlled variables are outside the fixed limit.

2 Important Features of the Proposed Automation System

- If necessary, the system can monitor and adjust parameters such as air, temperature and humidity [5].
- It extends the day according to the ambient illumination level.
- It measures the soil humidity, reacts accordingly and hence also controls wastage of water.
- It is sustainable, versatile and easy to use.
- It requires minimal changes to the existing greenhouse, so devices such as water valves, fans, humidifiers and whatever equipment might be installed in a common greenhouse should be interfaced [6].
- It uses active hardware–software interfacing and real-time monitoring [7].

Fig. 1 Arduino UNO board

3 Component Description

3.1 Arduino UNO

The Arduino UNO [8] is a widely used microcontroller. It has been developed by Arduino.cc and is based on the ATmega328P microcontroller.

The board is configured with a range of digital and analog input/output (I/O) pins that can be attached to different boards or other circuits. The machine has six analog pins and fourteen optical pins. The Arduino IDE can be configured with a type B USB cable. With a USB cable or an external 9 V battery, it can be conveniently powered. The Arduino UNO microcontroller is pre-programmed with a bootloader that allows easy uploading of new code to it without any specific hardware programmer being used (Fig. 1).

3.2 Soil Moisture Sensor

There are two tests used to assess the volumetric water content of the soil moisture sensor. The two samples cause the current to flow through the soil and then measure the resistance value and hence the moisture value. Soil will conduct more electricity which specifies that there is less resistance. Hence, the moisture level is high. Henceforth, the level of humidity will be higher. Less electricity is conducted by dry soil,

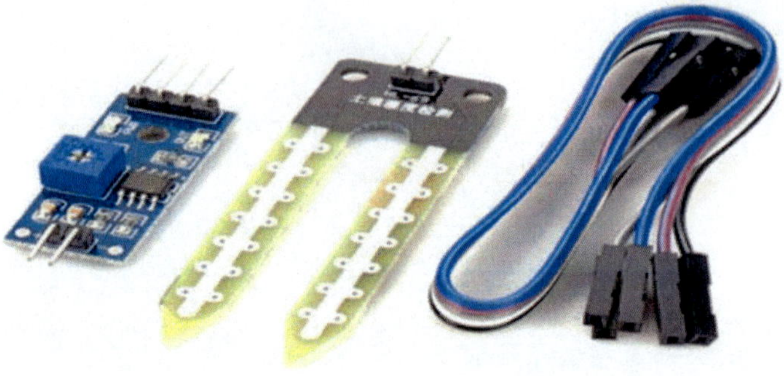

Fig. 2 Soil moisture sensor

so when there is less water, less electricity is conducted by soil, indicating high resistance. The moisture level will therefore be lower. It is possible to connect sensor in two modes: analog and digital modes (Fig. 2).

3.3 Temperature and Humidity Sensor

The DHT11 detects water vapor by measuring the electrical resistance between two electrodes. The humidity sensing tool is a substrate that keeps the electrodes applied to the surface moisture. The substratum produces ions when the substratum receives water vapor, which raises the electrode conductivity. The difference in resistance between the two electrodes is equal to the relative humidity. For measuring temperature, the DHT11 uses a NTC temperature sensor.

Higher relative humidity decreases resistance to the electrodes, while lower relative humidity raises resistance to the electrodes (Fig. 3).

3.4 Relay Module

Relays are based on the electromagnetism principle. When magnetizing the relay coil, it serves as a magnet and adjusts the switch position. The circuit driving the coil is entirely isolated from the switching portion ON/OFF. It provides good Fig. (4).

Insulation for electrical use. We can regulate a relay [9] that uses 5 V from an Arduino and link the other end to a 230 V appliance. The 230 V end of the 5 V Arduino circuit is completely isolated.

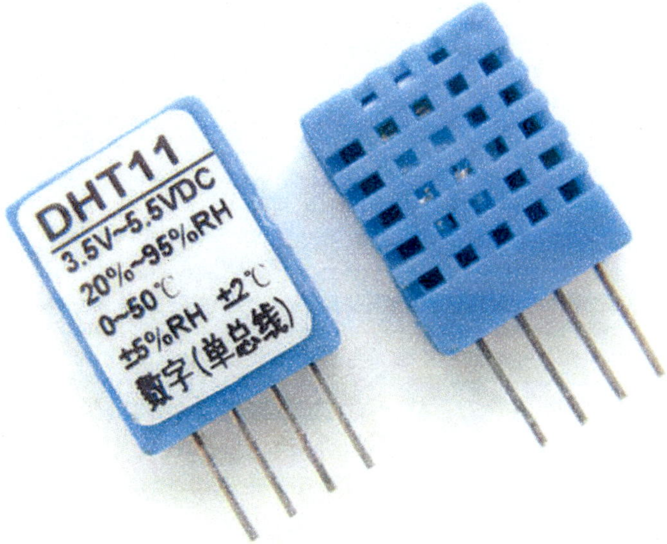

Fig. 3 Temperature and humidity sensor

Fig. 4 Relay module

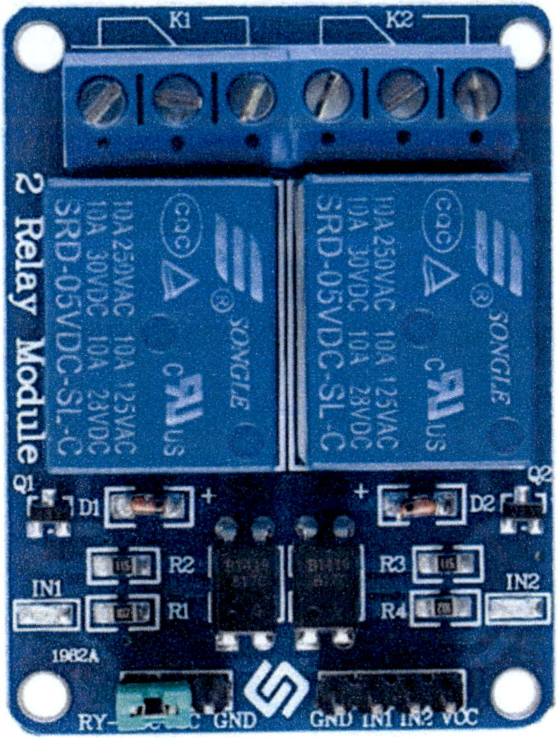

Fig. 5 Light-dependent resistor

3.5 Light-Dependent Resistor (LDR)

LDRs are made of materials with semiconductors that allow them to have their light-sensitive properties. Cadmium sulfide (CdS) is such a common material. LDRs or photoresistors are working on the "photoconductivity" concept. The theory of photoconductivity states that whenever light falls on the semiconductor surface (here LDR), the conductivity of the product increases or in other words, when the light falls on the LDR surface, the resistance of the LDR falls (Fig. 5).

3.6 Cooling Fan

For fast ventilation, a fan is used. It can relate to fans that extract cooler air from the outside environment, remove warm air from the inside or encourage air to flow through a heat sink to cool a specific area or entity. The cooling fans are used to maintain ambient temperature inside the greenhouse. The temperature sensor senses the temperature, then compares it with a preset desired temperature and accordingly switches the fan off and on in order to regulate the temperature (Fig. 6).

3.7 A.C. Water Pump

A pump is an electromechanical system that, by mechanical motion, transfers fluids (liquids or gasses). Generally, pumps are powered by electricity. An electric water pump uses electrical energy and converts it to mechanical energy by moving or displacing water. Pumps can be used to lift water from ground to a tank installed

Fig. 6 Cooling fan

Fig. 7 A.C. water pump

on terrace or even from underground water sources. Depending upon the source of water and our requirement, we can install any kind of pump that serves our purpose best (Fig. 7).

3.8 Heat Radiator

Radiators or convectors are devices that exchange heat. They are designed to transfer thermal energy from one medium to another medium for the purpose of heating a specific area.

According to our purpose and available sources of energy, we can use any kind of heating mechanism to regulate the temperature inside the greenhouse (Fig. 8 and Tables 1, 2 and 3).

Fig. 8 Heat radiator

4 System Architecture and Working

1. Block diagram.
2. Circuit.
3. Working.
4. Flowcharts.

   1. Block diagram (Fig. 9).
   2. Circuit (Fig. 10).
   3. Working.

   [1] To build this system, first we need an Arduino UNO board.
   [2] Then make the connections of different sensors required with Arduino UNO board. Such as:

   - LDR to $A0$.
   - Soil moisture to $A1$.
   - Temperature and humidity to $A2$.
   - Relay input 1 to $D5$.

Table 1 Pin description of Arduino UNO board

Pin category	Pin name	Details
Power	V_{in}, 3.3 V, 5 V, GND	V_{in}: input voltage to Arduino when using an external power source 5 V: regulated power supply used to power microcontroller and other components on the board 3.3 V: 3.3 V supply generated by on-board voltage regulator. Maximum current drawn is 50 mA GND: ground pins
Reset	Reset	Resets the microcontroller
Analog pins	*A*0–*A*5	Used to provide analog input in the range of 0–5 V
Input/output pins	Digital pins 0–13	Can be used as input or output pins
Serial	0 (Rx), 1 (Tx)	Used to receive and transmit TTL serial data
External interrupts	2, 3	To trigger an interrupt
PWM	3, 5, 6, 9, 11	Provides 8-bit PWM output
SPI	10 (SS), 11 (MOSI), 12 (MISO) and 13 (SCK)	Used for SPI communication
Inbuilt LED	13	To turn on the inbuilt LED
TWI	*A*4 (SDA), *A*5 (SCA)	Used for TWI communication
AREF	AREF	To provide reference voltage for input voltage

Table 2 Specification of Arduino UNO board

Parameter	Range
Microcontroller	ATmega328P-8 bit AVR family microcontroller
Operating voltage	5 V
Input voltage limits	6–20 V
Analog input pins	6 (*A*0–*A*5)
Digital I/O pins	14 (out of which 6 provide PWM output)
DC current on I/O pins	40 mA
DC current on 3.3 V pin	50 mA
Flash memory	32 KB (0.5 KB is used for bootloader)
SRAM	2 KB
EEPROM	1 KB
Frequency	16 MHz

Table 3 Components

SI. no.	Name	Function
1.	Arduino UNO	Provides a platform for development of different projects
2.	Soil moisture sensor	Detects soil moisture and sends digital/analog signals to the microcontroller
3.	Temperature sensor	Detects temperature and sends digital/analog signals to the microcontroller
4.	Humidity sensor	Detects humidity and sends digital/analog signals to the microcontroller
5.	Light sensor	Senses the amount of light present
6.	Relay module	To switch high-voltage appliances using low-voltage signal
7.	Cooling fan	Regulates the temperature
8.	A.C. water pump	To draw water from the source to the planted area
9.	Heat radiator	Maintains an ambient temperature

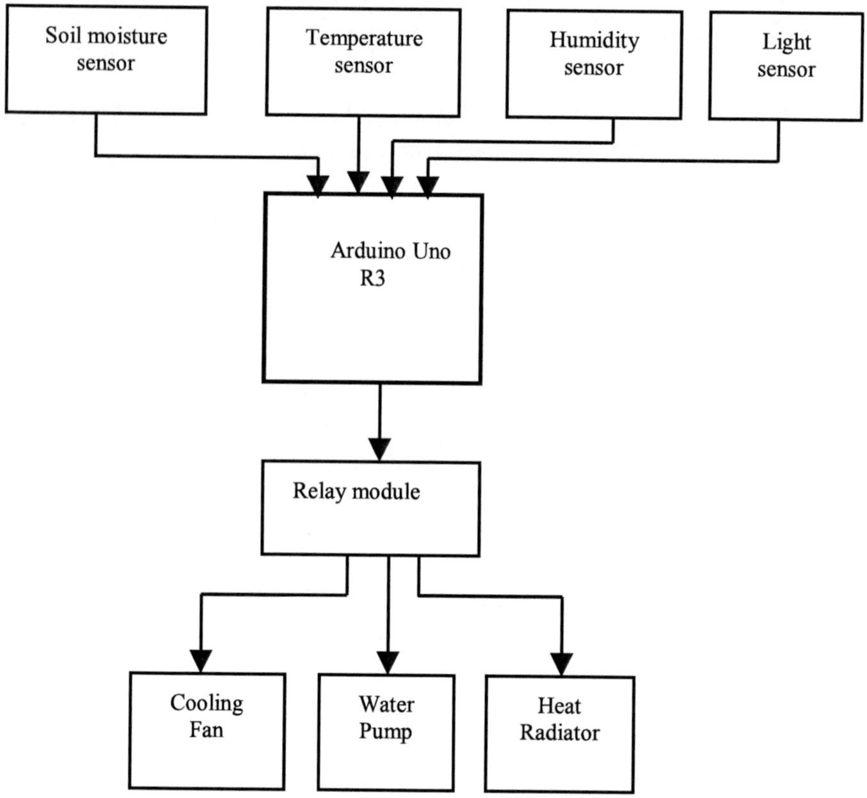

Fig. 9 Block diagram

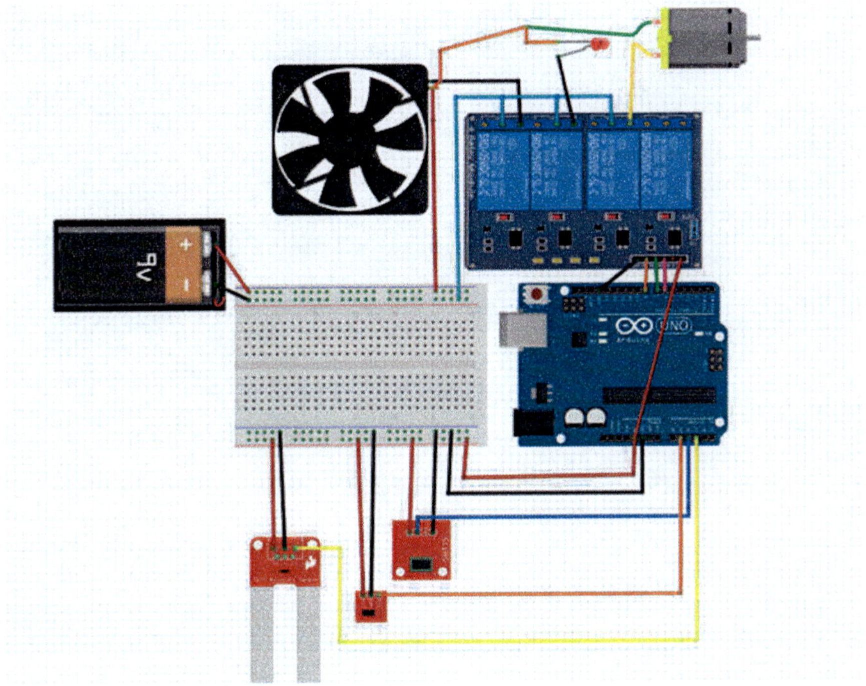

Fig. 10 Circuit diagram

- Relay input 2 to *D*6.
- Relay input 3 to *D*7.
- Relay output 1 to cooling fan.
- Relay output 2 to heat radiator.
- Relay output 3 to water pump or solenoid valve.
- LDR connected to analog pin *A*0 senses the light intensity and sends to Arduino UNO; thereafter, Arduino compares the value of data with standard set value and if the value crosses the set limit then it will activate the output connected to *D*5 pin, i.e., the light glows.
- Similarly, *A*1 senses the soil moisture and sends to the Arduino UNO. Arduino UNO compares it with the set value, and if the value crosses the limit, then it will activate the output connected to water pump to control the water level.
- Similarly, sensor such as temperature and humidity senses the temperature and humidity in the greenhouse and sends to Arduino UNO board. The board compares the value, and if the value crosses the limit, then it tries to balance the temperature and humidity by activating the exhaust fan.

Fig. 11 Uploading the program

[3] Then write the program according to the connections made, and then run the programs on Arduino using a PC.
[4] Supply power to Arduino with the help of a power source (Fig. 11).
[5] Our system is ready to work (Table 4).

Table 4 Data of humidity and temperature level

Sl. no	Fruits	Suitable Country	Required temperature	Required humidity
1.	Watermelon	China, Turkey, Iran	Higher than about 25 °C (77 °F) to thrive	Medium
2.	Kiwifruit	China, Italy, New Zealand	Average summer heat	Average
3.	Strawberry	USA, Turkey, Spain	15–27 °C	Low
4.	Raspberry	Russia, Poland, United States	Late summer and fall	High
5.	Orange	Brazil. United States, China, India	15.5 and 29 °C	Moderated

4. Flowchart.

Flowchart 1

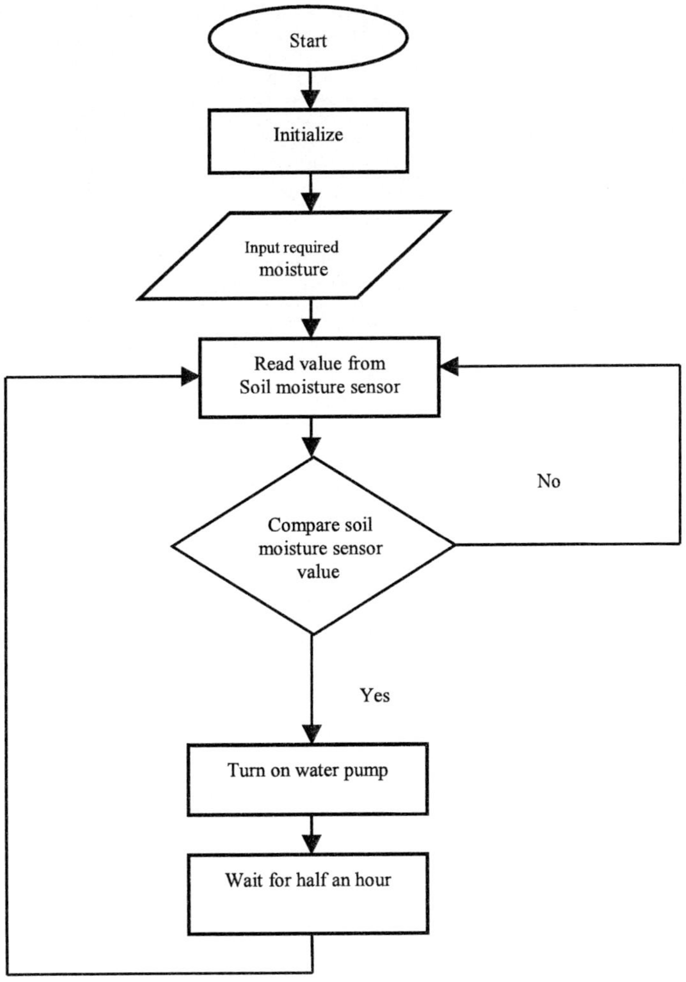

Flowchart 2

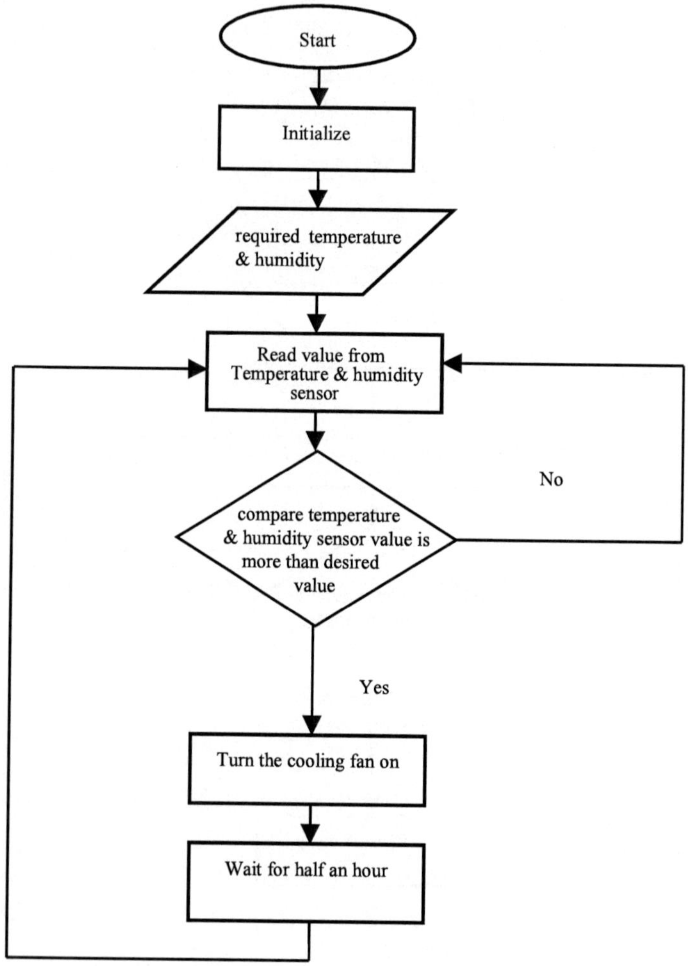
Start
Initialize
required temperature & humidity
Read value from Temperature & humidity sensor
No
compare temperature & humidity sensor value is more than desired value
Yes
Turn the cooling fan on
Wait for half an hour

Flowchart 3

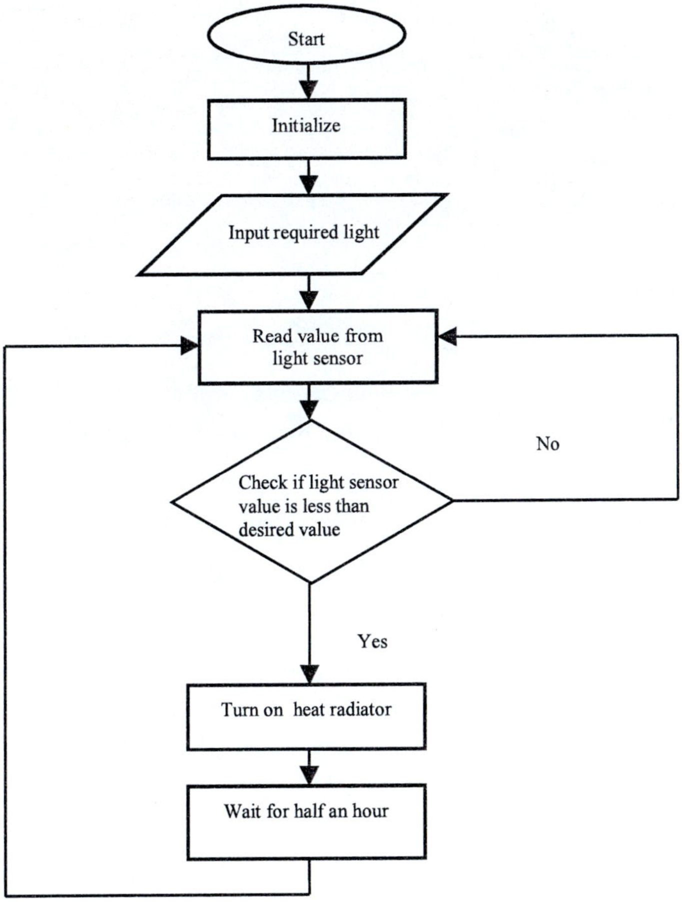

See Figs. 12 and 13.

5 Conclusion

The automatic monitoring system for the greenhouse comprises various sensors such as soil moisture, temperature and LDR light sensor. Sensors sense multiple restrictions such as soil moisture, temperature, humidity and light intensity, and data are direct to Arduino UNO and microcontroller control behavior to connect with current values. Automated greenhouse monitoring system eliminates the risk of greenhouse failure due to human error in specific environmental conditions. It is also possible to reduce labor costs. This program is most significantly cost-effective and environmentally friendly. The quantity and quality of yield also increase.

Fig. 12 Circuit

Fig. 13 Circuit

References

1. Fei W, Yan H, Qiu W, Dong L (2014) Design of intelligent greenhouse environment monitoring systembased on ZigBee and embedded technology. IEEE
2. Xu D; Li H (2008) Intelligent greenhouse control-system based on agent. In: 2008 international conference on intelligent computation technology and automation (ICICTA), vol 1, pp 390, 394, 20–22 Oct 2008
3. Mar´ıa Salom´e P´erez and Enrique V. Carrera Time synchronization in bluetooth-based sensornetworks. In: Proceedings of the 2014 IEEE central America and panama convention (CONCAPAN XXXIV)
4. Singh P, Saikia S Arduino-based smart irrigation using water flow sensor, soil moisture sensor, temperature sensor and ESP8266 WiFi module
5. Sripan M, Lin X, Petchlorlean P, Ketcham M (2012) Research and thinking of smart home technology. In: International conference on systems and electronic engineering (ICSEE'2012), Phuket (Thailand), 18–19 Dec 2012
6. Thengane VG, Gawande MB (2018) Arduino based tomato ripening stages monitoring system. Int J Innov Res Sci, Eng Technol 7(1)
7. Calder6n-C6rdova C, Bustamante B, Delgado J, Febres C, Montaño V, Saritama C, Ramírez C Wireless sensor network for real-time monitoring of temperature, humidity and illuminance in an orchid greenhouse
8. Parkhomenko A, Gladkova O, Ivanov E, Sokolyanskii A, Kurson S (2015) Development and application of remote laboratory for embedded systems design. Int J Online Eng 11(3):27–31
9. Adriansyah A, Dani AW Design of small smart home system based on Arduino. In: 2014 electrical power, electronics, communications, controls, and informatics seminar (EECCIS)
10. Rajalakshmi P, Mrs. Devi MS IOT based crop-field monitoring and irrigation automation
11. Vatari S, Bakshi A, Thakur T (2016) Green house by using IOT and cloud computing. In: IEEE international conference on recent trends in electronics information communication technology. India, 20–21 May 2016
12. Saha T, Jewel MKH, Mostakim MN, Bhuiyan NH, Ali MS, Rahman MK (2017) Construction and development of an automated greenhouse system using Arduino UNO. Int J Inf Eng Electron Bus 3: 1–8
13. Er. Poyen FB, Dr. Ghosh AK, Dr. Kundu P, Sen K (2017) Automatic shading of green house extension work. In: International conference on inventive systems and control (ICISC-2017)
14. Taras T, Pavlo D, Andriy K, Vasyl T (2015) Automated control system for Arduino and android based intelligent greenhouse. In: 2015 XI international conference on perspective technologies and methods in MEMS design (MEMSTECH)

An Involution Function-Based Symmetric Stream Cipher

Prashant Pranav, Sandip Dutta, and Soubhik Chakraborty

Abstract The use of cryptography meaning the art of secret writing for securing information shared over an open medium is an old methodology but its importance in the present scenario has increased tremendously. With the increase in the computing power of an eavesdropper, it has become unavoidable to look for a better and better cryptographic primitive that provides a level of security while keeping a check upon the time consumed to encrypt and decrypt the data. We have used an involution function to produce a symmetric stream cipher. Involution functions have the property that when functioned again they outputs the original variable. We have analyzed our work with respect to the execution time and compared it with the standard AES-128 algorithm. The result shows a much lesser time is required to encrypt data of same size with the proposed approach as compared to AES-128 and other block ciphers.

Keywords Cryptography · Symmetric key · Involution function · Algorithm analysis

1 Introduction

Cryptography is the art of writing something secretly; the first notable use of cryptography in writing dates back to circa 1900 B.C. when an Egyptian scribe used

P. Pranav (✉) · S. Dutta
Department of Computer Science and Engineering, Birla Institute of Technology, Mesra, Ranchi 835215, India
e-mail: prashantpranav19@gmail.com

S. Dutta
e-mail: sandipdutta@bitmesra.ac.in

S. Chakraborty
Department of Mathematics, Birla Institute of Technology, Mesra, Ranchi 835215, India
e-mail: soubhikc@yahoo.co.in

V. Nath and J. K. Mandal (eds.), *Proceedings of the Fourth International Conference on Microelectronics, Computing and Communication Systems*, Lecture Notes in Electrical Engineering 673, https://doi.org/10.1007/978-981-15-5546-6_5

non-standard hieroglyphs in an inscription. In data and telecommunications, the use of cryptographic algorithms is of utmost importance when communicating over any untrusted medium, such as Internet. The primary functions or objectives achieved by cryptographic algorithms are:

Privacy/Confidentiality It implies no one except the intended receiver can read the message. The sole communication must be done between the two involved communicating parties and not any eavesdropper.

Authentication It prevents the attacker from impersonating as a communicating party. It is the process of proving one's identity.

Integrity It assures the receiver that the delivered message has not been altered from the original on its way to the receiver. It prevents any third party from tampering the message without being noticed.

Non-repudiation It justifies the receiver that the sender has really sent the message and not anybody else.

Key Exchange The mechanism by which crypto keys are shared between the senders and the receivers. It is basically important for asymmetric key cryptography.

In cryptography, we start with a plain text which is in unencrypted format. The plaintext through some encryption technique is converted into cipher text, which is then again decrypted back into usable plain text. Mathematically, the process is written as:

$$C = E_k(P)$$

$$P = D_k(C)$$

where P = plain text, C = cipher text, E = the encryption protocol, D = the decryption protocol, k = key.

Cryptographic algorithms categorized on the basis of number of keys employed for both the cryptographic operations are discussed below.

1.1 Symmetric Key Cryptography or Secret Key Cryptography

In symmetric key or secret key cryptography, both the sender and receiver share the same or common key for encryption and decryption of message (Fig. 1).

There are two types of symmetric key cryptography:

Stream Cipher Stream cipher is an encryption algorithm that encrypts the plaintext (1 bit or byte) at a time.

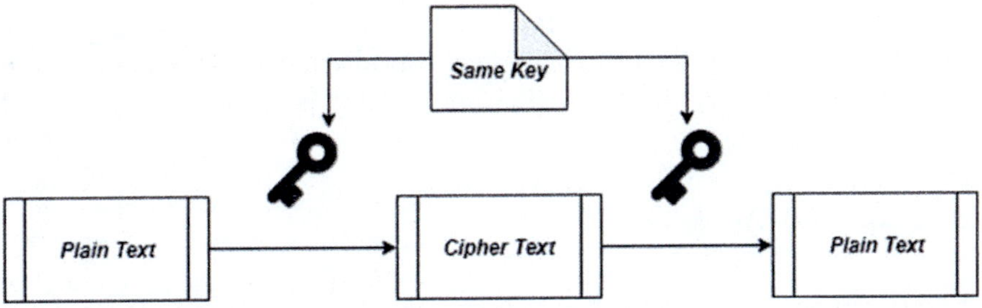

Fig. 1 Basic model of symmetric key cryptography

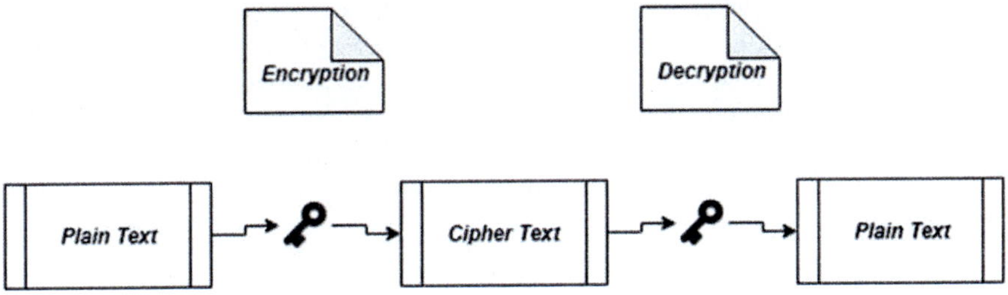

Fig. 2 Basic model of asymmetric key cryptography

Block Cipher Block cipher is an encryption algorithm that takes number of bits and encrypts them as a single unit, padding the plaintext so that it is a multiple of block size. The usual sizes of blocks are 64 bits, 128 bits, and 256 bits.

1.2 Asymmetric Key Cryptography or Public Key Cryptography

In asymmetric key or public key cryptography, two different key is used for encryption and decryption of message. Public key is used for encryption process and private key is used for decryption process (Fig. 2).

1.3 Hash Functions

The hash function is an algorithm that takes the input (message) of an arbitrary length and gives the output of fixed length of alphanumeric string. The alphanumeric string returned by the hash function is known as message digest. Hash function does not have any key that is why plaintext is not retransformed from cipher text.

In this paper, we have proposed the construction of a symmetric key stream cipher based on an involution function. An involution function is one whose inverse is itself, i.e., if *f(x)* is some function over a variable x, then $f(f(x)) = x$.

2 Proposed Work

The use of symmetric stream cipher is advantageous in many ways over the traditional block ciphers. The time consumed to encrypt and decrypt back a chunk of data using the stream-based technique is less as compared to block-based technique. As block ciphers works on a larger block of data at a time, they require more memory as compared to one bit/byte at a time of stream ciphers. Precisely, the space complexity of a steam cipher is constant. So, they are cheaper to implement in scenarios such as embedded devices and hardware.

We have used an involution function [1–4] to encrypt the data streams. Involution functions, as illustrated above, have the unique property of giving back the original when functioned again.

The proposed involution function is given below:

$$x = f(x) = \left(a - x^3\right)^{1/3}$$

When we compute $f(f(x))$, we get:

$$f(f(x)) = \left(a - \left(\left(a - x^3\right)^{1/3}\right)^3\right)^{1/3} = x$$

Now, fitting it into cryptographic aspect, we set

$a = K$, the shared secret key which can be exchanged through a key exchange algorithm such as Diffie–Hellman [5].

$$x = \text{Plaint text} = \text{PT}$$

$$x = \text{Plaint text} = \text{PT}$$

$$f(x) = \text{Cipher text} = \text{CT}$$

So, the equation for encryption of a plain text using the involution function and employing the Diffie–Hellman key exchange algorithm is:

$$\text{CT} = \left(K - \text{PT}^3\right)^{1/3}$$

And the corresponding decryption routine is:

$$\mathrm{PT} = \left(K - \left(\left(K - \mathrm{PT}^3\right)^{1/3}\right)^3\right)^{1/3}$$
$$= \left(K - \left((\mathrm{CT})^{1/3}\right)^3\right)^{1/3}$$

We have evaluated the discussed approach on MATLAB 2015a for examining the methodology on different parameters as compared to the existing standard AES-128. These are discussed in the below section.

3 Result and Discussion

We analyzed our work based on the execution time for varying input size data. We took three trials for each input length in order to diminish the execution time due to a cache hit. The results were compared to that of standard AES-128 [6, 7]. Execution time for our proposed involution function-based encryption and of AES-128 is shown in Tables 1 and 2, respectively.

From Tables 1 and 2 and the graphs in Figs. 3 and 4, it is clear that the proposed work takes much lesser time as compared to block ciphers. The execution time of AES-128 and hence other block ciphers (since all encrypts one block at a time) is

Table 1 Execution time: involution function-based symmetric stream cipher (encryption only)

Involution function-based				
Input (in bytes)	Trial 1	Trial 2	Trial 3	Execution time (mean)
16	0.002133	0.001410	0.001897	0.001813
32	0.002396	0.002630	0.002572	0.002533
48	0.002993	0.002236	0.003604	0.002944
64	0.003438	0.003510	0.003120	0.003356
80	0.004713	0.004655	0.004013	0.00446
96	0.004309	0.005403	0.006212	0.005308
112	0.005041	0.005973	0.006438	0.005817
128	0.005332	0.007261	0.008225	0.006939
144	0.005463	0.007669	0.005747	0.006293
160	0.009954	0.009345	0.008372	0.009224
176	0.011407	0.007338	0.011681	0.010142
192	0.011514	0.009660	0.009584	0.010253
208	0.013006	0.009621	0.010487	0.011038
224	0.010756	0.008492	0.013395	0.010881
240	0.019665	0.011644	0.012612	0.01464
256	0.014633	0.012793	0.012793	0.013406

Table 2 Execution time: AES-128 (encryption only)

AES-128				
Input (in bytes)	Trial 1	Trial 2	Trial 3	Execution time (mean)
16	0.048	0.0266	0.0222	0.032267
32	0.0739	0.059	0.0542	0.062367
48	0.0995	0.0843	0.0842	0.089333
64	0.1508	0.1271	0.1141	0.130667
80	0.1442	0.1464	0.1359	0.142167
96	0.1836	0.1605	0.1768	0.173633
112	0.2342	0.1896	0.2047	0.2095
128	0.243	0.2396	0.2663	0.249633
144	0.2516	0.2778	0.291	0.273467
160	0.2779	0.3353	0.2939	0.302367
176	0.3346	0.3306	0.3232	0.329467
192	0.3368	0.4291	0.359	0.374967
208	0.4466	0.4242	0.3949	0.4219
224	0.4473	0.3783	0.5521	0.459233
240	0.4734	0.5632	0.4454	0.494
256	0.5519	0.479	0.5316	0.520833

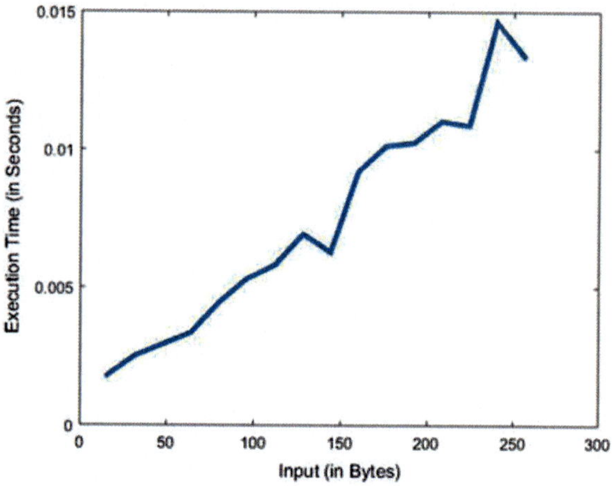

Fig. 3 Input versus execution time of the involution function-based symmetric stream cipher

in the worst case O(*n*), but the execution time of the proposed involution function based stream cipher do not show and pattern as can be seen from the graph in Fig. 3, and hence can be termed to be run in a constant time i.e. O(1).

The memory requirement of the proposed work is also less as compared to AES and other block ciphers. Our algorithm works on encrypting/decrypting one bit/byte at a time of data so requires much lesser memory as compared to AES-128 which

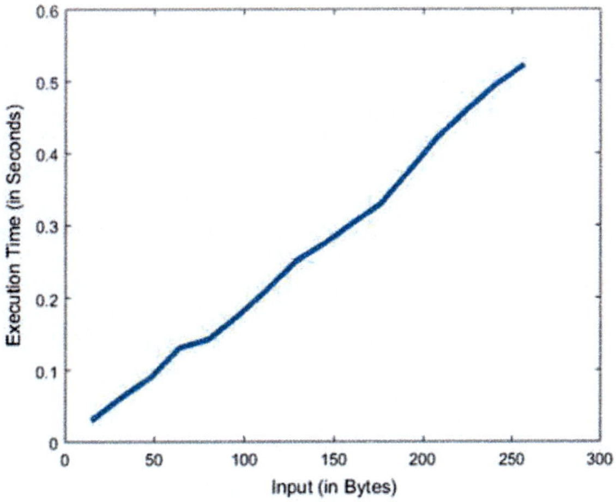

Fig. 4 Input versus execution time of AES-128

in turn works on a block of data at a time. This fact further increases the robustness of the algorithm as AES-128 and other block cipher are more prone to noise in the transmission, i.e., if one part in the data is messed up somehow, than it becomes practically impossible to recover the rest of the data. The proposed work and other symmetric stream ciphers encrypt bytes individually with no connection whatsoever with rest of the data and hence are less susceptible to noise in the transmission. The proposed work complies well with two goals of a cryptographic algorithm, viz. authentication and confidentiality.

4 Conclusion and Further Scopes

We have proposed an involution function-based symmetric stream cipher which provides security to confidential information with a very basic function whose inverse is the variable itself. The use of Diffie–Hellman key exchange algorithm further increases the versatility of the work by providing authentication. The cipher is much faster than any other block ciphers and takes much less memory. The disturbance during the transmission due to the noise is also minimum and recoverable as the process encrypts one bit/byte at a time of the data.

Because of the low memory usage of the cipher, its deployment in low storage devices such as embedded devices and mobile phones cannot be denied.

The work can be extended to be deployed in low storage devices with the use of a lightweight key exchange primitive and its correctness in terms of gate equivalent as indicated for a lightweight cipher [8] can be verified. Since involution functions are symmetric in nature, the work can be looked upon for exhibiting homomorphic encryption [9, 10] properties.

References

1. Wiener J, Watkins W (1988) A classroom approach to involutions. CollE Math J 19:247–250
2. Wiener J (1969) Differential equations with involutions. Differ Equs 5:1131–1137
3. John SR A course on group theory, p 10, Section 1.13
4. Ell TA, Sangwine SJ Quaternion involutions and anti involutions. Comput Math Appl 53(1): 137–143
5. Diffie W, Hellman M New directions in cryptography. IEEE Trans Inf Theory 22(6): 644–654
6. Announcing the ADVANCED ENCRYPTION STANDARD (AES), Federal Information Processing Standards Publication 197. United States National Institute of Standards and Technology (NIST). November 26, 2001. Retrieved October 2, 2012
7. Daemen J, Rijmen V (2003) AES proposal: Rijndael
8. McKay KA, Bassham L, Sönmez Turan M, Mouha N (2017) Report on lightweight cryptography. NISTIR 8114
9. Rivest R (2002) Lecture Note 15: Voting homomorphic encryption. Comput Netw Secur
10. Gentry C (2009) Fully homomorphic encryption using ideal lattices. In: The 41st ACM symposium on theory of computing (STOC)

Multi-objective Optimization of Block Size Based on CPU Power and Network Bandwidth for Blockchain Applications

Nikita Singh and Manu Vardhan

Abstract Distributed ledger technology (DLT) is fast growing as a solution that could address pertinacious challenges in the financial sector offering a more efficient and resilient approach to transaction record keeping. Its popularity is principally determined by its potential to transmit data in bulk securely over a peer-to-peer network. Block size optimization is an important issue in blockchain network as scalability bottlenecks prevent higher throughput and minimized latencies. Increasing adoption has raised concerns about its ability to scale and serve as a real-world usable network. An increased block size may cause a higher transmission time as the rate of transactions made will increase and may also cause the system to reach its maximum capacity to clear transactions. As compared to larger block size, small block size is more efficient, but creating too small a block might cause higher block building/creation time. An efficient blockchain-based application requires an optimal block size so that a justified performance is achieved. In the proposed approach, multi-objective optimization is performed, and 40 different solutions are obtained based on transaction selection time and block building time. The selection of a particular block size as an optimal solution is based purely on processing power of CPU, throughput and the available network bandwidth and latencies, where the application is to be deployed. The results establish that for various systems having 1.1 GHz CPU–3.0 GHz CPU, a 3.8 MB of block size selection optimizes the transaction selection and block building time.

Keywords P2P network · Distributed ledger technology · Network bandwidth · Multi-objective optimization · Blockchain · Consensus · IoT

N. Singh (✉) · M. Vardhan
Department of Computer Science and Engineering, National Institute of Technology, Raipur, India
e-mail: nsingh.cs@nitrr.ac.in

M. Vardhan
e-mail: mvardhan.cs@nitrr.ac.in

V. Nath and J. K. Mandal (eds.), *Proceedings of the Fourth International Conference on Microelectronics, Computing and Communication Systems*, Lecture Notes in Electrical Engineering 673, https://doi.org/10.1007/978-981-15-5546-6_6

1 Introduction

Distributed ledger technology (DLT) in a broader sense encompasses finance, banking, economics, cryptography and distributed technology. This technology has brought about a revolution and also leveraged its value in verticals such as finance, banking and other applications that were until now plagued with centralized control coupled with the threat of data highjack. The authors [1] observe that the immutable nature of blockchain makes it useful to various functionaries of the government. They further add that DLT has attracted significant investment by various banking giants and financial institutions who have deployed dedicated research and development teams to explore what this technology represents and its relevance to various financial institutions who have delved into joint ventures to propose industry standards. They also state the report published by the World Economic Forum in 2016 mentioning investments over $1.4 billion "to explore and implement application of DLT in the financial services industry."

Visa and other modern transaction management systems process transactions with the speed of 2400 transactions per second. It takes approximately 10 min for bitcoin to confirm a transaction providing user with a throughput of seven transactions per second [2]. The evident disparity between efficiency and throughput creates dilemma in minds of researchers who advocate the potential use of blockchain as a transaction processing system [3].

Scores of researchers are actively studying bitcoin transactions in order to estimate the size of bitcoin's transactional data. Studies have been conducted to study the rate at which the bitcoin network can confirm transactions and analyze some key metrics that have implications on the maximum size of the bitcoin block. The concern here is to arrive at an ideal block size. Blocks in bitcoin blockchain network have already reached their hard-coded limit of 1 MB which causes an increasing strain on the network and negatively influences the transaction confirmation time. The influence becomes more pronounced as the transaction throughput approaches the maximum. At certain times, an exponential growth in terms of number of transactions being processed has also been observed. Several users have complained about the blocks being filled up to 95% of the allowed limit (1 MB) resulting in network congestion [4]. There exist some systems such as Monero that allow dynamic block sizes; the impact of such dynamism is still being researched.

It is important to note that an ideal block size is one that is based on network transactional demand that avoids congestion as much as possible. Transactions that take place over the blockchain-based systems or the applications that run on DLT are placed in transaction pool sometimes referred to as the memory pool. Memory pool refers to the holding area or the aggregate size of the transactions waiting to be confirmed. When a transaction is transmitted to the network, it first gets verified by all the available nodes. After it successfully passes the verification, it goes inside the memory pool and waits until a miner picks it up to include it the next block. A certain number of these transactions are picked up to be stored into a block [5]. The transactions in this block are verified by the miner. This is only after the miner has solved the puzzle or the target value, thus entitling him to earn incentive in lieu of the verification of transactions. The challenge is to find a hash sum with the desired properties. Once transactions are verified, the miner signs the block and propagates it to the network. This is where impact of size of block gains importance.

The increase in block size becomes necessary especially if it aspires to become an alternative to the existing mainstream payment systems which have the capacity to handle thousands of transactions per second. Factors such as available computational power, network bandwidth, frequency of transaction generation and transaction validation time are all factors of the block size. Hence, the importance of this parameter in dictating transactional throughput motivates us to propose an optimized block size based on the above mentioned factors.

2 Related Work

In this section, we organize our discussion around the decomposition of the blockchain network by taking into consideration the hierarchy of layers present. First, we discuss DLT' potential for vast and pervasive changes in finance, health, supply chain, mobility and even government infrastructure. Later, in the section, we discuss the need for a different block size when the application to be deployed makes use of IoT devices that operate in a constrained environment. Finally, we delve into the relationship between block size, latencies, block intervals and the scalability bottlenecks they generate.

Guo and Lang [6] in their paper describe how blockchain technology is the combination of several other existing computer technologies, namely distributed data storage, peer-to-peer systems, distributed consensus mechanism and encryption algorithms. The paper discusses how major banking giants, such as Morgan Stanley and Goldman Sachs, have started to formulate transaction settlement system based on the blockchain technology. They discuss how the blockchain technology can revolutionize the existing banking industry by focusing on easy cross-border payments and point-to-point funds transfer.

Mathew and Abdul [7] discuss the impact of the collaboration between the Swiss firm WISeKey and the state government of Andhra Pradesh. The state of Andhra Pradesh, which is the seventh largest state of India, has announced a pilot project in collaboration with the cybersecurity firm WISeKey to securely store its state citizen data on a blockchain-based platform.

Dorri et al. [8, 9] recognize that Internet of Things encompasses numerous devices that gather, process and transmit huge amounts of sensitive data whose security is of utmost importance. The existing technologies to overcome these issues are centralized in nature and are also not well suited for these constrained devices in terms of energy consumption and processing overhead. The authors claim that these IoT devices demand decentralized, lightweight and distributed security mechanism which can be met by the use of blockchain technology.

Yuan and Wang [10] observe that a fork in bitcoin resulted in bitcoin cash. This also resulted in the increase in the block size of bitcoin cash from 1 to 8 MB. This permits more number of transactions to be placed into a single block during consensus competition. This has the benefit of improved transaction processing time and reduced time in transaction confirmation. They further express that the consensus mechanism ensures verification of transactions in transaction pool, and later, creation of a block in public blockchain is driven by consensus competitions. They further call this process as a crowd-sourcing task that involves miners and other participating nodes to contribute their computing power and farms. The monetary incentive gained calls for design of incentive compatible mechanisms so as to ensure revenue maximization aligned with the crucial target of guaranteeing a secured and trusted digital ledger technology-based system. These miners and other participating nodes verify transactions and create a new block. Certain amount of cryptocurrency or incentive is rewarded to the winning node. This ensures that entire network continually contributes to their efforts in transaction/data verification and block creation. Thus, the miners with huge computing power stand to gain as compared to weaker ones.

Eyal et al. [11] in their paper mention how the rate of transaction processing in a blockchain is capped by two parameters: the block size and the block interval. They outline a scalable bitcoin NG protocol which provides a significantly better throughput and minimized latencies in the network. In summary, their paper makes the following two contributions: propose a bitcoin NG protocol to show that blockchain scalability is possible and then propose some novel metrics to quantify the security and efficiency of the blockchain system.

Goswami [3] evaluates the scalability bottleneck of the present blockchain and defines parameters that directly or indirectly influence the block size and the throughput generated by the system. They also weigh the pros and cons of increasing and decreasing the block size. There are several proposals to increase the block size which is causing a heated debate in the bitcoin community. Croman et al. [12] suggest reparameterization of block size to be seen as a incremental and a major advancement in an approach to build the next generating of blockchains that are not only scalable but could also handle higher loads. They also highlight and enumerate various blockchain protocols ideas put forward by the research community along with the open challenges and the opportunities.

3 Motivation

The motivation for this work has been drawn from the scalability bottlenecks of the bitcoin blockchain. Bitcoin blockchain has reached their hard-coded limit of 1 MB. For any technology to be utilized as a real-world use case, it needs to compete with the already established mainstream technologies or payment processors that provide higher transaction rate, require lower bandwidth and as well require a lower transaction confirmation and block building time (Fig. 1).

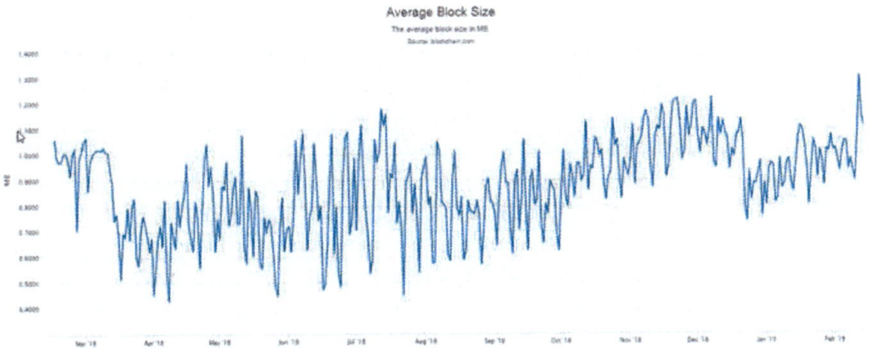

Fig. 1 Chart showing average block size in MB over a period of 1 year [19]

If a blockchain-based application is to positions itself for mainstream adoption with the real-world usage, it should be capable of handling higher throughput of transaction [11]. The number of transactions processed by the bitcoin blockchain is dwarfed as compared to the throughput of visa and other payment processing systems. Thus, the size of the block is of prime importance if the blockchain has to stay as a mainstream technology that could help build other applications on top of it.

Singh and Vardhan [13] propose a distributed and decentralized architecture that carries out analysis of the impact of query response time for digital ledger technology-based e-stamp and property registration system. This paper proposes fast parallel search algorithm in order to optimize the search mechanism. The proposed search algorithm searches the blocks by generating parallel threads for comparing the transactions that reduce the overall search time to 1/5 times as compared to the existing algorithms. This is done on a subset of blockchain against complete blockchain.

Singh and Vardhan [5] further propose a distributed and decentralized architecture that carries out analysis of the impact of query response time for different block sizes of blockchain. The authors observe that when the block size is increased by about 10 times, the time required to process and respond to a user query is reduced by 65–70%. They further add that if huge number of transactions takes place in a given time, the time taken for verification of transactions by the miners shall increase. Hence, the block size should be kept small so as to increase the visibility of the transactions.

So, the natural question is to understand the relationship between the CPU power, the query response time and the impact of block size so as to obtain optimal performance of the application under study. This is in fact optimization problems with multiple objectives that call for multi-objective optimization algorithms. These algorithms are significantly different from single-objective optimization algorithms.

Multi-objective optimization problem (MOOP) differs from single-objective optimization problem (SOOP) in the number of objective functions and the way solutions are compared [14]. SOOP typically has a single optimal solution in the solution space, and solutions are compared based on their fitness value. On the other hand, MOOPs have a set of optimal solutions which are known as Pareto front solutions

[14, 15]. This set of solutions cannot be dominated by other solutions in the search space and hence termed as non-dominated set of solutions. Generally, a Pareto front corresponds to a curve or an extremely complex hypersurface. MOOPs have many challenges like time complexity and dimensionality.

Generally, two major categories of approaches exist for solving the MOOPs, namely priori and posteriori approach [16]. The priori approaches solve MOOP by assigning weights of importance to the objective functions and finally reducing the approach into an SOOP. The weighted sum method, ϵ-constraint method, weighted metric method, weighted aggregation approach, value function method, Benson's method and goal programming methods fall in this category of classical priori methods [15]. The posteriori approach does not require any weight assignment. These approaches provide a Pareto front solution for MOOPs during simulation. A user can select any one optimal solution out of the set of solutions (Pareto front) based on their requirement and importance of objectives [17]. Hence, posteriori approaches are considered to be efficient between the block size and the incentive earned by the miners. The final objective to be achieved by the above discussion is adoption of DLT for constrained devices too so as to incrrease its wide spread applicability [20].

4 Implementation and Results

The motivation for this work has been drawn from the multi-objective optimization problem (MOOP) that handles two or more than two objectives. In this optimization problem, we have two minimization objectives: the first objective is to minimize transaction selection time of the block and the second objective is to minimize the block building time. Both objectives are based on the size block. Our overall objective is to find the block size for that transaction selection time, and block building time should be minimized. Higher the CPU power, transaction selection time shall be lower for a given block size. If block size is larger, the block building time is reduced. Both objectives are contradictory as seen by block size from the two equations. In the former equation, transaction selection time is directly proportional to block size, and in the later equation, block building time is inversely proportional to block size. Due to contradictory objectives, it has a set of solutions which is named as Pareto optimal solutions.

The formulation of the objective functions is:

$$\text{Transaction Selection Time (Block Size)} = \sum_{i=1}^{n} \frac{\text{Block Size}_i}{\text{CPU Power}_i}$$

$$\text{Block Building Time(Block Size)} = \frac{\text{Total Transaction}}{\text{Block Size}} \times \text{Per Block Overhead}$$

The problem has three input parameters—CPU power of miners, total number of transactions and per block building overhead. The size of the block varies between

Table 1 CPU power

M1	M2	M3	M4	M5	M6	M7	M8	M9	M10
1.1	2.3	2.7	1.8	1.5	2.1	1.9	1.4	3.0	2.6

Table 2 Other problem input parameters

Total number of transaction in pool of M1	10,000
Per block building overhead of M1	8
Number of miners (*n*)	10
Block size range	[0.5, 5]

[0.1, 5 MB]. The numerical value of CPU power of each miner is given in Table 1, and the numerical value of other input parameters is given in Table 2.

MOPSO is a widely used and well-established multi-objective version of PSO that is based on the Pareto envelope and grid-making technique [16]. In order to find a personal best particle/solution, it uses domination and probabilistic rules. Here, all the non-dominated particles are stored into a sub-swarm which is named as repository, and all the particles choose their global best value out of the particles that reside in the repository. The particles improve their position and velocity in a coordinated manner toward the food source [17].

Since we aim to find an optimal block size based on the conflicting objectives, the solution should be such that we get set of optimal block sizes which cannot be further improved. This is the principal behind non-dominated solutions as proposed by MOPSO.

The execution of MOPSO requires two types of parameter setting—common control parameters and algorithm-specific parameters. The common control parameters are listed in Table 3 and algorithm-specific parameters are listed in Table 4.

The obtained results are summarized in Table 5 after executing MOPSO in terms of minimum, maximum, range, standard deviation and mean. All fifty solution of Pareto front is illustrated in Fig. 2. The detailed graphical analysis of obtained solution is illustrated in Fig. 3.

Table 3 Common control parameters of MOPSO

Total number of transaction in pool of M1	10,000
Per block building overhead of M1	8
Number of miners (*n*)	10
Block size range	[0.5, 5]

Table 4 Algorithm-specific parameters of MOPSO

Total number of transaction in pool of M1	10,000
Per block building overhead of M1	8
Number of miners (*n*)	10
Block size range	[0.5, 5]

Table 5 Summary of obtained results

Objective 1. Transaction selection time	Objective 2. Block building time
Min = 2.6856	Min = 16,000
Max = 26.856	Max = 160,000
Range = 24.1704	Range = 144,000
St.D. = 8.237	St.D. = 47,380.43121
Mean = 11.5555	Mean = 64,848.2389

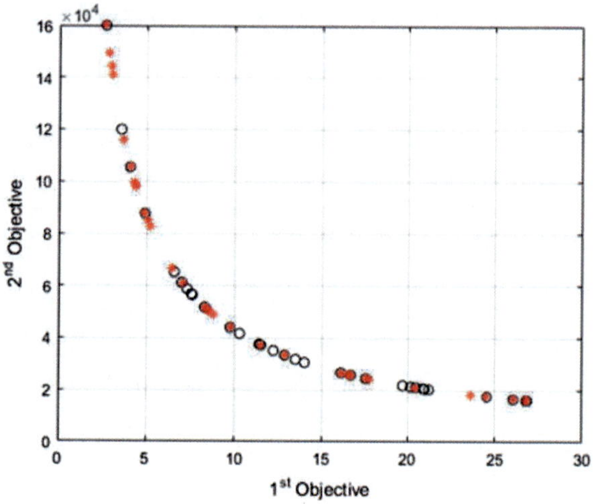

Fig. 2 llustration of obtained Pareto front by MOPSO

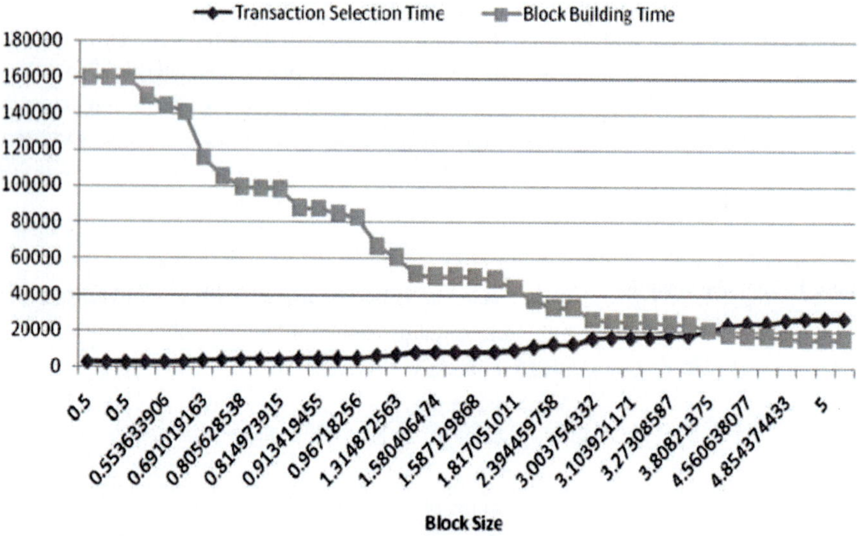

Fig. 3 Illustration of solutions obtained by MOPSO

MOPSO code is executed on MATLAB 2016a in a personal system with Windows 8.1 (64 bit) Intel(R) Core(TM) i7 CPU@3.60 GHz with 4 GB RAM. The code of MOPSO is freely available in open-source web page [18].

The obtained Pareto front shows the tradeoff between the objective 1 (transaction selection time) and objective 2 (block building time). It represents that if objective value of first function improves, then objective value of next function gets degraded.

Figure 3 illustrates all 40 blocks (obtained solutions) sizes and their respective transaction selection time and block building time. For example, if size of block is 0.5 MB, then value of transaction selection time is 2.68 s and block building time is 160,000 ms. The selection of a block size as a solution is purely based on the information available about the application deployment scenario. In this illustration, we have scaled the transaction selection time by 1000 for proper graphical illustration. Units

5 Conclusion and Results

Distributed ledger technology (DLT) has much potential in finance and banking sector. For mainstream adoption, it is imperative that block processing speed should match with the existing payment processing system. Ideal block size is the one that is based on network transactional demand that avoids congestion as much as possible. Factors such as CPU computation power of the miner, the available network bandwidth over which the DLT-based application is proposed to be deployed and frequency of transaction generation are all factors of the block size of the blockchain. The proposed approach finds optimal block size using multi-objective particle swarm optimization based on the deployment parameters such as CPU power of miners, total number of transactions and per block building overhead. The results establish that 3.8 MB of block size optimizes the transaction selection time and block building time. Thus, when IoT devices are proposed to be included in any blockchain-based application, the proposed approach can be efficiently used to arrive at an optimal block size as to ensure efficient working of the blockchain application.

References

1. Karan S, Singh N, Kushwaha DS (2018) An interoperable and secure E-wallet architecture based on digital ledger technology using blockchain. In: 2018 international conference on computing, power and communication technologies, India, pp 168–172.
2. How a Visa transaction works (2015) http://web.archive.org/web/20160121231718/http://apps.usa.visa.com/merchants/become-a-merchant/how-a-visa-transaction-works.jsp
3. Goswami S (2017) Scalability analysis of blockchains through blockchain simulation. UNLV Theses, dissertations, Professional Papers, and Capstones. 2976
4. https://www.coindesk.com/charts-determining-ideal-block-size-bitcoin. Accessed on 15.11.2018

5. Singh N, Vardhan M (2018) Digital ledger technology based real estate transaction mechanism & its block size assessment. Int J Blockchains Cryptocurrencies 1(1).
6. Guo Y, Liang C (2016) Blockchain application and outlook in the banking industry. Financ Innov 2(1):24
7. Mathew SA, Md AQ (2018) Evaluation of blockchain in capital market use-cases. Int J Web Portals 10(1):54–76
8. Drri A, Kanhere SS, Jurdak R, Gauravaram P (2017) Blockchain for IoT security and privacy: the case study of a smart home. In: 2017 IEEE international conference on pervasive computing and communications workshops (PerCom Workshops), pp 618–623
9. Eyal I, Gencer AE, Sirer EG, van Renesse R (2015) Bitcoin-NG: a scalable blockchain protocol. Technical report, CoRR
10. Yuan Y, Wang FY (2018) Blockchain and cryptocurrencies: model, techniques, and applications. IEEE Trans Syst Man, Cybern Syst 48(9): 1421–1428
11. https://bitfury.com/content/downloads/block-size-1.1.1.pdf
12. Croman K, Decker C, Eyal I, Gencer AE, Juels A, Kosba A, Miller A, Saxena P, Shi E, Gün E (2016) On scaling decentralized blockchains. In: Proceedings of 3rd workshop on bitcoin and blockchain research
13. Singh N, Vardhan M (2018) Blockchain based E-stamp procurement system with efficient consensus mechanism and fast parallel search. J Mech Contin Math Sci 13(4)
14. Deb K (2014) Multi-objective optimization. In: Search methodologies, Springer, pp 403–449
15. Deb K (2005) Multi-objective optimization using evolutionary algorithm. Wiley
16. Rao RV, Rai DP, Ramkumar J, Balic J (2016) A new multi-objective Jaya algorithm for optimization of modern machining processes. Adv Prod Eng Manag 11(4): 271
17. Fister Jr I, Yang XS, Fister I, Brest J, Fister D (2004) A brief review of nature-inspired algorithms for optimization. arXiv Prepr. arXiv1307.4186, 2013.Coello, In: Coello CAC, Pulido GT, Lechuga MS (eds) Handling multiple objectives with particle swarm optimization. IEEE Transactions on evolutionary computation, vol 8, Issue no 3, pp 256–279
18. Multi objective optimization [online] http://yarpiz.com/category/multiobjective-optimization/, Accessed on 10 Oct, 2018. Interface. IEEE Transl J Magn Japan 2: 740–741, August 1987 [Digests 9th Annual Conf. Magnetics Japan, p 301, 1982]
19. https://www.blockchain.com/charts/avg-block-size
20. Singh N, Vardhan M (2019) Distributed ledger technology based property transaction system with support for IoT devices. Int J Cloud Appl Comput (IJCAC) 9(2):60–78

Alamouti Code Generator in Optical Domain Using Mach–Zehnder Interferometer

Rajiv Kumar, Poonam Singh, and Niranjan Kumar

Abstract The paper describes the generation of Alamouti code in the optical domain. The proposed scheme can reduce the dimension of the device to a considerable amount. Code is implemented in optical domain and shows excellent resemblance with the original scheme. The scheme is useful in optical wireless communication system.

Keywords Alamouti code · Optical communication · Mach–Zehnder interferometer

1 Introduction

It is interesting to implement the optical digital devices using the Mach–Zehnder interferometer (MZI) structure. Desired amount of phase shift is produced by applying voltage across the arms of MZI, and excellent switching is observed across the arms. The MZI is implemented using the channel waveguides fabricated on the substrate of lithium niobate with two input and output ports which functions as a switch [1, 2]. MZI has been utilized to implement various modulation schemes in the optical domain [3–5]. The switching phenomena of MZI have also been utilized to implement various multiplexing schemes [6, 7]. A 1×2 digital optical switch is designed [8].Channel estimation is necessary for rapidly varying wireless channels [9]. The modern communication system involves the continuous effort to improve the efficiency of transmission and reception of signal. Alamouti code generator in

R. Kumar (✉)
Electronics & Communication Engineering, IIIT Ranchi, Ranchi, India
e-mail: rajivbit10@gail.com

P. Singh
Electronics & Communication Engineering, NIT Rourkela, Rourkela, India

N. Kumar
Electrical Engineering, NIT Jamshedpur, Jamshedpur, India

V. Nath and J. K. Mandal (eds.), *Proceedings of the Fourth International Conference on Microelectronics, Computing and Communication Systems*, Lecture Notes in Electrical Engineering 673, https://doi.org/10.1007/978-981-15-5546-6_7

the optical domain is discussed. The proposed methodology involves the concept of EO effect-based MZI structure. A simple diversity scheme with less complexity compared to maximum-ratio receiver combining is proposed [10]. It finds application in free-space optical communication [11]. An architecture of photonic Alamouti encoder is proposed [12]. Alamouti-type code is useful for the coherent detection over atmospheric turbulence channel [13]. Employing the phase-shift keying and Alamouti code the error rate can be calculated [14]. The Alamouti code generator in the optical domain using the particular arrangement of Mach–Zehnder interferometer is proposed. The layout is simulated on opti-bpm, and the simulated results are discussed.

Section 1 presents the introduction and the relevant research work carried out in the field of optical wireless communication. In Sect. 2, design of Alamouti code generator and the simulated results are discussed. Section 3 includes the conclusion.

2 Design of Alamouti Code Generator

The Alamouti space–time code sends symbols in two consecutive time slots. At the transmitter, two antennas are placed, and at the receiver end, one antenna is there, as shown in Fig. 1.

For a 2 × 1 Alamouti space–time code, there are two transmitter antenna and one receiver antenna.

The symbol pair X_1, X_2 and $-X_2^*$ and X_1^* are transmitted from antenna-1 and antenna-2 in two consecutive time slots as shown in Table 1.

MZI is a switching device the signal switch between the output terminals based upon the voltage applied at central electrode. The layout diagram is shown in Fig. 2.

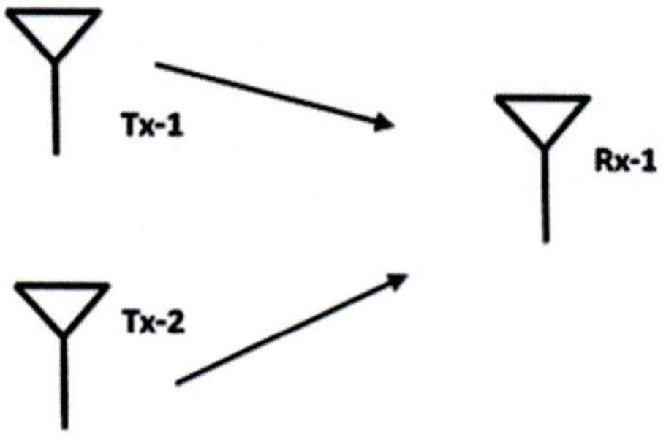

Fig. 1 Communication link of 2 × 1 antenna

Table 1 Symbol transmitted by the transmit antenna

Time	Antenna-1	Antenna-2
T	X_1	$-X_2^*$
t + T	X_2	X_1^*

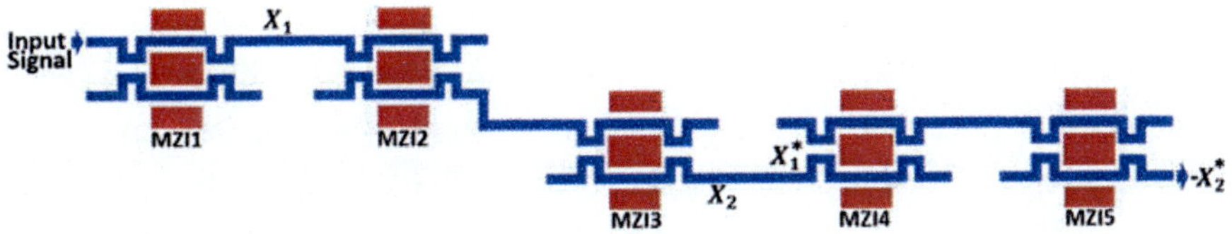

Fig. 2 Layout diagram of Alamouti code generator

The Alamouti code generator is proposed using the cascaded arrangement of 5 MZI structures. The output signal X_1, X_2, $-X_2^*$, and X_1^* is generated at different stages. The proposed layout is simulated using opti-bpm as shown in Fig. 3.

Simulated results of Alamouti code generator are shown in Fig. 4a Output Sequence X_1 and $X^*{}_1$ Fig. 4b Output Sequence X_2, $X^*{}_2$, and $-X^*{}_2$.

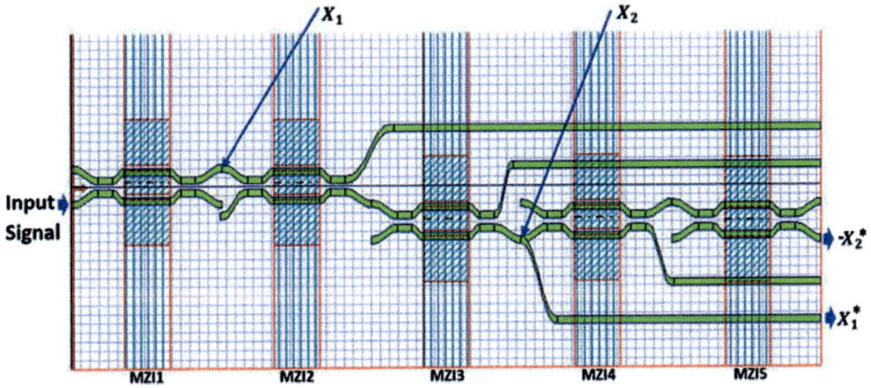

Fig. 3 Opti-bpm layout of Alamouti code generator

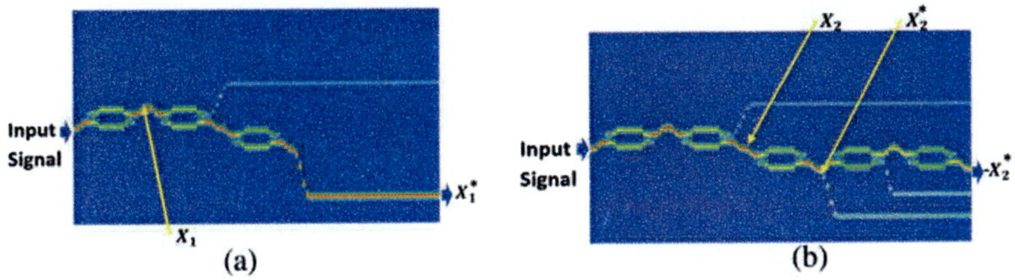

Fig. 4 **a** Output sequence X_1 and $X^*{}_1$, **b** Output sequence X_2, $X^*{}_2$ and $-X^*{}_2$

3 Conclusion

Alamouti code generator in optical domain is proposed. The proposed device is of smaller dimension and can be useful in the optical wireless communication system.

References

1. Nakamura S, Tajima K, Sugimoto Y (1994) Experimental investigation on high-speed switching characteristics of a novel symmetric Mach-Zehnder all-optical switch. Appl Phys Lett 65(3):283–285
2. Jun-qing L et al All-optical switch with switch-off time unrestricted by carrier lifetime. 1746
3. Yupapin PP, Teeka C (2010) OOK generation based on MZI incorporating a pumped nonlinear ring resonators system. Opt Express 18(10):9891–9899
4. Xu L, Wang BC, Baby V, Rand D, Glesk I, Prucnal PR (2002) All-optical data format conversion between RZ and NRZ based on wavelength converter. Conf Proc Lasers Electro-Optics Soc Annu Meet 1(2):308–310
5. Mishina K et al (2007) All-optical modulation format conversion from NRZ-OOK to RZ-QPSK using parallel SOA-MZI OOK/BPSK converters. Opt Express 15(12):7774–7785
6. Shi W et al (2013) Couplers for CWDM networks on silicon. 21(6):6733–6738
7. Huo L, Wang D, Wang Q, Xing Y, Jiang X, Lou C (2016) Optical time-division multiplexing signal processing using electro-optic modulators. Photonic Netw Commun 32(2):197–203
8. Singh G, Yadav RP, Janyani V (2010) Design and analysis of Ti: LiNbO3 1 × 2 digital optical switch with optimized electrode regions. J Opt Commun 31(2):78–80
9. Muneer P, Sameer SM (2016) Iterative joint carrier frequency offset and doubly selective channel estimation in high-mobility MIMO-OFDMA uplink using oblique projection. IEEE Trans Veh Technol 65(9):7110–7121
10. Alamouti SM (1998) STBC1.pdf. 16(8):1451–1458
11. Simon MK, Vilnrotter VA (2005) Alamouti-type space-time coding for free-space optical communication with direct detection. IEEE Trans Wirel Commun 4(1):35–39
12. Deng K, Wang BZ, Yao ZS (2014) Photonic Alamouti encoder for unifying hybrid FSO/RF system. Optik (Stuttg) 125(18):5051–5053
13. Niu M, Cheng J, Holzman JF (2014) Alamouti-type STBC for atmospheric optical communication using coherent detection. IEEE Photonics J 6(1)
14. Song X, Cheng J (2012) Alamouti-type STBC for subcarrier intensity modulated wireless optical communications. GLOBECOM—IEEE Glob Telecommun Conf 2936–2940

A Real-Time Road Congestion-Based Hybrid Approach for Finding the Optimized Route

Biru Rajak, Aprna Tripathi, and Dharmender Singh Kushwaha

Abstract As cities get overcrowded, commuters are spending most of their time in traveling and insane amount of time gets wasted in unnecessary traffic jams. A large number of countries are adopting various traffic management techniques to deal with traffic congestion. The objective of this research article is to develop a smart and intelligent road navigation system in a cost-effective way that dynamically adapts to the congestion of the route, and hence contribute to the development of smarter cities. Besides finding shortest route, congestion factor is computed to find an optimized route from source to destination. Congestion factor has been evaluated through numerical modeling in dense traffic scenario. Experiment results reveal that the travel time of a vehicle is reduced by up to 33% if it follows optimized route obtained from proposed approach. This can also be helpful for the emergency vehicles to identifying the best route in any emergency conditions. The proposed analytical model is further extended to identifying the optimal route for formation of on-demand green corridor in emergency situation.

Keywords Vehicle traffic routing · Traffic congestion · Optimal route finding · Travel time reduction

B. Rajak (✉) · D. S. Kushwaha
Department of Computer Science and Engineering, MNNIT Allahabad, Prayagraj, India
e-mail: birurajak@gmail.com

D. S. Kushwaha
e-mail: dsk@mnnit.ac.in

A. Tripathi
IET, Department of Computer Engineering and Application, GLA University, Mathura, India
e-mail: aprnatripathi@gmail.com

V. Nath and J. K. Mandal (eds.), *Proceedings of the Fourth International Conference on Microelectronics, Computing and Communication Systems*, Lecture Notes in Electrical Engineering 673, https://doi.org/10.1007/978-981-15-5546-6_8

1 Introduction

For any smart city, development of smart traffic management system is an important aspect because the traffic management system is the backbone of the faster movements in any modern society/city. It is also defining the affluence of the people who are residing in that society. In daily life, a person usually faces problems related to traffic such as traffic jam and vehicle parking problems. The problem intensifies with the rapid growth of vehicle on road. Efficient management, planning and adoption of smarter technologies can provide the better quality of life that converts cities into smart cities. Some country use on-demand green corridor for emergency vehicle clearance. Green corridor is not always feasible solution as it hampers the normal traffic flow and sometime green corridor set is not possible. Any system which works on real-time traffic data and provides an optimal path which is least congested without disturbing the normal flow is a better solution as compared to green corridor.

The remainder of this article is organized into five sections. Sections 2 summarized the literature survey. Section 3 proposed work and mathematical formulation has been done. Section 4 provides the simulation result and compares it with the shortest paths that do not consider congestion. Finally, Sect. 5 concludes this paper.

2 Related Work

Smart city traffic management is responsible to manage smooth traffic flow of the city. The task such as detection of rule violator, detection of congestion is the major role of traffic management. Among these various tasks, detection of the traffic congestion and managing the real-time traffic flow is challenging task. This is because the lack of integration of real-time traffic data and traffic control systems. There exist various techniques to detect the traffic congestion. Andrea and Marcelloni [1] proposed a GPS-based approach which tracks the location of the vehicles by using the GPS of the smartphones present in the particular area to identify the vehicle location. Based upon their location, traffic flow and congestion are estimated in real time. The limitation of this approach is that its faces serious issue of false congestion detection as GPS-enabled mobile is considered as a vehicle. Soufiene et al. [2] propose a smart traffic management system which is based on a communication-oriented mechanism (COM). Data is obtained from various devices like, GPS-enabled mobile phones, social media activity of users and other devices like camera. This data is further processed and sent to the users' smartphone to report the congestion incident that may have occurred at a particular route. Pan et al. [3] uses GPS or smartphone to collect the real-time traffic information and classifies the congestion in different categories denoted by different colors. High congestion is denoted by red color and green color is used to denote the less congestion. Another GPS-based traffic congestion detection approach proposed by Mingqi et al. [4] that uses the GPS-enabled undedicated mobile phone. This work requires continuously monitoring of the motion of the user mobile followed by map matching. Kumar and Kushwaha in [5] propose an approach for traffic congestion detection and its control. In this paper, image processing techniques are used to detect the congestion at the junction point. This congestion information is further classified into three different categories, namely high congestion, moderate congestion and low congestion. Real-time congestion information is later transferred to traffic control server. This information may be utilized for real-time traffic management. Mittal et al. in [6] have discussed the technique for emergency vehicle clearance called "Green Wave". Green wave makes all the signals green for an emergency vehicle at all post that lies between source and destination; subsequently stopping other vehicles. However, this leads to unnecessary traffic jams and hinders the passage on the entire route. Rajak and Kushwaha in [7] propose a green corridor based on RFID tag and reader. Every vehicle is attached with an RFID tag that is controlled by the central server. Each traffic junction is enabled with RFID reader that is used to sense the vehicle passing through it and reports to next traffic junction. Advance traveler information system (ATIS) is the part of intelligent transport systems [8, 9]. It recommends the alternate routes to improve the traffic congestion. Qi et al. in [10] have proposed a traffic model to reduce the congestion in cities by intimating vehicle drivers about the information of the real-time incidents which affect traffic on route. The limitation of this approach is that it does not classify the severity of congestion. Zhou et al. in [11] have used two-level hierarchical model to control the traffic network. At the parent level, large heterogeneous network is decomposed into a number of homogenous sub-networks; where every sub-network deals with demand balancing problems. Leaf level solves all the other traffic problems in the sub-network in a parallel fashion.

Prior prediction of congestion in traffic helps in reducing the traffic jams. Jia et al. in [12] have predicted the congestion pattern of urban transportation based on the historical data. Data consists of speed of the vehicle, direction of the vehicle where vehicle is moving and information related to traffic junction and this data is modeled to predict the location where the heavy congestion may exist. The drawback of the approach is that the result is based on the historical data instead of real-time traffic data. Kamal et al. in [13] have discussed the driving habits like slow or rash driving creates problems on road and leads to unnecessary jam. The authors have simulated the mathematical formulation for driving behavior of drivers and developed a smart driving system to improve traffic flow. Peng and Guanghan in [14] propose a lattice model of traffic flow where predictive individual driving behavior is considered. Based on the anticipated behavior, stability of traffic flow has been explored.

It is evident that many researchers are exploring various solutions to facilitate the society by providing on-demand optimal route. This required integration of real-time congestion information and the optimal route selection system. The optimal route may be based on shortest distance or may be based on least congestion route. There exist various approaches that propose optimal route based on shortest path. For shortest path, various well-defined algorithms are available such as Dijkstra's [15], A* search Floyd–Warshall, etc. Dijkstra's algorithm is the classical algorithm for finding the shortest route from source to destination. This algorithm does not consider congestion on the route, but in real-life scenario congestion on road is crucial factors for route optimization.

This paper proposes an approach for providing on-demand optimal route between source and destination based on the real-time congestion and shortest distance. For shortest distance path, lots of algorithms have been already developed; however, these algorithms only consider the distances from source to destination. This paper discusses the optimal solution for providing the least congested route considering issues such as congestion, numbers of vehicles, average flow of the vehicles and width of the lane. Congestion on road differs for different places at distinct times. The proposed hybrid model has represented the severity of congestion in numeric, where 0 and 1 denote the least and the most congested path, respectively. We simulate the proposed system using SUMO.

3 Proposed Work

Architecture of the proposed work includes various components such as traffic controller, data acquisition module, route finder, route optimizer, request handler and response generator. Data acquisition module is connected to all the sources which are installed at various junctions such as cameras or IoT devices. It is assumed that each traffic junction is equipped with the congestion detection system that is based on computer vision or IoT techniques. This congestion detection system captures real-time congestion density (C) and average velocity (V) at the specified junctions (Fig. 1).

The data acquisition unit obtains congestion density and traffic flow information for all the junctions in the city. This information is stored in a congestion matrix that is updated periodically. Each cell of this matrix stores a vector that contains

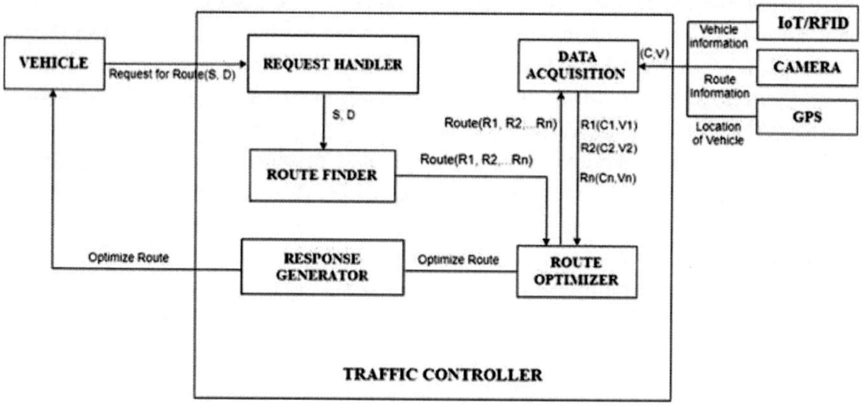

Fig. 1 Proposed model

Fig. 2 Example of congestion matrix

Junctions Name	A	B	C
A		(C, V)	
B			
C			

congestion parameters C and V between every two consecutive traffic junctions. For example, in Fig. 2, the cell highlighted with green color contains values *c*, *v*, i.e., at any time instance, the congestion density from junction *x* to junction *y* is *c* and average speed of traffic is *v*.

The request handler handles the all request obtained from the user needs to know optimal route for their journey. This request contains information about source and destination junctions. Request handler forwards the user's request to a route finder server. The route finder server has the city map in the form of adjacency matrix. The each cell of the matrix contains information such as distance and width of the road between two junctions points as shown in Table 1. The route finder server process the request obtained from request handler in order to find all the possible paths between the requested source and destination junction. The each path is comprises of all the intermediates junctions points including source and destinations points. Extracted routes information is forwarded to route optimizer server. The role of route optimizer server is to identify the best routes among all routes obtained from route finder server. The identification of best route is carried in such a manner that it identified route must be least congested and shortest as possible. To determine the best optimal route, this paper proposed a factor called congestion factor (CF) that is based on the distance, width of the path and average speed of traffic flow, congestion density on the path. The congestion factor is calculated by Eq. (1). Congestion factor (CF),

$$\mathrm{CF_{AB}} = \mathrm{CF}[((\mathrm{E_{AB}}(\mathrm{d}) \times \mathrm{E_{AB}}(\mathrm{C}))/((\mathrm{E_{AB}}(\mathrm{V}) \times \mathrm{E_{AB}}(\mathrm{W_{AB}}))] \tag{1}$$

The distance and width of the path are obtained from the route finder server and the real-time congestion density and average traffic flow are obtained from the data acquisition server. To obtain the optimal route, route optimizer uses the proposed algorithm that uses congestion factor in order to find best route. The route with minimum CF value is identified as an optimal route. Route optimizer server forwards this route information to response generator. Response generator sends this response to the respective users.

Table 1 CD and velocity adjacency matrix of city map (S-D)

	S	A	B	C	E	F	G	H	I	J	K	L	M	N	O	P	D
S	{0,0}	{0.6, 40}	{0.4, 50}	{0, 0}	{0, 0}	{0, 0}	{0, 0}	{0, 0}	{0, 0}	{0, 0}	{0, 0}	{0, 0}	{0, 0}	{0, 0}	{0, 0}	{0, 0}	{0, 0}
A	{0.6, 40}	{0, 0}	{0, 0}	{0.3, 54}	{0, 0}	{0, 0}	{0, 0}	{0, 0}	{0, 0}	{0, 0}	{0, 0}	{0, 0}	{0, 0}	{0, 0}	{0, 0}	{0, 0}	{0, 0}
B	{0.4, 50}	{0, 0}	{0, 0}	{0.6, 42}	{0, 0}	{0, 0}	{0.6, 40}	{0, 0}	{0, 0}	{0, 0}	{0, 0}	{0, 0}	{0, 0}	{0, 0}	{0, 0}	{0, 0}	{0, 0}
C	{0, 0}	{0.3,54}	{0.6,42}	{0, 0}	{0.4,48}	{0, 0}	{0, 0}	{0, 0}	{0, 0}	{0, 0}	{0, 0}	{0, 0}	{0, 0}	{0, 0}	{0, 0}	{0, 0}	{0, 0}
E	{0, 0}	{0, 0}	{0, 0}	{0.4,48}	{0, 0}	{0.65, 38}	{0, 0}	{0, 0}	{0, 0}	{0, 0}	{0, 0}	{0, 0}	{0, 0}	{0, 0}	{0, 0}	{0, 0}	{0, 0}
F	{0, 0}	{0, 0}	{0, 0}	{0, 0}	{0.65, 38}	{0, 0}	{0, 0}	{0, 0}	{0, 0}	{0.65, 38}	{0, 0}	{0, 0}	{0, 0}	{0, 0}	{0, 0}	{0, 0}	{0, 0}
G	{0, 0}	{0, 0}	{0.6,40}	{0, 0}	{0, 0}	{0, 0}	{0, 0}	{0.5,40}	{0, 0}	{0, 0}	{0, 0}	{0, 0}	{0, 0}	{0, 0}	{0, 0}	{0.8,32}	{0, 0}
H	{0, 0}	{0, 0}	{0, 0}	{0, 0}	{0, 0}	{0, 0}	{0.5,40}	{0, 0}	{0.3, 55}	{0, 0}	{0, 0}	{0.6, 38}	{0, 0}	{0, 0}	{0, 0}	{0, 0}	{0, 0}
I	{0, 0}	{0, 0}	{0, 0}	{0, 0}	{0, 0}	{0, 0}	{0, 0}	{0.5, 40}	{0, 0}	{0.4, 45}	{0.4,50}	{0, 0}	{0, 0}	{0, 0}	{0, 0}	{0, 0}	{0, 0}
J	{0, 0}	{0, 0}	{0, 0}	{0, 0}	{0, 0}	{0.65, 38}	{0, 0}	{0, 0}	{0.4, 45}	{0, 0}	{0, 0}	{0, 0}	{0, 0}	{0.4, 52}	{0, 0}	{0, 0}	{0, 0}
K	{0, 0}	{0, 0}	{0, 0}	{0, 0}	{0, 0}	{0, 0}	{0, 0}	{0, 0}	{0.4, 50}	{0, 0}	{0, 0}	{0.3, 55}	{0, 0}	{0, 0}	{0.4, 50}	{0, 0}	{0, 0}
L	{0, 0}	{0, 0}	{0, 0}	{0, 0}	{0, 0}	{0, 0}	{0, 0}	{0.6, 38}	{0, 0}	{0, 0}	{0.3,55}	{0, 0}	{0.4, 50}	{0, 0}	{0, 0}	{0, 0}	{0, 0}

(continued)

Table 1 (continued)

	S	A	B	C	E	F	G	H	I	J	K	L	M	N	O	P	D
M	{0, 0}	{0, 0}	{0, 0}	{0, 0}	{0, 0}	{0, 0}	{0, 0}	{0, 0}	{0, 0}	{0, 0}	{0, 0}	{0.4, 50}	{0, 0}	{0, 0}	{0.3, 55}	{0, 0}	{0, 0}
N	{0, 0}	{0, 0}	{0, 0}	{0, 0}	{0, 0}	{0, 0}	{0, 0}	{0, 0}	{0, 0}	{0.4, 52}	{0, 0}	{0, 0}	{0, 0}	{0, 0}	{0.4,40}	{0, 0}	{0.5, 30}
O	{0, 0}	{0, 0}	{0, 0}	{0, 0}	{0, 0}	{0, 0}	{0, 0}	{0, 0}	{0, 0}	{0, 0}	{0.4,50}	{0, 0}	{0.3, 55}	{0.5, 40}	{0, 0}	{0, 0}	{0, 0}
P	{0, 0}	{0, 0}	{0, 0}	{0, 0}	{0, 0}	{0, 0}	{0.8,32}	{0, 0}	{0, 0}	{0, 0}	{0, 0}	{0, 0}	{0, 0}	{0, 0}	{0, 0}	{0, 0}	{0.9, 28}
D	{0, 0}	{0, 0}	{0, 0}	{0, 0}	{0, 0}	{0, 0}	{0, 0}	{0, 0}	{0, 0}	{0, 0}	{0, 0}	{0, 0}	{0, 0}	{0.5,30}	{0, 0}	{0.9, 28}	{0, 0}

4 Optimal Path Finding Algorithm

This paper proposes a novel algorithm for finding optimal paths based on multiple parameters. Initial parameters such as source S and destination D are provided by the user. Route finder server extracts all possible routes ($P = \{R_1, R_2, R_3, \ldots R_n\}$) for the given source, and destination from the city map. Each route is defined as set of all edge vector (E = (d, w), where d is the distance, w is the width) that exist in intermediate junction in the route, i.e., $R_i = \{E1, E2, E3, \ldots\ldots. En\}$. After all possible routes are obtained from the route finder server, data acquisition server is invoked for the information such as congestion, average flow of the traffic distance and width of the lane for each edge. For example, for any route R = {E1, E2, E3, E4} at any instance of time t1, the value of C and V is {(c1, v1), (c2, v2), (c4, v4), (c5,v5)}. Data acquisition server periodically update the data, and hence at any other time t2, the value of C and V may be {(c6, v6), (c7, v7), (c8, v8), (c9, v9)}. Based on these above parameters, congestion factors (CF) for each route is calculated from Eq. (1) based upon the information that is acquired for the server. If the distance and traffic congestion is more, then congestion factor is high and when the velocity of vehicle and width of particular edge is more, then the value of congestion factor is low.

Algorithms: Optimal Path Finding (S, D, P)

```
while P has next
    R[i] = P.getRoute(i)
    while R has next
        [Dj, Wj]=R.getEdge(j)
        [Cj, Vj]=R. DataAcquisitionServer(Dj, Wj)
    end while
C.F[i]= [((Ei(d)*Ei(C))/((Ei(V)*Ei(W))]
end while
Optimal Route = function findMin(CF[])
return Optimal Route
```

Each route has a congestion factor and higher value of congestion factor is not suited for optimal path. The route which consists of the edges having least congestion factor is the optimal route. Thus, the above algorithm generates the optimized route.

5 Simulation and Result Analysis

To verify the performance of the proposed approach, we carried out simulation of various scenarios in SUMO [16]. SUMO is an open source, microscopic and multi-modal traffic simulation. It allows us to simulate the different scenarios pertaining to single vehicles moves through a given road network. The simulation allows us to address a large set of traffic management issues and has capacity to simulate 1000,000 vehicles.

A. *Simulation Scenario*

To simulate the performance of the proposed approach, we have imported the map of a locality of Allahabad city (https://www.openstreetmap.org/search?query=teliyarganj#map=18/25.49898/81.85891) in SUMO. We generate the city map of the exported map. To carry out simulation, we can define many vehicles in SUMO configuration and for a particular vehicle, source and destination is also generated randomly or can be defined manually. This city map comprises of the various junctions point. For the simplicity, junction point named as S, A, B, C, E, F, G, H, I, J, K, L, M and D, where in the proposed approach, it is considered that S is the source and D is the destination junctions as shown in Fig. 3.

The city map is represented in the form of adjacency matrix form having edges from S to A and a part of this is shown in Table 1. In this matrix, the weight of the two edges between junctions is defined by a set E {D, W} containing distance (D) and width (W). Value {0, 0} indicates that there is no direct route between said junctions or the two vertices are not adjacent.

When any vehicle demands for a least congested and optimal route between two junction by supplying its source junction (S) and destination junction D (S-D) pair,

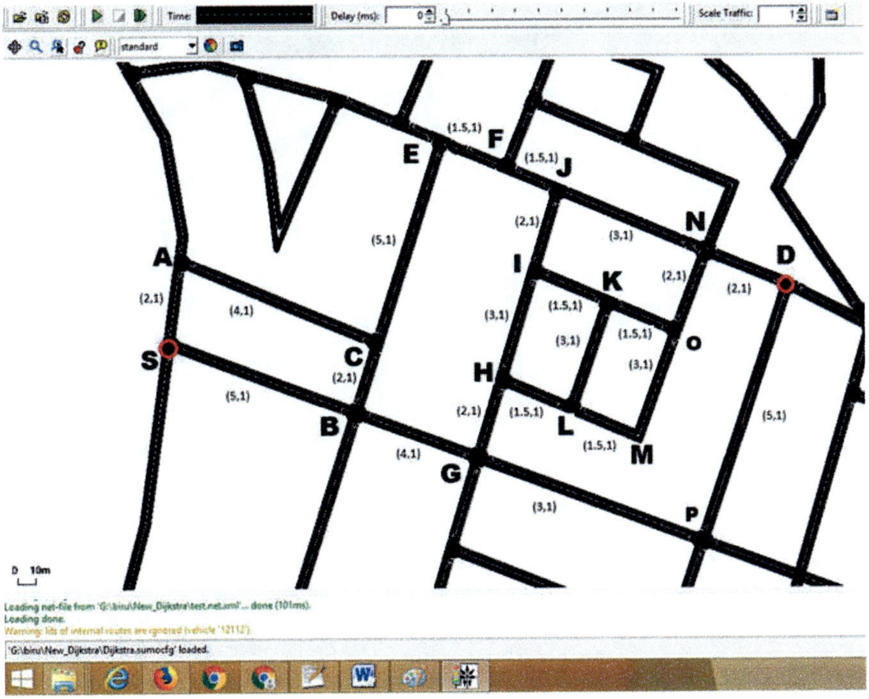

Fig. 3 City map with distance and width of each edge as a weight

the control server receives the requests and finds the all possible path between the (S-D). In this scenario, control server extracts four possible paths between (S-D). Let, these paths are defined by R1, R2, R3 and R4. R1 has staring edge S and passes through junctions A- > C- > E- > F- > J- > N- > D. Intermediate traffic junctions for R2 are B- > G- > P and for route R3 it passes through B- > C- > E- > F- > J- > N. Similarly, route R4 has the edges B- > G- > H- > L- > K- > O- > N. After extraction of all possible paths, route optimizer server obtains real-time congestion density and average speed of traffic flow between each edge of the all possible paths as shown in Fig. 4.

In this simulation, we have generated random traffic congestion by initializing 200 vehicles at random speed and random routes. Figure 4 shows that congestion map for the city at any time instance t for CD and V.

The congestion density and average traffic flow between the each junction of all possible routes R1, R2, R3, and R4 obtained from data acquisition server are shown in Table 2.

Now, route optimizer server finds and computes the congestion factors for all routes. The path with the minimum congestion factor is selected as an optimal path between (S-D). Initially, there existed four possible routes between (S-D), namely

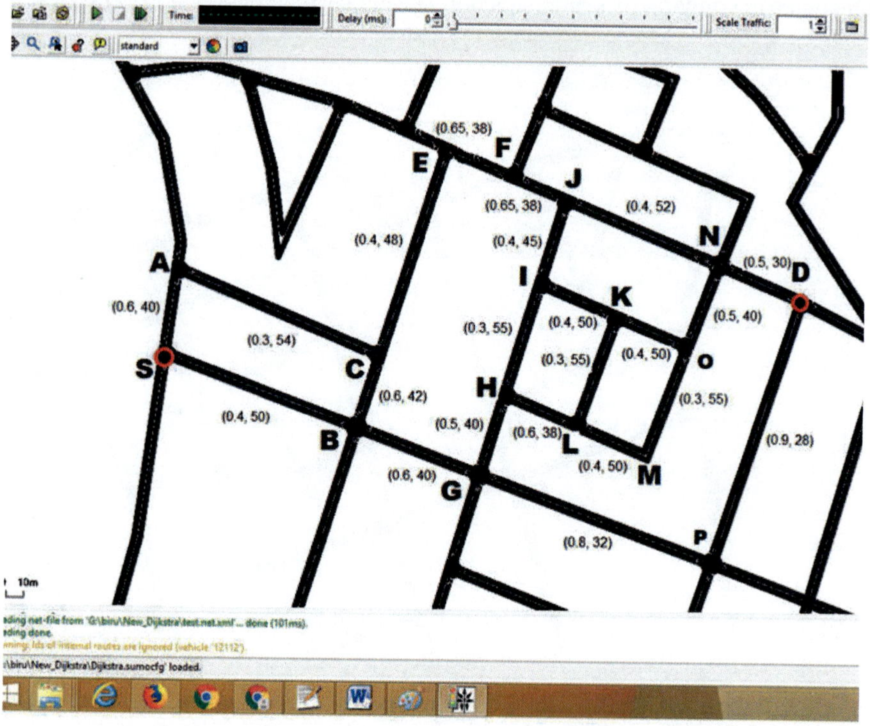

Fig. 4 City map with CD and velocity of each edge as a weight

Table 2 Details of each edge of possible route

Route R1			Route R2			Route R3			Route R4		
edge	C	V	edge	C	V	edge	C	V	edge	C	V
S-A	0.6	40	S-B	0.4	50	S-B	0.4	50	S-B	0.4	50
A-C	0.3	54	B-G	0.6	40	B-C	0.6	42	B-G	0.6	40
C-E	0.4	48	G-P	0.8	32	C-E	0.4	48	G-H	0.5	40
E-F	0.65	38	P-D	0.9	28	E-F	0.65	38	H-L	0.6	38
F-J	0.65	38	–	–	–	F-J	0.65	38	L-K	0.3	55
J-N	0.4	52	–	–	–	J-N	0.4	52	K-O	0.4	50
N-D	0.5	30	–	–	–	ND	0.5	30	O-N	0.5	40
–	–	–	–	–	–	–	–	–	N-D	0.5	30

R1, R2, R3 and R4 and their details are shown in Table 2. The control server computes the congestion factor for all above routes. Table 3 shows that, route R1 has minimum congestion factor value and this route R1 will be the optimal route based on the current congestion condition. The table also shows that route R2 has shortest distance, i.e., but its congestion factor value is greater the route R2, means congestion on this route is greater. Hence, the vehicle traveling on optimal route will reach to its destination in minimum travel time as compared to the vehicle traveling to shortest route with high congestion factor. The next simulation compares the travel time of the different vehicle on optimal route and shortest route (Table 3).

B. ***Analysis of travel time***

This simulation is carried out to study the impact of congestion factor on travel time. During this simulation, travel time of the two vehicles v1 and v2 is recorded for the different routes between the given source and destination junction (S-D) pair. The vehicle speed and start time are kept same during the various simulation. During each simulation, the shortest route between (S-D) pair is assigned to vehicle v1 and the optimal route found by the proposed approach is assigned to the vehicle v2. The congestion parameters such as congestion density and average speed traffic flow are generated randomly in each simulation. The congestion generation is simulated by generation of random routes for 300 vehicles in SUMO.

When a vehicle V1 demands route for source S and destination D, congestion factor is calculated by Eq (1) as:

$CF_{V1} = ((5 \times 0.4)/(1 \times 50)) + ((4 \times 0.6)/(1 \times 60)) + ((3 \times 0.8)/(1 \times 32)) + ((5 \times 0.9)/(1 \times 28)) = 0.3157$, and

$$\begin{aligned} CF_{V2} =&((2 \times 0.6)/(1 \times 40)) + ((4 \times 0.3)/(1 \times 54)) + ((5 \times 0.4)/(1 \times 48)) \\ &+ ((1.5 \times 0.65)/(1 \times 38)) + ((1.5 \times 0.65)/(1 \times 38)) \\ &+ ((3 \times 0.41)/\ (1 \times 52)) + ((2 \times 0.5)/(1 \times 30)) = 0.202. \end{aligned}$$

$$\text{Time taken(T)} = \text{Total distance} \times \text{Average speed} \times \text{CF}$$

Table 3 CF of possible route and optimal path

Source-destination	Possible route	Edges	Distance	CF.	Optimal path
S-D	R1	S- > A- > C- > E- > F- > J - > N- > D	19	0.202	R1
	R2	S- > B- > G- > P- > D	17	0.335	
	R3	S- > B- > C- > E- > F- > J- > N- > D	20	0.218	
	R4	S- > B- > G- > H- > L- > K- > O- > N- > D	21	0.235	

Table 4 Comparasion of CF and for shortest and optimal path

S-D	Vehicle	Route	Unit distance	Congestion factor	Time taken	Improved travel
S-D	V1	R2	17	0.315	0.1428	37.32%
	V2	R1	19	0.202	0.0895	
H-O	V1	R4	6	0.0520	0.0065	29.23%
	V2	R1	6	0.0403	0.0046	

Hence, for vehicle

$$\begin{aligned} V1_T &= 17/((50+40+32+28)/4)) * 0.315 \\ &= 0.1428(\text{in unit time}) \text{ and} \\ V2_T &= 19/((40+54+48+38+38+52+30)/7) * 0.202 \\ &= 0.0895(\text{in unit time}). \end{aligned}$$

Similarly, when source is G and destination is N:

$$\begin{aligned} CF_{V1}(A-G) &= (1.5*0.6)/(1*38)) + ((1.5*0.4)/(1*50)) \\ &\quad + ((3*0.3)/(1*55)) \\ &= 0.0520 \end{aligned}$$

$$\begin{aligned} CF_{V2}(A-G) &= ((3*0.3)/(1*55)) + ((1.5*0.4)/(1*50)) \\ &\quad + ((1.5*0.4)/(1*50)) \\ &= 0.0403 \end{aligned}$$

$$\begin{aligned} V1_T &= 6/((38+50+55)/3) * 0.0520 = 0.0065(\text{in unit time}) \\ V2_T &= 6/((55+50+50)/3) * 0.0403 = 0.0046(\text{in unit time}) \end{aligned}$$

Table 4 shows the results of the experiments that are carried out for analysis of travel time of vehicles. In Table 4, for the same source and destination, two vehicles start their journey at the same time. The average travel time of the vehicle is reduced up to 33% when it travels on optimal route identified by the proposed approach (Fig. 5).

6 Conclusion

Traffic congestion is a very common situation that arises frequently in urban areas. Inappropriate management of traffic congestion creates various problems such as traffic jams and air and sound population. Even valuable foreign currency is burnt

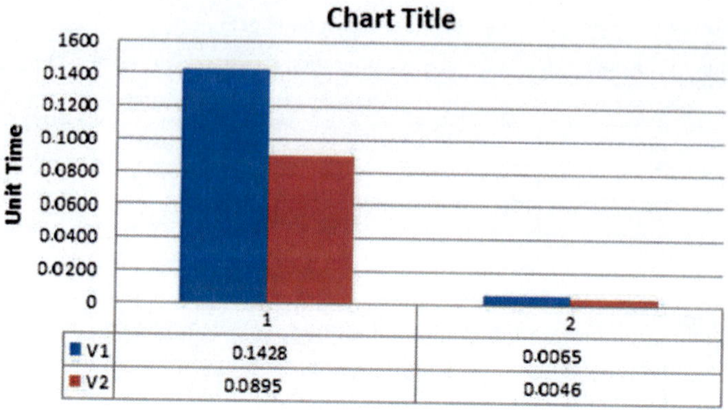

Fig. 5 Comparison graph

while traffic jams occur. This paper proposes an analytical model to manage real-time traffic by providing on-demand least congested route. The proposed approach utilizes various crucial parameters such as congestion density, average speed of traffic flow, distance and width of the road. The optimal route finding algorithm is also proposed in this paper. This proposed algorithm is very efficient as it has polynomial time complexity of O (n^2). Experiment results reveal that the travel time of a vehicle is reduced by up to 33% if it follows optimized route obtained from the proposed approach. This can be helpful for the emergency vehicles to identify the best route in any emergency conditions. The proposed analytical model is being further extended to identifying the optimal route for formation of on-demand green corridor in emergency situation.

References

1. D'Andrea E, Marcelloni F (2017) Detection of traffic congestion and incidents from GPS trace analysis. Expert Syst Appl 73:43–56
2. Soufiene D, Doolan R, Gabrial-Miro M, John M, (Nov 2013) A communications-oriented perspective on traffic management systems for smart cities: challenges and innovative approaches. IEEE Comm Surv Tutorials 10
3. Pan W, Chen Z, Li X(2014) Decryption method of baidumap'scoordinates based on artificial neural network. Comput Eng Appl 110–113
4. Mingqi L et al (2015) A road congestion detection system using undedicated mobile phones. IEEE Trans Intell Trans Sys 16.6:3060–3072
5. Kumar T, Kushwaha DS (2019) An approach for traffic congestion detection and traffic control system: proceedings of third international conference on ICTCS 2017
6. Mittal A, Bhandari D (2013) A novel approach to implement green wave system and detection of stolen vehicles. 2013(3rd IEEE international advance computing conference (IACC)
7. Rajak B, Kushwaha DS (2019) Traffic control and management overIoT for clearance of emergency vehicle in smart cities: proceedings of third intern-ational conference on ICTCS 2017. pp. 121–130. https://doi.org/10.1007/978-981-13-0586-3_12

8. Wang F-Y, Zheng N-N, Cao DP, Martinez CM, Li L, Liu T (2017) Parallel driving in CPSS: a unified approach for transport automation and vehicle intelligence. IEEE/CAA J Autom Sinica 4(4):577–587
9. Wang F-Y (2010) Parallel control and management for intelligent transportation systems: concepts, architectures, and applications. IEEE Trans Intell Transp Syst 11(3):630–638
10. Qi L, Zhou M, Luan W (2018) A two-level traffic light control strategy for preventing incident-based urban traffic congestion. IEEE Trans Intell Trans Sys 19(1):13–24. https://doi.org/10.1109/TITS.2016.2625324
11. Zhou Z et al (2017) Two-level hierarchical model-based predictive control for large-scale urban traffic networks. IEEE Trans Control Sys Technol 25.2: 496–508
12. Jia R et al (2018) Data driven congestion trends prediction of urban transportation. IEEE Internet Things J 5.2:581–591
13. Kamal MAS et al (2014) Smart driving of a vehicle using model predictive control for improving traffic flow. IEEE Trans Intell Trans Sys 15.2:878–888
14. Peng Guanghan (2013) A new lattice model of traffic flow with the consideration of individual difference of anticipation driving behavior. Commun Nonlinear Sci Numer Simul 18(10):2801–2806
15. Dijkstra EW (1959) A note on two problems in connexion with graphs. Numer Math 1:269–271
16. Krajzewicz D, Hertkorn G, Rössel C, Wagner P (2002) SUMO (Simulation of Urban MObility); An open-source traffic simulation. Proceedings of the 4th middle east symposium on simulation and modelling (MESM2002), Edited by Al-Akaidi A, pp. 183–187, SCS European Publishing House, ISBN 90-77039-09-0

Sentiment Analysis of Audio Diary

Debasmita Manna, Shaon Baidya, and S. Bhattacharyya

Abstract Natural language processing and sentiment analysis are a major field of research in this era. This paper explores the possibility of combining the presently available sentiment analysis algorithm in determining human emotions. This is done through diary case studies. The method is used on voice recordings of people over the period of one month. The result shows that presently available open-source sentiment analysis algorithms give satisfying results in the majority of cases but exceptional cases must be taken into consideration. Also, for effective sentiment analysis, we must consider the wide spectrum of human emotions and develop the algorithm accordingly. Spyder IDE is used for accessing IPython console to obtain results.

Keywords Depression · Natural language processing · Polarity · Sentiment analysis

1 Introduction

Everything that we speak has an emotion linked to it. Analyzing every word that we speak can give us an insight about emotions that a person is experiencing. Sentiment analysis is the task of categorizing words and linking them to different sentiments like positive, negative and neutral; or into an n-point scale, e.g., very good, good, satisfactory, bad and very bad [1, 2]. Through this, we can determine the general emotion reaction or polarity of a person. A thorough analysis for a long period of time can help determine the overall mental state of a person [3, 4].

D. Manna · S. Baidya (✉) · S. Bhattacharyya
Electronics and Communication Engineering, Dr. Sudhir Chandra Sur Degree Engineering College, 540 Dum Dum Road, Dum Dum, Kolkata 700074, India
e-mail: shaonbaidya@gmail.com

D. Manna
e-mail: debasmita.mn@gmail.com

S. Bhattacharyya
e-mail: swapanbhattacharyya@hotmail.com

V. Nath and J. K. Mandal (eds.), *Proceedings of the Fourth International Conference on Microelectronics, Computing and Communication Systems*, Lecture Notes in Electrical Engineering 673, https://doi.org/10.1007/978-981-15-5546-6_9

This paper presents the results of sentiment analysis of voice-based diary entries. Section 2 consists of historical background. Section 3 discusses with methodologies used in the project. Section 4 consists of results and case studies. Section 5 consists of conclusion.

2 Historical Background

Conventionally sentiment analysis is an algorithm is used by companies to estimate the extent of product acceptance. This also helps in determining strategies to improve product quality. It also facilitates policymakers or politicians to analyze public sentiments with respect to policies, public services or political issues. [1, 2, 5, 6] We are familiar with Twitter data sentiment analysis [7–12].

This paper has explored the possibility of using sentiment analysis algorithms to detect human emotions. We are living in this super busy twenty-first century. And we are living a sedentary life. We give more time to work and Internet. We give less time to people around us. And specially we give less time to ourselves. According to WHO around 300 million people of all ages are suffering from depression worldwide. And now depression is a modern-day epidemic [13]. It is well proven by research that depression is also linked with suicidal behavior, homicidal ideation and substance abuse [14–19].

Depression is ranked as the single largest contributor to non-fatal health loss. It accounts for 7.5% of global years lived with disability (YLDs) and 2.0% of global disability-adjusted life years (DALYs) in 2015 [20] (Fig. 1).

The number of people living with depression worldwide increased by an estimated 18.4% between the year 2005 and 2015 [3] (Fig. 2).

Most people with depression remain without treatment. Along with this, people often deny their illness and develop a reluctance to seek help. Early detection of

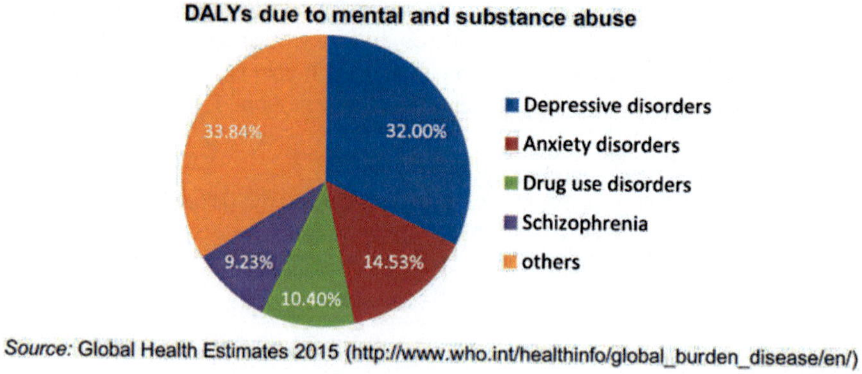

Fig. 1 DALYs due to mental and substance abuse. *Source* Global health estimates 2015 (http://www.who.int/healthinfo/global_burden_disease/en/)

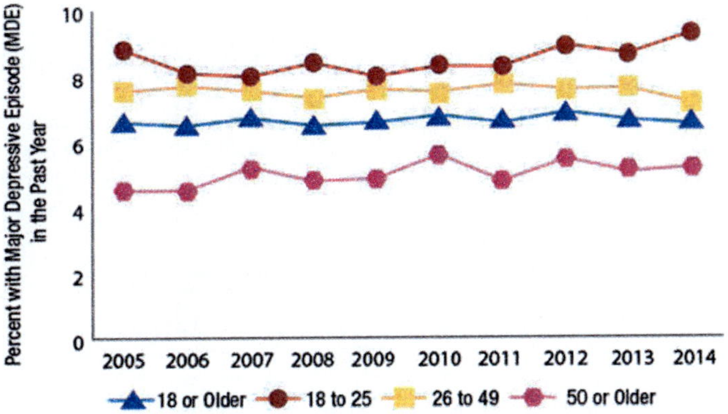

Fig. 2 Percent with Major Depressive Episode(MDE) in the Past Years. *Source* https://www.psycom.net/depression.central.html

depression is really beneficial. Data from NMHS revealed that the time gap between onset of depression symptoms and seeking professional help ranged between 1 month to 10 years [7].

This paper includes case studies of audio diary entries for sentiment analysis. Audio diary entries are used since it takes less time to voice record a diary entry than to write a diary. Also, recording diary entries gives a chance to emotion verbally which is more effective than writing thoughts.

3 Methodology

The following libraries are used for the project:

- TextBlob is a Python-based text processing library which provides a simple API for diving into natural language processing tasks.
- Googletrans is a Python library that implements Google translate API.
- Google speech recognition API is used for converting speech to text.
- Azlyrics is an API wrapper for extraction data from azlyrics.com.
- Seaborn and Matplotlib are Python-based libraries used for visual representation of data.
- Selenium is used for Web scrapping.

At first, sample audio is split into sixty-second frames. This audio is then converted into text. If the audio is in any other language than English, then it is first converted into English. The sentiment analysis is done with the help of a Python-based text processing library called TextBlob. The following flowcharts explain the workflow [3, 7, 15, 16, 21–34] (Figs. 3, 4).

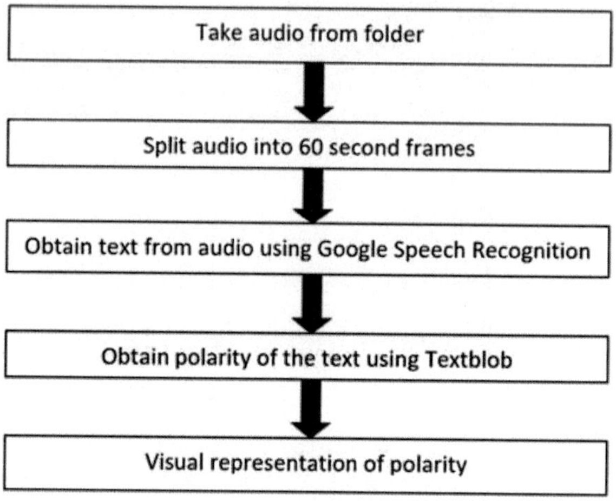

Fig. 3 Workflow of English audio diary data

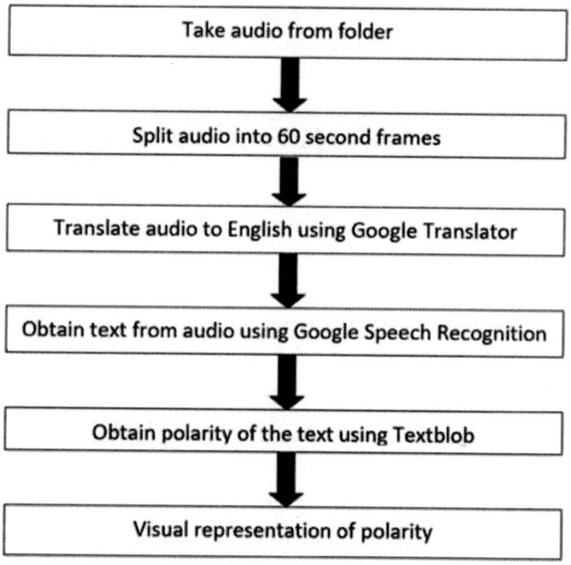

Fig. 4 Workflow of non-English audio diary data

4 Results and Case Studies

The following figures show the results obtained by sentiment analysis are audio diary data of a person for the month of February 2019.

The polarity or emotional reaction is in the range of -1 to $+1$.

The measure of polarity for an individual day is the average of the individual polarity of split parts of the audio for a particular day.

From Fig. 5, we can conclude that on 10th February the overall emotion experi-

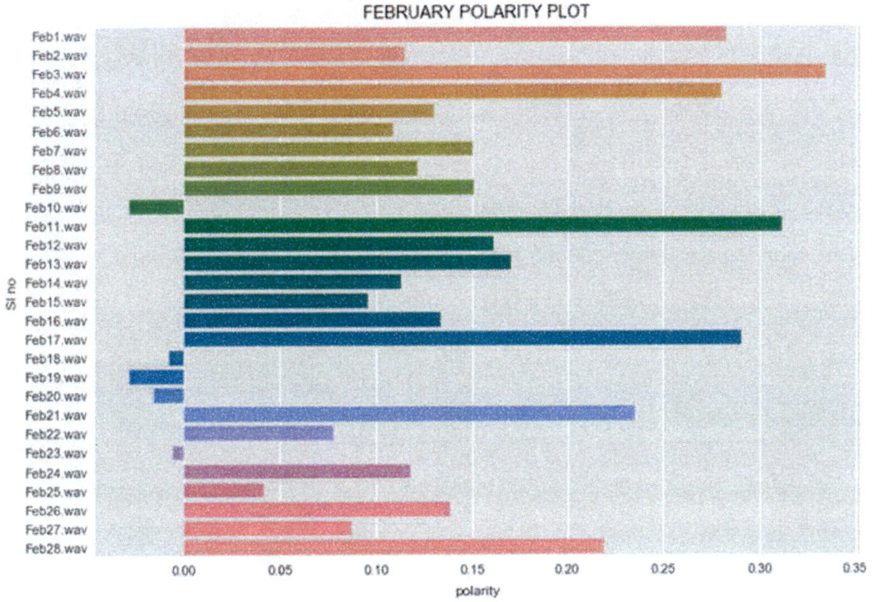

Fig. 5 Polarity Plot for month of February

enced is negative. Below is a detailed view of average of the individual polarity of split parts of the audio for 10th February (Figs. 6 and 7).

Similarly, from Fig. 5, it is concluded that on 3rd February the overall emotion experienced is positive. Below is a detailed view of average of the individual polarity of split parts of the audio for 3rd February (Fig. 8).

Figure 9 shows the comparative analysis of sentences with similar words is checked. The results drawn are shown in the following table.

From Table 1, we can clearly understand the effect of words on the overall sentiment of a sentence.

Apart from this, audio in Bengali language is also checked to explore the possibility of sentiment analysis of language other than English.

In Fig. 10, sentiment analysis of a Bengali song is shown. Google speech recognition does not work well on any audio with high background noise. So, to analyze song sentiment, we much use song audio with no background music.

Similarly, we can convert text from other languages too if it is supported by Google translator API.

The following link contains code for processing non-English audio.

https://drive.google.com/file/d/1-yScVRxjsNSftD3fy_XKMriWktKBhiHl/view?usp=sharing

The following link contains code for processing English audio

https://drive.google.com/file/d/14JCyzsbmBf2BSJ10IMk1LVvnUkp1yjdA/view?usp=sharing

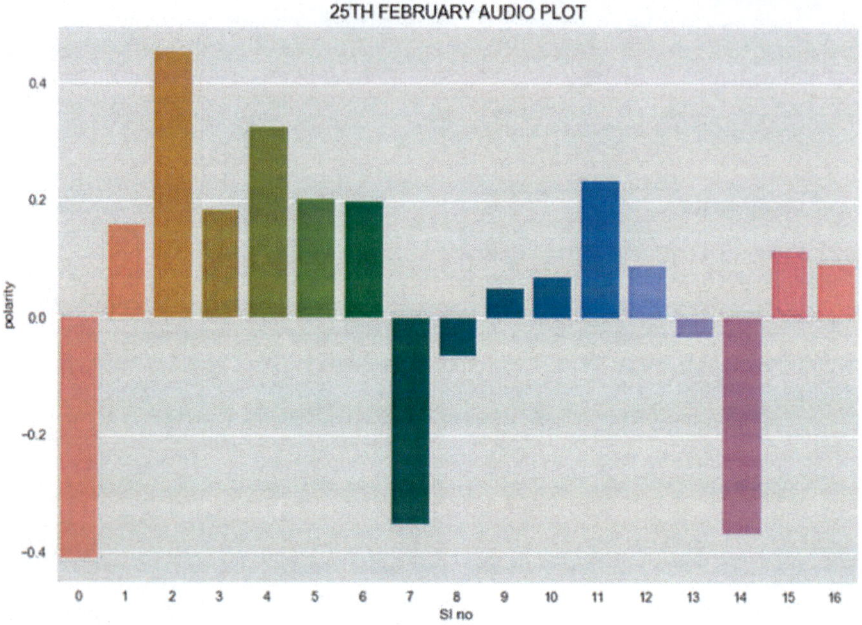

Fig. 6 Polarity Plot for 25th February

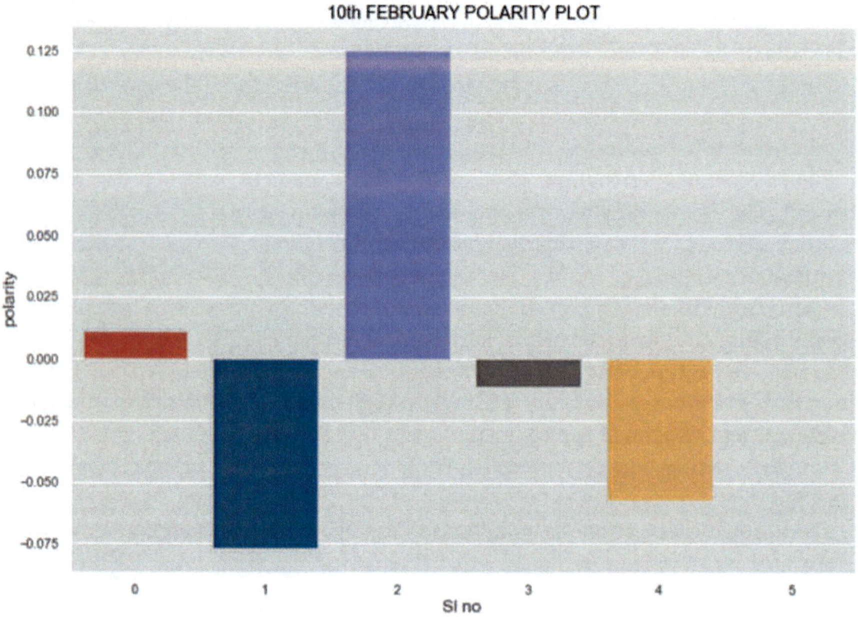

Fig. 7 Polarity Plot for 10th February

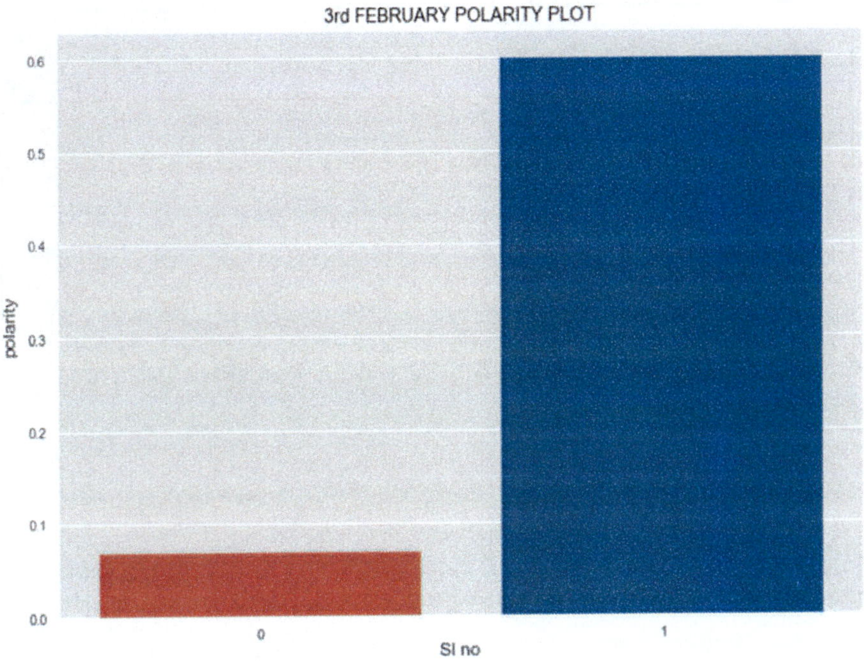

Fig. 8 Polarity plot for 3rd February. *Note* The number of bar plots for one audio depends on the duration in minutes of the audio

This kind of text processing can also be used to check the sentiment of music lyrics. Figures 11 and 12 show an example.

The following link contains code for processing outputs obtained in Fig. 9, 11 and 12.

https://drive.google.com/file/d/1GOQDZqvFyXhvQDOcYKr2qySbiQNUHNvY/view?usp=sharing

For obtaining English song lyrics, AZLyrics API is used.

5 Conclusion

The range of emotion spectrum is different from person to person. It is also found that positive polarity does not always mean positive emotions since magnitude of positive polarity and the emotion linked to it is also different and unequal from person to person. So, to implement this algorithm properly on human emotions, we need to make changes in the range of polarity according to individual differences of people. It should also be noted that predictive analysis is not possible here since emotions experience by a person is highly unpredictable. This is due to the fact that life events are uncertain.

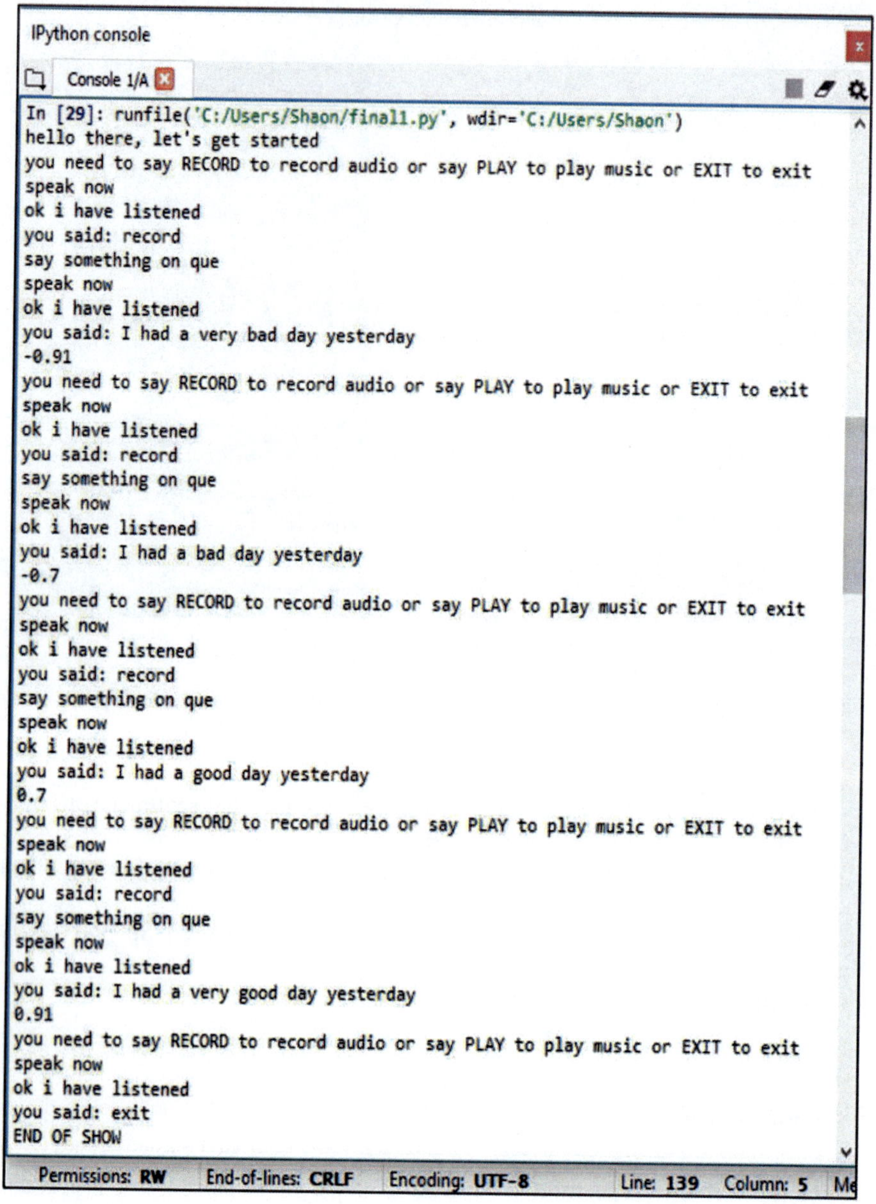

Fig. 9 Comparative analysis of sentences

Table 1 Comparative analysis of sentences

Sentence	Significant word	Polarity
I had a very bad day yesterday	Very bad	−0.91
I had a bad day yesterday	Bad	−0.7
I had a good day yesterday	Good	0.7
I had a very good day yesterday	Very good	0.9

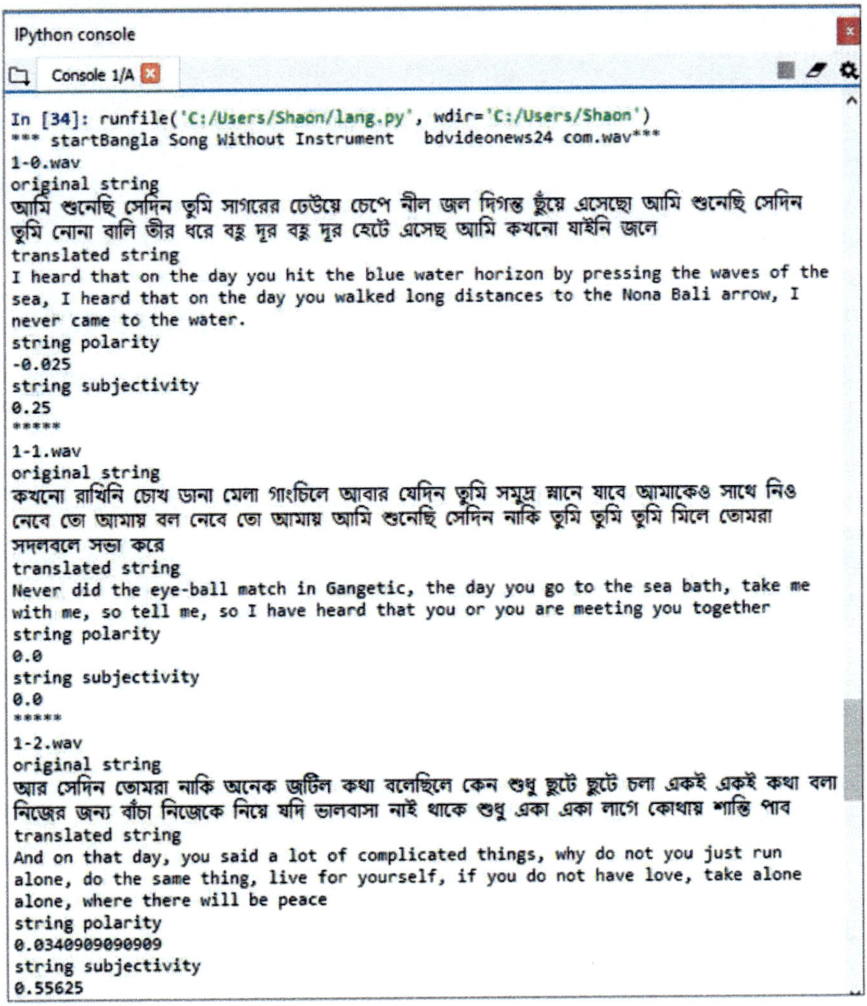

Fig. 10 Polarity calculation of Bengali song

```
IPython console
Console 1/A
In [31]: runfile('C:/Users/Shaon/final1.py', wdir='C:/Users/Shaon')
hello there, let's get started
you need to say RECORD to record audio or say PLAY to play music or EXIT to exit
speak now
ok i have listened
you said: play
music lyrics sentiment analysis
Give name of english songs only
speak song name
ok i have listened
you said: yellow
speak artist name
ok i have listened
you said: Coldplay


Look at the stars
Look how they shine for you
And everything you do
Yeah, they were all yellow

I came along
I wrote a song for you
And all the things you do
And it was called "Yellow"

So then I took my turn
Oh what a thing to have done
And it was all yellow

Your skin
Oh yeah your skin and bones
Turn into something beautiful
Do you know
You know I love you so
You know I love you so

I swam across
I jumped across for you
Oh what a thing to do

'Cause you were all yellow
```

Fig. 11 Polarity calculation of song lyrics(1)

Sentiment analysis has been around in the scientific community since long time. But this is a new way to manipulate it for analyzing human emotions. Keeping a track of polarity in case of human emotion is a field in which very little research has been done using conventional machine learning method. Deep learning will definitely help to improve the results of this paper. It will also allow predictive analysis.

Early detection of depression and other such mental illness is possible through this analysis. Building an app exclusive with this algorithm can help millions. One big problem of people with mental illness is that they do not discuss their emotions with others or are afraid to ask for help. If we can encourage them to confess their feelings

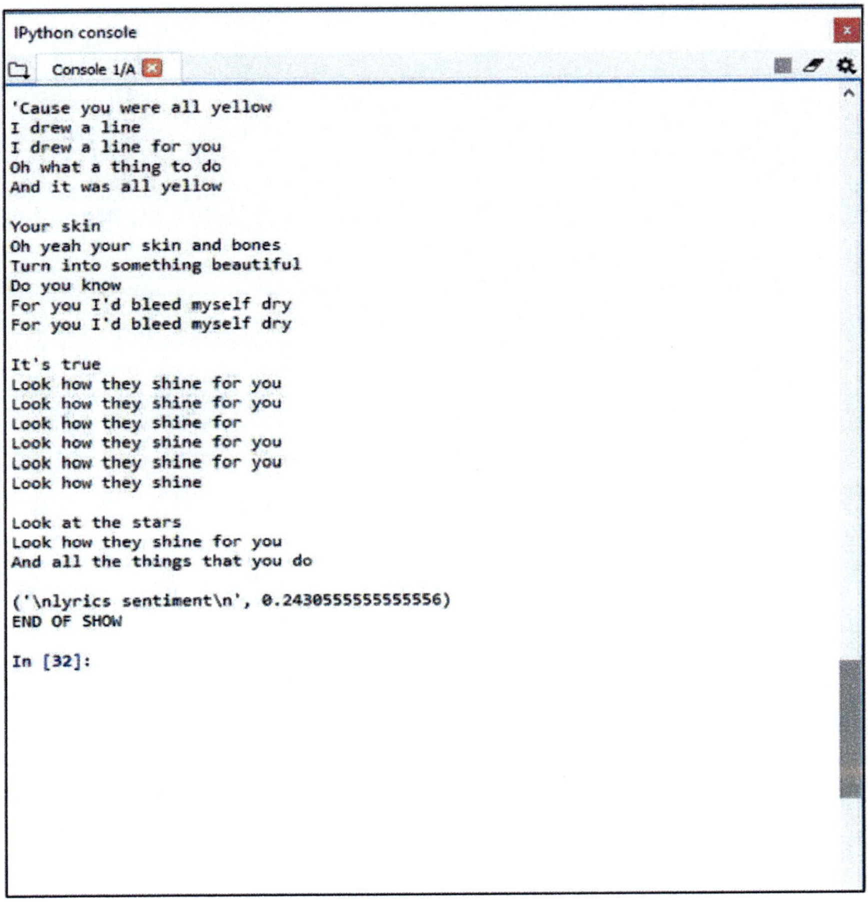

Fig. 12 Polarity calculation of song lyrics(2)

and voice record it, then there is a way to identify their magnitude of emotional reaction. Building a social media platform exclusive for people suffering from mental illness and giving them help can be fruitful too.

References

1. Liu B (2009) Handbook chapter: sentiment analysis and subjectivity. Handbook of natural language processing. Handbook of Natural Language Processing. Marcel Dekker, Inc., New York, NY, USA
2. Sobkowicz P, Kaschesky M, Bouchard G (2012) Opinion mining in social media: Modeling, simulating, and forecasting political opinions in the Web. Gov Inf Quart 29(4):470–479
3. Cambria E (2016) Affective computing and sentiment analysis. IEEE Intell Syst 31(2):102–107

4. Poria S, Cambria E, Howard N, Huang G-B, Hussain A (2016) Fusing audio, visual and textual clues for sentiment analysis from multimodal content. Neurocomputing 174:50–59
5. Zavattaro SM, French PE, Mohanty SD (2015) A sentiment analysis of US local government tweets: the connection between tone and citizen involvement. Gov Inf Quart 32(3):333–341
6. Jiang H, Lin P, Qiang M (2015) Public-opinion sentiment analysis for large hydro projects. J Construct Eng Manage 142(2):05015013
7. Rodrigues RG, das Dores RM, Camilo-Junior CG, Rosa TG (2016) SentiHealth-Cancer: a sentiment analysis tool to help detecting mood of patients in online social networks. Int J Med Inform 85(1):80–95
8. Ceron A, Curini L, Iacus SM, Porro G (2014) Every tweet counts? How sentiment analysis of social media can improve our knowledge of citizens' political preferences with an application to Italy and France. New Media Soc 16(2):340–358
9. Depression and other common mental disorders: global health estimates. Geneva: world health organization (2017) http://apps.who.int/iris/handle/10665/254610, Accessed 25 Mar 2017
10. GBD 2015 (2016) Disease and injury incidence and prevalence collaborators. Global, regional, and national incidence, prevalence, and years lived with disability for 310 diseases and injuries, 1990–2015: a systematic analysis for the Global Burden of Disease Study 2015. Lancet 388:1545–602
11. Pak A, Paroubek P (2010) Twitter as a corpus for sentiment analysis and opinion mining. In: LREc vol 10
12. National Mental Health Survey of India, 2015–16. Prevalence, pattern and outcomes. [website]. Bengaluru: national institute of mental health and NeuroSciences; 2016 http://www.nimhans.ac.in/, Accessed 25 Mar 2017
13. Is Depression at Epidemic Levels?. Mental health foundation of new zealand discussion paper. (January 2012)
14. John T, Cherian A (2001) Depressive disorders in the child andadolescent population (editorial). Indian Pediatr 38:1211–1216
15. Paltoglou G, Thelwall M (2012) Twitter, MySpace, Digg: unsupervised sentiment analysis in social media. ACM Trans Intell Syst Technol 3(4):66
16. Harzing A-W, Alakangas S (2016) Google scholar, Scopus and the web of science: a longitudinal and crossdisciplinary comparison. Scientometrics 106(2):787–804
17. Wildgaard L (2015) A comparison of 17 author-level bibliometric indicators for researchers in Astronomy, environmental science, philosophy and public health in web of science and google scholar. Scientometrics 104(3):873–906
18. Mongeon P, Paul-Hus A (2016) The journal coverage of web of science and scopus: a comparative analysis. Scientometrics 106(1):213–228
19. Adriaanse LS, Rensleigh C (2013) Web of science, scopus and google scholar: a content comprehensiveness comparison. Electron Libr 31(6):727–744
20. Agarwal A, Xi B, Vovsha I, Rambow O, Passonneau R Sentiment analysis of twitter data. Department of Computer Science Columbia University New York, NY 10027 USA
21. Poria S, Chaturvedi I, Cambria E, Hussain A (Dec. 2016) Convolutional MKL based multimodal emotion recognition and sentiment analysis. In: Proceedings IEEE 16th International Conference Data Mining (ICDM). pp. 439–448
22. Poria S, Cambria E, Gelbukh A (2016) Aspect extraction for opinion mining with a deep convolutional neural network. Knowl Based Syst 108:42–49
23. Yasin A, Ben-Asner Y, Menaeison A (Oct. 2014) Deep-dive analysis or the data analytics workload in cloudsuite. In: Proceedings IEEE international sympoisum workload characterization (IISWC). pp. 202–211
24. Chaturvedi I, Cambria E, Poria S, Bajpai R (Dec. 2016) Bayesian deep convolution belief networks for subjectivity detection. In: Proceedings IEEE 16th international conference data mining workshops (ICDMW). pp. 916–923
25. Cambria E, White B (2014) Jumping NLP curves: a review of natural language processing research. IEEE Comput Intell Mag 9(2):48–57

26. Cambria E, Schuller B, Xia Y, Havasi C (2013) New avenues in opinion mining and sentiment analysis. IEEE Intell Syst 28(2):15–21
27. Hsu C-W, Lin C-J (2002) A comparison of methods for multiclass support vector machines. IEEE Trans Neural Netw 13(2):415–425
28. Buczak L, Guven E (2016) A survey of data mining and machine learning methods for cyber security intrusion detection. IEEE Commun Surv Tuts 18(2):1153–1176. 2nd Quart
29. Khan MS, Wahab AWA, Herawan T, Mujtaba G, Danjuma S, Al-Garadi MA (2016) Virtual community detection through the association between prime nodes in online social networks and its application to ranking algorithms. IEEE Access 4:9614–9624
30. Hai Z, Chang K, Kim JJ, Yang CC (2014) Identifying features in opinion mining via intrinsic and extrinsic domain relevance. IEEE Trans Knowl Data Eng 26(3):623–634
31. Akhtar MS, Gupta D, Ekbal A, Bhattacharyya P (2017) Feature selection and ensemble construction: a two-step method for aspect based sentiment analysis. Knowl Based Syst 125:116–135
32. Addo-tenkorang R, Helo PT (2016) Big data applications in operations/supply-chain management: a literature review. Comput Ind Eng 101:528–543
33. Piryani R, Madhavi D, Singh VK (2017) Analytical mapping of opinion mining and sentiment analysis research during 2000–2015. Inf Process Manage 53(1):122–150
34. Poria S, Cambria E, Gelbukh A (2016) Aspect extraction for opinion mining with a deep convolutional neural network. Knowl-Based Syst 108:42–49

ISA-Based PI + FODN AGC of Multi-source System

Ravi Shankar, Utkarsh Raj, and Anjali Singh

Abstract This research work deals with the automatic generation control of two-area multi-source power system. Each control area comprises a reheat thermal, gas, and distributed generation (DG) generating units. Physical constraints like generation rate constraint (GRC) and governor dead-band (GDB) are considered for the thermal unit. PI + FODN is used for the secondary AGC control whose parameters are optimized using ISA optimization technique. 1% step load perturbation (SLP) is taken in area-1 to study the dynamics of the system. The performance of the proposed controller is compared with the ISA-based PIDN controller which shows the superiority of the proposed control scheme.

Keywords AGC · LFC · ISA · PI + FODN · PIDN

1 Introduction

Modern-day power system areas are interconnected with each other. It wants to maintain the frequency and tie-line power flow to its scheduled values automatically. This is achieved using the concept of automatic generation control (AGC) or load frequency control (LFC). The purpose of AGC is to maintain the balance between the load demand and the power generation automatically [1].

Researchers have focused on the AGC of conventional generating systems like thermal, hydro, and gas. Parmar et al. have considered a thermal, hydro, and gas

---

R. Shankar · A. Singh
Electrical Engineering Department, National Institute of Technology, Patna, India
e-mail: ravi@nitp.ac.in

A. Singh
e-mail: singhanjali296@gmail.com

U. Raj (✉)
Electrical & Electronics Engineering Department, Motihari College of Engineering, Motihari, India
e-mail: utkarshraj18@gmail.com

V. Nath and J. K. Mandal (eds.), *Proceedings of the Fourth International Conference on Microelectronics, Computing and Communication Systems*, Lecture Notes in Electrical Engineering 673, https://doi.org/10.1007/978-981-15-5546-6_10

system for the AGC of a two-area interconnected power system [2]. With the increased integration of non-conventional sources in the power system, authors have now started to investigate its effect on the dynamics of the system. Saha et al. have investigated the effect of DG and EV on the AGC of the multi-area power system [3].

For the effective AGC of the system, authors have suggested various control mechanisms. Conventional controllers like I, PI, and PID have been used extensively by the researchers for the secondary AGC of the system [4–6]. To overcome the drawbacks of the conventional controllers, the effectiveness of intelligent controllers like fuzzy controllers, ANN controllers, MPC controllers, etc., has been investigated by the authors [7–9]. Cascade controllers have the advantage of using the benefits of two controllers at the same time. Conventional controllers generally have integral parameters. To increase the effectiveness of the controller, parameter values between 0 and 1 are also allowed in fractional controllers.

For obtaining the best result from the controller, its parameters need to be properly optimized. For this purpose, various meta-heuristic techniques have been proposed by the researchers. Techniques like GA, PSO, WOA, and FOA have applied extensively for the optimization of the controller parameters [10–12]. Interactive search algorithm (ISA) is a new addition to the family of swarm-based optimization techniques proposed by Mortazavi et al. [13]. ISA is a hybrid of iPSO and TLBO algorithms, including the advantages of both while eliminating their limitations. The remaining part of the work is summarized as follows: The model of the proposed system is discussed in Sect. 2. Section 3 briefly explains the ISA algorithm. Section 4 gives the simulation results while Sect. 5 concludes the present research work.

2 Modeling of System

2.1 Proposed System

A linearized model of two-area multi-source power system is considered for the AGC analysis. Each area consists a thermal, DG, and gas generating unit. The reheat thermal generating unit incorporates the effect of physical constraints like governor dead-band (GDB) and generation rate constraint (GRC). The DG unit consists of geothermal power plant (GTPP), wind turbine system (WTS), solar photovoltaic system (SPV), and diesel engine generator (DEG) set. The transfer function model of the DG system is shown in Fig. 1 while that of the whole proposed system is shown in Fig. 2.

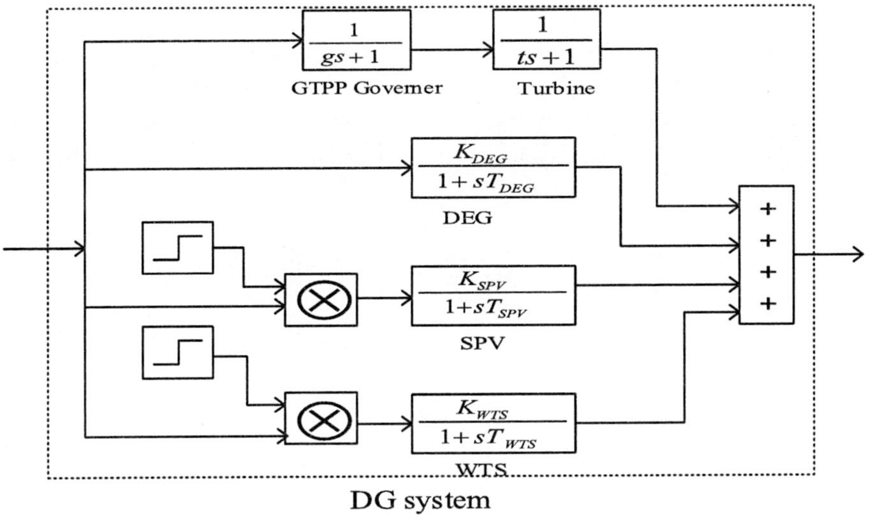

Fig. 1 Transfer function model of DG system

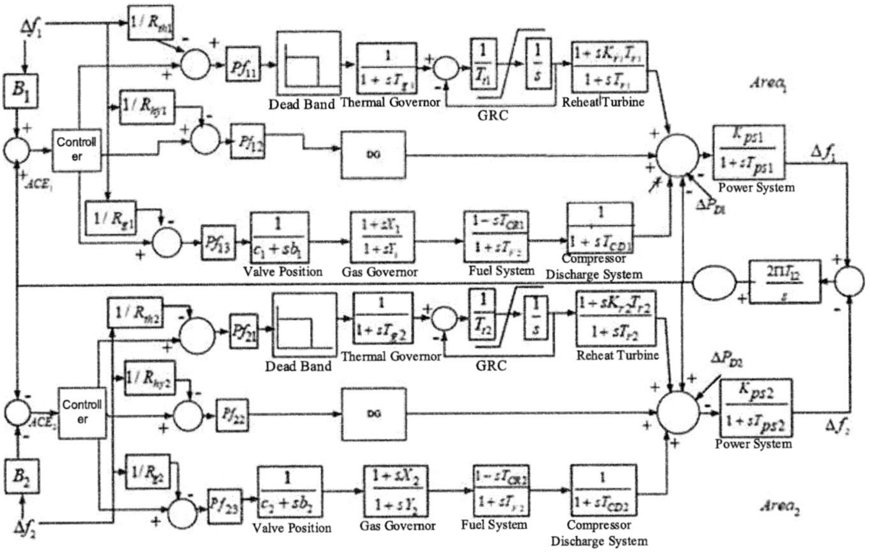

Fig. 2 Transfer function model of proposed two-area system

2.2 Proposed Controller

Conventional controllers suffer from various drawbacks, like slower speed of response and lesser noise rejection capability. The proposed controller is a sum of

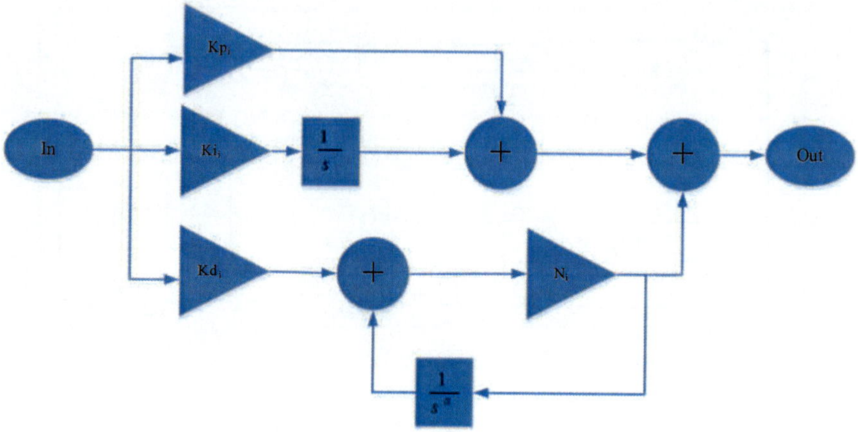

Fig. 3 Block diagram of proposed controller

proportional-integral (PI) and fractional-order derivative controller with filter (PI + FODN). Fractional-order controllers have the freedom to choose parameters values from 0 to N, while the conventional integral controllers can only choose integral values between 0 and N where $N \in (0, \infty)$. The various constraints applied to the controller parameters values based on practical experiences are as follows:

$$0 \le Kp_i, Ki_i, Kd_i \le 1 \tag{1}$$

$$0 \le \alpha_i \le 2 \tag{2}$$

$$0 \le N_i \le 100 \tag{3}$$

where $i = 1, 2$. The block diagram of the proposed controller is shown in Fig. 3.

3 Interactive Search Algorithm (ISA)

ISA combines the advantages of both the iPSO and TLBO. For complex problems, the best agent mostly gets stuck in the local optimum regions. To overcome this limitation of iPSO, the tracking and interacting paradigm of TLBO algorithm is assimilated within the iPSO algorithm to generate the ISA optimization technique. Also, in constrained optimization problems, an improved flyback (IFB) mechanism is used in the process of particle updation when there is a case of violation of constraint [13]. The algorithm of the ISA optimization technique is shown in Fig. 4.

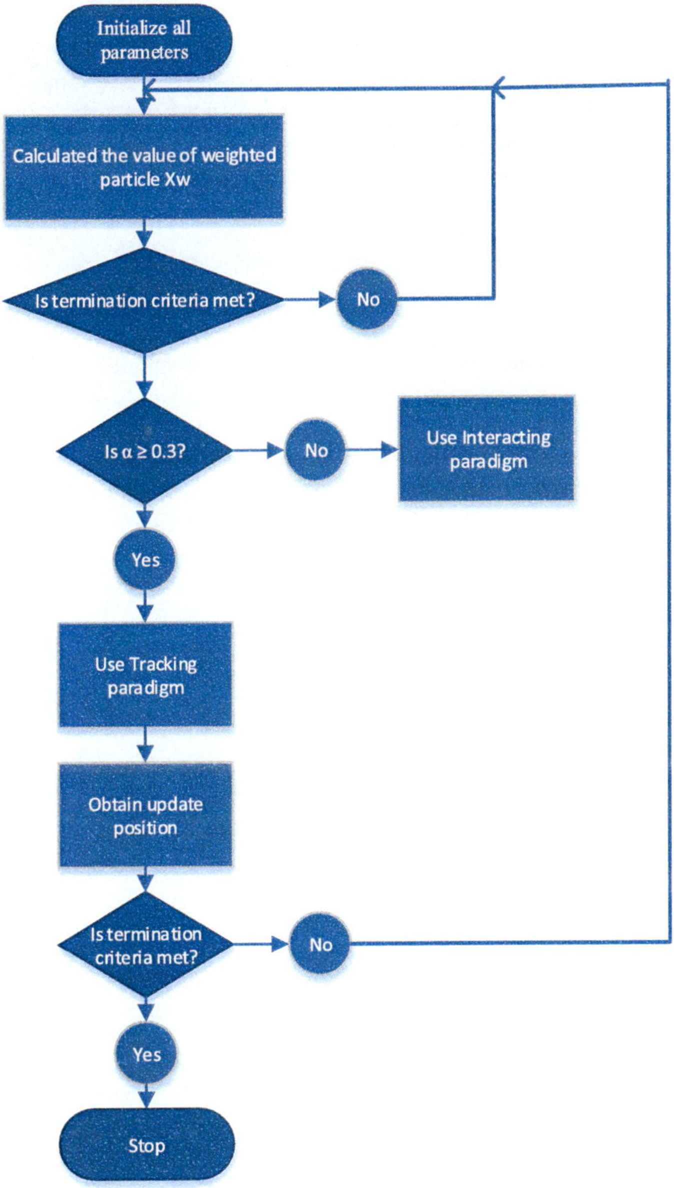

Fig. 4 Flowchart of ISA algorithm

4 Simulation Results

The proposed system model is simulated in the MATLAB® environment. A step load perturbation (SLP) of 1% is considered for the study of system dynamics. The objective function for the proposed study is the integral square error of the system which can be written in mathematical form as shown below:

$$ISE = \int_{0}^{t} (\Delta f_1^2 + \Delta f_2^2 + \Delta P_{tie}^2)dt \tag{4}$$

The proposed system is simulated to analyze the dynamics of the system. The parameters of the proposed controller and PIDN controller are optimized using ISA optimization technique and have been considered as the secondary controller for the system one at a time. The comparison between both the control methodologies is shown in Figs. 5, 6, and 7 while it is summarized in Table 1. The convergence profile for both the ISA optimized control schemes are shown in Figs. 8 and 9. The optimized parameter values of the proposed controller and PIDN controller are shown in Table 2 and Table 3, respectively.

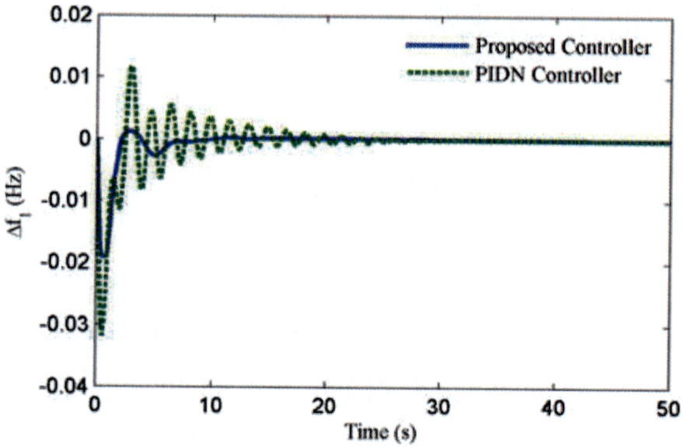

Fig. 5 Frequency deviation in area-1

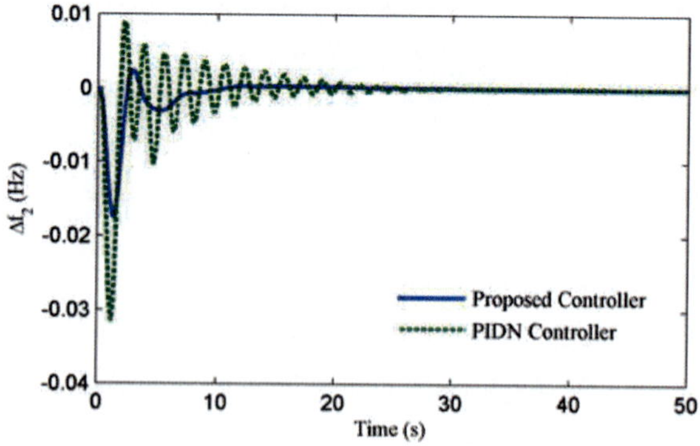

Fig. 6 Frequency deviation in area-2

Fig. 7 Tie-line power flow deviation

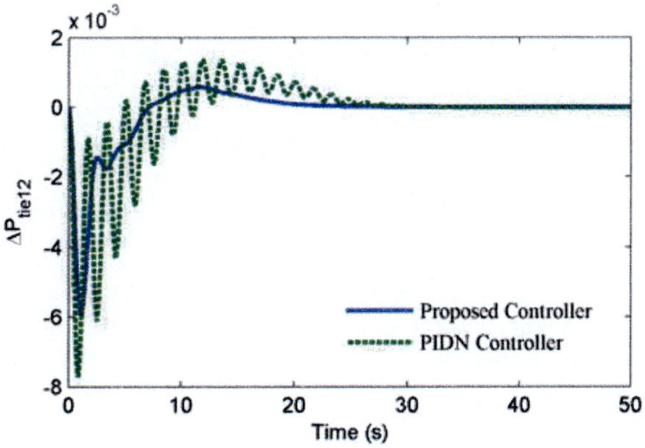

Table 1 Performance comparison of proposed controller with PIDN

	Proposed controller	PIDN controller
Δf_{1max}	−0.01,918	−0.03176
Δf_{2max}	−0.01741	−0.03141
$\Delta P_{tie,max}$	−0.00593	−0.00769

Fig. 8 Convergence profile of ISA-based proposed controller

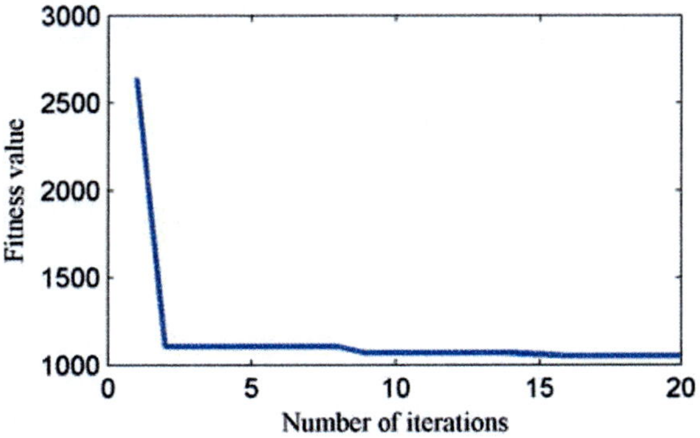

Fig. 9 Convergence profile of ISA-based PIDN controller

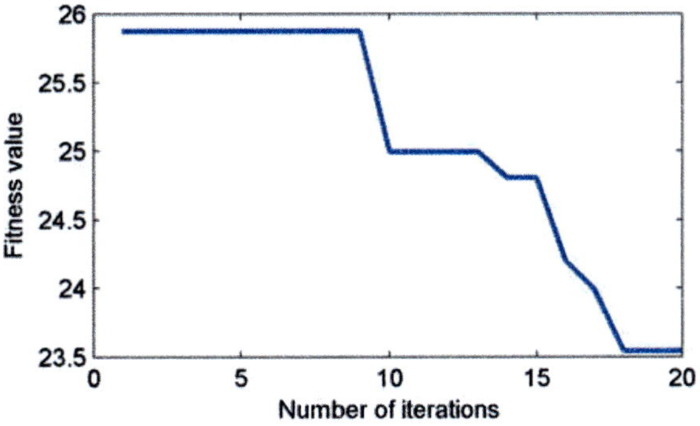

Table 2 Optimized parameters of proposed controller

Parameter	Optimized value
Kp_1, Kp_2	0.9388, 0.2359
Ki_1, Ki_2	0.2212, 0.6494
Kd_1, Kd_2	0.2240, 0.9388
N_1, N_2	13.4884, 93.8828
$\alpha_1. \alpha_2$	2.0000, 1.7312

Table 3 Optimized parameters of PIDN controller

Parameter	Optimized value
Kp_1, Kp_2	0.5123, 0.3276
Ki_1, Ki_2	0.8257, 0.2178
Kd_1, Kd_2	0.9394, 0.6619
N_1, N_2	99.8466, 78.8547

5 Conclusion

In the proposed work, a two-area multi-source power system has been frequency stabilized using ISA optimized PI + FODN controller. The simulation result of the proposed system for 1% SLP has been shown in the present work. The proposed control methodology shows improvement in settling time and peak deviation over the ISA optimized PIDN control scheme.

Appendix

Thermal power plant: $K_r = 0.3$, $T_r = 10\ s$, $T_t = 0.3\ s$, $R_{th} = 2.4$ Hz/pu, $T_g = 0.08\ s$, GRC = 0.0017pu/MW/s.

Gas power plant: $X = 0.6\ s$, $Y = 1.0\ s$, $b = 0.05\ s$, $c = 1$, $T_F = 0.23\ s$, $T_{CR} = 0.01\ s$, $T_{CD} = 0.2\ s$, $R_G = 2.4$ Hz/pu MW, $R_{th} = 2.4$ Hz/pu.

Power system: $K_{ps} = 120$ Hz/pu, $T_{ps} = 20\ s$;

DG: WTS: $K_{\mathrm{WTS}} = 1$ and $T_{\mathrm{WTS}} = 1.5\ s$; PV: $K_{\mathrm{PV}} = 1$ and $T_{\mathrm{PV}} = 1.8\ s$; DEG: $K_{\mathrm{DEG}} = 1/300$ and $T_{\mathrm{DEG}} = 2\ s$ EV: $K_{\mathrm{EV}} = 1$ and $T_{\mathrm{EV}} = 1$.

References

1. Shiva CK, Mukherjee V (2015) A novel quasi-oppositional harmony search algorithm for automatic generation control of power system. Appl Soft Comput J 35:749–765

2. Parmar KPS, Majhi S, Kothari DP (2014) LFC of an interconnected power system with multi-source power generation in deregulated power environment. Int J Electr Power Energy Syst 57:277–286
3. Saha A, Saikia LC (2018) Performance analysis of combination of ultra-capacitor and super-conducting magnetic energy storage in a thermal-gas AGC system with utilization of whale optimization algorithm optimized cascade controller. J Renew Sustain Energy 10(1)
4. Padhan DG, Majhi S (2013) A new control scheme for PID load frequency controller of single-area and multi-area power systems. ISA Trans 52(2):242–251
5. Mohanty B, Panda S, Hota PK (2014) Controller parameters tuning of differential evolution algorithm and its application to load frequency control of multi-source power system. Int J Electr Power Energy Syst 54:77–85
6. Shankar R, Bhushan R, Chatterjee K (2016) Small-signal stability analysis for two-area inter-connected power system with load frequency controller in coordination with FACTS and energy storage device. Ain Shams Eng J 7(2):527–541
7. Mohamed TH, Bevrani H, Hassan AA, Hiyama T (2011) Decentralized model predictive based load frequency control in an interconnected power system. Energy Convers Manag 52(2):1208–1214
8. Mortazavi A et al (2017) Maiden application of an sine–cosine algorithm optimised FO cascade controller in automatic generation control of multi-area thermal system incorporating dish-Stirling solar and geothermal power plants. Int J Electr Power Energy Syst 20(5):1–17
9. Guha D, Roy PK, Banerjee S (2016) Load frequency control of large scale power system using quasi-oppositional grey wolf optimization algorithm. Eng Sci Technol Int J 19(4):1693–1713
10. Shankar R, Kumar A, Raj U, Chatterjee K (2018) Fruit fly algorithm-based automatic generation control of multiarea interconnected power system with FACTS and AC/DC links in deregulated power environment. Int Trans Electr Energy Syst (August 2017):e2690
11. Roy R, Bhatt P, Ghoshal SP (2010) Evolutionary computation based three-area automatic generation control. Expert Syst Appl 37(8):5913–5924
12. Hasanien HM (2018) Whale optimisation algorithm for automatic generation control of inter-connected modern power systems including renewable energy sources. IET Gener Transm Distrib 12(3):607–614
13. Mortazavi A, Toğan V, Nuhoğlu A (2018) Interactive search algorithm: a new hybrid metaheuristic optimization algorithm. Eng Appl Artif Intell 71(October 2017):275–292

Assessing Overall Software Defect-Based Risk Using Analytic Hierarchy Process

Pooja Jha and K. S. Patnaik

Abstract Risk analysis and mitigation methods are gaining momentum among researchers and software professionals for preventing software project failures. It involves methods to deal with project uncertainties so that the project can become successful overcoming all threats. In this context, determining the potential risk factors at each phase of software development lifecycle and relating those risks to the defects they produce is important. Analytic Hierarchy Process (AHP) is used to determine the priority of defects that leads to risk. AHP helps in assessing risk for each phase of an agile project. As agile models primarily help projects to accommodate change requests, and with each change request, a new risk items is generated, and AHP-based risk analysis becomes extremely relevant to agile software development. This paper investigates use of AHP for effective decision making to identify and remove defects. Weighted criteria are used for evaluating the decision. To implement meaningful weighted criteria, a case study is conducted among 20 software professionals with various experience levels. Total 35 potential risk source factor-based questions were evaluated according to Saaty's scale.

Keywords Risk · Mitigation · Priority · Defect · Scale · Weight

1 Introduction

Software development projects are usually designed in stages, and each stage involves multiple activities. These activities are prone to risks. Several research papers on risk management in software projects have addressed the risk factors associated with the project [1]. If these factors are not identified early, then they can be held responsible

P. Jha (✉)
Department of CS/IT, Amity University Jharkhand, Ranchi 834001, India
e-mail: pooja.jha.ism@gmail.com

K. S. Patnaik
Department of CSE, BIT Mesra, Ranchi 835215, India
e-mail: kspatnaik@bitmesra.ac.in

V. Nath and J. K. Mandal (eds.), *Proceedings of the Fourth International Conference on Microelectronics, Computing and Communication Systems*, Lecture Notes in Electrical Engineering 673, https://doi.org/10.1007/978-981-15-5546-6_11

for the success or failure of the project. Various techniques of risk identification [2] and categorization [3–6] have been dealt along with analyzing the remedial approach. However, the fact that risk by nature is uncertain leaves the outcome of software projects to some element of chance.

The relationship between software project performance and risks involved in it is iteratively examined during different phases of SDLC. It is advisable for all project managers to apply different risk mitigation strategies in order to avoid software project failure. Nowadays, the client requirements are changing frequently, therefore agile software development which involves dividing the entire project into mini-projects and trying to solve most risky elements first. Risk analysis helps in combating threats before they actually cause a disaster.

Risks are encountered in each phase of SDLC. These risks can cause changes in important project parameters like cost, budget, and slip of schedule. Some risks become evident at later phases of SDLC. There are different types of risks, and each of them has separate severity level. Boehm coined the term risk exposure (RE) which is defined as the probability of risk happening and the loss it causes to affected parties [6]. It is important to know about the risks involved in various phases of SDLC and the impacts of these risks for software projects. If we are able to assess the risks involved in each phase separately, then we can investigate ways to reduce or minimize them in a more systematic way. Researchers have made a thorough study on risk using various methods and techniques. Several risk mitigation frameworks like continuous risk management (CRM), Chaos model [14], and earth moon model [10] are available. Software risk management provides feasible practice to improve the development process and control harmful elements in development processes. Its goal involves identifying and dealing with risks early in development, reducing long-term costs, and helping in preventing software disasters [7]. Each sprint of agile software projects gradually nullifies the involved performance and integration risks. But the popular expected values adopted in risk management cannot describe the uncertainty exactly, and it has caused some misunderstandings [8]. Software development risks depict the uncertainty which eventually creates a loss for the project. If the risks are identified and mitigated early, it becomes beneficial for the software project.

The Analytic Hierarchy Process (AHP) is a multi-criteria decision-making method, which was introduced by Saaty [9] in 1970s. AHP gained attention among researchers due to its mathematical construct and easy availability of input data. AHP helps in solving complex problems related to decision making. Comparative techniques are used for pairwise mapping in AHP. Weights of decision criteria are derived from these comparisons. A mechanism for improving consistency is followed if comparisons are not consistent. AHP helps in quantifying risks in each phase of SDLC. This paves frequent incremental release of working software.

Even though AHP is applied for effective decision making in various domains, it has been criticized from several perspectives. Rank reversal is a prominent problem regarding AHP [10]. Although AHP has seen major controversies [11–14], it has been practiced frequently. Weather forecasting, predicting election results, and fixing selection procedure in motor vehicle industry are several applications where AHP is widely used.

In this paper, we attempt to use AHP to determine the priority of defects that needs to be removed if they have been detected. Here, we have considered defects identified during requirement analysis, design, coding, and testing phase. We attempt to show how the priority of defect removal decision in each particular phase has influence when compared with other phases. Weighted parameters are used for evaluating the decision outcome. To build appropriate weighted criteria, we conducted a case study involving interviewing 20 software professionals with various experience level in CMM Level 5 multinational companies. Total 35 potential risk source factors were questioned, and answers were tabulated according to Saaty's scale. This case study is helpful in demonstrating the weights assigned to several criteria for decision making using AHP.

The paper is organized as follows. Section 2 describes the research procedure. Section 3 represents our survey results in statistical form where data is visualized by boxplot and summarized by ANOVA analysis. Section 4 concludes the study justifying several aspects of AHP usage considering large-scale projects and agile software development approach.

2 Research Method

The following steps are observed while taking a decision for defect removal using AHP:

1. ***Build decision-making model***: The decision is categorized into a series of goals, criteria, and alternatives. The goal is subdivided into intermediate goals for taking advantage of incremental development.
2. ***Assigning weights***: The weight of a criterion is **determined** by pairwise comparison with regard to a goal. A consistency index is then derived to ensure reasonable consistency with transparency.
3. ***Derived Priorities***: All the derived priorities are cumulated to obtain a weighted sum. The weight of individual criteria thus contributes to overall priority. This overall accumulated priority may be viewed as the ultimate goal.
4. ***Effect of change of weight determination***: Changes in weights of each criteria create an impact on overall priority. By adjusting the weight of each criteria, the reason behind taking a decision can be explained.

A judgment can be given based on the results of sensitivity analysis.

2.1 Developing a Model

The initial step in AHP analysis involves decision modeling. This step includes constructing a hierarchy to analyze the decision. Here, by decision means which

Fig. 1 Software project risk source

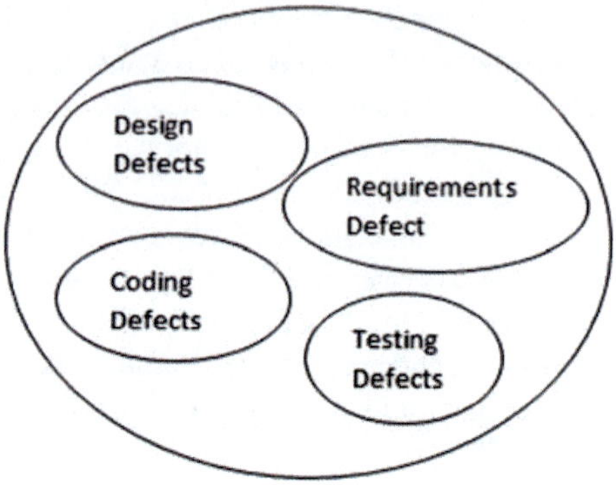

phase must be treated for defect removal on a priority basis so that the development process is not disturbed. Figure 1 shows the different source of risk factors.

2.2 *Assigning Weights to Criteria*

As every criterion has different importance level, the next step in AHP is the derivation of comparative priorities.

While developing software, several criteria contributes differently and so their weight age varies accordingly. For instance, a software programmer may allot more importance to coding rather than design and testing, while an end user may give more significance to the design aspect. Because of this relativity, we require to develop pairwise comparisons first, for each criterion using Saaty's scale [6] as shown in Table 1. Criteria comparison also helps in delivering incremental version of the implemented software where each version encompasses a subset of the requirements.

A comparison matrix is created in Table 2 after performing pairwise comparison of each criterion with respect to each other. This process of setting relative weight is also called the intensity judgment method. As we conducted the case study involving 20 software professionals, we obtain the average of the weights provided by them. AHP helps in improving process into projects by eliminating most risky elements first; hence, the study exemplifies practical implementation of AHP. Based on the

Table 1 Saaty's pairwise comparison scale

S. No	Judgment Categorization	Numeric Value
1	Extremely important	9.8
2	Very Strongly more important	7.6
3	Strongly more important	5.4
4	Moderately more important	3.2
5	Equally moderately more	1

Table 2 Pair-wise comparison matrix of criteria of overall software project risk with fractional values

Overall software Project risk	Requirement analysis defects	Design defects	Coding defects	Testing defects
Design defects	6.82/6.74	6.74/6.74	6.87/6.74	7.21/6.74
Coding defects	6.82/6.87	6.74/6.87	6.87/6.87	7.21/6.87
Testing defects	6.82/7.21	6.74/7.21	6.87/7.21	7.21/7.21
Requirement analysis defects	6.82/6.82	6.74/6.82	6.87/6.82	7.21/6.82

Table 3 Pair-wise comparison matrix of criteria of overall software project risk with decimal values

Overall software project risk	Requirement analysis defects	Design defects	Coding defects	Testing defects
Design defects	1.011	1	1.019	1.0697
Coding defects	0.9927	0.9810	1	1.0494
Testing defects	0.9459	0.9348	0.9528	1
Requirement analysis defects	1	0.9882	1.0073	1.0571

collected data, the calculated risk factor weights owing to requirement analysis defect is 6.82, design defect is 6.74, coding defect is 6.87, and testing defect is 7.21. We have depicted all the values in Table 2.

Table 3 shows the fractional values in decimal. For example, the ratio of the importance of design defects versus the importance of coding defects is 6.74/6.87 (design defects weight/coding defects weight).The importance of coding defects with respect to design defects will yield the reciprocal of this value (coding defects weight age/design defects weight age = 6.87/6.74) as shown in the coding defects-design defects cell in Table 2. In the comparison matrix of Table 3, it can be seen that when the credence of a criterion is compared with itself, for instance, coding defects versus coding defects, design defects versus design defects, requirement analysis defects versus requirement analysis defects or testing defects versus testing defects, then the input value is 1 which corresponds to equal importance as Saaty's scale. AHP methodology has its own advantage that it is very simple to use. This simplicity matches well with the simplicity of an agile project as an agile project is nothing but an iterative combination of mini waterfall models. The progress of an agile project is measured in terms of deliverable artifacts (SRS document, detail design document, test plan, code review, etc.). So it is important to use hierarchical process for ranking phase priorities. Even though many factors may be involved in making a decision, AHP methodology only requires two factors at a time. For calculating the overall priorities or weights of the criteria, we have selected the approximate method rather than the exact method due to its simplicity. The approximate method requires addition of the values in each column of Table 3.

Next, divide each cell value of Table 4 by the total of the column. The result is shown in Table 5.

This is followed by obtaining the overall priorities (Table 6) by simply calculating the average value of each row. For the first row: (0.256 + 0.256 + 0.256 + 0.2561)/4 = 0.256.

The comparison matrix is shown in Table 4 with the original judgments, and the calculated priorities are noted in Table 6. This is a practical way to see the judgments and priorities at the same time (Table 7). According to the results in Table 7, we give more importance to the design defect criterion (25.6%), followed by requirement analysis defect (25.3%). The coding defect factor priority attains third place (25.1%) in our overall software project risk decision. Testing defects priority attains the last position (23.9%). These priorities are derived based on our survey. As the priorities have a mathematical construct and the measurement values are derived from a ratio scale (Saaty's), AHP is quantifying risks figuratively rather than by intuition. Risk is

Table 4 Table column addition

Overall software Project risk	Requirement analysis defects	Design defects	Coding defects	Testing defects
Design defects	1.011	1	1.019	1.0697
Coding defects	0.9927	0.9810	1	1.0494
Testing defects	0.9459	0.9348	0.9528	1
Requirement analysis defects	1	0.9882	1.0073	1.0571
Sum	3.9496	3.904	3.9791	4.1762

Table 5 Normalized matrix

Overall software Project risk	Requirement analysis defects	Design defects	Coding defects	Testing defects
Design defects	0.256	0.256	0.256	0.2561
Coding defects	0.251	0.251	0.2513	0.2512
Testing defects	0.240	0.239	0.2394	0.2394

Table 6 Calculation of priorities (row averages)

Overall software project risk	Requirement analysis defects	Design defects	Coding defects	Testing defects	Priority
Design defects	0.256	0.256	0.256	0.2561	0.256
Coding defects	0.251	0.251	0.2513	0.2512	0.251
Testing defects	0.240	0.239	0.2394	0.2394	0.239
Requirement analysis defects	0.253	0.253	0.2531	0.2531	0.253

Table 7 Presentation of results (original judgments and priorities)

Overall software project risk	Requirement analysis defects	Design defects	Coding defects	Testing defects
Design defects	1.011	1	1.019	1.0697
Coding defects	0.9927	0.9810	1	1.0494
Testing defects	0.9459	0.9348	0.9528	1
Requirement analysis defects	1	0.9882	1.0073	1.0571

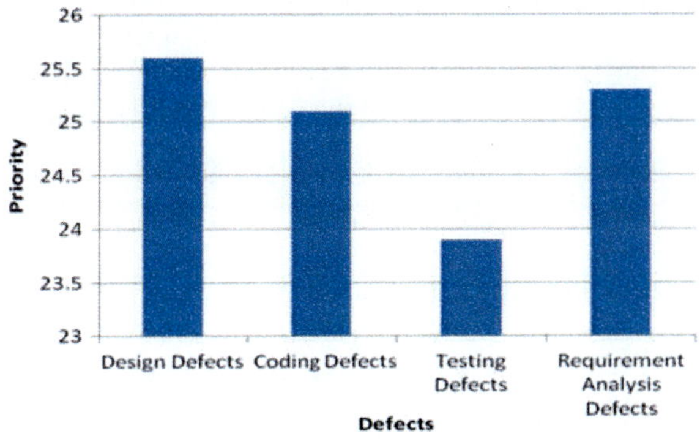

Fig. 2 Priority defects in different SDLC stages

an important iteration driver in case of agile projects. Assigning weights arbitrarily based on some unexercised guesswork creates misleading ranking. Toward this limitation, we conducted our survey for getting actual data from experienced people who regularly instrument risk management framework (Fig. 2).

2.3 Consistency

As the values of each weight are derived from individual's opinion, there will be some inconsistency in the final matrix of decision. But there is an acceptable limit to inconsistency. Consistency ratio (CR) is calculated with the help of derivation of consistency index (CI) of the matrix with reference to randomly generated matrix. A random matrix is highly inconsistent. Table 8 shows RI value for matrices of different dimension [9].

Table 8 Consistency indices for a random matrix

n	3	4	5	6
RI	0.58	0.9	1.12	1.24

The consistency ratio is CI/RI. A consistency ratio (CR) of 0.10 or less is acceptable for performing AHP analysis [6]. However, a consistency ratio which is greater than 0.10 indicates revision in the judgments is required to correct the inconsistency. The steps for estimating this value are as follows:

1. Begin with the matrix of decision comparison and derived priority (Table 6). It is restated in Table 9.
2. Priorities are utilized as factors (weights) for each column in Table 10.
3. Each value in the first column of the decision matrix in Table 10 is multiplied by the first column of Table 11. This process is continued for all the columns of the comparison matrix Table 11 shows the resulting matrix.
4. The values in each row are added to obtain weighted sum.
5. The weighted sum vector is divided by the corresponding priority of each criterion as shown in Table 12. The average of the values is called λ_{max}

$$\lambda_{max} = \frac{(3.996 + 4 + 4.002 + 3.9972)}{4} = 3.9975$$

6. The consistency index (CI) is calculated by:

$$CI = (\lambda_{max} - n)/(n - 1)$$

where n is the amount of compared elements (in our example $n = 3$). Therefore,

Table 9 Prioritization matrix

Overall software project risk	Requirement analysis defects	Design defects	Coding defects	Testing defects	Priority
Design defects	1.011	1	1.019	1.0697	0.256
Coding defects	0.9927	0.9810	1	1.0494	0.251
Testing defects	0.9459	0.9348	0.9528	1	0.239
Requirement analysis defects	1	0.9882	1.0073	1.0571	0.253

Table 10 Prioritization as factors

Overall software project risk	Requirement analysis defects	Design defects	Coding defects	Testing defects
Criteria weights >	0.253	0.256	0.251	0.239
Design defects	1.011	1	1.019	1.0697
Coding defects	0.9927	0.9810	1	1.0494
Testing defects	0.9459	0.9348	0.9528	1

Table 11 Calculation of weighted columns

Overall software project risk	Requirement analysis defects	Design defects	Coding defects	Testing defects	Weighted sum
Criteria weights >	0.2557	0.256	0.2557	0.2556	1.023
Design defects	0.2511	0.2511	0.2510	0.2508	1.004
Coding defects	0.2393	0.2393	0.2391	0.239	0.9567
Testing defects	0.2530	0.2529	0.2528	0.2526	1.0113

Table 12 Derivation of λ_{max}

Weighted sum/Priority	
1.023/0.256	3.996
1.004/0.251	4
0.9567/0.239	4.002
1.0113/0. 253	3.9972
Total	15.99
Divide total by 4 to obtain λ_{max} which is coming as 3.9975	

$$\mathrm{CI} = (\lambda_{max} - n)/(n - 1) = (3.9975 - 4)/(4 - 1) = 0.0025$$

7. The consistency ratio is calculated by

$$\mathrm{CR} = \mathrm{CI}/\mathrm{RI} = 0.\,0025/0.58 = 0.00431$$

As the value of 0.00431 for CR is less than 0.10, we can believe our judgment matrix is practically reliable and worthy of making decision making using AHP (Fig. 3).

3 Statistical Analysis

The boxplot data distribution indicates that the software professionals cumulatively provided highest weightage to testing phase (mean value of 7.21) of SDLC. After that comes the coding (mean value of 6.87), requirement analysis (mean value of 6.82), and design phase (mean value of 6.74), respectively. However, after applying AHP methodology, we notice that design defect gets highest attention (25.6% weightage) followed by requirement analysis defect (25.3%), coding defect (25.1%), and testing defect (23.9%). The difference in ranking of the phases before applying AHP and after applying AHP is true, but the notice worthy fact is that the difference in values is almost negligible. Because of popularity of agile methodology, both development

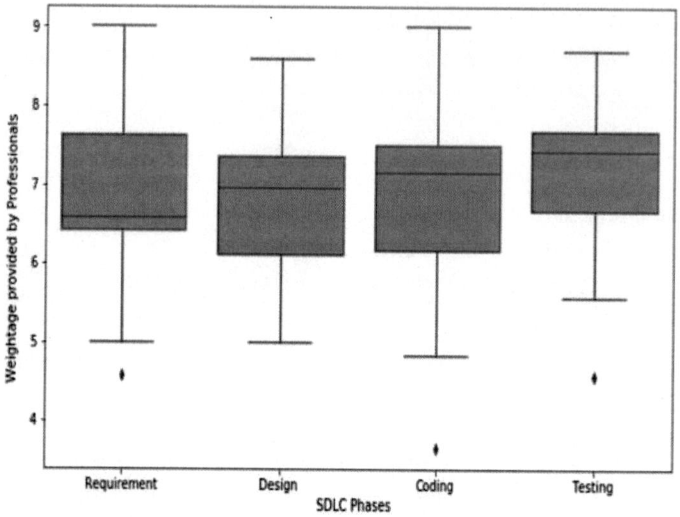

Fig. 3 Boxplot charting of collected data from software professionals for several SDLC phases

and testing activities are becoming concurrent and we are shifting toward TDD (Test Driven Development) where every iteration conducts its own testing phase. So an agile programmer, who is responsible for delivering artifacts every 2–4 weeks, may conduct coding and testing simultaneously from his viewpoint. Boxplot data visualization also supports the fact that there is very low difference among weightage values of several SDLC phases. As we are shifting toward agile paradigm where sprint-oriented delivery is getting popularity, all phases are starting right from day 1. So quite a lot of SDLC phases are becoming equally important to finish the project on time.

We also performed ANOVA analysis on our collected data from survey. One way ANOVA F test produces F value of 13.604 and p value of 2.29e−08. As p value is less than level of significance value (0.05) and F value is high, the null hypothesis (weightage values for all phases are equal) can be rejected

In this section, it is explained the results of research and at the same time is given the comprehensive discussion. Results can be presented in figures, graphs, tables, and others that make the reader understand easily [2–5]. The discussion can be made in several subchapters.

4 Conclusion

AHP methodology helps us to determine which defect produces more risk and needs to be mitigated first. Consistency index indicates the degree of consistency. We can conclude—Even though the difference in percentage values of several defects is small, the marginal difference becomes significant as software projects are mainly of budgets of multi million dollars. Large-scale projects demand the AHP methodology to manage risks. Our analysis based on the conducted case study indicates AHP facilitates multi-criteria decision making (MCDM) and bridges the gap between ordinal scale and ratio scale. According to AHP, design defects demand the highest priority followed by requirement analysis defects, coding defects, and testing defects.

We can also conclude that it is beneficial to fix a defect in earlier stages (design, feasibility study, etc.) of SDLC first, and then move toward fixing defects of later stages (testing, maintenance, etc.) of SDLC. The highest value of design defect priority and the lowest value of testing defect priority obtained from the AHP methodology case study data reinforce this fact. Complex decisions should not be based on intuition, and AHP helps in taking informed decision. The decision considers both qualitative and quantitative evaluation.

It becomes cumbersome to use AHP when number of alternatives or criteria is high. The very nominal difference in several defect priorities also indicates that if software projects are of small size; then, executing all the phases of SDLC in parallel (agile methodology) may be lucrative. Frequent changing customer requirements of agile software projects creates conflicting objectives/criteria, and AHP is best fit to sort out the confusion in these cases.

References

1. Schmidt R et al (2001) Identifying software project risks: an international delphi study. Sens Trans 17:5–36
2. Stoneburner G et al (2002) Risk management guide for information technology systems. In: Proceedings of ACM technical report SP 800-30. National Institute of Standards & Technology, Gaithersburg, MD, United States, pp 4–41
3. Artto KA (1997) Putting project risk management into perspective fifteen years of project risk management applications—where are we going? In: Proceedings of managing risks in projects Helsinki, Proceedings of IPMA symposium. Project Management, Finland, vol 1
4. Atkinson R, Crawford L, Ward S (2006) Fundamental uncertainties in projects and the scope of project management. Int J Proj Manag 24:687–698 (Elsevier)
5. Keil Mark, Tiwana Amrit, Bush Ashley A (2002) Reconciling user and project manager perceptions of IT project risk: a delphi study. Info Syst J 12:103–119
6. Boehm BW (1991) Software risk management—Principles and practices. IEEE Softw 8:32–41
7. Haimes YY (2015) Risk modeling, assessment, and management, 4th edn. Wiley, 720 p
8. Saaty TL (1980) The analytic hierarchy process. McGraw-Hill, New York
9. Belton V, Gear AE (1983) On a shortcoming of Saaty's method of analytic hierarchies. Omega 11(3):227–230 (Elsevier)
10. Dyer JS (1990) Remarks on the analytic hierarchy process. Manage Sci 36(3)
11. Saaty TL (1990) An exposition of the AHP in reply to the paper: remarks on the analytic hierarchy process. Manage Sci 36(3)
12. Harkar PT, Vargas LG (1990) Reply to "Remarks on the analytic hierarchy process" by Dyer JS. Manage Sci 36(3):269–273

13. Raccon L (1995) The chaos model and the chaos life cycle. Notes Softw Eng 20(1):55–66
14. Yu-Heng W, Jian J, Ying Q (2010) The earth-moon model on software project risk management. In: Proceedings of the ninth international conference on machine learning and cybernetics, Qingdao, July 2010;1999–2003

Reliable and Privacy Preserving Blockchain Based Medical Data Sharing Digital Ledger

Rene Thomas and V. S. Anitha

Abstract Secure sharing of health-care details along electronic media are challenging and yet an inevitable thing in the digital era. As valetudinarians pass between contributors, patients medical documents become disseminated across different organizations, resulting from drastic ethical issues to reliability and security. The decentralized and distributed mechanism of the blockchain gives a new stance to the digital frugality. Blockchain permits neither the entrant needs to know each other, and nor does it require third-party certification bodies to engage. It ensures data reliability and security through cryptographic algorithms and consensus used by the blockchain technology medical data sharing digital ledger using blockchain. In this paper, we use a private blockchain which makes use of ethereum and hyperledger fabric smart contract. Medical data services can be accessed by only those who have permission to ingress. The System uses a secure hash algorithm, proof of stake (PoS) consensus algorithm used for data provenance and thus it makes the block with tamper proof medical records than existing systems. Here the access granted for valid users only. Blockchain enhances security, anonymity, and privacy for medical assets.

Keywords Blockchain · Hyper-ledger fabric · Ethereum smart contract · Proof of stake (PoS) · Digital ledger · Cryptography · Anonymity

R. Thomas (✉)
Govt. Engineering College Wayanad, Kerala 670645, India
e-mail: reneemtk@gmail.com

V. S. Anitha
Dept. of CSE, Govt. Engineering College Wayanad, Kerala 670645, India
e-mail: anithavs@gecwyd.ac.in

V. Nath and J. K. Mandal (eds.), *Proceedings of the Fourth International Conference on Microelectronics, Computing and Communication Systems*, Lecture Notes in Electrical Engineering 673, https://doi.org/10.1007/978-981-15-5546-6_12

1 Introduction

1.1 General Background

As we have already known, the key to successful completion of any task performance is effective information gathering and processing of any complex functions. Nowadays the dissemination, public engagement, and exploitation of various digital records result in drastic ethical issues in the world of security. In this scenario, it is worth reasoning why we need the distribution of electronic records and how the information dissemination can be processed securely and tamper proof.

1.2 Introduction

Electronic medical records are the building blocks of emerging medical fields and their enterprises. As patients move between contributors, their medical records become disseminated across different organizations, and it lacks security. Heterogeneous data structures may preclude compatibility in the existing systems, while divergence in the medical records limits data comprehension. Even when the network could agree on the structures and semantics, security still compromises. Secure Data sharing [1] becomes more complicated as ossification of the internet architecture goes beyond limitations.

The idea of blockchain transpired into a big revolution in every industry, especially in medical data sharing. Nowadays dissemination of records becomes inevitable in medical fields. Blockchain-based medical data sharing digital ledger aims security in the medical record. Here the asset is transacted over the internet using hyperledger fabric and ethereum smart contracts and thereby promote faster and secure data record for medical information. In this proposed approach, instead of proof of work (PoW) algorithm, Proof of Stake (PoS) is used. So that the node is treated as validators used to mine the transactions which are more reliable than existing methods.

1.3 Blockchain

The blockchain [2] is a consecutively increasing series of records. It creates value, trust, and truth with suitable cryptography. Exponents of blockchain and its services considered it as a privacy-preserving indestructible block of records. Transactions can be easily verified with considerable clarity and trust to all the nodes involved.

The term blockchain technology was first elucidated in 2008 with the invention of bitcoin [3] to keep the past digital information safe and secure. The Hyperledger [4] project by IBM varies somehow, but similarly, this is geared more towards larger organizations than smaller companies or individuals. Ethereum is an open source blockchain computing platform featuring smart contracts.

1.4 Paper Organization

This paper deals with reliable and privacy preserving blockchain based data sharing digital ledger. Next section gives an overall idea of network selection, and brief descriptions on various consensus used in different data sharing medical records. Section 3 deals with the proposed approach and how the new variant for medical data sharing works. Section 4 gives an idea of design principles and methodology. Followed by comparing the performance based on different parameters. Finally concluded with future contributions.

2 Related Work

In this section, the focal point is on the network selection and identification of the best consensus used to securely share medical information among the system without compromising the certainty of the network. The core idea is to make medical asset transfer secure and faster in a tamperproof manner. The various steps evolved in the procedure of making blockchain based medical data sharing digital ledger includes network selection, finding the best suitable Consensus algorithm, ethereum smart contract implementation, and attachment of blocks and finally the analysis of medical data sharing blockchain system with blockbench.

2.1 Network Selection

Network identification is based on the requirements such as whether it is permission oriented [5] or permission-less. The difficulty in blockchain systems and cryptographic consensus algorithm is when each distributor asserts their blockchain platform is the best. The 2 criteria matters to the blockchain environment definition are:

1. *level of anonymity of validators*:
2. *level of trust in validators.*

Level of anonymity indicated the identity and trust means how inevitable is punishment for misbehaviour. Difference between permissoned and permissionless blockchain is explained with the help of the above Fig. 1. From the figure, it is clear that permissioned blockchain(private blockchain) is suitable for medical data sharing

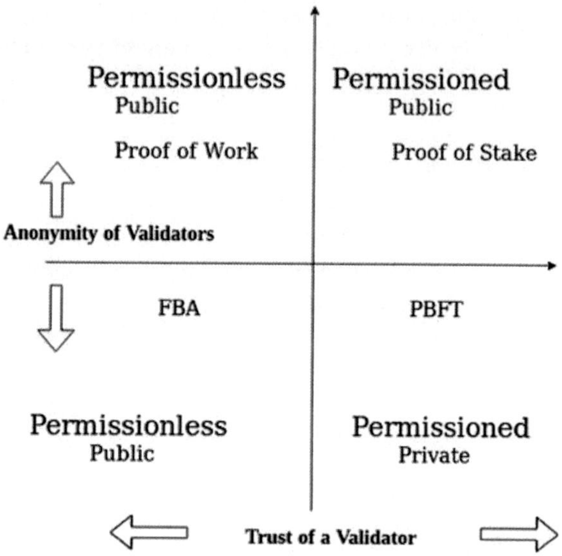

Fig. 1 Permissoned versus permission-less

since it meets the criteria and hence we can use hyperledger fabric with Proof of Stake as consensus algorithm.

2.2 *Consensus Algorithm*

After the successful selection of the network and studying the basics of permissioned blockchain, next is the selection of suitable consensus algorithms. Next we can discuss various medical data sharing systems.

1. *MedRec:* MedRec [6] is the first fully functional blockchain innovation to medical records. Medrec used for data access using blockchain technology and to provide secure data sharing. MedRec manages authentication, confidentiality, accountability, and data sharing. The Consensus used to secure the medrec is Proof of Work (PoW). Data transactions are done via smart contracts on an Ethereum Blockchain.

Pow is used for mining, the expensive mathematical calculation. The PoW is used to deterring cyber-attacks such as a distributed denial-of-service attack, (DDoS). The PoW can be used to confirm transactions. It will then produce new blocks to the chain. Here PoW, miners compete against each other to complete transactions on the network and get rewarded. This system is called proof of work because of the probability of mining of the block is increased with the amount of the work put in.

2. *PHIRchain* [7]: PHIRchain considers different methods to securely share medical information. This paper finds out all the possible and existing barriers to medical

data sharing. The system uses permissioned blockchain for clinical data sharing. Here the Trust is enforced through restricted access to transaction validation. Proof of Interoperability (PoI) is used as a consensus. PHIRchain architecture is customized with Decentralized App(DApp) to meet the ONC requirements. POI is the ability of health information systems to work together within and across organizational boundaries for effective delivery of demands.
3. *Secure ABS Scheme with Multiple Authorities* [8, 9]: The system uses Attribute-based Signature scheme for blockchain in Electronic Health Records system. Secure ABS protocol resists Collusion attack, where a node intentionally makes a secret agreement with an adversary. Rui Guo [8] analyze both security and performance analysis on the ABS scheme as well as MA(Multi-authority)-ABS scheme. Security of the ABS protocol is analyzed using a random oracle model. A random oracle analysis is an art of solving complex problems. Every query responded with an arbitrary output if those queries were distinctive. Oracle can be used as a substitution for every hash functions with high randomness. Security of the system is measured in terms of the random response to the queries.
4. *Blockchain: A Panacea for Healthcare Cloud-Based Data Security and Privacy* [10]: Proof of stake consensus is used to secure the blockchain. Proof of Stake (PoS) is cryptocurrency blockchain network aims to achieve distributed consensus. The creator of the next block is chosen via various combinations of random selection and wealth or age (i.e., the stake). The coins holders chosen in a deterministic way that is called staking.

Cloud-based EHR in blockchain is a new way of securing health care details. Consensus does not need to be confirmed by a trusted mediator. This system avoids performance bottleneck and a single point of failure. So the patients can even control their past records. Medical datas are stored in the blockchain network. Any changes to the blockchain are visible to all nodes in the network. It reduces the unauthorized modifications by trivially detecting them.

From these comparisons and analysis of consensus algorithms used in different blockchain based medical data sharing digital ledger, the suitable consensus algorithm for the proposed approach is proof of stake consensus algorithm and its modified version, proof of delegated stake also used in blockchain based system to boost the security and reliability in the network rather than existing proof of work consensus algorithm.

3 Reliable and Privacy Preserving Medical Blockchain

3.1 Overview

Blockchain integration is now almost spread all over the world with everlasting hopes in minds of millions. Now every programmer wants to convert or add their idea with blockchain in order to secure their idea with privacy as well as probity with maximum possible profit. The design part is mainly concentrated on the evolution, implementation, and analysis of blockchain based medical data sharing digital ledger via hyperledger fabric and ethereum.

To have better performance-based blockchain medical data sharing record system, here we have hyperledger fabric. This can be securely distributed in order to deal with current and future challenges and satisfy new service requirements, we need to contemplate the following design principles based on previous work on designing distributed network architecture, research new network technologies, and investigate new service requirements such as adaptability, high Availability, Fault Tolerance, performance, reliability, scalability and security. Ethereum smart contracts and proof of stake consensus algorithm is also added to ensure the guarantee and regularity in the system.

3.2 Problem Definition

Development, implementation, and analysis of medical data sharing record by integrating them into a secure and reliable blockchain based medical data sharing digital ledger via hyperledger fabric and ethereum smart contracts in django web framework using blockbench with the help of evaluation metrics.

4 Design Principles

The design methodology refers to the development of a system or a method for the current situation. The blockchain based medical data sharing digital record system is designed to secure the records all over the internet with no tamper proofs and the secure data is stored using interplanetary file system (IPFS). Figure 2 shows overall design of the proposed approach. Whenever a patient wants to consult a doctor, they are requested to generate identity card from the authority, or the hospital to check doctors availability based on patients preference. Once it is received, patient can undergo for diagnosis, after that his/her medical record will be electronically signed, cryptographically secure and added to the data store. At the time when someone requests for a medical record, the system then check the availability of requested data in the data store to confirm the validity of the request if the requester already registered with the system blockchain. If the request is valid then transaction event will be mined. The transaction event is then attached to the blockchain and listed as transaction details with the previous hash.

The blockchain in healthcare can be implemented with the help of hyperledger fabric as framework and ethereum smart contracts for the development of smart contracts in the blockchain and the architecture of medical blockchain is explained in Fig. 3. Here blockchain uses public key cryptography to create a tamperproof

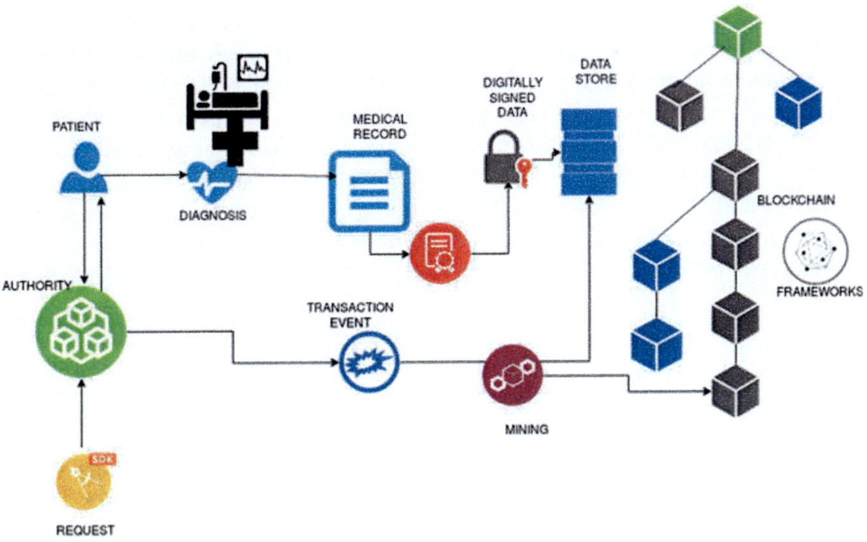

Fig. 2 Design overview

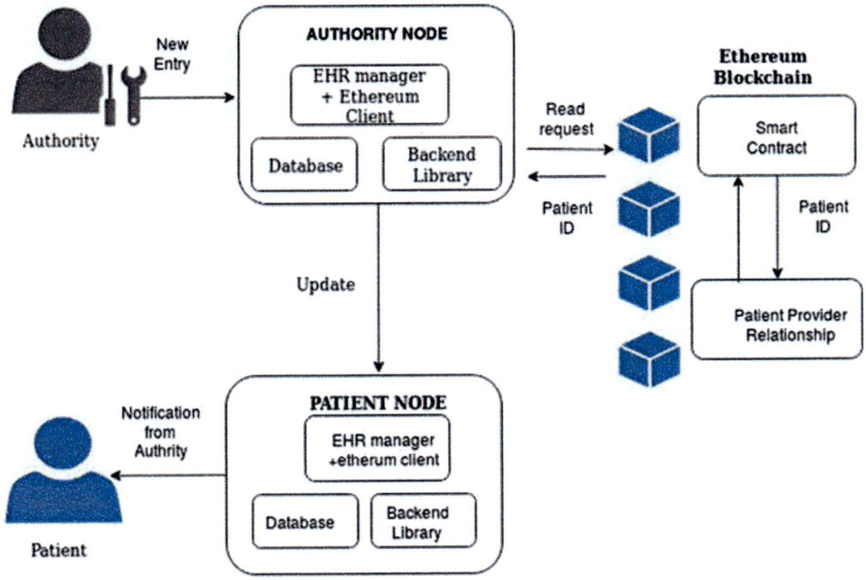

Fig. 3 Architecture of medical blockchain

chain of content, supported not by a central server, but a decentralized network of participating nodes. Each individual node works to solve a series of hashing exercises, which contribute to the formation of the chain a process known as mining.

Whenever a new entitiy is registered with the authority, node contains ehr manager, ethereum client, database and the backend library will be updated into the patient node. Patient node holds backend library details, manager, ethereum client and the database. The authority node request for the red request from ethereum blockchain. The smart contracts hold patient provider relationship smart contracts and other smart contracts. Once the patient id is fetched from the smart contract it then given to the authority node and therefore the details can be updated in the patient node. Once patient node get update the notification from the authority will be forwarded to the patient.

4.1 Ethereum Smart Contracts

Ethereum smart contract used to execute scripts on the blockchain of this system. Contracts are intelligent representation of that links where system does not store the record directly; rather encodes metadata that allows records to be accessed securely by patients, unifying access to data across disparate providers.

Contracts are built with the help of ERC-20 standard. Debugging the ethereum smart contract are done with truffle. We can get the truffle from internet and configure with the system. The system uses file structure layout, testing framework, console, and a debugging environment. Solidity is used here for contract coding. The contracts like a wrapper around a mapping (address = uint) object. The asserting equality with the hamcrest style.

After the contract is written and tested in the node, deployment into local blockchain can be completed. We can run the local blockchain with ganache. Ganache is then migrated with truffle by setting the port number on the ganache with the wallet.

4.2 Hyper-ledger Fabric

Hyperledger Fabric [11] is an framework for private blockchain. It is mainly used in the permissioned blockchain. Fabric provides membership services, chain code services, and other major services like consensus manager, ledger storage and the transaction details. On top of these modules, APIs, SDK, and CLI work.CLIs in the framework used to test the system.

The transaction is initiated by the peer node connected from the application client. The transaction is sent as a proposal to the Endorsing peers and each of the Endorsing peers validates the proposal (or the proposed transaction) against the World State database and sends the response back to the application client. The API in the blockchain are used to check the results. It checks whether the order of the transaction details added to the blockchain via ipfs are correct or not. SDK in the hyperledger fabric means software development kit which stores every software related to the fabric deployment [12].

5 System Implementation and Performance Analysis

In order to demonstrate the efficiency and security of blockchain based medical data sharing digital ledger, we can divide the overall implementation procedure into layers and blockbench uses these layer-wise analysis of the system. The implementation is divided into layers as GUI layer, execution layer, chain layer and the communication layer. The proposed system implementation layers are discribed in Fig. 4.

The bottom layer is implemented using solidity. Process synchronization of blocks in the blockchain are constructed in this layer. Here the transaction are executed in terms of ethereum smart contract which is visible for all the users. Smart contracts can interact with other contracts, make decisions, store data and send ether to others.

The chain layer contains consensus mechanism used for the blockchain. This layer hold protocols corresponding to blockchain part. The structure and semantics of the proposed blockchain systems are also supported by this layer. Both EHR and PHR and both of this layer employees PoS. Execution layer carries the details of how runtime interface will look like for example, where the codes can e executed and details of error checking and error correction. Finally the top most layer, application or GUI layer contains blockchain applications [13]. GUI layer uses Django web framework.

The analysis and comparison of proposed system can be done via formal analysis and formal verification [14]. Formal analysis means evaluating the possibility of attack on the specification of the protocol, products, or system by conducting some mathematical formalization of the security requirements whereas formal verification used to verify the correctness of the specification of the protocol, products, or by automated axiomatic theorem proving. Performance analysis can be done using tolerated malicious nodes and its efficiency is calculated based on the rate of transactions transferred successfully per seconds.

The formal analysis of different parameters can be done based on network comparison. In Table 1, throughput is analyzed in terms of number of successful transactions per second. Centralized and distributed systems can send higher amount of transactions in each seconds than decentralized systems and that is represented as high and low values. Asset ownership is depend upon the intermediator during the transaction and centralized network uses a trusted intermediator whereas the other two networks has direct transaction over the network. Since the centralized network needs a trusted authority the custodian risk is there and for the rest of the network have no custodian risk. Since the distributed and decentralized networks are completely integrated, centralized network integration is depends on platforms too. Hence the distributed and decentralized blockchain is selected for proposed system. Performance analysis in blockchain can be done using blockbench [13]. Blockbench is a framework for analyzing private blockbench [13]. Blockbench mainly use four parameters say application, execution engine, throughput, asset ownership, key management, integration and custodian risk and which are explained in Table 1 and also four different implementation layers. The layers explained in Fig. 5 can be used in the blockbench to analyze private blockchain.

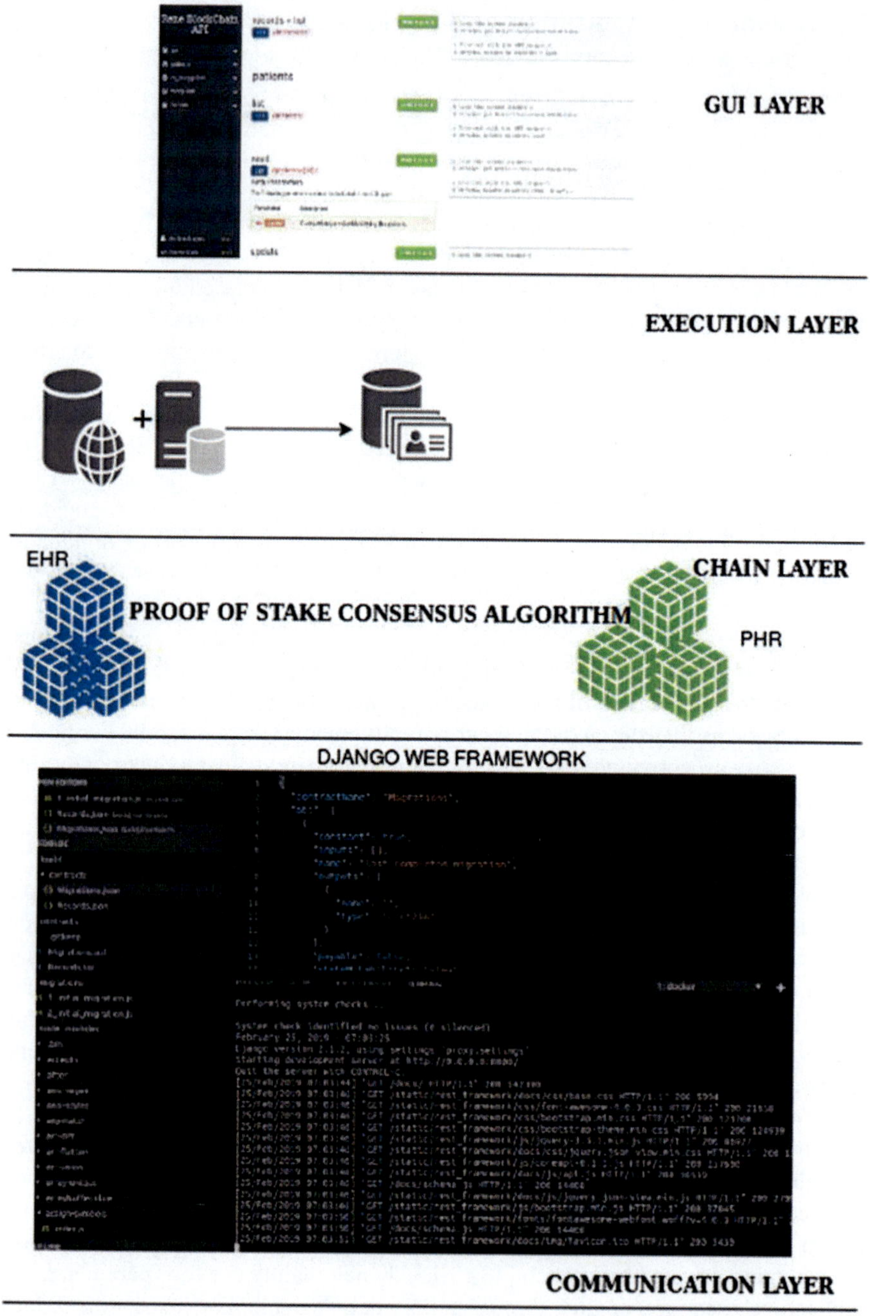

Fig. 4 Proposed system implementation layers

Table 1 Comparison of networks

	Centralized	Decentralized	Distirbuted
Execution engine	Centralized	Distributed	Distributed
Throughput	High	Very low	High
Asset ownership	Trusted	Directed	Directed
Custodian risk	Yes	No	No
PK management	Exchange side	Client	client
Blockchain integration	Depends	Full	Full

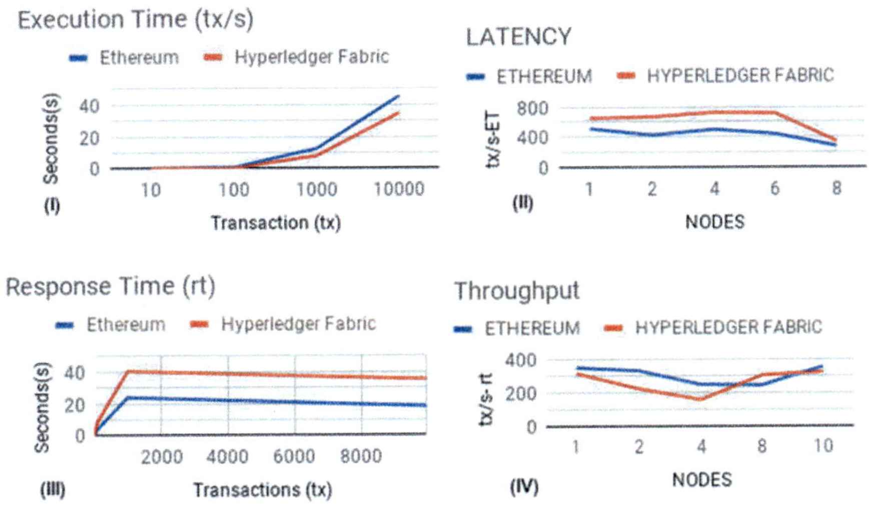

Fig. 5 Communication layer comparison

Blockbench measures performance in the communication layer in terms of throughput, latency, scalability and faulttolerance [13] and their result is shown in Fig. 5. First part (I) of Fig. 5 shows the execution time of both ethereum and hyperledger fabric end up with a linear increase after certain transactions are completed. Part (II) shows throughput where hyperledger has less successful transactions compared to ethereum blockchain. Response time represented in (III) and it shows that ethereum has less response time and part (IV) shows the latency of smart contracts where ethereum smart contracts have comparatively good performance due to the decreasing nature of latency. The third parameter used for performance analysis is scalability. Scalability is the difference in throughput and latency with respect to the nodes [13]. Here from Fig. 5, changes in throughput and latency result from better scalability for the ethereum smart contracts rather than hyperledger fabric [13]. The fourth and final parameter, the fault tolerance can be calculated as the changes occurred due to the node failure in latency and the throughput of both smart contracts. Since node failures and not added into our proposed system, our system has better performance compared to related approaches [8, 10].

Performance analysis based security can be analyzed using security metrics. Security of the system can be measured in terms of no of orphan blocks. Orphan or stake blocks are denoted as forks and which are created due to double spending and such vulnerabilities in the system. The ratio between no of blocks contained in the main chain of the blockchain and the no of stake blocks reportegives the security analysis. If the ratio is lower then we can say that the system is less vulnerable [13]. Since the ratio is lower for proposed system we can conclude that our system is secure where no other papers explained security analysis.

Finally, the analysis in the chain layer can be done with the help of comparative study on consensus its uses, major security concern, its blockchain type based on the blockchain network comparison, the tolerated malicious nodes and the performance efficiency. The detailed explanation of the comparative study is shown in Table 2, where BT means blockchain type, PE is the performance efficiency and TMN is the tolerated malicious nodes, PL indicates permissionless, P indicates permissioned, and finally C for consortium. Here tps is transaction per seconds. Pu and Pr are public and private.

From these comparisons, it is clear that Delegated Proof of stake(DPoS) is more good than other consensus but has difficulty in implementation. Proof of stake (PoS) also gives enough support and proof of work can be modified so it would be better to have more reliable system. Proof of Stake (PoS) consensus algorithm is reliable and suitable for medical data sharing systems using blockchain and it needs less effort from programmer side which is used in the proposed approach. The execution layer and application layer need not be analyzed since their results are already reflected in the previous comparisons.

Table 2 Comparison based on consensus efficiency

Authors	Consensus	BT	Purpose and security	TMN	PE
Azaria [6]	PoW	PL	Requirement to mining, against DDoS	50%	<20 tps
Zhang [7]	PoI	P	Interoperable, digitally signed and verified	Depends	≥50 tps
Guo [8]	MA-ABS	C	Secure via ROM and resists collision	Not Supported	>100 tps
Esposito [10]	PoS	C	No intermediator and avoid per formance bottleneck, single point of failure	50%	>300 tps
Liu [9]	DPoS	C/Pu/Pr	Advanced avoids 50% attack	Not needed	>300tps

6 Conclusion

Digital health records can't manage the complexities of the ledger storage and their security. As valetudinarian pass between medical centers, their data may be disbanded and thus losing easy access to past records. From the above comparisons, it is clear that blockchain based medical data sharing record system is the best way to rectify the problems with existing medical records. Each consensus promises security, data privacy in various factors. So we can conclude that Combination of hyperledger fabric along with the ethereum smart contract and Proof of Stake consensus algorithm will be the best option for blockchain based medical data sharing digital record. Since traditional Proof of Stake lacks in security and data privacy by compromising the decentralization; Delegated Proof of stake can also be implemented.

7 Future Works

Blockchain based systems act as a distributed decentralized digital ledger and its applications and features are increasing day by day. In this proposed approach even though hyperledger fabric is purely designed for private blockchain like blockchain based medical data sharing record system, it somehow results in delays in the network even if it is highly secured and ensure transparency. This type of delays in the network may cause less user friendly and it is time consuming. So the ethereum smart contracts used in the approach provide high throughput, scalability and avoid delay in the network. Future work in this system can be made by adding proof of delegated stake consensus along with ethereum protocol for smart contacts and other block addition by hyperledger fabric smart contracts makes the system comfortable, secure, transparent and reliable.

References

1. Krumholz HM, Waldstreicher J (2016) The yale open data access (yoda) projecta mechanism for data sharing. N Engl J Med 375(5):403–405
2. Underwood S (2016) Blockchain beyond bitcoin. Commun ACM 59(11):15–17
3. Nakamoto S (2008) Bitcoin: a peer-to-peer electronic cash system
4. Cachin C (2016). Architecture of the hyperledger blockchain fabric. Workshop on distributed cryptocurrencies and consensus ledgers, vol 310
5. Kravchenko P (2016) Ok, i need a blockchain, but which one?
6. Azaria A, Ekblaw A, Vieira T, Lippman A (2016) Medrec: Using blockchain for medical data access and permission management. Open and Big Data (OBD), international conference on IEEE, pp. 25–30
7. Zhang P, White J, Schmidt DC, Lenz G, Rosenbloom ST (2018) Fhirchain: applying blockchain to securely and scalably share clinical data. Comput Struct Biotechnol J 16:267–278
8. Guo R, Shi H, Zhao Q, Zheng D (2018) Secure attribute-based signature scheme with multiple authorities for blockchain in electronic health records systems. IEEE Access 776(99):1–12

9. Liu J, Li X, Ye L, Zhang H, Du X, Guizani M (2018) Bpds: a blockchain based privacy-preserving data sharing for electronic medical records. arXiv preprint arXiv:1811.03223
10. Esposito C, De Santis A, Tortora G, Chang H, Choo K-KR (2018) Blockchain: a panacea for healthcare cloud-based data security and privacy? IEEE Cloud Comput 5(1):31–37
11. Androulaki E, Barger A, Bortnikov V, Cachin C, Christidis K, De Caro A, Enyeart D, Ferris C, Laventman G, Manevich Y et al (2018) Hyperledger fabric:a distributed operating system for permissioned blockchains. Proceedings of the 13th European system conference, ACM, p. 30
12. Digest B Hyperledger fabric 101. blockdigest.com/courses/hlfabric101/
13. Dinh TTA, Wang J, Chen G, Liu R, Ooi BC, Tan K-L (2017) Blockbench: a framework for analyzing private blockchains. Proceedings of the 2017 ACM international conference on management of data. ACM, pp. 1085–1100
14. Matsuo S (2017) How formal analysis and verification add security to blockchain-based systems. Proceedings of the 17th conference on formal methods in computer-aided design, 1–4. FMCAD Inc
15. Xu X, Pautasso C, Zhu L, Gramoli V, Ponomarev A, Tran AB, Chen S (2016) The blockchain as a software connector. Software architecture (WICSA), 2016 13th Working IEEE/IFIP conference on IEEE, pp. 182–191

Survey on Lightweight Cryptography Algorithm for Data Privacy in Internet of Things

Monalisha Sinha and Sandip Dutta

Abstract Internet of things (IoT) has already made its way into every industry and has started helping us in making smart decisions. Two billion people using the Internet at the moment and that is set to grow to probably 2 billion in the next couple of years and devices will be linked with themselves together using networks. Conventional cryptography uses a huge amount of energy which results in cost and complexity as compared to lightweight cryptography. Lightweight cryptography has been implemented to overcome conventional cryptography. In this paper, we have done a survey among different lightweight cryptographic algorithm protocol in aspects of RAM, efficiency, encryption and decryption cycles.

Keywords IoT · Security · Cryptography · Encryption · Architectures · Attacks · Lightweight cryptography

1 Introduction

In this digital world, we can make everything talk to each other. The electronics objects that are connected to the Internet through their networks can exchange their data. Now billions and billions of objects are linked with the Internet. As these devices connected to each other will become the intelligent system and when they will exchange their data over the cloud they can transform our lives in a countless way whether it is in medical, house and so on. This results in consuming less cost and energy. IoT has brought evolution in this digital world by connecting to the Internet to extract meaningful information's.

M. Sinha (✉) · S. Dutta
Department of Computer Science and Engineering, Birla Institute of Technology, Mesra, Ranchi, Jharkhand, India
e-mail: monalisha530@gmail.com

S. Dutta
e-mail: sandipdutta@bitmesra.ac.in

V. Nath and J. K. Mandal (eds.), *Proceedings of the Fourth International Conference on Microelectronics, Computing and Communication Systems*, Lecture Notes in Electrical Engineering 673, https://doi.org/10.1007/978-981-15-5546-6_13

1.1 IoT Applications

The IoT is changing much about the world we lived in. Wireless sensor and chips that are embedded in the physical objects exchanges their data over the Internet. In IoT, the devices interact, collaborate and share experiences with each other. The house has become a smart house, the electronic things within the house communicate with each other. Research shows that by the year 2020, 50 billion objects will be connected to the Internet [1]. Each person approximately will use 7.1 objects online.

The IoT application is growing rapidly day by day [2]. Building electronics system using IoT is described in [3]. Most of the industries are showing their interest toward IoT. IoT applications in health care industries and improvement opportunities are discussed in [4]

In every application, IoT demand has increased whether it has converted home to smart homes, agriculture, smart retailer, manufacturing, education and many more [5].

1.2 IoT Architectures

IoT architecture establishes the connections between the physical object for data sharing [6]. Data sharing are done in real time. IoT architecture should provide quality of service, resistant to attack, security and data management.

IoT architecture comprises of four layers:

1. Layer I: Physical layer

The data present in the IoT are equipped with sensor and actuator to produce, accept and process the data. It contains sensors, actuator, cameras and many more.

2. Layer II: Network layer

Data acquisition system process the data received from the sensor and actuator. It contains all the hardware and software setup of the network.

3. Layer III: Application layer

Digitized data before sending to data store are processed through edge analytics where automated analytical computation is done on data. Application layer acts as a bridge between physical objects and the IT industry.

4. Layer IV: Perception layer

Data can be extracted from the data store through cloud analytics. It converts the analog signal to a digital signal (Table 1).

Table 1 IoT architecture

Layer I	Layer II	Layer III	Layer IV
Physical layer	Network layer	Application layer	Perception layer
Sensors, actuators, RFID tags	Wi-Fi, Bluetooth, 3G	Smart application and management	Barcode, RFID

1.3 IoT Attacks

Communication between the physical object and IoT devices is done through the Internet. Communication is done publicly, so attack occurrence can occur during communication.

1. Denial-of-service attack: The attack is done when it acts as distributed. System and data are made unavailable to the person who needs it. This results in jamming (unauthorized person occupy the channel) and spoofing (attacker send useless the message to the user to block the channel).
2. Man-in-middle: In this attack, the third party interrupts the channel by getting the key and pretends it to be the right person. Recipient thinks they are getting the right message. This type of attack is done in RFID technology.
3. Eavesdropping: The third party listens to the conversation between sender and receiver. This attack is in confidentiality [7].
4. Masquerading: The third party pretends to hold the identity of other people. It can result in a break down in the IoT resources.
5. Alternation: Information store in IoT devices can be modified or altered by the third party results in the integrity of data.

1.4 Security Challenges in IoT

In the IT industry, IoT has become the latest phase. Demands of IoT are increasing day by day, while a security challenge has brought fear regarding this technology. There are over 25 billion IoT devices connected to each other so it is impossible to test all the devices. Even yet not all the devices are updated in time. Uncertainty about security comes there when devices are not secure and they are prone to be hacked [8].

As the demand for IoT devices is increasing, IT manufacturer are producing more devices, so security challenges have become a major issue. Due to outdated software and hardware, it is more prone to be potential attacks. To prevent a potential attack, each device should be updated properly.

Data security and privacy are the two main challenges in this digital world. Rules should be provided as a lot of IT companies are transmitting a huge amount of data from IoT devices. Storage of data is not that much challenge but it is opened to be hacked.

A study discloses that 71% of the objects linked with IoT are opened to be attacked [9]. Man-in-the-attack can be performed by two parties who are directly communicating with each other. There is no solution regarding man-in-the-middle attacks. Encryption is done to minimize the data integrity.

1.5 Cryptography Overview for IoT

Cryptography is used for data security. Cryptography is the practice in which the original text is converted into unintelligent text for the security of data [10]. Symmetric and asymmetric cryptography are two parts of cryptography. The same key is used for encrypting and decrypting the message which results in faster and safer communication. The demerit of symmetric cryptography is the transfer of key between two parties. If the third party gains the key it can interrupt the communication. Confidentiality and integrity of data are assured in symmetric key but not validation of data. AES, DES, IDEA, etc., comes in symmetric key. In asymmetric two keys are used for communication, i.e., public and private key. It assures authentication, integrity and confidentiality of data. The public key is used to encrypt the message while the private key is used to decrypt the message which provides confidentiality. The main advantage of the asymmetric key is that it provides security of the key. If the key becomes large, then it results to slow down the encryption and complexity will be increased. Asymmetric key uses Rivest–Shamir–Adleman (RSA), Diffie–Hellman and elliptic-curve cryptography (ECC).

1.6 Lightweight Cryptography

Lightweight cryptography has the six-part standard: confidentiality, authentication, identification, non-repudiation and key exchange. Lightweight cryptography is implemented on resource-constraint devices like wireless sensor devices, embedded chip and so on. For a hardware implementation, sizes of chip and power consumption are measured. For a software implementation, smaller code and memory size are measured.

Advantages of lightweight cryptography over IoT:

1. Efficiency of continuous communication: To achieve end-to-end security, the symmetric key algorithm is used. For low-resource devices, limited energy consumption cryptographic algorithm should be used. The main advantage of a symmetric key algorithm is to reduce energy consumption.
2. Connection to low-constraint devices: Limited area of space is the main merit of lightweight cryptography. Limited area of space gives more chances to connect with low-constraint devices.

The needs of lightweight cryptography are discussed in [11].

1.7 Organization of Paper

This paper contains the following sections:

Section 2 is the survey on lightweight cryptography algorithm protocol for security in IoT. Section 3 is the comparison of different lightweight cryptographic protocols. Section 4 is the future scope of work. Section 5 concludes the conclusion of this paper.

2 Survey on Lightweight Cryptography Algorithm Protocol for Security in IoT

1. AES [12]:

Vincent Rijmen and Joan Daemen created AES. This algorithm is easy to implement and takes a sensible amount of time while running on a regular computer. AES is the advanced level of data encryption standard (DES). It comprises of exclusive-OR, octet substitution in S-box, row and column rotations and a mix column. It has key length of 128-bits, 192-bits or 256-bits and rounds of 10, 12 or 14. Rounds depend on the length of key.AES comprises of 128 blocks. For lightweight cryptography, AES-128 is the most suitable variant, due to rounds number and key length.AES is mainly designated for the software application. Based on the substitution-permutation network.

2. HIGHT [13]:

HIGHT stands for high security and lightweight. It is implemented in low-constraint devices. It is of the 64-bits block and 128-bits key length. This algorithm comprises of round functions, key schedule, encryption and decryption. Based on Feistel Network with 32-round iterative structure, it consists of XOR, addition-mod 2 to the power 8 and left-bitwise rotation and it is mainly used in hardware implementation.

3. SIMON [14]:

It is lightweight cryptography which performs better in hardware implementation. Key length is of 128 bits with block length of 128 bits.

4. TWINE [15, 16]:

It runs on small hardware. It performs better on computer software that is embedded. It is implemented on Feistel Network. TWINE 80 and TWINE 128 are of two types implemented according to key length. These types have 64 blocks with rounds of 36.

5. PRESENT [13]:

PRESENT is symmetric block cipher with 64 bits of block size with key length of 80 bits or 128 bits. 4-bit S-box which can be implemented on hardware optimizations. It is based on substitution-permutation network (SPN). It is used in low-constraint devices which use low power and high efficiency. Due to complexity estimation for the hardware implementation, it is most dense cryptographic algorithm. It is simple to implement with average security.

6. TEA [13]:

TEA is implemented on Feistel Structure with 64 rounds which are rounded by cycles. It has 32-bit unsigned integers that use 128-bit key. For each cycle, it has same the key. It is used in a low-constraint device.

7. CLEFIA [17, 18]:

CLEFIA is implemented on Feistel Network. It has a block length of 128 bits and key size of 128-bits, 192-bits and 256-bits. It gives better performance and security in hardware.

3 Comparison of Different Lightweight Cryptographic Algorithm Protocol

Table 2 shows that PRESENT, SIMON uses minimum gate that are acceptable to IoT [19]. The same hardware is used by Feistel Network for encrypt and decrypt of message. Using the same hardware, it minimizes the execution time and memory usage (Fig. 1).

ECC is the best known lightweight cryptography in asymmetric cryptography. It uses less key than RSA and uses less memory which increases the speed of execution. In wireless sensor network, ECC is the strongest among various attacks.

RSA uses two large prime numbers to create public key and private key which is more secure. To generate key, it uses discrete logarithm problem. Digital signature scheme uses asymmetric cryptography for data authentication. Table 3 shows differentiation between ECC and RSA in respect of data security and performance which shows that ECC is better than RSA. So, in the future, ECC will be used first than RSA for authentication.

4 Future Direction

By doing this survey and comparison analysis, we analyze that some research problems are found on which further research should be done.

Table 2 Comparison table of different lightweight cryptographic algorithm protocol [19]

Algorithms	Key length (bit)	Block length (bit)	Rounds	Structure	Tech. (μM)	Power (μM)	Area (Gate)	Throughput at 100 kHz (Kbps)	Attacks	Merit
AES-128	128	128	10	SPN	0.14	0.59	2411	57.61	Man-in-middle attack	Faster in hardware and software
HIGHT	128	64	32	FN	0.25	1.23	3091	191.20	Differential attack	High security
CLEFIA	128	128	18	FN	0.12	3.81	2410	41	Saturation attack	Faster processing
SIMON	128	1258	64	SPN	0.13	1.21	1327	23.8	Differential attack	Good in hardware
PRESENT	80	64	32	SPN	0.18	1.77	1045	13.5	Differential attack	Energy efficient
	128						1341	13.15		
TEA	128	64	32	FN	–	0.75	–	–	Related key attack	Same key
TWINE	80	64	36	FN	0.09	–	1876	181	Man-in-middle attack	Good for small hardware
	128									

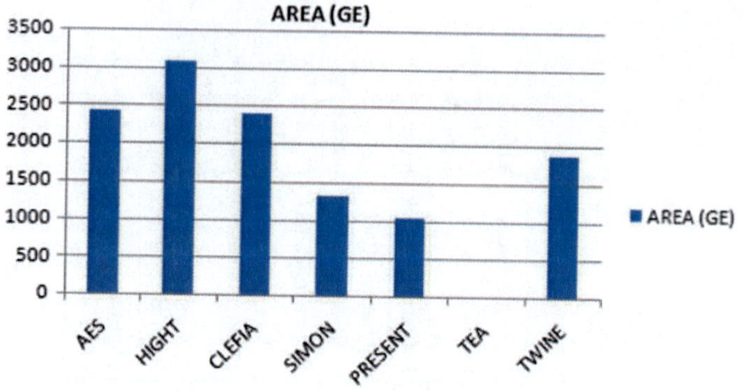

Fig. 1 Graph plot of gate

Table 3 RSA and ECC differentiation

Algorithm	Key length (bits)	Merit	Attacks
RSA	1024	Increases security	Man-in-middle
	15,360		
ECC	163	Increases speed, less memory, good security level	Side channel attack

1. Data security and data validation are the main part of IoT. So various techniques are implemented and a various hybrid model of encryption and authentication algorithm (AES and RSA techniques) are proposed. But this technique can causes increment in memory in devices. So CCM modes are used to counter the memory requirement for security and authentication of data.
2. Conventional cryptography provides more secure than lightweight cryptography. So round numbers are increased in lightweight cryptography for providing the same level of security which minimizes performance of algorithm. So, in the future research work, lightweight cryptography is implemented in such a way that it should provide faster performance in less number of rounds.

5 Conclusion

IoT connects the virtual world to the real world through technologies like sensor and embedded system. The technologies used in IoT makes our life more comfortable and easier. Lightweight cryptography has gained much attention from academic and IT industries as it uses low-power resources, limited memory which improves the performance enhances security and increases energy efficiency. IoT faces a lot of challenges regarding security, privacy and scalability, so cryptography is used for security. Conventional cryptography is not opted for IoT due to its constrained environment. In this survey, we have covered the different architecture of IoT, security challenges, privacy and lightweight cryptography algorithm to solve them.

References

1. Airehrour D, Gutierrez J, Ray SK (2016) Secure routing for internet of things: a survey. J Netw Comput Appl 66:198–213 (2016). Future Gener Comput Syst 29(7):1645–1660 (2013)
2. Gubbi J, Buyya R, Marusic S, Palaniswami M (2013) Internet of things (IoT): a vision, architectural elements, and future directions. Future Gener Comput Syst 29(7):1645–1660
3. Zhao P, Peffer T, Narayanamurthy R, Fierro G, Raftery P, Kaam S, Kim J (2016) Getting into the zone: how the internet of things can improve energy efficiency and demand response in a commercial building (2016)
4. Pang Z, Chen Q, Tian J, Zheng L, Dubrova E (2013) Ecosystem analysis in the design of open platform-based in-home healthcare terminals towards the internet-of-things. In: 2013 15th international conference on advanced communication technology (ICACT), IEEE, pp 529–534
5. Miorandi D, Sicari S, De Pellegrini F, Chlamtac I (2012) Internet of things: vision, applications and research challenges. Ad Hoc Netw 10(7):1497–1516
6. Borgia E The internet of things vision: key features, applications and open issues. Comput Commun 54:1–31
7. Xaio Q, Gibbons T, Lebru H (2009) RFID technology, security vulnerabilities, and countermeasures. Supply Chain the Way to Flat Organization, Intech, pp 357–382
8. Kumar SA, Vealey T, Srivastava H (2016) Security in internet of things: challenges, solutions and future directions. In: 2016 49th Hawaii international conference on system sciences (HICSS), IEEE, pp 5772–5781
9. Ban HJ, Choi J, Kang N (2016) Fine-grained support of security services for resource constrained internet of things. Int J Distrib Sens Netw 2016
10. Acharya K et al (2013) Analysis of cryptographic algorithms for network security. Int J Comput Appl Technol Res 3(2):130–135
11. Katagi M, Moriai S (2008) Lightweight cryptography for the internet of things. Sony Corporation, pp 7–10
12. Poettering B (2013) Rijndaelfurious AES-128 implementation for avr devices (2007)
13. Eisenbarth T, Gong Z, G¨uneysu T, Heyse S, Indesteege S, Ker-ckhof S, Koeune F, Nad T, Plos T, Regazzoni F, et al (2012) Compact implementation and performance evaluation of block ciphers in attiny devices. In: International conference on cryptology in Africa. Springer, pp 172–187
14. Beaulieu R, et al (2015) The SIMON and SPECK lightweight block ciphers. In: Proceedings of the 52nd Annual Design Automation Conference, pp 1–6
15. Suzaki T, et al. (2013) TWINE: a lightweight block cipher for multiple platforms. Selected areas in cryptography lecture notes in computer science. Springer, vol 7707, pp 339–354
16. Kumarkushwaha P et al (2014) A survey on lightweight block ciphers. Int J Comput Appl 96(17):1–7
17. Akishita T, Hiwatari H (2012) Very compact hardware implementations of the blockcipher CLEFIA. Selected areas in cryptography lecture notes in computer science. Springer, pp 278–292
18. Shirai T, et al. (2007) The 128-bit blockcipher CLEFIA (extended abstract). Fast Software Encryption. FSE 2007. Lecture notes in computer science. Springer, vol 4593
19. Kumar BA, Bansal M (2017) A review on lightweight cryptography algorithms for data security and authentication in IoTs. In: 2017 4th international conference on signal processing, computing and control (ISPCC). Solan, pp 504–509

nTunnel FET (nTFET) Reliability Study Against Positive Bias Temperature Instability (PBTI) for Different Device Architectures

Suman Das, Avik Chattopadhyay, and Suchismita Tewari

Abstract In this paper, for the first time, the nTFET reliability and stability have been studied against the adverse effect of positive bias temperature instability (PBTI) for three different device architectures, viz. elevated drain-(ED-), elevated source-(ES-) and elevated source–drain-(ESD-) TFETs, in terms of threshold voltage (V_T) shift, on-current/off-current (I_{ON}/I_{OFF}) deviation, off-current (I_{OFF}) deviation and minimum subthreshold swing (SS_{min}) deviation for different PBTI stress conditions. The channel lengths for all the devices have been considered as 70 nm. It is found that the ES-TFET is showing maximum stability against the PBTI effect in terms of I_{OFF} and I_{ON} with a V_T shift of 2.9% and SS_{min} shift of 27%, while maintaining measurable absolute values of the aforementioned device parameters.

Keywords Device architecture optimization · Elevated source–drain · PBTI · Interface trap charges · Tunnel FET · TFET reliability

1 Introduction

Continuous device miniaturization over the last few decades, in order to fulfill the increasing demand of "All-in-one" or "All-in-palm" concept, the need of devices, fit to work for next generation low-power electronics, has gained a lot of research attention.

S. Das (✉)
Department of Electronics and Communication Engineering, Sikkim Manipal Institute of Technology, Sikkim Manipal University, Sikkim, India
e-mail: das.suman194@gmail.com

A. Chattopadhyay · S. Tewari
Department of Radio Physics and Electronics, University of Calcutta, Kolkata, India
e-mail: avikjoy@yahoo.com

S. Tewari
e-mail: suchi_tewari@yahoo.co.in

V. Nath and J. K. Mandal (eds.), *Proceedings of the Fourth International Conference on Microelectronics, Computing and Communication Systems*, Lecture Notes in Electrical Engineering 673, https://doi.org/10.1007/978-981-15-5546-6_14

In this regard, Tunnel FET (TFET), due to its ability to overcome the 60 mV/decade subthreshold slope (SS) limitation at room temperature, of the conventional MOSFET devices, has emerged as a savior [1–5]. But Tunnel FET suffers from low I_{ON} value, which makes it difficult to use the device for analog application. But in order to overcome this shortcoming of the TFET device, alternate high mobility channel materials [6] as well as several novel device architectures have been proposed [7–15]. *L*-shaped and *U*-shaped gates [15, 16] are the examples of such innovative device architectures. Here, the vertical tunneling mechanism in addition to horizontal tunneling mechanism has been used, which increases the effective tunneling area, without affecting the device footprint and ultimately leading to an enhanced device current as well as a sharp subthreshold swing. However, in search for strict device performance improvement, the aspect of device stability and reliability against adverse effects, viz. bias temperature instability (BTI), hot carrier stress (HCS), etc., have been overlooked until the last decade. With device miniaturization, the BTI effects, e.g., negative BTI or NBTI for pTFETs and positive BTI or PBTI for nTFETs, become more and more significant from the perspectives of device dependability and consistency [17–20]. In the last few decades, although a few works have been found on the effect of BTI on the performance of the TFETs in the literature, most of them are restricted to the conventional TFET device architecture [21–23].

In this paper, for the first time, an extensive investigation of the nTFET device performance against the adverse effect of PBTI for three different architectural engineering, viz. elevated drain-(ED-), elevated source-(ES-) and elevated source–drain-(ESD-), has been reported. Based on the characteristic shifts for different PBTI stress conditions of the aforementioned nTFET devices, the most optimized device architecture has been found out with satisfactory device performance with a good reliability against the PBTI effect.

2 Device Structure and Simulation Framework

Figure 1 shows the schematic diagrams of ED-TFET, ES-TFET and ESD-TFET, considered in our study. The step-by-step fabrication process flows of the above-mentioned ED-TFET and ES-TFET on silicon-on-insulator (SOI) substrate are already demonstrated in Refs. [14, 15], respectively. The three structures are simulated with a fixed SOI layer thickness of 70 nm, for a channel length (L_{ch}) of 70 nm, and the corresponding equivalent oxide thickness (EOT) for gate insulator is taken as 1.6 nm following ITRS roadmap [24]. SILVACO ATLAS, version 5.18.3.R [25] as a numerical device simulator, has been used to perform all simulations. Non-local BTBT model, along with the CVT Lombardi model, the quantum density gradient and bandgap narrowing (BGN) models are employed in our simulations. The Shockley–Read–Hall (SRH) and Auger models are triggered on too to calculate carrier lifetime related to radiative and non-radiative recombination processes, respectively, and for carrier statistics, Fermi–Dirac statistical model has been invoked.

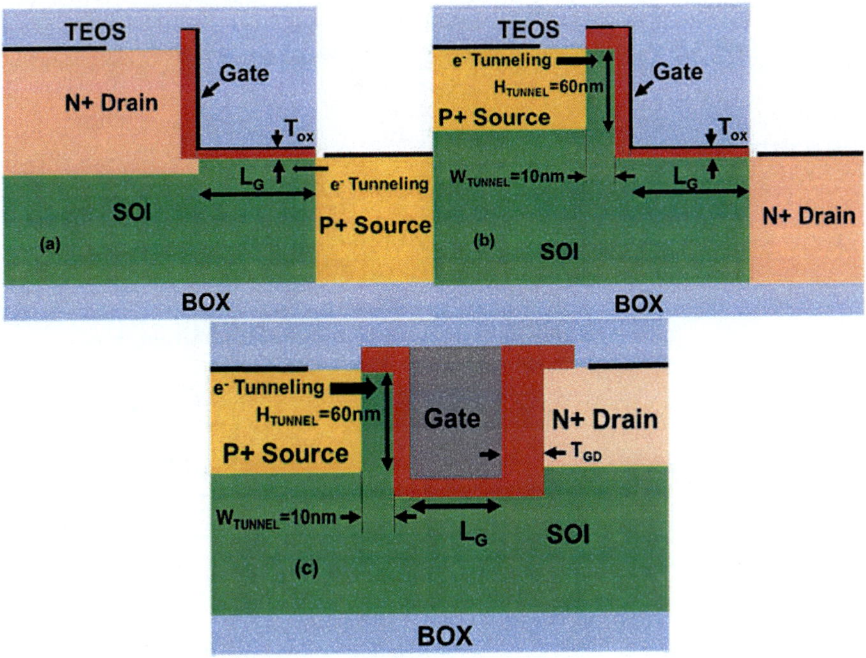

Fig. 1 Schematics of **a** ED-TFET, **b** ES-TFET and **c** ESD-TFET

3 Model Calibration

To confirm the validity of our simulated results, depicted in Fig. 2a, b, we first calibrated the transfer characteristics at different values of V_{DS} both for ED-TFET and ES-TFET to the already reported experimental data [14, 15], keeping all the device dimensions and parameters similar to those mentioned therein. Parameters related to non-local BTBT model and BGN model have also been tuned. A well agreement between the simulated and experimental results for both the types of TFETs, as can be evidenced from Fig. 2, proves the authenticity of our model calibrations.

4 Results and Discussion

To investigate the effect of PBTI on TFETs with different architectures, four different stress conditions are being chosen: 1. acceptor trap at room temperature, i.e., 300 K (PBTI-A300), 2. donor trap at room temperature, i.e., 300 K (PBTI-D300), 3. acceptor trap at 125 °C or 398 K (PBTI-A398) and 4. donor trap at 125 °C or 398 K (PBTI-D398). For the last two cases, the stress time corresponds to 1000 s [22]. For each case, the density of states is considered as 1×10^{12} $eV^{-1}Cm^{-2}$ [23] and the trap

Fig. 2 Comparison of transfer characteristics between **a** experimental data [14] and simulated data for ED-TFET, **b** experimental data [15] and simulated data for ES-TFET

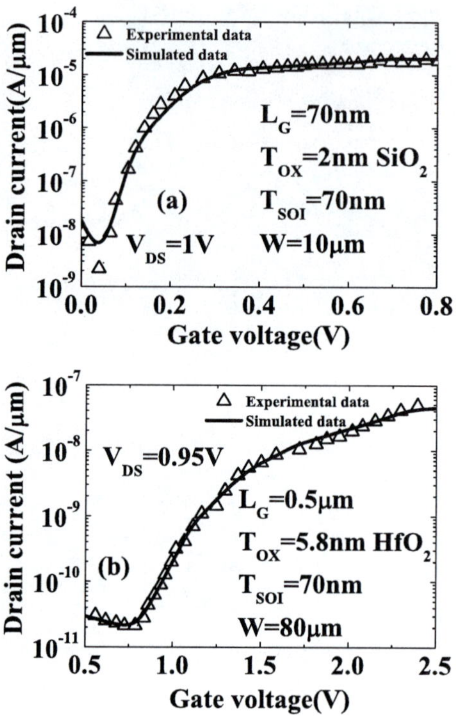

level is introduced at 0.45 eV below the conduction band (acceptor level) and 0.45 eV above the valence band (donor level) following the reference [23]. Figure 3 shows the transfer characteristics of ED-TEFT, with supply voltage (V_{DD}) = 0.4 V, for no PBTI effect and PBTI effect with aforementioned four stress conditions. It is to be noted that due to the *L*-shaped gate structure for ED-TEFT device, the trap level exists in channel, as well as in drain regions, but no such trap level is found in source region. It is clearly observed from Fig. 3 that the highest on-current (I_{ON}) and lowest

Fig. 3 Transfer characteristics of ED-TFET for no PBTI and different PBTI conditions at V_{DD} = 0.4 V

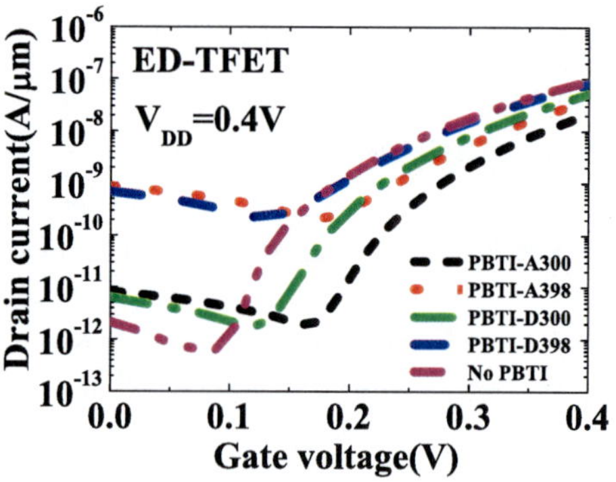

off-current (I_{OFF}) are obtained for no PBTI condition. From Fig. 3, it is also seen that for PBTI-A300, the knee point of the transfer characteristic shifts to extreme right, followed by the knee point for PBTI-D300 in the middle, keeping the knee point for no PBTI characteristic at extreme left for 300 K temperature. This phenomenon may be explained as follows. For PBTI-A300, the acceptor level (E_{TA}) exists above the Fermi level (E_F) for no bias condition, and hence, it always wants to remain empty. Even if electron comes to E_{TA} from source valance band through BTBT, it will get recombined immediately, resulting in least tunnel current through BTBT mechanism at source/channel junction, compared to the rest two cases. While in case of PBTI-D300, the donor trap level (E_{TD}) is always situated below E_F and hence always remains filled-up. But at source/channel junction, it has got no contribution in BTBT mechanism, as it is situated outside the tunnel window. However, the presence of this trap level reduces the effective bandgap (E_g) of the channel material, resulting in the flow of the tunnel current at a less $V_{GS,}$ and ultimately leading to a higher $I_{ON,}$ compared to the previous case. Now, for the case with no PBTI effect, the recombination probability is least, due to the absence of trap level in the forbidden energy gap, resulting in the left-most knee point of the corresponding transfer characteristic and the highest $I_{ON.}$ Additionally, It is to be noticed that the presence of trap level in drain and in channel reduces the bandgap (E_g) of the corresponding regions and hence increases the leakage current in subthreshold region, while the one with no PBTI effect has maintained its original E_g resulting low leakage current and hence least $I_{OFF,}$ compared to the rest two cases. For PBTI-A398 and PBTI-D398, the same trend has been followed for the elevated temperature, and the leakage increases, resulting in high I_{OFF} for both the cases. This entire phenomenon is supported by the corresponding electric field profiles, at source/channel and drain/channel junctions, as is observed from Fig. 4a, b, respectively, with highest (lowest) current corresponds to the highest (lowest) field.

Figure 5 shows the set of transfer characteristics for ES-TFET for no PBTI and PBTI with four stress conditions. In this structure, there is no trap level in source and/or drain region, but only in channel region. In case of PBTI-A300, as soon as capturing electrons maximizes and the corresponding accumulation of electrons in the E_{TA} screens the effect of tunnel field, resulting in low I_{ON}. No such screening effect takes place in case of PBTI-D300, and as a result, high I_{ON} is obtained, as is seen from Fig. 5. In case of no PBTI effect, the absence of the trap level reduces the leakage and ensures the maintenance of original $E_{g,}$ and original tunnel field, devoid of any screening effect, unlike the case with PBTI-A300. These combinations lead to the lowest leakage current and corresponding lowest I_{OFF} and a certain value of I_{ON}, which is higher than PBTI-A300 but lower than that of PBTI-D300. Similar trend is found for the cases with PBTI-A398 and PBTI-D398, and only, the high temperature is responsible for high current, throughout the entire range of V_{GS}, as is observed from Fig. 5. Figure 6a, b shows the corresponding electric field profile for source/channel junction and drain/channel junction, respectively. It is clearly witnessed from Fig. 6a, b that the nature of the transfer characteristics for different PBTI and no PBTI conditions is completely supported by the corresponding electric field profile. Figure 7 shows the set of transfer characteristics for ESD-TFET for

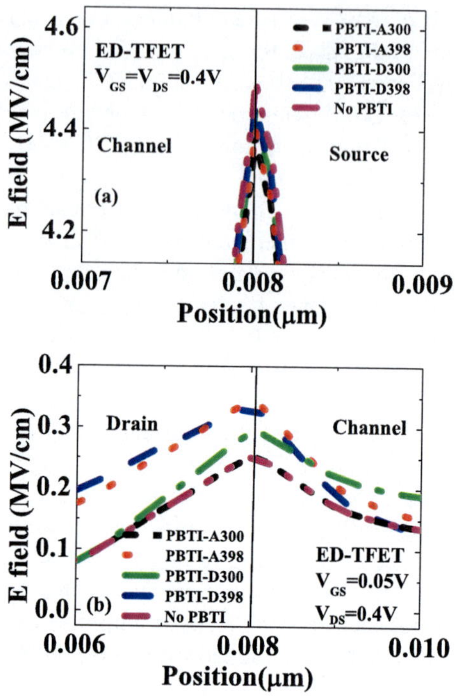

Fig. 4 Electric field profile of **a** source/channel **b** drain/channel junctions of ED-TFET for no PBTI and different PBTI conditions

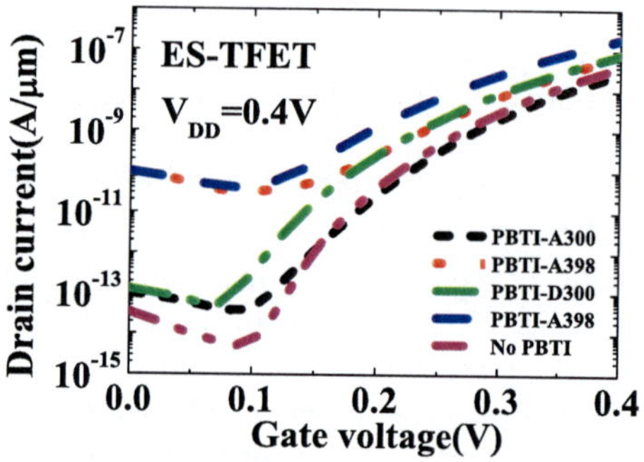

Fig. 5 Transfer characteristics of ES-TFET for no PBTI and different PBTI conditions at $V_{DD} = 0.4$ V

no PBTI and PBTI with four stress conditions. For this particular structure, there is trap level in drain and channel region only. But for this structure, there is no drain region below the horizontal arm of gate oxide, as is for the case of ED-TFET. The trend of the transfer characteristics is similar to that of ES-TFET, but only below the knee point, there is almost no change of I_{DS} with the decrease of V_{GS}. Figure 8a, b shows the corresponding electric field profile of ESD-TFETs with no PBTI and four aforementioned stress conditions, which supports the nature of the transfer characteristics, shown in Fig. 7.

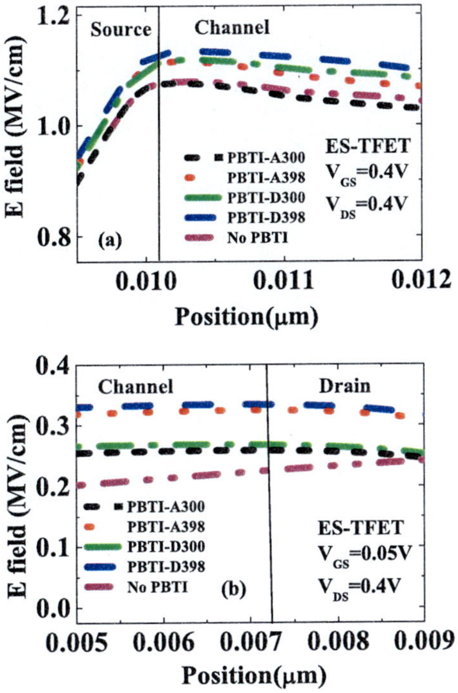

Fig. 6 Electric field profile of **a** source/channel **b** drain/channel junctions of ES-TFET for no PBTI and different PBTI conditions

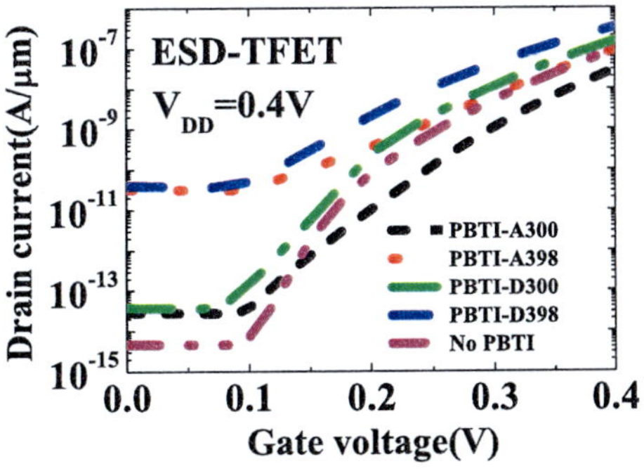

Fig. 7 Transfer characteristics of ESD-TFET for no PBTI and different PBTI conditions at V_{DD} = 0.4 V

The next different digital parameters, viz. I_{OFF}, I_{ON}/I_{OFF}, SS_{min}, are extracted from the corresponding characteristics and are given in Table 1. It is clearly observed from Table 1 that from the perspective of the absolute value of the parameters, viz. I_{ON} and I_{ON}/I_{OFF}, for no PBTI condition, ESD-TFET outperforms the rest two devices, while for SS_{min}, ED-TFET outperforms the rest two devices marginally, proving the overall superiority of the ESD-TFET device from the logic performance point of view. In order to investigate the reliability of the device against the adverse effect of PBTI, the change or shift of the parameters from its no PBTI condition is calculated and is given in Table 2. After critically scrutinizing, it is observed that although ED-TFET

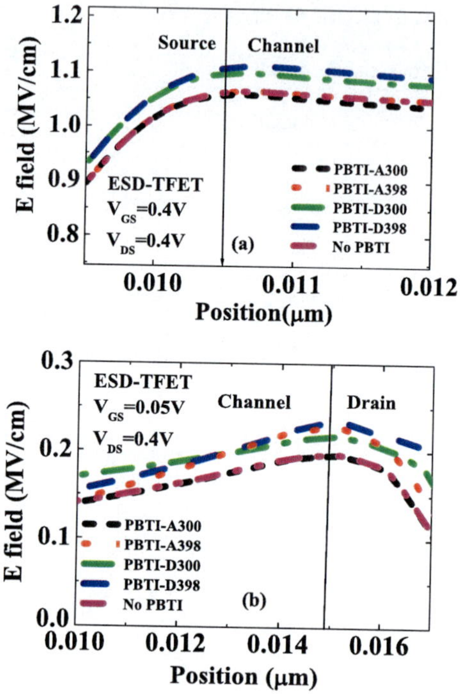

Fig. 8 Electric field profile of **a** source/channel **b** drain/channel junctions of ESD-TFET

produces the least SS_{min} out of the three TFET devices, it suffers from maximum V_T shift, hence maximum shift, due to PBTI effect. Another interesting fact that can be observed from Table 2 is that the superior device ESD-TFET with the lowest I_{OFF} and highest I_{ON}/I_{OFF} is suffering maximum from PBTI effect, producing highest changes in the aforementioned parameters due to PBTI effect, than that of the rest two devices, making it highly unstable device against the adverse effect of PBTI. Although when it comes to V_T shifting and SS shifting, it is producing the minimal change due to PBTI effect.

Interestingly, although the performance of ES-TFET in terms of the absolute value of $I_{OFF}$and I_{ON}/I_{OFF}, under no PBTI condition, remains worse than ESD-TFET, but better than that of ED-TFET, under PBTI conditions, it shows minimum parameters shift for I_{OFF} and I_{ON}/I_{OFF}, which is significantly low than the shifting of the aforementioned parameters for the rest two devices, as is observed from Table 2. Also, for V_T shift and SS_{min} shift, ES-TFET shows only marginal higher value than that of the ESD-TFET, against PBTI effect. Thus, after a thorough overall inspection, from device performance point of view, as well as considering reliability issue, ES-TFET has turned out to be the most optimized device of all. A performance comparison against PBTI effect, between our work and the nTFET, nMOSFET devices taken from Han et. al's work [26] in terms of V_T, SS_{min} and I_{ON}/I_{OFF} is made, and the corresponding values are tabulated in Table 3.

Table 1 Absolute value of digital parameters of different device architectures for no PBTI and PBTI with four stress conditions

Parameters	No PBTI	PBTI-A300	PBTI-D300	PBTI-A398	PBTI-D398
ED-TFET					
I_{ON} (A/μm)	9.2×10^{-8}	2.15×10^{-8}	5.24×10^{-8}	3.88×10^{-8}	8.15×10^{-8}
I_{OFF} (A/μm)	2.14×10^{-12}	8.2×10^{-12}	6.38×10^{-12}	8.66×10^{-10}	6.97×10^{-10}
I_{ON}/I_{OFF}	4.3×10^{4}	2.6×10^{3}	8.2×10^{3}	44.8	1.17×10^{2}
V_T (V)	0.3	0.39	0.325	0.375	0.31
SS_{min} (mV/decade)	13.8	25	20.8	62.5	50
ES-TFET					
I_{ON} (A/μm)	4×10^{-8}	2.7×10^{-8}	8.11×10^{-8}	7.26×10^{-8}	1.95×10^{-7}
I_{OFF} (A/μm)	3.44×10^{-14}	9.82×10^{-14}	1.3×10^{-13}	9.5×10^{-11}	9.96×10^{-11}
I_{ON}/I_{OFF}	1.16×10^{6}	2.75×10^{5}	6.24×10^{5}	7.64×10^{2}	2×10^{3}
V_T (V)	0.35	0.36	0.33	0.34	0.32
SS_{min} (mV/decade)	17.9	27.7	25	50	41.6
ESD-TFET					
I_{ON} (A/μm)	8.83×10^{-8}	2.94×10^{-8}	1.55×10^{-7}	9×10^{-8}	3.4×10^{-7}
I_{OFF} (A/μm)	4.6×10^{-15}	2.8×10^{-14}	3.8×10^{-14}	3.21×10^{-11}	4.1×10^{-11}
I_{ON}/I_{OFF}	1.9×10^{7}	1.1×10^{6}	4.1×10^{6}	2.8×10^{3}	8.3×10^{3}
V_T (V)	0.36	0.37	0.34	0.358	0.32
SS_{min} (mV/decade)	17.9	22.2	24.5	43.5	40.3

5 Conclusion

In this paper, we have reported a detailed investigation concerning the effect of PBTI on Si-nTFET for three different device architectures, viz. ES-, ED- and ESD-TFETs, followed by an optimization between the device performance and the reliability issue against the adverse effect of PBTI, to find the most suitable TFET device. After a thorough investigation, ES-TFET turns out to be the most optimized device in this regard. It suffers from minimum I_{OFF} and I_{ON}/I_{OFF} shifts from the no PBTI condition and shows significant low values for V_T shift and SS_{min} shift as well, maintaining a good device performance throughout, in terms of the device performance matrices.

Table 2 Percentage change in digital parameters of different device architectures for PBTI with four stress conditions with respect to no PBTI condition

Parameters	Deviation for PBTI-A300 over no PBTI condition	Deviation for PBTI-D300 over no PBTI condition	Deviation for PBTI-A398 over no PBTI condition	Deviation for PBTI-D398 over no PBTI condition
ED-TFET				
I_{OFF} (A/μm)	0.606 decade	0.424 decade	2.652 decade	2.483 decade
I_{ON}/I_{OFF}(%)	94	81	99.9	99.7
V_T (V) (%)	30	8.3	25	3.3
SS_{min} (mV/decade) (%)	81.2	50.7	352.9	262.3
ES-TFET				
I_{OFF} (A/μm)	0.638 decade	0.686 decade	3.606 decade	3.652 decade
I_{ON}/I_{OFF} (%)	76.3	46.2	99.9	99.8
V_T (V) (%)	2.9	5.7	2.9	8.6
SS_{min} (mV/decade) (%)	26.8	39.7	179.3	132.4
ESD-TFET				
I_{OFF} (A/μm)	0.82 decade	0.92 decade	3.86 decade	3.95 decade
I_{ON}/I_{OFF}(%)	94.2	78.4	100	100
V_T (V) (%)	2.8	5.6	0.5	11.1
SS_{min} (mV/decade) (%)	24	36.9	143	125

Table 3 Performance comparison between our work and the work of Han et al. [26]

Parameters	PBTI deviation for nTFET (Han et al.) (%)	PBTI deviation for nMOSFET (Han et al.) (%)	PBTI deviation for ES-TFET for A398 (our work) (%)	PBTI deviation for ES-TFET for D398 (our work) (%)
I_{ON}/I_{OFF}	99.9	99.9	99.9	99.8
V_T (V)	16.6	30	2.9	8.6
SS_{min} (mV/decade)	120	110	179.3	132.4

References

1. Mallik A, Chattopadhyay A (2012) The impact of fringing field on the device performance of a p-channel tunnel field-effect transistor with a high-gate dielectric. IEEE Trans Electron Dev 59(2):277–282
2. Mallik A, Chattopadhyay A (2012) Tunnel field-effect transistors for analog/mixed-signal system-on-chip applications. IEEE Trans Electron Dev 59(4):888–894

3. Guin S, Chattopadhyay A, Karmakar A, Mallik A (2014) Impact of a pocket doping on the device performance of a Schottky tunneling field-effect transistor. IEEE Trans Electron Dev 61(7):2515–2522
4. Chattopadhyay A, Mallik A (2011) Impact of a spacer dielectric and a gate overlap/underlap on the device performance of a tunnel field-effect transistor. IEEE Trans Electron Dev 58(3):677–683
5. Mallik A, Chattopadhyay A (2011) Drain-dependence of tunnel field-effect transistor characteristics: the role of the channel. IEEE Trans Electron Dev 58(12):4250–4257
6. Alian A, Mols Y, Bordallo CCM, Verreck D, Verhulst A, Vandooren A, Rooyackers R, Agopian PGD, Martino JA, Thean A, Lin D, Mocuta D, Collaert N (2016) InGaAs tunnel FET with sub-nanometer EOT and sub-60 mV/dec sub threshold swing at room temperature. Appl Phys Lett 109:243502
7. Bhuwalka KK, Schulze J, Eisele I (2004) Performance enhancement of vertical tunnel field-effect transistor with SiGe in the $\delta p+$ layer. Jpn J Appl Phys 43(7A):4073–4078
8. Patel N, Ramesha A, Mahapatra S (2008) Drive current boosting of n-type tunnel FET with strained SiGe layer at source. Microelectron J 39(2):1671–1677
9. Virani HG, Rao Adari RB, Kottantharayil A (2010) Dual-*k* spacer device architecture for the improvement of performance of silicon n-channel tunnel FETs. IEEE Trans Electron Dev 57(10):2410–2417
10. Saurabh S, Kumar MJ (2011) Novel attributes of a dual material gate nanoscale tunnel field-effect transistor. IEEE Trans Electron Dev 58(2):404–410
11. Mallik A, Chattopadhyay A, Guin S, Karmakar A (2013) Impact of a spacer–drain overlap on the characteristics of a silicon tunnel field-effect transistor based on vertical tunneling. IEEE Trans Electron Dev 60:935–943
12. Mallik A, Chattopadhyay A, Omura Y (2014) Gate-on-germanium source tunnel field-effect transistor enabling sub-0.5 V operation. Jap J Appl Phys 53:104201-1–104201-7
13. Madan J, Chaujar R (2016) Gate drain-overlapped-asymmetric gate dielectric-GAA-TFET: a solution for suppressed ambipolarity and enhanced ON state behaviour. Appl Phys A 122(11):973–981
14. Choi WY, Park BG, Lee JD, Liu TJK (2007) Tunneling field-effect transistors (TFETs) with subthreshold swing (SS) less than 60 mV/dec. IEEE Electron Dev Lett 28(8):743–745
15. Kim SW, Kim JH, King Liu TJ, Choi WY, Park BG (2016) Demonstration of *L* shaped tunnel field effect transistors. IEEE Trans Electron Dev 63(4):1774–1777
16. Shih PC, Hou WC, Li JY (2017) A *U*-gate InGaAs/GaAsSb heterojunction TFET of tunneling normal to the gate with separate control over ON- and OFF-state current. IEEE Electron Dev Lett 38(12):1751–1754
17. Jiao GF, Chen ZX, Yu HY, Huang XY, Huang DM, Singh N, Lo GQ, Kwong DL, Li MF (2010) Experimental studies of reliability issues in tunneling field-effect transistors. IEEE Trans Electron Dev 31(5):396–398
18. Beneventi GB, Gnani E, Gnudi A, Reggiani S, Baccarani G (2013) Can interface traps suppress TFET Ambipolarity? IEEE Electron Dev Lett 34(12):1557–1559
19. Gupta S, Nigam K, Pandey S, Sharma D, Kondekar PN (2017) Effect of interface trap charges on performance variation of heterogeneous gate dielectric junction less-TFET. IEEE Trans Electron Dev 64(11):4731–4737
20. Huang XY, Jiao GF, Cao W, Huang D, Yu HY, Chen ZX, Singh N, Lo GQ, Kwong DL, Li MF (2013) Effect of interface traps and oxide charge on drain current degradation in tunneling field-effect transistors. IEEE Electron Dev Lett 31(8):779–781
21. Han G, Yang Y, Guo P, Zhan C, Low KL, Goh KH, Liu B, Toh EH, Yeo YC (2012) PBTI characteristics of N-channel tunneling field effect transistor with HfO_2 gate dielectric: new insights and physical model. In: IEEE Proceedings of Technical Program of 2012 VLSI Technology, System and Application. Hsinchu, Taiwan
22. Ma MW, Chen CY, Su CJ, Wu WC, Wu YH, Kao KH, Chao TS, Lei TF (2013) Characteristics of PBTI and hot carrier stress for LTPS-TFT with high-*k* gate dielectric. IEEE Electron Dev Lett 29(2):171–173

23. Venkatesh P, Nigam K, Pandey S, Sharma D, Kondekar PN (2017) Impact of interface trap charges on performance of electrically doped tunnel FET with heterogeneous gate dielectric. IEEE Trans Dev Mater Reliab 17(1):245–252
24. International Technology Roadmap for Semiconductors (2005)
25. ATLAS User's Manual (2015) A device simulation package. SILVACO Int., Santa Clara, CA
26. Han G, Yang Y, Guo P, Zhan C, Low KL, Goh KH, Liu B, Toh EH, Yeo YC (2012) PBTI characteristics of N-channel tunneling field effect transistor with HfO_2 gate dielectric: new insights and physical model. In: Proceedings of Technical Program of VLSI Technology, System and Application

Three-Phase Space Vector Modulated Z-Source Inverter

Sweta Kumari, Rajib Kumar Mandal, and G. K. Choudhary

Abstract The simulation model and the control scheme of a three-phase space vector modulated (SVPWM) Z-source inverter (ZSI) is discussed in presented article. The result waveforms are exposed in the MATLAB-Simulink scope platform. This new topology of inverter is explored to defeat the all limited constraints of old VSI and CSI models. The output parameters of ZSI are controllable and adjust by the duty cycle of the switches used at the main bridge along with a primary switch at the source side. This variation in duty cycle is accomplished by SVPWM approach. Likewise the shoot-through state is worth in ZSI model.

Keywords Voltage-source inverter · Current-source inverter · Impedance-source inverter (ZSI) · Space vector pulse width modulation

1 Introduction

With technological advancements, semiconductor switches have become popular in control strategies. An inverter changes over DC to AC current [1]. Traditional voltage- or current-source converters are not buck–boost converter. They can be either buck converter or boost converter. Conventional voltage-source inverter is a DC-to-AC buck converter; i.e., the output AC signal is lower than the input DC signal. An extensive capacitor is connected between the DC input source voltages to maintain effectively the three-phase inverter output. Besides, an ordinary CSI is a DC-to-AC

S. Kumari (✉)
Electrical & Electronics Engineering Department, Darbhanga College of Engineering, Darbhanga, India
e-mail: eeesweta@gmail.com

R. K. Mandal · G. K. Choudhary
Electrical Engineering Department, National Institute of Technology, Patna, India
e-mail: rajib@nitp.ac.in

G. K. Choudhary
e-mail: girish@nitp.ac.in

V. Nath and J. K. Mandal (eds.), *Proceedings of the Fourth International Conference on Microelectronics, Computing and Communication Systems*, Lecture Notes in Electrical Engineering 673, https://doi.org/10.1007/978-981-15-5546-6_15

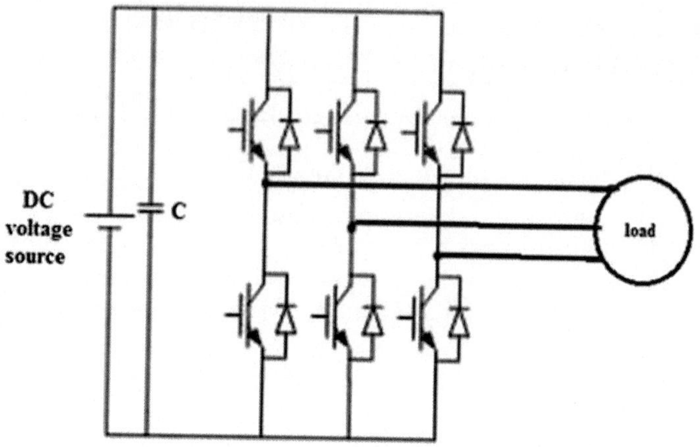

Fig. 1 Three-phase VSI model

boost converter. In comparison with the DC voltage provided to the bulky inductor circuit, the output AC voltage of CSI is better (Fig. 1).

The VSI has these limitations:

1. As the AC voltage signal cannot be superior than the input voltage (DC), the VSI behaves as a step-down inverter. Vice versa, it behaves a step-up converter. An extra boost chopper (DC–DC converter) is essential for those applications where the input dc voltage magnitude is constrained. [2].
2. For getting sinusoidal output, an additional LC channel is required. Due to this, the device gets bulky and complicated, resulting in more power losses.
3. To avoid the STS, which can happen due to electromagnetic interference, a dead time needs to be provided in both the top and bottom wings of the power switches. This leads to distortion in the output waveform.

The traditional CSI model is presented in Fig. 2. The input of the three-phase inverter circuit is a DC current source. The circuit comprises of six power semiconductor switches having reverse current blocking capability. Thus, it is a unidirectional bipolar semiconductor circuit.

The limitations of conventional CSI are:

1. Subsequently, the AC voltage magnitude is more than the source (DC) voltage; thus, an old CSI model behaves as a typical boost inverter. A dc voltage feeds a large inductor which performs as a constant current source. Thus, CSI is nothing more than a buck inverter since the magnitude of the output voltage is lower than that of input voltage at all the time.
2. Each of the topmost and bottom switches of bridge circuit should be gated simultaneously. Otherwise, an open circuit forms which can destroy the circuit. Thus, an overlap time is provided so that this condition does not occur. But, this overlap time results in the distortion of the output waveform.

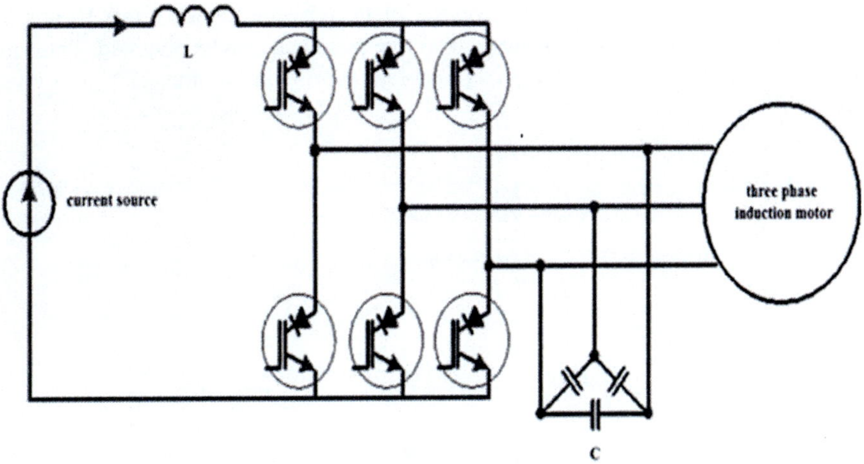

Fig. 2 Three-phase CSI model

3. For blocking the reverse voltage, a diode is connected across the fast-acting transistors, like IGBTs, of a CSI. This reduces the overall action of the circuit.

Accordingly, together VSI and CSI suffer from the succeeding issues:

1. They cannot be used as buck or boost converter simultaneously.
2. A VSI cannot act as CSI and vice versa. They are not compatible with each other.
3. The effect of EMI noise is pronounced in both the circuits.

Voltage Space Vector Modulation

The sinusoidal pulse width modulation (SPWM) technique is generally taken for modulation purpose in the traditional VSI models. The SPWM procedure is complicated, and furthermore, the usage proportion of DC voltage has reduced. In the present chapter, here is a discourse of space vector pulse width modulation (SVPWM) control technique with 3-phase ZSI. Fundamentally, SVPWM is a type of control calculation of PWM [3]. SVPWM is outstanding among other modulation technique to smother the sounds, uncommonly lower request music even in lower working frequencies [3]. Utilizing this regulation strategy additionally enhances the dynamic execution of the circuit. Another benefit of this strategy is that it gives 15% more fundamental component in the voltage output in comparison with PWM inverter circuits.

The three-phase ZSI is made up with 6 semiconductor power switches, each having binary states of "ON" and "OFF" in the switching circuit. Thus, there can be $2^3 = 8$ static space vectors which comprises of 2 zero vectors (000, 111) and 6 dynamic vectors $(001, 010, 011, 100, 101, 110)$ that can be obtained using switching cycles instantaneously. Thus, six divisions of activities are created. The (101) notation implies the top transistor control switch of leg 3 and leg 1 is ON and the lower transistor control switch of leg 2 is in ON condition. Using two dynamic vectors and one of the zero vector (111 or 000), reference vector (V_{ref}) is derived. According to

Fig. 3 V_{ref} is in area-1 in that condition it is planned by the stable vectors V_1 and V_2 and V_0 or V_7. Similarly, the voltage in rest five areas can also be obtained. Thus, the key function of SVPWM technique is to obtain approximate values of V_{ref} as given in Fig. 4 [4].

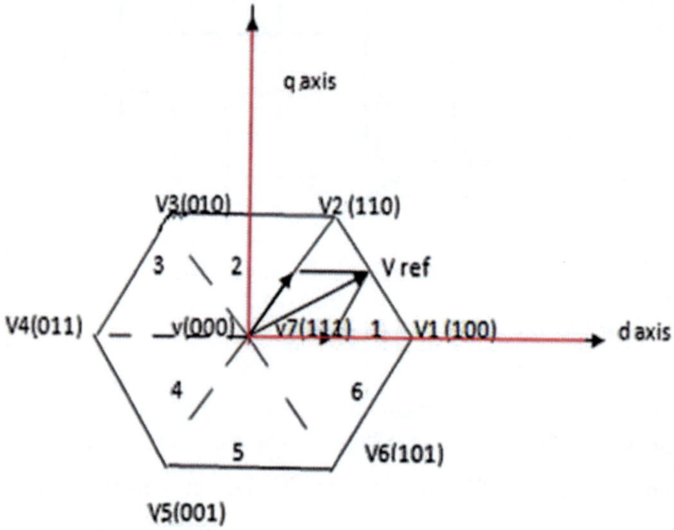

Fig. 3 Sector partition of voltage vectors

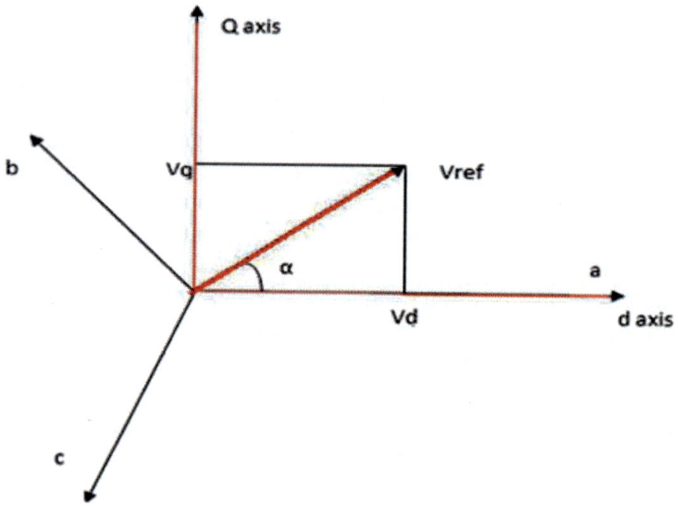

Fig. 4 V_{ref} space vector

2 Z-Source Inverter

For better control, another control scheme is presented that overcomes the constraints of old model VSI and CSI. Theoretically, the output parameter of a ZSI can vary from zero to up to infinite. The circuit model of a ZSI is presented in Fig. 5, while its equivalent circuit is shown in Fig. 6. Unlike conventional converters, shoot-through working condition is utilized in ZSI to improve output voltage. Controlling of buck–boost action of ZSI by shoot-through state (STS) and non-shoot-through state (NSTS) is explained in [2].

The general three-phase ZSI arrangement utilizes unprecedented impedance circuit as an association between DC source voltage and inverter circuit. For X-shaped impedance system, inductors and capacitors are utilized, as appeared in Fig. 5. In Fig. 6, a system of four terminals is demonstrated which consists of capacitors (C_1 and C_2) and two inductors (L_1 and L_2) that are joined in cross-structure. The cross-shape circuit gives rise to the linking between the main network and the DC source network. The values of reactance and capacitance should be chosen wisely so that no resonance can happen. On the other hand, the values of reactance and capacitance

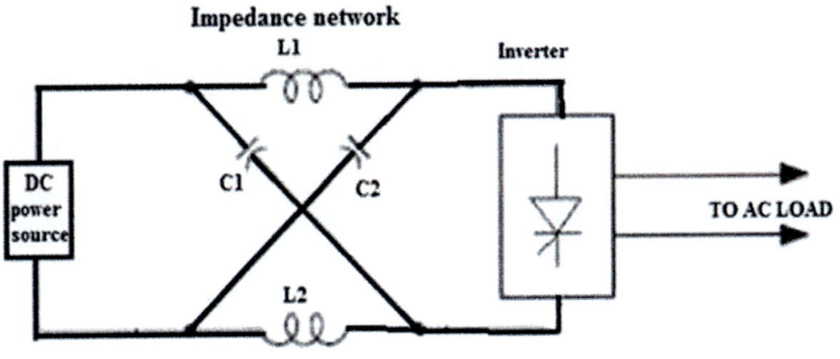

Fig. 5 Z-source converter

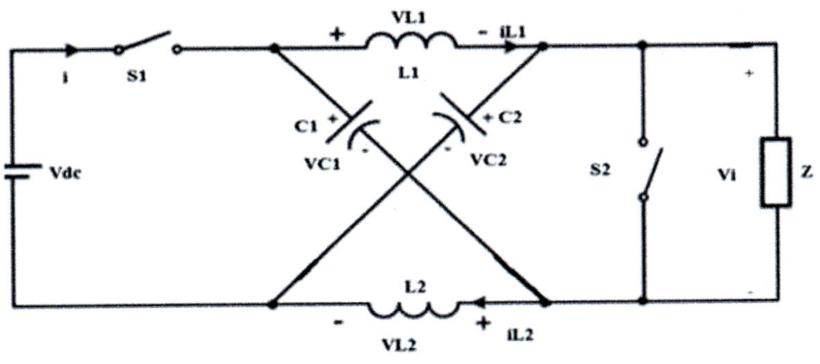

Fig. 6 ZSI equivalent circuit

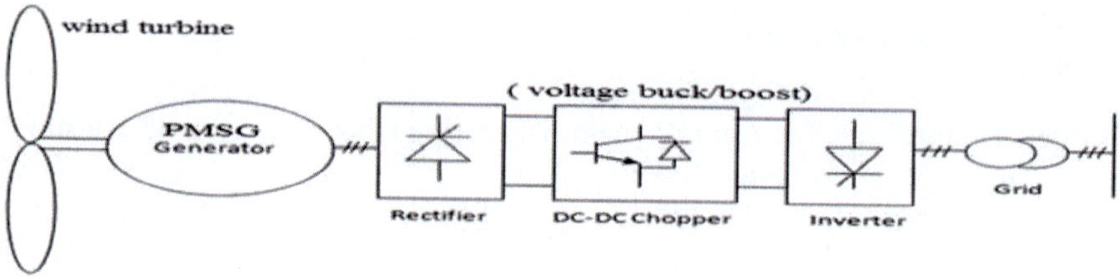

Fig. 7 Traditional inverter in general grid system

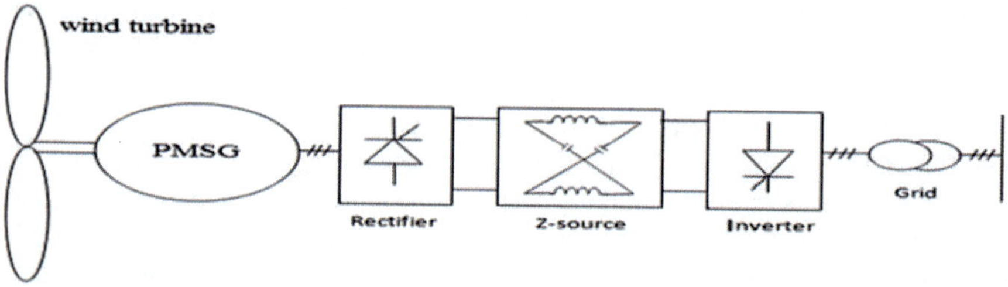

Fig. 8 ZSI-based grid interface

should not be large so that capacitor voltage and inductor current become negligible. The enhanced voltage is essentially the capacitor voltage in a ZSI. Similarly, switch S_1 of the ZSI is dealt with a bidirectional current for the buck–boost operation [5]. The diode should be connected in anti-parallel mode across the transistor. Figure 7 shows the popular power flow technique known as two-stage transformation for transfer of power from PMSG to grid [6, 7].

Since a typical VSI is unable to increase the yield voltage magnitude, ZSI is used for power transfer between PMSG and grid [8] as shown in Fig. 8. No DC booster is required since ZSI alone gives the required voltage magnitude of higher value than the input dc voltage [9].

3 Principal of Operation

The major unique feature of ZSI is that the yield AC potential difference can be varied between 0 and ∞, without any concern of the input value of potential difference. Thus, unlike VSI or CSI, it can give both buck and boost operation over a wide range of values. Essentially, the ZSI has three working conditions [2]. They are:

1. Normal state
2. Zero state
3. STS.

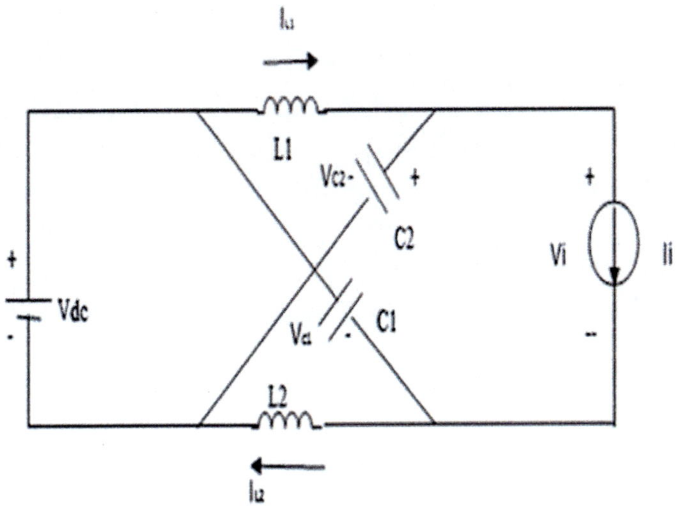

Fig. 9 Normal state

The "Normal State" and the "Zero State" of the Z-source resemble typical space-vector regulated inverter. To increase the magnitude of output voltage, the STS is applied. In ZSI, the use of STS mode can improve the reliability of the inverter, as the STS does not adversely affect the switching system. Thus, it gives diminished expense and extraordinary viable design for buck support activity. Figures 9 and 11 displays the binary methods of activity—STS and normal state. In normal operation, the AC voltage relies on the modulation index. Be that as it may, in the state of STS, the magnitude of voltage across of C_1 and C_2 gets step up. The capacitor voltage of the ZSI relies on the duty cycle of the STS. The following switching states are possible for the STS:

1. When the upper and lower switches are in function, i.e., when both the input and load are connected.
2. When there is short circuit of load terminals by the bottom and top switches of a same arms.
3. When there is short circuit of load terminals between the bottom and the top controls of a couple of phase arms (Fig. 10).

4 Investigation of ZSI Circuit

When both the uppermost and lowermost switches of same arms are triggered, then the circuit is in STS mode as in Fig. 11. At the input side, $S - 1$ is made reverse biased. In STS, C_1' and C_2' act as source of voltage for the ZSI and through the potential difference of C_1' and C_2' the charging of inductors L_1' and L_2' take place. So, for the inductor and capacitor potentials,

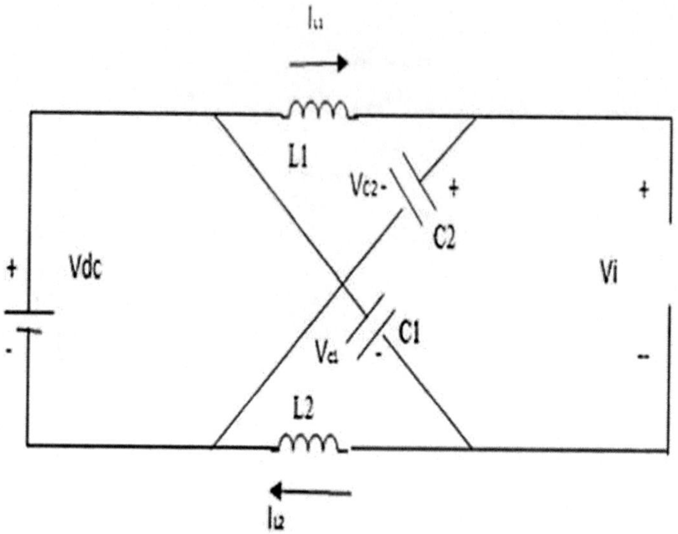

Fig. 10 Zero state

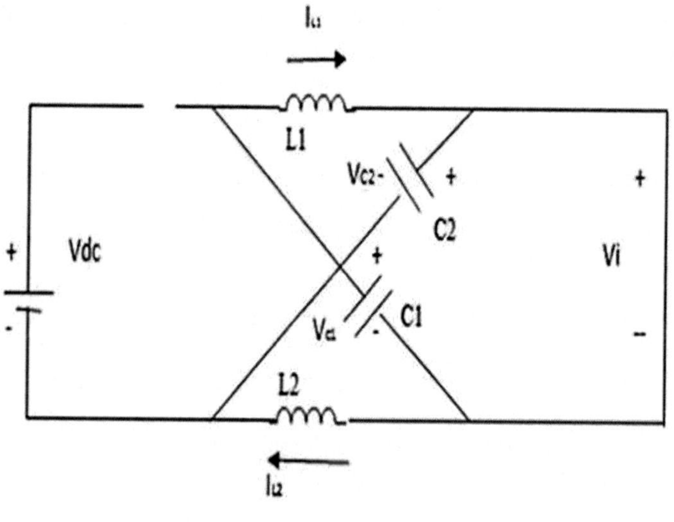

Fig. 11 STS

$$V'_{L'_1} = V'_{C'_1} \tag{1}$$

And

$$V'_{L'_2} = VV'_{C'_2} \tag{2}$$

If we assume that,

$$L' = L'_2 = L'_1$$

And

$$C' = C'_2 = C'_1$$

$$V'_{L'_1} = VV'_{L'_2} = V'_{L'} \tag{3}$$

And

$$V'_{C'_1} = VV'_{C'_2} = V'_{C'} \tag{4}$$

where

$$V'_{L'} = v'_{L'}\sin\left(w'_t + \theta''_{L'}\right) \tag{5}$$

And

$$V'_{C'} = v'_{C'}\sin\left(w'_t + \theta''_{C'}\right) \tag{6}$$

where

$\theta''_{L'}$ phase difference of inductor potential
$\theta''_{C'}$ phase difference of capacitor potential.

Therefore, the output potential difference can be written as,

$$V'_{\text{output}} = V'_0 = V'_{\text{DC}}\sin\left(w'_t + \theta''_0\right) \tag{7}$$

where

θ''_0 output AC voltage phase angle
V'_{DC} storage voltage.

In the STS from Fig. 11

$$V'_{L'} = V'_{C'} \tag{8}$$

And

$$I'_{L'_1} = I'_{L'_1} = I'_{L'_2} \tag{9}$$

At the time of STS (T_0), the potential difference across the dc link for the inverter bridge is

$$V'_j = 0 \tag{10}$$

Figure 9 displays the "Normal State," while Fig. 11 displays the "Zero State" ZSI. For the symmetrical mode of ZSI,

$$I'_{C'_1} = I'_{C'_2}$$
$$I'_{L'_1} = I'_{L'_2}$$

The potential difference between the inductor terminals is given by (11) when S_1 is turned ON at the source side,

$$V'_{L'} = V'_{\mathrm{DC}} - V'_{C'}$$

or

$$V'_{C'} = V'_{\mathrm{DC}} - V'_{L'} \tag{11}$$

During the normal and zero state (T_1), i.e., during NSTS,

$$\begin{aligned} V'_j &= V'_{C'} - V'_{L'} \\ V'_j &= V'_{C'} - (V'_{\mathrm{DC}} - V'_{C'}) \\ V'_j &= 2V'_{C'} - V'_{\mathrm{DC}} \end{aligned} \tag{12}$$

The duration for which DC input voltage V_{dc} is connected is taken as T, which is equal to the addition of duration of STS and NSTS, i.e., $T' = T_0 + T_1$.

At the steady state, the mean of inductor potential is,

$$\begin{aligned} V'_L . T' &= \left(V'_{L\,\text{shoot through}} . T_0\right) + \left(V'_{L\,\text{non shoot through}} . T_1\right) \\ V'_L . T' &= (V'_C . T_0) + [(V'_{\mathrm{DC}} - V'_C) . T_1] \\ V'_L &= 0 = [V'_C T_0 + (V'_{\mathrm{DC}} - V'_C) T_1]/T' \\ V'_C . T_0 &+ (V'_{\mathrm{DC}} - V'_C) . T_1 = 0 \\ V'_C &= [T_1/(T_1 - T_0)] V'_{\mathrm{DC}} \end{aligned} \tag{13}$$

Now, (12) can rearranged as,

$$
\begin{aligned}
V_i' &= 2V_C' - V_{DC}' \\
V_i' &= 2.[T_1/(T_1 - T_0)]V_{DC}' - V_{DC}' \\
V_i' &= [T_1/(T_1 - T_0)]V_{DC}' \\
V_i' &= B.V_{DC}'
\end{aligned} \tag{14}
$$

where

$$B = [T_1/(T_1 - T_0)] = \text{Boost factor}$$
$$B = [T_1/(T_1 - T_0)] \geq 1$$

Though, using space vector modulation, the maximum output voltage from inverter (per phase) is given by

$$V_{\text{output}} = M_{\text{i}} \frac{V_{\text{i}}}{2}$$

where M_{i} = index of modulation.

Thus, the output voltage using ZSI is

$$V_{\text{output}} = M_{\text{i}}.B\left(V_{\text{i}}'/2\right) \tag{15}$$

Thus, by varying the value of B (which in turn depends on T_0), the output voltage can be bucked or boosted. The design of impedance circuit using parameter estimation must be done when the current stress and B are large.

5 Simulation and Result

In this chapter, simulations are performed in the MATLAB environment. Figure 12 shows the circuit model, while for the input voltage = 100 V, the output is shown in Fig. 13. The impedance parameters for the ZSI are, $L' = L_2' = L_1' = 160\ \mu\text{H}$ and $C' = C_2' = C_1' = 1000\ \mu\text{F}$. The factor of modulation $(m) = 0.62$. The STS D_0 (duty cycle) is thus 0.38 (1 – M_{i}). The capacitor voltage is approximately expanded to V_C' = 220 V. The V_{phase} = 150 V (peak) at $f_{\text{switching}}$ = 10 kHz.

6 Conclusion

This chapter inspects the ZSI circuit arrangement and its activities. The traditional VSI and CSI have distorted waveforms along with decreased inverter voltage. In any case, the ZSI has a unique impedance, LC circuit for associating the applied voltage and inverter's circuit, which gives the improved output voltage. The ZSI awards the

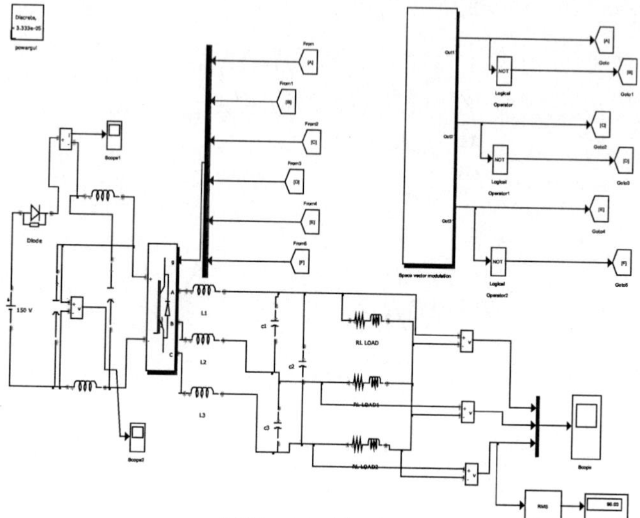

a) ZSI based simulation circuit

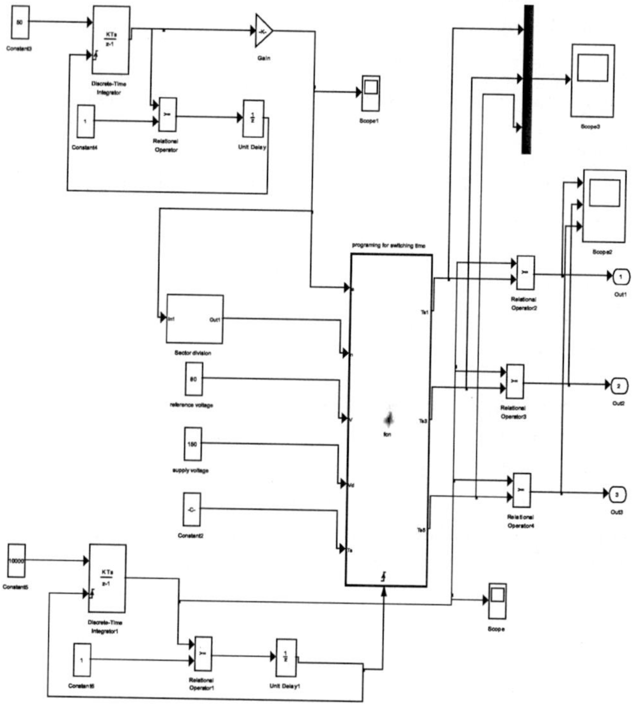

b) Space vector modulation

Fig. 12 SVPWM ZSI Simulink model

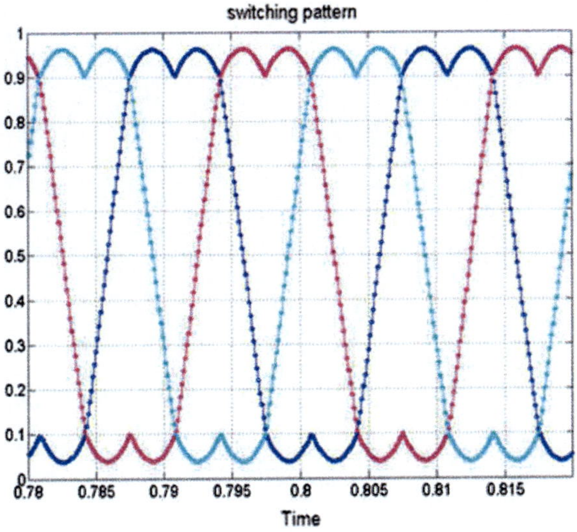

a) : Switching Pattern

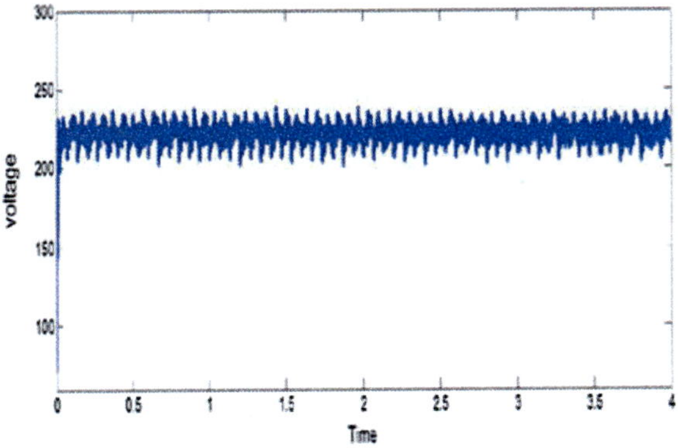

b) : Capacitor voltage

Fig. 13 Output of ZSI system

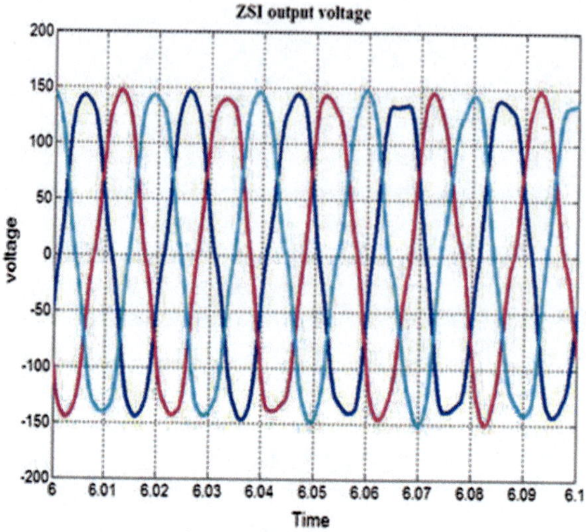

c) : AC Output waveform

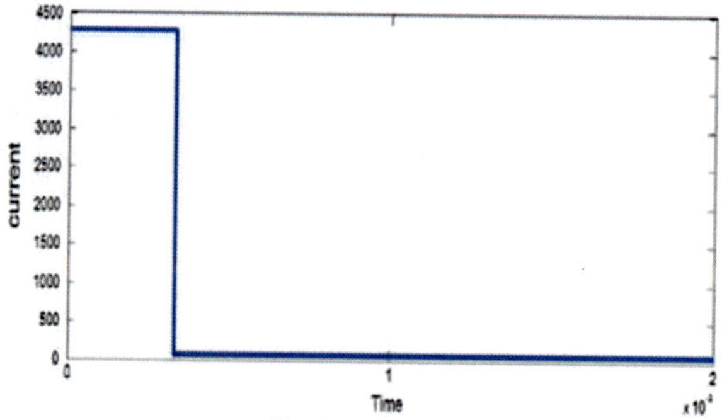

d) : Inrush current

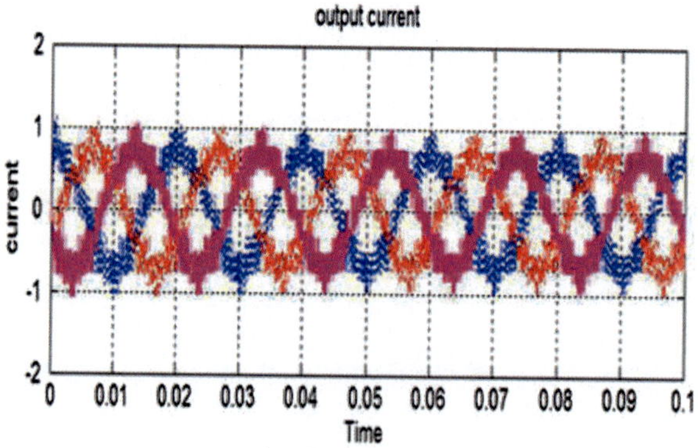

e) : Output current

Fig. 13 (continued)

usage of STS which conquers the prerequisite of overlap time for CSI and dead time for VSI, due to which the inverter circuit is in danger of harm. The proposed ZSI approach can be effectively used in a variable speed drive networks.

References

1. Bimal K (2002) Bose-modern power electronics and AC drives. Prentice Hall PTR
2. Peng FZ (2003) IEEE Trans Ind Appl 39(2):504–510
3. Xu W, Kaizheng H, Shijie Y, Bin X (2008) Simulation of three-phase voltage source PWM rectifier based on the space vector modulation. In: Chinese control and decision conference, CCDC 2008, pp 1881–1884
4. Ali US, Kamaraj V (2011) Z-source inverter with a new space vector PWM algorithm for high voltage gain. ARPN J Eng Appl Sci 6(6):9–13
5. Zhou ZJ, Zhang X, Xu P, Shen WX (2008) Single-phase uninterruptible power supply based on Z-source inverter. IEEE Trans Ind Electron 55(8):2997–3004
6. Amei K, Takayasu Y, Ohji T, Sakui M (2002) A maximum power control of wind generator system using a permanent magnet synchronous generator and a boost chopper circuit. In: Proceedings of power conversion conference, PCC-Osaka 2002, vol 3, pp 1447–1452
7. Chinchilla M, Arnaltes S, Burgos JC (2006) Control of permanent-magnet generators applied to variable-speed wind-energy systems connected to the grid. IEEE Trans Energy Convers 21(1):130–135
8. Chen Z, Guerrero JM, Blaabjerg F (2009) A review of the state of the art of power electronics for wind turbines. IEEE Trans Power Electron 24(8):1859–1875
9. Wang X-Y, Vilathgamuwa DM, Tseng K-J, Nguyen TD, Zhang S (2010) Design of a robust grid interface system for PMSG-based wind turbine generators. IEEE Trans Ind Electron 58(1):316–328

Optimum Cavity Length of a pTFET-Based Biosensor for Successful and Accurate Sensing of a Wide Range of Biomolecules

Sanu Gayen, Suchismita Tewari, and Avik Chattopadhyay

Abstract In this paper, for the first time, a rigorous analysis of determining minimum number of biomolecules and maximum length of the nanogap cavity, used to entrap biomolecules for the detection purpose, has been done. A pTFET-based biosensor (pTB-sensor) device is being proposed, and the corresponding detection sensitivity in terms of threshold voltage (V_{th})—sensitivity, threshold voltage (V_{th})—shift, subthreshold swing (SS) sensitivity, subthreshold swing (SS) shift, ON current/OFF current (I_{ON}/I_{OFF})—shift, leakage power (P_{leak})—sensitivity is determined. It is found that the minimum number of biomolecules or minimum detection limit is 2 for Myoglobin and Apomyoglobin, while the same is 3 for Protein-G, Ferricytochrome-C, and Ferrocytochrome-C, respectively. The maximum biomolecules required for successful detection are found to be 5 for Myoglobin and Apomyoglobin, while the same is 7 for Protein-G, Ferricytochrome-C, and Ferrocytochrome-C, respectively, with 78%, 67%, 41%, 55%, and 50% V_{th}—sensitivity and 83%, 74%, 53%, 63%, and 58% SS sensitivity for Apomyoglobin, Myoglobin, Protein-G, Ferrocytochrome-C, and Ferricytochrome-C, respectively. This leads to the minimum cavity length of 8 nm and maximum or optimum cavity length to be 20 nm, respectively, for the detection of all the five biomolecules successfully allowing space for scaling down of the channel length to sub-100 nm technology node.

Keywords Detection limit · Maximum cavity length · Nanogap cavity · pTFET-based biosensor

S. Gayen (✉) · S. Tewari · A. Chattopadhyay
Department of Radio Physics and Electronics, University of Calcutta, Kolkata, India
e-mail: sanugayen23@gmail.com

S. Tewari
e-mail: suchi_tewari@yahoo.co.in

A. Chattopadhyay
e-mail: avikjoy@yahoo.com

V. Nath and J. K. Mandal (eds.), *Proceedings of the Fourth International Conference on Microelectronics, Computing and Communication Systems*, Lecture Notes in Electrical Engineering 673, https://doi.org/10.1007/978-981-15-5546-6_16

1 Introduction

Over the last few decades, electronic biosensors or e-biosensors have gained a lot of popularity and became a topic of great interest in the interdisciplinary research field, due to their wide range of applications in different fields, starting from medical diagnosis to the detection of pollutants in environment, ultimately leading to the well-being of the human society [1–3]. Due to the label-free, hence economic detection process, easy scalability and compatibility with conventional CMOS process flow, the FET-based e-biosensors become attractive choice for the researchers nowadays [4–6]. From the detection sensitivity point of view, FET-based devices show maximum sensitivity in subthreshold regime of the device operation. Then again, due to the continuous device miniaturization over the last few decades in search for performance benefit, in terms of speed of operation, power dissipation, package density, etc., the need of the devices, suitable to work under highly scaled-down supply voltage, hence applicable for low-power electronics have increased to a great extent. To meet up the above-mentioned two points, tunnel FET with its sub-60 mV/decade subthreshold swing (SS) at room temperature [7–9] has emerged as the most appropriate device for e-biosensors.

A lot of work has been reported, discussing different architectural scheme, in order to increase the sensitivity of TFET-based biosensors [10–12]. But no detailed analysis has been reported, whatsoever regarding the optimum length of the nanogap cavity, formed in the device, in order to detect different biomolecules.

In this paper, for the first time, we have provided insight in determining the minimum number of biomolecules for detection purpose and have reported the corresponding optimum length of the nanogap cavity for a wide range of biomolecules detection, viz. Apomyoglobin, Myoglobin, Protein-G, Ferricytochrome-C, and Ferrocytochrome-C. The optimum cavity length gives the accurate vision of maximum scaling down of the e-biosensor device, maintaining the mechanical stability of the device, while providing high and precise measurement of detection sensitivity, with reduced device footprint, hence high package density.

2 Device Structure and Simulation Framework

Figure 1 represents the schematic of the pTFET-biosensor (pTB-sensor) device, considered for our work. The detailed fabrication process flow is given in the ref [13]. The gate oxide thickness of the proposed sensor device is chosen as 4 nm, in order to incorporate one monolayer of Apomyoglobin, whose diameter is 4 nm, highest amongst all the aforementioned biomolecules. A nanogap cavity of length 400 nm is formed by etching a portion of gate oxide, starting from the source/channel junction to the middle of the channel, maintaining the mechanical stability of the entire structure. TaN is used as the gate metal, having the work function of 4.15 eV

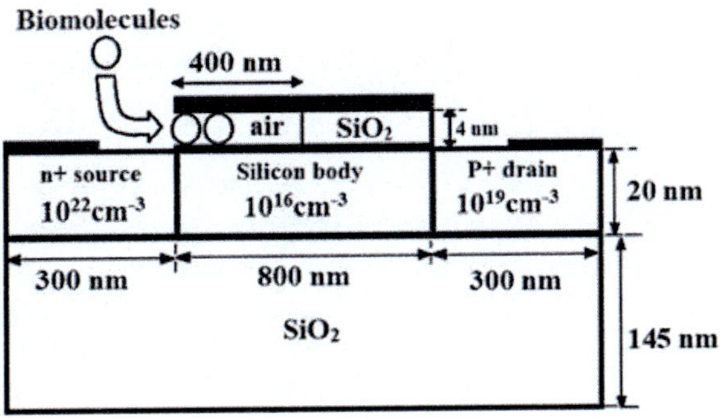

Fig. 1 Schematic of the pTFET-biosensor (pTB-sensor) device

[14]. The doping density of n-type source is considered as 10^{22} cm^{-3} and that of the p-type drain region is taken as 10^{19} cm^{-3}.

SILVACO ATLAS, a 2D numerical device simulator [15], has been used for the simulation of the proposed pTB-sensor device. In order to consider the spatial change in the energy band profile, the nonlocal band-to-band tunnelling (BTBT) model is invoked in our simulation, along with Shockley–Read–Hall (SRH) and Auger models to capture the essence of carrier generation and recombination. To capture the effects due to high doping concentration, band gap narrowing model is incorporated. Additionally, to incorporate the temperature, field (both lateral and longitudinal), and concentration dependency of carrier mobility, Lombardi CVT mobility model has been invoked. Throughout the simulation Fermi–Dirac (FD) statistics is used.

3 Model Calibration

Before proceeding with our proposed pTB-sensor device, we have validated our simulation scheme against the characteristic of an experimentally fabricated pTFET device of Ref. [13]. For the calibration purpose, the simulated device dimension, architecture and materials are kept exactly same as that of the fabricated pTFET, reported in [13]. Figure 2 shows a comparison of the transfer characteristics between the experimentally fabricated device [13] and our simulated pTFET device. From Fig. 2, it is pretty clear that both the characteristics tallied with each other in an excellent manner, confirming the authenticity of our simulation framework.

4 Results and Discussion

Biosensing mechanism through an electronic device stands on the successful translation of unique biological features of a biomolecule into some equivalent electronic or material parameter, followed by the shift of the output parameters of the

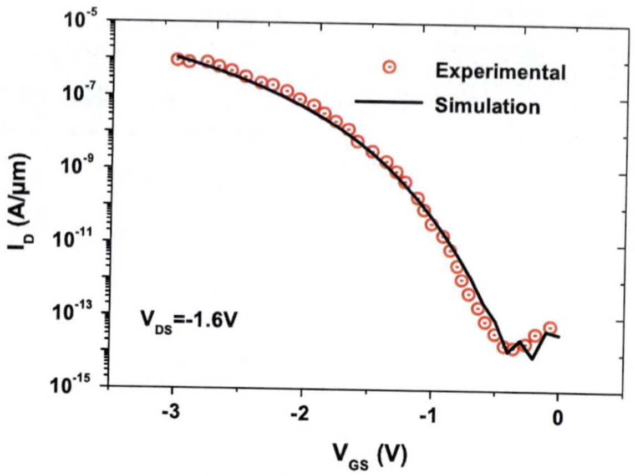

Fig. 2 Comparison of the transfer characteristics between the experimentally fabricated device [13] and simulated pTFET device

sensor device, due to the presence of biomolecules. Dielectric constant or K-value of the biomolecules is identified as such equivalent material parameters, and each biomolecule is represented as an insulator with its specified K-value and radius [16]. In our proposed pTB-sensor device, the presence of biomolecules in the nanogap cavity shifts the transfer characteristics of the said sensor device, hence reflecting its impact on the threshold voltage (V_{th}). Then again, noteworthy impact has been found in the subthreshold region of operation of the proposed pTB-sensor device, due to the presence of biomolecules. Hence, we have chosen the following performance parameters, in terms of which we have analysed different aspect of the proposed pTB-sensor device. The parameters are,

1. V_{th}—sensitivity or $S(V_{\text{th}})$
$$= \frac{(V_{\text{th}})\text{Empty Cavity} - (V_{\text{th}})\text{Biomolecules}}{(V_{\text{th}})\text{Empty Cavity}} \times 100\%$$

2. Subthreshold swing or SS—sensitivity or $S(\text{SS})$
$$= \frac{(\text{SS})\text{Empty Cavity} - (\text{SS})\text{Biomolecules}}{(\text{SS})\text{Empty Cavity}} \times 100\%$$

3. $I_{\text{ON}}/I_{\text{OFF}}$—shift or $\Delta I_{\text{ON}}/I_{\text{OFF}}$
$$= (I_{\text{ON}}/I_{\text{OFF}})\text{Empty Cavity} - (I_{\text{ON}}/I_{\text{OFF}})\text{Biomolecules}$$

4. Leakage power (P_{leak})—sensitivity or $S(P_{\text{leak}})$
$$= \frac{(P_{\text{leak}})\text{Empty Cavity} - (P_{\text{leak}})\text{Biomolecules}}{(P_{\text{leak}})\text{Empty cavity}} \times 100\%$$

Figure 3a shows the set of transfer characteristics curve for empty cavity and cavity with biomolecules having the number of the biomolecules as a parameter

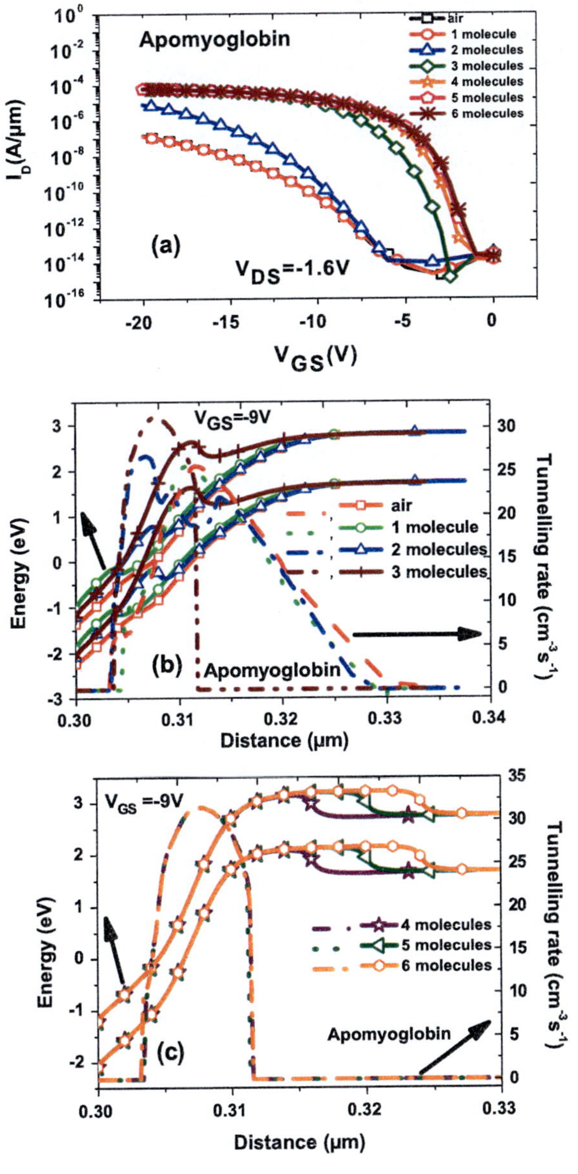

Fig. 3 **a** Transfer characteristics for empty cavity and the biomolecules Apomyoglobin with number the biomolecules as a parameter. Band diagram and hole tunnelling rate for **b** empty cavity and 1–3 molecules and **c** 4–6 molecules of Apomyoglobin

for Apomyoglobin. It is clearly witnessed from Fig. 3 that after adding just one molecule of Apomyoglobin, the transfer characteristic overlaps with that of the one with empty cavity. With the addition of two biomolecules, the V_{th}, hence the characteristic shifts towards right with sharper SS and better I_{ON}. But a maximum shift of V_{th}, hence that of characteristic is observed for the addition of three biomolecules from the source/channel junction side. After that the V_{th} shift is gradually decreasing and after five biomolecules, no such shift is found and the characteristics overlap with each other. The phenomena may be explained in terms of band diagram and tunnelling rate of the carriers. Figure 3b, c shows the corresponding band diagrams and hole tunnelling rates. It is clearly observed from Fig. 3b, c, after the capture of 1 biomolecule exactly at the source/channel junction, a slight notch is created both in the conduction band (E_C) and valence band (E_V), due to the increased K-value of Apomyoglobin, resulting higher electrostatic coupling at the junction. But this notch falls at the extreme low portion of the tunnelling window, where the carrier tunnelling probability is already low and producing negligible change of tunnelling width. Also the hole tunnelling rate for 1 Apomyoglobin molecule almost overlaps with that of the one with empty cavity, resulting no characteristic shift as is viewed from Fig. 3a. Now with the capture of more Apomyoglobin molecule, adjacent to the first one, the electrostatic coupling becomes stronger and the notch slowly moves towards right, i.e. more inside the channel, reducing the tunnelling width by significant amount in the middle portion of the tunnelling window, where the probability of the tunnelling is maximum. The tunnel width is found to be minimum for 3 Apomyoglobin molecules. From Fig. 3b, the corresponding hole tunnelling rate is observed to attain its peak value in the middle of the tunnelling window for two and three biomolecules of Apomyoglobin, resulting maximum current flow for the aforementioned two cases. As the number of Apomyoglobin molecules increases beyond 3, the effect of stronger electrostatic coupling spreads more inside the channel away from the source/channel junction and the notch moves towards the upper portion of the tunnelling window, i.e. towards the low-probability tunnelling zone, gradually reducing the impact of high K-value of the biomolecules on the tunnel junction and beyond five molecules, there is almost no effect of biomolecules on tunnel junction and the corresponding transfer characteristic overlaps with that of the case of five Apomyoglobin molecules, as is observed from Fig. 3a. Hence, the maximum cavity length, required for the successful detection of Apomyoglobin is $(5 \times 4\,\text{nm}) = 20\,\text{nm}$. And the lower detection limit for this particular biomolecule is 2.

Figure 4a demonstrates the set of transfer characteristics curve for empty cavity and cavity with biomolecules, with the number of the biomolecules as a parameter for Myoglobin. Here, also after the capture of one Myoglobin molecule, the characteristics does not shift but interestingly after the capture of two Myoglobin molecules, the transfer characteristic shift towards left, instead of right increasing the V_{th}. This anomalous behaviour may be explained in terms of corresponding energy band profile and tunnelling rate plot, as is shown in Fig. 4b. From Fig. 4b, it is clearly seen that the created notch, in the lower portion of the band diagram, due to the high K-value of two Myoglobin molecules, has actually increases the tunnelling width slightly at the middle portion of the band diagram, resulting increased tunnelling path and, reduced carrier tunnelling rate and ultimately increasing the V_{th} for this particular case. But as the number of Myoglobin molecules are increased from 2 to 3, the notch falls at the upper portion of the tunnelling window resulting high right shift of the transfer characteristics, like in the case for Apomyoglobin, discussed earlier. For this case also the impact of biomolecules in nanogap cavity is lost beyond five Myoglobin molecules. Hence, the maximum cavity length, required for the successful detection

Fig. 4 **a** Transfer characteristics for empty cavity and the biomolecules Myoglobin with number the biomolecules as a parameter. Band diagram and hole tunnelling rate for **b** empty cavity and 1–4 molecules and **c** 5–7 molecules of Myoglobin

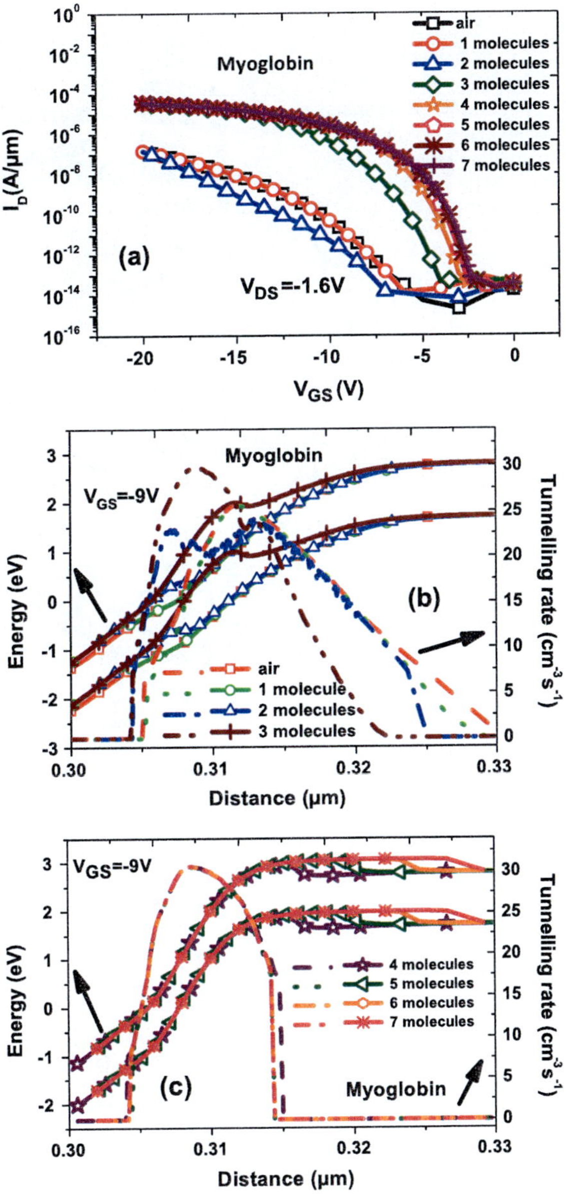

of Myoglobin is (5 × 4 nm) = 20 nm. And the lower detection limit for this particular biomolecule is 2. The same trends are being followed for the biomolecules Protein-G, Ferricytochrome-C, and Ferrocytochrome-C as is observed from the set of transfer characteristics and corresponding energy band diagrams along with tunnelling rate plots in Figs. 5a–c, 6a–c, and 7a–c, respectively. For Protein-G, Ferricytochrome-

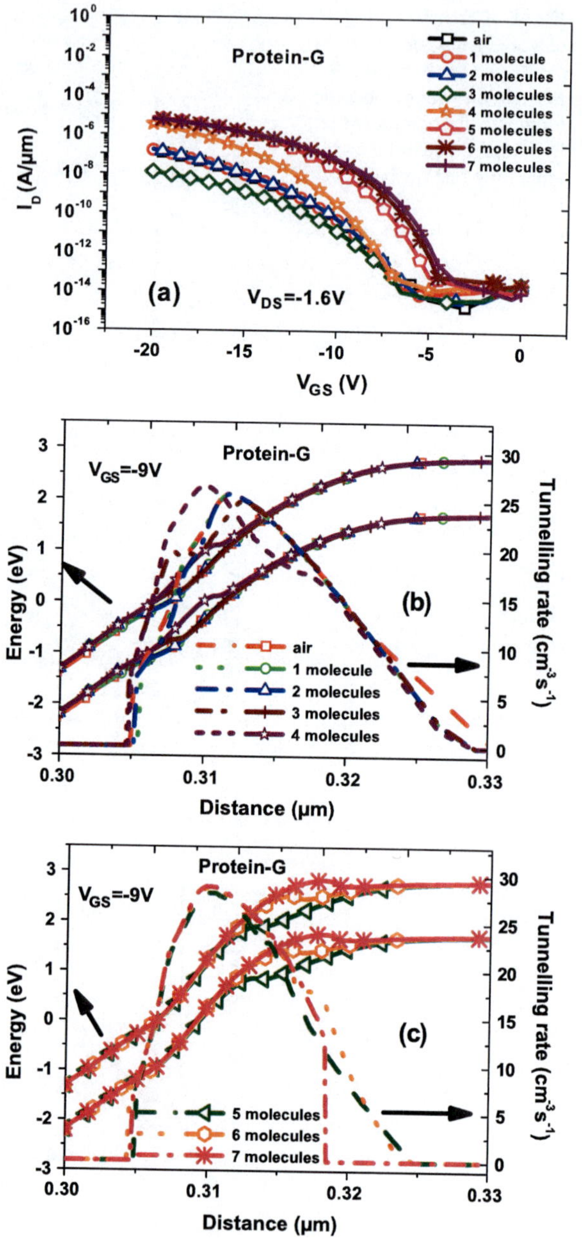

Fig. 5 **a** Transfer characteristics for empty cavity and the biomolecules Protein-G with number the biomolecules as a parameter. Band diagram and hole tunnelling rate for **b** empty cavity and 1–4 molecules and **c** 5–7 molecules of Protein-G

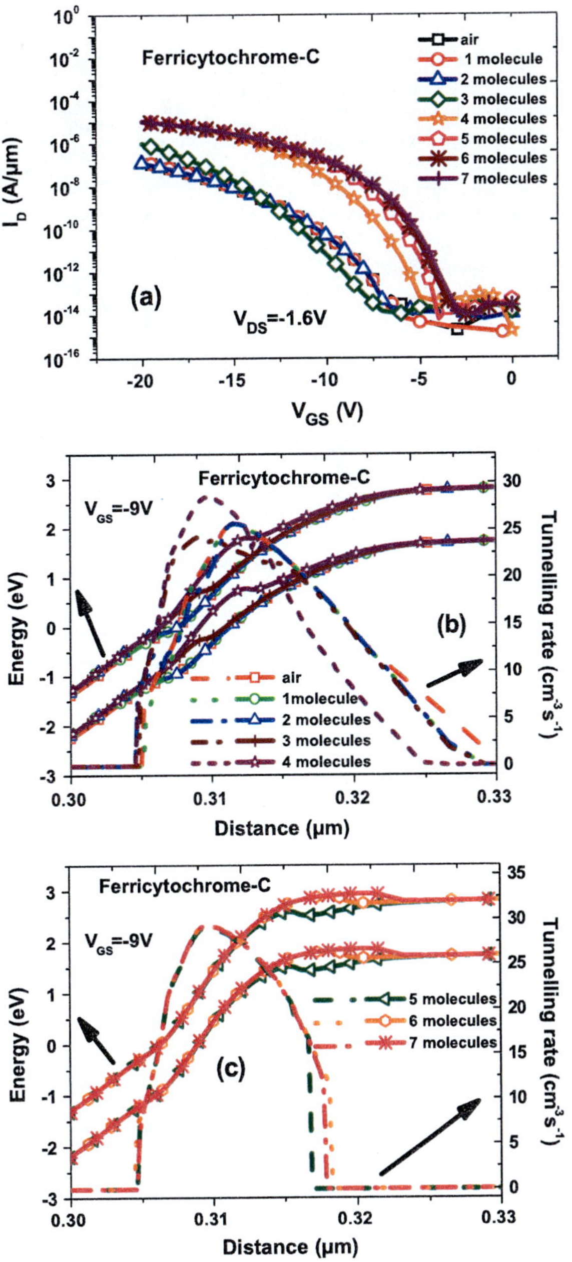

Fig. 6 **a** Transfer characteristics for empty cavity and the biomolecules Ferricytochrome-C with number the biomolecules as a parameter. Band diagram and hole tunnelling rate for **b** empty cavity and 1–4 molecules and **c** 5–7 molecules of Ferricytochrome-C

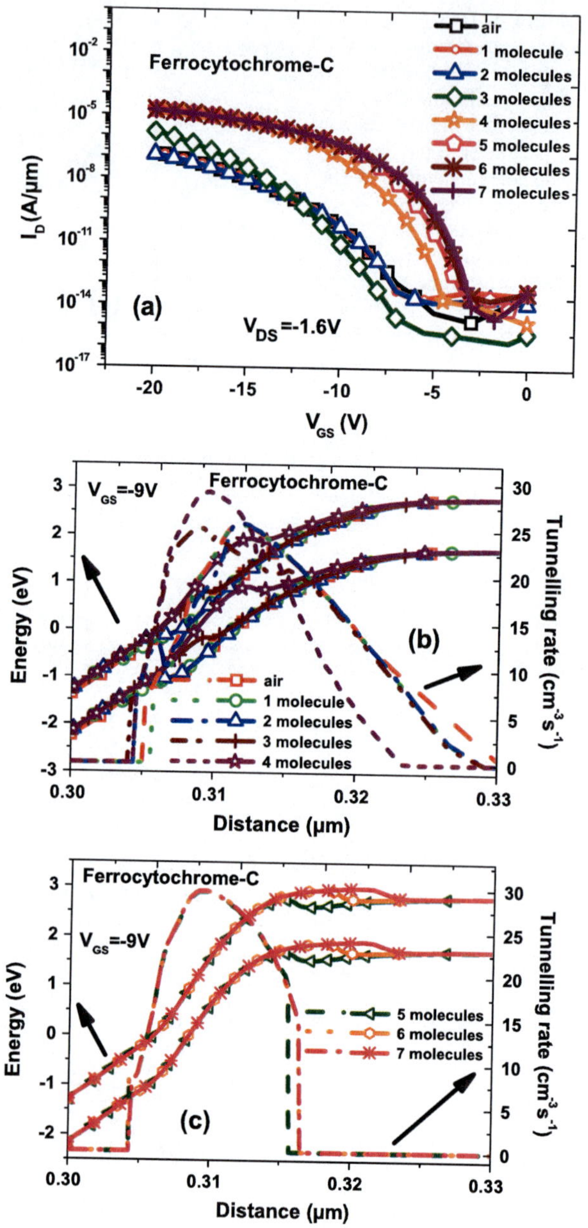

Fig. 7 **a** Transfer characteristics for empty cavity and the biomolecules Ferrocytochrome-C with number the biomolecules as a parameter. Band diagram and hole tunnelling rate for **b** empty cavity and 1–4 molecules and **c** 5–7 molecules of Ferrocytochrome-C

C, and Ferrocytochrome-C, the minimum detection limit is 3 as the radius of these molecules are smaller compared to Apomyoglobin and Myoglobin, respectively, and the maximum cavity length required for successful detection is (7×2.66 nm) = 18.62 nm, for Protein-G and (7×3.16 nm) = 22.26 nm for Ferricytochrome-C, and Ferrocytochrome-C each.

$S(V_{th})$, $S(SS)$, $\Delta I_{ON}/I_{OFF}$, and $S(P_{leak})$ are extracted and shown in Table 1, for all the five biomolecules. It is clearly observed from the Table 1 that the proposed pTB-sensor device becomes most sensitive after addition of third molecule in case of Apomyoglobin and Myoglobin, respectively, while capture of fifth molecule of Protein G and that of forth molecule in case of Ferrocytochrome-C and Ferricytochrome-C, respectively, trigger the maximum shift of sensitivity parameters from its just previous value, making the proposed sensor device most sensitive at that point. After critically, scrutinizing the data in the tables, we can clearly see the justification of fixing maximum cavity length as 20 nm for successfully detecting a wide range of biomolecules, without having any redundancy, whatsoever. This particular length of the cavity allows us to further scale down the device to sub-100 nm regime, which indefinitely will reduce the device footprint, and will offer high package density. A comparative study on the sensitivity parameters, viz. $S(V_{th})$ and $S(SS)$, has been shown in Table 2 between our work and the work of Kanungo et al. [17].

5 Conclusion

In this paper, we have reported a detailed investigation concerning the successful and accurate detection of a wide range of biomolecules, using a pTB-sensor device, and also determined an optimum nanogap cavity length for the said purpose, which avoids unnecessary lengthening of the device. After a rigorous investigation, we have reported 20 nm to be the optimum cavity length for the successful detection of a wide range of biomolecules, for the proposed pTB-sensor device. This enables us to scale down the sensor device up to channel length value of 40 nm, keeping 50% of the gate oxide space, reserved for cavity, which maintains the mechanical stability of the device while giving a promise of enhanced detection sensitivity in terms of the aforementioned performance matrices, along with being suitable for low-power electronics.

Table 1 V_{th}, SS, I_{ON}/I_{OFF}, P_{Leak}—sensitivities for five types of biomolecules

Name of biomolecules	No. of biomolecules	$S(V_{th})$ (%)	S(SS) (%)	$\Delta I_{ON}/I_{OFF}$	$S(P_{Leak})$ (%)
Apomyoglobin	0	0	0		0
	1	0	0	0	0
	2	6.54	25.89	0.48	29.24
	3	61.53	73.46	2.57	93.55
	4	72.62	75.49	3.05	95.21
	5	75.83	78.00	3.39	96.99
	6	77.61	82.71	3.99	99.11
Myoglobin	0	0	0		0
	1	0	0	0	0
	2	−16.29	−38.72	−0.35	−31.76
	3	43.24	60.23	1.51	77.65
	4	58.54	67.45	2.05	87.13
	5	63.41	71.67	2.37	91.88
	6	65.63	72.48	2.45	92.61
	7	67.07	74.27	2.45	94.25
Protein-G	0	0	0		0
	1	0	0	0	0
	2	0.55	2.27	0.02	2.35
	3	−8.31	−26.54	−0.25	−22.86
	4	5.76	34.82	0.58	41.18
	5	29.49	43.43	0.82	53.28
	6	37.14	48.86	0.96	61.34
	7	40.91	52.76	1.09	66.89
Ferrocytochrome-C	0	0	0		0
	1	0	0	0	0
	2	0.55	2.27	0.02	2.35
	3	−11.75	21.75	0.35	24.20
	4	36.36	52.19	1.14	66.05
	5	47.67	58.68	1.43	75.46
	6	52.22	61.93	1.62	80
	7	54.87	63.23	1.71	83.43
Ferricytochrome-C	0	0	0		0
	1	0	0	0	0
	2	0.33	2.27	0.02	2.35
	3	−12.97	8.44	−0.25	−22.86

(continued)

Table 1 (continued)

Name of biomolecules	No. of biomolecules	$S(V_{th})$ (%)	S(SS) (%)	$\Delta I_{ON}/I_{OFF}$	$S(P_{Leak})$ (%)
	4	30.82	47.56	0.58	41.18
	5	42.24	55.68	0.82	53.28
	6	46.89	56.57	0.96	61.34
	7	49.56	58.28	1.09	66.89

Table 2 Performance comparison between our work and the work of Kanungo et al. [17]

Parameters	Our work (%)	(Kanungo et al.) (%)
$S(V_{th})$	77.61	45.45
S(SS)	82.71	54.00

References

1. Luong JHT, Male KB, Glennon JD (2008) Biosensor technology: technology push versus market pull. Biotechnol Adv 26(5):492–500
2. Kim S, Baek D, Kim JY, Choi SJ, Seol ML, Choi YK (August, 2012) A transistor-based biosensor for the extraction of physical properties from biomolecules. Appl Phys Lett 101:073703-1-4
3. Barbaro M, Bonfiglio A, Raffo L (January, 2006) A charge-modulated FET for detection of biomolecular processes: conception, modeling, and simulation. IEEE Trans Electron Dev 53(1):158–166
4. Ribaut C, Loyez M, Larrieu JC, Chevineau S, Lambert P, Remmelink M, Wattiez R, Caucheteur C (2017) Cancer biomarker sensing using packaged plasmonic optical fiber gratings: towards in vivo diagnosis. Biosensors Bioelectron 92:449–456
5. Shin J, Choi S, Yang J-S, Song J, Choi J-S, Jung H-I (2017) Smart forensic phone: colorimetric analysis of a bloodstain for age estimation using a smartphone. Sens Actuators B Chem 243:221–225
6. Sheikhshoaie M, Karimi-Maleh H, Sheikhshoaie I, Ranjbar M (2017) Voltammetric amplified sensor employing RuO_2 nano-road and room temperature ionic liquid for amaranth analysis in food samples. J Mol Liq 229:489–494
7. Mallik A, Chattopadhyay A (2012) The impact of fringing field on the device performance of a p-channel tunnel field-effect transistor with a high-gate dielectric. IEEE Trans Electron Dev 59(2):277–282
8. Mallik A, Chattopadhyay A (2012) Tunnel field-effect transistors for analog/mixed-signal system-on-chip applications. IEEE Trans Electron Dev 59(4):888–894
9. Chattopadhyay A, Mallik A (2011) Impact of a spacer dielectric and a gate overlap/underlap on the device performance of a tunnel field-effect transistor. IEEE Trans Electron Dev 58(3):677–683
10. Kanungo S, Chattopadhyay S, Gupta PS, Rahaman H (March, 2015) Comparative performance analysis of the dielectrically modulated full-gate and short-gate tunnel FET-based biosensors. IEEE Trans Electron Dev 62(3):994–1001
11. Kalra S, Jagadesh Kumar M, Dhawan A (November, 2016) Dielectric-modulated field effect transistors for DNA detection: impact of DNA orientation. IEEE Electron Dev Lett 37(11):1485–1488
12. Narang R, Reddy KVS, Saxena M, Gupta RS, Gupta M (October, 2012) A dielectric-modulated tunnel-FET-based biosensor for label-free detection: analytical modeling study and sensitivity analysis. IEEE Trans Electron Dev 59(10):2809–2817

13. Migita S, Fukuda K, Morita Y, Ota H (2012) Experimental demonstration of temperature stability of Si-tunnel FET over Si-MOSFET. IEEE Silicon Nanoelectronics Workshop (SNW)
14. Lee BH, Choi R, Kang L, Gopalan S, Nieh R, Onishi K, Jeon Y, Qi WJ, Kang C, Lee JC (2000) Characteristics of TaN gate MOSFET with ultra thin hafnium oxide (8Å–12Å). IEDM Tech Dig 39–42
15. ATLAS User's Manual (2015) A device simulation package. SILVACO Int., Santa Clara, CA
16. Simson T, Brooks C (March, 1996) charge screening and dielectric constant of proteins: insights from molecular dynamics. J Am Chem Soc 8452–8458
17. Kanungo S, Chattopadhyay S, Gupta PS, Sinha K, Rahaman H (2016) Study and analysis of the effects of SiGe source and pocket-doped channel on sensing performance of dielectrically modulated tunnel FET-based biosensors. IEEE Trans Electron Dev 63(6):2589–2596

Spectrum Sensing in Cognitive Radio with SNR Wall Estimation

Chettiyar Vani Vivekanand, K. Bhoopathy Bagan, and G. Brenie Sekar

Abstract This paper suggests spectrum sensing method for the kind of noise which fluctuates with time in non-uniform manner. There are many methods available for spectrum sensing, but any of those methods do not show the property of stable false alarm in noisy environment. This shows there is a great need of stable false alarm condition for the sensing method to be robust, and it is achieved by Gershgorin disc method. In this method, the test statistic (GRCR—Gershgorin Radii and Centres Ratio) is the ratio of sum of the radii to the sum of the centres calculated from the covariance matrix of the received signal. A comparative analysis with few existing sensing methodologies is presented. But all the spectrum sensing techniques suffer with a problem called as SNR wall, a SNR at which the detector is not robust, even if many observations are made. In GRCR method, the difference between H_0 and H_1 becomes poor in lower SNR conditions and they overlap each other, which also happens with all the detectors. In this paper, SNR wall is estimated under correlated scenarios for the GRCR sensing method. The performance of this detector is evaluated using MATLAB simulations.

Keywords Cognitive radio · Gershgorin circle theorem · SNR wall · Probability Density Functions (PDFs)

C. Vani Vivekanand (✉)
RMK College of Engineering and Technology, Chennai, India
e-mail: vanivivekanand@rmkcet.ac.in

K. Bhoopathy Bagan
MADRAS Institute of Technology, Chennai, India

G. Brenie Sekar
RMD Engineering College, Chennai, India

V. Nath and J. K. Mandal (eds.), *Proceedings of the Fourth International Conference on Microelectronics, Computing and Communication Systems*, Lecture Notes in Electrical Engineering 673, https://doi.org/10.1007/978-981-15-5546-6_17

1 Introduction

In radio communication, static frequency allocation is adopted. But day o day increase in the demand of achieving high data rates due to increasing demands of human needs in wireless communication access with existing fixed frequency allocation leads to scarcity of spectrum. But FCC has observed the spectrum over wide range and observed the spectrum is underutilized, which leads to a conclusion that spectrum can be efficiently and effectively used by identifying the unused spectrum in the licensed spectrum range. This unused spectrum part is called as spectrum hole. The concept of dynamic spectrum access can give solution to the spectrum scarcity problem. This concept was proposed by Mr. Joseph Mitola, defined as Cognitive Radio. It is an intelligent radio with self-spectrum sensing capability and self-adjustment ability referred as dynamic spectrum access. There are mainly two participants of this dynamic spectrum access technique, one is the licensed user and the other one unlicensed user. The licensed user is the primary user (PU) who has higher priority in using the spectrum. The unlicensed user is called as secondary user (SU), and it can access the spectrum if it is not used by the primary user. One major condition imposed by FCC on secondary user is that it should not cause interference to the primary user. Hence, power used by SU should be less than the PU.

In order to use the spectrum effectively by dynamic spectrum access, detection of spectrum hole is necessary, which means that the spectrum is vacant and not used by PU. In this context, spectrum sensing plays a key role in detecting the spectrum hole. In today's wireless communication world, spectrum detection plays the vital role for the dynamic spectrum usage.

Spectrum sensing is an important task [2] in the cognitive radio network [3], for the efficient communication to take place. Spectrum sensing enables the secondary users (SU) to use the primary user's (PU) licensed spectrum, when the primary user (PU) channel is free. There are a number of spectrum detection techniques proposed to identify the presence of primary user. These techniques are energy detection, matched filter-based detection, cyclostationary detection and eigenvalue-based method. ED and eigenvalue-based method do not require the knowledge about the channel between the PU and SU. But ED is affected by noise uncertainty, and the eigenvalue method does not perform well for uncorrelated signals and complexity of eigenvalue method is more. Cyclic frequency information is required for implementing cyclostationary, and matched filter needs to know about the channel between the primary user and secondary user.

The goal of spectrum sensing is to decide between two hypotheses, the first of which states that the band of interest is free (H_0) and SUs can use the spectrum. The second hypothesis states that the band is occupied (H_1) so that SU should not interfere with the PU. Almost in all sensing algorithms, if enough samples are available, the probability density functions (PDFs) of the test statistics are well separable. But practically due to model uncertainties caused by certain parameters like non-stationary noise, non-ideal filters and imperfectly estimated parameters, these detection algorithms exhibit SNR walls-SNR values below which the detectors cannot robustly decide between H_0 and H_1. Hence. In addition to the spectrum sensing capability required by the CR, SNR estimation is essential in order to allow the spectral coexistence of primary and secondary users. This paper reflects the detection of the presence/absence of signals in uncertain low SNR environments. Small modelling uncertainties are unavoidable in communication systems and so robustness to them is a fundamental performance metric. The influence of these modelling uncertainties can be calculated by the position of the "SNR wall" below which a detector will fail to be robust, no matter how long it can observe the channel [7]. A simple mathematical model for the uncertainty in the noise and fading processes is proposed.

2 System Model

The cooperating sensing model is as shown in Fig. 1, in which "a" number of secondary users are present and each SU collects b samples, so totally ab samples are collected during sensing interval. In the spectrum sensing technique, samples of received signal are collected and sent to fusion centre, where matrix is calculated as given in (1)

$$\mathbf{Y} = \mathbf{AX} + \mathbf{N} \tag{1}$$

where $A \in \mathbb{C}^{a \times S}$ is the matrix representing the channel with elements a_{mn}, $m = 1, 2,$ 3, ..., a, $j = 1, 2, 3, \ldots, s$ represents the gain of channel between the nth PU and mth SU, e.g. a_{21} = is a channel gain between first primary user and second secondary user.

There are two assumptions made related to channel gain.

1. The gains are constant and identically distributed during sensing interval [4].
2. In real-time situation, the attenuation between PU transmitter and SU receiver is different; therefore, received power signals from SUs may not be constant.

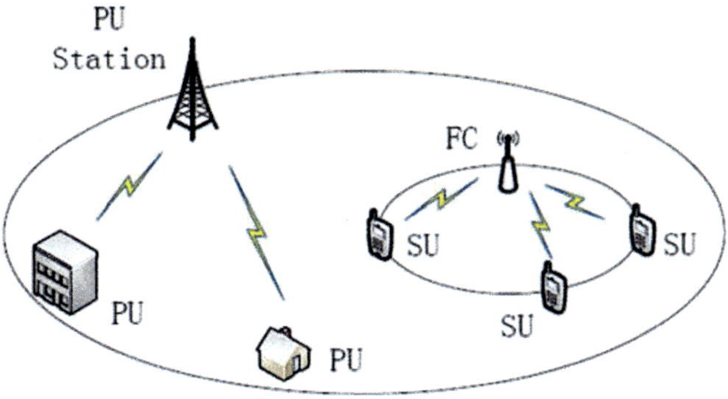

Fig. 1 Centralized cooperative sensing model

Based on these assumptions, the channel matrix is defined below:

$$\mathbf{A} = \mathbf{DC} \tag{2}$$

where $\boldsymbol{C} \in \mathbb{C}^{a \times S}$ is the channel matrix. The elements of this matrix are defined for two cases.

Case 1: For additive white Gaussian noise channel, $\alpha_{mn} = 1$

Case 2: For slow and flat Rayleigh's fading channels, α_{mn} is the Gaussian random variable.

The matrix $D \in \mathbb{R}^{a \times a}$ is a diagonal matrix in (2) given by

$$D = \text{diag}\left(\sqrt{\frac{\text{power}}{\text{power}_{\text{av}}}}\right) \tag{3}$$

where numerator represents received signal power in vector and power_{av} is the average received power. The expression for power_{av} is given by

$$\text{power}_{\text{av}} = \frac{1}{m}\sum_{m=1}^{a} pi \tag{4}$$

Each PU transmits constant power $\text{power}_{\text{avg}}$/s; therefore, overall channel gain is assumed to be unity.

The samples of each Pu signal are represented as row of the matrix $X \in \mathbb{C}^{S \times b}$ in (1). Gershgorin method is proposing a robust test statistics which is robust for both correlated signals and independent signals. Therefore, the elements of X are assumed with two conditions. The received signal is obtained from QPSK with μ samples per second, where μ is used for the differentiating between correlated signals and non-correlated signals [4].

SNR is an important parameter to be considered for analysing performance of GRCR method. power_{av} is the signal power, and the average noise variance is $\sigma_{\text{av}}^2 = \frac{1}{m}\sum_{i=1}^{a}\sigma_i^2$.

The overall SNR in SUs in terms of dB is given by:

$$\text{SNR} = 10\log_{10}\left(\text{power}_{\text{av}}/\sigma_{\text{av}}^2\right) \tag{5}$$

The performance metrics of spectrum sensing are:

1. **Probability of detection ($\boldsymbol{P_d}$)** also known as true positive detects the occupancy of channel correctly. It is represented as

   Probability of detection $P_d = P_{\text{ue}}(\text{decision} = H_1/H_1)$.

2. **Probability of false alarm ($\boldsymbol{P_{\text{fa}}}$)** also referred as false-positive corresponds to a case when the detector declares the frequency band occupied while it is actually vacant.

It is represented as $P_{\text{fa}} = P_{\text{ue}}(\text{decision} = H_1|H_0)$
where $P_{\text{ue}}(.)$ is the probability of the defined event.

3. **Probability of missed detection (P_m)** also refereed as false-negative means declaring the frequency band vacant when the PU signal is actually present. It can be derived from detection probability by using $P_m = 1 - P_d$.

If the primary user is detected, then Eq. (1) is satisfied. If PU is absent, then $X = 0$ in equation. The performance or this technique can be analyse by drawing receiver operating characteristics (ROC). This curve is plotted between P_d and P_{fa}, for varying SNR conditions.

3 Gershgorin Test Statistics

The Gershgorin test statistics is computed from the covariance matrix of the received signal at the fusion centre. The covariance is given as follows:

$$R = \frac{1}{b} Y Y^{\text{T}} \tag{6}$$

According to Gershgorin theorem, if R is an n × n matrix with complex entries a_{ij}, then the radius of the Gershgorin circle then $r_i(A) = \sum_{i \neq j} |a_{ij}|$ is defined as the sum of magnitudes of off diagonal entries. Then, Gershgorin disc $D(a_{ii}, r_i(a))$ cantered at a_{ii} with radius r_i. Eigenvalue of the matrix lies within minimum one disc. Applying theorem to the covariance matrix R, centre and radius of disc is given by,

$$C_{i=r_{ii}} \text{and } R_i = \sum_{j \neq \text{i}} |r_{ij}| \tag{7}$$

Based on this concept, from the covariance matrix R, sum of Gershgorin radius and sum of Gershgorin centres are calculated. It is interesting to observe that the ratio of both these values behaves differently under H_0 and H_1. Therefore, this ratio is used as test statistics for sensing the spectrum and given by:

$$\lambda_{\text{GRCR}} = \frac{\sum_{i=1}^{m} R_i}{\sum_{i=1}^{m} c_i} \tag{8}$$

This test statistics is having the characteristics of having different ratios for signal present and signal absent case, and the justification is given with the following example. In Fig. 2, Gershgorin circles for H_0 and H_1 are drawn by considering five secondary users and each SU having 5000 samples, with low SNR of −10 dB.

By considering the given conditions, observations from the Gershgorin discs can be made that radii changed from 2.19 (for H_0) to 6.83 (for H_1), but the variations in the centre are less from 49.92 (for H_0) to 52.42 (for H_1). So the difference between

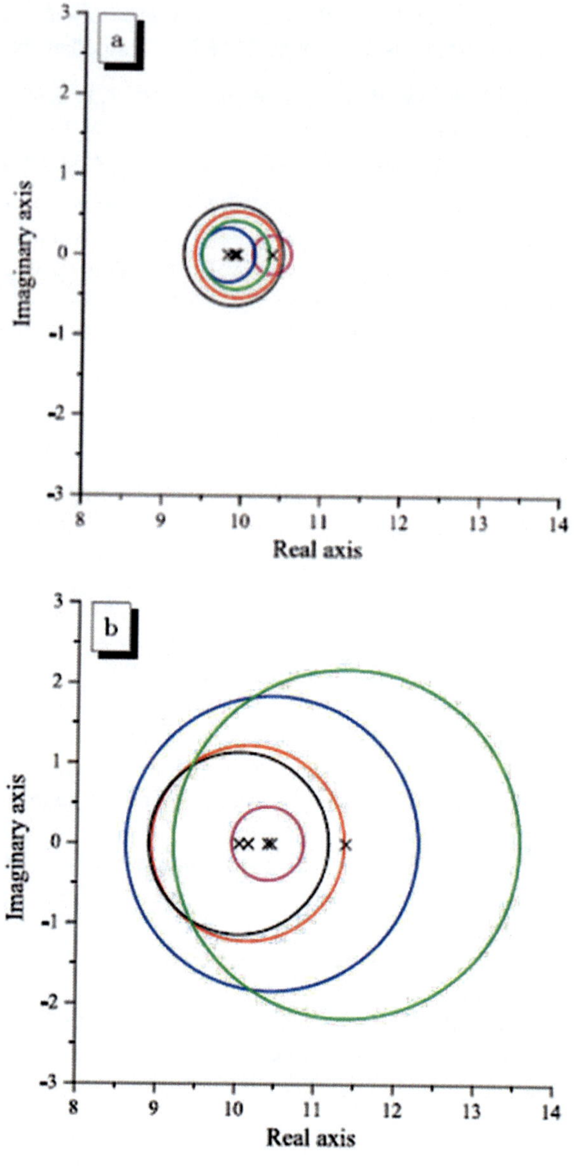

Fig. 2 Gershgorin circles drawn for hypotheses H_0 (**a**) and H_1 (**b**), for covariance matrix considering five secondary users and 5000 samples and lower SNR of −10 dB

signal present hypothesis and signal absent hypothesis is three times, i.e. there is a increase in test statistics result by three times. So, there is a considerable difference in the test statistics λ_{GRCR} between H_0 and H_1 as SNR increases [4]. This is the necessary and sufficient condition for any test statistics to work satisfactorily and efficiently. But the difference between signal present hypothesis (H_1) and signal absent hypothesis has very little difference when low SNR values are considered, and this property is same as other detectors. The following statistical results will confirm these statements. Spectrum sensing can also be done by only considering radius as test statistics, but the constant false alarm rate property is introduced by the denominator which is Gershgorin centre as given in (8).

4 Algorithm

Steps for the spectrum sensing:

1. Covariance matrix of the received signal is computed at the FC. Samples of received signal are taken during the spectrum sensing period in order to form the matrix.
2. By applying, Gershgorin theorem, sum of Gershgorin radius and sum of Gershgorin centres are computed. The ratio of former to later is formed as test statistics λ_{GRCR}.
3. This test statistics ratio is compared with the threshold value. If this ratio is greater than the threshold value, then primary user is detected and channel is occupied. But if it is less than the threshold, it indicates that PU is absent and channel is available for SU usage.

5 Comparison of Test Statistics

In this section, various popular methods are taken into consideration. The test statistics of these methods are compared with the Gershgorin model. Some of these methods are robust again dynamic noise, and some are not robust for non-uniform noise condition. But these detectors are considered for comparison as they are used more for PU detection [4].

5.1 Energy Detection Method

It is the simplest sensing technique, as it does not need any prior information about PU signal. It computes the power of the received signal and compares with a threshold. This threshold is fixed by the noise [1]. The decision statistics is computed by squaring the magnitude of the FFT averaged over N samples. This decision test statistics is given by:

$$T_{\text{ED}} = \frac{1}{\sigma^2} \sum_{i=1}^{m} \sum_{j=1}^{n} \left|y_{ij}\right|^2 \tag{9}$$

where y_{ij} is the element in the received signal matrix. For the case of non-uniform noise, test statistics is given by:

$$T_{\text{EDnu}} = \sum_{i=1}^{m} 1/\sigma_i^2 \sum_{j=1}^{n} |y_{ij}|^2 \tag{10}$$

The result of (9) is less that (10) as non-uniform noise is considered in the second case. The detection of PU signal is performed by comparing the energy detector output with the threshold. If TED > λ, then PU is present, but if $T_{ED} < \lambda$, then PU is absent. This threshold value λ depends on noise variance. As per this discussion, it is clear that SU does not require any knowledge about the PU signal and channel gain, which makes ED method very simple. Also, it has very simple circuitry. This method is widely used in secure cognitive radio and RADAR applications. This method performs better in low SNR ranges provided detection interval is high. It has the drawback that in order to fix threshold, it needs the knowledge about noise spectral density, which is an uncertain parameter. Hence, its performance deteriorates when noise has high uncertainty and high interference.

5.2 Cyclostationary Feature Detection

This is an effective detection method implemented based on the cyclostationary property of the received signal. As modulated signals are having periodicity property due to sine wave carriers, they can be classified as cyclostationary signal. As the noise signal is wide sense stationary without correlation, cyclostationary feature can be used for detecting the presence of primary user signal [8]. This method is also applicable for distinguishing between different types of transmissions with different modulation method. This detection method exploits the periodicity property of the primary user signal, where the parameters such as mean and autocorrelation of the received signal varies over time. Cyclostationary property of the signal can be extracted by computing spectral correlation density.

The two performance metrics of the cyclostationary is given as follows:

$$P_{\text{fa}} = \exp^{-\beta^2/2\sigma_x^2} \tag{11}$$

$$P_d = Q\left(\frac{\sqrt{2\eta}}{\sigma_y}, \frac{\beta}{\sigma_x}\right) \tag{12}$$

where $\sigma_x^2 = \frac{\sigma_y^2}{2N+1}$, the SNR is given by $= \frac{NA^2}{2\sigma_y^2}$. Q is given by $Q(a, b) = \int_b^\infty t\exp^{-(t^2+a^2)/2} I_0(\text{at})dt$ is the standard Gaussian complementary CDF, where I_0 is the zero-order Bessel function.

In this method, correlation between the spectral components is the key parameters in detecting primary user signal. This correlation measure is able distinguish between presence or absence of primary signal in less duration time. But it requires a large number of samples to distinguish between primary user, secondary user and noise. This demand of high sampling rate leads to increase in the computational complexity, sensing time, implementation of hardware cost.

The major advantage of this detection method is that it performs better in when SNR is low as compared to other detection techniques. This method has noise elimination capacity. So it performs better than other detection methods under low SNR conditions.

5.3 *Matched Filter Detection*

It is a coherent signal detection method, which is also called as optimum detection method. It can be used when the transmitted signal is known at the receiver side. It needs perfect knowledge of primary user signal features like bandwidth, type of modulation, etc. It maximizes SNR by reducing noise power at the output [8]. Probability of detection and probability of false alarm are as given below:

$$P_{\text{fa}} = Q\left(\frac{\beta}{\sqrt{\sigma_x^2 E_N}}\right) \tag{13}$$

$$P_d = Q\left(\frac{v^{-E_N}}{\sqrt{\sigma_x^2 E_N}}\right) \tag{14}$$

where $Q(a) = \frac{1}{\sqrt{2\Pi}} \int_a^\infty \exp^{-t^2/2} \text{d}t$ is Q function is the standard normal distribution. The energy at the output of the filter with N coefficients is given by $_N$, σ_x^2 is the noise power.

This method computes dynamic threshold with the knowledge about PU signal characteristics and the channel information between PU and SU. Because of dynamic threshold, spectral efficiency is improved. The major drawback of this method is that this method of spectrum detection needs the knowledge of the PU signal characteristics and the channel between the primary user to the secondary user. The advantage of this method is it performs better in the low SNR range, and it has good spectral efficiency. This method is commonly used in RADAR communication. The drawbacks of this method include prior signal characteristics knowledge, long observation time and complex computations involved.

5.4 *Generalized Likelihood Ratio Test (GLRT)*

GLRT spectrum sensing technique is a eigenvalue-based detection technique. Covariance of the received signal is computed, and the test statistics is formed as given in (15). This technique is mainly suitable for the situation when secondary users have less number of samples. GLRT technique is used to estimate the parameter when secondary users have few numbers of samples. Also, it does require knowledge about noise in estimating signal parameters, so it is classified as blind detection method. It is suitable for both flow and fast fading channels. The test statistics for GLRT test is given by:

$$T_{\text{GLRT}} = \frac{\lambda_1}{\sum_{i=1}^{m} \lambda_i} \tag{15}$$

where λ is the eigenvalue. Orthogonal frequency division multiple access and multiple input and multiple output techniques are using this spectrum sensing detection method. It requires the channel knowledge between PU and SU, and also it uses knowledge about modulation scheme. The drawback of this method is that it is having high computational complexity.

GLRT technique implements different spectrum sensing algorithm for different fading channels, in order to improve spectral efficiency. It performs better even in the presence of noise uncertainty.

5.5 *Hadamard Ratio Test*

In practical wireless communication scenario, SNR cannot be estimated accurately because of noise uncertainty. Hence, there is a need for a robust spectrum sensing technique to calculate SNR accurately. This method is suitable for uncalibrated receivers, and it performs better in multivariance analysis. In such conditions of uncertain environment [5], Hadamard ratio test is proposed aiming at accurate detection performance and robust spectrum sensing. Secondary users use multiple antennas to maximize the spectral efficiency and minimize the interference. Multiple antennas also detect the signal reliably in low SNR conditions. Energy detection and eigenvalue-based detection cannot provide reliable spectrum detection because both these methods assume the signal to be independent and identically distributed. The test statistics for the Hadamard ratio is given by:

$$T_{\text{HR}} = \frac{\det(R)}{\prod_{i=1}^{m} r_{ii}} \tag{16}$$

where determinant of covariance matrix is taken in the numerator in the above expression.

Hadamard ratio is developed from GLRT method in order to sense uncorrelated signals [6]. This method was further modified by calculating complementary covariance matrix instead of normal covariance matrix. This variant of Hadamard method also shows an improvement in the performance and robustness in presence of non-i.i.d signals. The performance of this method is theoretically analysed by computing detection probability. This method also performs better when non-uniform noise is present. It has very low computational complexity compared to other sensing techniques.

5.6 Volume-Based Detector

This detector has the property of maintaining stable false alarm rate for non-uniform noise. It is a non-blind spectrum technique. Generally, non-blind spectrum sensing techniques perform better for i.i.d. signals, but his method is not suitable for those signals as it exhibits poor performance for i.i.d. signals. The cyclostationary feature-based spectrum detector has very good sensing performance, and it is a robust technique in noisy environment. But it suffers with a drawback of frequency offset and synchronization errors. The detection statistics for this method is given by:

$$T_{\text{VD1}} = \log\left[\det\left(E^{-1}R\right)\right] \tag{17}$$

In this method, covariance matrix and inverse covariance matrix of received signal are calculated. This statistics shows better performance for non-uniform noise compared to other detectors. As in (17), the logarithm of determinant of the matrix is used to compute the test statistics to perform spectrum sensing. This method is the most complex compared to all other methods, but it performs better for non-uniform noise. By comparing all these methods, simulation results are shown in Fig. 3. It shows that GRCR performs better for both correlated and non-correlated signals, and also it is suitable for dynamic noise variances. It is robust detector with constant false alarm rate as compared to other techniques. But it also suffers with a problem of SNR wall, SNR value below which H_0 and H_1 overlaps and detector fails to perform better [9]. So, in the next section, SNR wall is estimated for GRCR, method and simulation results are shown to support the estimation.

6 SNR Walls in Spectrum Sensing

The receiver noise is modelled as AWGN because this approximation is close to reality for GRCR-based spectrum sensing. Here both the hypothesis of signal present ($\boldsymbol{H_1}$) and absent ($\boldsymbol{H_0}$) are differentiated by the correlation between the signal.

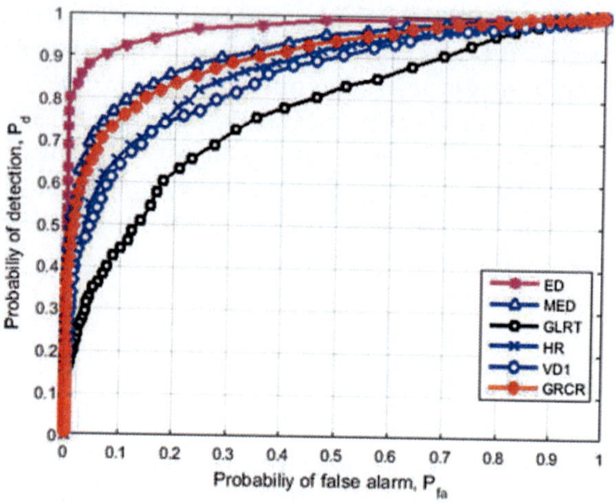

Fig. 3 Simulation results of comparison with all the test statistics

In this model, both noise and PU signal are assumed to have distributions with noise variances. The noise $\eta(n)$ is having variance σ_η^2, and PU signal $s(n)$ has the variance σ_s^2.

Based on this, the SNR is defined as

$$\mathrm{SNR} = \sigma_s^2/\sigma_\eta^2 \tag{18}$$

The performance metrics, false alarm and missed detection are given by the following formulae:

$$P_{\mathrm{fa}}(W, N) = P(\lambda_{\mathrm{GRCR}} \geq \varepsilon|\mathbf{H_0}, W),$$

$$P_{\mathrm{md}}(W, S, N) = P(\lambda_{\mathrm{GRCR}} \geq \varepsilon|\mathbf{H_1}, W, S) \tag{19}$$

These metrics are used to analyse the performance of proposed detector in terms of noise robustness [10]. For the given SNR values with larger values of samples (N), if the detector is not able to achieve a pair (P_{fa}, P_{md}), where $P_{\mathrm{fa}} \in [0, 0.5]$, $P_{\mathrm{md}} \in [0, 0.5]$, then it is declared as non-robust detector. The SNR wall is defined as

$$\mathrm{SNR_{wall}} = \sup\{\mathrm{SNR}_t, \text{ s.t. the proposed detector is not robust for all SNR} < \mathrm{SNR}_t\} \tag{20}$$

The detector is not robust if the sets of the medians of proposed threshold λ_{GRCR} under the two hypotheses overlap.

SNR Wall Lower Bound

The existence of SNR wall has to be proved with mathematical derivations. So, a lower bound on the SNR wall for GRCR detector is derived in this section. The

SNR value below which medians of test statistics under H_0 and H_1 overlap in the asymptotic condition ($N \to \infty$) is called as lower bound in SNR wall.

In order to derive SNR, lower-bound covariance matrices of signal and noise are required. These two parameters are represented by:

$\boldsymbol{R_s}$ = covariance matrix of signal

$\boldsymbol{R_\eta}$= covariance matrix of noise.

The correlation coefficients associated with $\boldsymbol{R_s}$ and $\boldsymbol{R_\eta}$ are denoted by ρ_{ij}^s and ρ_{ij}^η, respectively; these two parameters are used in further derivations.

Lower Bound on λ_{GRCR} (Test Statistics) Under H_0

In this condition, $\boldsymbol{R_x}$= $\boldsymbol{R_\eta}$, since no PU signal exists. For this case, the lower bound on test statistic can be found [10].

The considered covariance matrix has the following structure

$$\mathbf{R}_\eta = \sigma_\eta^2 \begin{bmatrix} 1 & \cdots & |\rho_{\max}^\eta| e^{j\phi} \cdots\cdots\cdots \\ \vdots & \ddots & \vdots \\ |\rho_{\max}^\eta| e^{-j\phi} & \cdots & \cdots\cdots \end{bmatrix}$$

$$\mathrm{SNR}_{\mathbf{H}_0}^{lo} = \frac{1 + |\rho_{\max}^\eta|}{1 - |\rho_{\max}^\eta|} \tag{21}$$

Detection becomes impossible for complete noise correlation as denominator of the equation becomes zero. Hence, correlation noise condition is excluded here.

Upper Bound on λ_{GRCR} (Test Statistics) Under H_1

In the H_1 case, $\boldsymbol{R_x}$= $\boldsymbol{R_s}$ + $\boldsymbol{R_\eta}$, upper bound can be calculated. r_{ij} is the element of ith row and jth column.

$$r_{ii} = \sigma_s^2 + \sigma_\eta^2 = (1 + \mathrm{SNR})\sigma_\eta^2,$$

$$r_{ij} = \sigma_\eta^2 \rho_{ij}^\eta + \sigma_s^2 \rho_{ij}^s = (\rho_{ij}^\eta + \mathrm{SNR}\rho_{ij}^s)\sigma_\eta^2 \tag{22}$$

The upper bound can be calculated from the following equation

$$\mathrm{SNR}_{\mathbf{H}_1}^{\mathrm{up}} = \frac{\mathrm{SNR} + 1 + \max \sum_{\substack{j=1 \\ j \neq 1}}^{g} \left|(\rho_{ij}^\eta + \mathrm{SNR}\ \rho_{ij}^s)\right|}{\mathrm{SNR} + 1 - \max \sum_{\substack{j=1 \\ j \neq 1}}^{g} \left|(\rho_{ij}^\eta + \mathrm{SNR}\ \rho_{ij}^s)\right|} \tag{23}$$

where g is the number of rows of $\boldsymbol{R_s}$. Combining Eqs. (21) and (23), SNR wall can be estimated as

$$\frac{\mathrm{SNR} + 1 + \max \sum_{\substack{j=1 \\ j \neq i}}^{g} \left|(\rho_{ij}^{\eta} + \mathrm{SNR}\ \rho_{ij}^{s}\right|}{\mathrm{SNR} + 1 - \max \sum_{\substack{j=1 \\ j \neq i}}^{g} \left|(\rho_{ij}^{\eta} + \mathrm{SNR}\ \rho_{ij}^{s})\right|} \leq \frac{1 + \left|\rho_{\max}^{\eta}\right|}{1 - \left|\rho_{\max}^{\eta}\right|} \tag{24}$$

The correlation coefficients can never be negative, so $\rho_{ij}^{s} \geq 0 \quad \forall i, j,$. By simplifying the above expressions, an expression for a lower bound on the SNR wall is derived as below:

$$\mathrm{SNR} \leq \frac{\alpha\max - 1}{p + \alpha\max(p - 2)} \tag{25}$$

where $\alpha\max = \frac{1+\left|\rho_{\max}^{\eta}\right|}{1-\left|\rho_{\max}^{\eta}\right|}$.

For $p = 2$ and $\rho_{\max}^{\eta} = 0.04$, the lower bound value on SNR $= 0.0456 = -13.41$ dB; it means that for SNR below -13.41 dB, the test statistics becomes, $\lambda\mathrm{GRCR}_{\mathbf{H}_1}^{\mathrm{asym}} < \lambda\mathrm{GRCR}_{\mathbf{H}_0}^{\mathrm{asym}}$ and the detector is not robust in this conditions. This condition is tested by receiver correlation as given below.

Receiver Correlation

In the receiver correlation, matrix multiplication method is used for generating coloured noise. White Gaussian noise with *P* dimension is considered for finding the correlation between receivers. This noise vector is represented as $w(n) \sim N\ (\mathbf{0}, \boldsymbol{R_W})$ for all the values of *n*, where $\boldsymbol{R_W}$= Identity matrix = $\boldsymbol{I}_{\mathbf{pxp}}$ and *p* is the number of receivers. From the *p*-dimensional zero mean Gaussian distribution, the noise vector $w(n)$ is generated. The identity matrix is considered to be its covariance matrix. The coloured noise vector $\eta(n)$ is measured at the receiver in the noise only hypothesis ($\boldsymbol{H_0}$) case. This measured noise is represented as $\boldsymbol{\eta(n) = Aw(n)}$, and $\mathbf{A} \in A \in \mathbb{C}^{\mathrm{pxp}}$, this A is selected such that covariance matrix of $\boldsymbol{\eta(n) = R_{\eta}}$.

This proposed method results into good results, as linear combinations of random variables are distributed as Gaussian distributions. The covariance matrix of $\boldsymbol{Aw(n)}$ is given by $\boldsymbol{A^H}\ \boldsymbol{R_W}\ \boldsymbol{A}$. Therefore, the matrix $\boldsymbol{A}$ can be easily found by calculating the Cholesky decomposition such that $\boldsymbol{R_{\eta}} = \boldsymbol{A^H}\ \boldsymbol{A}$.

Figure 4 shows the simulation results for the condition ($p = 2, Q = 0$) and different values of *N*. Noise correlation factor $\rho_{\max}^{\eta} = 0.04$. In this figure, continuous lines represent the histograms and dashed lines represent the means. The hypothesis of noise-only case ($\boldsymbol{H_0}$) result is shown in black, the hypothesis of signal-present case ($\boldsymbol{H_1}$) result is coloured, and range of SNR is considered as $\{-10;-11;\ldots;-15\ \mathrm{dB}\}$. It can be observed that with an increasing in the values of samples N, variance of test statistics decreases. The SNR wall for this scenario has been computed as -13.41 dB.

The simulation results also confirm this lower wall. When the SNR drops below the derived lower wall, the medians of the test statics under $\boldsymbol{H_1}$ are below the median

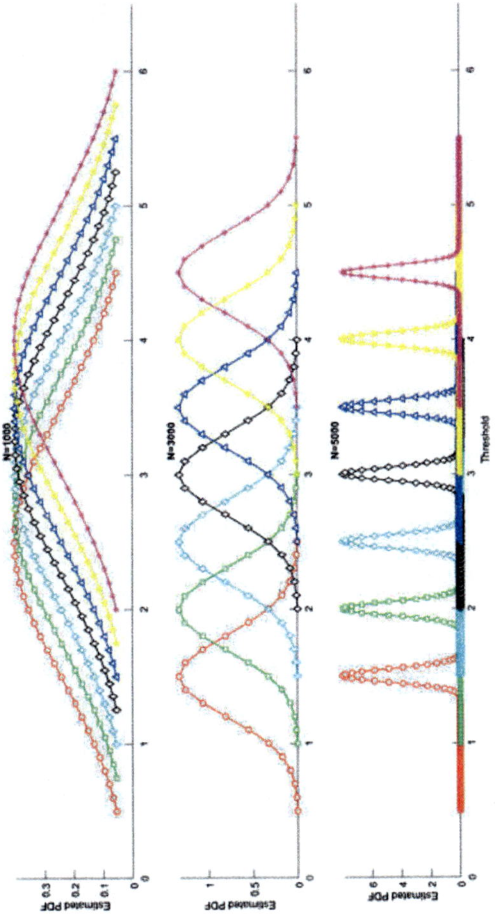

Fig. 4 Probability density function estimation for receiver correlation with different N

of the test statistic under $\boldsymbol{H_0}$ for all N, overlaps and the detector becomes non-robust at this lower bound SNR wall.

7 Conclusion

This paper provides the direct method of determining test statistics for the spectrum sensing. The test statistics is calculated from the covariance matrix of the signal at the receiver, and it is the ratio of the sum of the Gershgorin radii and the sum of the Gershgorin centres (GRCR). The specific features of this test statistics are that it is robust against varying noise, it is applicable to non-uniform noise and also it obeys constant false alarm property. This method is suitable for both types of signals which are correlated in time and not correlated in time. This method is suitable for Gaussian noise and fading channel.

The simulation results show the robustness of the method of compared to other detectors. The advantage is that GRCR method is simpler. It shows good performance in both uniform and non-uniform noise conditions. Also this method performs better with very less number of samples collected by secondary users. Hence, this method is suitable for applications having short sensing duration. For this method, SNR wall is estimated to find the lower bound of SNR, and the SNR value below which the method becomes non-robust is calculated. In this work, the presence of an SNR wall for GRCR spectrum sensing method is proven. The SNR wall is produced by uncertainty in the receiver. An expression for lower bound on the SNR wall, below which medians of H_1 and $\boldsymbol{H_0}$ overlap, is derived, and this derivation is supported by receiver correlation. SNR values below which the GRCR detector does not remain robust are shown with MATLAB simulations.

References

1. Semba Yawada P, Trung Dong M (2018) Performance analysis of new spectrum sensing scheme using multiantennas with multiuser diversity incognitive radio networks: research article. Received 13 June 2018; Accepted 15 August 2018; Published 3 September 2018
2. Akyildiz IF, Lo BF, Balakrishnan R (March, 2011) Cooperative spectrum sensing in cognitive radio networks: a survey. Phys Commun 4(1):40–62
3. Mitola J, Maguire GQ (April, 1999) Cognitive radio: making software radios more personal. IEEE Pers Commun 6(4):13–18
4. Guimaraes DA (2018) Robust test statistic for cooperative spectrum sensing based on the Gerschgorin circle theorem. Received November 6, 2017, accepted December 12, 2017, date of publication December 14, 2017, date of current version February 14, 2018
5. Kay SM (1993) Fundamentals of statistical signal processing: estimation theory. Prentice-Hall, Upper Saddle River, NJ, USA
6. Huang L, So H, Qian C (May, 2014) Volume-based method for spectrum sensing. Digit Signal Process 28(48):48–56. Available: https://doi.org/10.1016/j.dsp.2014.02.003

7. Ramirez D, Vazquez-Vilar G, Lopez-Valcarce R, Via J, Santamaria I (August, 2011) Detection of rank-*P* signals in cognitive radio networks with uncalibrated multiple antennas. IEEE Trans Signal Process 59(8):3764–3774
8. Diaz1 LMG, Marrero LM, Gómez JT (March, 20) Performance comparison of spectrum sensing techniques for cognitive radio networks. Conference paper: March 20
9. Tandra R, Sahai A (February, 2008) SNR walls for signal detection. IEEE J Sel Top Signal Proc 2(1)
10. Bollig A, Disch C, Arts M, Mathar R (2017) SNR walls in eigenvalue-based spectrum sensing. EURASIP J Wireless Commun Networking

Feature-Based Segmentation of Brain Tumor in MRI Images

Vijayshri Chaurasia, Princi Soni, Yashwant Kurmi, and Apoorva Modi

Abstract Image segmentation of magnetic resonance image (MRI) is a crucial process for visualization and examination of abnormal tissues, especially during clinical analysis. This paper presents an automatic lesion recognition method in MRI using image segmentation. In the proposed method, the intent region initialization is performed using low-level information by keypoint descriptor and the set of features and filter training data are analyzed to track tumor region. Further region growing and refinement process identifies consist boundary of tumors. The experimental outcomes on BrainWeb and BRATS 2013 datasets validate the effectiveness of the proposed method as that of existing techniques.

Keywords Brain tumor · MRI · Segmentation · Keypoint descriptor · Possibilistic model

1 Introduction

An uncontrolled growth of cells within the brain is called brain tumor. MR images are widely used in brain tumor detection and monitoring [1]. Segmentation of brain MRIs dispenses information about normal tissues, abnormal tissues, and anatomical structures. The brain itself is a complex structure and various types of tumor exhibit

V. Chaurasia · P. Soni · Y. Kurmi (✉)
Department of Electronics and Communication Engineering, Maulana Azad National Institute of Technology, Bhopal, India
e-mail: yashwantkurmi18@gmail.com

V. Chaurasia
e-mail: vijayshree21@gmail.com

P. Soni
e-mail: princi100ni@gmail.com

A. Modi
NSCB, Medical College, Jabalpur, India
e-mail: apoorvamodi2@gmail.com

V. Nath and J. K. Mandal (eds.), *Proceedings of the Fourth International Conference on Microelectronics, Computing and Communication Systems*, Lecture Notes in Electrical Engineering 673, https://doi.org/10.1007/978-981-15-5546-6_18

diversity in shape, size, location, appearance, and intensity. Innumerable efforts of researchers are continuing in the field of brain tumor detection, but there is still the need for accurate, precise, and reproducible characterization of abnormalities [2].

Numerous techniques for MRI segmentation have been investigated through the decades. Some of these are based on fuzzy C-means (FCM) [1, 3–5], *K*-means clustering [2, 5], tumor cut [6], Graph cut [7], grow cut [8], neural network [9] and patch base [10, 11]. In 1981, Bezdek et al. gave the first method for segmentation of MRI images, based on FCM algorithm [3].

In 2016, Pereira et al. presented multiple layers of convolutional neural network (CNN) with kernel weights [9]. The convolutional layers convolve an image with kernels to obtain the feature. In 2016, Cordier et al. [10] presented an advance patch-based voting technique with a probabilistic model that is convolution patch-based segmentation (CPBS) framework for MRI segmentation.

Recently, Gao et al. [11] in 2017, presented a tumor localization method based on Laplace regularization. Gaussian weighted Laplace prior (GWLP) method is a multimodal MRI segmentation technique without using morphological operations.

Outcomes of FCM-based methods are under segmented and time consuming while *K*-means clustering-based methods are faster but possess the limitation of poor intensity adjustment, low efficiency, and higher processing time. The image classification is investigated using karnel based approach [12], pixel based approach [13] and context based indexing [14].

The proposed method offers automatic segmentation of brain MRI for diagnosis of brain tumor and results in sharp edges, outlier, and consistent boundaries of the abnormal region. The proposal utilizes the concepts of max pooling [15], mid-level feature identification [16] and deep learning [17].

In rest of the paper, the second section describes the proposed system of brain MRI segmentation. The third section covers experimental results and comparative analysis. Finally, the fourth section draws the conclusion with frontrunners of a future framework.

2 Proposed Method

The proposed method provides brain tumor detection by segmentation with high accuracy and less complexity. Block diagram of the proposed method is shown in Fig. 1. Tumor detection is composed in the following steps: image enhancement, feature extraction, training section, region growing, and region refinement. Detailing of different steps is explained in following sub-sections.

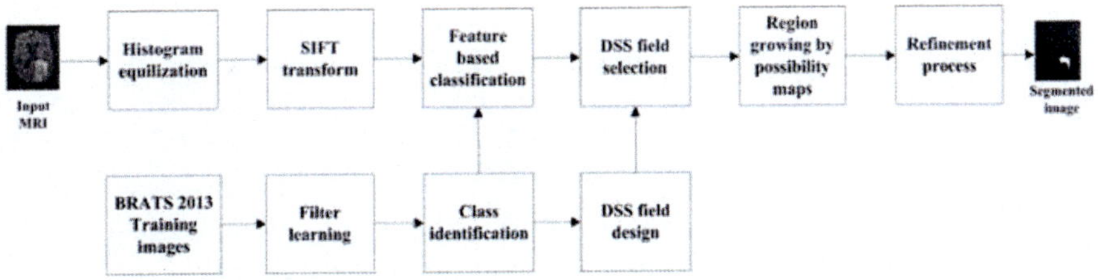

Fig. 1 Block diagram of the proposed method

2.1 *Image Enhancement*

Histogram equalization is performed for contrast and intensity enhancement using regional adaptive algorithm and high-pass filter [18]. For enhancement of the gray-scale image, each pixel is revaluated on the basis of gray-scale values of nearby pixels. Let input MRI image $f(x, y)$ composed of discrete gray levels in dynamic range of $[0, L - 1]$, then transform function is define by, e.g., (1).

$$s_l = C(r_l) = \sum_{i=0}^{l} P(r_i) = \sum_{i=0}^{l} \frac{n_i}{n_t} \tag{1}$$

where $l = 0, 1, 2, \ldots, L - 1$ is gray level, r_l represents an input gray level, n_l is number of pixel having gray level r_l, n_t is total number of pixels in input image, $P(r_l)$ probability density function (PDF) of gray level r_l,$C(r_l)$ is cumulative density function (CDF) and $0 \leq s_l \leq 1$.

The assignment of a new histogram level to each pixel is done with transform function s_l over a range $[0, L - 1]$. Histogram equalization improves the contrast of an area in an image by distributing an equal number of pixels across the range of intensities. It results a new enhanced image $I_e(x, y)$ using new pixel-intensity values.

2.2 *Feature Extraction*

Enhanced MR image $I_e(x, y)$ possess very high feature density. In this image, keypoint descriptors are used for extraction of abnormal tissues. The intent regions are defined as keypoint and their invariant features are considered as descriptor. In proposal, normalized scale-invariant feature transform (SIFT) [19] is employed to extract keypoint and create descriptors for the image region. The basic grid divides the image into regions.

SIFT descriptors perform keypoint localization of regions which are extracted from a set of training images. Gaussian pyramid $L(x, y, \sigma)$ is formed by iterative convolution of input image $I_e(x, y)$ and Gaussian kernel $G(x, y, \sigma)$, where σ represents standard deviation.

$$L(x, y, \sigma) = G(x, y, \sigma) * I_e(x, y) \tag{2}$$

$$G(x, y, \sigma) = \frac{1}{2\pi\sigma^2} e^{-(x^2+y^2)/2\sigma^2} \tag{3}$$

Feature location is determined by local extrema of difference-of-Gaussian function and given by (4) for a constant multiplication scalar β [20].

$$D(x, y, \sigma) = L(x, y, \beta\sigma) - L(x, y, \sigma) \tag{4}$$

Here, $I_e(x, y)$ is presented as set of regions which are expressed by$\{(x_1, y_1, w_1), (x_2, y_2, w_2), \ldots, (x_i, y_i, w_i), \ldots, (x_k, y_k, w_k)\}$. Where $w_i \in R^d$ is the ith normalized SIFT descriptor of an image region, (x_i, y_i) is the keypoint coordinate of a region in the image, R^d is d-dimensional real space and k is a total number of the local descriptor and $i = 1, 2, \ldots, k$. The magnitude of SIFT descriptor w is computed by (5).

$$|w_i| = \sqrt{(L(x_i + 1, y_i) - L(x_i - 1, y_i))^2 + (L(x_i, y_i + 1) - L(x_i, y_i - 1))^2} \tag{5}$$

The computed feature information is further used for classification.

2.3 *Training Section*

High-level image features are extracted using the local descriptors. Set of the linear filters are used for encoding of local statistical data. We have used random projection (RP) tree [21] for grounding of the filter bank (B) defined by $B \in R^{N \times d}$ where N denotes the number of cells in a tree structure and centered at mean value. The local descriptor and filter bank produce a feature response map [21] which is used for salient key detection.

The target region is defined inside discriminative spatial supporting (DSS) field. Every DSS field is surrounded by a relatively larger field called protecting the field [22]. If protecting field is too large only a few keypoints will be detected and if too small, then overlapping of the intent region will take place [23].

2.4 *Feature-Based Classification*

The largest filter response for each keypoint is taken and low-level descriptor vector W is defined:

$$W = [w_1, w_2, \ldots, w_k] \tag{6}$$

where $W \in R^{d \times k}$.

Further, linear filer responses F are obtained as:

$$\boldsymbol{F} = B \times W \tag{7}$$

Here, filter response matrix $\boldsymbol{F}$ with order $N \times k$ is defined as $\boldsymbol{F} \in R^{N \times k}$ and given by (8). In $\boldsymbol{F}$, largest value of a column shows a maximum similarity between a cell and local descriptor. Encoding of local descriptor is simply the largest inner product between local descriptor and cell.

$$\boldsymbol{F} = \begin{bmatrix} F_{11} & \cdots & F_{1k} \\ \vdots & \ddots & \vdots \\ F_{N1} & \cdots & F_{Nk} \end{bmatrix} \tag{8}$$

Set of the element of the ith column and filter response of ith descriptor is given by $F_{.i}$ and F_i, respectively. All lower inner product will set to zero as they cannot relate to the corresponding keypoint.

$$F_{.i} = \{F_{1i}, F_{2i}, \ldots, F_{ji}, \ldots, F_{Ni}\} \tag{9}$$

$$F_i = \max(F_{.i}) \tag{10}$$

Finally, a filter feature response F_i associated with the ith descriptor according to similarity parameter is computed.

2.5 DSS Field Selection

Let the size of the image I_e is $S \times S$, then the size of k number of DSS fields is given by a set $s \in \left[\frac{1}{4}, \frac{1}{3}, \frac{1}{2} \ldots\right] \times S$. Filter response according to ith keypoint is F_i and a DSS field are expressed by $D(s_i, i, L_i)$ corresponds to feature filter response map. Where s_i is the size and L_i is position of ith box. Figure 2 shows a typical DSS field $D(s_i, i, L_i)$ (dotted line box) and protecting field $D(2s_i, i, L_i)$ (solid line box). White stars denoted as $A_1, A_2, A_3, \ldots$ are keypoints with lower filter response and red stars denoted by K_1, K_2, these are the keypoints with largest filter responses in their related DSS field. If two DSS field exists in one protective filed and both have same keypoint, then DSS field is selected by discriminative score of both the fields.

The discriminative score $\xi(s_i, i, L_i)$ for filter response F_i is given by (11).

$$\xi(s_i, i, L_i) = F_i(L_i) \min_{\substack{L_n \in D(2s_i, i, L_i) \\ L_i \neq L_n}} |F_i(L_i) - F_i(L_n)| \tag{11}$$

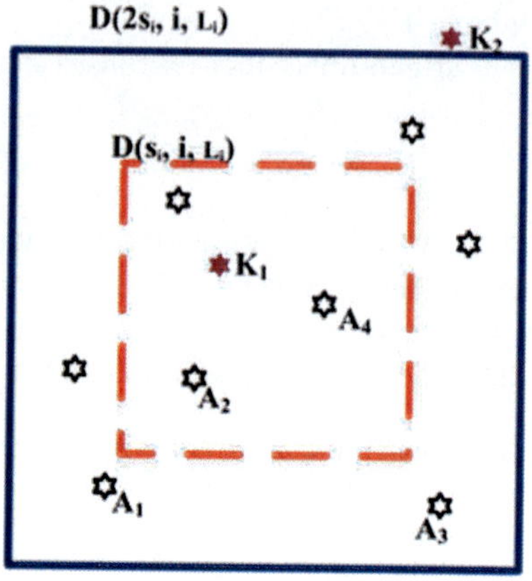

Fig. 2 Selection of DSS field

Every DSS field is defined according to one keypoint and $2N$ keypoint are selected and results an image J. All cells make a class set C and probability density functions (PDF) of classes derived from selected DSS field. Location of a keypoint is estimated by the DSS field scale and magnitude is defined by filter response, i.e., $f^n = |F_i|$, where $n = 1$ to $2N$. Therefore, the keypoint with a vicinity of s_n will be represented by quaternary (f^n, x^n, y^n, s^n), where $(x^n, y^n) = (x_i, y_i)$.

2.6 *Region Growing*

Proposed method uses possibilistic knowledge diffusion (PKD) method for region growing [24]. Class set C is delineated according to seed set ψ and neighborhood pixels are merged to the similar initial seed class regions by iteration process. Pixel set ψ for image J is defined by a set $\psi = \{\psi_0, \psi_1, \ldots, \psi_j, \ldots, \psi_N\}$ according to N semantic classes, $C = \{C_1, C_2, \ldots, C_j, \ldots, C_N\}$, where ψ_j is a bunch of pixels belongs to C_j class and $j = 1, 2, \ldots, N$. The probability density functions (PDF) are estimated for class set C with the kernel density estimation (KDE) technique [25]. If probability and possibility are represented by Pr and π, respectively, than $(Pr - \pi)$ transformation [26] and PDF for class C_j is $P(C_j)$, so possibility density of class C_j is defined by as:

$$\pi(C_j) = \sum_{t=1}^{N} \min[P(C_j), P(C_t)] \tag{12}$$

If maximum pixels of class C_j belong to intent region, then ψ_{j0} is characterized as initial sampled learning set of C_j. Further, ψ_{j0} is iterated m times for intent region growing. N possibility distributions correspond N possibility maps PM_{J,C_j} associated with image J. Different possibility maps (PM_{J,C_j}) are updated by possibilistic special information $\pi(K_n)$ to extract a new learning sample set (ψ_{01}) [26].

$$\mathrm{PM}_{J,C_j}(p_0) = \pi_{C_j}\{s(p_0)\} \tag{13}$$

where p_0 is a pixel of image J and $s(p_0)$ is observed feature such as gray level of pixel p_0. Nagao-Matsuyama filter [27] is an edge preserving, homogeneity controller, and smoothing filter is executed via 9 sub-windows (f_g), where g varies from 1 to 9 with 5×5 dimension. This leads to possibility distribution of pixel p_0, defined as

$$\pi_{p_0}(C_j) = \left[\pi_{p_0}(C_j, f_1), \pi_{p_0}(C_j, f_2), \ldots, \pi_{p_0}(C_j, f_9)\right] \tag{14}$$

where $\pi_{p_0}(C_j, f_g) = \frac{1}{9}\sum_{P\in f_g}\left\{\mathrm{PM}_{J,C_j}(P)\right\}$.

The window having the maximum value of the distribution $\pi_{p_0}(C_j)$ is select as wining window $\pi_{p_m}(C_j, f_g)$ in each possibilistic map.

$$\mathrm{PM}_{J,C_j}(p_0, m) = \underbrace{\Phi}_{p_0\in f_g} \mathrm{PM}_{J,C_j}(p_0, m-1) \tag{15}$$

where Φ is mean operator. This process will be terminated when no further pixels could be added to the class C_j and results in segmented image I_s with an intent abnormal region.

2.7 Region Refinement

Resultants of region growing process do not possess consistent boundaries. Image-guided disparity refinement method is applied to resolve boundary inconsistency [28] by computing constant time-weighted median (CTWM) filtering followed by hybrid super pixel technique [29]. If disparity of a pixel is much less than its adjacent pixels, it is marked as a hole. In image I_s, pixel p with nearest neighbor q will be a hole 'h' if,

$$h = \begin{cases} 1, & d_p < \frac{d_q}{2} \\ 0, & \text{otherwise} \end{cases} \tag{16}$$

where d_p and d_q are the disparities of pixel p and q, respectively. When any pixel with irregular disparity has a true condition for a hole, then that pixel will be labeled as a dark region. Correction in dark region depends on its neighbor. The corrected disparity is denoted by d_p^*.

$$d_p^* = \begin{cases} \min\left\{d_{q1}, d_{q2}\right\}, & \forall h(q_1)\|h(q_2) = 0 \\ \max\left\{d_{q1}, d_{q2}\right\}, & \forall h(q_1)\|h(q_2) = 1 \end{cases} \tag{17}$$

This refinement process is governed in horizontal as well as vertical direction. Since some regions, especially outlier regions are not well segmented even after refinement. Hybrid superpixel segmentation process undertakes for segmentation of in segmented region. Superpixel is a batch of pixels in proportion to a region and gray level. If bunch with the largest number of pixels contains less than 60% of the total pixel, than that superpixel is treated as an under segmented region. Under segmented region is processed further by mean-shift algorithm [30]. Finally, the process produces an MRI segmented image with an accurately extracted intent area.

3 Result and Discussion

The performance of the proposed MRI segmentation method is evaluated by using two benchmark data sets. First one is BrainWeb dataset that consists of the full 3D data volumes of normal and multiple sclerosis (MS) models with the variety of slice thickness, intensity levels, and noise levels [31]. This dataset has the extension of ".mnc" with a $T1$-weighted sequence. The second one is the BRATS 2013 database which contains three types of data sets; training, leader board, and challenge [32]. Both are specifically simulated with high-grade gliomas (HGG) and low-grade gliomas (LGG) tumor. For every patient in BRATS, there are four MRI sequences: $T1$ weighted ($T1$), $T1$ with gadolinium-enhancing contrast ($T1$c), $T2$-weighted ($T2$) and FLAIR. This dataset has an extension of ".mha".

The effect of number of training images, on segmentation, in term of accuracy and dice score is shown in Fig. 3. The performance of the proposed image segmentation method is investigated for different number of training images from BARTS 2013 training set for filter training. As shown, increase in the number of training images enhances the accuracy as well as the DSC score.

Figure 4 shows segmentation of the BrainWeb set in the first row and BRATS 2013 dataset in the next two rows. Simulation specification of the first dataset is severed MS lesions, $T1$ pulse sequence, 3 mm slice thickness, noise MS lesion

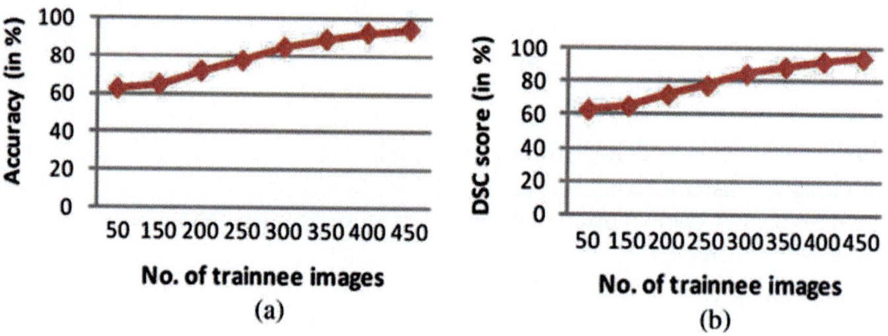

Fig. 3 Effect of number of training images on segmentation **a** accuracy and **b** DSC score

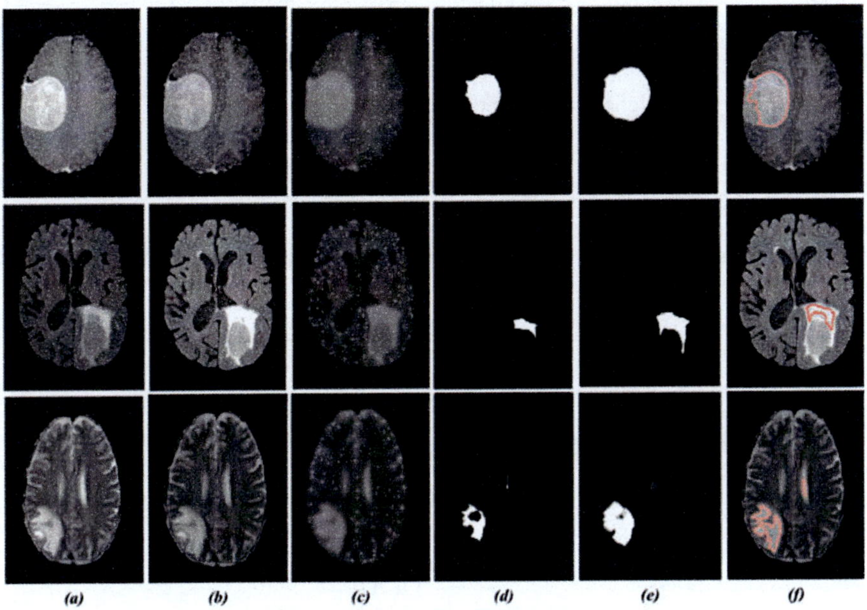

Fig. 4 Visual illustration of segmented images; **a** brain MRI slice, **b** enhanced MRI image, **c** image with keypoints sampling, **d** segmented image before refinement, **e** segmented image after refinement, and **f** intent region masked image

reference tissue and 5% noise level. Second and third rows are from leaderboard data set with $T1$-weighted, particularly HGG (subject id: 210) and LGG (subject id: 109), respectively. From Fig. 4, the effectiveness of the proposed method is visually justified.

The success of the proposed segmentation method is also examined by comparative analysis with some state-of-art methods. The methods considered for comparison are; the NFCM, enhanced nonlocal FCM algorithm [4], tumor cut, cellular automata-based method [6], PIGFCM, FCM with prior guide [1], KIFCM, hybridization of K-means classifier with FCM [5], CPBS, patched-based technique accomplish with neural network [10], two-tier KNN (K-nearest neighbor) with SOM [2], CNN, convolutional neural network [9], GWLP, Gaussian weighted Laplace prior [11]. The performance is investigated on the basis of precision, recall, F_1 − score, accuracy, and dice similarity coefficient (DSC).

Precision is the ratio of correctly predicted positive observations to the total predicted positive observations.

$$\text{Precision} = \frac{\text{TP}}{(\text{TP} + \text{FP})} \tag{18}$$

Recall is the ratio of correctly predicted positive observation to all observations in the actual class.

$$\text{Recall} = \frac{\text{TP}}{(\text{TP} + \text{FN})} \tag{19}$$

$F_1 -$ score is a weighted average of precision and recall.

$$F_1 - \text{Score} = \frac{2(\text{Recall} \times \text{Precision})}{(\text{Recall} + \text{Precision})} \tag{20}$$

Accuracy is a degree of closeness of result of measurement and calculation of a quantity to its true or actual value.

$$\text{Accuracy} = \frac{\text{TP} + \text{TN}}{\text{TP} + \text{TN} + \text{FP} + \text{FN}} \tag{21}$$

Dice similarity index is a measure of overlapping or spatial alignment between manual segmentation (by radiotherapist) and automatic segmentation. Its value comes in between 0 and 1 and higher value is desirable as it an indication of system efficiency.

$$\text{Dice} = \frac{2 \times \text{TP}}{(\text{FP} + \text{TP})(\text{TP} + \text{FN})} \tag{22}$$

The average comparative performance based on above-mentioned parameters using BRATS 2013 dataset is presented in Table 1. All quantities of Tables 1 and 2 are in percentage. The values mentioned in Table 1 are the average of simulation results drawn on 100 images of a dataset. As shown, the proposed method with region refinement provides most precise and accurate segmentation as compared to existing methods. For BRATS 2013 dataset, the proposed technique improves DSC by 4.27% and 5.45% as compared to KIFCM, and CNN methods, respectively. The accuracy of the proposed method is 7.79% and 10.36% higher as that of CNN and two-tier KNN methods, respectively. The inclusion of region refinement has improved the performance of the proposed method commendably. It has shown an average increase of 12.68% and 10.47% in $F_1 -$ score and DSC, respectively. In terms of precision,

Table 1 Average comparative performance analysis of different MRI image segmentation methods using BRATS 2013 dataset

Method	Precision	Recall	$F_1 -$ score	Accuracy	DSC
Tumor cut [6]	79	81	79.99	78	89
KIFCM [5]	86	91.8	89.23	93	89
Two-tier KNN [2]	85	87.8	86.38	84	83
CNN [9]	88	79	83.26	86	88
CPBS [10]	87	84	85.47	88	90
Proposed before refinement	81	85	81.95	83	84
Proposed	93	92.6	92.34	92.7	92.8

Table 2 Average comparative performance analysis of different MRI image segmentation methods using BrainWeb dataset

Method	Precision	Recall	F_1 − score	Accuracy	DSC
NFCM [4]	78	84	80.89	87	82
PIGFCM [1]	89	78	83.14	91	92
KIFCM [5]	88	90.5	88.64	95	91
GWLP [11]	88	94	90.90	91	74
Proposed before refinement	83	87	84.95	81	85
Proposed	92	91.7	91.84	94.4	93.2

recall, DSC, and F_1 – score, the proposed method outperforms all companions listed in Table 1.

Table 2 shows the average comparative performance analysis of BrainWeb dataset. The quantities shown in the table are average of simulation results drawn on 80 images of BrainWeb dataset. Here also, the proposed MRI image segmentation technique with region refinement produces the most efficient performance as that of existing methods considered for comparison.

The proposed method has 6.81% higher precision level than GWLP [11] and KIFCM [5] methods. GWLP shows 94 as recall value, that is best in overall methods included proposed method. Proposed method provides 3.73% higher accuracy than GWLP and PIGFCM [1] techniques. Our method is also computationally efficient technique with leading DSC value 92.8 which is 0.86% higher than Dice index of PIGFCM. In terms of precision, DSC, and F_1 – score, the proposed method outperforms all companions listed in Table 2 with highest values of 92, 93.2, and 91.84, respectively. As F_1 – score is a mirror of error rate, so the proposed method signifies low error rate.

4 Conclusion

MRI is a very functional modality for diagnosis of brain tumor. A novel feature-based MRI segmentation technique has been proposed in this paper. In proposal, firstly, the image enhancement is performed using histogram equalization followed by intent region identification using SIFT and feature-based classification with DSS field. Here, the keypoint descriptor sampling is used to transform low-level information into high-level image features by a set of linear filters. Further, the intent region is extracted by possibilistic model and Nagao-Matsuyama filter. Finally, the region refinement identifies overall the tumor structure. The proposed method shows better F_1-score and DSC as compared to state-of-art methods. MRI segmentation by the proposed method reports an average accuracy of 93.46%, average F_1-score of 92.12% with average DSC 92.98%.

References

1. Mokhtar EM (2014) Tumor segmentation in brain MRI using a fuzzy approach with class center priors. EURASIP J Image Video Process 2014(1):21
2. Anitha V, Murugavalli S (2016) Brain tumour classification using two-tier classifier with adaptive segmentation technique. IET Comput Vis 10(1):9–17
3. Peizhuang W (1983) Pattern recognition with fuzzy objective function algorithms (James C. Bezdek) SIAM Rev 25(3):442–442
4. Ahmed MN, Yamany SM, Mohamed N, Farag AA, Moriarty T (2002) A modified fuzzy *C*-means algorithm for bias field estimation and segmentation of MRI data. IEEE Trans Med Imaging 21(3):193–199
5. Abdel-Maksoud E, Elmogy M, Al-Awadi R (2015) Brain tumor segmentation based on a hybrid clustering technique. Egypt Inf J 16(1):71–81
6. Hamamci A, Kucuk N, Karaman K, Engin K, Unal G (2012) Tumor-cut: segmentation of brain tumors on contrast enhanced MR images for radiosurgery applications. IEEE Trans Med Imaging 31(3):790–804
7. Boykov YY, Jolly MP (2001) Interactive graph cuts for optimal boundary & region segmentation of objects in N-D images. Proc Eighth IEEE Int Conf Comput Vision (ICCV) 2001(1):105–112
8. Vezhnevets V, Konouchine V (2005) Grow-cut—interactive multi-label N-D image segmentation by cellular automata. Graphicon 150–156
9. Pereira S, Pinto A, Alves V, Silva CA (2016) Brain tumor segmentation using convolutional neural networks in MRI images. IEEE Trans Med Imaging 35(5):1240–1251
10. Cordier N, Delingette H, Ayache N (2016) A patch-based approach for the segmentation of pathologies: application to glioma labelling. IEEE Trans Med Imaging 35(4):1066–1076
11. Gao Y, Wang K, Jiang S, Liu Y, Ai T, Tian J (2017) Bioluminescence tomography based on gaussian weighted laplace prior regularization for in vivo morphological imaging of glioma. IEEE Trans Med Imaging 36(11):2343–2354
12. Comaniciu D, Ramesh V, Meer P (2003) Kernel-based object tracking. IEEE Trans Pattern Anal Mach Intell 25(5):564–577
13. Alsahwa B, Solaiman B, Almouahed S, Bosse E, Gueriot D (2016) Iterative refinement of possibility distributions by learning for pixel-based classification. IEEE Trans Image Process 25(8):3533–3545
14. Vailaya A, Figueiredo MAT, Jain AK, Zhang HJ (2001) Image classification for content-based indexing. IEEE Trans Image Process 10(1):117–130
15. Yu X, Yang J, Wang T, Huang T (2015) Key point detection by max pooling for tracking. IEEE Trans Cybern 45(3):444–452
16. Boureau Y, Bach F (2010) Learning mid-level features for recognition 2559–2566
17. Lecun Y, Bengio Y, Hinton G (2015) Deep learning. Nature 521(7553):436–444
18. Chen Y, Ding J, Hsiao C, Chang H (2014) A region adaptive encoding algorithm for simple image compression 2014(3):595–598
19. Zhou H, Yuan Y, Shi C (2009) Object tracking using SIFT features and mean shift. Comput Vis Image Underst 113(3):345–352
20. Pani A, Shende P, Dhumal M, Sangle K, Shiravale S (2013) Advantages of using sift for brain tumor detection. Int J Students Res Technol Manag 1(3):327–338
21. Freund Y, Dasgupta S, Kabra M, Verma N (2007) Learning the structure of manifolds using random projections. Adv Neural Inf Process Syst 473–480
22. Yang J, Yu K, Gong Y, Huang T (2009) Linear spatial pyramid matching using sparse coding for image classification. 2009 IEEE Comput Soc Conf Comput Vis Pattern Recognit Work CVPR Work 2009:1794–1801
23. Collins RT, Liu Y (2005) On-line selection of discriminative tracking features 27(10):1–14
24. Kaur A, Singh N (2014) Region growing and object extraction techniques 3(10):2012–2015
25. Rudemo M (1982) Empirical choice of histograms and kernel density estimators. Scand J Stat 9(2):65–78

26. Dubois D, Foulloy L, Mauris G, Prade H (2004) Probability-possibility transformations, triangular fuzzy sets, and probabilistic inequalities. Reliab Comput 10(4):273–297
27. Nagao M, Matsuyama T (1979) Edge preserving smoothing. Comput Graph Image Process 9(4):394–407
28. Jiao J, Wang R, Wang W, Li D, Gao W (2017) Color image-guided boundary-inconsistent region refinement for stereo matching. IEEE Trans Circuits Syst Video Technol 27(5):1155–1159
29. Ma Z, He K, Wei Y, Sun J, Wu E (2013) Constant time weighted median filtering for stereo matching and beyond. In: Proceedings of the IEEE international conference on computer vision, pp 49–56
30. Comaniciu D, Meer P (2002) Mean shift: a robust approach toward feature space analysis. IEEE Trans Pattern Anal Mach Intell 24(5):603–619
31. Brain Web: Simulated Brain Database. http://brainweb.bic.mni.mcgill.ca/brainweb/
32. Virtual Skeleton, BRATS 2013 Sep. 30, 2015 Available: https://www.virtualskeleton.ch/BRATS/Start2013

A Synthesis of Reversible Digital Circuits to Solve the Boolean Satisfiability

Martin Stava

Abstract The paper presents a methodology of synthesising the reversible digital circuits that solve the Boolean satisfiability (SAT). There is a full range of applications of such circuits, for example, accelerating SAT solvers, generating compacted tests for the sequential circuits designed for testability (with a scan chain), producing a set of results for the relation inverse to combinational functions, generating input vectors of a circuit from output ones, etc. Further, compacted representation of functions and the straightforward and reverse approaches to obtain responses are discussed. The methodology comprises a few rules to synthesise a combinational function (circuit) into a reversible circuit in a way to implement them into FPGAs. A method of shortening the truth table for implementation in an FPGA is presented, and the architecture of reversible circuits is described. The experimental data and features of the physical implementation of the circuits reverse to the ISCAS'85 benchmarks are presented for Xilinx FPGA Spartan 2E—the area overhead and computing complexity of generating the first valid input vector. The experiments show that both the features have asymptotically polynomial complexity.

Keywords Boolean satisfiability · Reversible circuits · VLSI · FPGA · Combinational circuit

Research described in this paper was partially financed by the BUT grant MA90901008, partially by the grant LO1401 under the National Sustainability Program, and partially by the author himself. For the research, the infrastructure of the SIX Center was partially used. I would especially like to thank my prime employer, InvaSys, a. s., namely Dr Kyrre Sletsjoee, for the individual flexible working schedules and for understanding of my academic activities, which has allowed me to carry out scientific research in my spare time.

M. Stava (✉)
Department of Microelectronics, Faculty of Electrical Engineering and Communication, Brno University of Technology, Brno, Czech Republic
e-mail: stavam1@feec.vutbr.cz

V. Nath and J. K. Mandal (eds.), *Proceedings of the Fourth International Conference on Microelectronics, Computing and Communication Systems*, Lecture Notes in Electrical Engineering 673, https://doi.org/10.1007/978-981-15-5546-6_19

1 Introduction

The Boolean satisfiability problem (SAT) has a remarkable impact on a wide range of digital circuit domains, such as testing, verification, backtracking, backtracing, etc. The majority of existing professional and academic automatic test pattern generation (ATPG) systems uses a path sensitisation technique at the logic level of circuit representation. The technique consists of three steps: fault sensitisation, path sensitisation, and line justification. During the second and third steps, there is a high probability of conflict occurrence between the signal assignments—currently required and previously done. It forces the ATPG algorithm to backtrack. Since the number of backtracks plays the key role in the test generation speed, a few heuristics and techniques to limit the number of backtracks have been developed [1], especially for speeding up the test generation process [2, 3].

There are two approaches usable for the deterministic test generation (DTG). The first approach maps the problem of DTG onto a finite space search, where algorithms run on a structural system of gates. Roth [4] firstly developed such a structural algorithm, so-called D-algorithm, employable also in computer-aided processes. After the D-algorithm had been introduced, a lot of its followers appeared along with new terminology and a variety of heuristics. For example, the structural algorithm PODEM [5] brought a special term backtrace—a novel technique setting the appropriate inputs of a combinational function/circuit for the objective to be achieved. PODEM laid the foundations to other techniques improving the DTG process—RAPS [6] or FAN [7]. RAPS introduced a new kind of backtracing—random backtrace. FAN focused on the parallel processing of a few objectives and constituted another backtracing technique—multiple backtrace that decides on those internal lines always-justifiable. Other improved ATPG techniques and systems are explored in [8, 9].

The second approach is based on Boolean satisfiability solving algorithms. Here, the D-algorithm problem formulation is translated into a characteristic equation that is solved using a branch-and-bound search. In recent years, SAT solvers have been optimised for computation time [10–12] and a number of algorithms employing them have been proposed, e.g. [13, 14]. The main issues (features) of the state-of-the-art hardware-based SAT solvers can be summarised as follows: (i) find only one matching assignment of an input vector to an output vector (or few such assignments); (ii) no direct support for either XOR logic gates or large combinational blocks; (iii) run at relatively low clock frequencies.

Novelty statement: This paper presents the synthesis of a hardware SAT solver dealing with the issues (i) and (ii). That is our solver is able to find all matching assignments of input vectors to an output vector and directly supports XOR logic gates and large combinational blocks. Moreover, it can be incorporated in hardware accelerators as an IP core; for example, in a SoC or MPSoC under Xilinx Vivado with the AXI4 interconnecting hierarchy.

The presented methodology of synthesising the reversible digital circuits that solve the Boolean satisfiability is based on our previous research [15]. The methodology is shown on an application to digital circuit testing, especially test pattern generation.

2 Motivation

When there is a need to test a structured design-for-testability sequential circuit, the most popular technique is a scan design [16]; Fig. 1 shows an example of such a sequential circuit. The symbols CP_1 through CP_3 mean combinational parts, PIs/POs primary inputs/outputs, S_{in} and S_{out} are a serial input and a serial output of the scan chain R_s, respectively. The sequential circuit is divided by the scan chain into a few combinational parts. Thus, any of the algorithms generating deterministic tests for combinational circuits (mentioned in Sect. 1) can be employed.

Further, it is assumed to have a list of test vectors, marked as L. Consider a simple task: Applying L to CP_2, whose inputs and outputs are directly connected to R_s, so that CP_2 make a feedback to R_s. Let the number of test vectors in L be t and the length of the scan chain R_s be k. The simplest way to apply L is to perform the well-known procedure (described, e.g., in [6]): Selection of a vector p of the length k from L and its scanning into R_s during k clock cycles, loading a response during the $(k + 1)$st clock cycle; it is repeated until L is empty and finally, the last response is scanned out from R_s. The computing complexity of the procedure is $\{t * (k + 1) + k\}$ clock cycles. A reduction of that complexity may lead to getting both the whole time to apply L and the power consumption lower.

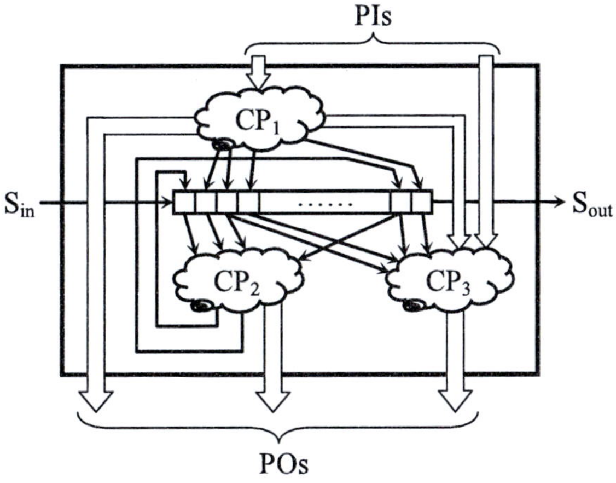

Fig. 1 A scan-based sequential circuit to exploit a Boolean satisfiability algorithm

Fig. 2 **a** Truth table; **b** primitive cubes of combinational functions $f1(x2, x1, x0)$ and $f0(x2, x1, x0)$ after compacting by Algorithm 1

x_2	x_1	x_0	f_1	f_0
0	0	0	1	1
0	0	1	1	1
0	1	0	0	1
0	1	1	1	1
1	0	0	1	0
1	0	1	1	0
1	1	0	0	0
1	1	1	0	0

(a)

Alg. 1 ⇒

x_2	x_1	x_0	f_1	f_0
1	1	x	0	0
1	0	x	1	0
0	1	0	0	1
0	0	x	1	1
0	x	1	1	1

(b)

A reduction of the computing complexity lies in having such a test-vector sequence that a response to the currently processed vector will fulfil one of the following conditions:

1. The response to the currently processed vector is equivalent to a next test vector (the best case).
2. The next test vector is achievable from the response by scanning as few values as possible into R_s.

For simplicity, let it be considered that only the scan cells of R_s that are directly influenced by the feedback through CP_2 (see Fig. 1) and that form a specific part of the test vectors, marked as SPTV.

There are two approaches to achieve the test length reduction—*straightforward* and *reverse*. Before describing them, a process of generating a truth table is worth recalling. If a truth table is not given, an exhaustive table for a combinational circuit (i.e., that consists of 2^n entries) is generated—see Fig. 2a. Such a table is possible to shorten by applying Algorithm 1. Then, the straightforward as well as reverse implications are given by searching through the table. This approach is rather useful for solving the test reduction by software.

In the former approach, the SPTV is considered as an input vector applied to the input signals of CP_2. The responses, obtained from the output signals of CP_2, to input vectors are specified either by the CP_2 structure description at the gate level or by the CP_2 truth table at the functional level. The left and right parts of the truth table represent the values of inputs and responses, respectively. In that case, any input value determines one response value. This straightforward determination is unique but non-surjective and non-injective. The table in Fig. 2a is static and preferable to that in Fig. 2b with respect to the search speed. However, in the case of the responses specified by the CP_2 structure description, those are dynamic and computed from input vectors, so that it is not necessary to know the corresponding exhaustive truth table.

In the latter approach, the SPTV is considered as an input vector applied to the output signals of CP_2. The responses, obtained from the input signals of CP_2, to input vectors are, like in the former approach, specified either by the CP_2 structure description at the gate level or by the CP_2 truth table at the functional level. Unlike in the former approach, the right and left parts of the truth table represent the values of inputs and responses, respectively. In this case, one input value can correspond to more than one response value. Therefore, the determination of a response, marked as reverse determination, is non-surjective and non-injective. Both the tables in Fig. 2a, b are easily implementable in FPGAs. However, owing to an area overhead, the table in Fig. 2b is preferred. In the case of the responses specified by the CP_2 structure description, it is necessary to transform this structure into a reversible circuit. Then, if requested, the responses are computed from input vectors—the computation is similar to the operation backtrace (as in the ATPG algorithms).

The properties of the straightforward and reverse determination approaches can be now summarised. The main property is observed: The straightforward implication strictly determines only a one-element set of responses to an input vector while the reverse implication can result in an empty or one-element or multi-element set of responses to an output vector. Thus, in the case of the multi-element set, there is a possibility to select such a response that best fits a need of the test pattern compaction/compression. While the employment of reversible circuits for the compacted/compressed test generation seems to be advantageous as indicated above, its implementation in an FPGA is ambiguous (see the experimental results in Sect. 4).

3 Synthesis of Reversible Digital Circuits

The synthesis of reversible digital circuits assumes a description of a combinational function in a form of an algebraic equation or a structure of gates. The following subchapters introduce the theory of combinational functions, a reversible circuit structure, a controller architecture, and transformation techniques for the structural parts of combinational circuits, namely fan-out stems and more-input gates.

3.1 *Combinational Function Representation*

The simplest way to represent a combinational circuit is by its truth table, such as that in Fig. 2a. Let an n-tuple $(x_{n-1}, \ldots, x_0)$ of input variables be marked as an input vector $\mathbf{X}$. Assume binary input values, a circuit performing functions $f_{m-1}(x_{n-1}, \ldots, x_0), f_1(x_{n-1}, \ldots, x_0), \ldots, f_0(x_{n-1}, \ldots, x_0)$ of n input variables requiring a truth table with 2^n entries. Let an m-tuple $(f_{m-1}, \ldots, f_0)$ of output variables be marked as an output vector $\mathbf{F}$ and a one-dimensional array V of size $m * 2^n$ be a data structure representing output variables of the truth table. Arrange 2^n input combinations in their increasing binary order. Then $V(0) = (f_{m-1}(0,0, \ldots, 0), \ldots, f_0(0, 0, \ldots, 0)), \ldots, V(2^n - 1) = (f_{m-1}(1, 1, \ldots, 1), \ldots, f_0(1, 1, \ldots, 1))$. A typical procedure to determine the value of variables defined by $\mathbf{F}$ works as follows:

1. Concatenate the input values in proper order to form one binary word.
2. Let i be an integer value of this word.
3. The value of the vector **F** is given by $V(i)$.

For a circuit with m outputs, an entry in the array V is an m-bit vector defining the output values. Hence, truth tables are often represented by read-only memories (ROMs).

The cube representing a common prime implicant of **F** is called a *primitive cube* [6]. Primitive cubes provide a compact representation of a function. To find values for a subset of **X** to produce a corresponding value for **F** is the desired goal. Clearly, the solutions are given by one of the separate groups of primitive cubes of **F**. For example, Fig. 2b shows that either $(x_2, x_1) = (0, 0)$ or $(x_2, x_0) = (0, 1)$ is needed to set $\mathbf{F} = (1, 1)$.

The cube associated with the input vector **X** is a $(v_{n-1} \ldots v_1\ v_0)$ vector of corresponding variable values. The cube of the **F** function vector has a form of $(v_{n-1} \ldots v_1\ v_0|v_{fm-1} \ldots v_{f0})$ marked as a **W** vector of values, where $v_{fi} = f_i\ (v_{n-1}, \ldots, v_1, v_0)$ and $v_{fi} \in \{0, 1\}, 0 \leq i \leq m-1$, $v_j \in \{0, 1, x\}, 0 \leq j \leq n-1$ and the value x denotes an unspecified or "don't care" value. Thus, the cube of **F** can represent an entry in its truth table. A g implicant of **F** is represented by a cube, which can be constructed by Algorithm 1. An example of a truth table and its primitive cubes is shown in Fig. 2.

Algorithm 1 *Construct_Cubes*

1. Find a G set of common prime implicants for the same values of **F**. That is to say common prime implicants of all functions $f_i(\mathbf{X})$ whose value is 1, and $\overline{f_j}(\mathbf{X})$ whose value is 0 are looked for. Naturally, those implicants are not needed to be prime for individual $f_i(\mathbf{X})$ or $\overline{f_j}(\mathbf{X})$ but are needed to be common prime, i.e. as big as possible, for the whole vector **F**.
2. Set $v_i = 1$ (0) if x_i ($\bar{x}_i$) appears in $g \in G$.
3. Set $v_i = x$ if neither x_i nor $\bar{x}_i$ appears in g.
4. Evaluate and set v_{fi}.
5. Go to Step 2 if G is not empty.
6. Go to Step 1 if not all values of **F** have been used to construct cubes.

Example 1 Consider two functions $f_1(x_2, x_1, x_0)$ and $f_0(x_2, x_1, x_0)$ given by the truth table in Fig. 2a. Find their primitive cubes by applying Algorithm 1.

Solution $n = 3$, $m = 2$, $\mathbf{X} = (x_2, x_1, x_0)$, $\mathbf{F} = (f_1(\mathbf{X}), f_0(\mathbf{X}))$, $\mathbf{W} = (v_2\ v_1\ v_0|v_{f1}\ v_{f0})$. Karnaugh's maps can be used to find common prime implicants.

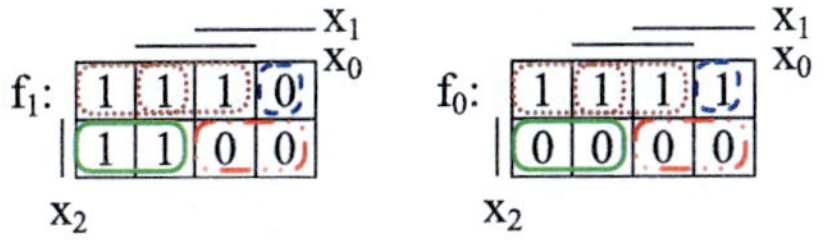

for $\mathbf{F}_0 = (0, 0)$: $\overline{f_1} = \overline{f_0} = x_2 x_1$ → $\mathbf{W}_0 = (1\ 1\ \mathrm{x} \mid 0\ 0)$

for $\mathbf{F}_1 = (1, 0)$: $f_1 = \overline{f_0} = x_2 \bar{x}_1$ → $\mathbf{W}_1 = (1\ 0\ \mathrm{x} \mid 1\ 0)$

for $\mathbf{F}_2 = (0, 1)$: $\overline{f_1} = f_0 = \bar{x}_2 x_1 \bar{x}_0$ → $\mathbf{W}_2 = (0\ 1\ 0 \mid 0\ 1)$

for $\mathbf{F}_3 = (1, 1)$:

$f_1 = f_0 = \bar{x}_2 \bar{x}_1 + \bar{x}_2 x_0$ → $\mathbf{W}_3 = (0\ 0\ \mathrm{x} \mid 1\ 1)$

→ $\mathbf{W}_4 = (0\ \mathrm{x}\ 1 \mid 1\ 1)$

The final primitive cubes are shown in Fig. 2b. (*End of the example*)

If the cube q can be obtained from the cube p by replacing one or more x values in p by 0 or 1, it means that p covers q, e.g., the cube 00x|11 covers the cubes 000|11 and 001|11. The primitive cubes of the **F** vector of functions given by the truth table are shown in Fig. 2.

3.2 A Reversible Circuit Structure

A combinational circuit has to be converted into a corresponding reverse structure to perform the Boolean satisfiability, as shown in Fig. 3. The symbol **CP** stands for a combinational circuit or its part, the symbol **Cs** denotes constraints (e.g., fixed logic values assigned to given signals) and **TCP** stands for a matching transformed combinational part (circuit). The symbol **CL** denotes control logic. Let the whole circuit (Cs + TCP + CL) be defined as a *reversible circuit* (*RC*). The reversible circuit produces one input vector (*InpVector*) per one clock cycle for an applied output vector (*OutVector*). The RC can generate all possible InpVectors to the applied OutVector.

Each generated InpVector carries a validity flag—the Valid signal. As soon as all the possible input vectors have been computed for the applied OutVector, a signal Done is set to 1 otherwise to 0. The signals Valid and Done are of two-value logic.

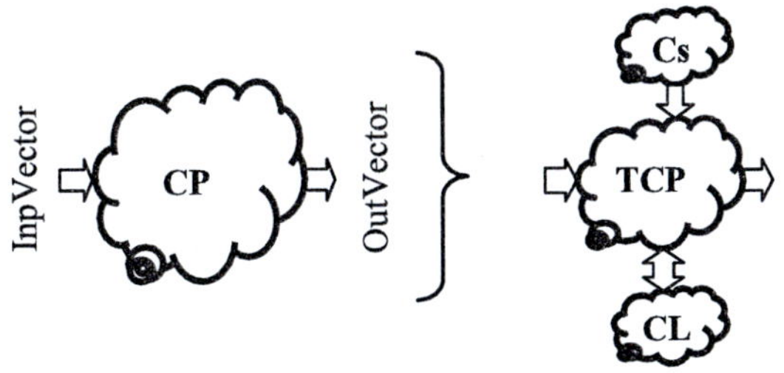

Fig. 3 Combinational circuit transformation into the corresponding reversible circuit

The vectors InpVector and OutVector are of three-value logic. The control logic uses two-value signals.

3.3 Gate Transformation

Within transforming a given circuit into a reversible one, each gate is transformed to a Logic Element (LE), implementing the matching reverse implication table (RIT) of the gate. For example, the reverse implication table for the 2-AND gate and the appropriate LE are shown in Fig. 4.

Further, a way of selection a line from the RIT (see Fig. 4) is explained. If there is requested gate output $c = 0$, one of the two matching RIT lines ($k0$, $k1$) can be chosen. Once the reset signal (rst_i) has come, the first matching line is always selected, i.e. the $k0$ line, and the status signal $NLExist_i$ is set to 1. $NLExist_i$ signalises that a next line in the RIT exists and can be chosen for the given output *c*. If a conflict occurs or a next (*a, b*) vector is demanded during a subsequent run, the second matching line will be selected by the control signal $NextLine_i$, i.e. the line from the $k1$ address. The signal $NLExist_i$ will be set to 0 if there is no other (*a, b*) vector for the given output *c*. The addresses $k0$ and $k1$ are kept in the register $Reg0$.

LE can represent a greater combinational unit (circuit, function) than only a gate RIT—the idea is identical.

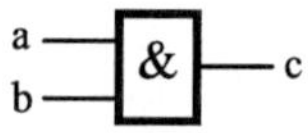

c	a	b	Address
0	0	x	k0
0	x	0	k1
1	1	1	k2
x	x	x	k3

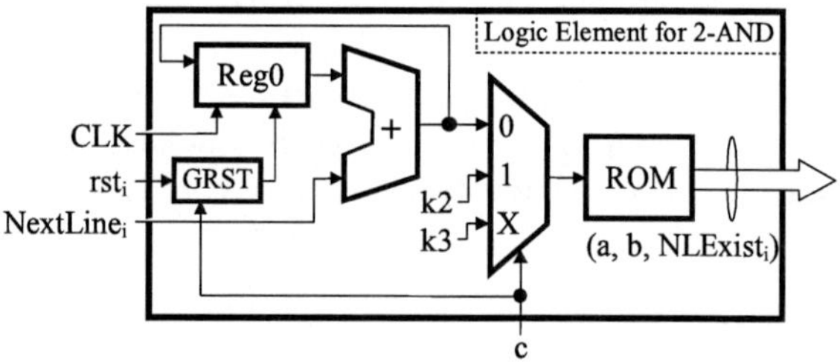

Fig. 4 A reverse implication table (RIT) of the 2-AND gate and the relevant Logic Element

3.4 Fan-Out Stem Transformation

∩	0	1	x
0	0	*err*	0
1	*err*	1	1
x	0	1	x

Each fan-out has to be replaced by a *Cond* block, performing a guarding function with the binary intersection operator according to the table ∩. The *err* item in the table means that there is a conflict between two logic values (0 and 1)—it is signalised to the controller as a status signal Conflict_i. Note that an input vector generated is valid if and only if no conflict has occurred during solving the Boolean satisfiability.

3.5 The Controller Architecture

The controller architecture is shown in Fig. 5. The *Conflicts* signal, consisting of the Conflict_i signals from the Cond block, represents a vector of conflicts (see the *err* items in Sect. 3.4). The block *Controlled Nullity* resets the *NLExists* signals that are influenced by the *Conflicts* signal. The function of the other blocks is comprehensible from their names.

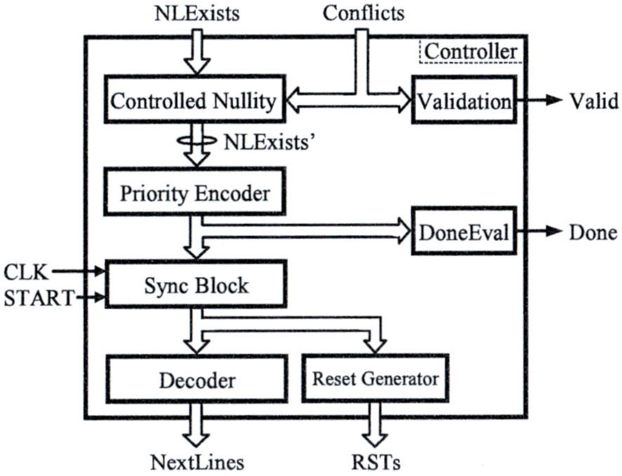

Fig. 5 Controller architecture

4 Experimental Results

The transformation method was applied on the ISCAS'85 benchmarks. The experimental results are presented in Table 1. Its Column 3 shows an area overhead in the thousands of equivalent gates (GE) for an original benchmark. Columns 4 up to 7 show an area overhead in the thousands of GE, a clock speed, an average number of clock cycles and an average time for producing the first valid input vector (VIV) by the reversible circuits for the benchmarks. To calculate the average number of clock cycles, all of the three-value-logic output vectors were applied and the first occurrence of a valid input vector, if any, for each output one was observed.

The average time for the first VIV to occur can be simply calculated. It takes $\frac{10^{-6}}{CLK\_speed} * VIV$ seconds. Column 2 shows the number of gates involved in the benchmarks.

To get the area overhead, CLK and VIV features of reversible circuits, our RevCiGen tool automatically transformed the benchmarks into the corresponding reversible circuits, described in VHDL'93. As a target platform for getting the area overhead and CLK features, the Xilinx FPGA of the Spartan 2E series, namely xc2s600e, was utilised and all the circuits were synthesised in Xilinx ISE of version 14.7. The VIV features were found out by simulation in Mentor's ModelSim of version 10.5b.

The values in Table 1 present data for the graphs below. As shown in Figs. 6 and 7, the complexities of the area overhead and the average time to produce the first valid input vector for the reversible circuits to the benchmarks are approximately polynomial of the sixth order. Figure 8 shows an exponentially falling curve of the maximal clock (CLK) frequency of the reversible circuits. The exponential dependence is given mainly by the logic included in the Controlled Nullity and Priority Encoder

Table 1 Results for the reversible circuits with the ISCAS'85 benchmarks

Benchmarks			Reversible circuits			
Name	#Gates	a-o in #GE (10^3)	a-o in #GE (10^3)	CLK speed (MHz)	avg_#cycles for 1st VIV	avg_time for 1st VIV (μs)
c17	6	<1	1	82	2	<0.1
c432	160	3	28	12	48	4
c499	202	4	34	10	32	3
c880	383	5	60	6	67	11
c1355	546	4	76	4	139	30
c1908	880	3	99	3	207	68
c2670	1193	12	158	2	244	107
c3540	1669	6	211	1	311	188
c5315	2307	17	296	1	432	359
c6288	2406	7	329	1	780	675
c7552	3512	18	406	< 1	697	872

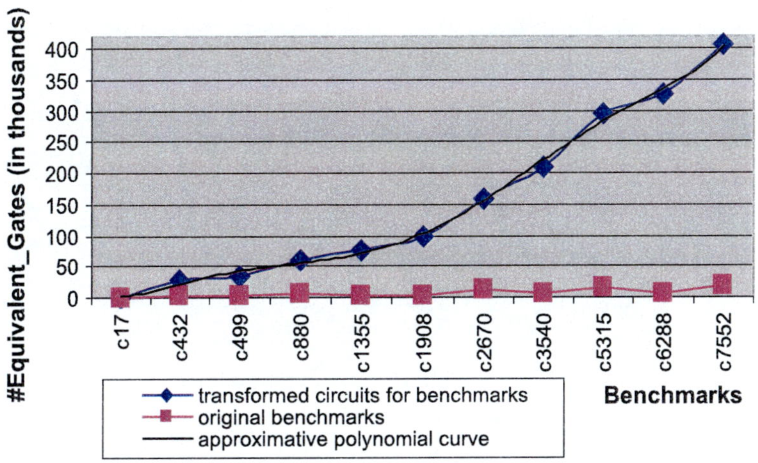

Fig. 6 Area overhead for the benchmarks and their reversible circuits

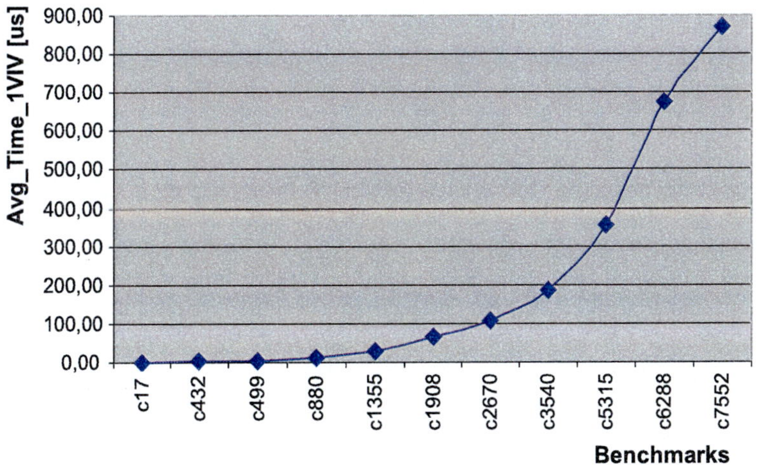

Fig. 7 Average time to produce the first valid input vector by the reversible circuits to the benchmarks

based on priority evaluation, thus remarkably lengthening the critical combinational path (i.e. the maximal delay).

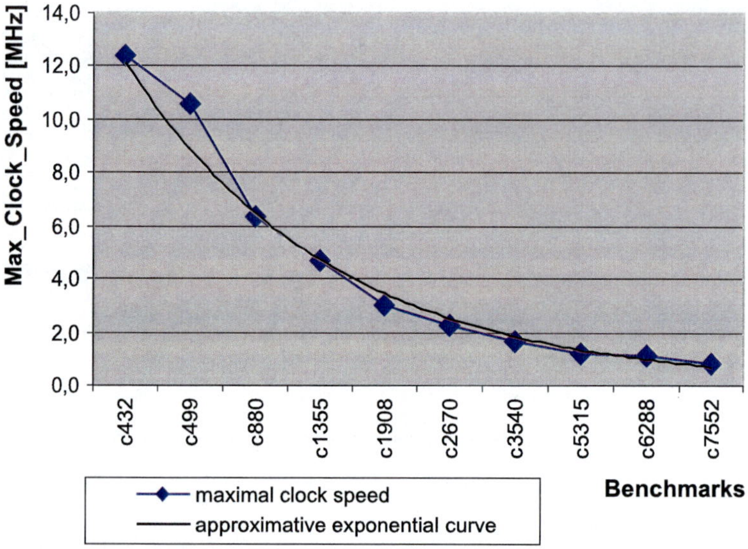

Fig. 8 Maximal clock speed for the reversible circuits to the benchmarks

5 Conclusions

We introduced the methodology of synthesising the reversible digital circuits that solve the Boolean satisfiability problem (SAT). The reversible circuits can be used in various applications; for example, a backtracking accelerator co-operating with a software automatic test pattern generation (ATPG) tool, a computing unit for compacted test generation for sequential circuits designed for testability, accelerating SAT solvers, generating input vectors of a circuit from output ones, etc.

The complexities of the area overhead as well as the computing time for the circuits reversible to the set of the ISCAS'85 benchmarks are polynomial of the sixth order. It can be further generally assumed that this dependence trend continues with the increasing complexity of straightforward combinational circuits (in the number of gates).

We succeeded in proposing a methodology of synthesising the reversible circuits, describing them in VHDL'93, implementing them into an FPGA, and finding out their features. Unlike the other SAT solvers, the reversible circuits according to the proposed methodology are able to produce not only one solution but also all possible solutions; i.e. for an output vector of a straightforward combinational circuit to find all the input vectors leading to the given output one.

Table 2 sums up and compares the properties of representative SAT solvers and our reversible circuits. The first representative [10] uses configurable hardware to speed up Boolean satisfiability with a direct mapping of Boolean operations to logic

Table 2 Comparison of SAT solvers

Property	SAT solvers [10, 12]	Our reversible circuits
# Clock cycles to find one solution	Several (mostly thousands)	One
Clock frequency	Low	Low
Supporting XOR or larger blocks	No	Yes
Possible to find all inputs to an output	No	Yes
Area overhead	Middle	Large

gates and the massive parallelism in processing implications. The second representative [12] performs backtracing all goals along all possible lines and the concurrent assignments of a few variables in a massively parallel way.

In our future work, we shall focus on the reduction of the asymptotically exponential time complexity of reversible circuits and on a stream processing of SAT, which will result in getting clock frequency higher. In general, the high-performance data processing is under our actual research activity.

References

1. Montanaro A (2018) Quantum-Walk speedup of backtracking algorithms. Theory Comput 14(1):1–24
2. Venkatasubramanian M, Agrawal VD (2016) Failures guide probabilistic search for a hard-to-find test. In: Proceedings of 25th IEEE North Atlantic Test Workshop, pp 18–23
3. Wang C, Reddy SM, Pomeranz I, XJ, Rajski J (2002) Conflict driven techniques for improving deterministic test pattern generation. Digest of Technical Papers of IEEE/ACM international conference on CAD-02, pp 87–93
4. Roth JP (1966) Diagnosis of automata failures: a calculus and a method. IBM J Res Dev 10:278–291
5. Goel P (1981) An implicit enumeration algorithm to generate tests for combinational logic circuits. IEEE Trans Comput 215–222
6. Abramovici M, Breuer MA, Friedman AD (1990) Digital systems testing and testable design. IEEE Press, 652 p
7. Fujiwara H, Shimono T (1983) On the acceleration of test generation algorithms. IEEE Trans Comput:1137–1144
8. Hamzaoglu I, Patel JH (1998) New techniques for deterministic test pattern generation. In: Proceedings of IEEE VLSI test symposium, pp 446–452
9. Tsai KH, Tompson R, Rajski J et al (1999) STAR-ATPG: a high speed test pattern generator for large scan designs. In: Proceedings of International Test Conferrence, pp 1021–1030
10. Zhong P, Ashar P, Malik S, Martonosi M (1998) Using reconfigurable computing techniques to accelerate problems in the CAD domain: a case study with Boolean satisfiability. In: Proceedings of the design automation conference, pp 194–199
11. Zhang L, Malik S (2002) Conflict driven learning in a quantified Boolean satisfiability solver. In: Proceedings of International conference on computer aided design, pp 442–449

12. Abramovici M, de Sousa JT, Saab D (1999) A massively-parallel easily-scalable satisfiability solver using reconfigurable hardware. In: Proceedings of the conference on design automation, pp 684–690
13. Li JW, Pu GG, Zhang LJ, Vardi MY, He JF (2018) Accelerating LTL satisfiability checking by SAT solvers. J Logic Comput 28(6):1011–1030
14. Larrabee T (1992) Test pattern generation using boolean satisfiability. IEEE Trans Comput Aided Design 4–15
15. Stava M, Novak O (2006) HW implementation of the Backtrace algorithm with conflict-driven dynamic reconfiguration. In: Proceedings of workshop on design and diagnostics of electronic circuits and systems, pp 250–252
16. Bushnell ML, Agrawal VD (2000) Essentials of electronic testing for digital, memory and mixed-signal VLSI circuits. Kluwer Academic Publishers, 690 p

An Improved Method for Detection of Laryngeal Cancer and Its Stages

T. M. Inbamalar, T. Jayashre, D. Jayashre, and V. S. Jayakrishnan

Abstract According to the fact sheet of GLOBOCAN 2018, the number of people suffering from oral cancer is increased to 1,19,992 in 2018 from 56,000 cases in 2012 which is about 11.4% rise in 6 years. It is a known fact that if laryngeal cancer can be detected in the preliminary stages, probability of cure is more and survival rate is also very high. When a person is diagnosed having cancer, it is vital to identify the cancer stages which will be useful for the treatment to be carried out. In this paper, an improved method has been presented for detection of laryngeal cancer and its stages. Here, PET scan image is used as input and image processing concepts have been employed to detect cancer and its stages. MATLAB software has been used for implementation. The proposed method has been tested on various inputs, and it has been proved that this method has improved performance compared to the methods in the literature.

Keywords Cancer · Digital image processing · Laryngectomy · PET/CT image · Thresholding

T. M. Inbamalar (✉) · T. Jayashre · D. Jayashre · V. S. Jayakrishnan
Electronics and Communication Engineering, RMK College of Engineering and Technology, Chennai, India
e-mail: inbamalarece@rmkcet.ac.in

T. Jayashre
e-mail: jaya15055.ec@rmkcet.ac.in

D. Jayashre
e-mail: jaya15056.ec@rmkcet.ac.in

V. S. Jayakrishnan
e-mail: jaya15054.ec@rmkcet.ac.in

V. Nath and J. K. Mandal (eds.), *Proceedings of the Fourth International Conference on Microelectronics, Computing and Communication Systems*, Lecture Notes in Electrical Engineering 673, https://doi.org/10.1007/978-981-15-5546-6_20

1 Introduction

Oral cancer accounts for around 30% of all cancers in India. New Global Cancer Data: GLOBOCAN during the year 2018 by IARC at Geneva, Switzerland, estimate 1,19,992 new throat cancer cases in India. It is expected to double in the year 2030. Laryngeal cancer is the ninth and seventh most common cause of cancer in males in Asia and India, respectively. Especially, in India, laryngeal cancer contributes to approximately 3–6% of all cancers in men. In 2012, 17,560 Indians lost their lives from laryngeal cancer. Throat cancer refers to abnormal cancer cells present in the voice box, the vocal cords and other parts of the throat, such as the tonsils and oropharynx. It is often grouped into two categories: pharyngeal cancer and laryngeal cancer. Here, in our proposed work, only laryngeal cancers have been considered.

Around 7 lakh people in this world, i.e., approximately 0.01%, suffer from laryngeal cancer, and around 2 lakh people suffer death due to laryngeal cancer. Though the percentage of people in this world having laryngeal cancer may appear less, their impact is more. The people having last stage of laryngeal cancer have the impact of removing the larynx. Normally, such persons undergo a surgery named laryngectomy. They lose their speech after surgery and suffer from various humiliations, suffocations, health issues related to breathing and also different challenges. These people strive to attain rehabilitation. To provide remedy for such people, this work has been initiated. Here, the objective is providing an algorithm for detection of laryngeal cancer which can mention about stages also after examination. When the cancer stage is 0 or 1, then the cancer is benign and it can be cured via medicines. But if the stage is 2, 3 or 4, then it is said to be malignant and at the last stage, it is necessary to remove the larynx. Hence, only when stages are known, further treatment can be done effectively. Thus, an algorithm has been proposed to identify the cancer stages in a simple and easy way.

2 Review of Related Works

There are various methods available, contributed by various scientists for the diagnosis of laryngeal malignancy. Works have been done before recurrence surgeries. Such methods are described in this section.

Becker et al. [1] have done performance analysis of FDG-PET/DWIMRI for identification of carcinoma by squamous cells in skull and neckline. The person undergoes hybrid PET/MRI test with morphological MRI, DWI and FDG-PET. In this, the performance parameters are reported as sensitivity—93.0%, specificity—93.5% and PPV—90.9% per lesion. Another scientist Allegra et al. [2] used PET/CT methods for the laryngeal cancer analysis. They identified neighborhood cancer recurrence subsequent to surgical treatment. Accuracy is a parameter used for analysis and has been evaluated in doubtful patients for repetition. From the results, it has been identified that PET/CT imaging is a valuable means for alleged reappearance in laryngeal tumors where traditional image processing methods are imprecise. Moreover, the disease discovery has been done after radiotherapy. So, clinical procedures are also involved in this technique. The performance metrics such as specificity, sensitivity and accuracy of this method has been reported as 88%, 100% and 93.3%, respectively. Corina et al. [3] proposed a technique to identify or locate the lesion area. Also, segmentation of blood vessels is included in the method. It uses anisotropic as well as matched filter. In this method, lesion has been categorized based on numerical investigation of thickness, tortuosity and density of blood. Based on the experimental results, the accuracy has been reported as 84.3%. Kraft et al. [4] used image processing in narrowband endoscopic image for the diagnosis of malignant laryngeal cells. This system produced same value of 97% for sensitivity as well as accuracy and specificity value 96%. Watanabe et al. [5] used imaging techniques in narrowband in the analysis of laryngeal lesions. Narrowband image (NBI) is used as inputs, and the abnormal intra-epithelial micro-vessel changes were considered. The diagnostic criterion of lesion was the occurrence of isolated area in brown color or spots sprinkled in brown color. Sensitivity and specificity were 91.3% and 91.6%, respectively. But, the cost is very high due to the use of filters, processors and the endoscopic equipment. Geets et al. [6] described a scheme for analyzing FDG-PET descriptions using watershed transform and hierarchical cluster analysis. Here, gradient and thresholding are used for segmentation. Volumes and radii of the spheres are calculated, and later, they are compared for mismatches. But, the major disadvantage is that the accuracy of the source-to-background ratio is uncertain and also it is a time-consuming process. Schwartz et al. [7] scrutinized the application of 18-F-FDG-positron emission tomography for staging of the carcinoma cells in head and neck prior to radiotherapy. FDG-PET/CT descriptions were shaped using image registration algorithm. But, the accuracy is uncertain as the contrast among the blood vessels is poor and also, the cancer lesions in image obtained are less visible.

From the literature, it is understood that there are methods to detect laryngeal cancer before recurrence surgeries. But there are no methods for the laryngeal cancer stage identification. And the existing methods involve complex clinical assessment techniques.

3 Proposed Innovation

In the proposed system, laryngeal cancer in a person is to be identified using PET scan images as input. Thus, the first step is the acquisition of PET scan images of the larynx. PET scans are generally color images with the cancerous part of the larynx bright than the other parts. Hence, the color image is converted to grayscale counterpart. Then, it is again transformed to binary image. This is done by set value which is fixed for segmenting only the brightest cancer part using thresholding concept. After this, the cancer part is separated from the other areas where the white pixels correspond to the cancer cells and the black pixels correspond to the other larynx part. Then, the number of white pixels (cancer part) is counted and based on this count the cancer

Table 1 Relation between size, pixel and stages

Size (cm)	Pixel count	Stage
$0 < \text{size} \le 1$	$0 < \text{pixels} \le 30$	0
$1 < \text{size} \le 2$	$30 < \text{pixels} \le 119$	1
$2 < \text{size} \le 4$	$119 < \text{pixels} \le 475$	2
$4 < \text{size} \le 6$	$475 \le \text{pLiels} \le 1069$	3
Size > 6	pixels > 1069	4

and its stages are detected which is done as follows. Every stage of laryngeal cancer can be classified using the size of the cancer cells. Though cancer stage is classified based on the spreading of cancer to the lymph nodes and the adjacent organs, it is also evident that in every stage of cancer, the dimension of the cancer cells varies and they are standardized for each stage. Hence, size of the tumor has been considered as a feature and feature extraction is done. For the feature extraction, the dimension of the cancer cell is determined. This determined dimension is compared with the already fixed standard size of each stage by which each stage is identified and displayed as output. Feature extraction, i.e., the determination of size of cancer part in the larynx, is done by counting the white pixels. But, the already set values of the cancer size are in centimeters and here measurement has been done in terms of pixels since conversion between centimeters and pixels can be done. Table 1 shows the standard values of each stage and their corresponding pixel count. The flow diagram of the technique to be implemented is shown in Fig. 1.

4 Results

The proposed method has been tested on various inputs, and results are given in Table 2. A sample output is shown in Fig. 2. The figure shows original PET input image with laryngeal cancer shown in the left and the segmented cancer cell part of the PET scan displayed on the right. From the results obtained, the standard performance metrics that are useful in comparative studies and validation of the proposed schemes have been evaluated. The values obtained for true positive ($\mathbf{T}_+$), false negative ($\mathbf{F}_-$), true negative ($\mathbf{T}_-$) and false positive ($\mathbf{F}_+$) are given in Table 3. With the aim of ascertaining the supremacy of the above-explained method, the evaluation parameters such as accuracy, sensitivity, specificity and positive precision are evaluated and the values obtained are given in Table 4.

To establish the validity of the proposed method, the results obtained are compared to the existing methods in the literature. Table 5 shows the comparison of accuracy of the proposed scheme with the existing schemes in the literature.

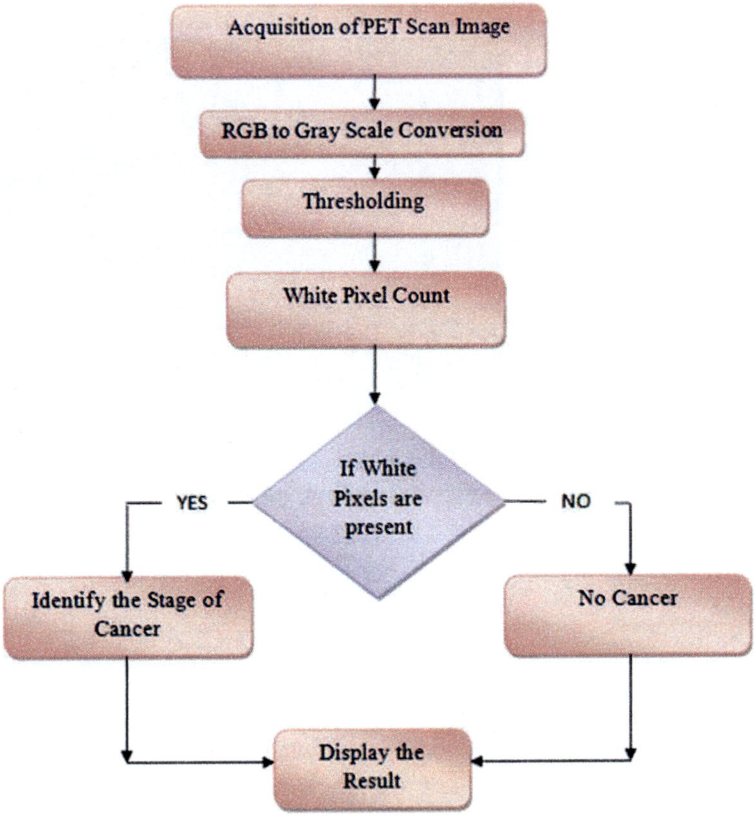

Fig. 1 Flowchart for the proposed algorithm

Table 2 Inputs and test results

Input	Actual stage	Test results
Sample 1	Stage 3	Stage 3
Sample 2	Stage 3	Stage 3
Sample 3	Stage 2	Stage 2
Sample 4	Stage 4	Stage 4
Sample 5	*Stage 0*	*No cancer*
Sample 6	Stage 1	Stage 1
Sample 7	Stage 4	Stage 4
Sample 8	Stage 2	Stage 2
Sample 9	No cancer	No cancer
Sample 10	No cancer	No cancer

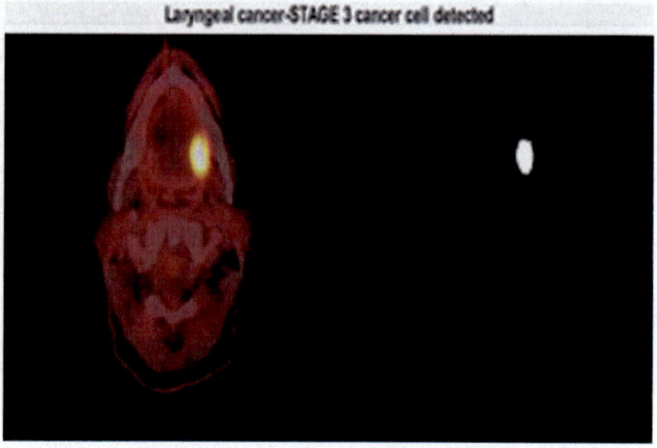

Fig. 2 Sample input and output of laryngeal cancer PET image

Table 3 T_+, T_-, F_+ and F_- values obtained

T_+	T_-	F_+	F_-
7	2	0	1

Table 4 Parameters obtained for the proposed method

Sn (%)	Sp (%)	Ac (%)	Pp (%)
87.5	100	90	100

Table 5 Comparison of parameters

Method	Sn (%)	Sp (%)	Ac (%)	Pp (%)
Proposed	87.5	100	90	100
Becker et al.	93	93.5	–	90.9
Allegra et al.	88	100	93.3	–
Barbalata et al.	–	–	84.3	–
Kraft et al.	97	96	97	–
Watanabe et al.	91.3	91.6	–	–

5 Conclusion

Thus, a method to detect laryngeal cancer and its stages has been proposed, tested and validated. This method will have a great response because of its improved performance. As the identification of the stages of laryngeal cancer can be done at a faster rate with greater accuracy, the further consultation for the laryngeal cancer stage detection is not required and also the treatment can be taken quickly without any delay. The proposed method helps the medical practitioners to learn more about cancer. Oncologists will then be able to design treatment strategies targeted directly to each specific stage of cancer.

References

1. Becker M, Varoquaux AD, Combescure C, Rager O, Pusztaszeri M, Burkhardt K, Delattre BMA, Dulguerov P, Dulguerov N, Katirtzidou E, Caparrotti F, Ratib O, Zaidi H, Becker CD (2018) Local recurrence of squamous cell carcinoma of the head and neck after radio(chemo)therapy: Diagnostic performance of FDG-PET/MRI with diffusion-weighted sequences. Eur Radiol 28(2):651–663
2. Allegra E, Saita V, De Natale M, Marino N, Trapasso S, Tamburrini S, Alessio Caterina, Ippolito M (2017) Use of PET/CT to detect local and regional laryngeal cancer recurrence after surgery. Rep Med Imaging 10:31–36
3. Barbalata C, Mattos LS (2016) Laryngeal tumor detection and classification in endoscopic video. IEEE J Biomed Health Inf 20(1):222–232
4. Kraft M, Fostiropoulos K, Gürtler N, Arnoux A, Davaris N, Arens C (2016) Value of narrow band imaging in the early diagnosis of laryngeal cancer. Head Neck 38(1):15–20
5. Watanabe A, Taniguchi M, Tsujie H, Hosokawa M, Fujita M, Sasaki S (2009) The value of narrow band imaging for early detection of laryngeal cancer. Eur Arch Otorhinolaryngol 266(7):1017–123
6. Geets X, Lee JA, Bol A, Lonneux M, Grégoire V (2007) A gradient-based method for segmenting FDG-PET images: methodology and validation. Eur J Nucl Med Mol Imaging 34(9):1427–38
7. Schwartz DL, Ford E, Rajendran J, Yueh B, Coltrera MD, Virgin J, Anzai Y, Haynor D, Lewellyn B, Mattes D, Meyer J, Phillips M, Leblanc M, Kinahan P, Krohn K, Eary J, Laramore GE (2005) FDG-PET/CT imaging for preradiotherapy staging of head-and-neck squamous cell carcinoma. Int J Radiat Oncol Biol Phys 61(1):129–136

IoT-Based Secure Communication to Enhance Blockchain Model

Nayancy, Sandip Dutta, and Soubhik Chakraborty

Abstract Internet of things (IoT) privacy and security remain a most important, challenging factor in this era, mostly due to distributed nature and immense scale of IoT networks. The blockchain technology provides decentralized security, public digital ledger and yet occupies significant delay, computational overhead, and high energy that is not fit for low-resource IoT devices. In this paper, our proposed work will analyze possibilities for enhancement of the blockchain in IoT without sacrificing security. However, the traditional blockchain technology cannot be applied to IoT devices with limited computing resource and network resources. Hence, we will solve the verification problem by generating random number to verify the correct transaction in the blockchain by using Keccak-256 hash function in IoT.

Keywords Internet of things (IoT) · Blockchain · Hash function · Security

1 Introduction

Internet of things (IoT) is an entity used to attach many devices in network to transfer and receive immense amounts of secure data as well as privacy-perceptive information through the Internet which is practiced in different applications, namely sensor network, health care, embedded systems, artificial intelligence, and RFID [1]. Nowadays, IoT is the most important development field in the area of information technology [2]. The fast growth in the study of IoT causes the appearance of different

Nayancy (✉) · S. Dutta
Department of Computer Science and Engineering, Birla Institute of Technology, Mesra, Ranchi, Jharkhand, India
e-mail: nayancy1992@gmail.com

S. Dutta
e-mail: sandipdutta@bitmesra.ac.in

S. Chakraborty
Department of Mathematics, Birla Institute of Technology, Mesra, Ranchi, Jharkhand, India
e-mail: soubhikc@yahoo.co.in

V. Nath and J. K. Mandal (eds.), *Proceedings of the Fourth International Conference on Microelectronics, Computing and Communication Systems*, Lecture Notes in Electrical Engineering 673, https://doi.org/10.1007/978-981-15-5546-6_21

problems; one of the most important targets was the exposure to various cyberattacks [2]. Sicari et al. show the challenges and their active solutions in the area of IoT security and privacy, identify open issues with signifying few hints for further research. According to [3] to formulate IoT devices stay secure be a challenging thing to make out as various security policies are contravene to construct IoT devices, low-cost, comfortable to use, and small in size [4]. One of the ways to strengthen IoT security can be done by using blockchain technology [3–5].

Nowadays, blockchain technology is rising in haste and is progressively more being used as in term of bitcoin, and it was introduced by Santoshi Nakamoto [6]. Initially, concept of blockchain was only used in transaction to avoid double spending by Santoshi Nakamoto. After a while, blockchain technology is used for several purposes, namely in the area of IoT, transportation, health care, and other fields.

The blockchain technology has the features of decentralization, tamperproof, and high security, which enables all users to recognize value interconnection and communication at a very low cost. Dorri et al. [7] showed that adopting blockchain in IoT was not straightforward and entails various major challenges that are low latency for the transaction confirmation, large resource demand for solving the proof-of-work (PoW) and small scalability that is the result of the broadcasting transactions, and all the existing block to the entire network. In this paper, we will perform research on implementation of blockchain for IoT security. We will solve the verification problem by generating random numbers to verify the correct transaction in the blockchain by using Keccak-256 hash function in IoT.

The remaining paper is planned as follows: In Sect. 2, we briefly introduced the related works, while Sect. 3 presents the methodology of the proposed enhanced blockchain technology in it. Section 4 describes the result of the proposed design. Section 5 concludes the work of this paper and summarizes our future work.

2 Related Work

In order to construct an efficient blockchain in IoT for intelligent devices, in this paper, we summarized the related research into the following perspectives.

2.1 Blockchain

The concept of blockchain was initially introduced by Satoshi Nakamoto [6] through bitcoin. In the beginning, bitcoin was created with blockchain structure to avoid the problem of double spending. However, in this era, blockchain technology is being used for several purposes such as IoT, embedded system, and health care.

Blockchain is a decentralized computation and information/data sharing platform that enables various authoritative domains, without trust on each other, to coordinate, collaborate, and cooperate in the rational decision-making process. Blockchain is

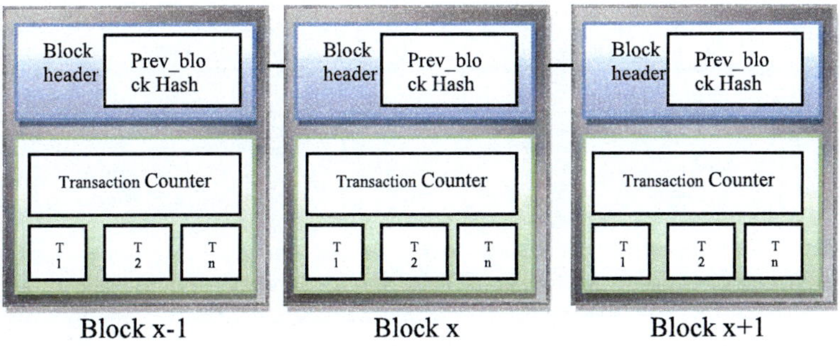

Fig. 1 Blockchain

an open distributed public ledger to protect recorded transactions in the untrusted network environments [8]. Blockchain is a collection of blocks arranged in a linear sequence, where every block consists of timestamp and transactions. In blockchain, every block is linked with the preceding block and contains a hash value from the preceding block. Hence, the transaction record stored in the blockchain cannot be altered or removed without totally changing the content of the blockchain [9]. This is the reason that causes blockchain which is secure from intruder/hackers.

Figure 1 shows the structure of blockchain. In Fig. 2, a block is divided into two parts, namely (a) a block header with metadata and (b) set of associated transactions.

The several characteristics of blockchain are as follows:

- ***Decentralization***: No need to verify the transaction that means it does not depend on the central authority. Consent algorithms require to retaining the data reliability on the blockchain networks.
- ***Anonymity***: Each person in the blockchain network can communicate among each other with the help of generating address. Hence, user actual identity is not shown in the network during communication.
- ***Persistency***: Transactions are easily validated, and invalid transactions will be removed by miners in the blockchain. Therefore, it is challenging to erase the existing transactions from the blockchain.
- ***Auditability***: In blockchain, every transaction can be easily verified and tracked because every transaction will refer to the previous block transaction.

2.2 Hash Function in Blockchain

The hash function is a cryptographic method to map the data of random length to a fixed-length string. The main few security requirements that are one way and collision resistance are generally essential for hash functions. The former ensures that it is very difficult to find same hash value of two inputs. With n-bits length output for a hash function, the complexities of flouting one way and finding the collision are correspondingly bounded by Ó($2^{n/2}$) birthday attack and Ó(2^n) brute force attack.

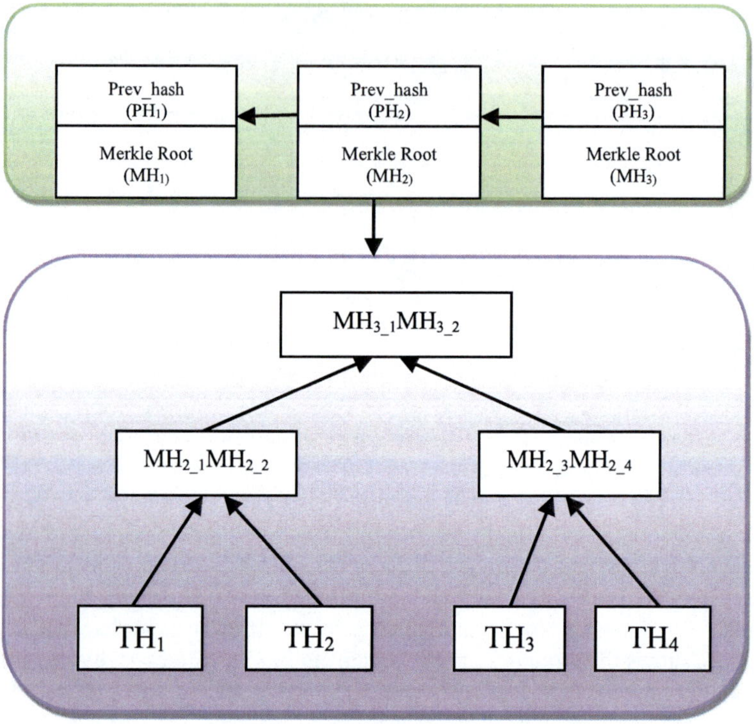

Fig. 2 Representation of datastore and associated block

Hence, for ensuring at least 80-bits security, the output size of hash function must be at least 160-bits.

SHA-256 cryptographic hashing algorithms are mainly applied in the blockchain, and SHA-256 is a family of hash functions called secure hash algorithms (SHA). SHA is a US Federal Information Processing Standard (FIPS), and algorithms which belong to the SHA family including SHA0 (proposed in 1993), SHA1 (proposed in 1995), and SHA2 (proposed in 2001) are published by the US National Security Agency (NSA). Though Bertoni et al. [10] proposed SHA3 (published in 2014), the padding function is customized by the National Institute of Standards and Technology (NIST). According to the requirements of the existing security, SHA2 and SHA3 are suggested for using in cryptocurrencies and blockchains.

2.3 IoT–Blockchain

According to Reyna et al. [11], in IoT–blockchain, all transactions get through blockchain, enabling immutable records of interactions. This approach makes sure that all the selected interactions are distinguishable as their details can be queried in

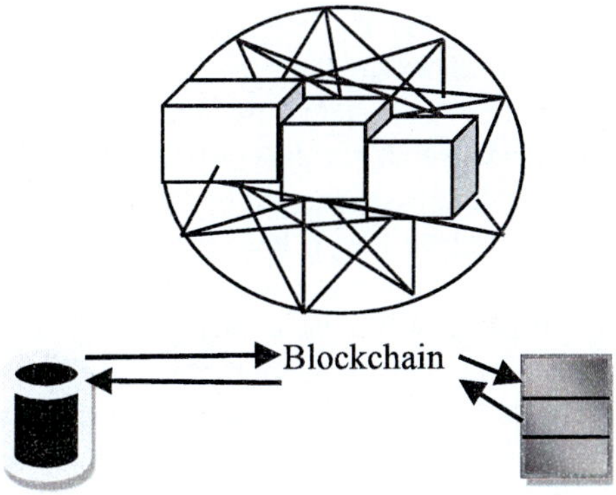

Fig. 3 IoT–blockchain

blockchain, and also, it increases the autonomy of the IoT devices. Recording every valid interaction in blockchain would engage an increase in data and bandwidth, which is one of the main known challenges in blockchain (Fig. 3). Also, all data on IoT related to these transactions should also be stored in the blockchain.

2.4 Security in IoT

Internet of things (IoT) is an entity used to tie many devices in the network to transfer data/information through the Internet which is practiced in different applications using network sensors, embedded systems, artificial intelligence, and RFID [1]. Security into IoT is challenging because of the limited resource capability of the huge amount of devices, heterogeneity between the ubiquitous computing devices, lack of standardization, and immense scale. Moreover, various IoT devices share and bring a very huge quantity of data from private spaces, thus opening up major privacy concerns. To defend user's data privacy, the researchers in [12] proposed an ample range of off-the-shelf low-resource devices require primary security considerations.

In summary, regardless of the current proposals for providing the privacy and security in IoT, few challenges still require to be addressed:

- ***Privacy***: Protecting the client confidentiality while edifying various types of information.
- ***Resource optimization***: In IoT, low-resource devices are not appropriate used for complex and high level of security methods.
- ***Centralization***: Centralized methods are unsuitable for IoT and get the challenges many-to-one traffic, single point of failure, and reduced scalability.

3 Proposed Work

Blockchain technology uses the concept of digital signature for validation purpose. The sender sends the data or transaction details with his secrete key, and the receiver receives data with his public key, and the user needs to know the digital signature and the private key, if he wants to access the data. In traditional blockchain, every block has the previous hash value only, while in this paper, random number that is initial value is used as a unique key for the verification process. Keccak-256 is used as the lightweight cryptographic hash function because it supports IoT device.

In this part, we will show the design of enhanced blockchain technology, so as to resolve the problem of trust among the user and valuators. Figure 4 shows the method of enhanced blockchain technology. In this method, transaction T_1 is hashed using Keccak-256 and then that hash value is XOR with random number. The output of this is called the previous hash value that is used in the next block, and after nth transactions, the final hash value is generated.

3.1 Verification

The transactions must be "validate" in order to imitate in a public ledger. In the verification process, verifier takes the final value and transactions record. With the help of final value and transactions record, verifier generates initial value and if the initial value is same as the final value that means transactions in the blockchain is verified and there are no changes occurred.

Figure 5 shows the process of the block structure. Block N is the present block of the blockchain of ongoing communication. The present block contains the previous

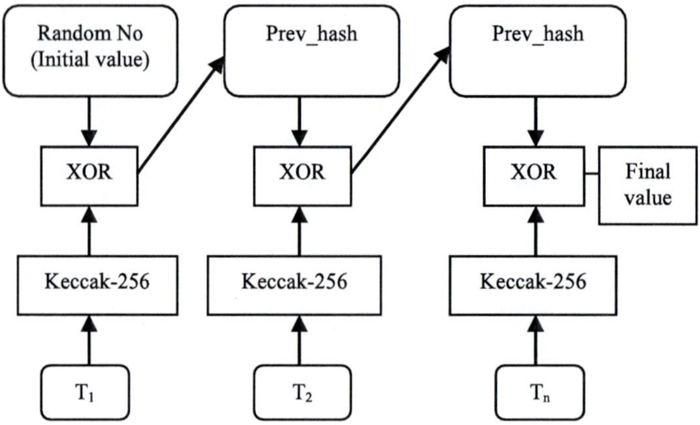

Fig. 4 Block diagram of enhanced blockchain method

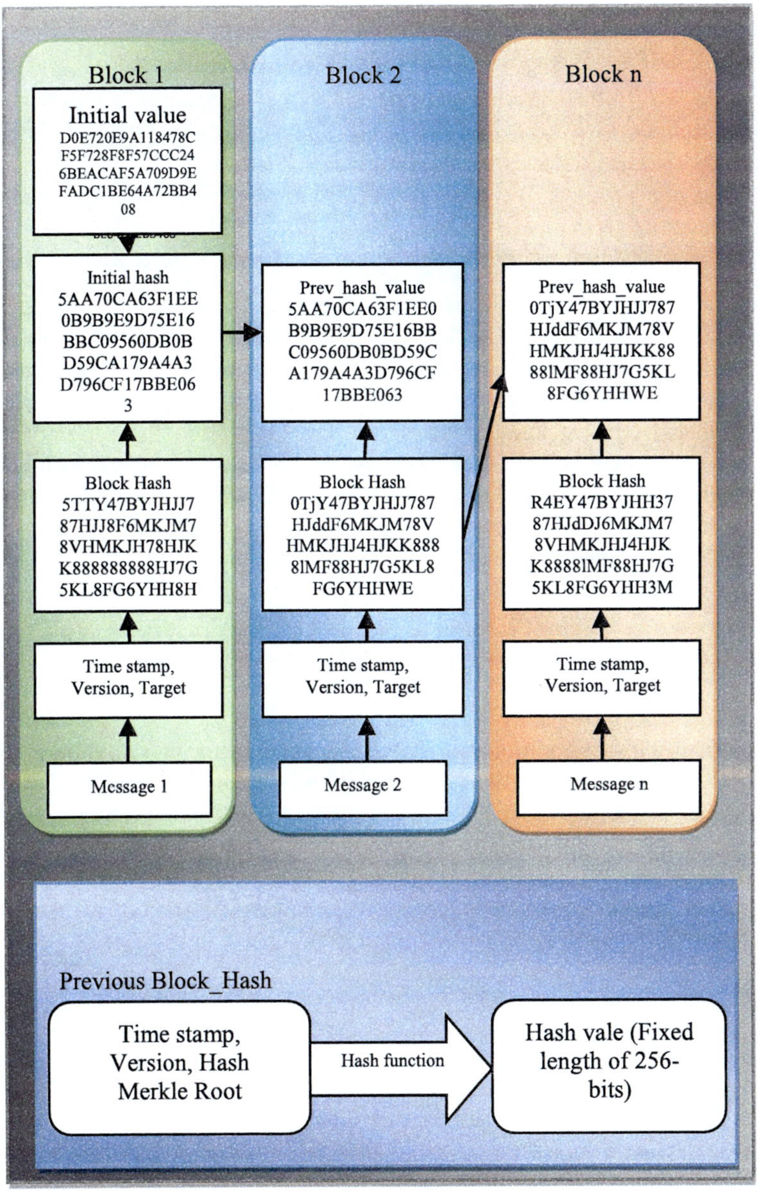

Fig. 5 Process of block structure

hash value, the block hash, time stamp, target address, and block version including messages.

- ***Random No***: Initial value of the first block that is used as a unique key for verification.
- ***Block hash***: The hash value of the current block's message.
- ***Previous block hashes***: It can be one or more if the blockchain is split and all are appended.
- ***Timestamp***: The timestamp is the time at which block was generated.

Each block contains a hash value (unique identifier), time-stamped batches of the current valid transactions, and hash value (Prev_hash) of the preceding block. The hash of the preceding block connects the blocks simultaneously and prevents every block from being changed or a new transaction/block being added in between two existing blocks. In this technique, each succeeding block strengthens the verification of the previous block and thus the complete blockchain. The blockchain contains verified proof of transactions. However, while enhanced blockchain basically serves as a secure database for the transactions, its benefits expand far beyond than the traditional blockchain.

4 Results

In comparison with the traditional blockchain technology, the most favorable motivation of utilizing random number as initial value and Keccak-256 hashing algorithm is that it is extensively more adaptable for IoT devices. The enhanced blockchain technology advocates the idyllic decentralized architecture for secure transaction transfer using IoT devices. The information on every transaction is added to the enhanced blockchain by the random number and mining technology. The proof-of-work (PoW) insists the high protection of the blockchain itself.

According to Fig. 6, it represents that the hash function Keccak-256 has better avalanche effect as compared to SHA-256 and SHA-128. Therefore, Keccak-256 has a high security level than the SHA-256 and SHA-128.

5 Conclusion

In this paper, we proposed the analyzed possibilities designed for enhancement of blockchain in IoT without sacrificing security. One of the major problems in blockchain technology is verification processes that slow down the transactions. In this work, we will solve the verification problem by generating random numbers to verify the correct transaction in the blockchain.

In the future, we will work on the implementation and deployment of the proposed method for lightweight encryption cryptography using blockchain in IoT device.

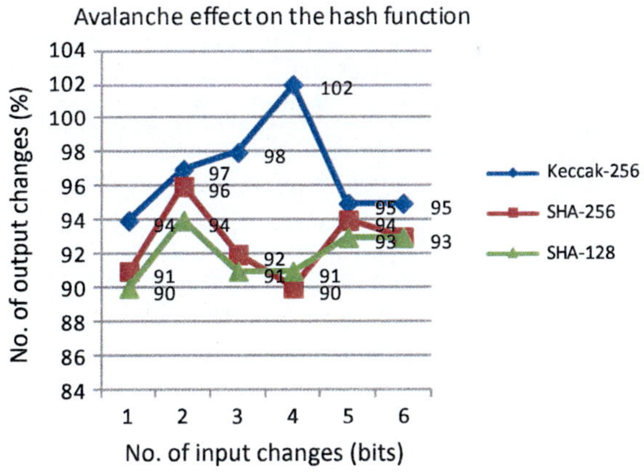

Fig. 6 Avalanche effect

Acknowledgements This work is carried out under the guidance of Dr. Sandip Dutta and Dr. Soubhik Chakraborty is gratefully acknowledged at Birla Institute of Technology, Mesra, Ranchi.

References

1. Madakam S, Ramaswamy R, Tripathi S (2015) Internet of Things (IoT): a literature review. J Comput Commun 3(05):164
2. Yue Z, Sun W, Li P, Ur Rehman M, Yang X (2015) Internet of things: architecture, technology and key problems in implementation. In: 2015 8th International Congress on Image and Signal Processing (CISP), pp 1298–1302. IEEE 2015
3. Singh M, Singh A, Kim S (2018) Blockchain: a game changer for securing IoT data. In: IEEE 4th World Forum on Internet of Things (WF-IoT), pp 51–55. IEEE
4. Sicari S, Rizzardi A, Grieco LA, Coen-Porisini A (2015) Security, privacy and trust in Internet of Things: the road ahead. Comput Netw 76:146–164
5. Conoscenti M, Vetro A, De Martin JC (2016) Blockchain for the Internet of Things: a systematic literature review. In: IEEE/ACS 13th international conference of computer systems and applications (AICCSA), pp 1–6. IEEE
6. Pulkkis G, Karlsson J, Westerlund M (2018) Blockchain-based security solutions for IoT systems. Wiley, USA
7. Dorri A, Kanhere SS, Jurdak R, PGauravaram R (2017) Blockchain for IoT security and privacy: the case study of a smart home. In: IEEE international conference on pervasive computing and communications workshops (PerCom Workshops), pp 618–623. IEEE
8. Hari A, Lakshman TV (2016) The internet blockchain: a distributed, tamper-resistant transaction framework for the internet. In: Proceedings of the 15th ACM workshop on hot topics in networks, pp 204–210. ACM
9. Xu X, Weber I, Staples M, Zhu L, Bosch J, Bass L, Pautasso C, Rimba P (2017) A taxonomy of blockchain-based systems for architecture design. In: 2017 IEEE international conference on software architecture (ICSA), pp 243–252. IEEE
10. Bertoni G, Daemen J, Peeters M, Van Assche G (2009) Keccak sponge function family main document. Submission to NIST (Round 2) 3(30)

11. Reyna Ana, Martín Cristian, Chen Jaime, Soler Enrique, Díaz Manuel (2018) On blockchain and its integration with IoT. Challenges and opportunities. Fut Gener Comput Syst 88:173–190
12. Sivaraman V, Gharakheili HH, Vishwanath A, Boreli R, Mehani O (2015) Network-level security and privacy control for smart-home IoT devices. In: IEEE 11th international conference on wireless and mobile computing, networking and communications (WiMob), pp 163–167. IEEE

Tuberculosis Bacteria Segmentation in Acid Fast Stained Images

Yashwant Kurmi, Vijayshri Chaurasia, Aditya Goel, Deepti Joshi, and Neelkamal Kapoor

Abstract This paper presents a novel method for tuberculosis bacteria segmentation in microscopic images. Manual identification of the bacterial cell is a very difficult process. The automation in tuberculosis bacteria detection is the object of this article using microscopic image processing. In the proposed segmentation method, firstly image enhancement is done followed by the bacteria region masking. Further, the marking of bacteria points is performed by the marked point process model. Finally, the complete bacteria are identified by the superellipse and supervised variational contour models. MATLAB simulation results confirm the superiority of the proposed method as compared with the state of the art methods, on the basis of segmentation accuracy, F_1-score, and Dice similarity coefficient.

1 Introduction

Tuberculosis (TB) identification is a global problem. The MultiDrug-Resistant TB (MDR-TB) made TB a critical health issue [1]. The WHO's report says annually 480,000 new cases of MDR-TB occur globally out of them, and approximately, 9.0% are extensively drug-resistant TB cases [2]. Early diagnosis and effective anti-TB treatment (ATT) is necessary to control this epidemic. The automatic detection process may be used to decrease technician participation in screening for TB and is also very much useful in laboratories of countries which have high burden of TB. TB treatment is directed for a long duration, generally minimum of six months. During this long course of treatment, mostly the specific drugs become resistant into the patient's body for future purpose [3].

Y. Kurmi (✉) · V. Chaurasia · A. Goel
Maulana Azad National Institute of Technology, Bhopal, India
e-mail: yashwantkurmi18@gmail.com

D. Joshi · N. Kapoor
All India Institute of Medical Sciences, Bhopal, India

V. Nath and J. K. Mandal (eds.), *Proceedings of the Fourth International Conference on Microelectronics, Computing and Communication Systems*, Lecture Notes in Electrical Engineering 673, https://doi.org/10.1007/978-981-15-5546-6_22

Many researchers have devoted their valuable concentration to address this domain. Vargasa et al. presented two segmentation methods [4] as chromatic information-based segmentation and segmentation using morphological operation on green channel's gray levels of stained tuberculosis images. The chromatic channel also used by Forero et al. [5] with by shape features characterized using Gaussian mixture model. They have further minimized the segmentation error using Bayesian classification. It reduces the large amount of debris and concluded that the pattern-recognition in image-processing techniques is appropriate tools to improve the manual screening of samples. Khutlang et al. have implemented a two stage classifiers [6]. The first stage is used as a pixel classifier, and the second one is used as an object classifier. The combination of a mixture of Gaussian in both stages performs well as compared to the other combinations.

Li et al. presented an energy framework for object segmentation [7]. It uses intensity distribution and local spatial constraints-based background and object descriptors. Performance of Li's method is governed by initiation of appropriate contour which is very crucial and limits the segmentation capacity. Mozos et al. designed an automatic real-time screening system [8]. They have also used the chromatic information and SVM classifier in the screening system. Mozos's technique offers high accuracy with pretty low dependability. Hence, whatever cases the system is resulting positive outcome they definitely confirm as positive by manual testing. But it may miss some of the TB positive cases and regard them as TB negative cases. Sadanandan et al. in [9] presented segmentation using curvature-based approach. The tracking of bacteria is performed by the Viterbi algorithm followed by post tracking as the refinement process for error removal. This approach is suitable for high contrast images only. The performance of Sadanandan's method is image dependent and not fits in all the datasets of tuberculosis bacteria. Most of the existing methods require human intervention for appropriate segmentation and affects the detection accuracy on the basis of personnel expertise.

This proposal presents an automatic method to segment the bacteria image using proper initialization of the bacteria points. The proposed method consists of the four steps. The first step is image contrast enhancement [9, 10]. The image color space [8] is analyzed [11] in the second step to form a bacteria region mask using Hessian of Laplacian of Gaussian (HLoG). The third step marks the specified bacteria, by applying the marked point process (MPP) model [12] on the outcome second step. In the fourth step, firstly complete bacteria are extracted using superellipse model [9] followed by application supervised variational contour model [13] for separation of overlapped bacteria. The rest of the paper is organized as the second section explains the proposed method. Third section covers the simulation results and discussion. The fourth section concludes the paper with future possibilities.

2 Proposed Method

Let us define an image element as $I(x, y)$ with (x, y) being the spatial coordinates. The objects to be segmented are assumed clearly distinguishable from the background of the image, and their contours form approximately dark long rice shape. Figure 1 shows the flowchart of the proposed method, which segments tuberculosis bacteria in four steps.

The step by step method is explained in the following subsections.

2.1 Contrast Enhancement

The Ziehl Neelsen Sputum smear microscopy images [6] are the considered as input that contain the tuberculosis bacteria cells appear as dark long rice-shaped objects on an illuminating background. This means that any pure intensity-based segmentation will result an incorrect output. To overcome the above problem, we attempted the separation of the principal curves of the intensity surface. Minimal curvature-based regions are separated using differential geometry-based techniques [10].

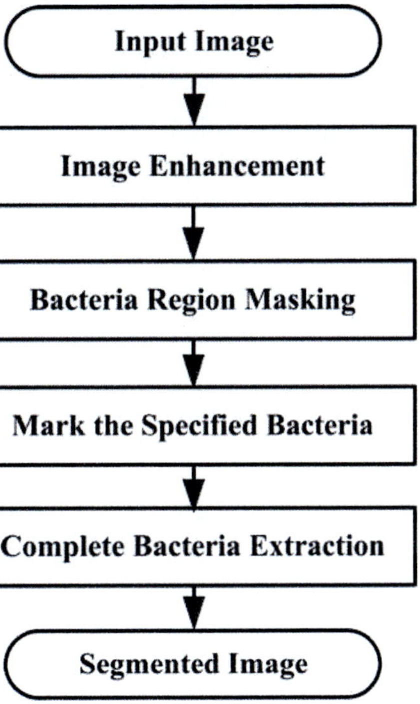

Fig. 1 Flowchart of the proposed method

Consider a grayscale image with spatial coordinates (x, y) in image surface and gray level intensity I. The image surface is supposed to be continuous up to second order of partial derivatives. Here, image smoothening is made first by using the Gaussian kernel. The standard deviation (SD) of Gaussian is set to 1.4 pixels, which is approximately $1/10$th the width of the tuberculosis bacteria cell. The specific point on image surface has an infinite number of curves passing through it. There is an infinite number of curves near the specific point on the image surface. In these curves, two mutually orthogonal curves with maximum curvature and minimum curvature exist. These specified curvatures are indigenous to the Hessian matrix [14]. The second derivative matrix of the image is called Hessian, which is calculated for each pixel.

$$H = \begin{bmatrix} I_{xx} & I_{xy} \\ I_{yx} & I_{yy} \end{bmatrix} \tag{1}$$

where I_{xx} and I_{yy} denote the second derivatives of the image matrix along x- and y-directions, and I_{xy} denotes the derivative of the matrix first along x-direction and then along y-direction, using discrete approximations [15]. The eigenvalues are computed using trace $\mathrm{tr}(H)$ and determinant $|H|$ as follows

$$P_{C1,2} = \frac{\mathrm{tr}(H) \pm \sqrt{\mathrm{tr}(H)^2 - 4 \times |H|}}{2} \tag{2}$$

The notations P_{C1} and P_{C2} are the principal curvatures with $P_{C1} < P_{C2}$. In phase contrast images of tuberculosis bacteria, consider two long rice-shaped cells lying nearby, the cells are dark, and the region between the cells is bright. The cell curvatures are extracted using the orientation of the cell. One curvature is taken as along the major axis of the cell, and the other one is taken orthogonal to the major axis of the cell. Taking the lowest value of the two and apply this value to bacteria enhancement (i.e., lowering the bacteria intensity up to this lowest value) in the image. The enhancement is avoided outside the region of bacteria cells.

2.2 *Bacteria Region Masking*

Images have different color and textural variation in different object regions and background. The color space and the HLoG are described in the following subsections.

2.2.1 Color Space

The acid fast stained images from the various datasets of tuberculosis bacteria are analyzed. It is found that (i) the green color component value of the tuberculosis bacteria is very low as compared to the other constituents of the image. All the color components and their counterparts are computed for the analysis of the image. (ii) The green to red color ratio and (iii) green to intensity ratio [8] are very informative for the bacteria separation in the microscopic images of tuberculosis bacteria, since, these values are very low at the bacteria point's positions.

2.2.2 Hessian-Based Laplacian of Gaussian (HLoG)

The application of the Hessian function is performed at the pre-segmentation stage of the bacteria region. The theoretic base of the Hessian analysis assures that the pre-segmentation recognizes all the bacteria including noise [11].

The scale space depiction $L(x, y; t)$ of the image matrix $I(x, y)$ at the point (x, y) with scale parameter t can be given as

$$L(x, y; t) = I(x, y) * G(x, y; t) \tag{3}$$

where the symbol asterisk (*) denotes the convolution operator and $G(x, y, t)$ is a Gaussian kernel defined as

$$G(x, y; t) = \frac{1}{2\pi t^2} \exp\left(-\frac{x^2 + y^2}{2t^2}\right) \tag{4}$$

The Laplacian of $L(x, y, t)$ is defined in the Expression (5), as given below

$$\nabla^2 L(x, y; t) = \nabla^2 I(x, y) * G(x, y; t) \tag{5}$$

By using the differentiation transforms, the commutes performed through the convolution operator, G can be presented as $\nabla^2 G$, and hence, the Eq. (5) can be rewritten as

$$\nabla^2 L(x, y; t) = \nabla^2 I(x, y) * \nabla^2 G(x, y; t) \tag{6}$$

The Laplacian of Gaussian (LoG) decreases with increase the value of t, and LoG is the maximized without the convolution operation [14]. The LoG optimization is performed using a compensatory normalizing factor α as

$$LoG(x, y; t) = I(x, y) * t^{\alpha} \nabla^2 G(x, y; t) \tag{7}$$

The value of α is tuned to maximize the LoG with respect to the bacteria sizes [7]. Since the bacteria studied in this article have different sizes, the scale parameter value has an essential role. A smaller value of t may cause over-segmentation, whereas for large t, may lead to under segmentation and several small structures may be left unidentified. Hence, optimal t is essential to avoid over as well as under segmentation and identification of different bacteria size in the images. The suitable value of t is chosen as per the bacteria size for efficient extraction of the bacteria. The application of Hessian analysis preserves the geometric shape of the bacteria. The eigenvalues of the Hessian matrix of bacteria structure describe the structure's geometry [11]. The utilization of the LoG provides a smoothness to the image. It is taken at any pixel (x, y) in image $LoG(x, y; t)$ at scale t, and Hessian matrix for that point is given as

$$H(x, y; t) = \begin{pmatrix} \frac{\partial^2 LoG(x,y;t)}{\partial x^2} & \frac{\partial^2 LoG(x,y;t)}{\partial x \partial y} \\ \frac{\partial^2 LoG(x,y;t)}{\partial x \partial y} & \frac{\partial^2 LoG(x,y;t)}{\partial y^2} \end{pmatrix} \tag{8}$$

The pixel geometric organization [16] and specific shapes alignment [17] are taken corresponding to the pixel. Both eigenvalues of $H(x, y; t)$ are negative for concave elliptic pixel (x, y) [11]. These results in a need of transformed bacteria computation since every pixel of transformed-bright bacteria have a negative definite Hessian [11].

Let bacteria T in LoG space is an eight-connected component of set U. This set may be defined as U which is equal to $\{(x, y)|(x, y) \in LoG(x, y, t), I(x, y, t) = 1\}$.

where

$$I(x, y, t) = \begin{cases} 1, & \text{for Negative Definite Hessian} \\ 0, & \text{otherwise} \end{cases} \tag{9}$$

The Hessian certainty is evaluated by the leading principal minors instead of computing the eigenvalues [11]. Let

$$\kappa = \begin{bmatrix} \kappa_{11} & \kappa_{12} \\ \kappa_{21} & \kappa_{22} \end{bmatrix}, \tag{10}$$

If κ_{11} is negative and $|\kappa|$ is positive, then κ be negative definite. The multiple thresholding [9] is used to estimate the bacteria region using (i), (ii), and (iii) from Sect. 2.2.1 with the Hessian of Laplacian of Gaussian matrix. All these parameters provide an estimate of the bacteria region. It is used to mask the specified bacteria region in the image. This mask is used to apply the MPP model for bacteria points marking.

2.3 *Marking the Specified Bacteria*

The MPP is used to indicate the basic bacteria points in the image. MPP model is based on the stochastic framework. The objects in the images are random in number and localized randomly. The geometry of these objects is defined using random parameters [12]. MPP is defined as the Poisson measure-based density function [18, 19]. A set of random parameterized marked points is chosen to configure MPP. These parameters explain some geometrical characteristics of a considered object. The density term contains prior information of geometrical constraints of the underlying object. The point process is defined by considering some point's configuration $s = \{s_1, \ldots, s_n\}$ in $K \subset R^v$. Then, the density is defined with respect to the Poisson measure as

$$\mathrm{d}P(s) = f(s)\mu(\mathrm{d}s), \tag{11}$$

where

$$\mu(\mathrm{d}s) = \mathrm{e}^{-\beta(K)}\left\{1[\varphi \in \mathrm{d}s] + \sum_{n=1}^{\infty} \frac{\mu_n(\mathrm{d}s)}{n!}\right\} \tag{12}$$

with

$$\mu_n(\mathrm{d}s) = \int 1[s_1, \ldots, s_n \in \mathrm{d}s]\beta(\mathrm{d}s_1)\ldots\beta(\mathrm{d}s_n), \tag{13}$$

The parameter β is an intensity extent on K. An MPP is a point process in which each point s_i is a part of a mark $m_i \in M$, where M is the mask. Then, density is the function of points and its corresponding masking points as $u_i = (s_i, m_i)$. Global constraints on the whole configuration are modeled by neighborhood relations and associated interactions [12].

2.4 Complete Bacteria Extraction

The bacteria extraction is performed in two stages; firstly, the superellipse model identifies all the bacteria which regard overlapped bacteria as a single one. Since diagnosis and treatment of TB are dependent on bacteria count, hence segmentation of overlapped bacteria is very important. In the second stage, the supervised variational contour model is used to segment overlapped bacteria.

2.4.1 Bacteria Segmentation

The segmentation of bacteria is performed using a repeated thresholding approach [9]. The threshold is computed on the normalized image by quantizing it into 256 intensity levels for simplicity. The image bacteria are fitted with an ellipse using the multi-threshold. The ellipse parameters are found using moments [20] in matrix W as follows:

$$W = \begin{bmatrix} w_{02} & w_{11} \\ w_{11} & w_{20} \end{bmatrix}$$

$$\text{major axis} = 4 \times \sqrt{\frac{\lambda_1}{w_{00}}} \tag{14}$$

$$\text{min or axis} = 4 \times \sqrt{\frac{\lambda_2}{w_{00}}} \tag{15}$$

where λ_1 and λ_2 represent the eigenvalues of moment matrix W, w_{pq} is the pqth central moment along x- and y-directions, respectively. w_{00} is 0th central moment (corresponding to area of bacteria). The ellipse parameters are analyzed for individual bacteria. The bacteria were filtered based on the minor and major axis length to remove very small and very large regions. The minor and major axes lengths are given as a parameter to the algorithm. The weight W_{gt} is calculated corresponding to each bacteria and assigned to the bacteria [9] as follows:

$$W_{\text{gt}} = 0.5 \times \{\text{RAR} + \text{convexity}\} \tag{16}$$

where RAR: Residual Area Ratio is defined in terms of area (A) and ellipse area $\left(A_{\text{ellipse}}\right)$ as

$$\text{RAR} = \frac{\min\left(A, A_{\text{ellipse}}\right)}{\max\left(A, A_{\text{ellipse}}\right)} \tag{17}$$

and convexity is defined as

$$\text{convexity} = \frac{A}{\text{area of convex hull of object}} \tag{18}$$

The ellipse area is found as

$$A_{\text{ellipse}} = \frac{\pi \times \text{major axis} \times \text{min or axis}}{4} \tag{19}$$

The basic bacterial image characteristics, as lower intensity is the bacterial region and high intensity is the background region, are used. The lower valley in bimodal histogram shows the bacteria region. The threshold level was computed experimentally. The bacteria image range is assigned [0, 1] to the segment, and the threshold is set to 0.75 [9]. The bacteria those do not obey the linear path, overlapped each other or have curved trajectory, create difficulty in detection. The solution to this problem is to find the contour using the supervised variational model [13].

2.4.2 Supervised Variational Contour Model

An image u: $\Omega \rightarrow \Re$ is defined in continuous domain (Ω). It is segmented into two parts; objects Ω_O and background Ω_B, here $\Omega_O \cup \Omega_B = \Omega$ and $\Omega_O \cap \Omega_B = \emptyset$. The intensity and region features-based energy function is formulated with the guidance of labeled bacterial points. The coupled statistical and graph (CSG) supervised variational formulation is used for labeling of all the pixels of containing a unique descriptors for both; objects and background.

$$F^{CSG}(s) = \int_{\Omega_O} T^O(s)O^O(u(s))ds + \int_{\Omega_B} T^B(s)O^B(u(s))ds \tag{20}$$

where $O^O(\cdot)$ and $O^B(\cdot)$ are the statistical inference descriptors and $T^O(s)$ and $T^B(s)$ are the weights of bacterial used to compute the probability density of pixels. Morphological erosion [21] of two pixels is applied to remove the overlapping of bacteria. The multi-model distribution persists in medical images. It is easily described by the mixture-of-mixtures Gaussian model [22]. It is assumed that the pixel value $s \in \Omega$ and is derived from intensity distributions. The multimodal intensity distribution of the objects and the background is given as

$$p(u(s)|\Theta) = \sum_{\gamma=1}^{N} c_\gamma p_\gamma(s) = \sum_{\gamma=1}^{N} c_\gamma \left(\sum_{m=1}^{M_{\text{Modes}}} c_m^\gamma G_m^\gamma(u(s)) \right) \tag{21}$$

The above background descriptor is defined through the mixture-of-mixtures Gaussian model. Here, Θ represents the full set of Gaussian parameters containing N number of segments with M_{Modes} number of Gaussian components. The mixing coefficients of segments are c_y, and modes are c_y^m.

$$G_m^\gamma(u(s)) = G_m^\gamma\left(u(s)|\mu_m^\gamma, \sigma_m^\gamma\right) \tag{22}$$

Single bacteria segmentation by energy function is described in the following section for a given image. Two segments as $N = 2$ in the mixture-of-mixtures model simplify the object (single bacteria) and background approximation. The mixing coefficients probability for two segments is assumed to be equal and here by c_γ is taken as 0.5. The discrete presentation is converted into continuous level set formulation through level set function φ: $\Omega \rightarrow \Re$, with signed values corresponding to the partitions. The implicit contour model φ separates the objects $\Omega_O = \{s|\varphi(s) \geq 0\}$ and background $\Omega_B = \{s|\varphi(s) < 0\}$. The statistical model of (21) has also been reformulated through variational formulation:

$$F(\varphi, \Theta) = \int_{\Omega} (p(u(s)\Theta_O)H(\varphi) + p(u(s)\Theta_B)(1 - H(\varphi)))ds$$

$$F(\varphi, \Theta) = \int_{\Omega} \sum_{\chi=1}^{M_{\mathrm{O}}} c_{\chi}^{\mathrm{O}} G_{\chi}^{\mathrm{O}}\left(u(s)|\mu_{\chi}^{\mathrm{O}}\sigma_{\chi}^{\mathrm{O}}\right) H(\varphi)\mathrm{d}s + \int_{\Omega} \sum_{\rho=1}^{M_{\mathrm{B}}} c_{\rho}^{\mathrm{B}} G_{\rho}^{\mathrm{B}}\left(u(s)|\mu_{\rho}^{B}\sigma_{\rho}^{B}\right)(1 - H(\varphi))\mathrm{d}s \quad (23)$$

where H represents the Heaviside function, given as

$$H(\varphi) = \begin{cases} 1, & \varphi \geq 0 \\ 0, & \varphi < 0 \end{cases};$$

Let M_{O} and M_{B} represent the number of Gaussian components of objects and background, respectively; Θ_{O} and Θ_{B} are the complete set of Gaussian parameters. where $\Theta = \{\Theta_{\mathrm{O}}, \Theta_{\mathrm{B}}\}$; c_{χ}^{O} and c_{ρ}^{B} show the mixing coefficients of the modes for segments, in which $\sum_{\chi=1}^{M_O} c_{\chi}^{O} = 1$ and $\sum_{\rho=1}^{M_{\mathrm{B}}} c_{\rho}^{\mathrm{B}} = 1$; also $G_{\chi}^{O}\left(u(s)|\mu_{\chi}^{O}\sigma_{\chi}^{O}\right)$ and $G_{\rho}^{\mathrm{B}}\left(u(s)|\mu_{\rho}^{\mathrm{B}}\sigma_{\rho}^{\mathrm{B}}\right)$ is a Gaussian density at any pixel x for the segments, where $\mu_{\chi}^{\mathrm{O}}\sigma_{\chi}^{\mathrm{O}}$ and $\mu_{\rho}^{\mathrm{B}}\sigma_{\rho}^{\mathrm{B}}$ denote set of mean and standard deviation of Gaussian components for mixture model. The maximum likelihood of the Gaussian mixture model is derived for both segments Ω_{o} and Ω_{B} by the Heaviside function. It provides the objects (bacteria) segmented binary image as 1 for object or bacteria region and zero for the background region. The fast radial symmetry (FRS) transform [23] is then applied to find the center point locations and each bacteria size and counting of the bacteria in images.

3 Simulation Results and Discussion

The images taken for result analysis are from the tuberculosis dataset [24]. The dataset taken here as Data 1: Single_or_few_Microscope-1_set1, it is a set of 99 images of size 800 × 600 × 3 in JPEG format. Data 2: Without_bacilli_Microscope-1_set1, it is a set of 50 images of size 800 × 600 × 3 in JPEG format. Data 3: Overlapping_Microscope-1_set1, containing set of 50 images of size 800 × 600 × 3 in JPEG format. Data 4: Ocluded_Microscope-1_Set-1 consist a set of 200 images of size 800 × 600 × 3 in JPEG format. Figure 2 shows a sample input image in (a), its HLoG feature in (b), and (c) the segmentation result.

The qualitative analysis by a visual illustration of the segmented bacteria is shown in Fig. 3. Figure 3a, c, e is the input images, and (b), (d), and (f) are the corresponding segmentation result images. Following parameters are used for quantitative result analysis of the proposed method with available state of the art methods.

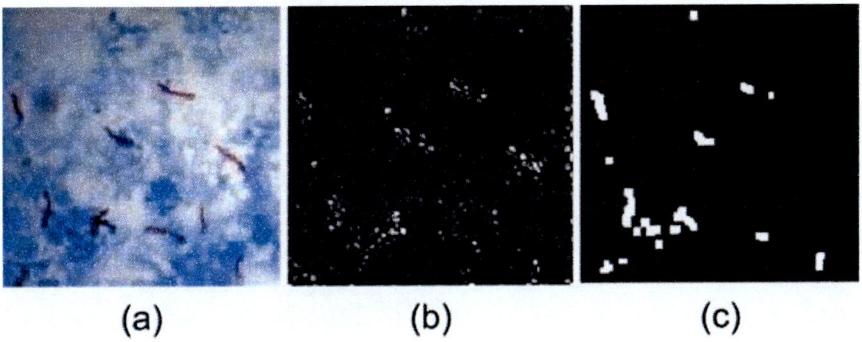

Fig. 2 Sample illustration of tuberculosis bacteria image **a** input image, **b** HLoG feature **c** segmentation of tuberculosis bacteria image

3.1 Segmentation Evaluation Parameters

The segmentation performance [9] was evaluated using precision, recall, F_1-score, and accuracy. The F_1-score is harmonic mean of precision and recall. These were found as follows

$$\text{recall} = \frac{\text{TP}}{\text{TP} + \text{FN}} \tag{24}$$

$$\text{precision} = \frac{\text{TP}}{\text{TP} + \text{FP}} \tag{25}$$

$$F_1\text{-score} = \frac{2 \times \text{precision} \times \text{recall}}{\text{precision} + \text{recall}} \tag{26}$$

$$\text{Accuracy} = \frac{\text{TP} + \text{TN}}{\text{P} + \text{N}} \tag{27}$$

where $\text{P} = \text{TP}+\text{FN}$ and $\text{N} = \text{FP}+\text{TN}$, TP are true positive pixels, FP is false positive pixels and TN is true negative pixels, and FN is false negative pixels. The perfect segmentation figured out 100% segmentation accuracy. The boundaries touching objects are ignored in the result. The comparative analysis of the proposed method with state of the art methods is done by computing the performance parameters using the segmented images and ground truths.

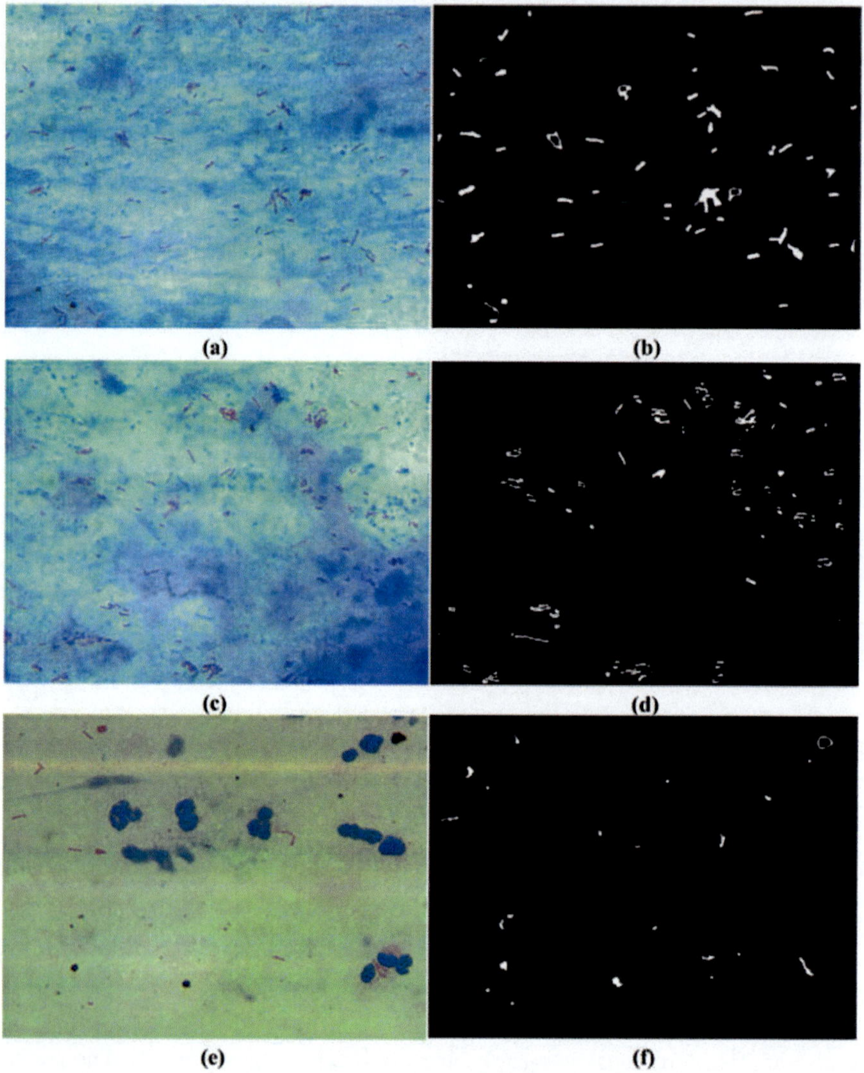

Fig. 3 Segmentation results of the acid fast stained tuberculosis bacterial images. **a, c, d** sample images, **b, d, f** segmentation outcome by the proposed method

3.2 Dice Similarity Coefficient (DSC)

Dice similarity coefficient (DSC) is a coefficient that measures the overlap between ground truth and segmentation result [25]. The ground truth is used for segmentation accuracy validation. The segmentation capability is accessed using DSC of the segmentation results for dataset images [13, 24]. Using segmented image I_s and its ground truth I_m, the DSC is defined as follows:

$$\mathrm{DSC}(I_s, I_m) = 100\% \times \frac{2|I_s \cap I_m|}{|I|_s + |I_m|} \tag{27}$$

Table 1 shows the average precision value comparison. The proposed method has average precision value of 0.861. The boldface is used to highlight the best result in table for a better look. It is 6.8 and 7.4% higher in comparison with [8, 9], respectively. The average recall or sensitivity of the proposed method is 0.890 as shown in Table 2. It is 7.8 and 11% higher as compared to [8, 9], respectively.

Tables 3 and 4 show the F_1-score and accuracy performance comparison, respectively. As shown, accuracy and F_1-score are reported the highest value in proposed technique. Average F_1-score of the proposed work is 0.875. The average accuracy of the proposed method also marks an improvement of 10% than that of [9].

The DSC is compared in Table 5. The DSC value for [5] with data 2 has good performance, but for all other datasets, the proposed method performs better. The average DSC performance of the proposed method is 0.84% better than [5]. The average performance of the proposed method for the mixed dataset is shown in Fig. 4, which confirms the superiority of the proposed segmentation method over other existing method.

Table 1 Precision value for four datasets

Methods	Data 1	Data 2	Data 3	Data 4	Average
[4]	0.727	0.761	0.693	0.689	0.718
[5]	0.743	0.758	0.757	0.789	0.762
[6]	0.723	0.786	0.802	0.786	0.774
[7]	0.695	0.753	0.793	0.782	0.756
[8]	0.769	0.787	0.819	0.834	0.802
[9]	0.703	0.784	0.817	0.818	0.781
Proposed	**0.814**	**0.861**	**0.876**	**0.893**	**0.861**

Table 2 Recall value for four datasets

Methods	Data 1	Data 2	Data 3	Data 4	Average
[4]	0.757	0.791	0.647	0.647	0.711
[5]	0.771	0.788	0.805	0.795	0.790
[6]	0.723	0.816	0.845	0.795	0.795
[7]	0.705	0.785	0.839	0.798	0.782
[8]	0.790	0.816	0.848	0.846	0.825
[9]	0.733	0.828	0.857	0.828	0.812
Proposed	**0.842**	**0.897**	**0.916**	**0.903**	**0.890**

Table 3 F_1-score value for four datasets

Methods	Data 1	Data 2	Data 3	Data 4	Average
[4]	0.742	0.776	0.669	0.667	0.714
[5]	0.757	0.773	0.780	0.792	0.776
[6]	0.723	0.801	0.823	0.790	0.784
[7]	0.700	0.769	0.815	0.790	0.769
[8]	0.779	0.801	0.833	0.840	0.813
[9]	0.718	0.805	0.837	0.823	0.796
Proposed	**0.828**	**0.879**	**0.896**	**0.898**	**0.875**

Table 4 Accuracy performance on different dataset

Method	Data 1	Data 2	Data 3	Data 4	Average
[4]	72.67	76.07	60.67	63.70	68.28
[5]	74.34	75.76	76.47	78.47	76.26
[6]	69.34	78.62	80.46	78.46	76.72
[7]	67.53	75.53	79.85	78.75	75.42
[8]	76.86	78.63	80.87	83.63	80.00
[9]	70.27	79.81	81.73	81.78	78.40
Proposed	**81.37**	**86.67**	**87.56**	**89.34**	**86.24**

Table 5 Average DSC for different methods on four datasets

Methods	Data 1	Data 2	Data 3	Data 4	Avg.
[4]	91.78	89.08	90.78	93.80	91.36
[5]	94.34	**95.87**	95.48	95.98	95.42
[6]	87.34	87.72	86.47	88.74	87.57
[7]	86.53	85.58	88.75	87.85	87.18
[8]	89.77	93.73	90.78	93.73	92.00
[9]	89.28	90.71	89.83	91.87	90.42
Proposed	**96.38**	95.18	**97.17**	**96.14**	**96.22**

4 Conclusion

An automatic system for identification of tuberculosis bacteria has been established using ZN-stained sputum smear microscopic images segmentation. Entire image segmentation of the acid fast stained images is performed in four steps; image contrast enhancement, bacteria region map creation, marking of the specified bacteria, using MPP model, and complete bacteria extraction using multiple thresholding with the supervised variational contour model. Proposed method offers better segmentation

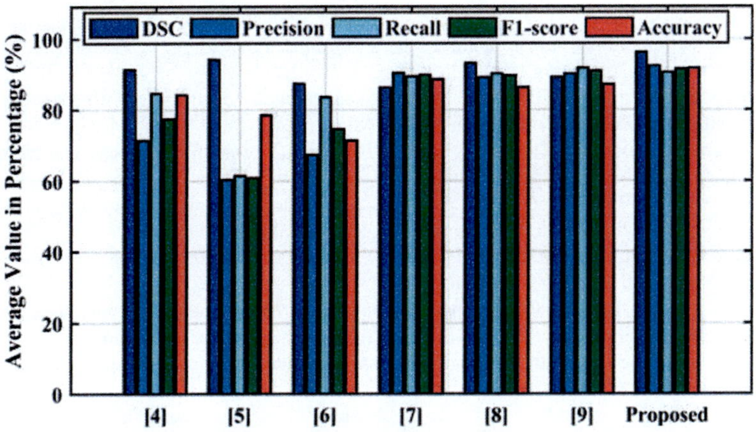

Fig. 4 Average parameter comparison of the proposed method with different state of the art methods for the combined dataset of data 1 to data 4

accuracy, F_1-score, and DSC as compared to the state of the art methods. The average accuracy and F_1-score of the proposed method are 91.75 and 91.5, respectively. The proposed technique provides the average value of DSC as 96.2. In future, segmentation outcome of the proposed method may be used for TB diagnosis and grading by applying classifiers.

Declarations Ethics approval and consent to participate All the ethics are kept in mind regarding the submission of the manuscript.

Consent for publication: Authors are agreed as per requirement of the journals requirement.

Availability of data and material: It will be available as per the journals requirement.

Competing interests: There are no competing interests in the manuscript.

Funding: No fund is available for this research work.

Authors' contributions

1. Corresponding author introduced the methodology and perform the simulation.
2. Second and third authors support in the improvement of simulation result and the preparing the manuscript.
3. Fourth and fifth authors support to understand the dataset and also support to know about the suitable object segment in microscopic images.

References

1. Kanade S, Nataraj G, Ubale M et al (2016) Fluorescein diacetate vital staining for detecting viability of acid-fast bacilli in patients on antituberculosis treatment. Int. J. Myobacteriol. 5:294–298
2. World Health Organization: Global tuberculosis report 2015, WHO/HTM/TB/2015-22
3. World Health Organization: Treatment of Tuberculosis: Guidelines, 4th edn. WHO, Geneva (2010). WHO/HTM/TB/2009.420

4. Vargasa, M. F., Sroubekd, F., Borregoc, J. A., et al.: Segmentation, auto focusing and signature extraction of tuberculosis sputum images. In: SPIE Proceedings of Photonic Devices and Algorithms for Computing IV, 4788 (2002)
5. Forero MG, Cristóbal G, Desco M (2006) Automatic identification of Mycobacterium tuberculosis by Gaussian mixture models. J. Microsc. 223(2):120–132
6. Khutlang R, Krishnan S, Dendere R et al (2010) Classification of mycobacterium tuberculosis in images of ZN-stained sputum smears. IEEE Trans. Info. Tech. Biomed. 14(4):949–957
7. Li C, Wang X, Eberl S et al (2013) A new energy framework with distribution descriptors for image segmentation. IEEE Trans Image Process 22(9):3578–3590
8. Mozos, R. S., Cruz, F. P., Madden, M. G., *et al.:* An automated screening system for tuberculosis. IEEE J. Biomed. Health Inf. 18(3):855–862
9. Sadanandan SK, Baltekin O, Magnusson KEG et al (2016) Segmentation and track-analysis in time-lapse imaging of bacteria. IEEE J. Sel. Top. Signal Process. 10(1):174–184
10. Willmore TJ (1959) An introduction to differential geometry. Oxford University Press, Oxford, U.K.
11. Zhang M, Wu T, Bennett KM (2015) Small blob identification in medical images using regional features from optimum scale. IEEE Trans Biomed Eng 62(4):1051–1062
12. Descombes, Zerubia, J.: Marked point process in image analysis. IEEE Signal Process. Mag. **19**(5):77–84
13. Li C, Wang X, Eberl S et al (2015) Supervised variational model with statistical inference and its application in medical image segmentation. IEEE Trans. Biomed. Eng. 62(1):196–207
14. Lindeberg T (1998) Feature detection with automatic scale selection. Int. J. Comput. Vis. 30(2):79–116
15. Woodford C, Philips C (2012) Numerical methods with worked examples, Matlab edn. Springer, New York
16. Salden, A.H., Romeny, B.M.T.H., Viergever, M.A., et al.: Differential geometric description of 3D scalar images. Internal Report 3 DCV, 91, (5) (1991)
17. Frangi, A., Niessen, W., Vincken, K., et al.: Multiscale vessel enhancement filtering. Proceedings of International Conference on Medical Image Computing and Computer-Assisted Intervention, vol 1496, pp. 130–137 1998
18. Ripley, B.D.: Modelling spatial patterns. J. R. Stat. Inst. ser. B **39**, 172–212 (1997)
19. Stoyan D, Kendall WS, Mecke J (1987) Stochastic geometry and its applications. Wiley, New York
20. Burger W, Burge MJ (2009) Principles of digital image processing: core algorithms. Springer, New York
21. Haralick, R.M., Zhuang, S., Lin, C., et al.: The digital morphological sampling theorem. IEEE Trans. Acoust. Speech Signal Process. **37**(12), 2067–2090 (1989)
22. Browne RP, McNicholas PD, Sparling MD (2012) Model-based learning using a mixture of mixtures of Gaussian and uniform distributions. IEEE Trans. Pattern Anal. Mach. Intel. 34(4):814–817
23. Loy G, Zelinsky A (2003) Fast radial symmetry for detecting points of interest. IEEE Trans. Pattern Anal. Mach. Intell. 25(8):959–973
24. ZNSM-iDB: Ziehl Neelsen Sputum smear microscopy image data base. Available: http://www.tbdb.org/cgi-bin/data/downloadData.pl
25. Foster B, Bagci U, Xu Z et al (2015) Segmentation of PET images for computer-aided functional quantification of tuberculosis in small animal models. IEEE Trans. Biomed. Eng. 61(3):711–724

Condition Monitoring of PV-Wind System Using Stockwell Transform

Debopoma Kar Ray, Debarati Das, Om Prakash Shah, Sanu Kumar Singh, and Surajit Chattopadhyay

Abstract This paper deals with the fault and abnormality analysis in PV array, monitoring the grid-tied hybrid micro-grid current. Firstly, a hybrid system has been built-in software, and a series arc fault has been inserted. The micro-grid current has been recorded and analyzed using Stockwell transform-based statistical parameter monitoring, and optimized parameter has been selected for fault analysis in the system. Depending on the variation in the statistical parameter values, features have been extracted. A logic has been proposed at the end for abnormality assessment in the system with practical validation of the work using field data.

Keywords Hybrid micro-grid · Kurtosis · Practical validation · Series arc fault · Skewness · Stockwell transform

1 Introduction

Hybrid micro-grid systems can be operated in islanded as well as grid-connected modes [1]. With the uncertain future of the conventional sources of energy, renewable

---

D. K. Ray (✉) · D. Das · O. P. Shah · S. K. Singh
Department of Electrical Engineering, MCKV Institute of Engineering, Liluah, Howrah, India
e-mail: debopoma86@gmail.com

D. Das
e-mail: debaratidas113@gmail.com

O. P. Shah
e-mail: shah.vishal178@gmail.com

S. K. Singh
e-mail: sanu90m@gmail.com

S. Chattopadhyay
Department of Electrical Engineering, Ghani Khan Choudhury Institute of Engineering and Technology, Malda, India
e-mail: surajitchattopadhyay@gmail.com

V. Nath and J. K. Mandal (eds.), *Proceedings of the Fourth International Conference on Microelectronics, Computing and Communication Systems*, Lecture Notes in Electrical Engineering 673, https://doi.org/10.1007/978-981-15-5546-6_23

has become inevitable in recent times. Hence, condition monitoring of islanded and grid-tied renewable energy conversion system is of prime importance in recent days. Numerous works have been seen for assessing faults in micro-grids. A 300–50 kW wind solar system analysis has been seen for power flow, fault analysis using buck converter and SPWM technique with power control and amplitude control methods [2]. Maximum power point tracking technique-based optimization has been done for hybrid systems [3]. Optimal sizing of a PV-wind system has been seen where enhanced genetic algorithm has been used for minimizing the total cost of the system [4]. An algorithm has been seen for calculating the number of generating units in PV-wind system for periodically invariant battery charging and optimum cost-effective generation [5]. Fuzzy logic control-based optimization and distributed energy generation has been seen to be developed for a PV-wind system [6]. An integrated biogas–wind–solar hybrid micro-grid formation has been seen for catering base and peak loads of the connected habitable area at low cost of electricity [7]. Optimum design can be done for PV-wind system using genetic algorithm and particle swarm optimization techniques [8]. A new PV-wind system proposition reduces carbon emission in the environment [9]. A static compensator-based simulation and control strategy improves load voltage and current for PV-wind systems [10]. Charge controller and renewable energy converter designs have been effective for a standalone PV-wind system [11]. ETAP-software-based PV-wind system stability, and fault analysis is done for minimizing system outage and maximizing system reliability [12].

Fault detection analysis has been viewed monitoring the grid current of a PV-wind system using MRA of DWT-based statistical parametric optimization [13–15]. Energy storage unit-based supercapacitor has been observed to be used in a wind-PV form for stability improvement [16], wherein a proportional integral derivative (PID)-supplementary damping controller (SDC) has been seen to be used for enhancing the damping characteristics of low-frequency oscillations. A pattern search-based optimization in conjunction with sequential Monte Carlo simulation has been seen for cost optimization and reliability analysis in wind-PV hybrid system [17]. Implementation of a low-pass filter and hysteresis control-based supervisory control strategy can be used for versatile power transfer in a PV-wind system [18]. Implementation of copula-based buffered power incorporated probability analysis has been seen for cost optimization and stability analysis of a PV-wind system [19]. Hybrid optimization model for electric renewable (HOMER)-based lowest net present cost (NPC) and cost of energy (COE) analysis can be used for a remote PV-wind-battery-diesel system for reducing the operating cost of the system [20].

However, none of the analysis seen so far has observed to deal with the fault estimation in PV-wind system, monitoring the Stockwell transform-based optimization technique in time-frequency domain. Also series arc fault analysis has also not been seen under attention in any of the researches available in literature.

Thus, this work deals with the fault and abnormality assessment for PV-wind system, monitoring the system current Stockwell transform-based Skewness and Kurtosis values. Firstly, a PV-wind system has been developed in software, a series arc fault has been simulated at the PV array, and the system current assessment has been done computing the Stockwell transform-based statistical values. Based on the optimized value, patterns have been extracted for effective series arc fault in the micro-grid. An algorithm has been proposed at the end to identify these types of faults in the network. Practical authentication of the algorithm has been done with offline data obtained from industry with satisfactory outcome.

2 System Development

To proceed with the analysis, firstly, a PV-wind system has been designed (MATLAB software, R2018b), and data acquisition has been data at the load side for normal and fault conditions of the system. The schematic diagram of the system has been shown in Fig. 1.

The system comprises a 30 kVA, 440 V, 50 Hz alternator, static exciter of 400 kW relay (operating time 10 ms, percentage voltage sag of 63% supply voltage), circuit breaker (resistance 0.01 Ω, transition time of 0.2 s), solar cell (nominal voltage 24 V, 280 W peak, module efficiency 16.7%), inverter (5000 Ω snubber resistance), solar charge controller, battery, LC filter (2 mH, 6.85 F), domestic load (5 kW), commercial load (10 kW).

In this system, the micro-grid current has been tracked at series arc fault in the PV array. The acquired current signature has been assessed using ST-based parametric optimization algorithm and features have been extracted for effective fault and assessment in the system.

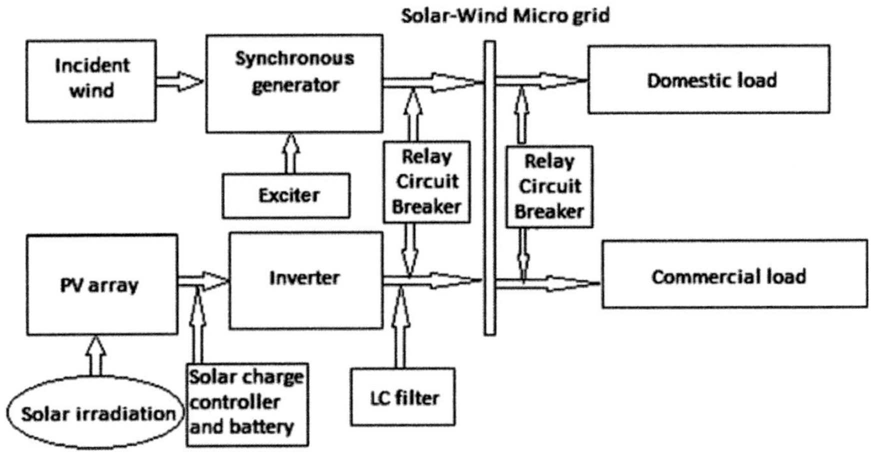

Fig. 1 PV-wind system schematic

3 Theoretical Backgrounds

The Stockwell transform (ST) can be derived using Gaussian

$$g(t) = \frac{|f|}{\sqrt{2\pi}} e^{\frac{t^2 f^2}{2}} \tag{1}$$

as

$$s(\tau, f) = \mathrm{STFT}(\tau, f) = \int_{-\infty}^{\infty} h(t) \frac{|f|}{\sqrt{2\pi}} e^{\frac{(\tau - t)^2}{2}} e^{-j2\pi f t} dt \tag{2}$$

Wider windowed ST provides better frequency resolution for lower frequency. However, narrower windowed ST provides better time resolution for higher frequency [21].

The ST coefficients of the micro-grid current at normal and fault have been analyzed using statistical parameter value monitoring. Skewness is a measure of symmetry, or more precisely, the lack of symmetry. Kurtosis is a parameter that describes the shape of a random variable's probability distribution.

4 Fault Analysis in PV Array

The PV-wind system current has been tracked for different percentage faults in the PV array as presented in Fig. 2. The recorded currents have been assessed using the ST program, and the Skewness and Kurtosis values of the ST coefficients have been calculated. Depending on the deviation of the Skewness and Kurtosis values, best-fit parameter selection has been done for effective percentage arc faults occurring in the PV array. The ST output frame (two-dimensional) has been presented in Fig. 3. The variation in the Skewness and Kurtosis values has been presented in Table 1.

It is very difficult from the above figures to disintegrate any percentage arc faults in the PV array. Thus, ST algorithm has been used to extract features.

Monitoring Fig. 3, it has been observed that each signal has a distinctive nature of frequency distribution for different percentages of series arc faults in the PV array. But monitoring only the nature of the frequency distribution, fault assessment in a micro-grid is very difficult. Thus, Skewness–Kurtosis-based monitoring is necessary.

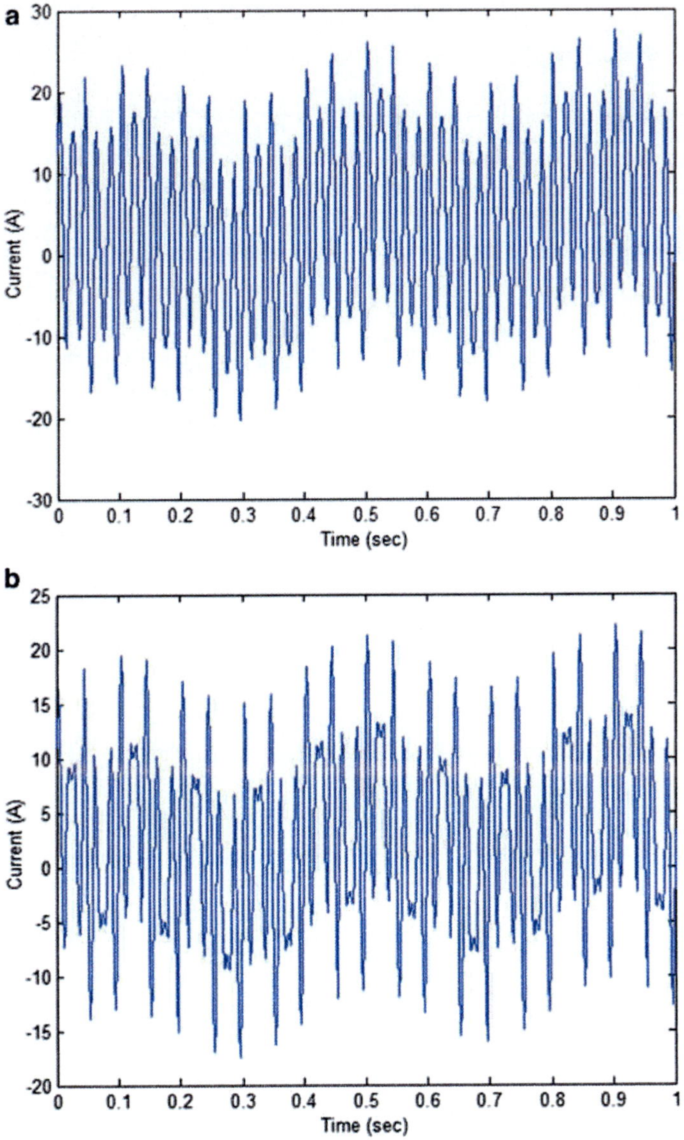

Fig. 2 **a** Micro-grid current at healthy condition, **b** micro-grid current at 1% series arc fault, **c** micro-grid current at 2% series arc fault, **d** micro-grid current at 3% series arc fault, **e** micro-grid current at 4% series arc fault

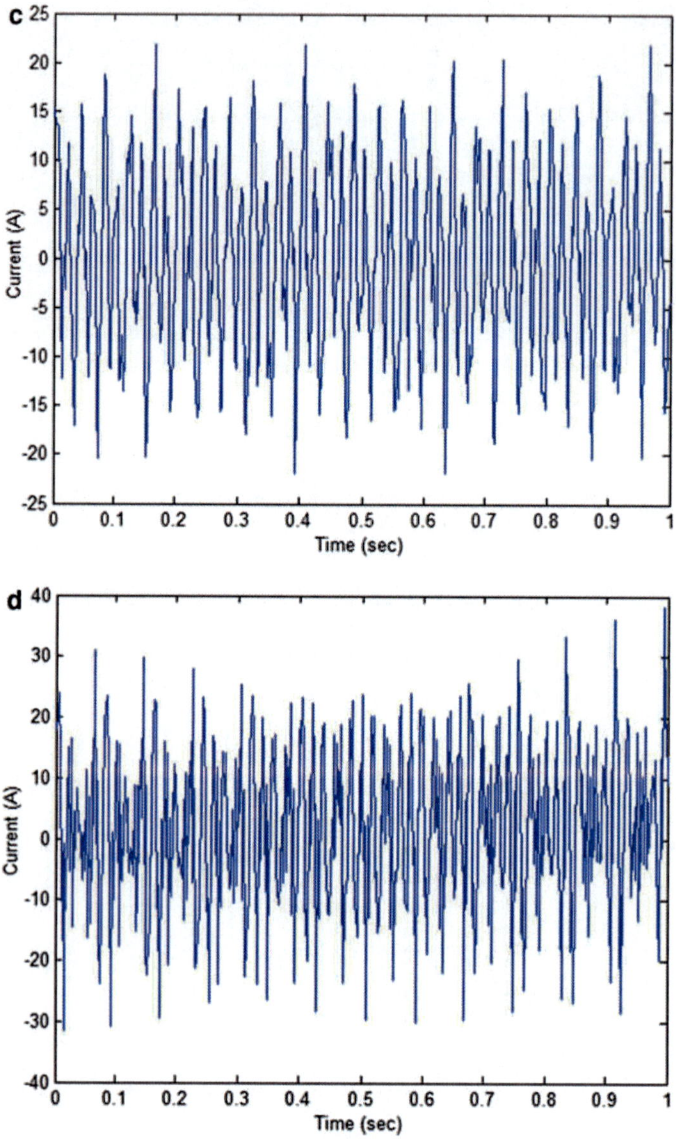

Fig. 2 (continued)

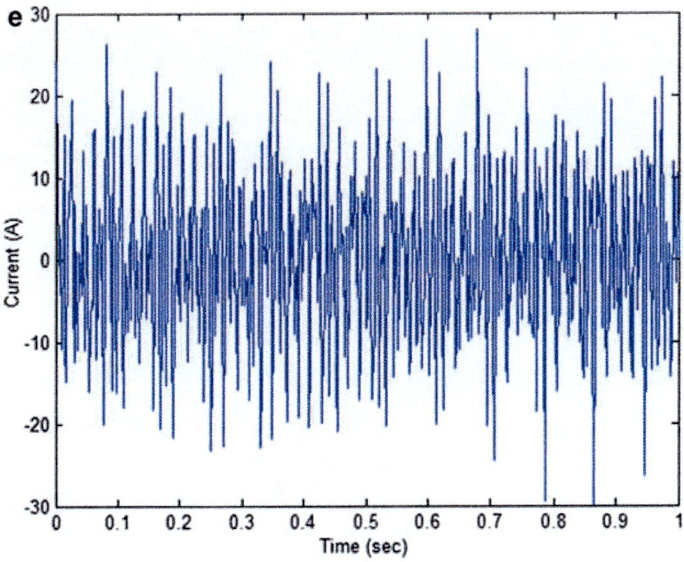

Fig. 2 (continued)

5 Feature Extraction

The Skewness and Kurtosis values of the responses obtained in Fig. 3 have been computed, and the variation in each value has been monitored computing the covariance [17] of the responses. The respective covariance values for different conditions of the system have been presented in Table 1.

Assessing Table 1, it has been inferred that the variation in the covariance values is distinctively different and can be used as a feature for percentage series arc fault estimation in the system. It has also been observed that the covariance value for 3% fault is the highest, which may be due to the pattern of harmonic generation in the system or due to the change in instantaneous nature of current signature at fault. Depending on the results obtained from his analysis, logic has been developed for percentage series arc fault detection in the PV array. The logic has been presented below:

Step 1: Record grid current
Step 2: Perform ST analysis
Step 3: Calculate Skewness and Kurtosis values of ST coefficients.
Step 4: Calculate statistical parameter covariance as in Table 1.
Step 5: Determine fault percentage.

The algorithm takes around 1.2 s, which has been seen to be satisfactorily less.

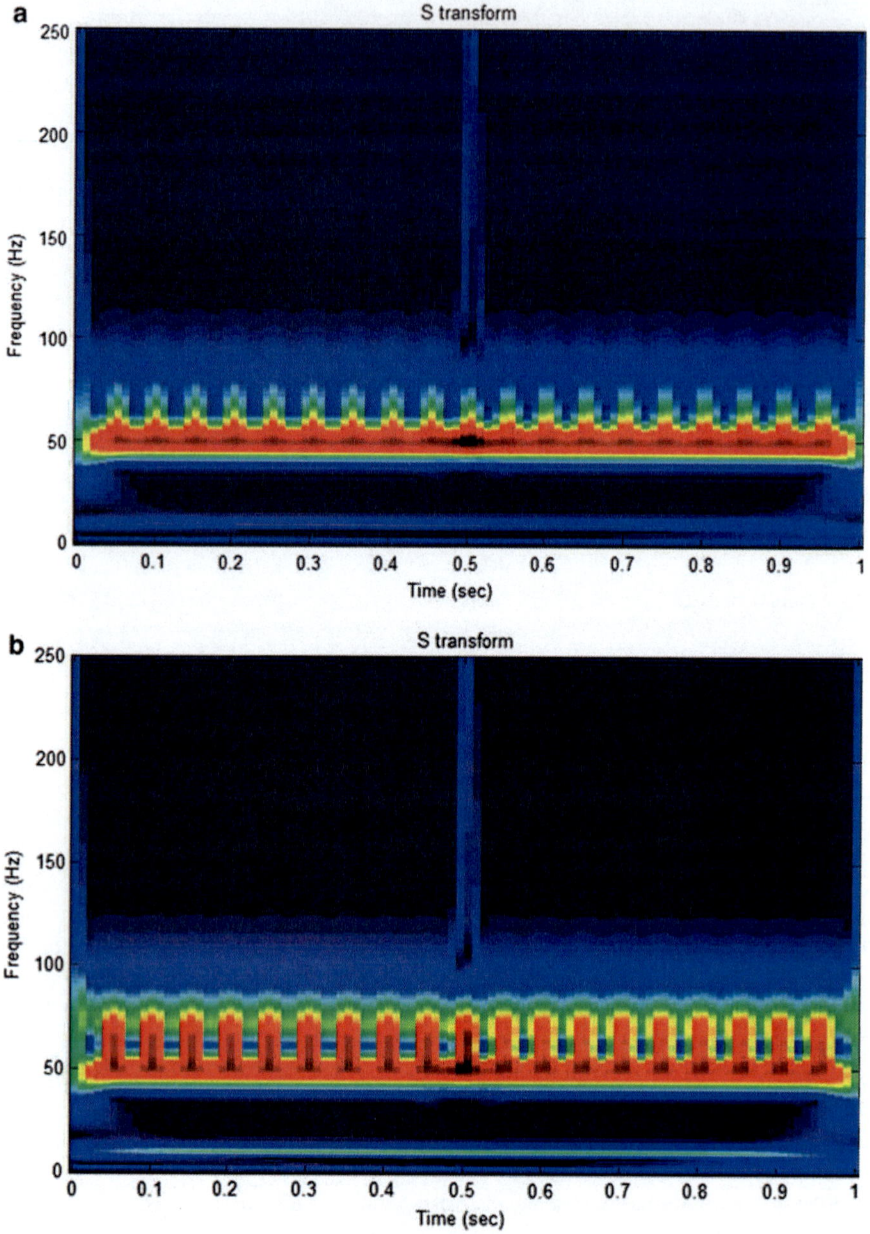

Fig. 3 **a** Healthy condition ST frame, **b** ST output frame for 1% fault, **c** ST output frame for 2% fault, **d** ST output frame for 3% fault, **e** ST output frame for 4% fault

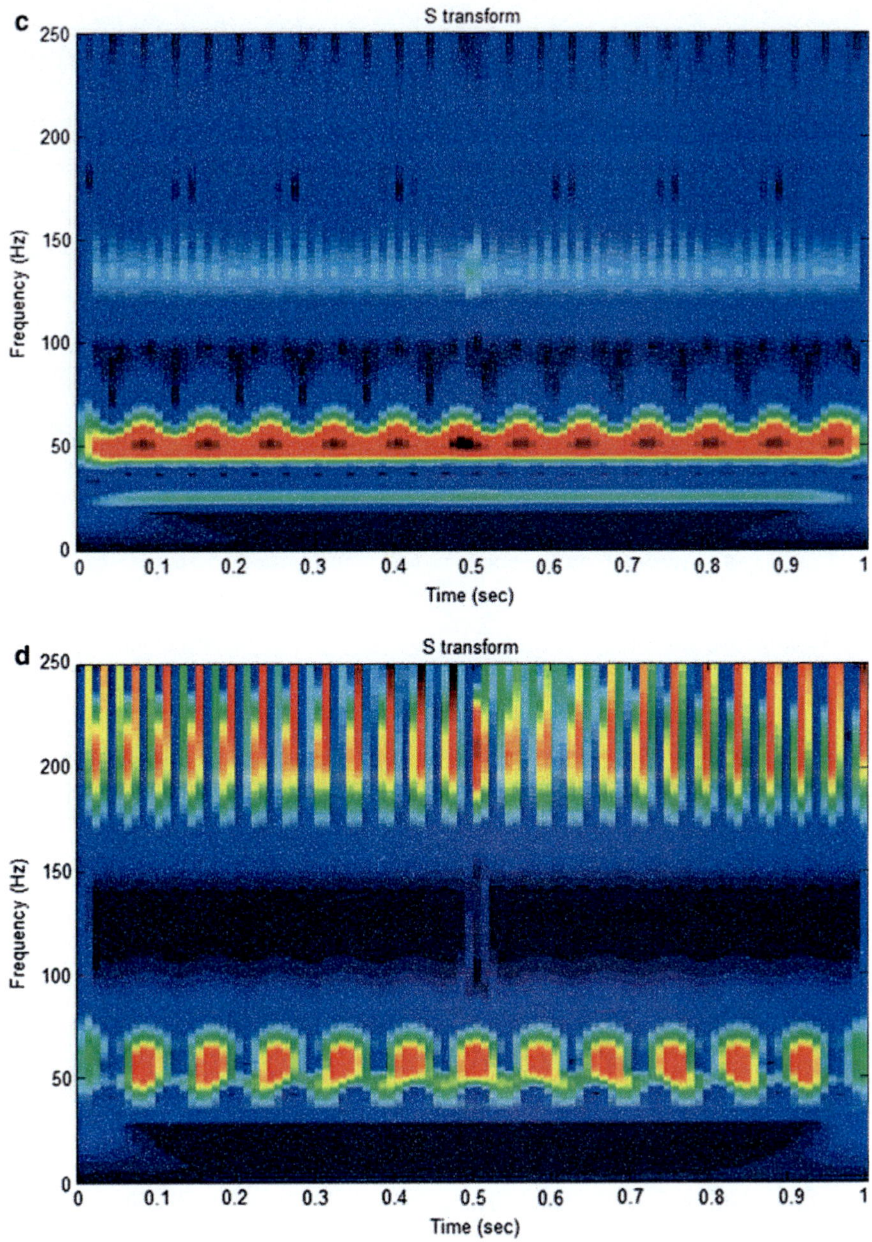

Fig. 3 (continued)

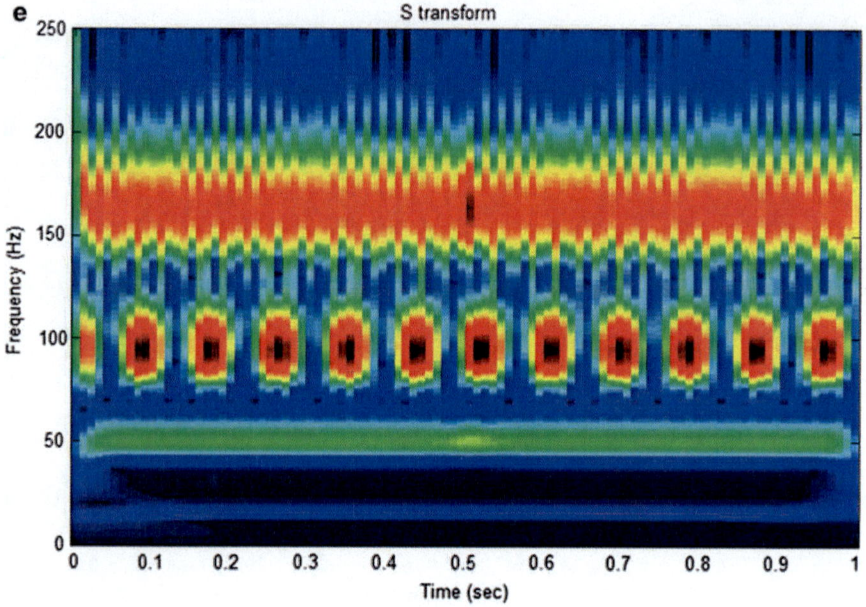

Fig. 3 (continued)

Table 1 Covariance assessment from Skewness–Kurtosis values

System condition	Skewness value covariance	Kurtosis value covariance
Healthy	0.125	3.69
1% fault	0.029	0.63
2% fault	0.038	0.297
3% fault	0.111	0.811
4% fault	0.05	0.059

6 Practical Validation

To authenticate the algorithm proposed in Sect. 5, data has been collected from field study for analysis. Offline data has been collected from an Indian 15.12 kWp SPVRT system, wherein the current signature has been assessed using the developed algorithm. Here analysis has been advanced considering the idea that all renewable and non-renewable energy generations are ultimately fed to the grid for connecting to the load centers in modern days. The single line diagram of the system has been presented in Fig. 4 [22].

The system has been seen to be built up with 48 numbers of solar panel with 315 Wp each, one solar inverter of 15 kW, 415 V, mounted on 15.12 kWp GI mounting structure. Data has been collected at an instant, and the current signature has been

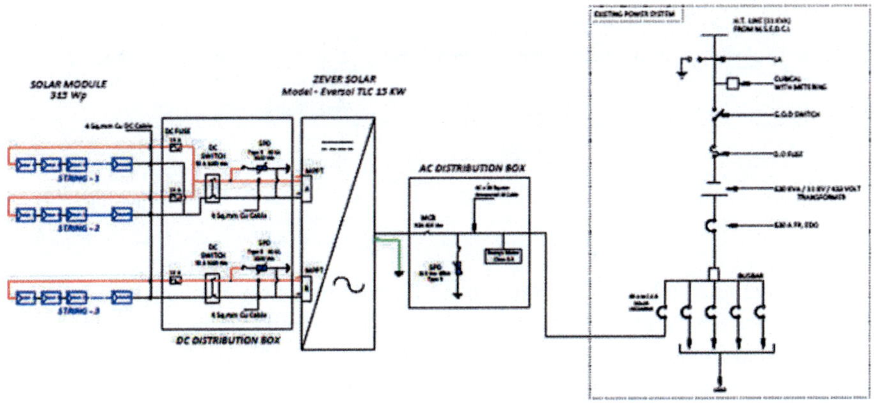

Fig. 4 Single line diagram of the grid-tied PV generating system

analyzed using the developed algorithm as in Sect. 5. The ST output frame of the collected data has been presented in Fig. 5. The ST coefficient statistical parameter covariance has been calculated, and a comparative analysis between the theoretical and practical analysis has been presented in Table 2.

In Table 2, *SC* and *KC* denotes the covariance values of Skewness and Kurtosis coefficients, respectively. Monitoring Table 2, it has been inferred that, the data has

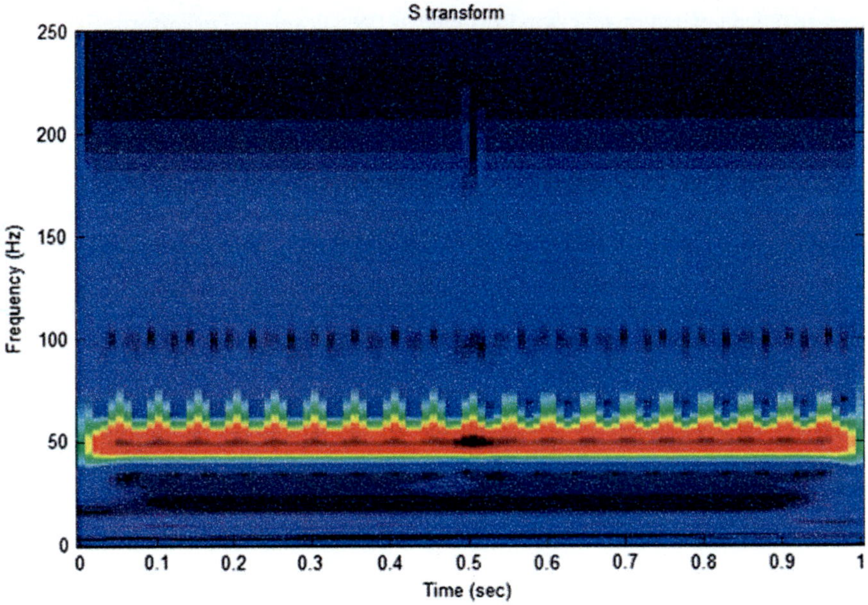

Fig. 5 ST output frame for data collected from field study

Table 2 Feature extraction and data analysis

Skewness value covariance	Kurtosis value covariance	Percentage error w.r.t Table 1		Data acquisition to decision-making operating time	Remarks
0.04	0.312	*SC*	*KC*	1.42 s	2% series arc fault in PV module
		5%	4.8%		

been obtained for 2% series arc fault in the PV module of the system. This has been inferred since the data obtained from Table 2 has been seen to closely match with the data obtained in Table 1 for 2% series arc fault in the PV array. Also the algorithm operating time has been observed to be 1.42 s from data acquisition to decision making, which is satisfactorily less. Also percentage error of the algorithm has been observed to be within 5% which is within tolerable limits.

7 Conclusion

Renewable energies are of great use in modern days due to the fact that after 20–30 years, the conventional energy resources will deplete. Keeping this in mind, researches on renewable energy generations with condition monitoring of the generation systems is of vivid concern nowadays. In this context, the present work has been advanced, wherein a PV-wind system has been developed in software. The grid current has been recorded at healthy and for different percentages of series arc faults in the PV array of the micro-grid. The recorded current has been analyzed using the developed ST algorithm, and the ST coefficients have been assessed computing Skewness–Kurtosis-based covariance values. Depending on the variation in these values, the fault percentages have been calculated. An algorithm has been proposed at the end for determining various fault percentages in the system. The operating time of the algorithm form data acquisition to decision making takes around 1.2 s, which is satisfactory. In this approach, analysis has only been done in software domain. Practical authentication of the work has been done with data obtained from Sterlite Power, India, with a vivid comparative study of the theoretical and practical approaches with percentage error of $\leq 5\%$ which is acceptable. The operating time of the decision-making algorithm from data acquisition to solution is <1.5 s.

Acknowledgements We are grateful to Mr. Raktim Basak, Associate Manager, Sterlite Power, for providing the necessary data for analysis purpose.

References

1. Jayachandran M, Ravi G (2017) Design and optimization of hybrid microgrid system. Energy Proceedings, vol 117. Elsevier, pp 95–103. Available: https://www.sciencedirect.com/sciencedirect.com/science/article/pii/S1876610217323184
2. Kabalci E, Bayindir R, Hossain E (2017) Hybrid microgrid testbed involving wind/solar/fuel cell plants: a design and analysis test bed. Presented at ICRERA. Available: https://ieeexplore.ieee.org/document/7016511
3. Aisheh AA (2018) Hybrid solar/wind micro-grid systems. Presented at conventional and sustainable energies. Available: https://www.omicsonline.org/conference-proceedings/2090-4541-C1-050-005.pdf
4. Traore A, Elgothamy H, Zohdy MA (2018) Optimal sizing of solar/wind hybrid off-grid microgrids using an enhanced genetic algorithm. J Power Energy Eng 6:64–77. Available: https://file.scirp.org/pdf/JPEE_2018053014523053.pdf
5. Li J, Wei W, Xiang J (2012) A simple sizing algorithm for stand-alone PV/wind/battery hybrid microgrids. Energy 5:5307–5323. Available: https://www.mdpi.com/1996-1073/5/12/5307/energies-05-05307.pdf
6. Ruban MAM, Rajasekaran GM, Rajeswari N (2015) Implementation of energy management system to PV-wind hybrid power generation system for DC microgrid applications. Int Res J Eng. Tech 2(8):204–210. Available: https://www.irjet.net/archives/V2/i8/IRJET-V2I835.pdf
7. Habibullah M, Mahmud K, Kocar G, Islam AKMS (2017) Economic challenges of hybrid microgrid: an analysis and approaches for rural electrification. Presented at AIP. Available: https://aip.scitation.org/doi/10.1063/1.4984672
8. Sawlw Y, Gupta SC, Bore AK (2016) PV-wind hybrid system: a review with case study. Cogent Eng 3(1):1–31. Available: https://www.tandfonline.com/doi/pdf/10.1080/23311916.2016.1189305?needAccess=true
9. Priyadharshini E, Meena S (2015) Optimization of hybrid solar wind power by implementing microgrid control and controlling pollution by reducing. Int J Emerg Technol Comput Sci Electron 13(1):203–208. Available: http://www.ijetcse.com/wp-content/plugins/ijetcse/file/upload/docx/646OPTIMIZATION-OF-HYBRID-SOLAR-WIND-POWER-BY-IMPLEMENTING-MICROGRID-CONTROL-AND-CONTROLLING-POLLUTION-BY-REDUCING-CO-pdf.pdf
10. Acharya S (2016) Simulation and control of PV-wind-hydro hybrid renewable energy system and possibility I Nepal. Int J Sci Eng Res 7(8):1540–1555. Available: https://www.ijser.org/researchpaper/Simulation-and-Control-of-Pv-Wind-HydroHybrid-Renewable-Energy-System-and-Possibility-in-Nepal.pdf
11. Lauria D, Lamb S, Aisheh AA (2016) Designing stand-alone microgrid hybrid solar/wind energy systems and grid-connected smartgrid hybrid solar/wind energy systems. Presented at IAJC-ISAM. Available: http://cd16.iajc.org/wp-content/uploads/Camera-ready-papers/070-x-16___Designing%20a%20Microgrid%20Hybrid%20Solar-Wind%20Energy%20System%20_REVISDED2_Abu_Aisheh_Lamb_Lauria.pdf
12. Sharif MSE, Khan MMZ, Moniruzzaman M, Bose A (2017) Design, simulation and stability analysis of wind-PV-diesel hybrid power system using ETAP. Am J Mod Energy 3(6):121–130. Available: http://www.sciencepublishinggroup.com/journal/paperinfo?journalid=363&doi=10.11648/j.ajme.20170306.12
13. Kar Ray D, Chattopadhyay S, Sengupta S (2018) Fault diagnosis in isolated renewable energy conversion system using skewness and Kurtosis assessment. In: Advances in intelligent systems and computing, pp 57–71. Available: https://link.springer.com/chapter/10.1007/978-3-319-74808-5_7
14. Kar Ray D, Chattopadhyay S, Sengupta S (2017) Multi-resolution analysis based line to ground fault discrimination in standalone wind energy conversion system in harmonic environment. In: Lectures on modelling and simulation; a selection from AMSE, pp 139–150. Available: http://amsemodelling.com/publications/lectures_on_modeling_and_simulation/calcutta_14.pdf

15. Ray DK, Sarkar G, Paul P, Haldar R, Ghoshal S, Sadhukhan S, Chattopadhyay S (2017) Fault detection in wind energy conversion system using multi-resolution analysis of discrete wavelet transform. C+ CA Program Eng Sci 42(1), pp 26–31. ISSN: 0045–6152
16. Wang L, Vo QS, Prokhorov AV (2018) Stability improvement of a multimachine power system connected with a large-scale hybrid wind-photovoltaic farm using a supercapacitor. IEEE Trans Ind App 54(1), pp 50–60. ISSN: 1939–9367
17. Arabali A, Ghofrani M, Amoli ME, Fadali MS (2014) Stochastic performance assessment and sizing for a hybrid power system of solar/wind/energy storage. IEEE Trans Sustain Energy 5(2):363–371. ISSN: 1949–3037
18. Kim SK, Jeon JH, Cho CH, Ahn JB, Kwon SH (2008) Dynamic modeling and control of a grid-connected hybrid generation system with versatile power transfer. IEEE Trans Ind Elec 55(4), pp 1677–1688. ISSN: 1557–9948
19. Feng L, Zhang J, Li G, Zhang B (2016) Cost reduction of a hybrid energy storage system considering correlation between wind and PV power. Prot Cont Mod Power Syst 1-9. https://doi.org/10.1186/s41601-016-0021-1
20. Adefarati T, Bansal RC, Justo JJ (2017) Techno-economic analysis of a PV-wind-battery-diesel standalone power system in a remote area. J Eng IET 740–744. ISSN: 2051–3305
21. George NV (2009) S-transform: time frequency analysis and filtering. M.Sc. thesis, EE., NIT., Rour., Orissa
22. Technical data: 15.12 kWp PVRT system, Sterlite Power, India

An Investigation on Photocatalytic Dye Degradation of Rhodamine 6G Dye with Fe- and Ag-Doped TiO_2 Thin Films

B. K. Nahak, T. S. K. Subudhi, L. K. Pradhan, A. Panigrahi, R. Roshan, S. S. Mahato, and S. Mahata

Abstract Transparent Fe- and Ag-doped titania thin films were developed on fluoride-doped tin oxide (FTO) glass substrate via sol-gel dip coating process. Structural, basic morphological as well as optical attributes of the films were studied by X-ray beam diffraction (XRD), field emission SEM (FESEM) and UV–visible spectroscopy. XRD result depicts prevalent formation of anatase crystalline phase of TiO_2. In UV–visible spectroscopy, optical properties of the film like transmission, reflectance and absorption were studied over wavelength range of 300–800 nm. Some important parameters like band gap (E_g), thickness (t) and refractive index (n) of the developed films were calculated from the spectroscopic data. It was found that the optical and structural properties of the film were reliant on the type of doped material incorporated in the film. Acquired band gap of 2.78 eV and 2.85 eV, respectively, for Ag- and Fe-doped films reflects higher strained Ag-doped film compared to that of Fe-doped one. Higher activity of Ag assimilation over Fe toward Rhodamine 6G degradation is credited to Ag deposition on TiO_2 surface creating electron trap center which reduces the recombination probability and improves charge transfer process on the matrix.

Keywords TiO_2 · Dye degradation · Rhodamine 6G · XRD · Anatase · FESEM · UV–Vis analysis · Band gap

B. K. Nahak · T. S. K. Subudhi · L. K. Pradhan · A. Panigrahi · R. Roshan · S. S. Mahato
Department of Electronics and Communication Engineering, National Institute of Science and Technology, Palur Hills, Berhampur 761008, India

S. Mahata (✉)
Department of Chemistry, National Institute of Science and Technology, Palur Hills, Berhampur 761008, India
e-mail: shrabani.mahata@gmail.com

V. Nath and J. K. Mandal (eds.), *Proceedings of the Fourth International Conference on Microelectronics, Computing and Communication Systems*, Lecture Notes in Electrical Engineering 673, https://doi.org/10.1007/978-981-15-5546-6_24

1 Introduction

In recently developed technology, among the widely used semiconducting materials titanium dioxide (TiO_2) has attracted exceptional interest due to its outstanding physicochemical properties as well as high stability. Since the discovery of titanium dioxide (TiO_2) as photocatalyst by Honda and Fujishima, it has brought in-depth research in the field of photocatalytic application due to its advantages like low cost, small reactor systems, environment-friendly and renewable way of producing hydrogen [1, 2]. Anatase phase TiO_2 is able to consume only 4% of the radiation coming to the earth's surface. It has been seen that the application of TiO_2 is mostly done where UV rays are sufficiently available, e.g., in outdoor applications or in the presence of any artificial UV source in order to initiate the photocatalytic applications [3]. Therefore, to enhance photocatalytic efficiency of TiO_2 researchers from all over the world are trying to extending the spectral response toward the visible light. Henceforth, it is highly indispensable to explore correlation between structural and optical properties of the sample with their electronic and electrical properties.

Decrease in the intrinsic band gap of TiO_2 makes it more efficient for the utilization of either solar or artificial light. This goal can be achieved by some of the most popular methods which are (i) addition of carbonate salts/sacrificial agents, (ii) noble metal doping, (iii) dye sensitization [4, 5]. In the first approach, added sacrificial reagents react with holes of valance band which significantly increase the electron–hole separation giving rise to soaring quantum efficiency. The main downside of this procedure is to sustain the process that continuous addition of electron donors is essential [6]. Li et al. stated that addition of organic chemicals like formaldehyde, formic acid and oxalic acid effectively increases the competence of photocatalytic H_2 generation. In metal doping method, it has been reported that addition of Au, Ni, Rh, Pd, Pt, Ag, Cu, noble metals, increases the TiO_2 photocatalysis significantly [7–13]. Owing to the fact that Fermi level of TiO_2 being higher than above-mentioned noble metals, excited conduction band electrons readily transferred to metals incorporated on titania film surface, while valance band holes continue to stay on the TiO_2. In this method, most extensively used technique is used to consume visible light for energy transformation. For achieving the aim, dye having sensitivity toward the visible light and having redox property can be used in photocatalytic applications as well as in solar cell fabrication [4, 6, 9, 10, 14–16].

This photocatalytic mechanism promotes to the degradation of many harmful industrial dyes such as phenols, haloaromatics and halogenated biphenyls. Out of those, Rhodamine 6G belongs to the class of xanthene dyes and these are mostly used in laser materials. The present paper explores the photocatalytic dye degradation of Rhodamine 6G dye using Ag-/Fe-doped TiO_2 materials. To compare the photocatalytic efficiency, the experiment was conducted for different time periods. In order to validate structural, basic morphological and optical properties of 0.01% Fe and Ag doped titania films were characterized by XRD, FESEM, UV–Vis spectroscopic techniques.

2 Experimental Details

TiO_2 sol was developed by sol-gel process and subsequently deposited on FTO-coated glass by dip coating technique. All chemicals used were from Sigma-Aldrich and of analytical grade. In a typical procedure, titanium isopropoxide (TIP) was used as precursor to prepare the sol. In first step, TIP was mixed with 1-propanol in a mole ratio of 1:9. The solution was stirred for 10 min at normal room temperature followed by gradual addition of acetyl acetone maintaining 0.02 mol ratio to that of Ti. In the next step, a mixture of 0.01% doped material containing DI water and 1-propanol was added dropwise to the prepared solution. Obtained solution was stirred for 30 min on magnetic stirrer producing red solution and black solution for Fe and Ag doping, respectively. Finally, a mixture of polyethylene glycol-600 and acetic acid gradually was mixed to the obtained solution, continued to stir on a magnetic stirrer for another 2 h and kept for 24 h (Table 1).

Deposition of titania on FTO glass was done by dip coating method. Cleaning process of glass slide is imperative for the suitable adherence of films. Glass substrates were ultrasonicated in DI water for about 30 min followed by subsequent 5 min ultrasonication in each of acetone and isopropanol. Finally, the glass slides were dried in pure nitrogen atmosphere. Glass slides were coated with the prepared sol-gel solution at a lifting speed of 20 cm/min (maintaining delay time of 10 s) at room temperature (28 °C). The obtained slides were subjected to heating up to 470 °C at 3 °C/min heating rate and with a soaking time of 30 min in an oxidation furnace.

To investigate the crystallite size and to detect the structural properties of the developed titania films, X-ray beam diffraction analysis was done with Cu-Kα radiation ($\lambda = 1.54$ Å) using X-ray powder diffraction from PAN analytical in the range of 20–50°. Optical characterization was done using UV-1800 Shimadzu, Japan UV–Vis spectrophotometer. Morphological study of the developed film was explored by field emission SEM (Supra 40) from Carl Zeiss, Germany.

Table 1 Molecular structure and chemical formula of Rhodamine 6G

Molecular formula	Chemical formula $C_{28}H_{31}N_2O_3Cl$ Molar mass 479.02 g/mol Appearance dark red, purple or brown
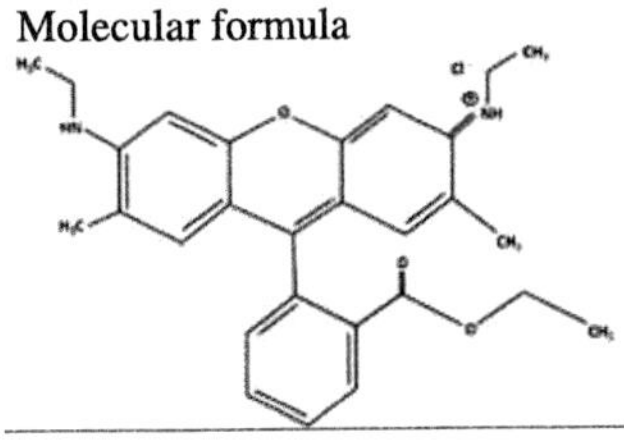	

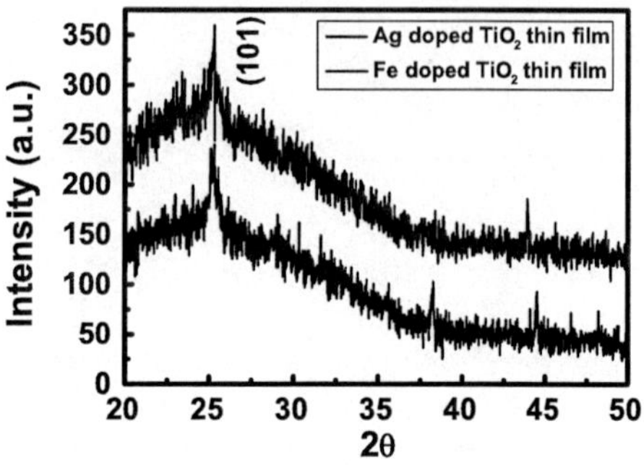

Fig. 1 X-ray beam diffraction pattern of Fe- and Ag-doped TiO_2 thin films

3 Results and Discussion

3.1 Structural Characterizations

Powder X-ray beam diffraction was carried out to find out crystallinity and crystal phase of Fe- and Ag-doped films. Owing to polycrystalline nature of the films, phase contents as well as crystallite size of the created films were analyzed with monochromatic X-ray. Obtained XRD patterns are shown in Fig. 1.

Diffraction intensity was recorded under varying diffraction angle of 20–50°. Formation of anatase phase in Ag- and Fe-doped calcined (at 470 °C) samples was confirmed from appearance of (101) diffraction peak at 2θ values of 25.3°, assigned to characteristics of anatase phase of TiO_2 as shown in Fig. 1. Observation established predominant formation of crystalline anatase phase only in both cases.

Oxide coating on substrate surface developed from doped sol can provide suitable sites for enhancing nucleation of the phase formation during initial heat treatment of the film. In addition, doping also significantly supports nucleation within the layer and enhances crystallinity [17, 18]. Observation in the present article showed no transformation of phase from anatase to rutile takes place, even after doping which is expected result at this temperature in such diluted systems. Width of XRD (101) peak becomes wider demonstrating low crystallinity of the developed film.

3.2 Crystallite Size

Average crystallite size of TiO_2 particles was calculated from Scherrer's equation as below Eq. (1):

$$D_{\text{Avg}} = \frac{k\lambda}{\beta \cos\theta} \tag{1}$$

Table 2 Calculated crystalline size, average strain, dislocation density of Fe- and Ag-doped TiO_2 thin films

Sample name	Crystallite size (nm)	Average strain (ε_{str})	Dislocation density
TiO_2–Fe	15.02	1.46	0.004
TiO_2–Ag	13.42	1.59	0.005

Here, D_{Avg} is the avg. size of the crystallite, **λ** represents X-ray beam wavelength, k is a constant (0.90), β signifies X-ray peak broadening measured as full breadth at half maxima in radian (FWHM) and θ denotes ensuing angle of diffraction.

Calculated crystallite size of anatase titania was obtained as 15.02 nm and 13.42 nm for Fe- and Ag-doped films, respectively. Generally, bond between ligands and TiO_2 does not break easily below temperature of 400 °C. Thus, this metal–ligand bond may remain in the final structure of TiO_2 at the time of calcinations and affect the morphology and structure like crystallite size of materials.

3.3 Average Strain

To enhance the device performance, average strain analysis of materials is significantly important. Exploration of structural parameters like average strain has impending significance in microelectronic devices. Improvement in performance is attained by regulating strain in the transistor channel to enhance electron–hole mobility which significantly affects overall conductivity throughout the channel.

Owing to the lattice strain, unit cells of nanocrystalline materials deviate from their normal positions resulting in very common X-ray broadening. Twinning, distortion, surfaces, domain boundaries, etc., defects result in different types of displacements. During spontaneous growth of the crystal, there is scope for development and extension of defects. The average strain ($\boldsymbol{\varepsilon}_{\mathbf{str}}$) was calculated for Fe- and Ag-doped TiO_2 thin films by applying Stokes–Wilson equation Eq. (2).

$$\varepsilon_{str} = \frac{\beta \cos\theta}{4} \tag{2}$$

where θ denotes angle of diffraction and β represents X-ray peak broadening, i.e., full width (in radian) at half maxima of the diffraction peak. The calculated values of average strain are given in Table 2.

3.4 Dislocation Density

Dislocation density (δ) was obtained with the help of Eq. (3), and the values have been mentioned in Table 1.

$$\delta = \frac{1}{D_{\text{Avg}^2}} \tag{3}$$

where D_{Avg} is the average size of crystallite. Results of dislocation density (δ) well demonstrate extent of defects in samples.

3.5 Morphological Characterizations

Figure 2a–d illustrates FESEM images of Fe- as well as Ag-doped TiO_2 thin films (morphology and cross section). It has been observed that the films exhibit an almost uniform spherical grain-like structure. Shape of particles is quite similar to each other and dense. Compact morphology can be identified from Fig. 2. Figure 2c, d shows cross section of Fe- and Ag-assimilated titania films presenting thickness of 509 nm and 319 nm for Fe- and Ag-doped films, respectively.

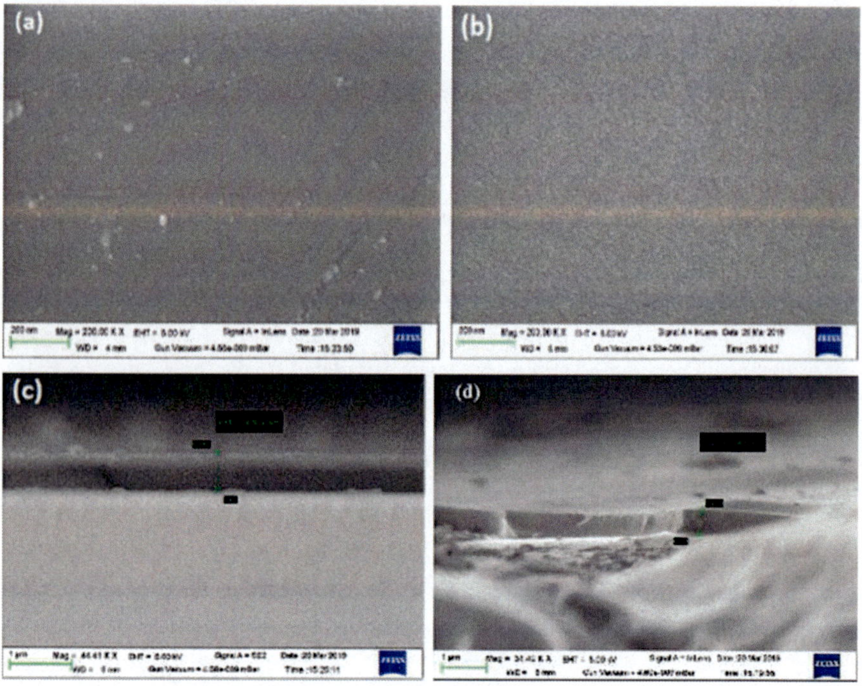

Fig. 2 **a**, **b** are the FESEM surface images of Fe- and Ag-doped thin films. **c**, **d** are the FESEM cross-sectional images of Fe- and Ag-doped thin films

3.6 Optical Characterizations

For the global development of science, technology and industry, the study of optical behavior of thin films is inevitably significant. Photoactivity of TiO_2 thin film can be studied from optical performance of the film in visible and ultraviolet ranges. Photocatalytic performance of the developed titania film attributes to its UV–visible light absorption efficiency. Depending on specific applications, color of thin film is highly important; i.e., film must be transparent and conductive.

Effect of Fe and Ag doping on film behavior was studied with UV–Vis spectroscopic measurement. Quantification of absorption, light transmission and reflection to obtain electronic properties of material at the microscopic level is exceedingly essential. In order to determine optical parameters, transmission spectra of different modified titania films on glass slide were recorded in 300–800 nm range. The transmission spectra of the single layer TiO_2 coatings on glass are shown in Fig. 3. UV–visible spectra of the developed film demonstrate higher extent of transmittance in visible light region as depicted in Fig. 3.

Interference phenomenon observed in transmission spectrum of the coated glass indicates a good surface quality (smoothness and flatness) of the developed film.

Figure 3 also demonstrates film surface is reflecting without much scattering/absorption in the bulk of the film. Fabrication techniques, composition and structure have noteworthy influence on optical properties of titania film. The width of the absorption band increases with increase in film thickness (Fig. 4). The original peak in the UV region was shifted to higher wavelength region with increase in coating thickness.

Index of refraction (n) was calculated using the formula Eq. (4):

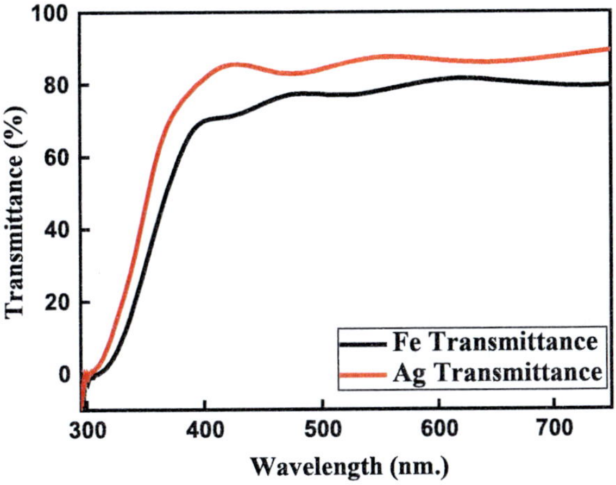

Fig. 3 UV–Vis transmittance spectra of Fe- and Ag-doped TiO_2 thin films

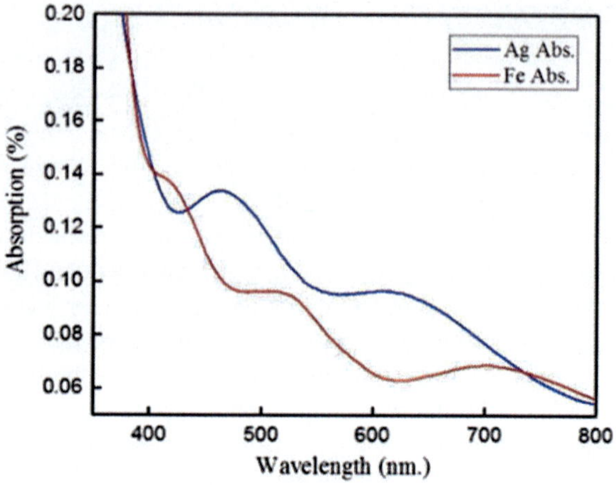

Fig. 4 UV–Vis absorbance spectra of Fe- and Ag-doped TiO_2 thin films

$$n = \left(\left(\frac{N(\lambda_1 \times \lambda_2)c}{2(\lambda_1 - \lambda_2)t} \right)^2 + \sin^2 a \right)^{1/2} \tag{4}$$

where

λ_1 = maximum wavelength, λ_2 = wavelength minimum
N = fringe numbers, n = index of refraction
a = incidence angle, c = angstrom conversion factor.

Obtained E_g for Fe- and Ag-doped films were 2.85 eV and 2.78 eV, respectively. A number of layers as well as variation of dopant have substantial effect on band gap of developed film [19]. Decrease in band gap (E_g) in case of doped material compared to that of undoped one evidenced that Fe and Ag cations are successfully integrated into titania lattice [20, 21]. The interaction of conduction band e^- with dopant e^- results in decreasing the band gap by shifting the valence band (VB) and lowering the conduction band (CB). Figure 3 displays spectral response of 0.01% Fe- and Ag-doped titania films. Spectra revealed that optical characteristics of both the doped films were quite good. Developed Fe- and Ag-doped titania films showed significant transmittance in the visible region but low transmittance in UV range [22, 23]. Absorption peak at 465 nm and 525 nm for Ag-doped and Fe-doped films, respectively, may be endorsed to inherent band–band transition (Table 3).

The absorption maxima in visible range for Fe- and Ag-incorporated film were obtained at 643 nm and at 710 nm, respectively. Observed absorption in this region for the doped samples may be attributed to the metal electronic transition. It was

Table 3 Calculated thickness (nm), refractive index and band gap of the Fe- and Ag-doped TiO_2 thin films, respectively

Sample name	Thickness (nm)	Refractive index	Band gap (E_g in eV)
TiO_2–Fe	509.5	1.5	2.85
TiO_2–Ag	314.6	2.1	2.78

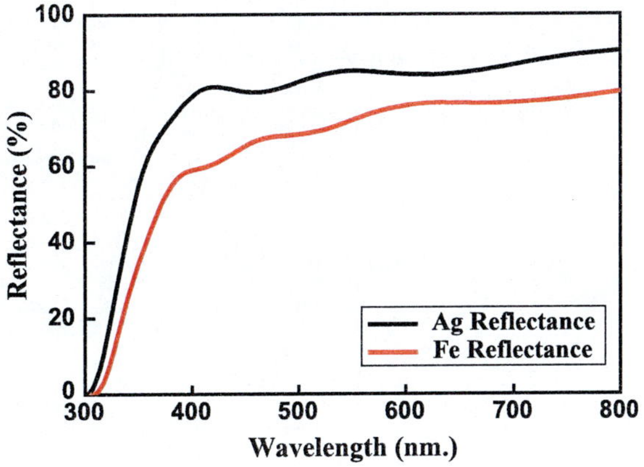

Fig. 5 Reflectance of Fe- and Ag-doped TiO_2 thin films

noted that the index of refraction for developed titania films varies with change in wavelength (Fig. 5).

Introduction of new states of energy in forbidden zone of TiO_2 was accountable for absorption peak shifting. Electrons efficiently excited from VB to CB when photon energy equals to that of energy band gap of the material.

These electrons are efficiently arrested by the external circuit in photovoltaic devices and generate current in the circuit. From the graphical plot of $h\nu$ against $(\alpha h\nu)^{1/2}$, energy band gap (E_g) of doped TiO_2 film was calculated (Figs. 8 and 9).

Structure of energy band gap of the developed material can be analyzed from the study of absorption property of the material. Optical band gap energy of semiconductor plays a major role in fabrication of photovoltaic cells. Variation of optical band gap (E_g) and refractive index (n) with thickness (nm) for TiO_2 thin film is shown in Table 2. Good optical property of the developed TiO_2 thin film was observed from its sharp absorption edge. Calculated band gap for both Ag- and Fe-doped films was found to be 2.78 and 2.85, respectively, as presented in Figs. 8 and 9, respectively.

3.7 Photocatalytic Activity Study

Photocatalytic dye degradation efficiency of the developed film was explored by testing its degradation test toward Rhodamine 6G dye [24]. Figure 6 shows UV–Vis spectra of dye degradation study before and after 2 h visible light irradiation in the presence of Ag- and Fe-incorporated films. Obtained spectra predominantly demonstrate high efficiency of Ag-doped film over Fe-doped one. Higher activity credited to the effect of Ag deposition on TiO_2 surface acting as electron trap center reducing the recombination probability of electron–hole pair on the oxide surface and resulting in improved charged transfer efficiency on the matrix result likewise corresponds to strain and lower band gap in Ag-doped film that prompts to higher absorption in the visible region.

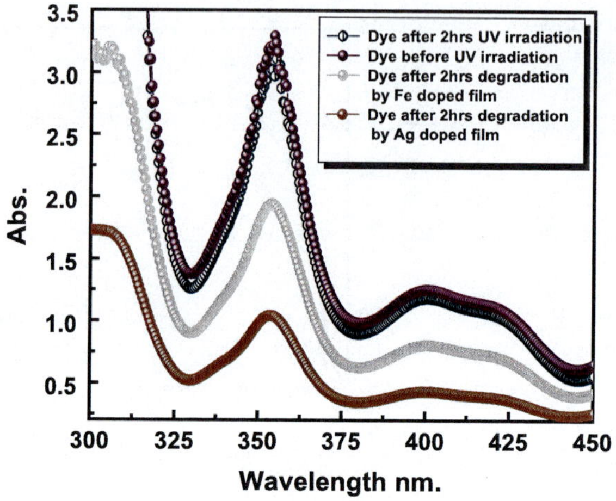

Fig. 6 Dye degradation study by Fe- and Ag-doped TiO_2 thin films

Figure 7 shows the degradation mechanism of photocatalytic dye degradation using Fe/Ag–TiO_2 materials. In an aqueous environment, the holes created under UV radiation are scavenged by the hydroxyl groups preset on the surface, generating OH. radicals which promote oxidation of organics. This promotes the degradation of many harmful industrial dyes. As shown in Fig. 7, the color of the dye gets faded when it is being exposed to sunlight. Figures 8 and 9 show the obtained band gap of Ag- and Fe-doped TiO_2 thin films, respectively, from the spectroscopic data.

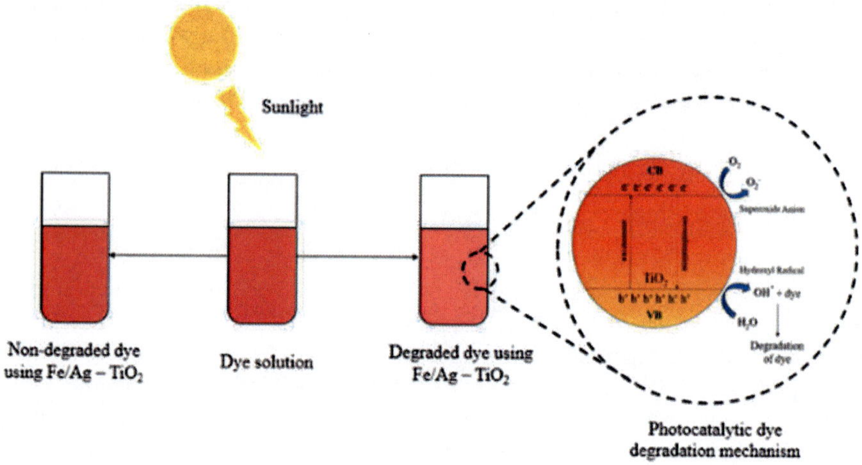

Fig. 7 Dye degradation mechanism

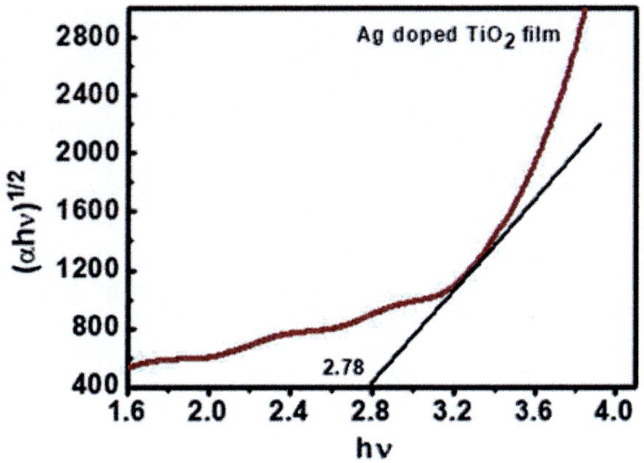

Fig. 8 $(\alpha h\nu)^{1/2}$ versus $h\nu$ plot of Ag-doped TiO_2 thin film

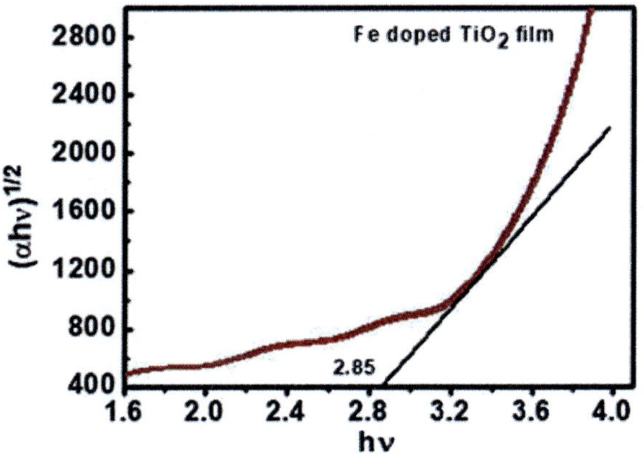

Fig. 9 $(\alpha h\nu)^{1/2}$ versus $h\nu$ plot of Fe-doped TiO_2 thin film

4 Conclusion

Highly uniform doped titania film was efficaciously developed from a different metal-doped sol by dip coating process. Effect of metal integration on optical, structural and morphological behavior of the developed films was investigated in detail. X-ray beam diffraction study showed prevalent formation of anatase crystalline phase.

Crystallite size and dislocation density of Fe- and Ag-doped films were calculated as 15.02 nm, 13.42 nm and 0.004, 0.005, respectively. Refractive index of the film was calculated as 1.5 and 2.1 for Fe- and Ag-doped films, respectively. Decreased band gap from 3.2 eV to 2.78 and 2.85, respectively, for Ag- and Fe-doped films attributed to the increased strain in the film as a result of doping. Ag doped films are more strained compared to Fe doped due to smaller band gap. Higher photocatalytic dye degradation of Ag-doped film over Fe-doped one is credited to Ag deposition on TiO_2

surface creating electron trap center which reduces the recombination probability and results in improved charged transfer process on the matrix.

References

1. Finegold L, Cude JL (1972) One and two-dimensional structure of alpha-helix and beta-sheet forms of poly (L-alanine) shown by specific heat measurements at low temperatures (1.5–20 K). Nature 238:37–38
2. Khan SUM, Al-Shahry M, Ingler WB (2002) Efficient photochemical water splitting by a chemically modified n-TiO_2. Science 224
3. Hagfeldt A, Gratzel M (1995) Light-induced redox reactions in nanocrystalline systems. Chem Rev 49–68
4. Ni M, Leung MKH, Leung DYC, Sumathy K (2007) A review and recent developments in photocatalytic water-splitting using TiO_2 for hydrogen production. Renew Sustain Energy Rev 11:401–425
5. Sen S, Mahanty S, Roy S, Heintz O, Bourgeois S, Chaumont D (2005) Investigation on sol-gel synthesized Ag-doped TiO_2 cermet thin films. Thin Solid Films 474(1–2):245–249
6. Liao C, Huang C, Wu JCS (2012) Hydrogen production from semiconductor-based photocatalysis via water splitting. Catalysts 490–516
7. Hoffmann MR, Martin ST, Choi W, Bahnemannt DW (1995) Environmental applications of semiconductor photocatalysis. Chem Rev 69–96
8. Linsebigler AL, Lu G, Yates JT (1995) Photocatalysis on TiO_n surfaces: principles, mechanisms, and selected result. Chem Rev 735–758
9. Soc JE, Natarajan C, Nogami G (1996) Cathodic electrodeposition of nanocrystalline titanium dioxide thin films. J Electrochem Soc 143(5):1547–1550
10. Kenry J, Yeo C, Lim CT (2016) Emerging flexible and wearable physical sensing platforms for healthcare and biomedical applications. Microsyst Nanoeng 2:16043
11. Viana MM, Mohallem NDS, Miquita DR, Balzuweit K, Silva-Pinto E (2013) Preparation of amorphous and crystalline Ag/TiO_2 nanocomposite thin films. Appl Surf Sci 265:130–136
12. Dholam R, Patel N, Adami M, Miotello A (2009) Hydrogen production by photocatalytic water-splitting using Cr- or Fe-doped TiO_2 composite thin films photocatalyst. Int J Hydrogen Energy 34(13):5337–5346
13. Kim NJ, La YH, Im SH, Ryu BK (2010) Optical and structural properties of Fe–TiO_2 thin films prepared by sol–gel dip coating. Thin Solid Films 518(24):e156–e160
14. Gao Y, Masuda Y, Koumoto K (2004) Micropatterning of TiO_2 thin film in an aqueous peroxotitanate solution. Chem Mater 16:1062–1067
15. Kang M, Kim SW, Park HY (2018) Optical properties of TiO_2 thin films with crystal structure. J Phys Chem Solids 123:266–270
16. Nasralla N, Yeganeh M, Astuti Y, Piticharoenphun S, Shahtahmasebi N (2013) Structural and spectroscopic study of Fe-doped TiO_2 nanoparticles prepared by sol–gel method. Sci Iran 20(3):1018–1022
17. Saini KK, Sharma SD, Chanderkant, Kar M, Singh D, Sharma CP (2007) Structural and optical properties of TiO_2 thin films derived by sol-gel dip coating process. J Non-Cryst Solids 353(24–25):2469–2473
18. Bally AR et al (1998) Structural and electrical properties of Fe-doped TiO_2 thin films. J Phys D Appl Phys 1149
19. Mahata S, Mahato SS, Nandi MM, Mondal B (2012) Synthesis of TiO_2 nanoparticles by hydrolysis and peptization of titanium isopropoxide solution. AIP Conf Proc 1461(1):225–228
20. Lin CYW, Channei D, Koshy P, Nakaruk A, Sorrell CC (2012) Effect of Fe doping on TiO_2 films prepared by spin coating. Ceram Int 38(5):3943–3946

21. Karuppuchamy S, Nonomura K, Yoshida T, Sugiura T, Minoura H (2002) Cathodic electrodeposition of oxide semiconductor thin films and their application to dye-sensitized solar cells. Solid State Ionics 151:19–27
22. Fuchs TM et al (2002) Deposition of titania thin films from aqueous solution by a continuous flow technique. J Mater Chem 1597–1601
23. Gao Y, Masuda Y, Peng Z, Yonezawa T, Koumoto K (2003) Room temperature deposition of a TiO_2 thin film from aqueous peroxotitanate solution. J Mater Chem 4:1–6
24. Mondal B, Usha K, Mahata S, Kumbhakar P, Nandi MM (2011) Synthesis and characterization of nanocrystalline TiO_2 thin films for use as photoelectrodes in dye sensitized solar cell application. Trans Indian Ceram Soc 70(3):173–177

MMaSA—A Modified Matrix Synthesis Approach for Thermal-Aware Gate Array Placement

Somnath Roy Choudhury, P. Sai Srinivas, and Sambhu Nath Pradhan

Abstract This work presents a thermal-aware gate array placement based on a stochastic thermal model and the simulated annealing (SA) algorithm. The proposed placement algorithm modified matrix synthesis approach (MMaSA) optimizes the peak on-chip temperature of gate array placement. The thermal characterization of a placement by MMaSA involves a modified model of the combinatorial optimization problem, the MSP with the inclusion of the adiabatic influence of the die boundary. Experiments have been conducted by including the characteristics of gate arrays on MCNC benchmark circuits. Results show that MMaSA gives improved results in comparison with the MSP approach.

Keywords SA · On-chip temperature · MCNC · Optimization · Matrix synthesis problem (MSP)

1 Introduction

Processors with large functionality and very high operating frequencies often become 'thermally sick', due to the formation of hotspots and temperature gradients. Hotspots and temperature gradients lead to phenomenon such as electro-migration [1, 2], generate thermal stress and lead to clock skew and logical errors [3]. At the physical design stage of the VLSI design process, placement and floorplan play a very vital role in the hotspot minimization and hence designing of efficient floorplan or placement algorithms becomes very important.

Some of the efforts made towards the realization of thermal-aware placement and floorplan are discussed as follows. Paper [4] presents a GA-based thermal-aware floorplan algorithm for optimizing the chip area and on-chip temperature of slicing floorplan. Paper [5] presents a hybrid PSO-GA algorithm for minimizing chip area,

S. R. Choudhury (✉) · P. Sai Srinivas · S. N. Pradhan
Department of Electronics and Communication Engineering, National Institute of Technology Agartala, Agartala, India
e-mail: som.vlsi@gmail.com

V. Nath and J. K. Mandal (eds.), *Proceedings of the Fourth International Conference on Microelectronics, Computing and Communication Systems*, Lecture Notes in Electrical Engineering 673, https://doi.org/10.1007/978-981-15-5546-6_25

wire length and temperature of non-slicing floorplan. The authors in [6] present a model for the thermal-aware placement for gate arrays by implementing a combinatorial optimization, the matrix synthesis problem (MSP). Moreover, the authors in [7] also present a technique for the thermal-aware placement of gate arrays. Our work is one such effort towards the mitigation of hotspots for gate array placement.

Contribution of our work: (a) Development of a new thermal model and a thermal metric, viz. 'modified critical threshold (MCT)' for thermal characterization of the gate array placement. (b) Development of a thermal-aware algorithm MMaSA, for gate array placement. (c) Development of a reference placement algorithm based on the simple approximation algorithm (SAA) [6] for gate arrays. (d) Comparative study between the reference algorithm SAA and the proposed placement algorithm MMaSA by implementing them on MCNC benchmarks arranged in gate array placement.

Motivation of the work: MSP [6] considers the placement of the gate array cells in the form of a square matrix and divides the placement into thermal zones of txt sub-matrices. Parameter t is an integer related to the heat transfer in the die. For poor heat transfer, $t = 2$ gives good approximation. The target of MSP is to minimize the maximum of the sum of heat (or heat dissipation per unit time, i.e. power) amongst all the txt sub-matrices or thermal zones. The solution to the MSP can be obtained by the rearrangement of the gate array cells and the SAA one such solution to the MSP.

The SAA first arranges the cells in descending order of power values (or heat) and then divides the cells into four groups (Gi) of equal numbers of cells, viz. G0 = {L0}, G1 = {L1}, G2 = {L2} and G3 = {L3}. The power (or heat) of the cells is such that power of group G0 > G1 > G2 > G3. It then constructs the gate array placement with the repeating pattern of the txt (with $t = 2$) sub-matrices with cells (L0, L1, L2, L3) by considering one cell from each of the four groups. This always ensures that a high power cell is always surrounded by low and medium power cells and as a result the hotspot minimization occurs. MSP defines a thermal metric μ_t (which may be called as the critical threshold) and is the maximum of the aggregate of power (or heat) of the cells amongst all the txt zones. So according to SAA, placements having equal μ_t values are equally thermally fit. The SAA targets towards achieving the minimum critical threshold value in order to minimize the hotspots. In order to form the groups (Gi) of equal sizes (with total number of elements of the placement in multiples of t^2), sometimes dummy cells of zero power need to be added to the placement.

As an example of MSP, we consider a random arrangement of gate array cells (with $t = 2$) as shown in Fig. 1a. The numbers in the cells denote the power dissipation value (in watt) of the corresponding cells. The indexes L0, L1, L2 and L3 denote the cell placement according to SAA in a txt zone. The dotted red boundary denotes the zone of critical threshold. Placement 1 in Fig. 1a is improper with respect to SAA since the cells are not arranged according to the (L0, L1, L2, L3) pattern. It is a bad solution with high critical threshold. Placement 2 in Fig. 1b is in agreement with SAA arrangement pattern and is the best possible placement with respect to SAA with the least possible critical threshold. Placement 3 in Fig. 1c has equal μ_t or critical

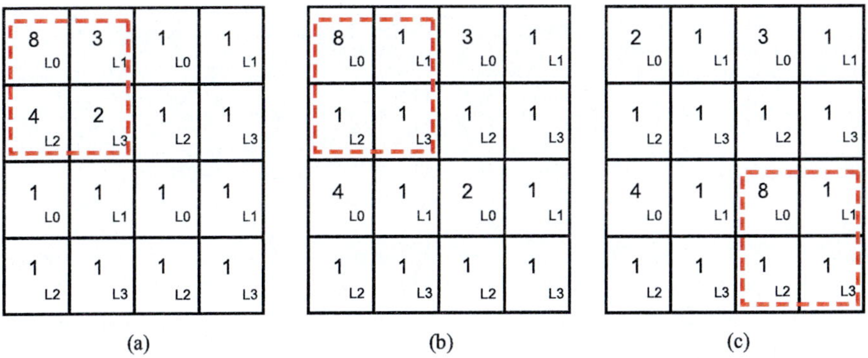

Fig. 1 Example placement of MSP with $m = 4$ (rows) and $n = 4$ (columns). **a** Placement 1 bad solution. **b** Placement 2 optimal solution. **c** Placement 3 optimal solution

threshold value, as Placement 2 and hence SAA may not consider the formation of Placement 3 while determining a better placement solution than Placement 2.

For experimental implementation of the example, the height and width of each block are considered to be 0.003 m and 0.0015 m, respectively. Experimental results in Fig. 2a–c represent the thermal maps of Placement 1, Placement 2 and Placement 3, respectively.

Table 1 and Fig. 2 show that: (i) Placement 1 has the highest peak temperature, and it is in agreement with the high μ_t value predicted by SAA. (ii) Placement 2 has lesser peak temperature due to the lower (best) μ_t value predicted by SAA. (iii) Placement 3 (having same μ_t value as Placement 2) has lesser peak temperature than Placement 2, and this temperature improvement is not supposed to happen according to SAA. The peak temperature has reduced in Placement 3 because the high power

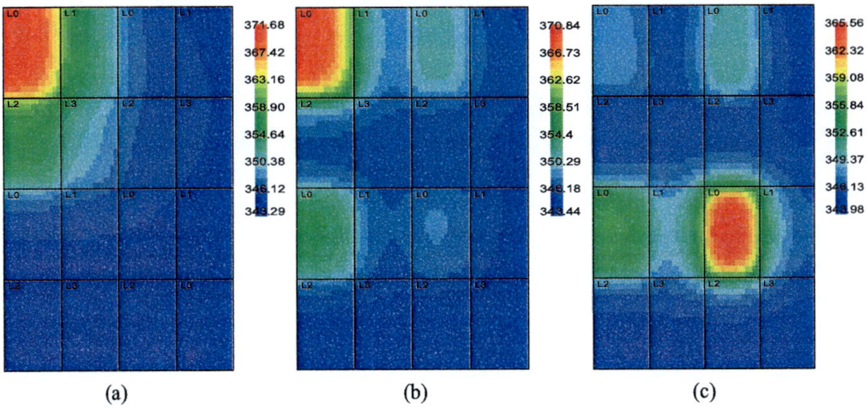

Fig. 2 Thermal map of the placements illustrated in Fig. 1. **a** Placement 1 bad solution. **b** Placement 2 optimal solution. **c** Placement 3 optimal solution

Table 1 Critical threshold value and peak temperatures of the placements in Fig. 1

Solution	Critical threshold (μ_t) [W]	Peak temperature [K]
Placement 1	17	370.29
Placement 2	11	368.79
Placement 3	11	365.55

block of 8 W (a critical block) sitting at the die corner in Placement 2 has been moved to a relatively distant location from the die boundary in Placement 3.

Hence, the MSP is not enough to guarantee a thermal-aware placement of gate arrays since it considers the (txt) region with maximum sum to be having the maximum temperature without accounting for the relative position of the functional blocks from the die boundary. As shown in Fig. 2 of [8], it can be observed that there are no lateral resistive heat paths available along the die boundary. This is because the die boundary behaves adiabatically to the heat flux released from a functional block. Hence in this work, we have modified the MSP by including the adiabatic influence of the die boundary during the thermal metric evaluation.

2 Definitions Related to the Proposed Gate Array Placement

(a) Dimension of a gate array cell: In a gate array, all the gate array cells are considered as rectangular envelopes, having the same dimensions, i.e. identical height and width. It has been considered that the height and width of a gate array cell are equal to the highest height and highest width of all functional blocks in the circuit so that a gate array cell can encompass any functional block in the circuit within itself. A circuit gets arranged in a gate array, such that one gate array cell gets assigned with each functional block.

(b) Dimension of a functional block: The functional blocks of a circuit are considered to be rectangular in shape. The height and width of a functional block may be lesser than or equal to the height and width of a gate array cell. A functional block within a gate array cell is so positioned that the lower left corners of both the functional block (rectangle) and the gate array cell (rectangle) coincide.

The height and width dimensions of both a gate array cell and a functional block lie on the plane of the die and also stay parallel to the edges of the die.

(c) Critical functional block: Functional blocks whose power density dp_i is such that $dp_i \geq dp_{avg} + Md$ are the critical blocks, and they are highly probable of forming the hotspots. Here, dp_{avg} is the average power density and Md is the mean deviation in power density of the functional blocks of a circuit.

(d) Power density of gate array cell: Power density of a gate array cell Pd_i is value of the power of the functional block contained in the gate array cell divided by the area of the gate array cell.

3 The Proposed Thermal Model

Assumptions of the model are as follows:

1. Instead of considering the power (or heat) of the gate array cells for defining the critical threshold of a txt thermal zone, the proposed model considers the power density of gate array cells for defining the critical threshold.
2. Instead of focusing on all the possible txt zones, to reduce the computational complexity, the model focuses only on those txt zones which contain the critical functional blocks, since hotspots are caused by critical blocks. Such zones are termed as critical zones. A critical zone is considered to contain only one critical functional block.
3. It redefines the critical threshold function by considering two terms which are defined as follows.

Aggregate power density term (μ): It is the sum of the power densities (Pd_i) of the gate array cells '*i*' which belong to a critical zone and is given by

$$\mu = \sum_{i=1}^{(t \times t)} \mathrm{Pd}_\mathrm{i}. \tag{1}$$

With the increase in the power density aggregate term, the probability of rise in temperature of a critical zone gets enhanced.

Heat reflection term (ϕ): This term is related only to the critical functional block in a critical zone. The model considers the die boundary to be adiabatic, due to which the heat flowing from a critical functional block (along *x* and *y* directions) to the boundary gets reflected back to the functional block. As the critical block approaches closer to the die boundary, the reflected heat component goes on increasing resulting in an increase in temperature of the critical functional block and the critical zone. It has been considered that: (a) $\phi \propto L$, where L is the height or width of the critical functional block which ever faces the boundary. (b) $\phi \propto \mathrm{dp_i}$, where dp_i is the power density of the critical functional block. (c) $\phi \propto 1/(s + \Delta)$, where '*s*' is shortest perpendicular distance (along *x* or *y* direction) between the die boundary and the side of the functional block facing the boundary. Parameter Δ ensures that the division value does not become infinity when '*s*' becomes zero. The new modified critical threshold (MCT) function for a critical zone becomes:

$$\text{MCT} = \frac{\mu}{\mu_{\max}} + \frac{\phi}{\phi_{\max}} \tag{2}$$

Parameter Φ includes the sum of the heat reflection components along x and y directions. Parameters $\mu_{\max}$ and $\phi_{\max}$ are the highest possible values of μ and heat reflection component (along x or y), respectively. The thermal metric has been normalized since the two terms are dimensionally different. Now the objective of the placement problem is to obtain an optimal placement for the gate array arrangement such that the new modified critical threshold (MCT) function is minimized.

4 Implementation, Results and Analysis

In this work, two gate array placement algorithms have been developed in C programming language, based on the simulated annealing (SA) algorithm [9]. (i) The first algorithm uses the simple approximation algorithm (SAA) [6] for solving the MSP of distributing the functional blocks in the gate array. For further improving the placement, the perturbations adopted are (a) swapping of entire two random txt sub-matrices and (b) randomly swapping the corresponding L0 or L1 or L2 or L3 cells between two random txt sub-matrices. Perturbations occur while maintaining the SAA principle intact. The first algorithm is used as a reference.

(ii) The second algorithm (MMaSA) uses the proposed thermal model (discussed in Sect. 3). For further improvement of the placement, the perturbation used is swap of random cells between two randomly selected txt sub-matrices.

For finding the peak temperature value of the optimal placements obtained from the placement algorithms, hotspot tool [10] has been used. For the experimentation purpose, the functional blocks of the MCNC benchmark circuits [11] have been arranged in a gate array placement distribution. Power densities of MCNC benchmark modules have been chosen randomly between 0.022 and 2.4 W/mm^2 such that the average power density of benchmark apte is 0.593831 W/mm^2, xerox is 0.729076 W/mm^2, hp is 0.72433 W/mm^2, ami33 is 0.729749 W/mm^2, and ami49 is 0.728304 W/mm^2. From the experimental results of Table 2 and Fig. 3, we observe that the proposed MMaSA algorithm gives −0.39%, 3.4%, 0.26%, 2.67% and 3.44% improvement in the peak temperature (considered in °C), respectively, for apte, xerox, hp, ami33 and ami49 over SAA. The proposed algorithm MMaSA gives an average improvement of 1.88% in peak temperature (considered in °C) over SAA.

5 Conclusion

This work depicts a thermal-aware placement for gate arrays, the algorithm has been developed in C language, and experimentation has been done in a Linux environment running on a 3.0 GHz Intel Core i5 processor. The proposed algorithm modifies the

Table 2 Attributes of the best placement solutions with respect to peak temperature for SAA and MMaSA

Circuit	Gate array cell size		Number of functional blocks	Number of dummy cells	Matrix array size [row × column]	Area [10^{-6} m^2]	Peak temperature [K]	
	Height [10^{-3} m]	Width [10^{-3} m]					SAA [6]	MMaSA
apte	1.832	3.186	9	7	4 × 4	93.388	359.69	360.03
xerox	2.569	1.295	10	6	4 × 4	53.2297	365.71	362.54
hp	0.7	3.304	11	5	4 × 4	37.0048	334.11	333.95
ami33	2.982	3.360	33	3	6 × 6	360.703	324.32	322.95
ami49	25.872	24.640	49	15	8 × 8	40799.109	352.31	349.58

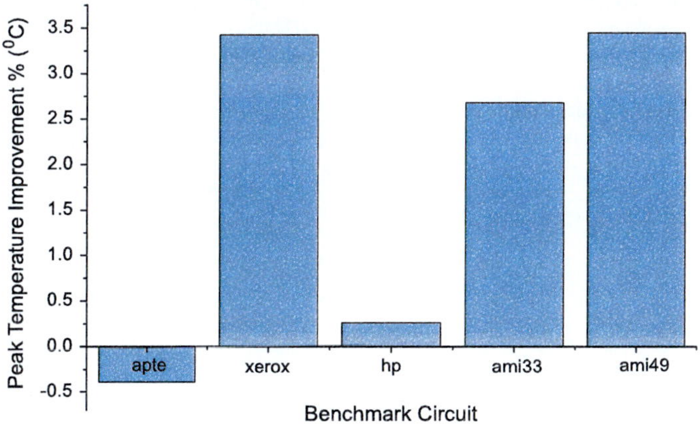

Fig. 3 Percentage improvement in peak temperature (in °C) of MMaSA over SAA

critical threshold thermal metric of the MSP by including the adiabatic influence of the die boundary. Experimental results show that the proposed placement algorithm is efficient in terms of minimizing the peak on-chip temperature of gate arrays. As a future scope, it has a good possibility of implementation for gate array placement in 3D IC.

Acknowledgements This work has been supported by SMDP-C2SD project sponsored by MeitY, Government of India.

References

1. Black JR (1969) Electromigration—a brief survey and some recent results. IEEE Trans Electron Devices 16(4):338–347

2. Ho PS (1982) Basic problems for electromigration in VLSI applications. In: Proceedings of the IEEE 20th international reliability physics symposium, pp 288–291
3. Cheng YK, Raha P, Teng CC, Rosenbaum E, Kang S-M (1998) ILLIADS-T: an electrothermal timing simulator for temperature sensitive reliability diagnosis of CMOS VLSI chips. IEEE Trans Comput-Aided Des Integr Circuits Syst 17(8):668–681
4. Hung W-L, Xie Y, Vijaykrishnan N, Addo-Quaye C, Theocharides T, Irwin MJ (2005) Thermal-aware floorplanning using genetic algorithms. In: Proceedings of IEEE international symposium on quality electronic design (ISQED), pp 634–639
5. Sivaranjani P, Senthil Kumar A (2015) Thermal-aware non-slicing VLSI floorplanning using a smart decision-making PSO-GA based hybrid algorithm. Circuits Syst Signal Process 34(11):3521–3542
6. Chris C, Chu N, Wong DF (1998) A matrix synthesis approach to thermal placement. IEEE Trans Comput-Aided Des Integr Circuits Syst 17(11):1166–1174
7. Ghosal P, Samanta T, Rahaman H, Dasgupta P (2008) Thermal-aware placement of standard cells and gate arrays: studies and observations. In: IEEE computer society annual symposium on VLSI
8. Huang W, Stan MR, Skadron K (2005) Parameterized physical compact thermal modeling. IEEE Trans Compon Packag Technol 28(4):615–622
9. Kirkpatrick S, Gelatt CD Jr, Vecchi M (1983) Optimization by simulated annealing. Science 220(4598):671–680
10. http://lava.cs.virginia.edu/hotspot
11. MCNC benchmark netlists. Avalaible at https://s2.smu.edu/~manikas/Benchmarks/MCNC_Benchmark_Netlists.html

Computation of Absorption Coefficient for Pöschl–Teller Potential in Double Quantum Well Structure for Photodetector Applications

Suporna Bhowmick, Debarati Chakraborty, Dayita Guha, Bijoya Chakraborty, Pampa Debnath, and Arpan Deyasi

Abstract Absorption coefficient due to lowest quantum state of double-well-triple-barrier structure is analytically investigated for Pöschl–Teller potential configuration. Kane-type conduction band nonparabolicity of first order is considered for evaluation of eigenstates for realistic computation, and Lorentzian lineshape is taken into account. Tuning parameters of the Pöschl–Teller geometry is varied along with structural parameters and results are compared with earlier obtained rectangular, parabolic and Gaussian potentials. Corresponding blueshift/redshift is obtained for each specific case, and effect of applied field along the direction of quantum confinement is analyzed compared with the situation when field is absent. Comparative study speaks in favor of the present structure both in terms of higher value of absorption as well as lower linewidth spectrum (wavelength scale). Result speaks in favor of the configuration compared to the idealized potential well for photodetector and sensor applications.

Keywords Absorption coefficient · Pöschl–Teller potential · Lorentzian lineshape · Eigenstates · Redshift/blueshift · Structural parameters

---

S. Bhowmick · D. Chakraborty · D. Guha · B. Chakraborty · P. Debnath · A. Deyasi (✉)
Department of Electronics and Communication Engineering, RCC Institute of Information Technology, Kolkata, India
e-mail: deyasi_arpan@yahoo.co.in

S. Bhowmick
e-mail: bhowmicksuporna@gmail.com

D. Chakraborty
e-mail: chakrabortydebarati6991@gmail.com

D. Guha
e-mail: dayitaguha25@gmail.com

B. Chakraborty
e-mail: bijoyachakraborty1410@gmail.com

P. Debnath
e-mail: poonam.4feb@gmail.com

V. Nath and J. K. Mandal (eds.), *Proceedings of the Fourth International Conference on Microelectronics, Computing and Communication Systems*, Lecture Notes in Electrical Engineering 673, https://doi.org/10.1007/978-981-15-5546-6_26

1 Introduction

Optical application of quantum devices is the subject of research since the last thirty years, and several quantum-confined structures are proposed [1–3] for novel optical performance. Published data suggests the focuses are made on different material configurations [4, 5] as well as complex geometries [6, 7], and mathematical computation becomes more complex for the later sections for the theoretical researchers due to inability of analytical methods to solve Schrödinger equations. Several numerical techniques are proposed [8–12] for solving those structures in order to investigate the fundamental electronic property, namely the eigenenergies. With the advances of photonics, optical properties of these structures are also investigated, and absorption coefficient gets the prime attention owing to its importance of applying the device as optical emitter/detector [13].

Effect of size quantization on optical absorption coefficient of quantum well is analyzed a decade ago [14] in presence of electric field where ideal orthonormalization conditions of wavefucntions are assumed. Work is extended by considering distribution of carriers [15] in the devices, and simultaneously gain and recombination coefficients are estimated. Experimental results are obtained for MQW structures, and modeled with theoretical data for lower magnitude of electric field [16]. Combined effect of electro-optic field [17] is calculated on quantum well structures, later, which is a consequence of the previous findings of suitability of GaAs material for fastest holography [18] based on absorption coefficient analysis. Effect of nonparabolicity is included a few years back [19], preceeded by work on strain engineering [20]. Redshift in double quantum well structure is reported in recent past [21] at 1550 nm, and work is applied on Morse quantum well [22] as well as inverse parabolic well [23]. Very high absorption coefficient is recently reported [24] for lattice matched structures.

Pöschl–Teller potential has one unique advantage compared with any other potential well geometry is that it exhibits Bose–Einstein condensation of noninteracting 39 K atoms [25]. Also, it shows higher frequency of operation when designed for photodetector applications. Combining these two effects, authors simulated the absorption profile of Pöschl–Teller potential in presence of field for various structural parameters, and also in a specified tuning range of design parameters. Simulation is performed using MATLAB® software. The reasons for choosing Pöschl–Teller potential over other configurations are established with supremacy of the findings through comparative study, which clearly established novelty of the work. Results are represented in graphical form, and key data are presented in tabular format.

2 Mathematical Formulation

Optical absorption coefficient for arbitrary quantum well device can be written in the following form

$$\alpha(\omega) = \frac{q^2 m^*}{2\varepsilon_0 \varepsilon_r c n_r \hbar d m_0^2} \frac{M^2}{\hbar\omega} \times \sum_{n,m} \langle \zeta_v^m | \zeta_c^n \rangle \Theta(E_{mn} - \hbar\omega) \quad (1)$$

where $\langle \zeta_v^m | \zeta_c^n \rangle$ is the integral function denotes the overlap between the envelope functions of valence and conduction bands, respectively, 'd' denoted the quantum well width, Θ represents Heaviside step function, 'M' is defined as

$$M = N \int_{\text{cell}} \psi_{\kappa_c}^*(\vec{r}') p_A \psi_{\kappa_v}(\vec{r}') \mathrm{d}v \quad (2)$$

In Eq. (1), the integral function gives the selection rule between two adjacent quantum states. It may be written in the form

$$\langle \zeta_v^m | \zeta_c^n \rangle = \delta_{\kappa\kappa'} \int_{-d/2}^{d/2} \chi_{\mathrm{m}}^h(z) \chi_n^e(z) \mathrm{d}z \quad (3)$$

where it may be noted that we have considered z-axis is the direction of applied field. Here, the wavefunctions χ_i is selected from generalized solution of Schrödinger wave equation.

Substituting the values of wavefucntions with some rearrangement, we get

$$\alpha(\omega) = \frac{2\pi q^2 \hbar}{\varepsilon_0 \varepsilon_r c n_r m^{*2} V} \frac{1}{\hbar\omega} \times \sum_{i,j} \left| \langle i | p_z | j \rangle^2 \right| \delta(E_j - E_i - \hbar\omega)(f_{\mathrm{FD}}^i - f_{\mathrm{FD}}^j) \quad (4)$$

where 'i' and 'j' are the initial and final states, and f_{FD} is the Fermi–Dirac distribution function. Kronecker delta function is introduced to conserve the momentum.

Introducing Lorentzian lineshape function and the concept of oscillator strength, we get

$$\alpha(\omega) = \frac{n_s \pi q^2 \hbar}{\varepsilon_0 \varepsilon_r c n_r m^{*2} d} \frac{1}{\hbar\omega} \left[\frac{8\hbar}{3d} \right]^2 \delta(\Delta E - \hbar\omega) \quad (5)$$

where 'Γ' is half-width at half of the maximum.

3 Results and Discussion

Using Eq. (5), absorption coefficient is computed and plotted for Pöschl–Teller potential configuration of the double quantum well system in presence of electric field applied along the direction of dimensional confinement. Here, we consider those eigenenergies, obtained when band nonparabolicity is taken into account. Also, we invoked Lorentzian lineshape function in order to compute realistic coefficient profile. Full-width half-maximum is taken in order to fit the data with published results of rectangular quantum well. Result is also calculated in absence of bias, and peak position is measured w.r.t. wavelength of incident light. For validation of the result, simulated behavior is compared with that obtained for ideal rectangular quantum well structure and corresponding blueshift/redshift is measured.

Figure 1 shows the variation of absorption coefficient for Pöschl–Teller quantum well configuration, and result is compared with well-known rectangular potential structure with identical configuration and under identical electric field. From the comparative study, it is found out that ideal potential profile gives higher magnitude of absorption, but position of the peak appears at higher wavelength. This speaks in favor of Pöschl–Teller potential as this structure is more suitable for photodetection application at optical spectrum.

Figure 2 shows the variation of absorption profile for different well width. It is seen from the plot that reduction of quantum confinement drastically reduces the absorption profile. This is due to the fact that lowering the quantum well dimension enhances quantum confinement, and this, in turn, increases the eigenvalue. Thus, the resonant tunneling condition can be achieved at higher energy states and makes very high peak profile. Sudden reduction of ground state eigenvalue lowers the absorption profile also. A few results are provided in Table 1.

Similar simulation is carried out for different contact and middle barrier widths also, which are represented in Figs. 3 and 4, respectively. A comparison between the two figures show that peak position remains unaltered with change of contact

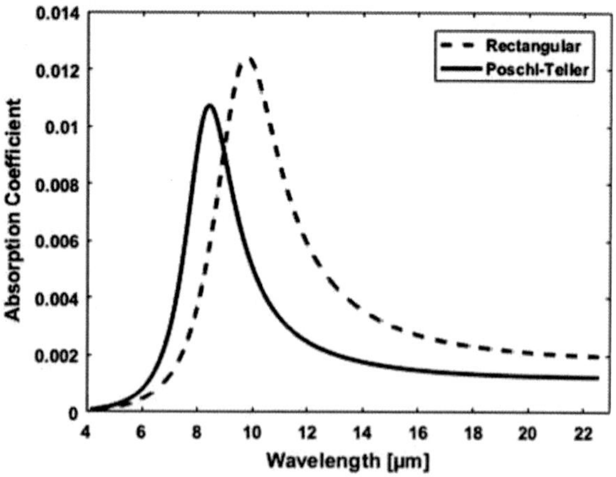

Fig. 1 Comparative study of absorption coefficient between Pöschl–Teller potential with rectangular well

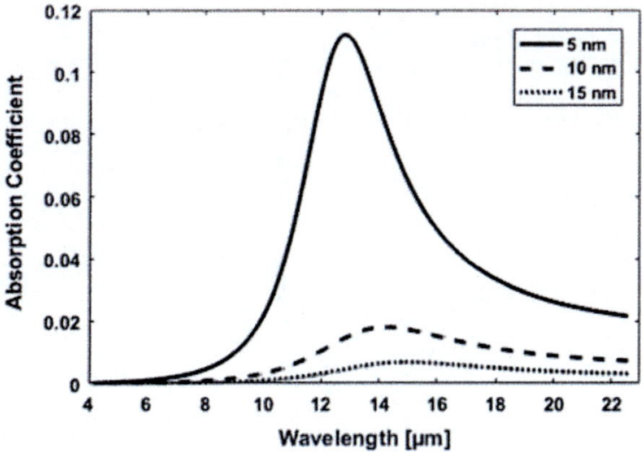

Fig. 2 Absorption coefficient of Pöschl–Teller potential for three different well dimensions

Table 1 Absorption coefficient for different well widths

Well width [nm]	E_1 [eV]	E_2 [eV]	Peak value of absorption coefficient
2	0.05503	0.164	0.7582
3	0.04503	0.1505	0.3255
4	0.04203	0.1435	0.1791
5	0.04203	0.1405	0.111
6	0.04203	0.1375	0.07548
7	0.04003	0.137	0.0518
10	0.04253	0.1325	0.02372

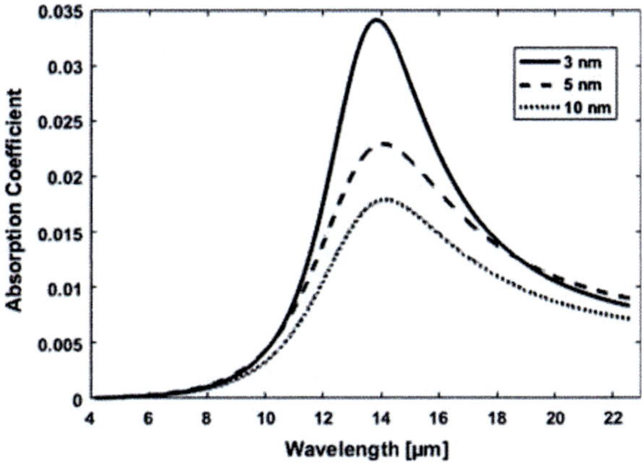

Fig. 3 Absorption coefficient of Pöschl–Teller potential for three different contact barrier widths

potential width, whereas change is taken place when middle barrier width is varied. For Fig. 3, higher contact barrier width makes higher quantum energy levels, which causes change of peak magnitude. But change of sandwich barrier layer width affects

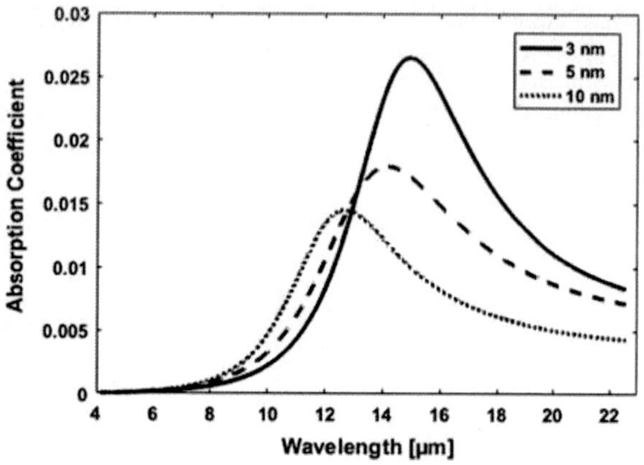

Fig. 4 Absorption coefficient of Pöschl–Teller potential for three different middle barrier widths

resonant tunneling. Thus, this effect is more influential in determining optical candidature of the device. High middle barrier width hinders the resonant condition by increasing the quantum confinement, and thus energy difference between lowermost and first excited state increases. Increase in subband transition energy reduces the peak value of absorption wavelength; simultaneously, it appears at lower wavelength. Results are summarized in Tables 2 and 3, respectively.

Table 2 Absorption coefficient for different contact barrier widths

Contact barrier width [nm]	E_1 [eV]	E_2 [eV]	Peak value of absorption coefficient
2	0.08853	0.18	0.03622
3	0.08303	0.174	0.03402
4	0.07703	0.168	0.03402
5	0.07153	0.162	0.03034
6	0.06553	0.156	0.03034
7	0.06003	0.15	0.02739
10	0.04253	0.1325	0.02739

Table 3 Absorption coefficient for different middle barrier widths

Middle barrier width [nm]	E_1 [eV]	E_2 [eV]	Peak value of absorption coefficient
2	0.04603	0.128	0.02794
3	0.04503	0.1295	0.02648
4	0.04353	0.131	0.02497
5	0.04253	0.1325	0.02372
6	0.04153	0.1335	0.02271
7	0.04053	0.1345	0.02174
8	0.03953	0.1355	0.02084

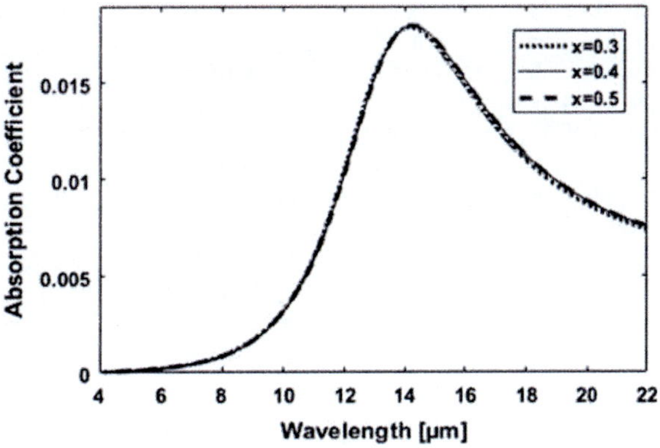

Fig. 5 Absorption coefficient of Pöschl–Teller potential for three different barrier material compositions

Table 4 Absorption coefficient for different barrier material compositions (x)

Mole fraction (x)	E_1 [eV]	E_2 [eV]	Peak value of absorption coefficient
0.2	0.04853	0.1945	0.01092
0.3	0.04253	0.1325	0.01789
0.4	0.1265	0.216	0.018
0.5	0.21	0.3	0.01788
0.6	0.2935	0.3835	0.01788
0.7	0.377	0.467	0.01788

Figure 5 shows the effect of material composition on absorption coefficient, which is very insignificant as evident. As change of eigenvalue with barrier material composition is relative small for Pöschl–Teller potential configuration, so the effect is also seen in the corresponding plot. Results are given in tabular in Table 4. The effect of applied electric field is plotted in Fig. 6a. As electric field enhances the tunneling phenomenon between adjacent quantum wells, henceforth, corresponding intersubband energy decreases. This also simultaneously makes redshift of the spectrum. Figure 6b indicates the biasing effect where progressive redshift is observed with increase of external bias. In this context, it may be noted that bias is applied in the direction of quantum confinement.

The Pöschl–Teller potential is configured using two design parameters: 'α', the width parameter, and 'λ', the depth parameter. Variation of these parameters change the well geometry, and thus it becomes essential to study the effect of these parameters for best possible optical application. Figure 7 indicates the depth parameter effect, whereas Fig. 8 shows the width parameter effect. Higher the 'λ', lower the peak and lower the operating wavelength. But the variation is reversed for width parameter, as revealed from Fig. 8. A closer inspection of Fig. 8 gives the data that just slight tailoring of width parameter makes a very large shift of absorption profile. Thus, from design point-of-view, width parameter gets the priority.

Effect of electron concentration inside the well is analyzed in Fig. 9. Higher the

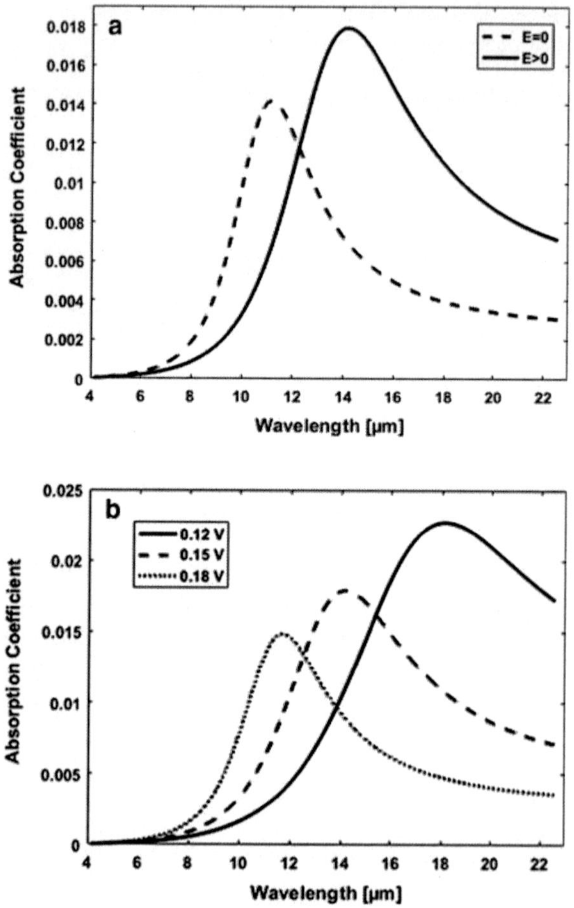

Fig. 6 **a** Comparative study of absorption coefficient of Pöschl–Teller potential in presence and absence of electric field, **b** absorption coefficient of Pöschl–Teller potential for three different electric fields

concentration, lower the subband energy. This ensures larger peak of absorption profile. One point may be noted in this context that we have taken the average electron concentration of the adjacent quantum wells, and the variation is insignificant w.r.t. the average value. This data is very important as the concentration ultimately determines the current in the device.

In Table 5, we have introduced one comparative study with published data in literatures. Result suggests that for identical composition, dimensions and biasing conditions, the present potential configurations provide better peak value. Also, the biggest advantage lies here that without changing the dimensions of the well, the structure can be modified by suitably tuning two critical design parameters, 'α' and 'λ'. This facility is not present in other potential geometries. Tuning the shape of potential well changes the device shape prior to fabrication. Also, design parameters can tune the blueshift/redshift occurrence in spectrum scale. Another very important point may be noted in this context that the present potential configuration provides

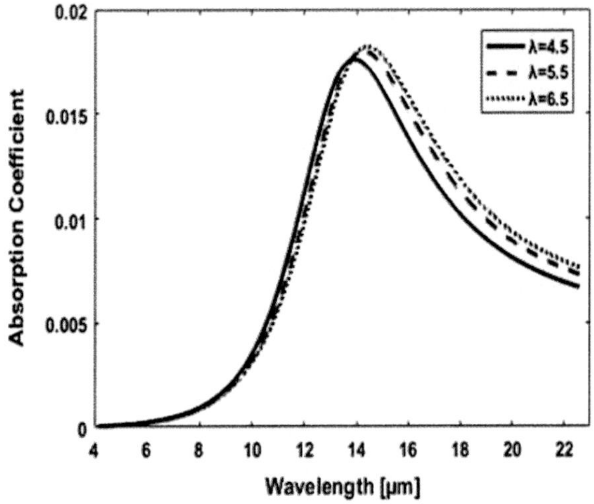

Fig. 7 Absorption coefficient of Pöschl–Teller potential for three different depth parameters

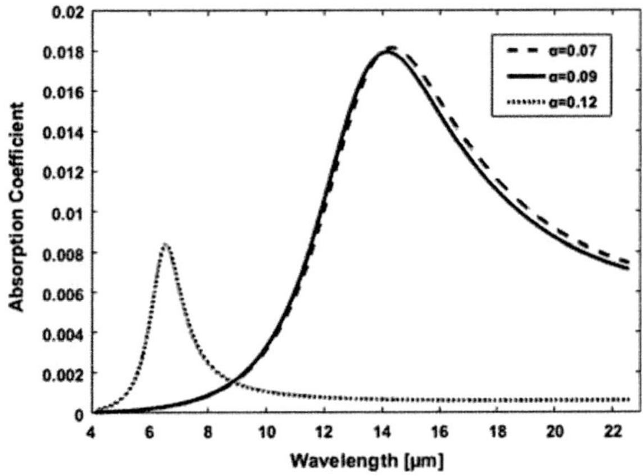

Fig. 8 Absorption coefficient of Pöschl–Teller potential for three different width parameters

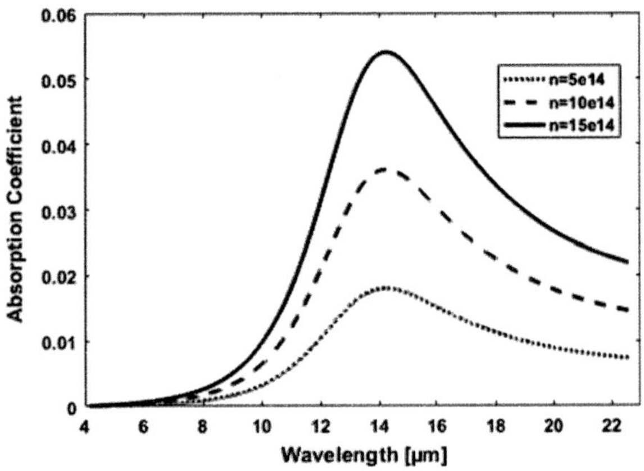

Fig. 9 Absorption coefficient of Pöschl–Teller potential for three different electron concentrations

Table 5 Comparative study with other potential geometries

Type of potential	Peak absorption coefficient	Approximate linewidth [μm]
Parabolic [13]	0.017	38
Gaussian [13]	0.015	30
Pöschl–Teller [present work]	0.018	20

reduced linewidth than the other known potentials, which gives a special feature for application as sensor.

4 Conclusion

Absorption coefficient of double quantum well structure is computed for Pöschl–Teller potential in presence of electric field for various structural and internal parameters, where design constraints are tailored within the fabrication limit to optimize the absorption profile. Result is compared with rectangular geometry, which reveals that the present configuration is more suitable for high frequency applications. Also, comparative study is carried out for both parabolic and Gaussian profiles, which speaks in favor of the present work both in terms of peak of absorption coefficient and linewidth. Lorentzian lineshape is included in the computation along with band nonparabolicity to make the results closer to practical importance. The present structure has the potential advantage of tailoring the shape of profile without changing the dimensional parameters, which novelty is totally absent in all other configurations. Redshift/blueshift is measured by tuning the material and dimensional parameters, which are a key to maximize absorption for case-specific applications.

References

1. Ashalley E, Gryczynski K, Wang Z, Salamoc G, Neogi A (2019) Plasmonically-powered hot carrier induced modulation of light emission in a two-dimensional GaAs semiconductor quantum well. Nanoscale 11:3827–3836
2. Wan Y, Inoue D, Jung D, Norman JC, Shang C, Gossard AC, Bowers JE (2018) Directly modulated quantum dot lasers on silicon with a milliampere threshold and high temperature stability. Photonics Res 6(8):776–781
3. Alves RA, Costa JC, Gomes M, Silva NA, Guerreiro A (2017) Quantum wires as sensors of the electric field: a model into quantum plasmonics. In: 25th optical fiber sensors conference
4. Lu F, Bhattacharya I, Sun H, Tran TTD, Ng KW, Malheiros-Silveria GN, Chang-Hasnain C (2017) Nanopillar quantum well lasers directly grown on silicon and emitting at silicon-transparent wavelengths. Optica 4(7):717–723
5. Dong M, Winful HG, Cundiff ST (2018) Quantum-well diode lasers for frequency comb generation. In: Frontiers in optics/laser science. OSA technical digest, paper LW6F.2

6. Vinasco JA, Radu A, Kasapoglu E, Restrepo RL, Morales AL, Feddi E, Mora-Ramos ME, Duque CA (2018) Effects of geometry on the electronic properties of semiconductor elliptical quantum rings. Sci Rep 8. Article number 13299
7. Lebiadok YV, Shalayeva AA (2018) Interface defects and silicon impurities in $GaAs/Al_{0.3}Ga_{0.7}As$ heterostructures. IOP Conf Ser J Phys Conf Ser 961:012009
8. Knyazev SY, Shcherbakova EE (2017) Method for numerical solution of the stationary Schrödinger equation. Russ Phys J 59(10):1616–1622
9. Belov PA, Khramtsov ES (2017) The binding energy of excitons in narrow quantum wells. IOP Conf Ser J Phys Conf Ser 816:012018
10. Dubrovskiy SV, Zubkov VI (2017) Self-consistent solution of Schrödinger and Poisson equations by means of numerical methods in the LabVIEW development environment. In: IEEE conference of Russian young researchers in electrical and electronic engineering
11. Zhang Z, Ding J, Wang HT (2017) Efficient numerical solution of excitation number conserving quantum systems. AIP Adv 7:085225
12. Lô B, Gueye SB (2016) Numerical verification of transition's energies of excitons in quantum well of ZnO with the finite difference method. J Mod Phys 7(3):329–334
13. Sarkar D, Deyasi A (2016) Comparative analysis of absorption coefficient for parabolic and gaussian quantum wells for photodetector application. Adv Ind Eng Manag 5(2):197–201
14. Sinyak VA (1996) Optical absorption in a quantum well with an applied electric field. In: International semiconductor conference
15. Blood P (2000) On the dimensionality of optical absorption, gain, and recombination in quantum-confined structures. IEEE J Quantum Electron 36(3):354–362
16. Kvietkova J, Siozade L, Disseix P, Vasson A, Leymarie J, Damilano B, Grandjean N, Massies J (2002) Optical investigations and absorption coefficient determination of InGaN/GaN quantum wells. Phys Status Solidi A 190(1):135–140
17. Oziurk E, Sari H, Sokmen I (2004) Intersubband optical absorption in quantum wells under applied electric and intense laser fields. Surf Rev Lett 11(3):297–303
18. Ramos-Garcia R, Chiu-Zarate R, Nolte D, Melloch MR (2003) Measurements of absorption coefficient and refractive index changes without the Kramers–Kronig relation in photorefractive quantum wells of GaAs. J Opt A Pure Appl Opt S514
19. Dey A, Neogi B, Maiti B, Chandra D (2012) A simple analysis of the interband absorption coefficient of bulk and quantum well of nonparabolic semiconductors with application to $Hg_{1-x}Cd_xTe$ material. J Optoelectron Adv Mater 14(3–4):210–218
20. Lever L, Hu Y, Myronov M, Liu X, Owens N, Gardes FY, Marko IP, Sweeney SJ, Ikonic Z, Leadley DR, Reed GT, Kelsal RW (2011) Modulation of the absorption coefficient at 1.3 μm in Ge/SiGe multiple quantum well heterostructures on silicon. Opt Lett 36(21):4158–4160
21. Dakhlaoui H (2015) Linear and nonlinear optical absorption coefficients and refractive index changes in $GaN/Al_xGa_{1-x}N$ double quantum wells operating at 1.55 μm. J Appl Phys 117:135705
22. Sakiroglu S (2016) Linear and nonlinear optical absorption coefficients and refractive index changes in Morse quantum wells under electric field. Int J Mod Phys B 30(30):1650209
23. Kadadra E, Fellaoui K, Abouelaoualim D, Oueriagli A (2016) Optical absorption coefficients in GaN/AlGaN double inverse parabolic quantum wells under static external electric field. Mod Phys Lett B 30(25):1650165
24. Park SH, Ahn D, Park CY (2017) Intersubband transition in lattice-matched BGaN/AlN quantum well structures with high absorption coefficients. Opt Express 25(4):3143–3152
25. Sarath R, Vinodkumar PC (2015) Bose–Einstein condensation in generalized Pöschl–Teller potential. PRAMANA J Phys 85(1):77–89

QoS-Based ACO Routing Protocols in MANETs: A Review

Priyanka Kumari and Sudip Kumar Sahana

Abstract A mobile ad hoc network (MANET) is an assortment of mobile nodes that exchange information with non-mounted framework and centralized organization. All node behaves as a router when have to find out the routes from the source node to the destination node and maintain those routes within the communication network. But routing in MANETs becomes complex due to the dynamic environment of network and bandwidth limitation. Swarm intelligence (SI) primarily nature-based algorithms have pulled in a great deal of consideration of the researchers to solve this problem. SI-based ant colony optimization (ACO) algorithms appeared to be fantastic for creating routes in a network due to the issue of finding the shortest paths by ant is similar to the issue of routing in networks. This paper shows an exhaustive review and evaluation of numerous ACO-based routing protocols in MANETs.

Keywords MANET · Swarm intelligence (SI) · Ant colony optimization (ACO) · Routing protocols

1 Introduction

MANET is a continuously self-configuring connected mobile and wireless node through radio communication where every node acts as host as well as the router. Due to self-configured and lack of infrastructure and centralized administration, it becomes a tedious task to develop routing protocols in MANETs. Since the 1990s, many routing protocols have been developed such as ad hoc on-demand distance vector (AODV) [1], dynamic source routing (DSR) [2], optimized link state routing (OLSR) [3], and distance-sequenced distance vector routing [4]. But these routing

P. Kumari (✉) · S. K. Sahana
Department of Computer Science and Engineering, Birla Institute of Technology, Mesra, Ranchi, Jharkhand 835215, India
e-mail: priyankakumari42@gmail.com

S. K. Sahana
e-mail: sudipsahana@bitmesra.ac.in

V. Nath and J. K. Mandal (eds.), *Proceedings of the Fourth International Conference on Microelectronics, Computing and Communication Systems*, Lecture Notes in Electrical Engineering 673, https://doi.org/10.1007/978-981-15-5546-6_27

protocols do not fulfill the requirements of MANETs such as self-organized manner, self-adaptation due to dynamic topology and improving QoS parameters through low end-to-end delay, high packet delivery ratio (PDR), and low routing overhead. Therefore, there is a need to develop a protocol which is adaptive, robust, and self-healing. Recently, a new class of routing protocols has developed which is grounded on "swarm intelligence (SI)." SI is nature-inspired from biological systems such as ant colony optimization (ACO), particle swarm optimization (PSO), bird flocks, bee colonies, and fish schools which are capable to tackle issues related to self-organization and self-adaptation. Many researchers have proved that ACO is an excellent algorithm for searching the optimal paths in MANETs. The main source of motivation behind ACO is the foraging behavior of ants [5].

The aim of this paper is to provide a survey of various ACO-based algorithms for MANETs. Rest of the paper is described as: basic ant colony algorithm is explained in Sect. 2. Section 3 describes the taxonomy of ACO-based MANETs routing protocols. Section 4 explains the various QoS-aware ACO-based routing algorithms for MANETs. Section 5 provides a table summary to describe simulation parameter comparative analysis of all reviewed protocols based on ACO. Finally, we summarized the whole study work with the conclusion in Sect. 6.

2 Basic Ant Colony Algorithm

ACO is a famous swarm intelligence approach. Based on the food-seeking behavior of ants, Dorigo [6] introduced ant colony optimization (ACO). It is based on the inspiration by the behavior of real ants which are wondering around their nests to forage for the search of food. Ants deposit a particular quantity of a chemical substance (named as pheromone) which serves as a route mark that ants have taken. Initially, ants select a random path between nest and food. When finding food they come back to their nests and at the same time, they discharge trails on the ways. On the premise of deposited secretion quantity, the newer ant selects the optimal path having higher secretion concentration that is the shortest path among all alternative remaining paths. The concentrations of the pheromone of the shortest path are more than that of all other remaining paths. Finally, the ants get the shortest path from the nest to food. Figure 1 shows how the ants search the shortest route between ant's nest and source food.

Algorithm: Basic flow of ACO (adapted from [7])

1. Represent the problem by means of a creation graph.
2. Initialize the ACO parameters.
3. Generate ant solutions from each ant's casual walk on the construction graph.
4. Update pheromone intensities.
    (a) Local updation.
    (b) Global updation.

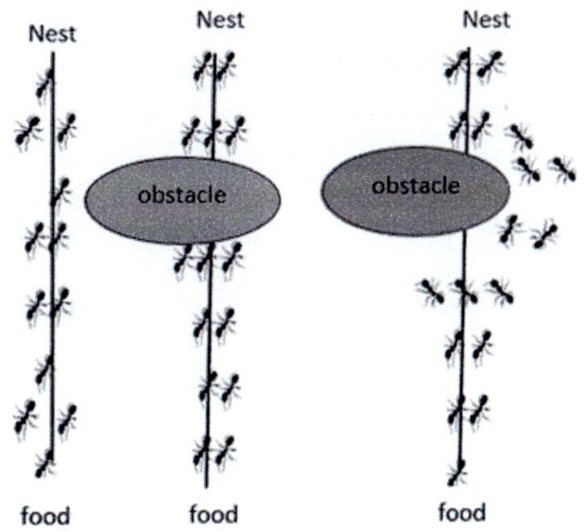

Fig. 1 Ant colony optimization

5. Move to Step 3 and repeat till the termination condition is met.

The third step of the above basic ACO algorithm is to generate ant solutions by using the probability rule or state transition state (Eq. 1). The probability of choosing the next node *j* from node *i* is:

$$P(ij) = \left(\left[\tau_{ij}\right]^{\alpha} \cdot \left[\eta_{ij}\right]^{\beta}\right) / \left(\Sigma s \in \text{allowed}[\tau_{is}]^{\alpha} \cdot [\eta_{is}]^{\beta}\right) \quad (1)$$

where
τ_{ij} = pheromone intensity of edge (i, j).
η_{ij} = heuristic information of edge (i, j).
α = importance of the pheromone trail.
β = importance of heuristic visibility.

There are two phases of Step 4: local updation and global updation. The pheromone, τ_{ij}, related with edge joining node *i* and *j* updated locally by using Eq. (2). After that to enhance pheromone concentration on edge, updated globally by using Eq. (3).

$$\tau_{ij} = (1 - \rho) \cdot \tau_{ij} + \rho \cdot \tau_0 \quad (2)$$

where
τ_{ij} = pheromone value associated with edge (i, j).
τ_0 = initial pheromone value.
ρ = evaporation rate.

$$\tau_{ij} \leftarrow \tau_{ij} + (1 - \rho) \cdot \tau_{ij} + \Sigma^{m} k - 1 \Delta\tau_{ij}^{k} \quad (3)$$

where
ρ = evaporation rate.
$\Delta\tau_{ij}^{k}$ = pheromone amount laid on edge (i, j) by ant *k*.

Table 1 Similarities between networks and ants

S. No.	Parameters	Networks	Ants
1.	Physical structure	Unstructured, dynamic, distributed	-Do-
2.	Source of route	Requests are sent from the source to get neighboring information and then build new routes	Pheromones are used to construct new routes
3.	Multipath support	Single path, partially multipath	-Do-
4.	Basic structure	Self-configuring and self-organizing	-Do-
5.	Goal	To find the shortest path	-Do-

$$\tau_{ij}^{k} = \begin{cases} \frac{q}{L_k} & \text{If } k\text{th ant uses edge } (i, j) \text{ in its tour} \\ 0 & \text{Otherwise} \end{cases}$$

q is constant, and L_k is the length of trip of kth ant. The ants behave like data control packets, whose task is to discover a path connecting the source and destination pairs and gather information about it. There are a lot of similarities among networks and ants as shown in Table 1. Thus, the general characteristics of ants make them suitable for routing in MANETs.

3 Taxonomy of ACO-Based MANETs Routing Protocols

We categorize ACO-based routing protocols into various categories with regard to their services and design purposes such as QoS-based ACO routing protocols, location-based routing protocols, security-based ACO routing protocols, and power-based ACO routing protocols. Figure 2 shows the taxonomy of ACO-based routing protocols in detail. Quality of services is defined as the parameters which affect the performance of any networks. The basic parameters of QoS constraints are:

1. Delay (s)
2. Jitter (s)
3. Bandwidth
4. Throughput (bit/s)
5. Routing overhead
6. Packet delivery ratio (PDR).

To maintain QoS [8], a service can be characterized by a set of significant requirements while sending data packets from one node to another node such as maximum throughput, minimum bandwidth, minimum delay, minimum jitter, maximum PDR, and less routing overhead. To support real-time application, it is necessary to support all these above QoS parameters in network environments. In this paper, we discuss various QoS-based ACO routing protocols in MANETs.

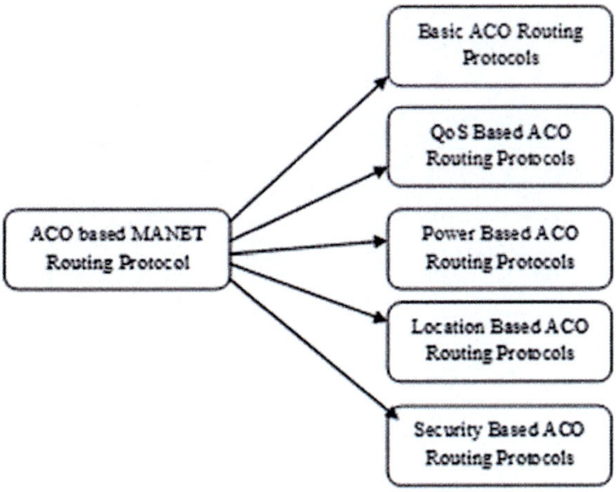

Fig. 2 Taxonomy of ACO-based MANETs routing protocols

4 Review of QOS-Aware ACO Routing Protocols

Due to the dynamic nature of topology and the limitation of bandwidth in MANETs, it is also necessary to maintain QoS constraints while searching the shortest paths. Conventional methods like DSDV [9], DSR [10], AODV [11], OLSR [3], and ZRP [12] are used to search the shortest path. But these methods take more time to obtain the optimal path. Many authors get inspired from the pheromone concept of ant colony algorithm to choose the paths on the basis of QoS constraints [13–19]. Figure 3 shows the types of routing protocols based on ACO.

AntNet was the first ACO-based routing protocol proposed by Di Caro and Dorigo [20, 21]. In this method, FANTs and BANTs are used to keep routing information and network status in routing tables. However, the routing information in AntNet takes more time to propagate. Ant colony-based routing algorithm (ARA) [22] is introduced by Gunes et al. which depends on swarm intelligence methods. It generates similar results as DSR along with reduces routing overhead. But it is not true in case of high mobility.

"Ant routing algorithm for mobile ad hoc networks (ARAMA)" [23] is a biological inspired proactive routing protocol proposed by Hussein and Saadawi where FANTs collect information about networks and BANTs keep informed the routing information. This algorithm considers delay and hop counts as QoS constraints. AntHocNet is a versatile nature-inspired algorithm which is proposed by Di Caro et al. [24]. It is a combination of reactive and proactive routing multipath algorithm having six types of ants: PFANTs, RPANTs, RBANTs, RPFANTs, and RPBANTs. This routing algorithm beats AODV but generates high routing overhead. ADSR is proposed by Asokan et al. [25] to support QoS parameters such as bandwidth, delay, and residual energy. ADSR produces improved results over DSR for PDR, delay, and residual energy at the node. However, routing overhead of ADSR is to some extent higher than DSR. PACONET routing algorithm is based on the principles of ACO algorithm

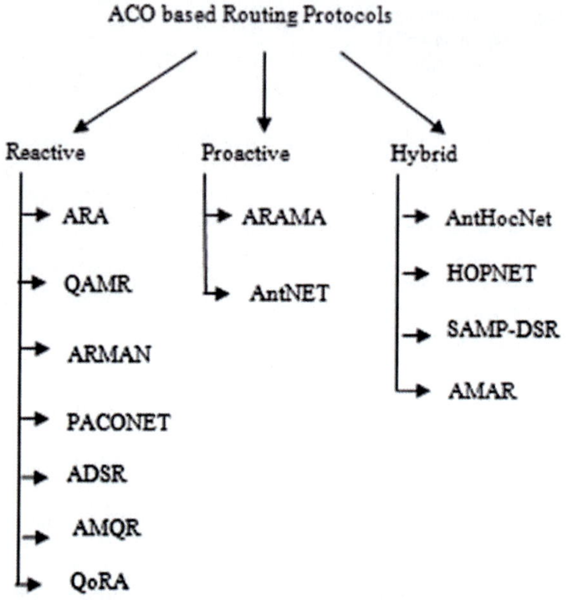

Fig. 3 Types of ACO-based routing protocols

proposed by Osagie et al. [26]. In PACONET, FANTs explore the routes and BANTs also update pheromone value in the reverse direction of FANTs. The performance of PACONET is improved than AODV but the PDR is the same as compared to AODV. HOPNET is introduced by Wang et al. [27] which is based on ACO and also used the combined features of ZRP and DSR. It uses ZRP framework which outperforms AODV. However, routing overhead of HOPNET decreases as the network size increases. ARMAN [28] is an ant-based routing protocol which improves QoS in terms of bandwidth and delay in high mobility cases. Khosrowshahi-Asl et al. [29] introduced switched ACO-based multi-path dynamic source routing (SAMP-DSR). It is based on ACO algorithm as well as conventional DSR algorithm. It is having two components: ACO-based component and path discovery component. It improves the results of the dynamic network through switching approach. AMQR [30] based on ACO is a reactive routing protocol proposed by Deepalakshmi and Radhakrishnan. It considers PDR, delay, and routing overhead as QoS constraints. The AMQR outperforms AODV and AntHocNet but generates high routing overhead. QAMR [31] is introduced by Krishna et al. which is based on ACO where the route is selected based on the stability of the nodes and the path preference probability. Routing overhead of QMAR is higher than ARAMAN and AODV. AMAR [32] is a hybrid multipath ACO-based routing protocol proposed by Samadi and Beigy. It considers PDR, delay, and routing overhead as QoS constraints. The AMAR outperforms AODV, OLSR, and AntHocNet but generates high routing overhead. QoRA [33] is proposed by Al-Ani and Seitz which is based on ant colony optimization and Simple Network Management Protocol (SNMP). The QoRA outperforms AODV and CLWPR but consumes more energy and bandwidth.

5 Comparative Analysis

A simulation parameter comparative analysis of the reviewed protocols is presented in Table 2. From Table 2, we conclude that NS2, OMNET++, GloMoSim, and QualNet are used for the implementation of ACO-based routing protocols. Many researchers considered three QoS parameters such as delay, packet delivery ratio (PDR), and routing overhead (RO) for the evaluation of the performance of ACO-based routing protocols.

6 Conclusion

Because of the dynamic and self-organizing characteristics of MANETs, to find out the optimal route becomes a tedious challenge. Many researchers used ACO-based routing algorithm to solve this problem due to two reasons: (i) There are a lot of similarities between message packets and ants. (ii) ACO is used for optimization problems especially, dynamic problems. In this paper, we have figured out to show how various ACO-based routing protocols whose primary aim is to discover the efficient path between the source node to the destination node. This research infers that ACO approaches are very promising for route optimization in MANETs. After reviewing it was concluded that using ACO algorithm of MANET protocols leads to good results in end-to-end delay, PDR, and routing overhead. The performance of ACO-based routing algorithm varies due to the initialization of the pheromone value.

Table 2 Simulation parameters-based comparison of ACO routing protocols

S. No.	Protocols	Year	Simulator	Compared with	No. of nodes	Considered QoS parameters				Remarks
						Delay	Jitter	PDR	RO	
1.	AntNET [20]	1999	C++	OSPF, SPF, BF, Q-R, PQ-R, Daemon	8, 14, 57	✓	✓	✓	✓	1. Improve delay and PDR 2. Huge delay in propagating the routing information
2.	ARA [22]	2002	NS2	DSDV, AODV, DSR	50	✓	X	✓	X	1. Similar performance as DSR 2. Generate very less overhead because there are no routing tables 3. Not adaptive mechanism
3.	ARAMA [23]	2003	OPNET	Without	30	X	X	✓	X	1. Reduce route discovery 2. Reduce route maintenance
4.	AntHocNet [24]	2005	QualNet	AODV	100	✓	X	✓	✓	1. Improve PDR, delay, and jitter 2. Generate high routing overhead
5.	ADSR [25]	2008	NS2	DSR	100	✓	✓	X	X	1. Improve PDR, delay and residual energy at node 2. Generates higher routing overhead as compared to DSR

(continued)

Table 2 (continued)

S. No.	Protocols	Year	Simulator	Compared with	No. of nodes	Considered QoS parameters				Remarks
						Delay	Jitter	PDR	RO	
6.	PACONET [26]	2008	GloMoSim	AODV	50	✓	X	✓	✓	1. Improve delay and control overhead 2. PDR is same as compared to AODV 3. Since each node maintains its own routing table so routing overhead is high
7.	HOPNET [27]	2009	GloMoSim	AODV, ZRP, AntHocNet	50	✓	X	✓	✓	1. Scalable for large networks 2. Improve delay and packet delivery ratio 3. Routing overhead decreases as network size increases
8.	ARMAN [28]	2009	NS2	AODV	50	✓	X	✓	X	1. Highly adaptive 2. Scalable 3. Reduces delay
9.	SAMP-DSR [29]	2011	OMNET++	MP-DSR, EMP-DSR, AntHocNet, AODV	100	✓	X	✓	✓	1. Improve the performance of the dynamic network through switching approach

(continued)

Table 2 (continued)

S. No.	Protocols	Year	Simulator	Compared with	No. of nodes	Considered QoS parameters				Remarks
						Delay	Jitter	PDR	RO	
10.	AMQR [30]	2011	NS2	AODV, AntHocNet	20, 40, 60, 80, 100	✓	X	✓	X	1. Provides good PDR 2. Reduces delay 3. High routing overhead
11.	QAMR [31]	2012	NS2	ARAMAN, AODV	50	X	X	✓	✓	1. Higher routing overhead
12.	AMAR [32]	2012	QualNet	AntHocNet, OLSR, AODV	100	✓	X	✓	X	1. Provides good PDR 2. Reduces delay 3. High routing overhead
13.	QoRA [33]	2015	SUMO	AODV, CLWPR	100	✓	X	✓	X	1. Improve bandwidth, delay and packet loss 2. QoRA consumes more energy and bandwidth

References

1. Perkins C, Belding-Royer E, Das S (2003) Ad hoc on-demand distance vector (AODV) routing. Tech. Rep. rfc3561. IETF, Fremont
2. Johnson DB et al (2001) DSR: the dynamic source routing protocol for multi-hop wireless ad hoc networks. Ad Hoc Netw 5:139–172
3. Clausen T, Jacquet P, Laouiti A, Muhlethaler P, Qayyum A, Viennot L (2001) Optimized link state routing protocol for ad-hoc networks. In: Proceedings of IEEE INMIC, Dec 2001, pp 62–68
4. Perkins CE, Bhagwat P (1994) Highly dynamic destination-sequenced distance-vector routing (DSDV) for mobile computers. ACM SIGCOMM Comput Commun Rev 24(4):234–244
5. Gupta AK, Sadawarti H, Verma AK (2012) MANET routing protocols based on ant colony optimization. Int J Model Optim 2(1):42
6. Dorigo M (1992) Optimization, learning and natural algorithms. Ph.D. dissertation, Politecnico di Milano, Milan
7. Dorigo M, Stützle T (2004) Ant colony optimization. MIT Press, Cambridge. ISBN 978-0-262-04219-2
8. Junhai L, Lie X, Danxia Y (2008) Research on multicast routing protocols for mobile ad hoc networks. Comput Netw 52:988–997
9. Perkins CE, Bhagwat P (1994) Highly dynamic destination-sequenced distance-vector (DSDV) for mobile computers. In: Proceedings of the ACM conference on communications architectures, protocols and applications, London, Aug 1994, pp 234–244
10. Maltz DA, Johnson DB (1996) Protocols for adaptive wireless and mobile computing. IEEE Pers Commun 3(1)
11. Perkins CE, Royer EM (1999) Ad-hoc on-demand distance vector routing. In: Proceedings of the 2nd IEEE workshop on mobile computing systems and applications, Feb 1999, pp 90–100
12. Haas ZJ, Perlman M (2001) The performance of query control schemes of the zone routing protocol. IEEE/ACM Trans Netw 9(4):427–438
13. Zhibo G, Xinquan J (2003) An algorithm of QoS unicast routing based on the principle of ant system. Sig Process 19(5):432–436
14. Lu G, Liu Z, Zhou Z (2001) A distributed QoS routing algorithm based on ant algorithm. J China Inst Commun 22(9):34–42
15. Marwaha S, Tham CK, Srinavasan D (2002) A novel routing protocol using mobile agents and reactive route discovery for ad-hoc wireless networks. In: Toward network superiority proceedings of IEEE international conference on networks 2002(ICON 2002), 27–30 Aug 2002, pp 311–316
16. Guoying L, Zemin L, Zheng Z (2000) Multicast routing based on ant algorithm for delay-bounded and load-balancing traffic. In: Proceedings 25th annual IEEE conference on LCN 2000, 8–10 Nov 2000, pp 362–368
17. Lu G, Liu Z (2000) Multicast routing based on ant-algorithm with delay and delay variation constraints. In: IEEE APCCAS 2000. 2000 IEEE Asia-Pacific conference on circuits and systems. Electronic communication systems, 4–6 Dec 2000, pp 243–246
18. Subing Z, Zemin L (2001) A QoS routing algorithm based on ant algorithm. In: ICC 2001. IEEE international conference on communications, vol 5, 11–14 June 2001, pp 1581–1585
19. Rajeshwar S, Singh DK, Kumar L (2009) Swarm intelligence based routing in mobile ad hoc networks: an overview. Natl J Technol PSG Tech 5(2). ISSN-0973-1334
20. Di Caro G, Dorigo M (1999) AntNet: distributed stigmergetic control for communications networks. J Artif Intell Res 9:317–365
21. Di Caro G, Dorigo M (1998) Two ant colony algorithms for best-effort routing in datagram networks. In: The proceedings of 10th international conference on parallel and distributed computing and systems (PDCS'98), Las Vegas
22. Gunes M, Sorges U, Bouazizi I (2002) ARA—the ant-colony based routing algorithm for MANETs. In: Proceedings of international conference on parallel processing workshop. IEEE, pp 79–85

23. Hussein O, Saadawi T (2003) Ant routing algorithm for mobile ad-hoc networks (ARAMA). In: Proceedings of the IEEE international performance, computing, and communications conference, Apr 2003, pp 281–290
24. Di Caro G, Ducatelle F, Gambardella LM (2005) AntHocNet: an adaptive nature-inspired algorithm for routing in mobile ad hoc networks. Eur Trans Telecommun 16(5):443–455
25. Asokan R, Natarajan AM, Venkatesh C (2008) Ant-based dynamic source routing protocol to support multiple quality of service (QoS) metrics in mobile ad hoc networks. Int J Comput Sci Secur 2(3):48–56
26. Osagie E, Thulasiraman P, Thulasiram RK (2008) PACONET: improved ant colony optimization routing algorithm for mobile ad hoc networks. In: Proceedings of 22nd international conference on advanced information networking and applications (AINA), pp 204–211
27. Wang J, Osagie E, Thulasiraman P, Thulasiram RK (2009) HOPNET: a hybrid ant colony optimization routing algorithm for mobile ad hoc network. Ad Hoc Netw 7(4):690–705
28. Deepalakshmi P, Radhakrishnan DS (2009) Ant colony based QoS routing algorithm for mobile ad hoc networks. Int J Recent Trends Eng 1(1):459–462
29. Khosrowshahi-Asl E, Noorhosseini M, Pirouz AS (2011) A dynamic ant colony based routing algorithm for mobile ad-hoc networks. J Inf Sci Eng 27(5):1581–1596
30. Deepalakshmi P, Radhakrishnan S (2011) An ant colony-based multi-objective quality of service routing for mobile ad hoc networks. EURASIP J Wirel Commun Netw 1(153):1–12
31. Krishna PV, Saritha V, Vedha G, Bhiwal A, Chawla AS (2012) Quality-of-service-enabled ant colony-based multipath routing for mobile ad hoc networks. IET Commun 6(1):76–83
32. Samadi S, Beigy H (2012) An adaptive multipath ant routing algorithm for mobile ad hoc networks. Int J Comput Electr Eng 4(2):175–180
33. Al-Ani AD, Seitz J (2015) QoS-aware routing in multi-rate ad hoc networks based on ant colony optimization. Netw Protoc Algorithms 7(4):1–25

A Novel NOR Gate-Based Dynamic Power Gating Technique in SRAM

Abhishek Nag, K. Ruchira Reddy, Nilanjana Majumder, Elizabeth Debbarma, and Sambhu Nath Pradhan

Abstract An autonomous power gating technique is implemented in static random-access memory (SRAM) in this paper to minimize leakage power dissipation. SRAM cells are the most predominantly used programming technology which is used for storing configuration bits in programmable devices like FPGA. The proposed approach extracts information from the operation of the SRAM and automatically generates the criteria for gating. The design and simulation of the SRAM have been carried out in Cadence Virtuoso spectrum and resulted in a maximum of 74.24% power savings. The access time associated with our technique is minimal. A comparison with prior works indicates the effectiveness of our technique.

Keywords Power gating · Leakage power · SRAM · FPGA · Access time

1 Introduction

In the present generation, the implementation of special-purpose processors [1, 2] is one of the main applications of field-programmable gate arrays (FPGA). However, the excessive power dissipation of FPGAs compared to ASIC creates a barrier to its

A. Nag (✉) · K. Ruchira Reddy · N. Majumder · E. Debbarma · S. N. Pradhan
Department of Electronics and Communication Engineering, NIT Agartala, Agartala, Tripura, India
e-mail: abhi14379@gmail.com

K. Ruchira Reddy
e-mail: ruchirareddykonda@gmail.com

N. Majumder
e-mail: nilanjanamajumdar14@gmail.com

E. Debbarma
e-mail: elizabethdebbarma0412@gmail.com

S. N. Pradhan
e-mail: sambhupradhan.ece@nita.ac.in

V. Nath and J. K. Mandal (eds.), *Proceedings of the Fourth International Conference on Microelectronics, Computing and Communication Systems*, Lecture Notes in Electrical Engineering 673, https://doi.org/10.1007/978-981-15-5546-6_28

popularity. Static random-access memory (SRAM) [3] is used to store the data in a programmable device and allow the end user to read from it as well as write into it. SRAM-based field-programmable devices constitute a significant number of SRAM cells to store the configuration bits. Hence, along with the internal blocks of the programmable devices, reducing the power dissipation associated with SRAM will further increase the efficiency of the device. Moreover, with the scaling of technology, low-power techniques in VLSI circuits [4] have become the need of the hour.

A gate leakage current reduction technique in SRAM cell is presented in [5]. Authors in [5] achieved the leakage reduction by implementing a dynamic voltage scaling technique, presented in Fig. 1. The basic logic in the concept of NC-SRAM structure in [5] is the enabling of multiple ground levels to the memory cell by virtue of the pass transistors NC1 and NC2. Gate leakage currents of the cell are reduced by the effect of positive voltage (virtual ground). During the transition of the SRAM cell to idle mode from active mode, word line (WL) changes from ‘1’ to ‘0’ causing the gate voltage of NC1 to become 1, and thus NC1 becomes active and the source voltage of NC1 becomes VS, which is greater than VSS.

As a result, voltage difference and consequently Gate leakage current is reduced. The design and analysis of two different low-power SRAM cell structures are presented in [6]. Gate voltages for the NMOS pass transistors are reduced in the first cell structure (IWL-VC SRAM CELL), hence lowering the gate leakage current. The ground level is increased during the idle (inactive) mode, correspondingly reducing the sub-threshold leakage current. The IWL-VC SRAM CELL is presented in Fig. 2. A new pass transistor ‘P3’ is introduced in the earlier NC-SRAM cell with a view to reducing gate voltages of the N3/N4 pass transistors. As a result, there is a simultaneous reduction of gate tunneling as well as sub-threshold currents. In order to further minimize the gate leakage current, PMOS pass transistors are introduced in the second cell structure (PP–SRAM). The concept of PMOS pass transistor SRAM structure is introduced in this technique, which results in lower gate leakage current in comparison with a normal SRAM cell.

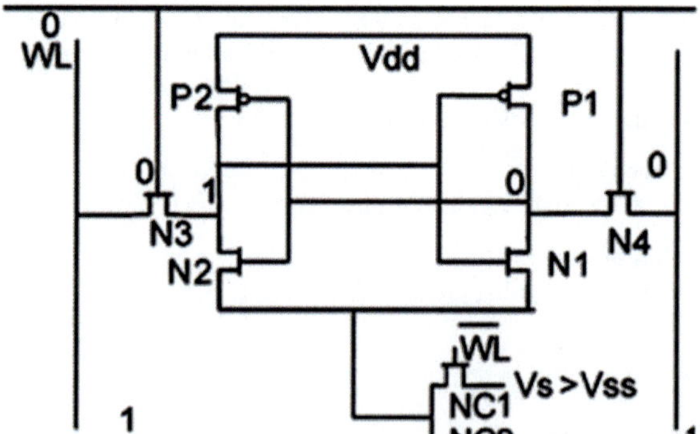

Fig. 1 Low-power NC-SRAM [5]

Fig. 2 Low-power IWL-VC SRAM [6]

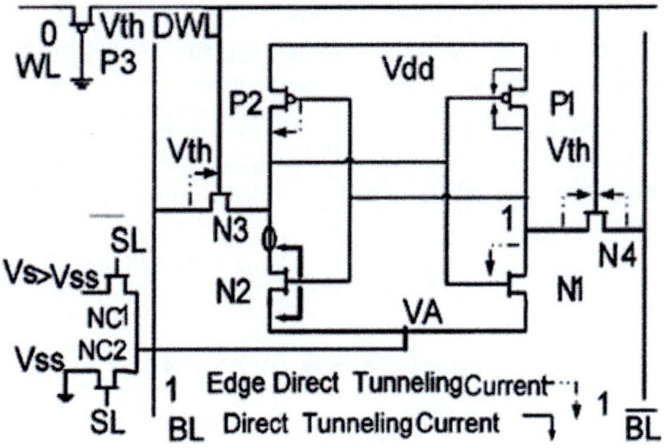

The PMOS transistors P3 and P4 replace the NMOS transistors N3 and N4 of Fig. 2, resulting in reduced gate leakage current, since the PMOS transistor's gate tunneling current is lowering than that of NMOS [5].

In [7], authors proposed a design of SRAM cell (TG8T) based on transmission gates that consisting of 8 transistors. The proposed design in [7] is presented in Fig. 3.

A concept of differential operation is introduced in TG8T technique, where a PMOS is introduced in the architecture of a conventional 6T SRAM cell, parallel with each of the access NMOS [3] represented by M5 and M6 in Fig. 3.

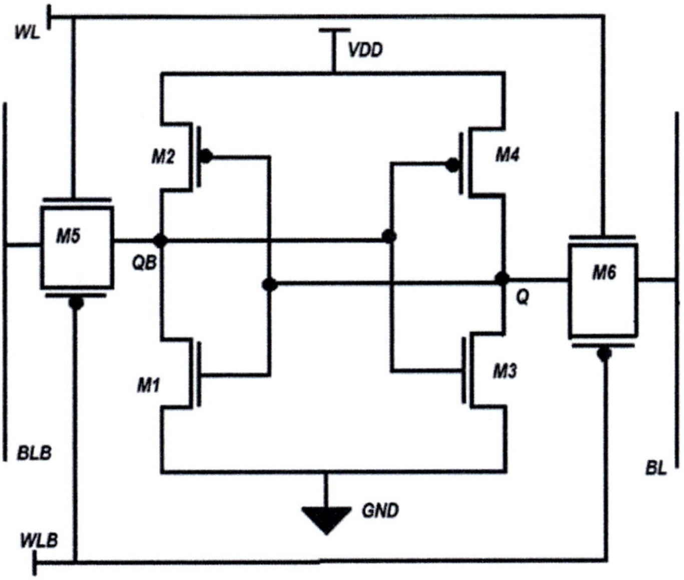

Fig. 3 High-performance SRAM architecture in [7]

A new approach of implementing supply power gating [8] is proposed in this work, which aims at reducing leakage power dissipation as well as retains the data stored in the SRAM. A NOR gate-based power gating logic is introduced which effectively reduces the total power consumption of the SRAM by approximately 74.24% with a minimal increase in delay. The information required for gating is obtained from the knowledge of bit lines and word lines of the SRAM and automatically generates the criteria for gating the supply voltage.

2 Operation of 6T SRAM Cell

The basic 6T SRAM cell uses the idea of two inverters cross-coupled to each other by virtue of 4 transistors, as shown in Fig. 4. In this particular structure, there are two stable states equating to the logic '1' and '0' states. 2 more transistors are needed in the SRAM structure, which acts as the access control for the memory cell. As a result, 6 transistors are required to form the SRAM structure, and hence, it is termed as a 6T memory cell. In figure, WL is the word line used to control the transistors M5 and M6. The bit lines are 'BL' and 'BLB' which are used as I/O lines. M1, M2, M3, and M4 form the memory part, and the data stored in it is accessed through M5 and M6 transistors. The BL and BLB lines are connected to a sense amplifier, which reads the final stored value (either 1 or 0).

Normally, SRAM operation is governed by three modes, namely standby mode, read operation, and write operation.

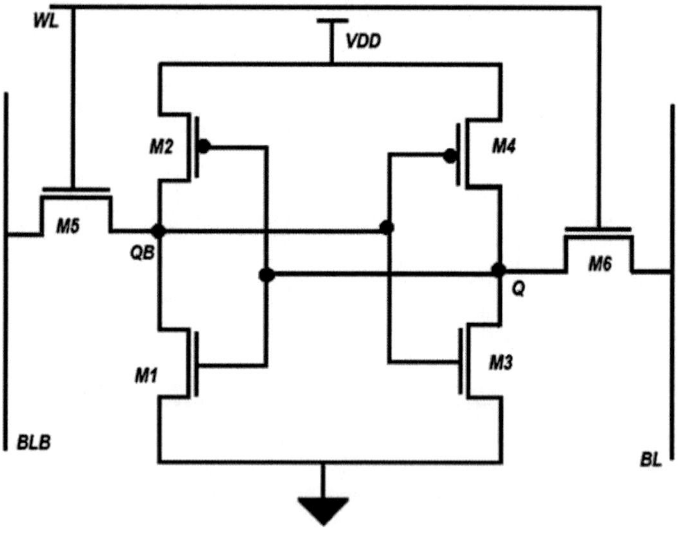

Fig. 4 Transistor-level architecture of 6-T SRAM cell

Standby Mode In this mode, the access transistors M5 and M6 are shut down by virtue of a low 'WL'. Hence, the cross-coupled inverters are separated from the bit lines, but still keep on reinforcing each other until the power supply is disconnected.

Read Operation During read operation, WL is asserted 1. Bit lines BL and BLB act as output lines. The word line WL now allows the connection between cell and bit lines through the 2 transistors M5 and M6, which are known as access transistors. The bit lines 'BL' and 'BLB' have been pre-charged to VDD. Initially, we assume Q at VDD and QB at 0. As QB and BLB are at different potential, charge at BLB discharges through M5. Sense amplifier (comparator) gives output as 1 when BLB is less than BL. Similarly, sense amplifier gives output as 0 when BL is less than BLB.

Write Operation During write operation, WL remains 1. BL and BLB now act as input lines. BLB is now forced to GND. Let values stored at QB is assumed initially to be VDD and Q is 0. As 'QB' and 'BLB' are at different potential, charge stored at QB discharges through M5 and charges the BLB node to VDD; thus, gate voltage of M3 decreases (lower than the threshold voltage of M3), M3 becomes OFF, and M4 becomes ON. Thus, Q gets charged to VDD. Similar procedure is followed for storing 0.

3 Proposed Approach

The method of power reduction method is based on introducing 'NOR' gates in 6T-SRAM structure (NOR6T), resulting in lower power consumption when compared to a normal 6T-SRAM cell. In order to decrease power consumption of the SRAM cell, NOR gates NOR1 and NOR2 are connected and followed by two PMOS sleep transistor, M7 and M8, as shown in Fig. 5.

M7 and M8 are connected to the 2 inverters of the SRAM to make it work normally when the word line (WL) is 1. When the SRAM is in standby mode, i.e., when WL = 0, the inverter which is having output as 0 will be cut off from the power supply so that unnecessary leakage power consumption will be minimized.

The detailed working principle of the proposed technique is explained below:

Read and Write Operation
During write and read operation, word line (WL) is asserted 1; bit lines BL and BLB act as input and output lines, respectively. The memory cell remains connected to the bit lines. NOR1 and NOR2 will have 1 as input in one terminal and either 1 or 0 as input in the other terminal (depending on the value at BL and BLB), leading to outputs of both NOR1 and NOR2 as 0. As gate terminal input of M7 and M8 is 0, the two inverters will operate normally as conventional 6T-SRAM.

Standby Mode Operation
During the standby mode, value '0' is asserted to 'WL'. Hence, the two inverters in Fig. 5 get disconnected from bit lines, as the access transistors M5 and M6 remain

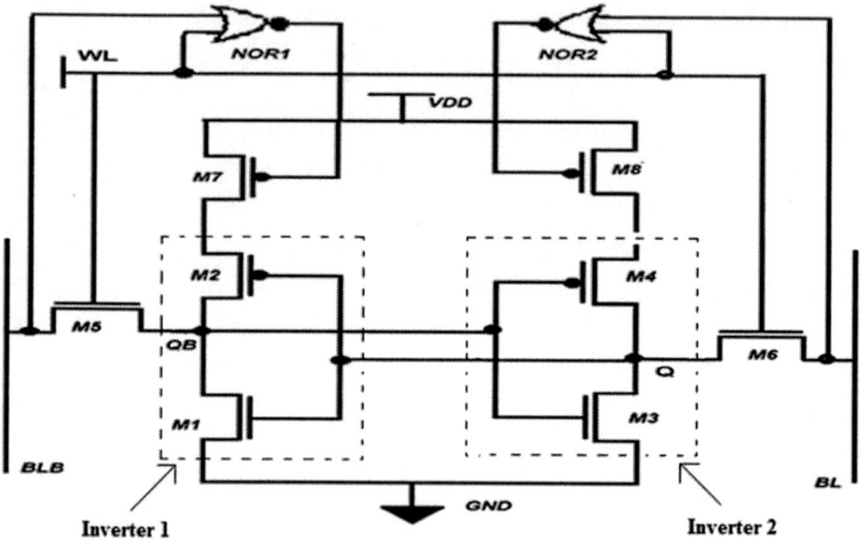

Fig. 5 Proposed low-power NOR6T SRAM cell

switched off. NOR1 and NOR2 will have 0 as input in one terminal having input WL and either 1 or 0 as input in the other terminal.

- When WL is 0 and BL is 1, NOR2 will have input as 0 and 1, leading to output as 1, and as a result, gate terminal input of the PMOS (M8) will also be 0, and hence, the inverter 2 in Fig. 5 will remain connected to supply voltage VDD.
- At the same time BLB is 0, NOR1 will have input as 0 and 0, leading to output as 1 and as a result gate terminal input of the PMOS (M7) will also be 1, and hence, the inverter 1 will be disconnected from the supply voltage VDD, and the inverter will be OFF.
- When WL is 0 and BLB is 1, NOR1 will have input as 0 and 1, leading to output as 1, and as a result, gate terminal input of the PMOS (M7) will also be 0, and hence, the inverter 1 will remain connected to supply voltage VDD.

At the same time, BL is 0, and NOR2 will have input as 0 and 0, leading to output as 1, and as a result, gate terminal input of the PMOS (M8) will also be 1, and hence, the inverter 2 will be disconnected from the supply voltage VDD and the inverter will be OFF.

Thus, at a particular instant only one of the two inverters will be connected to the supply voltage, leading to a significant amount of power savings during standby mode. The NOR6T approach will result in a significant increase in area overhead for a single SRAM. However, when a row and column of SRAM cells are considered in an actual programmable device, multiple SRAM cells can be governed by the same NOR gates and sleep transistors for gating, which will result in an acceptable power–area trade-off. This is represented in Fig. 6.

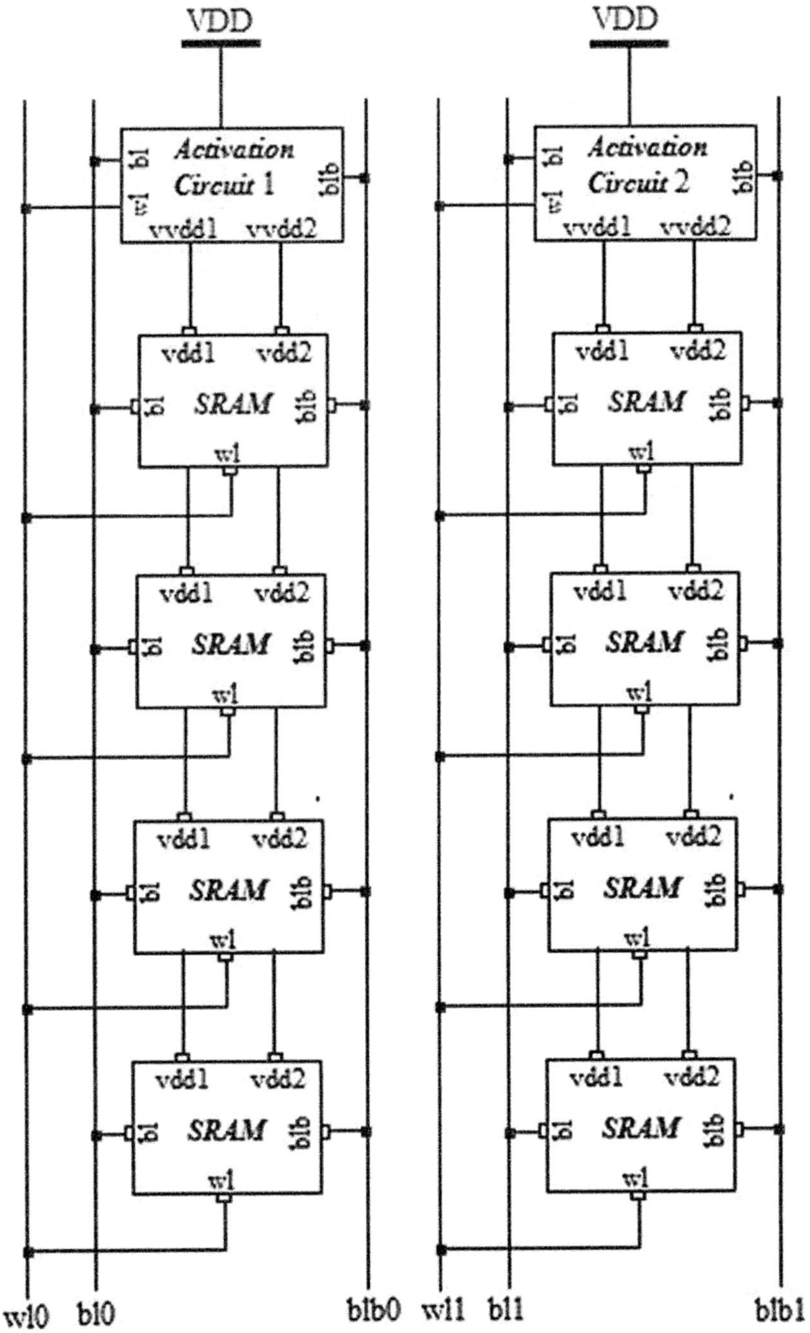

Fig. 6 Proposed low-power NOR6T technique in an array of 4 × 4 SRAM cells

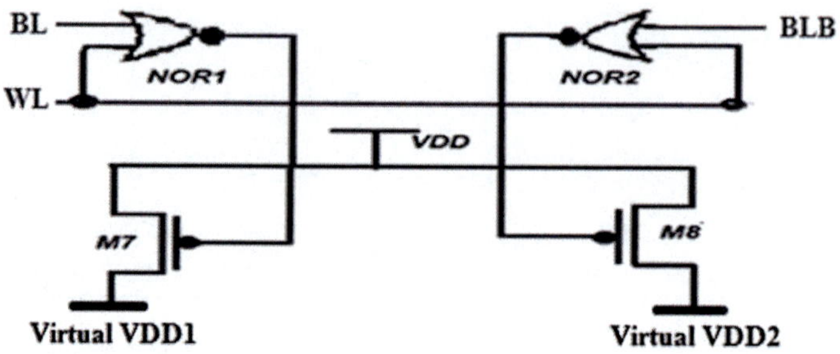

Fig. 7 Activation circuit of the proposed approach

The activation circuit in Fig. 6 is shown in Fig. 7. A single activation circuit can control the operating mechanisms of all the SRAM cells that are fed with the same WL, BL, and BLB.

As can be observed from Fig. 6, all the SRAM cells of the first column are controlled by activation circuit 1, whereas the SRAM cells of second column are controlled by activation circuit 2. Although the internal architecture of both the activation circuit will be the same, only separate inputs will be considered (different WL, BL, and BLB) for the generation of virtual VDD (VVDD).

4 Simulation Results

The design and simulation of the SRAM architecture are carried out in Cadence Virtuoso spectrum at 45 nm technology. Figure 8 presents the simulation waveform of a conventional 6T-SRAM.

In Table 1, the detail analysis of 6T SRAM and the proposed approach are presented. In Table 1, average power, number of transistor, and write access time are presented. It is observed from the table that significant power savings have been achieved in our proposed technique.

On an average, around 74.24% power savings is achieved in our technique. The delay associated with our technique is measured between inputs (BL and BLB) and outputs (Q and QB), resulting in a minimal increase from normal 6T SRAM. The delay analyzed is in terms of 'write access time' [3].

The NOR6T approach results in a significant increase in area overhead for a single SRAM. However, when a row and column of SRAM cells are considered in an actual programmable device, multiple SRAM cells can be governed by the same NOR gates and sleep transistors for gating, which will result in an acceptable power–area trade-off.

The detailed simulation waveform of our proposed technique is presented in Fig. 9.

In order to evaluate the effectiveness of our proposed approach, a comparison with prior techniques is being made and presented in Table 2. The prior works are also being developed and simulated in Cadence virtuoso spectrum, to obtain an accurate

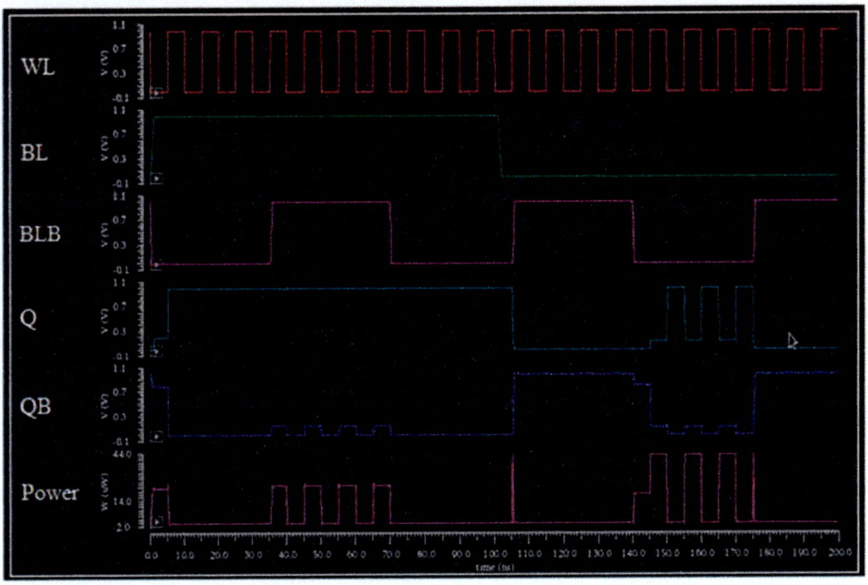

Fig. 8 Simulation waveforms of 6T SRAM

Table 1 Power, area, and delay analysis

Parameters		6T SRAM	NOR6T
Average power (μW)		6.6	1.7
Area (number of transistors used)		6	16
Write access time (ns)	Q & BL	4.599	4.614
	QB & BLB	4.915	4.938

comparison. From Table 2, it is observed that our proposed techniques have resulted in better power savings than all of the prior techniques.

The proposed approach in this paper is mainly low-power–low energy-oriented and results in an increase in area overhead. Reduction of the excess area is kept as a scope for future works.

5 Conclusion

An autonomous low-power technique is implemented in an SRAM cell in this paper, with a view to minimizing the total power dissipation. The supply power gating proposed in this work aims at reducing leakage power dissipation as well as the retention of the stored data in SRAM.

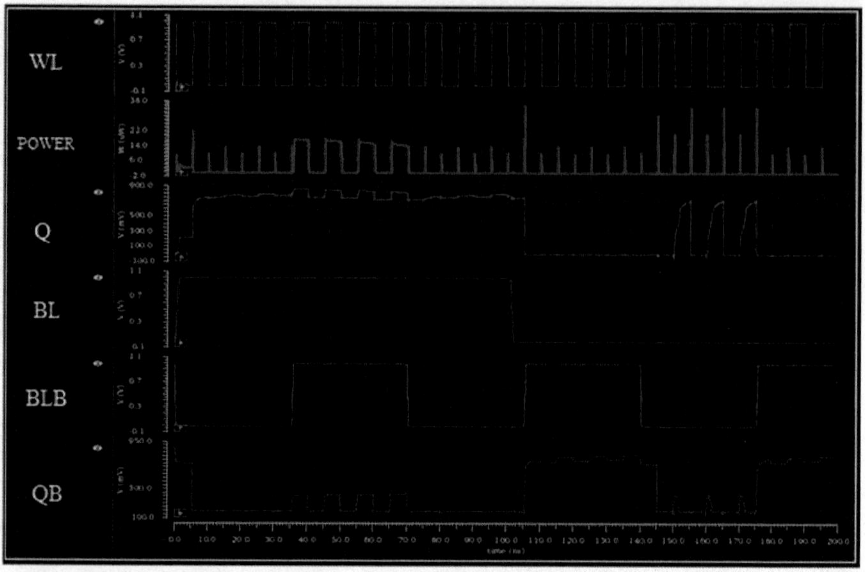

Fig. 9 Simulation waveforms of proposed NOR6T SRAM

Table 2 Comparison with prior works

Parameters		Prior works				Proposed approach
		NC-SRAM	IWL-VC SRAM	PP-SRAM	TG8T SRAM	NOR6T
Average power (μW)		5.594	11.04	16.03	11.78	1.7
Area (number of transistors used)		8	9	8	8	16
Write access time (ns)	Q & BL	5.594	4.789	4.599	4.594	4.614
	QB & BLB	5.610	4.795	4.789	4.915	4.938

The proposed theory of the low-power strategy is to automatically disconnect the power supply to the SRAM cell during periods of inactivity. The gating criterion is automatically extracted from the knowledge of bit lines and word line of the SRAM. By virtue of the proposed gating technique, an approximate of 74.24% of the total power consumption of the SRAM has been reduced, with an increase in delay of only 0.46%. A comparison with prior works indicates the effectiveness of our approach.

Acknowledgements The work presented in this paper has the support of SMDP-C2SD project, sponsored by the deity, Government of India.

References

1. Liu Zhiheng, Duan Xiongying, Liao Minfu, Zou Jiyan (2017) A FPGA-based digital synchronous methodology for IEC 61850-9-2 process bus. AEU Int J Electron Commun 76(10):137–145
2. Shah Divya K, Chaurasiya Rohit B, Vyawahare Vishwesh A, Pichhode Khushboo, Patil Mukesh D (2017) FPGA implementation of fractional-order chaotic systems. AEU Int J Electron Commun 78:245–257
3. Gupta V, Anis M (2009) Statistical design of the 6T SRAM bit cell. IEEE Trans Circ Syst I 57(1):93–104
4. Ojha P, Rana C (2015) Design of low power sequential circuit by using adiabatic techniques. Int J Intell Syst Appl (IJISA) 8:45–50
5. Elakkumanan P, Thondapu C, Sridhar R (2004) A gate leakage reduction strategy for sub-70 nm memory circuit. In: Proceedings of IEEE Dallas/CAS workshop, pp 145–148
6. Razavipour G, Afzali-Kusha A, Pedram M (2009) Design and analysis of two low-power SRAM cell structures. IEEE Trans Very Large Scale Integr (VLSI) Syst 17(10):1551–1555
7. Sharma J, Khandelwal S, Akashe S (2015) Implementation of high performance SRAM cell using transmission gate. In: 2015 Fifth international conference on advanced computing & communication technologies, pp 257–260
8. Nag A, Nath D, Pradhan SN (2017) Leakage reduction of SRAM based look up table using dynamic power gating. J Circ Syst Comput (JCSC) 26(3):1750041-1 to 1750041-12

Multi-level Encryption Based on Fractional Fourier Transform, Double Random Phase Encoding Combined with Chaos, and Arnold Transform

Aarushi Shrivastava and Janki Ballabh Sharma

Abstract A novel scheme for colour image encryption using fractional Fourier transform, random phase encoding, and chaotic map combined with Arnold transform is proposed using software MATLAB R2016a. We use the logistic chaotic map for diffusion. Scrambling along each layer is processed with replacement matrices by chaotic system and Arnold transform. The three scrambled images are converted into one-encrypted image by using double random phases. Proposed method provides multi-layer security of the colour image in both intermediate (fractional) and time domains. Simulation result verifies its efficiency and superiority in comparison to the other related schemes. The various analysis results display that the keys used are highly sensitive to their values, which is the indication to the high level security of the algorithm. Moreover, additional keys used in the proposed scheme provide more flexibility. While, statistical analysis result verifies its robustness against the statistical attacks.

Keywords Fractional Fourier transform · Double random phase encoding · Arnold transform

1 Introduction

With the growth of the computational and Internet technology, multimedia data is more susceptible to the unauthorized distribution and manipulation. With the problems dealing with illegal data access over Internet and due to the high speed for processing the data in parallel, the optical image encryption techniques have played an important role in data security. The double random phase encoding (DRPE) was proposed first [1] to encrypt the data using two different planes. Conventional DRPE has been enhanced further to improve the security level by applying it with the fractional Fourier transform (FRFT) [2], Fresnel transform [3], etc. However, some more

A. Shrivastava (✉) · J. B. Sharma
Electronics Department, Rajasthan Technical University, Kota, Rajasthan, India
e-mail: srivastava.aarushi11@gmail.com

V. Nath and J. K. Mandal (eds.), *Proceedings of the Fourth International Conference on Microelectronics, Computing and Communication Systems*, Lecture Notes in Electrical Engineering 673, https://doi.org/10.1007/978-981-15-5546-6_29

studies have been shown to chaos-based encryption methods in [7] which gives us more randomness, security, sensitivity to initial conditions. Some amount of work has been done on colour image encryption schemes. In [4], single channel colour image encryption has been proposed by using discrete cosine transform in YCbCr space. In [5], the colour image encryption is processed by using the multiple-order discrete fractional Fourier transform and chaotic scrambling. In [6], the colour image is encrypted by using logistic map and double random phase encoding but with limited encryption keys.

Therefore, in demand to boost the security of transmitting image, we introduce the sufficient amount of additional keys to it. In this paper, a multi-layer colour image encryption scheme using double random phase encoding, fractional Fourier transform, combined with chaos, and Arnold transform is presented. The main objective of this paper is to propose a colour image encryption technique for the transmission over unsecure or secure network. Firstly, we applied the fractional Fourier transform (FRFT) on each component. Secondly, the chaotic-based scrambling is performed along with Arnold transform with the different iterations which severs as an additional key. Finally, the double random phase encoding is used to obtain one encrypt image through three scrambled images. The comparative results show its resistivity through exhaustive attack and the secure transmission of data in both time and frequency domains.

2 Preliminaries for Proposed Work

2.1 2D-Coupled Logistic Map

The two-dimensional-coupled Logistic map [7] is described mathematically as follows:

$$\begin{cases} X_{m+1} = \mu_1 X_m(1 - X_m) + \gamma_1 Y_m^2 \\ Y_{m+1} = \mu_2 Y_m(1 - Y_m) + \gamma_2(X_m^2 + X_m Y_m) \end{cases} \tag{1}$$

The system is chaotic when $2.75 < \mu_1 \leq 3.4$, $0.15 < \gamma_1 \leq 0.21$, $2.7 < \mu_2 \leq 3.45$ and $0.13 < \gamma_2 \leq 0.15$ and generate chaotic sequences X, Y in the interval (0, 1).

2.2 Double Random Phase Encoding

This property comprises of the two random phase masks which are in the input and Fourier planes [1].

Let the (x, y) and (u, v) denote the space and Fourier plane coordinates. The original image $f(x, y)$ is multiplied by the random phase function $\psi(x, y)$ and is afterward Fourier transformed. In following step, the output obtained is for the second time multiplied by the another phase mask $\varphi(u, v)$ and inverse Fourier transform is accomplished on the image. Both the sequences are statistically independent to each other.

2.3 Fractional Fourier Transform

The fractional Fourier transform [2] is a simplification of the ordinary Fourier transform with an order of parameter 'n'. The nth order FRFT, $f_n(x_n)$ of a function $f(x)$ is defined as

$$f_n(x_n) = F_n\{f(x)\}(x_n) = \int_{-\infty}^{\infty} K_n(x, x_n) f(x) \mathrm{d}x$$

The kernel is known as

$$K_n(x, x_n) = \begin{cases} \chi_\varphi \exp\{j\pi(x^2 \cot\varphi - 2xx_n + x_n^2 \cot\varphi)\}, & 0 < |n| < 2 \\ \delta(x - x_n), & n = 0 \\ \delta(x + x_n), & n = \pm 2 \end{cases} \quad (2)$$

$$\chi_\varphi = \exp\left[-j\pi(\mathrm{sgn}(\sin\varphi))/4 + j\varphi/2\right], \quad \varphi = a\pi/2$$

Here, x and x_n signify the input and the output (nth order) fractional domains.

2.4 Arnold Transform

It is used to shuffle the pixel position of two-dimensional-squared database. Arnold transform of an image $f(x, y)$ with dimensions of $N \times N$ pixels which is mathematically defined as:

$$\begin{pmatrix} x' \\ y' \end{pmatrix} = \mathrm{ART}[f(x, y), N] = \begin{pmatrix} 1 & 1 \\ 1 & 2 \end{pmatrix} \begin{pmatrix} x \\ y \end{pmatrix} (\mathrm{mod}\, N) \quad (3)$$

where (x, y) denotes the pixel position of original image and (x', y') denotes the pixel position of the scrambled image after the ART operation is done.

3 Proposed Work

The proposed color image encryption scheme has been depicted in Fig. 1 and the brief steps are summarized as follows:

Step 1: First, apply the FRFT to the original image I on each component of it with the different fractional orders given as (α_1, α_2) for red, (β_1, β_2) for green, $(\vartheta_1, \vartheta_2)$ for blue as shown in Fig. 1.

Step 2: Now, to generate the random numbers, we iterate the logistic map of (1) for $3N \times N$ times. For every single iteration, we obtain the x_i and y_i, then these values are treated to acquire the sequences P, L, Y, Z as shown below

$$\begin{aligned} P &= 10^K x_i - \text{floor}(10^K x_i), \\ L &= 10^K y_i - \text{floor}(10^K y_i), \\ Y &= \text{floor}\big((10^K x_i - \text{floor}(10^K x_i))10^K\big) \bmod 256 \\ Z &= \text{floor}\big((10^K y_i - \text{floor}(10^K y_i))10^K\big) \bmod 256 \end{aligned} \tag{4}$$

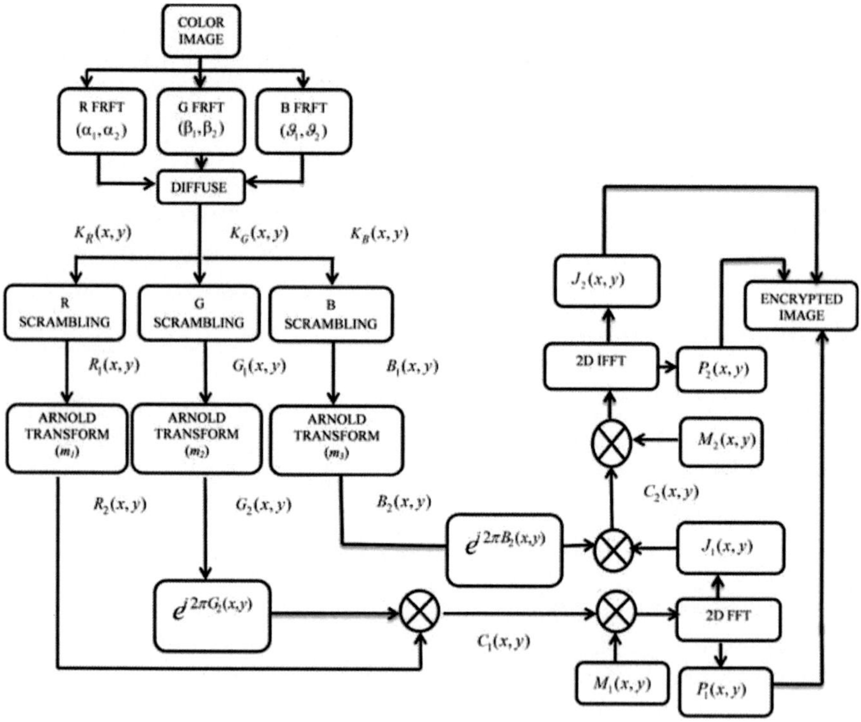

Fig. 1 Proposed diagram of encryption scheme

Step 3: Now, to obtain the scrambling matrices through random numbers, we take the sequence P and L and arrange them in ascending order, that gives us the new vector as P' and L' now divide vector P by P' and L by L' So, through which we obtain the six scrambling matrices, i.e., $A_1, A_2 \ldots A_6$.
Step 4: By taking the sequences P and L obtained from Step 1, we obtain two random phase masks of size $N \times N$, namely Q and V as shown below:

$$M_1(x, y) = \mathrm{e}^{\mathrm{i}2\pi Q} \tag{5}$$

$$M_2(x, y) = \mathrm{e}^{\mathrm{i}2\pi V} \tag{6}$$

Step 5: Now, the sequences Y and Z are used to construct H and B matrices which are used for diffusion of colour image as demonstrated below:

$$K(m, n, k) = ((I(m, n, k) + H(m, n, k)) \bmod 256) \oplus B(m, n, k) \tag{7}$$

Here, $m, n = 1, 2, \ldots 256$ and $k = 1, 2, 3$. The exclusive OR operation bit by bit is indicated by the symbol $\oplus$.
Step 6: The matrix $K(m, n, k)$ is converted into RGB components and scrambling is done by using the scrambling matrices separately.

$$\begin{cases} R_1 = A_1 \times K_R \times A_2 \\ G_1 = A_3 \times K_G \times A_4 \\ B_1 = A_5 \times K_B \times A_6 \end{cases} \tag{8}$$

Step 7: Normalization is done on to each component.
Step 8: To enhance the security, all three components are scrambled using Arnold transform with iteration numbers as m_1, m_2, m_3 and are converted into R_2, G_2, B_2.
Step 9: Now, the components in Step 8 are combined as the amplitude and phase part so as to achieve the complex image C_1 as shown below:

$$C_1 = G_2 \mathrm{e}^{\mathrm{i}2\pi R_2} \tag{9}$$

Step 10: Now, C_1 is multiplied by the random phase mask M_1 with the 2D fast Fourier transform (2D-FFT) on it which separates into two parts as presented below:

$$\begin{aligned} J_1 &= \text{amplitude}\{F[C_1 M_1]\} \\ P_1 &= \text{phase}\{F[C_1 M_1]\} \end{aligned} \tag{10}$$

where F [] represent 2D-FFT.
Step 11: By repeating again the Step 9, combine $J_1(x, y)$ as amplitude and $B_2(x, y)$ as phase; thus, we obtain $C_2(x, y)$, now repeat the Step 10 by $M_2(x, y)$, after which

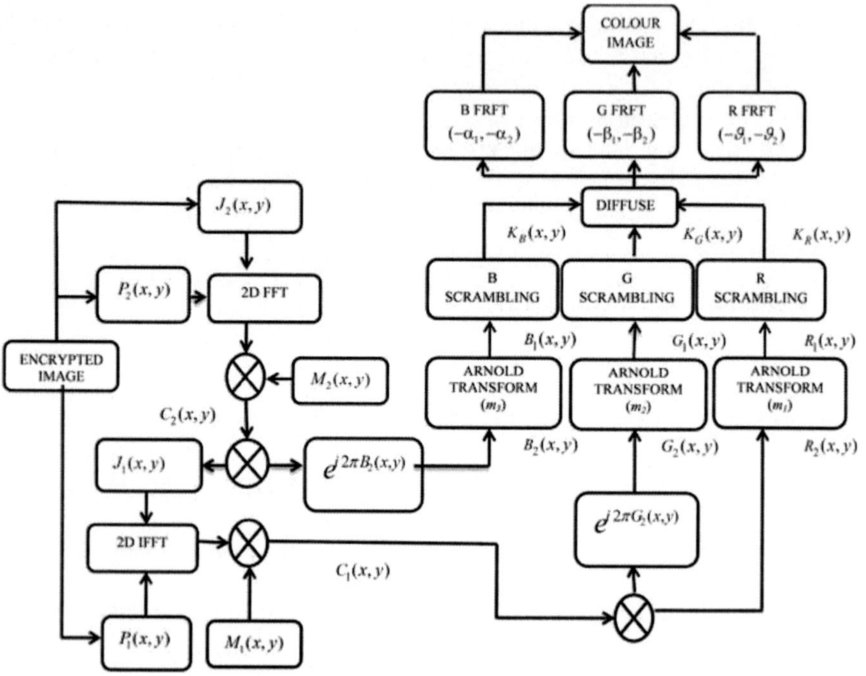

Fig. 2 Proposed diagram of decryption scheme

we subsequently perform 2D inverse fast Fourier transform to obtain amplitude $J_2(x, y)$ and phase part as $P_2(x, y)$.
Step 12: Finally, concatenate the matrices $P_1(x, y) P_2(x, y) J_2(x, y)$ to achieve the encrypted image.

Decryption is achieved by reversing the above procedure in which we have encrypted the image. Moreover, by performing the inverse of FRFT with the fractional orders $(-\alpha_1, -\alpha_2)$ for red, $(-\beta_1, -\beta_2)$ for green, $(-\vartheta_1, -\vartheta_2)$ for blue as shown in Fig. 2 to get the approximated image.

4 Computational Results

We have proposed and stimulated the encryption algorithm on colour image Lena with 256 × 256 pixels as shown in Fig. 3. For convenience, the initial conditions used in logistic map are set to $x_0 = 0.0215$, $y_0 = 0.57$, $z_0 = 0.3$, $\mu_1 = 2.93$, $\mu_2 = 3.17$, $\gamma_1 = 0.19$, $\gamma_2 = 0.139$, $\lambda = 3.99$. Experiments are performed for better performance of proposed scheme for different fractional orders of (α_1, α_2), (β_1, β_2), $(\vartheta_1, \vartheta_2)$ as (0.5, 1.5), (0.6, 1.4), (0.7, 1.3). The iteration numbers of the Arnold transform $m_1 = 5$, $m_2 = 6$, $m_3 = 4$ which are used for scrambling. By the use of these keys, the

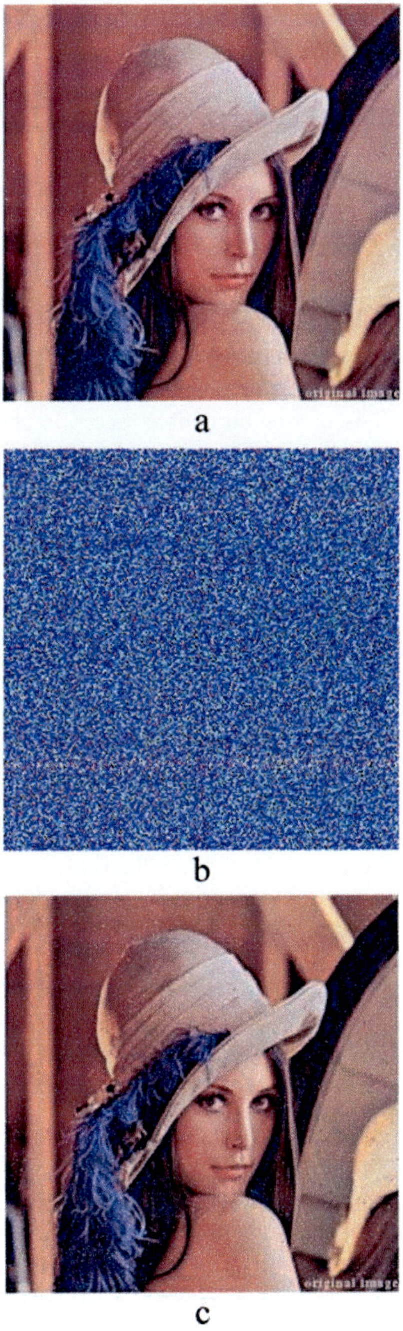

Fig. 3 Encryption and decryption of respective Lena image of the proposed scheme with its original image (**a**), encrypted image (**b**), decrypted image (**c**)

original image is encrypted as shown in Fig. 3b. The decryption is done by using (−0.5, −1.5), (−0.6, −1.4), (−0.7, −1.3) fractional orders and are processed to get the decrypted image as displayed in Fig. 3c.

4.1 Key Space Analysis

A good algorithm should include the large key space which is big enough to make the brute force attack ineffective. In our scheme, the fractional orders, Arnold transform parameters with chaotic scrambling matrices are used as the encryption keys. Here, the key parameters used are given as $(x_0, y_0, z_0, \mu_1, \mu_2, \gamma_1, \gamma_2, \lambda)$, (α_1, α_2), (β_1, β_2), $(\vartheta_1, \vartheta_2)$ as fractional orders and m_1, m_2, m_3 as Arnold transform iterations, which are all used as 64 bit double precision key. So, the key space is $2^{(64)17}$, i.e., 2^{1088} which is good enough to protect the data from the brute force attacks when compared to the advanced encryption standard (AES) encryption algorithm with the key space as 2^{128} (Fig 4).

4.2 Sensitivity Analysis

Here, the sensitivity analysis for Lena color image is performed by altering the fractional order of red component (0.5, 1.5) by (0.2, 1.3), keeping other orders correct the image is not decrypted correctly as shown in Fig. 5a. Similarly, for green and blue, if we change (0.6, 1.4) by (0.7, 1.7) and (0.7, 1.3) by (0.8, 1.5), we are unable to get the correct image shown in Fig. 5b, c. Now, if we use wrong Arnold keys at its inverse as $m_1 = 6$, $m_2 = 3$ and $m_3 = 4$, then the decrypted image is shown in Fig. 5d. So, here, in the sensitivity of the cipher keys results in good and secure cryptosystem.

The proposed technique has been numerically evaluated. The mean square error of the proposed technique is 4.8×10^{-04} between plain and decrypted image. Table 1 shows the computed PSNR value of the proposed technique is better than the algorithms used in [4–6] which truly indicates the quality approximation between the recovered image and the original image.

4.3 Statistical Analysis

1. *Histogram Analysis*

It is observed from Fig. 4a–f that the histogram of the encrypted image and original football image is much different from each other. Hence, it is difficult for an unapproved user to obtain any valid information by unfair means from its statistical property. Figure 4 shows the comparison of histograms before and after Lena image

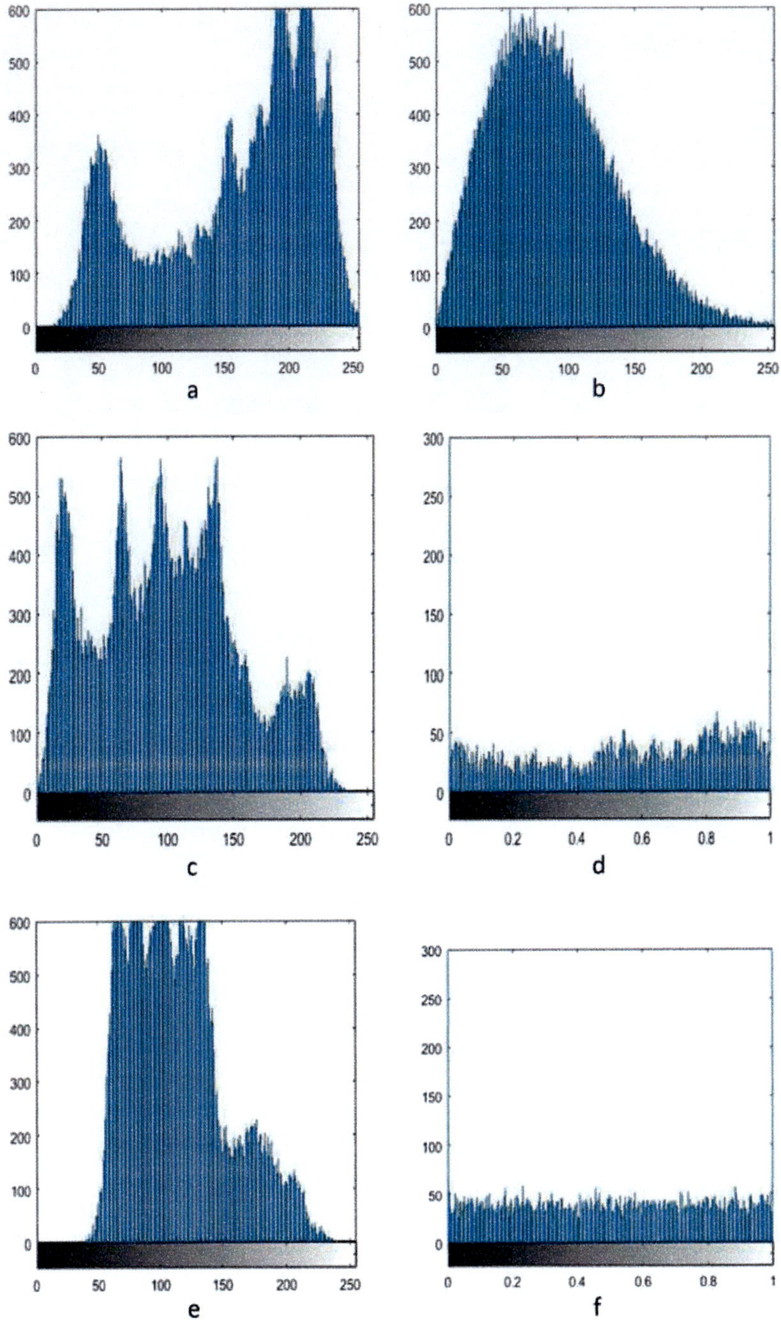

Fig. 4 Histograms of **a** red, **c** green, **e** blue components, of original image, **b** Red, **d** green, **f** blue component of encrypted image

Fig. 5 Image decrypted using incorrect keys **a** when the fractional order of red is incorrect keeping all others same. Similarly for **b** green and **c** blue, **d** incorrect Arnold keys

Table 1 Comparison of peak signal-to-noise ratio between decrypted and original images

	Proposed algorithm	Luo et al. [5]	Wu et al. [4]	Huang et al. [6]
PSNR	43.2 dB	37.11 dB	37.023 dB	–

encryption. It can be seen from the figure that the encrypted histogram is relatively uniform, effectively masking the distribution of the pixels in the plaintext image, making the statistical analysis even more difficult.

Table 2 Correlation coefficients of original and ciphered images

Scheme			Horizontal	Vertical	Diagonal
Proposed	Original image	*R*	0.9993	0.9996	0.9989
		G	0.9480	0.9991	0.9974
		B	0.9980	0.9994	0.9985
	Encrypted image	*R*	−0.0080	0.0190	0.0143
		G	−0.4887	−0.0259	0.0105
		B	−0.0091	−0.0259	0.0105
Huang et al. [6]	Original image	*R*	0.9566	0.9812	0.9295
		G	0.9432	0.9695	0.9199
		B	0.9269	0.9586	0.9020
	Encrypted image	*R*	0.0027	−0.0013	0.0039
		G	0.0034	−0.0034	0.0021
		B	0.0046	0.0038	0.0031

2. *Correlation Analysis*

To test it, we have randomly selected the 10,000 pairs of two adjacent pixels from an image. Calculated correlation coefficients of images are shown in Table 2 in comparison with [6] which demonstrates that nearby pixels in input images are strongly matched and are near to 1. However, for the encrypted images, correlation coefficient are low and approximately near to 0 which defines that our scheme is more secure and nearby pixels are weakly matched.

5 Conclusion

This paper demonstrates a secure colour image encryption technique using fractional Fourier transform, double random phase encoding and each component undergoes scrambling by using chaotic system and Arnold transform separately. There exists not only the single layer of security but provides the multi-layer security of the colour image. Moreover, if there is no information of keys used at different layers its unruly to decrypt the image because all keys and positions used are very sensitive to its own values. So, the presented technique is sensitive to encryption keys, resistible to the statistical attacks, and the value of peak signal-to-noise ratio is good enough to have a secure transmission.

References

1. Refregier P, Javidi B (1995) Optical image encryption based on input plane encoding and Fourier plane random encoding. Opt Lett 20:767–769
2. Liu Z, Li S, Liu W, Wang Y, Liu S (2013) Image encryption algorithm by using fractional Fourier transform and pixel scrambling operation based on double random phase encoding. Opt Lasers Eng 51(1):8–14
3. Singh H, Yadav AK, Vashisth S, Singh K (2015) Optical image encryption using devil's vortex toroidal lens in the Fresnel transform domain. Int J Opt 13. Article ID 926135
4. Wu J, Guo F, Zhou N (2013) Single channel colour image encryption using reality preserving fractional discrete cosine transform in YCbCr space. J Comput 8(11)
5. Luo X, Fan J, Wu J (2012) Single-channel color image encryption based on the multiple-order discrete fractional Fourier transform and chaotic scrambling. In: 2012 IEEE international conference on information science and technology, Wuhan, 23–25 Mar 2012
6. Huang H, Yang S (2017) Colour image encryption based on logistic map and double random phase encoding. IET Image Process 11(4):211–216
7. Wu Y, Noonan JP, Yang G, Jin H (2012) Image encryption using the two-dimensional logistic chaotic map. J Electron Imaging 21:013014

SEPIC-Boost CSCCC-Based SPWM Inverter with MPPT Technique

Debarati Banerjee, Sarbojit Mukherjee, and Rajarshi Dhar

Abstract In this paper named 'SEPIC-Boost CSCCC-based SPWM Inverter with MPPT Technique' a new design is proposed which uses PV panel to extract power from solar energy. Its main aim is to extract power from single solar PV panel for MLI. It also uses a SIMO-SEPIC-CSC combination converter in between the PV panel and the MLI. As the conventional methods, two-staged power conversion is implemented and along with this, it has a SIMO-SEPIC-CSC combination converter to generate inverter input. To improve efficiency and make it more reliable, MPPT tracking is employed which helps in drawing of maximum power from the PV panel. Simulation is done in MATLAB/Simulink 2016A. Results are displayed as a verification of the proposed idea.

Keywords PV panel · MLI · MPPT · SIMO-SEPIC-CSC combination converter

1 Introduction

In terms of power generation compared to any other forms of renewable energy resource, solar panels are more frequently used. This popularity of the solar energy owes to the fact that it is plenty and hence is freely available also that it is void of pollution and noise. There is also a high increase of the rate of usage [1].

There are three stages for generation of power, i.e., solar energy is sent for the DC-to-DC conversion which is in turn sent for the DC-to-AC conversion [2]. Other

D. Banerjee (✉) · R. Dhar
Electronics and Communication Engineering, RCC Institute of Information Technology, Kolkata, India
e-mail: debarati.rcc26@gmail.com

R. Dhar
e-mail: rajarshidhar1997@gmail.com

S. Mukherjee
Electrical Engineering, RCC Institute of Information Technology, Kolkata, India
e-mail: sarbo.1234@gmail.com

V. Nath and J. K. Mandal (eds.), *Proceedings of the Fourth International Conference on Microelectronics, Computing and Communication Systems*, Lecture Notes in Electrical Engineering 673, https://doi.org/10.1007/978-981-15-5546-6_30

proposals include mainly two stages of conversion which uses DC-to-AC conversion from solar energy with a DC-link capacitor used in between the two stages [3, 4]. This has a disadvantage in which the equipment is made expensive and heavy which is unsuitable for any types of applications. In a multilevel inverter from only one solar panel, seven levels can be developed. An inverter consisting of nine levels can also be developed but there is a disadvantage of asymmetrical configuration leading to uneven switch stress [5]. As the total number of levels of the voltage increases the number of solar panels required also increases which make the design heavy and expensive added to which MPPT algorithm is used to all the panels which create several problems. Unbalance in current and voltage occurs when distributed MPPT algorithm is implemented [6, 7].

For step up or step down of the voltage or current, DC-to-DC converter is used which naturally includes boost converter or buck-boost converter or SEPIC converter and many others. Combination of any two converters can also be used for applications of high-level devices. Dual input and dual output topologies are also available but if implemented with combination of converters then the complexity is increased.

This paper which is presented is successful in overcoming the above mentioned disadvantages and it also extends from three levels to higher voltage levels while drawing power from a single solar panel.

2 Analysis and Design of the Proposed System

This system consists of a PV panel, a SIMO-SEPIC-CSC combination converter and a DC-to-AC converter which is a SPWM inverter [8].

2.1 MPPT Algorithm

Using the MPPT algorithm as in Fig. 1, system efficiency of extraction of maximum power from PV panel can be enhanced. Among the different methods that can be used, the algorithm of perturb and observe is being implemented here for reliability and enhancement of efficiency.

As the name suggests, the controller changes voltage of the panel each time and the power is measured and further adjustments are done if the power increases again till the power becomes constant. Hence, in perturb and observe method, observation of the change in power is done each time. This is the most commonly used method but oscillations in the output power can be observed.

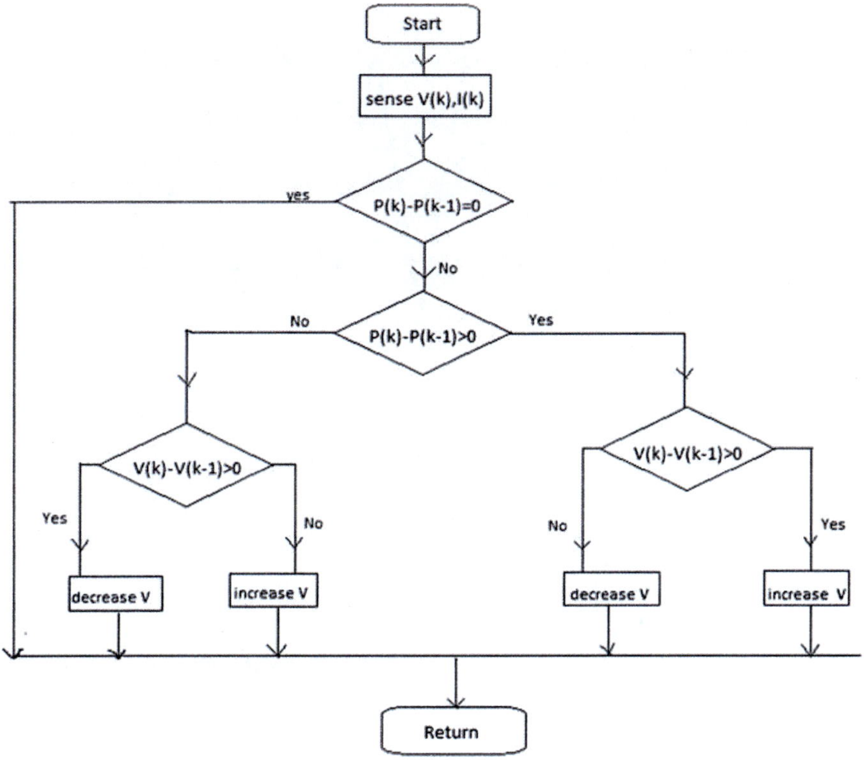

Fig. 1 Perturb and observe algorithm

2.2 *Control Mechanism*

Sinusoidal pulse width modulation (SPWM) technique is used for controlling the switches of the inverter as in Fig. 2.

2.3 *SIMO Converter*

Here, we are using non-isolated single switch DC-to-DC converter. Two output-type DC-to-DC converter configurations are obtained using combination of non-isolated single inductor DC-to-DC converter like boost [9, 10], canonical switching cell (CSC) [11, 12], Zeta converter [13, 14] using two-inductor topology.

There are two different modes of the operation of DC-to-DC converters:

(i) Continuous Conduction Mode (CCM)

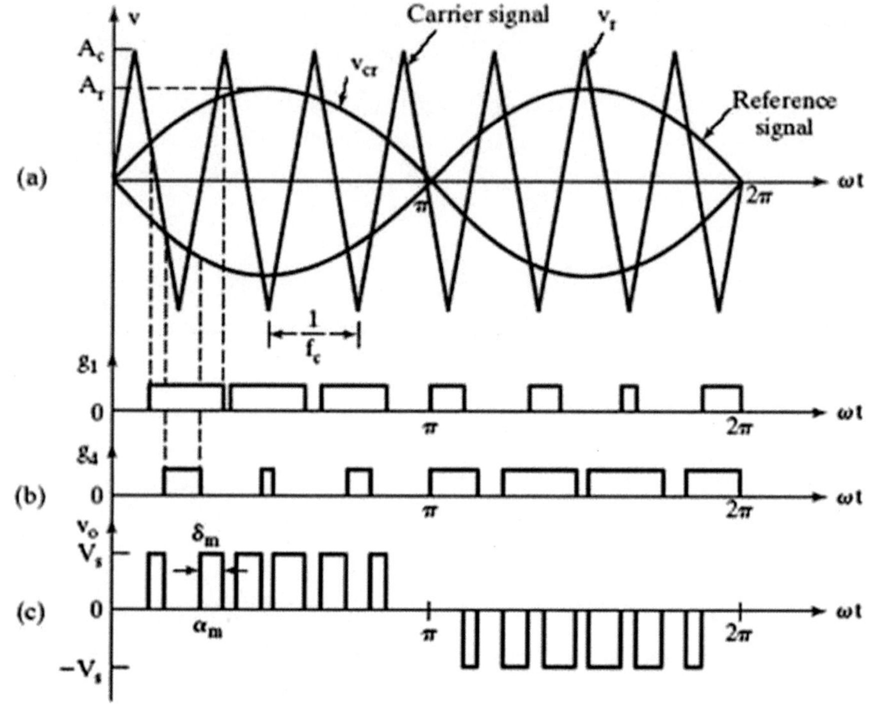

Fig. 2 Sinusoidal pulse width modulation

Here, current through the inductor of the converter circuit is continuous. In CCM, it enables load regulation as there is no load dependency observed in the DC conversion ratio (Fig. 3).

(ii) Discontinuous Conduction Mode (DCM)

Here, current through the inductor of the converter circuit is discontinuous. In the basic and derived topologies for the same output power, peak currents and root-mean

Fig. 3 Boost converter

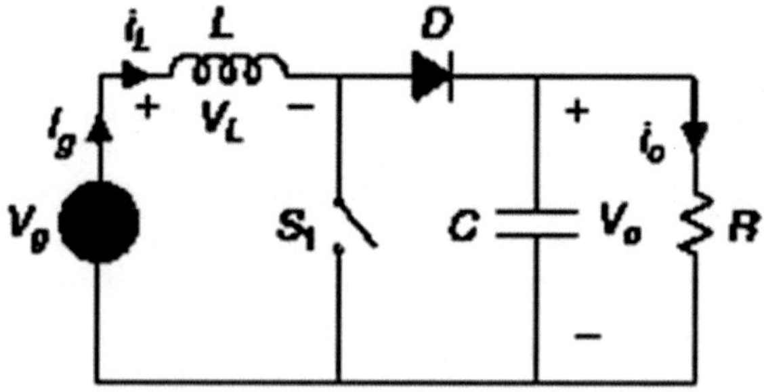

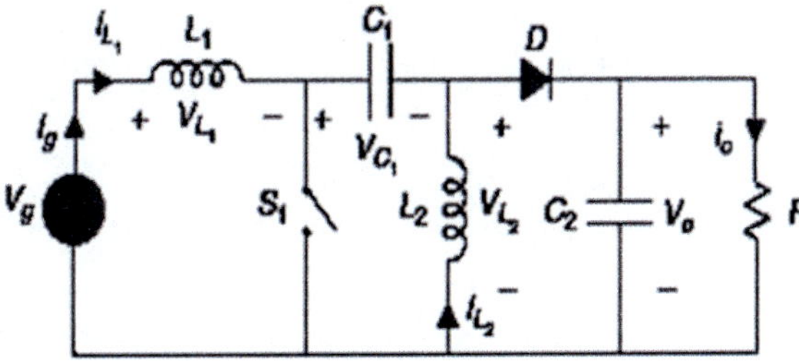

Fig. 4 SEPIC converter

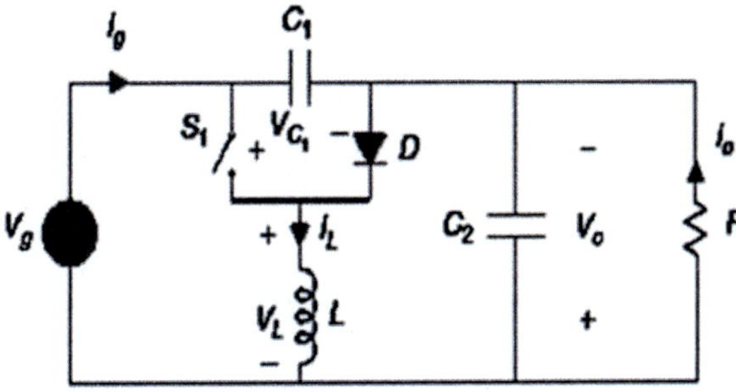

Fig. 5 CSC converter

square (RMS) currents are larger in DCM than in CCM. As a result of this, higher conduction path power losses is observed and also a high ringing.

A SEPIC-Boost CSE combination converter (SE-Bo-CS-CCC) is the combination of boost converter (as in Fig. 6), SEPIC converter (as shown in Fig. 4), and a CSC converter (as shown in Fig. 5). The conversion relations are described as the ratio of the output DC voltage (V_o) to the input DC voltage (V_g) in steady-state conditions. Ripples magnitude with small and lossless switching are taken as analogues to their respective components, where R is the load resistance attached to the converter. The tendency of a converter to operate in DCM is given by the dimensionless parameter K. CCM results from large values of K while DCM results from small values of K (Table 1).

This is a converter configuration with three outputs. These combined converters provide voltage which is a DC and bipolar one and an higher output voltage than the corresponding input voltage. The simulation diagram of the SIMO SEPIC-Boost-CSC combination converter is shown as in Fig. 7.

2.4 *SIMO-Based SPWM Inverter*

The output from the DC links behaves as the inverter source which is cascaded according to the requirements. With this model, output voltage of three levels has been shown as in Fig. 8. SPWM is used for generating the gate signal.

Figure 9 shows the proposed converter's SIMULINK model. Hence, the figure

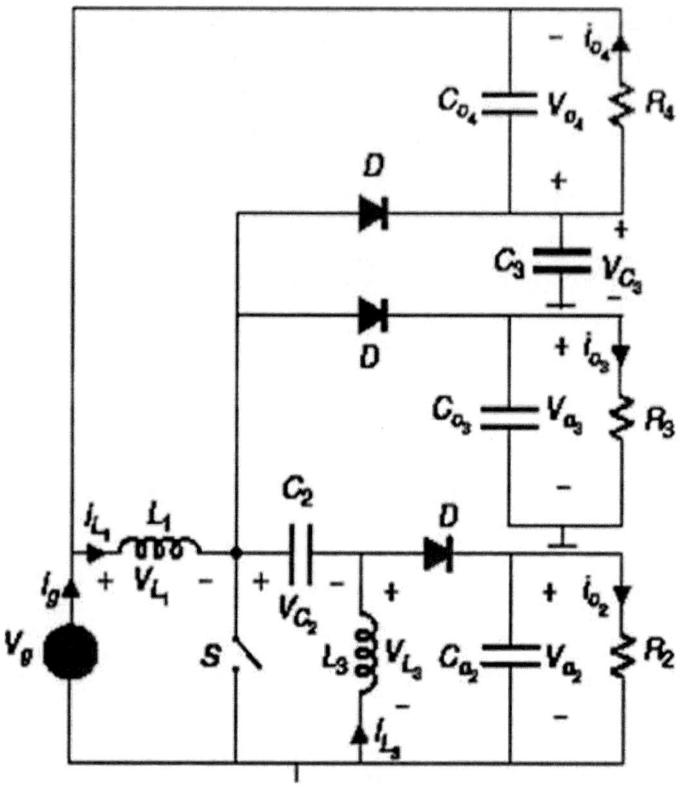

Fig. 6 SIMO SEPIC-boost-CSC combination converter

Table 1 A summary of characteristics of basic non-isolated DC–DC converters

Converter	Output type	Features
Boost	Step-up and non-inverted output	• High efficiency • Simplicity • Discontinuous input current • Continuous output current
SEPIC	Non-inverted and step-up output	• High efficiency • Continuous input current • Discontinuous output current • High peak output current in power components
CSC	Step-up-down inverted output	• Less number of passive components • High electrical stresses on the switch • Continuous input current

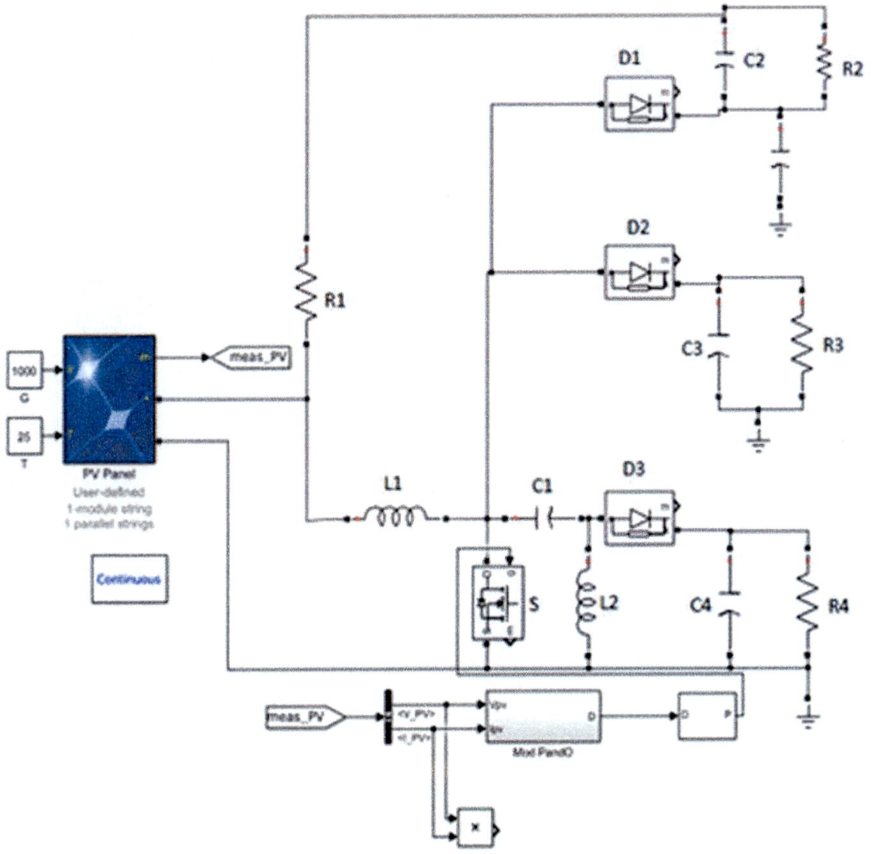

Fig. 7 Simulation diagram of SIMO-SEPIC-CSC combination converter

shows the model for a three-level inverter using SIMO-SEPIC-CSC combination converter, a PV panel, and the MPPT algorithm [15, 16].

3 Simulation Results

The simulation results of different characteristics are presented to support and validate the ideas presented in this paper as shown below.

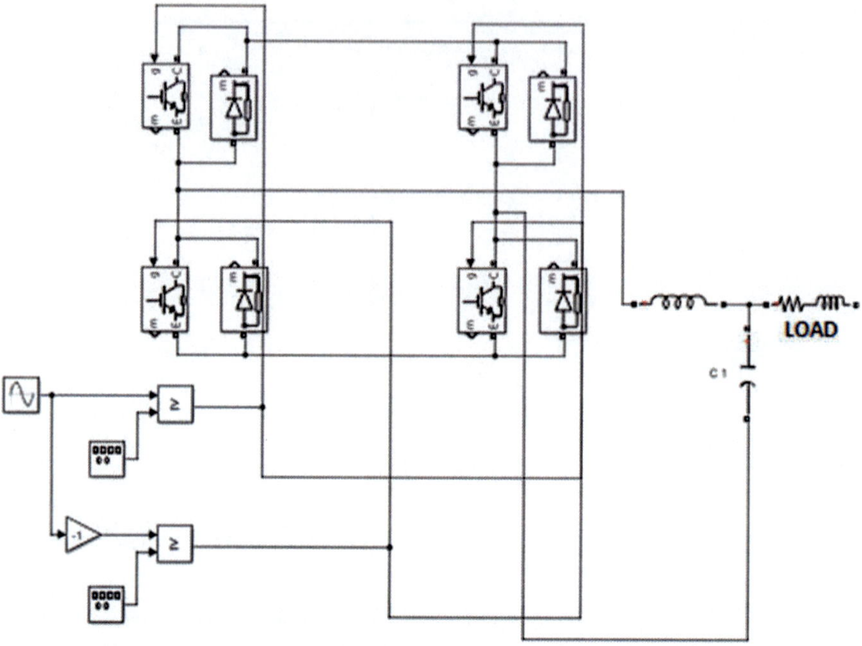

Fig. 8 SPWM Inverter of three levels based on SIMO

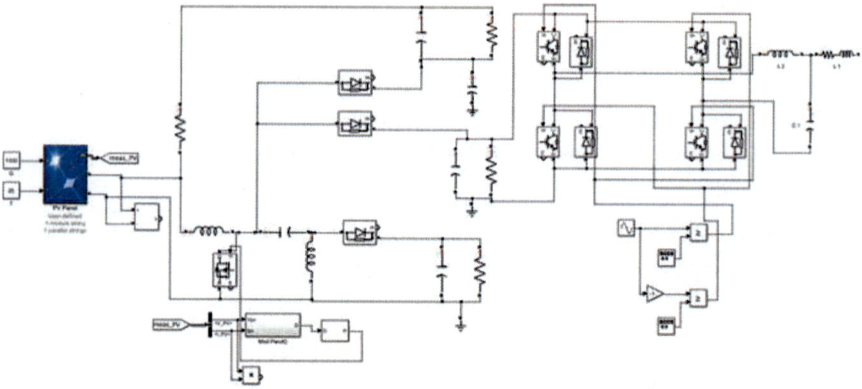

Fig. 9 Simulink model for SIMO SEPIC-Boost-CSC combination converter

3.1 MPPT Pulse Input

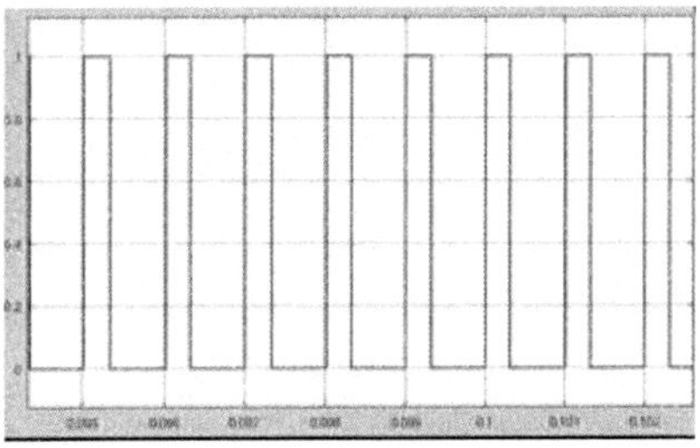

3.2 SEPIC Converter Output

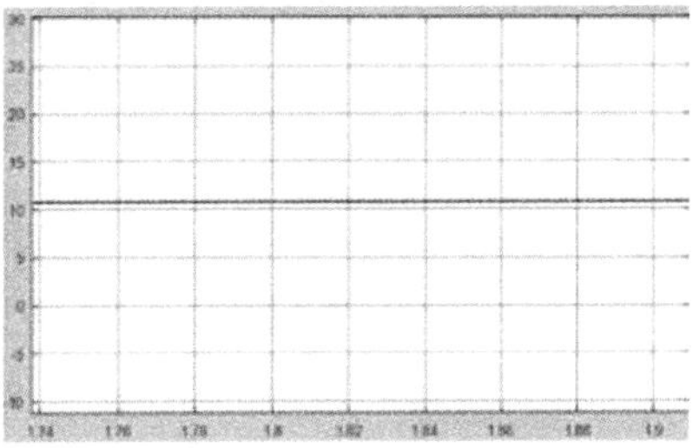

3.3 Boost Converter Output

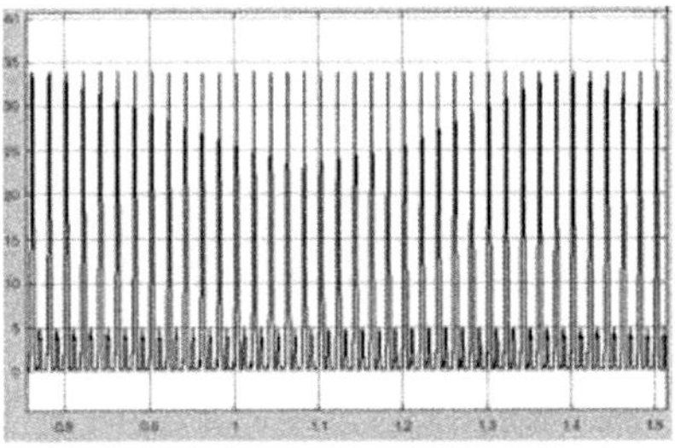

3.4 CSC Converter Output

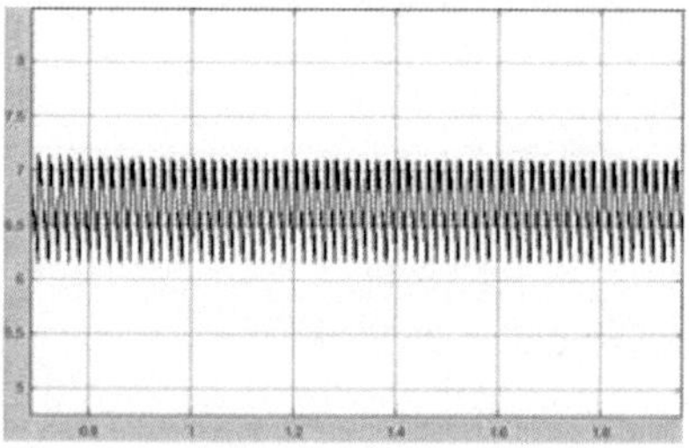

3.5 Voltage Across Inductor L1

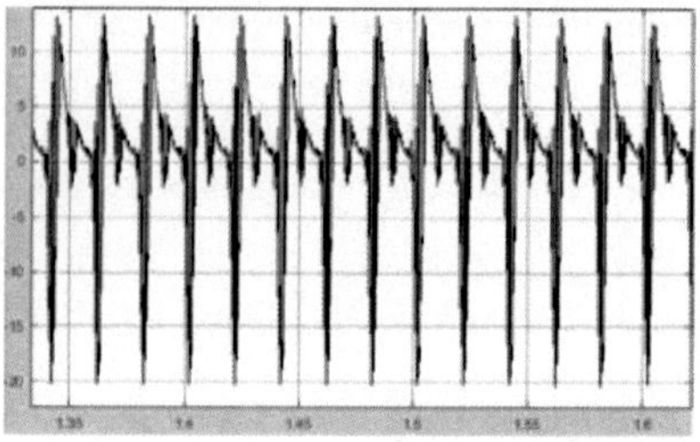

3.6 Voltage Across Inductor L2

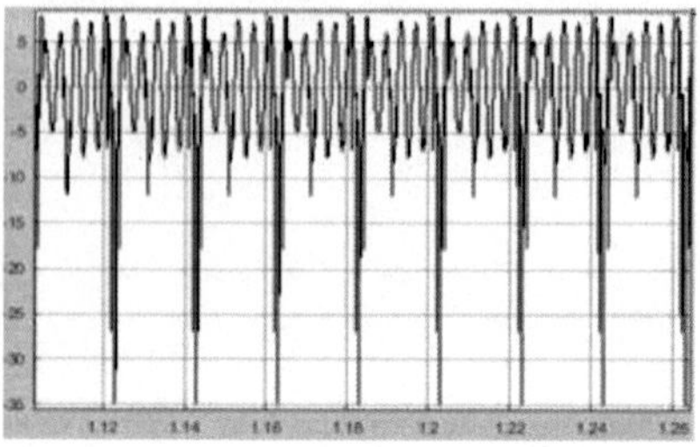

3.7 *Voltage Across Inverter Load*

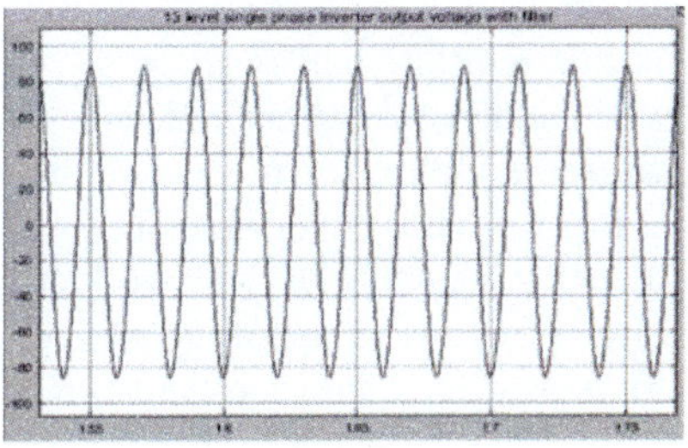

3.8 *SPWM Pulse*

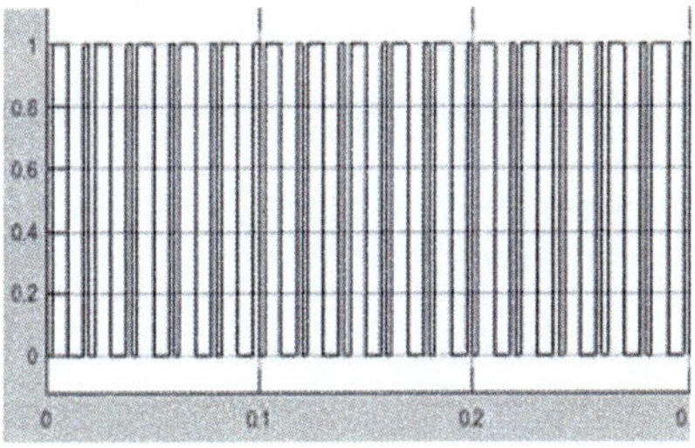

After the simulation of the circuit in MATLAB, we find DC–DC SEPIC converter output voltage is 11 V. The peak voltage of the boost converter output is 34 V. From the DC–DC CSC converter we are getting 7.1 V. The Inductor L_1 is getting charge up to 14 V. The peak voltage across Inductor L_2 is 8 V. The voltage across the Inverter load is 85 V.

4 Conclusion

Multiport converters are capable of interfacing or connecting to different sources and ports. In Fig. 6 the circuit diagram shows the multiple outputs of the converter in CCM mode of operation where the three different outputs are in three different modes. The first terminal voltage (across R_4 load) is in buck mode, the second terminal voltage (across R_3 load) is in the boost mode and the third terminal voltage (across R_2 load) is in buck-boost mode. Thus, from a single input, we get three different outputs and thus can be used to drive three different circuits. The paper presented here proposes a system with an inverter connected to one of the outputs of the SIMO system and thus AC loads can also be driven with this system.

In conventional [17] and modified [18, 19] SEPIC converters, the topology is constructed such that the systems have single input single output (SISO). In comparison to that system, our proposed topology has single input and multiple outputs (SIMO). The main advantage of this system is that multiple loads can be driven from the same source keeping all the properties of the boost converter constant. As presented in [18], the system mentioned uses voltage doubling circuits which uses only a single inductor or normal boost converter logic, whereas in our topology, we have used a SEPIC converter with double inductor to have a stable DC with workable amount of current at the output. Thus, the novelty of our circuit mainly focuses on the multiple output system with different load-driving capacity with constant current output.

References

1. Carrasco JM, Franquelo LG, Bialasiewicz JT, Galvan E, Guisado RCP, Prats MAM, Leon JI (2006) Power electronic systems for the grid integrated renewable energy sources: a survey. IEEE Trans Indust Electron 53(4):1002–1016
2. Lu X, Sun K, Ma Y, Huang L, Igarashi S (2009) High efficiency hybrid cascade inverter for photovoltaic generation. In: Proceedings of IEEE region 10 conference TENCON. https://doi.org/10.1109/tencon.2009.5396210
3. Chithra M, Dasan B (2011) Analysis of H-bridge multilevel inverters with photo voltaic arrays. In: Proceedings of international conference on emerging trends in electrical and computer technology, Nagercoil, India, pp 441–447
4. Marquez A, Leon JI, Vazquez S, Franquelo LG (2014) Advanced control of multi level h-bridge converter for pv applications. In: Proceedings of the 40th annual conference of the IEEE industrial electronics society, Dallas, TX, USA, pp 4548–4553
5. Manoharan MS, Ahmed A, Kim HW, Park JH (2015) A single source photo voltaic power conditioning system using asymmetric cascaded multi level inverter. In: Proceedings of the 9th international conference on power electronics and ECCE Asia (ICPE-ECCE Asia), Seoul, South Korea, pp 21–26
6. Patil SN, Prasad RC (2015) Design and development of MPPT algorithm for high efficient DC-DC converter for solar energy system connected to the grid. In: Proceedings of international conference on energy systems and applications, Indore, India, pp 228–233
7. Xiao B, Hang L, Mei J, Riley C, Tolbert LM, Ozpineci B (2014) Modular cascaded H-bridge multilevel PV inverter with distributed MPPT for grid-connected applications. IEEE Trans Indus Appl 51(2):1722–31
8. Huang J, Shi H (2014) Reducing the common-mode voltage through carrier peak position modulation in an SPWM three-phase inverter. IEEE Trans Power Electron 29(9):4490–4495
9. Hwu KI, Peng TJ (2012) A novel buck–boost converter combining KY and buck converters. IEEE Trans Power Electron 27(5):2236–2241
10. Wu H, Mu T, Ge H, Xing Y (2016) Full-range soft-switching-isolated buck-boost converters with integrated interleaved boost converter and phase-shifted control. IEEE Trans Power Electron 31(2):987–999
11. Jha A, Singh B (2006) Power quality improvement using CSC converter for high power LED driver. In: Proceedings of the 6th IEEE international conference on power systems (ICPS), New Delhi, India, pp 1–6
12. Singh B, Bist V (2015) A BL-CSC converter-fed bldc motor drive with power factor correction. IEEE Trans Industr Electron 62(1):172–183

13. Murthy-Bellur D, Kazimierczuk MK (2011) Isolated two-transistor zeta converter with reduced transistor voltage stress. IEEE Trans Circuits Syst II Express Briefs 58(1):41–45

14. Khatab AM, Marei MI, Elhelw HM (2018) An electric vehicle battery charger based on zeta converter fed from a PV array. In: Proceedings of IEEE international conference on environment and electrical engineering and IEEE industrial and commercial power systems Europe (EEEIC/ I&CPS Europe), Palermo, Italy, pp 1–5
15. Chen C, Fayed A (2015) A low-power dual-frequency SIMO buck converter topology with fully-integrated outputs and fast dynamic operation in 45 nm CMOS. IEEE J Magaz 50:2161–2173
16. Raghamadhuri SV, Subramanian K (2017) SIMO based inverter with MPPT technique. In: Proceedings of international conference on computation of power, energy information and communication (ICCPEIC), Melmaruvathur, India, pp 506–509
17. Tidke A, Savakhande V, Chewale M, Sonawane P, Wanjari R (2018) MPPT based sepic converter with voltage doubler cell for PV systems. In: Proceedings of international conference on control, power, communication and computing technologies (ICCPCCT), Kannur, India, pp 195–199
18. Ranjana MSB, Reddy NS, Kumar RKP (2014) A novel sepic based dual output DC-DC converter for solar applications. In: Proceedings of power and energy systems: towards sustainable energy, Bangalore, India, pp 1–5
19. Kaouane M, Boukhelifa A, Cheriti A (2015) Design of a synchronous sepic DC-DC converter for a standalone photovoltaic system. In: Proceedings IEEE Canadian conference on electrical and computer engineering, Halifax, Canada, pp 870–874

A Fuzzy Logic Approach for Software Cost Prediction

Farha Masroor, Vandana Bhattacharjee, and Debjani Mustafi

Abstract Software cost prediction is the technique of accurately evaluating the amount while developing the software. Estimation involves the total time required for the completion of the software, effort required that is measured in terms of person per month (PM), and the total cost to complete the activity. Accuracy and duration are the two desirable criteria in the software estimation process. In software estimation process, there are several inputs that are being fed to the system and these inputs are used for the generation or calculation of the set of outputs. The important work of the software project managers in the present scenario is the computation of cost or effort before the absolute advancement of any particular software. There are several methods applied for software cost estimation but we will focus on the fuzzy logic which is a soft-computing method. We feel that model which is based on fuzzy logic for the software cost estimation should be able to give the uncertain values rather than other models which give precision and certain values and the results based on fuzzy logic model will be more accurate than other models. In this research paper, we have developed the fuzzy logic model by taking several inputs and membership functions, also the fuzzy rules have been deduced by expert-knowledge with the help of MATLAB's Fuzzy logic toolbox and the results are then compared with the multiple regression model using SPSS Tool upon desharnais data set which gives less accuracy than fuzzy-based model.

Keywords Fuzzy logic · Fuzzy inference system · Effort · Estimation · Desharnais data set · Linear regression

F. Masroor (✉) · V. Bhattacharjee · D. Mustafi
Department of Computer Science and Engineering, Birla Institute of Technology, Mesra, Ranchi, Jharkhand, India
e-mail: farhamasroor01@gmail.com

V. Bhattacharjee
e-mail: vbhattacharya@bitmesra.ac.in

D. Mustafi
e-mail: debjani.mustafi@bitmesra.ac.in

V. Nath and J. K. Mandal (eds.), *Proceedings of the Fourth International Conference on Microelectronics, Computing and Communication Systems*, Lecture Notes in Electrical Engineering 673, https://doi.org/10.1007/978-981-15-5546-6_31

1 Introduction

Accuracy and duration are the most important tasks for the software projects. In the past history, several researchers have already worked in the field of software computation to advance the efficiency and several techniques have been developed. This paper deals with soft-computing technique which are used for approximate models. Soft computing replaces the hard computing technique as it is less time consuming with more intelligent processing systems. Fuzzy-based model is one of the soft-computing techniques which we will focus in this paper.

1.1 Fuzzy Logic

It is a path for computing based on "degree of truth."It follows the concept of truth that is partial. It is not based on modern computer that gives values either "true" or "false"(1 or 0). Fuzzy systems have the edge of better performance, great productivity, simplicity, and lesser cost. Membership function plays vital role in fuzzy system as it represents the "degree of truth" in fuzzy logic. It defines whether all the information and elements in the fuzzy set are discrete or continuous. Basically, there are four membership functions: Triangular, Gaussian, Trapezoidal, and Singleton. It is very difficult to choose the correct set of membership functions for a particular fuzzy model. In this research work, we will use trapezoidal and triangular membership functions and have taken five membership values as very low, low, medium, high, and very high.

1.2 Fuzzy Rules

Fuzzy rules are the rules that are produced by human expertise brains. It may vary from person to person as different person will think differently and with their own brains, they can generate fuzzy rules accordingly. Fuzzy rules are basically 'IF','THEN' rules.

1.3 Fuzzy Logic System

It has four important parts:

1. Fuzzification: It is the process of converting the crisp value that is non-fuzzy input values to fuzzy data.
2. Rule-Based System: Rule-based system mean the rules that are used for the process (IF-THEN rules). Fuzzy IF-THEN rules basically connect m-conditional

variables to n-consequent variables in the form: IF (X_1 is A_1 and … X_m is A_m) THEN (Y_1 is B_1 and … Y_n is B_n) Where: Linguistic terms are A_1 …, A_m and B_1 …, B_n and Linguistic variables are (X_1 …, X_n and Y_1 …, Y_n) "Ancedent" is the IF part, and "Consequent" is THEN part.

3. Inference Rule: These are the IF-THEN rules, which are carried out in fuzzy inference system and then the rules that will be fired according to the data set are the final fired rules which will be applied in the inference system to get the result.
4. Defuzzification: It is the technique to convert the fuzzy values to original values. The rest of the paper is coordinated as such: Segment 2 as Literature Review, Segment 3 consists of Proposed Framework, followed by Segment 4 as Experimental Results, Conclusion as a part of Segment 5 and Segment 6 as References.

2 Literature Review

Bedi and Singh [1] have compared the fuzzy logic approach with the COCOMO Models and have shown the accuracy by evaluating the MRE% and MMRE%. The data set that is used in this research paper is the Promise data set. The fuzzy inference system is based on Mamdani approach. Five input parameters and one output parameter are taken which is used for prediction of software effort. With different input/output parameters, different linguistic variables are taken. Based on linguistic variables and parameters, four different fuzzy rules are considered. The result with the COCOMO Model is MMRE% of 25.604 which has much higher % than fuzzy logic approach of MMRE% of 17.613. Dizaji and Gharehchopogh [2] have evaluated the price of software projects as per the meta-heuristic algorithms and the data set of NASA has been taken for further research. At first, the classification of the projects is done according to the project type. After the classification process, the ACO and the COA algorithm are used for the cost estimation of software projects. In ant colony optimization (ACO) technique, the ants are proficient enough to find the shortest path with the help of concentration level of pheromones. When an ant moves, it leaves some concentration level of pheromones in the ground, and reaches its final destination. The other ants when moves forward follow the same path with the help of pheromones and moves in the direction which has more concentration level of pheromones. This way they find the shortest path and is called ACO technique. Chaos optimization algorithm (COA) is done with the help of iterative maps. The experimental results say that the performance increased when ACO algorithm is combined with COA. Also, MARE for COCOMO Model is 0.29% and 0.078% is the MARE with proposed technique(ACO combined with COA) which proves that there is an improvement in the estimated costs of software projects. An optimized fuzzy logic framework for the research work has been used by Sharma and Verma [3]. Many research works have proved that techniques involving COCOMO Models give poor results but in this paper, the framework is built upon COCOMO-81 and intermediate COCOMO is used. To get the exact size of software projects is very difficult in COCOMO Model as it doesn not consider the projects not lying exactly in any of the three categories(organic, semi-detached, and embedded), so fuzzy logic approach is used for improvement in the results. Gaussian membership function is used for software development mode and effort. For the development of fuzzy rules, the basic components of COCOMO Model are used. Triangular and Trapezoidal membership functions are used to fuzzify the cost drivers having linguistic values such as very low, low, and nominal high. For every cost drivers, independent fuzzy inference system is built. From each effort multiplier, the defuzzified value is collected by individual FISs after matching, aggregation, and subsequent defuzzification. By multiplying them all together, the total EAF is obtained. The comparison has been done of nominal effort prediction on actual real project data by FIS and COCOMO Model. Also another comparison of the comprehensive effort predicted by FIS and COCOMO Model with the addition of effort multipliers is done. The results show that the nominal effort prediction by the FIS has less than 50% error for most of the projects. Shivakumar et al. [4] have worked with the concept of adaptive neuro-fuzzy logic for the effort estimation [5] which helps in the improvement of accuracy and reliability. Ninety three instances of NASA project data were considered and 30 projects from different case studies and experiments were gathered consisting actual effort and 15 attributes together with field of work, size, and domain. After the collection of data and attributes, these 15 attributes were converted into three index values. The adaptive neuro-fuzzy technique is then built [4]. Artificial neural networks are made up of neurons which are connected in parallel. This model has an input of six-grouped attributes which leads to the development effort. The neural network and fuzzy logic principles are combined in this ANFIS framework. The MRE and MMRE are calculated and then the results have been compared with the proposed ANFIS model and other algorithmic models. Kumar et al. [6] deal with the fuzzy logic technique for software prediction taking two inputs as lines of code and adjustment difficulty level and effort as an output. The data set has been taken from BIT MCA students. Three membership functions [3] as low, medium, high have been taken. Also, according to the data set, the parameters have been taken for the inputs and output, respectively. Fuzzy inference system is built and fuzzy IF-THEN rules are applied. The predicted effort is obtained and with the help of it, the MRE and MMRE are calculated. The same data set is used for the testing in multiple regression method and the results are than compared by both the methods. By fuzzy logic technique, the results are more accurate as the MMRE is 0.1762% whereas by multiple regression, 0.5358% is the MMRE which is much higher. So fuzzy logic gives much more accurate results but the limitation in this paper is that it is difficult to determine a correct pair of M.F when the dimensionality and volume of data are broad. An approach based on fuzzy logic and optimization process to evaluate software project effort has been given by Ganesh et al. [7]. At first, Fuzzy logic approach is used for both categorical and numerical data which are specified by fuzzy sets and while generating fuzzy rules, the grouping of the optimization is done by means of particle swarm optimization(PSO), so that the rules could function better. The fitness for the optimization function is assumed to be the effort of the software. The outcome of these fitness functions that is the fitness values are further carried for optimization of fuzzy rules. The comparison by Bhatnagar et al. [8] has been done between fuzzy logic and neural network models for the development of software effort estimation. Radial basis neural network (RBNN), FFNN, and fuzzy logic models are built and the results are evaluated based upon parameters like MRE, MMRE, BRE, and prediction. Neural network as a tool and backpropagation (learning algorithm) method for training the networks are used. At first, the fuzzy inference system is built with fuzzy rules having Gaussian membership functions, mode, and size as inputs and effort as output. For each cost drivers, fuzzy inference system is defined. EAF is calculated by multiplying the values of each cost drivers. The final effort is then evaluated by combining the two components, i.e., nominal effort and EAF. After the designing of the fuzzy system, the FFNN and RBNN with ten hidden layers are designed and these neural networks are trained with 50 randomly chosen projects and 2500 epoch value is taken and are saved as FFNN and RBNN. The evaluation of these saved networks with the value of effort is done. At last, the comparison between both the neural networks, i.e., FFNN and RBNN and fuzzy logic are done on the basis of MMRE and prediction. Kumar and Chopra [9] have focussed on the literature of fuzzy logic and other algorithmic models. The basic purpose of this paper is to take a review on the studies of software estimation using fuzzy and other models in the past years to improve the accuracy. The theoretical part is discussed in this paper, so that this could help for the development of new framework for estimation or to do the changes in the existing one. The comparison between different techniques for software effort/cost prediction is also shown in the paper. Singh and Sahoo [5] have introduced the ANN structure and the performance analysis of different ANNs for software effort estimation is done. ANN is structured between the independent (cost drivers) variables and dependent (effort)variable. Four types of ANN is assumed, MATLAB 10NNTool is used using NASA data set. Kushwaha and Suryakant [10] have developed fuzzy logic technique for software cost estimation and the comparison in the performance has been done with the COCOMO Model in this research paper. The basic difference in this paper with the others is in the membership function. Generally, in the fuzzy logic approach, many researchers have used triangular MF's but in this research work, Gaussian mfs are used in fuzzy logic technique. According to the results, 13 GMF gives better and is closer to the actual effort than 11 GMF and COCOMO Model. Also, by analyzing the results, it is clear that higher the membership functions better will be the results.

3 Proposed Framework

3.1 Data Analysis

The proposed model is validated on desharnais data set. The data set consists of several parameters/metrics from which four inputs and one output parameter/metrics have been used which are important for software modelling and contains 81 program data values.

The metrics that are used are Transactions, Entities, PointNonAdjusted, and Language which are inputs to the fuzzy system and effort as the only output to the fuzzy system.

- Transactions—It is the number of necessary transactions in the data model and is measured in the range of 0–1000.
- Entities—Entities are the total number of objects to represent the software or systems and is measured in the range of 0–400.
- PointNonAdjust—It is used to measure the size of the project in adjusted function points and is measured in the range of 0–1200.
- Language: How many programming languages are used in the scheme and are expressed as 1, 2, or 3 measured in the range of 1–3.
- Effort—Actual effort is evaluated in person per hour and is deliberated in the range of 500–24,000.

3.2 Fuzzy Rules

Fuzzy Rules are the IF-THEN rules which is used for constructing fuzzy model. The high expertise knowledge in oral form is being converted to a set of IF-THEN Rules. The membership functions and weights of the rules are coordinated with the help of input and output data. This paper consists of 14 fuzzy rules:

1. If Transaction is LOW, Entity is MEDIUM, and PointNonAdjust is LOW, then Effort is LOW.
2. If Transaction is LOW, Entity is HIGH, and PointNonAdjust is LOW, then Effort is LOW.
3. If Transaction is LOW, Entity is LOW, and PointNonAdjust is MEDIUM, then Effort is LOW.
4. If Transaction is LOW, Entity is MEDIUM, and PointNonAdjust is MEDIUM, then Effort is LOW.
5. If Transaction is MEDIUM and PointNonAdjust is MEDIUM, then Effort is LOW.
6. If Transaction is LOW, Entity is LOW, and PointNonAdjust is LOW, then Effort is VERY LOW.

7. If Transaction is LOW and PointNonAdjust is HIGH, then Effort is VERY LOW.
8. If Transaction is MEDIUM, Entity is MEDIUM, and PointNonAdjust is MEDIUM, then Effort is LOW.
9. If Transaction is LOW and PointNonAdjust is HIGH, then Effort is MEDIUM.
10. If Transaction is MEDIUM, Entity is MEDIUM, PointNonAdjust is MEDIUM, and Language is MEDIUM, then Effort is HIGH.
11. If Transaction is MEDIUM, Entity is LOW, PointNonAdjust is LOW, and Language is MEDIUM, then Effort is HIGH.
12. If Transaction is MEDIUM, Entity is HIGH, and PointNonAdjust is MEDIUM, then Effort is VERY HIGH.
13. If Transaction is LOW, Entity is HIGH, and PointNonAdjust is HIGH, then Effort is VERY HIGH.
14. If Transaction is HIGH, Entity is HIGH, and PointNonAdjust is MEDIUM, then Effort is VERY HIGH.

- All the membership functions are triangular in the inputs but for the output, we have used both triangular and trapezoidal membership functions as shown in Tables 1 and 2 with all the scalar parameters(a, b, c)-input and (a, b, c, d)-output

Table 1 Membership function characteristics (input)

Variable name	Range	MF	Parameters		
Tranasactions	0–1000		*a*	*b*	*C*
		L	0	250	500
		M	250	500	750
		H	500	750	1000
Entities	0–400	L	0	100	200
		M	100	200	300
		H	200	300	400
PointNonAdjust	0–1200	L	0	300	600
		M	300	600	900
		H	600	900	1200
Language	1–3	L	1	1.5	2
		M	1.5	2	2.5
		H	2	2.5	3

Table 2 Output

Variable name	Range	MF	Parameters			
Effort	500–24,000		*a*	*b*	*c*	*D*
		VL	500	500	2000	3500
		L	2000	3500	5000	
		M	4000	8000	12,000	
		H	10,000	14,000	18,000	
		VH	16,000	18,000	24,000	24,000

4 Evaluation Criteria

4.1 Magnitude of Relative Error (MRE)

It is the common criteria for the evaluation of software effort models.

$$\mathrm{MRE} = \frac{\text{Actual Effort} - \text{Predicted Effort}}{\text{Actual Effort}}$$

MRE will be calculated for each data value whose effort is predicted. So for the given data set, there are 81 data values and for each data values, the MRE will be calculated. The cumulative of MRE for all the observations(M) can be calculated through Mean MRE(MMRE).

$$\mathrm{MMRE} = 1/\mathrm{M}\sum_{i}^{\mathrm{M}} \mathrm{MRE}$$

4.2 Multiple Regression

With four independent variables, multiple regression can be expressed as:

$$y = a + b_1x_1 + b_2x_2 + b_3x_3 + b_4x_4$$

where y is the dependent variable; a, b_1, b_2, b_3 and b_4 are constants and x_1, x_2, x_3 and x_4 are the four independent variables. We can deduce the values of constants by solving these equations for multiple regression:

Table 3 Predicted effort with MRE using fuzzy logic technique

Project ID	Actual effort	Predicted effort using fuzzy logic	MRE
1	5152	1803.077549	0.650024
2	5635	2051.513135	0.635934
3	805	1882.352941	1.338327
4	3829	1970.810351	0.485294
5	2149	1743.221877	0.188822
–	–	–	–
77	1400	2844.228408	1.031592
78	2800	1688.431373	0.396989
79	9520	3380.234657	0.644933
80	5880	4863.416623	0.172888
81	23,940	1852.796053	0.922607

$$\sum x_1 y = b_1\left(\sum x_1^2\right) + b_2\left(\sum x_1 x_2\right) + b_3\left(\sum x_1 x_3\right) + b_4\left(\sum x_1 x_4\right)$$
$$\sum x_2 y = b_2\left(\sum x_2^2\right) + b_1\left(\sum x_1 x_2\right) + b_3\left(\sum x_2 x_3\right) + b_4\left(\sum x_2 x_4\right)$$
$$\sum x_3 y = b_3\left(\sum x_3^2\right) + b_1\left(\sum x_1 x_3\right) + b_2\left(\sum x_2 x_3\right) + b_4\left(\sum x_4 x_3\right)$$
$$\sum x_4 y = b_4\left(\sum x_4^2\right) + b_1\left(\sum x_1 x_4\right) + b_2\left(\sum x_2 x_4\right) + b_3\left(\sum x_3 x_4\right)$$

5 Experimental Results

To proclaim the feasibility of the proposed framework, the experimentation has been done with multiple regression and fuzzy logic (proposed methodology) methods by taking the large amount of data from the data set.

The snapshot of predicted effort and MRE using fuzzy logic is shown in Table 3.

Table 4 shows the snapshot of predicted effort along with MRE through linear regression.

Table 5 shows the comparison between both the techniques and the screenshot of the final output has been attached.

6 Conclusion

For every data from the given model, we have analyzed the results of the actual and the predicted efforts and then evaluated the mean relative error (MRE) of each

Table 4 Predicted effort with MRE using linear regression technique

Project ID	Actual effort	Predicted effort using linear regression	MRE
1	5152	5915.0551	0.148109
2	5635	6427.91261	0.140712
3	805	2465.08731	2.06222
4	3829	6377.22194	0.665506
5	2149	4851.76832	1.257687
–	–	–	–
77	1400	3941.1818	1.81513
78	2800	5900.69183	1.10739
79	9520	11187.66875	0.175175
80	5880	8153.78173	0.386698
81	23,940	20,490.10925	0.144106

Result showing MMRE for both the techniques

Table 5 Comparitive results of both the techniques in terms of MRE and mean MRE

	Multiple regression	Fuzzy logic
Min (MRE)	0.0171	0.0049
Max (MRE)	3.4712	2.2984
MMRE	0.5743	0.5089

project and mean MRE (MMRE). The same data set has also being tested for multiple regression model and evaluated the results in the same manner. Table 4 shows the comparison where MMRE% of 0.5089 by proposed technique is much superior than MMRE% of 0.5743 by multiple regression. So, after analyzing the outcomes, we came to the conclusion that the multiple regression technique gives less accuracy than the proposed fuzzy logic method.

References

1. Bedi RPS, Singh A (2017) Software cost estimation using fuzzy logic technique. Indian J Sci Technol 13(10)
2. Dizaji ZA, Gharehchopogh FS (2015) A hybrid of ant colony optimization and chaos optimization algorithms approach for software cost estimation. Indian J Sci Technol 8(2)
3. Sharma V, Verma HK (2010) Optimized fuzzy logic based framework for effort estimation in software development. IJCSI Int J Comput Sci Issues 7(2)
4. Shivakumar N, Balaji N, Ananthakumar K (2016) A neuro fuzzy algorithm to compute software effort estimation. Glob J Comput Sci Technol 16(1)
5. Singh J, Sahoo B (2011) Software effort estimation with different artificial neural network
6. Kumar S, Pal J, Bhattacherjee V (2013) Software effort prediction-a fuzzy logic approach. Glob J Comput Sci Technol 13(10)

7. Ganesh MKS, Thanushkodi K (2015) An efficient software cost estimation technique using fuzzy logic with the aid of optimization algorithm. Int J Innov Comput Inf Cont 11(2)
8. Bhatnagar R, Bhattacharjee V, Ghose MK (2010) A proposed novel framework for early effort estimation using fuzzy logic techniques. Glob J Comput Sci Technol 10(14)
9. Kumar S, Chopra V (2013) To design and implement neural network and fuzzy logic for software development effort prediction. Int J Comput Appl 84(11)
10. Kushwaha N, Suryakant (2014) Software cost estimation using the improved fuzzy logic framework. In: 2014 conference on IT in business, industry and government (CSIBIG). IEEE, pp 1–5
11. Ochodek M, Nawrocki J, Kwarciak K (2011) Simplifying effort estimation based on use case points. Inf Softw Technol 53(3):200–213
12. Thamarai I, Murugavalli S (2016) An evolutionary computation approach for project selection in analogy based software effort estimation. Indian J Sci Technol 9(21)
13. Singh H, Singh G, Bhatia N (2012) Election results prediction system based on fuzzy logic. Int J Comput Appl 53(9)
14. Mittal A, Parkash K, Mittal H (2010) Software cost estimation using fuzzy logic. ACM SIGSOFT Softw Eng Notes 35(1):1–7
15. Sheta A, Rine D, Ayesh A (2008) Development of software effort and schedule estimation models using soft computing techniques. In: 2008 IEEE congress on evolutionary computation (IEEE world congress on computational intelligence). IEEE, pp 1283–1289
16. Pahariya JS, Ravi V, Carr M (2009) Software cost estimation using computational intelligence techniques. In: 2009 world congress on nature and biologically inspired computing (NaBIC). IEEE, pp 849–854
17. Verma HK, Sharma V (2010) Handling imprecision in inputs using fuzzy logic to predict effort in software development. In: 2010 IEEE 2nd international advance computing conference (IACC). IEEE, pp 436–442
18. Ghose MK, Bhatnagar R, Bhattacharjee V (2011) Comparing some neural network models for software development effort prediction. In: 2011 2nd national conference on emerging trends and applications in computer science. IEEE, pp 1–4
19. Johanyák ZC, Tikk D, Kovács S, Wong KW (2006) Fuzzy rule interpolation Matlab toolbox-FRI toolbox. In: 2006 IEEE international conference on fuzzy systems. IEEE, pp 351–357
20. Idri A, Hosni M, Abran A (2016) Improved estimation of software development effort using classical and fuzzy analogy ensembles. Appl Soft Comput 49:990–1019
21. Capers JT (2007) Estimating software costs. McGraw-Hill Inc, New York
22. https://www.kaggle.com/toniesteves/desharnais-dataset

Cost-Effective Resource Provisioning in Cloud Using Cooperative Coevolutionary Genetic Algorithm

Monika Kumari and Gadadhar Sahoo

Abstract Nowadays, cloud is widely used platform for intensive computing, bulk storage, and networking. Elastically scalable on demand pay as per used model is one of the major reasons for the acceptability of the cloud-based computing. Data centers with huge amount of computing devices are distributed over the cloud service provider as the back end to provide services. Computing devices are arranged bombastically in data centers to ensure 24/7 availability. Inspired from Pantriyagin's maximization principle for optimal control and self-balancing population with predator–prey interaction in ecological environment, a model for resource provisioning with cost-effective approach and efficient resource utilization scheme is presented in this paper. We present an optimal resource provisioning model by taking into account the matter of sustainability and profitability. Cooperative coevolutionary genetic algorithm is implemented in CloudSim simulation tool to find the optimal cost per provisioning.

Keywords Resource provisioning · Profit-scaling model · Optimization · Resource utilization · Profit

1 Introduction

In recent years, various organizations, viz. Amazon, Google, etc., provide platforms to avail computational and storage facilities. To avail various facilities, these organizations charge some amount as per the need. Here, optimality in allocation of desired resource is highly desirable to achieve optimal usage of resource available with cloud service providers. Resource allocation, a way to assign available resource to the needed cloud request, as per demand is possible only by employing an efficient

M. Kumari (✉) · G. Sahoo
Department of Computer Science and Engineering, BIT Mesra, Ranchi, Jharkhand, India
e-mail: monika.kit@gmail.com

G. Sahoo
e-mail: gsahoo@bitmesra.ac.in

V. Nath and J. K. Mandal (eds.), *Proceedings of the Fourth International Conference on Microelectronics, Computing and Communication Systems*, Lecture Notes in Electrical Engineering 673, https://doi.org/10.1007/978-981-15-5546-6_32

load sharing technique [1]; else, the utilization of scarce resource and satisfaction of cloud consumer within the limit of cloud is not possible [1]. In this mechanism, a set of resource is supplied for execution of task and keeping data reserve. Assigning user's requirement to the suitable resource distributed geographically over the internet is a tedious job. For this resource is divided into several execution environments known as virtual machines (VM) and provides a complete system to execute the user's request [1].

Resource provisioning is one of the crucial and important tasks in the cloud. In resource provisioning, we have to make a sequence of optimal decisions, such as identification and selection of resources, capacity determination of resource for allocation, acquiring and releasing of resources and methods for resource utilization dynamically at runtime. Cloud provider offers mainly two types of resource provisioning scheme to the consumer: on-demand plan and reservation plan. However, the cost of resource provisioning is cheaper in reservation plan than on-demand plan. As in reservation plan consumer has to pay in advance to the provider [2]. Due to uncertainty in demand of resources both plans cannot put the best provisioning scheme. With reservation plan two types of problem occurred frequently: *under-provisioning problem* occurs as unpredictable demand pattern unfit to fulfill the request on available reserved resources and *over-provisioning problem* occurs if the reserved resources are over the actual demand and part of reserved resource will be unutilized. So it is important for cloud consumers to minimize the on demand cost and oversubscribed cost of under-provisioning and over-provisioning [2]. As the arriving pattern of request is not predictable and finding the optimal or near optimal solution is NP-hard because of the large size data center, the resource provisioning is very critical and complex task [3]. Also minimization of cost per reservation as input request by the suitable scheduling algorithm of resource provisioning is NP-hard.

1.1 Resource Provisioning Schemes

Resource provisioning schemes are broadly classified into two parts: sustainable provisioning (continuous) and seasonal provisioning (peak time):

i. **Sustainable Provisioning**: This is the initial provisioning of resource based on previous collocated data as input, done during system start-up or in offline mode [4]. After the primary calculation, the allocation may not be re-estimated for a sustainable amount of time, such as several months [5]. This provisioning type is used by the cloud provider for reservation plan, which is based on non-live migration. When a new request comes to the cloud provider, it sends this request to the data center controller. On the basis of arrived set of request information A such as CPU, memory (RAM), storage, network, etc., data center controller fulfill the request by sending request for data center status information to the data center managers. The data center managers have their respective data center's resource information. The data center managers distribute the resource information to the data center controller. By studying the request type and available information about different data centers, it maps the VM request to suitable data center controller by using optimal scheduling algorithm [4].

ii. **Seasonal Provisioning**: In seasonal provisioning used in on-demand resource provisioning plan which alters the assignment of tasks onto resources periodically at runtime on data center for effective performance satisfying the service level agreements (SLAs) for different purpose of services [4]. In view to acheive the optimum provisioning from current state dynamic allocation is implemented through live migration technique. The main objective of the dynamic resource provisioning problem is to achieve the maximum arrangement of current assignment of task at minimum cost. The optimal decision varies on multiple constraints, i.e., maximization of resource utilization to minimization of reservation cost or a combination of multiple parameters [4]. The dynamic resource provisioning can be further subdivided into two parts: sensitive and dedicated [3].
iii. **Sensitive resource provisioning**: This is the provisioning scheme used to reach from an undesired state into desire state from an initial placement for improving the performance, optimal power utilization, and load distribution.
iv. **Dedicated resource provisioning**: This provisioning adjusts the initial allocation request assigned to the resource aforetime the system meet a satisfied state.

We basically motivated to find out the optimal control mechanism for resource provisioning to get the best performance of the system as well as the limit and design of some acceptable control parameter. We used here Pantriyagin's maximum principle (PMP) for profit maximization. To evaluate the performance, we used cooperative coevolutionary genetic algorithm (CCGA).

2 Related Work

Resource Provisioning Optimization Approach: The resource provisioning problem studied in three different cloud architectures: Single cloud, i.e., only one cloud, multi-cloud, i.e., many cloud computing data centers from one or more cloud service providers, and federated clouds. A different modeling and solution approach depends on specific system assumptions and objectives. The established optimization techniques are broadly categorized and have been shown in Fig. 1 as follows [6]:

1. Single-objective optimization problem (SOP).
2. Multi-objective solved as single objective (MAS)
3. Pure multi-objective optimization (PMO).

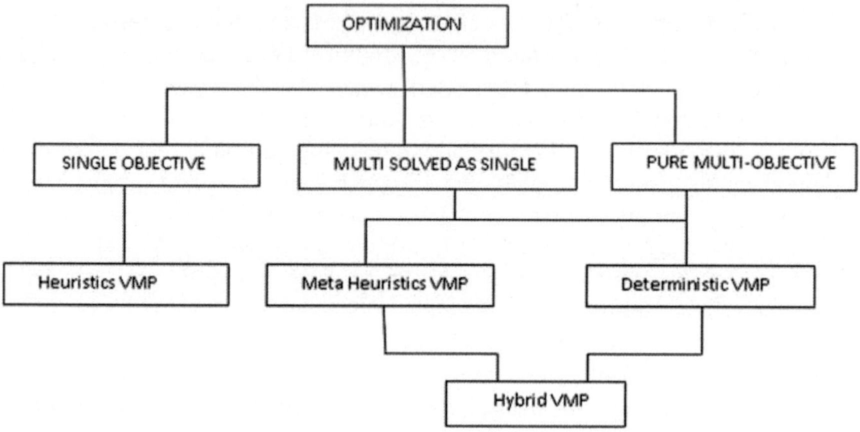

Fig. 1 Classification of optimization technique

Single-objective function is evaluated at once in a single-objective approach. Most of the virtual machine problem has been solved as a single-objective optimization approach [6]. Two or more functions are taken for evaluation in multi-objective optimization. When two or more objective functions are combined to form a single-objective function and solved, it is known as multi-objective solved as single-objective optimization. This approach uses a trade-off rule between goals to decide one of several goals. Fuzzy logic is one of the best rules for combining multiple objectives. This approach was used in [7], where Junwei Cao et al. presented two goals: efficient power distribution and efficient job assignment for multiple multi-core severe processors with different sizes and certain loads considering only one at a time.

2.1 Resource Provisioning Techniques

Resource provisioning problems are presented using various modeling techniques for optimization. The effectiveness and applicability of different resource provisioning techniques depend on the analytical and modeling techniques used. Different models have assumptions and goals of their own system. Modeling techniques are widely divided into four groups:

1. Deterministic methods.
2. Heuristics methods.
3. Meta-heuristics methods.
4. Approximation methods.

1. **Deterministic Methods**: Under deterministic approach to modeling the resource provisioning problem, algorithms are adopted.

Constraint Programming: The limitations on associated variables should be complied with in this programming result. It comprises three fundamental strategies [5] first to declare the domains of variables, second to determine the constraints on the declared variables, and third to search the domain. The resource supply solved by this technique is a two-stage supply process [8]. Local Decisions—it pertains to each application environment. Global Decisions—It takes the local decision from all applications as an input and then concerted efforts to maximize the global utility function.

Bin Packing: The bin packing is a NP-hard combinatorial problem that is convexly optimized. In it, the physical machines of various amounts/quantities (thought about/believed as bin) must be packed with a limited number of tasks in a course that reduces the number of bins adopted [5]. The set of computer instructions should be divided into three phases [9]: creation of past demand patterns, The future demand forecasting based on the past demands and Measure-Forecast-Remap (MFR) for mapping or remapping of tasks to resources. The purpose of the algorithm is to minimize the number of resources required for tasks if the likelihood of over-burdening the server is fixed in the interval. The final algorithm stage uses bin packing. Request in the following interim is anticipated in the previous interim dependent on interest

Stochastic Integer Programming [1]: To model optimization problems that involve uncertainty stochastic programming is used. Because of some unknown parameters, most of the real-world problems cannot be captured by deterministic integer programming. To estimate the prediction of uncertainty stochastic programming models take advantage of probability distributions. Stochastic Integer Programming effective for problems in which the distribution of demands is known or can be estimated but exact demands are unpredictable. The objective of this approach is to minimize cost simultaneously fulfill the demands of users. So this method allocates the resource in three phases [10].

2. **Heuristic Methods**: It is a static method based on prior knowledge of problem domain a heuristic is developed to solve the problem. The algorithm comes under this modeling technique are as follows: First Fit, Next Fit, Best Fit, Least Full First, Most Full first, Best Fit Decreasing and First Fit Decreasing.
3. **Metaheuristics Method**: A metaheuristic is a high-level search procedure that applies some form of rule or set of rules based on some source of knowledge, in order to analyze the fishing space effectively. The inspiration is Darwin's survival of the fittest principle used in metaheuristics methods. Most bio-inspired metaheuristics are stochastic search techniques (such as all evolutionary algorithms viz. artificial bee colony, Firefly algorithm, particle swarm optimization, League Championship Algorithm, Ant Colony Optimization, BAT Algorithm etc.). The unified framework of metaheuristics method involves four operators: the initial Transition, Evaluation and Determination. The resource provisioning problem can solve using the following algorithms come under metaheuristics method: Genetic Algorithm, Ant Colony Optimization, Partial Swarm optimization, Simulated Annealing, Tabu search.

Genetic Algorithm: A static approach based on set of heuristics, particularly useful in dynamically changing objective function. It works with population of solution and application of genetic operator will generate optimal solution. Liu et al. [11] proposed an improved genetic algorithm called NS-GGA designed to incorporate fast, non-dominated sorting NSGA-II to Grouping Genetic Algorithms. In NS-GGA fast, non-dominated sorting combined with grouping of genetic operators are used to approach the Pareto-optimal front [1].

Cooperative Coevolutionary Genetic Algorithm (CCGA): The CCGA was introduced by Potter and De jong [12], which consolidates the essential thoughts of the decomposition– coordination enhancement of vast scale framework with developmental improvement. The fundamental thought is to loosen up the framework wide limitations of the basic issue and structure the double issue, at that point to break down the double issue into various subproblems. A subgradient-based stochastic improvement strategy is utilized to take care of the double issue. For the low-level subproblems, which are non-raised, discrete, nonlinear and with lower measurement, the GA is utilized. In Table 1, presents a comparative study of existing methodology in from various research papers.

3 Research Methodology

3.1 Profit-Scaling Model

After exhaustive study of different research paper the methodology for resource provisioning problem in cloud environment can be suggested and proposed as shown in Fig. 2.

The step by step procedure for proposed resource provisioning methodology is discussed as follows:

1. At the very beginning of resource provisioning mechanism user submit their job to cloud broker.
2. Now cloud broker analyzed the job requirement with various registered resource provider and configure the virtual machine accordingly.
3. Each resource provisioner comprised with a Data center. At Data center Data Center Controller(DCC) uses CCGA for mapping VMs with suitable resources(PMs) and choose the best mapped resource on the basis of proposed profit-scaling fitness function (Eq. 5).
4. To overcome from the resource under-provisioning and over-provisioning issue auto-scaling mechanism will be proposed inspired from Ecological environment of predator–prey relationship for self-balancing consumer-resource requirement (Table 2).

and the cost of job per unit time, $Cb(t)$ is

Table 1 Comparative study of existing optimization technique

Author	Optimization methodology	Constraints	Objective
Cao et al. [7]	Used queuing model (M/M/m). Multi-objective solved mono objective comes under deterministic resource provisioning technique	Power and performance constraint	Optimal load distribution
Chaisiri et al. [2]	Used stochastic and deterministic integer programming multi-objective solved as mono objective comes under deterministic resource provisioning technique	Uncertain demand and resource price	Minimizing the resource provisioning cost
Sun et al. [13]	Used mixed integer programming multi-objective solved as mono objective comes under deterministic resource provisioning technique	Resource capacity and location	Minimize the total power consumption, i.e., network communication power and server power consumption
Tang et al. [14]	Alternating optimization	Flexible task arrival rate	Optimal request allocation and optimal service capacity
Chiang et al. [11]	Queuing model with N policy(multi-objective solved as mono objective) constraint programming comes under deterministic resource provisioning technique	Minimum power consumption with controlled service rate and minimum response time without violating SLAs	Minimization of operational Cost

Author	Features	Limitation
Cao et al. [7]	Multicore heterogeneous server with multi clouds, optimal speed of sever for different core speed	Tradeoffs of optimization has been developed by fixing one constraint constant
Chaisiri et al. [2]	Multistage resource in multiple cloud	No proper probability distribution through stochastic programming also needs separate provisioning planning at each stage, Benders cut needs extra computational time

(continued)

Table 1 (continued)

Author	Features	Limitation
Sun et al. [13]	Designed a heuristic algorithm for power efficient VN provisioning	Only optimizing the power consumption of VN request
Tang et al. [14]	Cost aware dynamic request allocation policy and service capacity scaling policy	Redirection policy is for single dispatcher not consider the case of multiple dispatcher within a datacenter
Chiang et al. [11]	To overcome the idle power consumption cost three different power saving policies with optimal decision-making process and mode switching control	Higher response time with low arrival rate

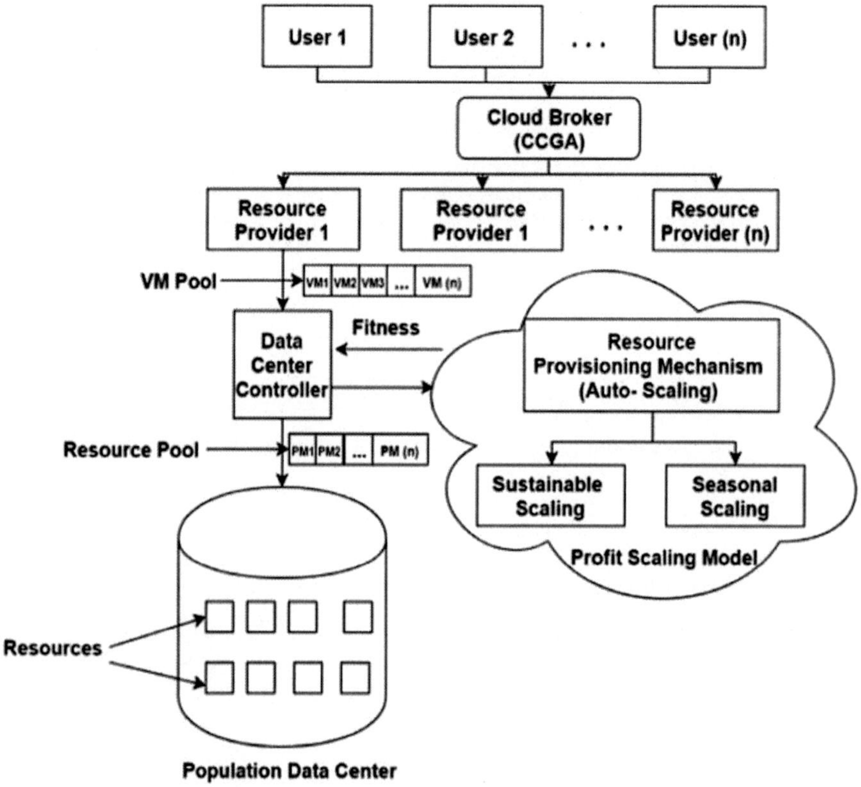

Fig. 2 Framework of resource provisioning in cloud

Table 2 Variables and parameters

Variables
X (t) = the population of available resources at time t
$b(t)$ = amount of job per resource operational at time t
$H(t)$ = the scaling rate (units of resource scaled per unit time)
Parameters
C_B= the overhead cost of allocating one job per unit of time
n = the mean number of broker per resource
p = the reservation rate per unit of resource
w = per broker's wage per unit of time

$$Cb(t) = C_B + nw\, b(t)$$

This give the profit per unit time, denoted as $P\ (t)$, as

$$P(t) = pH(t) - Cb(t) \tag{1}$$

Let us include an interest (discount) rate $\delta\ > 0$ which assume to be constant. Using Eq. (1), the present value of the expected profit at some time t, $E\ (P(t))$, is given by

$$E(P(t)) = \mathrm{e}^{-\delta t}(pH(t) - Cb(t)]$$

Integrating $E\ (P(t))$, the true return or total profit rate in present dollars, denoted as J, is expressed as

$$J = \int \mathrm{e}^{-\delta t}[pH(t) - Cb(t)]\mathrm{d}t \tag{2}$$

This is the sustainable profit function to be maximized. A severe weakness in this model is that the interest rate δ is assumed to apply to everything, yet C_B, p and w are assumed to be constant. In reality, p will probably grow with or even faster than general inflation, and C_B are influenced by factor such as technological advancements, union negotiations, government policies and taxation rates (all comes under SLAs). It is possible to include stochastic fluctuations in numerical simulation in order to arrive at more realistic predictions. However, for a first analysis we will proceed with the given unrealistic assumptions.

The Provisioning rate $H\ (t)$ is taken as:

$$H(t) = qX(t)b(t)$$

i.e., proportional to the job per resource and the population of resource with a proportionality constant q known as resource selection rate. For limited (constant) job per resource, therefore

$$H(t) = qU\,X(t)$$

The function (2) becomes

$$J(b) = \int_{0}^{\infty} \mathrm{e}^{-\delta t} b(t)[pqx(t) - C]\mathrm{d}t\angle$$

which is a functional of resource pool size $b = b(t)$. It is easy to include a modification where the reservation price of resource and the cost per job, p and C, both increase with time. Suppose for example that they both grow at the same rate α such that

$$p = P(t) = P_0 \mathrm{e}^{\alpha t} \tag{3}$$

$$C = C(t) = C_0 \mathrm{e}^{\alpha t} \tag{4}$$

We then obtain an objective functional

$$J(b) = \int_{0}^{\infty} \mathrm{e}^{(\alpha-\delta)t} b(t)[P_0 qx(t) - C_0)\mathrm{d}t \tag{5}$$

which is well defined if $\delta > \alpha$ (if $\delta \leq \alpha$, the functional will in general no longer be finite).

4 Conclusion

In this paper a novel idea for maximizing the profit of resource provider in cloud environment using CCGA has been proposed. The result and sensitive analysis of the proposed methodology will further be enhanced in the extended version of the paper.

References

1. Vinothina V et al (2012) A survey on resource allocation. Strat Cloud Comput IJACSA 3(6)
2. Chaisiri S et al (2012) Optimization of resource provisioning cost in cloud computing. IEEE Trans Serv Comput 5(2):164–177

3. Masdari M et al (2016) An overview of virtual machine placement schemes in cloud computing. J Netw Comput Appl 66:106–127
4. Paul AK (2015) Dynamic virtual machine placement in cloud computing. M.Tech thesis guided by Dr. Sahoo B, NIT Rourkela, Odisha, India, May 2015
5. Shankar A (2010) Virtual machine placement in computing clouds. Guided by Prof. Bellur U, Technical report, IIT Bombay 10 April 2010
6. Lopez-Pires F et al (2015) Virtual machine placement literature review. arXiv:1506.01509v1 [cs.DC], 04 June 2015
7. Cao J et al (2014) Optimal power allocation and load distribution for multiple heterogeneous multicore server processors across cloud and data centers. IEEE Trans Comput 63(1):45–58
8. Van HN, Tran FD, Menaud JM (2009) Autonomic virtual resource management for service hosting platform. In: ICSE workshop on software engineering challenges of cloud computing
9. Bobroff N, Kochut A, Beaty K (2007) Dynamic placement of virtual machines for managing SLA violations, pp 119–128
10. B-S Lee, Chaisiri S, Niyato D (2009) Optimal virtual machine placement across multiple cloud providers 2009. In: IEEE Asia Pacific services computing conference, pp 103–110. APSCC
11. Chiang Y-J et. al (2015) An efficient green control algorithm in cloud computing for cost optimization. IEEE Trans Cloud Comput 3(2)
12. Potter MA, De Jong KA (1994) A cooperative coevolutionary approach to function optimization. In: Proceedings of the 1999 IEEE international conference on parallel problem solving from nature (PPSNIII), Jerusalem, Israel, pp 249–257
13. Sun G et al (2015) Power-efficient provisioning for online virtual network requests in cloud-based data centers. IEEE Sys J 9(2)
14. Tang J et al (2014) Dynamic request redirection and elastic service scaling in cloud-centric media networks. IEEE Trans Multimedia 16:1434–1445

Design of 2–7 GHz Voltage-Controlled Oscillator Using Multi-stage Current Starved Configuration

Niraj Prasad Gupta and Srikanta Pal

Abstract In this article, a three-stage and five-stage voltage-controlled oscillator (VCO) using current starved configuration has been proposed. Among the two, only a three-stage VCO gives the required bandwidth of 5 GHz. Current starved VCO is designed using complementary metal oxide semiconductor (CMOS) technology. The oscillating output frequency bandwidth (2–7 GHz) is considered as the main objective for the proposed circuit. The proposed three-stage VCO circuit uses low supply voltage of 1 V, low power of 29.93 uW. Transient analysis and periodic steady state (PSS) analysis are performed, and after simulation, a wide frequency bandwidth of 5 GHz was obtained. The phase noise for the three-stage VCO at 1 GHz is −132 dBc/Hz. The circuit has been designed in Cadence Virtuso using gpdk 45 nm CMOS technology with supply voltage of 1 V.

Keywords CMOS VCO · Current starved VCO · Ring oscillator

1 Introduction

The voltage-controlled oscillators are used very widely in wireless transceiver and other electronic components such as clock data recovery circuits (CDR), phase locked loop (PLL), and frequency translation in radio circuits [1, 2]. Oscillators are tuned at certain frequency, and the oscillators whose frequency can be varied electronically are termed as voltage controlled oscillator. VCO is one of the important components of PLL which is a deciding factor for the power consumption by PLL and also responsible for deciding area occupied by PLL [3, 4]. VCO also finds its place in RF transceivers where it is used to drive the up-conversion and down-conversion mixer

N. P. Gupta (✉) · S. Pal
Department of Electronics and Communication Engineering, Wireless Communication, Birla Institute of Technology, Mesra, Ranchi, India
e-mail: nirajgupta1717@gmail.com

S. Pal
e-mail: spal@bitmesra.ac.in

V. Nath and J. K. Mandal (eds.), *Proceedings of the Fourth International Conference on Microelectronics, Computing and Communication Systems*, Lecture Notes in Electrical Engineering 673, https://doi.org/10.1007/978-981-15-5546-6_33

[5]. The design architecture of VCO can be divided under two categories: first being the ring oscillator and the LC oscillator being the second one. LC oscillators are made by using inductors and capacitors with an amplifier. A ring oscillator consists of a odd number of inverter stages where the output of the last stage of inverter is fed back to the input of the first stage. Figure 1 shows an n-stage CMOS ring oscillator. The ring oscillators work by controlling the time taken in charging and discharging of the gate capacitance of the next inverter. We know that for any circuit to oscillate the total phase shift has to be 2π and a gain of unity at the oscillation frequency. A VCO can also be designed using differential ring configuration [6], but the power consumption is higher compared to this work. Although the LC oscillators have found to have superior phase noise performance as compared to ring oscillators, still ring oscillators have numerous advantages over LC oscillators [7]. The area requirement of the ring oscillators is comparatively less than the LC oscillators and thus can be easily integrated with CMOS circuit, which reduces fabrication cost. The second reason why they are used over LC oscillators is their wide tuning range, and this makes them robust over temperature variation and process variation [8]. The basic equation for a VCO is given as [9]:

$$f_{\mathrm{VCO}} = f_0 + K_{\mathrm{VCO}} V_{\mathrm{ctrl}} \tag{1}$$

Here of is the frequency at which VCO will oscillate without any control voltage. V_{ctrl} is the external electronic input given to the VCO that helps in changing or tuning the oscillating frequency as per requirement. In this paper, we have designed a three-stage and five-stage and CMOS VCO using current starved configuration. Among

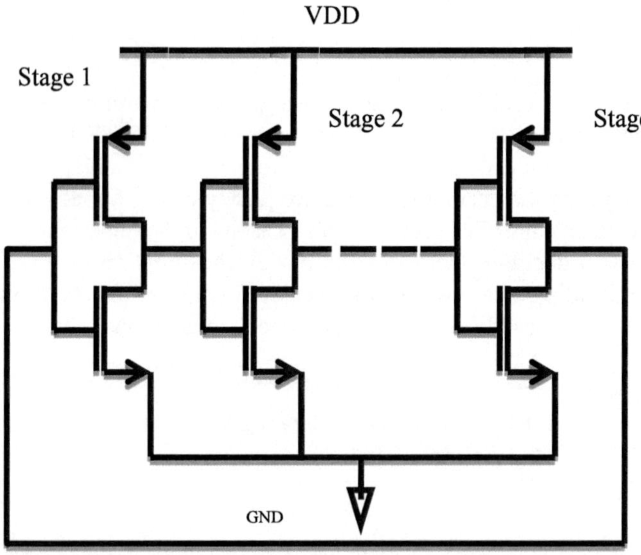

Fig. 1 N-stage CMOS ring oscillator

these two VCOs, the three-stage VCO has obtained the desired output oscillating frequency of 2–7 GHz.

This paper consists of four sections, Sect. 2 describes the basics of current starved configuration in ring oscillator, Sect. 3 discusses about the design of VCO, Sect. 4 consists of simulation results and plots, Sect. 5 is the conclusion section.

2 Current Starved VCO

A current starved voltage-controlled oscillator as the name suggests is a configuration or arrangement made in the circuit which makes the circuit to starve or struggle for current flow. The schematic diagram for a current starved VCO has been given in Fig. 2. The circuit has two parts one being the inverter, and other is current starving circuit. From the figure, it can be observed that every inverter stage is covered or enveloped by a *n*-type MOSFET at the lower end and by a *p*-type MOSFET at the upper end of the *N*-stage ring oscillator. Now, this whole circuit is guided by a pair of *n*-type MOSFET and *p*-type MOSFET, and this *p*-type MOSFET allows the other *p*-type MOSFET (excluding the inverter stage *p*-type MOSFET) to mirror the current across them depending on their W/L ratio. As the drain terminal of the MOSFET M6 is shorted with its gate terminal, MOSFET M6 acts as main current source for all the upper *p*-type MOSFET (such as MOSFET M4 and that MOSFET M4, M3, M2, M1 are connected in series, so the current flow across all these MOSFETs must be same. Hence the current I_{D4} and current I_{D1} must have same value. The current flow across MOSFET M6 can be varied by varying the supply voltage, but this is not acceptable. The other way to vary the current flow is to keep varying the channel width of MOSFET M5, and this allows the current across it to increase. Now,

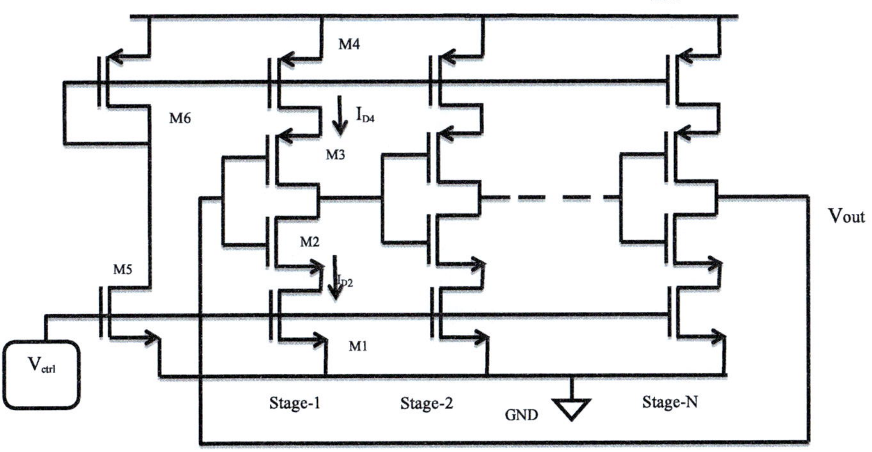

Fig. 2 *N*-stage CMOS VCO

changing the internal device parameters again and again during the circuit operation is not possible. The relation of drain current with the width of channel in saturation region is given as

$$I_D = \frac{1}{2}\mu_n C_{ox}\frac{W}{L}(V_{GS} - V_{th})^2 \quad (2)$$

$$I_D = \frac{1}{2}\mu_p C_{ox}\frac{W}{L}(V_{GS} - V_{th})^2 \quad (3)$$

The above two equation gives the relation between the W/L ratio and drain current for *n*-type MOSFET and *p*-type MOSFET in saturation region, respectively. At the gate terminal of MOSFET M5, a DC voltage source is attached, which helps to vary the its gate to source voltage.

3 Design Of VCO

For an '*N*' stage ring oscillator, the oscillating frequency is given as [10, 11]:

$$f_o = \frac{1}{N t_d} \quad (4)$$

where td is the delay

$$t_d = \frac{Q}{I} = \frac{C_{total} V_{DD}}{I} \quad (5)$$

Q = charge, I = current, C_{total}= total capacitance, V_{DD} =supply voltage

From (3) and (4), the oscillating frequency can be given as

$$f_o = \frac{I_D}{N C_{total} V_{DD}} \quad (6)$$

From Eq. (4), it can be observed that the oscillating frequency is inversely proportional to number of stages. As the number of stages increases, the oscillating frequency decreases, for three-stage VCO, the output oscillating frequency is higher than the five-stage VCO. The total capacitance for single inverter stage is the summation of output capacitance and input capacitance and is given as [12, 13]:

$$C_{total} = C_{out} + C_{in}$$

$$C_{total} = C_{ox}(W_p L_p + W_n L_n) + \frac{3}{2}C_{ox}(W_p L_p + W_n L_n)$$

Table 1 Design parameters of current starved VCO

Parameters	Value (nm)
Width of current starved PMOS (*WPCS*)	800
Width of current starved NMOS (*WNCS*)	200
Width of PMOS in inverter (W_P)	200
Width of NMOS in inverter (W_N)	100
Length of every NMOS $L_{NCS} = L_{PCS} = L_N = L_P = L$	45

$$C_{\text{total}} = \frac{5}{2} C_{\text{ox}} (W_p L_p + W_n L_n) \tag{7}$$

where C_{ox} is the oxide capacitance, W_p and W_n are the width of *p*-type MOSFET and *n*-type MOSFET, respectively, L_p and L_n are length of *p*-type MOSFET and *n*-type MOSFET, respectively.

The oscillating frequency f_0 is the center frequency without any external voltage given by Eq. (6).

The value of oscillating frequency varies with the external voltage V_{ctrl} as given in Eq. (1). Also, from Eq. (6), it is evident that the total capacitance or the width of the inverter stages also decides the oscillating frequency. The supply voltage V_{DD} is fixed at 1 V. The only parameter that can be varied externally is the current running through the various stages of the VCO circuit. We can now vary this current with help of our control input connected externally to the *n*-type MOSFET M5 (Fig. 2). This external voltage should be above the threshold voltage of the device, otherwise the device stays in cutoff region and hence no operation. Table 1 gives the detail of the parameters. Figure 3 shows the schematic diagram of the three-stage VCO designed in cadence software.

4 Simulation Results

The displays the oscillation frequency obtained for the change in control voltage. The average power for three-stage was obtained to be 29.93 uW. Frequency of oscillation is being controlled by the external control voltage applied at the MOSFET M5 (Fig. 2). The input control voltage was varied from 300 mV to 1 V for the three-stage VCO. The oscillating frequency without any external voltage was obtained to be 5.2 GHz. Table 2 displays variation of oscillating frequency (Figs. 4 and 5).

Figure 6 gives the power dissipation in the circuit, and mostly, the power is consumed during the charging and discharging process. Figure 7 gives the phase noise plot for three-stage VCO.

The plot of (Fig. 8) a oscillation frequency variation with control voltage and power dissipation (Fig. 9) is obtained for five-stage VCO.

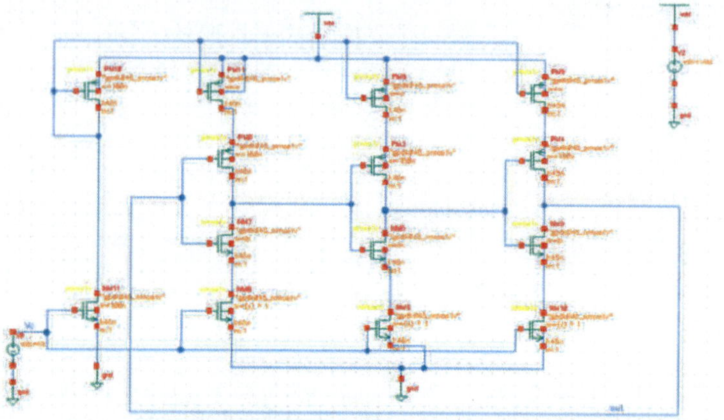

Fig. 3 Design schematic of three-stage VCO

Table 2 Variation of frequency with control voltage

Control Voltage (V)	Oscillation frequency (GHz)
0.3	1.59
0.312	2.00
0.328	2.509
0.344	3.03
0.359	3.50
0.374	4.01
0.390	4.52
0.405	5.01
0.421	5.50
0.432	6.025
0.4507	6.50
0.4661	7.00
0.4829	7.5
0.500	8.03

5 Conclusion

In this article, we have observed that for a VCO, as the number of stage increases, there is a decrease in the output oscillating frequency. For a three-stage circuit, the linearity (Fig. 5) exists from 1 to 11 GHz, whereas in case of five-stage circuit, the linearity exists from 2.8 to 5.93 GHz. Also, the power dissipation in the five-stage

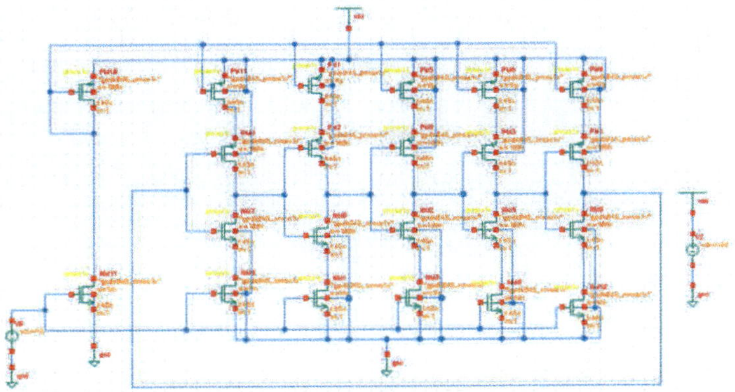

Fig. 4 Design schematic of five-stage VCO

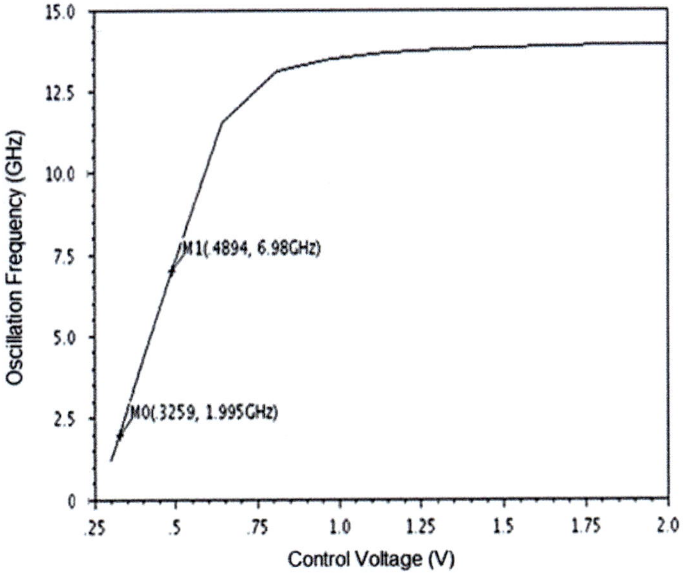

Fig. 5 Oscillation frequency variation with control voltage for three-stage VCO

VCO is more as compared to three-stage VCO. A comparison table for three-stage and five-stage for various parameters has been given below (Tables 3 and 4).

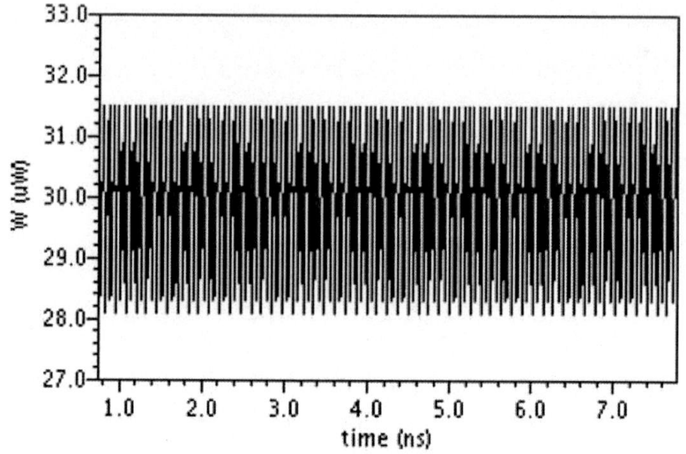

Fig. 6 Power estimation in cadence

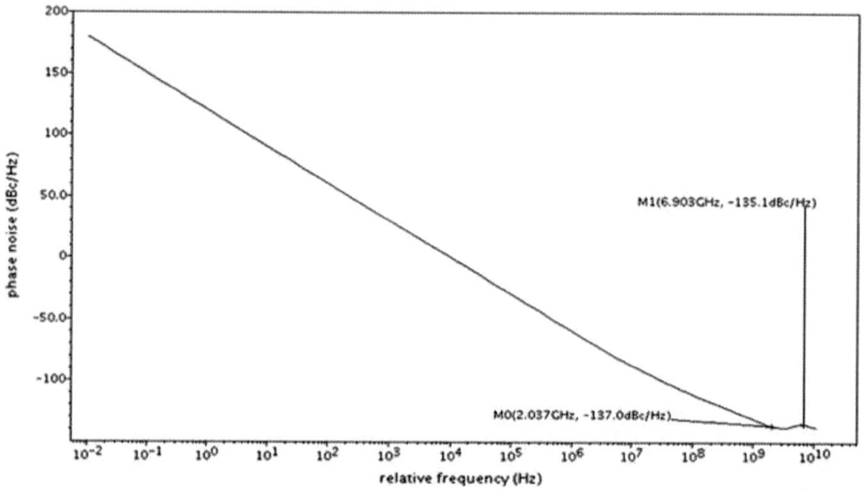

Fig. 7 Phase noise plot for three-stage VCO

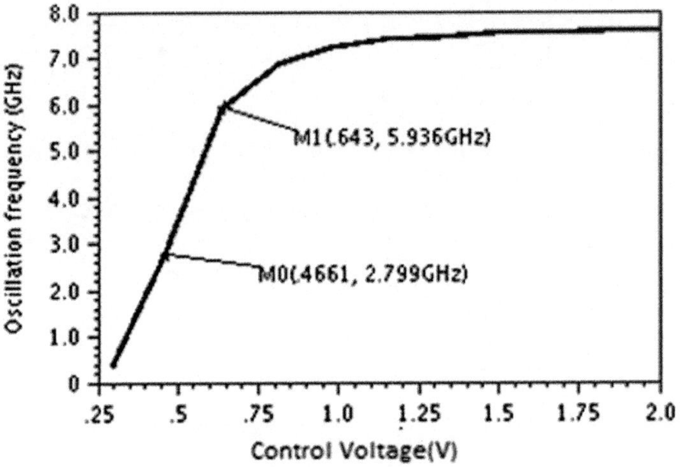

Fig. 8 Oscillation frequency variation with control voltage for five-stage VCO

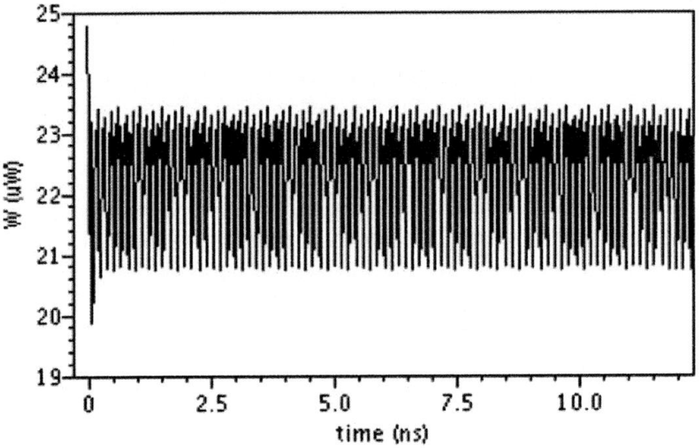

Fig. 9 Power estimation in cadence for five-stage VCO

Table 3 Parameter comparison for three-stage and five-stage VCO

Parameters	Three-stage	Five-stage
Supply voltage (V)	1	1
Average power (uW)	3.461	9.477
Maximum frequency (GHz)	13.5	8
Technology used (nm)	45	45
Phase noise @1 GHz dBc/Hz	−132	−129

Table 4 Comparison analysis with recent papers

Parameters	This work (three-stage)	Reported [7]	Reported [12]	Reported [6]
Technology (nm)	45	45	90	180
Center frequency (GHz)	4	2	1	6.6
Power	3.461 uw	−185.8 dBm (power gain)	386.64 uW	6.01 mW
Supply voltage (V)	1	1	1.8	1.8
Frequency tunable range (GHz)	1–11	–	0.023–1.940	3.1–10.1

References

1. Yang YC, Lu SS (2010) A single –VCO fractional –*N* frequency synthesizer for digital TV tuners. IEEE Trans Electron 57(9):3207–3215
2. Thacker T, Boroyevich D, Burgos R, Wang F (2011) Phase-locked loop noise reduction via phase detector implementation for single-phase systems. IEEE Trans Ind Electron 58(6):2482–2490
3. Loveless TD, Massengill LW, Holman WT, Bhuva BL (2007) Modeling and mitigating single-event transients in voltage controlled oscillators. IEEE Trans Nucl Sci 54(6):2561–2567
4. Kulkarni M, Kalmeshwar N (2013) Design of a linear and wide range current starved voltage controlled oscillator for PLL. Int J Cyber Inf 2(1)
5. Oleg N, Pokharel RK, Kanaya H, Yoshida K (2009) Low noise wide tuning range quadrature ring oscillator for multi-standard transceiver. IEEE Microw Wirel Comp Lett 19(7)
6. Kamran S, Ghaderi N (2017) A novel high speed CMOS pseudo-differential ring VCO with wide tuning control voltage range. In: 25th Iranian conference on electrical engineering
7. Saw S, Nath V (2015) An ultra low power and low phase noise current starved CMOS VCO for wireless application. In: International conference on industrial instrumentation and control (ICIC), COE Pune
8. Ghai D, Mohanty SP, Kougianos E (2009) Design of parasitic and process variation aware Nano-CMOS RF circuits: a VCO case study. IEEE Trans VLSI Syst 17(9)
9. Razavi B. RF microelectronics. Prentice Hall Communications Engineeering and Emerging Technology Series
10. Rout PK, Acharya DP, Panda G (2014) A multi-objective optimization based fast and robust design methodology for low power and low phase noise current starved VCO. IEEE Trans Semicomd Manuf 27(1)
11. Docking S, Sachdev M (2004) An analytical equation for the oscillation frequency of high-frequency ring oscillators. IEEE J Solid-State Circ 39(3):533–537
12. Muddi V, Shinde KD, Shivprasad BK (2015) Design and implementation of 1 GHz current starved voltage controlled oscillator (VCO) for PLL using 90 nm CMOS technology. In: International conference on control, instrumentation, communication and computational technologies 2015
13. Baker RJ, Li HW, Boyce DE (2002) CMOS: circuit design layout and simulation. IEEE Press, Piscataway

Rectenna for Wireless Power Transmission

Rupam Bharati and Janardan Sahay

Abstract Rectenna (rectifying antenna) is designed for better conversion efficiency. The obtained efficiency is 57.8%. The return loss obtained is −25.043 dB. The working characteristics of the rectenna are determined by the characteristic curves which depend on the input power. The power varies from −20 to 20 dBm. This paper focus on the microwave power transmission and conversion efficiency and how efficiently the conversion takes place. The rectifier circuit was designed and simulated in advanced design system (ADS) software while the antenna was designed using high-frequency structure simulator (HFSS).

Keywords Microwave power transmission · Conversion efficiency · Rectenna

1 Introduction

The rectenna (rectifying antenna) was first demonstrated by William C. Brown in 1962. Recently, wireless power transmission (WPT) is an important solution for providing power for wireless or non-contact-based devices. WPT can be achieved by different methods such as near-field inductive coupling, magnetic resonant coupling, and microwave power transmission (MPT) [1–3]. Among these, MPT has much longer transmission range. DC voltage used to drive the terminals should be followed by a rectification circuit which transfers the DC voltage at microwave frequency from receiving antenna to DC voltage output. To increase the beam efficiency over long distance, the frequency should be increased, but due to increased frequency, the efficiency of the system decreases. Wireless power transfer is discussed by diffrent authors in various papers [4–14] keeping main focus to improve the conversion

R. Bharati · J. Sahay (✉)
Department of Electronics & Communication Engineering, Birla Institute of Technology, Mesra, Ranchi, JH 835215, India
e-mail: jsahay@bitmesra.ac.in

R. Bharati
e-mail: rupambharati1993@gmail.com

V. Nath and J. K. Mandal (eds.), *Proceedings of the Fourth International Conference on Microelectronics, Computing and Communication Systems*, Lecture Notes in Electrical Engineering 673, https://doi.org/10.1007/978-981-15-5546-6_34

efficency. In this paper, the focus is on the 2.45 GHz frequency which leads to smaller aperture as well as better conversion efficiency. Different input power is given to the rectenna from −20 to 20 dBm. The simulated result has been shown with different conversion efficiency. The main objectives of designing the rectenna were that rectenna will be able to receive microwave energy at 2.45 GHz frequency and converting that to DC power, better return loss so that maximum output can be obtained, increase the overall efficiency so that RF-DC conversion is more efficient.

1.1 What Is Rectenna?

The main function of a rectenna is to receive the RF waves and convert them to the DC power. The block diagram of recenna has been shown in Fig. 1. The receiving power function is done by the antenna, while the conversion is done by the rectifier. Higher efficiency rectifying system is required because WPT is an energy-oriented system. The rectifier circuit consists of a diode Zero-biased Schottky diode HSMS285X family. The rectifier consists of low pass filter for harmonic rejection properties and a capacitance for power conversion and storage, load resistor, and microstrip lines. Different types of rectifiers like half wave rectifier, full wave rectifier, voltage doubler, and voltage multipliers can be used. In this paper, half wave rectifier with a zero bias single series diode HSMS2852 is used which is a surface mount diode.

The antenna is designed on printed circuit board (PCB) on FR4 substrate which is the cheapest. The ε_r is 4.4, and tan δ is 0.018. The understanding of working of antenna is most important step, the frequency at which it works and the rectifier topology it uses for greater efficiency and most important matching between the rectifier and the antenna. The characteristics of efficient rectenna is: high power conversion efficiency, wider frequency band, higher gain, maximum output DC voltage, low input power for higher efficiency.

Fig. 1 Block diagram of the rectenna

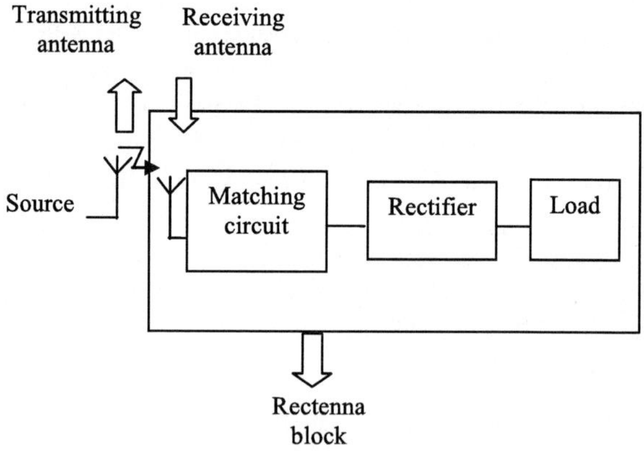

1.2 Power Transmission

The frequency bands 869 MHz and 2.45 GHz are used for wireless power transmission. They are used for telemetry and RFID applications. It is allocated for ISM bands (industry, science, and medicine) The Friss transmission equation is used to calculate the power received from one antenna (when another antenna transmits at a distance R). The Friss power transmission equation is used for wireless power transfer which is

$$P_r = P_t \frac{G_t G_r}{(4\Pi R/\lambda)^2} \tag{1}$$

P_r power received
P_t power transmitted
G_r receiver antenna gain
G_t transmitter antenna gain
λ free space wavelength
R transmission distance.

Assuming no losses due to propagation effect.
For 2.45 GHz, more directive antennas with smaller structure can be framed.

1.3 Power Conversion Efficiency

This is the parameter through which the performance of the rectenna could be measured. This is the ratio of DC power at the receiver to the AC input power sensed by the rectenna. Power conversion efficiency is given by:

$$\eta = \frac{V_{dc}^2}{R_{load} + P_d + A_{eff}} \tag{2}$$

$$\Gamma(P_{in}) = \frac{(Z_{in} - Z_{rect})}{(Z_{in} + Z_{rect})} \tag{3}$$

$$P_d = P_{in}[1 - \Gamma(P_{in})] \tag{4}$$

P_{in} = input power, V_{dc}^2 = utput DC voltage, R = load resistance, P_d = power density, A_{eff} = effective area, Z_{in} = input impedance, Z_{rect} = impedance of the rectifier.

The maximum power is collected by the receiving antenna and delivered to the rectifying diode. For maximum power transfer, there must be proper matching done between the source and the load. Impedance of the source, the output power of the source, and the input impedance varying with the net input power are given which is used to calculate the input power according to Eqs. (2)–(4).

2 Design Parameters

2.1 *Antenna*

The GCPW antenna is simulated in 1.6 mm thick FR4 substrate with relative permittivity of 4.4. The configuration of CPW antenna has been shown in Fig. 2 and its dimensions are shown in Table 1. The dimension of the substrate is 90 * 70. The

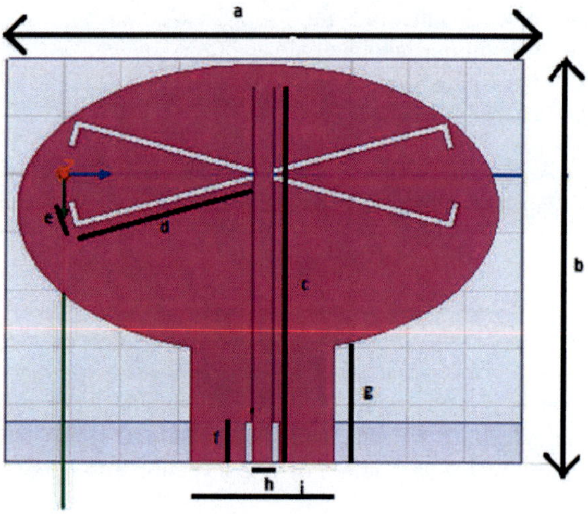

Fig. 2 Configuration of the CPW antenna

Table 1 Dimensions of CPW antenna

Symbol	Dimension (mm)
a	90
b	70
c	35
d	30
e	4
f	3.5
g	27
h	3.5
i	25

feeding is done by CPW. The beginning of the structure is fed by GCPW. The input impedance is matched to 50 Ω by properly moving the length of the CPW stub. The center frequency 2.45 GHz depends on the length *d*, and *e*.$|S_{11}|$ is found to be below −10 dB at 2.45 GHz frequency. The −10 dB bandwidth has been found to be 26.9%. The VSWR is 1.58. VSWR is the ratio of maximum voltage to minimum voltage in a standing wave. The 3D gain pattern has been simulated.

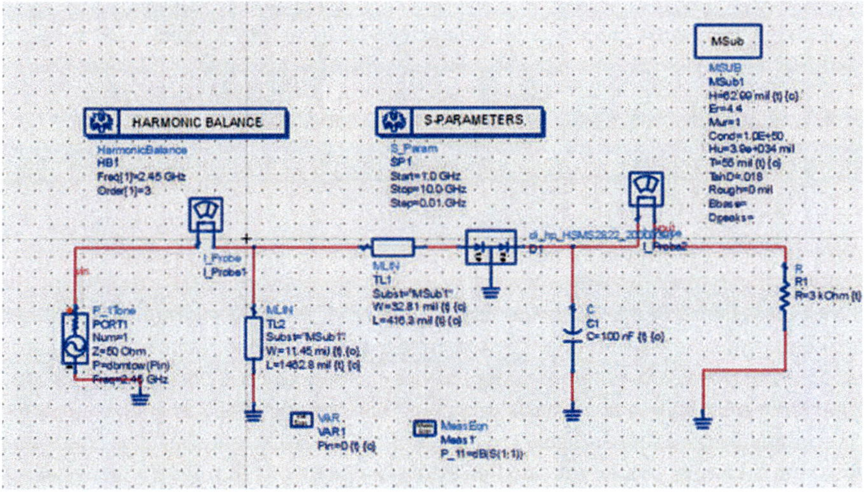

Fig. 3 Simulated circuit of the rectifier

Fig. 4 $|S_{11}|$ parameter of the CPW antenna

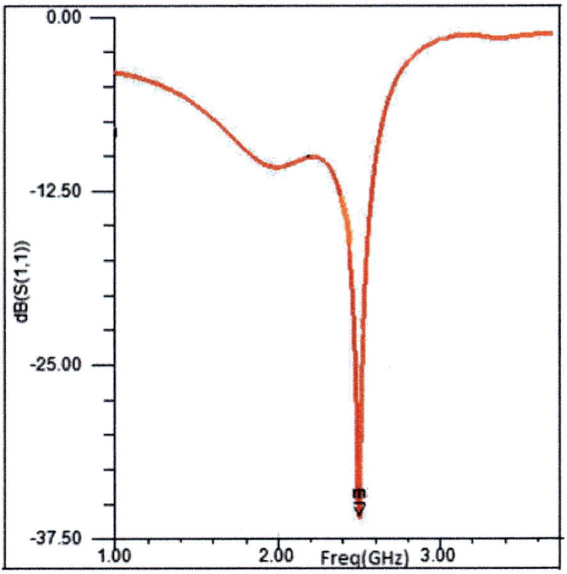

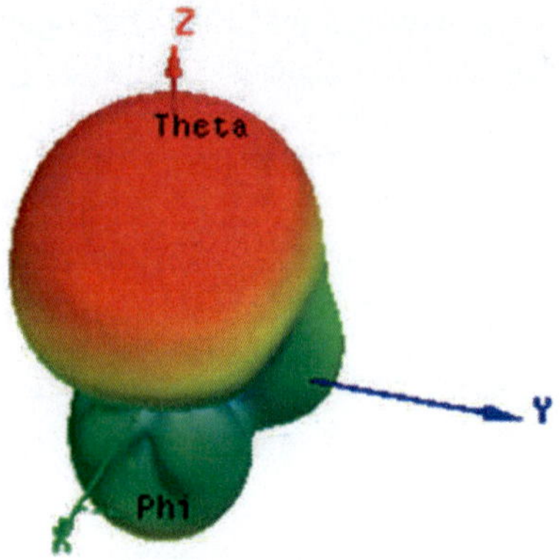

Fig. 5 3D polar plot of radiation pattern of antenna

2.2 *Rectifier Circuit*

The rectifier is used as AC voltage to DC voltage converter. The purpose of the conversion is for efficient transmission of power. The rectifier circuit has a diode, capacitor, and transmission line. The impedance of the diode is obtained by simultion as (93.905-j219.421) and thier results has been shown in Figs. 6 and 7. Single stub impedance matching is used for the matching between the 50 Ω antenna source and the rectifier circuit which is discussed later. **HSMS2852** diode is used. The equivalent linear circuit model is used for calculating the impedance of the diode which is based on the spice parameters of the data sheet of the diode.

The impedance matching gives maximum power transfer. The rectifier circuit consists of load resistor, transmission line, storage capacitors. Single series diode is used in the circuit. The load resistor (R1) is tuned to different values for better output voltage and conversion efficiency. The capacitor C1 (100 nF) acts as a high-pass filter also stores energy. Only the positive half cycle of the wave can be rectified by the diode, and the negative half cycle is rejected since it is a half wave rectifier. The return loss obtained is −25.043 dB. In Fig. 3, the simulated circuit of the rectifier is given.

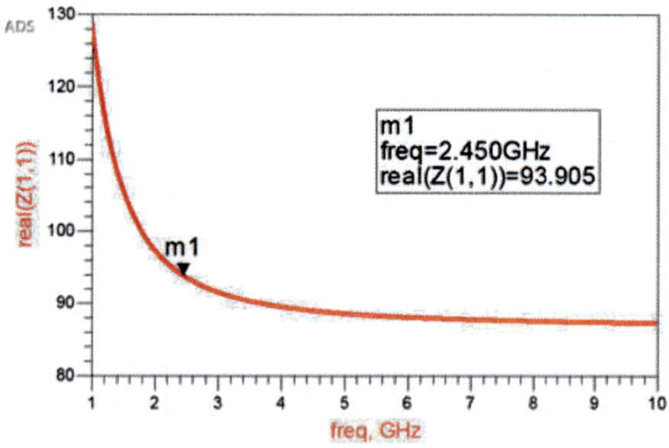

Fig. 6 Real impedance of the diode

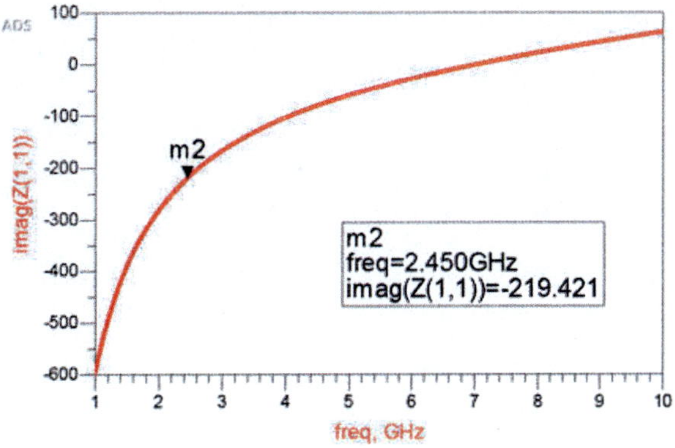

Fig. 7 Imaginary impedance of the diode

The length of the stub is 1462.8 mil, and the width is 11.45 mil. The length of the main transmission line is 416.3 mil, and the width is 32.81 mil. Harmonic balance is used which is a frequency domain-based simulator for simulating nonlinear circuits and their distortions.

3 Results and Discussion

Figure. 4 depicts the return loss S11 of the antenna having value −35 dB at 2.45 GHz frequency. Return loss determines the amount of power lost during transmission. Figure 8. shows the return loss of the rectifier circuit and it is −25.043 dB.

The 3D polar plot shows the radiation pattern of the antenna in E plane as well as H plane and directivity at 2.45 GHz is shown in Figs. 5, 6 and 7.

The efficiency of the rectenna is determined through the rectifier circuit. Fig. 9. shows the efficiency vs input power graph, and the maximum conversion efficiency is 57.8% at −5dBm.

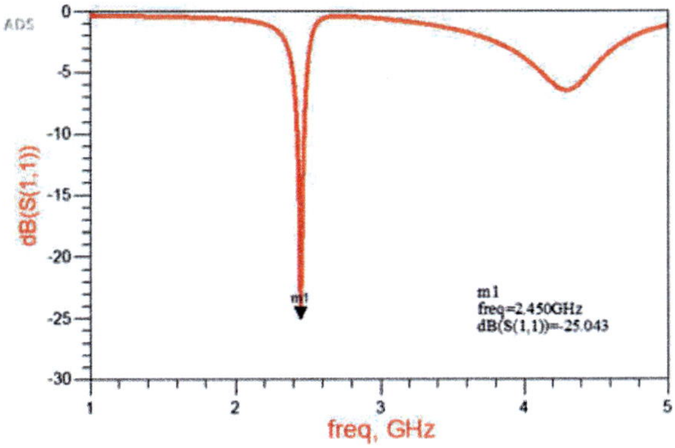

Fig. 8 S_{11} of the rectifier circuit

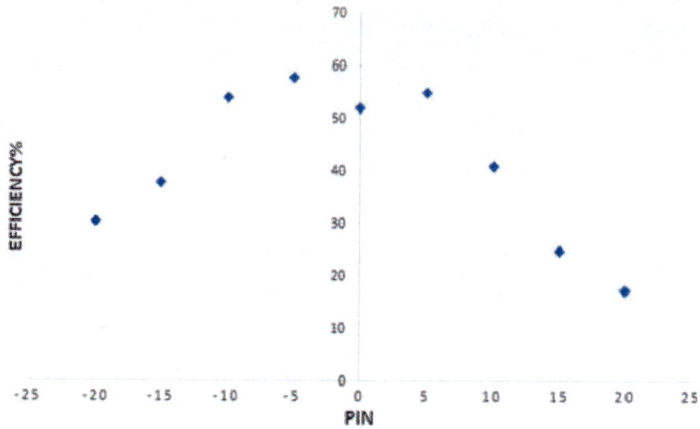

Fig. 9 Simulated efficiency versus input power in dBm of the rectifying circuit at 2.45 GHz

Table 2 Comparison between this work and related work

Refs.	Freq. (Ghz)	Dimension	Input power (dBm)	Conversion efficiency (%)
[1]	2–3.1	35 × 50	5	70
[2]	2.45	110 × 90	−17.2	50
[3]	2.45 Ghz	120×115	13	72
This work	2.45	90 × 70	−5	57.8

4 Conclusion

This paper shows the working of the rectenna at input power of −20 to 20 dBm. The conversion efficiency was found out to be 57.8% at −5dBm. Different parameters like frequency band, matching between antenna impedance, and rectifier impedance has been taken care while doing the simulations. The bandwidth efficiency is 26.9%, and the peak gain is found to be 9.85 dBi. The size of the substrate is also reduced with a better bandwidth and return loss. Table 2 shows the comparison between the proposed rectenna and some relevant works. The result measured shows that the rectenna has a broad bandwidth, and higher conversion efficiency is also found. Further, the fabrication of the rectenna is required and it should be compared with the measured results.

References

1. Shi Y, Fan Y, Li Y, Yang L, Wang M (2018) An efficient broadband slotted rectenna for wireless power transfer at LTE band. IEEE Trans Antennas Propag 1:814. https://doi.org/10.1109/tap.2018.2882632

2. Sun HC, Guo YX, He M, Zhong Z (2012) Design of a high-efficiency 2.45-GHz rectenna for low-input-power energy harvesting. IEEE Antennas Wireless Propag Lett 11:929–932
3. Nie MJ, Yang XX, Tan GN, Han B (2015) A compact 2.45-GHz broadband rectenna using grounded coplanar waveguide. IEEE Antennas Wireless Propag Lett 14:986–989
4. Heikkinen J, Kivikoski M (2001) Performance and efficiency of planar rectennas for short-range wireless power transfer at 2.45 GHz. Microw Opt Technol Lett 31:86–90
5. Haboubi W et al (2014) An efficient dual-circularly polarized rectenna for RF energy harvesting in the 2.45 GHz ISM band. Prog Electromagnet Res 148:31–39
6. Sun H, Geyi W (2016) A new rectenna with all polarization-receiving capability for wireless power transmission. IEEE Antennas Wireless Propag Lett 15:814–817
7. Chen YS, Chiu CW (2017) Maximum achievable power conversion efficiency obtained through an optimized rectenna structure for RF energy harvesting. IEEE Trans Antenna Propag 65(5):2305–2317
8. Lu P, Yang X, Li J, Wang B. A compact frequency reconfigurable rectenna for 5.2- and 5.8-GHz wireless power transmission. IEEE Trans Power Electron
9. Song C et al (2017) Matching network elimination in broadband Rectennas for high-efficiency wireless power transfer and energy harvesting. IEEE Trans Ind Electron 64(5)
10. Kim S et al (2014) Ambient RF energy-harvesting technologies for self-sustainable standalone wireless sensor platforms. Proc IEEE 102(11):1649–1666
11. Song C, Huang Y, Zhou J, Zhang J, Yuan S, Carter P (2015) A high-efficiency broadband rectenna for ambient wireless energy harvesting. IEEE Trans Antennas Propag 63(8):3486–3495
12. Lu P, Yang X-S, Li J-L, Wang B-Z (2016) Polarization Reconfigurable broadband rectenna with tunable matching net-work for microwave power transmission. IEEE Trans Antennas Propag 64(3):1136–1141
13. Suh YH, Chang K (2002) A high-efficiency dual-frequency rectenna for 2.45-and 5.8-GHz wireless power transmission. IEEE Trans Microwave Theory Techn 50(7):1784–1789
14. Kuhn V, Lahuec C, Seguin F, Person C (2015) A multi-band stacked RF energy harvester with RF-to-DC efficiency up to 84%. IEEE Trans Microw Theory Techn 63(5):1768–1778

An Energy-Efficient LEACH Routing Protocol for Wireless Sensor Networks

Rahul Priyadarshi and Ravi Ranjan Kumar

Abstract Sensor network is one of the leading exploration fields recently that may be used in almost every field. The outdated LEACH routing protocol is still reflected for the various research field, but in this paper, upgraded LEACH routing protocol is presented by way of an energy-efficient cluster head selection allowing distance and energy as main factors. The existing probability function has been reformed according to the distance and energy parameters. Therefore, distance and energy parameter select the energy-efficient cluster head for communication of data. The simulation outcomes display that the selection of cluster head in modified LEACH protocol is improved compared to the existing LEACH protocol. Better cluster head selection further advances the overall performance of network along with system lifespan compared to current LEACH protocol.

Keywords Sensor · Network · Cluster · Nodes · Energy · Probability · Distance

1 Introduction

Sensor nodes in the wireless sensor network (WSN) are grouped together and form a system to collect the data from the different location where human intervene is not possible all the time like in very dense area of forest and many more. Generally, these types of networks are installed in military application where collected data play an important role in battlefield. As these types of the network is wireless in nature where energy is the main concern of all the kind of data gathering process. There are various routing protocol have been proposed to save the energy and make overall system more efficient. One of the leading routing protocols is LEACH protocol that

R. Priyadarshi (✉) · R. R. Kumar
Department of Electronics and Communication Engineering, National Institute of Technology Patna, Patna, Bihar 800005, India
e-mail: rahul.glorious91@gmail.com

R. R. Kumar
e-mail: kravirrk@gmail.com

V. Nath and J. K. Mandal (eds.), *Proceedings of the Fourth International Conference on Microelectronics, Computing and Communication Systems*, Lecture Notes in Electrical Engineering 673, https://doi.org/10.1007/978-981-15-5546-6_35

is presented by author in [1–3]. This protocol is planned with two phases. In first phase, selection of nodes as cluster head (CH) is done by threshold value. Selection of CH is done by using generated random number between 0 and 1, and further, it is compared with threshold value which is given by Formula (1)

$$T(n) = \begin{cases} \frac{P}{1-P(r \bmod 1/P)} & \text{if } n \in G \\ 0 & \text{otherwise} \end{cases} \tag{1}$$

When the selected value is smaller than the threshold value, in that case, node will be selected as CH or else node will behave as simple member node. This iteration is done in each and every round for the selection of CH. Once selection of CH is done, then TDMA technique [4, 5] is planned to send the data from member nodes to the Base Station (BS). In next phase, i.e., steady phase, nodes will receive TDMA schedule first, and then, in that schedule time, all the member nodes communicate data to their respective CH. Once the data are collected by CH from the member nodes, then it forwards gathered data to BS.

This process of gathering the data from member node to the CH is done unceasingly till scheduled time. Afterward, completion of each and every TDMA: setup phase is worked for the selection of CH for succeeding round [6, 7]. The node selected as CH for proceeding round will not be selected as CH for the succeeding round in order to maintain energy level of nodes that further extend lifetime of network. LEACH method follows solitary hop routing in which all the node can connect to the CH and data transmission can done straight away [8, 9]. Consequently, the progresses are much needed for determining consumption of energy for dynamic clustering is existing research development. Figure 1 displays ordinary network topology for WSN. In this paper, an enhanced LEACH routing protocol has been considered for dynamic clustering [10]. Selection of CH is done according to the energy of each node and distance of each node from the neighboring node in addition to distance

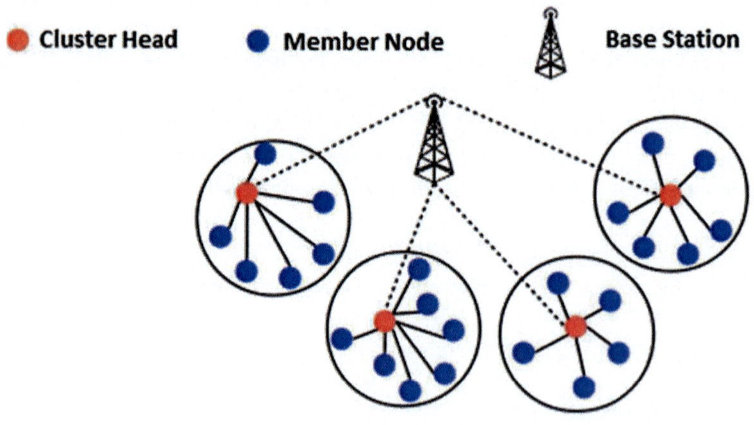

Fig. 1 Network topology in WSN

from BS. These enhancements give an improved result in balancing the load that displays an enhanced network lifetime as well as network scalability.

2 Related Work

In [11, 12] authors proposed a new kind of method which basically deals with the bulk transmission of data or one can say data accumulation between the nodes and BS. They proposed a new kind of scheme named as "efficient cluster head replacement scheme" where the selected CH will remain unaffected till it has adequate amount of energy [13, 14]. By doing so, this method saves energy while selecting the CH in each round. Lifetime of the node is also increased when furthering CH energy to the advanced level while transmitting data to the BS. Also, power intensification for the inter cluster has been introduced by doing so.

As we know in the LEACH protocol [15, 16] formation of CH and selection of CH takes place randomly. The main work of CH is not only to collect the data from member node and send it to the BS but also synthesis of data takes place. The consumption of energy while data transmission is smaller than data fusing. For reducing energy intake of CH, LEACH protocol is upgraded by author in [17, 18] by using two levels CH. The selection process of CH is similar as of LEACH protocol. Energy of each node as well as distance between BS and CH is computed which are further compared with average distance and average energy between BS and CH, respectively. If node energy and distance of CH from BS are greater than the average distance and average energy, then CH starts doing transmission of data. In case of node energy and distance of CH from BS are smaller than the average distance and average energy, then instead of primary CH secondary CH starts transmission of data in order to maintain the lifetime of primary CH at some extent. Main concept that has been developed in this method is that selection of primary and secondary user, and also, selection process of CH has to be notified to all other nodes in the network which is responsible for energy intake. Author in [19] is created on static clustering which decreases complication of the network as well as the power dissipation of system. It is realized by unraveling total network area into smaller area to reduce energy consumption of node while transferring data between bulky networks. Basically, LEACH protocol that is proposed by author in [8] is grouping of DEEC [20] as well as LEACH [21]. DEEC routing protocol is basically select CH based on the residual energy. DEEC method that employed in LEACH verifies that it advances network lifetime by selecting accurate CH. For time being, Ad-LEACH protocol is showing the development in data fusing method with the cost of more energy consumption of node.

Author in [22–24] projected a new type of LEACH protocol which gives preference to the energy constraint and how to diminish energy consumption of node and hence formation of CH properly. Main idea is to keep CH in static way until and unless CH loses all its energy. This process certainly prolongs the lifetime of system by dropping energy consumption in each round of CH selection [25, 26]. One of the major problems with this method is that it works for only in static node scheme. In case of dynamic node, nodes move dynamically which lead to the formation of new cluster after certain period to reduce the cluster weight by adding some extra nodes to the cluster.

3 Proposed Routing Protocol

In proposed routing protocol, distance and energy are two foremost limitations that are considered in network. Consequently, the necessity for enhancements established on such limitations always displays significance role in arena of exploration. Selection of CH is totally done by probability function for every round. Instead of probability function, a new kind of function (let us consider notch function, NF) has been introduced which is computed in every round and rationalized with probability function which further used to classify the better selection of CH in order to extend the network lifetime. Based on the proposed function, residual energy of each node is computed as given below Eq. (2)

$$\mathrm{NF}_1 = \frac{\mathrm{EG}_{\mathrm{Current}}}{\mathrm{EG}_{\max}} \tag{2}$$

where $\mathrm{EG}_{\mathrm{Current}}$ signifies residual energy of the node and $\mathrm{EG}_{\max}$ signifies the maximum energy of node. The NF_1 is computed for every node with CH to evaluate residual energy of the node. Similarly, distance between nodes also plays an important role in WSN to save the energy. The consumption of energy will be more in case of selected CH is far away from the BS during communication takes place between CH and BS. As we know main work of CH is to collect the data from its respective member nodes and transfer data to BS. As a result, distance is one of the major factors that must be taken into consideration between the CH and BS while selecting the CH. The distance between BS and CH is calculated using by given below Eq. (3)

$$\mathrm{NF}_2 = \frac{D_{\mathrm{BS}}}{D_{\mathrm{FAR}}} \tag{3}$$

where D_{BS} is distance between BS and the node and D_{FAR} is the distance between furthermost node in the cluster to the BS. Merging the two notch functions NF_1 and NF_2 and multiplying both with existing probability function gives new function which enhances the network lifetime. The merging of notch functions with existing probability function is given by Eq. (4)

$$\mathrm{NF} = \sum_{i=1}^{2} w_i \times \mathrm{NF}_i \tag{4}$$

Above Eq. (4) can be expanded and rewritten as given below equation.

$$\mathrm{NF} = \left[w_1 \left[\frac{\mathrm{EG}_{\mathrm{Current}}}{\mathrm{EG}_{\max}} \right] + w_2 \left[\frac{D_{\mathrm{BS}}}{D_{\mathrm{FAR}}} \right] \times \frac{P}{1 - P(r \bmod 1/P)} \right]$$

where sum of total $w_1 + w_2$ is taken as one and considered any value for w_1 and w_2 according to the requirement of distance and energy of any network.

4 Simulation and Result Analysis

The notch function NF is performed in each and every round, and selection of node as CH is taken on the basis of NF. Table 1 displays simulation parameters and their value.

Figure 2 displays the plot between total numbers of alive node versus total number of rounds. The below figure reflects that total number of alive node in improved

Table 1 Simulation parameters

Parameters	Values
Area	150 m × 150 m
Sensor nodes	500 nodes
Node's energy	0.5 J
BS location	(50 m, 50 m)
Number of rounds	1000

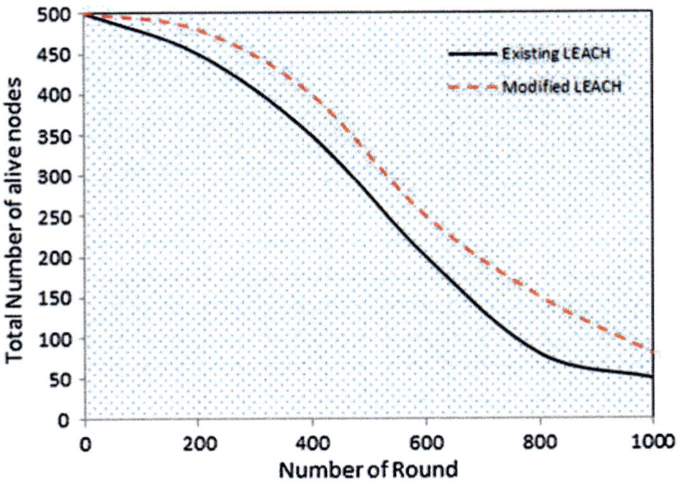

Fig. 2 Total number of alive nodes

LEACH is more compared to the existing LEACH. It is almost 30% improvement in alive nodes in improved LEACH compared to the existing LEACH, and this can be better understood by using Fig. 3 which is shown below. Thus, both figures further reflect the improvement in lifetime of network.

Figure 4 displays the plot between average residual energy verses number of rounds which is achieved on the basis of distance parameter which is taken considered in improved LEACH protocol, while it was not in existing LEACH routing protocol. It is clear from above graph is that proposed routing protocol shows better performance in view of energy saving compared to the existing protocol.

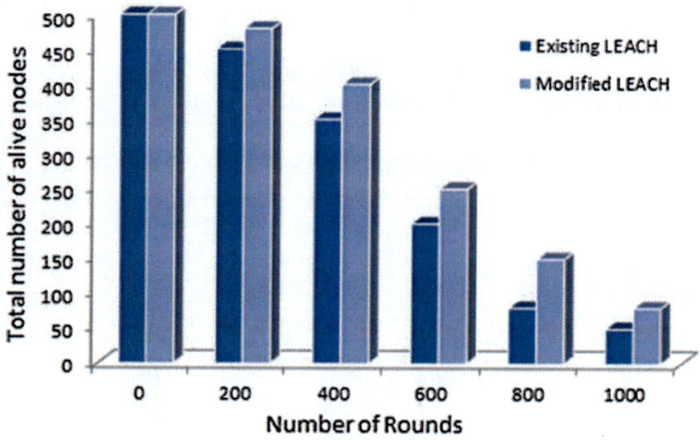

Fig. 3 Total number of alive nodes increased by 30%

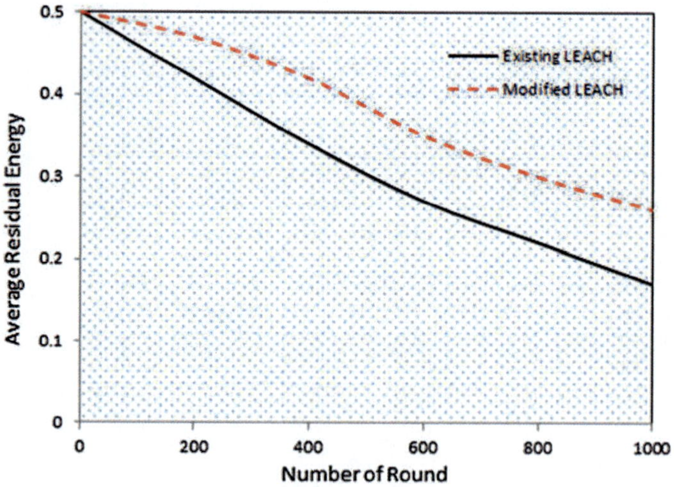

Fig. 4 Average residual energy

5 Conclusion

The modified LEACH protocol which is proposed selects CH based on energy and distance parameters. Simulation results reflect that proposed algorithm shows better performance in terms of alive nodes and residual energy compared to the existing algorithm. Advance way for selecting CH advances the total performance of the network in terms of network lifetime. Combination of notch function provides better results, and further, it can be modified with different parameter like distance between member nodes and CH as well as establishing cluster with identical number of nodes might also advance the network lifetime.

References

1. Heinzelman WR, Chandrakasan A, Balakrishnan H (2000) Energy-efficient communication protocol for wireless microsensor networks. In: Proceedings of the 33rd annual Hawaii international conference on system sciences. IEEE, p 10
2. Heinzelman WB, Chandrakasan AP, Balakrishnan H (2002) An application-specific protocol architecture for wireless microsensor networks. IEEE Trans Wirel Commun 1(4):660–670
3. Akyildiz IF, Su W, Sankarasubramaniam Y, Cayirci E (2002) Wireless sensor networks: a survey. Comput Netw 38(4):393–422
4. Wang CF, Shih JD, Pan BH, Wu TY (2014) A network lifetime enhancement method for sink relocation and its analysis in wireless sensor networks. IEEE Sens J 14(6):1932–1943
5. Mahmood D, Javaid N, Mahmood S, Qureshi S, Memon AM, Zaman T (2013) MODLEACH: a variant of LEACH for WSNs. In: 2013 Eighth international conference on broadband and wireless computing, communication and applications (BWCCA). IEEE, pp 158–163
6. Priyadarshi, R., Soni, S. K., & Sharma, P. (2019). An enhanced GEAR protocol for wireless sensor networks. In: Lecture notes in electrical engineering. doi:10.1007/978-981-13-0776-8_27
7. Priyadarshi R, Singh L, Kumar S, Sharma I (2018) A hexagonal network division approach for reducing energy hole issue in WSN. Eur J Pure Appl Math 118
8. Priyadarshi R, Rawat P, Nath V (2018) Energy dependent cluster formation in heterogeneous wireless sensor network. Microsyst Technol. https://doi.org/10.1007/s00542-018-4116-7
9. Priyadarshi R, Bhardwaj A (2017) Node non-uniformity for energy effectual coordination in WSN. Int J Inf Technol Secur 9(4):3–12
10. Priyadarshi R, Soni SK, Nath V (2018) Energy efficient cluster head formation in wireless sensor network. Microsyst Technol 24(12):4775–4784. https://doi.org/10.1007/s00542-018-3873-7
11. Priyadarshi R, Singh L, Sharma I, Kumar S (2018) Energy efficient leach routing in wireless sensor network. Int J Pure Appl Math 118(20):135–142
12. Priyadarshi R, Tripathi H, Bhardwaj A, Thakur A (2018) Performance metric analysis of modified LEACH routing protocol in wireless sensor network. Int J Eng Technol. https://doi.org/10.14419/ijet.v7i1.5.9146
13. Heinzelman WB, Chandrakasan AP, Balakrishnan H (2002) An application-specific protocol architecture for wireless microsensor networks. IEEE Trans Wirel Commun 1(4):660–670
14. Priyadarshi R, Yadav S, Bilyan D (2019) Performance analysis of adapted selection based protocol over LEACH protocol. In: Luhach AK, Hawari KBG, Mihai IC, Hsiung P-A, Mishra RB (eds) Smart computational strategies: theoretical and practical aspects. Springer Singapore, Singapore, pp 247–256. https://doi.org/10.1007/978-981-13-6295-8_21

15. Priyadarshi R, Yadav S, Bilyan D (2019) Performance and comparison analysis of MIEEP routing protocol over adapted LEACH protocol. In: Luhach AK, Hawari KBG, Mihai IC, Hsiung P-A, Mishra RB (eds) Smart computational strategies: theoretical and practical aspects. Springer Singapore, Singapore, pp 237–245. https://doi.org/10.1007/978-981-13-6295-8_20
16. Bani Hani RM, Ijjeh AA, A survey on LEACH-based energy aware protocols for wireless sensor networks.
17. Lindsey S, Raghavendra CS, PEGASIS: power-efficient gathering in sensor information systems. In: Proceedings, IEEE aerospace conference, vol 3, pp 3-1125–3-1130
18. Smaragdakis G, Matta I, Bestavros A, SEP: a stable election protocol for clustered heterogeneous wireless sensor networks *
19. Ali MS, Dey T, Biswas R (2008) ALEACH: advanced LEACH routing protocol for wireless microsensor networks. In: 2008 international conference on electrical and computer engineering, pp 909–914
20. Priyadarshi R, Singh L, Singh A, Thakur A (2018) SEEN: stable energy efficient network for wireless sensor network. In: 2018 5th international conference on signal processing and integrated networks, SPIN 2018. https://doi.org/10.1109/spin.2018.8474228
21. Priyadarshi R, Singh L, Randheer, Singh A (2018) A novel HEED protocol for wireless sensor networks. In: 2018 5th international conference on signal processing and integrated networks, SPIN 2018. https://doi.org/10.1109/spin.2018.8474286
22. Kumar V et al (2018) Multi-hop communication based optimal clustering in hexagon and voronoi cell structured WSNs. AEU—Int J Electron Commun
23. Rehman O, Javaid N, Manzoor B, Hafeez A, Iqbal A, Ishfaq M (2013) Energy consumption rate based stable election protocol (ECRSEP) for WSNs. Procedia—Procedia Comput Sci 19:932–937
24. Loscri V, Morabito G, Marano S (2005) A two-levels hierarchy for low-energy adaptive clustering hierarchy (TL-LEACH). In: 2005 IEEE 62nd vehicular technology conference VTC-2005-Fall, vol 3, pp 1809–1813
25. Barati H, Movaghar A, Rahmani AM (2015) EACHP: energy aware clustering hierarchy protocol for large scale wireless sensor networks. Wirel Pers Commun 85(3):765–789
26. Fu C, Jiang Z, Wei WEI, Wei A (2013) An energy balanced algorithm of LEACH protocol in WSN. Int J Comput Sci 10(1):354–359

Power Management of Pedaling System with SPV System: An Alternate Option of Power Generation Without Emission

Ambarisha Mishra

Abstract An environment-friendly electric power generation without emission is considered here with pedal system, and SPV system is presented here. The rural and urban areas are facing the situation of electricity shortage and outage. The present work proposes a system to generate electric power at a miniature stage. A gym bicycle pedal system (GBPS) integrated with SPV is to overcome the above situation. The system is developed for power management to channelize energy output without harmful emission. It is a straightforward methodology of electricity generation and found better especially for those areas where power outage is an issue. The efficacy of power management presented is practically demonstrated on the GBPS, and its output is unified with PV system to meet the total load. To generate sufficient voltage at good speed, a gearbox assembly of 1:30 ratio is provided. A 50 kWp solar-based system with GBPS is taken into consideration here. The forecasted load is supplied with power management in peak and off-peak hours. This work is an alternate choice of electric power without harmful discharge. Proposed work of GBPS is developed at the prototype stage.

Keywords SPV system · Pedal system gym bicycle · Generation scheduling · Energy

1 Introduction

The power demand is increasing nowadays. Everyone wants to live a lavish life with the application of different electrical and electronics accessories, but just think the situation of life with no electric power. A large number of countries (developing) are in power crisis. More than 75% populations are not getting electricity that is approximately 1.2 billion people, (World Energy Outlook 2015). The problem of power cut is not limited to the rural locations but in the urban areas too. Electric power

A. Mishra (✉)
Department of Electrical Engineering, National Institute of Technology Patna, Patna, India
e-mail: ambrishee@gmail.com

V. Nath and J. K. Mandal (eds.), *Proceedings of the Fourth International Conference on Microelectronics, Computing and Communication Systems*, Lecture Notes in Electrical Engineering 673, https://doi.org/10.1007/978-981-15-5546-6_36

generated by fossil fuel produces a lot of unfavorable effect to environment (SO_x, NO_x and CO_2) and other harmful emissions. There is a need of alternative method of power generation to fulfill load demand without the emission of greenhouse gases.

The objective of this work is to integrate the energy generated by gym bicycle pedaling system located at the gym point and SPV system. In literature, many works have been reported on use of pedaling system for power generation. A bicycle lighting approach is developed in [1] where 4 W energy at 5–10 kmph is produced. To obtain a usable voltage, a gearbox assembly was employed with pedals [2]. A system is proposed with pedaling mechanism for power generation [3] using physical exercise. Cost of power generation and transmission is not affordable or too much high cost for few developing countries. In such cases, distributed generation like renewable energy sources or its integration with other sources is a better solution [4, 5].

The generation of electricity for rural locations is reported [6, 7] using human beings. A pedal rickshaw with power was established to reduce power burning up at Chittagong, Bangladesh [8]. An emission-free, ecological and a reliable power generation method is presented in [9].

This work focuses on gym center where solar photovoltaic system is provided at roof top and gym bicycle is used for physical exercise purpose. These two systems combined together to provide electrical supply to nearby areas at an economical rate. An integration of thermal and SPV generation is reported in [10]. The scenario of PV/battery system connected to grid is presented in [11].

There is sufficient amount of literature published regarding the power generation problem. The SPV is not completely considered among these. A hybrid vehicle powered by the manual pedaling, electric and SPV is presented [12].

The application of gym bicycle along with SPV for power generation is less focused in reported literature and that is vital area of research to solve electricity crisis issue for regions with more power cut or power outage [13]. The paper mainly focused the energy generation by the GBPS along with SPV on to fulfill the total demand. The major influences of the present work are as follows:

1. The amount of electrical power generation and cost analysis of the GBPS on daily, monthly and yearly basis.
2. Power contribution with and without GBPS.
3. Cost analysis of on-site and off-site SPV-based power generation.
4. Management of power with both the systems in peak and off-peak hours.

2 Theory

The objective of present work is to generate electrical power from man power efforts with SPV integration that is independent of fuel-based power generation to satisfy load demand without harmful emission. Theoretical analysis is presented next here.

2.1 GBPS

The gym bicycle pedal system is used at gym stations for physical exercise. The paddle rotation used for rotation of alternator generates electrical power. To increase the speed of alternator, 1:30 ratio gearbox is used. The converter is used.

For speeding up the rotation of the alternator, a gearbox is arranged with 30 multi-turn ratios. A convertor is for converting alternator output into appropriate form along with battery and inverter arrangement. The block diagram of the GBPS is shown in Fig. 1.

The cycle used here in this work is rated 90–110 rpm connected to front wheel axel and chain. A gearbox of turn ratio 1:30 is used to connect it to alternator for production of sufficient voltage even at less bicycle speed. The setup system parameters are given in Table 1.

It is obvious that paddle speed is always constant, so to maintain the output voltage a charge controller is mandatory that also improves the life of used battery value. The typical charging time of battery capacity is around 6–8 h. Here a 1080 Wh of battery taken that will be charged in approximately 8 h. At the end point, an inverter is used to get the required AC supply.

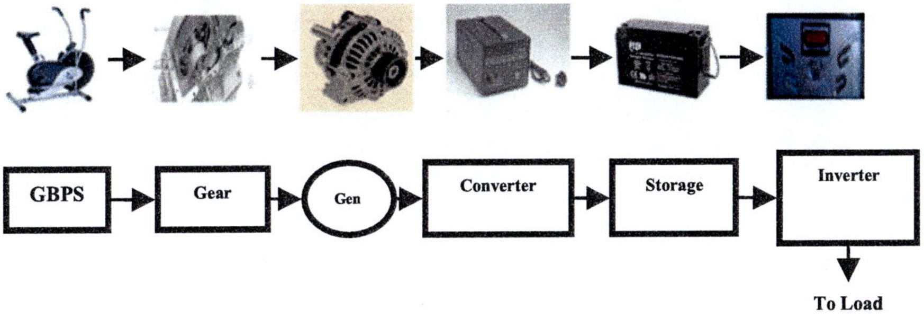

Fig. 1 Block diagram of GBPS

Table 1 Parameters of GBPS system

S. No.	Parameter	Value unit
1	Gym bicycle	90–110 rpm
2	Ratio of gearbox	1:30
3	Battery capacity	12 V 90 Ah
4	Generator output	12 V 50 A
5	Inverter rating	1 kW
6	Power convertor	12 V 10 A

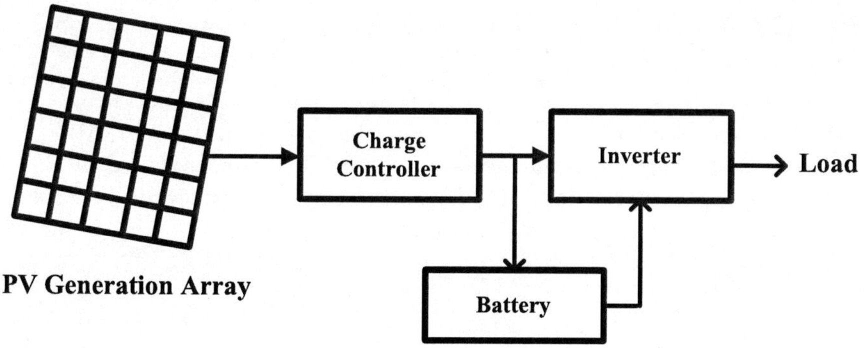

Fig. 2 SPV-based power generation

2.2 SPV System

The solar modules are used to convert sunlight to electric power. A solar cell does not have any movable parts and needs less maintenance and long life [14]. To reduce the land requirement, roof can be used for installation of solar panels.

A typical SPV-based power generation system is shown in Fig. 2. At gym center rooftop, 50 kW solar modules can be installed that is considered here in this work.

2.3 On-Site and Off-Site Solar Power Generation

The on-site plant here means solar system considered with land required for installation of solar panels, and off-site means land and battery is not considered. The performance evaluation of solar plant is done with and without GBPS.

In on-site solar photovoltaic power plant, the land cost is considered while in off-site SPV plant there is no land cost. Off-site SPV power plants can be used without battery and land. In this work, the effects before and after installing the GBPS with SPV system are evaluated.

The combined power developed by GBPS and solar plant is used to fulfill the load demand. During the weekend and night, low power output of GBPS is considered here that is more in peak hours more output is a better option to meet load demand. The power capacity of overall system can be increased as per demand.

3 Experimental Investigation

It is obvious here that pedal system is operated by human efforts. Due to gearbox arrangement, alternator speed will vary linearly with pedal speed. The speed of

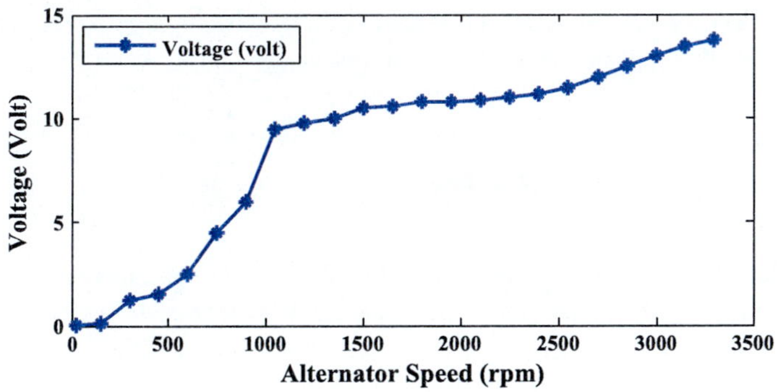

Fig. 3 Voltage developed with speed

alternator will be 1500 rpm at pedal speed of 50 rpm that can be increased by putting more effort and consequently more voltage can be produced.

3.1 Electrical Response of GBPS

The voltage produced by alternator is directly proportional to speed of padel, and there is a chance of small fluctuations. The charging time of battery can be reduced at the highest speed of alternator around 3000 rpm. The variation of output voltage with speed is shown in Fig. 3.

The emf developed by alternator can be expressed as given in Eq. (1)

$$e = B \times l \times v \tag{1}$$

3.2 Electrical Power from GBPS System

The normal timing of exercise at gym centers is 05:00 am to 09:00 am and 05:00 pm to 09:00 pm. Total time duration of above 8 h is sufficient to charge the battery fully. The electrical power output can be doubled by connecting two batteries, and they take same time to charge by GBPS system.

This power stored in the batteries is to be used for 24 h in case of power unavailability from grid. The total electrical power developed by GBPS is 918 W by one battery and 1836 W with two batteries considering 15% loss in battery. The cost of one unit of GBPS system is shown in Fig. 1 is $900.

The amount of energy generated by such five numbers of GBPS units is 9.18 kWh/day, i.e., 3350.7 kWh/year and by ten numbers of units is 18.36 kWh/day, i.e., 6701.4 kWh/year. If such 100 units are combined, the amount of electrical energy generated per day will be 183.60 kWh and per year 67,014.0 kWh.

3.3 Development of Solar Plant

For the design and development of a solar plant, solar modules (panel), battery, inverter, controller and space (land) are required. To design of a 50 kWp solar power plant, total amount of energy needs to be extracted from modules is 1560 kWh per day and peak watt rating of modules is 249.6 kW.

The solar modules designed by TATA power of TP 250 series are selected for solar plant design. The solar cells are monocrystalline Si wafer, and 998 modules are required for 50 kWp solar plant.

The design of inverter mainly depends on peak demand, and as per requirement batteries are selected considering loss in battery. Total 53 PV arrays are used for that 3850 m^2 land is needed [15]. Total cost for 50 kWp solar plant is given in Table 2.

4 Result Analysis

In the present work, electrical power generation without emission is presented to fulfill the load demand in the region suffering from power cut or power outage. The performance of solar and GBPS integrated power generation system is analyzed.

In the present work, a 50 kWp solar plant with 20 numbers of GBPS systems is considered. The load forecasted for 24 h is shown in Fig. 4.

In the integration, the contribution of power generated by GBPS is in the peak hours with high load demand. Here SPV system is analyzed with and without GBPS.

Table 2 Cost of equipments

S. No.	Items	Rate ($)
i	SPV module	197,255
ii	Installation frames	13,774
iii	Installation labor cost	469
iv	Hardware packing and freight	1719
v	Inverters	8125
vi	Design and management cost	313
vii	Electrical items	10,000
	Overall cost for off-site SPV plant	231,655
viii	Land required	240,625
ix	Storage battery	36,125
	Overall cost for on-site SPV plant	508,405

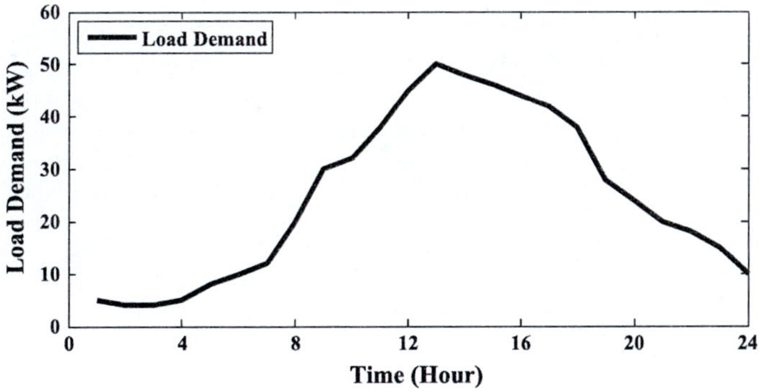

Fig. 4 Forecasted load for 24 h

4.1 SPV System Without GBPS for Load

From the forecasted load chart, it is clear that peak load is 50 kW at 12:00 pm to 01:00 pm, and minimum is 4 kW at 04:00 am.

The variation of power output of a 50 kWp solar plant along the load variation is shown in Fig. 5. From the graph, it is clear that in noon time power generation by SPV is good and enough to meet the demand. The load demand is highest at 01:00 pm. At certain time, SPV generation is more than load demand (10:00 am), and during this period extra power can be used to charge the battery for future use. To meet the peak load, GBPS is integrated with SPV and performance is shown in Fig. 6.

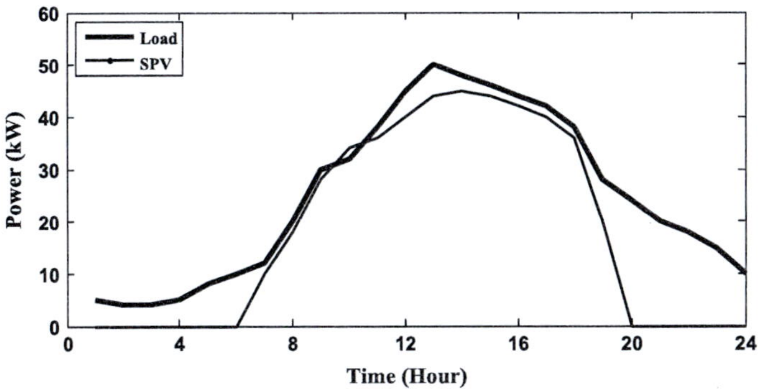

Fig. 5 Variation of SPV output with load for 24 h

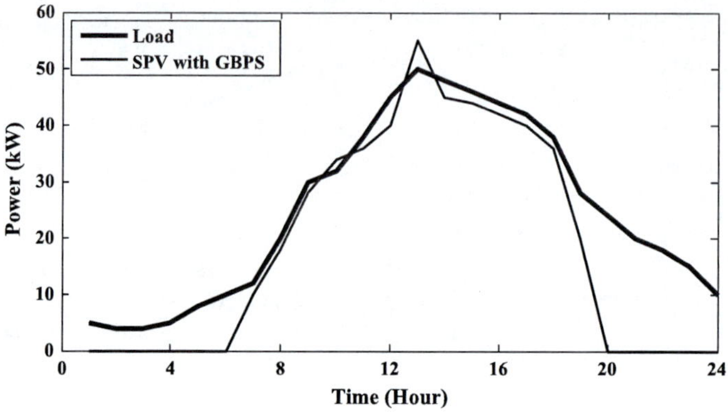

Fig. 6 SPV + GBPS power output with load demand

4.2 SPV System with GBPS

With addition of GBPS, the power demand from SPV is reduced in peak hours. Figure 6 shows the variation of power generated by SPV + GBPS (50 kWp SPV and 20 GBPS units) with load during 24 h.

The extra power generated by GBPS can be utilized in off-peak hours. So the integrated SPV GBPS system is capable of meeting load demand in peak and off-peak hours.

4.3 Power Generation of SPV with GBPS in Off-Peak Hours

The availability of sunlight is changing from season to season and also depends on geographical region. In North India, it is available from 07:00 am to 06:30 pm during summer and 07:30 am to 05:00 pm during winter. SPV generation is best in noon time. Figure 7 shows individual power generation by SPV and GBPS with load demand in off-peak hours marked.

The energy stored in batteries with GBPS system is utilized in off-peak hours (12:00 am to 06:30 am and 06:00 pm to 12:00 am). Figure 8 shows the demand and power generated between 6:30 am and 06:00 pm by SPV and remaining demand fulfilled by GBPS stored power. This analysis is for one system, and similar multiple systems can be developed for load demand of a town, village, etc. This is a power solution without emission.

This idea is useful in urban area also for reducing electricity bill and can be connected to grid also. Still there are some villages facing lack of power supply as per requirement that is also affecting the education of children residing there.

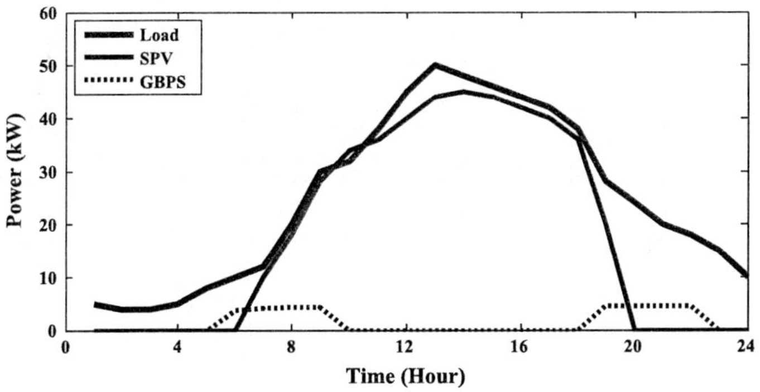

Fig. 7 Variation of individual power SPV and GBPS with load demand

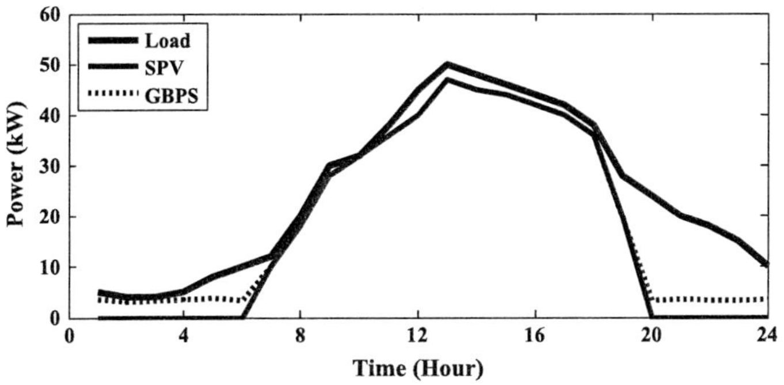

Fig. 8 Utilization of surplus power of GBPS in off-peak hours

5 Conclusions

This work presented here is methodology with a bicycle pedal-based system with the SPV system for generation of electric power. The generated power here is safe and reliable and harmless for the civilization. The integration of GBPS and SPV system is providing a solution to power crisis in rural areas that too without emission. A solar PV system 50 kWp is with GBPS integrated. The total load is managed in peak and off-peak hours by combining GBPS with SPV and saving the surplus power of GBPS whenever it is available. It can easily be used with other renewable energy sources too. The arrangement is good for the rural areas. The hardware setup for the presented work is designed and developed at prototype level.

References

1. Schwaller E (1999) Bicycle lighting system and generator. United States Patent
2. Hicks RB (2000) Electricity generation for pedalled vehicles. United States Patent
3. Gagne JP (2001) Pedal operated power generating system. United States Patent
4. Bhusal P, Zahnd A, Eloholma M, Halonen L (2007) Energy efficient innovative lighting and energy supply solutions in developing countries. Int Rev Electr Eng 2(5):665–670
5. Podmore R, Larsen R, Louie H, Waldron B (2011) Affordable energy solutions for developing communities. IEEE, pp 1–8
6. Mok SC (2011) New village gadgets synergy of human-powered generation and ultra-efficient electrical appliances for poverty eradication. IEEE Computer Society, pp 25–29
7. Megalingam RK, Nair LM, Viswanath M, Sugathan S (2012) In: IEEE global humanitarian technology conference pedalite: lighting up lives in un-electrified villages, pp 12–17
8. Barua S, Hasan MM (2014) Savings of electricity consumption through solar based recharging station for power assisted pedal rickshaw in Chittagong, pp 1–4
9. Kumar J, Garg N, Ali A (2016) Energy generation using paddling system with solar system. MIT Int J Mech Eng 6(1):5–8. ISSN 2230-7680
10. Marwali MKC, Haili MA, Shahidehpour SM, Abdul-Rahman KH (1998) Short term generation scheduling in photovoltaic-utility grid with battery storage. IEEE Trans Power Syst 13(3):1057–1062
11. Lu B, Shahidehpour M (2005) Short-term scheduling of battery in a grid-connected pv/battery system. IEEE Trans Power Syst 20(2):1053–1061
12. Hsu T-R, Yee RK, Le TT (2010) Hybrid-powered zero-emission vehicle using human, electric and solar power inputs. United States Patent
13. Mishra A, Kumar J (2019) Generation scheduling and power management of pedaling system with solar photovoltaic system. In: International conference on sustainable development, ICSD-2019 IJIERT, pp 12–17
14. Kalogirou S (2009) Solar energy engineering. Library of congress cataloging in publication data
15. Chandel M, Mathur S, Mathur A, Aggarwal GD (2014) Techno-economic analysis of solar photovoltaic power plant for garment zone of Jaipur city. Case Stud Thermal Eng 2, pp 1–7

Calculating Threshold Voltage Shift for Shallow Implanted Short-Channel MOSFET in Presence of High-K Dielectric

Rajarshi Dhar, Payel Halder, and Arpan Deyasi

Abstract Shift of threshold voltage is analytically measured for shallow implanted short-channel MOSFET considering HfO_2 as dielectric material. With suitable design parameters, depletion width is made greater than implant depth, and Poisson's equation is solved for threshold voltage computation under inversion condition. Simulated findings are compared with the ideal results obtained for sheet charge approximations and for different structural and doping parameters. For every design consideration, corresponding implant depth is also computed as a function of substrate ion concentration. Results are important for design of transistors under such doping condition in the submicron region.

Keywords Threshold voltage shift · Shallow implant · Implant depth · Depletion width · High-K dielectric

1 Introduction

In present day, ion implantation processed is favored for fabrication of low-dimensional devices [1, 2]; but with the shrinking in the dimension of devices [3], problems related to transistors falsely trigger themselves with unwanted/noise signals at the gate terminal. Many different techniques have been applied [4, 5] to prohibit the undesired behavior of the transistors when incorporated in the VLSI circuit. In present VLSI systems, local oxidation of silicon process (LOCOS) is being applied, but for the future processes like the ULSI or the GLSI where the dimension is less

R. Dhar · A. Deyasi (✉)
Department of Electronics and Comm Engineering, RCC Institute of Information Technology, Kolkata, India
e-mail: deyasi_arpan@yahoo.co.in

R. Dhar
e-mail: rajarshidhar1997@gmail.com

P. Halder
Department of Electronics and Comm Engineering, Dream Institute of Technology, Kolkata, India

V. Nath and J. K. Mandal (eds.), *Proceedings of the Fourth International Conference on Microelectronics, Computing and Communication Systems*, Lecture Notes in Electrical Engineering 673, https://doi.org/10.1007/978-981-15-5546-6_37

than 1 μm, LOCOS technique is not going to work satisfactorily [6]. Therefore, the need of novel techniques becomes the need of the hour [7, 8]. The trench isolation process is one of those processes already reported in various literatures [9, 10] which can satisfactorily overcome this problem so far.

For isolation of very high-voltage devices at the terminals/junctions, we have to create larger separation between two consecutive devices, which is impractical from the miniaturization of chip size point of view. In this context, the use of deep trench isolation [11, 12] significantly reduces the device size, hence reducing the device cost. Even though transistors are isolated in a circuitry by a deep trench structure, they start to become connecting between themselves by virtue of inter-capacitance property and thus greatly degrade the electrical performance of the circuit. The quality of the trench sidewall (i.e., the silicon/trench interface) is also extremely crucial in determining the reliability and leakage characteristics at interface [13, 14]. Interface quality is therefore experimentally evaluated by measuring the state density. This density is directly dependent on the implant profile, whether shallow depth or deep implant. In this context, it may be mentioned that if depletion region thickness (approximate) is greater than the depth of implant, then it will be called shallow implant; otherwise, it is called deep implant. Normally, deep implant is preferable in order to increase the drain current density [15, 16], and henceforth, shallow implant is less investigated.

The present paper examines the shift of threshold voltage in short-channel MOSFET for shallow implant profile in presence of high-K dielectric. Corresponding implant depth is also measured with variation of structural and design parameters. Computation is carried out using MATLAB® software. Novelty of the work lies in the fact that deep implant is generally considered for MOSFET design in order to achieve higher current density, but this also causes a possibility of burnout of the devices. The present work nullifies the effect where shallow implant exhibits better device control over threshold voltage. A few works on shallow implants are previously reported on bulk structure, and in the present work, effect is investigated on micron domain. Result is compared with the approximated profiles obtained using sheet charge density estimation.

2 Mathematical Formulation

Computation of implant depth begins with 1D Poisson's equation where field and potential profiles may be represented in the following forms

$$\xi_{R1}(z) = \frac{qN_A}{\varepsilon_{\text{sub}}}(W_{\text{dep}} - W_I) + \frac{qN_I}{\varepsilon_{\text{sub}}}(W_I - z) \tag{1}$$

$$\phi_{R1}(z) = -\frac{qN_A}{\varepsilon_{\text{sub}}}(W_{\text{dep}} - W_I)z - \frac{qN_I}{\varepsilon_{\text{sub}}}\left(W_I - \frac{z}{2}\right) \tag{2}$$

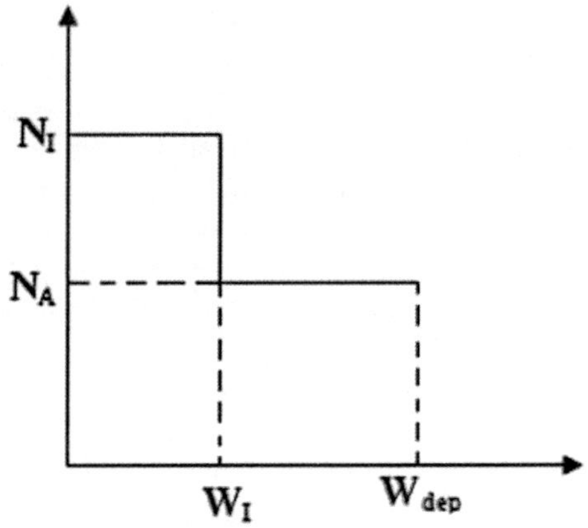

Fig. 1 Implant profile with depth

where W_{dep} denotes the width of depletion region with corresponding doping concentration N_A, and W_I indicates implant depth with N_I as doping concentration. These equations are for region 1 as indicated in Fig. 1.

In the region 2, corresponding electrical functions are written as

$$\xi_{R2}(z) = \frac{qN_A}{\varepsilon_{\text{sub}}}(W_{\text{dep}} - z) \tag{3}$$

$$\phi_{R2}(z) = \frac{qN_A}{2\varepsilon_{\text{sub}}}(W_{\text{dep}} - z)^2 \tag{4}$$

After computation of surface potential, final expression of threshold voltage may be written as

$$V_T = V_{FB} + 2\phi_F + \frac{1}{C_{\text{ox}}} \times \left[qD_I + \sqrt{2q\varepsilon_{\text{sub}}N_A\left(2\phi_F - \frac{qD_IW_I}{2\varepsilon_{\text{sub}}}\right)}\right] \tag{5}$$

Hence, the shift of threshold is given by

$$V_T = \frac{q(N_I - N_A)W_I}{C_{\text{ox}}}\left(1 - \frac{W_I}{2W_{\text{dep}}}\right) \tag{6}$$

3 Results and Discussions

Using Eq. (5), threshold voltage shift is computed as a function of substrate concentration for different internal and structural parameters. All the results are compared with the ideal characteristics obtained for sheet charge estimation.

In Fig. 2, shift of threshold is plotted for different dielectric materials, and the result is compared with that obtained for conventional SiO_2. Simulated findings showed that shift of threshold voltage increases with electrical permittivity and also increased with substrate concentration. This is due to the fact that higher permittivity increases the barrier potential, which enhances input voltage requirement to get drain current at linear condition. This enhances shift. It is also noted that for SiO_2 material, simulated results are almost similar to the ideal behavior obtained for sheet charge approximation, but the result differs for high-K dielectrics and with higher concentration.

Threshold voltage shift is also measured with different dielectric thickness, and the result is plotted in Fig. 3. From the plot, it is seen that lower the thickness, the shift decreases. From the result, it is found that for higher dielectric thickness, shift is relatively large as oxide capacitance decreases, whereas for lower thickness, reverse phenomenon takes place. Computations are carried out considering HfO_2 as the material.

Shift is also measured for different implant depth, and corresponding result is plotted in Fig. 4. This is important in the context that this depth is taken in such a

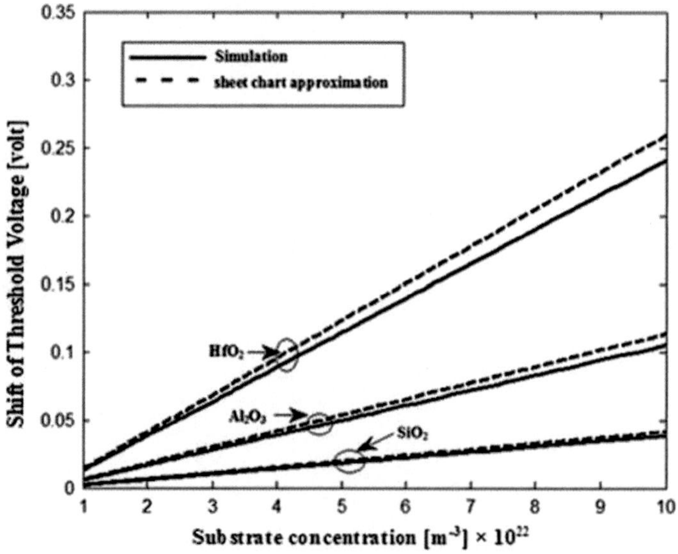

Fig. 2 Shift of threshold voltage with substrate concentration for different dielectrics

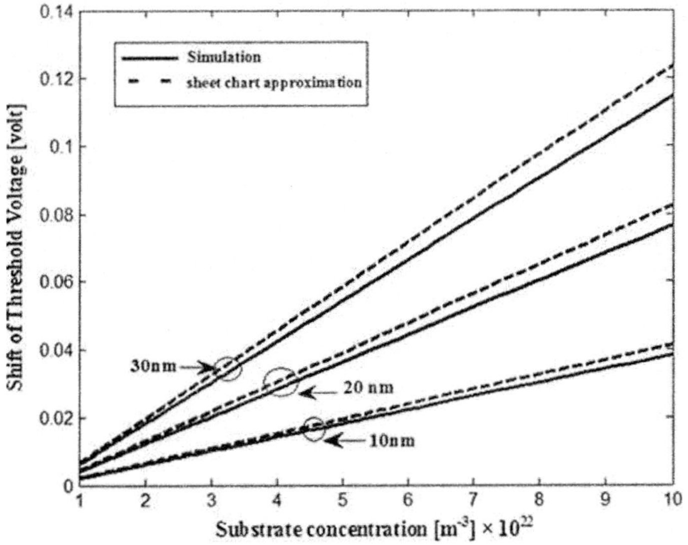

Fig. 3 Shift of threshold voltage with substrate concentration for different dielectric thickness

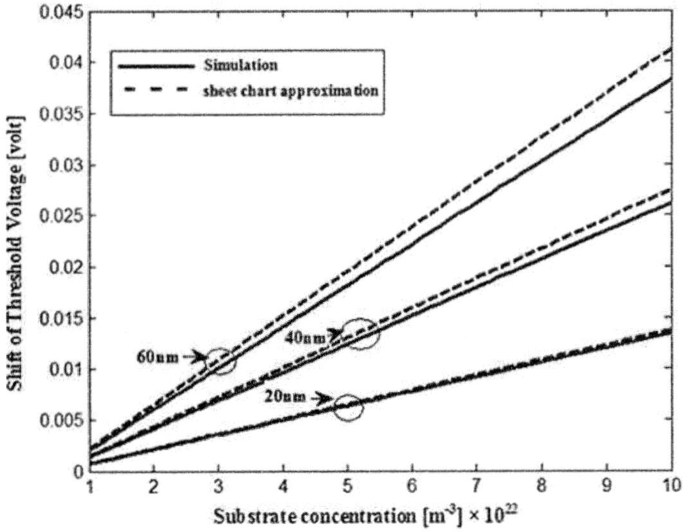

Fig. 4 Shift of threshold voltage with substrate concentration for different implant profile

way that it becomes less than the depletion depth. Design parameters are chosen in that way.

Higher implant depth makes larger shift of threshold as it becomes comparable with depletion width. This requires the need of depth profile measurement for different structural and internal design parameters. Figure 5 shows the implant profile with substrate ion concentration, and it is independent of the permittivity of the oxide material.

Figure 6 shows maximum depletion depth profile as a function of implant depth.

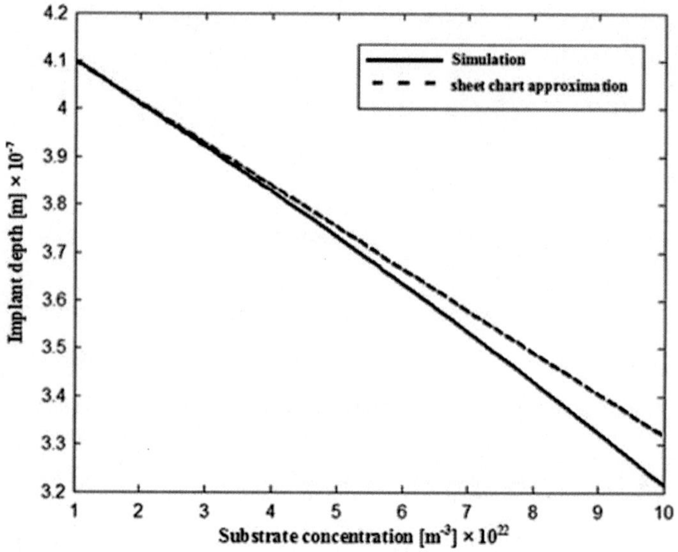

Fig. 5 Implant depth profile as a function of substrate ion concentration

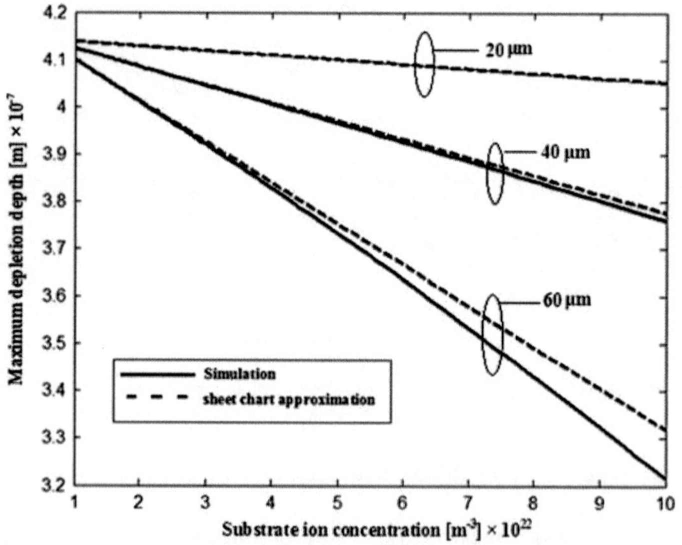

Fig. 6 Maximum depletion depth as a function of different implant depth

From the plot, it is seen that the higher the implant depth, the depletion depth decreases. This is a turnout form shallow implant to deep implant, as reduction of depletion depth indicates that it will become comparable at a particular magnitude. Also at larger depth, the value deviates from the ideal value obtained from sheet charge approximation.

Figure 7 shows the transition from hi-lo to lo-hi structure with change of doping concentration in the depletion region. Once it becomes comparable to the concentration of hi structure, then depletion becomes almost negligibly small, and also becomes constant with substrate concentration variation. Lowering depletion width indicates that the structure is now at the threshold of deep implant.

Table 1 shows a comparative study for implant depth where works are carried out for shallow implant profile. Table 2 contains the comparison for depletion depth.

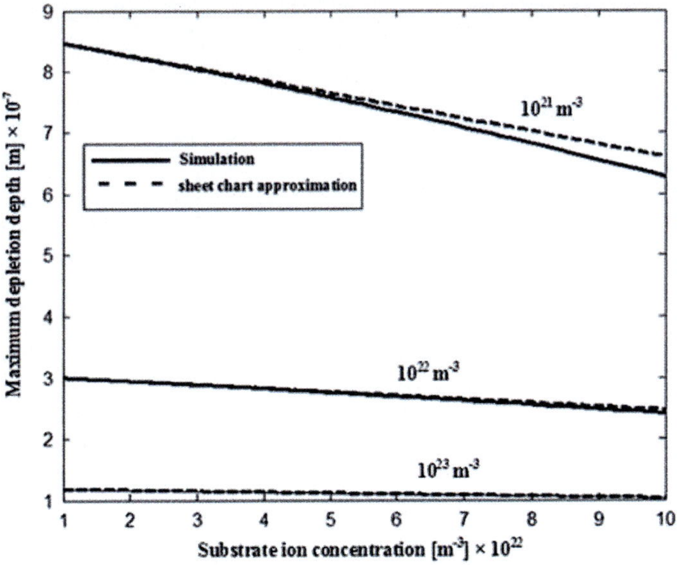

Fig. 7 Maximum depletion width for different implant concentrations

Table 1 Comparative study for implant depth

Data for implant depth	Substrate concentration (m^{-3})	Implant depth (μm)
Arora [17]	7×10^{22}	0.351
Present work	7×10^{22}	0.355

Table 2 Comparative study for depletion depth

Data for implant depth	Substrate concentration (m^{-3})	Implant depth (μm)	Depletion depth (μm)
Arora [17]	7×10^{22}	60	0.361
Present work	7×10^{22}	60	0.363

4 Conclusion

Simulated findings show that threshold voltage shift is higher for high-K dielectric material compared to conventional material when shallow implant is considered with hi-lo configuration. Also higher implant depth can turn the device from hi-lo to lo-hi structure with suitable choice of design parameters. Thickness of the dielectric material and background concentrations will play important role in tuning the device characteristics.

References

1. Upadhyay S, Mandal A, Chavan V, Subrahmanyam NBV, Bhagwat P, Chakrabarti S (2018) Impact of phosphorus ion implantation on the material and optical properties of InAs/GaAs quantum dots. In: Proceedings volume 10543: quantum dots and nanostructures: growth, characterization, and modeling XV; 1054309
2. Chee SW, Kammler M, Graham J, Gignac L, Reuter MC, Hull R, Ross' FM (2018) Directed self-assembly of ge quantum dots using focused Si_2 + ion beam patterning. Sci Rep 8:9361
3. Li WQ, Xiao XH, Stepanov AL, Dai ZG, Wu W, Cai GX, Ren F, Jiang CZ (2013) The ion implantation-induced properties of one-dimensional nanomaterials. Nanoscale Res Lett 8:175
4. Hu Y, Jiang H, Lau KM, Li Q (2018) Chemical vapor deposited monolayer MoS_2 top-gate MOSFET with atomic-layer-deposited ZrO_2 as gate dielectric. Semicond Sci Technol 33(4):045004
5. Hong MH, Perng DC (2017) Effects on selective epitaxial growth of strained-SiGe p-MOSFETs on various (001) Si recess structures. J Theor Appl Phys 11(4):313–317
6. Lee M, Cho S, Lee N, Kim J (2018) Radiation-tolerance analysis of I-gate n-MOSFET according to isolation oxide module in the CMOS bulk process. Microelectron Eng 200:45–50
7. Imamoto T, Ma Y, Muraguchi M, Endoh T (2015) Low-frequency noise reduction in vertical MOSFETs having tunable threshold voltage fabricated with 60 nm CMOS technology on 300 mm wafer process. Jpn J Appl Phys 54(4S):04DC11
8. Chang WH, Irisawa T, Ishii H, Uchida N, Maeda T (2019) Physical mechanisms of mobility enhancement in ultrathin body GeOI pMOSFETs fabricated by hetero-layer-lift-off technology. IEEE Trans Electron Devices 66(3):1182–1188
9. Berens J, Rasinger F, Aichinger T, Heuken M, Krieger M, Pobegen G (2019) Detection and cryogenic characterization of defects at the SiO_2/4H-SiC interface in trench MOSFET. IEEE Trans Electron Devices 66(3):1213–1217
10. Chahar S, Rather GM, Hakim N (2019) The effect of shallow trench isolation and sinker on the performance of dual-gate LDMOS device. IEEE Trans Electron Devices 66(1):585–591
11. Zhang J, Zou H, Zhao Q, Wang K, Li Z, Li Z, Zhang B (2018) High-voltage SOI deep trench LDMOS with quasi vertical super junction structure. In: 14th IEEE international conference on solid-state and integrated circuit technology, pp 1–3

12. Heider F, Janeschitz C, Haberjahn M, New CE, Chow KL, Roberts JW (2016) Poly recess depth in two different trench MOSFETs measured with scatterometry: AM: advanced metrology. In: 27th annual SEMI advanced semiconductor manufacturing conference, pp 50–53
13. Zhang CM, Jazaeri F, Borghello G, Faccio F, Mattiazzo S, Baschirotto A, Enz C (2019) Characterization and modeling of gigarad-TID-induced drain leakage current of 28-nm bulk MOSFETs. IEEE Trans Nucl Sci 66(1):38–47
14. Ferhati H, Djeffal F (2018) Graded channel doping junctionless MOSFET: a potential high performance and low power leakage device for nanoelectronic applications. J Comput Electron 17(1):129–137
15. Baishya S, Mallik A, Sarkar CK (2005) A subthreshold drain current model for deep submicron pocket implanted MOSFETs. In: International semiconductor device research symposium, pp 360–361
16. Kim HW, Seok O, Moon JH, Bahng W, Jo J (2017) Fabrication of 4H-SiC lateral double implanted MOSFET on an on-axis semi-insulating substrate without using epi-layer. Jpn J Appl Phys 56:120305
17. Arora ND (1987) Semi-empirical model for the threshold voltage of a double implanted MOSFET and its temperature dependence. Solid State Electron 30(5):559–569

Healthcare Recommendation System

Rakesh Raja, Indrajit Mukherjee, and Bikash Kanti Sarkar

Abstract In the era of digitalization of information, big data plays an important role in the field of healthcare industry. The reason is that it generates a huge amount of digital data due to the digitalization of healthcare system. This digital data can be compared with big data according to its characteristics and can be characterized as 'healthcare big data'. On analysing healthcare big data with various tools and techniques, it helps healthcare providers to provide precise and relevant predictions to the patients based on their medical history. The proposed model is an effort to develop a personalized healthcare recommendation system to predict the relevant recommendations to the healthcare providers for making healthcare decisions from the big data generated by the healthcare industries.

Keywords Big data · Health care · Personalized health care · Collaborative filtering · Prediction model · Linear regression · Decision tree · Bayesian classifier · Survival analysis

1 Introduction

In the past few years, healthcare system gives rise to a tremendous amount of data considered as healthcare big data as well as focuses on value-based treatment and personalized health care due to the digitalization of the world. In general, healthcare data means data related to health risks which incorporates electronic health records (EHRs) such as digital version of patient's information and medical history, doctor's prescription, diagnostic reports, biometric data and other health-related data. Due to the volume and diversity of medical data, big data in health care is overwhelming. By employing analytical tools on big data, it helps healthcare providers for healthcare

R. Raja (✉) · I. Mukherjee · B. K. Sarkar
Department of Computer Science and Engineering, Birla Institute of Technology, Mesra, Ranchi, India
e-mail: rajarakeshchauhan@gmail.com

V. Nath and J. K. Mandal (eds.), *Proceedings of the Fourth International Conference on Microelectronics, Computing and Communication Systems*, Lecture Notes in Electrical Engineering 673, https://doi.org/10.1007/978-981-15-5546-6_38

decision-making with the aim of promoting health and preventing disease. To be more specific, using big data analytical tools and techniques in health care has the potential to improve the quality of care, lifestyle habits and reduce costs by identifying associations and relationships between healthcare data [1].

The term 'health' refers to the condition of complete emotional and physical well-being. Healthcare services have been characterized as a large number of administrations rendered to people, families or networks by well-being administration experts for advancing, keeping up or re-establishing well-being. The quality of healthcare service helps people to maintain this optimal state of health. The data stored in healthcare systems has enhanced during the last ten years, driving to be considered as big data. There is an abundance of data prepared to be analysed [2].

The concept of healthcare big data refers to the electronic medical records (EMRs) created by digitization of information-too large and complex for traditional software and/or hardware to make sense of [3]. According to author in [3], the healthcare system has traditionally generated massive amounts of information by maintaining medical records, compliance and regulative needs and patient care. This large amount of knowledge holds the promise of supporting an excellent range of medical and aid features, as well as clinical decision support, health management, monitoring of diseases, health condition based on sensors, etc. [4–6]. For example, cancer diagnosis requires petabytes of data from different sources to identify the patient's condition and potential for survival. In addition, the use of analytical tools and techniques on big data in health care today reduces healthcare costs while improving the quality of health care by underlining more customized and preventive care and ongoing monitoring [7]. As according to the study of [8], big data analytics consists of a set of analytical methodologies and tools like correlations, cluster analysis, regression analysis, decision trees, neural network analysis, Bayesian analysis and so on.

Health information, however, needs to be open and accessible to everyone concerned within the healthcare system to satisfy the above-mentioned healthcare services. In this regard, according to the study of [9] high-level data integration, interoperability and sharing among various healthcare providers are essential to deliver high-quality, secure healthcare services to the patients they serve [9]. Big data-driven methodology plays a critical role in the development of personalized healthcare services. There are many diseases with risk factors that can be prevented or have a minimum of risk indicators. Enlightening these characteristics of the disease can help to reduce the burden of disease in personalized health care. However, the possible aggregate of threat factors is so complicated and it is far impossible for a medical doctor to completely examine it in real time at the time of patient interaction. Stakeholders associated with healthcare system take careful medical history and perform particular research facility testing to work outpatient health and risk of further disease.

2 Motivation

In the era of digitalization of information, healthcare service shifts from disease-centric version to a patient-centric version. Health practitioner's decision-making focused across the clinical expertise and information from medical proof and varied tests in a disease-centric version. Alternatively, in a patient-centric version, patient effectively takes an interest in their own consideration and acquires service focused on individual desires and preferences knowledgeable by the usage of advice and supervision of their healthcare providers. The proposed work is driven by patient-centric version developing a customized risk profile for disease as well as health plan and wellness plan for a personal.

3 Problem Statement

The three principle questions emerge in order to develop a patient-centric version for personalized healthcare system:

- What are my prospects for disease development?
- How do I deal with them?
- Which health strategies can work best for me?

4 Methodology

This drawback can be addressed by collaborative filtering methodology, according to the research study of Su et al. [10]. The fundamental rule of collaborative filtering is that patients having similar health conditions are possible to possess similar characteristics on different diseases. This statistic is then used to develop personalized management system for health care in addition to generate predictions on different diseases supported a group of other similar patients.

5 Proposed Model

Collaborative filtering approach is used for the development of a patient-centric prediction model. The goal is to develop a general prediction healthcare recommendation system to offer a recommendation for acting a specific interest with a purpose to improve the health of patient, based totally on his health condition and information from another patients' history. The point of the proposal is to find which activities individually affect the value of each parameter of health. The recommendation algorithm proposed depends among the estimations of the health parameters and also the

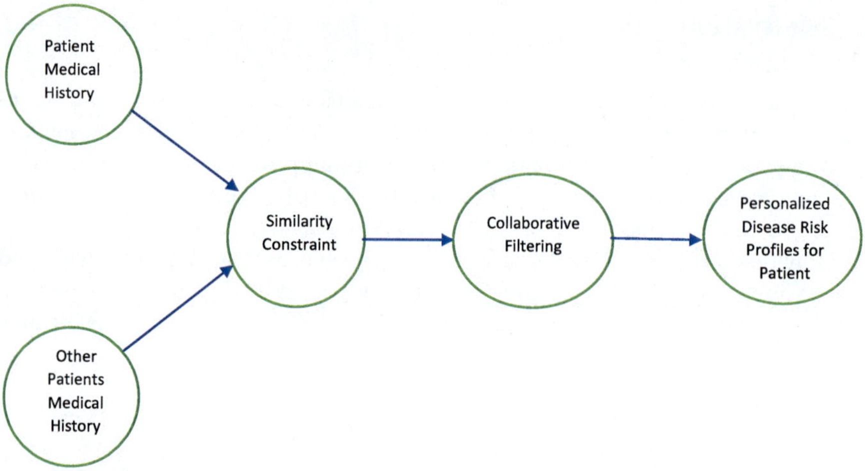

Fig. 1 Recommendation system for personalized healthcare

physical activities of the clients. The essential thought is to discover which physical activities influence changes in the estimation of health parameters. The algorithm continues to use this dependence to acknowledge the identical or similar health situations that are found in any other person with similar characteristics. If there is information in the clients' history that once doing some physical activity their health condition has progressed, the algorithm accepts this information and suggests the activity to other clients with similar health issues. In order to accomplish this, collaborative filtering and classification techniques are carried out so as to cluster clients with similar characteristics in the proposed recommendation algorithm. Such utilization of classified knowledge provides significant recommendations based on prior information of patients with similar health conditions. As a result, this recommendation system influences on enhancing the health condition, resulting in a better lifestyle for the person. Figure 1 shows the recommendation system for personalized disease prediction process.

6 Collaborative Filtering Methodology

The collaborative filtering approach is used to construct a model from the past medical history of a patient as well as similar health scenario of other patients with similar characteristics. This model is then used to predict the disease based totally on a group of other comparable patients through constructing a personalized recommendation system on the patient-centric version. Additionally, collaborative filtering approach uses content-based filtering for communication between different patients and generating totally different recommendation and suggestion to the clients. This system requires all time to monitor the condition of the patients and recommending possible emergency services for their medical conditions. Also, the framework empowers the patient to contact other individuals with comparative health condition and trade their experience.

7 Recommendation Algorithm

The motivation behind recommendation algorithm is to present a suggestion for playing out a particular activity so that it will enhance patient's prosperity, in perspective on his given health circumstances and medical records of another patients' history. In the scenario of health care, it also searches out which carrying activities have an impact on the worth of every parameter of health care. The recommendation algorithm presented belongs to the collaborative model of healthcare system [11]. Figure 2 shows the various stages of recommendation algorithm.

Phases of Recommendation Algorithm The recommendation algorithm for personalized healthcare system comprises of four essential stages:

1. Classification of patients as per their diagnosis and separating all clients that do not have a place with a similar class with the active patient.
2. Selection of patients closest to the active patient as per the health record with the use of collaborative filtering.
3. Calculate the utility of the behaviour of active patient with his comparable patients while using their health profiles.
4. Recommendation generated by measuring the effectiveness of utility activities.

1. **Categorization of patients according to diagnosis**: The first phase of the recommendation algorithm is to classify the patients according to diagnosis. Patients having the similar permissible activities with the same diagnosis are classified to cluster clients. We need expert knowledge in this stage in order to characterize the classes and the membership conditions in those training. This knowledge might be utilized by some classification rules as a training set so as to assign unlabelled samples to classes.
2. **Choice of the most comparable clients**: The second stage of recommendation algorithm is utilized to choose clients who have a place with a class similar to the active client and are very similar in terms of their health state to the active

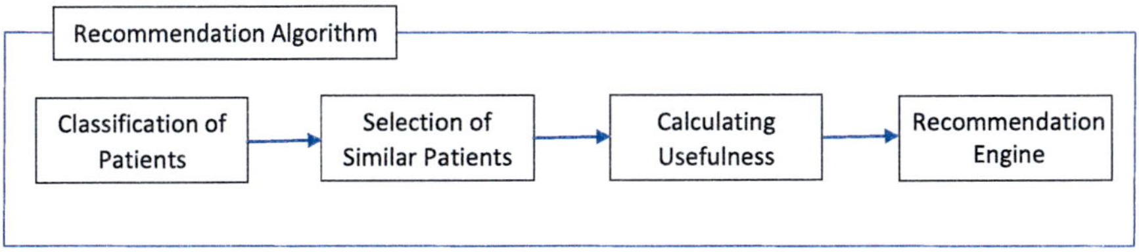

Fig. 2 Phases of recommendation algorithm

clients. The health information of the selected client would assist to obtain better recommendations.

3. **Estimate the value of the health parameters**: The third stage of recommendation algorithm is the most critical stage where we determine the utility of the activities. During this stage for each comparable patient, we tend to calculate the utility of every lifestyle activity of the active patient based on its present health condition so that we can improve the value of the health parameters as well as health care of an individual.
4. **Generation of recommendations**: This stage is concerned with recommending the activity that most of the comparable patients consider to be the most valuable.

8 Clinical Prediction Model for Survival Analysis

Clinical prediction model is one of the most critical components of healthcare analytics that has a massive effect on disease diagnosis and treatment. Survival analysis [12] aims at modelling the time to event data. The observation starts from a particular starting time and will continue until the occurrence of a certain event. The starting point for the assessment in the healthcare context is typically a formal medical intervention such as referral to hospitalization, starting to take a specific medication or a diagnosis of particular disease. The event of interest might be death, discharge from hospitalization or any other interesting incident that can happen during the observation period. The survival model in statistics can be applied to predict the patient's survival outcomes [13]. These outcomes can fall into any one of these categories:

- Continuous outcomes
- Binary outcomes
- Categorical outcomes
- Ordinal outcomes
- Survival outcomes

These outcomes are shown in Table 1.

Basic Statistical Prediction Model
The well-known basic statistical models that are widely used in clinical data analysis include:

Linear Regression
In clinical data analysis, linear regression is often employed in clinical cost prediction and the estimation of some medical inspection.

Generalized Additive Model
To model the continuous outcomes in regression, the generalized additive model is used. It is a variant of linear regression that can handle nonlinear distribution.

Table 1 Categorization of survival outcome

Outcome	Method	Application
Continuous outcome	• Linear regression • Generalized additive method	• Medical cost prediction • Estimation of some medical inspection
Binary outcome	• Logistic regression • Decision tree • Bayesian model	• Disease diagnostic • Prediction of patient's death or risk • Medical image segmentation
Categorical outcomes	• Polytomous logistic regression	• Multiple disease diagnostics
Ordinal outcomes	• Bayesian classifier	• Predict the grade/severity of illness
Survival outcomes	• Survival analysis	• Predict the time to event of interest

Logistic Regression
Logistic regression is one of the most popular binary classification methods which is widely adopted for clinical prediction tasks.

Bayesian Classifier
The main function of Bayesian classifiers is modelling probabilistic relationships between the attributes set and the class variable. This approach is commonly studied in the context of clinical prediction.

Decision Tree
Decision tree is a general learning technique that is utilized in a wide range of healthcare diagnosis systems.

Polytomous Logistic Regression
Polytomous logistic regression is an extension of the basic logistic regression, which is designed to handle multiclass problems. Polytomous logistic regression is used when there is no predefined order among the categories.

9 Conclusions and Future Research

The aim of this study was to highlight the big data in health care which plays a crucial role to improve the healthcare services. The systematic analytics of big data in health care help healthcare providers to provide precise and relevant recommendations to the patients based on their medical history. The proposed model is an effort to develop a personalized healthcare recommendation system to predict the relevant recommendations to the healthcare providers for making healthcare decisions from the big data generated by the healthcare industries. In the era of digitalization of information, healthcare service shifts from disease-centric version to a patient-centric version. The enormous data in health care has many challenges such as aggregating unstructured healthcare data from disparate data sources, privacy concern and security of healthcare big data. These challenges can be the future studies to be addressed.

References

1. Sarkar BK (2017) Big data for secure healthcare system: a conceptual design. Complex Intell Syst 3:133–151
2. Mohapatra S, Murarka S (2016) Improving patient care in hospital in India by monitoring influential parameters. Int J Healthc Manag
3. Raghupathi W, Raghupathi V (2014) Big data analytics in healthcare: promise and potential. Health Inf Sci Syst 2(3):2047–2501
4. Raghupathi W (2010) Data mining in health care. In: Kudyba S (ed) Healthcare informatics: improving efficiency and productivity, pp 211–223
5. G Dembosky A (2012) Data prescription for better healthcare. Financ Times 2012:19–22
6. Feldman B, Martin EM, Skotnes T (2012) Big data in healthcare hype and hope. Dr. Bonnie 360. http://www.west-info.eu/files/bigdata-inhealthcare.pdf
7. Fernandes L, O'Connor M, Weaver VJ (2012) Big data, bigger outcomes. AHIMA 2012:38–42; 7. CCC2011c (2011) Smart health and wellbeing. Computing Community Consortium. Springer, Berlin
8. Davis CK (2014) Beyond data and analytics. Commun ACM 57:39–41
9. Alnuem M, EL-Masri S, Youssef A, Emam A (2011) Towards integrating national electronic care records in Saudi Arabia. In: International conference on bioinformatics and computational biology, Monte Carlo Resort, Las Vegas, Nevada, 18–21 July
10. Su X, Khoshgoftaar TM (2009) A survey of collaborative filtering techniques. Adv Artif Intell 4:421–425
11. Kulev I et al (2013) Recommendation algorithm based on collaborative filtering and its application in healthcare. CIIT
12. Rupert G, Miller Jr (2011) Survival analysis, vol 66. Wiley
13. Marubini E, Valsechhi MG, Analysing survival data from clinical trials and observational studies. Wiley, 414 pages. ISBN 0-971-93987-0

Comparative Study of Logic Performance of Hybrid CMOSFETs at Deca-Nanometer Regime

Suchismita De, Suchismita Tewari, and Abhijit Biswas

Abstract A comparative analysis is carried out to study the digital circuit behavior of hybrid p-Ge/n-Si CMOSFETs, p-Si/n-InGaAs CMOSFETs as well as conventional Si CMOS devices at channel length of 60, 30, and 20 nm. Extensive numerical investigation is used to investigate the performance of the different inverter circuits in terms of noise margin low and high, rise and fall times alongside propagation delay. Our studies show that hybrid CMOSFET comprising Si nMOS and Ge pMOS devices performs the best with respect to noise margin high (NM_H), noise margin low (NM_L), and rise time (t_r) compared to the hybrid CMOSET comprising InGaAs nMOS and Si pMOS devices and the conventional CMOSFET. However, n-InGaAs/p-Si CMOSFET yields the lowest value of fall time (t_f) and time delay per inverter (t_d) in relation to the values found in other two inverters. Our investigation reveals that all the time parameters for both hybrid CMOSFETs show reduced value compared to the conventional Si counterpart while noise margins show improvement for Si CMOSFETs.

Keywords Fall time · Hybrid CMOSFET · Logic performance · Noise margin · Propagation delay · Rise time

1 Introduction

Researchers all over the world are continuously working on the improvement of the performance of integrated circuits (ICs) over the past few decades. Considerable

S. De (✉) · S. Tewari · A. Biswas
Department of Radio Physics and Electronics, University of Calcutta, Kolkata, India
e-mail: suchismitade@ymail.com

S. Tewari
e-mail: suchi_tewari@yahoo.co.in

A. Biswas
e-mail: abiswas5@rediffmail.com

V. Nath and J. K. Mandal (eds.), *Proceedings of the Fourth International Conference on Microelectronics, Computing and Communication Systems*, Lecture Notes in Electrical Engineering 673, https://doi.org/10.1007/978-981-15-5546-6_39

improvement has been achieved as a result of aggressive scaling down of device dimensions as well as exploration of novel materials with high channel mobility for the construction of CMOSFET inverters. Stacks of various new materials are also being experimented with, apart from the conventional CMOS device with Si channel nMOS and Si channel pMOS, in order to achieve performance improvement in CMOS devices which will also bring about further extension of Moore's law.

Among the high mobility channel materials that are being explored, Ge with a hole mobility of around 1900 cm^2/Vs and InGaAs with an electron mobility of around 8450 cm^2/Vs present themselves as an attractive choice for *p*-channel and *n*-channel materials, respectively. Work has also been done with the combination of these materials using InGaAs channel nMOSFET and Ge channel pMOSFET so as to bring about a considerable enhancement in their performance [1, 2]. The digital behavior of hybrid n-InGaAs/p-Ge CMOS inverters has been reported in [3, 4]. Notably, in such hybrid CMOSFETs, the electron-to-hole mobility ratio is almost 10. As a result, in order to attain the VTC curve with symmetry in such inverters, the optimum width ratio of pMOS and nMOS devices ($W_p/W_n = r$) would be equal to 10. This design leads to a larger chip area when compared with conventional Si CMOS inverters in which the optimum ratio r is 3. This problem has been addressed in our previous work [5] by constructing a hybrid CMOS device incorporating a Ge pMOSFET and a Si nMOSFET where $r = 1$ and hence occupying the least chip area possible for improved logic performance. The advantage of outstanding electron mobility in InGaAs has also been taken into account by constructing a CMOS amplifier consisting of InGaAs nMOSFET and Si pMOSFET in our previous work [6] as a contender to the conventional CMOSFET. In this case, Si has been used for the pMOS device instead of Ge due to its considerable suppression of short-channel effects like lower DIBL in contrast to Ge, which leads to a higher output resistance and also low leakage. Although there have been many reports regarding digital performance of hybrid CMOSFETs, a comprehensive investigation on various hybrid CMOSFETs at different gate lengths have not been reported.

In this paper, the digital performance of such hybrid devices is investigated, and a comparative analysis of two different hybrid CMOSFETs comprising Si channel nMOS and Ge channel pMOS devices, InGaAs channel nMOS and a Si channel pMOS devices is performed and compared to a Si CMOSFET at three different channel lengths, e.g., 60, 30, and 20 nm. The performance analysis is conducted in terms of noise margins, rise and fall times as well as propagation delay of a single-stage inverter.

2 Device Structure and Simulation Setup

A comparative analysis of the logic performance of a hybrid CMOSFET consisting of Si nMOSFET and Ge pMOSFET, a hybrid CMOSFET comprising InGaAs nMOS and Si pMOS device, and a conventional CMOSFET comprising Si nMOS as well as pMOS device is carried out. The corresponding device schematics are illustrated in

Fig. 1a–c, respectively. The detailed process flow for the Ge pMOSFET may be found in [7] and the elaborate process flow for realization of the InGaAs MOSFETs on Si substrate with Al_2O_3 box layers is described in [8]. The Si CMOSFETs may also be realized using the direct wafer bonding technique, as was demonstrated successfully by Tagaki et al. in [3]. In this study, three channel lengths viz 60, 30, and 20 nm are considered.

ATLAS from SILVACO [9] is used for simulation of the three different MOSFETs. In the simulation, the interface-trapped charge density (D_{it}) has been considered to be 1×10^{12} eV^{-1} cm^{-2} for Ge/high-k interfaces as per experimental findings reported in Ref. [10], as well as for the $In_{0.7}Ga_{0.3}As$/high-k interface, as per experimental findings reported in Ref. [11], whereas in case of high-k/Si interface the D_{it} value

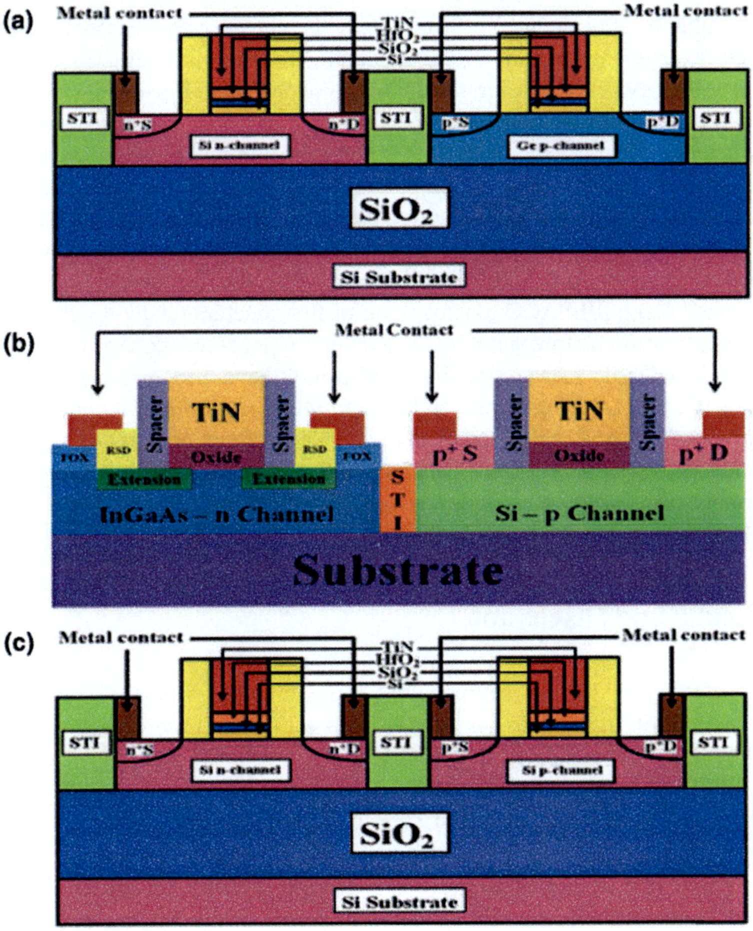

Fig. 1 Cross-sectional view of CMOSFETs comprising **a** Si nMOSFET and Ge pMOSFET, **b** InGaAs nMOSFET and Si pMOSFET, and **c** Si nMOSFET and Si pMOSFET

is considered as 1×10^{11} eV^{-1} cm^{-2}. The source/drain (S/D) contact resistance is considered as 82 Ω μm [12] for Ge pMOS device, 93 Ω μm [13] for $In_{0.7}Ga_{0.3}As$ nMOS device, and 30 Ω μm [14] for Si nMOS and pMOS devices, respectively. In order to take into account carrier recombination events in Ge, InGaAs, and Si channels, the Shockley-Read-Hall (SRH) and Auger recombination models have been employed. The inversion carrier mobility is a function of multiple parameters which include carrier concentration in the channel, temperature, horizontal and vertical electric fields, surface roughness between oxide and semiconductor, etc. Importantly for low band gap semiconductors such as Ge and InGaAs, band-to-band tunneling and trap assisted tunneling current play vital role in determining the total drain current. To capture all these physical effects, the relevant models are invoked in the present simulation. Since the channel thickness of both Ge and InGaAs MOSFETs is comparable with the exciton Bohr radius of the corresponding material, quantum-mechanical effects impact the current voltage characteristics of such devices. Hence, quantum effects are included while simulating Ge and InGaAs channel MOS devices. However, the quantum-mechanical effects are not significant for Si channel MOSFETs having channel thickness above 5 nm. Hence, in the present study quantum effects are not taken into account while simulating Si devices. For the computation of carrier concentration distributions, Fermi–Dirac (FD) statistics has been used. An important phenomenon in case of InGaAs MOSFETs is the velocity overshoot effect and has been included by invoking energy balance model for InGaAs nMOS device simulation. At the circuit-level investigation, mixed-mode simulation is performed using SILVACO to extract various parameters related to digital circuit applications for three aforesaid CMOS inverters in push–pull configuration. All the nMOS and pMOS devices comprising CMOSFETs have the capacitance equivalent thickness (CET) of 2.3 nm, 1 nm, and 0.8 nm for $L_g = 60$ nm, 30 nm, and 20 nm, respectively, following ITRS roadmap [15]. The work functions of the gate materials of n- and p-MOSFETs are tuned such that the threshold voltage values of 0.2 and −0.2 V may be set for n- and pMOS devices, respectively.

3 Results and Discussion

In order to study the logic circuit behavior of the three different CMOS devices, our device simulation framework has already been validated in our earlier publications [5, 16] by making a comparison of the transfer characteristics of Ge device that has been simulated, with the experimental characteristic curve at $L_g = 30$ nm [6], as well as that of the InGaAs device at $L_g = 60$ nm [17]. The three different devices are named as follows for future reference: CMOS1—hybrid CMOSFET consisting of a Si nMOS and a Ge pMOS devices, CMOS2—hybrid CMOSFET comprising InGaAs channel nMOS and a Si channel pMOS devices, and CMOS3–CMOSFET comprising Si channel nMOS as well as pMOS devices.

First of all, the voltage transfer characteristic (VTC) curves of the three different devices are studied, and the corresponding results for the devices with channel length

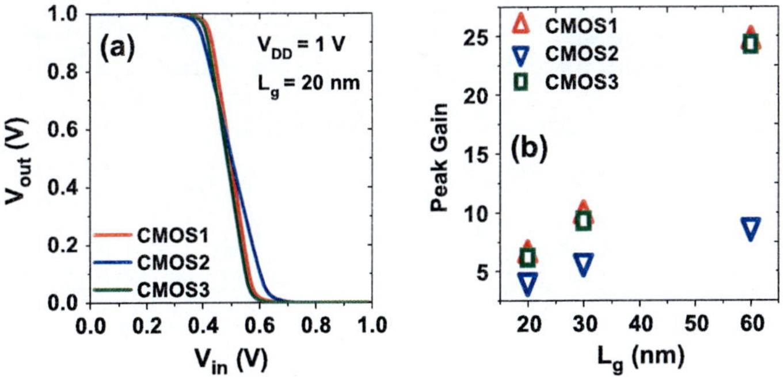

Fig. 2 Comparison of **a** VTC curves and **b** peak gain of CMOS1, CMOS2, and CMOS3 inverters at a channel length of 20 nm

of 20 nm have been plotted as shown in Fig. 2a. Also the peak gain obtained from the corresponding VTC curves for the three different devices has been plotted as shown in Fig. 2b. The most symmetric VTC curves for all the three devices are plotted which has been obtained at a device-width ratio (W_p/W_n) of 1 for CMOS1, 5 for CMOS2, and 3 for CMOS3, as the symmetry of a VTC curve is dependent on the effective carrier mobilities in the nMOS and pMOS devices, respectively, constituting the inverter which has been calculated to obtain the above W_p/W_n ratios. It can be observed from Fig. 2 that the sharpest VTC curve is obtained for CMOS1 followed by CMOS3, and finally, the VTC curve of CMOS2 has the least sharpness. The gain of the inverter is expressed as $[(g_{mn} + g_{mp})/(g_{dn} + g_{dp})]$ [18], where $g_{dn/p}$ and $g_{mn/p}$ denote the output conductance and transconductance of n/pMOS devices, respectively. This gain is an indication to the amount of steepness of the VTC curve. Due to the higher mobility of holes in the Ge pMOS device, it yields a larger transconductance (g_{mp}), which in turn results in the gain being enhanced and hence the high sharpness of VTC curve in CMOS1 as compared to CMOS3. On the contrary, in case of CMOS2, a very high electron mobility in the InGaAs nMOS device resulting in a high value of g_{mn} is however compensated by the high output conductance (g_{dn}) of nMOS device, thus producing a reduced gain and hence less sharpness than both CMOS1 as well as CMOS3. This trend has been followed by the devices having channel lengths of 30 nm and 60 nm as well. The low noise margin (NM_L) and high noise margin (NM_H) of all the three devices for the three different channel lengths have been compared as plotted in Fig. 3a, b, respectively. Here, it can be observed that since the devices having the most symmetric VTC curves pertaining to their corresponding optimum W_p/W_n ratios are considered, the values of NM_L are almost equal to the values of NM_H for all the three devices. It can be seen that the noise margins follows the same trend as explained above with CMOS1 possessing the highest value followed by CMOS3 and CMOS2. Next, a train of pulse with a frequency 5 GHz and amplitude 1 V in case of devices with $L_g = 20$ and 30 nm and 1.3 V in case of the device with $L_g = 60$ nm is fed to the input of the different

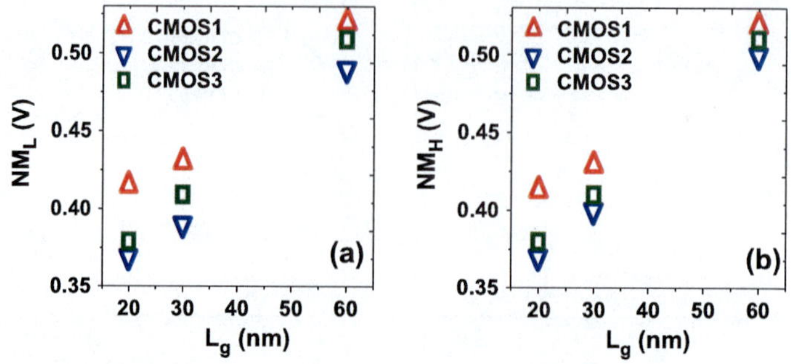

Fig. 3 Comparison of **a** noise margin low (NM_L) and **b** noise margin high (NM_H) of CMOS1, CMOS2, and CMOS3 inverters for channel lengths of 20, 30, and 60 nm

inverters, and the corresponding output waveforms have been obtained using mixed-mode simulation. The comparison of rise time (t_r) as well as fall time (t_f) between the three different devices for a channel length of 20 nm has been plotted as shown in Fig. 4a, b, respectively.

As the rise times are controlled by the pMOS devices, it can be seen clearly that CMOS1 has the least t_r followed by CMOS2 and CMOS3, which is attributed to the higher hole mobility of Ge in the pMOS device in CMOS1 as compared to the Si pMOS devices in CMOS2 and CMOS3, respectively. However, it can be observed that t_r of CMOS2 is lower than that of CMOS3. This occurs due to the reduction in t_r with the increase in W_p/W_n ratio which is 5 in case of CMOS2 and 3 for CMOS3. Since the width of the pMOS device increases, the channel resistance reduces resulting in a reduction in RC time constant and hence the obtained results. The fall times on the other hand are controlled by the nMOS devices, and it can be clearly seen that the lowest t_f is possessed by CMOS2 which can be attributed to the higher mobility of InGaAs in the nMOS device, whereas both CMOS1 and CMOS2 has similar values of t_f as the width of both the Si nMOS devices are same. The comparison of the rise times as well as the fall times for the different devices for all the three channel

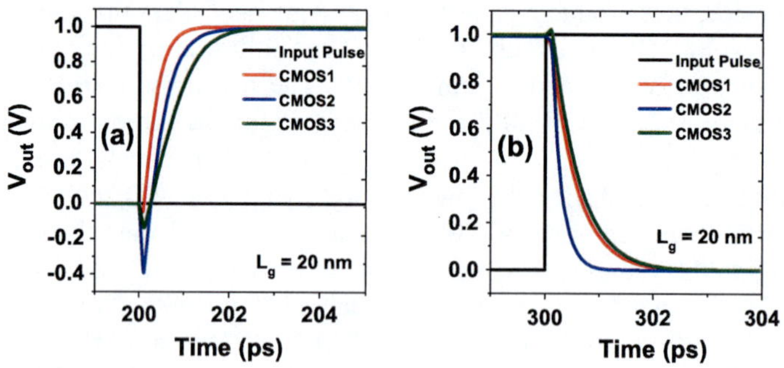

Fig. 4 Comparison of **a** rise time and **b** fall time of CMOS1, CMOS2, and CMOS3 inverters at a channel length of 20 nm

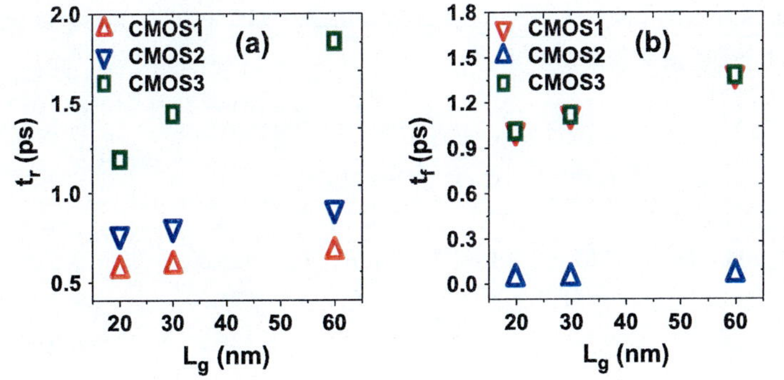

Fig. 5 Comparison of **a** rise time and **b** fall time of CMOS1, CMOS2, and CMOS3 inverters for channel lengths of 20, 30, and 60 nm

lengths have been plotted as shown in Fig. 5a, b, respectively. It can be seen that the above trend has been maintained for the devices of all the channel lengths. Also the propagation delay per inverter (t_d) is calculated which is given by ($t_r + t_f$)/2. The values of t_d for the different devices for three different channel lengths are shown in Fig. 6. One can easily observe that t_d follows the same trend as found for t_r and t_f. Finally, a comparative study of the logic parameters of CMOSFETs in our work is done with those of other research groups for channel length of 30 nm as shown in Table 1.

Here, Ref. [16] reports the logic parameters of a CMOSFET comprising of an InGaAs nMOSFET and a Ge pMOSFET, whereas Ref. [19] shows the logic parameters of a TFET inverter as well as a CMOS inverter. It is evident from Table 1 that our proposed hybrid devices perform better compared to earlier reported devices in terms of digital device parameters.

Fig. 6 Comparison of time delay per inverter of CMOS1, CMOS2, and CMOS3 inverters for channel lengths of 20, 30, and 60 nm

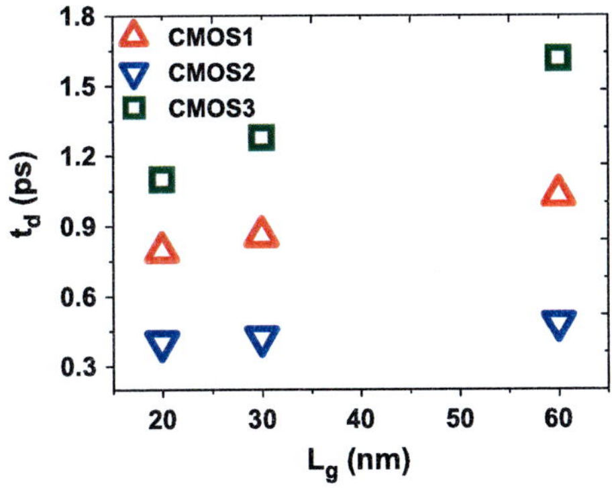

Table 1 Comparative study of logic parameters of different CMOSFETs at channel length of 30 nm

Parameters	Different CMOSFETs					
	This work			Ref. [16]	Ref. [19] TFET inverter	Ref. [19] CMOS inverter
	CMOS 1	CMOS 2	CMOS 3			
NM_L (V)	0.43	0.39	0.409	0.283	1	0.17
NM_H (V)	0.429	0.4	0.41	0.281	0.6	0.16
t_r (ps)	0.596	0.81	1.44	1.6	125.00	600.00
t_f (ps)	1.12	0.041	1.12	0.5	1524.00	900.00
t_d (ps)	0.858	0.426	1.28	1.05	824.5	750.00

4 Conclusion

The comparative study of two different hybrid devices comprising Si nMOSFET and Ge pMOSFET (CMOS1) and InGaAs nMOSFET and Si pMOSFET (CMOS2) with reference to a conventional Si CMOSFET (CMOS3) has been presented at gate lengths of 60, 30, and 20 nm. Our analysis reveals that the highest values of noise margin low, noise margin high as well as peak gain are obtained for CMOS1 in relation to those found for CMOS3 and CMOS2 with all the three channel lengths. While computing the rise time of the inverters, it has been observed that CMOS1 possesses the least rise time followed by CMOS2 and CMOS3, whereas in case of fall time, CMOS2 has the least value while CMOS1 and CMOS3 have similar values which are higher than that of CMOS2. Furthermore, with respect to the time delay per inverter, it is found that CMOS2 outperforms the rest followed by CMOS1 and CMOS3. Our comparative analysis suggests that hybrid CMOSFETs comprising high mobility channel materials prove themselves as promising candidates for improved digital circuit performance for deca-nanometer technology nodes.

References

1. Xue F, Jiang A, Zhao H, Chen YT, Wang Y, Zhou F, Lee J (2012) Sub-50-nm In0.7Ga0.3As MOSFETs with various barrier layer materials. IEEE Electron Device Lett 33(1):32–34
2. Lin D, Brammertz G, Sioncke S, Fleischmann C, Delabie A, Martens K, Bender H, Conard T, Tseng WH, Lin JC, Wang WH, Temst K, Vatomme A, Mitard J, Caymax M, Meuris M, Heyns M, Hoffmann T (2009) Enabling the high-performance InGaAs/Ge CMOS: a common gate stack solution. In: Proceedings of international electron devices meeting, pp 327–330
3. Takagi S, Yokoyama M, Kim S-H, Zhang R, Suzuki R, Taoka N, Takenaka M (2013) III-V/Ge CMOS device technologies for high performance logic applications. ECS Trans 53(3):85–96
4. Irisawa T, Oda M, Kamimuta Y, Moriyama Y, Ikeda K, Mieda E, Jevasuwan W, Maeda T, Ichikawa O, Osada T, Hata M, Tezuka T (2013) Demonstration of InGaAs/Ge dual channel CMOS inverters with high electron and hole mobility using stacked 3D integration. In: Symposium on VLSI technology digest of technical papers, pp T56–T57

5. De S, Tewari S, Biswas A, Mallik A (2017) Impact of channel thickness and spacer length on logic performance of p-Ge/n-Si hybrid CMOSFETs for ULSI applications. Superlattices Microstruct 109:316–323
6. Tewari S, De S, Biswas A, Mallik A (2017) Impact of sidewall spacer on n-InGaAs devices and hybrid InGaAs/Si CMOS amplifiers in deca-nanometer regime. Microsystem Technologies
7. Hutin L, Royer CL, Damlencourt JF, Hartmann JM, Grampeix H, Mazzocchi V, Tabone C, Previtali B, Pouydebasque A, Vinet M, Faynot O (2010) GeOIpMOSFETs scaled down to 30-nm gate length with record off-state current. IEEE Electron Device Lett 31(3):234–236
8. Kim SH, Yokoyama M, Nakane R, Ichikawa O, Osada T, Hata M, Takenaka M, Takagi S (2013) Strained extremely-thin body $In_{0.53}Ga_{0.47}As$-on-insulator MOSFETs on Si substrates. In: Proceedings of VLSI symposium, pp T50–T51
9. ATLAS User's Manual (2012) A 2D device simulation software package. SILVACO Int, Santa Clara
10. Grandchamp B, Jaud M-A, Scheiblin P, Romanjek K, Hutin L, Le Royer C, Vinet M (2011) In-depth physical investigation of GeOI pMOSFET by TCAD calibrated simulation. Solid State Electron 57(1):67–72
11. Wang LS, Xu JP, Liu L, Lu HH, Lai PT, Tang WM (2015) Interfacial and electrical properties of InGaAs metal-oxide-semiconductor capacitor with TiON/TaON multilayer composite gate dielectric. Appl Phys Lett 106(12):123504–1–123504–4
12. van Dal MJH, Vellianitis G, Duriez B, Doornbos G, Hsieh C-H, Lee B-H, Yin K-M, Passlack M, Diaz CH (2014) Germanium p-channel FinFET fabricated by aspect ratio trapping. IEEE Trans Electron Dev 61(2):430–436
13. Yonai Y, Kanazawa T, Ikeda S, Miyamoto Y (2011) High drain current (>2A/mm) InGaAs channel MOSFET at $V_D = 0.5$ V with shrinkage of channel length by InP anisotropic etching. In: Proceedings of 2011 IEEE international electron device meeting, pp 13.3.1–13.3.4
14. Agrawal A, Lin J, Barth M, White R, Zheng B, Chopra S, Gupta S, Wang K, Gelatos J, Mohney SE, Dutta S (2014) Fermi level depinning and contact resistivity reduction using a reduced titania interlayer in n-silicon metal-insulator-semiconductor ohmic contacts. Appl Phys Lett 104:112101-1-3
15. The International Technology Roadmap for Semiconductors 2009. http://www.itrs2.net
16. Tewari S, Biswas A, Mallik A (2015) Investigation on high performance CMOS with p-Ge and n-InGaAs MOSFETs for logic applications. IEEE Trans Nanotechnol 14(2):275–281
17. Sun Y, Majumdar A, Cheng C-W, Kim Y-H, Rana U, Martin RM, Bruce RL, Shiu K-T, Zhu Y, Farmer D, Hopstaken M, Joseph EA, de Souza JP, Frank MM, Cheng S-L, Kobayashi M, Duch EA, Sadana DK, Park D-G, Leobandung E (2012) Self-aligned III–V MOSFETs: towards a CMOS compatible and manufacturable technology solution. In: IEEE electron devices meeting, pp 2.7.1–2.7.4
18. Allen PE, Holberg DR (2002) CMOS analog circuit design, 2nd edn. Oxford University Press, New York
19. Aswathy M, Biju NM, Komaragiri R (2013) Comparison of a 30 nm tunnel field effect transistor and CMOS inverter characteristics. In: 2013 third international conference on advances in computing and communications, Cochin, pp 149–152

Student Performance Prediction Using Classification Algorithms

Akash Ranjan, Rohit Raj, Aman Deep, and Kishore Ku Senapati

Abstract Education has become an important medium for every individual to achieve excellence in their life. A better and consistent performance in study makes the journey to the excellence easy and achievable. So, in today's competition-driven world, consistent performance has become a big concern for every student out there. Students who are performing poorly in their courses need to get proper guidance at an early stage. To aid in identifying such students, we have proposed a framework based on classification algorithms that classifies and predicts the performance of the students based on their past academic records. The algorithm first learns about the previous performances of the students and subsequently, uses it to predict the future performance. We first train the algorithm through the dataset comprising of the performance records of more than 100 students and then use the model to predict the future performance. Automatic feature selection method has been used to process the data so as to provide input to the classification algorithms in order to predict the student performance. On applying the classification algorithms, it was observed that C4.5 showed a better accuracy of 86% than Naive Bayes algorithm, that had an accuracy of 70%. This proposed model will help in the early detection of weak students, decreasing dropout rates, reducing educational pressure on students and increasing the skill set of students.

Keywords Data mining · Classification algorithm · C4.5 · Naive Bayes

1 Introduction

In every student's life, the most important phase is education. New technology helps in effective learning and provides teachers with the tools to support each and every student with their performance even when the number of students is large. Grades

A. Ranjan · R. Raj · A. Deep · K. K. Senapati (✉)
Department of Computer Science and Engineering, Birla Institute of Technology, Mesra, Ranchi, Jharkhand 835215, India
e-mail: kksenapati@bitmesra.ac.in

V. Nath and J. K. Mandal (eds.), *Proceedings of the Fourth International Conference on Microelectronics, Computing and Communication Systems*, Lecture Notes in Electrical Engineering 673, https://doi.org/10.1007/978-981-15-5546-6_40

reflect the overall performance of students and represents how well a student has grabbed the knowledge comprehended the and lesson taught in a course. So it is very important for any students to get a proper guidance and support to pass with good marks and do well in class.

In this paper our aim was to predict the grades of the students and identify those who need proper guidance and support at an early stage. To achieve this, previous performance data of the students were retrieved and then algorithms were applied. Data mining, which is an important machine learning tool used for identifying patterns and inferring knowledge by training on datasets, has been used to extricate the relevant data from a vast set and use it for further processing. In educational data mining, prediction techniques are used to speculate the performance of the student. Processes such as regression, classification and estimation of density have been used in the course of predicting the student performance.

Classification is an information mining process that groups objects from a collection into labelled target classes and categories. The objective of this function is to precisely predict the objective class for each case in the dataset. The classification method is divided into two phases-training and testing based on a bifurcated dataset datasets model: first is for training the model and the second is for testing the model. The model is trained by comparing the predicted class against the actual class in the training dataset. Classification has numerous applications in client division, business demonstrating, showcasing, credit examination, and biomedical and sedate reaction displaying.

One of the most basic and essential classification algorithms in machine learning is *k*-nearest neighbours. The *k*-nearest neighbour belongs to supervised learning model and has a great application in different fields like pattern recognition, data mining and intrusion detection. Another type of supervised learning algorithm is "support vector machine" (SVM) used for both regression and classification purposes. In this computation, every datum is point in m space dimension (where m is equal to total features) and is plotted with the estimation of each part being the estimation of a specific request [1].

Bayesian classifiers are factual classifiers that are represented as chart structure. A few Bayes calculations have been created, among which Bayesian and gullible Bayes are the two fundamental procedures.

A computational model known as artificial neural network (ANN) is based on the structure and components of natural neural frameworks. Information that travels through the framework impacts the structure of the ANN due to the fact that a neural framework changes—or learns, based on the data fed and yields accordingly [2].

For numerical prediction, the most important methodology is regression. The goal of this methodology is to accomplish an element of the free factors that permits processing the contingent desire for a reliant variable for expectation and anticipating practices subject to the minimization of a particular sort of error by means of an iterative technique. Generally classification and regression trees (Truck) abide by these undertakings.

2 Background Study and Related Works

The accomplishment of machine learning is affected by numerous factors on a given task. The portrayal and nature of the model information are above all else [3]. Feature subset determination is the way towards recognizing and expelling however much insignificant and repetitive data as could reasonably be expected. At times, precision on future classification can be improved; in others, the outcome is an increasingly conservative, effectively interpreted portrayal of the target concept. Decision tree calculations, for example, C4.5 can once in a while over fit preparing information (data), bringing about vast trees. Most of the time, evacuating insignificant and excess data can result in C4.5 creating smaller trees. C4.5 produces decision trees which can be utilized for classification. It is an augmentation of ID3 algorithm. To choose the attributes for splitting in C4.5 calculations, the ideas of information gain and entropy are utilized. It is the standardized (normalized) data gain (contrast in entropy) that helps in selecting the feature for split. The attribute having the highest normalized information gain is selected. The C4.5 calculations at that point continue by considering the remaining sub-records having the next highest standardized (normalized) information gain.

The performance prediction of students has been a popular area in the research field of recent years. Mustafa Agaolglu conducted the research on predicting the performance of instructor. They have used different analysis algorithms for predicting the instructor performance. Their results show classifier which is the best algorithm to predict the performance of instructor is C4.5 [4]. Galit gave a contextual analysis that utilizes student information to break down their learning conduct to foresee the outcomes and to caution understudies in danger before their end of the year tests [5].

Halees depicted that data mining is also helpful in improving the learning process which can focus on perceiving and isolating factors relating to the learning system of students [6].

Shaleena and Shaiju Paul directed an examination on anticipating the student performance by utilizing decision trees, class irregularity and cost touchy arrangement strategy. They found the significant elements and connections that lead to a student to pass or fall flat. They said a few elements are connected with the student disappointment [7].

Pandey and Pal [8] examined 600 students from different schools and applied different classification algorithms which can predict that the new enrolled students will score good marks or not.

3 Proposed Methodology

The implementation of the proposed framework was separated into five phases. In the first phase, the database, that include the information about marks and previous performance was collected from the students. After collecting the data, extraneous

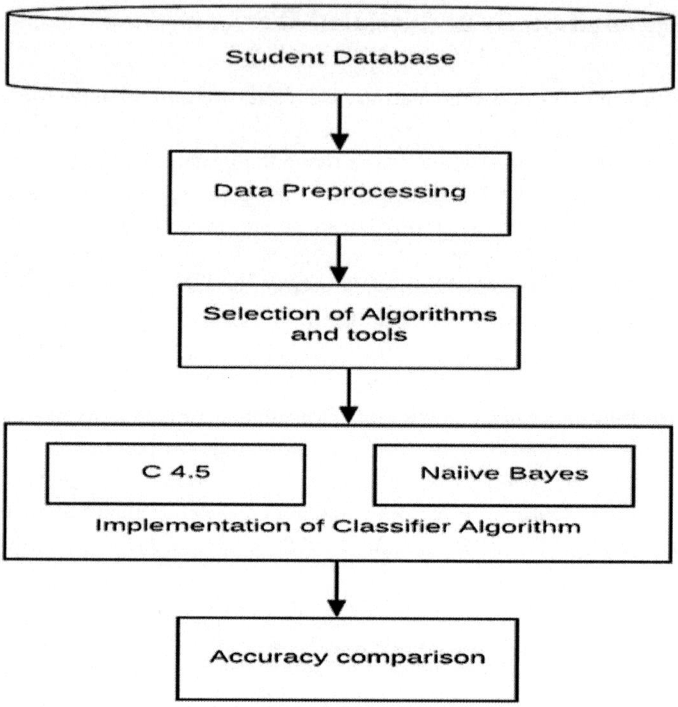

Fig. 1 Processing model

information was removed and different algorithms were applied to extract the relevant attributes present in the given dataset. These were the second and third phases respectively. In the fourth stage various, classification algorithms were used to classify and predict the future performance of the students. In the last stage, their accuracies were compared for better analysis in future work. The flowchart of these stages is shown in Fig. 1.

3.1 Data Processing

The database contained more than 100 records, each having attributes like name, state, gender, 10th marks, 12th marks, Jee mains, 1st SGPA marks, 2nd SGPA marks, label of 3rd sem, attendance. The sample of original datasets is shown in Table 1.

Co-relation-based attributes selection algorithm is used to select the relevant attributes for the classification methods. A good component subset contains components which are strongly associated with the class, but should not be correlated with each other. CFS evaluates the relationship of nominal qualities, so numerical attributes are discretized. The final construct of correlation-based feature selection does not rely upon any explicit information transformation—all that needs to be provided is a method for estimating the connection between any two factors. So, the technique could also be applied to many supervised classification issues, together with numeric class.

The temporal minimum redundancy (TMRMR), which is maximum relevance feature selection approach, deals with multivariate transient information without any prior past information. The proposed technique is assessed on datasets dependent on maximum and minimum relevance redundancy criteria. The algorithm for its implementation is as follows.

***Algorithm*: TMRMR-M / TMRMR-C**

Input: dataset $D = \{(X_i, c_i)\}_{i=1}^{N} \in \mathbb{R}^{G \times T} \times \{1, \ldots, K\}$, $algorithm \in \{TMRMR-M, TMRMR-C\}$; number of features to select m, α

Output: Feature set S

1. $S_{ALL} \leftarrow \{1, 2, \ldots, G\}$
2. $S \leftarrow \varnothing$
3. $S_\alpha \leftarrow \varnothing$
4. **for each** $g_j \in \mathbb{R}^{N \times T}$ **do**
5. $\quad F(g_j, c) = temporalRelevance(g_j, c)$ // Eqs. (1)-(2)
6. **end for**
7. **for** $i = 1$ **to** $round(\alpha G)$ **do** // function $round(\alpha G)$ rounds αG to nearest integer
8. $\quad S_\alpha \leftarrow S_\alpha \cup \underset{j \in S_{ALL} \setminus S_\alpha}{\arg\max} F(g_j, c)$
9. **end for**
10. $S \leftarrow \underset{j \in S_{ALL}}{\arg\max} F(g_j, c)$
11. **while** $length(S) < m$ **do** // function $length(S)$ returns the number of elements in S
12. $\quad$ **for each** $k \in S_\alpha \setminus S$ **do**
13. $\quad\quad S' \leftarrow S \cup k$
14. $\quad\quad W_{dm}(g_k) = \frac{1}{length(S')^2} \sum_{i,j \in S'} R(g_i, g_j)$ // $R(g_i, g_j) = R_c(g_i, g_j)$ if $algorithm = TMRMR-C$ (Eq. (3))
// $R(g_i, g_j) = R_m(g_i, g_j)$ if $algorithm = TMRMR-M$ (Eq. (4))
15. $\quad\quad V_F(g_k) = \frac{1}{length(S')} \sum_{i \in S'} F(g_i, c)$
16. $\quad$ **end for**
17. $\quad S \leftarrow S \cup \underset{k}{\arg\max} \left(\frac{V_F(g_k)}{W_{dm}(g_k)} \right)$
18. **end while**
19. **return S**

After applying co-relation-based feature selection algorithm, the datasets containing only relevant attributes are shown in Table 2.

Table 1 Sample dataset before processing

Name	State	Gender	Att. %	10th %	12th %	Jee mains marks	1st sem GPA	2nd sem GPA	Label
Ankit	Bihar	Male	92	96	94	160	7.28	7.3	H
Sumit	Jharkhand	Male	97	95	98.50	150	7.62	7.27	H
Ajay	Up	Male	93	95	91.2	174	8.03	8.73	H
Shreya	Jharkhand	Female	86	96.7	90.6	151	8	7.6	M
Sameer	Delhi	Male	70	95	92.3	113	7.45	6.8	L

Table 2 Sample dataset after processing

Attendance	10th	12th	Jee mains marks	1st sem GPA	2nd sem GPA	Label
92	96	94	160	7.28	7.3	H
97	95	98.5	150	7.62	7.27	H
93	95	91.2	174	8.03	8.73	H
86	96.7	90.6	151	8	7.6	M
70	95	92.3	113	7.45	6.8	L

Table 3 Comparative study of different publication with their accuracy

No.	Authors	Algorithm used	Accuracy (%)
1.	Adhatrao et al. [12]	C4.5	75.145
2.	Osmanbegovic and Suljic [13]	Naive Bayes	76.65
3.	Pierrakeas et al. [14]	C4.5	83.75
4.	Pierrakeas et al. [14]	Naive Bayes	82.89
5.	Al-Shawakfa et al. [15]	C4.5	87
6.	K. K. Senapati, Akash Ranjan, Aman and Rohit Raj	C4.5	87.4

3.2 *Implementation of Classification Algorithms*

After selection of the relevant attributes, the filtered data was classified through Naive Bayes and C4.5 classification algorithms.

Based on the Bayes' theorem, Naive Bayes is very efficient in classifying the model. The datasets are mutually independent but Naive Bayes performs very well under this assumption [9, 10].

C4.5 develops the classification model as a decision tree. This procedure is implemented recursively [11].

After passing the dataset containing relevant attributes as an input to C4.5 algorithm, we get a decision tree as output which can be used for predicting the performance of the testing dataset. Fig. 2 shows the decision tree returned by C4.5 after

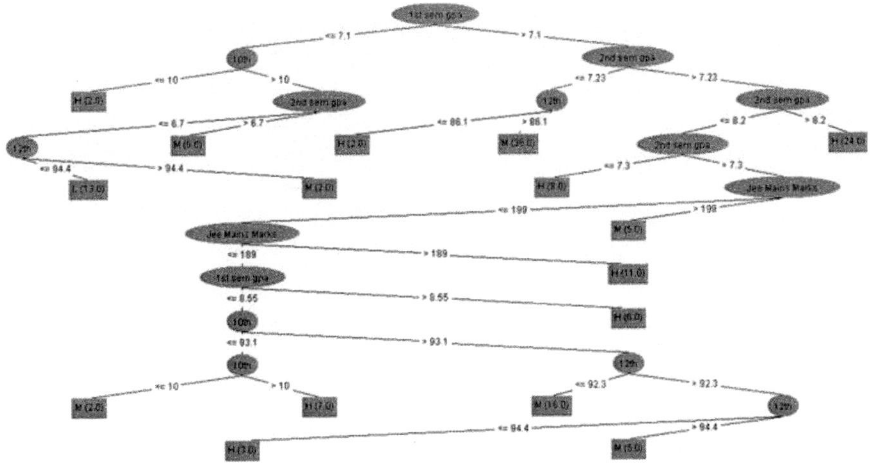

Fig. 2 Decision tree generated by C4.5

processing the input dataset.

Algorithm 1.1 C4.5(D)

Input: an attribute-valued dataset D

Tree = {}
if D is "pure" OR other stopping criteria met **then**
    terminate
end if
for all attribute $a \in D$ **do**
    Compute information-theoretic criteria if we split on a
end for
a_{best} = Best attribute according to above computed criteria
Tree = Create a decision node that tests a_{best} in the root
D_v = Induced sub-datasets from D based on a_{best}
for all D_v **do**
    Tree$_v$ = C4.5(D_v)
    Attach Tree$_v$ to the corresponding branch of Tree
end for
return Tree

---

4 Experiments and Results

The dataset for our experiment includes the details of undergraduate students enrolled in 2016 and 2017. The dataset contains the performance of all students in the final semester examination. The dataset also contains the previous academic records such as 10th marks, 12th marks, and Jee main marks. In each semester there are 5 theoretical and 3 sessional papers. Combining the grades of all papers, the final GPA is

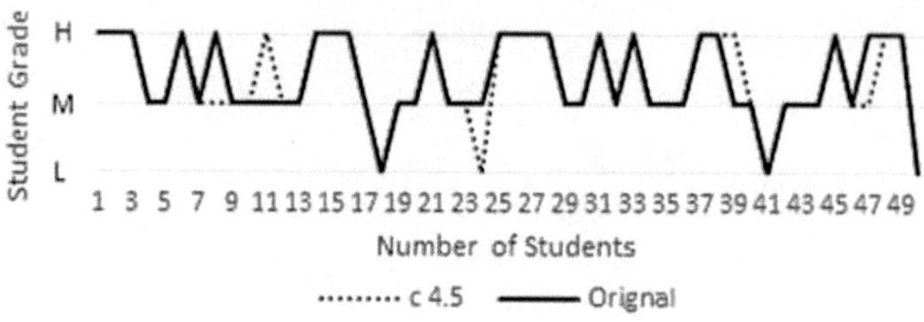

(a)

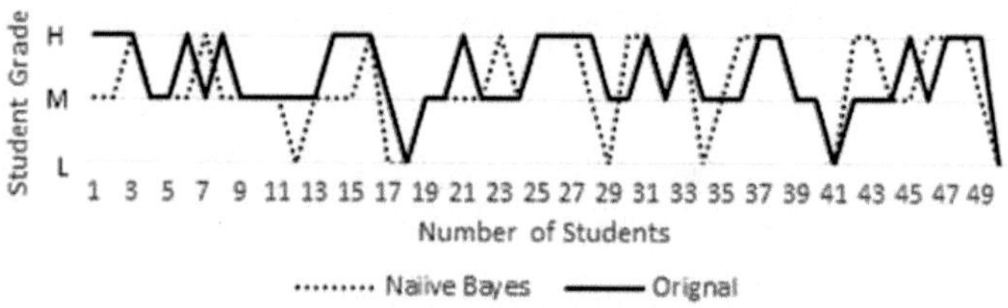

(b)

Fig. 3 **a** Result analysis of C4.5. **b** Result analysis of Naive Bayes

calculated. For training purpose, we have used the dataset of 2016 batch students and predicted the results of 2017 batch students. After getting the actual final grade of 2017 batch students, it was compared with predicted result. The comparison graph of original versus predicted for the sample of the first 50 students is shown below in Fig. 3a, b respectively.

We evaluated the accuracy of both the algorithms and compared their efficiency it was found that C4.5 has classified 87.4% of the instances correctly, whereas Naive Bayes has classified only 70.6% of the instances correctly, as shown in Table 4. According to this result, C4.5 can help in a better way to predict the student performances. Through this result, instructor can easily get the predicted grade which will help them to give proper support and guidance to weak students.

Our algorithm was implemented in JAVA 1.8 and runs on Intel Core i3 1.70 GHz CPU, 4 GB RAM with AMD Radeon HD 8750 M GPU having 4 GB of memory. The machine learning simulation tool which we have used is WEKA with version 3.7.0 written in Java developed at the University of Waikato, New Zealand [3].

Table 4 Predictive performance of classifiers for dataset containing 150 instances

Algorithm	Correctly classified instances	Prediction accuracy
Naive Bayes	106	70.6
C4.5	132	87.4

5 Conclusion

This paper proposes the implementation of different classification algorithms for predicting students' performance. In most of the cases, attendance and GPA played an imperative role in accomplishing students' performance.

The implementation of our model in universities is manageable because it is easy to collect marks of the student rather than their socio-economic data from an administrative point of view. Our result shows the decision tree which can help in the identification of the major factors affecting the performance. By identifying these factors, students can be warned earlier in the degree program. In future, we plan to gather more training samples and consider other factors which can affect the performance of the students.

References

1. Zhang L, Li KF (2018) Education analytics: challenges and approaches. In: 2018 32nd international conference on advanced information networking and applications workshops (WAINA), pp 193–198
2. Amonkar S, Gangan T, Doddihal N, Pasad C, Bhatia G (2017) Design and implementation of tourism enhancement through use of web mashup. In: 2017 fourteenth international conference on wireless and optical communications networks (WOCN), pp 1–7
3. http://weka.sourceforge.net/doc.stable/
4. Agaoglu M (2016) Predicting instructor performance using data mining techniques in higher education, vol 4. IEEE
5. Ben-Zadok G (2007) Examining online learning processes based on log files analysis: a case study. In: ICT 2007
6. El-Halees AM (2009) Mining students data to analyze e-learning behavior: a case study
7. Paul S, Shaleena KP (2015) Data mining techniques for predicting students. In: International conference on engineering and technology (ICETECH). IEEE 2015
8. Pandey UK, Pal S, Data mining: a prediction of performer or underperformer using classification
9. Rish I (2001) An empirical study of the Naive Bayes classifier. In: IJCAI 2001
10. Domingos P, Pazzani M (1997) On the optimality of the simple Bayesian classifier under zero-one loss. Mach Learn 29(2–3):103–130
11. http://www.otnira.com
12. Jha R, Gaykar A, Adhatrao K, Dhawan A, Honrao V (2013) (IJDKP), Sept 2013
13. Osmanbegovic E, Suljic M (2012) Data mining approach for predicting student performance
14. Pierrakeas CJ, Kotsiantis SB, Pintelas Educational Software Development Laboratory, Greece
15. Al-Shawakfa EM, Al-Radaideh QA, Al-Najjar MI (2006) Mining student data using decision trees. ACIT, Jordan
16. http://www-gth.die.upm.es/partners/sony/Chapter5.doc

Effect of Integrated Competency Management and Human Resource Development at the Level of Efficiency of Workforce

Thaya Madhavi and Rajesh Mehrotra

Abstract The successful running of an establishment is only dependent on trust and competence infused in its employees. In this context, competence management has a key function to take part in with the business front of any establishment. For the viability of any organization, the latter as the ultimate responsibility to pour in enough confidence and competence among these employees going on various sections of the institutions might finally contribute to success of the organization in the business field. The operational and behavioral facets of employees have to be satisfied by implementing competency management by providing training and updating their skills. Human resource policies and practices as well as competencies of human resource play a tangible role to reap competitive advantage in the field. The concept of competency is a vital organ of human resource management presenting a groundwork to analyze workforce practices and bring them near to the management attention in the organization. This report discusses an innovative, vibrant and proactive approach to the consolidation of human resource development, competency management and knowledge management to attain the best execution for an efficient organization's business performance in combination with strategic business needs at the right direction for its winner. Further, it may also form one of the success factors for the organization by contemplating enhancement of a workforce sufficiently through the development of human resources and management of competencies.

Keywords Competency management · Human resource management · Knowledge management · Integration

---

T. Madhavi (✉) · R. Mehrotra
School of Business and Management, Jaipur National University, Jaipur, Rajasthan 302017, India
e-mail: madhavimbabit@gmail.com

R. Mehrotra
e-mail: rajeshmehrotra7@rediffmail.com

V. Nath and J. K. Mandal (eds.), *Proceedings of the Fourth International Conference on Microelectronics, Computing and Communication Systems*, Lecture Notes in Electrical Engineering 673, https://doi.org/10.1007/978-981-15-5546-6_41

1 Introduction

In the present world class, business has emerged on account of trade, the principle of service oriented and commercial organization with the concept of gaining ground faster in the world trade and in the development of updated technology. Unlike a business trend prevailed upon in the market two decades ago, the present-day business management has been looking forward for short-term earnings and thus becomes more flexible, efficient and customer friendly. In their attempt to implement modern technology, they are in one away, setting apart their workforce [1].

Nohria et al. [2] have surveyed over two hundred companies at a span of ten years and made viable suggestions for their successful survival. They have noticed that they do possess aligned culture, implementing plans successfully and have concentrated strategy. In addition to the above, they are found to be having, committed leadership, the industry growth oriented and quick to transform through mergers. The authors suggested the partnership package in the industry that would advance the organization well. By the end of 1980, competence management has been debated exhaustively in the instruction of its successful implementation in the governing body.

It is a fact that human competence seeks of intellectual and human capital [3]. To differentiate between the two, the human capital is the capability of the employees in any arrangement for self-promotion of skills, whereas human-based capital, societal capital and structural capital are involved in intellectual capital. The human capital research and its consideration of knowledge management go hand in hand while solving competency approach-related problems. The skills, the stimulus and emotional states are the three components of human capital, as presented in Fig. 1.

Some of the researchers like Ulrich, Beeker and Gerhart, and Lake have researched on by trusting main emphasis on skills through forecasting, coordinating, monitoring and training of workforce. Further, they have found that blending of the conventional employee developmental activities with training methods and approach is a confrontation set by the organization asks for varying key position of human resource management.

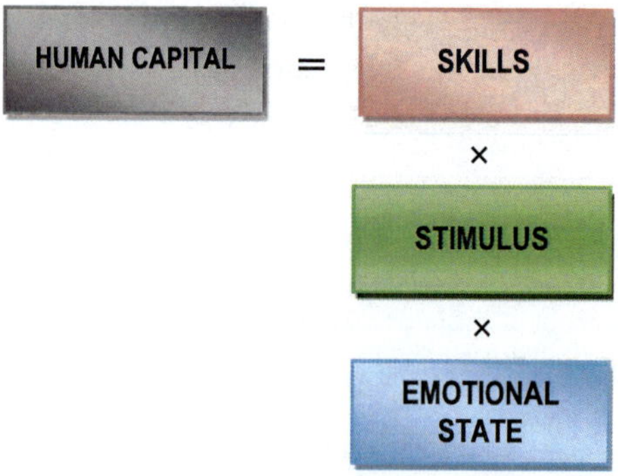

Fig. 1 Origin of human capital

2 Viable Integration of Development of Human Resource and Management of Competency

A wide research work was carried out by organizations in particular on the effect of competencies and new theories adopted to streamline the prototype in the company management. Structure included was described with reference to its visualization, strategy and proficiency management of a company that gives limitations and guidance for the development of employees in the establishment.

An integrated and streamlined system of competent management and human resource development has been reflected in Fig. 2. This model provides focus areas, essential theories combined into a rational unit.

2.1 Management of Competency

What does competency mean? The word competency takes its origin from a Latin word 'competentia' meaning 'having the right to speak'. The competency management concept was traced back to 1970 in human resource management. The main thrust for competency management is the combination of strategic vision with human resource planning in an organization. It is used mainly to analyze the quantitative and qualitative competencies of working employees at present in the organization. Its main aim is to compare the levels of requisite competencies of employees to achieve set targets.

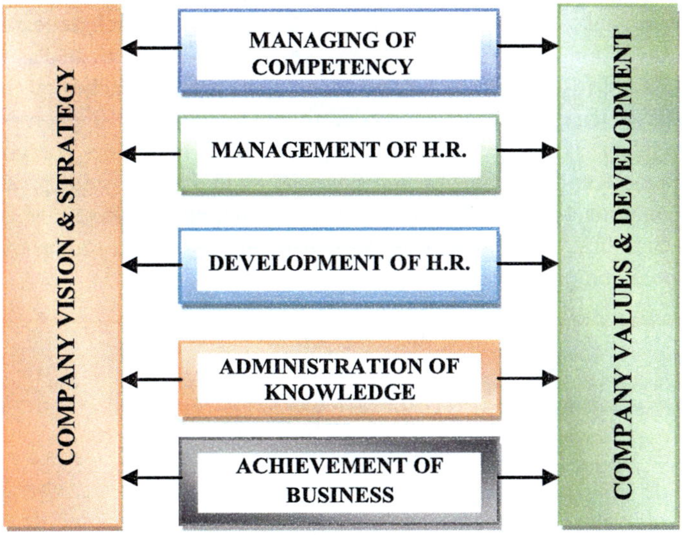

Fig. 2 Efficient organization of competency management and human resource development

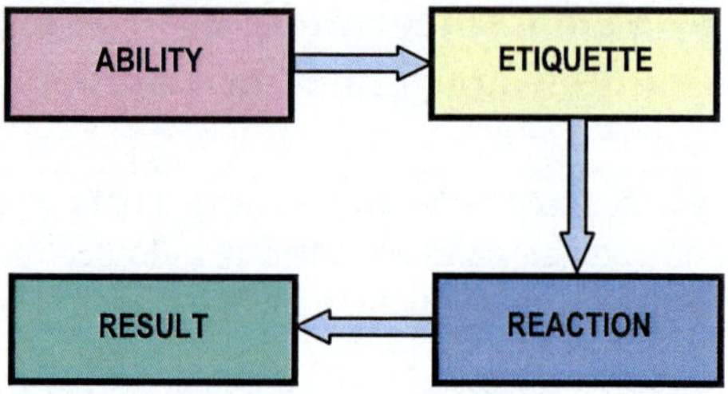

Fig. 3 Behavioral skill

The competition involves public rather than produce, and the core of the issue is the capability of concerning the organization. Competence refers to an ability of employees to do a job, whereas competency refers to the skill and knowledge of an employee to accomplish a work fruitfully. Competence is termed as a task-related concept. Figure 3 shows that competency speaks of the etiquette of an employee [4].

2.2 Competency Types

(a) *Specific Competencies*
These are the attributed skilled employees who are to carry out while performing a work. They are the person's work obligations on specific functions. They are company's employees with outstanding accomplishments or knowledge. These capabilities are otherwise identified as hard skills.

(b) *General Competencies*
To achieve coveted result, the employee is needed to act within the framework of the organizational surroundings [5]. Competencies such as team making, decision making, leadership issues and communication are treated as soft skills. McClelland [6] postulated behavioral competency approaches of employees. Boyatzis [5] in 1982 has provided a definition and publicized the word behavioral competency. The work he has carried out shows that many components are accountable to differentiate much successful and fewer thriving presentations. The components were individual character, experience, motives and behavioral characteristics. The author had distinguished performance competencies from threshold competencies.

2.3 Outline of Competency

It is a mix of human resource strategies and competencies which forms a support for evolving different processes of human resources. Tools of the framework are band matrix, competency dictionary, assessment tools and competency profile [7].

Referring to the glossary of competency, it is a deposit of competencies for an organization. It is depicted in image 4. This type of competency is essential for executive

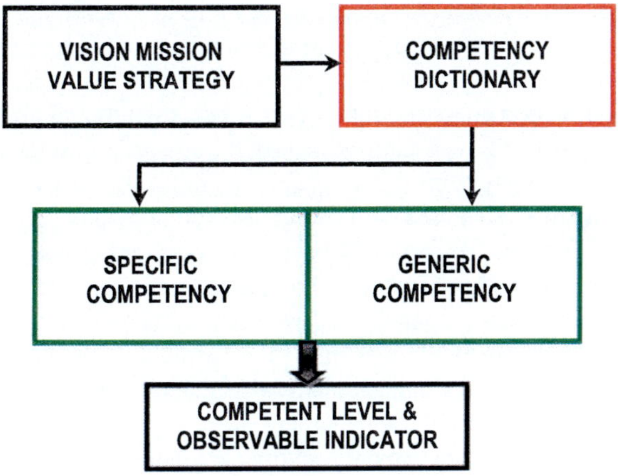

Fig. 4 Dictionary of competency

level employees and operative level employees at diverse actions commonly called as competent level, whereas observable action indicates the level of competencies needed at that particular juncture.

A further step in competency mapping is to trace back the needs of competency is otherwise called band matrix mapping. Profile of competency dictates the range of human resource processes involving career management, performance management, selection and attributes of training. Every work has its competency profile fixed with certain proficiency scale to be identified with the sale. A practical assessment method has been evolved for every line of work through competition management practices for employees, which would prove for its practical approach, reliability for its implementation and viability. Assessment of efficiency or suitability through competency is a part and parcel while selecting a prospect for a job through performance management or developmental process. In this, skill gaps can be identified in the process of assessment, steps taken to fulfill the openings through the conduct of training program to its employees. Further the development in that area can be done through training.

3 Competent Human Resource Management

The strategies allocated through human resource management are a lengthy process for attaining results for the evolution of business in an organization in a right way. It supplies all the required conditionalities for achieving laid down policies over the form of personnel involved, their capability to handle the task, right step in recruitment observance of safeguards while recruitment, rewards and engage outsourcing management. Human resource management and organizational strategy work hand in hand. Involvement of top management affairs in organizational strategy is a must in which human resource strategy will take sufficient lead. The success of any organization needs involvement of human resource management to act as an important role in the business front. Further, human resource management requires the full obligation on its shoulders as it affects the integral processes for the ultimate winner of the business organizations by the merger of all the subsystems. Bringing closure of the subsystems based on competency and subsystem based on performance under one roof paved the path for achieving the results [2].

Bratton and Gold [8] have said that human resource management has aimed at two principles of which one is an enhancement of employee performance and the other organizational level development. Human resource management is confronted with certain challenges in the marketplace. The challenges being confronted by the human resource management in the market are strategically based on human resource management, the management anticipating frequent changes in its establishment and finally management seeking personal guidance and support of human resources management. While reviewing traditional practices of human resource management, it has been emphasizing main thrust on quantity if the workforce and the nature of people being recruited. The workforces so taken in have the obligation to meet the labor requirement. The organization's success is primarily dependent on the impositions of competency development over the targets fixed and the prerequisite proposed for its achievement at the organizational stage.

3.1 Human Resource Development as a Base for Knowledge Management

The experience and competencies of the present and past employees have a bearing on the company's functioning. The human resource development programs have a direct impact on employees to further their personal accomplishments. Besides that, the employees have a moral duty to raise their capabilities and update their knowledge and know-how through practical experience for the welfare of the system they work for [2]. It is equally significant that it develops core competency of the system that would be a vision of the company. Human resource development programs have an important part to play in providing valuable information to the organization on business matters based on ability.

3.2 Impact of Performance Management

The concept of performance management has come to be realized in the business market since 1970. The exponents of the performance management have suggested that the latter has been felt at different levels of functioning, such as organizational level, process and job level [9]. Personal level performance is reflected in the job level, achievement of the work suggests the process layer, and the whole system is concerned with organizational level. Different levels of organization acquire varied levels of performance output. It is a mix of various factors contributed to the success of the organization. Ives et al. [10] have pointed out certain contributory factors. This limits the growth of organization, roles to be played, processes to be watched over, the nature of culture, the physical environment prevailing in the system and finally human factors and guidance they seek from the institution. The standard to be employed stipulates the nature of competencies and motivation of employees in the workplace. In the corporate sector, top management has the final say on the performance indices and empowerment of management tools in sub-areas which influence productivity and growth. Kaplan and Norton's scorecard is an indicative of utilization of monitoring tool that enhances the efficiency of employee which introduced balanced scorecard which is utilized as a monitoring tool. It identifies the targets and work out the problems, easily. In this scorecard, two types of indicators are pronounced. One is result indicators and the other performance indicators. In the wake of strategic planning, the analysis of core competencies has been undertaken that helps in bettering the overall competency and growth strategies of the corporation.

3.3 Personal Level Competence

Mapping of personal competencies has a strong character taking on. Companies are assessing the primary competencies of employee keeping in view of their earlier performance, so that estimation of individual competencies has become a critical issue in the performance front. Adult learners have masterminded the ownership trait through assessment process. In fact, they have better access to competencies and planning of performance which require a strict point of concentration and individual point of exertion. Competencies and job profile are thoroughly investigated by probing individual targets set for them. To summon an example, the sales manager is not simply to oblige to possess the complete story of the merchandise and its price tag, but he should likewise possess the technique to convince the customers through business tactics which help to sell away the merchandise. Competency mapping is always at the receiving end to face the challenges in the market and redress the same on time if it is so. Competency management can also resolve problems through deliberation with the help of identifying techniques or discussions keeping in view of the company's vision and the prevailing strategies. Personal competencies are studied in various ways. Further, mapping of individual bases competences can claim propriety of the organization to have a review of its key areas of strength and weakness as a whole. The importance of mapping and the result realized have a prominent role to play in the business field [3].

Competencies are not only evaluation of personal accomplishments and abilities, but more of deep involvement in the operation. The process is undermined by drawing of current and future plan for HRD. The process starts with self-assessment of accomplishments by employees' current level, but it may not be so up to the level of the change of outplay.

Company's vision and strategy, assessment, company level competence, the scope of competence and development plans can be visualized both at the level of individual and organization [3]. Top management has an important part to play in the competency assessment for which the tools, systems and principles have been prepared to maintain a grip over competencies at corporate level and to achieve further advancement in the area.

4 Human Resource Development

Development of human resource constitutes the main framework to benefit the employee while enhancing their skills, attitudes and abilities. HRD has an obligation to increase the share of better quality employees so that the institute will achieve their commitment toward the customers. The HRD practitioner has the committed obligation to get the employee multi-skilled in the order of existing competency and make the employee play a vital part in the welfare of the system [11]. It is quite a combined phenomenon of competence building even though it may not include all facts and figures while implementing competency management. The introduction of human resource development includes notification of scope of performance, such as performance of task, implementation of competency decision, focus on character improvement, promotion of new products, innovation of new fields of surgical procedure and scope for enhancement of employment prospects for prospective nominees.

Boudreau and Ramstad [12] have visualized that the investment in human resource strategy might lead to establishment of organization on firm footing. Human resource development activity if accomplished at right perspective employees would be able to improve their quality work life (QWL) that ultimately increases the business productivity of the organization. The enhancement of quality work life for employees can be judged by the collective competencies of employees, which includes teamwork, leadership and various prospects. The organization's environment is guiding factor in which position and strategy-based competency could vest the evaluation while reviewing the operation of the employee. It is therefore important for creating workplace order. An interlinked relation among QWL, human intercourse, human relation development and business realization of objects exists [12, 13].

4.1 Strategic Planning for Human Resource Development

HRD is otherwise considered for extension of knowledge. McClelland [14] has undertaken research into the issues of the topic and is of the opinion that the organization usually does not involve items related to developmental activities while attributing competency strategies. The primary function of human resource management is to assist strategic management. In this, the ability of employee knowledge will be assessed reported and well met. The HRD mechanisms are confined to many areas and some involves interaction, part play, assessment of achievements prospective development, change of tariffs, net worth and enrichment of job, etc. These would increase the competency of employees. These mechanisms may vary according to the prevalence of internal environment, the organization size, the commitment and guidance extended by the top level management. The significant role played by HRD has been realized very fast in the business area in the existing circumstances which is now difficult to draw a strategic plan on HRD for a long period.

4.2 HRD Criterion and Tendency

The employees learn and gain experience through HRD and prepare the grounds for better prospects for future sustenance in the business field. Burgoyne and Hodgson [15] have tabulated the above concept. Therefore, an employee selects such a task that would give good result in the future. Vast changes have taken place in the field of HRD in the past two decades. Some, such, changes attributed to conduct internal training programs in the organization focused primarily on the strategy-oriented topics. These topics include management and leadership development, performance consultation, learning design, proactive learning, career and performance management and training assessment.

4.3 Developmental Activities Through HRD

It is a system that promotes the capability among employees of an organization. It includes strategy competence, determination, development needs, identification and assessment of competencies and discussion on development as appearing in Fig. 5. The levels of need analysis include development areas, competence region, levels of competence required, current competencies, its results and summary obligations [2]. A wide range of assessment of competence areas has led to tactful planning for enhancement of functioning. HRD works on diverse planning processes, work review, change of work allotment, review of competencies and present competence areas. The backing and cooperation of top management extended are necessary to continue blending in the developmental activities in the system. The organizational activities are limited to support extended by the team rather than individual-based performance [16].

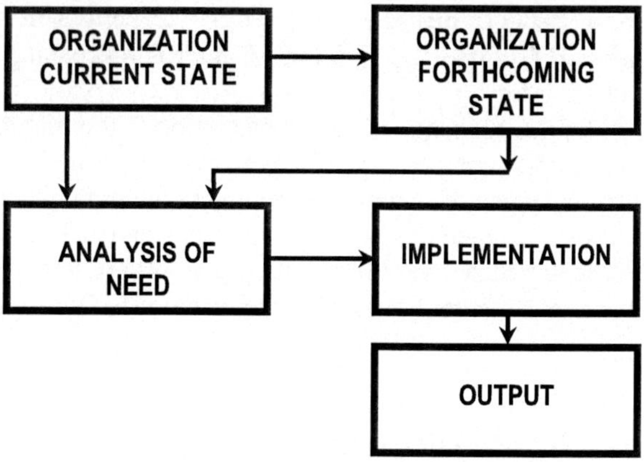

Fig. 5 Developmental stages of human resource

5 Interrelationship Among Human Resource Development, Competence and Knowledge Management

Knowledge-based management is correlated with competency management. It has two areas of operation. One dealing with well-planned knowledge of management and other the spread of knowledge through the competence process. Further knowledge management can be acquired through codification and personalized strategy. Codification is meant for information storage in computer and utilized several times, to produce faster and cheaper products and services. Communication through people network and advice to specific problem resulting in revision of premium services are included in personalization. The strategy of codification targets at information system, personalization strategy, deals with shifting of skills and experts constantly in the purview of information between two individuals [2].

Worthy organization generates, promotes and deals in knowledge-based issues. Active teams work in this field and encourage staff in this field of competencies by involving new strategies. This sustains an impact on the behavior of employees. It identifies the required competency to bring out strategies with success. Specific viewpoints will be taken up in competence management supported by the right cause. In this problem set, organizational level will be addressed and discussed in HRD. The specialists in human resource development are termed as practitioners who are motivated by educational prospective, professional knowledge and external networks. The top management has the real challenge to identify the skills of employees, enhance and build a core competency for the welfare of the organization.

6 Conclusion

In HRD, field competency mapping helps in hiring of talented person for a job through filtration among core competencies and elimination of the unqualified applicants is another attribute. It also helps in placing the qualified candidate in the suitable department from the date of recruitment. HRD helps the management to survive in the industry with efficient and well-qualified staff and manager. Though competency is incorporated into HRM with the basic aim of acquisition, vision and communication strategies and goals, it could be sufficiently applied in the field. It is seen that in the present-day scenario many organizations are looking for a selection of subsystem based on competencies. It is also opined that the blending of competence management, knowledge management and human resource development better results has emanated for successful business performance compared to the age old practices adopted earlier in much better business performance as compared to the age old traditions followed. Despite all the above requirements possessed by an organization, it needs employees sincerity and commitment toward business which are the successful factors for the development of business with the organization.

References

1. Vakola M, Soderquist KE, Gregory PP (2007) Competency management in support of organizational change. Int J Manpow 28(3/4):260–275. https://doi.org/10.1108/01437720710755245
2. Nohria N, Joyce W, Robertson B (2003) What really works. Harvard Bus Rev 81(7):43–52
3. Laakso-Manninen R, Viitala RR (2007) Competence management and human resource development—a theoretical framework for understanding the practices of modern finnish organizations. Haaga-Helia University of Applied Science Publication
4. Woodruffe C (1992) What is meant by a competency? In: Boam R, Sparrow P (eds) Designing and achieving competency. McGraw-Hill, Maidenhead, pp 16–29
5. McClelland DC (1973) Testing for competence rather than for intelligence. Am Psychol 28(1):1–14. https://doi.org/10.1037/h0034092
6. Boyatzis RE (1982) The competent manager: a model for effective performance. Wiley, London
7. Pande S, Basak S (2014) Human resource management text and cases. Pearson, Dorling Kindersley (India) Pvt. Ltd., New Delhi
8. Bratton J, Gold J (1999) Human resource management. Theory and practice. Palgrave, New York
9. Rummler GA, Brache AP (1990) Improving performance, how to manage the white space on the organisation chart. Jossey-Bass Publishers, USA
10. Ives W, Torrey B, Gordon C (2000) Knowledge sharing is a human behavior. Knowledge management: classic and contemporary works. The MIT Press, Cambridge, pp 99–131
11. Barney JB, Ketchen DJ, Wright M (2011) The future of resource-based theory: revitalization or decline? J Manag 37(5):1299–1315
12. Boudreau J, Ramstad P (1999) Human resource metrics: can measures be strategic? Res Person Hum Resour Manag Suppl 4:75–98
13. Kesti M (2012) Organization human resources development connection to business performance. Proc Econ Financ 2:257–264. https://doi.org/10.1016/S2212-5671(12)00086-X
14. McClelland S (1994) Gaining competitive advantage through strategic management development. J Manag Dev 13(5):4–13

15. Burgoyne JG, Hodgson VE (1983) Natural learning and managerial action: a phenomenological study in the field setting. J Manag Stud 20(3):387–399
16. Leibman M, Bruer R, Maki BR (1996) Succession management: the next generation of succession planning. Hum Resour Plann 19(3):16–29

Predicting Type 2 Diabetes Using Logistic Regression

Neha Prerna Tigga and Shruti Garg

Abstract Type 2 diabetes is possibly going to be the most significant plague in the history of mankind. If we bring all the diabetic people from around the world together, it can form the third-largest nation in the world. Lately, the predominance of diabetes has altogether expanded in India by 13%. While diet and change in the art of living are the foundation to control type 2 diabetes, most of the patients in the long run require medications to control glucose and related health problems. Therefore, early diagnosis and treatment are important to restrain from future complications. This study uses logistic regression, a popular machine learning classification algorithm to predict the risk of type 2 diabetes among individuals. The aim of this study is to improve prediction so that the logistic regression algorithm can be used on any dataset to give result with good accuracy. The Pima Indian Diabetes dataset is taken for analysis, and RStudio is used to process and visualize the result. Our model is showing pretty good prediction with an accuracy of 75.32%. This study will help the future researchers to develop new interventions to reduce the prevalence of diabetes.

Keywords Type 2 diabetes · Machine learning · Logistic regression · Risk prediction

1 Introduction

Diabetes is a condition that weakens the body's capacity to process blood glucose, also called as blood sugar. Without continuous and watchful administration, diabetes can prompt development of sugars in the blood, which can lead to life-threatening complications, including stroke and heart-related disease. Various types of diabetes

N. P. Tigga · S. Garg (✉)
Department of Computer Science and Engineering, Birla Institute of Technology, Mesra, Ranchi, Jharkhand 835215, India
e-mail: gshruti@bitmesra.ac.in

N. P. Tigga
e-mail: tigganeha4@gmail.com

V. Nath and J. K. Mandal (eds.), *Proceedings of the Fourth International Conference on Microelectronics, Computing and Communication Systems*, Lecture Notes in Electrical Engineering 673, https://doi.org/10.1007/978-981-15-5546-6_42

can arise, and dealing with the condition relies upon the type. Not all types of diabetes come from an individual being overweight or living an unhealthy way of life. In fact, some can have diabetes from a young age. Currently, over 425 million people are living with diabetes [1].

Diabetes is quickly picking up the status of a possible plague in India with nearly 72.9 million diabetic individuals presently determined to have the disease. According to the reports from Indian Diabetes Federation (IDF) Atlas (7th Edition) 2017, India (72.9 million) is the second-largest country with diabetes prevalence after China (114.3 million) bagging the first place [2]. India right now faces a risky future in connection to the potential load that diabetes may put upon the nation. Numerous impacts influence the growth of this ailment all over India, and finding of those variables is important to further change when confronting health challenges. So what are the variables at present influencing diabetes in India that are making this issue so outrageous? The cause of diabetes in India is based on multiple factors and incorporates hereditary variables combined with natural impacts, like obesity related with rising comfort in quality of life, stable urban relocation and change in the standard of living. Rajadhyaksha in [3] said that there is a need for building a robust disease management system by engaging all medical and related health experts, patients and caretakers to triumph over this epidemic. Diabetes prevalence by wealth bracket is shown in Fig. 1. However, even with the prevalence of diabetes in India, there are no national but a few multi-driven studies being carried out on the predominance of diabetes and its associated problems [5].

Different types of diabetes that can develop are:

- Type I diabetes: This is also known as juvenile diabetes, and this type happens when the body is unable to build insulin. Individuals with type I diabetes depend

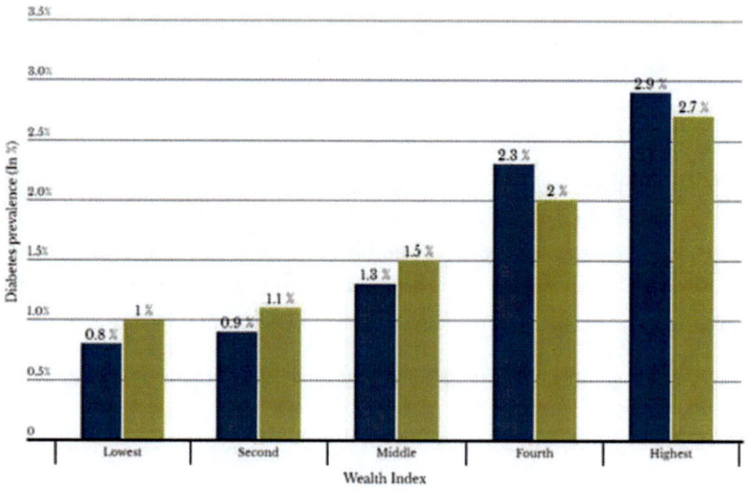

Fig. 1 Diabetes in India, by wealth bracket [4]

on insulin, which implies they should take insulin shots every day to manage their blood sugar level.
- Type 2 diabetes: This is also called as hyperglycemia. Type 2 diabetes affects the manner in which the body utilizes insulin. While the body still makes insulin, different from that in type I, the cells in the body do not react to it as productively as they once did. This is predominantly the most acknowledged kind of diabetes, as per the National Institute of Diabetes Digestive and Kidney Diseases, and it has solid connections with being obese.
- Gestational diabetes: This type manifests in females amid pregnancy when the body can turn out to be less susceptible to insulin. Gestational diabetes does not happen in all females and settles down subsequently after childbirth [1].

Some common symptoms of diabetes are:

- Frequent urination.
- Feeling thirstier.
- Always hungry—despite the fact that you are eating.
- Extreme tiredness.
- Blurred vision.
- Slow healing of cuts and wounds.
- Unplanned weight loss—Despite the fact that you are eating more (Type 1).
- Pain or numbness of hands or legs (Type 2).

From the past few years, researchers have been creating and using different algorithms and toolkits for analysis. Machine learning and data mining are the most popular and tremendously growing research areas in the field of diabetes and health care. Wu [6] applied improved *k*-means cluster algorithm, and logistic regression algorithm got 95.42% accuracy, which was 3.04% higher accuracy than the rest. Patil [7] proposed Hybrid Prediction Model (HPM) which is based on basic *k*-means clustering algorithm and seeks to validate the chosen class label of provided data and obtained an accuracy of 92.38%.

Meng [8] did a study comparing the accuracy of decision tree, logistic regression and artificial neural network, out of which decision tree algorithm gave the best accuracy of 77.87%. Tapak [9] compared the traditional classification methods (logistic regression and Fisher linear discriminant analysis) and four machine learning classifiers (neural networks, support vector machines, fuzzy *c*-mean and random forests), among which the performance of support vector machine was the highest with an accuracy of 98.6%. Kavakiotis [10] reviewed different machine learning and data mining algorithms and tools used for research in the area of diabetes. The study concluded that 85% of research were on supervised learning and rest 15% research were on unsupervised learning. Out of all, support vector machine (SVM) was the most popular algorithm used in the field of diabetes. Wang [11] proposed an improved electromagnetism-like mechanism (IEM) algorithm and compared it to the nine most widely used feature selection and classification methods and determined that the IEM algorithm is superior to all when it comes to accuracy and kappa index. Orabi [12] applied decision tree technique to predict age as it relates to diabetes incidents with an accuracy of 85%. Perveen [13] did a study comparing the ensemble techniques (AdaBoost and bagging) using J45 decision tree as the base learner to predict the risk of an individual getting diabetes. The result showed that AdaBoost is better than bagging as well as standalone J45 decision tree.

Although many of the researchers worked in this field and found accuracy more than 90%, there is a chance of overfitting because of high accuracy rate. Overfitting occurs when the model tries to adapt too much to the training data. In other words, the model tries to learn the training data and fits really well to the dataset. The problem with overfitting is that when a new value is added to the dataset, the model misbehaves and is unable to classify because it is so well fitted to the training data [14].

The description of upcoming sections in this paper is as follows: Sect. 2 includes a brief description of logistic regression. Result and discussion are discussed in Sect. 3. And lastly, in Sect. 4, conclusion is presented.

2 Methodology

Logistic regression is a technique of machine learning which has borrowed its concept from statistics. It is the most common machine learning algorithm after linear regression. In various numbers of ways, linear regression and logistic regression are comparable. But the major distinction lies in the way they are used. Linear regression is used for prediction of values, whereas logistic regression is used in task related to classification. Logistic regression is named after logistic function (also called as sigmoid function) which is the base of this method. Figure 2 shows architecture of logistic regression [15].

where 1, x_1, x_2, x_3, ..., x_n are the input variables. c, w_2, w_2, w_3, ..., w_n are the model coefficients. y is considered to be the predicted output.

Logistic regression computes the weighted sum of the input variables and runs it through a special function known as logistic function or sigmoid function which produces an output y. Here, the output is taken in binary or in terms of 0/1 or −1/1.

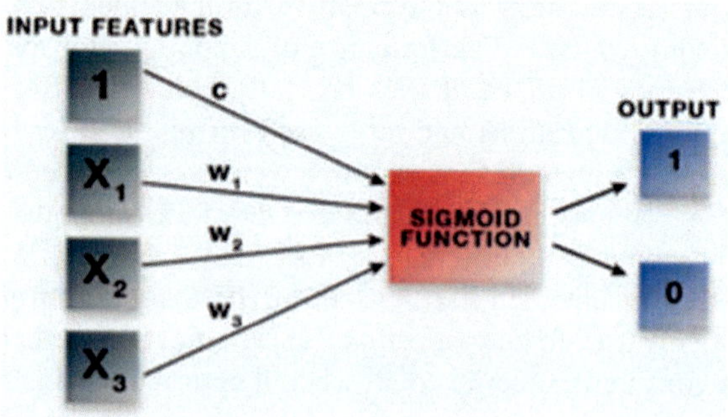

Fig. 2 Architecture of logistic regression [15]

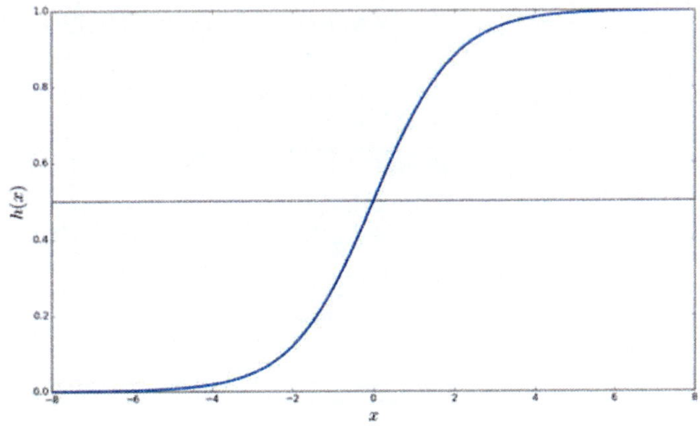

Fig. 3 Logistic regression [12]

$$y = \text{logistic}(c + x_1 * w_1 + x_2 * w_2 + x_2 * w_2 + \ldots + x_n * w_n) \tag{1}$$

$$y = \frac{1}{\left(1 + \left(\mathrm{e}^{-(c + x_1 * w_1 + x_2 * w_2 + x_3 * w_3 + \cdots + x_n * w_n)}\right)\right)} \tag{2}$$

The sigmoid/logistic formula is as follows:

$$y = \frac{1}{1 + \mathrm{e}^{-x}} \tag{3}$$

Figure 3 illustrates *S*-shaped curve, and this symbolizes that the increase in the input variable above 0 brings it the closer to 1 and the decrease in input variable below 1 brings it closer to 0. So, if the value is greater than 0.5, then the outcome is said to be positive (or 1); otherwise, the outcome is said to be negative (or 0).

The flowchart of work present in this paper is presented in Fig. 4.

3 Result and Discussion

3.1 Research Objective

Applying logistic regression algorithm to predict type 2 diabetes and verifying its accuracy using test set results.

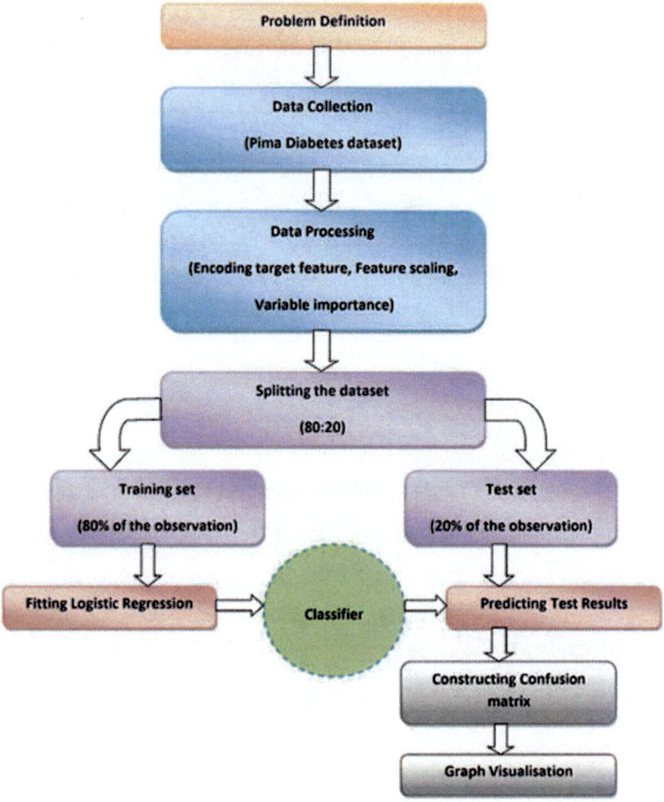

Fig. 4 Flowchart of the study

3.2 Research Methodology

- Data Acquisition: The dataset used in this study is the very popular Pima Diabetes dataset [16]. This dataset is from the National Institute of Diabetes and Digestive and Kidney Disease. The target of this dataset is to determine whether a patient has diabetes or not, considering certain estimations into the dataset. All patients here are females of at least 21 years of age. There are total 769 patients with nine parameters in this dataset. The parameters are shown in Fig. 5.
- Design and implementation: Tool used to construct this implementation is RStudio. For coding and data visualization, R programming language is used.
- Description of algorithm: The point of logistic regression is to find the best fitted line that fits the parameters we have in our dataset. Logistic regression is used to predict probability. Here, the outcome parameter is considered to be the dependent variable, and the rest are taken as independent variables. The outcome parameter is in binary form of 0 or 1, where 0 means the person does not have diabetes and 1 means the person has diabetes. Here, in this study, we are predicting the possibility

	A	B	C	D	E	F	G	H	I
1	Pregnanci	Glucose	BloodPres	SkinThick	Insulin	BMI	DiabetesF	Age	Outcome
2	6	148	72	35	0	33.6	0.627	50	1
3	1	85	66	29	0	26.6	0.351	31	0
4	8	183	64	0	0	23.3	0.672	32	1
5	1	89	66	23	94	28.1	0.167	21	0
6	0	137	40	35	168	43.1	2.288	33	1
7	5	116	74	0	0	25.6	0.201	30	0

Fig. 5 Parameters used in the dataset

of an individual having type 2 diabetes based on certain variables. In order to train the dataset, the entire sample is randomly divided into training set and test set in the ratio 80:20, respectively. Logistic regression is fitted to the training set which is then used to predict the test set results. Confusion matrix is shown in Fig. 6. The following are the parameters for assessment of quality of classification. The measured values of accuracy rate, error rate, sensitivity, specificity and precision are shown in Table 1, and the formulas are as follows:

$$\text{Acurracy Rate} = \frac{\text{TP} + \text{TN}}{\text{TP} + \text{TN} + \text{FN} + \text{FP}} \tag{4}$$

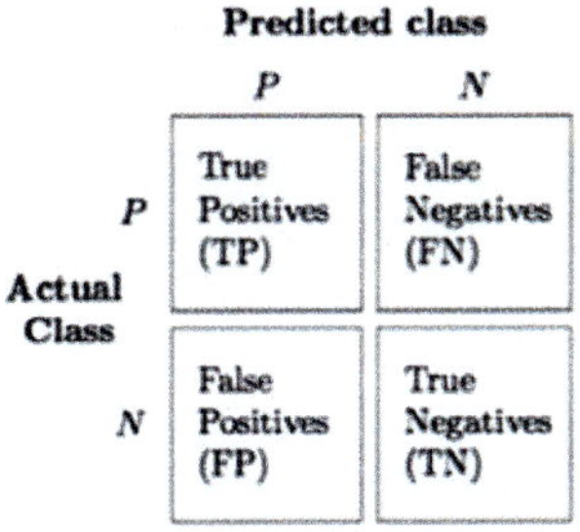

(a) Confusion matrix Terminology

	FALSE	TRUE
0	86	14
1	24	30

(b) Resultant Confusion matrix

Fig. 6 Confusion matrix

Table 1 Analysis of confusion matrix

Measuring rates	Percentage (%)
Accuracy rate	75.32
Error rate	24.68
Sensitivity	86
Specificity	55.56
Precision	78.18

```
Coefficients:
                         Estimate Std. Error z value Pr(>|z|)
(Intercept)               -0.8643     0.1083  -7.982 1.44e-15 ***
Pregnancies                0.4464     0.1228   3.636 0.000277 ***
Glucose                    1.1850     0.1365   8.684  < 2e-16 ***
BloodPressure             -0.3060     0.1110  -2.758 0.005816 **
SkinThickness              0.1165     0.1268   0.919 0.358146
Insulin                   -0.2626     0.1213  -2.164 0.030434 *
BMI                        0.6770     0.1319   5.134 2.84e-07 ***
DiabetesPedigreeFunction   0.2245     0.1122   2.002 0.045297 *
Age                        0.1726     0.1226   1.408 0.159132
---
Signif. codes:  0 '***' 0.001 '**' 0.01 '*' 0.05 '.' 0.1 ' ' 1
```

Fig. 7 Variable significance

$$\text{Error Rate} = \frac{\text{FN} + \text{FP}}{\text{TP} + \text{TN} + \text{FN} + \text{FP}} \tag{5}$$

$$\text{Sensitivity} = \frac{\text{TP}}{\text{TP} + \text{FN}} \tag{6}$$

$$\text{Specificity} = \frac{\text{TN}}{\text{TN} + \text{FP}} \tag{7}$$

$$\text{Precision} = \frac{\text{TP}}{\text{TP} + \text{FP}} \tag{8}$$

3.3 Discussion

In present study, logistic regression is applied to predict type 2 diabetes. For successful application of the method, we need to know the significant importance of each variable in dataset which is calculated by R function and given in Fig. 7.

Figure 7 shows that the variables Pregnancies, Glucose and BMI have the highest significance value as compared to the rest. So, it can be stated that Pregnancies, Glucose and BMI are highly significant in this dataset and have a greater impact on predicting diabetes. Because Glucose and BMI are found most significant, graphs are plotted in Fig. 8 for each training set and test set results. Here, the green region predicts patients with diabetes, and the red region predicts patients without diabetes. The green points represent patients with diabetes, and red points represent patients without diabetes based on the Pima Diabetes dataset. It can be observed that majority of red points are within the red region and majority of the green points are within the green region. Therefore, it can be determined that this study has good predictions.

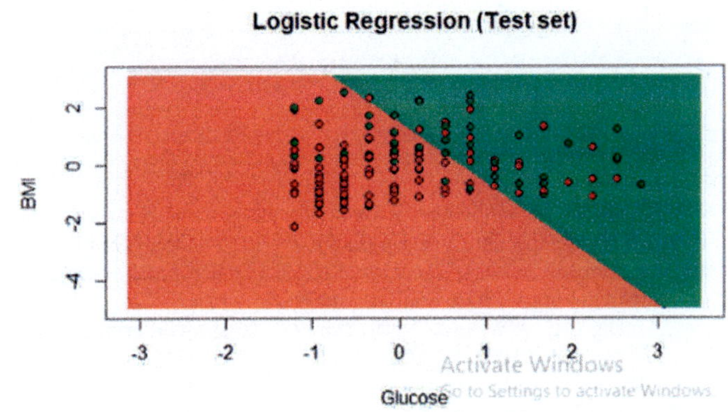

(a) Visualisation of Training set results.

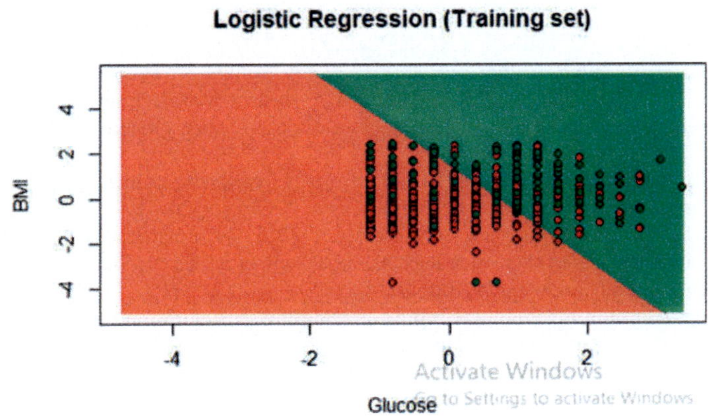

(b) Visualisation of Test set results.

Fig. 8 Graph visualization

4 Conclusion

The aim of this study was to use logistic regression to predict diabetes for an appropriate result. From the result, it can be said that the prediction is pretty good with an accuracy of 75.32% and error rate of 24.68%. Therefore, logistic regression can be considered as a good predictor of diabetes. The accuracy of logistic regression will improve with the increase in the number of independent variable to an extent that it does not lead to overfitting. For future work, it is important to take hospital's latest data to predict diabetes for a more accurate result. The larger the dataset, the better will be the prediction. Various researches are being carried out in the field of health care and computer science as it relates to diabetes. This study is useful for people to not only care for their health but also to establish a healthy way of life.

References

1. https://www.nhp.gov.in/world-diabetes-day-2018_pg
2. https://diabetesatlas.org/
3. Rajadhyaksha V (2018) Managing diabetes patients in India: Is the future more bitter or less sweet? Perspect Clin Res 9(1):1–3
4. https://www.firstpost.com/india/diabetes-is-indias-fastest-growing-disease-72-million-cases-recorded-in-2017-figure-expected-to-nearly-double-by-2025-4435203.html
5. Kaveeshwar SA, Cornwall J (2014) The current state of diabetes mellitus in India. Aust Med J 7(1):45–48. https://doi.org/10.4066/amj.2013.1979
6. Wu H, Yang S, Huang Z, He J, Wang X (2018) Type 2 diabetes mellitus prediction model based on data mining. Inform Med Unlock 10:100–107
7. Patil BM, Joshi RC, Toshniwal D (2010) Hybrid prediction model for type-2 diabetic patients. Expert Syst Appl 37(12):8102–8108
8. Meng XH, Huang YX, Rao DP, Zhang Q, Liu Q (2013) Comparison of three data mining models for predicting diabetes or prediabetes by risk factors. Kaohsiung J Med Sci 29(2):93–99
9. Tapak L, Mahjub H, Hamidi O, Poorolajal J (2013) Real-data comparison of data mining methods in prediction of diabetes in Iran. Healthcare informatics research 19(3):177–185
10. Kavakiotis I, Tsave O, Salifoglou A, Maglaveras N, Vlahavas I, Chouvarda I (2017) Machine learning and data mining methods in diabetes research. Comput Struct Biotechnol J 15:104–116
11. Wang KJ, Adrian AM, Chen KH, Wang KM (2015) An improved electromagnetism-like mechanism algorithm and its application to the prediction of diabetes mellitus. J Biomed Inform 54:220–229
12. Orabi KM, Kamal YM, Rabah TM (2016) Early predictive system for diabetes mellitus disease. In: Industrial conference on data mining. Springer, Cham, pp 420–427
13. Perveen S, Shahbaz M, Guergachi A, Keshavjee K (2016) Performance analysis of data mining classification techniques to predict diabetes. Proc Comput Sci 82:115–121
14. https://en.wikipedia.org/wiki/Overfitting
15. https://www.quantinsti.com/blog/machine-learning-logistic-regression-python
16. https://www.kaggle.com/uciml/pima-indians-diabetes-database/version/1

FPGA Implementation of PICO Cipher

Nigar Ayesha and Bibhudendra Acharya

Abstract The rise in ubiquitous computing connecting resource-constrained devices had led to the emergence of a sub-domain in cryptography called lightweight cryptography. The algorithms in this domain target moderate security while satisfying the constraints set for a desired application. PICO is one of the lightweight SPN ciphers suitable for low-area and low-power applications. In this paper, two hardware architectures ideal for low-area footprint and low power are proposed. This comprises a serial architecture and a reduced datapath architecture which are designed specifically for reduced area and low-power applications while choosing trade-offs from latency and performance. The results are evaluated for Xilinx FPGA platforms like Spartan 3, Virtex 4 and Virtex 6. The synthesis and implementation of the proposed architectures are performed using Verilog HDL in Xilinx ISE Design Suite 14.6. The area footprint is estimated in terms of the slices, flip flops and LUTs. The power consumption is estimated using XPower Analyzer for a clock frequency of 10 MHz. The simulations for the proposed designs are verified using ISim Simulator. Further, the area in terms of slices occupied is compared with the existing ciphers for Spartan 3 and Virtex 4 families.

Keywords PICO cipher · Security · Lightweight cryptography · Low-area · Low-power · Verilog · IoT · FPGA · SPN

1 Introduction

With the advent of Internet of Things (IoT), communication between devices and wireless networks has reached unprecedented levels. The interaction of numerous

---

N. Ayesha · B. Acharya (✉)
Department of Electronics and Telecommunications, National Institute of Technology, Raipur, Raipur, India
e-mail: bacharya.etc@nitrr.ac.in

N. Ayesha
e-mail: nigarayesha11@gmail.com

V. Nath and J. K. Mandal (eds.), *Proceedings of the Fourth International Conference on Microelectronics, Computing and Communication Systems*, Lecture Notes in Electrical Engineering 673, https://doi.org/10.1007/978-981-15-5546-6_43

tiny computing devices for pervasive computing is essential for various applications like smart homes, smart grid, smart cities, etc. [1]. Due to the huge data exchange between these devices, security is essential at various stages of communication against possible security threats, hacks and breaches. However, there are no fixed industry-accepted standards due to the heterogenous systems involved in the interconnection. This has further led to challenges in imparting security to the IoT systems which are compatible with various industry applications. The level of security depends largely on the application and the significance of the communicating network [2].

These security-sensitive applications also have constraints on the resources like hardware footprint, power, energy, latency, performance, memory, etc. The same can be said to be true for other resource-constraint devices like RFID tags, contactless smart cards, sensor nodes, etc., where lack of security can lead to data infiltration with severe consequences [3].

The conventional algorithms like Advanced Encryption Standard (AES) and Data Encryption Standard (DES) are suitable ideally for satisfying security objectives in desktop/server environments [4]. However, these algorithms have high power consumption, memory requirements, processing speed and large area footprint. As a result, they are not practical solutions for applications where resources are a major constraint.

This resulted in a sub-domain of cryptography called lightweight cryptography. The insatiable demands set on the design metrics, and security objectives have resulted in making this domain a hot topic of research [5]. Each design is application specific with choices in the selection of the type of cipher, number of rounds, input plaintext size, key size, memory requirements, etc. A good lightweight cipher should ideally be memory efficient, resource efficient, less complex operations, rich security standards, high throughput, lower latency, low power/energy consumption and resistance to popular attacks like meet-in-the-middle attack, biclique attack, etc. [6]. However, it is not possible to satisfy all the desirable features and is often required to choose trade-offs by prioritizing the desired features in an application. Hardware-based security solutions are gaining popularity in recent times to meet the security challenges with ultralow-area and power requirements [7].

Lightweight ciphers are generally classified as block ciphers or stream ciphers. Further, block ciphers are classified as substitution–permutation networks (SPN) and Feistel networks. PRESENT [8], RECTANGLE [9], LED [10], PICO [11], etc., are examples for SPNs which typically involve a chain of connected mathematical operations. This comprises a combination of substitution layer and permutation layer with key-scheduling operations in a single round. The Feistel networks include ciphers like SIMON, SPECK [12], LiCi [13], PICCOLO [14], etc., which tend to operate on one half section of the data per round and thus require additional rounds for improving security in comparison with SPNs.

PICO is an SPN lightweight cipher which is ideal for low-area and low-power applications with good performance in hardware as well as software. Setting low area and low power as the chief constraints and trade-offs from latency and throughput, two lightweight hardware architectures are proposed for this cipher. This includes a serial architecture and a reduced datapath architecture with memory-based key scheduling.

In general, lightweight cryptography targets application-specific integrated circuits (ASICs) which are followed by huge costs and long time to market. In recent times, field-programmable gate arrays (FPGAs) are gaining popularity due their reconfigurable fabric with scope for resource optimization. These are ideal for testing prototypes and reduce the time to market with scope for possible upgradation with the changing industry standards [15].

FPGAs contain arrays of programmable logic blocks with reconfigurable interconnects which facilitate the blocks to be connected with other logic gates. The logic blocks can be programmed to perform various combinational functions based on truth tables. In addition, logic blocks are composed of memory elements and flip flops [16].

Structure of the Paper

The rest of the paper is organized as follows. Section 2 gives a brief overview of the algorithm for PICO cipher. Section 3 describes the proposed hardware architectures. Section 4 presents the hardware comparison of the results. Section 5 deals with the discussion of the implemented results. Section 6 is the conclusion of the work.

2 Algorithm Overview

Figure 1 shows the block diagram of PICO cipher [11]. It is an SPN cipher with a total of 32 rounds. It has an input of 64-bits plaintext and 128-bits key. Here input plaintext,

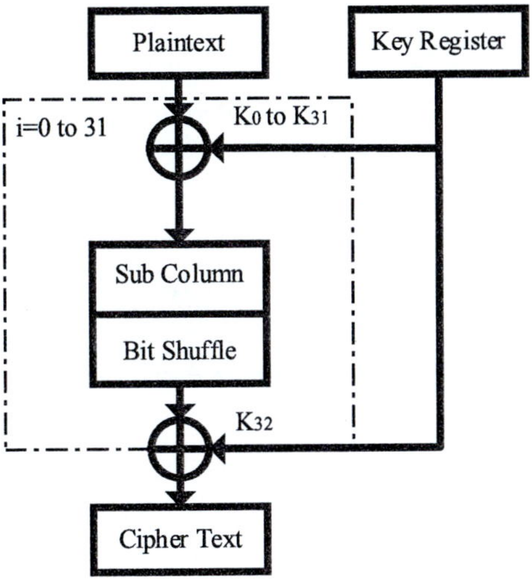

Fig. 1 Block diagram of PICO algorithm [11]

$$P = p_{63} p_{62} p_{61} \dots p_0.$$

Add Round-Key
This performs the XOR operation with the 64-bits plaintext and the 64-bits sub-key.

SubColumn
The substitution is performed as per the 4×4 S-box values [11]. This is performed similar to the column-wise substitution in RECTANGLE cipher.

Permutation Layer
This performs bit-wise permutation. In hardware, this can be accomplished by simple wiring. Permutation in a cipher imparts diffusion and randomness [11].

Key Scheduling
The key scheduling is similar to the design in SPECK cipher. It is compact in memory requirements and quite robust against attacks [12]. A total of 33 sub-keys are generated of which 64-bits from the 128-bits are used.

For a given key,

$$K = k_{127} k_{126} k_{125} \dots k_2 k_1 k_0$$
$$L_0 = k_{31} k_{30} k_{29} \dots k_2 k_1 k_0$$
$$M_0 = k_{63} k_{62} k_{61} \dots k_{34} k_{33} k_{32}.$$

The subkeys are generated as,

for j=0 to 31
$L_1=((M_j)\oplus RCS(L_0,3))\oplus(L_0)$
$M_{j+1}=((L_1)\oplus LCS(M_j,7))\oplus j$
$L_0=L_1$
End for

The subkeys generated are arranged as an array of 4×16 bits and used in the add round-key operation.

3 Proposed Architectures

3.1 Proposed Serial Architecture

Figure 2 shows the proposed serial architecture for an input datapath of 4-bits. The initial 32 clock cycles are needed for loading of the 64-bits plaintext and 128-bits key. The loading of the plaintext bits is done such that the bits to be substituted are stored together. This is accomplished using proper comparators and FSM states as

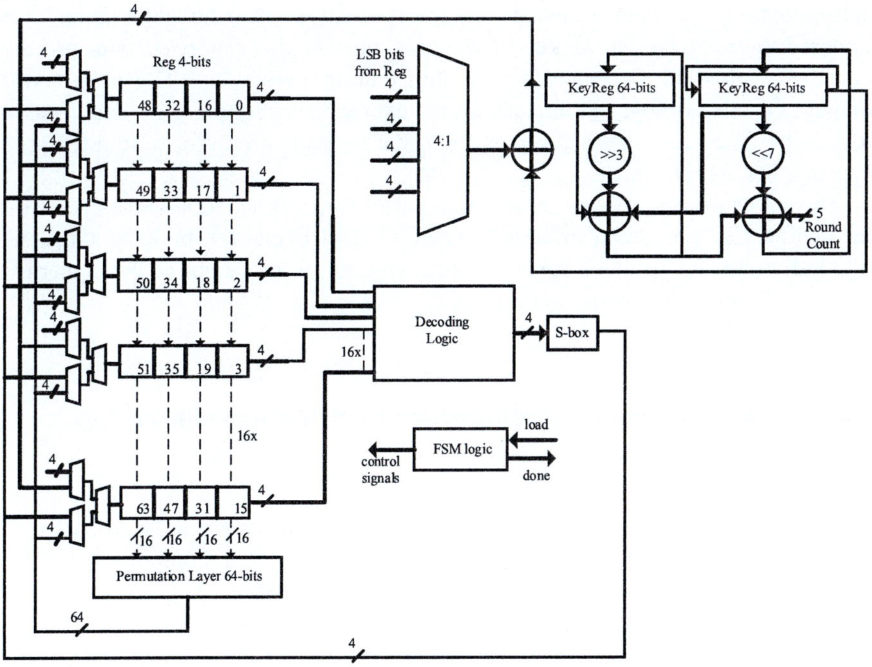

Fig. 2 Proposed serial architecture

shown in Fig. 3 to control the routing through the S-box. This accounts for the first state S_0. In the state S_1, the XOR operation is performed with the plaintext bits and the key. The required key bits are shifted with each clock cycle for a total of 16 clock cycles. The desired plaintext bits are selected using decoding logic, and the XORed result at each clock cycle is stored back into the registers.

In the next state S_2, the substitution is performed by choosing the input from the 16, 4-bit registers using a decoding logic for a total of 16 clock cycles. The substituted outputs are stored in the same registers. In the final state S_3, the permutation of the

Fig. 3 FSM for the proposed serial architecture

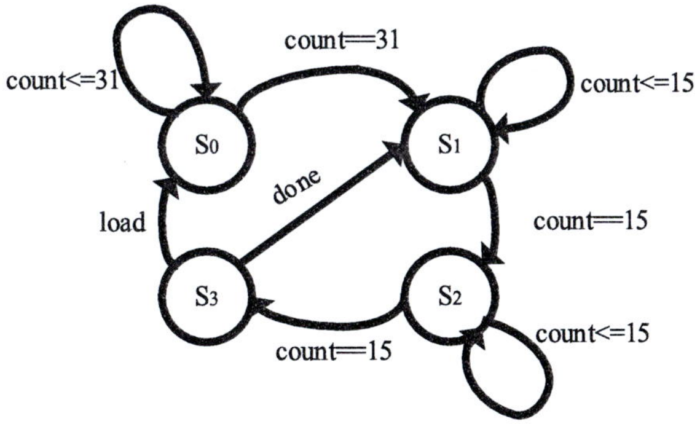

substituted outputs is performed and the resultant output from the permutation layer is stored. In this state, the subkey for the next round is also generated. This process repeats for a total of 32 rounds to generate the desired ciphertext. The ciphertext is output as 4-bits serially, and the total ciphertext requires additional 16 clock cycles. It is to be noted that in all the case, the same registers are used in an attempt to decrease the area overhead.

The whole operation requires 32 clock pulses for loading, 1056 clock cycles for the intermediate operations in the 32 rounds, 16 clock cycles for the XOR at the end of 32nd round and 16 clock cycles for receiving the serial output of the ciphertext bits. Hence, the total process requires 1120 clock cycles.

3.2 Reduced Datapath Architecture with Memory-Based Key Scheduling

Figure 4 shows the proposed hardware architecture targeting a reduced datapath. The datapath width is reduced from 64-bits to 16-bits input and output width. This design is thus suitable for FPGAs with a smaller number of IOBs. The key-scheduling process involves a RAM module in which the key along with the subkeys is to be initially stored in memory.

This is useful for resource-constraint devices with moderate emphasis on security. Due to the absence of additional architecture to address the key scheduling, the area

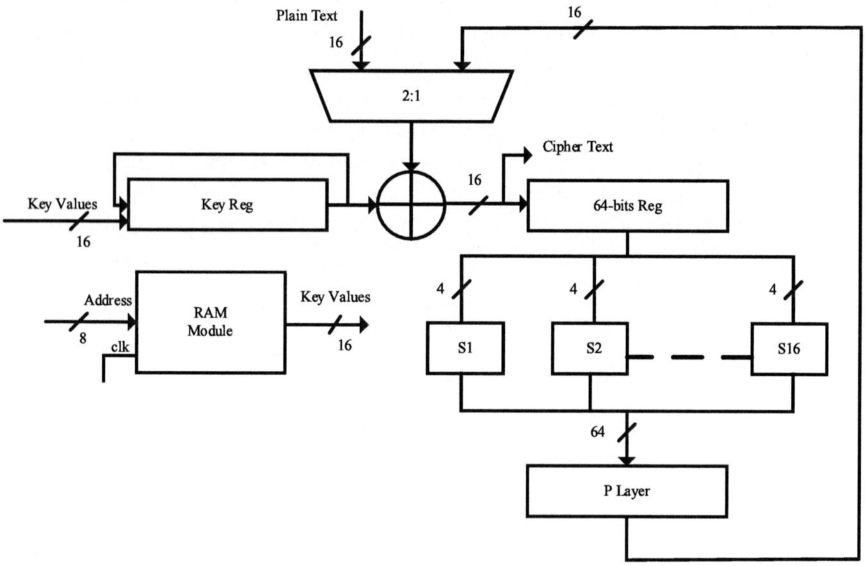

Fig. 4 Proposed reduced datapath architecture with memory-based key scheduling

is drastically reduced [17]. However, the design is vulnerable to side-channel attacks and is thus limited to applications with moderate to low security levels [18].

The hardware architecture consists of a multiplexer to choose from the input plaintext bits to be loaded and the updated plaintext for the next round. This is followed by the XORing of the corresponding key bits and the plaintext bits. The results are stored in a 64-bits register which can operate as a shifter register/parallel register. The substitution is performed by the 16 S-boxes in a column-wise manner as per the original algorithm [11]. The substituted output is passed through the P-layer which serves as the input for the next round.

In this design, a RAM module is incurred which has the respective key values initialized in the memory. The desired values can be accessed through a counter which serves as the address for the corresponding key values. An alternative design for the same architecture can be designed using a ROM module or the inbuilt BRAM resources in an FPGA.

4 Hardware Evaluation

The proposed architectures were designed using Verilog in Xilinx ISE Design Suite 14.6. The architectures were implemented for Post-Place and Route and the simulations were observed in ISim Simulator. The power consumption was estimated for a clock frequency of 10 MHz using Xilinx XPower Analyzer. The results for hardware utilization and power consumption for the proposed architectures are provided in Table 1.

The results for the proposed architectures are presented for three FPGAs chosen as the implementation platforms. The FPGAs considered are Spartan 3 (xc3s50-5pq208), Virtex 4 (xc4vlx25-12ff668) and Spartan 6 (xc6vlx75tff484-3). The synthesis of the proposed architectures was set with focus on area as the optimization goal and optimization effort being high.

The hardware evaluation for the chosen FPGAs is analyzed based on area, performance and power consumption. For an FPGA, slices are the chief metrics for evaluating area. Performance is analyzed in terms of throughput and latency for the designs under consideration. The values for throughput-per-slice give the relation between the area and the performance.

Table 2 provides the area estimation of the FPGA implementation of some existing ciphers. Figure 5 gives a comparison of area in terms of slices for Virtex 4 and Spartan 3 families. The comparison of dynamic power consumption for Virtex 6 family at 10 MHz frequency is presented in Fig. 6. The designs are referred as, This Work(S) for the proposed serial architecture and This work (Rd) for the proposed reduced datapath architecture in Figs. 5 and 6.

Table 1 Resources occupied and performance evaluation

Design	FF	LUTs	Slices	F_{max} (MHz)	Latency (cycles)	Thr (Mbps)	Thr/slice (Mbps)	Static power (mW)	Dynamic power (mW)	Total power (mW)
Device xc6vlx75-3ff484										
Serial	210	259	77	350.51	1120	20.1	0.26	1292.81	2.79	1295.6
Reduced datapath	120	51	37	193.44	136	91.03	2.46	1292.81	2.72	1295.53
Device xc4vlx25-12ff668										
Serial	216	454	234	312.53	1120	17.85	0.08	233.10	19.91	253.02
Reduced datapath	138	125	87	227.37	136	107	1.23	233.01	15.07	248.08
Device xc3s50-5pq408										
Serial	216	455	233	127.38	1120	7.3	0.03	27.34	0.03	27.38
Reduced datapath	138	125	87	188.21	136	88.56	1.02	27.34	0.06	27.4

Table 2 Area utilization of some existing ciphers

Work	Slices	F_{max} (MHz)
Spartan 3 xc3s50		
SIMON [19]	399	135
XTEA [20]	254	62.6
Camellia [21]	318	125.8
F-FCSR-16 [22]	473	134
Virtex 4 xc4vlx25		
ICEBERG [23]	575	247
SEA [23]	438	241
PRESENT [24]	192	284.3

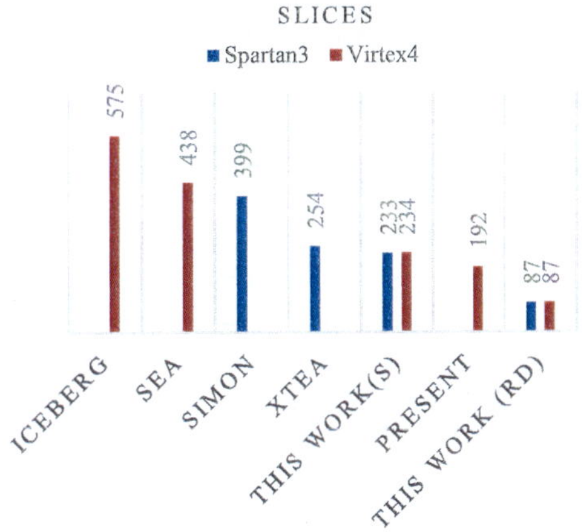

Fig. 5 Resource usage comparison of the proposed architectures

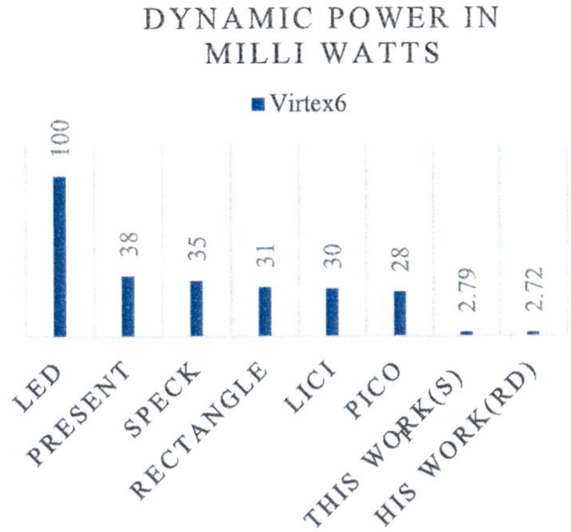

Fig. 6 Comparison of dynamic power consumption [11, 13]

5 Discussion of Results

It can be observed that the area footprint is for the proposed architectures, thus suiting lightweight applications. It can be inferred from Fig. 5 that the slices occupied are almost equal for the two families as they both consist of four input LUTs. The area reduction for the reduced datapath architecture is significant due to the absence of key scheduling compared to the serial architecture. The serial architecture needs an FSM to control the states and the processing of the intermediate states requires registers to perform the column-wise and row-wise operations. Due to the non-uniformity in the processing of bits, area overhead increased to some extent. However, in both the proposed architectures, there is a drastic reduction in dynamic power, thus suiting low-power applications. The performance is a trade-off for the serial architecture due to the substantial increase in latency. Security is a trade-off for the reduced datapath architecture due to the vulnerability to side-channel attacks.

6 Conclusion

In this paper, two lightweight hardware architectures for the PICO cipher were proposed. The proposed serial architecture and reduced datapath architecture were designed primarily for minimal-power and low-area footprint applications. The principal trade-offs chosen for serial architecture included high latency and low-performance, while the reduced datapath architecture had moderate latency and security trade-offs. Further, there is scope for increasing the performance or decreasing the area of the architectures by choosing a wider datapath for serial architecture and utilizing the FPGA resources like BRAM. Alternative designs for the reduced datapath architecture can be proposed with emphasis on performance, embedding the key in ROM modules or BRAM modules for a given application.

References

1. Silva BN, Khan M, Han K (2018) Internet of things: a comprehensive review of enabling technologies, architecture, and challenges. IETE Techn Rev 35(2):205–220
2. Xu T, Wendt JB, Potkonjak M (2014) Security of IoT systems: design challenges and opportunities. In: Proceedings of the 2014 IEEE/ACM international conference on computer-aided design. IEEE Press, pp 417–423
3. Sehrawat D, Gill NS (2018) Lightweight block ciphers for IoT based applications: a review. Int J Appl Eng Res 13(5):2258–2270
4. NF Pub (2001) 197: Advanced Encryption Standard (AES), Federal Information Processing Standards Publication 197, US Department of Commerce/NIST, 26 Nov 2001. Available from the NIST website, National Institute of Standards and Technology. FIPS 197: Advanced Encryption Standard. Available via: https://csrc.nist.gov/publications/fips/, 12 Mar 2019
5. Hammad BT, Jamil N, Rusli ME, Z'aba MR, Ahmed IT (2017) Implementation of lightweight cryptographic primitives. J Theoret Appl Inf Technol 95

6. McKay KA, Bassham L, Turan MS, Mouha N (2017) NISTIR 8114 report on lightweight cryptography. The National Institute of Standard and Technologies
7. Eisenbarth T, Kumar S, Paar C, Poschmann A, Uhsadel L (2007) A survey of lightweight-cryptography implementations. IEEE Des Test Comput 522–533
8. Bogdanov A, Knudsen LR, Leander G, Paar C, Poschmann A, Robshaw MJ, Seurin Y, Vikkelsoe C (2007) PRESENT: an ultra-lightweight block cipher. In: International workshop on cryptographic hardware and embedded systems, pp 450–466
9. Xiang Z, Zhang W, Bao Z, Lin D (2015) RECTANGLE: a bit-slice lightweight block cipher suitable for multiple platforms. Sci China Inf Sci 58:1–15
10. Guo J, Peyrin T, Poschmann A, Robshaw M (2011) The LED block cipher. In: Lecture notes in computer science (including subseries lecture notes in artificial intelligence and lecture notes in bioinformatics): preface, vol 6917, pp 326–341
11. Bansod G, Pisharoty N, Patil A (2016) PICO: an ultra lightweight and low power encryption design for ubiquitous computing. Def Sci J 66(3):259–265
12. Beaulieu R, Treatman-Clark S, Shors D, Weeks B, Smith J, Wingers L (2015) The SIMON and SPECK lightweight block ciphers. In: Design automation conference (DAC), 2015 52nd ACM/EDAC/IEEE, pp 1–6
13. Patil J, Bansod G, Kant KS (2017) LiCi: a new ultra-lightweight block cipher. In: 2017 international conference on emerging trends and innovation in ICT (ICEI), pp 40–45
14. Shibutani K, Isobe T, Hiwatari H, Mitsuda A, Akishita T, Shirai T (2011) Piccolo: an ultra-lightweight blockcipher. In: International workshop on cryptographic hardware and embedded systems, pp 342–357
15. Yalla P, Kaps J (2009) Lightweight cryptography for FPGAs. In: 2009 international conference on reconfigurable computing and FPGAs, Quintana Roo, pp 225–230
16. Sbeiti M, Silbermann M, Poschmann A, Paar C (2009) Design space exploration of present implementations for FPGAS. In: Proceedings of the Southern conference on programmable logic (SPL), pp 141–145
17. Lara-Nino CA, Morales-Sandoval M, Diaz-Perez A (2016) Novel FPGA-based low-cost hardware architecture for thc PRESENT block cipher. In: Proceedings of the Euromicro conference on digital system design, pp 646–650
18. Moradi A, Barenghi A, Kasper T, Paar C (2011) On the vulnerability of FPGA bitstream encryption against power analysis attacks: extracting keys from Xilinx Virtex-II FPGAs. In: Proceedings of the 18th ACM conference on computer on communications security (CCS). New York, NY, USA, pp 111–124
19. Aysu A, Gulcan E, Schaumont P (2014) SIMON says: break area records of block ciphers on FPGAs. IEEE Embed Syst Lett 6(2):37–40
20. Kaps JP (2008) Chai-tea, cryptographic hardware implementations of xtea. In: INDOCRYPT 2008. Lecture notes in computer science, vol 5365. Springer, pp. 363–375
21. Yalla P, Kaps J-P (2009) Compact FPGA implementation of Camellia. In: Daněk M, Kadlec J, Nelson B (eds) FPL 2009. IEEE, pp 658–661
22. Hwang D, Chaney M, Karanam S, Ton N, Gaj K (2008) Comparison of FPGA-targeted hardware implementations of eSTREAM stream cipher candidates. In: Proceeding state of the art of stream ciphers workshop, SASC 2008, Lausanne, Switzerland, pp 151–162
23. Mace F, Standaert F-X, Quisquater JJ (2008) FPGA implementation(s) of a scalable encryption algorithm. IEEE Trans Very Large Scale Integr (VLSI) Syst 16(2):212–216
24. Lara-Nino CA, Diaz-Perez A, Morales-Sandoval M (2017) Lightweight hardware architectures for the PRESENT cipher in FPGA. IEEE Trans Circuits Syst I Regul Pap 64:2544–2555

VLSI Implementation of ESF and QTL Lightweight Ciphers

Nivedita Shrivastava, Bibhudendra Acharya, and Ajay Singh Raghuvanshi

Abstract In order to provide security in resource scare environment, various lightweight ciphers are proposed. Many optimized architectural designs for these ciphers are presented by several authors to enhance the compatibility with lightweight domain. In this work, in order to provide security in lightweight environment, three optimized hardware architectures of eight-sided fortress (ESF) and QTL ciphers are proposed. Two architectures for ESF cipher are presented based on different S-box generation techniques. RAM-based hardware design of QTL cipher is also presented. Designs are evaluated and compared in various FPGA platforms, on the basis of resource consumption, performance and power requirement for their implementation. ESF architecture with RAM-based S-box requires least resource consumption for its implementation.

Keywords Lightweight cryptography · Security · SPN network · QTL · ESF · Security

1 Introduction

In recent years, there is huge growth in various computing devices like smartphones, smart cards, radio-frequency identification tags (RFID), etc. With the growing usage of these ubiquitous devices and rapid advancement in Internet of things (IoT), there is constant threat to security and privacy of the information and data of end users. These devices are equipped with low-end microcontrollers, which have small word length

N. Shrivastava · B. Acharya (✉) · A. S. Raghuvanshi
Department of Electronics and Telecommunication, National Institute of Technology, Raipur, Raipur, India
e-mail: bacharya.etc@nitrr.ac.in

N. Shrivastava
e-mail: nivedita.shrivastava01@gmail.com

A. S. Raghuvanshi
e-mail: asraghuvanshi.etc@nitrr.ac.in

V. Nath and J. K. Mandal (eds.), *Proceedings of the Fourth International Conference on Microelectronics, Computing and Communication Systems*, Lecture Notes in Electrical Engineering 673, https://doi.org/10.1007/978-981-15-5546-6_44

[1]. They also use slow oscillators. There are limited power generation resources available. As a result for enhancing performance various power optimization techniques [2], novel transmission techniques [3] and many energy harvesting techniques [4] are used. This all leads to the requirement for lightweight primitives to provide proper privacy and security in this constrained-resource environment.

A wide range of ciphers is introduced by various authors to combat the issue of security and privacy. Along with traditional ciphers like advanced encryption standards (AES) [5] and data encryption standard (DES) [6] many other ciphers are proposed in the field of lightweight cryptography. Lightweight ciphers are mainly distinguished as Feistel network-based ciphers and substitution-permutation (SP) network-based ciphers. In SP network-based ciphers, diffusion and confusion of bits are performed. Due to the property of diffusion when a minor modification is made in plaintext, huge variation is observed in cipher text. As a result of property of confusion, there is increased complexity between the key and the plaintext. Some of the popular SP network-based ciphers are PRESENT [7], KLEIN [8], RECTANGLE [9], light encryption device (LED) [10], mCrypton [11], etc. Another type of lightweight ciphers is known as Feistel ciphers. In these ciphers, encryption is performed on half of the data, thus they require more rounds of encryption in order to provide proper security. Some of the famous Feistel ciphers are PICCOLO [12], LBLOCK [13], CLEFIA [14], SIMON and SPECK [15], TWINE [16], etc.

Various design strategies are proposed by different authors in order to establish an optimum trade-off between resource requirements, power consumption as well as to enhance performance of the designs. Some of the famous design strategies are pipelined architecture, serial architectures, iterative designs, parallel architecture and many more.

The major disadvantage of application-specific integrated circuits (ASIC) is the need for special memory generators, which leads to increased cost of design. In the case of field programmable gate array (FPGA), this problem is avoidable as block random access memory (BRAM) is already present in the device. For applications with massive computational requirement, registers are used for storing data. This leaves BRAM unused. Block ciphers are considered perfect for such applications [17].

In this work, a new hardware implementation of lightweight ciphers using BRAM of FPGA is proposed. BRAM is used for storing data and subkey. It is also used for designing substitution-boxes (S-boxes) of the architectures. With this approach slices, look-up table (LUTs) and flip-flops are free to be used for any other applications. For designing hardware architectures, target algorithms are eight-sided fortress (ESF) [18] and QTL [19] cipher. These are basically 64-bits Fiestel cipher with SP network-based round function. For ESF cipher two designs are proposed based on S-box implementation techniques. A change in resource consumption and performance with the change in the implementation technique of S-box is observed. A BRAM based architecture is proposed for the QTL cipher also. Two FPGA platforms are used for implementation, comparison and evaluation of the designs.

This work is divided into six sections. Section 2 describes the basic algorithm of QTL and ESF cipher. Section 3 describes the proposed hardware architecture of the both the ciphers. Section 4 presents the experimental methodology used in this work. Section 5 illustrates results obtained and presents short discussion of the results. Section 6 summaries the work in a short conclusion.

2 Algorithm Overview

2.1 *ESF Cipher*

ESF is a 64-bits Fiestel cipher with a key length of 80-bits. It performs 32 rounds of encryption to generate cipher text. Figure 1 shows the basic algorithm of ESF cipher. In each round of encryption the following operations are performed:

- Input of 64-bits is divided into set of 32-bits. Leftmost 32-bits of data are shifted left with an offset of 7-bits. Rightmost 32-bits are processed via a round function 'F' which is described further.
- These rightmost and leftmost 32-bits are then XORed. Thereafter, the output of this XOR operation is swapped with un-processed rightmost 32-bits.
- After completion of all the rounds, the final ciphertext is obtained. Least significant bits (LSB) are obtained through right part, whereas most significant bits (MSB) are obtained through left part of the architecture.

Round Function Figure 2 shows the steps involved in the generation of round function, which is used to process rightmost 32-bits of ESF cipher. It is described as below:

- Rightmost 32-bits of intermediate data are XORed with 32-bits of the subkey.

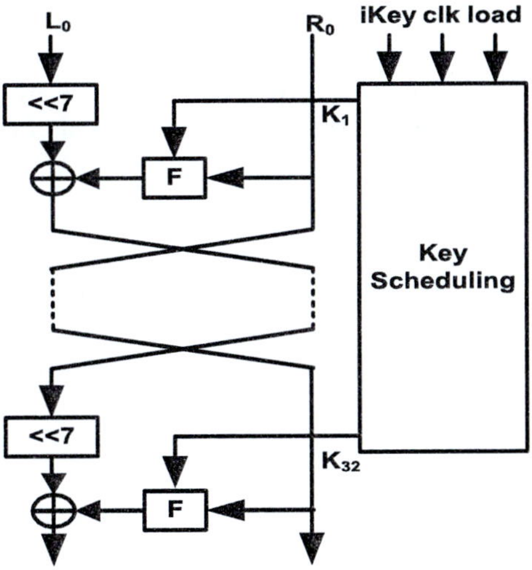

Fig. 1 Encryption process of ESF cipher

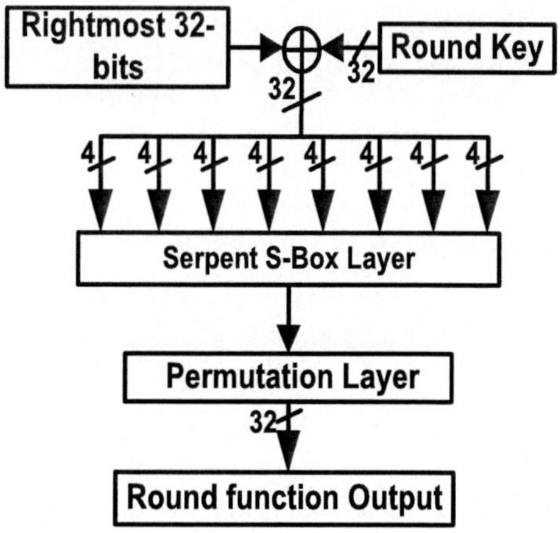

Fig. 2 Round function of ESF cipher

- In these 32-bits, substitution is performed by means of SERPENT S-boxes. These are basically 4 × 4 S-boxes, i.e., they produce an output of 4-bits and they also accept input of 4-bits. They are basically a set of eight different S-boxes, which process data in parallel. There are various methods of designing S-boxes like RAM-based S-box, Boolean S-box or LUT-based S-box.
- After substitution, permutation is performed by means of a permutation layer (P-layer). ESF performs bit-wise permutation. It follows the principle used in designing the permutation layer of PRESENT cipher.

2.2 QTL

QTL is an ultra-lightweight block cipher. It has a block length of 64-bits and can work for a key length of both 64-bits and 128-bits referred as QTL-64 and QTL-128. For the key size of 64-bits, 16 rounds are required, while for 128-bits, 20 rounds of encryption are required. It is devoid of any key scheduling mechanism. Figure 3 shows the basic encryption process of QTL cipher. For a single round of encryption, the following steps are followed:

- Master key of 64-bits is divided in set of 16-bits each and is represented as $\{K_0, K_1, K_2, K_3\}$.
- For each set of 32-bits of plaintext, separate Fiestel transformation is applied. For a single Fiestel transformation structure, rightmost 16-bits of plaintext are XORed with subsequent leftmost 16-bits. These leftmost bits are first processed through round function F1/F2. Thereafter, bits are swapped in the end of each Fiestel transformation. The process is again repeated for a single round.
- At the end of single encryption round, round transposing is performed as shown in Fig. 3.

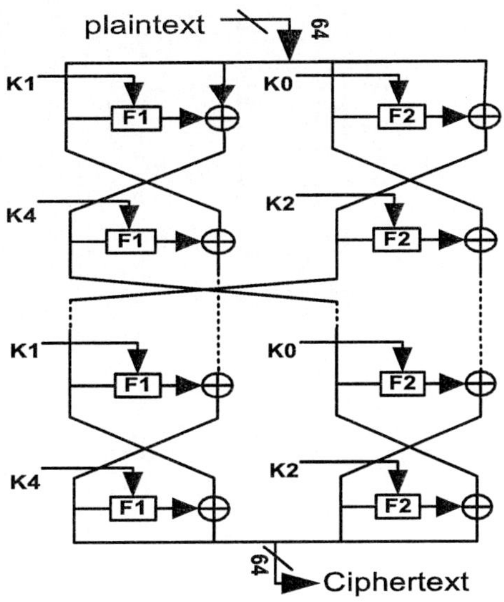

Fig. 3 Encryption process of QTL cipher

- For processing rightmost 32-bits of data, round function F1 is used, while for processing leftmost 32-bits, F2 round function is used. These round functions have separate sets of round constants and different S-boxes as well.

3 Proposed Hardware Architecture of ESF and QTL Cipher

In this section, the proposed hardware architectures for both ESF and QTL ciphers are discussed.

3.1 Eight-Sided Fortress

In this architecture, BRAM of FPGA is used to store processed intermediate data. BRAM is also used to design S-box layer. ESF uses SERPENT S-boxes, which are basically a set of eight different S-boxes. As shown in Fig. 4, two BRAMs are used to store the data. Two clock cycles are required to read and write data from BRAM. This leads to a total latency of 63 clock cycles. For each round, subkeys are pre-generated and stored in a separate RAM module. Subkeys are extracted and XOR operation is performed with rightmost 32-bits of data. The result is then sent to another BRAM for performing substitution.

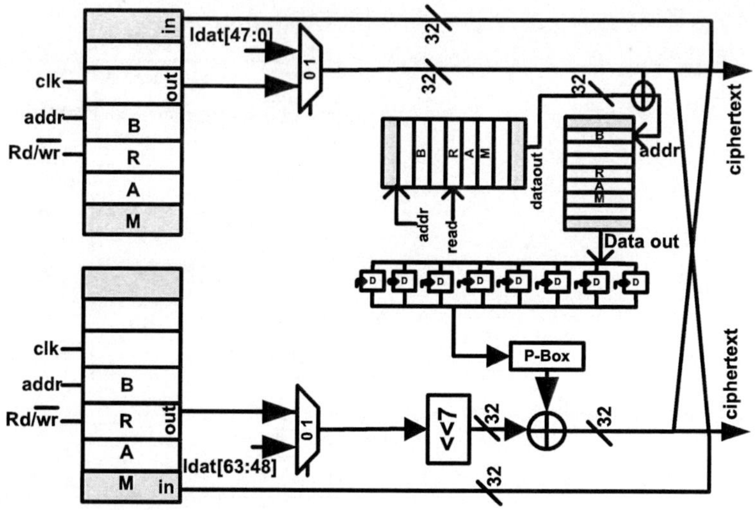

Fig. 4 Hardware architecture of ESF cipher with RAM-based S-box

Thereafter, diffusion of bits is performed using permutation layer. Leftmost 32-bits are shifted by 7-bits and result is XORed with processed rightmost bits. After completion of all the rounds of encryption, the ciphertext is obtained.

The architecture of ESF cipher with a different technique of implementing S-boxes is shown in Fig. 5. Separate BRAMs are used to store 64-bits of data and 32-bits of the key. There are various ways of designing S-boxes. In this work, Boolean expressions for all the eight S-boxes are derived and used to design the S-boxes. As a result, S-boxes can be easily designed using basic logic gates, i.e., AND, OR and NOT. XOR operation is performed with pre-generated subkey and rightmost 32-bits of data. Confusion of bits is performed using a set of Boolean S-boxes, followed by diffusion through P-box. Again, XOR operation is performed between these processed rightmost bits and seven-bits shifted leftmost bits. Thereafter, data is swapped and again send to BRAM for storage.

3.2 QTL

As shown in Fig. 6, 64-bits master key is divided in the set of 16-bits subkey and fed directly to the round functions F1 and F2. RT is a function used to perform round transposition. In this function, 64-bits of input is divided in set of 16-bits, and permutation is performed. After that bit-sets are again concatenated into set of 64-bits. Figure 7 shows round function for QTL cipher. 8-bits round constants for both functions F1 and F2 are stored in a separate ROM module.

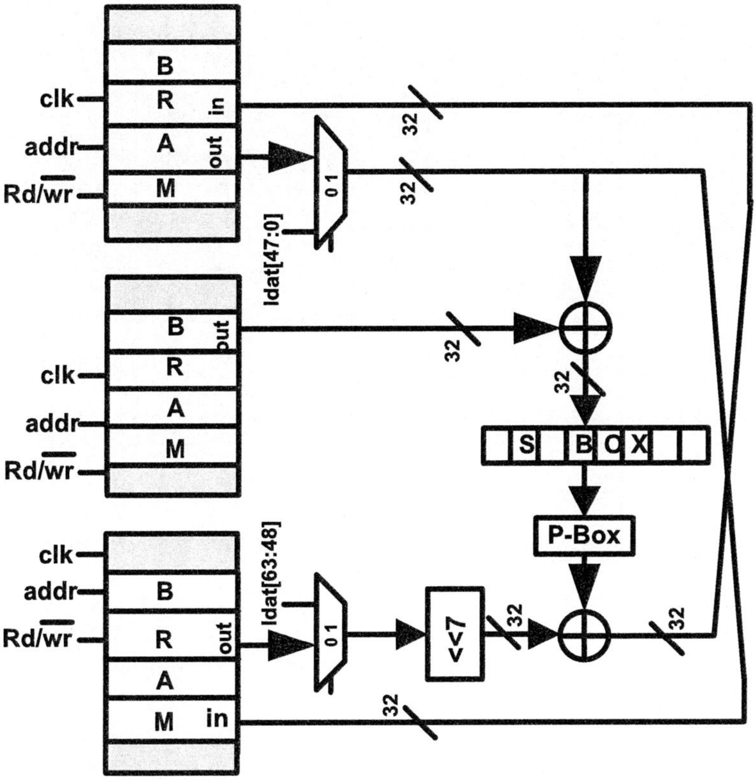

Fig. 5 Hardware architecture of ESF block cipher with Boolean S-box

They are retrieved from ROM and XORed with rightmost 8-bits of intermediate data. Then, another XOR operation is performed with the input key. Thereafter, substitution is performed followed by permutation again followed by substitution. For both the functions F1 and F2, a different set of S-boxes and round constants are used.

4 Simulation Platform

4.1 Metrics

Various metrics are discussed to evaluate ciphers. For analysis of resource consumption by the architectures; number of slices, LUTs and flip-flops are considered. For power analysis dynamic, static and total power is considered at the operational frequency of 13.56 MHz.

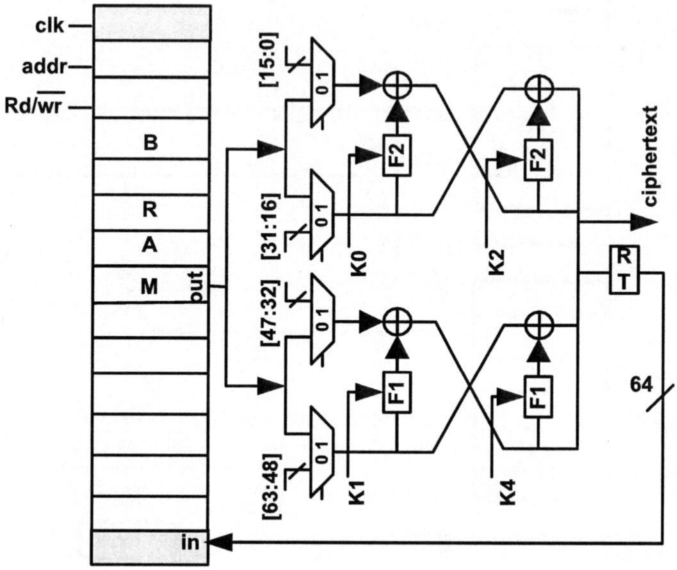

Fig. 6 Hardware architecture of QTL cipher

Fig. 7 Round function of QTL cipher

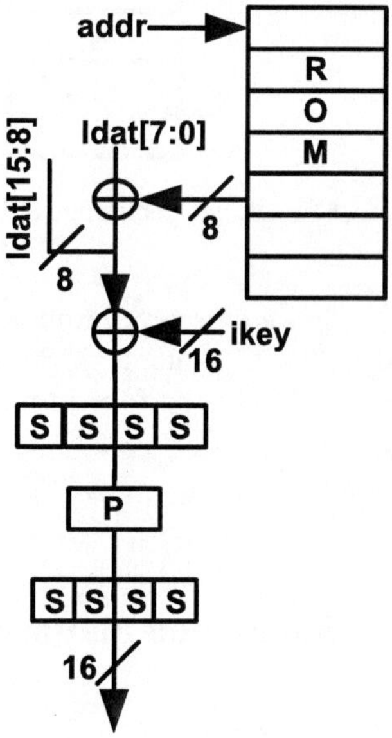

Performance is evaluated using both latency and throughput of the architectures. Throughput at the maximum operational frequency is calculated according to Eq. (1):

$$\text{Throughput} = \frac{\text{Block Size} \times \text{FMax}}{\text{Latency}} \tag{1}$$

Low power devices do not work at the maximum operational frequency. They work at some lower operational frequency. For RFID tags, a frequency of 13.56 MHz is used [20]. Throughput at 13.56 MHz (Thr*) is calculated according to Eq. (2):

$$\text{Thr}^* = \frac{\text{Block Size} \times 13.56\,\text{MHz.}}{\text{Latency}} \tag{2}$$

A derived metric throughput per slice (Thr*/slc) is used to analyze the relationship between performance and resource consumption of the architecture. It is obtained by dividing the throughput achieved at 13.56 MHz with the number of slices used in the implementation of the respective design.

4.2 *Environments*

All the proposed architectures are designed and implemented in ISE design tool 13.6. To analyze the architectures LUT-4 and LUT-6 FPGA platforms are used.

For LUT-4 FPGA, Virtex-4 (xc4vlx25-12ff668) is considered. While for analysis in LUT-6 FPGA, Virtex-5(xc5vlx50t-3ff1136) is considered. Power is estimated using Xilinx XPower Analyzer tool. This tool gives static, dynamic and total power consumed by the architectures.

4.3 *Configurations*

Three designs which are implemented and evaluated are summarized below:

A1 A RAM-based design of ESF cipher. This design stored subkeys and intermediate data in a separate BRAM module. S-boxes are also designed using BRAM available in FPGA.

A2 Second RAM-based design of ESF cipher. BRAM is used to store processed intermediate data only. S-boxes are designed using Boolean logic.

A3 RAM-based design of QTL cipher. This design uses BRAM for storing intermediate data. Boolean S-boxes are used to perform substitution operation.

5 Results and Discussion

Different FPGA platforms are used to implement and evaluate different designs. The obtained results are shown in Table 1. Proposed architectures are compared with hardware implementation of PRESENT cipher in the respective FPGA platform. A4 shows the result obtained by the implementation of the base hardware design of PRESENT cipher. A5 shows result obtained for optimized implementation of PRESENT cipher.

It is observed that the number of flip-flops remains consistent for both the platforms. Their number is reduced significantly on using BRAM for storage; as a result, unused resources can be used for some other applications. Least number of LUTs is required by A1 for Virtex-4 FPGA. On changing the implementation technique of S-box from Boolean (A2) to RAM-based (A1), the number of LUTs is reduced by 19.28%, while the number of slices is reduced by 20.67% in Virtex-4 platforms. In the Virtex-5 platform, the number of slices is reduced by 14.67%. For Virtex-5 platforms, both A1 and A2 require nearly the same number of LUTs, followed by A3, A5 and A4, respectively.

From Fig. 8, it is seen that QTL hardware architecture requires more area for its implementation as compared to ESF hardware design. For both FPGA platforms, the best throughput at a frequency of 13.56 MHz is given by QTL cipher. The best result of throughput at the maximum operational frequency is given by A4 followed by A3 for both the FPGA platforms.

Figure 9 shows a graphical illustration of Thr*/slc obtained for Virtex-4 and Virtex-5 platform. For Virtex-4 platform best result is given by ESF architecture with RAM-based S-box (A1). For Virtex-5 FPGA, RAM-based design of QTL cipher (A4) gives the best trade-off between performance and resource consumption of the architecture. As compared to A5, Thr*/slc is improved in A4 by 40.78%. Thr/slc* for A1 is nearly 2.42 times as compared to A4 for Virtex-4 FPGA.

It is observed that the static power remains nearly constant for both the FPGA platforms. Dynamic power shows variation as it depends on switching activity of the architecture. Total power is simply the addition of static and dynamic power. Among the proposed designs, for Virtex-4 FPGA, A3 shows the least power requirement, while for Virtex-5 design, A1 shows least power requirement. Power analysis shows that there is a significant change in the value of power consumption with the change in the device used for the implementation of architectures.

6 Conclusion

In this work, hardware architectures of ESF and QTL ciphers are designed and implemented in FPGA platforms. Results are evaluated and compared on the basis of resource consumption, power consumption and performance of the designs in different FPGA platforms.

Table 1 Presents results obtained for different architectures in various computing platforms

Conf.	State (bit)	Key (bit)	Latency (cycle)	Slice	LUT	Flip-flop	FMax. (MHz)	Static power (mW)	Dynamic power (mW)	Total power (mW)	Thr (Mbps)	Thr* (Mbps)	Thr*/slc (Mbps/slc)
Virtex-4 (xc4vlx25-12ff668)													
A1	64	80	63	69	134	6	179.79	233.66	49.17	282.83	182.64	13.77	0.199
A2	64	80	63	87	166	6	116.17	233.59	45.60	279.19	118.02	13.77	0.158
A3	64	64	31	173	324	5	99.15	233.63	47.68	281.31	204.69	27.99	0.161
A4 [7]	64	128	55	192	382	200	284.33	333.44	15.44	348.88	330.86	15.78	0.082
A5 [1]	64	128	136	152	265	201	364.56	233.01	15.01	248.02	171.56	6.38	0.041
Virtex-5 (xc5vlx50t-3ff1136)													
A1	64	80	63	75	135	6	184.89	560.15	13.81	573.96	187.82	13.77	0.184
A2	64	80	63	64	134	6	118.02	560.41	39.99	600.40	119.89	13.77	0.215
A3	64	64	31	111	234	5	111.63	560.41	40.60	601.02	230.46	27.99	0.252
A4 [7]	64	128	55	88	283	200	271.67	560.04	3.47	563.51	316.12	15.78	0.179
A5 [1]	64	128	136	73	239	201	431.78	560.04	2.63	562.67	203.19	6.38	0.087

*Results obtained using frequency of 13.56 MHz

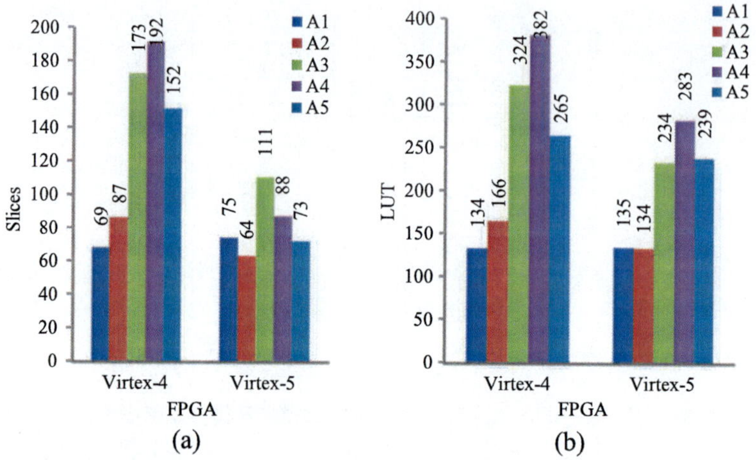

Fig. 8 Resource consumption for different architectures (a) slices and (b) LUT

Fig. 9 Throughput/slice for different designs as 13.56 MHz

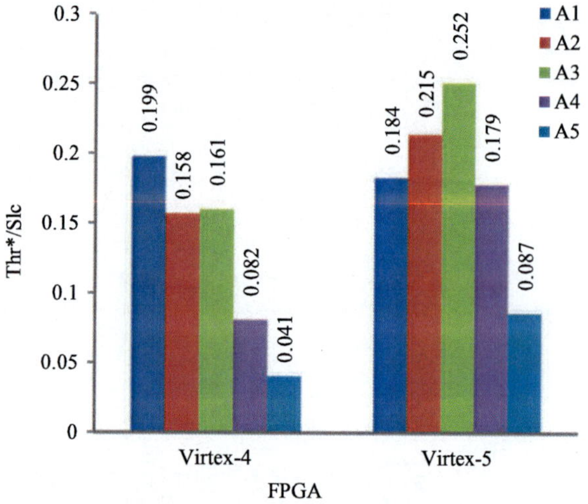

BRAM of FPGA is used for storing intermediate processed data. This makes registers free to be used for any other purpose. In the case of ESF cipher, the impact of changing S-box implementation technique in performance and resource consumption is also presented. Least resource consumption is required by ESF architecture with RAM-based S-box implementation. Best performance is given by RAM-based QTL cipher in the Virtex-4 platform.

Further studies are required to design and compare other architecture of QTL cipher with different S-box implementation techniques. There is also scope for designing of serial and pipelined architectures of both QTL and ESF cipher.

References

1. Lara-Nino CA, Diaz-Perez A, Morales-Sandoval M (2017) Lightweight hardware architectures for the PRESENT cipher in FPGA. IEEE Trans Circuits Syst I Regul Pap 64:2544–2555
2. Haeri AAR, Karkani MG, Sharifkhani M, Kamarei M, Fotowat-Ahmady A (2017) Analysis and design of power harvesting circuits for ultra-low power applications. IEEE Trans Circuits Syst I Regul Pap 64(2):471–479
3. Zou Z et al (2011) A low-power and flexible energy detection IR-UWB receiver for RFID and wireless sensor networks. IEEE Trans Circuits Syst I Regul Pap 58(7):1470–1482
4. Alippi C, Galperti C (2008) An adaptive system for optimal solar energy harvesting in wireless sensor network nodes. IEEE Trans Circuits Syst I Regul Pap 55(6):1742–1750
5. Pub NF (2001) 197: advanced encryption standard (AES). Fed Inf Process Stand Publ 197:0311
6. DE Standard (1997) Federal information processing standards publication 46, vol 23. National Bureau of Standards, US Department of Commerce
7. Bogdanov A, Knudsen LR, Leander G, Paar C, Poschmann A, Robshaw MJ, Seurin Y, Vikkelsoe C (2007) PRESENT: an ultra-lightweight block cipher. In: International workshop on cryptographic hardware and embedded systems, vol 4727, pp 450–466
8. Gong Z, Nikova S, Law YW (2011) KLEIN: a new family of lightweight block ciphers. In: International workshop on radio frequency identification: security and privacy, pp 1–18
9. Xiang Z, Zhang W, Bao Z, Lin D (2015) RECTANGLE: a bit-slice lightweight block cipher suitable for multiple platforms. Sci China Inf Sci 58:1–15
10. Guo J, Peyrin T, Poschmann A, Robshaw M (2011) The LED block cipher. In: Lecture notes in computer science (including subseries Lecture notes in artificial intelligence and lecture notes in bioinformatics): preface, vol 6917, pp 326–341
11. Lim CH, Korkishko T (2005) mCrypton—a lightweight block cipher for security of low-cost RFID tags and sensors. In: International workshop on information security applications, pp 243–258
12. Shibutani K, Isobe T, Hiwatari H, Mitsuda A, Akishita T, Shirai T (2011) Piccolo: an ultra-lightweight block cipher. In: International workshop on cryptographic hardware and embedded systems, pp 342–357
13. Wu W, Zhang L (2011) LBlock: a lightweight block cipher. In: International conference on applied cryptography and network security, pp 327–344
14. Shirai T, Shibutani K, Akishita T, Moriai S, Iwata T (2007) The 128-bit block cipher CLEFIA. In: International workshop on fast software encryption, pp 181–195
15. Beaulieu R, Treatman-Clark S, Shors D, Weeks B, Smith J, Wingers L (2015) The SIMON and SPECK lightweight block ciphers. In: Design automation conference (DAC), 52nd ACM/EDAC/IEEE, pp 1–6
16. Suzaki T, Minematsu K, Morioka S, Kobayashi E (2011) Twine: a lightweight, versatile block cipher. In: ECRYPT workshop on lightweight cryptography
17. Kavun EB, Yalcin T (2011) RAM-based ultra-lightweight FPGA implementation of PRESENT. In: Proceedings of the international conference on reconfigurable computing and FPGAs (ReConFig), pp 280–285
18. Xuan LIU, Zhang WY, Liu XZ, Feng LIU (2014) Eight-sided fortress: a lightweight block cipher. J China Univ Posts Telecommun 21:104–128
19. Li L, Liu B, Wang H (2015) QTL: a new ultra-lightweight block cipher. Microprocess Microsyst 45:45–55
20. Identification cards—contactless integrated circuit cards—proximity cards—part 2: radio frequency power and signal interface, document ISO/IEC 14443-2, Aug 2010

Optimal Allocation of AVR and DGs in Distribution Systems Using HSA

K. R. K. V. Prasad and Kollu Ravindra

Abstract In recent years, distributed generation (DG) penetration in distribution networks has increased at a high rate due to advancement in electricity markets. Apart from serving loads locally, installation of DG offers several technical benefits such as enhancement of voltage profile and reduction in network loss. Application of automatic voltage regulator (AVR) along with DGs simultaneously placed will further increase technical benefits. In this work, the impact of DG along with AVR placement is analyzed with an objective to minimize network losses and voltage deviation. The solution to the optimization problem is determined using harmony search algorithm (HSA). Proposed approach's efficacy is verified on a standard IEEE 69-bus distribution network.

Keywords Distributed generation · Automatic voltage regulator · Harmony search algorithm

1 Introduction

Improved system operation with good voltage profile and minimum energy losses can benefit both utilities and customer. Enhancement of voltage profile can be done by placing AVRs, DGs, capacitor banks and resorting to reconfiguration and optimal conductor selection [1–4]. The optimal placement of distributed generation (DG) units is pivotal for planning engineers in distribution networks as improper allocation of DG results in the higher fault currents, losses, and over-voltages. So it is essential to examine the consequence of DG on distribution systems [5, 6]. Several researchers

K. R. K. V. Prasad (✉)
EEE Department, Aditya College of Engineering and Technology, Surampalem, AP, India
e-mail: krvprasad219@gmail.com

K. Ravindra
EEE Department, University College of Engineering Kakinada, JNTUK, Kakinada, AP, India
e-mail: ravikollu@gmail.com

V. Nath and J. K. Mandal (eds.), *Proceedings of the Fourth International Conference on Microelectronics, Computing and Communication Systems*, Lecture Notes in Electrical Engineering 673, https://doi.org/10.1007/978-981-15-5546-6_45

used evolutionary methods such as simulated annealing [4], particle swarm optimization [7], genetic algorithms (GA) [4], and artificial bee colony to solve optimization problems in electrical networks [4]. In [8], GA is employed to optimally allocate AVRs in a distribution network. Benefits can be maximized if DGs and other devices like capacitors, AVRs are correctly integrated into the system.

Aim of this work is to find optimal sizes and corresponding locations for DGs and optimal tap setting of AVRs and their locations in a radial distribution system by employing HSA. Optimal allocation problem is solved subjected to the constraints of maximum DG penetration limit, branch current limits, tap position of AVRs, injected active and reactive power limits of DG units, and permissible bus voltage limits [9].

2 Problem Formulation

In this paper, the objective is to minimize distribution network active power loss and enhance bus voltages. Generally, loss can be minimized by optimizing the currents flows in the lines. Hence, DG and AVRs are placed optimally to mitigate power loss and reduce voltage deviation.

2.1 Objective Function

In this work, minimization of loss index (f_p) is chosen as an objective function which is formulated as ratio of power loss with DG and/or AVR to the base case power loss, where base case power loss means network power loss without DG and/or AVR

$$f_p = \left(\sum_{i}^{\text{Nbr}} PL_i^{\text{with\_ DG/AVR}}\right) / \left(\sum_{i}^{\text{Nbr}} PL_i^{\text{base}}\right) \tag{1}$$

Optimization problem is solved adhering to the below constraints:

(i) Voltage constraint

$$V_{\min} \leq V_i^{\text{with\_ DG/AVR}} \leq V_{\max} \tag{2}$$

(ii) Branch current limits (thermal limits)

$$I_i \leq I_{\text{cap}} \tag{3}$$

(iii) DG power capacity constraint

$$P_{\text{DG\_ min}} \leq P_{\text{DG}i} \leq P_{\text{DG\_ max}} \tag{4}$$

(iv) Tap settings constraint for AVR

$$T_{\min} \leq T \leq T_{\max} \quad (5)$$

where $\mathrm{PL}_i^{\text{with\_ DG/AVR}}$ represents the total real power loss with DG and/or AVR, $\mathrm{PL}_i^{\text{base}}$ corresponds to the total real power loss with base case, $V_{\min}$ is the minimum voltage, $V_{\max}$ is the maximum voltage, $P_{\text{DG\_ min}}$ is the minimum DG power capacity, $P_{\text{DG\_ max}}$ is the maximum DG power capacity, $T_{\min}$ is the minimum tap setting, and $T_{\max}$ is the maximum allowable tap setting.

Bus voltage deviation index (f_v) of the system is computed as given in (6). Minimization of f_v can be chosen as objective.

$$f_v = \left(\sum_{i=2}^{\text{nbus}} \left(V_i^{\text{with\_ DG/AVR}} - 1\right)^2\right) / \left(\sum_{i=2}^{\text{nbus}} \left(V_i^{\text{base}} - 1\right)^2\right) \quad i = 1, 2, \ldots, \text{nbus} \quad (6)$$

where nbus represents a total number of nodes, f_v denotes voltage deviation index, which is formulated as the ratio of voltage deviation with DG/AVR to the voltage deviation without DG/AVR.

2.2 *Weighted Sum Approach*

In this, multiple objectives of minimization of f_p and f_v are combined to form a single objective as shown in (7). Weights w_1 and w_2 are chosen according to distribution network planner priorities. The overall objective function is optimized and a near-optimal solution is found for DG and AVR placement problem.

$$f = w_1 f_p + w_2 f_v \quad (7)$$

2.3 *Automatic Voltage Regulator (AVR) Model*

A model is defined to know the impact of AVR in a distribution network [10]. An AVR is an autotransformer with tap setting control mechanism on its windings and it is used to control distribution network voltages. AVR placed between buses A and B with its initial tap setting as t and internal admittance (Y) is shown in Fig. 1. AVR is modeled as series and parallel admittances as represented in Fig. 2.

Modified AVR model when autotransformer tap setting is changed from t to $t + \Delta t$ is shown in Fig. 3. An alternate way to replicate tap position variation is to adjust the model shown in Fig. 3 by including imaginary injection currents I_A and I_B as shown in Fig. 4. The fictitious injected currents are given by (8) and (9).

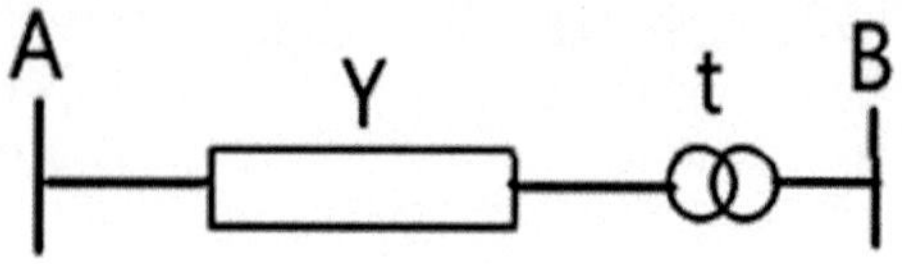

Fig. 1 AVR model in branch

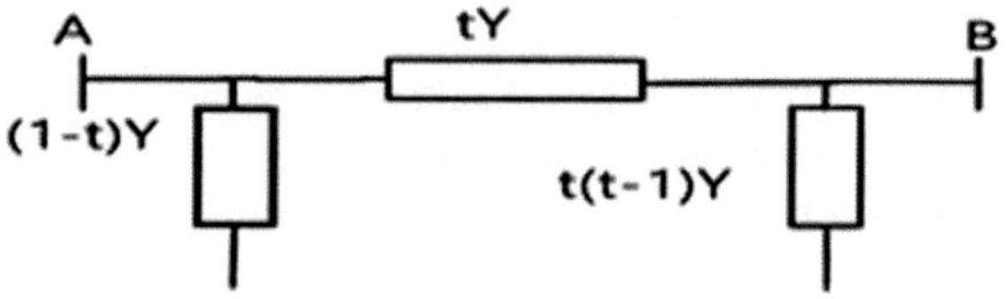

Fig. 2 π model representation of AVR

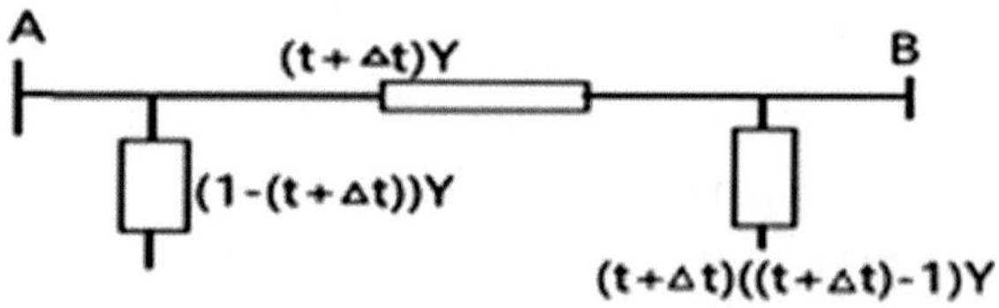

Fig. 3 AVR model including tap position variation

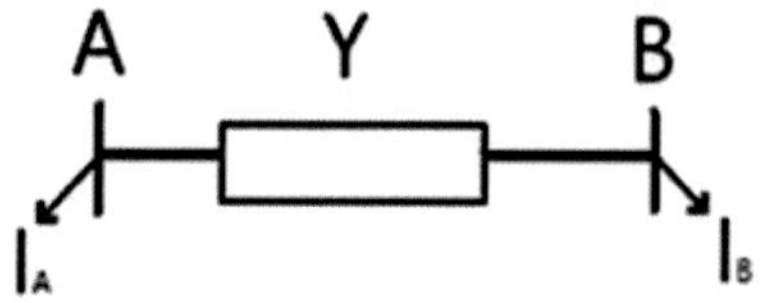

Fig. 4 AVR equivalent with fictitious current injections

$$I_A = (1 - (t + \Delta t)) * Y * V_B \tag{8}$$

$$I_A = (1 - (t + \Delta t)) * Y * V_B + \left((t + \Delta t)^2 * Y * V_B\right) \tag{9}$$

2.4 Harmony Search Algorithm

HSA is inspired by natural musical process searching to get an intact harmony. This step-by-step efficient search process uses a stumble search. The workflow for HSA is as follows [11, 12].

Step 1: Choose the objective function and identify decision variables subjected to boundary conditions of the variables. Specify the algorithm parameters, viz. harmony memory size (HMS), the memory consideration rate (HMCR), bandwidth (BW), the pitch adjusting rate (PAR), and the maximum number of iterations required ($iter_{max}$).

$$\begin{gathered}\text{Minimise } f(x)\\ \text{Subject to } x_{iL} \leq x_i \leq x_{iu}\ (i = 1, 2, 3, \ldots, N)\end{gathered} \tag{10}$$

where x_{iL} and x_{iu} correspond to minimum and maximum boundaries for decision variables, respectively.

Step 2: Initialization the harmony memory (HM) using the following equation:

$$x_i^j = x_{iL} + \text{rand} * (x_{iU} - x_{iL}) \quad j = 1, 2, \ldots, \text{HMS} \tag{11}$$

where rand corresponds to an arbitrary number generated between 0 and 1.

Step 3: In this step, a new harmony is improvised from *HM*. A new harmonic improvised based on three norms: memory deliberation, pitch adjustment, and random generation. Firstly, a uniform random number r is originated in the boundary [0, 1]. If r is lower than the HMCR, the decision variable x_i^{new} is originated by the memory deliberation; otherwise,x_i^{new} is randomly generated. Then, pitch adjustment for x_i^{new} is done with a probability of PAR if it is originated by the memory deliberation. The pitch adjustment norm is shown below:

$$x_i^{\text{new}} = x_i^{\text{new}} + r * bw \tag{12}$$

Step 4: After originating x_i^{new}, the harmony memory will be upgraded as follows. If the fitness of the $x_i^{\text{new}} = \left(x_1^{\text{new}}, x_2^{\text{new}}, \ldots, x_n^{\text{new}}\right)$ is better than the fitness of the least fit harmony vector in the HM, subsequently it is incorporated in HM. Otherwise, it is discarded.

Step 5: Repeat steps 3 and 4 for $\text{iter}_{\max}$ number of times.

3 Proposed Approach

Step 1: Define loss as objective function, DG sizes and locations, tap changers and locations as decision variables. Initialize HSA parameters and specify the IEEE 69-bus test system line and load data, constraints.

Step 2: Initialize harmony memory with solution vectors (DG sizes and locations, tap changers and locations).

$$\text{SV} = \left[\text{DG}_1^1\ \text{DG}_2^1 \ldots \text{DG}_N^1\ L_1^1\ L_2^1\ L_3^1 \ldots L_N^1\ T_1^1\ T_2^1\ T_3^1 \ldots T_N^1\ TL_1^1\ TL_2^1\ TL_3^1 \ldots TL_N^1\right] \tag{13}$$

Each solution vector in HM is evaluated by finding the objective function value by running a load flow. Backward–forward sweep method is used to determine load flows. Sort the solution vectors in HM based on their ascending order of functional values.

Step 3: Extemporize a new harmony.

Step 4: Upgrade the HM.

Step 5: Examine the termination criteria and stop.

4 Results

To verify the efficacy of the proposed approach, it is tested on an IEEE 69-bus system which is a 12.66 kV system whose data is taken from [13]. HSA parameters chosen for this work are: HMS = 50, $\text{iter}_{\text{max}} = 500$, par = 0.3, HMCR = 0.95, $w1 = 0.5$, and $w2 = 0.5$.

Four scenarios are simulated in this work.

Scenario I: System without DGs and AVRs (base case)
Scenario II: System with AVRs only
Scenario III: System with DGs only
Scenario IV: System with DGs and AVRs (proposed approach 1)
Scenario V: Multi-objectives with DGs and AVRs (proposed approach 2 based on weighted sum approach)

DG penetration is limited to 50% in this work and all the DG units are assumed to be operating at 0.8 pf lagging. Results obtained for the four scenarios are tabulated in Table 1. Loss in the network for scenario I is 224.98 kW. The real power losses are observed to be 205.52 kW for scenario II, 11.23 kW for scenario III, and 10.68 kW for scenario IV using HSA (proposed approach 1). From Table 2, it is observed that power loss is 11.87 kW for scenario IV solved using adaptive genetic algorithm (AGA) [14]. Comparing scenario IV, i.e., HSA and AGA; it is clear that power loss mitigation with HSA is greater than obtained with AGA. Results show that combining DG with AVR results in the highest power loss reduction. From Table 2, it is also interpreted that power loss in scenario IV (AGA) is 11.87 kW when 5 DGs and 15 AVRs are deployed in the network. But with scenario IV (HSA), 5 DGs and 4 AVRs are deployed in the network reducing the loss to 10.68 kW. It can be concluded that proposed approach 1 based on HSA is superior when compared to the method based on AGA. Further, the proposed approach 1 lead to the highest percentage of loss reduction which is 95.25 compared to other scenarios. From Fig. 5, it is evident that voltage profile of the network has considerably improved due to simultaneous placement of AVRs with DGs (scenario IV). Convergence characteristic of the algorithm is shown in Fig. 6.

Table 3 shows that when multiple objectives are chosen for optimization, then both loss reduction percentage and voltage deviation minimization have increased when AVR is employed. Similarly, Table 4 shows that when multiple objectives are chosen for optimization, then both loss reduction percentage and voltage deviation minimization have increased when DG is employed. Based on the results from Tables 1, 3, and 4, it is clear that an approach of simultaneous optimal placement of both DG and AVR using a multi-objective approach will give better results.

Table 1 Results of IEEE 69-bus test system with loss minimization as objective

Planning cases	DGs placement and rating		Total DG apparent power and penetration (MVA)	AVR locations and tap ratio				Total Real power loss (kW)	Minimum bus voltage (p.u)	Percentage loss reduction w.r.t. scenario I	Voltage deviation
	Bus no.	Ratings (MVA)		Branch No.	Tap ratio	Branch no.	Tap ratio				
Scenario I	–	–	–	–	–	–	–	224.98	0.9091	–	0.0993
Scenario II	–	–	–	16 22 06 62	1.1000 0.9063 1.0313 1.0938	–	–	205.52	0.9118	8.65	0.0881
Scenario III	61 17 59 65 62	0.823 0.432 0.187 0.282 0.605	2.329	—	—	–	–	11.23	0.9880	95.01	0.0119
Scenario IV	60 21 63 62 61	0.455 0.456 0.606 0.244 0.555	2.316	02 61 67 63	0.9188 1.0625 1.0938 1.0813	–	–	10.68	0.9878	95.25	0.0121

Table 2 Comparison of results: Scenario IV with AGA and HSA with loss minimization as objective

Planning cases	DGs placement and rating		Total DG size (MVA)	AVR locations and tap ratio				Total loss (kW)	Minimum bus voltage (p.u)	% loss reduction w.r.t. scenario I	Voltage deviation
	Bus no.	Ratings (MVA)		Branch no.	Tap ratio	Branch no.	Tap ratio				
Scenario IV (AGA) [15]	21	0.441	2.328	5	1.0989	48	1.0637	11.87	0.9953	94.739	0.005
	61	0.572		6	1.0938	49	1.098				
	62	0.348		7	1.0984	55	0.9237				
	63	0.539		8	0.9014	57	1.063				
	64	0.428		9	1.0906	59	0.9098				
				10	1.0983	66	0.9305				
				11	1.0958	68	0.9092				
				44	1.0509						
Scenario IV (HSA)	63	0.271	2.33	68	1.0875	09	1.0875	11.28	0.9936	94.98	0.005
	61	0.449		11	0.9188	13	0.9063				
	62	0.859		55	0.9188	59	1.0875				
	65	0.348		39	1.0438	10	0.9563				
	18	0.403		24	1.0750	03	0.9125				

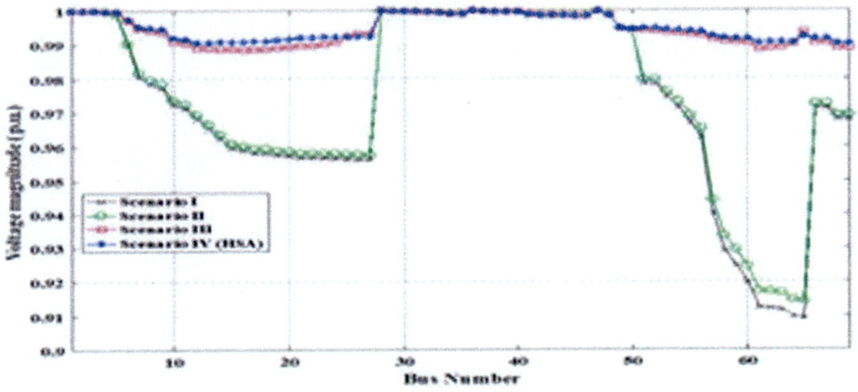

Fig. 5 Voltage profile of 69-bus system

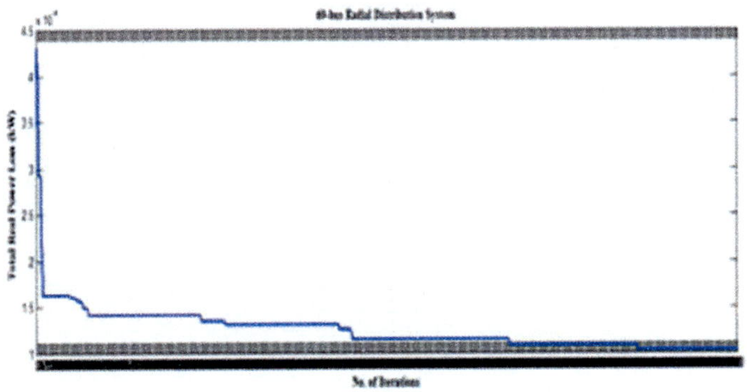

Fig. 6 Convergence characteristic using HSA for scenario IV

Hence, the same is implemented (proposed approach 2—scenario V) and the results are tabulated in Table 5 which vindicate the efficacy of proposed approach 2 giving the best compromised solution in terms of loss and voltage deviation reduction.

5 Conclusion

In this work, AVRs along with DGs placement are done on IEEE 69-bus system with an objective to minimize energy loss. Optimal DG allocation problem, optimal AVR allocation problem, and DG with AVR allocation problem are solved using harmony search algorithm. Power loss reduction using the proposed approach 1, i.e., scenario IV (HSA) is maximum when compared to scenario I, II, and III. Voltage profile has considerably improved due to simultaneous deployment of DGs and AVRs in the

Table 3 Results of IEEE 69-bus test system with AVR

Planning cases	DGs placement and rating		Total DG size (MVA)	AVR locations and Tap ratio				Total loss (kW)	Minimum bus voltage (p.u)	% loss reduction w.r.t. base case	Voltage deviation
	Bus no.	Ratings (MVA)		Branch no.	Tap ratio	Branch no.	Tap ratio				
Base case	–	–	–	–	–	–	–	224.98	0.9091	–	0.0993
Single loss minimization objective (f_p)				16 22 06 62	1.1000 0.9063 1.0313 1.0938			205.52	0.9118	8.65	0.0881
Single Vdev minimization objective (f_v)				64 67 33 20	1.0938 1.0688 0.9625 0.9125			207.55	0.9123	7.75	0.0876
Multi objectives (weighted sum method)				63 61 12 15	0.9313 1.0625 0.9063 1.0813			206.75	0.9119	8.10	0.0880

Table 4 Results of IEEE 69-bus test system with DG

Planning cases	DGs placement and rating		Total DG size (MVA)	AVR locations and tap ratio				Total loss (kW)	Minimum bus voltage (p.u)	% Loss reduction w.r.t. scenario I	Voltage deviation
	Bus no.	Ratings (MVA)		Branch no.	Tap ratio	Branch no.	Tap ratio				
Base case	–	–	–	–	–	–	–	224.98	0.9091	–	0.0993
Single loss minimization objective (f_p)	61 17 59 65 62	0.823 0.432 0.187 0.282 0.605	2.329					11.23	0.9880	95.01	0.0119
Single Vdev minimization objective (f_v)	62 63 26 65 61	0.461 0.659 0.443 0.534 0.233	2.33					15.33	0.9892	93.19	0.0107
Multi-objectives (weighted sum method)	23 17 65 61 60	0.246 0.142 0.503 0.678 0.761	2.33					13.56	0.9882	93.97	0.0117

Table 5 Results of IEEE 69-bus test system with DG and AVR

Planning cases	DGs placement and rating		Total DG size (MVA)	AVR locations and tap ratio				Total loss (kW)	Minimum bus voltage (p.u)	% loss reduction w.r.t. scenario I	Voltage deviation
	Bus no.	Ratings (MVA)		Branch no.	Tap ratio	Branch no.	Tap ratio				
Base case	–	–	–	–	–	–	–	224.98	0.9091	–	0.0993
Single loss minimization objective (f_p)	60 21 63 62 61	0.455 0.456 0.606 0.244 0.555	2.316	02 61 67 63	0.9188 1.0625 1.0938 1.0813	–	–	10.68	0.9878	95.25	0.0121
Single Vdev minimization objective (f_v)	60 64 62 18 65	0.252 0.492 0.734 0.396 0.449	2.323	62 18 58 25	0.9437 0.9000 1.0688 0.9125	–	–	14.19	0.9906	93.69	0.0093
Multi-objectives (weighted sum method)	57 15 61 64 60	0.262 0.354 0.702 0.542 0.458	2.318	63 25 12 26	0.9500 0.9000 0.9063 0.9625	–	–	12.36	0.9882	94.5	0.0120

network. Multi-objective approach of minimizing voltage deviation and loss reduction (proposed approach 2) to solve optimal simultaneous DG and AVR placement problem gave good compromised solution.

References

1. Milosevic MB, Begovic M (2004) Capacitor placement for conservative voltage reduction on distribution feeders. IEEE Trans Power Deliv 19(3):1360–1367
2. Vu H, Pruvot P, Launay C, Harmand Y (1996) An improved voltage control on large-scale power systems. IEEE Trans Power Syst 11(3):1295–1303
3. Ramalinga Raju M, Ramachandra Murthy KVS, Ravindra K, Srinivasa Rao R (2010) Optimal conductor selection for agricultural distribution system—a case study. In: 2010 international conference on intelligent and advanced systems, Manila, pp 1–6
4. Ng HN, Salama MMA, Chikhani AY (2000) Classification of capacitor allocation techniques. IEEE Trans Power Deliv 15(1):387–392
5. Ackermann T, Andersson G, Sder L (2001) Distributed generation: a definition. Electr Power Syst Res 57:195–204
6. Lopes AP, Hatziargyriou N, Mutale J, Djapic P, Jenkins N (2007) Integrating distributed generation into electric power systems: a review of drivers, challenges and opportunities. Electr Power Syst Res 77:1189–1203
7. Naresh G, Ramalinga Raju M, Narasimham SVL, Kollu R (2011) Tuning of power system stabilizers in multi-machine systems using particle swarm optimization. Int Rev Model Simul 4:2470–2479
8. Mendoza J, Morales D, Lopez R, Lopez M, Lopez E, Vannier JC (2005) Optimal location of voltage regulators in radial distribution networks using genetic algorithms. In: Proceedings of 15th power systems computation conference, Belgium
9. Keane A (2013) State-of-the-art techniques and challenges ahead for distributed generation planning andoptimization. IEEE Trans Power Syst 28(2):1493–1502
10. Alarcon-Rodriguez A, Ault G, Galloway S (2010) Multi-objective planning of distributed energy resources: a review of the state-of-the-art. Renew Sustain Energy Rev 14:1353–1366
11. Niknam T, Ranjbar AM, Shirani AR (2005) An approach to Volt/Var control in distribution networks with distributed generation. ScientiaIranica 12(1):34–42
12. Kollu R, Rayapudi SR, Sadhu VLN (2014) A novel method for optimal placement of distributed generation in distribution systems using HSDO. Int Trans Electr Energy Syst 24(4):547–556
13. Geem ZW, Kim JH, Loganathan GV (2001) A new heuristic optimization algorithm: harmony search. Simulation 76(2):60–68
14. Savier JS, Das D (2007) Impact of network reconfiguration on loss allocation of radial distribution systems. IEEE Trans Power Deliv 22(4):885–8977
15. Ganguly S, Samajpati D (2017) Distributed generation allocation with on-load tap changer on radial distribution networks using adaptive genetic algorithm. Appl Soft Comput J 59:45–67

ARM Microcontroller Based Safety and Surveillance System

Jayendra Kumar, S. V. S. Gowtham Reddy, P. N. V. Shiva Krishna, and G. Anjan Kumar

Abstract Nowadays, the entire world is getting into automation of devices which has eased the life of people in many ways. Moving further into the world of automation and building up an environment which will help the people in many ways such as controlling appliances providing security and making home free from thefts and fire accidents has been discussed in this paper. Here, a camera has been used for the identification of the person and a gas sensor which senses the gas leakages if any in the home and alerts the owner of the house regarding this. A temperature detecting sensor has been used which detects any abnormal change in temperature and alerts the user. ARM LPC2148 microcontroller is used here for controlling the sensor, and it also sends an alert message to user. For connectivity between the user and system, GSM module has been used in this process. PIR sensor has also been used to detect any intrusion and LDR sensor for detecting any thefts.

Keywords ARM microcontroller · GSM module · PIR sensor · Temperature sensor · LDR sensor

J. Kumar · S. V. S. Gowtham Reddy (✉) · P. N. V. Shiva Krishna · G. Anjan Kumar
Department of ECE, NIT Jamshedpur, Jamshedpur, Jharkhand 831014, India
e-mail: gowthamsvsr@gmail.com

J. Kumar
e-mail: jkumar.ece@nitjsr.ac.in

P. N. V. Shiva Krishna
e-mail: pnvshivakrishna@gmail.com

G. Anjan Kumar
e-mail: anjan4136@gmail.com

V. Nath and J. K. Mandal (eds.), *Proceedings of the Fourth International Conference on Microelectronics, Computing and Communication Systems*, Lecture Notes in Electrical Engineering 673, https://doi.org/10.1007/978-981-15-5546-6_46

1 Introduction

The main aim of this system is to make home a safer place away from fire accidents and thefts. Security is one of most concerned aspects of everyone. This system enables the user to identify any unwanted intrusion that occurs at home or any fire accident that may arise due to short circuit or gas leakage. The mishap can be identified, and it is reported to the user within minimum time to the user so that he can take necessary measures to avoid any major circumstances. These days we have developed much advancement in technology on IOT [1–3] platform through which we can control any of the appliances be it at home, office or industries with the help of our smart devices like phones and laptops even from large distances. This system is fully automated with minimal manual instructions which make it fast, efficient and reliable. The components used in this system are widely available and are of low cost which makes it low cost effective and available to large part of people.

This system consists of ARM microcontroller [4, 5], GSM module [4, 6], PIR sensor, temperature sensor, gas detecting sensor, camera and relays [7]. ARM LPC2148 microcontroller is user friendly and can be easily interfaced and programmed with various output devices. GSM module is preferred over technologies like Bluetooth [6] and Wi-Fi [1, 5] due to its long-range connectivity with better speeds and efficiency due to its widespread coverage. Long-range connectivity is not possible with Bluetooth and Wi-Fi.

Here, sensors are connected to microcontroller. GSM modem is connected to the microcontroller which acts as a bridge between the user and the system at home. Whenever any person tries to enter home, PIR sensor [8] senses it and camera captures his face and sends it to the user. Based on the reply from the user whether the person is known person or unknown, the system takes necessary actions. If the person is identified as unknown person, then the system raises the alarm and alerts the neighbours and it also sends message to police through GSM module. Thus, it also provides surveillance along with safety from any fire accidents. GSM system has high security infrastructure, which gives us maximum reliability where people cannot monitor information sent or received.

2 Literature Review

Kodali and Jain [1] designed a system which is used for home automation and intruder detection. If any human movement is sensed near the entrance of home, the system sends an alert message to user and also the user can raise the alarm to warn the intruder and prevent him. Here, author implemented home automation in the same system through which the user can control the appliances of the home like lights, fan and AC through his mobile. Here, they sent the alert messages to user through voice calls which can be replaced with SMS which can also include the picture of intruder captured through camera installed in home. Kumar and Kumar [2] designed a system which is used to control theft by using Raspberry Pi integrating with webcam and motion sensor. Whenever any intrusion is detected or wrong pin is entered at door lock, the webcam takes the image of the person and sends it to the user and the user can take necessary actions. In other paper [3], they designed a low cost-effective system which saves large amount of data and power. The system remains idle when there is no motion, and it records only when there is an intrusion; thus, it saves more data and consumes less power which makes it efficient and better. When motion is detected, the captured video sensor data are stamped on image using image processing techniques on Python platform. Data stored in local storage device, after certain threshold, will be uploaded into the cloud along with data log created during processing stage.

Sunehra [4] used PIR sensor and IR sensor with ARM7 microcontroller to design a surveillance system to detect intrusion. They used GSM for connectivity between user and system. Keshamoni and Hemanth [5] used MQ-2 (gas sensor) and LM35 (temperature sensor) to detect any gas leakages in the surroundings. When any change is detected by one of the sensors, the siren is triggered and the user is warned. Vaidya and Vishwakarma [6] provided a comparative analysis of the technologies like GSM, Bluetooth and Wi-Fi. After ample research, they provided the comparative analysis of each technology, its advantages and disadvantages. Author concluded that GSM is preferable over other technologies like Wi-Fi and Bluetooth due to its long-range connectivity and widespread coverage which enables us to control the devices even when we are far away.

Recently, Gupta and Kumar [7] designed a system for the purpose of monitoring and irrigation of plants using sensors. The data about the temperature and humidity can be obtained from anywhere by using the Android mobile application as data is uploaded to cloud. To avoid the events such as burglary or intrusion, Sahoo [8] designed a system using PIR sensor to detect intruder and ESP8266 module to send data to remote server. Here, GSM module is used to send text alerts to user when an intrusion occurs.

3 Proposed System

The block diagram of our proposed safety and surveillance system is given in Fig. 1.

The sensors are connected to ARM LPC2148 microcontroller shown in Fig. 2. ARM LPC2148 is a 32-bit RISC microcontroller. It has two UART ports, and it works on a 5 V output DC supply. It also includes many features like I2C serial interfaces and two SPI serial interfaces along with pulse width modulation (PWM) unit, a real-time clock and can reach up to a speed of 60 MHz.

It has a vectored interrupt controller with configurable priorities and vector addresses.

PIR sensor used to detect intrusion is shown in Fig. 3. PIR sensor is used to detect intrusion by an unknown person. PIR sensor [9] senses the IR light radiating from objects, and it also senses the movement of people in its range, and whenever an

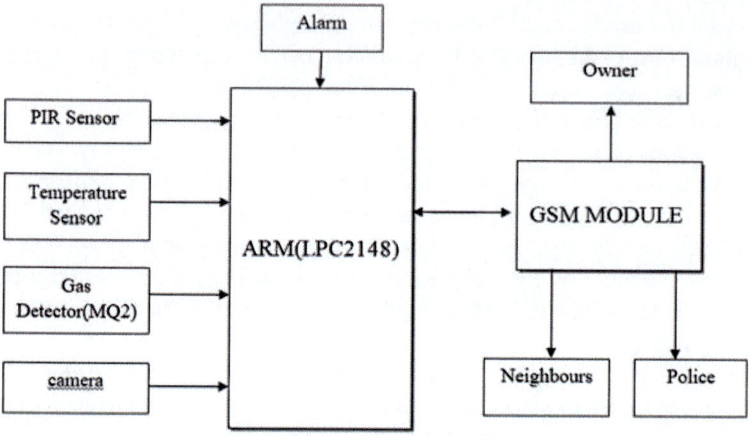

Fig. 1 Block diagram of ARM-based safety and surveillance system

Fig. 2 ARM LPC2148 microcontroller

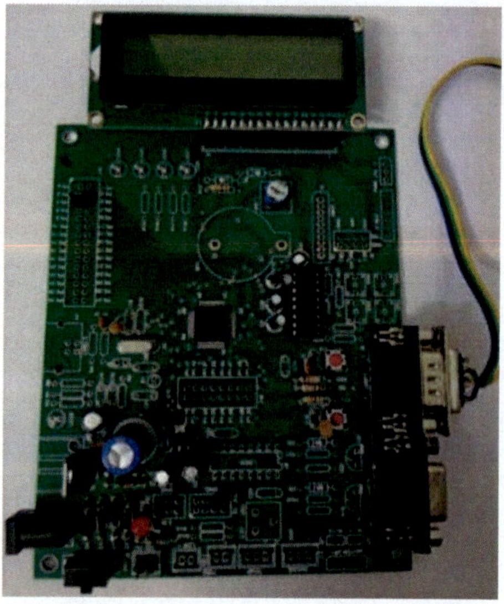

intrusion occurs, it sends a signal to microcontroller which in turn sends an alert message to the user about the intrusion. Depending upon the amount of IR radiation that is emitted by body, it can differentiate between human and animal, so that it will not send an alert message when any small animal like cat enters the house.

Here, a camera is installed in the system which is used to take the picture of the intruder and send it to the user through microcontroller. If the intruder is a known person to the user, then by instructions, he tells the system that he is a known person and there is no harm. Then, the system which is initially given with pictures of user's family adds the picture of this person to the list of known persons. Due to this, if the

Fig. 3 PIR sensor for intrusion detection

same person tries to enter the house afterwards, the system will not warn the user which is of much convenience.

To avoid fire accidents and gas leakages, temperature sensor is employed which is shown in Fig. 4 and gas detecting sensor in Fig. 5. Initially, threshold value of temperature is given to the system to which the system compares the recorded value. The temperature sensor senses the temperature timely and sends the value to controller. If the present value is higher than the threshold value, then it indicates an abnormal increase in temperature which may be due to fire accidents like short circuit. The smart home [10] system then immediately sends a warning message consisting of detected temperature to the user. The user will send a message through GSM [11]

Fig. 4 Temperature sensor

Fig. 5 MQ-2 gas detecting sensor

after which the neighbours are alerted through the alarm [12], and it will send a message also to fire department about the situation which may allow them to curb the fire accident within scope.

A MQ2 gas detecting sensor is installed in the kitchen. If at any instant gas leakage occurs, then the sensor sends signal to microcontroller which immediately sends an alert message to user about the gas leakage, and also, it alerts the neighbours about leakage.

GSM modem is used for connectivity between the user and microcontroller. GSM module is connected to microcontroller using an RS232 interface. RS232 voltage levels are at ±12 V, whereas microcontroller input and output operate at 0–5 V; hence, RS232 [13] is not compatible to microcontroller. Here, MAX232 [14] which is shown in Fig. 6 is used to enable the communication between GSM module and microcontroller. MAX232 converts the RS232 level signals to TTL level signals. Hence, when any mishap occurs at house, the ARM microcontroller [15] then sends an immediate message though GSM module to the user using IOT [16–20]. The outgoing message contains information regarding mishap whether it is intrusion, fire accident or gas leakage. It waits for user's instruction to take necessary action. The GSM module used is shown in Fig. 7.

GSM module consists of SIM800A for connectivity. It is a quad-band GSM/GPRS module that works on frequencies GSM 850 MHz; EGSM 900 MHz, PCS 1900 MHz and DCS 1800 MHz. SIM800A can meet most of the space requirements in user's applications, such as M2M, smartphone and other mobile devices and supports one pulse width modulation (PWM) and pulse coded modulation(PCM). Our proposed architecture is shown in Fig. 9.

4 Proposed System Architecture

Our working procedure is depicted in Fig. 8. The PIR sensor continuously checks for the intruder when it is connected to the microcontroller. When the PIR sensor

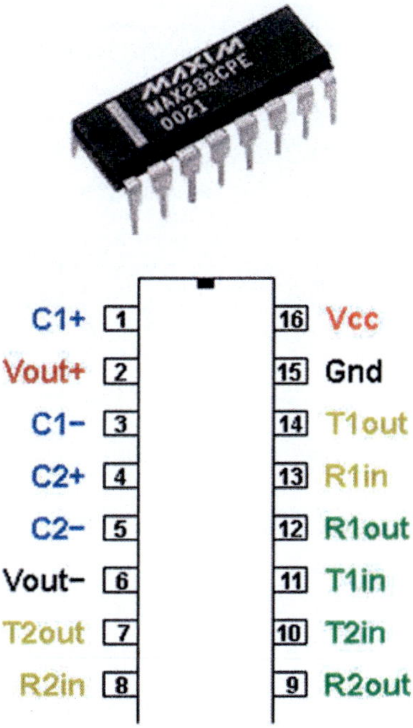

Fig. 6 MAX232

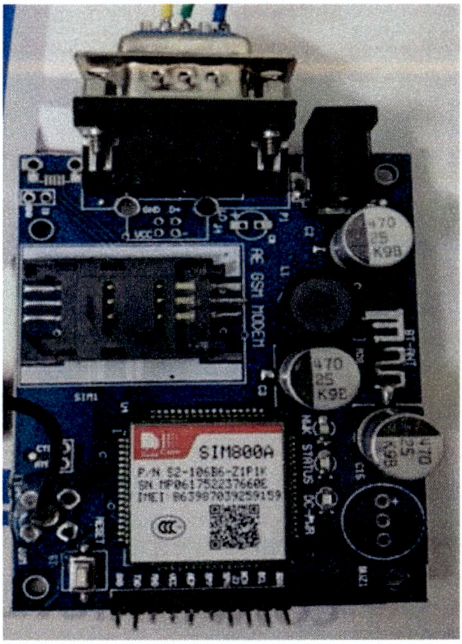

Fig. 7 GSM module consisting of SIM800a

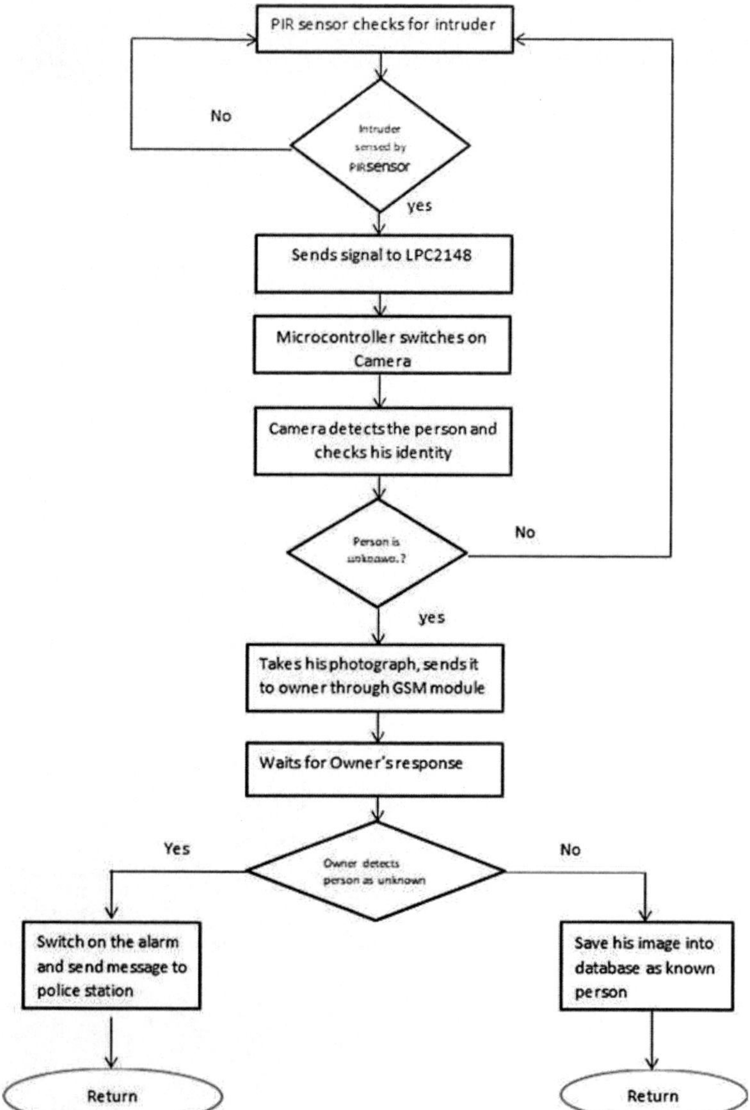

Fig. 8 Flowchart for detecting intrusion using PIR sensor and surveillance camera

detects any infrared radiation which is of any human beings (i.e. the threshold value of infrared radiation is assigned such that it is of humans), it sends a signal to ARM LPC2148 microcontroller which in turn on the camera. The camera then takes snap of the person and checks whether the person is actually an intruder or known person using data which is already assigned given to the microcontroller. If the person is an intruder, the microcontroller sends the image of the person using GSM module.

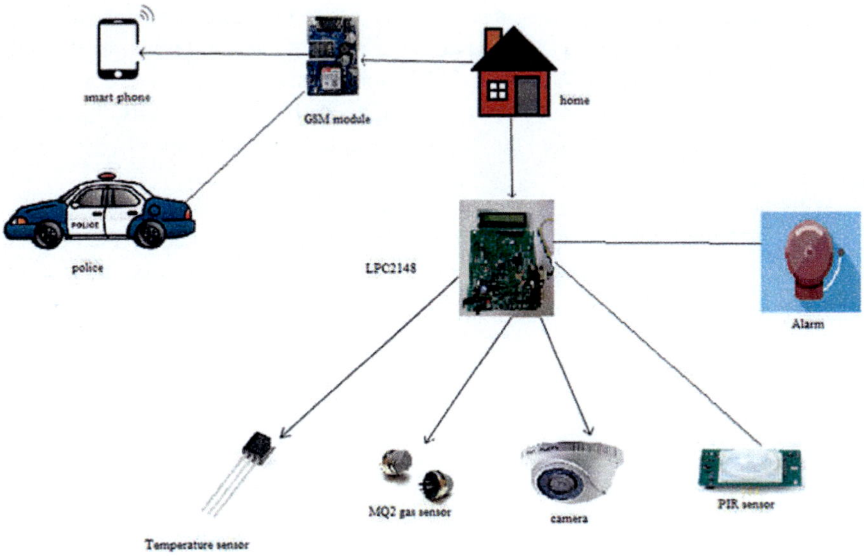

Fig. 9 Architecture of the proposed system

It waits for the owner's instruction. If the owner recognizes the person, he sends a message to add the person's image to database, and if the person is an intruder, the microcontroller then sends message to police and neighbours, and it also switches on buzzer alarm. The same conditions and procedure are followed for fire detection and gas leakage (Fig. 9).

5 Advantages

Compared to the existing models, this system provides more features which are as follows

- Provides more security.
- Includes safety and surveillance of home under same system.
- Contains minimal manual instructions making it more efficient and reliable.
- Can also be used for home automation of appliances.

6 Conclusion

We designed a cost-effective system based upon ARM LPC2148 microcontroller which can be used for safety of the home from fire accidents and gas leakages. It also provides theft control by informing the user whenever any intrusion occurs. The system can differentiate between a known person and an unknown person which prevents sending an alert message when the person entered is known to the user. The fast and efficient connectivity of GSM module enables the system to send an immediate alert message to user even when he is far from home. With the help of cloud storage, we can store the known person's image in it and also store the video recorded by camera during surveillance.

With the advent of the latest technology like LTE, we can get greater speeds and better connectivity, which can be used in system for better efficiency. We can also use the system to control appliances like light, fan, AC and other devices making the system a fully home automated and security system.

References

1. Kodali RK, Jain V, Bose S, Boppana L (2016) IoT based smart security and home automation system. In: IEEE international conference on computing, communication and automation (ICCCA), Noida, India, 29–30 Apr 2016
2. Kumar J, Kumar S, Kumar A, Behera B (2019) Real-time monitoring security system integrated with Raspberry Pi and e-mail communication link. In: 9th international conference on cloud computing, data science & engineering confluence 2019, Noida, India, 10–11 Jan 2019
3. Kumar J, Ramesh PR (2018) Low cost energy efficient smart security system with information stamping for IOT networks. In: 2018 3rd international conference on internet of things: smart innovation and usages (IoT-SIU), Bhimtal, India, 23–24 Feb 2018
4. Sunehra D, Bano A (2014) An intelligent surveillance with cloud storage for home security. In: 11th annual IEEE India conference (INDICON-2014) on emerging trends and innovations in technology, Pune, India, 11–13 Dec 2014
5. Keshamoni K, Hemanth S (2017) Smart gas level monitoring, booking & gas leakage detector over IoT. In: IEEE 7th international advance computing conference (IACC), Hyderabad, India, 5–7 Jan 2017
6. Vaidya VD, Vishwakarma P (2018) A comparative analysis on smart home system to control, monitor and secure home, based on technologies like GSM, IOT, Bluetooth and PIC microcontroller with Zigbee modulation. In: IEEE international conference on smart city and emerging technology (ICSCET), Mumbai, India, 5 Jan 2018
7. Gupta N, Kumari A, Kumari S, Kumar J (2019) Automatic plant watering and monitoring system using Nodemcu. In: 9th international conference on cloud computing, data science & engineering confluence 2019, Noida, India, 10–11 Jan 2019
8. Sahoo KC, Pati UC (2017) IoT based intrusion detection system using PIR sensor. In: 2017 2nd IEEE international conference on recent trends in electronics, information & communication technology (RTEICT), Bangalore, India, 19–20 May 2017
9. Moghavvemi M, Seng LC (2004) Pyro electric infrared sensor for intruder detection. In: IEEE region 10 conference TENCON 2004, Chiang Mai, Thailand, 24 Nov 2004
10. D'souza M, Wilfred N, Pereira R, Rayen T, Telgote A (2016) Home automation using internet of things. In: IEEE 7th annual ubiquitous computing, electronics & mobile communication conference (UEMCON), New York, USA, 20–22 Oct 2016
11. Gu G, Peng G (2010) The survey of GSM wireless communication system. In: IEEE international conference on computer and information application, Tianjin, China, 3–5 Dec 2010
12. Mrinal M, Priyanka L, Saniya M, Poonam K, Gavali AB (2017) Smart home—automation and security system based on sensing mechanism. In: 2017 second international conference on electrical, computer and communication technologies (ICECCT), Coimbatore, India, 22–24 Feb 2017

13. Teymourzadeh R, Ahmed SA, Chan KKW, Hoong MK (2013) Smart GSM based home automation system. In: IEEE conference on systems, process & control (ICSPC2013), Kaula Lumpur, Malaysia, 13–15 Dec 2013
14. Loganthurai P, Shalini M, Vanmathi A, Veeralakshmi M, Vivitha V (2017) Smart energy meter billing using GSM with warning system. In: IEEE international conference on intelligent techniques in control, optimization and signal processing (INCOS), Srivilliputhur, India, 23–25 Mar 2017
15. Vidya KS, Manikanthan SV (2015) An android based secure access control using ARM and cloud computing. In: 2015 2nd international conference on electronics and communication systems (ICECS), Coimbatore, India, 26–27 Feb 2015
16. Gupta KK, Shukla S (2016) Internet of Things: security challenges for next generation networks. In: IEEE international conference on innovation and challenges in cyber security (ICICCS-INBUSH), Noida, India, 3–5 Feb 2016
17. Vikram N, Harish KS, Nihaal MS, Raksha U, Kumar SAA (2017) A low cost home automation system using Wi-Fi based wireless sensor network incorporating Internet of Things (IoT). In: 2017 IEEE 7th international advance computing conference (IACC), Hyderabad, India, 5–7 Jan 2017
18. Lee HR, Lin CH, Kim WJ (2016) Development of an IoT-based visitor detection system. In: International SoC design conference (ISOCC), Jeju, South Korea, 23–26 Oct 2016
19. Beltran V, Martinez JA, Skarmeta A, Julia PM (2016) An ARM-compliant IoT platform: security by design for the smart home. In: 2016 IEEE 5th global conference on consumer electronics, Kyoto, Japan, 11–14 Oct 2016
20. Mandula K, Parupalli R, Murty CHAS, Magesh E, Lunagariya R (2015) Mobile based home automation using Internet of Things (IoT). In: 2015 international conference on control, instrumentation, communication and computational technologies (ICCICCT), Kumaracoil, India, 18–19 Dec 2015

Attenuation of Millimeter Wave in Storm Layers with Spherical and Non-spherical Dust Particles

Swastika

Abstract In the present paper, an attempt has been made to estimate the attenuation of millimeter wave from the layers of spherical and non-spherical dust particles. It is observed that attenuation of millimeter wave is very much related to the extinction cross section of dust particles. The attenuation caused by the layers of spherical and non-spherical dust particle increases with increasing frequency of operation for different values of visibility. It is also found that the scattering cross section, absorption cross section and extinction cross-section have got important role to estimate the loss due to scattering and absorption of millimeter wave.

Keywords Millimeter wave attenuation · Absorption cross section · Extinction cross section

1 Introduction

"Effects of Sand and Dust Storms on Microwave Propagation." S. C. Gupta [1] "Theory of Microwave Remote Sensing." Leung Tsang [2]. "Depolarization of MM W Back Scattered from Sand and Dust Storms." C. S. Rai and B R Vishvakarma [3].

2 Method

In order to consider the scattering effects of main constituents of sand and dust storms such as sand, silt and clay, the consideration of outline structure of these components of atmosphere should be observed. So considered length of communication link be l, which contains the layers of different constituents of storms blown to the height of

Swastika (✉)
Department of Electronics & Communication Engineering, Netaji Subhas Institute of Technology-Bihta, Near IIT Patna, Patna, Bihar, India
e-mail: swastikaanuj54@gmail.com

V. Nath and J. K. Mandal (eds.), *Proceedings of the Fourth International Conference on Microelectronics, Computing and Communication Systems*, Lecture Notes in Electrical Engineering 673, https://doi.org/10.1007/978-981-15-5546-6_47

link because of dust storms. As the storm is assumed to contain three constituents, the entire section is represented in the form of three layers in cascade. Further, the layer with sand particle extends from Z = 0 to Z = l_1, the layer of silty particle extends from Z = l_1, to Z = l_2, and layer of clay particle extends from Z = l_2 to Z = l_3. If ε_1, ε_2, ε_3 and σ_{1ext}, σ_{2ext}, σ_{3ext} denote the permittivity and extinction cross section of sand, silt and clay, respectively, then,

$$\sigma_{ext}(Z) = \begin{matrix} \sigma_{1ext} \text{ for } 0 \le Z < l_1 \\ \sigma_{2ext} \text{ for } l_1 \le Z < l_2 \\ \sigma_{3ext} \text{ for } l_2 \le Z < l_3 \end{matrix}$$

where $l_1 + l_2 + l_3 = l$.

3 Scattering from Dust Particles

In order to study dust particle's scattering behavior which is considered as a dielectric sphere, here energy balance concept over the surface of sphere surrounding the scattering volume is utilized. H_{inc} and E_{inc} denote the incident magnetic and electric field, and H_{sca} and E_{sca} denote the corresponding scattered fields. Total amplitude of the fields at any point on the surface of the sphere can be given as

$$\begin{aligned} E &= E_{inc} + E_{sca} \\ H &= H_{inc} + H_{sca} \end{aligned} \tag{1}$$

As the problem is concerned with the flow of energy, Poynting vector must be used. In this case, only the radial component of Poynting vector actually crosses the surface of the sphere. In spherical coordinate system, if H (θ, Φ) and E (θ, Φ) denote the magnetic and electric fields, then the radial component of the average Poynting vector will be in the radial direction, the value of which may be mentioned as follows:

$$\begin{aligned} S_R &= 1/2\, R_e\big(E_\theta H_\Phi^* - E_\Phi H_\Phi^*\big) \\ S_R &= 1/2\, R_e\big[\big(E_{\theta inc} H_{\Phi inc}^* - E_{\Phi inc} H_{\theta inc}^*\big) \\ &\quad + E_{\theta sca} H_{\Phi sca}^* - E_{\Phi sca} H_{\theta inc}^* \\ &\quad + E_{\theta inc} H_{\Phi sca}^* - E_{\theta sca} H_{\theta inc}^* \\ &\quad + E_{\Phi inc} H_{\theta sca}^* - E_{\Phi sca} H_{\Phi inc}\big)\big] \end{aligned} \tag{2}$$

From the observation of Eq. (2), it is found that the integral of first term on right hand of equation is zero as it gives the net flow of energy in incident plane wave, whereas second term when integrated gives the total scattered power (P_{sca}) out of the incident wave. The third term if integrated, fields ($-P_{ext}$) and the integral of (S_R) yield the outflow of energy from the sphere.

3.1 Extinction Cross section

From Eq. (2), it is evident that to maintain the energy balance, third term of the equation must be equal in magnitude of the sum of absorbed and scattered energy. Therefore,

$$P_{\text{ext}} = P_{\text{abs}} + P_{\text{sca}} = 1/2\text{Re}\int_{o}^{\pi}\int_{o}^{2\pi}(E_{\theta\text{inc}}H^{*}_{\phi\text{sca}} - E_{\phi\text{inc}}H^{*}_{\theta\text{sca}} + E_{\theta\text{sca}}H^{*}_{\phi\text{inc}} - E_{\phi\text{sca}}H^{*}_{\theta\text{inc}}) R^{2}\sin\theta\,\text{d}\theta\,\text{d}\phi \tag{3}$$

Hence, the extinction cross section may be defined by

$$\sigma_{\text{ext}} = -\left(2\pi/\text{k}^{2}\right)\text{Re}\sum_{n=1}^{\infty}(2n+1)\left(a_{n}^{s} + b_{n}^{s}\right)\text{cm}^{2} \tag{4}$$

where

$$a_{n}^{s} = -\frac{\mu_{1}J_{n}(X)[mXJ_{n}(mX)]' - \mu_{2}J_{n}(mX)[XJ_{n}(X)]}{\mu_{1}h_{n}^{(2)}(X)[mXJ_{n}(mX)]' - \mu_{2}J_{n}(mX)[Xh_{n}^{(2)}(X)]'}$$

$$b_{n}^{s} = -\frac{\mu_{2}m^{2}J_{n}(mX)[XJ_{n}(X)]' - \mu_{2}J_{n}(X)[mXJ_{n}(mX)]'}{\mu_{1}m^{2}J_{n}(mX)[Xh_{n}^{(2)}(X)]' - \mu_{2}h_{n}^{(2)}(X)[mXJ_{n}(mX)]'}$$

Jn	Bessel functions of first kind.
$h_{n}^{(2)}$	spherical Hankel function of second kind.
$m = \sqrt{\epsilon^{*}}$	Complex refractive index
μ_{1}	permeability of internal medium
μ_{2}	permeability of external medium
X	$2\ a/\lambda$
K	$2/\lambda$

In this equation, a_{n}^{s} and b_{1}^{s} indicate the amplitudes of the field on the surface of the sphere which are proportional to the amplitudes of the magnetic and electric multi-poles induced by the incident wave, respectively. More specifically, a_{1}^{s} quantifies the field distribution corresponding to induced magnetic dipole and a_{1}^{s} is related to magnetic quadrupole. Similarly b_{1}^{s} signifies the field distribution corresponding to induced electric dipole, and b_{2}^{s} corresponds to electric quadrupole. Under this condition, it is quite logical that higher order multiples oscillations will be weakly excited. All the higher order terms could be neglected without significant loss in accuracy, excepting the terms a_{1}^{s} and b_{1}^{s}. Since 2 a/λ is small, compared to unity, Bessel functions can be expanded in powers of (*X*) to find out the a_{1}^{s} and b_{1}^{s} coefficients.

$$J_n(X) = 2^n \frac{n!}{(2n+1)!} X^n \left(1 - \frac{n+1}{(2n+1)(2n+3)} X^2 + \cdots\right) \quad (5)$$

The resulting value of a_n^s and b_n^s can be written as

$$a_n^s = \left\{-j \frac{2^{2n} n!(n+1)!}{(2n)!(2n+1)!} \frac{x^{2n+1}}{\mu_1^n + \mu_2^{n+1}}\right\} \times \{\mu_2 - \mu_1\}$$
$$\left\{\frac{\mu_2\left[(n+1) + (n+3)m^2\right] - \mu_2\left[(n+3) + (n+1)m^2\right]}{(2n+2)(2n+3)} X^2\right\} \quad (6)$$

$$b_n^s = \left\{J \frac{2^{2n} n!(n+1)!}{(2n)!(2n+1)!} \frac{X^{2n+1}}{\mu_1(n+1) + \mu_2 m_n^2}\right\} \times \left\{\mu_1 - \mu_2 m^2\right\}$$
$$- \left\{\frac{\mu_1[(n+1) + (n+3)m^2] - \mu_2 m^2 [n+3) + (n+1)m^2] x^2}{(2n+2)(2n+3)} x^2\right\} \quad (7)$$

In the above equation, $X = 2\,a/\lambda \ll 1$, and thus, all the terms having X^7 or higher power can be neglected. The final expression for σ_{ext} can be obtained as

$$C_{\text{ext}} = \left(4\pi/k^2\right)(X)^3$$
$$\left[\text{Im}\left(\frac{\varepsilon - 1}{\varepsilon + 2}\right) + 2/3(X)^3 \text{Re}\left(\frac{\in -1}{\in +2}\right)^2\right] \quad (8)$$

The calculated value of σ_{ext} is shown in Fig. 1.

4 Non-spherical Particle

The scattering coefficients for particle which is not spherical can be obtained by applying scattering amplitude matrix of small ellipsoid which approaches to spheroid under the condition $a = b$ where a, b and c are ellipsoid axes. We can find out scattering and absorption cross section for the dust particle which is spheroidal are[2]

$$\sigma_{\text{sca}} = \left(k^4/6\pi\right) \int_o^\infty \frac{\sin\beta}{2} P(\beta) \mathrm{d}\beta \left\{|a_\theta|^2\right.$$
$$+ \left[1/2 \cos^2\theta' \sin^2\beta + \sin^2\theta' \cos^2\beta\right] |a_r - a_\theta|^2$$
$$+ 2\left[\sin^2\theta' \cos^2\beta + 1/2 \cos^2\theta' \sin^2\beta\right]$$
$$\left.\text{Re}[a_\theta(a_r - a_\theta)]\right\} \quad (9)$$

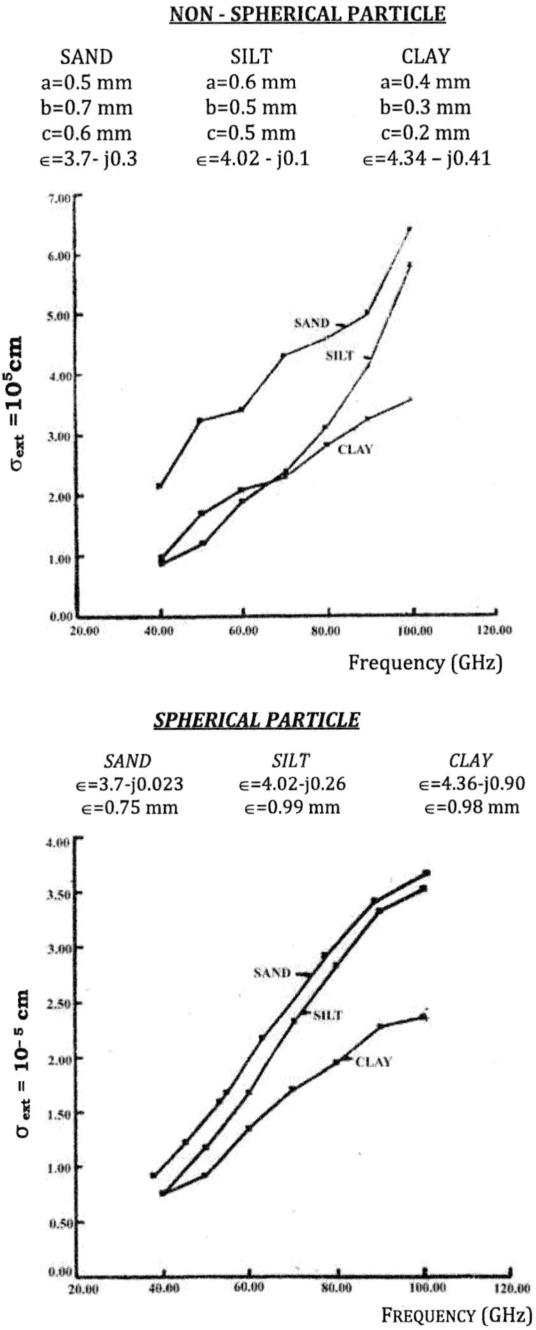

Fig. 1 Extinction cross section against frequency for non-spherical dust particles

$$\sigma_{\text{abs}} = \int\limits_{o}^{\pi} \frac{\sin\beta}{2} \mathrm{d}\beta p(\beta)$$

$$\left\{ b_\theta + \left[\frac{\cos^2\theta' \sin^2\beta}{2} + \sin^2\theta' \cos^2\beta \right] (b_r - b_\theta) \right\} \tag{10}$$

where

$$a_r \frac{\varepsilon_s - 1}{1 + v_d A_2}, a_\theta = \frac{\varepsilon_s - 1}{1 + v_d A_1}, b_r = \text{klm}(a_r)$$
$$b_\theta = \text{klm}(a_\theta)$$

and

$$\int\limits_{o}^{\pi} \frac{\sin\beta \mathrm{p}(\beta)}{2} \mathrm{d}\beta = 1, \ \mathrm{A}_1 = \int\limits_{o}^{\infty} \frac{\mathrm{ds}}{(\mathrm{s} + \mathrm{a}^2)} \mathrm{Rs}$$

$$A_2 = \int\limits_{o}^{\infty} \frac{\mathrm{d}s}{(s + c^2)} \mathrm{Rs}$$

where, $\mathrm{Rs} = [(s + a^2)(s + b^2)(s + c^2)]^{1/2}$ and

$$v_d = \frac{abc}{2}(\varepsilon_s - 1)$$

ε_s relative permittivity of spheroidal dust particle.
θ' —θ
θ $0 \leq \theta \leq -$
β $0 \leq \beta \leq -$

and,

$$\sigma_{\text{ext}} = \sigma_{\text{sca}} + \sigma_{\text{abs}} \tag{11}$$

5 Analytical Treatment of Attenuation

Waves will be attenuated when the microwave and millimeter wave pass through sand and dust particles. This will be caused by two phenomena, scattering and absorption. Considered $N(a)\mathrm{d}a$ is dust particles number per unit volume of storm where radii in interval a to $a + \mathrm{d}a$. Here extinction cross section of dust particles is σ_{ext}, and then, total removed power with incident Poynting vector S_r of dust particle volume of single cross-sectional area is mentioned as

$$\frac{dS}{dl} = -S_r \int_o^\infty \sigma_{ext} N(a) da$$

By denoting the integral in this equation by ⌊ (attenuation)

$$\alpha = \int_o^\infty \sigma_{ext} N(a) da \text{ and putting } N(a) = nP(a) \tag{12}$$

where particle's number in unit volume of stock is n and $P(a)$ probability density function.

The attenuation for spherical dust particle is

$$\alpha = n \int_o^\infty \sigma_{ext}\, a^3\; P(a) da \text{ nep/cm} \tag{13}$$

Similarly, for non-spherical dust particle (Fig. 2)

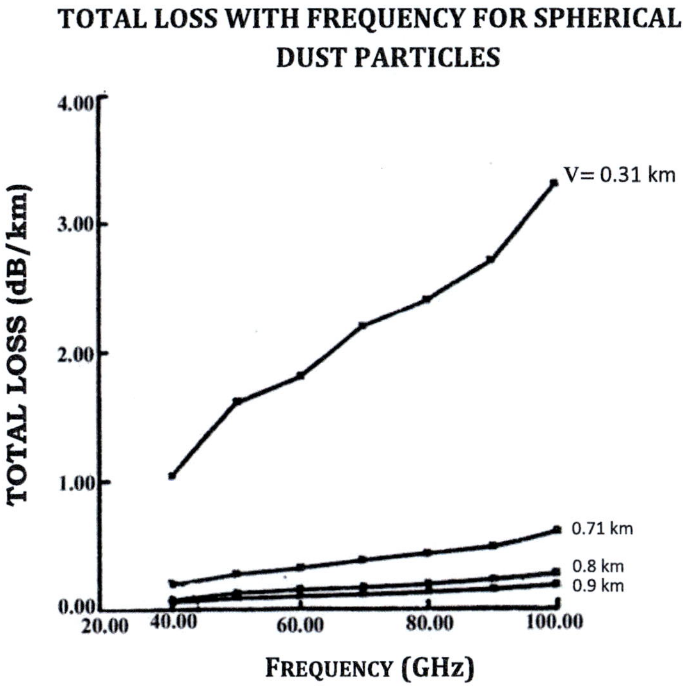

Fig. 2 Total loss with frequency for spherical dust particles

$$\alpha = n \int_{o}^{\infty} \sigma_{\text{ext}} abc\, P(s) \mathrm{d}s \text{ nep/cm} \tag{14}$$

where parameters *a*, *b* and *c* are denoted by *s*. The particle's number *n* in unit value of the storm is given by

$$n = \frac{9.43 \times 10^{-9}}{V^{y}\left(\frac{4}{3}\pi a^{3}\right)} \tag{15}$$

where *V* is the visibility in km, *a* radius of the dust particle, γ a constant = 1.07.

The value of ⌊ for spherical and non-spherical dust particle can be obtained by combining Eqs. (12)–(15) (Figs. 3, 4 and 5).

$$\alpha_{\text{sph}} = \frac{9.43 \times 10^{-9}}{V^{r}\left(\frac{4}{3}\pi a^{3}\right)} \int_{o}^{\infty} \sigma_{\text{ext}} a^{3} p(a) \mathrm{d}a \tag{16}$$

$$\alpha_{\text{non-sph}} = \frac{9.43 \times 10^{-9}}{V^{r}\left(\frac{4}{3}\pi \overline{a}^{3}\right)} \int_{o}^{\infty} \sigma_{\text{ext}}\, abc\, p(s) \mathrm{d}s \tag{17}$$

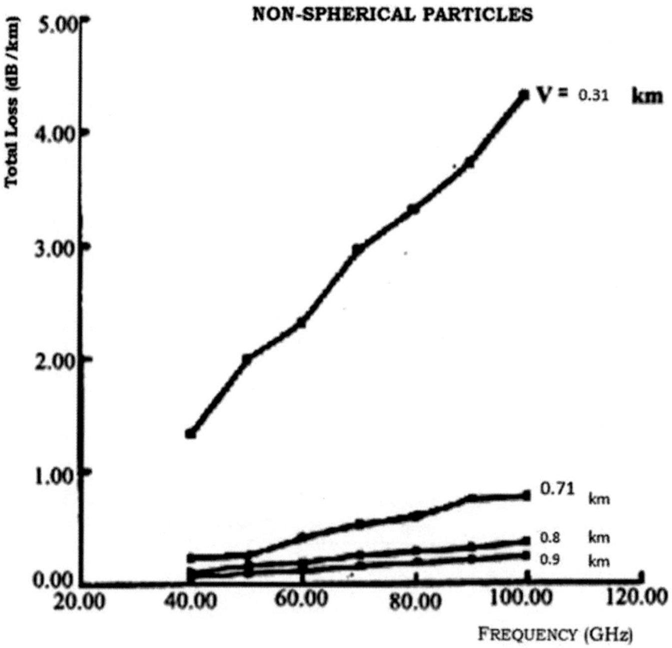

Fig. 3 Total loss against frequency for non-spherical dust particles

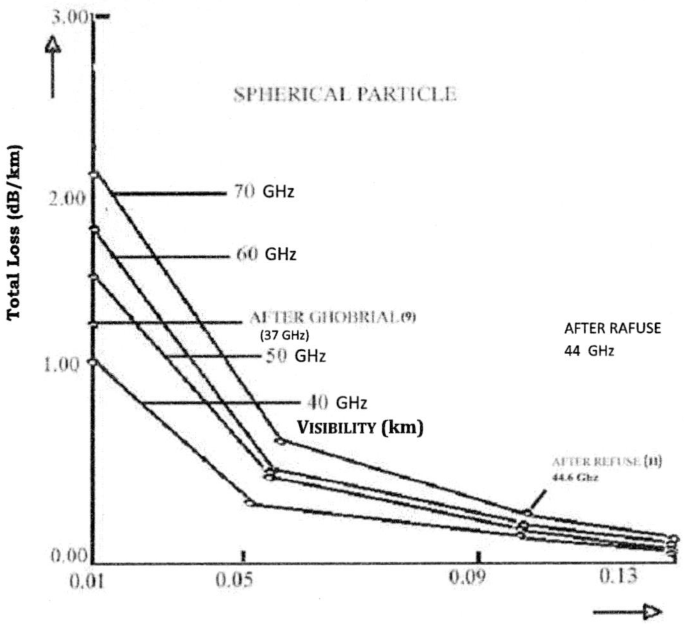

Fig. 4 Total loss against visibility for spherical dust particles

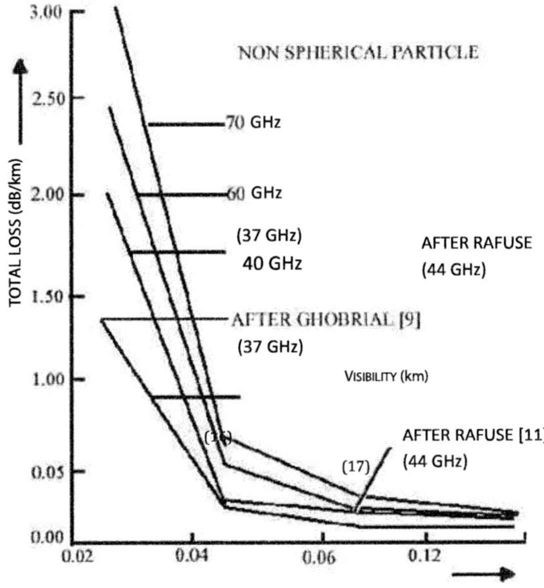

Fig. 5 Total loss against visibility for non-spherical dust particles

The attenuation (⌊) taken as dB/km and Eqs. (16) and (17) are

$$\alpha_{\text{sph}} = \frac{434 \times 9.43 \times 10^{-9} \times 3}{V^r\left(4\pi a^3\right)} \int_o^\infty \sigma_{\text{ext}} a^3 \, p(a)\, da \text{ dB/km} \tag{18}$$

$$\alpha_{\text{non-sph}} = \frac{434 \times 9.43 \times 10^{-9} \times 3}{V^r\left(4\pi \, a\bar{b}c\right)} \int_o^\infty \sigma_{\text{ext}} \, abc \, p(s) ds \text{ dB/km} \tag{19}$$

6 Conclusion

Frequency for sand, silt and clay increases with increase in extinction cross section. The value of extinction cross section, σ_{ext}, for non-spherical dust particle is found to be higher than the particle which is spherical for specified frequency and complex dielectric constant. This is one new evidence that the extinction cross section directly depends on the particle size and number of small particles in the atmosphere. The suspended larger particles are ellipsoidal or spheroidal. Computational work is done to observe the variation in attenuation with respect to frequency and visibility. The attenuation calculations are based on the Rayleigh scattering. The attenuation increases almost linearly with frequency, both for spherical and non-spherical dust particles. The magnitude of attenuation increases with decreasing value of visibility. This is because of the fact that decreasing visibility enhances the number of dust particles in the storm. It has also been observed that non-spherical dust particles offer enhanced attenuation as compared to spherical ones. The non-spherical model for attenuation evaluation is in close agreement with the other reported work as compared to spherical model. In study, the author has also referred Ghobrial and Rafuse.

References

1. Gupta SC et al (1981) Proceeding on international symposium on microwave and communication, IIT Kharagpur, pp 104–109
2. Tsang L et al (1985) Wiley, p 307
3. Rai CS, Vishvakarma BR (1992) J Inst Eng (India) 73(ET 3):111–114

VLSI Implementation of Tunable Band-Pass Notch IIR Filter for Localization of Hot spots in Proteins

Vikas Pathak, Satyasai Jagannath Nanda, Amit Mahesh Joshi, and Sitanshu Sekhar Sahu

Abstract A tunable band-pass notch (BPN) IIR digital filter (including zero phase filtering) is proposed by Ramachandran et al. in 2009 to detect the hot spot regions in proteins. The hot spots are the locations of amino acids at which proteins communicate with each other to achieve biological functions. In this paper, the tuning technique of above BPN filter is modified as per the characteristics frequency of protein functional group. The VLSI architecture of this tuned filter is developed and synthesized using Artix-7 family FPGA. The implemented architecture performance is compared with that obtained by MATLAB for FGF protein family. It is observed that the hardware provides approximately 51 times faster results than MATLAB run time.

Keywords IIR digital filter · IEEE-754 floating point standard · Proteomics · Protein hot spot detection

1 Introduction

Proteins are the fundamental elements of any living organism, which are generated by combinations of different amino acids [1]. They are formed by a linear chain of 20-amino acids. Every amino acid in any protein sequence is denoted by a character. The protein linear chains of amino acids are folded in a specific way to form complex 3-D structures. Proteins carry out their organic functions with the help of these 3-D structures by making interactions with other proteins molecules called as targets.

V. Pathak (✉) · S. J. Nanda · A. M. Joshi
Malaviya National Institute of Technology, Jaipur, Rajasthan 302017, India
e-mail: 2013rec9567@mnit.ac.in

V. Pathak
Swami Keshvanand Institute of Technology, Jaipur 302017, India

S. S. Sahu
Birla Institute of Technology, Mesra, Ranchi 835215, India

V. Nath and J. K. Mandal (eds.), *Proceedings of the Fourth International Conference on Microelectronics, Computing and Communication Systems*, Lecture Notes in Electrical Engineering 673, https://doi.org/10.1007/978-981-15-5546-6_48

These protein-target relations are performed at specific active sites [2–4]. These active sites are stabled using the various group of amino acids surrounding that site, commonly known as hot spots. Alternatively, hot spots are also called as group of few amino acids locations, responsible for the stability of protein-target interactions. Finding the locations of these hot spots is very important in understanding the protein functionality. Hence a reliable and efficient technique is required for detections of these hot spot locations. In this paper, the hardware implementation of tunable band-pass notch (BPN) digital filter is carried out, which is used for localization of hot spot regions.

Literature review reveals that many researchers have explored several techniques for detections of hot spot locations in the protein family. Initially, Short-Time Discrete Fourier Transform (STDFT) [5] is used for analysis. But the computational complexity of the transform based method is higher. Then Ramachandran et al. [6, 7] improved the computational complexity by using IIR narrow band-pass digital filters. The inverse-Chebyshev IIR filters used by [6, 7] require higher filter order (6–8), which results in lower selectivity. To achieve high selectivity with low filter order, second-order band-pass notch (BPN) IIR filter is employed in [8, 9]. In these papers, an automatic tuning approach for adjusting the filter coefficients of BPN filter is discussed, which required several steps (iterations) of any optimization technique.

In this paper, the iteration steps are reduced by directly designing the filter in the form of an all-pass filter and computational complexity is further reduced by the hardware implementation of BPN.

The paper is summarized as follows: Sect. 1 introduces the basic concept and literature review of hot spot localization problem in protein. The RRM model and filter based hot spots detection technique are detailed in Sect. 2. VLSI implementation details of proposed BPN IIR digital filter system is described in Sect. 3. Various results are given and discussed in Sect. 4. Finally, Sect. 5 concludes the paper.

2 RRM Model and Filter Based Hot Spot Detection

2.1 Resonance Recognition Model

The hot spots in any protein family is determined by resonance recognition model (RRM) of protein-target interactions [10, 11]. According to the RRM model, protein character sequence is first converted to a numerical sequence using EIIP (electron-ion interaction potential) method. Various EIIP values of 20-amino acids are shown in Table 1 [4–6], which represent the typical energy of the valence electrons in that particular amino acid.

Then DFT (Discrete Fourier Transform) of these protein numerical sequences belonging to same biological function is calculated, known as consensus spectrum ($Y(k)$) and is represented by:

Table 1 Amino acids of protein sequence denoted by EIIP values

Amino acid	EIIP value	Amino acid	EIIP value
Aspartic acid (D)	0.1263	Alanine (A)	0.0373
Arginine (R)	0.0959	Lysine (K)	0.0371
Phenylalanine (F)	0.0946	Histidine (H)	0.0242
Threonine (T)	0.0941	Proline (P)	0.0198
Cysteine (C)	0.0829	Glutamic (E)	0.0058
Serine (S)	0.0829	Valine (V)	0.0057
Methionine (M)	0.0823	Glycine (G)	0.0050
Glutamine (Q)	0.0761	Asparagine (N)	0.0036
Tryptophan (W)	0.0548	Isoleucine (I)	0.0000
Tyrosine (Y)	0.0516	Leucine (L)	0.0000

$$Y(k) = |X_1(k)X_2(k)\ldots X_M(k)| \quad (1)$$

where X_1, X_2, …, X_M represents the DFTs of M-proteins related to that protein family.

A distinct peak in the plot of $Y(k)$ versus amino acids locations, signifies the characteristics frequency (ω_0) of that functional group of proteins. Figure 1 shows the Consensus spectrum of FGF (fibroblast growth factor) protein family, which shows the normalized $\omega_0 = 0.9$. All proteins of the same biological functional group interact with their targets at ω_0. Hot spots are the locations in a protein sequence (denoted by EIIP values), where the ω_0 is leading.

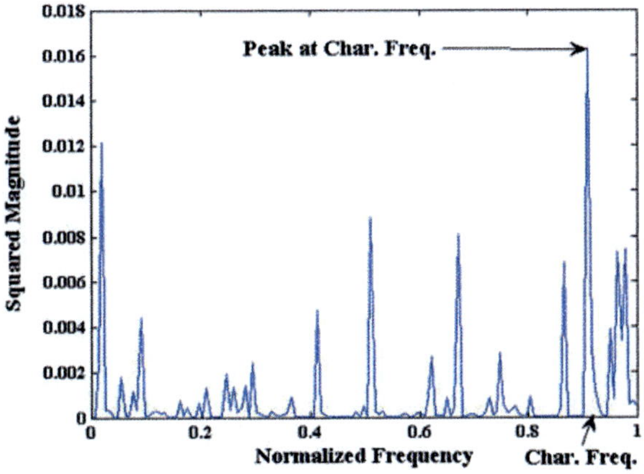

Fig. 1 Consensus spectrum of FGF protein family

2.2 Hot spot Region Detection Using Digital Filters

As described in the earlier section according to the RRM model, once the characteristics frequency of protein family is determined, then hot spots locations are the regions in the equivalent numerical sequence of any protein where ω_0 is prevailing. So to determine the hot spot locations in any protein sequence, those characteristics frequency components are to be selected among the various irrelevant frequencies present in protein numerical sequence. This can be easily obtained by passing the protein sequence through narrow band-pass or band-pass notch digital filter by tuning that filter for characteristics frequency. Then peaks in the plotted filtered output will signify the hot spot location of that particular protein of interest.

Higher selectivity (for effective localization of hot spots) and lower filter order (offers fewer computations for effective hardware implementation) of any digital filter are desired. Out of the two choices of FIR or IIR digital filters, IIR digital filtering approach is preferred in this paper because it requires low filter order to achieve higher selectivity. A delay is introduced at the filtered output due to the protein sequence. Computation of this delay is very important to facilitate the correct output. This computation is not very simple for IIR digital filter, and it will be very Large for FIR filters. Zero phase filtering (detailed in Sect. 12.5 of [12]) is the simple solution for removing this delay.

The complete digital filter based system (including zero phase filtering) for hot spot region identification is shown in Fig. 2. The character sequence of proteins is first of all converted to numerical sequence $S[n]$ using the EIIP method. Then according to zero phase filtering, this numerical sequence is passed to $F(z)$ filter, and then the filtered output is reversed (using R) and applied again to same filter $F(z)$ to eliminate the delay introduced by the first filter. The second filter output is then again reversed to get the final filter output $P[n]$. The energy of $P[n]$ is computed as follows:

$$E_P[n] = |P[n]|^2 \tag{2}$$

The peaks in this energy plot represent the hot spot location in that particular protein for its interaction with the target of that protein functional group.

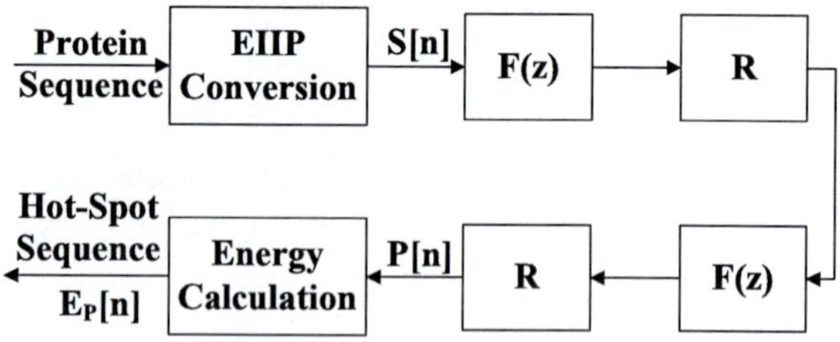

Fig. 2 Filter based hot spot detection system

2.3 Tuning of Band-Pass Notch IIR Digital Filter

Initially, an inverse-Chebyshev IIR narrow bandpass digital filter was used by [6, 7] and then subsequently second-order bandpass notch (BPN) filter was proposed in [8]. The transfer function $F(z)$ of BPN is derived from all-pass filter $A(z)$ as follows:

$$F_{\mathrm{BPN}}(z) = \frac{1}{2}\left[\frac{(1-d_0)\left(1-z^{-2}\right)}{1+d_1 z^{-1}+d_0 z^{-2}}\right] \tag{3}$$

where, d_0 and d_1 are the filter coefficients. The determination of these filter coefficients is the main concern here. In [8] d_0 and d_1 were calculated using 1-dimensional optimization approach like the golden section search, which requires 32-iterations. These calculations were further simplified using a second-order least-squares polynomial in [9]. All these efforts made by [8, 9] requires more computational time and more hardware for calculation of these filter coefficients. This computational time and hardware requirement is reduced in this paper.

Vaidyanathan [13] proposes an all-pass filter based anti-notch filter (ANF) (which is equivalent to BPN filter) for exon region identification in DNA sequence, whose filter response is given by:

$$F_{\mathrm{ANF}}(z) = \frac{1}{2}\left(\frac{\left(1-R^2\right)\left(1-z^{-2}\right)}{1-2R\cos\theta z^{-1}+R^2 z^{-2}}\right) \tag{4}$$

Any standard IIR digital filter is represented by:

$$F(z) = \frac{b_0+b_1 z^{-1}+b_2 z^{-2}}{1+a_1 z^{-1}+a_2 z^{-2}} \tag{5}$$

By comparing Eqs. (3), (4) and (5),

$$b_0 = -b_2 = \frac{1-R^2}{2} = \frac{1-d_0}{2}, b_1 = 0 \tag{6}$$

$$a_1 = d_1 = -2R\cos\theta = -\left(1+R^2\right)\cos\omega_0, a_2 = d_0 = R^2 \tag{7}$$

where R is the pole radius of $F(z)$ transfer function and ω_0 (in radian) is the desired normalized frequency (char. Freq. of protein family group) of which filter has to be designed.

Here R determines the selectivity of the BPN (ANF) filter, which should be closer to unity for good selectivity. Assuming $R = 0.992$, the filter coefficients are given by:

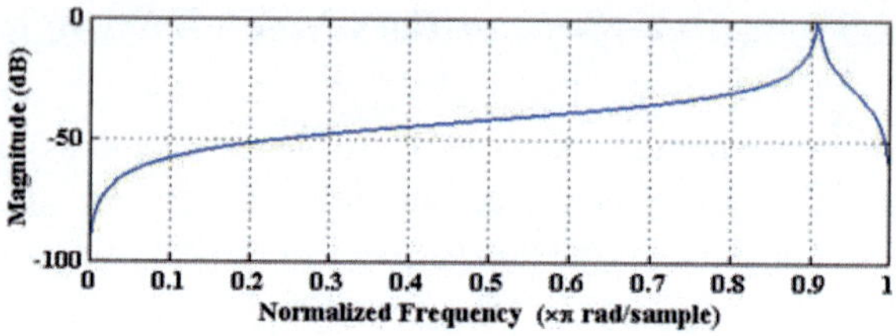

Fig. 3 The magnitude response of BPN filter tuned for the FGF protein family

$$\begin{vmatrix} b \\ a \end{vmatrix} = \begin{vmatrix} 0.007968 & 0 & -0.007968 \\ 1 & 1.984064 \cos \omega_0 & 0.984064 \end{vmatrix} \tag{8}$$

It is clear from these filter coefficients that here all coefficients except d_1 (a_1) are constants and a_1 is only depending on $\omega_{0,}$ which can be easily calculated using finding the 'cos' function rather than longer calculations in [8, 9]. Hence BPN (ANF) filter can be easily designed and tuned just by varying the ω_0 for any particular protein family functional group and pass the protein sequence of that protein family for hot spot region detection. The magnitude response of BPN filter tuned at $\omega_0 = 0.9$ for the FGF protein functional group is shown in Fig. 3.

3 Hardware Implementation of Band-Pass Notch Digital IIR Filter

3.1 Implementation of Band-Pass Notch IIR Filter System

Block diagram (Data-path) of the system used for hot spot region identification is shown in Fig. 4, in which two anti-notch filters are used and tuned according to characteristics frequency of particular protein family. For reading the filter output in

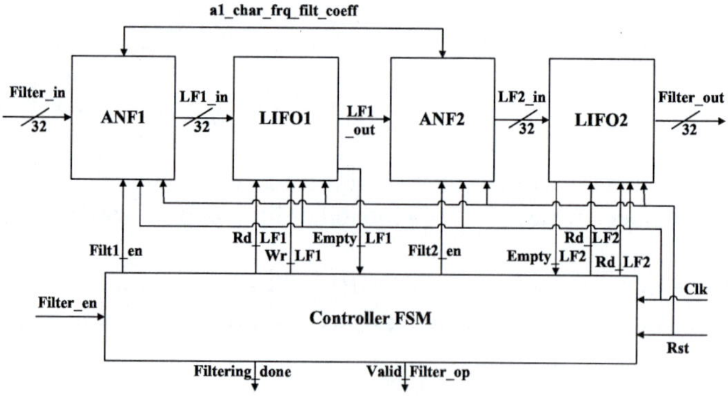

Fig. 4 Block diagram of ANF including zero phase filtering

reverse order two last in first out (LIFO) blocks are used. In LIFO block one RAM and two counters are used for writing to and reading from the block RAM.

Controller FSM is used for providing various control signals to LIFO and ANF. External filter input is directly applied to ANF1 and its output is written to LIFO1. Then after reading from LIFO1, its output is applied to ANF2. LIFO2 is written by the ANF2 output. Final band-pass notch filter output is supplied after reading from LIFO2. Two control output signals 'valid_filter_op' and 'filtering_done' are generated from controller FSM.

3.2 Controller FSM

Various control signals for reading and writing into LIFO and enabling and disabling of ANF are supplied from controller FSM (as shown in Fig. 5). Here S_1, S_3, S_5, S_7 states are the main states and S_2, S_4, S_6, S_8 states are the extra states used for providing synchronization between LIFO read & write and ANF enabling signals. S_1 is the power on reset state in which all internal control signals like filt_en to ANF1 & ANF2, Rd_LIFO & Wr_LIFO to LIFO1 & LIFO2 and external signals like valid_filt_op & filtering_done are cleared to zero. This is a Mealey FSM diagram, in which inputs are written on transition arrow and various outputs are written side

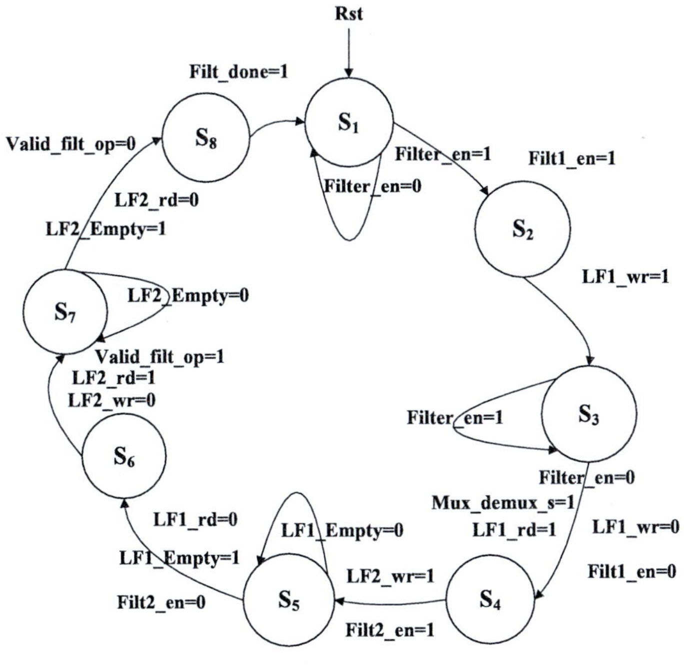

Fig. 5 State diagram of controller FSM

by side of the transition line. Proposed hot spot filter is activated using external filter_en signal and FSM comes out of reset state to S_2 for filter_en = '1'. During this transition filt1_en signal is set to '1' and then FSM transit to S_3 state. LIFO1 is enabled for writing after S_2 state. There is one state difference between filt1_en and wr_LIFO1 signals due to the latency of one clock pulse of ANF.

FSM remains in S_2 state until filter_en='1' and comes to extra S_4 state before which writing to and reading from LIFO1 is disabled and enabled respectively and ANF1 is disabled. Then FSM unconditionally comes to state S_5. During this transition, ANF2 is selected and LIFO2 is enabled for writing. In S_5 state ANF1 output values (written to LIFO1) are read in reverse order from LIFO1 and again applied to ANF2 & filter output is written into LIFO2, until LIFO1 is empty. Then FSM comes to S_6 extra state. During S_5 to S_6 transition, reading from LIFO1 is stopped and ANF2 is disabled. After this FSM comes to state S_7. In S_7 state final hot spot filter output values are supplied after reading (in reverse order) from LIFO2, which is indicated by valid_filt_op = '1'. For this LIFO2 is disabled and enabled for writing and reading respectively before coming into S_7 state. The FSM remains in S_7 state until LIFO2 is empty, after which it comes into state S_8. LIFO2 is stopped from reading, valid_filt_op = '0' (for indicating that output on the output data bus is not valid) and filt_done = '1' (for representing that hot spot filtering is complete) before coming into S_8 state. This filt_done signal remains '1' for single clock pulse, then FSM comes to reset state, and all control signals are reset again and then hot spot filter FSM wait for another set of filter inputs.

3.3 VLSI Realization of IIR ANF

Band-pass notch IIR digital filter can be realized either by direct form-I structure or by direct form-II structure according to filter coefficients of Eq. 8. In this paper, direct form-II structure is considered for hardware implementation because it requires only two delay elements as compare to four delay elements of direct form-I structure.

Here there is a requirement of real data type because filter input (Protein EIIP numerical sequence) and filter coefficients are of real values. For representing these real values, fixed point or floating point arithmetic can be used. But in this paper 32-bit single precision IEEE 754 floating point standard [14] is used due to its higher range and more precision as compared to fixed point arithmetic.

VLSI architecture of ANF IIR filter (according to direct form-II structure) is shown in Fig. 6. In this architecture, the b_1 filter coefficient is skipped because in Eq. 8 $b_1 = 0$. The main components of IIR digital filter like adders and multipliers (represented by AD & M respectively in proposed architecture) are implemented according to the IEEE 754 floating point standard [15]. Two delay elements (registers) D_1 and D_2 are used. Then finally these components are combined to make the VLSI architecture according to direct form-II structure of IIR ANF.

The overall clock frequency of any hardware DSP system is determined by critical path delay, which is the delay provided by particular longest path in that system. The

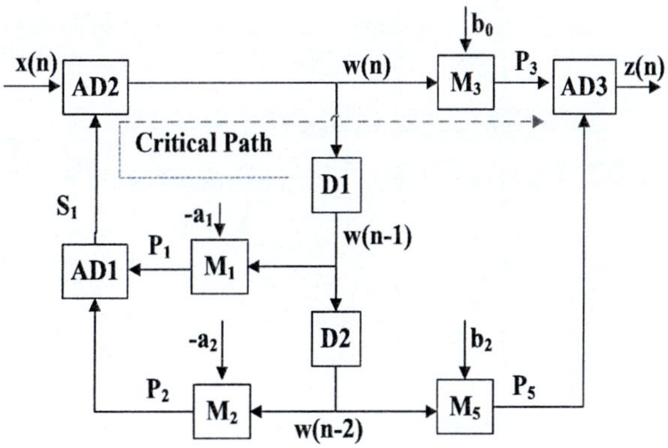

Fig. 6 VLSI architecture of IIR ANF

maximum clock frequency will be equal to a inverse of this delay and responsible for performance of any VLSI system. The critical path of IIR filter is shown by red dotted line.

Theoretical Critical path delay (T) is equal to $3T_A + 2T_M$, where T_A and T_M are propagation delay of adders and multipliers units respectively. So the maximum operating clock frequency is equal to $1/(3T_A + 2T_M)$.

4 Results and Discussion

4.1 Simulation Result

Protein benchmark datasets are easily available on various online databases like PDB (protein data bank) [16], and Swiss-Prot [17]. The online alanine-scanning energetics database (ASEdb) [18] of protein hot spots locations according to alanine-scanning mutagenesis (ASM) practical method is prepared. In order to validate the proposed design, human basic fibroblast growth factor ('bFGF') protein (with PDB ID '4fgf' and Swiss-Prot ID 'P09038') of FGF protein family is used for hot spot region identification. The filter coefficient a1 is calculated and applied to the ANF hardware system according to char. freq. $(\omega 0) = 0.9$ for FGF family. The proposed design is simulated using Xilinx Isim simulator, whose waveform for "bFGF" protein is shown in Fig. 7. This waveform clearly signifies that filter output can be sampled for valid_filt_op = '1' and after that filtering_done = '1' to indicate filtering is complete. Various LIFO control signals are also shown.

Filter input passes three times through S_2, S_3 and S_4 states plus some extra states. Therefore the total number of clock cycles (M) required to complete this filtering process is equal to $3 * N + 8$, where N is the length of the filter input sequence. So total computational time of hardware filter is equal to $M * T$, where T is the critical path delay of the filter system. To validate the design, the complete hot spot ANF

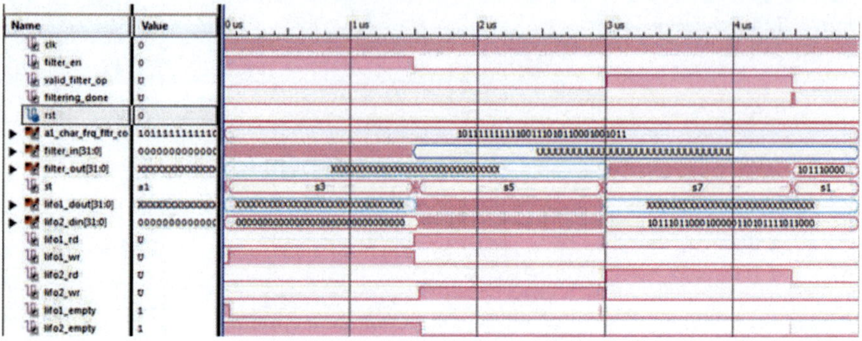

Fig. 7 Simulation waveform for FGF protein functional group

system is also implemented on MATLAB software. Then MATLAB computational time is also calculated by evaluating the average CPU time over 100 runs for each protein example sequence using ‘tic’–‘toc’ commands. Several assumptions are made during MATLAB simulations like a fresh MATLAB session is started, an intel (R) core I3@2.53 GHz processor with 4 GB RAM is used and no other work is done on the CPU during this simulation

For “bFGF” protein sequence ($N = 146$ and critical path delay $T = 39.817$), computational times of hardware and MATLAB simulation (of BPN filter) are 17.84 μs and 910 μs respectively (as detailed in Table 2). Therefore hardware implementation of BPN IIR digital filter speeded up the computation for hot spot region detection by 51-times compared to its MATLAB simulation.

Hot spot locations are detected for ‘bFGF’ protein example, which are shown as similar plot for MATLAB and hardware implementation in Fig. 8a, b respectively. Same hot spot locations (24, 96 and 103) as of reported in literature [5, 7] and ASEdb data base [18] are detected (detailed in Table 2). Some other new hot spot locations like 5, 16, 37, 48, 59, 70, 80, 113 and 124 are also identified except those well known locations of ASEdb data.

Table 2 Comparison of the proposed hardware architecture of BPN filter method with previous literature for ‘bFGF’ data

Parameter	Hot Spot Locations	Computational Time (μs)
MATLAB Simulation of STDFT Transform based technique [5]	24, 96, 103	3764.3
MATLAB Simulation of Filter based technique [7]	24, 96, 103	393.7
MATLAB Simulation of IIR BPN Filter based technique (proposed in this paper)	24, 96, 103	910
Hardware simulation of IIR BPN Filter (proposed in this paper)	24, 96, 103	17.84
ASEdb Data [18]	24, 96, 103, 140	–

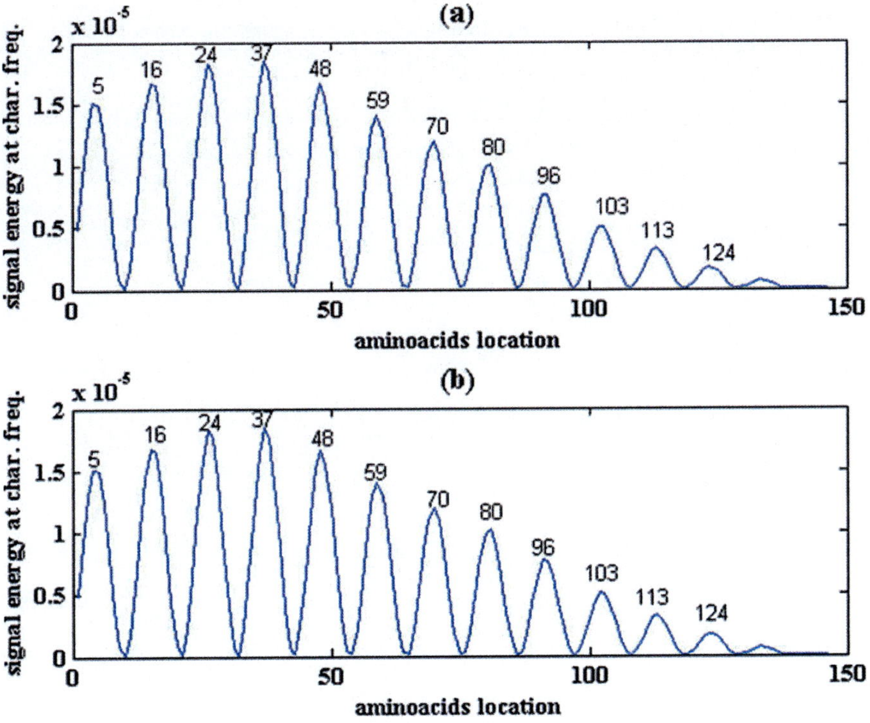

Fig. 8 Detection of Hot spot locations for human bFGF protein using **a** MATALB, **b** hardware implementation

4.2 Synthesis Result

The proposed BPN filter system is synthesized using Xilinx ISE 14.4 by considering the Artix-7 FPGA family (xc7a100t-2-csg324 device). FPGA resource utilization of the proposed design is shown in Table 3, which clearly indicates very less hardware is required. Summary of various FPGA hardware blocks used are detailed in Table 4,

Table 3 FPGA resource utilization of proposed design

Parameter	Used	Available	% Utilization
Slice registers	291	126,800	0
Slice LUTs	3264	63,400	5
Fully used LUT FFpairs	113	3442	3
Bonded IOBs	101	210	48
BUFB/BUFCTRLS	1	321	3
DSP48E1 Blocks	16	240	6
Block RAM/FIFO	2	135	0

Table 4 Summary of FPGA hardware blocks for proposed design

Name of H/W unit	Number of H/W blocks
Multipliers (24 * 24 bit)	8
Adders/subtractors	48
Registers	8
Latches	38
Comparators	18
Multiplexers	278
XORs	14
FSMs	1
RAM (256 * 32 bit)	2

Fig. 9 RTL Schematic of the proposed design

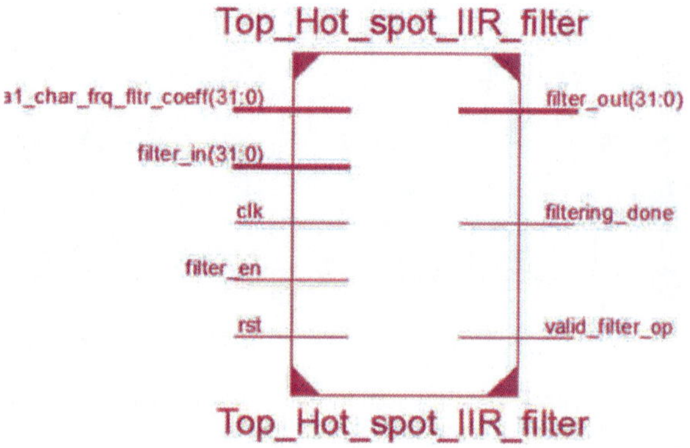

which clearly shows that two block RAM and 1-FSM is used (as per our design).

Minimum clock period and maximum clock frequency of the proposed design is 39.817 ns and 25.11 MHz respectively. Top level RTL schematic of the hot spot filter is shown in Fig. 9, which indicates the various inputs and outputs signals of the proposed design.

5 Conclusion

Hardware implementation of tunable BPN IIR digital filter system (including zero phase filtering) is carried out in this paper, which is used for localization of hot spot regions. Hardware filter results are validated through MATLAB implementation. Hardware BPN filter requires 17.84 us computational time as compared to 910 us of MATLAB software implementation; i.e., hardware filter is 51 times faster compared to its MATLAB simulation with similar hot spot locations for 'bFGF' protein sequence.

References

1. Alberts B, Bray D, Johnson A, Lewis J, Raff M, Roberts K, Walter P (1998) Essential cell biology. Garland Publishing, New York
2. Bogan AA, Thorn KS (1998) Anatomy of hot spots in protein interfaces. J Mol Biol 280:1–9
3. Clackson T, Wells JA (1995) A hot spot of binding energy in a hormone-receptor interface. Science 267(5196):383–386
4. Kortemme T, Baker D (2002) A simple physical model for binding energy hot spots in protein-protein complexes. PNAS 99(22):14116–14121
5. Ramachandran P, Antoniou A, Vaidyanathan PP (2004) Identification and location of hot spots in proteins using the short-time discrete Fourier transform. In: Thirty-eighth Asilomar conference on signals, systems, and computers, Pacific Groove, CA, Nov 2004, pp 1656–1660
6. Ramachandran P, Antoniou A (2006) Localization of hot spots in proteins using digital filters. In: Proceedings of IEEE international symposium signal processing and information technology, Vancouver, BC, Canada, Aug 2006, pp 926–931
7. Ramachandran P, Antoniou A (2008) Identification of hot-spot locations in proteins using digital filters. IEEE J Sel Top Sig Process 2(3):378–389
8. Ramachandran P, Lu WS, Antoniou A (2008) Improved hot-spot location technique for proteins using a band-pass notch digital filter. In: Proceedings of IEEE international symposium on circuits and systems, Seattle, pp 2673–2676
9. Ramachandran P, Lu W-S, Antoniou A (2009) Tuning technique for the location of hot spots in proteins using a bandpass notch digital filter. In: 2009 IEEE international workshop on genomic signal processing and statistics. IEEE
10. Cosic I (1994) Macromolecular bioactivity: is it resonant interaction between macro-molecules?—theory and applications. IEEE Trans Biomed Eng 41(12):1101–1114
11. Cosic I (1997) The resonant recognition model of macromolecular bioactivity—theory and applications. Birkhauser Verlag, Basel
12. Antoniou A (2005) Digital signal processing: signals, systems, and filters. McGraw-Hill, New York
13. Vaidyanathan PP, Yoon BJ (2002) Gene and exon prediction using all pass-based filters. In: Proceedings of IEEE workshop on general signal processing and statistics
14. Kahan W (1996) IEEE standard 754 for binary floating-point arithmetic. Lect Notes Status IEEE 754(94720–1776):11
15. Bharade P, Joshi Y, Manthalkar R (2016) Design and implementation of IIR lattice filter using floating point arithmetic in FPGA. In: Proceedings of IEEE conference on advances in signal processing (CASP), Pune, pp. 321–326
16. Protein Data Bank (PDB), Research Collaboratory for Structural Bioinformatics (RCSB) (Online). Available http://www.rcsb.org/pdb/
17. Swiss-Prot Protein Knowledgebase. Swiss Institutes of Bioinformatics (SIB) (Online). Available http://us.expasy.org/sprot/
18. Alanine Scanning Energetics Database (ASEdb) (Online). Available http://nic.ucsf.edu/asedb/index.php

Smart Data Logger for Solar and Wind Power Generation

Sujoy Mondal, Ashoke Mondal, and Shilpi Bhattacharya

Abstract As solar and wind power generation systems are becoming more and more popular owing to the depleting fossil fuels, smarter monitoring systems with precision data logging are an active area of development. The data accumulated are very useful if analyzed properly, and hence, the data logged so far need to be accurate without significant error. In this paper, a smart design and implementation of a microcontroller-based data acquisition system are presented to monitor solar irradiance and speed of wind as these are the primary parameters deciding power generation in a solar–wind hybrid power plant. A pyranometer is used as a solar radiation sensor, and an anemometer senses wind speed. Temperature (solar insolation) and generated power have also being logged. A relationship has been tried to develop among the key parameters and dependent variables.

Keywords Anemometer · Data logging · Pyranometer · Solar insolation

1 Introduction

Energy is indispensable for the survival of surrounding life. The conventional sources of energy like coal, oil or gas are being used commonly to provide energy. Depletion of these conventional fossil fuels as projected in major studies and surveys predicts reserves will run short in 50 years, gas reserves are only enough for 70 years, and coal reserves are consumed away completely in 200 years. Exhaustion of the energy reserves from conventional sources has led to the rise in exploitation of renewable

S. Mondal
Department of ECE, RCCIIT, Kolkata, India
e-mail: sujoy.mondal@rcciit.org

A. Mondal · S. Bhattacharya (✉)
Department of EE, RCCIIT, Kolkata, India
e-mail: shilpibhatt2013@gmail.com

A. Mondal
e-mail: ashoke.mondal@gmail.com

V. Nath and J. K. Mandal (eds.), *Proceedings of the Fourth International Conference on Microelectronics, Computing and Communication Systems*, Lecture Notes in Electrical Engineering 673, https://doi.org/10.1007/978-981-15-5546-6_49

energy sources primarily solar and wind. Both these renewable energy availabilities depend on geographical and meteorological conditions. The PV power scenario is expanding at a rapid pace and is considered one of the major sources of renewable energies because of its abundance. Also, because of the increasing popularity of micro-grid systems and the clean power being easily fed to electrical grid, PV-based power plants are gaining utmost importance [1–10]. So, the profitability of such a power plant and their implementation in various locations are an important area of research [10–14]. PV and wind power installations however operate in outdoor locations which are not confined to definite environmental conditions. PV modules of different makes or technologies work differently and deliver different irradiance [15–17].

Quantity of solar power generation in a PV plant depends on solar insolation which generally is guided by the geographical location of the PV unit. On the other hand, wind energy-based power plant also depends on the wind speed which definitely is governed by the coordinates of the region selected for such a plant. Generally, nowadays, hybrid power plants based on solar and wind power are set up for more reliability. In remote areas, these hybrid power systems, generally based on renewable energy sources such as wind and solar, play significant roles. The optimum use of these renewable sources depends on several climatic factors of a region. The solar irradiation and wind speed are the primary parameters governing the energy output from a hybrid wind–solar power plant. The availability of power continuously from such a system also depends on the said parameters. In view of that, a smart data acquisition system to monitor these parameters continuously in real time to operate such a hybrid power system effectively is an active area of interest owing to the growth in power generation through renewable sources.

Owing to these factors, online measurement and recording of solar irradiance and wind speed are of utmost importance to study the feasibility for setting up or extension of a solar, wind or solar–wind hybrid power plant at a certain geographical location. This paper presents the development and testing of a smart data logger to be used in a solar–wind hybrid power plant. The proposed data acquisition system is based on an AVR processor taking inputs from pyranometer and anemometer for solar irradiation (W/mt^2) and wind speed (m/s). The board is programmed to store the logged data in an external as well as internal memory. Microcontroller-based data acquisition systems have also been attempted in [18]. Several data acquisition systems were developed to record and constantly monitor the important meteorological data of a specific location such as wind speed, solar irradiation, humidity and temperature. Moreover, important electrical parameters of significance such as voltage and current generated from photovoltaic and wind turbine generator-based power systems [19–25].

2 Proposed System

Figure 1 shows the block diagram of the proposed smart data logging system. An ARM processor is central to the data acquisition system. The system consists of two

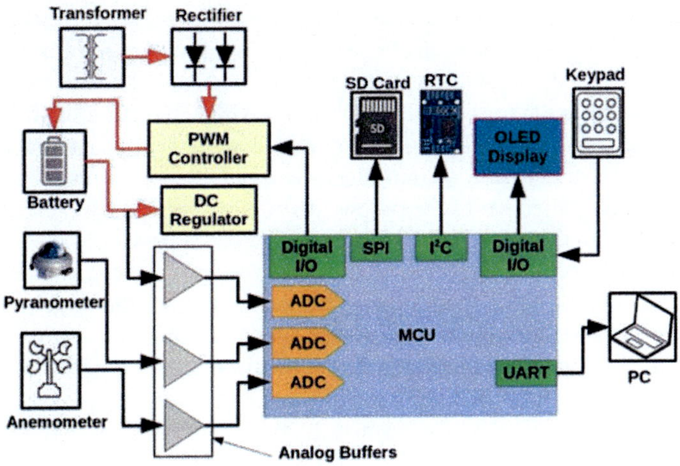

Fig. 1 Block diagram of the proposed system

sensors, solar insolation and wind speed. A calibrated pyranometer shown in Fig. 2 has been used to convert the solar irradiance into corresponding analog signal.

The built-in ADC of the processor samples and converts the electronic signal into digital word and stores in the memory (SD card) with time stamp. A real-time clock module is used to keep track of the current time. The measured data of solar insolation and wind speed need to be logged with respect to the respective current time they were captured. The real-time clock is being generated from RTC (Dallas). Similarly, output of anemometer which is proportional to the wind speed is also logged through another built-in ADC and stored in the memory. Both the samplings are done at a slow rate of 1 sample/min. This rate is quite adequate to have an idea about the environmental conditions. In order to continuously record the data through the day without any interruption in backup, power supply has been provided with 12 V 7AH power supply battery. The charging of the battery has been controlled through the

Fig. 2 Pyranometer module

processor and transformer and a switch (MOSFET) to prevent overcharging and deep discharge. A keypad and OLED display are used to obtain information about the present date, time, value of the measured parameters, solar insolation and wind speed. The data from SD card could be dumped into PC/laptop through UART (serial port). The data obtained so far could be drawn as plot using proper software. The system consists of an AVR microcontroller (ATmega328) along with other peripheral circuits. The main objective of the proposed system is to read two sensor data and store it to a SD card memory. However, there is another 1kByte of in system memory is available as backup memory. Pyranometer and anemometer sensors are used to measure solar irradiance and wind speed, respectively. Both the sensors output are voltage type. The sensor output voltages are measured by microcontroller (MCU) ADC unit. A RTC module is used to capture the time instant which also finds data sampling interval. The status about system time, date, memory, sensor data all are displayed through a GLCD. A SD card module attached to MCU through SPI protocol to store the data in micro SD card (Fig. 3).

3 System Algorithm

The system works in three different modes: Normal mode, system date/time update mode and PC interface mode. System Date/Time Update Mode: If the menu button (M) is pressed during power-on reset (PWR-RST), this mode will be activated wherein the date and time settings can be updated. PC Interface Mode: If the up (U) and down (D) keys of the keypad are pressed during PWR-RST, the processor works in PC interface mode. In this mode, the content of SD card is transferred directly to PC through a PC data logger software (software implemented in C++). Normal Mode: Unlike previous modes on PWR-RST, MCU operates in the normal mode. In the normal mode, the system starts with initialization of the MCU ports, registers, counter and flag values. At first, MCU reads the battery voltage to control the MOSFET-based buck converter operated in PWM mode. To prevent deep discharge and overcharging of the battery, two threshold voltages, respectively, 11.8 V and 14.75 V are used. A real-time clock module (RTC) is constantly monitoring to keep track of sampling intervals of sensor data. In this prototype module, MCU gets triggered after each one minute of interval. Figure 3 depicts the complete work flow of the system software. First, the sensor data in terms of analog voltages are read by MCU, and then, it is converted by proper scaling factor associated to each sensors. The scaling factors provided by the makers are specified in Table 1. Thereafter, the values are written to SD card in the data format as in Fig. 4. All the sensor data, time, date, SD card status are updated to OLED display after each second.

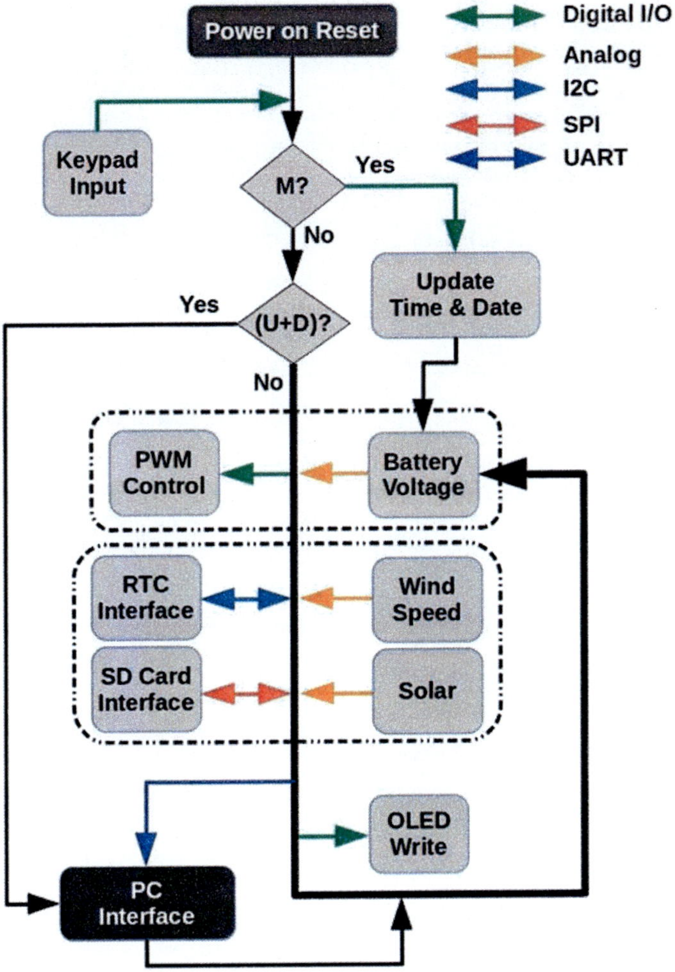

Fig. 3 System software flow diagram

4 Results and Discussions

The results obtained are as depicted in Figs. 5 and 6. Figure 5 indicates the voltage obtained from the solar panel which is correlated with the solar irradiance in watt/mt^2. The generated voltage and the solar irradiance are recorded during a span of 8 am to around 4 pm. This sample data were recorded on March 1, 2018. Figure 6 shows the battery voltage correlated with solar irradiance (watt/mt^2) and wind speed (m/s) recorded continuously on 23rd and 24th of August 2018 consecutively. The experimental microcontroller unit and electronic controller unit are shown in Figs. 7 and 8 respectively.

Table 1 System components

Parameters	Specifications
Pyranometer	Input: +12 V (Bias) Output range: 0–2 V Sensitivity: 1 mV = 1 W/mt^2
Anemometer	Input: +12 V (Bias) Output range: 0–2000 mV @ 0–100 mt/s Sensitivity: 20 mV @ 1 mt/s
Battery	12 V, 7 AH, C-10
RTC	Dallas-3231
OLED display	128 × 64 pixel (0.96 in.)
Keypad	1. RESET 2. MENU 3. UP 4. DOWN
Microcontroller	ATmega328
MOSFET	IRF540/9540
Micro SD card	4 GB

DATE	TIME	SOLAR (Watt/m^2)	WIND (m/s)	BAT VOLT(V)

Fig. 4 Data format for memory card

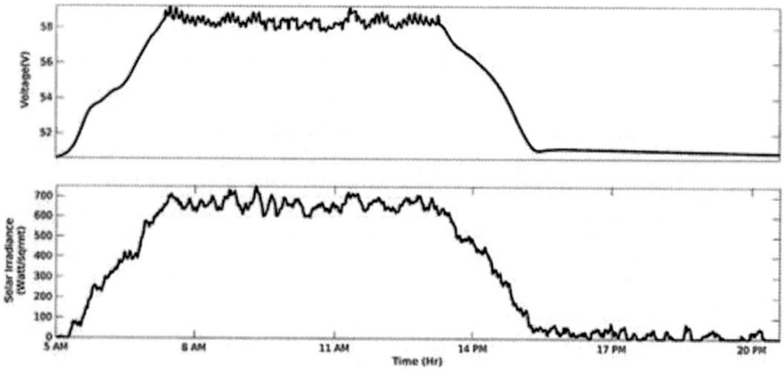

Fig. 5 Solar irradiation (W/mt^2) and voltage (V)

The blue and orange colored plots represent solar irradiance and wind speed, respectively.

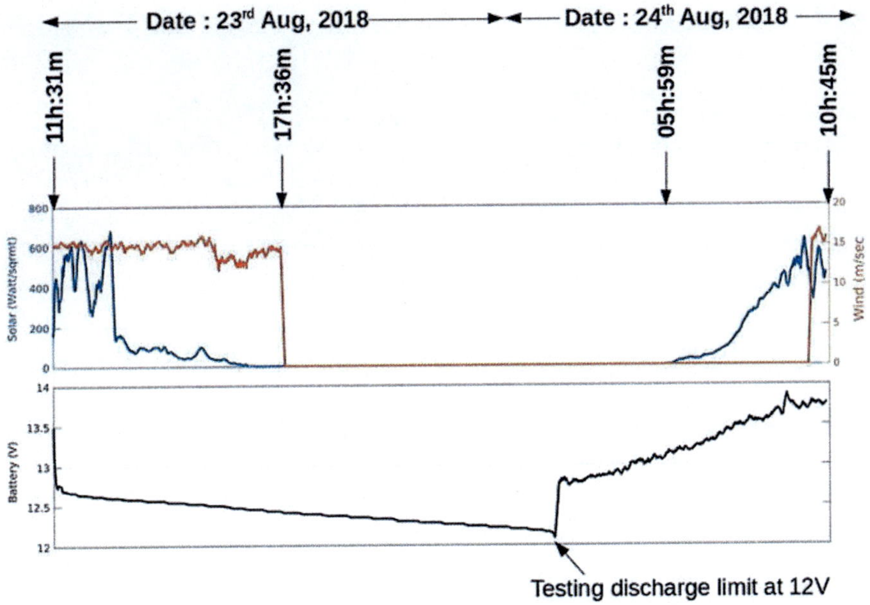

Fig. 6 Solar irradiation (W/mt^2) and wind speed (mt/s) battery voltage (V)

Fig. 7 Microcontroller unit

Fig. 8 Hardware setup

5 Conclusion

As a hybrid power plant based on solar and wind is gaining popularity in renewable energy-based power generation sector, a data acquisition system is designed and implemented here for continuous monitoring of environmental parameters. These parameters are measured and displayed as well as recorded and stored in a SD card or PC for further analysis. This smart data logger is capable of monitoring specifically the solar irradiance and wind speed and displays the same on an OLED display. The data acquisition system is so designed to operate with power from the mains, whereas it is also capable of working successfully during power outage.

References

1. Chen T, Yang JM (2009) Research on energy management for wind/PV hybrid power system. In: 2009 3rd international conference on power electronics systems and applications (PESA). IEEE, pp 1–4
2. Liu J, Fang W, Yang Y, Yang C, Lei S, Fu S (2013) Increasing wind power penetration level based on hybrid wind and photovoltaic generation. In: 2013 IEEE international conference of IEEE region 10 (TENCON 2013). IEEE, pp 1–5
3. Kiymaz Ö, Yavuz T (2016) Wind power electrical systems integration and technical and economic analysis of hybrid wind power plants. In: 2016 IEEE international conference on renewable energy research and applications (ICRERA). IEEE, pp 158–163
4. Kanathipan K, Lam J (2017) A new modified maximum power extraction technique for wind and hybrid renewable energy systems. In: 2017 IEEE electrical power and energy conference (EPEC). IEEE, pp 1–6
5. Daut I, Irwanto M, Irwan Y, Gomesh N, Rosnazri NA (2011) Potential of solar radiation and wind speed for photovoltaic and wind power hybrid generation in Perlis, Northern Malaysia. In: 2011 5th international power engineering and optimization conference. IEEE, pp 148–153
6. Hao W, Sun C, Leilei Z (2016) The research on adaptability evaluation of wind power integration capacity in power grid considering wind-solar hybrid complementary characteristics and flexible loads interference. In: 2016 IEEE PES Asia-Pacific power and energy engineering conference (APPEEC). IEEE, pp 1856–1860
7. Rather ZH, Chen Z, Thøgersen P, Lund P (2015) Dynamic reactive power compensation of large-scale wind integrated power system. IEEE Trans Power Syst 30(5):2516–2526
8. Tomasson E, Söder L (2018) Generation adequacy analysis of multi-area power systems with a high share of wind power. IEEE Trans Power Syst 33(4):3854–3862

9. Gu M (2009) Research on strategic importance of non-grid-connected wind power for us energy development. In: 2009 World non-grid-connected wind power and energy conference. IEEE, pp 1–5
10. Kim J-h, Yang H-J, Ko S-y, Baatarbileg A, Kwon H, Oh S-B, Lee G-m (2018) Contribution of large scale PV plants to electric power during winter peak times in Jeju Island. In: 2018 5th international conference on renewable energy: generation and applications (ICREGA). IEEE, pp 41–45
11. Rahman MA, Al Awami AT (2015) Decentralized wind-PV-diesel based hybrid power generation. In: 2015 international conference on electrical engineering and information communication technology (ICEEICT). IEEE, pp 1–5
12. Lukutin B, Shandarova E, Fuks I (2016) Energy efficiency of photovoltaic power plants in stand-alone power supply systems. In: 2016 2nd international conference on industrial engineering, applications and manufacturing (ICIEAM). IEEE, pp 1–5
13. Kumar A, Verma A (2016) Optimization and implementation of hybrid energy sources in remote and grid active microgrids—a case study for Indian scenario. In: 2016 IEEE 7th power India international conference (PIICON). IEEE, pp 1–6
14. Freire MC, Pica CQ, Martins MAI, da Silva AL (2015) Analysis of relevant technical and economic aspects to support the choice of feasible locations for photovoltaic power plants in Brazil. In: 2015 IEEE PES innovative smart grid technologies Latin America (ISGT LATAM). IEEE, pp 258–263
15. Osterwald C, Anderberg A, Rummel S, Ottoson L (2002) Degradation analysis of weathered crystalline-silicon PV modules. In: Conference record of the twenty-ninth IEEE photovoltaic specialists conference, 2002. IEEE, pp 1392–1395
16. Carullo A, Vallan A (2012) Outdoor experimental laboratory for long-term estimation of photovoltaic-plant performance. IEEE Trans Instrum Meas 61(5):1307–1314
17. Del Cueto J (2002) Comparison of energy production and performance from flat-plate photovoltaic module technologies deployed at fixed tilt. In: Conference record of the twenty-ninth IEEE photovoltaic specialists conference, 2002. IEEE, pp 1523–1526
18. Mukaro R, Carelse XF (1999) A microcontroller-based data acquisition system for solar radiation and environmental monitoring. IEEE Trans Instrum Meas 48(6):1232–1238
19. Koutroulis E, Kalaitzakis K (2003) Development of an integrated data-acquisition system for renewable energy sources systems monitoring. Renew Energy 28(1):139–152
20. Soetedjo A, Nakhoda YI, Suryadi D (2013) Development of data acquisition system for hybrid power plant. In: 2013 international conference on QiR. IEEE, pp 197–201
21. Benghanem R, Maafi A (1997) Measurement system for solar radiation. In: IEEE instrumentation and measurement technology conference sensing, processing, networking. IMTC proceedings, vol 2. IEEE, pp 932–935
22. Itagaki A, Okamura H, Yamada M (2003) Preparation of meteorological data set throughout japan for suitable design of PV systems. In: Proceedings of 3rd world conference on photovoltaic energy conversion, 2003, vol 2. IEEE, pp 2074–2077
23. Gomes JM, Ferreira PM, Ruano AE (2011) Implementation of an intelligent sensor for measurement and prediction of solar radiation and atmospheric temperature. In: 2011 IEEE 7th international symposium on intelligent signal processing. IEEE, pp 1–6
24. Lita I, Cioc IB, Visan DA, Dorobantu I (2009) Technologies for wireless data acquisition and monitoring of environmental parameters. In: 2009 32nd international spring seminar on electronics technology. IEEE, pp 1–5
25. Schuss C, Eichberger B, Rahkonen T (2018) Impact of solar radiation on the output power of moving photovoltaic (PV) installations. In: 2018 IEEE international instrumentation and measurement technology conference (I2MTC). IEEE, pp 1–6

Deep Learning-Based Alzheimer Disease Detection

Swathi S. Kundaram and Ketki C. Pathak

Abstract Deep learning methods have gained more popularity recently in medical image analysis. This work proposes a deep convolutional neural network (DCNN) for Alzheimer's disease classification using magnetic resonance imaging (MRI) samples. Alzheimer disease (AD) is an irreversible neurological brain disorder; its early symptoms are memory loss and losing thinking abilities called cognitive functions. The accurate diagnosis of Alzheimer's disease at an early stage is very vital for patient care and conducting future treatment. Deep learning techniques are capable of learning high-level features from dataset compared to hand-crafted feature learning methods such as machine learning techniques. The proposed method classifies the disease as Alzheimer's disease (AD), mild cognitive impairment (MCI) and normal control (NC). Spyder software obtained from anaconda bundle with Keras library and Tensorflow backend on GPU is used to model DCNN. Experiments are conducted using ADNI dataset and output classification result showed 98.57% accuracy compared to other studies. Our approach also enables us to expand this methodology to predict for more stages of disease classification.

Keywords Alzheimer's disease · Convolutional neural network · Deep learning · MRI · Neurological disorder

1 Introduction

Alzheimer's disease (AD) is a progressive neurological brain disease, which is caused due to the damage of nerve cells in parts of the brain [1]. It begins with the loss of memory, difficulty in speaking language and other cognitive functions making a

S. S. Kundaram (✉) · K. C. Pathak
Department of Electronics and Communication, Sarvajanik College of Engineering and Technology, Gujarat Technological University, Surat, India
e-mail: lilly.kundaram@gmail.com

K. C. Pathak
e-mail: ketki.joshi@scet.ac.in

V. Nath and J. K. Mandal (eds.), *Proceedings of the Fourth International Conference on Microelectronics, Computing and Communication Systems*, Lecture Notes in Electrical Engineering 673, https://doi.org/10.1007/978-981-15-5546-6_50

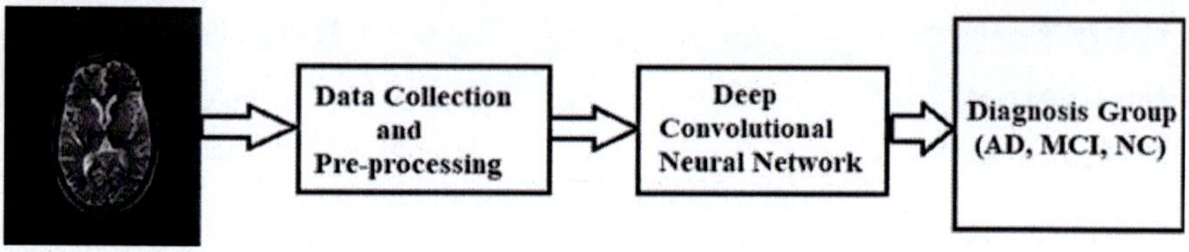

Fig. 1 Proposed deep learning flow for classification of Alzheimer's into AD, MCI and NC

patient unable to perform day-to-day life activities. In particular, [1] researchers found that AD is not only common cause of dementia but eventually leading to death of people, which become a remarkable focus in research (Fig. 1).

According to Alzheimer's association, it is the sixth leading cause of death in the USA[2]. A survey [3] stated that there will 131.5 million people living with dementia worldwide and most of them with age greater than 65 has higher rate of risk with this disease. The brain region including thinking ability, memory, reasoning of the patient wrinkle up and shrinks in the hippocampus area. This is the main cause of suffering from AD. The visualization of AD and healthy brain shown in Fig. 2 gives the idea that memory and language muscles have diminished. Genetic mutation is another cause for AD; estimated to affect about 1% or less than 1% people [4].

An early diagnosis of this disease becomes crucial and requires good clinical assessment based on patient's medical history, several neuropsychological tests such as mini-mental state examination (MMSE), neuropsychiatric inventory questionnaire, clinical dementia rating and other pathological evaluations [6]. In addition to these clinical methods, there many other techniques to detect AD such as biomarkers, cerebrospinal fluid (CSF) analysis, brain imaging include magnetic resonance imaging (MRI)/positron emission tomography (PET), analysing proteins in blood.

Fig. 2 Brain cross-sectional to visualize difference between healthy brain and Alzheimer brain [5]

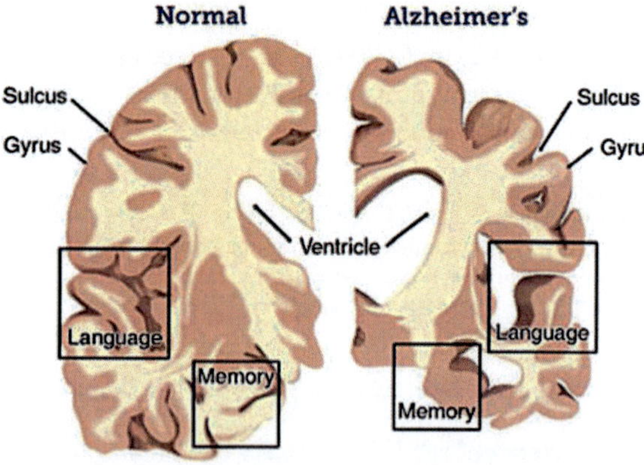

Traditional approach for classification of AD to assist diagnosis was to generate feature wavelets using discrete wavelet transform (DWT) method. This does not give detection of disease; machine learning algorithms are required to do further processing [7].

Machine learning approaches are best for accurate classification of AD [8]. Most popular among these approaches is support vector machine (SVM). Rathore et al. [9] proposed a framework uses feature-based ranking method with SVM to classify the disease as AD and healthy controls (HC). SVM is used to build predictive classification models, which extracts informative features, high dimensional from the MRI [9]. However, this requires handcrafted features of brain structures, which is laborious and time consuming leading to need of experts advice.

An alternative family of machine learning methods is deep learning algorithms. Deep learning algorithms perform automatic feature extraction without human intervention. Due to the availability of more number of hidden layers, deep learning approaches learn the high-level representation from the raw data. This is the reason behind its popularity in the computer vision domain. Ortiz et al. [10] discussed many deep learning architectures for early diagnosis of AD. Convolutional neural network (CNN) are inspired from human visual cortex and learns the features from simple edges to more complex edges from the dense hierarchical structure. It is building block of convolution and pooling layers. Convolutional layer provides the feature maps by multiplying the input image with the kernel, and pooling layer down samples the image keeping the similarity features [11]. This stimulated many neuroscience researchers to find their solution to the problem associated with neuroimaging. Shi et al. [12] described multimodal classification of AD with four classes. Staked autoencoders (SAEs) are used for feature learning for both MRI and PET. These features are fused and trained using SVM, which achieved very less accuracy compared to other available multimode classification. Cui et al. [13] addressed sequential analysis of MRI image along time axis by measuring the longitudinal progression. Multi-layer perceptron (MLP) is used for spatial features, and to train these features, recurrent neural network (RNN) is used. However, such algorithm requires rigid segmentation as a preprocessing task. The accuracy achieved is 89.69% in two-way classification as AD and NC. Islam et al. [14] proposed a DCNN model for four classes. In this, five DCNN models have trained and output features are fused to get the prediction of disease. The uniqueness of this approach is that every model gives various features different from one another making the model generalized for unseen data prediction with accuracy 93.18%.

There are many works available on CNN method for detection of AD. Gunawardena et al. [15] addressed the problem for pre-detection of AD for three classes with accuracy achieved is 84.4%. Combination of CNN and RNN is a new approach for AD diagnosis proposed by Liu et al. [16]. 3D PET images sliced into 2D images which trained by CNN and RNN are used to classify the CNN features with accuracy 91.2% for one-vs-all of three classes. Khvostikov et al. [17] used fusion of structural MRI and mean diffusivity-diffusion tensor imaging (MD-DTI) on hippocampal

ROI region for AD diagnosis. Wang et al. [18] proposed an eight layer CCN for with different types of activation functions and pooling operations for two classes, achieved accuracy is 97.65%.

In this work, a framework-based DCNN is implemented for detection of AD, which uses three layer of CNN model for three classes—AD, MCI and NC. This work does not use any rigid segmentation for grey matter (GM) of the data. Experimental evaluation is performed in terms of accuracy and loss on both training as well as testing dataset. The rest of the papers describes the methodology and implementation in Sect. 2, experiments and results discussion in Sect. 3 and conclusion in Sect. 4.

2 Proposed Methodology

Figure 1 shows the flow of proposed methodology which consists of two steps, preprocessing and network training. The detail description is in the following sections.

2.1 Data Preprocessing

Medical images when acquired from any imaging equipment are in Digital Imaging and Communications in Medicine (DICOM). After acquisition, they need to be converted into proper format (JPG, PNG, TIFF, etc.) for further processing. In our work, we have converted MRI samples into JPEG slices in MATLAB tool. Pixel size of each sample is reduced to 8-bit from 14-bit size by rescaling to 255. Processed slices for each class are shown in Fig. 3.

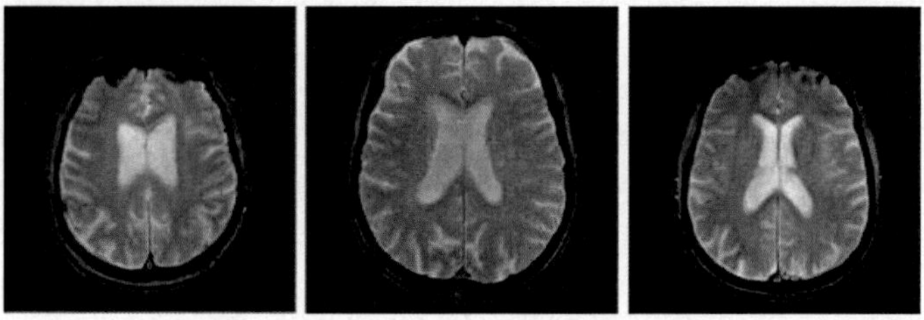

Fig. 3 JPEG slices for each diagnosis class after preprocessing (left: *AD*, centre: *MCI*, right: *NC*)

2.2 Network Architecture

Convolutional neural networks (CNN) are inspired from human visual system. The visual system has small number of neuron cells sensitive to a specific field, i.e., some neurons in the brain fired only in the presence of edges in particular orientation. Such operation is depicted in CNNs. The functioning of convolution layer is to automatically extract features maps from the input images by using element-wise multiplication with filter along entire image. Pooling layer is generally used to avoid overfitting problem, i.e., when network memorizes the data instead of generalization. Rectified linear unit (ReLU) activation is used to fire the neuron or to determine the output of neural network. Feature maps are extracted with combination of several such Conv-ReLU-Pool operations and reach the final single or multiple fully connected layers. The detail operation of proposed architecture is as follows.

Our proposed model shown in Fig. 4 is an ensemble of three blocks; each of the individual blocks has several layers performing three basic operations, which are:

- Convolution
- ReLU activation
- Max pooling

The architecture consists of three sets of convolutional and max pooling layers, followed by a flattening convolutional layer features, then two fully connected layers and finally a softmax/sigmoid classifier. Output has three classes, which are Alzheimer disease (AD), normal control (NC) and mild cognitive impairment (MCI). The input for architecture is a 256×256 gray scale image, which passes through the first convolutional layer with 32 feature maps with filters having size 3×3, a stride of one, and pooling is made zero with ReLU activation function. The image dimensions change from $256 \times 256 \times 1$ to $254 \times 254 \times 32$, according to the dimension formula given below:

$$n^{[l]} = \left(n^{[l-1]} + 2p^{[l-1]} - f^{[l]}\right)/s^{[l]} + 1 \tag{1}$$

where n is the size of input image or previous layer image, p refers to padding and s refers to stride, l refers to current layer, and F refers to the filter size. Then, the network applies max pooling layer with a filter size 3×3 and a stride of one. The resulting

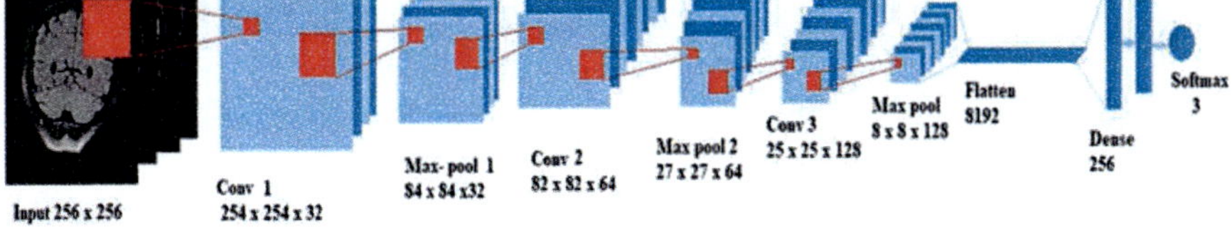

Fig. 4 Architecture of proposed model

Table 1 Specification of proposed model

Layer name	Output shape	Parameter number
Conv2D_1	254, 254, 32	896
ReLU	254, 254, 32	0
Max Pooling_1	84, 84, 32	0
Conv2D_2	82, 82, 64	18,496
ReLU	82, 82, 64	0
Max Pooling_2	27, 27, 64	0
Conv2D_3	25, 25, 128	73,856
ReLU	25, 25, 128	0
Max Pooling_3	8, 8, 128	0
Flatten_1	8192	0
Dense_1	256	2,097,408
ReLU	256	0
Drop out	256	0
Dense	3	771
Softmax	3	0
Total Params	0	2,191,427
Trainable	0	2,191,427
Non trainable	0	0

image dimension reduced to $84 \times 84 \times 32$. Next, there is a second convolutional layer with 64 feature maps having size 3×3 and a stride of 1, so image dimensions reduced to $82 \times 82 \times 64$. Then again max pooling with filter size 3×3, dimension reduces to $27 \times 27 \times 64$. Next, third convolutional layer with 128 feature maps having size 3×3 and stride 1, dimension reduces to $25 \times 25 \times 128$, and with max pooling, the dimension reduced to $8 \times 8 \times 128$. Then, total parameters obtained are 8192 by flattening.

The fifth layer is a fully connected convolutional layer with 256 feature maps each of size 1×1. Each of the 256 units is connected to all the 8192 ($8 \times 8 \times 128$) in the fourth layer. The sixth layer is a fully connected layer with 256 units. Finally, there is a fully connected softmax output layer with 10 possible values corresponding to the digits from 0 to 9. Specification of the proposed model shown in below Table 1.

3 Experiments and Results Discussion

3.1 Dataset

The dataset used in this work obtained from Alzheimer's disease neuroimaging initiative (ADNI) [19]. Dataset consists of 110 AD, 105 MCI and 51 NC subjects, where each subject contains 44 to 50 sample of images. Out of which 110 AD subjects are collected from Horizon Imaging Centre [20]. There are total of 9540 images used for training the network and 4193 images for testing. Data augmentation on images is done with rescale operation.

3.2 Hyper Parameters

Deep learning approaches advocate solving the problem end to end rather breaking the problem into different parts. Due to which it fails to interpret the reasoning behind the result obtained; because of the unknown work done by the collective neurons behind the dense structure of network. Hence, to improve the performance of the model, the following parameters are available for tuning.

- Loss function
- Optimizers
- Layers
- Drop out
- Epochs
- Augmentation

The loss function is the guide to the terrain, telling the optimizer when it is moving in the right or wrong direction. Optimizers shape and mould the model into its most accurate possible form by futzing with the weights. Dropout is used to prevent the network from overfitting by disabling the neurons on purpose; which is done with some probability; normally 0.2 probability drop is preferred. Data augmentation artificially increases the size of training set to avoid regularization of the network. By tuning the number of layers and epochs, accuracy of the model can be increased.

3.3 Experimental Setup and Evaluation

Proposed model is implemented with the Keras library with Tensorflow backend. The experiments are conducted on laptop with 8 GB RAM of Dell Intel Corei7. Model is trained on NVIDIA Ge Force 540 M GPU with 8 GB memory. Relu activation is applied for each neuron of CNN. Output is classified as AD, MCI and NC. There are total of 9540 images are used for training the network and 4193 images for testing. Loss function used is binary cross-entropy. Batch size taken is 10. Optimizer used are Adam, SGD (lr-0.0001, decay-0.00001), Adagrad, Nadam, SGD, Adadelta, Rmsprop. Drop out is kept 0.2 probability. Dense activation function used is softmax. Network is trained for 10 epochs.

Table 2 shows the results of the proposed DCNN model. Performance is evaluated in terms of accuracy and loss for training as well as validation set. Loss gives the best knowledge of how fit is the model. Out of all the optimizers, Adagrad proved to

Table 2 Performance of proposed framework

Optimizer	Training accuracy	Training loss	Validation accuracy	Validation loss
Adagrad	0.9857	0.0386	0.8772	0.3907

give best accuracy with less loss because it does not need manual tuning of learning rate as it makes small updates for frequent parameters and big updates for infrequent parameters. The accuracy versus epoch and loss versus epoch graph for both training and validation set is shown in Fig. 5. It is seen that for training set as the accuracy reached 98.57% loss is dropped down to zero. This gives the measure of progression during training period of the model. While the validation set gives the measure of the quality of the model. Validation accuracy has reached 87.72%, which describes that with 87.72% accuracy model can predict the detection on new data.

We have experimented our dataset on different optimizers with same setup discussed at the beginning of section C. Though the highest accuracy is achieved by adam, adadelta and nadam which can be depicted from Table 3, but the validation loss is constantly increasing; i.e., the model is starting to fit on noise and is beginning to overfit. This ultimately loses the model's ability to predict on new data.

Performance comparison of proposed model with other approaches along with techniques and data modalities is explained in Table 4. Among all the approaches, our proposed model has achieved accuracy as high as 98.57% without any pre-learnt features.

A sample illustration of Conv-ReLU-Maxpool operations on single jpeg image with size of 320 × 240 × 3 is shown in Fig. 6. Single feature with filter size of 3 × 3 is extracted using convolution operation shown in Fig. 6b. Relu activation

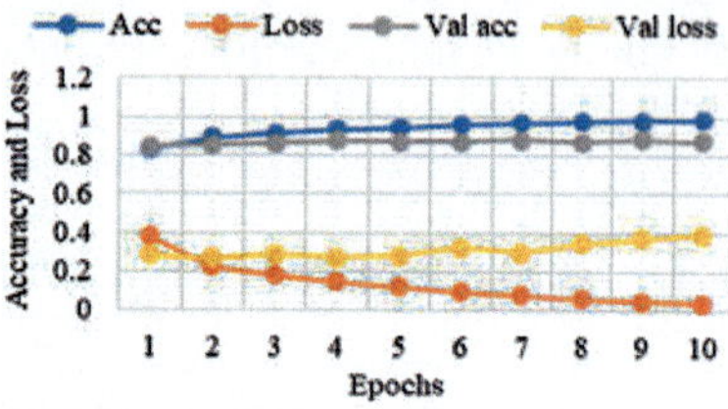

Fig. 5 Accuracy and loss for training set and validation set for 10 epochs with Adagrad optimizer using proposed model is shown

Table 3 Performance of proposed framework for different optimizers

Optimizer	Training accuracy	Training loss	Validation accuracy	Validation loss
SGD	0.9156	0.1853	0.8637	0.2698
Adam	0.9870	0.0384	0.8881	0.6613
Adadelta	0.9947	0.0145	0.8739	0.8607
Nadam	0.9877	0.0353	0.8691	0.8733
Rmsprop	0.9742	0.0808	0.8894	0.4834
SGD-0001	0.7238	0.7550	0.7550	0.5714

Table 4 Performance comparison of proposed framework with other approaches

Approach	Technique	Classification	Modalities	Accuracy (%)	Dataset
Islam et al. [14]	DCNN	4 class (AD/cMCI/MCI/NC)	MRI	93.18	OASIS
Liu et al. [16]	DCNN-RNN	3 class (AD/MCI/NC)	FDG-PET	91.2 with AD versus NC 89.9 with MCI versus NC	ADNI
Shi et al. [12]	SAE & SVM	4 class (AD/ncMCI/cMCI/NC)	MRI + PET	0.53 ± 0.47 with SAE 0.47 ± 0.18 with SVM	ADNI
Ruoxuan et al. [13]	MLP-RNN	2 class (AD/NC)	MRI	89.69	ADNI
Proposed	DCNN	3 class (AD/MCI/NC)	MRI	98.57	ADNI

(a) Jpeg image size $320 \times 240 \times 3$ (b) Convolution operation $318 \times 238 \times 1$ (c) Relu activation $318 \times 238 \times 1$ (d) Max pooling $158 \times 118 \times 1$ (e) Conv-maxpool-Relu-max pool $34 \times 25 \times 1$

Fig. 6 Illustration of Conv-ReLU-Maxpool operations on sample jpeg image

and size reduction of image using maxpool operation is emphasized in Fig. 6c, d. Another interpretation shown in Fig. 6e is the whole operation of Conv-Maxpool-ReLU-Maxpool. From the illustration, one can visualize how the feature maps are produced by the CNN model and also how feature sizes are reduced in every layer of operation. This experiment is conducted on Spyder software with Keras library. Images are visualized with the help of computer vision library.

4 Conclusion

In this work, we have presented a framework based on deep convolutional neural network for Alzheimer's disease detection in terms of accuracy. We have achieved 98.57% accuracy on our dataset without using any handcrafted features for training the network. Validation accuracy achieved is 87.72%. Experimental data is obtained from ADNI, and total 13,733 images from 266 subjects are used. Further studies will focus on achieving performance parameters such as specificity, sensitivity, recall and F1-score by improving the DCNN model. Therefore, another experiment can be implemented to test the prediction of disease by tuning hyper parameters. As a future enhancement, we will use support vector machine to classify the DCNN features.

Acknowledgements We express our gratitude towards Horizon Imaging Centre for providing us the offline raw dataset for AD.

References

1. Alzheimer's Association (2018) 2018 Alzheimer's disease facts and figures. Alzheimer's Dement 14(3):367–429
2. Alzheimer's Association (2016) 2016 Alzheimer's disease facts and figures. Alzheimer's Dement 12(4):459–509
3. Prince MJ (2015) World Alzheimer report 2015: the global impact of dementia: an analysis of prevalence, incidence, cost and trends. Alzheimer's Disease International
4. Bekris LM, Lutz F, Chang-En Yu (2012) Functional analysis of APOE locus genetic variation implicates regional enhancers in the regulation of both TOMM40 and APOE. J Hum Genet 57(1):18
5. The difference between a healthy brain and a brain affected by Alzheimer's (online). Available: https://www.brightfocus.org
6. McKhann GM, Knopman DS, Chertkow H, Hyman BT, Jack CR Jr, Kawas CH, Klunk WE et al (2011) The diagnosis of dementia due to Alzheimer's disease: recommendations from the National Institute on Aging-Alzheimer's Association workgroups on diagnostic guidelines for Alzheimer's disease. Alzheimer's Dement 7(3):263–269
7. Valenzuela O, Román BS, Ortuño Guzman FM, Villamor JLB, Saéz-Lara MJ, Rojas F, Rojas I (2014) Development of soft-computing techniques capable of diagnosing Alzheimers disease in its pre-clinical stage combining MRI and FDG-PET images. In: IWBBIO, pp 1644–1650
8. de Bruijne M (2016) Machine learning approaches in medical image analysis: from detection to diagnosis, pp 94–97
9. Rathore S, Habes M, Iftikhar MA, Shacklett A, Davatzikos C (2017) A review on neuroimaging-based classification studies and associated feature extraction methods for Alzheimer's disease and its prodromal stages. NeuroImage 155:530–548
10. Ortiz A, Munilla J, Gorriz JM, Ramirez J (2016) Ensembles of deep learning architectures for the early diagnosis of the Alzheimer's disease. Int J Neural Syst 26(07):1650025
11. Milletari F, Navab N, Ahmadi S-A (2016) V-net: fully convolutional neural networks for volumetric medical image segmentation. In: 2016 fourth international conference on 3D vision (3DV). IEEE, pp 565–571
12. Shi J, Zheng X, Li Y, Zhang Q, Ying S (2018) Multimodal neuroimaging feature learning with multimodal stacked deep polynomial networks for diagnosis of Alzheimer's disease. IEEE J Biomed Health Inform 22(1):173–183
13. Cui R, Liu M, Li G (2018) Longitudinal analysis for Alzheimer's disease diagnosis using RNN. In: 2018 IEEE 15th international symposium on biomedical imaging (ISBI 2018). IEEE, pp 1398–1401
14. Islam J, Zhang Y (2018) Brain MRI analysis for Alzheimer's disease diagnosis using an ensemble system of deep convolutional neural networks. Brain Inform 5(2):2
15. Gunawardena KANNP, Rajapakse RN, Kodikara ND (2017) Applying convolutional neural networks for pre-detection of Alzheimer's disease from structural MRI data. In: 2017 24th international conference on mechatronics and machine vision in practice (M2VIP). IEEE, pp 1–7

16. Liu M, Cheng D, Yan W (2018) Classification of Alzheimer's disease by combination of convolutional and recurrent neural networks using FDG-PET images. Front Neuroinform 12:35
17. Khvostikov A, Aderghal K, Benois-Pineau J, Krylov A, Catheline G (2018) 3D CNN-based classification using sMRI and MD-DTI images for Alzheimer disease studies. arXiv preprint arXiv:1801.05968
18. Wang S-H, Phillips P, Sui Y, Liu B, Yang M, Cheng H (2018) Classification of Alzheimer's disease based on eight-layer convolutional neural network with leaky rectified linear unit and max pooling. J Med Syst 42(5):85
19. ADNI—Alzheimer's disease neuroimaging initiative (Online). Available: http://adni.loni.usc.edu/
20. Horizon Imaging Centre, Surat, Gujarat

Novel Low-Complex 4 × 4 and 16 × 16 Intra-prediction Architecture for Error Concealment for H.264

Ketki C. Pathak, Anand D. Darji, and Jignesh N. Sarvaiya

Abstract There are various advanced video-displaying instruments which are responsible for producing high-resolution video data. Due to damaged disk like DVD or other media player, there is huge probability of distortion occur in reproduced video data. Error concealment technique is at decoder side, which has been developed to recover the damaged or lost region by utilizing temporal/spatial redundant information using intra/inter-mode prediction. This phase of intra-prediction mode decision in H.264/AVC intra-frame coder consumes more time with high computational complexity due to iterative process of prediction. Now, for real-time application the novel VLSI architecture for intra-prediction was developed which reduces redundancy and high number of memory access. This proposed architecture is implemented for all nine modes, and novel architecture is processing for 4 × 4 as well as 16 × 16 block size for intra-prediction modes. The proposed hardware design is implemented in VHDL with target device vitex6 (xc6vlx75t-3ff484). This proposed method provides rearrangement of intra-prediction equations, which reduced computational complexity by reducing gate count, and also minimizes iterative process by 29 clock cycle for one macroblock retrieval. Compared to state of the art, the proposed architecture reduces computational complexity.

Keywords Intra-prediction modes · Error concealment · Spatial prediction · Computational complexity · Block size

K. C. Pathak (✉)
Sarvajanik College of Engineering and Technology, Surat, India
e-mail: ketki.joshi@scet.ac.in

A. D. Darji · J. N. Sarvaiya
Sardar Vallabbhai National Institute of Technology, Surat, India

V. Nath and J. K. Mandal (eds.), *Proceedings of the Fourth International Conference on Microelectronics, Computing and Communication Systems*, Lecture Notes in Electrical Engineering 673, https://doi.org/10.1007/978-981-15-5546-6_51

1 Introduction

Video compression system utilized in many applications like consumer electronics and cellular phones, but it is very difficult to apply in video teleconferencing or M-Health system. Technological development makes high-throughput and low computational complexity and highly efficient video coding standard. Due to some advanced features like spatial and temporal predictions, the compression efficiency is better compared to other preceding standard like JPEG 2000.

There are three basic building blocks, namely intra/inter-prediction, transform, and entropy coding/decoding which are responsible for bit stream compression/decompression as shown in Fig. 1. Intra-prediction is used to find residue error within the same frame, while inter-prediction is used to find the residue from previously encoded frame. Most efficient prediction mode calculation is challenging task as it requires various intra/inter-prediction mode calculation in H.264.

Due to almost similar value of neighboring pixel along the local edges within the frame, it is easy to reconstruct or predict the lost pixel or erroneous pixel using intra-prediction mode calculation. The extrapolation method is used to perform intra-prediction between previously encoded pixel and pixel that is to be reconstructed. Bit rate can be reduced by accurate prediction of current pixel from neighboring pixels. Prediction error can be minimized if we can able to predict the lost or erroneous pixel exactly or nearer to original pixel. This error is called prediction error or residual error [2], which can be further minimized using reconstruction loop.

There are various hardware architectures implemented and suggested for H.264 intra-prediction block [3, 4]. Due to high-resolution video sequences, the reproduction of same quality video at decoder side in real-time scenario for various $N \times N$ intra-prediction algorithms is one of the challenging issue = 4, 8, 16, 32… modes of intra-prediction. The FPGA-based VLSI architecture suggested by Abhilash et al. [5] utilize parallel processing of three modes without planar mode calculation for 720p

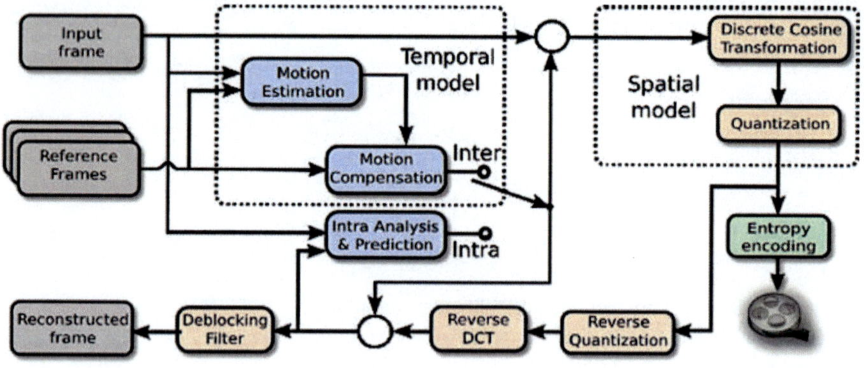

Fig. 1 Block diagram of H.264 vocoder [1]

Fig. 2 Luma 4 × 4 block to be predicted

tl	t1	t2	t3	t4	t5	t6	t7	t8
l1	p1	p2	p3	p4				
l2	q1	q2	q3	q4				
l3	r1	r2	r3	r4				
l4	s1	s2	s3	s4				
l5								
l6								
l7								
l8								

(1280 × 720) and 1080i (1920 × 1080) HD resolution videos, as planar mode calculation required more area and time to process planar mode. Hsia et al. suggested central-based calculation of planar mode, and they obtained 239 clock cycles to process entire 16 × 16 intra-prediction mode calculation [6]. Xu et al. introduced a power-efficient architecture which emphasized on data reused and decomposition techniques to implement intra-prediction [3]. Nadeem et al. [4] have evaluated some modified equations of planar mode calculation and suggested some partial data path which was overlapped between 4 × 4 and 16 × 16 mode calculation. Thus, they were able to achieve area reduction due to decrement in redundancy in addition operation in real-time HD video sequences of 1080p and 4k × 2k resolution videos. Milca et al. has suggested 1080 p for 4k × 4k resolution with 94 MHz operating frequency with 629 Mb/s throughput [2]. There are certain architecture suggested for optimization in data flow mechanism in VHDL programming for baseline intra prediction architecture [7]. There are some optimization of parallel mechanism [8] for intra mode calculation for FPGA implementation is also suggested [7–10]. Some of research papers based on Fast mode intra prediction architecture by skipping mode formulation have been suggested to reduce delay to improve speed of operation as well as complexity reduction [11–15].

The organization of paper is as follows. Basics of intra-frame coding for H.264 video standard is covered in Sect. 2, which is followed by proposed intra-prediction architecture implementation procedure covered as Sect. 3. Synthesized results and performance evaluation are described in Sect. 4. This section ensures the high performance of proposed VLSI architecture in terms of complexity reduction as no of clock cycle is reduced to 29. Section 5 is concluded with summary of proposed architecture with its achievement in performance improvement facts.

2 Basics of Intra-frame Coding

Inter-frame and intra-frame coding are two methods of redundancy compression that have been used to good effect in the development of video processing [16]. It is also possible to extract information about differences between spatially nearby pixels at a given instant in time. Spatial redundancy between predicted pixel and original pixel is reduced by means of intra-frame coding which is applicable within the single frame. This process of intra-frame coding makes large areas. If successive frames are highly correlated, then there is lack of temporal redundancy; at that time, intra-mode prediction is best suitable choice. The intra only uses spatial prediction, and inter uses temporal prediction that is in between the frames. The prediction taken within the frame is called I frame. The prediction taken from past is called P frame, and if from future frame is called B (bidirectional frame).

2.1 Detail Analysis of 4 × 4 Prediction Modes

The smallest macroblock in H.264 standard is 4 × 4 luma macroblock (MB). Prediction accuracy is inversely proportional to size of macroblock, smaller the size of MB gives better prediction accuracy. There are basically three regions in any images like smooth region, edges, and texture regions, and small MBs are most suitable for texture- and edge-based regions.

To predict any mode the basic arrangement is shown in Fig. 2. The region indicated as p1 to s1 till p4 to s4 is predicted block (PB). The samples above this PB is called above reference samples which are t1 to t8 samples. The samples left to PB left samples which are l1 to l8. DC mode and planar mode equation is formed based on the availability of left most reference samples or tom most reference samples. Remaining angular diagonal modes are calculated as per the availability of neighboring pixels. While calculation of any prediction mode among 9 modes, if there is unavailability of any samples, this vacant samples can be replaced by available samples by directly copying samples at that position. In each 4 × 4 sub macro block is predicted using total eight directional prediction modes and pure DC mode. The directional prediction modes are known as pure vertical mode, pure horizontal mode, diagonal down left mode, diagonal down right mode, horizontal down mode, vertical left mode, horizontal up mode and vertical right mode. Weighted average of the prediction samples can be used for calculation of direction mode. The mean of all other directional samples is used to calculate DC mode. There are total nine angular intra modes are available in 4 × 4 luma blocks [1]. Overall architecture to calculation intra prediction modes is shown in Fig. 3.

The equations of 4 × 4 sub macro block for few prediction modes like Pure vertical mode, pure Horizontal mode, Diagonal down left mode with reference to Fig. 2 is shown in Eqs. (1), (2) and (3) respectively [1]:

(1) Vertical

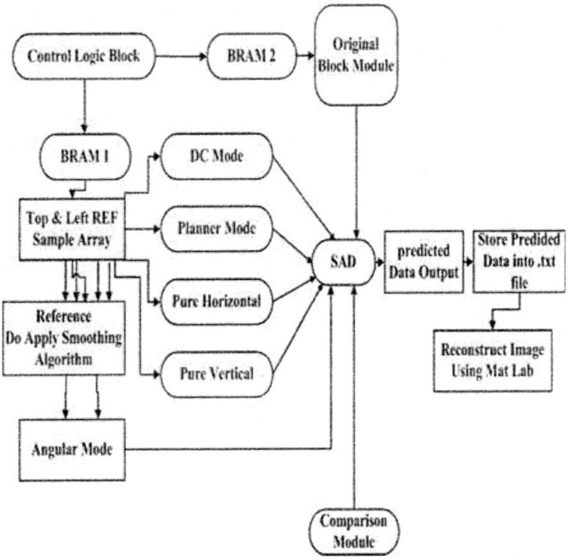

Fig. 3 Overall block diagram for intra-prediction mode

$$\begin{aligned} p1 = q1 = r1 = s1 = t1 \\ p2 = q2 = r2 = s2 = t2 \\ p3 = q3 = r3 = s3 = t3 \\ p4 = q4 = r4 = s4 = t4 \end{aligned} \quad (1)$$

(2) Horizontal

$$\begin{aligned} p1 = p2 = p3 = p4 = l1 \\ q1 = q2 = q3 = q4 = l2 \\ r1 = r2 = r3 = r4 = l3 \\ s1 = s2 = s3 = s4 = l4 \end{aligned} \quad (2)$$

(3) Diagonal down left

$$\begin{aligned} p1 &= (t1 + 2t2 + t3 + 2) >> 2 \\ p2 = q1 &= (t2 + 2t3 + t4 + 2) >> 2 \\ p3 = q2 = r1 &= (t3 + 2t4 + t5 + 2) >> 2 \\ p4 = q3 = r2 = s1 &= (t4 + 2t5 + t6 + 2) >> 2 \\ q4 = r3 = s2 &= (t5 + 2t6 + t7 + 2) >> 2 \\ r4 = s3 &= (t6 + 2t7 + t8 + 2) >> 2 \\ s4 &= (t7 + 3t8 + 2) >> 2 \end{aligned} \tag{3}$$

2.2 Intra 16 × 16 Prediction Modes

Larger macroblock size is best suitable for smooth and flat areas which do not contain much of the information, and the variation in pixel intensities is very gradual. Compared to 4 × 4 MB, it requires total four directional prediction modes which can be calculated with the help of top reference samples and left reference samples of current MB. Compared to 4 × 4 MB, few bits are needed to encode 16 × 16 MB, and it produces large value of residuals. Again compared to 4 × 4, there are some common modes like horizontal mode, vertical mode, DC mode, and planar mode, which are shown in Fig. 4 .

As depicted in Fig. 4, we can evaluate vertical mode from upper most samples PB (x; −1) with x = 0,…, 15 and horizontal mode from left most neighboring samples PB (−1; y) with y = 0,…, 15. DC mode can be calculated by using the available left most samples and upper most samples which is calculated as shown below Eq. (4).

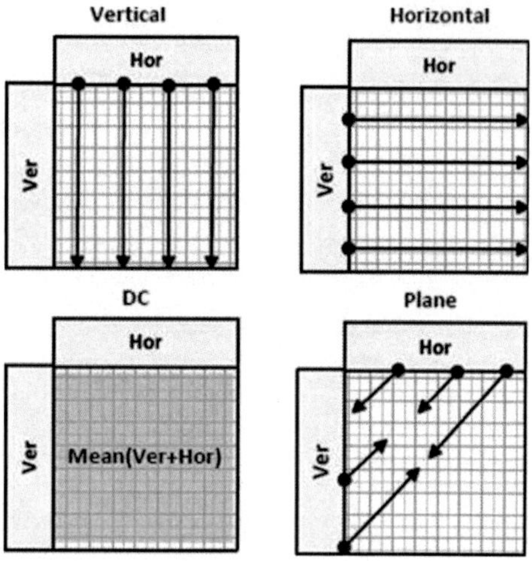

Fig. 4 16 × 16 prediction modes (8)

$$\mathrm{DC} = \sum_{x=0}^{15} \mathrm{PB}(x, -1) + \sum_{y=0}^{15} \mathrm{PB}(-1, y + 16) >> 5 \tag{4}$$

The planar mode is calculated using following Eqs. (2–4).
Pure horizontal mode is calculated as Eq. (5)

$$H \text{ mode} = \sum_{x=0}^{7} (x + 1) \times (P(8 + x, -1) - P(6 - x, -1)) \tag{5}$$

Pure vertical mode is calculated as Eq. (6)

$$V \text{ mode} = \sum_{y=0}^{7} (y + 1) \times (P(-1, 8 + y) - P(-1, 6 - y)) \tag{6}$$

Angular modes are calculated as following series of Eqs. (7)–(10). The parameter A1 is computed by summing input prediction pixels PB (15; −1) and PB (−1; 15) and by performing right shift on the sum as presented in Eq. (7):

$$A1 = 16 \times \mathrm{PB}\,(-1,\ 15) + \mathrm{PB}\,(15, -1));\tag{7}$$

While parameters $A2$ and $A3$ are calculated as Eq. (8) and Eq. (9) by referring to parameters H and V:

$$A2 = (5 \times H \text{ mode}) \gg 6;\tag{8}$$

$$A3 = (5 \times V \text{ mode}) \gg 6.\tag{9}$$

Three parameters $A1$, A2, and A3 and Clip1 function are used for computing planar mode prediction pixels as Eq. (10).

$$\mathrm{PB}(x, y) = \mathrm{Clip1}(A1 + A2(x - 7) + A3(y - 7) \gg 5 \tag{10}$$

2.3 *Selection Criteria for Mode Selection*

For accurate prediction, some of the criteria have been decided according to availability of the neighborhood macroblock for precision of mode decision of intra-prediction. Following Table 1 suggests the selection criteria for mode selection for 4 × 4 macroblock size. Table 2 shows the selection criteria for mode selection of 16 × 16 block size.

Table 1 Availibility of 4 × 4 luma prediction mode (8)

Prediction mode for 4 × 4 luma block	Available reference samples to predict the mode
DC mode predicted	None of the reference samples available
Horizontal mode predicted DC mode predicted Horizontal up mode predicted	Left reference samples available and top reference samples not available
Vertical right mode predicted, vertical left mode predicted, DC mode predicted, diagonal down left mode predicted	Top reference samples available, left reference samples not available
All modes can be predicted	Both top reference samples and left reference samples available

Table 2 Availibility of 16 × 16 luma prediction mode (8)

Availability of neighboring MBs	Available 16 × 16 luma prediction modes
None of the reference samples available	DC mode predicted
Left reference samples available, top reference samples not available	Horizontal mode predicted, DC mode predicted
Top reference samples available, left reference sample not available	Vertical mode predicted, DC mode predicted
Both reference samples available	All modes predicted

2.4 General Architecture for Intra-predication

The general architecture for intra-prediction is shown in Fig. 5. There are two buffers used. One of them is used for storing top reference samples for vertical mode prediction and other one is used for storing left reference samples for horizontal mode prediction. Both these buffers are considered as Block RAM (BRAM). For different mode calculation, any of these BRAM can be extracted as per the equation described in Section C of Sect. 2. The best mode is selected through multiplexer as shown in Fig. 5. Left register array and upper register array are used to store previously coded and reconstructed neighboring pixels of 4 × 4 luma block in current MB.

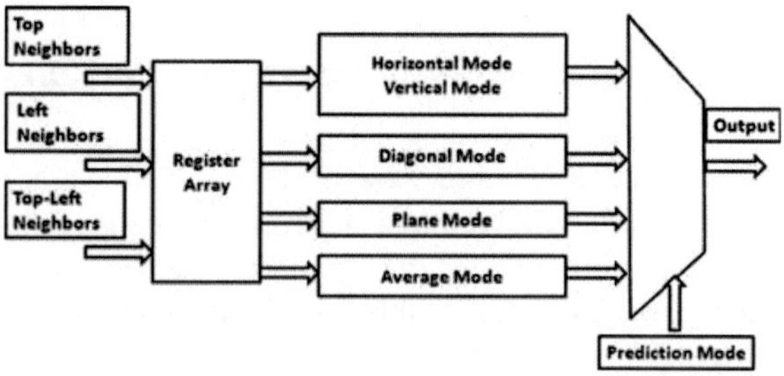

Fig. 5 General architecture for intra-prediction

3 Proposed Architecture

In this proposed architecture shown in Fig. 6, the prediction of lost macroblock is correctly predicted using available prediction mode by considering available left or upper most reference samples. We have considered two macroblock sizes 4 × 4 MB and 16 × 16 MB in this proposed architecture. this architecture, we have utilized the concept of block-wise prediction in which reconstructed pixels are responsible for prediction of the next blocks. So, in this proposed hardware architecture, the input neighboring pixels are applied through demultiplexer to array registers which store the initial values. As in (see Sect. 2), common parts calculations are stored in register and using adder and shifting operations. All the related equations are implemented and stored in registers. In mode calculation, some of the equations are identical that are directly used from array registers.

According to the availability of both the left reference samples and top reference samples, 4 × 4 luma block can be predicted using 161 clock cycles. Clock cycles required for preprocessing and performing available 4 × 4 luma predictions based on the availability of neighboring 4 × 4 luma blocks for a 4 × 4 luma block are given in Tables 2 and 3. As described above, 4 × 4 having nine modes from that calculations of horizontal, vertical planar and DC is same as in intra 16 × 16. They are implemented using simple adders and shifter. Here, compared to existing architecture [2] proposed architecture used D-ff which automatically reduced delay and hardware.

Planar mode implementation is different in both 4 × 4 MB and 16 × 16 MB. Planar mode implementation for 16 × 16 MB required data path planning to avoid multiplication operation to minimize the computation time and prediction process complexity. The organized planar mode prediction equations for initial block 0 in a MB are shown in Fig. 7.

Fig. 6 Data path for proposed 4 × 4 and 16 × 16 luma prediction modes

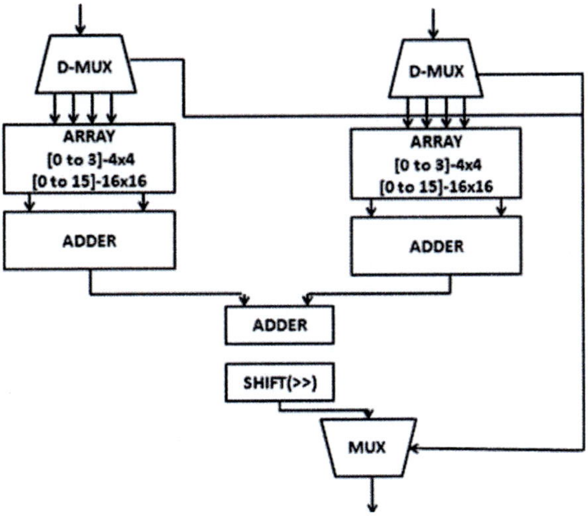

Table 3 Clock cycle analysis one macroblock for 4 × 4 modes

Available modes	No. of clock cycle for mode operation			
	Processed	Compute + Output	[16]	[10]
Horizontal (0)	1	17	17	17
Vertical (1)	1	17	17	17
DC (2)	3	19	20	21
Diagonal Down left (3)	2	18	18	24
Diagonal down right (4)	3	19	18	24
Vertical right (5)	2	18	19	23
Horizontal down (6)	2	18	19	23
Vertical left (7)	2	18	19	22
Horizontal up (8)	2	18	18	20
Total		162 CLK cycles	165 CLK cycles	191 CLK cycles

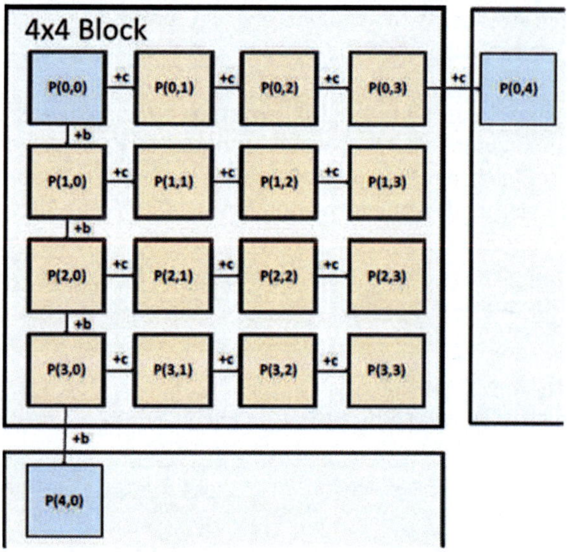

Fig. 7 Planar mode calculation for one 4 × 4 block (9)

To perform the planar mode prediction for 16 × 16 MB, the total block size is divided into 4 × 4 MB. Then, the further calculation of common equation like horizontal, vertical, and DC equation implementation is done with available left and top reference samples, and predicted pixel values are stored in temporary registers. For each row, pixels are predicted, adds the value c, and stores the result momentary registers. Then, it follows for next consecutive rows to predict the pixels and always adds the value *c*. This process continues for whole 16 × 16 MB.

4 Proposed Intra-prediction Selection Method

In this proposed algorithm, the novel approach is implemented in such a way that the delay is reduced and complexity is minimized by demultiplexer usage. In this proposed algorithm, the common mode equations are stored in register array, and then, it is fetched from array using adder and shifter. DC and planar modes are implemented using Fig. 7 by considering availability of neighboring pixels for both kinds of intra-prediction mode calculation like 4 × 4 mode and 16 × 16 mode. In error concealment algorithm, the lost pixel can be predicted by considering best mode among all the predicted modes, and that mode is replaced as a concealed block in the image. This can be performed by Sum of Absolute Difference (SAD) calculation which predicts optimized best mode as shown in Fig. 8 and Eq. (9).

4.1 Sum of Absolute Differences (SAD)

In H.264 standard, there is prediction unit (PU) before transformed unit (TU). The main purpose of prediction unit is to calculate best prediction mode based on all directional mode, DC mode, and planar mode evaluation. This best mode calculation is done as SAD which is the absolute difference between available reference samples and samples which is predicted. The minimum value of SAD is suggested as a best mode for that particular angular and DC as well as planar modes. As a startup point, we can consider any an arbitrary mode, then there is a comparator unit which is used to compare the absolute difference as computational valued for the selected arbitrary mode with current mode which is to be predicted.

The decision of best mode depends on this computational cost. If this absolute difference is minimum, then current mode is to be considered as one of best modes among all the modes. This procedure is performed as a parallel processing for all modes, and then, final comparator makes a decision of final best mode among all other modes. Equation for calculating SAD for particular block is described in Eq. (11). In this equation *F(i,j)* is original pixel value and *P(i,j)* is the predicted pixel value. SAD architecture is shown in Fig. 8.

Fig. 8 SAD architecture

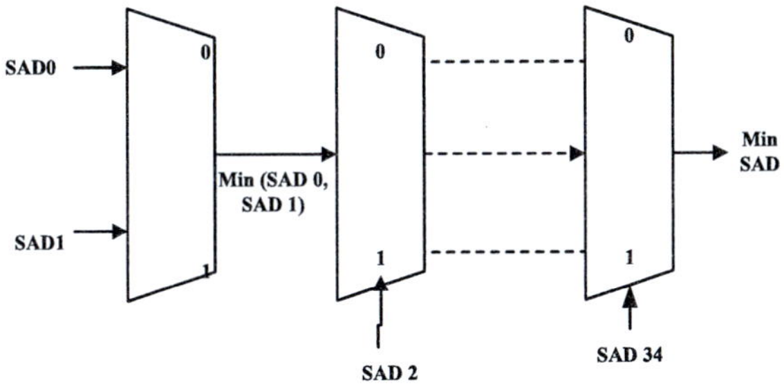

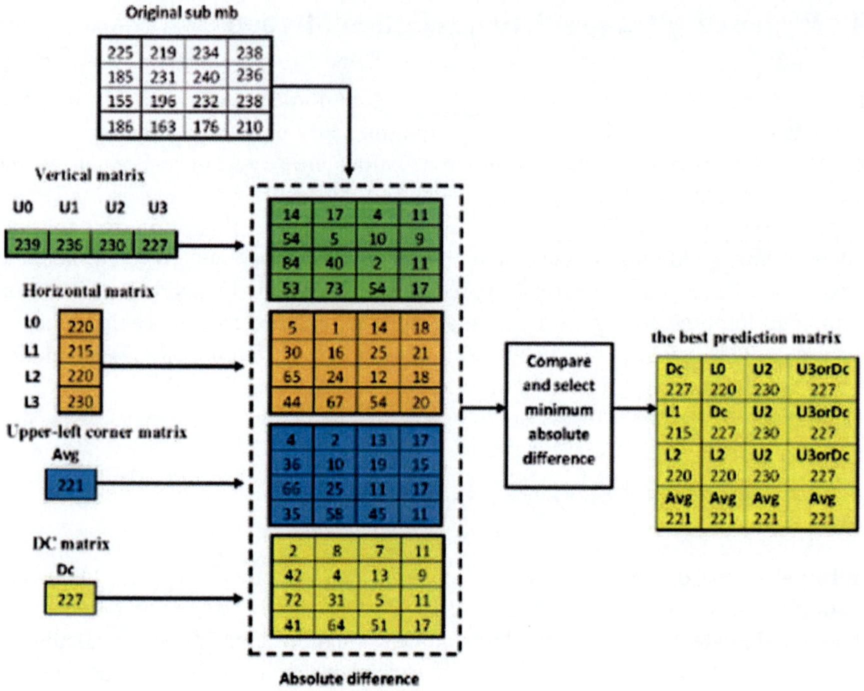

Fig. 9 Example of SAD (10)

$$\text{SAD} = \sum_{i=0}^{N} \sum_{j=0}^{N} |F(i, j) - P(i, j)| \tag{11}$$

Here, in Fig. 9 small example of working algorithm is described in which predicted mode values are compared with original MB, and then, absolute difference should be taken and one of the mode is selected which has minimum distance. Main advantage of SAD is that it is required only addition and difference between the pixel values. Algorithm is proposed which optimizes design metrics such as performance and quality.

5 Experimental Analysis

5.1 Result Analysis of Implementation

From careful analysis in Table 3, it is observed that in intra-prediction for 4 × 4 macroblock size, if both neighbors are available then for one MB total 162 CLK cycles are needed to implement the proposed architecture. Here, For MATLAB analysis,

Table 4 Clock cycle analysis for one marcoblcok for 16 × 16 modes

Available modes	No. of clock cycle for mode operation		
	Processed	Compute + output	[16]
Horizontal (0)	1	257	257
Vertical (1)	1	257	257
DC (2)	4	260	273
Plan (3)	52	308	340
Total	57 CLK cycles	1082 CLK cycles	1127 CLK cycles

coastguard frame sequence is considered in which one frame [144 × 176] has total 1584 MBs from which 1504 MBs have both neighbors, so 1584 × 162 = 2,56,608 CLK cycles are needed and for intra 16 × 16 has total 99 MBs from which 79 MBs have both neighbors, so 79 × 1082 = 85,478 CLK cycles are needed to implement the whole frame. As shown in Tables 3 and 4, intra 4 × 4 algorithm used just 1% LUTs and slice registers, while in intra 16 × 16 has 3% utilization of LUTs, respectively. So, hardware utilization is reduced with increase in speed from the overall architecture shown in Fig. 3; all the related modes with respect to their block sizes and their mathematical equations modes for 1 MB are implemented.

VHDL simulation results are shown in Fig. 10 for horizontal mode, Fig. 11 for vertical mode, Fig. 12 for DC mode and Fig. 13 for planar mode for 4 × 4 intra predictions for H.264 standard.

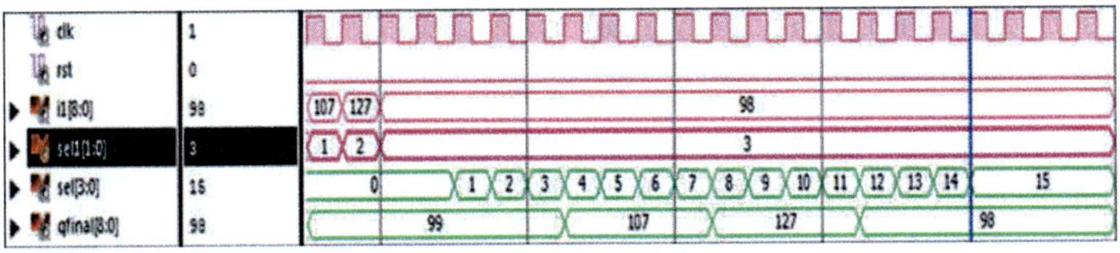

Fig. 10 Output waveform for horizontal mode (4 × 4)

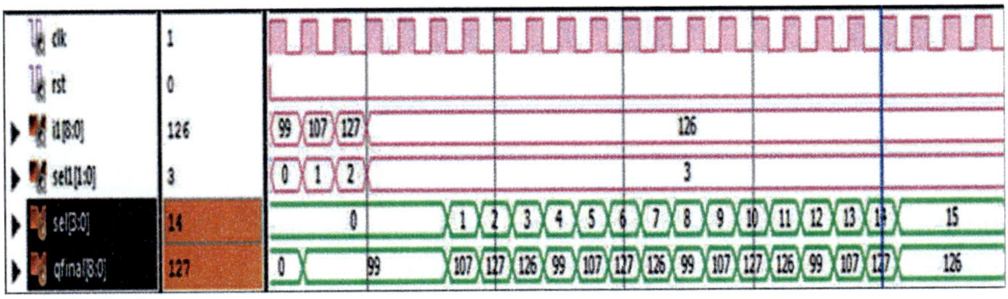

Fig. 11 Output waveform for vertical mode (4 × 4)

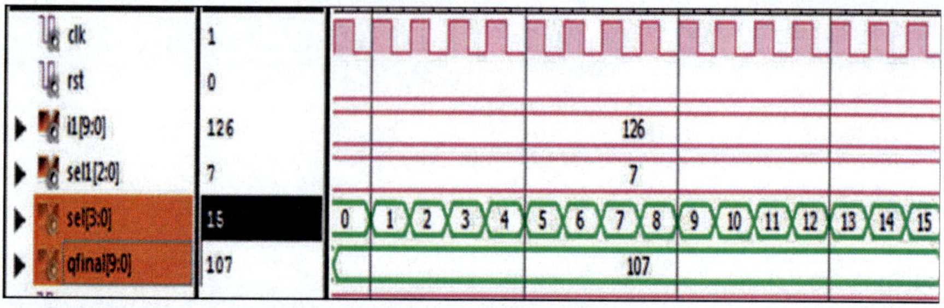

Fig. 12 Output waveform for DC mode (4 × 4)

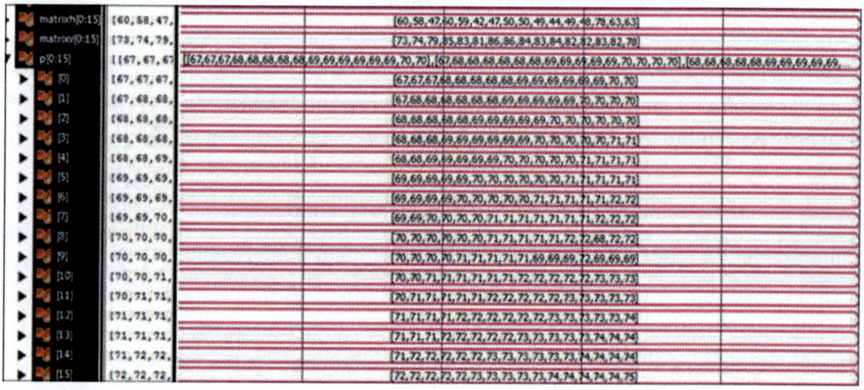

Fig. 13 Waveform for planar mode

Table 5 Device utilization for vertex 6 for 4 × 4 block size

Available modes	LUTs available-(46,560)	Slice register available-(93,120)	Global CLK buffer available-(32)
Horizontal	14 (0%)	72(0%)	2
Vertical	14 (0%)	72(0%)	2
DC	86 (0%)	87(0%)	2
Diagonal down left (3)	85 (0%)	94(0%)	2
Diagonal down right (4)	211 (0%)	156(0%)	2
Vertical right	220 (0%)	191(0%)	2
Horizontal down	220 (0%)	191(0%)	2
Vertical left	134 (0%)	127(0%)	2
Horizontal up	127 (0%)	120(0%)	2
Total	1111 (1%)	1110 (1%)	18 (5%)

Table 6 Device utilization for 16 × 16 block size

Available modes	LUTs available-(46,560)	Slice register available-(93,120)	Global CLK buffer available-(32)
Horizontal	89 (0%)	384(0%)	2
Vertical	91 (0%)	384(0%)	2
DC	412 (0%)	422(0%)	2
Plan	3707(7%)	2113(2%)	2
Total	4229 (9%)	3303 (3%)	18 (5%)

Table 7 Comparison of performance parameter of proposed architecture

Available modes	All modes (4 × 4) (proposed)	All modes (16 × 16) (proposed)	[7]	[2]
LUT	1111	4229	4358 (4 × 4) 1951 (16 × 16)	4465 (4 × 4)
Max frequency (MHZ)	113	111	130	118.03
Throughput	2270	772	122 (4x4) 177 (16 × 16)	629
Clock cycle	162	57	272 (4 × 4) 188 (16 × 16)	48

Clock cycle analysis for One Macroblock (MB) is demonstrated for 4x4 Modes in Table 3 and 16 × 16 Modes in Table 4. It shows that, we need less clock cycle compared to architecture suggested in reference [16] and [10].

Device utilization of one Macroblock (MB) is demonstrated for 4 × 4 modes and 16 × 16 modes in Tables 5 and 6 respectively.

6 Conclusion

We have implemented 4 × 4 and 16 × 16 block sizes for intra-prediction architecture. By comparing device utilization and memory organization of architecture, 4 × 4 block architecture is more memory efficient compared to 16 × 16 block architecture. The hardware design is based on a novel organization of the intra-prediction equations which reduced computational complexity and reduced iterative process by reducing processing 29 cycles compared to suggested architecture in [16]. We have also proposed optimization in throughput improvement and operating frequency compared to [2, 7] and [15]. Compared to other architectures like [11–13], we have also maintained the reconstruction of image without skipping any mode of operation. It also supports real-time video encoding sequences for higher resolution like 4k $\times$ 4k for 30–60 fps also.

References

1. Richardson IE (2010) The H.264 advanced video compression standard, 2nd edn. Vocodex Limited, Willey
2. Orlandic M, Svarstad K (2013) A high-throughput and low-complexity H. 264/AVC intra 16 $\times$ 16 prediction architecture for HD video sequences. In: 2013 21st Telecommunications forum (TELFOR). IEEE, pp 529–532
3. Xu K, Choy CS (2008) A power-efficient and self-adaptive prediction engine for H. 264/AVC decoding. IEEE Trans Very Large Scale Integr (VLSI) Syst 16(3):302–313
4. Nadeem M, Wong S, Kuzmanov G (2011) An efficient hardware design for intra-prediction in h. 264/AVC decoder. In: Electronics, Communications and Photonics Conference (SIECPC), 2011 Saudi International. IEEE, pp 1–6
5. Atitallah AB, Loukil H, Masmoudi N (2011) Fpga design for H. 264/avc encoder. Int J Comput Sci Eng Appl 1(5)
6. Hsia S, Hsu W, Chou Y (2012) Fast low-complexity computation and real-time architecture for h.264/avc intra-prediction. J Real time Image Process
7. Atitallah AB, Loukil H, Masmoudi N (2011) FPGA design for h. 264/avc encoder. Int J Comput Sci Eng Appl 1(5):119
8. Meng B, Au OC, Wong C-W, Lam H-K (2003) Efficient intra-prediction mode selection for 4 $\times$ 4 Blocks in H.264. In: IEEE conference on multimedia and expo, vol 3, pp III-521–4
9. Shafique M, Bauer L, Henkel J (2009) A parallel approach for high performance hardware design of intra prediction in H. 264/AVC video codec. In: Proceedings of the conference on design, automation and test in Europe. European Design and Automation Association, pp 1434–1439
10. Wang X, Cui X, Yu D (2009) A parallel intra prediction architecture for H.264 video decoding. In: IEEE international conference on ASIC, pp 859–862
11. Chiang K-C, Wu MF, J.J.-J. Shann, Modification and implementation of an edge-based fast intra prediction mode decision algorithm for H.264/AVC high resolution real-time systems. J Visual Commun Image Representation 23(2), 245–253 (ISSN 1047-3203)
12. Shrivastava VK, Muralidhar P, Rama Rao CB (2013) Architecture for H.264 intra prediction fast mode decision algorithm. Int J Comput Appl 68(3):0975–8887
13. Kannangara CS, Richardson IEG, Bystrom M, Zhao Y, Solera J, MacLennan A, Cooney R (2006) Low complexity skip prediction for H.264 through lagrangian cost estimation. IEEE Trans Circ Syst Video Technol 16(2):202–208
14. Kim D-Y, Han K-H, Lee Y-L (2010) Adaptive single-multiple prediction for H.264/AVC intra coding. IEEE Trans Circ Syst Video Technol 20(4):610–615
15. Lee W, Jung Y, Lee S, Kim J (2007) High speed intra prediction scheme for H.264/AVC. IEEE Trans Consum Electron 53(4):1577–1582
16. Sahin E, Hamzaoglu I (2007) An efficient hardware architecture for H. 264 intra prediction algorithm. In: Design, automation & test in Europe conference & exhibition, 2007. DATE'07. IEEE, pp 1–6
17. Huang H, Cao T-Y, Zhang X-W (2008) The enhanced intra prediction algorithm for H.264. In: IEEE conference on image and signal processing, vol 2, pp 161–165

18. Shah J, Raghunandan KS, Varghese K (2012) HD resolution intra prediction architecture for H.264 decoder. In: IEEE international conference on VLSI design, pp 107–112
19. Hamdy S, Ibrahim MEA, Abdelhalim Z (2013) VHDL realization of efficient H.264 intra prediction scheme based on best prediction matrix mode. Int J Comput Appl (IJCA) 77:1–7

Genetic Algorithm-Based Image Watermarking Using Multiple Location

Divya Paikaray and Abhijit Mustafi

Abstract The process of image watermarking is used to digitally mark the image before storing or transmitting to other agents. For preservation of the original content, sometimes the log data of the image or document being used is also required. This paper mainly aims at the development of an algorithm which embeds a watermark to an image which can be used for its copyright preservation issues or patent issues. The paper aims at development of an algorithm which remains unaltered in most adverse situations. An image is made of pixels and anything which is being embedded into an image will invariably be made of pixels. Any bulk embedding process which aims at a particular section of an image can be easily targeted or altered by several filtering techniques. But in this paper, the concept that the fundamental entity of any image is pixel is being used. This helps us to embed the required image with much more secured process. Moreover, the process of locating the sections for embedding is also determined by the machine itself which makes it much safer reason being if allocation of the section would have been done by use of any algorithm, there are chances of decoding that algorithm, which would eventually reveal the position of the watermark. But since this entire process is automated by machine itself, the probability of any such decoding becomes nil. Platform used for this is MATLAB. Wavelet transform is used for development of the program because of its higher efficiency in case of digital domain.

Keywords Invisible watermark · Embedding · Genetic algorithm · Haar transformation · Heuristic approach

D. Paikaray (✉) · A. Mustafi
Department of Computer Science and Engineering, Birla Institute of Technology, Mesra, Ranchi, Jharkhand, India
e-mail: divyaprasi@rediffmail.com

A. Mustafi
e-mail: abhijit@bitmesra.ac.in

V. Nath and J. K. Mandal (eds.), *Proceedings of the Fourth International Conference on Microelectronics, Computing and Communication Systems*, Lecture Notes in Electrical Engineering 673, https://doi.org/10.1007/978-981-15-5546-6_52

1 Introduction

The process of watermarking is used extensively for digital signature of the documents of any nature. With the increase in advancement of technology, the security issues pertaining for preserving the ownership of documents have also increased. There are cases of copyright violation with documents and many other products. To counter these cyber-insurgencies, digital watermarking was introduced. But most of the algorithms are prone to alteration and most of the cases can be either decoded or back-engineered to decode the embedding technique and further removal of the watermark from the document. Then, there is another issue of noise and filters that are either intentionally applied to the document or image while transmission or get interfered while transmission. These noise and filter also tend to alter the original watermarking of the document which makes the decoding or retrieval of the watermark at receiver side problematic, in some cases impossible. Therefore, there is need for a robust watermarking which is immune to most of the attacks, noise and filters. The main purpose of embedding is not just to embed the image with a signature but also its successful retrieval at the receiver end when needed. In this paper, therefore an approach to robust watermarking is made, so as to make it immune to such unwanted interference in the transmission medium along with higher immunity to the cyber-attacks which aim to alter the content intentionally.

2 Background

For digital images, many watermarking algorithms are designed to get a robust image which will be immune to attacks. Many transformations are used for embedding process. Historically, for the embedding process, discrete Fourier transformation (DFT) was used, but due to some drawbacks, discrete wavelet transformation (DWT) is being used for more good results. The previous papers are described as follows:-

Dehghan proposed in [1] an image watermarking scheme which is based on wavelet coefficients and variance. The paper provides two methods, namely *METHOD 1* and *METHOD 2*. M1 uses wavelet coefficients in low pass scale to embed watermark into the image, whereas in M2 variance of each block is calculated and first N blocks with higher variances are selected for watermarking. In decoding process, both original image and received images are transformed and corresponding coefficients are divided. The ratio of majority is compared, and if the value exceeds the set threshold, then value encoded is '1' else it is '0'. Strength factor $a = 1.01$ and $a = 1.025$ is found out by experiments.

Liu proposed in [2] a scheme of watermarking which is based on distributed discrete wavelet transform (DDWT) for image. In this, it involves the transfer of original data using DDWT technique from spatial domain to frequency domain and after that the image which is to be embedded in the four sub-bands is in the frequency domain. The embedded data in this is assigned to the spatial coefficients due to which it prevents attack that is cropping, etc. In this, the performance is improved so that the image remains robust and immune against several geometric attacks that are scaling or rotation and several non-geometric attacks that are Gaussian noise, contrast adjustment, and sharpening.

Ellinas proposed in [3] an algorithm which is based on embedding of image in wavelet domain and edge detection. In this, the embedding image is embedded into the sub-band coefficients that are present on the edge. The watermark image is then embedded on the coefficients that are selected around edges, and for good watermark strength, a different scale factor is chosen. Kim developed in [4] a watermarking algorithm using wavelet transform. In this approach, coefficients of all sub-bands were utilized to embed the watermark into whole image. For the selection of coefficients, level adaptive threshold scheme is used. In this, scale factor is selected on basis of decomposition level of selected coefficients. Sridhar proposed in [5] a technique based on watermarking in wavelet domain using merge image. The watermarking is performed on the mixed-up image. It has separated the original image into two matrices for the intermixing of images, and then, watermarking has been done on the shuffled image. Khanna has proposed in [6] a method to watermark the image which is proposed to be immune to most of the attacks listed in the paper. The common attacks which are listed in the paper are transformation like rotation, shearing, scaling and translation. Other than transformation, filtering, such as Gaussian filter blur and image sharpening, is also one of the major attacks on the watermark of the image. Other than geometrical transformation there is noise that either intentionally or unintentionally gets added to the image which may affect the watermarking of the image. Basic noises are like duplicating and removing lines or column of the image or cropping. Compression of image also sometimes results in the interference with the watermark of the original image. The paper has extensively studied major areas which affect the watermark decoding process. One of the attacks that neutralizes most of the watermarking algorithms is that chopping or decomposing the image in small fragments and then rearranging them which results in retrieval of the original image, but since the image was broken into various pieces, there are chances of breakdown of the watermarking too, which would not have got rearranged the same way it was embedded. This method is known as content modification. Watermarking is not only used for copyright issues, but can be used to store certain other information which may be required at the receiver end. The proposed method has two levels of discrete wavelet transformation unlike other watermarking algorithms which use only one layer of transformation. The proposed method uses singular value decomposition because any change in their values is invisible to the naked human eyes. Firstly, a second-level DWT is applied to get approximation of the detailed component. Then singular value decomposition is performed on those components. Similar operation is performed on the watermarking image and embedding of the watermarking is done on the main image. The paper also proposes the use of genetic algorithm for obtaining the optimal value of the scaling factor of the watermark. The values of the parameters are obtained experimentally. The algorithm has been used to watermark three images of 512×512 size each, with a watermark of 128×128 size. The efficiency of the watermarking is tested using the peak signal-to-noise ratio value for each of the images. The paper concludes with major focus on the usage of the watermark and its efficient embedding into the host image with minimal deviation from the original content. Genetic algorithm was used for optimization of the entire process which makes a trade-off between image quality and the peak signal-to-noise ratio of the original image. The image thus obtained from this algorithm is supposed to be immune to most of the geometrical attacks and noise while transmissions. Hurrah has proposed in [7] a method to watermark the image using DCT and DWT for different purposes. In this image watermarking method, the image is embedded into all three components that are red–green–blue (RGB) of the colour image using the transformation that is discrete cosine transform (DCT) domain. In this (DCT), discrete wavelet transform is then used for fragmenting the image into the sub-bands from each of the sub-band of the three colour image. For ensuring better security of the watermark, multiple encryptions are used. Experimental results show that even after subjecting the image to two simultaneous attacks, the affect on image of this attack is very low, which is confirmed from the several parameters such as peak signal-to-noise ratio, BCE, NCC and BER. According to the proposed algorithm, first the watermark image is encrypted by using Arnold transform and then by the chaos encryption methods. This adds to double layer of the encryption before embedding. Data embedding is done firstly by dividing the original image into (RGB) red, green and blue components. Now to extract the four frequency bands, we apply the DWT. Now we use the component that is the LL components of each discrete wavelet transform transformed, and for embedding 4×4 sub-block is used. Then, the block-level DCT is used to transform each 4×4 sub-block. The data extraction involves similar process by applying an extraction algorithm which gives us about the sector where embedding was done. After extraction, the techniques that are used for retrieving the image are chaos encryption and inverse Arnold. The paper claims to have developed a watermarking technique that is robust for colour images in hybrid transform domain. The ratio of security is enhanced by using multiple encryptions. The results were compared for all geometrical and signal processing attacks.

3 Proposed Framework

A robust image watermarking algorithm has been proposed. The watermarking has been done in wavelet domain. The watermarked image has been embedded in multiple locations in the original image. The watermarking done is invisible to the human naked eyes so that it cannot be easily destroyed. For the multiple locations, genetic algorithm is used to get the optimized location and with minimum root-mean-square error (RMSE) of the image. So that there will be less distortion in image and the resulted image will be of good quality. The whole process is explained further:

3.1 Embedding of Watermark Image

In this, the watermarked image is to be watermarked in the host image in a different location. For this, the wavelet decomposition is done on both the images and the image that is to be embedded is divided into specific parts, and embedded into different locations.

3.2 Extraction of Watermark Image

After the invisible watermarking, the original image becomes immune to different attacks. To extract the watermark image, we need to find out the entire wavelet coefficients which are located in different location fulfilling all the conditions. By the process of concatenating the entire wavelet coefficient which is at a different location and resizing it, we apply inverse wavelet for reconstruction, so as to get the watermark image.

3.3 Embedding at Different Multiple Location

For embedding or inserting the watermark image on the original image, the watermark image is being divided into no. of parts that the user wants to. So that much different location is also required for the embedding process. The multiple locations are found through genetic algorithm keeping few of the constraints in mind.

- Firstly, the different parts of image should not overlap each other while getting embedded into the original image/host image.
- Secondly, the watermark image wavelet coefficients should not exceed the last portion of the host image wavelet coefficients.

3.4 Genetic Algorithm

In this for the good result of finding locations, a heuristic method is used known as genetic algorithm. Genetic algorithm is used here to find different locations where the different parts of watermark image are to be embedded. It gives the better results by giving the root-mean-square error (RMSE) minimum. So that there will be minimum distortion after embedding the watermark image. Finally, the result gives a good image with less distortion and minimum root-mean-square error (RMSE).

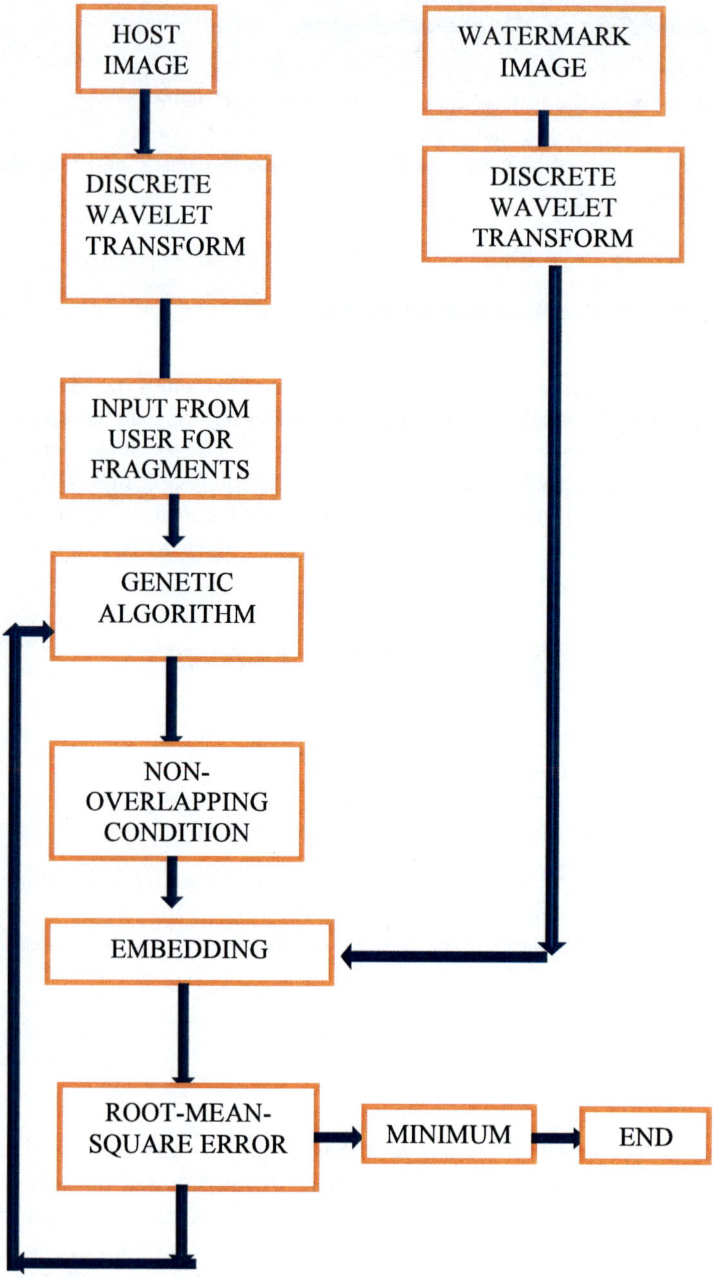
HOST IMAGE
WATERMARK IMAGE
DISCRETE WAVELET TRANSFORM
DISCRETE WAVELET TRANSFORM
INPUT FROM USER FOR FRAGMENTS
GENETIC ALGORITHM
NON-OVERLAPPING CONDITION
EMBEDDING
ROOT-MEAN-SQUARE ERROR
MINIMUM
END

Fig. 1 Before embedding

4 Experimental Results

The result of the proposed algorithm is efficient enough and gives a very effective result. By applying this algorithm better, genetic algorithm has given better locations as per requirements by keeping the root-mean-square error minimum and a good quality of image. Here, Fig. 1 represents the original image. After the embedding process keeping all the constraints or condition in the algorithm, the image produced is represented in Fig. 2. The extracted image is shown in Fig. 3 which is extracted from the original image after embedding. Figure 4 represents the process of finding an appropriate location. The results are as follows (Fig. 5).

A tabular comparison for the algorithm by varying scaling factor has been given below. The algorithm has been executed for a range of scaling factor, and corresponding root-mean-square error value has been generated. The trend of varying RMSE value can be used to find optimum scaling factor (Tables 1 and 2).

5 Conclusion

An approach is made to develop an algorithm which implements the watermarking of an image with providing immunity with respect to the geometrical and filtering

Fig. 2 After embedding

Fig. 3 Extracted watermark image

attacks. The robustness of the watermarking is of major concern along with optimized use of the watermark, which is ensured using genetic algorithm to reduce net error in the image. The variation between host image and embedded image is being compared using PSNR and is confirmed to be within acceptable limits.

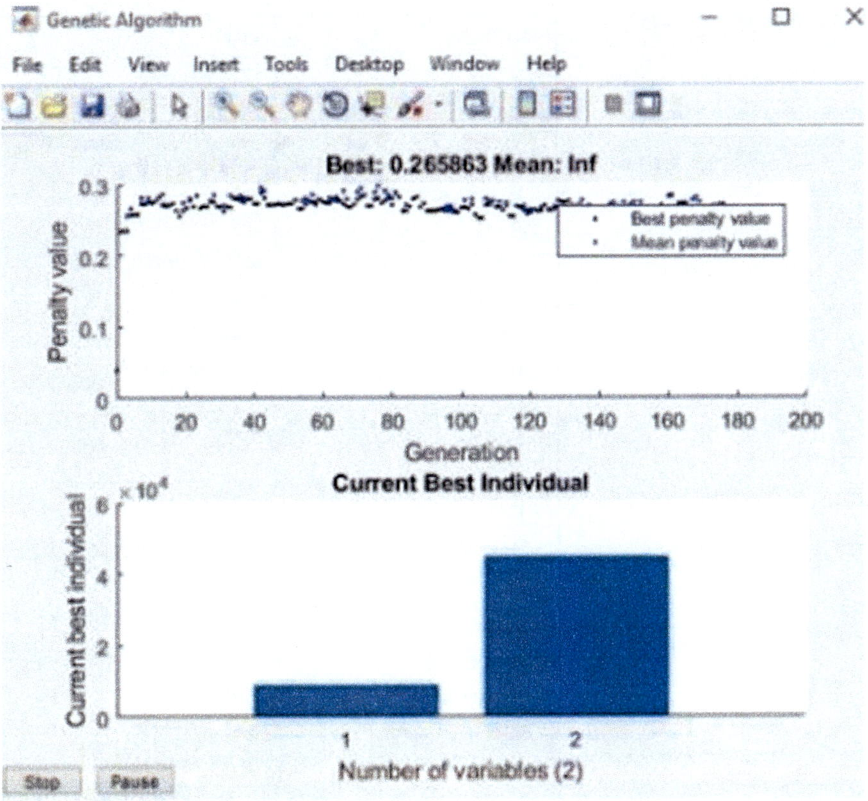

Fig. 4 Genetic algorithm finding location

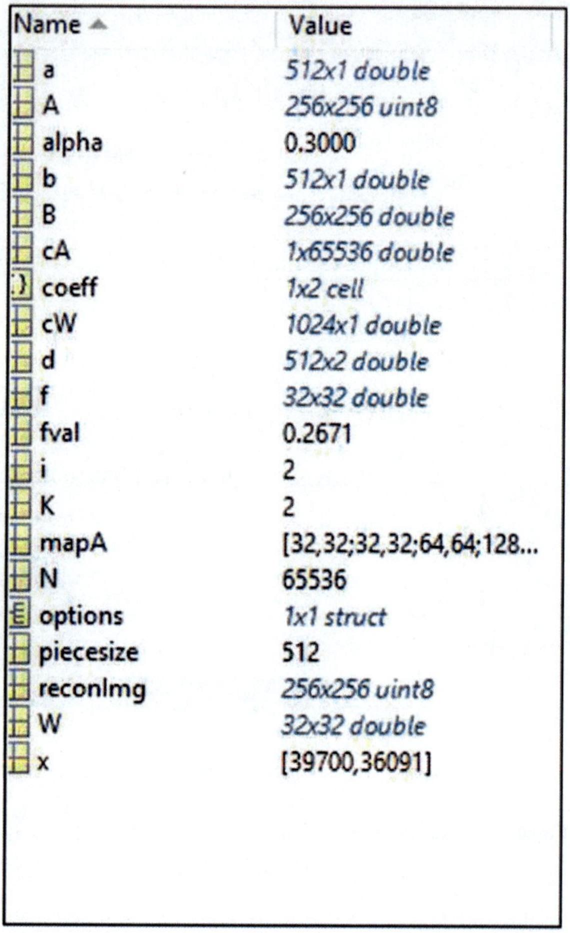

Fig. 5 Values of variables used

Table 1 Fragmentation $k = 2$

Scaling factor (α)	Root-mean-square error
0.2	0.2953
0.4	0.2678
0.6	0.2641
0.8	0.2556

Table 2 Fragmentation $k = 4$

Scaling factor (α)	Root-mean-square error
0.2	0.3235
0.4	0.2613
0.6	0.2592
0.8	0.2542

References

1. Dehghan H, Safavi SE (2006) Robust image watermarking in the wavelet domain for copyright protection. IEEE Trans Image Proc 15(9):2831–2842
2. Liu J-C, Lin C-H, Kuo L-C (2006) A robust full-band image watermarking scheme. 1-4244-0411-8/06© 2006 IEEE
3. Ellinas JN (2007) A robust wavelet-based watermarking algorithm using edge detection. World Acad Sci Eng Technol 34:291–296
4. Kim JR, Moon YS (1999) A robust wavelet-based digital watermarking using level-adaptive thresholding. 0-7803-5467-2/99 ©1999 IEEE
5. Sridhar B, Arun C (2013) A wavelet based image watermarking technique using image sharing method. In: 2013 International conference on information communication and embedded systems (ICICES), Chennai, pp 629–633
6. Khanna AK (2016) Digital image watermarking and its optimization using genetic algorithm. Galgotias University, UP, India. 978-1-5090-1666-2/16 ©2016 IEEE
7. Hurrah NN, Loan NA, Parah SA, Sheikh JA (2017) A transform domain based robust color image watermarking scheme for single and dual attacks. In: 2017 Fourth international conference on image information processing (ICIIP), Shimla, pp 1–5

Fuzzy Supervisory Expert Tuner for PID Controller

Payel Pramanick, Sayanika Bandyopadhyay, and Chanchal Dey

Abstract Goal of this paper is to study the role of fuzzy rule-based supervisory auto-tuner designing for proportional-integral-derivative (PID) controller. Initial tuning of the conventional PID controller is performed through standard relay tuning methodology. The proposed fuzzy supervised auto-tuner works toward the performance enhancement during closed-loop operation through continuous updating of the PID tuning parameters. This online gain modification scheme is so designed that it attempts to provide desired transient and steady-state behavior. Proposed fuzzy auto-tuner is designed by a fuzzy rule set which is developed from the understanding of an experienced process operator. Performance of the reported controller is substantiated through simulation experiment and compared with the classical PID controller with the help of different performance measures—settling time, integral absolute error, percentage overshoot, and integral-of-time multiplied absolute error. Closed-loop responses and the performance indices clearly justify the superiority of the proposed controller compared to conventional PID controller.

Keywords PID controller · Fuzzy auto-tuner · Fuzzy rule base · Performance measure

P. Pramanick (✉) · S. Bandyopadhyay · C. Dey
Department of Applied Physics, University of Calcutta, Kolkata, India
e-mail: payel256pramanick@gmail.com

S. Bandyopadhyay
e-mail: sayanikabanerjee880@gmail.com

C. Dey
e-mail: cdaphy@caluniv.ac.in

V. Nath and J. K. Mandal (eds.), *Proceedings of the Fourth International Conference on Microelectronics, Computing and Communication Systems*, Lecture Notes in Electrical Engineering 673, https://doi.org/10.1007/978-981-15-5546-6_53

1 Introduction

To control complex industrial process loops, a number of control methodologies have been developed over the last 50 years. But, till date, the majority of the industries use simple proportional-integral-derivative (PID) controller for plant control applications [1]. Due to its simple structure and satisfactory performance, PID controller is still widely accepted in industrial applications. Easy understanding and clear functionality along with its cost-effective performance make it indispensible for closed-loop process control applications. Due to its simple structure, PID controller is usually suitable for first- and second-order processes. However, most of the industrial plants deal with the higher-order processes and nonlinearities [2] are also present in their behavior. Due to change in ambient condition and aging, controller parameters get changed over prolonged operation. Presence of dead time also worsen the efficacy of the controller by delaying its control action, and hence dead time is considered as one of the most arduous dynamic elements that naturally occur in any physical system.

Hence, the conventional PID controller with their linear functionality is not capable enough to regulate such processes with their fixed settings. As the process operating condition changes, PID controller's tuning parameters should be updated so that desirable performance can be ensured throughout the transient and steady-state operating phases. In this context fuzzy logic controller [3] seems to be quite appropriate as it can incorporate desired amount of nonlinearity in its behavior with a competently designed rule base. A number of research findings are reported for designing fuzzy controller for controlling processes with nonlinear and dynamic nature. Nomura et al. [4] proposed a technique where a controller is tuned iteratively by minimizing a cost function calculated from the difference of the fuzzy controller output and desired output provided by set of training data. This method of tuning may be effective for control system with time varying in nature, but its application is restricted due to unavailability of a particular set of training data. According to Yoshida et al. [5] gain tuning methodology is suitable for first-order processes with dead time where empirical relations involving process parameters are used to calculate the membership functions (MFs) for input and output variables. For higher-order processes, control performances provided by this technique are not adequate. In order to eradicate overshoot due to accretion of control action in a PI type fuzzy controller Lee [6] proposed two intensified versions of classical PI type fuzzy controller. Transient response of a second-order linear system including integrating element is remarkably improved by this scheme. But authors in [7] have clearly showed that the controller proposed by Lee [6] is almost similar to a conventional fuzzy PD type controller.

Here, in the proposed methodology, classical PID controller is not been replaced by a fuzzy controller. Rather a fuzzy rule-based expert tuner is designed to modify the tuning parameters of a conventional PID controller during closed-loop operation depending on the instantaneous process operating conditions. Thus, a supervisory fuzzy expert tuner is incorporated along with the classical PID controller in the control loop. PID controller is initially tuned by using standard relay tuning methodology [8]. Depending on the knowledge of an experienced process operator two fuzzy tuners are reported here consisting of 25 and 9 rules. Depending on the instantaneous process operating condition fuzzy tuner provides a gain updating parameter which will modify all three parameters of the PID controller so that desirable closed-loop response can be achieved during set point tracking as well as load recovery phases.

Rest of the paper has three sections. In Sect. 2, the proposed fuzzy supervisory PID tuning scheme is described with its various designing parameters (scaling factors, membership functions, and determination of rule base). In Sect. 3, simulation results for two second-order dead time processes are presented and the conclusion is provided in Sect. 4.

2 Proposed Fuzzy Supervised PID Tuning Scheme

Standard mathematical form of PID control action is given by Eq. (1)

$$u(t) = k_p\left[e(t) + \tau_d\frac{\mathrm{d}e(t)}{\mathrm{d}t} + \frac{1}{\tau_i}\int_0^t e(\tau)\mathrm{d}\tau\right]. \tag{1}$$

where

$e(t) = y_{sp(t)} - y(t)$ is the process error
$u(t) =$ Control signal
$k_p =$ Proportional gain
$\tau_d =$ Derivative time
$\tau_i =$ Integral time.

Equation (1) can be written as,

$$u(t) = k_p e(t) + k_d\frac{\mathrm{d}e(t)}{\mathrm{d}t} + k_i\int_0^t e(\tau)\mathrm{d}\tau \tag{2}$$

where $k_d = k_p \times \tau_d$, $k_i = k_p/\tau_i$.

PID tuning is nothing but fixing three parameters k_p, k_i, and k_d so that one can achieve desired process response during set point change and load varying phases. Moreover, the controller should exhibit robust behavior in case of process model uncertainties and presence of measurement noise. Most of the available tuning relations are designed either targeting toward improved set point tracking or enhance load rejection performance. In addition, if they are considerably robust then performance is not found to be satisfactory and vice versa.

To avoid this incompetence up to certain extent, fuzzy rule-based auto-tuner is reported here which will provide a gain updating parameter α at every sampling instant for updating of the proportional-integral-derivative terms of a classical PID

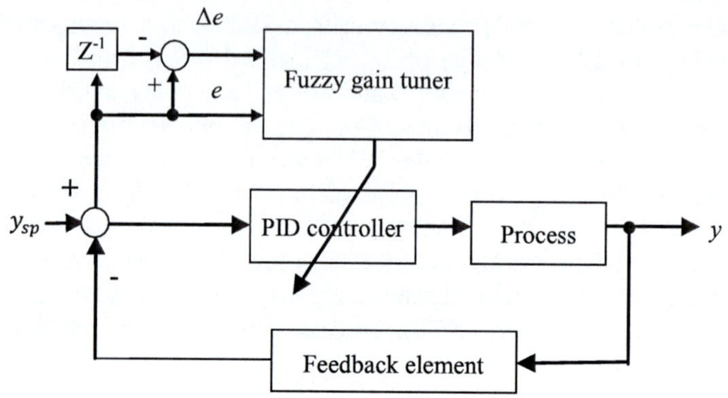

Fig. 1 Block diagram of Fuzzy supervisory tuner for PID controller

controller. Schematic diagram of the proposed fuzzy tuned PID (F-PID) controller is depicted in Fig. 1.

Here, our goal is to achieve better performance during closed-loop operation by fuzzy tuned PID controller (F-PID) compared to classical PID controller. To ascertain the requirement, following gain updating relations are defined for individual gain parameters of PID controller

$$k_p' = k_p(1 + k_1|\alpha|), \tag{3}$$

$$k_{\mathrm{i}}' = k_{\mathrm{i}}(0.3 + k_2\alpha), \tag{4}$$

$$k_d' = k_d(1 + k_3|\alpha|). \tag{5}$$

k_{p}', k_{i}', k_d' are the modified tuning parameters of F-PID controller. Here, the gain updating parameter a is obtained from the fuzzy rule base involving 9 rules and 25 rules, respectively. Here, the fuzzy tuner is developed based on the knowledge of a process expert.

In Fig. 1, input to the fuzzy gain tuner is error (e) and change of error (Δe) and depending on the value of e and Δe expert designed rule base is employed to provide the gain updating parameter, i.e., α which further modifies the tuning parameters of the PID controller as given by Eqs. (3)–(5). Here, 3 {Negative(N), Zero(Z), Positive(P)} and 5 {Negative Big(NB), Negative Small(NS), Zero(ZE), Positive Small(PS), Positive Big(PB)} fuzzy sets are considered for defining input variables— error (e), change of error (Δe), and gain updating parameter α. Inputs to the fuzzy gain tuner as shown in Fig. 1 are normalized error and normalized change of error instead of their physical values.

A. *Scaling factor*

Here, the values of error (e) and change of error (Δe) are mapped within the interval $[-1, 1]$ with the help of scaling factors. Suitable values are to be chosen based on the knowledge about the system under control and sometimes through trial and error for achieving the desired process response. Other than optimization-based approach there is no standard method for selection of the scaling factor. Output of the fuzzy tuner is the gain updating parameter α which is fed to the conventional PID controller.

B. *Membership function*

Gaussian type membership function is used for error (e), change of error (Δe), and gain updating factor (α) within the interval $[-1, +1]$ for both the cases with 3 and 5 membership functions (MFs) as depicted in Figs. 2 and 3, respectively.

Rule base: The fuzzy rule bases consisting of 25 and 9 rules is designed based on the expert's knowledge. To have a clear understanding regarding the rule base

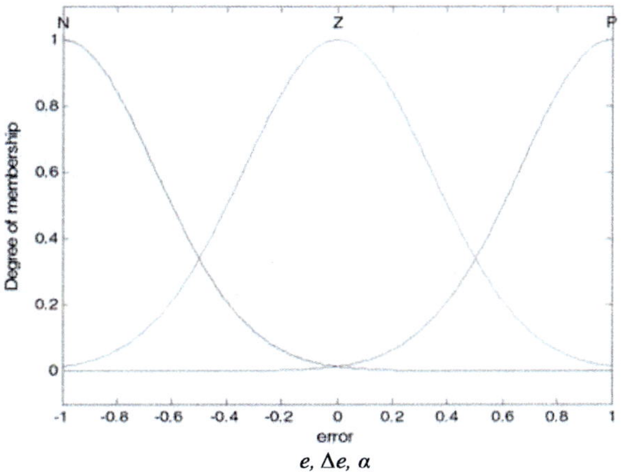

Fig. 2 Membership functions with three fuzzy sets

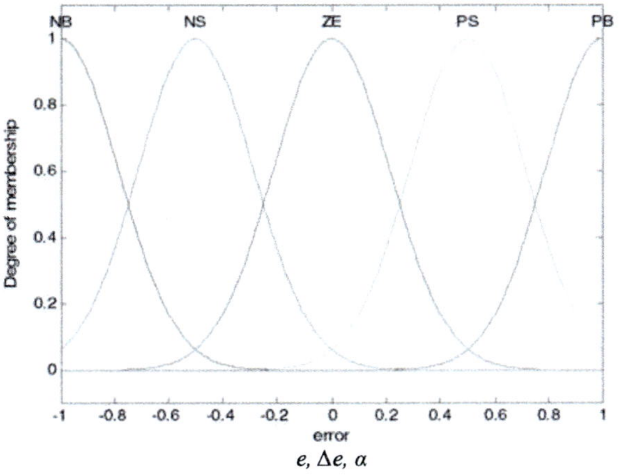

Fig. 3 Membership functions with five fuzzy sets

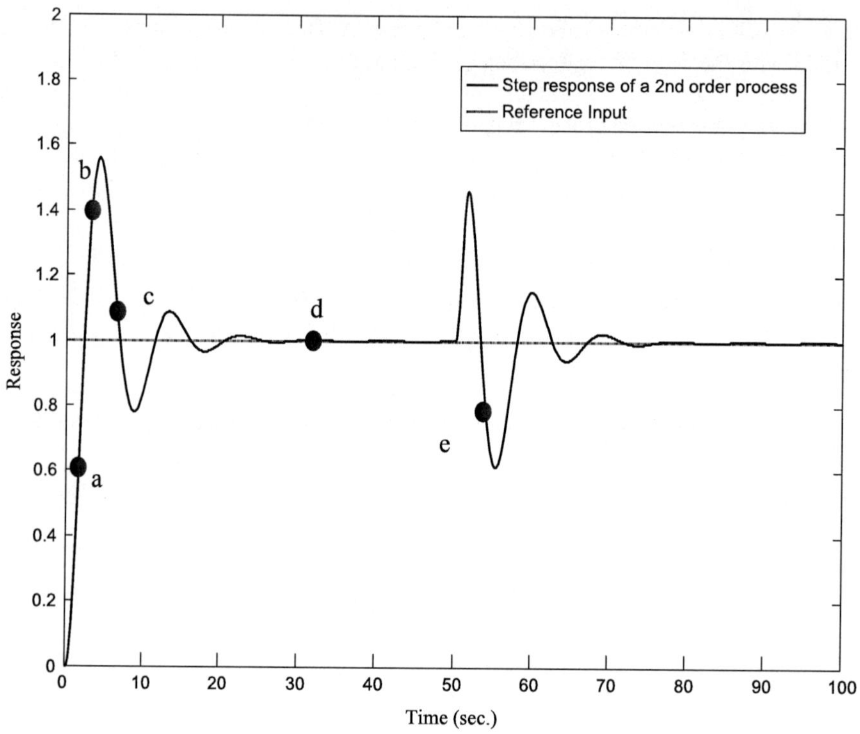

Fig. 4 Step response of a second-order system

designing strategy, a typical under-damped response of a second-order process is provided here as shown in Fig. 4. Formations of the rule bases for some specific operating points are discussed in the subsequent section:

1. For operating point 'a', to have lower overshoot and smaller settling time, the controller gain is required to be small. For this operating phase, where the error (e) is positive big (PB), but change of error (Δe) is negative big (NB). To obtain satisfactory response under such situation controller gain is required to be reduced, and hence the gain updating parameter is resulted with the property NB, i.e., α is negative big. So, the rule has the form—*If e is PB and Δe is NB then α is NB.* Such gain alteration is expected to provide improved performance.
2. For operating point 'b', where the process output is producing an overshoot, i.e., error (e) is negative big (NB) and change of error (Δe) is negative big (NB). To restrict the process, overshoot controller gain should be large but in opposite direction, and hence the value for gain updating factor α should be positive big (PB) as depicted in the rule base as shown in Fig. 9. The corresponding gain modification rule is given by—*If e is NB and Δe is NB then α is PB.*
3. At point 'c', error (e) is negative small (NS) and change of error (Δe) is positive small (PS), that means the process is moving toward the set point and under

such operating condition one can reduce the value of control action to avoid any possible undershoot in the subsequent operating phases. As a result, the value for gain updating parameter is chosen as negative small (NS) which signifies that there will be reduced control action as given by the rule of the form—*If e is NS* and *Δe is PS then α is NS*.

4. At steady-state condition, i.e., at point 'd' when both error (e) and change of error (Δe) are almost close at zero, controller gain should be small to avoid any possible chattering behavior. Hence, under such operating phase, the gain updating parameter must have zero value as defined by the rule—*If e is* ZE *and* *Δe is* ZE *then α is* ZE as depicted in the rule base of Fig. 6.

Hence, in similar way, the entire rule base (Figs. 5 and 6) for providing the appropriate gain updating factor (α) for the conventional PID controller can be framed so that one can expect an improved closed-loop response may be achieved due to F-PID compared conventional PID controller.

The nature of the gain surface with 9 and 25 rules for all possible operating phases is shown in Figs. 7 and 8, respectively. This surface is found to be highly nonlinear in nature. Depending on the process operating condition, gains of the PID parameters get modified as per the nature of this gain surface.

Fig. 5 Fuzzy rule base with 9 rules for computation of α

Δe/e	**N**	**Z**	**P**
N	P	Z	N
Z	Z	Z	Z
P	N	Z	P

Fig. 6 Fuzzy rule base with 25 rules for computation of α

Δe/e	**NB**	**NS**	**ZE**	**PS**	**PB**
NB	PB	PS	ZE	NS	NB
NS	PS	PS	ZE	NS	NS
ZE	ZE	ZE	ZE	ZE	ZE
PS	NS	NS	ZE	PS	PS
PB	NB	NS	ZE	PS	PB

Fig. 7 Surface view for α with 9 rules

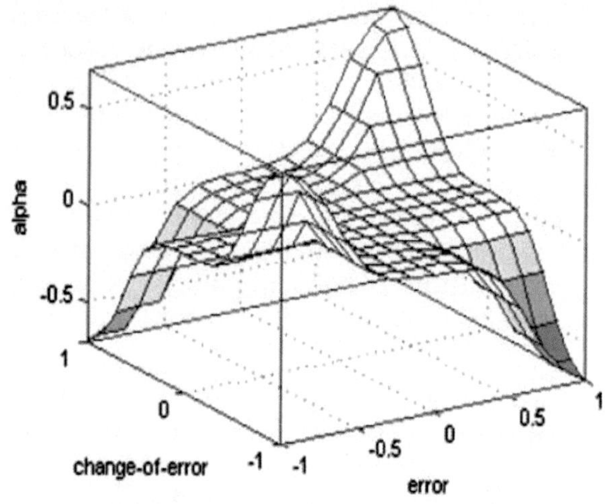

Fig. 8 Surface view for α with 25 rules

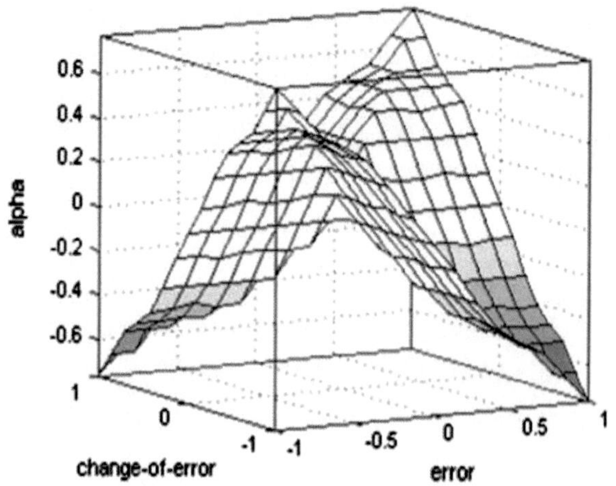

3 Simulation Results

Here, result of simulation studies will be represented with two second-order processes for F-PID, i.e., fuzzy expert-tuned PID controller along with classical PID controller. For clear assessment among two controllers, performance indices, such as percentage overshoot (%OS), settling time (t_s), integral time absolute error (ITAE), and integral absolute error (IAE), are evaluated and their values are provided in Tables 1 and 2. Here, *IAE* and *ITAE* are considered because only by observing response curves are always not sufficient to assess various responses. Robustness of the proposed scheme is evaluated with the same rule bases for two different process models where dead time is perturbed by +50% from its nominal value.

Table 1 Performance analysis for Model I

Type of controller	Performance measurer					
	Dead time (s)	Rule base	Settling time (t_s s)	% Overshoot	IAE	ITAE
PID	0.2		5.20	35.70	1.81	6.92
F-PID		9	6.80	1.00	1.57	6.80
		25	4.50	0.00	1.50	6.68
PID	0.3 (+50% perturbed)		5.50	48.66	2.09	8.33
F-PID		9	7.20	4.62	1.64	8.00
		25	3.00	0.00	1.53	6.80

Table 2 Performance analysis for Model II

Type of controller	Performance measurer					
	Dead time (s)	Rule base	Settling time (t_s s)	% Overshoot	IAE	ITAE
PID	0.2		10.20	63.16	4.26	41.92
F-PID		9	14.50	25.90	3.50	37.06
		25	14.90	22.17	3.37	34.09
PID	0.3 (+50% perturbed)		14.60	77.31	5.13	53.46
F-PID		9	14.20	30.38	3.65	38.73
		25	15.70	22.69	3.41	34.76

For simulation study, two second-order process models are considered as described by the following relations

$$\text{Model I}: \quad G_1(s) = \frac{e^{-\tau s}}{(s+1)^2}. \tag{4}$$

$$\text{Model II}: \quad G_2(s) = \frac{e^{-\tau s}}{S(S+1)}. \tag{5}$$

PID controller is initially tuned using relay tuning methodology [8] and tuning parameters for Model I and Model II are given by

$$k_{p1} = 4.15, k_{i1} = 3.19, K_{d1} = 1.34 \quad \text{(Model I)} \tag{6}$$

$$k_{p2} = 2.08, K_{i2} = 1.16, K_{d2} = 0.92 \quad \text{(Model II)} \tag{7}$$

For both, the process models as described by Eqs. (4) and (5), i.e., for Model I and Model II, dead time is considered as $\tau = 0.2$ s. Response of Model I with PID and F-PID consisting of rule bases with 9 and 25 rules along with the control action

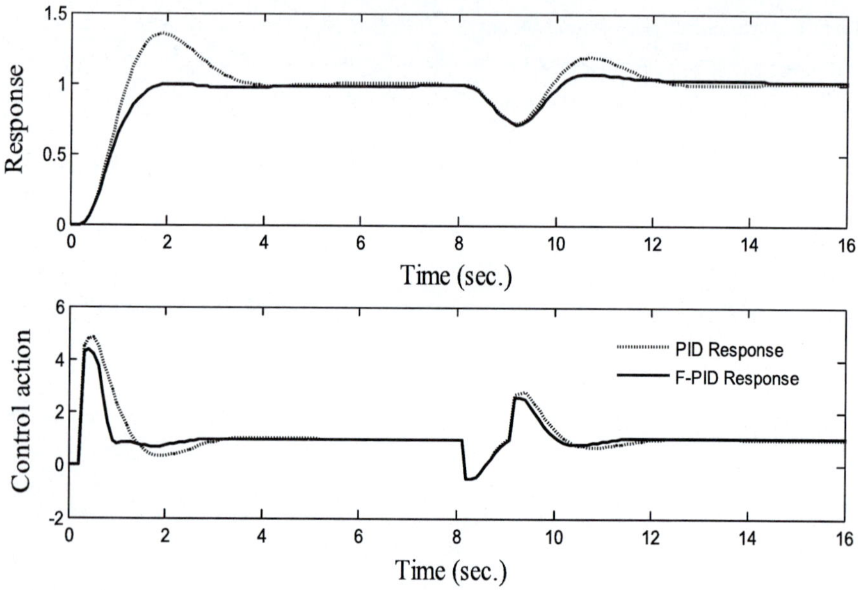

Fig. 9 PID and F-PID responses of Model 1 (nominal) using 9 fuzzy rules

are depicted in Figs. 9 and 10. To study the robust feature of the reported controller under parametric uncertainty, dead time of the Model I is considered to be 50% more from its nominal value. Corresponding closed-loop responses for Model I with 9 and 25 fuzzy rules are depicted in Figs. 11 and 12, respectively. During simulation study, at the beginning, unit step set point change is applied and after reaching the desired steady state, load disturbance is applied at $t = 8$ s for Model I and $t = 17$ s for Model II.

Once load disturbance is applied, PID takes longer time to recover but more aggressive recovery is found for F-PID resulting significant impact in the performance measures as the value of %OS, t_s, IAE, ITAE are more for PID compared to that of the F-PID (shown in Table 1). Similar nature of performance improvement can also be found from the closed-loop responses provided by the perturbed model. In addition, it is also observed that performance of supervisory fuzzy PID controller (F-PID) is improved by increasing the number of fuzzy rules from 9 to 25 and it is reflected by the values of performance measures that are described in Table 1.

The second-order process represented by Eq. (5), i.e., Model II has a pole at $s = 0$, this leads to oscillatory behavior of the process and existence of dead time makes the system more difficult to control. Initial setting of PID controller for this model is given by relation (7). Conventional PID controller fails to perform satisfactorily for this kind of oscillatory process. Figures 13 and 14 show the responses of the Model II for dead time $\tau = 0.3$ s. In case of F-PID with rule bases consisting of 9 and

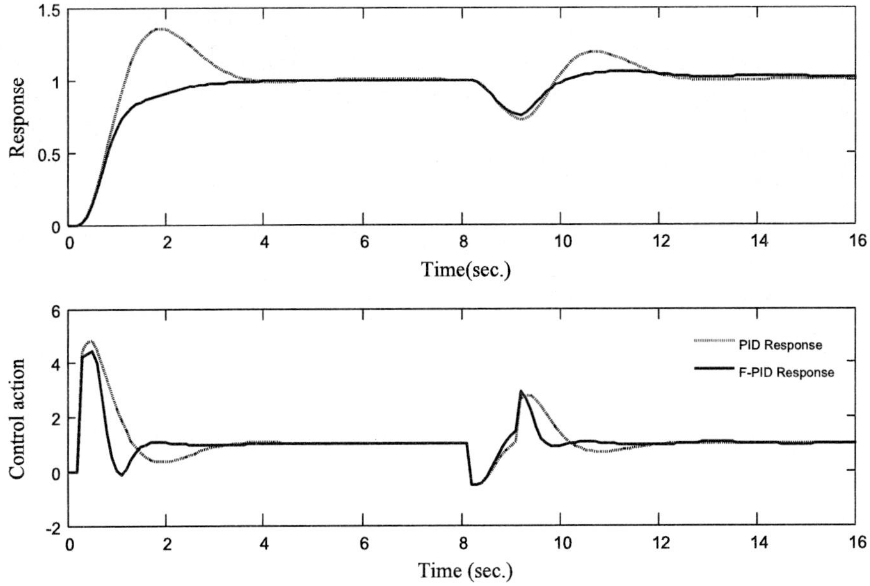

Fig. 10 PID and F-PID responses of Model 1 (nominal) using 25 fuzzy rules

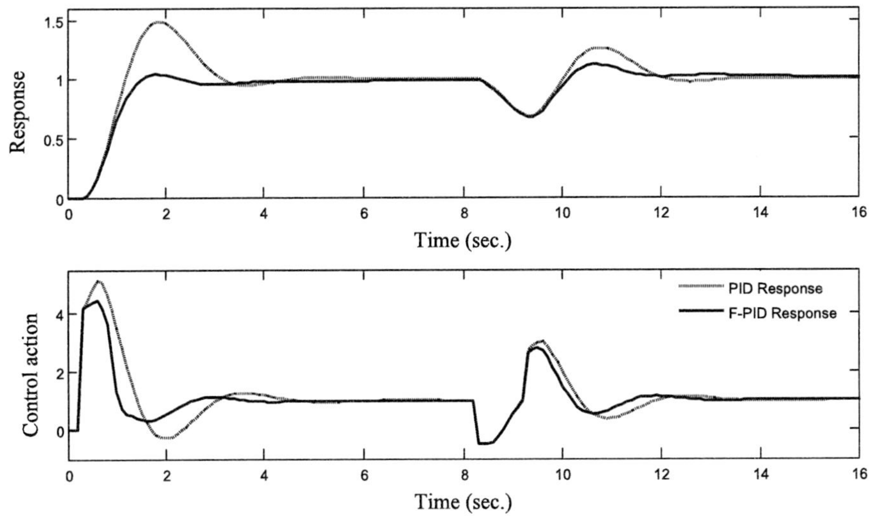

Fig. 11 PID and F-PID responses of Model I (perturbed) using 9 fuzzy rules

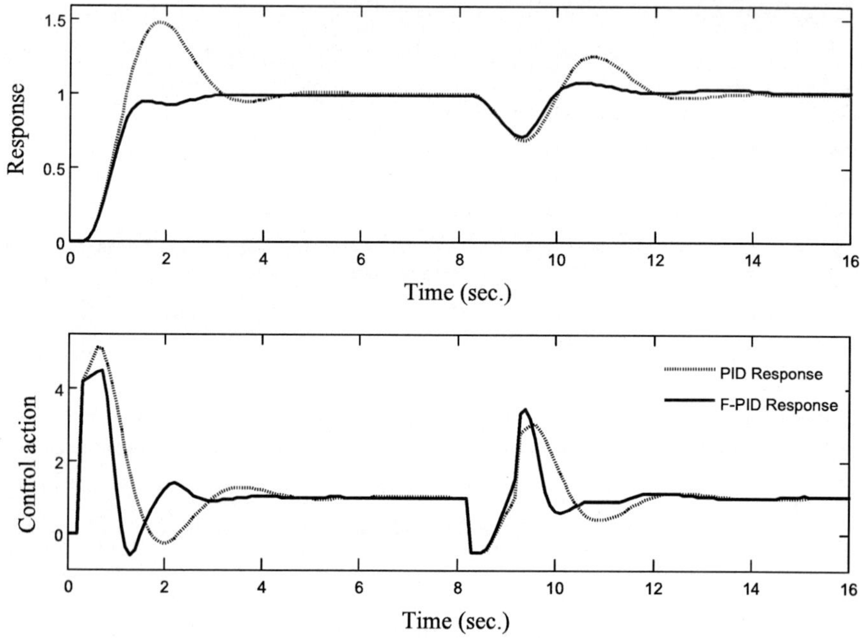

Fig. 12 PID and F-PID responses of Model I (perturbed) using 25 fuzzy rules

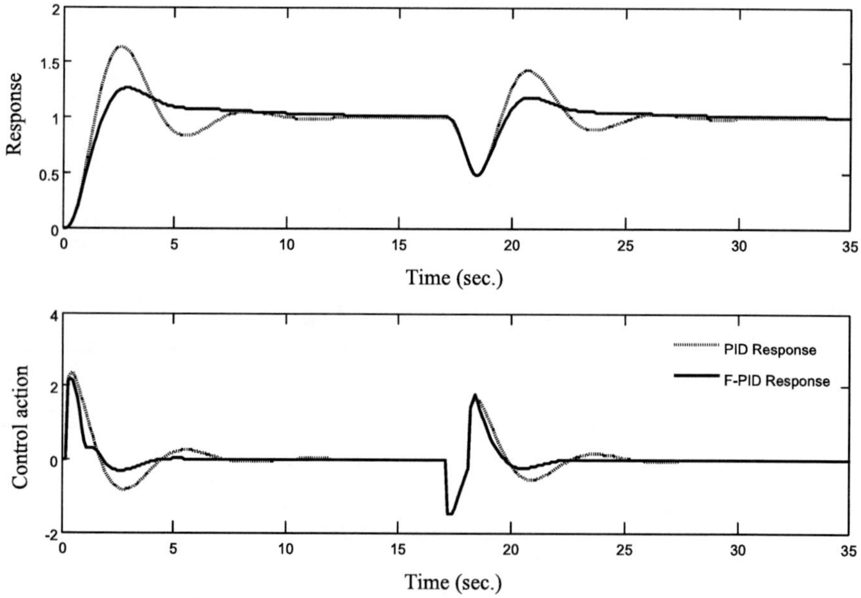

Fig. 13 PID and F-PID responses of Model II (nominal) using 9 fuzzy rules

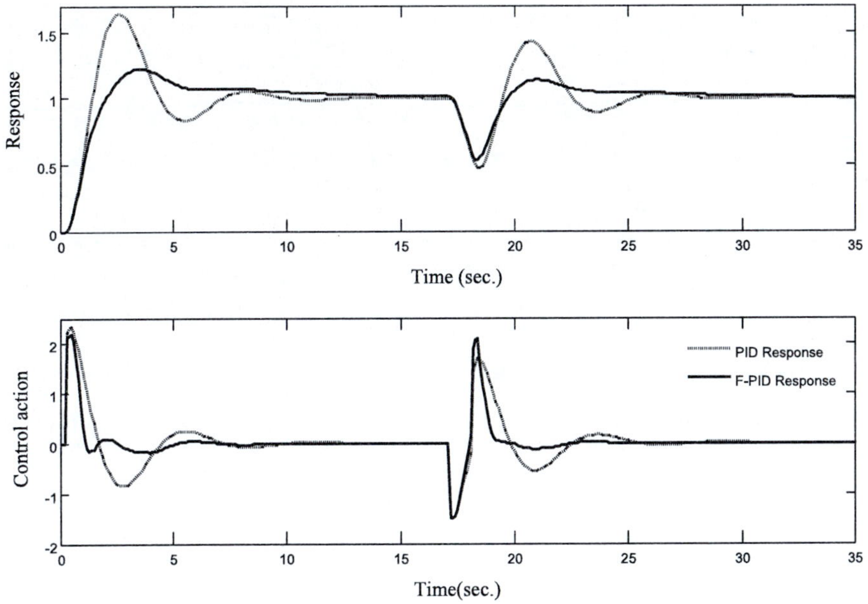

Fig. 14 PID and F-PID responses of Model II (nominal) using 25 fuzzy rules

25 rules responses get improved considerably compared to classical PID controller. Performance enhancement of F-PID is demonstrated from the value of performance indices as shown in Table 1.

Figures 15 and 16 show the responses of the perturbed Model II with the dead time $\tau = 0.3$ s (i.e., +50% perturbed). Performance of PID and F-PID is evaluated using rule bases consisting of 9 and 25 rules, and the corresponding performance measures for this perturbed model are shown in Table 2. From the responses, it is clearly visible that the proposed supervisory fuzzy tuned PID controller (F-PID) provides considerable improvement with +50% perturbed dead time, i.e., justifying its stronger robust feature compared to conventional PID controller.

4 Conclusion

In this paper, fuzzy supervisory expert tuners are designed with 9 and 25 fuzzy rules for online updating of the PID tuning parameters. Limitation of the conventional PID controller is augmented with the help of fuzzy expert tuner where the rules are framed based on the knowledge of experienced human operator. From the simulation results, it is clearly visible that the proposed F-PID controller provides superior performance to achieve improved set point tracking of input and faster load rejection compared to the conventional PID controller. It is further observed that the proposed fuzzy tuner

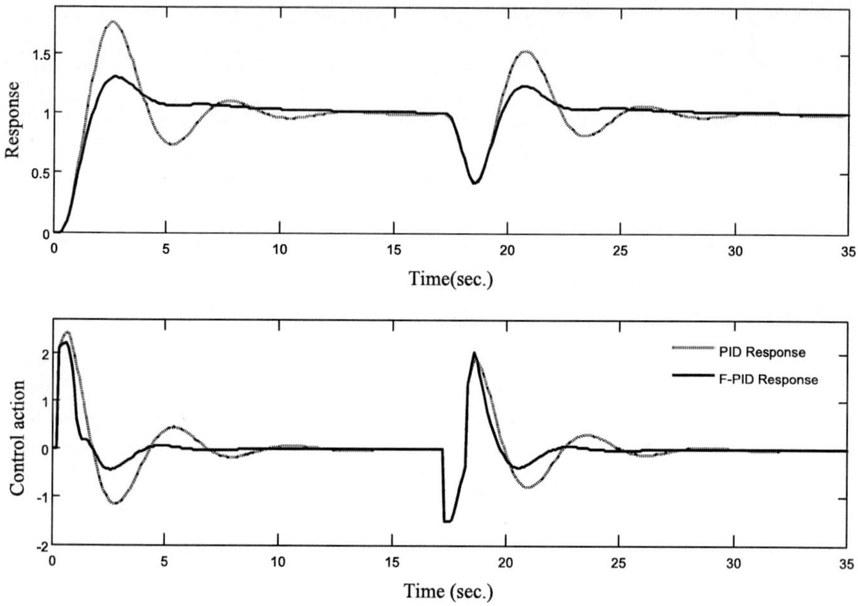

Fig. 15 PID and F-PID responses of Model II (perturbed) using 9 fuzzy rules

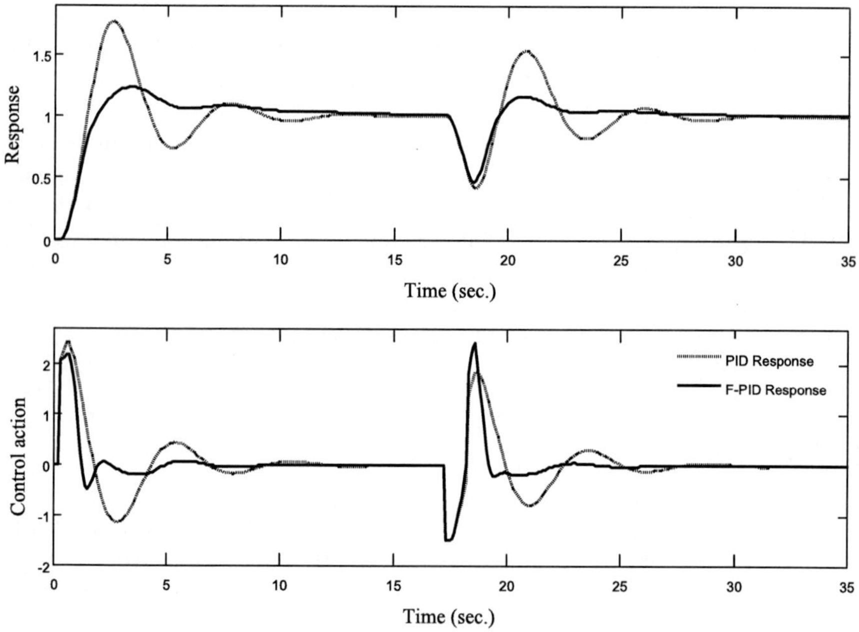

Fig. 16 PID and F-PID responses of Model II (perturbed) using 25 fuzzy rules

performs considerably better with larger rule base. Moreover, their performance is found to be more robust under model uncertainty. In the future, there is scope to verify the efficacy of the reported scheme on other process models and to evaluate its performance in real time.

References

1. Yu J, Liu C (2015)Design of self-tuning PID controller with fuzzy variable parameters based on LabView. In: 2015 IEEE international conference on information and automation, IEEE, pp 2586–2591
2. Visioli A (2001) Tuning of PID controller with fuzzy logic. EIE Proc Control Theor Appl 148(1):1–8
3. Mudi R, Pal NR (1999) A robust self-tuning scheme for PI- and PD-type fuzzy controllers. IEEE Trans Fuzzy Syst 7(1):2–16
4. Zheng L (1992) A practical guide to tune of proportional and integral (PI) like fuzzy controllers. In: Proceedings of Fuzzy IEEE, San Diego, CA, pp 633–641
5. Yoshida M, Tsutsumi Y, Ishida T (1990) Gain tuning method for design of fuzzy control systems. In: Proceedings of international conference fuzzy logic neural networks, Fukuoka, Japan, pp 405–408
6. Lee J (1993) On methods for improving performance of PI-type fuzzy logic controllers. IEEE Trans Fuzzy Syst 1:298–301
7. Mudi RK, Pal NR, A note on fuzzy PI-type controllers with resetting action. Fuzzy Sets Syst, communication
8. Bequette BW (2002) Process control modelling, design and simulation. Prentice Hall, PTR

Smart Racking and Retailing Using IOT

Neha Mishra and Anup Kumar Keshri

Abstract This paper describes about smart racking and retailing using IOT. Hence, it focuses on retailers to understand their challenges and the way IOT can help them overcome these challenges. Nowadays, we find that inventory management has become a very complex problem. Not only the store owner as well as the customer sometimes faces a lot of problems in searching products in store. To solve this problem, we use IOT technology which will resolve both the retailer's and customer's problem. With the help of this system, the owner can maintain inventory information smartly. The system will also be helpful for the customers to search products easily in store and may know the details of the product such as product's availability in store, its price, and similar items. Hence, this will increase more and more customer to the store owner.

Keywords T-slot · Actuator · Arduino · Load cell · HX711module · Creo

1 Introduction

Internet of things includes devices that can connect to the Internet serving the purpose of sending as well as receiving information or data. Thus, devices which can be monitored from a remote location are known as an IOT device. Generally, in general stores or shopping mall, the retailer needs a round for checking the stock position or the quantity of products left. Hence, it requires man power. Thus, with the inclusion of IOT, this work can be done automatically through smart racking and retailing, as IOT is a system of interconnecting computing devices. IOT includes various types of sensors. Each sensors includes an individual IP address, which are then connected to a network for sending and receiving information or data. By adopting IOT retailers

N. Mishra (✉) · A. K. Keshri
Department of Computer Science, BIT MESRA, Ranchi, India
e-mail: nehanisha1214@gmail.com

A. K. Keshri
e-mail: anup_keshri@bitmesra.ac.in

V. Nath and J. K. Mandal (eds.), *Proceedings of the Fourth International Conference on Microelectronics, Computing and Communication Systems*, Lecture Notes in Electrical Engineering 673, https://doi.org/10.1007/978-981-15-5546-6_54

may become experience-based retailers and can do business smartly. It combines the online and in-store experience of the customers to the retailers.

Many of the times when a customer requires a product and he is not able to search the product, he inquires the retailer and then, finally, the retailer after searching informs the product is out of stock. So, this can be considered as a loss in time for customer as well as to the store owner.

This research is all about designing an automated system using IOT that helps in doing business smartly. To design a Smart Rack which will automatically replace an empty rack by a filled rack. Further, to regularly update stock position as per the customer picks up or put a product back to the rack. And when the rack goes out of stock a warning message will be sent to the retailer as well as to the main branch from where the products are supplied. So that, after receiving warning message, the main branch can immediately send the stock of products to the required store.

2 Methods and Material

A common problem faced by the store owner nowadays is increased sales and reduced operational cost. Whereas customers also face problems such as searching products in the store, and all on a sudden occurrence of unavailability of the product may make a loss to the customer according to the profit he would earn on final bulk shopping. Suppose, in a city, there are ten general stores under same branch and the main branch announces offers on any particular day or on any festive seasons. To avoid rush in stores, smart rack should be implemented so that, as soon as the front rack gets empty, it could be automatically replaced by the filled rack behind. To further improve the technique, weight sensors are used on the racks which may regularly update the quantity of the product to the store owner. If the product is reduced to threshold value then immediately a warning message is generated to the retailer's desk as well as to the main branch to refill the products (Fig. 1).

Materials or the hardware components required for implementing the system includes t-slot, actuator, Arduino Uno, load cell, HX711 load cell amplifier module.

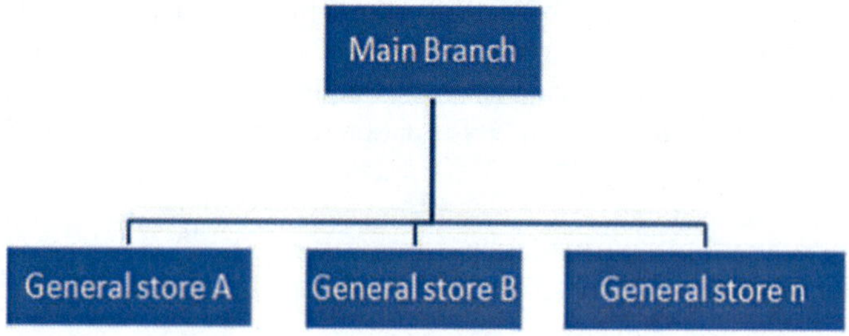

Fig. 1 System architecture

T-slot: A t-slot is a framing system which consists of lengths of square or rectangular extruded aluminum profile and is great for assembly. It has t-shaped groove on each side (Fig. 2).

Actuator: An actuator acts as a motor which converts energy into torque to move or to control a mechanism or a system to which it is incorporated. Thus, it runs on electric or pressure (Fig. 3).

Arduino Uno: It is an open-source microcontroller board which controls whole process (Fig. 4).

Load cell: A transducer that transforms a pressure into electrical output. A load cell consists of strain gauge which deforms when pressure is applied on it, generating an electric signal on deformation. The signal generated is in few millivolts (Fig. 5).

Fig. 2 T-slot aluminum profile

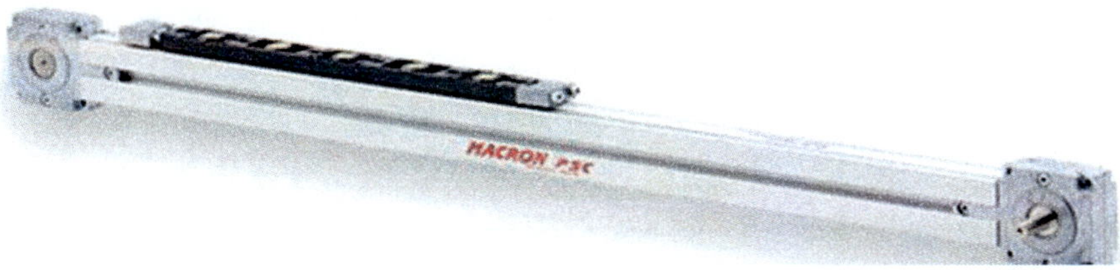

Fig. 3 Linear actuator

Fig. 4 Arduino Uno

Fig. 5 Load cell

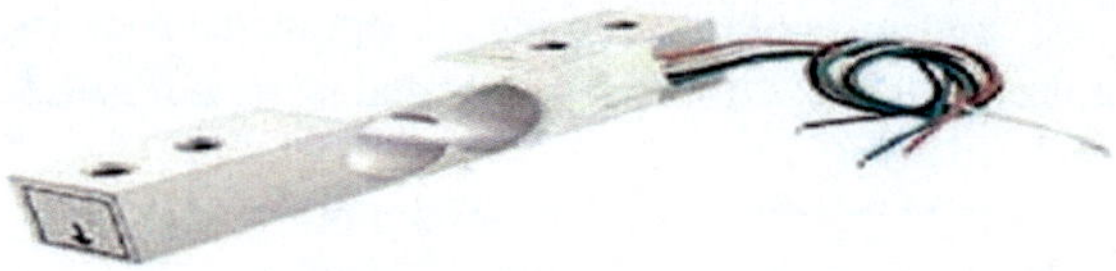

Fig. 6 HX711

HX711module: Acts as an amplifier to amplify the value obtained from load cell. Analog to digital converter. Hence, known as hx711 load cell amplifier module (Figs. 6 and 7).

A customer enters a store and starts shopping.

The customer picks up a product.

As soon as he picks up a product left is calculated and is counter checked with the set threshold value.

If he put back the product, then again the stock position is updated.

As soon as the product left goes below the threshold value, then immediately and automatically the empty rack is replaced by the filled rack arranged behind.

And, at the same time, a warning message is generated to the retailers desk and to the main branch as well.

To further enhance the working of the rack, a module is designed on Creo software for proper functioning of the rack (Figs. 8 and 9).

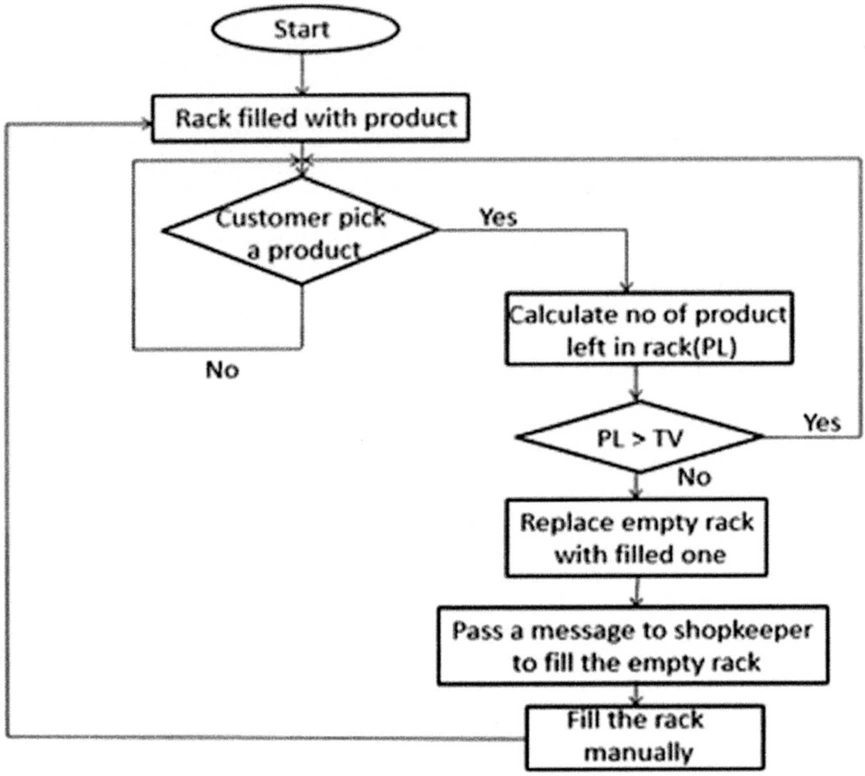

Fig. 7 Flow diagram

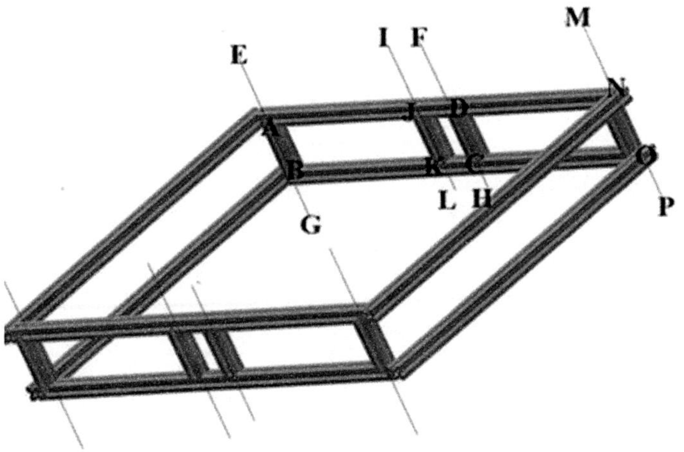

Fig. 8 Designed on Creo

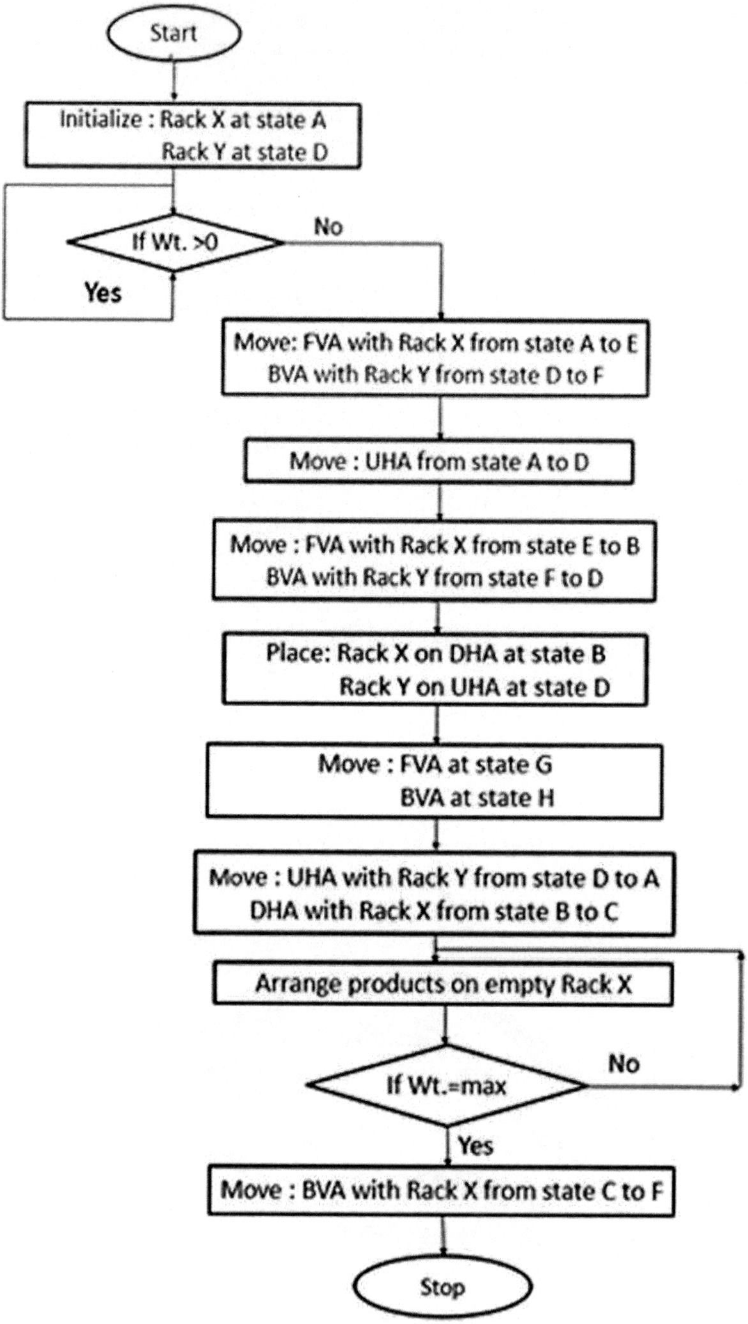

Fig. 9 Flow diagram representing rack movement on above module

1. Start
2. initialize: rack X at state A
   rack Y at state D
3. check weight
4. if weight > 0
   Continue shopping
5. else
6. move: FVA with rack X from state A to E
   BVA with rack Y from state D to F
7. Now move UHA from state A to D
8. Move: FVA with rack X from state E to B
   BVA with rack Y from state F to D
9. place: rack X on DHA at state B
   rack Y on UHA at state D
10. move: FVA at state G
    BVA at state H
11. move: UHA with rack Y from state D to A
    DHA with rack X from State B to C
12. arrange products on empty rack X
13. check wt. on X
14. if wt. on X = maximum
    move BVA from state H to C
15. lift and take rack X from state C to F with BVA
16. check weight on Y
17. End

3 Discussion

For the implementation of this paper objective, a proper algorithm is made to define the automatic movements of the rack. Size and weight of all the hardware required are also kept in mind for smooth functioning of the system so that it would be quite useful to the retailers as well as the customers. Features of all the hardware components will be kept in mind according to the retailers' perspective and also customers view point. Therefore, such materials will be used so that efficiency of the system will be maintained for a longer period of time.

Kalange et al. [1] In this paper, the author has made use of RFID tags on the product, i.e., radio frequency identification tag which gives information about the product to RFID reader and it gives the information to the arduino. The author has also used load cell which gives data to Hx711 and it further provides data to the arduino. Arduino further processes these information and through wifi module saves it to cloud. Smart phone used by the customer can easily access the cloud to retrieve information of the product as the RFID tag provides all the information related to the product such as expiry date, availability in store, nutrition, price, and profit. This application automatically manages the inventory system which is very efficient to the store owner as well as to the customer in economic manner.

Hwangbo et al. [2] In this era of modernization, online retail has influenced business very deeply. Even though online retail has influenced the customers view, offline retail is still most important for a large proportion of popularity. Thus, the author has made use of technologies such as sensors, indoor positioning, vision, augmented reality, and interactive interface which drastically improved the retail shops service quality. Indoor positioning system (IPS) is a system similar to a sensor which observes the natural position of objects and people around the building. Facial recognition is a technique for identifying a person from a digital image or video frame from a video source. It is a part of biometrics as it has the advantage of mass identification in public places.

Author has introduced various IOT products, including smart hanger working on sensor technologies, smart CCTV based on indoor positioning and facial recognition, and smart mirror based on augmented reality and digital signage. Thus, the term "Internet of Everything" is a perfect definition to Internet of things. Dave Evans, who originated the concept of IoE, defined IoE as "the intelligent connection of people, process, data and things." Thus, Iot focuses on machine-to-machine (M2M) communications. The author has tried to investigate persuasive marketing and customer's experience on use of various technologies. Hence, the author has gone through a case study. Therefore, smart retailing will change customer behavior.

4 Conclusion

Smart rack can be very useful to retail industry as it can store all perishable items as well as non-perishable items. Regular updation of stock position will be also very helpful for the retailer as well as the customer through proper invoicing. Thus, smart rack will reduce man power resulting in the reduction of effort as well as time of the customers and retailers. Therefore, it will be feasible to use and so user friendly. Searching the products in the store will get easier. Also the product details can be easily observed through proper invoicing. Web-based application will handle multiple branches of shops all together.

References

1. Kalange SH, Kadam DA, Mokal AB, Patil AA (2017) Smart retailing using IOT. Int Res J Eng Technol (IRJET) 4:263–268
2. Hwangbo H, Kim YS, Cha KJ (2017) Use of the smart store for persuasive marketing and immersive customer experiences: a case study of Korean apparel enterprise. Copyright © 2017
3. Khanna A, Tomar R (2016). IoT based interactive shopping ecosystem. https://doi.org/10.1109/ngct.2016.7877387

4. Farooq MU, Waseem M, Mazhar S, Khairi A, Kamal T (2015) A review on internet of things (IoT). Int J Comput Appl (0975 8887) 113(1), 1–7
5. Ding K, Jiang P (2018) RFID-based production data analysis in an IoT-enabled smart job-shop. IEEE/CAA J Automatica Sinica 5(1)
6. Chen R, Peng L, Qin Y (2010) Supermarket shopping guide system based on internet of things. School of IoT Engineering, Jiangnan University, Wuxi, Jiangsu 214122
7. Vengurlekar PN, Kulkarni VS, Kulkarni PR, Jadhav SZ (2016) Cloud based online reatil management framework: a different aspect. Imperial J Interdisc Res 2
8. Xiaohui X (2013) Study on security problems and key technologies of the internet of things. In: Computational and information sciences (ICCIS), pp 407–410
9. Elmustafa SAA, Zeinab KAM (2017) Internet of things applications, challenges and related future technologies. World Sci News 2(67):126–148
10. Satapathy R, Prahlad S, Kaulgud V (2015) Smart shelfie—internet of shelves: for higher on-shelf availability. In: 2015 IEEE region 10 symposium, Ahmedabad, pp 70–73
11. Rosado L, Gonçalves J, Costa J, Ribeiro D, Soares F (2016) Supervised learning for out-of-stock detection in panoramas of retail shelves. In: 2016 IEEE international conference on imaging systems and techniques (IST), Chania, pp 406–411
12. Mitchell A (2011) Improving on-shelf availability. [White paper] SymphonyIRIGroup
13. Kejriwal N, Garg S,Kumar S (2015) Product counting using images with application to robot-based retail stock assessment. In: IEEE international conference on technologies for practical robot applications, pp 1–6
14. Higa K, Iwamoto K (2018) Robust estimation of product amount on store shelves from a surveillance camera for improving on-shelf availability. In: 2018 IEEE international conference on imaging systems and techniques (IST), Krakow, pp 1–6
15. Satapathy R, Prahlad S, Kaulgud V (2015) Smart shelfie internet of shelves: for higher on-shelf availability. In: 2015 IEEE region 10 symposium (TENSYMP), pp 70–73
16. Kejriwal N, Garg S, Kumar S (2015) Product counting using images with application to robot-based retail stock assessment. In: 2015 IEEE international conference on technologies for practical robot applications (TePRA), pp 1–6
17. Perona P, Malik J (1990) Scale-space and edge detection using anisotropic diffusion. IEEE Trans Pattern Anal Mach Intell 12:629–639

Fuzzy Rule-Based Supervisory PID Auto-Tuner for TRMS Process

Sayanika Bandyopadhyay, Payel Pramanick, and Chanchal Dey

Abstract The main concern of this paper is to study a strategy of fuzzy rule-based supervisory auto-tuner designing for proportional integral derivative (PID) controller and its performance evaluation on a two rotor multi-input multi-output system (TRMS). TRMS is considered to be a simplified helicopter model with two dynamic motions, i.e., the movements of yaw and pitch. Here, we have designed a fuzzy auto-tuned PID controller toward achieving desired response during yaw and pitch movement for TRMS model. Controlling of MIMO process is always a challenging task due to the interaction between its individual degree of freedom. So, the proposed controller design should be in such a way as if any drift in one degree has minimum effect on other, and overall, the system should provide stable behavior. Hence, conventional PID controller with defined tuning parameters usually fails to execute a satisfactory close-loop response in such cases. To overcome this limitation of PID controller, two fuzzy rule-based gain tuners are incorporated for two separate PID controllers responsible for controlling the direction and speed of the motors responsible for yaw and pitch motion. From simulation results, it can be clearly substantiated that the proposed scheme is superior to conventional PID controller to track the desired trajectory and reject the disturbance.

Keywords TRMS · Model · MIMO system · PID controller · Nonlinear control · Fuzzy auto-tuner

---

S. Bandyopadhyay (✉) · P. Pramanick · C. Dey
Department of Applied Physics, University of Calcutta, Kolkata, India
e-mail: sayanikabanerjee880@gmail.com

P. Pramanick
e-mail: payel256pramanick@gmail.com

C. Dey
e-mail: cdaphy@caluniv.ac.in

V. Nath and J. K. Mandal (eds.), *Proceedings of the Fourth International Conference on Microelectronics, Computing and Communication Systems*, Lecture Notes in Electrical Engineering 673, https://doi.org/10.1007/978-981-15-5546-6_55

1 Introduction

In recent years, interest in designing intelligent controller for aeronautical systems has steadily increased. Modern control schemes which provide satisfactory performance for aerospace problems are not found to be enough fruitful in various process control applications. Ordinary PID controllers still dominate this region for its simple and straightforward tuning strategy. Due to strong dynamics and nonlinearity of TRMS, conventional PID controllers cannot provide acceptable control performance during desired trajectory tracking and disturbance rejection phases. TRMS is considered to be a miniaturized and simplified model of a helicopter which is a highly coupled nonlinear system [1]. It is also regarded as a benchmark MIMO system for performance evaluation of control laws. Conventional PID tuning with fixed setting is not found to be adequate for controlling the system having such complex dynamics resulting in unacceptable control performance [2].

To overcome such limitation related to the static nature of conventional PID controller with linear behavior, the controllers based on fuzzy logic have been extensively used in such purposes [3, 4]. Desired extent of nonlinearity may be easily incorporated in a fuzzy controller to counteract the process nonlinearity. Moreover, fuzzy gain scheduling [4] technique is quite effective for controlling processes like TRMS with strong interaction. By combining the benefits of evolutionary computing tools with fuzzy controller [5], their performance can be made further superior than their conventional approaches. In [6], nonlinear cascade-based control technique is used for controlling TRMS. Here, we have designed fuzzy supervisory expert autotuner for conventional PID controller. This fuzzy rule base expert tuner modifies the PID tuning parameters with the help of a gain tuning parameter during close-loop operation, i.e., online in nature. Fuzzy rule base comprising of 49 rules is arranged based on the knowledge of an expert in terms of desired performance of a TRMS process. As a result what is expected that the online gain tuning feature of the PID controller should be capable to provide satisfactory trajectory tracking and unwanted disturbance rejection performance compared to conventional PID settings. The superiority of the designed tuning scheme as expressed by its controller performance indices is clearly substantiated in the detailed simulation study.

2 TRMS Description

Twin rotor multiple input multiple output system (TRMS) is an electro-mechanical setup in laboratory scale designed by Feedback Instruments Ltd for performing control experiments. The mechanical section of TRMS consists of two rotational components (main or pitch rotor which enables vertical positioning and tail or yaw rotor responsible for horizontal positioning) placed on a beam along with a counterbalance. The beam is pivoted on its base so that it is free to rotate in both its horizontal

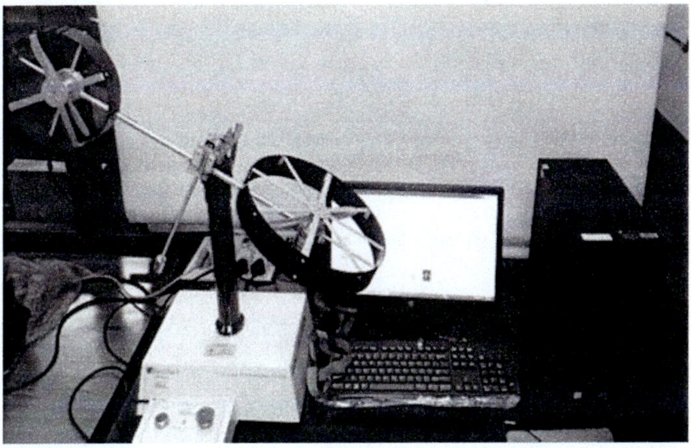

Fig. 1 TRMS control unit

and vertical planes. The voltage input to TRMS is limited to the range ±10 V. The entire assembly is attached to the tower allowing for safe experimentation (Fig. 1).

TRMS setup comprising of four main components:

1. PC with control system algorithms
2. DAQ device acting as an interface between the PC and external environment
3. The process under control, i.e., TRMS
4. Sensors for angular speed and position measurement.

3 TRMS Schematic

TRMS is a plant which is termed as TITO process as it is a two-input two-output system. Figure 2 presents a simplified schematic of the TRMS.

The control action u_1 is applied to the pitch motor responsible for motion in vertical plane, and the control action u_2 is applied to the yaw motor resulting motion in the horizontal plane. There is always significant amount of coupling effects between these two motions and its one of the key features of TRMS as shown in Fig. 2. So, to avoid the coupling effect, a decoupler is required to be introduced in the control loop. Decopuler will be responsible to reduce the coupling effect, i.e., the interaction between yaw and pitch motion of the TRMS.

4 Block Diagram Analysis of MIMO process

A typical multivariable system block diagram is shown in Fig. 3. Each of the variables shown in this diagram is a vector, and each block represents matrix transfer function.

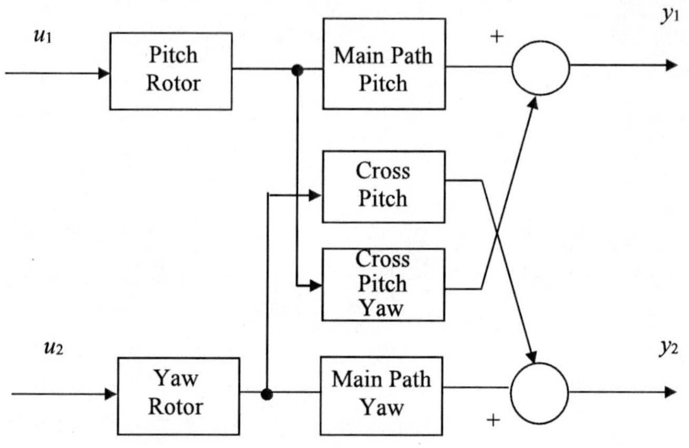

Fig. 2 TRMS simplified system schematic

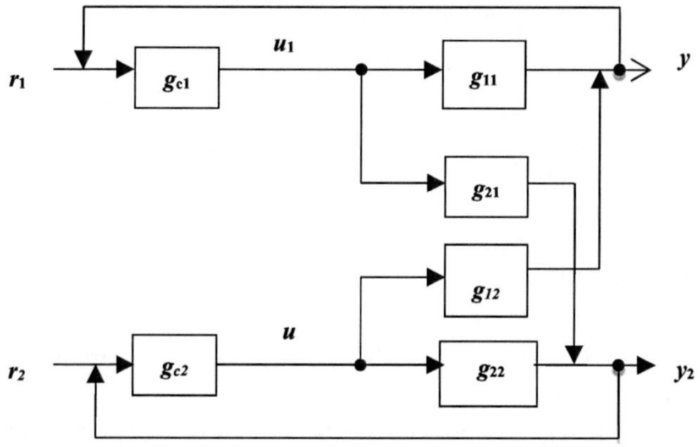

Fig. 3 Feedback connection for a two-input two-output process

So, there are two inputs and two outputs. Set points, outputs, manipulated inputs, and the disturbance input are

$$r = \begin{bmatrix} r_1 \\ r_2 \end{bmatrix}; y = \begin{bmatrix} y_1 \\ y_2 \end{bmatrix}; u = \begin{bmatrix} u_1 \\ u_2 \end{bmatrix}; l = [l]$$

The controller, process, and disturbance matrices are

$$G_c(s) = \begin{bmatrix} g_{c11}(s) & g_{c12}(s) \\ g_{c21}(s) & g_{c22}(s) \end{bmatrix},$$

$$G_{\rm p}(\rm s) = \begin{bmatrix} g_{11}(\rm s) & g_{12}(s) \\ g_{21}(\rm s) & g_{22}(s) \end{bmatrix}, G_d(s) = \begin{bmatrix} g_{d1}(s) \\ g_{d2}(s) \end{bmatrix}$$

It is to note that if there is no coupling effect, then two SISO controllers will be capable to ensure the desired response from such a system, and hence, the controller transfer function will be

$$G_c(s) = \begin{bmatrix} g_{c1}(s) & 0 \\ 0 & g_{c2}(s) \end{bmatrix}$$

where $g_{\rm c}$(s) and $g_{\rm c2}$s) are two controllers responsible for independent yaw and pitch motion. The relationship between inputs (both manipulated and disturbance) and output is

$$y(s) = G_p(s)u(s) + G_d(s)l(\rm s) \tag{1}$$

But, in reality, there will be a strong interaction between the yaw and pitch motion, and hence, we need a decopuler to minimize it. Hence, the choice of an appropriate decoupler is an important aspect for MIMO process. The relationship between the output vector and the process input vector is

$$y(s) = G_p(s)D(s)u^*(\rm s). \tag{2}$$

For a TITO process,

$$\begin{bmatrix} y_1(s) \\ y_2(s) \end{bmatrix} = G_p(s)D(s)\begin{bmatrix} u_1^* \\ u_2^* \end{bmatrix}$$

where G_p(s)D(s) is a 2 × 2 transfer function matrix (Fig. 4).

5 Proposed Control Approach

Here, a supervisory fuzzy expert tuner consisting of 49 rules is designed which will modify the tuning parameters of the two PID controllers involved for controlling TRMS process. The fuzzy tuner will modify the individual tuning parameter for PID controller with the help of a gain updating parameter α based on the process operating condition. Schematic diagram along with the designed fuzzy-tuned PID controller is shown in Fig. 5.

In our proposed scheme, inputs are given to the fuzzy inference system are error (e) and change of error (Δe), and depending upon the values of e and Δe, some

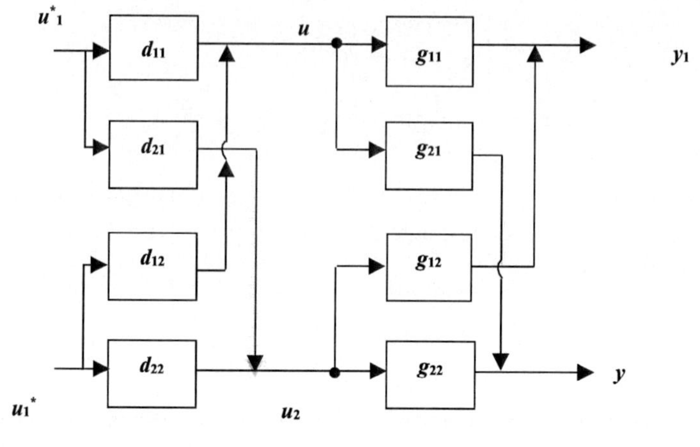

Fig. 4 Block diagram of a general multivariable system with decoupler

Fig. 5 Block diagram of proposed fuzzy-tuned PID controller

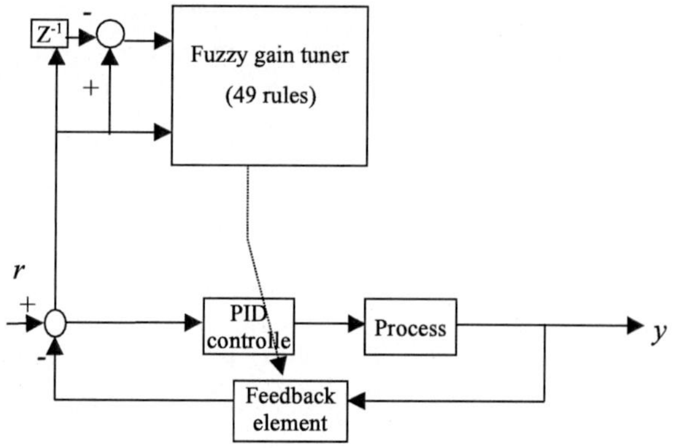

rules are defined by expert that will provide the gain updating parameter, i.e., α which further modifies the parameters of PID controller. Here, seven fuzzy sets (NB NM NS ZE PS PM PB) are considered for error (e), change of error (Δe), and gain updating parameter (α) where ZE represents zero, NS is negative small, NM is negative medium, NB is negative big, PS is positive small, PM is positive medium, and PB represents positive big. To make this scheme independent of process, the values of error (e) and change in error (Δe) are normalized according to their maximum and minimum values.

Here, the main concern is to achieve better performance in system response for fuzzy-tuned PID controller (F-PID) compared to conventional PID controller. To ascertain the requirement, the following gain updating relations are defined for individual gain parameters of PID controller.

$$k_p^{'} = k_p(1 + k_1\alpha) \tag{3}$$

$$k_{\mathrm{i}}^{'} = k_{\mathrm{i}}(1 + k_2\alpha) \tag{4}$$

$$k_d^{'} = k_d(1 + k_3\alpha) \tag{5}$$

where $k_{\mathrm{p}}^{'}$, $k_{\mathrm{i}}^{'}$, $k_d^{'}$ are the modified tuning parameters of F-PID controller. Here, the gain revising parameter α is obtained from 49 fuzzy rule bases.

Membership function: The two inputs of TRMS, error (e), and change of error (Δe) are defined by some membership functions (MF) which lie within a common interval $(-1, 1)$ as shown in Fig. 6. Similarly, the gain updating parameter, α, is also defined by some membership functions, in the interval of (0, 1) as depicted in Fig. 7.

Scaling factor: Here, the values of actual outputs, i.e., error (e) and change of error (Δe) are mapped within the interval $[-1, 1]$ with the help of scaling factor G_{e} and $G_{\Delta\mathrm{e}}$. Based on the knowledge about the system under control, some suitable values are chosen and sometimes through trial and error for achieving the desired control response.

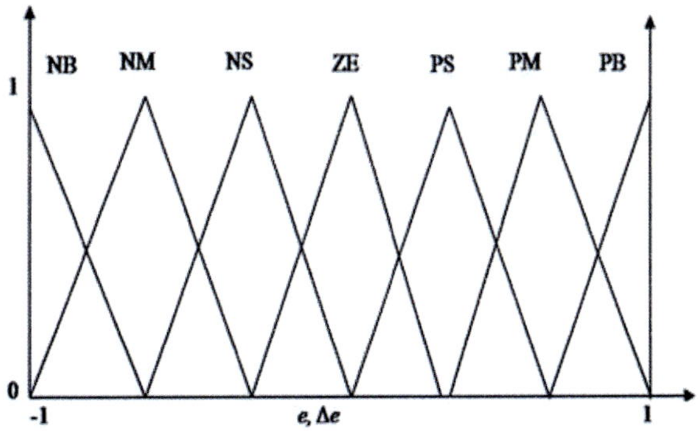

Fig. 6 Membership functions for error (e) and change of error (Δe)

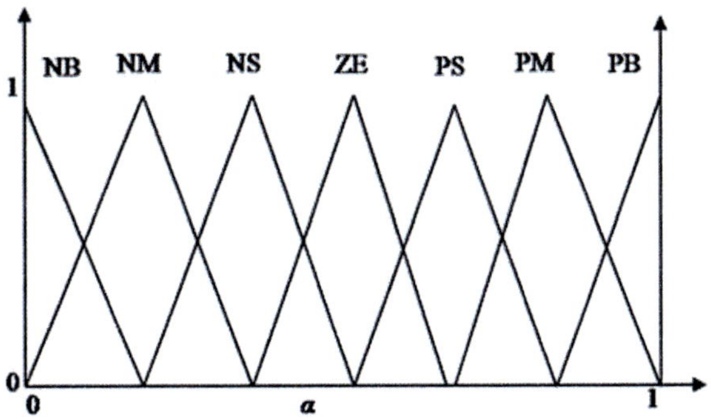

Fig. 7 Membership functions for gain updating factor (α)

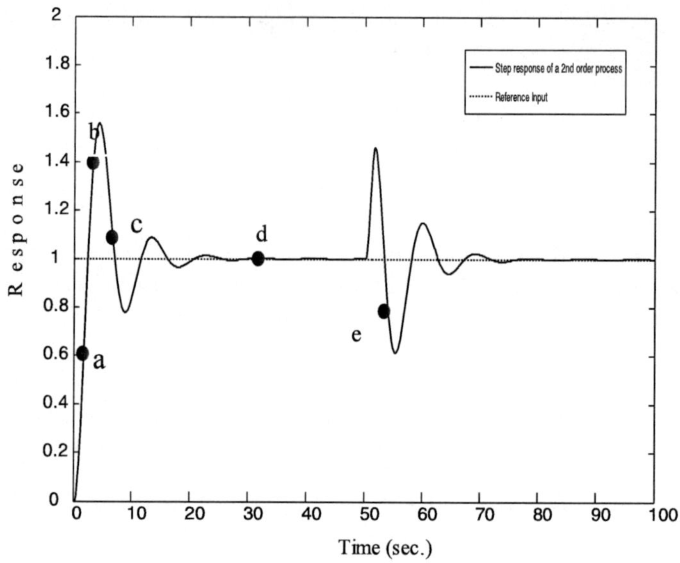

Fig. 8 Step response of a second-order system

Rule base: To construct the fuzzy rule base, 49 rules are designed based on the human knowledge. To have a clear understanding regarding the rule base designing strategy, a typical under-damped response of a second-order process is provided here as shown in Fig. 8. Some rules for formation of such rule base for some specific operating points are discussed in the subsequent section:

1. For operating point 'a', to have a lesser overshoot and smaller settling time, the gain of the controller is required to be small. For this operating phase where the error (e) is positive medium is (PM) but change of error (Δe) is negative big (NB). To obtain satisfactory response, controller gain is required to be reduced, and hence, the gain updating parameter is resulted with the property NS, i.e., α is negative small. So, the rule has the form—*If e is PM and Δe NB is then α is NS*. Such gain variation is expected to provide improved performance.
2. In case of the operating point 'b' where an overshoot is observed in the process output, i.e., error (e) is negative medium (NM) and change of error (Δe) is negative big (NB). The way to lower the process overshoot, controller gain should be large but in opposite direction, and hence, the value for gain updating factor α should be negative big (NB) as shown in the rule base Fig. 9. The corresponding gain modification rule is given by—*If e is NB and Δe is NB then α is NB.*
3. At point 'c', error (e) is negative medium (NM), and change of error (Δe) is positive medium (PM) that means the process is moving toward the set point, and under such operating condition, we can maintain the same value of control action to avoid any possible undershoot in the subsequent operating phases. As a result, the value for gain updating parameter is chosen as zero (ZE) which

Δe/e	NB	NM	NS	ZE	PS	PM	PB
NB	NB	NB	NB	NM	NS	NS	ZE
NM	NB	NM	NM	NM	NS	ZE	PS
NS	NB	NM	NS	NS	ZE	PS	PM
ZE	NB	NM	NS	ZE	PS	PM	PB
PS	NM	NS	ZE	PS	PS	PM	PB
PM	NS	ZE	PM	PM	PM	PM	PB
PB	ZE	PS	PM	PM	PB	PB	PB

Fig. 9 Fuzzy rule base with 49 rules

signifies that there will be no change in control action as given by the rule—*If e is NM and Δe is PM then α is ZE.*

4. At steady-state condition, i.e., at point 'd', when both error (e) and change of error (Δe) are almost zero, the controller gain should be small to avoid any possible chattering behavior. Hence, under such operating phase, the gain updating parameter must have zero value as defined by the rule—*If e is ZE and Δe is ZE then α is ZE* depicted in the rule base of Fig. 9.

Hence, in similar way, the entire rule base (Fig. 9) for providing the appropriate gain upgrading factor (α) for the PID controller can be framed so that an improved close-loop response can be expected from F-PID compared to conventional PID controller.

In Fig. 10, the nature of the gain surface for all possible operating phases is shown. It is observed that this surface is highly nonlinear in nature. Depending on the process operating condition, gains of the PID parameters get modified as per the nature of this gain surface. So, for TRMS process through close inspection of its behavior,

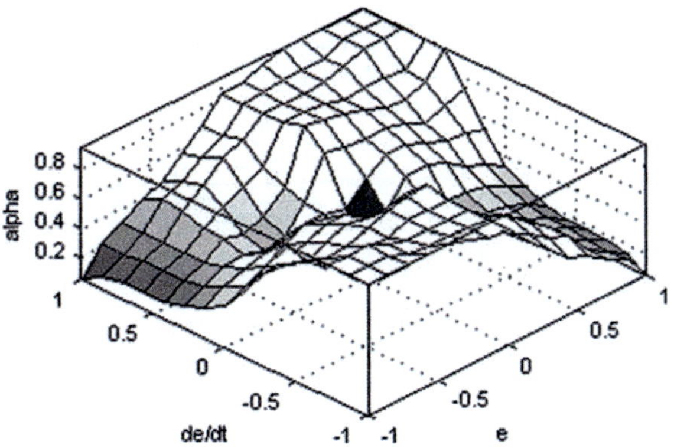

Fig. 10 Variation of alpha (α) with error (e) and change of error (Δe)

the desired amount of nonlinearity can easily be incorporated in designing the gain surface to ensure improved trajectory tracking and load rejection behavior.

6 Simulation Results

In simulation study, performance of our proposed F-PID, i.e., supervisory fuzzy expert-tuned PID controller along with conventional PID controller is evaluated on the TRMS model. Here, the performance of F-PID and PID has been studied for yaw and pitch control separately, i.e., for one degree of freedom. Simulation study is also performed to justify the efficacy of the proposed scheme on MIMO process, for both yaw and pitch simultaneously, i.e., for two degrees of freedom motion control. In case of two degrees of freedom experimentation, the performance of the controller in presence as well as absence of decoupler has also been studied. From the responses as well as performance indices, it is clearly visible that F-PID is capable to perform more satisfactorily during trajectory tracking and load rejection phases compared to convention PID tuning. Moreover, interaction between yaw and pitch motion is found to be less sensitive in case of F-PID compared to conventional PID controller.

Pitch and yaw responses along with control action due to step excitation for PID and F-PID are depicted in Figs. 11 and 12, respectively, with a dead time of 0.1 s. Here, during performance evaluation, no decoupler is used for both PID and F-PID. To study the robust feature of the controller under parametric uncertainty, the dead time of the process is taken to be 50% more from its nominal value, and the corresponding responses for yaw and pitch are shown in Figs. 13 and 14, respectively.

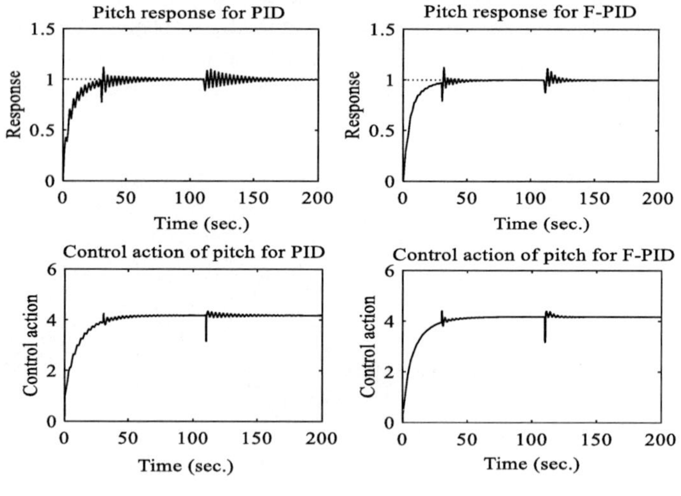

Fig. 11 Comparative study of pitch response for PID and F-PID for dead time 0.1 s without decoupler

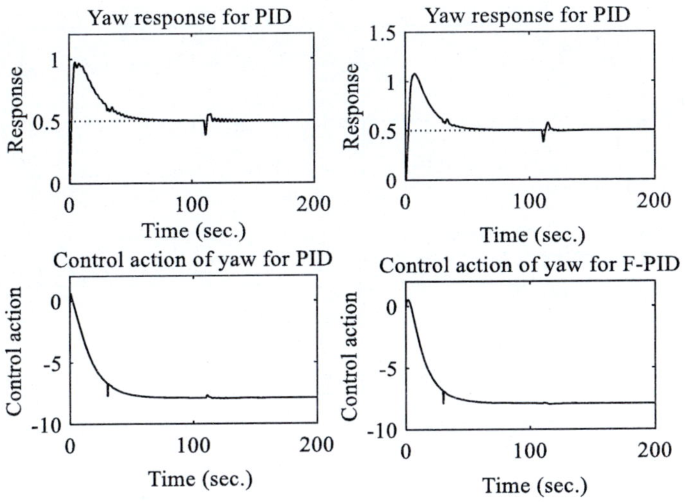

Fig. 12 Comparative study of yaw response for PID and F-PID for dead time 0.1 s without decoupler

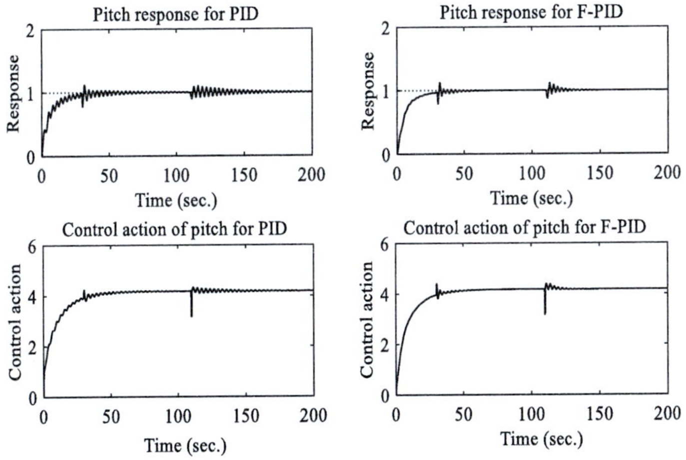

Fig. 13 Comparative study of pitch response for PID and F-PID for dead time 0.15 s (+50% perturbed) without decoupler

From Figs. 11 and 12, it is found that during load changing phases of yaw and pitch motion, PID takes longer time to recover than F-PID, and thus, there is a significance impact in the performance measures as the values of ITAE is more for PID compared to that of F-PID for both pitch and yaw control (shown in Table 1). In the presence of decoupler, yaw and pitch responses are shown in Figs. 15 and 16. With +50% perturbations in dead time responses are shown in Figs. 17 and 18. The subsequent performance indices are listed in Table 2.

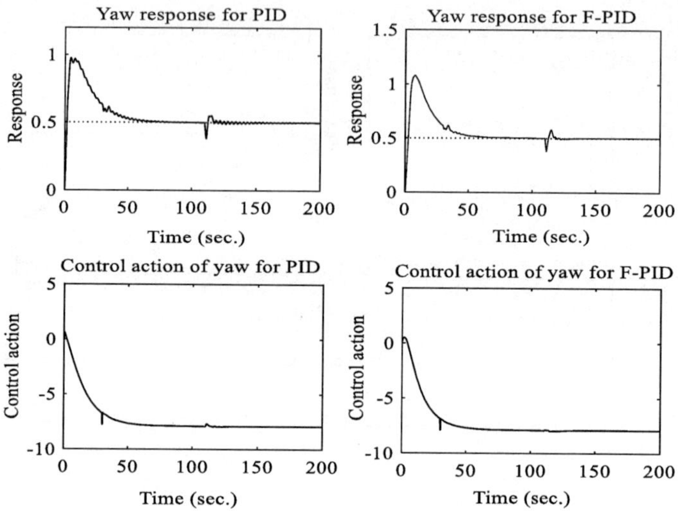

Fig. 14 Comparative study of yaw response for PID and F-PID for dead time 0.15 s (+50% perturbed) without decoupler

Table 1 Performance indices during yaw and pitch motion without decoupler

Dead time (s)	Performance index			
	System	Controller	IAE	ITAE
0.10	Pitch	PID	9.70	357.30
		F-PID	7.87	147.80
	Yaw	PID	11.88	259.50
		F-PID	11.46	240.30
0.15 (+50% perturbed)	Pitch	PID	9.99	398.10
		F-PID	7.97	157.90
	Yaw	PID	11.51	266.40
		F-PID	11.02	241.70

7 Conclusion

This paper addressed the designing of a supervisory expert fuzzy-tuned PID controller for TRMS during both one and two degrees of motion control. Performance of the reported scheme is also evaluated in the presence of perturbation in process dead time value. To evaluate the load recovery feature, disturbance signal is applied during yaw and pitch motion separately. Initially, the system identification toolbox helped to understand the dynamics of the twin rotor MIMO system (TRMS). Identified model is found to be a highly coupled nonlinear system. Both the yaw and pitch dynamics are identified separately, i.e., one degree of freedom structure. Later,

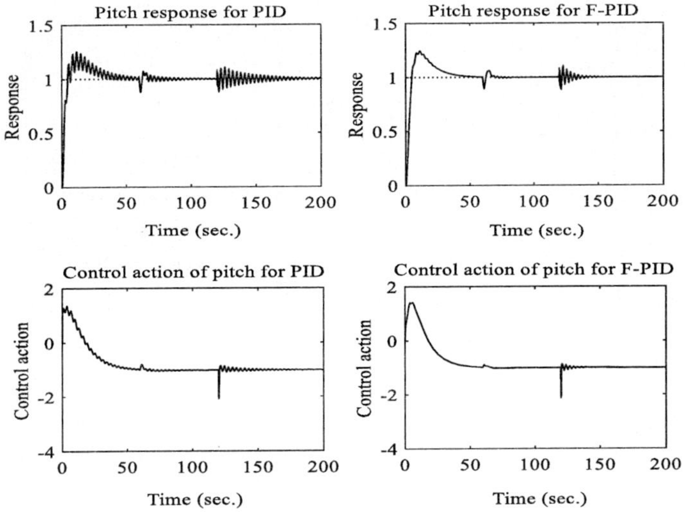

Fig. 15 Comparative study of pitch response for PID and F-PID for dead time 0.1 s with decoupler

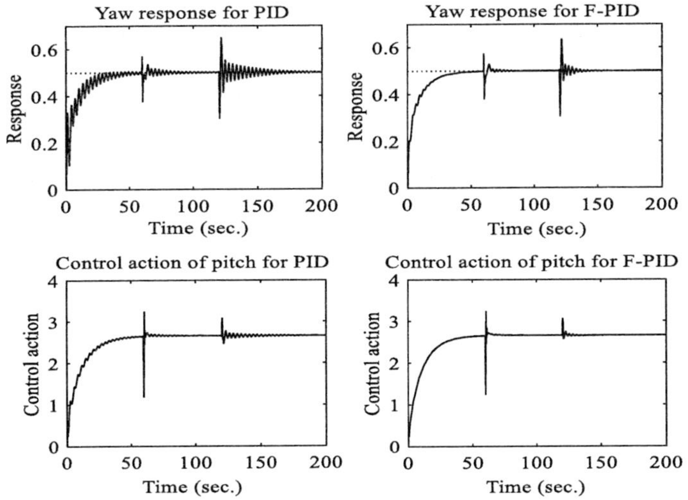

Fig. 16 Comparative study of yaw response for PID and F-PID for dead time 0.1 s with decoupler

yaw and pitch model with their interacting behavior is also identified for studying two degree of freedom structure. To reduce the interaction between yaw and pitch motion, decoupler is designed. From the simulation results, it is clearly visible that the proposed F-PID controller provides superior performance to achieve stable tracking of inputs compared to the conventional PID controller. In future scope, we have a plan to verify the performance of the reported scheme on a real-time TRMS setup.

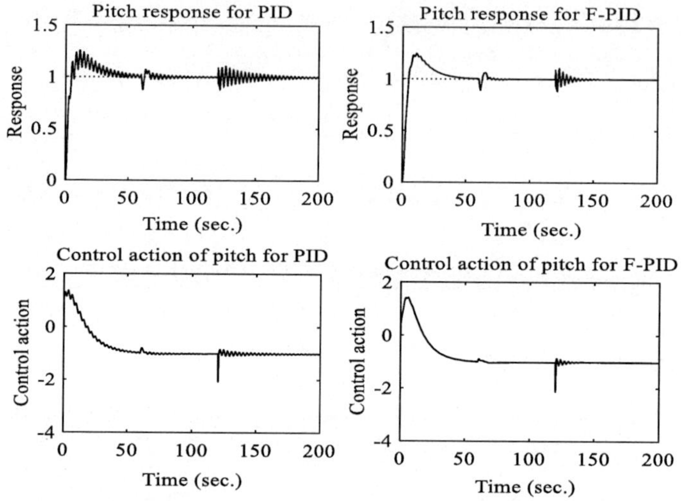

Fig. 17 Comparative study of pitch response for PID and F-PID for dead time 0.15 s (+50% perturbed) with decoupler

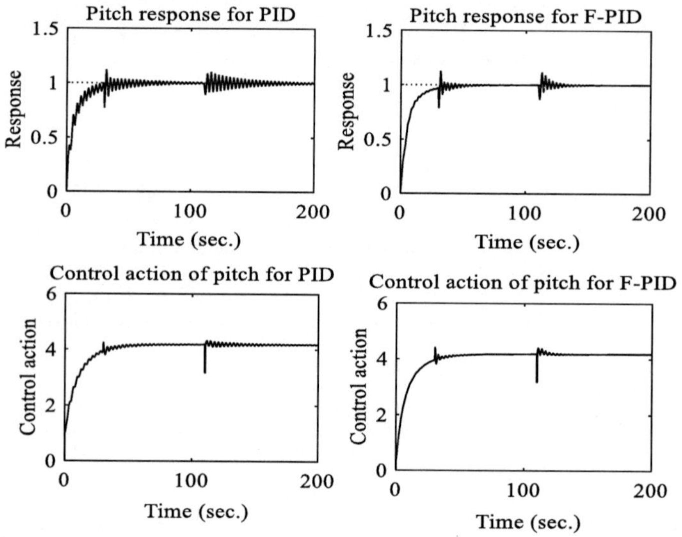

Fig. 18 Comparative study of yaw response for PID and F-PID for dead time 0.15 s (+50% perturbed) with decoupler

Table 2 Performance indices during yaw and pitch motion with decoupler

Dead time (s)	Performance index			
	System	Controller	IAE	ITAE
0.10	Pitch	PID	7.98	348.00
		F-PID	7.65	180.30
	Yaw	PID	4.74	204.80
		F-PID	3.90	94.98
0.15 (+50% perturbed)	Pitch	PID	8.40	398.50
		F-PID	7.75	187.30
	Yaw	PID	5.02	238.10
		F-PID	3.93	99.18

References

1. Chalupa P, Prilkryl J, Novak J (2014) Modelling of twin rotor MIMO system. In: 25th DAAAM intelligent manufacturing and automation, DAAAM 2014
2. Patel AA, Pithadiya PM, Kannad HV (2015) Control of twin rotor MIMO system (TRMS) using PID controller. Int J Adv Eng Res Dev (IJAERD)
3. Islam B, Ahmed N, Bhatti DL, Kha S (2004) Controller design using fuzzy logic for a twin rotor MIMO system. In: IEEE Transactions conference paper
4. Ghellab MZ, Zeghlache S, Bouguerra A (2015) Real time implementation of fuzzy gain-scheduled PID controller for twin rotor MIMO system
5. Hashim HA, Abido MA (2015) Fuzzy controller design using evolutionary techniques for twin rotor MIMO system: a comparative study, vol 2015. Hindawi Publishing Corporation
6. Juang J-G, Tu K-T (2013) Design and realization of a hybrid intelligent controller for a twin rotor MIMO system. J Mar Sci Technol 21:333–341

Utilizing Cooperative-MIMO for an Event-Based Data Transmission in a Wireless Sensor Network

Monica Kumari, Sanjeet Kumar, and Sarah Asheer

Abstract An efficient use of energy is a challenging task in battery-constrained WSN. Cooperative-MIMO (CMIMO) where the nodes cooperatively transmit data packets to the base station (BS) is an efficient way that leads to huge energy savings. But still there exists a problem of selection of receiver node by the multiple transmitters. This can be overcome by using an effective selection strategy which ensures end-to-end connectivity as well as lifetime enhancement. In this paper, an Event-driven data transmission Routing Protocol (EDTRP) is proposed for routing and central node (CN) selection. In EDTRP, only the nodes which detect an event are involved in the communication process. If the data sensed by the individual node is above a threshold value of the particular physical attribute in concern, then only the transmitter of that node is switched on. This avoids unnecessary data transmission leading to energy savings. The proposed work shows a significant improvement when compared to the non-cooperative-MIMO scheme such as LEACH and Improved Vice-LEACH (V-LEACH). EDTRP shows an improvement of almost 211% compared to LEACH and 171.30% compared to Improved V-LEACH in terms of half-lifetime.

Keywords Wireless sensor network · Multiple-input multiple-output · Clustering · Aggregation · Event driven · Routing

M. Kumari · S. Kumar · S. Asheer (✉)
Department of Electronics and Communication Engineering, Birla Institute of Technology, Mesra, Ranchi, India
e-mail: sarah.bitrnc@gmail.com

M. Kumari
e-mail: kumari.monica21@gmail.com

S. Kumar
e-mail: sanjeet@bitmesra.ac.in

V. Nath and J. K. Mandal (eds.), *Proceedings of the Fourth International Conference on Microelectronics, Computing and Communication Systems*, Lecture Notes in Electrical Engineering 673, https://doi.org/10.1007/978-981-15-5546-6_56

1 Introduction

A wireless sensor network (WSN) has varied application in industrial as well as commercial domain. A WSN is defined as a network of small battery powered devices called as nodes that can communicate the information gathered from a monitored field to each other as well as distant sink through wireless links. It provides a user the capability to implement, perceive, and respond to the events and circumstances in a defined environment. A sensor node is the core module of WSN. It is a device which senses the environment and converts the physical attributes into a form which is suitable for analysis and interpretation of the environment [1]. As the nodes of a WSN have strict energy constraints, design of an energy-efficient WSN has always been a challenge to the researchers. The nodes perform three basic operations: sensing, data processing, and communication. Of the three basic operations, the communication process is responsible for maximum energy consumption, thus in order to make optimal use of energy, efficient communication techniques should be used. Many researches were proposed to achieve high stability in network along with preserving the energy for communication in network.

The Cooperative-MIMO (CMIMO) is one of the approaches that can be used to improve the energy efficiency of a WSN without compromising other performance metrics. In a CMIMO system, several nodes each equipped with an antenna cooperate among themselves to form a multi-antenna node [2]. In Fig. 1, the basic steps involved in CMIMO have been explained. In first step, all the nodes at the transmitting end communicate cooperatively by synchronizing and exchanging their information among themselves. The second step represents the transmission of information using a MIMO technique. The nodes at the transmitting end transmit distinct symbols at the same time interval with the help of a Vertical-Bell Laboratories Layered Space-Time (V-BLAST) transmission. Basically, in a general transmission, a single symbol

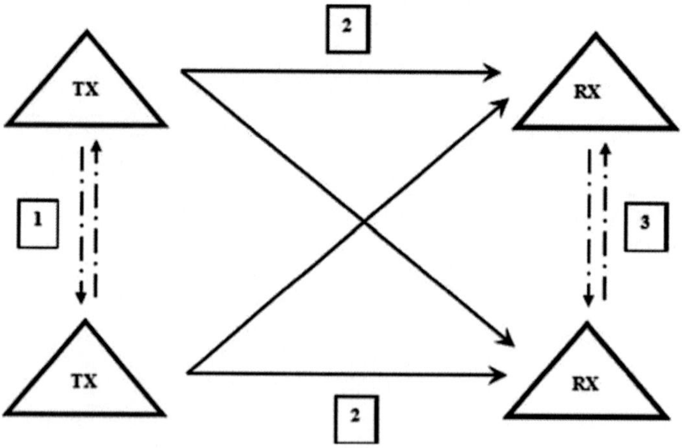

Fig. 1 Communication steps in CMIMO

would be transmitted over the channel at each interval of time. But, in the case of V-BLAST, each transmitting antenna transmits its own independent coded substream of data and the symbols or data are grouped into frames. Finally, in the third step, the nodes at the receiving end communicate cooperatively by exchanging the received information or messages. This is done in order to decode the received symbol.

The rest of this paper is organized as follows. In Sect. 2, related work relevant to our work has been discussed. In Sect. 3, the proposed model is discussed which aims in improving the lifetime of the network by introducing data transmission based on event detection. Section 4 discussed the simulation results of this model. Finally, Sect. 5 concludes the paper along with an insight on directions and scope for future work.

2 Related Work

Till date, an immense research has been done which aims at enhancing the performance of WSN by implementing the concept of CMIMO. In this section, we have discussed the models and algorithms which have been extended and applied to cooperative-MIMO systems.

In [2], the authors have proposed an Event Clustering Routing Protocol (ECRPC) for WSN, which modifies the clustering process based on the energy. The authors have compared the energy requirement of the nodes in ECRPC with that of Adaptive Routing Protocol with Energy-Efficiency and Event Clustering for WSN (ARPEES) and it is found that ECRPC outperforms ARPEES.

Clustering is a technique that facilitates healthy cooperation among the nodes of a WSN. Various clustering techniques such as Low Energy Adaptive Clustering Hierarchy (LEACH) [3], Hybrid Energy-Efficient Distributed Clustering (HEED) [4], and Threshold sensitive Energy-Efficient sensor Network (TEEN) [5] have been proposed. Several variants of the LEACH protocol have also been proposed which shows enhanced performance. The authors in [1] have proposed an improved version of LEACH called as ….V-LEACH protocol, in which instead of choosing the CH periodically, it selects vice CH during the selection of CH and saving a large amount of energy. Dhankhar [3] has presented a detailed analysis of LEACH, LEACH C and HEED protocol. Comparison between them is done in terms of cost of sensor nodes, in order to determine the best energy saving routing protocol and has been concluded that the performance of HEED is better compared to the other two. The authors in [4] propose a model for cooperative multiple-input single-output (CMISO) scheme and analyze the cluster lifetime by considering the transmission distances of each cluster. Jain and Pachouri [5] have proposed an algorithm in which if a base station (BS) fails, then a new BS is established with least computation cost by employing a LEACH and Threshold sensitive Energy-Efficient sensor Network (TEEN) algorithm. Chen and Zhao [6] have proposed a greedy approach in order to maximize the network lifetime by taking residual energy under consideration. They have developed a model which utilizes both channel state information and residual energy information.

The LEACH protocol operates into stages and each stage comprises of two phase: setup phase and steady-state data transmission phase. In the setup phase, clusters are organized in which the CH selection is done on a random basis. In steady-state data transferral phase, each CH gathers the packets from its cluster members, aggregates them, and then transmits the aggregated data to the BS. In Improved V-LEACH protocol, there is an alternate CH which will work after the main CH has lost all its energy and is dead. It considers minimum distance and maximum energy for the selection of CH as well as the alternate CH. But, this protocol does not consider the concept of performing aggregation on the redundant data generated from the closely spaced sensor nodes.

To overcome the shortcomings of the non-cooperative-MIMO scheme such as LEACH, Improved V-LEACH, there exists a need to propose a system which effectively selects the CHs and improves the network lifetime by introducing the concept of event detection with hard and soft thresholds during communication in network. In our proposed protocol called the Event-driven data transmission Routing Protocol (EDTRP), the CH is the central node (CN) whose selection is based on a parametric approach which considers the parameters:

- Probability
- Residual energy.

The packet transmission to the sink is based on event detection. The node's transmitter is switched on only if the value of the sensed data is greater than the hard threshold. The concept of hard and soft thresholds is discussed in detail in Sect. 3. This restricts the transmission of unnecessary data which leads to energy savings.

3 Network Operation and the Energy Model of the Proposed Protocol

3.1 Network Operation

The proposed model is based on event-driven data transmission scheme. As shown in Fig. 2, 100 nodes are distributed randomly over an area of 100 m $\times$ 100 m. The base station (BS) is located at the center of the network area.

The nodes are categorized based on their functionality. The CNs perform aggregation and also participates in the intercluster communication in addition to the sensing, and the remaining nodes which perform the basic sensing operation are known as the normal nodes (NN). The nodes are organized into clusters with each cluster having a central node (CN). The CN selection is done on the basis of probability and residual energy level. The cluster members are chosen based on the minimum

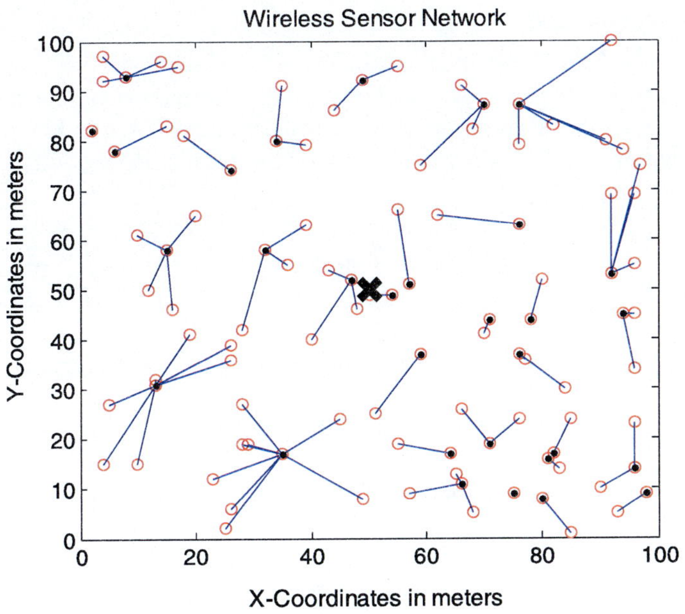

Fig. 2 Wireless sensor network with 100 nodes and BS

distance between each normal node to CN. The number of bits sensed by each node is denoted by "b_{data}". All the nodes as well as the BS have a single antenna and both the node as well as the BS are assumed to be static. The sensing parameter used in this model is temperature.

The operation of the proposed scheme takes place into rounds. Each round consists of CN selection, data gathering, data aggregation, data broadcasting, and data transmission phase. The communication between the nodes as well as the BS is divided into intracluster, intercluster, and long haul communication phase. During intracluster communication phase, after sensing an event, the data collected by the normal nodes is shared among the cluster members and the nominated CNs perform aggregation. During the intracluster communication phase, the sharing of data takes place only with its cluster members. Now, if the distance of the CN to BS is within the threshold distance, then only the data is communicated directly to the BS. But, if CNs are located at a far distance from BS, then communication takes place cooperatively via several CNs to the BS. This takes place during the intercluster communication.

3.2 Energy Model

We have considered the energy model used in [2]. The energy consumption during communication in wireless sensor network includes three parts: transmission, reception, and aggregation. The energy consumption for transmitting b_{data} bits is given by Eq. (1):

$$E_{\text{Trx}}(b_{\text{data}}, d) = \begin{cases} b_{\text{data}} \times E_{\text{elec}} + b_{\text{data}} \times E_{fs} \times d^2, \; d < d_0 \\ b_{\text{data}} \times E_{\text{elec}} + b_{\text{data}} \times E_{mp} \times d^4, \; d \geq d_0 \end{cases} \tag{1}$$

The energy consumption during reception of b_{data} bits by CN is given by Eq. (2):

$$E_{\text{Rcx}}(b_{\text{data}}) = b_{\text{data}} \times E_{\text{elec}} \tag{2}$$

The amount of energy required by CN in order to aggregate the data in each cluster is:

$$E_{\text{DataAgg}}(b_{\text{data}}) = b_{\text{data}} \times E_{\text{DAgg}} \tag{3}$$

Here, E_{elec} is the amount of energy demanded by sensor node to transmit or receive per bit data, d_0 is the threshold distance, E_{DAgg} indicates the energy required for per bit aggregation, E_{fs} and E_{mp} depend on the transmitter amplifier model used and d is the distance between the transmitter and the receiver and they are related by the following equation:

$$d_0 = (E_{\text{fs}}/E_{\text{mp}})^{1/2} \tag{4}$$

The probability equation considered for CN selection in the proposed model is same as LEACH [3], which is given by (5):

$$T(n) = \begin{cases} p/((1-p) \times (r \bmod (1/p))), & \text{if } n \in G \\ 0, & \text{otherwise} \end{cases} \tag{5}$$

In each round, all the nodes generate a random number between 0 and 1. All those nodes having the maximum remaining energy level and generated number less than the threshold values $T(n)$ are elected as CN. After CN selection, the cluster formation will be done on the basis of minimum distance between CN and normal nodes. Each of the nodes is going to sense the event data continuously and store those data on the memory of the sensor. The nodes having the sensing value greater than the hard and soft threshold are only allowed to cooperatively transmit the data to CN and the number of such nodes is denoted by n_{Th}.

Table 1 Simulation parameters

Symbol	Parameters	Values
$m*m$	Network field	100*100
b_{data}	Data packet size	500 bytes
d_0	Threshold distance	87.7058 m
p	Probability of a node to become CN	0.2
n	Total number of nodes	100
	BS location	(50, 50)
$E_{initial}$	Initial energy of sensor node	0.5 J
E_{elec}	Energy dissipated per bit to run the circuit	50 nJ/bit
E_{fs}	Propagation through free space	10 pJ/bit/m
E_{mp}	Propagation through multipath	0.0013 pJ/bit/m^4
rmax	Maximum number of rounds	7000

Hard Threshold—It is a threshold value for the sensed attribute, beyond which NN only transmits data information to CN by turning on its transmitter, only if the sensed value is in the range of interest.

Soft Threshold—This is a small variation in the value of the sensed attribute, which activates the node to turn on its transmitter.

The other non-triggered nodes will remain in sleep mode. The CN aggregates the data received by the triggered normal nodes of their cluster. Each of the CN broadcasts their residual energy, location information, and its identity among the nodes. The CN whose location is closer to BS, further forwards the aggregated data to BS. But, if the CN are located at a far distance, i.e., beyond the threshold distance defined in Table 1, then communication to BS takes place via several CN. The shortest path selection is done only after the formation of the CN and the same path will be used for further event data communication. In this model, there is no need to wait for the next transmission period.

4 Results and Discussion

The results obtained are based on the energy model discussed in the previous section. MATLAB (version 15.0) has been used as the simulation tool which provides an efficient way to compare the performance of the network with LEACH and Improved V-LEACH. The result obtained here depends upon certain parameters listed in Table 1.

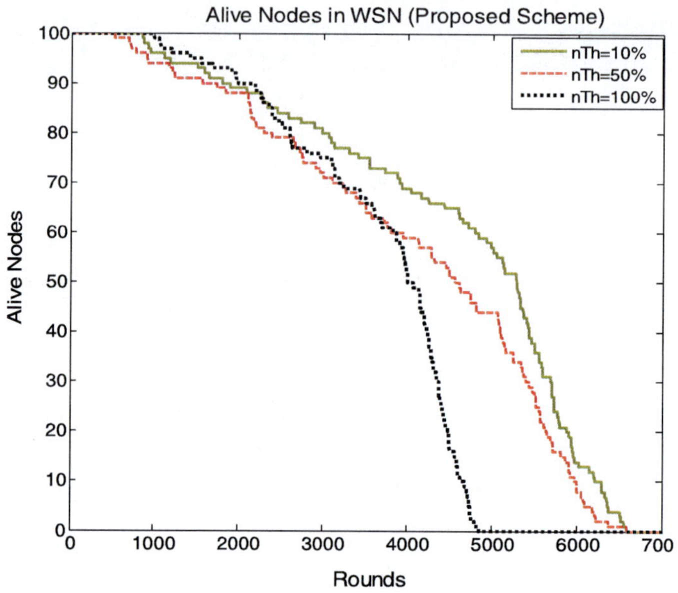

Fig. 3 Comparison of alive nodes per rounds for different percentage of n_{Th}

Based on the discussions of the previous section, the following results are obtained. Lifetime of the proposed protocol is evaluated and is compared to that of LEACH and Improved V-LEACH. Here, lifetime is defined in terms of number of rounds till which half of the nodes becomes dead viz. also called as the half-lifetime of the network. The total number of rounds till all the nodes become dead has also been shown and compared graphically.

Figure 3 provides comparison of alive nodes per rounds for different percentages of n_{Th}. Figure 4 indicates comparison of dead nodes per rounds for different percentages of $n_{\mathrm{Th}.}$ The EDTRP model with n_{Th} is 10% has the highest network lifetime as compared to other two cases. In terms of half-lifetime, when n_{Th} is 50% show significant improvement of 12.7% over the case where n_{Th} is 100%. But, n_{Th} is 10% shows an improvement of 29.93 and 15.30% when compared to the half-lifetime of n_{Th} is 100% and n_{Th} is 50%. The case with n_{Th} is 10% saves a large amount of energy by eliminating the transmission from unwanted nodes.

Figures 5 and 6 provides the comparison of the proposed model (EDTRP) with n_{Th} is 10% to the existing non-cooperative-MIMO scheme (LEACH and Improved V-LEACH). The above result indicates that the EDTRP model provides better lifetime as compared to LEACH and Improved V-LEACH. The EDTRP model with n_{Th} is 10% has the highest network lifetime as compared to all other cases.

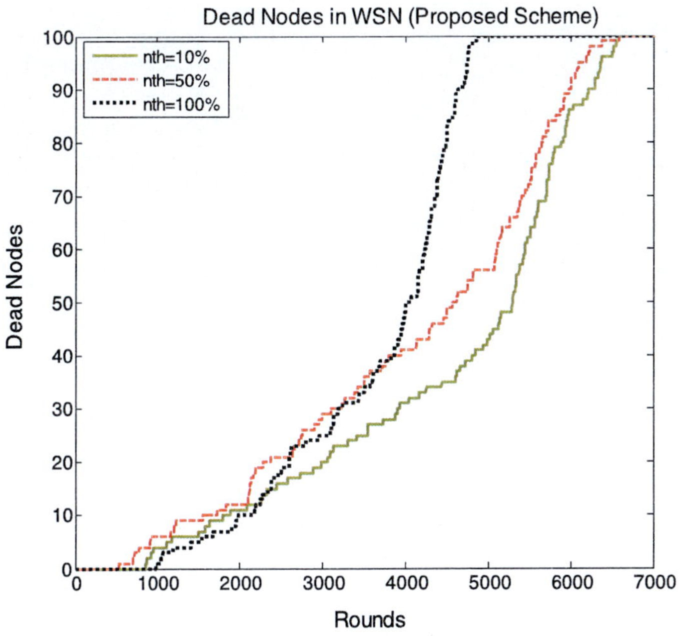

Fig. 4 Comparison of dead nodes per rounds for different percentages of n_{Th}

In Table 2, we have compared the half-lifetime of EDTRP with n_{Th} is 10% with that of LEACH and Improved V-LEACH. And, it shows an improvement of 211 and 171.30% when compared to the half-lifetime of LEACH and Improved V-LEACH respectively.

5 Conclusion and Future Work

In this paper, the lifetime of the proposed scheme has been compared to LEACH and Improved V-LEACH. In EDTRP, only the sensor nodes which have significant information after the preliminary phase of sensing are involved in the communication process. Thus, the proposed model supports an event-based packet selection and transmission. From the above results, certain conclusion has been drawn. First, he proposed a model which provides an improvement in the network lifetime as compared to LEACH and Improved V-LEACH by incorporating event-based node selection as well as residual energy and minimum distance criteria for selection of the CN. Second, the EDTRP model with n_{Th} equal to 10% has the maximum lifetime compared to the other two cases with n_{Th} 50% and 100%.

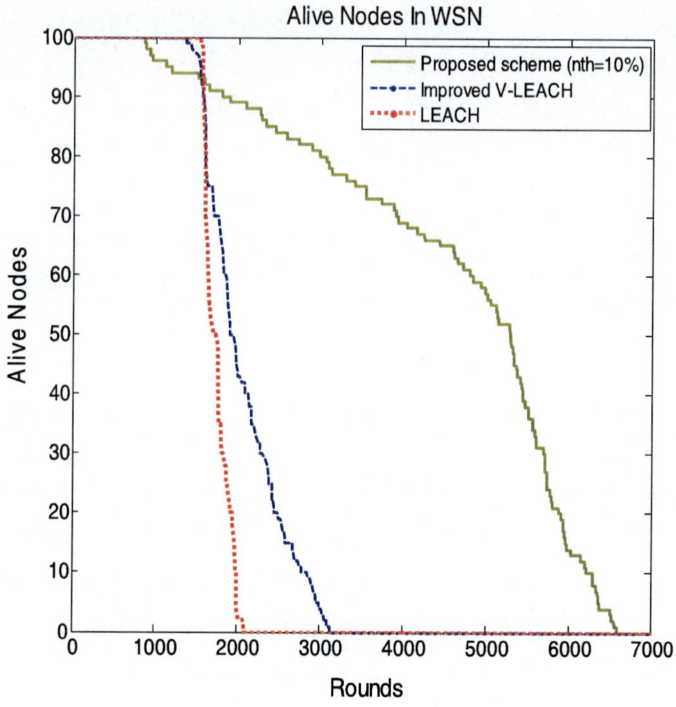

Fig. 5 Comparison of lifetime in terms of alive nodes

The above work can be extended to large network area. In a large network area, the hard threshold value has to be set for each cluster separately since the event detected in each part of a network has a significant difference. For smaller network area as in EDTRP, only a single threshold value is considered which is same for the entire network.

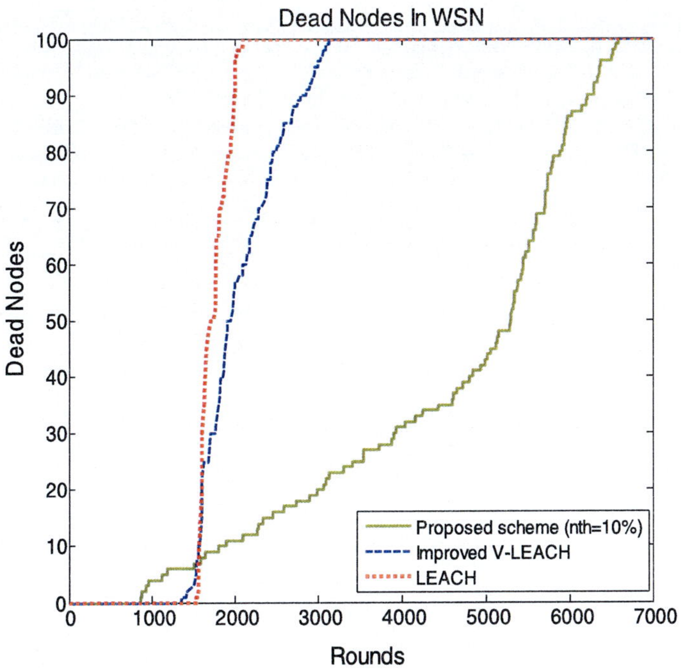

Fig. 6 Comparison of lifetime in terms of dead nodes

Table 2 Improvement of proposed scheme over LEACH and improved V-LEACH

Protocol	Half-lifetime (rounds)	All dead nodes (rounds)
LEACH	1701	2098
Improved V-LEACH	1950	3131
$n_{\mathrm{Th}} = 100\%$	4072	4889
$n_{\mathrm{Th}} = 50\%$	4589	6590
$n_{\mathrm{Th}} = 10\%$	5291	6640

References

1. Ahlawat A, Malik V (2013) An extended vice-cluster selection approach to improve v leach protocol in WSN. In: Third international conference on advanced computing and communication technologies (ACCT). IEEE, pp 236–240
2. Chelbi S, Duvallet C, Abdouli M, Bouaziz R (2016) Event-driven wireless sensor networks based on consensus. In: IEEE/ACS 13th international conference of computer systems and applications (AICCSA), pp 1–6
3. Dhankhar S, Singh ES (2016) Performance comparison of LEACH & HEED clustering protocols in WSN using MATLAB—a review. Int J Techn Res 5(1):167–170

4. Huang Z, Okada H, Kobayashi K, Katayama M (2012) A study on cluster lifetime in multi-hop wireless sensor networks with cooperative MISO scheme. J Commun Netw 14(4):443–450
5. Jain P, Pachouri R (2017) Wireless sensor network node to node mapped on base station with energy consumption. Int J Comput Sci Trends Technol 5(4)
6. Chen Y, Zhao Q (2005) On the lifetime of wireless sensor networks. IEEE Commun Lett 9(11):976–978

Rapid Modeling of a Fast-Steering Mirror Assembly from Time Response Data

Mahua Pal, Dhiman Biswas, Anusree Das, Kumardeb Banerjee, and Bivas Dam

Abstract A systematic approach for a rapid modeling of a fast-steering mirror (FSM) assembly from its time response data is presented here. The FSM assembly comes with integrated piezo-actuators and pick-off sensors for a two-axis tip-tilt motion of the mirror. The step response data is acquired in a digital storage oscilloscope and analyzed to generate a mathematical model of the FSM assembly. The modeling method is compared with results obtained by existing methods reported in literature for its correctness. The frequency response of the model is then generated in MATLAB and compared with the actual frequency response of the FSM assembly to verify the correctness of the model.

Keywords Fast-steering mirror · Frequency response · Time response analysis · System modeling · Validation

1 Introduction

Two-axis FSM assemblies driven by piezo-actuators are used widely in laser beam steering applications [1–4]. Design of an FSM controller for precise pointing of a

M. Pal (✉)
Techno Main Salt Lake, Kolkata, India
e-mail: mahuapal@yahoo.com

D. Biswas · A. Das · K. Banerjee · B. Dam
Jadavpur University, Kolkata, India
e-mail: dhiman.biswas91@gmail.com

A. Das
e-mail: anusreedas2013@rediffmail.com

K. Banerjee
e-mail: kumardebb@rediffmail.com

B. Dam
e-mail: b_dam@rediffmail.com

V. Nath and J. K. Mandal (eds.), *Proceedings of the Fourth International Conference on Microelectronics, Computing and Communication Systems*, Lecture Notes in Electrical Engineering 673, https://doi.org/10.1007/978-981-15-5546-6_57

laser beam in presence of atmospheric jitter requires an accurate mathematical model of the FSM assembly. This paper proposes a rapid modeling of an FSM assembly by driving the piezo-actuators by a low frequency square wave and acquiring angular displacement of the mirror assembly from the integrated position pick-off sensors. The input drive comes from an Arbitrary Waveform Generator from *M/S Agilent Technology*; the response is acquired in a DSO from *M/S GWINSTEK Inc.*

Experimentally determined mathematical models of one-axis and two-axis FSM assemblies are as documented in [5] and [6], respectively. Here, the fast-steering mirrors are the heart of a Laser Jitter Control (LJC) test bed. Ignoring the negligible cross-coupling between the two axes, the functional block diagram of the FSM assembly is illustrated in Fig. 1.

The transfer function of the FSM and PSD is reported as:

$$\mathrm{GFSM}(s) = \left(\frac{\omega_n^2}{s^2 + 2\xi\omega_n s + \omega_n^2}\right)$$

$$\mathrm{GPSD}(s) = \frac{1}{s\tau_d + 1}$$

The position sensing devices (PSDs) are modeled as first-order systems to account for the time lag τ_d in the command circuitry to the mirror and the detection circuitry. As stated in [5], value for τ_d was determined by comparing the step response of the experimental system to a simulated step response using the mathematical model, and adjusting τ_d until the simulated and actual step responses matched.

The FSM model is thus characterized by a pair of complex conjugate poles resulting in a damped oscillatory response. As reported in [5, 6], the transfer function coefficients are determined from the frequency response of the combined FSM-PSD modules.

The work reported in [7] uses the experimentally obtained frequency response of the FSM module to obtain a MIMO linear time invariant model of the same using subspace-based multivariable system identification algorithm [8]. However, the method is computationally intensive and not suitable for any fast modeling process.

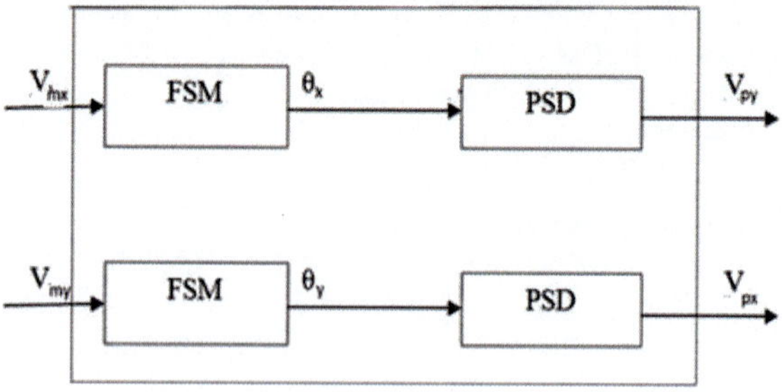

Fig. 1 Block diagram of two-axis FSM

Another approach [9] proposes a linear least square estimator that uses the frequency response data to estimate the system parameters. Once again the computational complexity involved does not make it suitable for any fast modeling process.

The FSM module [10] with built-in piezo-actuators and integrated position pick-off sensors reported in this work comes with a piezo-actuator driver from *M/S piezosystem jena* [11]. Analysis of the time response data shows that the FSM response matches with those documented in [5] and [6]. However, it encounters a dead time that reduces the stability and limits the speed of response of the system. Instead of adopting the computationally intensive system identification method described in [8], a relatively simple method for parameter estimation of the FSM assembly from time response data is presented in this paper. Existing modeling methods for second-order systems with dead time from their time reaction curves [12–14] are later used to validate the model obtained by using the method proposed in this work. The proposed algorithm can be embedded in the FSM controller unit for a fast in-system modeling of the FSM assembly.

2 Experimental Setup

The experimental test bench comprising of an FSM assembly, piezo-driver (d-Drive), and a laser source are mounted on an optical bread board shown in Fig. 2.

The test-bench components are:

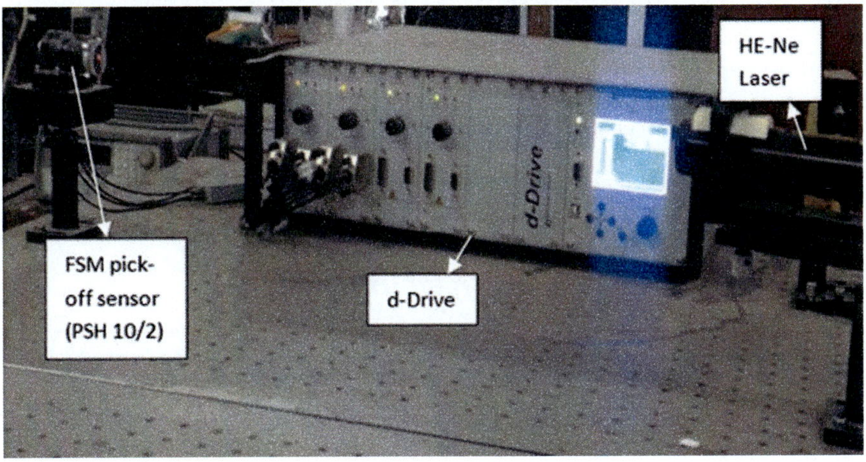

Fig. 2 Experimental Setup

2.1 Fast-Steering Mirror Assembly

A two-axis FSM of PSX x/2 series [10] is used in the experimental setup. It is characterized by a maximum angular displacement of ±4 mrad and a resonant frequency of 3500 Hz.

2.2 Digital Drive (D-Drive) Module

The FSM assembly is driven by a piezo-actuator digital drive module (*d-Drive*) from *M/S piezosystemjena* [11] with in-built function generator, noise injector, digital PID controller, a display module and a serial interface for PC connectivity. The d-Drive has the MOD input and the MON output with special features mentioned below:

2.3 Modulation Input: MOD

The motion of the actuator can be remotely controlled using this input. The control signal applied through the MOD input must be in the range of 0 to +10 V. In this test procedure, signal from a function generator is used to feed the MOD input.

2.4 Monitor Output: MON

With a specific command, different signals can be taken out from the MON output. The voltage range of the signal taken from MON is scaled to 0 to +10 and can be monitored in an oscilloscope. In this test procedure, the signal from the integrated pick-off sensor of the FSM is monitored.

2.5 Arbitrary Waveform Generator (AWG)

An AWG, Agilent 33220A from *M/S Agilent Technology* is used to generate the input drive for the MOD input of the d-Drive unit.

2.6 *Digital Storage Oscilloscope (DSO)*

A 70 MHz, two Gs/s, two-channel DSO, GDS-2072A from *M/S GWINSTEK Inc* is used to acquire the response data from MON output of the d-Drive unit for storage and analysis.

The FSM assembly is first driven by an unipolar square wave of amplitude 2 V and frequencies 5 Hz, 10 Hz, and 15 Hz, respectively, to generate and acquire the step response data. Later, it is driven by a sinusoid whose frequency is varied from 1 Hz to 1000 Hz, whence the frequency response is determined.

3 Modeling Methodology

To generate the time response data of the FSM assembly, we first program the d-Drive to work in open-loop mode, thereby bypassing the in-built digital PID block. The block diagram of the working setup is illustrated in Fig. 3.

From the acquired time response data in the DSO, we obtained the values of dead time, peak voltage, and peak time using the on-screen cursors and from these values, we calculate the system parameters (ξ and ω_n) of an equivalent under-damped second-order system. The model so obtained is simulated in MATLAB and its step response is compared with the actual response to check the correctness of the model. The model data are then used to estimate the system parameters based on two-points and three-points of the step response as reported in [13] and [14] to corroborate the results obtained by our approach.

The actual open-loop frequency response is obtained by driving the FSM assembly by a sinusoid whose frequency is varied from 1 to 1000 Hz, and acquiring the angular tilt of the mirror position from the MON output of the d-Drive, and then comparing it with drive input on the DSO for amplitude and phase measurement. The acquired frequency response is again compared with the MATLAB generated frequency response of the second-order mathematical model of the FSM for model validation.

Fig. 3 Functional block diagram of the d-Drive setup

4 System Identification

The time response performance indices (PIs) for a square wave drive at three different frequencies are listed in Table 1, for an x-axis drive. The y-axis drive yields an almost identical result.

The time response plots stored for measuring of the PIs for a 15 Hz input drive are illustrated in Figs. 4 and 5.

From the time response data, the model of the FSM assembly turns to be second-order plus dead time system with a transfer function of:

Table 1 Time response data

Freq (Hz)	Peak response V_{pp} (Volt)	Steady-state voltage V_{ss}(Volt)	Dead time t_d (ms)	Peak time t_p (ms)	% Peak overshoot % M_p
5	2.32	2	0.53	2.45	16
10	2.32	2	0.53	2.45	16
15	2.32	2	0.53	2.45	16

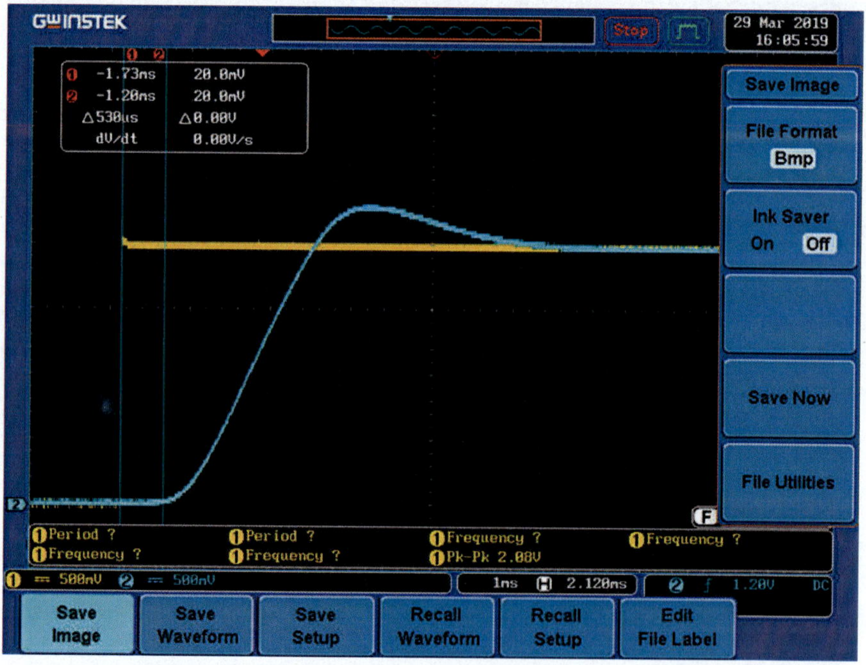

Fig. 4 Open-loop FSM output with an input of 0–2 V square wave of freq. 15 Hz, in finding t_d

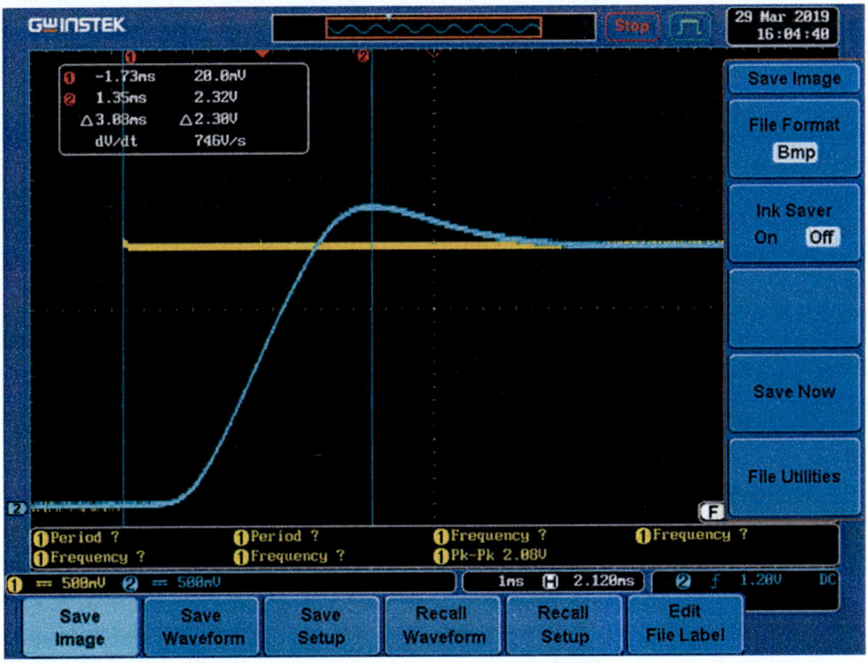

Fig. 5 Open-loop FSM output with an input of 0–2 V square wave of freq. 15 Hz, in finding t_p

$$G_{\text{cl}}(s) = \text{K}e^{-\tau s}\left(\frac{\omega_n^2}{s^2 + 2\xi\omega_n s + \omega_n^2}\right) \tag{1}$$

From the % peak overshoot data:

$$M_p = 100\frac{V_{\text{pp}} - V_{\text{ss}}}{V_{\text{ss}}} = 0.16 = e^{\frac{-\pi\xi}{\sqrt{1-\xi^2}}} \tag{2}$$

whence, the damping ratio is

$$\xi = 0.5 \tag{3}$$

From Fig. 4, the dead time,

$$t_d = 0.53 \text{ ms} \tag{4}$$

As observed from Fig. 5,

$$t_p + t_d = 3.08 \text{ ms} \tag{5}$$

Hence, the peak time is

$$t_p = 2.45 \text{ ms} \tag{6}$$

As $t_p = \frac{\pi}{\omega_n (1-\xi^2)^{1/2}}$
we get

$$\omega_n = 1451 \text{ rad/sec} \tag{7}$$

The DC gain

$$K = \frac{V_{\text{ss}}}{V_{\text{in}}} = 1 \tag{8}$$

Hence, the overall system transfer function takes the form:

$$G_{\text{FSM}}(s) = \left(\mathrm{e}^{-0.00053\text{s}}\right)\left(\frac{2.1052 \times 10^6}{s^2 + 1451\text{s} + 2.105 \times 10^6}\right)$$

$$G_{\text{FSM}}(s) = \left(\mathrm{e}^{-0.00053\text{s}}\right)\left(\frac{1451^2}{s^2 + 1451\text{s} + 1451^2}\right) \tag{9}$$

Step response of the model as given in (9) is MATLAB platform gives a plot as illustrated in Fig. 6. The parameters obtained from the plot are: dead time of 0.5 ms, peak time of 2.5 m, and %peak overshoot of 16%. The model time response is a very close match to the actual time response shown in Figs. 4 and 5.

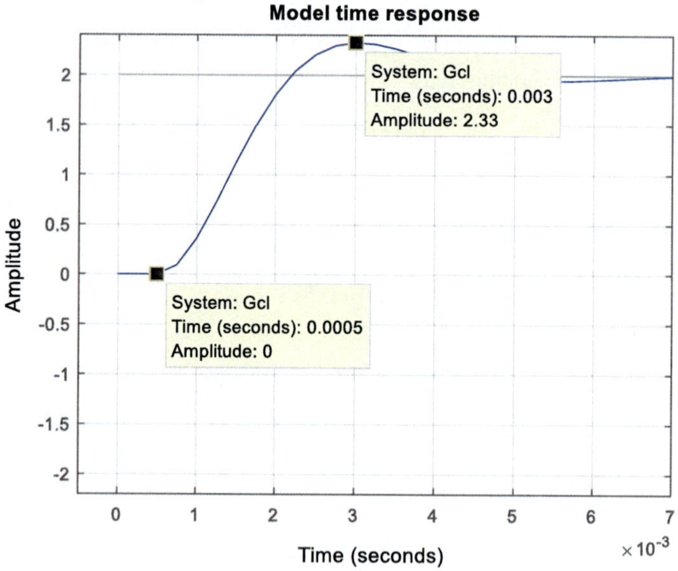

Fig. 6 Model step response

In line with [13] parameter estimation of any second-order model using two-points or three-points of step response is tried out from the MATLAB generated step response data of (9). The same parameters along with the dead time are also estimated from the same response data using the method documented in [14].

4.1 Parameter Estimation Based onTwo- or Three-Point of Step Response [13]

From the step response curve, we find t_1, t_5, t_9 which are the time past the dead time at which the step response attains 10%, 50%, and 90% of its final (steady-state) value, respectively, for the first time. The dead time is taken to be 0.53 ms.

$$t_9 = (2 - 0.53) = 1.47 \text{ ms} \tag{10}$$

$$t_5 = (1.42 - 0.53) = 0.89 \text{ ms} \tag{11}$$

$$t_1 = (0.866 - 0.53) = 0.336 \text{ ms} \tag{12}$$

Next, we calculate α and β which are both functions of damping ratio ξ:
For three-point method,

$$\alpha = \frac{t_9 - t_1}{t_5} = f(\xi) = 1.2472 \tag{13}$$

$$\text{For two pont method, } \beta = \frac{t_9}{t_1} = f(\xi) = 4.375 \tag{14}$$

The inverse correlations give ξ as functions of α and β as:
For three-points,

$$\xi = \left(0.4646 \times 10^{-8}\right)e^{6.4075\alpha} + 0.605\alpha - 0.2 = 0.55 \tag{15}$$

For two-points,

$$\xi = 0.1354\beta + \left(2.1586 \times 10^{-4}\right)e^{0.462\beta} = 0.58 \tag{16}$$

For three-points, the time constant T is found out by taking average of T_{t1}, T_{t5}, T_{t9} given as follows:

$$T_{t1} = \frac{t_1}{0.0137\xi^2 + 0.07267\xi + 0.4445} = 6.87 \times 10^{-4} s \tag{17}$$

$$T_{t5} = \frac{t_3}{0.03922\xi^2 + 1.09678\xi + 0.548} = 7.65 \times 10^{-4}\text{s} \quad (18)$$

$$T_{t9} = \frac{t_9}{-0.0469\xi^2 + 5\xi - 0.983} = 8.3 \times 10^{-4}\text{s} \quad (19)$$

Hence,

$$T = 7.63 \times 10^{-4}\text{s and } \omega_n = \frac{1}{T} = 1310\text{rad/s} \quad (20)$$

$$T\,.\,F_{3\_\text{pt}} = \frac{1.71 \times 10^6}{s^2 + 1441s + 1.7 \times 10^6}\text{e}^{-0.00053} \quad (21)$$

Similarly for two-point method, we obtain

$$T_{t1} = \frac{t_1}{0.0137\xi^2 + 0.07267\xi + 0.445} = 6.84 \times 10^{-4}\text{s} \quad (22)$$

$$T_{t9} = \frac{t_9}{-0.0469\xi^2 + 5\xi - 0.983} = 7.73 \times 10^{-4}\text{s} \quad (23)$$

Hence,

$$T = 7.28 \times 10^{-4}\text{s and } \omega_n = \frac{1}{T} = 1373\,\text{rad/s} \quad (24)$$

$$T\,.\,F_{2-\text{pt}} = \frac{1}{T^2s^2 + 2T\xi s + 1}\text{e}^{-\theta s} \quad (25)$$

$$T\,.\,F_{2-\text{pt}} = \frac{1.89 \times 10^6}{s^2 + 1593s + 1.89 \times 10^6}\text{e}^{-0.00053} \quad (26)$$

Estimated model parameters as shown in (21) and (26) are a near match of those reported in (9).

4.2 *Parameter Estimation Based on Two- or Three-Point of Step Response [14]*

From the step response curve, we find $y_{p,}$ the first peak, and get a direct estimate of ξ using the formula:

$$\xi = \sqrt{\frac{[\ln(yp - 1)]^2}{\pi^2 + [\ln(yp - 1)]^2}} = 0.5 \quad (27)$$

It then computes the inflection point coordinates (y_i, t_i) and a second point (y_2, t_2), where y_2 and t_2 are correlated to ξ as:

$$y_2 = 1.8277 - 1.7652\xi + 0.6188\xi^2 \quad (28)$$

$$t_2 = 3.4752 - 1.3702\xi + 0.1930\xi^2 \quad (29)$$

From these data, the three-points estimates of T and θ are:

$$T = 6.8259 \times 10^{-4}\text{s and } \omega_n = \frac{1}{T} = 1465\text{rad/s} \quad (30)$$

$$\theta = 0.532 \text{ ms} \quad (31)$$

The two-points method uses the points (y_i, t_i) and (y_p, t_p) and the corresponding estimates are:

$$T = 7.0337 \times 10^{-4}\text{s and } \omega_n = \frac{1}{T} = 1421\text{rad/s} \quad (32)$$

$$\theta = 0.545 \text{ ms} \quad (33)$$

Estimated model parameters as given in (30–31) and (32–33) are even a closer match of those reported in (9).

Table 2 gives a comparative list of the parameter estimates in line with methods given in [13] and [14] and the method proposed in this work. However, our method has a limitation that it cannot be applied for critically damped and over-damped systems.

Table 2 Comparison list of parameter estimates

Methods [13]	ω_n (rad/s)	ξ	θ (ms)
Two-points method	1373	0.58	–
Three-points method	1310	0.55	–
methods [14]			
Two-points method	1421	0.5	0.545
Three-points method	1451	0.5	0.532
Our method	1451	0.5	0.53

5 Model Validation

As discussed in the modeling methodology, we generate the model bode plot in MATLAB and compare it with the bode plot generated from the actual frequency response of the FSM assembly in x-axis.

The FSM assembly is driven by a fixed amplitude (0 to +2 V) sinusoid whose frequency is varied from 1 to 1000 Hz and the steady-state output is compared with that of the input in the DSO to ascertain the input to output amplitude ratio and phase relationship. The amplitude ratio is then converted to dB, and along with the phase information, are used to generate the bode plot (Fig. 7).

The −3 dB bandwidth (BW) for the actual response is 251 Hz and that for the FSM model is 270 Hz.

To have a closer look at the magnitude and phase errors between the actual response and FSM model response, the magnitude and phase errors are listed in Table 3. It is observed that up to 100 Hz, the magnitude errors are within 0.5 dB; however, the phase error gradually increases to about 6.5°. Since standard laser beam steering loops aim at a BW of around 50 Hz, the FSM model obtained is a good approximation of the actual FSM transfer function.

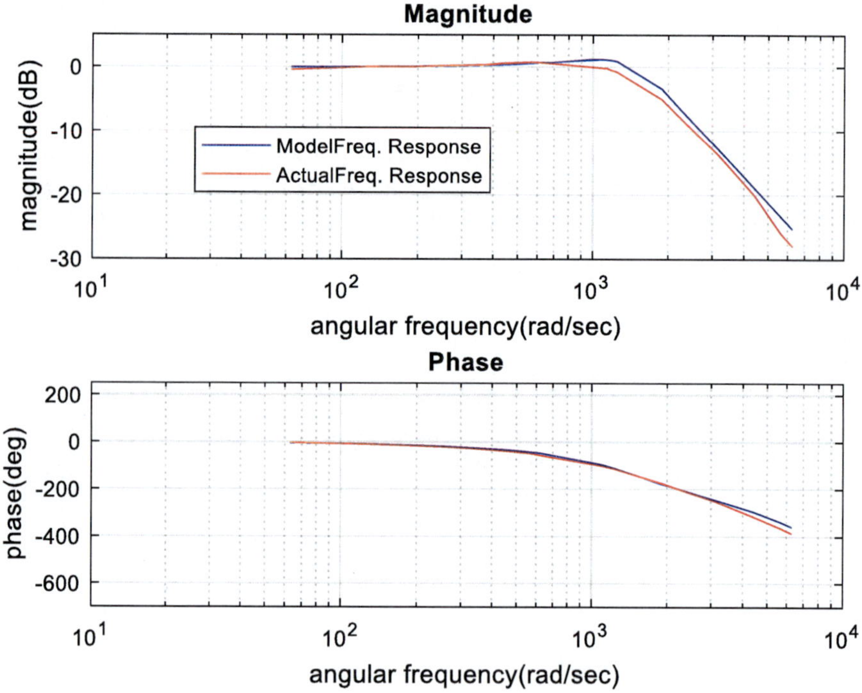

Fig. 7 Bode plots of FSM model and actual FSM unit

Table 3 Error between actual and model responses

Freq. (Hz)	Magnitude error	Phase error
10	0.5	0.8
20	0.0	1.3
30	0.1	2.9
40	0.0	3.8
50	0.1	4.7
60	0.1	5.4
70	0.2	6.0
80	0.2	6.4
90	0.1	6.5
100	0.0	6.5
175	1.4	6.7
180	1.4	6.1
182	1.3	5.8
185	1.5	5.4
190	1.6	4.6
195	1.6	3.9
200	1.7	3.1
300	1.7	2.3
500	0.9	7.0
700	1.2	21.8
900	2.7	26.5
1000	2.7	28.1

6 Conclusion

This paper presents a simple, fast, and accurate methodology for modeling a two-axis FSM assembly from its time response data. The transfer function parameters for the FSM model and actual system closely match each other.

The magnitude and phase errors in frequency responses of the actual FSM and its model at higher frequencies are due to the fact that we have identified the FSM as a second-order plus dead time model. Fitment of a higher order module from the frequency response data, as reported in [8], might have resulted in a more accurate model of the FSM. However, keeping in mind the computational complexity involved in the process and the control bandwidth requirement in typical laser beam steering systems, the second-order plus dead time model obtained from the time response is fairly accurate for designing the controller.

References

1. Mitchell PV, Technology development, newport corporation (2001) Fast steering mirror technology: active beam stabilization
2. Claeyssen F et al, Cedrat technologies (2018) Beam steering mirrors: from space applications to optronic applications
3. Portillo AA, Ortiz GG, Racho C (2001) Fine pointing control for optical communications. In: IEEE aerospace conference, BigSky, Montana
4. Optics in motion. Standard fast steering mirrors (2008) https://www.opticsinmotion.net/OIM100%20Series%20Users%20Manual%20rev%20D.pdf
5. Watkins RJ, Agarwal Bn (2007) Use of least means squares filter in control of optical beam jitter. J Guidance, Control Dynamics, 30(4):1116–1122
6. Watkins RJ, Chen H-J, Agrawal BN, Shin YS, Optical beam jitter control, free-space laser communication technologies XVI. In: GS Mecherle, CY Young, JS Stryjewski, (ed), Proceeding of SPIE, vol 5338
7. Chen N, Potsaid B, Wen JT, Barry S, Cable A (2010) Modelling and control of fast steering mirror in imaging application. In: 2010 IEEE international conference on automation science and engineering, pp 27–32
8. McKelvey T, Akcay H, Ljung L (1996) Subspace-based multivariable system identification from frequency response data. IEEE Trans Autom Contr 41(7):960–979
9. Cheres E, Eydelzon A (2000) Parameter estimation of a second order model in the frequency domain from closed loop data. Trans IChemE 78(Part A):293–298
10. PiezoSystmjena PSH x/2 Series (2017) https://www.piezosystem.com/fileadmin/redakteure/datenblatt/Piezo%20Aktoren%20Nano-Positioniersysteme/Spiegelkippsysteme/PSH%20x%202/PSH_5
11. Digital piezo-controller d-Drive (2011) https://www.piezosystem.com/products/piezo_controller/digital_system_d_drive/
12. Huang CT, Chou CJ (1994) Estimation of the underdamped second-order parameters from the system transient. Ind Eng Chem Res 33:174–176
13. Huang C-T, Clements Jr WC (1982) Parameter estimation for the second-order-plus-dead-time model. Ind Eng Chem Process Des Dev 601–603
14. Rangaiah GP, Krishnaswamy PR (1996) Estimating second-order dead time parameters from underdamped process transients. Chem Eng Sci 51(7):1149–1155

RF Front-End Band-Pass Filters for GPS Receiver

Prashant Kumar Singh, Rajesh Kumar Patjoshi, and Anjini Kumar Tiwary

Abstract The RF front end for Global Positioning System (GPS) receiver generally composed of an antenna, band-pass filter, and low noise amplifier (LNA). This chapter illustrates the filter parameters and provides comprehensive review of GPS filter with design methods. As the GPS system works on L-band frequencies, so it is difficult to employ lumped or discrete components in the designing of the RF filters, due to their inability to work in microwave frequency. So, these filters are realized by using transmission line components. In this chapter, microstrip technology is used for the realization of band-pass filter, which is good for low-power application as well as the advantages like onboard integration with active component, easier fabrication, and low cost. Full-wave EM simulators are used to validate the results of band-pass filter structures.

Keywords GPS receiver · Band-pass filter · Microstrip · Coupled line · Radial stub

1 Introduction

Today, Global Positioning System (GPS) became a vital part of our daily life, which is used for location positioning and navigation. The Navigation Satellite Time and Ranging (NAVSTAR) popularly known as GPS is a satellite-based system, which provides the location and time information over the earth to the GPS receiver. Beside the US-based GPS system, there are various satellite navigation systems operating

P. K. Singh (✉) · R. K. Patjoshi
Department of ECE, NIST, Berhampur, Odisha, India
e-mail: pksingh@nist.edu

R. K. Patjoshi
e-mail: rajeshpatjoshi1@nist.edu

A. K. Tiwary
Department of ECE, BIT Mesra, Ranchi, Jharkhand, India
e-mail: aktiwary@bitmesra.ac.in

V. Nath and J. K. Mandal (eds.), *Proceedings of the Fourth International Conference on Microelectronics, Computing and Communication Systems*, Lecture Notes in Electrical Engineering 673, https://doi.org/10.1007/978-981-15-5546-6_58

around the globe and some are under development, such as Global Navigation Satellite System (GLONASS, Russia), Galileo Positioning System (European Union), BeiDou Navigation Satellite System (China), Quasi-Zenith Satellite System (QZSS, Japan), and Indian Regional Navigation Satellite System (NAVIC, India) [1].

The GPS includes three segments, namely space, control, and user segments. The space segment comprises of the orbiting GPS satellites in space, which transmits the signals having the ranging and data information. These signals are received and processed by the user segment (i.e., GPS receiver) for navigation, timing, surveying, and other uses. Nowadays, a vast range of users like civilians, farmers, scientists, pilots, surveyors, and the military are using GPS receivers. However, the control segment looks on the monitor, command, and control of the GPS satellite constellation.

The GPS satellites receive the signals from the command segment in S-band frequency and transmit the navigational signals in L-band frequency. The GPS user segment (GPS receiver) is designed to receive and process these L-band signals. The ionospheric delay limits the use of lower frequency signals; however, rain (or ice or fog) attenuation limits the use of higher frequencies. Hence, L-band signals are preferred for the satellite navigation systems. The frequencies allocated for the use of radio navigation satellite system in L-band are 1164–1300 MHz (lower L-band) and 1559–1610 MHz (upper L-band) [2]. The lower L-band comprised of L5 (1164–1189 MHz), L2 (1215–1239.6 MHz), E5 (1164–1215 MHz), E6 (1260–1300 MHz), G3 (1189–1214 MHz), G2 (1237–1254 MHz), B2I (1179–1203 MHz), and B3 (1256–1280 MHz). The upper L-band comprised of L1 (1567–1587 MHz), E1 (1559–1591 MHz), G1 (1593–1610 MHz), and B1I (1553–1569 MHz). Out of these, L1, L2, and L5 bands are allocated for GPS.

Each GPS satellite transmits data on two carrier frequencies at 1575.42 MHz (L1) and 1227.60 MHz (L2). The GPS L1 signal is used for the civilians; however, GPS L2 signal is reserved for the military receivers. The modernized GPS signals include three more civilian signals, namely L1C, L2C, and L5. The career frequency for L1C and L2C signals are the same as is for L1 and L2, respectively. The difference is in coding scheme. However, an extra carrier frequency at 1176.45 MHz is allotted for L5 signal.

The GPS transmission at multiple bands needs multiband receiver at ground. Nowadays, many of the systems like mobiles, vehicles, laptops have inbuilt GPS receiver system in addition with many more different applications like GSM (Global System for Mobile Communication), GPRS (General Packet Radio Service), WLAN (Wireless LAN), WIMAX (Worldwide Interoperability for Microwave Access), Bluetooth. Even, the system combining the GPS and other navigational systems like GLONASS constellations is advantageous [3] to cope with the difficulties found due to the terrain variations (urban/mountainous areas), satellite visibility, interference, etc. This discussion leads to the requirement of dual or multiband antenna and filter at the receiver unit.

This chapter covers a detailed literature overview of the single and multiband bandpass filter (BPF) for RF front-end of the GPS receiver. Initially, a fundamental GPS receiver architecture is discussed with its block diagram and components. Further, microstrip BPFs for RF front-end is discussed followed by the filter parameters.

2 GPS Receiver

2.1 *Receiver Architecture*

With the escalation of technology, various receiver designs are developed for GPS receivers. One of the basic receiver structural designs is shown in Fig. 1. Fundamental GPS receiver comprised of antennas, RF front-end (low-noise amplifier (LNA), RF Filter, and Mixer), intermediate frequency (IF) section, demodulator, microcontroller, power supply, memory, and data display.

The L-band signal transmitted from the GPS satellites reaches to the receiver end with extremely low power. These signals are right-hand circular polarized (RHCP). Accordingly, the receiver's antennas must be capable to process these fainted signals and also be RHCP. The miniaturization, multiple bands, and reduction in higher-order modes are essential for emerging and rapidly developing communication system. To achieve these requirements, microstrip antenna [4, 5] is a good choice. After the reception of the GPS signal through the antenna, these signals required amplification as the available signal strength at the GNSS receiver is about 0.14 μV peak [2] buried in noise. This needs amplification up to 1 V with minimum distortion, which can be achieved by using LNA.

Even enhanced signal-to-noise ratio (SNR) is considered in receiver; however, the in-band and out-of-band interferences deteriorate the receiver performance. The in-band interference is due to the third-order harmonic of the high gain LNA. However, the out-of-band interference is due to the second- or third-order harmonic of signals of RADAR, cellular base stations, etc. Hence, low gain LNA is used after the antenna to minimize the in-band interference; however, the out-of-band interference can be rejected by using RF filter. There are two different plans to connect the RF filter in the

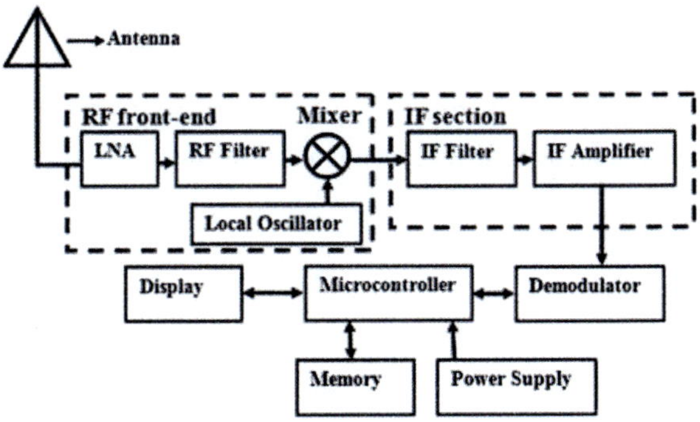

Fig. 1 Basic building block of GPS receiver

receiver unit. This filter may be connected before or after the LNA having different pros and cons.

The RF filter connected before the LNA improves the selectivity of the receiver; on the other hand, this configuration also increases the undesired noise figure. Filter connected after LNA depicts good selectivity with less degradation in noise figure. Further, the RF output from the filter will be mixed with locally generated signal to down-convert the frequency in single or multiple stages. This mixing and down-conversion are further explained in the next section. Thereafter baseband extraction and data display are done through the demodulation and microcontroller units.

2.2 *Superheterodyne Receiver*

There are several receiver architectures available, out of which superheterodyne is most common. The block diagram, shown in Fig. 2, depicts the working of a super-heterodyne or superhet receiver. It works on the heterodyning (mixing) principle and gives a superior performance compared to its counterparts like tuned radio frequency receiver (TRF Receiver) which gives its name as superheterodyne receiver.

The tuned circuit picks up the desired band-pass signal centered on carrier frequency f_{RF} and outputs it to the mixer. A local oscillator, which generates a signal of frequency f_{LO}, is fed to the mixer along with the received signal (f_{RF}) and produces a set of different frequencies like $f_{\mathrm{LO}} + f_{\mathrm{RF}}$ and $f_{\mathrm{LO}} - f_{\mathrm{RF}}$. However, the choice of oscillator frequency has to be carefully taken as the intermediate frequency (IF) at mixer's output is pre-decided. There are two choices available for oscillator frequency which are either $f_{\mathrm{LO}} = f_{\mathrm{RF}} + f_{\mathrm{IF}}$ (preferred) or $f_{\mathrm{LO}} = f_{\mathrm{RF}} - f_{\mathrm{IF}}$. For a received signal of frequency f_{RF}, mixer produces two frequency components of f_{IF} and $f_{\mathrm{IF}} + 2f_{\mathrm{RF}}$. Of these, the signal of frequency f_{IF} is passed and the higher frequency is rejected by the IF filter. However, if an undesired signal at frequency $f_{\mathrm{RF}} + 2f_{\mathrm{IF}}$ is received and is passed through mixer, then it produces two signals at f_{IF} and $2f_{\mathrm{IF}} + 3f_{\mathrm{RF}}$, where the filter will stop higher frequency and pass the lower frequency leading to reception and processing of an unwanted signal. This interference is called as adjacent channel interference (ACI), and the frequency $f_{\mathrm{RF}} + 2f_{\mathrm{IF}}$ is called image frequency. To get rid of the problem of image frequency, there is a need of BPF centered at f_{RF}, before the mixer stage to reject the image frequency.

The discrete or lumped inductor and capacitor cannot be used for the designing of the above-said BPF, as above 1 GHz frequency, the component size becomes comparable with the signal wavelength. Hence, these filters need distributed components

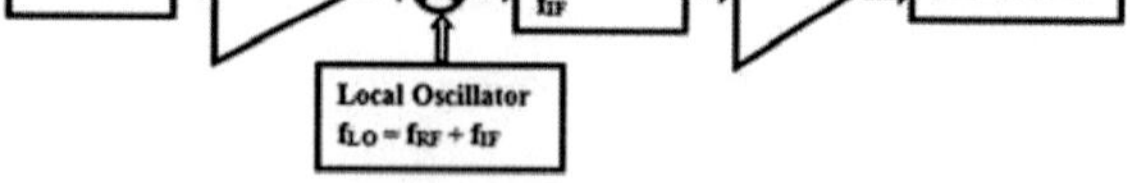

Fig. 2 Superheterodyne receiver

like waveguide and transmission lines. The filters designed for the microwave/RF frequencies are known as microwave RF filters. The RF filters for GPS application are discussed in the next section.

3 RF BPF for Front-End GPS Receiver

The applicability of BPF in receiver is already discussed in the last section. The increasing demand of the handheld system requires the miniaturize components with low cost and lightweight. The GPS receiver system comes into various forms like solo GPS system, cell phone unit (multiple applications). This needs the designing of single narrowband BPF as well as wideband or multiband BPF.

As the signal power available at the GPS receiver unit is very low, the components used at the front-end should be low-power device. The microstrip technology gives a better solution to overcome these limitations as is useful for low-power application and provides onboard integration with active component, easier fabrication, and low cost. To achieve above-said requirements, numerous design methods are proposed [6–17].

The microstrip dual-mode BPF [6] is designed using rectangular ring resonator coupled structure. The narrow pass-band region is controlled by coupling mechanism; however, the out-of-band performance is improved by electronic band gap (EBG) effect of the multiple defected ground structure (DGS). In order to improve the selectivity and miniaturize, the size of the BPF for GPS receiver application, a portion within the square patch resonator, is etched out [7]. This perturbation is used to improve the selectivity of the filter by generating transmission zeros in stop-band area. A dual-band band-pass filter [8] with low loss and compact size is developed for dual application GPS and wireless local area network (WLAN). This filter utilizes diverse quarter-wave resonator with short at one end using via and open at other end. In order to achieve compactness and good out-of-band performance, some more configurations developed using stepped impedance resonators (SIR) [9–11] and DGS structures [10]. Incorporating the GPS band few compact diplexers [13, 14] is also investigated. Recently, an advanced multiband BPF [15] is developed using the concentric resonators, which is also reconfigurable. A narrowband BPF [16] for GPS receiver application is implemented using DGS resonator. The designed miniaturized BPF rejects the second and third harmonics using stepped impedance DGS resonator. In addition, open stubs are used for the improvement of rejection level and widening of stop-band by creating two independently adjustable transmission zeros. These band-pass filters with technology used, applications, operating frequencies, fractional bandwidth (FBW), out-of-band region, and size is shown in Table 1.

Here, the design methods for two different planar BPF configurations [17, 18] are discussed in details. The first BPF is designed using the concept of coupled microstrip lines and J-inverter [17] for the operational frequency of 2.5 GHz with 1 GHz bandwidth. In this chapter, the same configuration is used for designing of the

Table 1 Some GPS BPFs available in the literature

S. No.	Operating frequency (GHz)	FBW (%)	Stop-band (GHz)	Size (mm × mm)
[6] 2008	1.575	5.0	>4.8	–
[7] 2008	1.575	3.3	>2.4	19 × 19
[8] 2008	1.575/5.7	4/4	>8	25 × 14
9] 2009	1.575/2.4	6/3.8	>8	21.16 × 30.14
[10] 2010	1.57/2.4	3/7	>4.5	–
[11] 2011	1.57/5.35	3.2/2.9	>8	–
[12] 2012	1.59/1.82/2.44	6.6/4.4/3.8	>4.5	19 × 21.8
[13] 2013	1.575/5.8	6.35/3.45	>11	–
[14] 2015	1.575/2.4	4.4/3.5	>4.5	62 × 32
[15] 2018	1.2/1.5 (GPS)	4.5/9 (GPS)	>3.5	30 × 18
[16] 2014	1.615	18	>5.5	33 × 27
[18] 2015	1.58	3.2	>11	16.7 × 17.5

BPF for GPS. However, the next discussed narrowband BPF configuration for GPS application [18] utilizes the combination of high-pass section and low-pass section.

3.1 BPF Using Coupled Lines

The FR-4 substrate (thickness, $h = 1.6$ mm; dielectric constant, $\varepsilon r = 4.4$; and loss tangent, $\tan\delta = 0.016$) is chosen for this BPF configuration. The simulation is done using IE3D software for the verification of the design.

The conventional asymmetric parallel-coupled line filter is shown in Fig. 3. Here, the modified compact symmetrical coupled band-pass filter [17] configuration is explained. The compactness is achieved by means of shifting the lower right half portion of the conventional structure to the upper half right half as shown in Fig. 4.

The FBW of 15%, center frequency of 1.57 GHz, and pass-band ripple of 0.1 dB are chosen specifications for the Chebyshev-type BPF design. Utilizing the concept

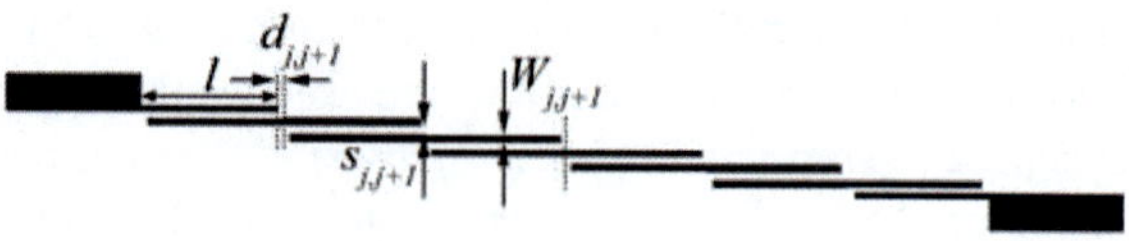

Fig. 3 Conventional asymmetric parallel-coupled line BPF

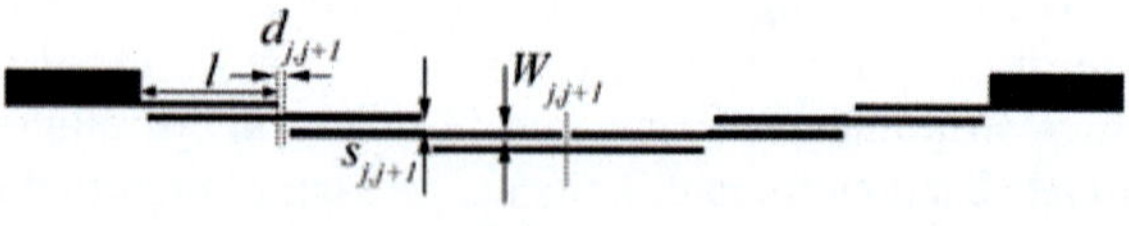

Fig. 4 Modified compact symmetrical BPF [17]

of J-inverter LPF prototype is transformed into BPF using Eq. (1–5).

$$\frac{J_{01}}{Y_0} = \sqrt{\frac{\Pi \times \Delta}{2 \times g_0 \times g_1}} \tag{1}$$

$$\left.\frac{J_{j,j+1}}{Y_0}\right|_{j=1 \text{ to } n-1} = \frac{\Pi \times \Delta}{2 \times \sqrt{g_j g_{j+1}}} \tag{2}$$

$$\frac{J_{n,n+1}}{Y_0} = \sqrt{\frac{\Pi \times \Delta}{2 \times g_n \times g_{n+1}}} \tag{3}$$

$$\left.(Z_{oe})_{j,j+1}\right|_{j=0 \text{ to } n} = \frac{1}{Y_0}\left[1 + \frac{J_{j,j+1}}{Y_0} + \left(\frac{J_{j,j+1}}{Y_0}\right)^2\right] \tag{4}$$

$$\left.(Z_{oo})_{j,j+1}\right|_{j=0 \text{ to } n} = \frac{1}{Y_0}\left[1 - \frac{J_{j,j+1}}{Y_0} + \left(\frac{J_{j,j+1}}{Y_0}\right)^2\right] \tag{5}$$

where $J_{j,j+1}$ and Y_0 are admittance inverter parameters and characteristic admittance of input/output lines, respectively. The element values of Chebyshev low-pass prototype for the chosen specifications are $g0 = g6 = 1.0$, $g1 = g5 = 1.1468$, $g2 = g4 = 1.3712$ and $g3 = 1.9750$. Knowing the values of the even-mode (Z_{oe}) and odd-mode (Z_{oo}) impedance of the strips (by using Eqs. (1–5)), the dimensions of the strip width $W_{j,j+1}$ and gap $s_{j,j+1}$ can be obtained from the given graph [19].

The correction factor $d_{j,j+1}$ (0.165 h; h is the thickness of the substrate) is incorporated in the calculation of the length of the microstrip lines, which is quarter wavelength at center frequency. The design parameters of the filter are shown in Table 2. The simulation result of this filter is shown in Fig. 5. The simulations result shows the band-pass region centered at 1.575 GHz frequency.

Table 2 Design parameters of BPF shown in Fig. 4

J	0	1	2
$J_{j,j+1}/Y_0$	0.4533	0.1879	0.1432
$(Z_{oe})_{j,j+1}$	82.93 Ω	61.16 Ω	58.18 Ω
$(Z_{oo})_{j,j+1}$	37.60 Ω	42.37 Ω	43.86 Ω
$W_{j,j+1}$	0.512 mm	0.688 mm	0.704 mm
$s_{j,j+1}$	0.104 mm	0.32 mm	0.416 mm
$d_{j,j+1}$	0.264 mm	0.264 mm	0.264 mm

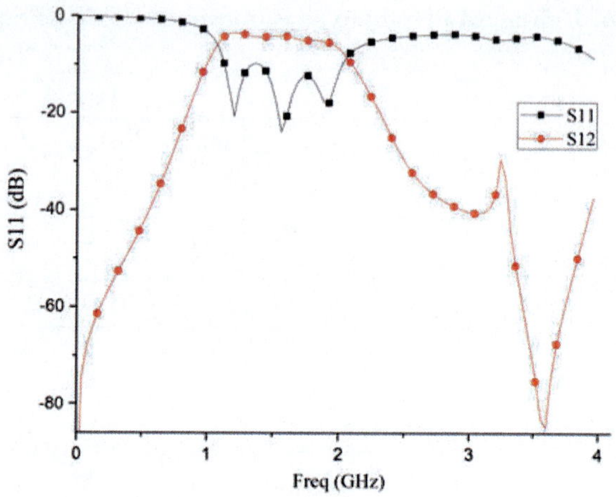

Fig. 5 S-parameter characteristic of the modified parallel-coupled BPF

3.2 *BPF Using Shorted Stub*

In this configuration, RT/Duroid RO5880 is used as the substrate material; however, same IE3D simulator is used for the verification. The substrate used has the dielectric constant of 2.2, height 1.57 mm, and the loss tangent 0.0009.

The conventional designs of the BPF aim for good in-band characteristics (pass-band) and ultra-wide stop-band along with compact sizes. The use of parallel-coupled resonators and coupled branch resonators for the implementation of BPF increases the circuit size. In this section, a compact BPF is designed using a low-pass filter (LPF) along with two short-circuited stubs embedded within the 50 μm line as shown in Fig. 6. A radial stub loaded structure is used in order to improve the out-of-band characteristic of the filter. The radial stub can be analyzed by a series combination of inductor (L_r) and capacitor (C_r) given by:

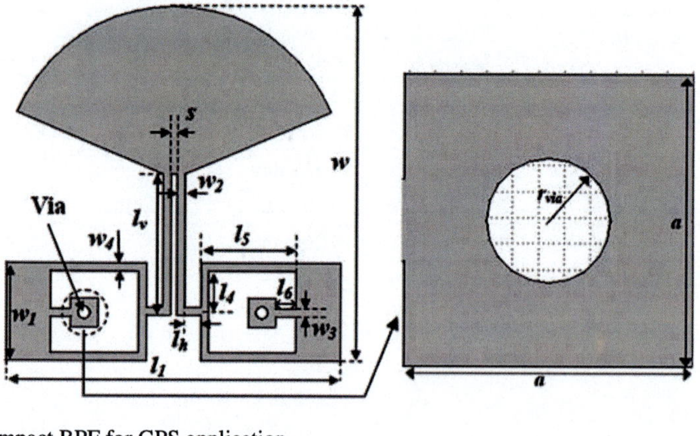

Fig. 6 Compact BPF for GPS application

Table 3 Parameteric values of BPF in Fig. 6

S. No.	Parameters	Length (mm)
1	r_i	0.6
2	r_o	8.5
3	w_1	4.84
4	w_2	0.4
5	l_v	7
6	l_2	2.42
7	l_3	3
8	l_h	0.8
9	s	0.3

$$L_r = \left[\frac{120\pi h}{\theta_r c}\right]\left[\ln\frac{r_o}{r_i}\right] \tag{6}$$

$$C_r = \frac{\theta_r r_o^2 \varepsilon_{\text{eff}}}{240\pi hc} \tag{7}$$

where h = dielectric thickness, β = phase constant, θ_r = Spanning angle of radial stub, c = speed of light, ε_{eff} = effective dielectric constant, r_i = inner radius of stub, and r_o = outer radius of radial stub, respectively. Initially, a LPF is designed using radial stub loaded coupled line structure using the design equations [20–23] for θ_r = 135°. The dimensional parameters [18] are as shown in Table 3.

Further, some portions of the 50 μm transmission line are etched out and a short-circuited transmission line implanted into the etched portion of on both sides of the LPF section as shown in Fig. 6. This shorted transmission line (via) attenuates some specific frequencies from DC to some frequencies of pass-band, which makes its characteristic look like that of as band-pass filter.

The stop-band has been enhanced by positioning transmission zeros in the stop-band while a transmission pole in pass-band allows its operation as a BPF with ultra-wide stop-band. Figure 7 shows a narrowband BPF characteristic for GPS receiver with 3 dB bandwidth (pass-band) from 1.559–1.610 GHz. This filter shows a very good operability of up to 11 GHz.

4 Conclusion

In this chapter, initially GPS system is introduced. Further, after the discussion of the applicability of microstrip filter in GPS receiver system, two different methods of implementation of BPF for GPS are elaborated. The first design comprises of parallel-coupled line and shows a wide pass-band bandwidth of 15%, and the second configuration shows narrow pass-band of only 3.2%. Comparing the two configurations each has their advantages and disadvantages. In terms of compactness and

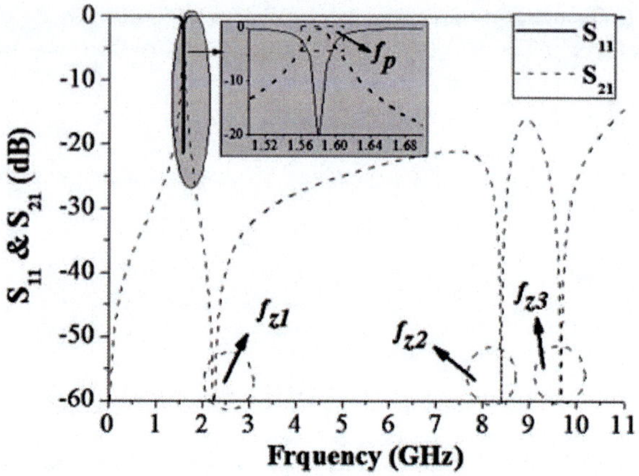

Fig. 7 S-parameter characteristic of the BPF shown in Fig. 6

wideband operability, obviously second configuration is better. However, the fabrication is much easier in case of the first configuration. A performance table of different available BPFs for GPS receiver is compiled in Table 1.

References

1. El Khoury F, Zgheib A (2018) Building a dedicated GSM GPS module tracking system for fleet management: hardware and software. Taylor and Francis Group, CRC Press, New York
2. Yegin K (2018) Antennas and front-end in GNSS, multifunctional operation and application of GPS. IntechOpen. https://doi.org/10.5772/intechopen.74971
3. Tamazin M, Karaim M, Noureldin A (2018) GNSSs, signals, and receivers, multifunctional operation and application of GPS. IntechOpen. https://doi.org/10.5772/intechopen.74677
4. Basilio LI, Chen RL, Williams JT, Jackson DR (2007) A new planar dual-band gps antenna designed for reduced susceptibility to low-angle multipath. IEEE Trans Antennas Propag 55(8)
5. Dierck A, Rogier H, Declercq F (2013) A wearable active antenna for global positioning system and satellite phone. IEEE Trans Antennas Propag 61(2)
6. Hon K, Heng-Yau P (2008) Design of a dual-mode bandpass filter with wide stopband performance for GPS application. Microwave Opt Technol Lett 50(2):445–447
7. Yan-Kuin S, Jau-Rung C, Min-Hang W, Cheng-Yuan H (2008) Design of a miniature and harmonic control patch dualmode bandpass filter with transmission zeros. Microwave Opt Technol Lett 50(8):2161–2163
8. Hung-Wei W, Yan-Kuin S, Min-Hang W, Ru-Yuan Y (2008) Design of dual-band bandpass filter using diverse quarter wavelength resonators for GPS/WLAN applications. Microwave Opt Technol Lett 50(10):2694–2696
9. Yu-Chi C, Chia-Hsiung K, Min-Hang W, Ru-Yuan Y (2009) Design of the compact dual-band bandpass filter with high isolation for GPS/WLAN applications. IEEE Microwave Wireless Compon Lett 19(12):780–782
10. Deng H-W, Zhao Y-J, Zhang L, Zhang X-S, Zhao W (2010) Dual-band BPF with DSIR and TSIR. Electron Lett 46(17)
11. Deng H-W, Zhao Y-J, Zhang X-S, Chen W, Wang J-K (2011) Compact and high selectivity dual-band dual-mode microstrip BPF with single stepped-impedance resonator. Electron Lett 47(5)

12. Wang L, Guan B-R (2012) Compact and high selectivity tri-band BPF using nested DDGSRs. Electron Lett 48(7)
13. Deng H-W, Zhao Y-J, Fu Y, Zhou X-J, Liu Y-Y (2013) compact and high performance microstrip diplexer for GPS and WLAN application. Microwave Opt Technol Lett 55(6):1233–1237
14. Yang R-Y, Hsiung C-M, Hung C-Y, Lin C-C (2010) Design of a high band isolation diplexer for GPS and wlan system using modified stepped-impedance resonators. Prog Electromagnet Res 107:101–114
15. Kingsly S, Kanagasabai M, Mohammed GNA, Subbaraj S, Panneer Selvam Y, Natarajan R (2018) Multi-band reconfigurable microwave filter using dual concentric resonators. Int J RF Microwave Comput-Aided Eng e21290. https://doi.org/10.1002/mmce.21290
16. Liu H, Ren B, Xiao X, Zhang Z, Li S, Peng S (2014) Harmonic-rejection compact bandpass filter using defected ground structure for GPS application. Hindawi Publishing Corporation, active and passive electronic components, vol. 2014, Article ID 436964, https://doi.org/10.1155/2014/436964
17. Tiwary AK, Gupta N (2011) Design of compact coupled microstrip line band pass filter with improved stopband characteristics. Prog Electromagnet Res C 24:97–109
18. Singh PK, Tiwary AK, Gupta N (2015) Design of radial microstrip band pass filter with wide stop-band characteristics for GPS application. Prog Electromagnet Res C 59:127–134
19. Mattaei G, Young L, Jones EMT (1980) Microwave filters, impedance-matching networks, and coupling structures. Artech House, Norwood, MA
20. Ma K, Yeo KS (2011) New ultra-wide stopband low-pass filter using transformed radial stubs. IEEE Trans Microwave Theory Technol 59(3):604–611
21. Ma K, Mou S, Wang K, Yeo KS (2012) Radial loaded transformed radial stub for LPF stopband extension. Prog Electromagnet Res Lett 30:125–132
22. Singh PK, Tiwary AK, Gupta N (2014) Design and development of radial stub low-pass filter with improved characteristics. 10th Int. Conference on Microwaves, Antenna, Propagation and Remote Sensing (ICMARS) Jodhpur, India, pp 13–16
23. Kwon H, Lim H, Kang B (2007) Design of 6–18 GHz wideband phase shifters using radial stubs. IEEE Microwave wireless Comput Lett 17(3):205–207

Comparative Study of Parameter Estimation Methods for a Single Diode PV Module

Kollu Ravindra and S. M. Sulthana

Abstract In recent years, deployment of solar photovoltaic generation units has increased due to advancement in PV technologies and dwindling costs. Modeling of a PV module is important to know the impact of a PV system both economically and technically from maximum power point tracking perspective. A single diode model is used which is accurate enough and simple to model PV module behavior. In this work, meta-heuristic harmony search algorithm (HSA) is used to estimate the PV module parameters under standard test conditions (STC). Comparison of HSA-based extraction method with analytical and iterative-based Newton–Raphson, Gauss Seidal methods is done. Root mean square error (RMSE) is used as an accuracy indicator for the parameter extraction methods. Two mathematical models are used to estimate parameters under varying irradiance and temperature conditions. Results show that HSA-based extraction method accurately extracts parameters of a PV module.

Keywords PV module · Parameter extraction · Harmony search algorithm

1 Introduction

Depleted fossil fuels, pollution due to fossil fuels burning and emissions led to the increase in renewable energy sources-based generation worldwide. Solar PV energy is being used extensively to serve the increased demand of electrical energy. Photovoltaic (PV) generation is an important renewable source of energy in the world due to several merits such as

- Emission and pollution free
- Abundantly available with no cost on fuel

K. Ravindra (✉) · S. M. Sulthana
Electrical and Electronics Engineering Department, University College of Engineering Kakinada, JNTUK, Kakinada, Andhra Pradesh, India
e-mail: ravikollu@gmail.com

V. Nath and J. K. Mandal (eds.), *Proceedings of the Fourth International Conference on Microelectronics, Computing and Communication Systems*, Lecture Notes in Electrical Engineering 673, https://doi.org/10.1007/978-981-15-5546-6_59

- Less maintenance and related costs
- Feeding the loads locally there by reducing T&D losses
- Easy installation, no noise
- Unattended operation.

For evaluation of the performance, better understanding of the I-V and PV characteristics is necessary. Therefore, for optimization of the PV system during planning, accurate modeling is an important tool. This accurate modeling results in optimal dimensioning of PV power plants maximizing benefits. The accurate modeling can help in describing specifications for the power conditioning equipment [1]. Analytical method by Kishore [2] estimated single diode model parameters using simplified equations, whose accuracy is questionable due to the assumptions made in simplifying equations. Kashif et al. [3] used Newton–Raphson method to extract single diode model parameters and Chatterjee et al. [4] used Gaussian iteration method to extract parameters. Since the I-V equation is nonlinear finding explicit solutions is difficult. Brano and Ciulla [5] used an analytical method based on generalized reduced gradient for PV module parameter extraction. Kebir [6] provided a review of various parameter extraction methods such as analytical, numerical and meta-heuristic artificial bee colony (ABC) algorithm and concluded that ABC algorithm convergence speed is fast and it accurately estimates the PV module parameters compared to numerical and analytical methods. Shongwe and Hanif [7] compared performance of iterative method and method-based maximum power point matching for parameter extraction. It is shown that iterative method based on Gaussian iteration performs better than other methods. Also, five models are compared to account for variation in irradiance and module temperature on PV module performance. Ishaque and Salam [8] used differential evolution to extract single diode module parameters.

2 Problem Formulation

The objective of this work is to extract the parameters of a PV module accurately for accurate modeling, simulation and control studies. Single diode and two diode-based models commonly used to model a solar PV module. In this work, one diode-based equivalent circuit model is used because it is simple and accurate. The basic circuit of one diode model is shown in Fig. 1.

Where I_{ph} is the solar photocurrent, I_s corresponds to diode reverse saturation current, R_p is the shunt resistance current, I output current, V is the PV module output voltage, and R_s is the series resistance.

By connecting N_s number of cells in series a module is formed the I-V combination of nonlinear equation acquired as

$$I = I_{\text{Ph}} - I_s\left\{\exp\left(\frac{q(V + R_s * I)}{\text{AKTN}_s}\right) - 1\right\} - \left(\frac{V + I * R_s}{R_p}\right) \quad (1)$$

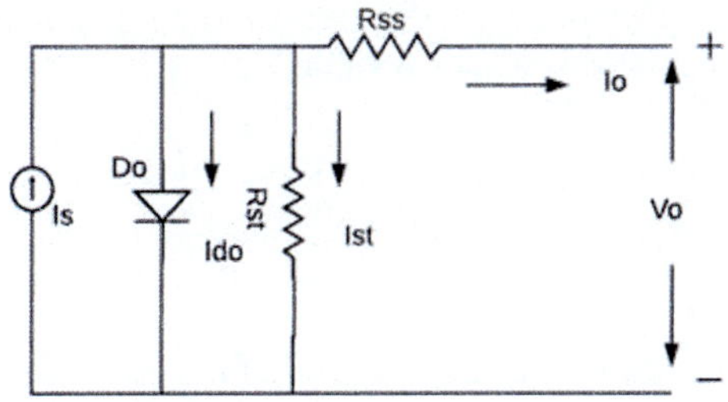

Fig. 1 Basic circuit of solar module

where I_s reverse saturation current of diode, q corresponds to electron charge (1.602 $\times$ 10^{-19} Coloumbs), A represents diode ideality factor, K is Boltzmann constant and its value is (1.38 $\times$10^{-23} J/K), and T is temperature in Kelvins.

2.1 Analytical Method

Analytical method [2] is discussed in this section where five unknown parameters of single diode model are extracted. These parameters can be found from points depicted on the curve shown in Fig. 2.

By considering known points from the characteristic curve and substituting in (1) and solving the following equations are written. Under open-circuit point condition and short-circuit conditions (1) is rewritten as (2) and (3).

$$0 = I_{\text{ph}} + I_s.\left(\exp\left(\frac{V_{\text{oc}}}{V_t\,.\,A}\right) - 1\right) + \frac{V_{\text{oc}}}{R_p} \tag{2}$$

$$I_{\text{sc}} = I_{\text{ph}} + I_s.\left(\exp\left(\frac{R_s\,.\,I_{\text{sc}}}{V_t\,.\,A}\right) - 1\right) + \frac{R_s\,.\,I_{\text{sc}}}{R_p} \tag{3}$$

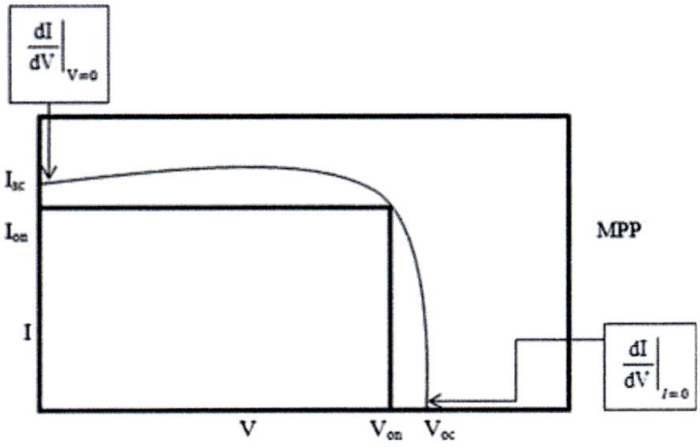

Fig. 2 Characteristic curve—analytical method

Derivate of (1) with respect to V is taken and considering open-circuit condition, below equation is obtained

$$\left.\frac{dI}{dV}\right|_{I=0} = -I_s\left(\frac{1}{V_t \,.\, A}\left(1 + \left.\frac{dI}{dV}\right|_{I=0} R_s\right)\exp\left(\frac{R_s V_{oc}}{V_t \,.\, A}\right)\right) - \frac{1}{R_p}\left(1 + \left.\frac{dI}{dV}\right|_{I=0} R_s\right) \tag{4}$$

Under short-circuit condition, the equation is written as

$$\left.\frac{dI}{dV}\right|_{V=0} = -I_s\left(\frac{1}{V_t A}\left(1 + \left.\frac{dI}{dV}\right|_{V=0} R_s\right)\exp\left(\frac{R_s I_{sc}}{V_t \,.\, A}\right)\right) - \frac{1}{R_p}\left(1 + \left.\frac{dI}{dV}\right|_{V=0} R_s\right) \tag{5}$$

Since the derivate at maximum power point is zero, an equation is obtained as

$$-\frac{I_{mp}}{V_{mp}} = -I_s\frac{1}{AV_t}\left(1 + \frac{I_{mp}}{V_{mp}} R_s I_o \exp\left(\frac{V_{mp} + I_{mp}R_s}{V_t A}\right)\right) - \frac{1}{R_p}\left(1 - \frac{I_{mp}}{V_{mp}} R_s\right) \tag{6}$$

Making some approximations leads to simplified expressions which are given below

$$R_s = R_{so} - \frac{AV_t}{I_s}\left(\exp\left(-\frac{V_{oc}}{V_t \,.\, A}\right)\right) \tag{7}$$

$$R_p = R_{po} \tag{8}$$

$$I_{ph} = I_s\left(\exp\left(\frac{R_s \,.\, I_{sc}}{V_t \,.\, A}\right) - 1\right) + I_{sc}\left(1 + \frac{R_s}{R_p}\right) \tag{9}$$

$$I_s = \left(I_{sc} - \frac{V_{oc}}{R_p}\right)\left(\exp\left(-\frac{V_{oc}}{V_t \,.\, A}\right)\right) \tag{10}$$

$$A = \frac{\left(V_{mp} + I_{mp}R_{so}\right) - V_{oc}}{V_t\left\{\ln\left(I_{sc} - \frac{V_{mp}}{R_{po}} - I_{mp}\right) - \ln\left(I_{sc} - \frac{V_{oc}}{R_p}\right) + \frac{I_{mp}}{\left(I_{sc} - \frac{V_{oc}}{R_{po}}\right)}\right\}} \tag{11}$$

where R_{so} and R_{po} are series resistance and shunt resistance under open-circuit conditions, respectively, I_{mp} and V_{mp} are maximum current and maximum voltage, respectively, and V_t is equal to *KT/q*.

2.2 Newton–Raphson Method

Equation (1) is transcendent in nature, hence, iterative methods such as Newton–Raphson method [3] is used to solve the equation and find parameters I_{ph}, I_s, A, R_s, and R_p of the single diode model. The initial values for I_{ph}, I_s, A, and R_p are given by (12) to (15)

$$I_{\text{ph}} = (R_s + R_p)/R_p I_{\text{sc}} \tag{12}$$

$$I_s = \frac{I_{\text{sc}} - \frac{V_{\text{oc}}}{R_p}}{\exp\left(\frac{\frac{V_{\text{oc}}}{V_t}}{\frac{A}{N_s}}\right) - 1} \tag{13}$$

$$A = \frac{K_v - \frac{V_{oc}}{\text{Tn}}}{N_s V_{\text{tn}}\left(\frac{K_i}{I_{\text{ph}}} - \frac{3}{T_n} - \frac{E_{\text{gap}}}{kT_n^2}\right)} \tag{14}$$

$$R_p = V_{\text{mp}} \frac{V_{\text{mp}} + I_{\text{mp}} R_s}{V_{\text{mp}} I_{\text{mp}} - V_{\text{mp}} I_s \exp\left(\frac{\frac{V_{\text{mp}} + I_{\text{mp}} R_s}{V_t}}{\frac{N_s}{A}}\right) + V_{\text{mp}} I_s - P_{\text{mp}}} \tag{15}$$

The flowchart of iterative N-R method is shown in Fig. 3.

2.3 Harmony Search Algorithm

HSA is inspired by natural musical process searching to get an intact harmony. This step by step efficient search process uses a stumble search. The workflow for HSA is as follows [9, 10].

Step 1: Formulate the objective function and identify decision variables subjected along with boundary conditions for the variables. Specify the algorithm parameters, viz. harmony memory size (HMS) or the numbers of solution vectors, the memory consideration rate (HMCR), bandwidth (BW), the pitch adjusting rate (PAR), and the maximum number of iterations required (iter_{max}).

$$\begin{aligned} &\text{Minimise } f(x) \\ &\text{Subject to } x_{iL} \le x_i \le x_{iU} \ (i = 1, 2, 3 \ldots N) \end{aligned} \tag{16}$$

where x_{iL} and x_{iU} correspond to minimum and maximum boundaries for decision variables, respectively.

Minimization of RMSE [13] is chosen as objective in this work which is given by

Fig. 3 Flowchart of N-R method

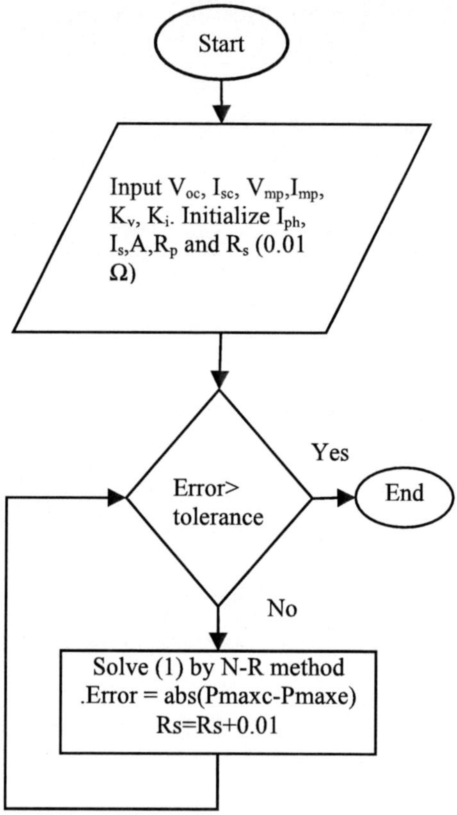

$$\text{RMSE} = \sqrt{\left(\frac{1}{N}\right)\sum_{i=1}^{N}(f_i(V, I, x))^2} \tag{17}$$

Objective function for single diode model is formulated using (1) where x is a solution vector given by $[I_{\text{ph}}, I_s, R_s, R_p, A]$.

$$f(V, I, x) = \left(I - I_{\text{ph}} + I_s\left(\exp\left(q\left(\frac{V + (IR_s)}{AN_skT}\right)\right) - 1\right) + \left(\frac{V + (IR_s)}{R_p}\right)\right) \tag{18}$$

where N is the number of experimental I-V values.

Step 2: Initialization the harmony memory (HM) using the following equation:

$$x_i^j = x_{iL} + \text{rand} * (x_{iU} - x_{iL}) \quad j = 1, 2, \ldots, \text{HMS} \tag{19}$$

where rand corresponds to a random number generated between 0 and 1.

Each solution vector in HM is evaluated by finding the objective function value (18). Sort the solution vectors in HM based on their ascending order of functional values.

Step 3: In this step, a new harmony is improvised based on three norms: memory deliberation, pitch adjustment, and random generation. If r the random number generated between 0 and 1 is lower than the HMCR, the decision variable x_i^{new} is originated using memory deliberation; otherwise, x_i^{new} is randomly generated. Then, pitch adjustment for x_i^{new} is done with a probability of *PAR* if it is originated using memory deliberation which is shown below:

$$x_i^{\text{new}} = x_i^{\text{new}} + r * bw \tag{20}$$

Step 4: After originating x_i^{new}, the harmony memory will be upgraded as follows. If the fitness of the $x_i^{\text{new}} = \left(x_1^{\text{new}}, x_2^{\text{new}}, \ldots, x_n^{\text{new}}\right)$ is superior to the worst vector in memory, then replace it with the new harmony in HM. Otherwise, it is discarded.

Step 5: Replicate steps 3 and 4 for $\text{iter}_{\max}$ number of times.

A. Models to account for variations in irradiance (G) and temperature (T)

Model 1 [11]:

$$I_s = I_{sn}\left(\left(\frac{T}{T_n}\right)^3\right)\exp\left(\left(\frac{qE_{\text{gap}}}{nk}\right)\left(\frac{1}{T_n} - \frac{1}{T}\right)\right) \tag{21}$$

$$I_{\text{ph}} = (I_{\text{phn}} + K_i dT)\left(\frac{G}{G_n}\right) \tag{22}$$

where I_{sn}, T_n, and I_{phn} correspond to I_s, T, and I_{ph} values at STC condition.

Model 2 (proposed model):

In the proposed model, apart from variations in I and I_{ph} are given by (21) and (22), variations in R_p are also considered given by (23).

$$R_p = \frac{V_{\text{mp}}\left(V_{\text{mp}} + I_{\text{mp}}R_s\right)}{V_{\text{mp}}I_{\text{ph}} - V_{\text{mp}}I_s\exp\left(\frac{V_{\text{mp}}+I_{\text{mp}}R_s}{N_sV_tA}\right) + V_{\text{mp}}I_s - P_{\max}} \tag{23}$$

3 Results

With the aim of verifying the proposed method based on HSA for parameter extraction and models, KC200GT module [12] is used to evaluate and test whose parameters are given in Table 1. Analytical, Newton–Raphson and HSA-based extraction

Table 1 Datasheet parameters of KC200GT

Parameters	KC200GT
I_{sc}	8.21 A
V_{oc}	32.9 V
V_{mp}	26.3 V
I_{mp}	7.61 A
K_I	0.00318 A/°C
K_V	−0.123 V/°C
N_s	54

methods are used to extract parameters of single diode model. Then, the extracted parameters are used to plot the I-V curve for the PV module. The lower and upper ranges of the five parameters chosen for the HSA-based extraction method are tabulated in Table 2.

The comparison of current vs. voltage curves at STC obtained using analytical, iterative, and meta-heuristic methods is shown in Fig. 4. Table 3 shows the RMSEs calculated and parameters extracted from analytical, iterative, and meta-heuristic methods. From Table 3, based on calculated RMSEs and parameters extracted by using analytical, iterative, and meta-heuristic methods, it is concluded that proposed method based on HSA (meta-heuristic technique) is the best method which gives least error at STC.

Results with G = 1000 W/m² and T = 50 °C:

Table 2 Upper and Lower range of solar cell parameter

Parameters	Lower range	Upper range
R_s (Ω)	0	0.5
R_p (Ω)	0	500
I_s (nA)	0	0.1
A	1	2
I_{ph} (A)	0	10

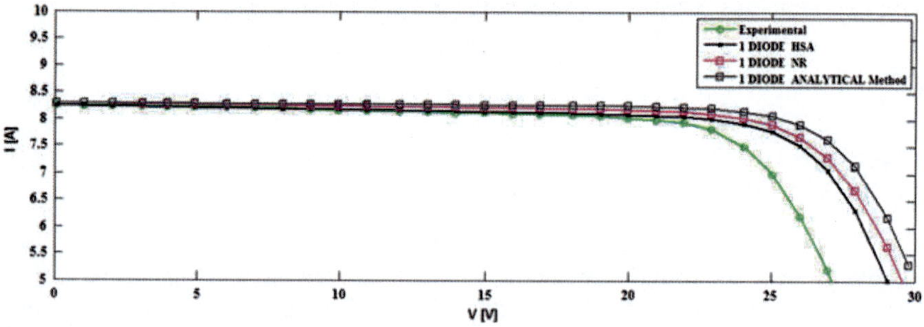

Fig. 4 I-V characteristics of KC200GT at STC

Table 3 Parameter extraction of KC200GT using different methods

Parameters	Analytical	N-R	HSA
RMSE	0.560	0.2381	**0.0215**
$R_s (\Omega)$	0.2278	0.309	0.25731799
R_p (Ω)	136	194.003059	196.02646
I_o (A)	2.096e-9	2.1143e-09	0.7637e-009
A	1.075	1.075559	1.035175476
I_{ph} (A)	8.22	8.223281	8.1809206

Table 4 Extracted parameters and RMSE of single diode model (1 and 2) at $S = 1$ kW/m^2, $T = 50$°C

Models	HSA model 1	HSA model 2
RMSE	1.4573	**1.3446**
R_s (Ω)	0.25731799	0.25731799
R_p (Ω)	196.026460	49.1150
I_o (A)	9.72482e-010	9.72482e-010
A	1.035175476	1.035175476
I_{ph} (A)	8.260421	8.260421

As it is concluded in the previous section that HSA gives more accurate results when compared to analytical and iterative N-R technique; hence, proposed model 2 and model 1 are used along with meta-heuristic technique HSA and results are compared which are presented in Table 4. The I-V characteristics are shown in Fig. 5.

A convergence characteristic of the algorithm is shown in Fig. 6.

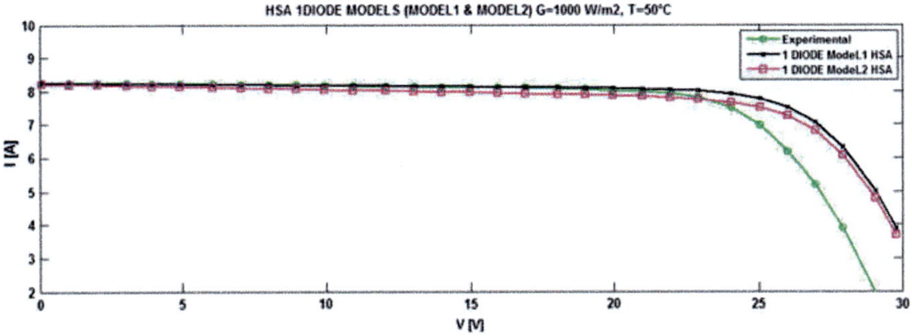

Fig. 5 I-V characteristics of KC200GT with $G = 1000$ W/m^2 and $T = 50$ °C

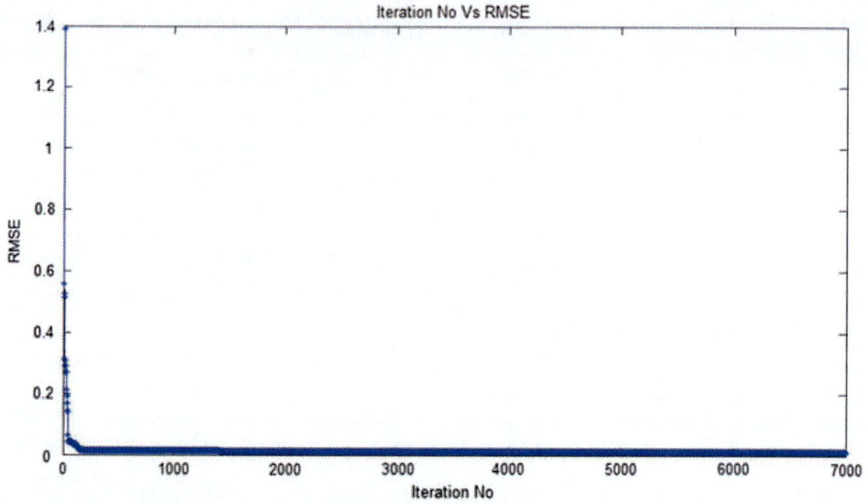

Fig. 6 Convergence characteristic of HSA algorithm for parameter extraction

4 Conclusion

The work has been carried out to extract the parameters by using analytical, iterative (N-R), and HSA (meta-heuristic)-based extraction techniques. From this work, it is concluded that HSA is more accurate technique compared to analytical and iterative techniques as the RMSE error between the simulated I-V curve and experimental I-V curve is small. Proposed model accounting for variations in irradiance and temperature leads to less RMSE error compared to the existing model.

References

1. Vun JC, Salam Z, Ishaque K (2015) Cell modelling and model parameters estimation techniques for photovoltaic simulator application: a review. Appl Energy154:500–519
2. Kishore R (1989) Accurate analytical expressions for the parameters of the single exponential model. Solid-State Electron. Pergamon Press plc
3. Ishaque K, Salam Zl, Taheri H (2011) Simple, fast and accurate two-diode model for photovoltaic modules. Sol Energy Mater Sol Cells 95:586–594
4. Chatterjee A, Keyhani A, Kapoor D (2011) Identification of photovoltaic source models. IEEE Trans Energy Convers 26(3):883–889
5. Brano VL, Ciulla G (2013) An efficient analytical approach for obtaining a five parameters model of photovoltaic modules using only reference data. Appl Energy 111:894–903
6. Kebir ST, Haddadi M, Ait-Cheikh MS (2015) An overview of solar cells parameters extraction methods. In: 2015 3rd International conference on control, engineering & information technology (CEIT), Tlemcen, pp 1–7
7. Shongwe S, Hanif M (2015) Comparative analysis of different single-diode PV modeling methods. IEEE J Photovoltaics 5(3):938–946

8. Ishaque Kashif, Salam Zainal (2011) An improved modeling method to determine the model parameters of photovoltaic (PV) modules using differential evolution (DE). Sol Energy 85(9):2349–2359
9. Kollu R, Rayapudi SR, Sadhu VLN (2014) A novel method for optimal placement of distributed generation in distribution systems using HSDO. Int Trans Electr Energy Syst 24(4):547–556
10. Geem ZW, Kim JH, Loganathan GV (2001) A new heuristic optimization algorithm: harmony search, simulation 76(2):60–68
11. Islam MA, Merabet A, Beguenane R, Ibrahim H (2013) Modeling solar photovoltaic cell and simulated performance analysis of a 250 W PV module. In: Proceedings of electrical power and energy conference, pp. 1–6, 21–23 Aug 2013
12. KC200GT High Efficiency Multicrystal Photovoltaic Module Datasheet Kyocera 2000 http://www.kyocera.com.sg/products/solar/pdf/kc200gt.pdf
13. Nassar-eddine I, Obbadi A, Errami Y, El Fajri A, Agunaou M (2016) Parameter estimation of photovoltaic modules using iterative method and the Lambert W function: A comparative study. Energy Convers Manag 119:37–48

Estimation of MMW Attenuation in Dust Storms Under Rayleigh Approximation

Swastika

Abstract It is found that frequency under Rayleigh approximation the scattering cross section dominates over absorption cross section. It is clearly found that scattering cross section, absorption cross section and extinction cross section increase with increasing frequency. It is further observed that the loss due to scattering and absorption increases with increasing frequency and decreases with increasing visibility; this is corroborated with the fact that increasing visibility decreases the number of particles present in the unit value of storm. It may therefore be concluded that with the advent of communication engineering and due to enhance superiority of millimeter wave over infrared. The millimeter wave has finite contribution in communication engineering. It is imperative for us to investigate the other useful characteristics of millimeter wave under adverse atmospheric condition which will be considered as the milestone in the area of communication.

1 Introduction

“Electromagnetic Theory” Sarition [1],”Electromagnetic Scattering and its Application” Bryvel and Jones [2]. The all Relation in the Calculation of Rain Attenuation IEEE Transaction LORoderic, David and Daniel [3].”Propagation of Short Radio Waves” Kerr [4] “High Scattering by small particles” Van de Hulei [5] ‘Effect of sand and Dust Storm on Terrestrial Microwave link’ Alrizzo et al. [6] “Microwave/Millimeter Wave Propagation in Sand and Dust Storms” Rai [7].”Millimeter Wave Propagation in Smoke” Knox [8], “The Effect at Low Attitude Nuclear Burst on Millimeter Wave Propagation” Ahschular [9].

Swastika (✉)
Department of Electronics and Communication Engineering, Netaji Subhas Institute of Technology-Bihta, Near IIT Patna, Patna 801106, India
e-mail: swastikaanuj54@gmail.com

V. Nath and J. K. Mandal (eds.), *Proceedings of the Fourth International Conference on Microelectronics, Computing and Communication Systems*, Lecture Notes in Electrical Engineering 673, https://doi.org/10.1007/978-981-15-5546-6_60

Sand and dust causing attenuation is a problem in utilization of millimeter wave. They arise problems in terrestrial and space communication. The signals get attenuated when it passes through the medium which has precipitations as sand and dust particles. These particles cause absorption of energy and also cause scattering of energy out of beam which are two phenomena for attenuation.

2 Method

Absorbed energy gets converted into heat. Dust particles do oscillatory movement as ions and energy will be radiated in all directions when it comes under the influence of incident wave. This causes attenuation which depends on size and particles concentration. In terms of scattering and absorption cross section of a single particle, the attenuation theory can be explained.

3 Scattering from Dust Particles

Here, scattered fields and incident fields are indicated as E_{sca} H_{sca} and E_{inc} H_{inc}. Total amplitude fields amplitude at some point on the surface of the sphere which can be mentioned as

$$\begin{aligned} E &= E_{\text{inc}} + E_{\text{sca}} \\ H &= H_{\text{inc}} + H_{\text{sca}} \end{aligned} \tag{1}$$

If the problem is related with the energy flow, then Poynting vector must be used. So only radial component of Poynting vector crosses sphere surface. The magnetic and electric fields is indicated as H (θ, Φ) and E (θ, Φ) in spherical co-ordinate system then the radial component of the average Poynting vector will be in radial direction, the value of which may be given by

$$\begin{aligned} S_R &= 1/2R_e\left(E_\theta H_\Phi^* - E_\Phi H_\theta^*\right) \\ S_R &= 1/2R_e[\left(E_{\theta\text{inc}} H_{\Phi\text{inc}}^* - E_{\Phi\text{inc}} H_{\theta\text{inc}}\right) \\ &\quad + \left(E_{\theta\text{sca}} H_{\Phi\text{sca}} - E_{\Phi\text{sca}} H_{\theta\text{sca}}^*\right) \\ &\quad + (E_{\theta\text{inc}} H_{\Phi\text{sca}} + E_{\theta\text{sca}} H_{\Phi\text{inc}} + E_{\Phi\text{inc}} H *_{\theta\text{sca}} - E_{\Phi\text{sca}} H *_{\Phi\text{sca}})] \end{aligned} \tag{2}$$

From Eq. (2), the integral of first term on right-hand side of the equation is zero as it gives the net flow of energy in incident plane wave, whereas second term when integrated gives the total scattered power (P_{sca}) out of incident wave. The third term if integrated yields ($-P_{\text{ext}}$), and the integral of (S_R) yields the out flow of energy from the sphere.

4 Absorption Cross Section

The ratio of total absorbed energy per second to the energy density of the incident wave is defined as the absorption cross section, and it can be written as

$$\sigma_{\text{abs}} = \frac{-\text{Re} \int_o^{\pi} \int_o^{2\pi} S_R\, R^2 \sin\theta d\theta d\phi}{\frac{E_o^2}{2\eta_o}} \tag{3}$$

where ½ $\frac{E_o^2}{\eta_o}$ is mean energy flow of incident wave per unit area.

5 Scattering Cross Section

It is defined as the ratio of the total scattered energy per second to the energy density of incident wave. The total scattered power of spherical dust particle is given by

$$P_{\text{sca}} = {}^{1}\!/\!{}_{2}\text{Re} \int_o^{\pi} \int_o^{2\pi} (E_{\theta\text{sca}} H_{\phi\text{sca}} - E_{\phi\text{sca}} H^{*}_{\theta\text{sca}}) \\ R^2 \sin\theta\, d\theta\, d\Phi \tag{4}$$

and scattering cross section σ_{sca} is

$$\sigma_{\text{sca}} = \frac{2\pi}{K^2} \sum_{n=1}^{\infty} (2n+1)|a_n^s|^2 + |b_n^s|^2 \text{cm}^2 \tag{5}$$

where a_n^s and b_n^s indicate the amplitude of the field on the surface of the sphere which are proportional to the amplitudes of the magnetic and electric multiples induced by the incident wave.

6 Extinction Cross Section

From Eq. (2), it is evident that to maintain the energy balance, third term of the equation must be equal in magnitude of the sum of absorbed and scattered power. Therefore,

$$\begin{aligned} \mathrm{P_{ext}} &= \mathrm{P_{abs}} + \mathrm{P_{sca}} \\ &= -1/2\mathrm{Re} \int\limits_{o}^{\pi} \int\limits_{o}^{2\pi} E_{\theta\mathrm{inc}} H^*_{\phi\mathrm{sca}} - E_{\phi\mathrm{inc}} H^*_{\theta\mathrm{sca}} \\ &\quad + E_{\theta\mathrm{sca}} H^*_{\phi\mathrm{inc}} - E_{\phi\mathrm{sca}} H^*_{\theta\mathrm{inc}}) R^2 \sin\theta d\theta d\phi \end{aligned} \tag{6}$$

Similarly the extinction cross section may also be defined as

$$\sigma_{\mathrm{ext}} = R_e \frac{2\pi}{K^2} \sum_{n=1}^{\infty} (2n+1)(a_n^s + b_n^s)\mathrm{cm}^2 \tag{7}$$

and the resulting value of a_n^s and b_n^s can be written as

$$a_n^s == j \frac{2^{2n} n!(n+1)! X^{2n+1}}{(2n)!(2n+1)\mu\ 1^{n+\mu} 2^{n+1}}$$

$$\left\{ \mu_2 - \mu_1 - \frac{\mu_2\{(\mathrm{n}+1) + (\mathrm{n}+3)\mathrm{m}^2\} - \mu_1(\mathrm{n}+3) + (\mathrm{n}+1)\mathrm{m}^2}{(2\mathrm{n}+2)(2\mathrm{n}+3)} \right\}$$

$$\begin{aligned} b_n^s = j \frac{2^{2n} n!(n+1)!}{(2n)!(2n+1)!} \frac{X^{2n+1}}{\mu_1(n+1) + \mu_2 m^2 n} |\mu_1 - \mu_2| \\ - \frac{\mu[(n+1) + (n=3)m^2] - \mu_2 m^2[(n+3) + (n+1)m^2]}{(2n+2)(2n+3)} X^2 \end{aligned}$$

In the above equation, $X = 2\pi a/\lambda$ is a very small quantity as compared to unity, and therefore, all the terms having X^7 or higher power can be neglected. The final expression for σ_{ext} can be obtained as

$$\sigma_{\mathrm{ext}} = \frac{\lambda^2}{2\pi} x^3 (C_1 + C_2 X^2 + C_3 X^3) \tag{8}$$

where

$$C_1 = \frac{\sigma \varepsilon_2}{(\varepsilon_1 + 2)^2 + \varepsilon_2^2}$$

$$C_2 = \frac{\sigma}{5} \frac{\varepsilon_2 (7\varepsilon_1^2 + 7\varepsilon_2^2 - 4\varepsilon_1 - 20)}{[(\varepsilon_1 + 2)^2 + \varepsilon_2^2]^2} + \frac{\varepsilon_2}{15} + \frac{5}{3} \frac{\varepsilon_2}{2_1^{\varepsilon} + 3)^2 + 4\varepsilon_2^2}$$

$$C_3 = \frac{4}{3} \frac{(\varepsilon_1 - 1)^2 (\varepsilon_2 + 2)^2 + \varepsilon_2^2 [2(\varepsilon_1 - 1)(\varepsilon_1 + 2) - 9] + \varepsilon_2^4}{[(\varepsilon_1 + 2)^2 + \varepsilon_2^2]^2}$$

And $\varepsilon = \varepsilon_1 - j\,\varepsilon_2$

The derivation of Eq. (8) is based on the concept of Rayleigh scattering from dielectric sphere where dust particle has been assumed to be a dielectric sphere with small circumference. The method may be adopted to derive the expression for n_{ext} which will be similar to Eq. (8). The slab of particles, a model followed by Vande Hulst and accordingly the forward scattering amplitude are written as $[S(O, D)]$ extinction cross section (σ_{ext}) of dust particle is given by

$$\sigma_{\text{ext}} = \frac{4\pi}{K^2} R_e[S(O.D)]$$
$$\text{where } K = \frac{2\pi}{\lambda} \tag{9}$$

The scattering amplitude $S\ (O, D)$ can be defined as

$$S(O, D) = JX^3 \left(M_1 + M_2 X^2 + M_3 X^3 + M_4 X^4 + M_5 X^5 + M_6 X^6 + M_7 X^7 \ldots \right) \tag{10}$$

where

$$M_1 = \frac{m^2 - 1}{m^2 + 2}$$

$$M_2 = M_1 \left[\frac{3}{5} \left(\frac{m^2 - 2}{m^2 + 2} \right) + \frac{1}{30} (m^2 + 2) + \frac{1}{6} \left(\frac{m^2 + 2}{2m^2 + 3} \right) \right]$$

$$M_3 = -j \frac{2}{3} M_1^2$$

$$M_4 = M_1 \left[\frac{3}{350} \frac{(m^6 + 20m^4 - 200m^2 + 200)}{(m^2 + 2)^2} + \frac{2}{225} \frac{(m^2 + 2)}{(3m + 3)} \right] + \frac{1}{315} (m^2 + 2)(m^2 - 2) - \frac{5}{42} \left(\frac{(m^2 + 2)}{(2m^2 + 3)^2} \right)$$

$$M_5 = -j \frac{4}{5} M_1^2 \left(\frac{m^2 - 2}{m^2 + 1} \right)$$

and $m = n_r - jn_1$ is complex refracting index of the spherical particle which is related to complex dielectric constant by the relation,

$$\varepsilon_1 = n_r^2 - n_i^2, \varepsilon_2 = 2n_r n; \tag{11}$$

Neglecting the terms having X^7 and higher power of X in Eq. (10), one has

$$S(O, D) = jX^3 M_1 + jX^5 M_2 + jX^6 M_3 \tag{12}$$

The real part of S (O, D) is given by

$$\text{Re}\,[S(O, D) = X^3 I_m(M_1) + X^5(M_2) + X^6 I_m(M_3) \tag{13}$$

The imaginary parts M_1, M_2 and M_3 are found to be

$$I_m M_1 = \frac{3\varepsilon_2}{(\varepsilon_1 + 2)^2 + \varepsilon_2^2}$$

$$I_m(M_2) = \frac{3}{5}\frac{\varepsilon_2(7\varepsilon_2^2 + 7\varepsilon_2^2 + 4\varepsilon_1 - 20)}{[(\varepsilon_1 + 2)^2 + \varepsilon_2^2]^2} + \frac{\varepsilon_2}{30} + \frac{5}{6}\frac{\varepsilon_2}{(2\varepsilon_1 + 3)^2 + 4\varepsilon_2^2}$$

and

$$I_m(M_3) = \frac{2}{3}\frac{(\varepsilon_1 - 1)(\varepsilon_1 + 2)^2 + \varepsilon_2^2[2(\varepsilon_1 - 1)(\varepsilon_1 + 2 - 9) + \varepsilon_2^4}{[(\varepsilon_1 + 2)^2 + \varepsilon_2^2]^2} \tag{14}$$

Combining Eqs. (12), (13) and (14)

$$\sigma_{\text{ext}} = \frac{4\pi}{K^2}X^3\frac{3\varepsilon_2}{(\varepsilon_1 + 2)^2 + \varepsilon_2^2} + X^2\frac{3}{5}\frac{\varepsilon_2(7\varepsilon_1^2 + 7\varepsilon_2^2 + 4\varepsilon_1 - 20)}{[(\epsilon_1 + 2)^2 + \epsilon_2^2]^2}$$
$$+ \frac{5}{6}\frac{\varepsilon_2}{(2\varepsilon_1 + 3)^2 + 4\varepsilon_2^2} + \frac{\varepsilon_2}{15} + X^3\frac{2}{3} \tag{15}$$

Solving Eq. (15)

$$\sigma_{\text{ext}} = \frac{\lambda^2}{2\pi}x^3(C_1 + C_2X^2 + C_3X^3) \tag{16}$$

where C_1, C_2 and C_3 have the same values as in Eq. (8). It can be observed that the value of extinction cross section (σ_{ext}) obtained by using the concept of Rayleigh scattering and Van de Hulst's slab model gives the same value.

7 Scattering Cross Section Based on Dipole Radiation Concept

Apart from the two concepts discussed earlier, the evaluation of extinction cross section (σ_{ext}) can also be made using the dipole radiation concept. It is well known that for $a \ll \lambda a$ the far zone scattered field from the dielectric sphere is the same as that radiated by a small electric dipole of total strength P_0. Thus, for such cases, a spherical dust particle with a radius (a) much smaller than the wave length of incident plane wave and characterized as a dielectric sphere with complex dielectric constant $\square = \square_i - j\,\square_2$ is considered. If the incident electric field is considered as $E_1 = E_{o}{}^{-jk_x}_{e0}$, then uniform electric field over the extent of dust particle can be given by $E_O a_Z$. It may be emphasized that polarization produced by uniform static field in dust particle is same as would be produced in dielectric sphere. Therefore, the dipole polarization P per unit volume in the sphere is obtained from the solution of Laplace equation and is given by the relation.

$$P = 3\left|\frac{\varepsilon - 1}{\varepsilon + 2}\right|\varepsilon_o E_o a_z \tag{17}$$

The total dipole moment P_O of dust particle is obtained as obtained as

$$\begin{aligned} P_o &= \frac{4}{3}\pi a^3 P \\ &= 4\pi a^3\left(\frac{\varepsilon - 1}{\varepsilon + 2}\right)\varepsilon_o E_o a_z \end{aligned} \tag{18}$$

Now the value of scattered field E_x of a dielectric sphere is attributed to the radiation of small electric dipole P_o; therefore, it will be imperative to obtain the radiation field of a small dipole which as follow.

The vector potential A is given by

$$A = \mu_o I dl \frac{e^{-jk_o r}}{4\pi r} a_z \tag{19}$$

Further in terms of components in spherical coordinates the vector potential A is modified as

$$A = \frac{\mu_o I dl}{4\pi r} e^{-jk_o r}(a_r \cos\theta - a_\theta \sin\theta) \tag{20}$$

Then, the electric field in terms of vector potential A can be found by the relation

$$E = -j\omega A + \frac{\Delta\Delta\,.\,A}{j\omega\mu_o \in_o} \tag{21}$$

Combining Eqs. (20) and (21), one has

$$E = \frac{jZ_o Idl\cos\theta}{2\pi k_o}\left(\frac{jk_o}{r^2} + \frac{1}{r^3}\right)e^{-jk_o r}a_r - \frac{jZ_o ldl}{4\pi k_o}$$

$$\sin\theta\left(\frac{k_o^2}{r} + \frac{jk_o}{r^2} + \frac{1}{r^3}\right)e^{-jk_o r a_\theta} \tag{22}$$

Now for r larger than the wave length $\lambda_{O,}$ Eq. (22) is modified to

$$E = \frac{jZ_o Idl\sin\theta e^{jk_o r}k_o}{4\pi r}a_\theta \tag{23}$$

It is observed from Eq. (23) that the current element *Idl* is equivalent to the time derivative of the dipole moment such that

$$q = \frac{jl}{\omega}or\,Idl = -j\omega P_o \tag{24}$$

Thus *Idl* may be replaced by $j\omega P_O$ for the far zone radiated field. Thus, the far zone scattered field (E) can be written as

$$E_s = \frac{-\omega Z_o k_o P_o\sin\theta e^{-jk_o r}}{4\pi r}a_\theta \tag{25}$$

Since far zone-radiated field of small electric dipole is same as scattered pattern of a small dielectric sphere, Eq. (25) must characterize the scattered field of a dielectric sphere.

Now in sand and dust storms where the particles are in far zone of each other, the far field approximation of the scattered field is valid. Therefore, Eq. (25) can be utilized to obtain the scattered power of dust particles which are characterized as a dielectric sphere. Hence, total scattered power (P_s) for dust particles can be obtained by using Eq. (25) as

$$P_s = \frac{1}{2}\gamma_o^2\int_o^\pi\int_o^\pi |E_s|^2 r^2\sin\theta d\theta d\phi = \frac{\omega k_o^2 Z_o|P_o|^2}{12\pi} \tag{26}$$

Substituting the value of P_O from Eq. (18), (26) one has

$$P_s = \frac{4\pi}{3}a^3(K_o\alpha)^4\gamma_o|E_o|^2\left|\frac{\varepsilon-1}{\varepsilon+2}\right|$$

The scattering cross section (σ_s) is defined as

$$\sigma_s = \frac{P_s}{P_1} = \frac{P_s}{\frac{1}{2}\gamma_o|E_o|^2} = \frac{8}{3}\pi a^2 (k_o a)^4 \left|\frac{\varepsilon - 1}{\varepsilon + 2)}\right| \quad (27)$$

where

$$P_1 = \frac{1}{2}\gamma_o |E_o|^2$$

In the similar fashion, the time average absorbed power is given by

$$P_a = \frac{1}{2} R_e \int_o^a \int_o^{2\pi} \int_o^{\pi} E J_p^* r^2 \sin\theta d\phi = \frac{2}{3}\pi a^3 R_e E J_p^*$$

$$= 6\pi a^3 K_o \gamma_o \left|\frac{\varepsilon - 1}{\varepsilon + 2}\right|^2 \frac{\varepsilon_2 |E_o^2|}{(\varepsilon - 1)^2 + \varepsilon_2^2} \quad (28)$$

Dividing Eq. (28) by incident power density, one can obtain the absorption cross section as follows:

$$\sigma_a = \frac{P_a}{P_1} = 12\pi a^2 (k_o a) \left|\frac{\varepsilon - 1}{\varepsilon_1 + 1}\right|^2 \frac{\varepsilon_2}{(\varepsilon - 1)^2 + \varepsilon_2^2} \quad (29)$$

Now, the extinction cross section (σ_{ext}) which is the sum of scattering cross section (σ_s) and absorption cross section (σ_a) can be given by

$$\sigma_{ext} = \sigma_a + \sigma_s$$

$$= [8/3\pi a^2 (k_o a)^4 + 12\pi a^2 (k_o a) \frac{\varepsilon_2}{(\varepsilon_1 - 1)^2 + \varepsilon_2^2}] \left|\frac{\varepsilon - 1}{\varepsilon + 2}\right|^2 \quad (30)$$

This extinction cross section can be utilized to quantify the attenuation of the signal while propagating through the sand and dust storms.

8 Analytical Treatment of Attenuation

The millimeter wave gets attenuated if its medium contains dust and sand particles. If N (a) da be the number of dust particles per unit volume of storm with radii in the interval a to $a +$ da. Let dust particle's extinction cross section is σ_{ext}, then the total power removed from the wave with incident Poynting vector Sr by the dust particles in volume element of unit cross-sectional area and thickness d is given as

$$\frac{dS_r}{dl} = -S_r \int\limits_{o}^{\infty} \sigma_{\text{ext}} N(a)\text{da} \quad (31)$$

If the integral in this equation is denoted by □ which signifies the attenuation,

$$\text{then} \int\limits_{o}^{\infty} \sigma_{\text{ext}} N(a)\text{da} = \alpha$$

and Eq. 7. (31) modifies to

$$\frac{dSr}{dl} = -S_r\alpha \quad (32)$$

Integrating Eq. (31), one gets.

$$Sr = S_o e^{-\int \alpha dl}$$

where S_0 indicates the amplitude of incident Poynting vector. In general the dust particles are never uniformly distributed over an extended region of storm. Therefore, N (a) and (□) are the function of distance l along the path. When the particles are very small in comparison to wave length as they are in sand and dust storms, it turns out that the attenuation is directly proportional to concentration, irrespective of particle size distribution. Therefore, the attenuation can be written as

$$\alpha = \int\limits_{o}^{\omega} \sigma_{\text{ext}} N(a)da = N\sigma_{\text{ext}}\text{nep/cm} \quad (33)$$

where mono-dispersed particle size distribution is assumed; i.e., all the particles are of the same size. Also, it is assumed that there are no mutual interactions among the particles due to their wide separations. Now combining Eqs. (16) and (33), one has based on Rayleigh scattering

$$\alpha = \frac{N\lambda^2}{2\pi} X^3 (C_1 + C_2 X^2 + C_3 X^3) \quad (34)$$

The total volume of particles in unit volume of the storm is defined by the following relation suggested by Samin and Ghobrial [9].

Volume of particles per unit volume of storm $= \frac{9.43x10^{-9}}{V^r}$

Then number of spherical dust particles per unit volume of storm can be given by

$$N = \frac{9.43x10^{-9}}{V^r x \frac{4\pi}{3} a^3} \quad (35)$$

where V = visibility in km. a = radius of the particle and $\gamma = a$ constant = 1.07.

The value of attenuation can be obtained by combining Eqs. (34) and (35). Thus,

$$\alpha = \frac{9.43x10^{-9}}{V^{r}\frac{4\pi}{3}a^{3}}\frac{\lambda^{2}}{2\pi}X^{3}(C_{1} + C_{2}X^{2} + C_{3}X^{3})N_{p/cm} \tag{36}$$

If the attenuation is taken in dB/km, then Eq. (36) will be modified to,

$$\alpha = \frac{4.34 \times 9.43x10^{-9} \times 3\lambda^{2}X^{3}}{V^{\gamma} \times (4\pi)\alpha^{3}}\left(C_{1} + C_{2}X^{2} + C_{3}X^{3}\right)dB/km \tag{37}$$

The value of attenuation based on the dipole radiation concept can also be obtained by combining Eqs. (30), (33) and (35). The resulting equation is given by

$$\alpha = \frac{434x9.43 \times 10^{-9} \times 3}{V^{\gamma} \times (4\pi)a^{3}} 8/_{3}\pi a^{2}(k_{o}a)^{4} + 12\pi a^{2}(k_{o}a)\frac{\varepsilon_{2}}{(\varepsilon_{1}-1)^{2}+\varepsilon_{2}^{2}}\left|\frac{\varepsilon - 1}{\varepsilon + 1}\right| dB/km \tag{38}$$

9 Numerical Computation

Using Eqs. (3), (5) and (8), the values of σ_{ab}, σ_{sca} and next have been calculated based on Rayleigh scattering as a function of particle size, permittivity and frequency. The data thus obtained are shown in the form of graphs in Figs. 1. The data on various cross sections based on dipole radiation concept, Eqs. (27), (29) and (30) are also shown in these figures. Similarly, using Eqs. (37) and (38), the value of attenuation has been calculated as a function of frequency and visibility for different sample (sand, silt and clay). The calculated values are shown in the form of graphs in Figs. 1, 2 and 3.

10 Discussion of Results

In order to examine the variation of absorption cross section, scattering cross section and extinction cross section with frequency calculations were made using Eqs. (3), (5), (8), (27), (29) and (30) for sand, silt and clay which are supposed to be the main constituents of the storms. The calculated data for σ_{abs}, σ_{sca} and n_{ext} are shown in Fig. 1a–c for different particle size of three constituents. It is observed that values of various cross sections calculated from both the methods of analysis show a good agreement.

Fig. 1 Variation of cross section with frequency

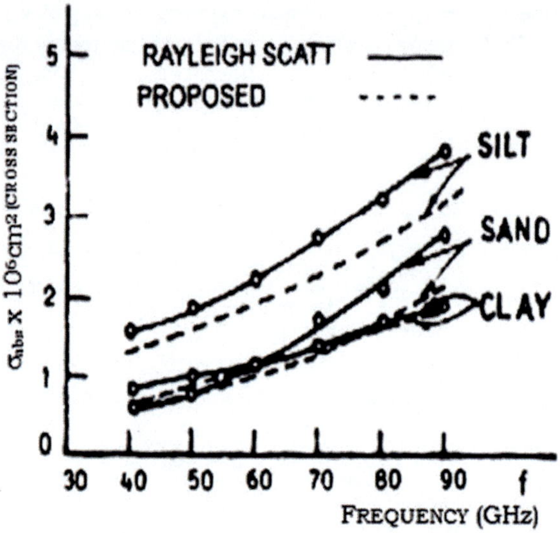

Fig. 2 Variation of cross section with frequency

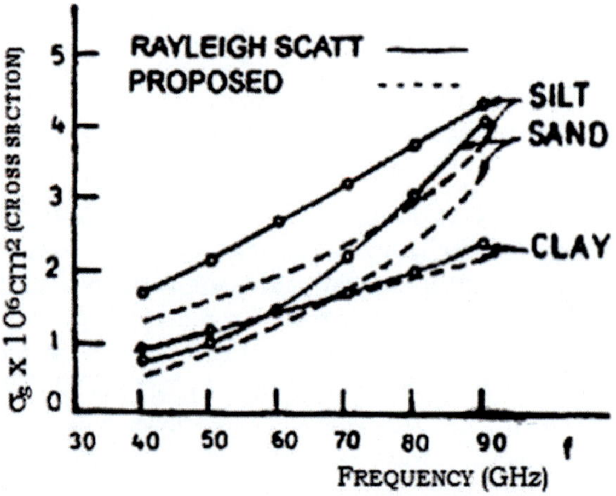

Some computational works are done to find the variation in attenuation with respect to frequency, visibility and particle size and expressed it by using Eqs. (37) and (38). The concept of Rayleigh scattering ($a \ll \lambda$) is used for developing equations for estimating the attenuation. Fig. 2 shows the attenuation variation with frequency for different visibilities; it is observed as frequency increases for given values of visibility attenuation increases linearly.

The change in higher frequency is said to be due to deviation from the assumption ($a \ll \lambda$) on which the Rayleigh scattering is being derived from. The emphasis is given that particle's scattering cross section depends $(2\square a/\lambda)^4$. At lower frequency

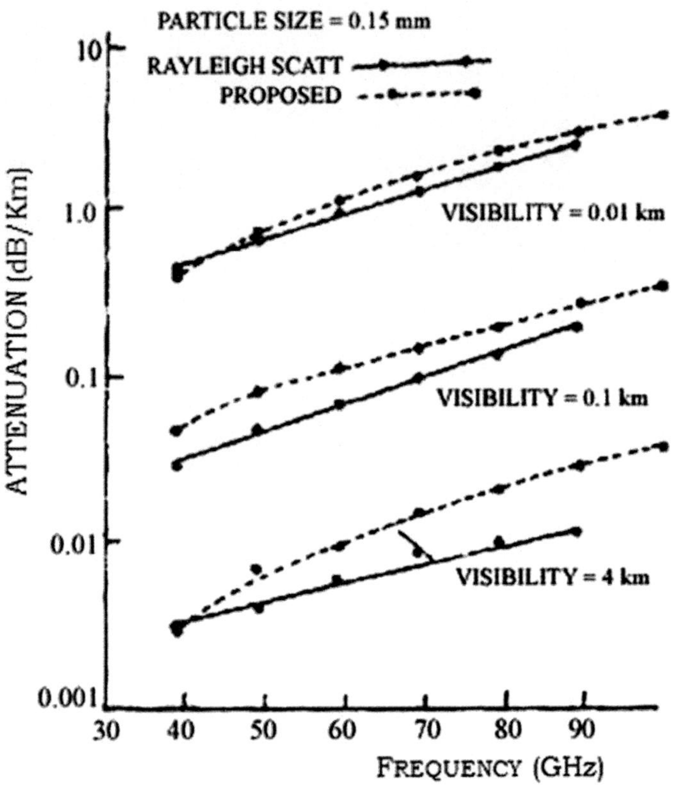

Fig. 3 Variation of attenuation with frequency for different visibilities

that is at longer wavelength, scattering is very weak. This is supported from the data shown in Fig. 1.

According to Rayleigh scattering theory, the particles radius is very much less than the free wavelength. If frequency is increased, then the wavelength reduces. Thus, at higher frequency, the radius of the particle may be comparable to free space wave length and the back scattering cross section approaches the geometrical cross section (□a^2). This may cause enhanced back scattered power resulting into significant increase in attenuation which has been observed in this paper. This is further supported by the radiation concept in which the particles are supposed to radiate as dipole when they are exposed to incident electromagnetic radiation. In this case, the intensity of radiation from particle (dipole) depends directly on the frequency of exciting field. Thus, the increase in frequency is found to increase the scattered reduction on which ultimately enhances the overall loss of energy.

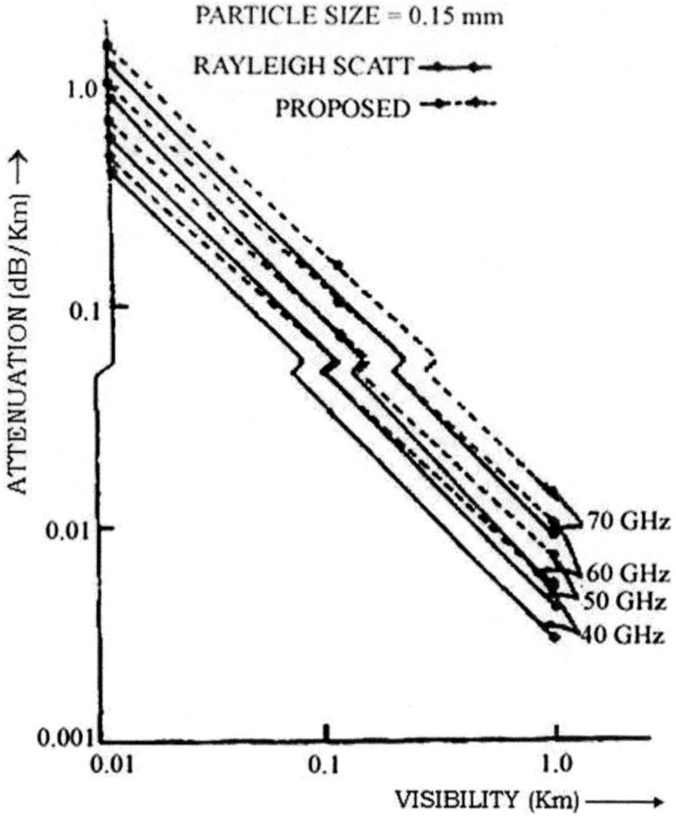

Fig. 4 Variation of attenuation with visibility for different frequencies

So in general the attenuation due to sand and dust particle depends heavily in the maximum particle radius, frequency The concept of Rayleigh scattering as well as dipole radiation could be used to predict the loss due to attenuation at lower frequency side the prediction will be associated with a large deviation due to inherent limitation of the assumptions used in the Rayleigh scattering.

References

1. Sarition JA (1941) McGraw Hill, New York, p 568
2. Bryvel LP, Jones AP (1981) Applied Science Publication, p 19
3. Roderic LO, David VR, Daniel BH (1978) Transaction, vol 20, pp 318–320
4. Kerr DE (1951) McGraw Hill, New York, p 676

5. Van de Hulei HC wilex Chapters 4 and 6, M Alrizzo, HT Al Hafid, BR Vishvakaram'
6. Hof (1993) vol 74, ETI May, pp 26–30
7. Rai CS (1992) Ph.D. Thesis submitted to Banaras Hindu University (India)
8. Knox JE (1979) In: IEEE Conference 79, conference record, vol 2, pp 357–361
9. Ahschular EE (1983) AUARD conference Proceedings, Norway

Solar Power and Water Quality Monitoring Using Wireless Sensor Network with ZigBee

V. Leela Vathi and V. Aravind

Abstract Nowadays, renewable energy power generation is playing a main role in which solar power is widely used. Maximum power point tracking controller (MPPTC) improves the efficiency of solar power system. Providing better quality drinking water to public is a challenge due to pollution in the ground water and contamination even during distribution. Hence, water quality monitoring (WQM) is required. This paper mainly concentrates on developing a flexible, reliable wireless sensor network (WSN)-based WQM across a wireless sensor zone fed by MPPT-controlled solar PV system.

Keywords MPPT · WSN · WQM

1 Introduction

Solar power prevents maintenance and is pollution free, but installation cost is high [1], and a power conditioner is also required for load interface. MPPTC operation [2] maximizes the output power, irrespective of irradiation, temperature, and load characteristics.

WQM [7, 8, 10] is essential for the planet. It consists of spatially distributed autonomous sensor to monitor water parameters [14, 15]. A survey on WQM is given in [11, 12], WSN concept in [4, 8], and its application for real-time WQM is considered in [3, 6, 9]. WSN based on PV energy harvesting system is given in [5]. The application of ZigBee is given in [13].

In Sect. 2, WSN technology is considered, and in Sect. 3, Zigbee technology is explained. Section 4 demonstrates proposed block diagram of the system. Results and discussions are given in Sect. 5. Conclusions are described in Sect. 6.

V. Leela Vathi (✉) · V. Aravind
Department of ECE, GRIET Hyderabad, Hyderabad, Telangana, India
e-mail: leel44d@gmail.com

V. Nath and J. K. Mandal (eds.), *Proceedings of the Fourth International Conference on Microelectronics, Computing and Communication Systems*, Lecture Notes in Electrical Engineering 673, https://doi.org/10.1007/978-981-15-5546-6_61

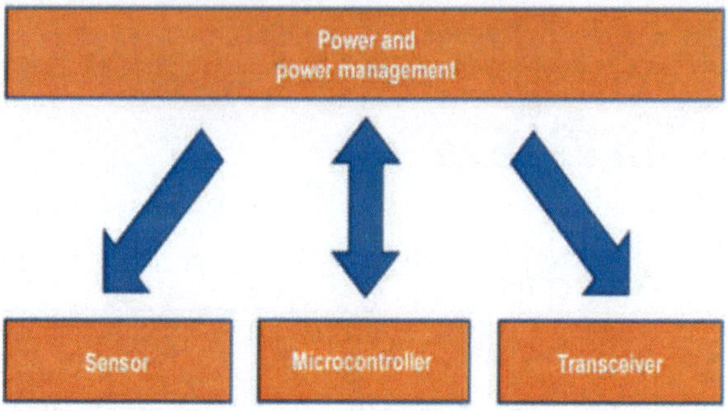

Fig. 1 Structure of a sensor node

2 WSN Technology

The structure of WSN sensor node is shown in Fig. 1; it consists of base station, Arduino microcontroller, a transceiver, and sensors for WQM. Power is supplied from solar panels.

3 Zigbee Technology

Figure 2 shows the block diagram of ZigBee module and various layers of Zigbee. It consists of three nodes: Routers, coordinators, and end device.

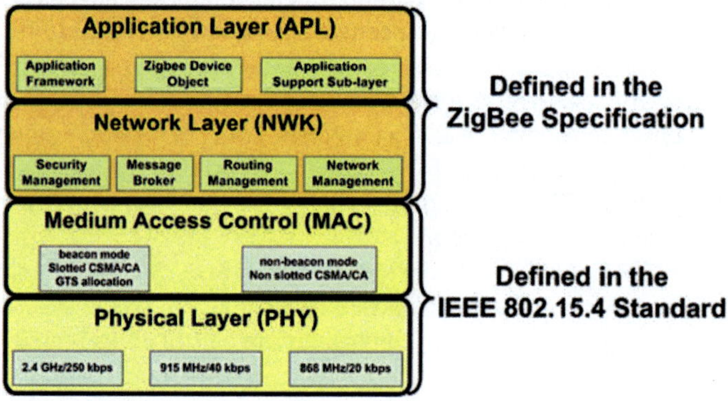

Fig. 2 Block diagram of ZigBee technology

4 Proposed Block Diagram for Domestic Applications

The block diagram of sensor node fed by solar panel is as shown in Fig. 3, which includes three sensors, Arduino microcontroller, solar power module and XBee block. The data processed by the Arduino is transmitted through these modules to coordinator [9] as given in Fig. 4.

Solar PV panel converts light energy available during sunlight into electrical energy to charge the battery and also simultaneously supply voltage to the system. When there is no sunlight, battery supplies voltage to the connected system. For the sensing of solar panel voltage and current, battery voltage, current and system voltages, and current sensors are used.

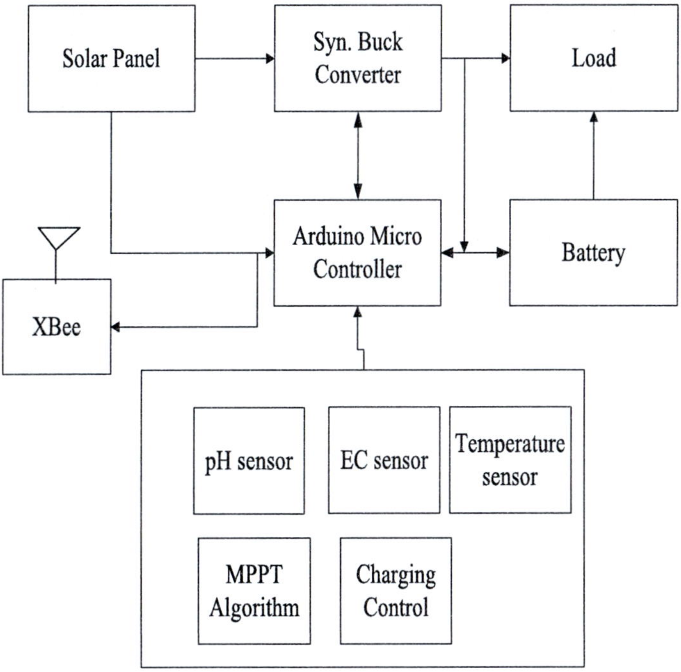

Fig. 3 Block diagram for MPPT solar charge controller and sensor node

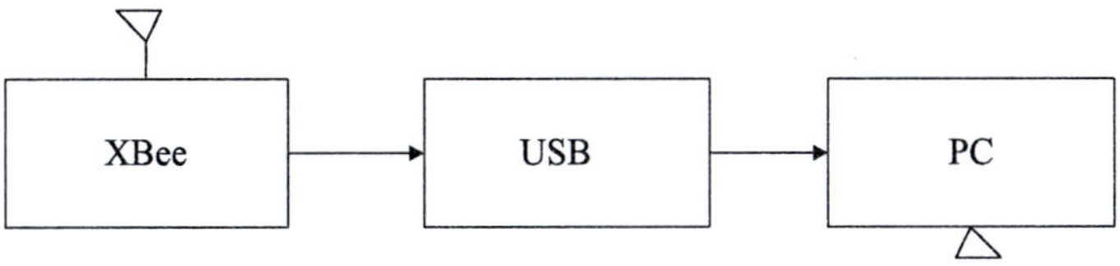

Fig. 4 Block diagram of coordinator

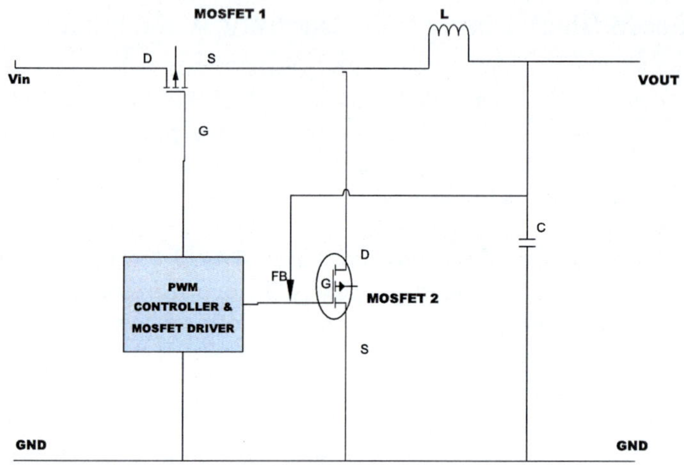

Fig. 5 Synchronous buck converter

Lead acid battery contains sulfur acid coagulated electrolyte so that it cannot spill out. It is made of plates, lead, and lead oxide solution with 35% of sulfur acid to 65% of water ratio.

Charge controller is used for limiting the rate of electric current that is added to or drawn from the batteries. It is used from preventing overcharge and protecting against over voltage as they reduce the battery performance and life span that leads to risk and is also used for preventing deep discharging. MPPT algorithm is used to match closely PV voltage with battery voltage.

The buck converter is a DC-DC converter used in power management. Figure 5 shows the block diagram of synchronous buck converter. MOSFET 1 conducts to charge the inductor by transferring energy from the input. When the switch control is off, MOSFET 2 switch turns on, and it provides a current path for the inductor when discharging. MPPTC with buck-type DC-DC converter controlled by Arduino microcontroller is used.

The pin diagram of AT mega328 microcontroller is shown in Fig. 6. XBee PRO module and its pin diagram used for wireless transmission of data are shown in Fig. 7.

5 Results and Discussion

Figure 8 shows the MATLAB simulation results of the voltages of PV module, inverter, LC filter, and transformer output.

Hardware configuration is illustrated in Figs. 9 and 10 shows the voltage waveform from solar panel and gate pulses to MOSFET is given in Fig. 11.

The energy collected from solar panels is stored in battery with the help of charge controller. To use power from solar to domestic loads, inverter is necessary.

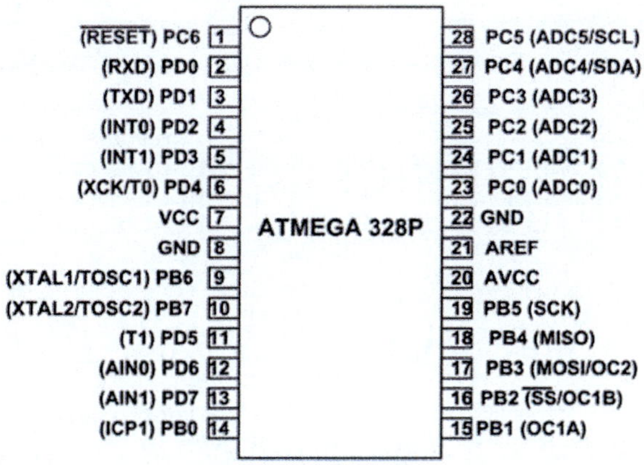

Fig. 6 Pin configuration of AT mega328 microcontroller

Fig. 7 XBee PRO module and its pin configuration

Transformer of 230v/12-0-12v is used in reverse connection to boost the voltage to drive the loads and isolate the inverter circuit from load fluctuations

Sensor nodes placed at different locations collect data and transmit to base station. Figure 12 shows hardware setup of sensors. Figures 13 and 14 illustrate nodes placed on overhead tanks.

Electrical conductivity, pH, and temperature values which represent the water quality are displayed in the serial monitor. Figure 15 shows the values obtained at one sensor node and data collected from various nodes is transmitted through Internet.

After switching ON the power, sensors placed inside the water collect the data. Data center receives the water quality data and processes the data.

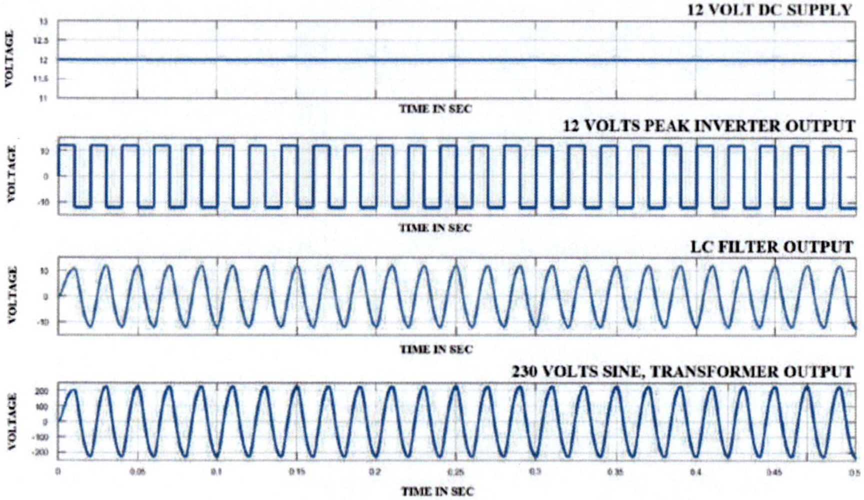

Fig. 8 Simulink waveforms of voltages at different units

Fig. 9 Hardware configuration

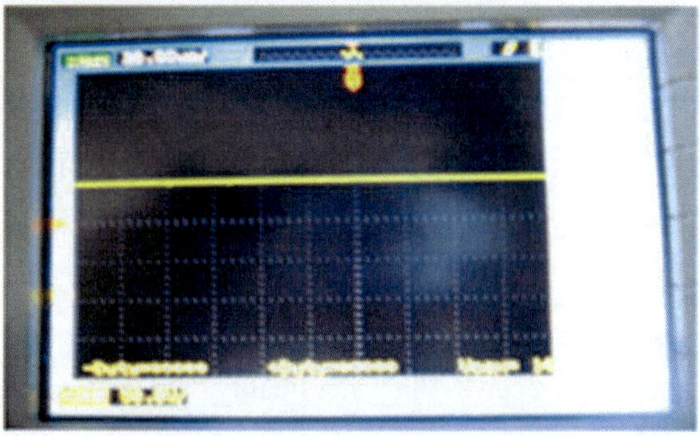

Fig. 10 O/p voltage waveform from solar panel

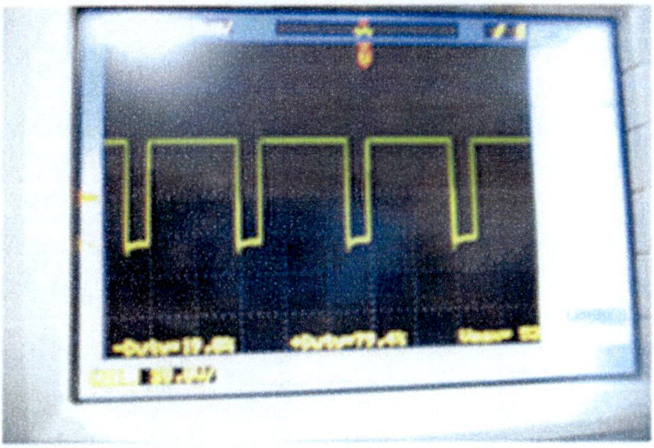

Fig. 11 Gate pulses to MOSFET

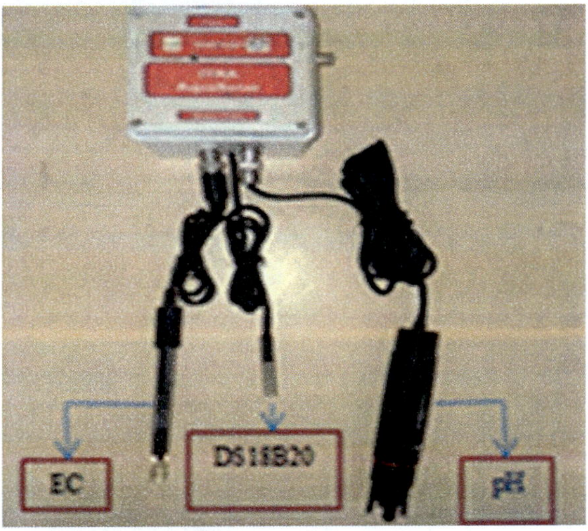

Fig. 12 Implementation of sensor node

In the map view, two locations, Pragathi Nagar and Bachupally, are displayed along with the water quality parameters as shown in Fig. 16. The obtained values are compared with the minimum permissible values. If the values fall beyond the permissible levels, water is not recommended for drinking. The values of water quality parameters obtained in this work are approximately equal to values in the literature [9], but the performance of the system is improved with solar MPPTC.

6 Conclusions

WSN is developed for WQM. Various water quality parameters are analyzed by applying the above model. ZigBee technology is used for data transfer between base

Fig. 13 Node1 placed on overhead tank

Fig. 14 Node 2 placed on overhead tank

station and sensor nodes. Power to the system is supplied from solar power module through MPPTC. Inverter circuit and transformers are used to give supply to AC appliances. It is useful for domestic and agriculture applications.

Fig. 15 Arduino IDE serial monitor

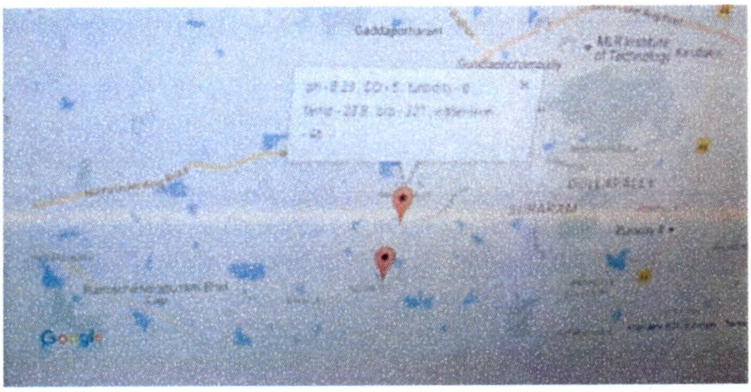

Fig. 16 GUI Web page displaying real-time data

References

1. Trishan E, Patrick LC (2007) Comparison of photovoltaic array maximum power point techniques. IEEE Trans Energy Convers 22(2):439–449
2. Anshul G, Srinivas NVNS (2010) Implementation of MPPT solar charge controller with power Psoc. J Cypress Perform Document No. 001–54121, 1–14
3. Menon KAU, Divya P, Ramesh MV (2012) Wireless sensor network for river water quality monitoring in India. In: 2012 IEEE third international conference on computing, communication and networking technologies (ICCCNT'12), Coimbatore, pp 1–7
4. Amruta MK, Satish MT (2013) Solar powered water quality monitoring system using wireless sensor network. In: 2013 IEEE international mutli-conference on automation, computing, communication, control and compressed sensing (iMac4s), Kottayam, pp 281–285
5. Li Y, Shi R (2015) An intelligent solar energy-harvesting system for wireless sensor networks. Springer EURASIP J Wireless Commun Netw 179:1–12
6. Wong BP, Kerkez B (2016) Real-time environmental sensor data: an application to water quality using web services. Elsevier Environ Modell Software 84:505–517

7. Malika Haroon, Szwilskib Anthony (2016) Towards monitoring the water quality using hierarchal routing protocol for wireless sensor networks. Procedia Comput Sci 98:140–147
8. Kageyama T, Miura M, Maeda A, Mori A, Lee SS (2016) A wireless sensor network platform for water quality monitoring. In: IEEE international conference on sensors, pp 1442–1444
9. Sowmya C, Naidu CD, Prasad R (2017) Implementation of wireless sensor network for real time overhead tank water quality monitoring. IEEE 7th international advance computing conference (IACC), pp 546–551
10. Mane V, Medsinge P, Chavan A, Patil S (2017) Water quality measuring system using wireless sensor network. Int Res J Eng Technol 4(2):1926–1930
11. Bokingkito Jr PB, Llantos OE (2017) Towards Monitoring the water quality using hierarchal routing protocol for wireless sensor networks. Procedia Comput Sci 124:698–705
12. Pule M, Yahya A, Chuma J (2017) Wireless sensor networks: a survey on monitoring water quality. J Appl Res Technol 15(6):562–570
13. Suryawanshi V, Khandekar PM (2018) Design and development of wireless sensor network (WSN) for water quality Monitoring using Zigbee In: 2018 IEEE second international conference on intelligent computing and control systems (ICICCS), Madurai, India, pp 862–865
14. Srivastava S, Vaddadi S, Sadistap S (2018) Smartphone based system for water quality analysis. Springer Appl Water Sci 8(130):1–13
15. Chen Y, Han D (2018) Water quality monitoring in smart city: a pilot project. Elsevier Autom Constr 89:307–316

Designing and Implementing a Fully Differential Amplifier Using Current Conveyor of Third Generation

Pooja Gupta, Rajeev Kumar Ranjan, and Vijay Kumar Verma

Abstract The emerging current mode devices are being familiar because of their absolute linearity and larger constant bandwidth as compared to the presently available voltage mode devices. This paper emphasizes the designing of a two-stage differential amplifier of which the high gain stage constitutes a current mirror converting the differential current wisely to an output current which is single ended. The advantage of having higher output impedances in filter with active current mode elements can facilitate easier cascading of additional filters. These cascading on outputs of active devices can indeed improve filter responses to a greater extend which is not possible while using passive elements. The idea of designing a differential amplifier using active current conveyor blocks is pronounced widely because of their low component count, lower output impedances, higher input impedances, higher bandwidth, gain, and accuracy. Moreover, this differential amplifier with current mode devices exhibits greater reduced voltage excursion in sensitive nodes, lesser power usage, better accuracy and bandwidths, lower harmonic distortions, and excellent phase matching output gains. The simulations here are carried out using Cadence/PSpice.

Keywords Analog signal processing · Current conveyors · Voltage and current source · Amplifier · CMOS device

P. Gupta (✉)
Departtment of ECE, KIIT Bhubaneshwar, Bhubaneshwar, India
e-mail: poojajune241992@gmail.com

R. K. Ranjan
Department of ECE, IIT-ISM Dhanbad, Dhanbad, India

V. K. Verma
ISRO Bangalore, Bangalore, India

V. Nath and J. K. Mandal (eds.), *Proceedings of the Fourth International Conference on Microelectronics, Computing and Communication Systems*, Lecture Notes in Electrical Engineering 673, https://doi.org/10.1007/978-981-15-5546-6_62

1 Introduction

The current mode devices have paved its way after the undoubted rule of voltage mode circuits for more than four decades in designing and innovation of linear and nonlinear circuits. Potential advantages of simpler circuitry, higher slew rates, inherently wider bandwidths, and lower power consumptions have made the current mode circuits to receive a considerable attention [1–3]. The operational transconductance amplifier, the four-terminal floating nullor [4], and the current feedback operational amplifier are popular realization of current mode devices employing current conveyors.

Moreover, the enhancements in analog circuit and systems are due to the improving quality of the active elements. Recent years have shown tremendous increase in the growth of electronically controllable active elements. Signal processing applications have also found many advantages [3, 5–8]. The current mode devices also exhibit single-parameter control of input resistances, voltage gains [9, 10], transconductance, and trans-resistance. A fully differential amplifier is basically a balanced output difference amplifier which specializes in providing lower distortions in output and higher voltage swing. These are widely employed in many of the high-performing analog integrated circuits. These fully differential amplifiers are seen to have various advantages over the amplifiers with single-ended outputs. These amplifiers provide larger output voltage swing and are very less susceptible to common mode noises.

The use of CCI was prevalent and then came the second-generation current conveyor, namely the CCII where one of the input terminals named "Y" has high input impedance, and this factor turned to be a disadvantage once the CCII was used in an instrumentation amplifier to handle the differential signals. This disadvantage overpowered the advantages of higher dynamic ranges, greater bandwidth, better linearity, and lower power consumption with respect to the voltage mode counterparts [11, 12]. Most of the current conveyor applications were also limited because of the use of current mirrors which have an inaccurate current following action which was then resolved on introducing floating current source (FCS) [4]. Then came the third-generation current conveyor (CCIII) [13] which was implemented much later [13, 14] and this was again followed by controlled current conveyor III (CCCIII) for realizing the floating resistance using the intrinsic resistance of the input X terminal which is controlled by the DC bias current.

This paper is precisely organized into section where the second section emphasizes on the prevailing literature and the studies accounting the proposed work. The third section comprises of the proposed method and well describes the operation of these circuits and the active devices. The fourth section of the paper discusses the simulation results, parameters of the proposed design, and the performance of the proposed approach. The last section well summarizes all the main results of this paper.

2 Proposed Work

A CMOS-based differential current controlled CCII [15] can be easily integrated conveniently as it comprises of CMOS transistors. But the CMOS translinear circuit is difficult to fabricate and it has restricted practical applications [16]. The current controlled differential voltage current conveyor (CCDVCC) can be used to fabricated numerous analog circuits [17], which retain the benefits of electronic tunability and availability of the differential signal at the input port.

Here, we try proposing a novel design of both current- and voltage-controlled fully differential amplifier using third-generation current conveyors. Further, the characteristics of this proposed design are studied and employed for implementing differential amplifier using computer-based simulation software. These requests are accomplished using either by using propelled advances or adapting new design procedures [18]. These days the voltage buffers are integrated with differential amplifier and flipped voltage follower [19] to achieve higher input impedance. Cascading of differential amplifiers helps achieving a much higher gain.

In the research work done by [12, 20–23], both second- and third-generation current conveyors were presented incorporating CMOS in translinear loop, and it was optimized on the basis of static and dynamic operation. Herencsar et al. [24] and Alaybeyoglu et al. [25] developed a novel application of gain-variable third-generation current conveyor (GVCCIII) and Z-Copy current differencing buffered amplifier (ZC-CDBA), respectively, using third-generation current conveyor, and they further suggested designing of useful proposed active element in analog signal processing besides the other topologies in the literature [26]. However, the researchers recommend the design of filters which can operate at higher frequencies in microwave range. Previous studies further suggested that the enhancement techniques have to be developed to further increase the bandwidth, to reduce the output offset current errors and static power dissipation by designing of low biasing currents, and to operate in low supply voltages [27]. In this regard, the researcher attempts to design and implement a fully differential amplifier in both current- and voltage-controlled mode using current conveyors.

3 Current Conveyors and Its Three Generations

The "current conveyors" are one of the inventions that happened to be by chance, and these have today benefited the electronic circuits and systems widely. As the history prevails that in late sixties when A. S. Sedra working on the design of programmable instruments under the supervision of Prof K. C. Smith, discovered a circuit which could precisely convert voltage to current using a bipolar transistors. This circuit again was employed in designing of voltage-controlled waveform generator. However, simultaneously they came up with the realization of a new three-port network which was then termed as the "current conveyor." It is basically a current follower which has a defined unity gain, along with the added advantage that here that the voltage at one of the input terminals is even conveyed to the other input terminal. This invention initially did not receive popularity but later on it was realized that the use of current instead of voltage as active parameter finds better accuracy, higher usable gain, and better bandwidth due to the reduction of voltage excursion at sensitive nodes. Further, a lot many numbers of modifications and improvements were made to the newly discovered current conveyor for bringing up new generations and variations.

The various generations of current conveyors, namely CCI, CCII, and CCIII are being used for performing different applications like the oscillators, multifunction and universal filters [17], in differentiators, integrators, and many more. The absence of a dominant pole in low-frequency region makes its fundamental frequency limit much higher [17]. This is again an added advantage over voltage mode. Subsequently the designing of current mode devices encounters difficulties as the count of the diode-connected transistors probably increases. Further the PMOS current mirrors in the CC cannot be used directly because of the current following action which then results in an extremely wide response in frequency [14]. Therefore, the research these days focuses in improving techniques in order to achieve lesser power consumption along with lower supply voltages and adaptable features of high frequency and slew rate [28].

The basic third-generation current conveyor (CCIII) [16] is a current-sensing unit in floating or grounded terminal. Later, a number of CCIII and its variants were developed in the literature [29, 30]. A third-generation CC is typically represented as (Fig. 1).

The matrix equation for an ideal third-generation current conveyor is

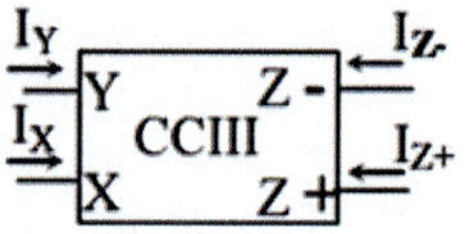

Fig. 1 Ideal third-generation current conveyor. *Source* Adopted from Feki and Masmoudi [23]

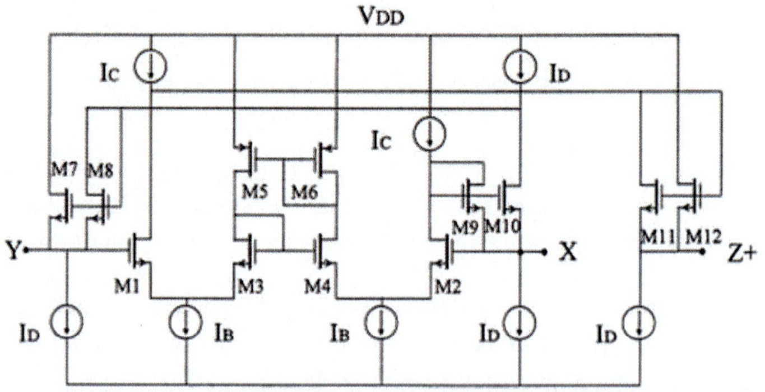

Fig. 2 Circuit configuration of the CCIII. *Source* Adopted from Ismail and Soliman [14]

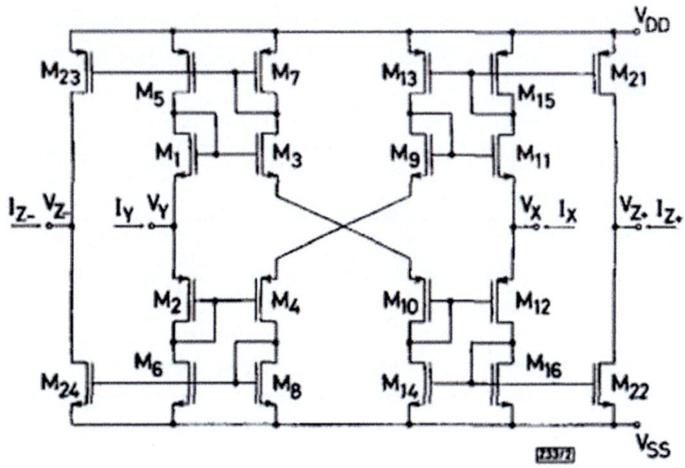

Fig. 3 CMOS implementation of CCIII

$$\begin{pmatrix} I_Y \\ V_X \\ I_Z \end{pmatrix} = \begin{pmatrix} 0 & -1 & 0 \\ 1 & 0 & 0 \\ 0 & \pm 1 & 0 \end{pmatrix} \begin{pmatrix} V_Y \\ I_X \\ I_Z \end{pmatrix}$$

Here, the positive I_z denotes the non-inverting CCIII+ and negative current I_z denotes the inverting CCIII-, respectively, where the currents are all flowing into the nodes. The port X and Y carries a negative current gain that enables the use of the circuit as an integrated floating current-sensing device. The figure below shows the circuit configuration of CCIII+. CCIII as shown in Fig. 3 can be obtained using the same principle. The wider bandwidth and better performance of these implementations have made this new building block to be very demanding for many applications [14] (Fig. 2).

3.1 Advantages of Third-Generation Current Conveyors

The advantages of CCIII include high linearity, low gain errors, and wider frequency responses. Moreover, the cascading of CCIII is easier with other circuits because of its high output impedance at the Z-terminal devoid of additional active elements. In addition, the impedances at X and Y terminals of the CCIII are low. CCIII can be used as input cell of probes and current-sensing units. They have shown significant appearances in many applications as they can provide the current in the floating terminal as well.

The third-generation current conveyor-based inductances and simulators are being widely used in designing of current mode circuits like active filters and can be used further in the realization of equivalent circuits. Furthermore, CMOS current conveyors are more suitable in many high-frequency applications with wide operating range and excellent frequency responses [14]. CCIII can be made to realize equivalent biquad as well.

3.2 CMOS Implementation of CCIII

The figure below shows the CMOS implementation of CCIII.

All the transistors in the figure are biased in saturation region for precise circuit operation. The performance of the circuit is adversely affected whenever there is a decrease in the device threshold voltage and supply voltage of CMOS. This brings into picture the reduced dynamic range, more low noise margins, and increased propagation delay. This usefulness and simpler circuit make the incorporation of CMOS current conveyors as an emerging class of effective performance analog circuit design based upon current mode approaches. This study emphasizes the designing of fully differential amplifier using third-generation conveyor, and the proposed design works toward increasing the gain–bandwidth product and reduction in power consumption in analog circuit. Additionally, this paper deals with current conveyor-based design criteria for the low power operational transconductance amplifier. Here, the CCIII is used for the control on replacing the input bias current (I_{bias}). Cadence simulation platform is used for the performance analysis on Microwind layout which shows the high-frequency application of the proposed circuit.

3.3 Current and Voltage Mode Design

Attainment of high bandwidth and impedance levels can be achieved appropriately by using current and voltage mode design in order to develop high performing analog circuits. The proposed design of CCIII can be evaluated in terms of its performance by calculating the current and voltage levels at each node. In order to achieve a better signal-to-noise ratio, we need to use a higher supply voltage [31]. The parasitic capacitance in transistors results in lower power consumption and improved speed due to its charging and discharging mechanism at the nodes. This mechanism starts once the supply voltage is distributed to the nodes. Many others like the log domain filter and switched current filters can make use of this model. But anyhow this model is not found to be useful in areas which require higher linearity. The current mode operations need to be incorporated with linearization techniques to decrease the nonlinearity introduced due to device mismatching and other phenomenas.

In order to obtain high performance in analog circuits, the current and voltage mode design is an appropriate approach toward attaining high bandwidth and impedance levels. The performance of the proposed FDACV-CCIII design is on the basis of the voltage and current levels at every node that includes input and output node. Commonly, the supply voltage is considered to be higher for attaining suitable signal-to-noise ratio (SNR) [31]. Also, the parasitic capacitance of the transistors utilized is charged and then discharged once the supply voltage gets distributed to the nodes, which decreases the power consumption and improves the speed. In this manner, it is possible to obtain greater speed and lower power consumption in the voltage and current mode approaches [32]. This model can be applied in switched current filters, log domain filters and also in nonlinear current and voltage mode circuits. However, in case of device mismatches, which leads to the nonlinear operation and generates distortion, they cannot be used in applications which require high linearity. Hence, linearization technique is introduced in current mode operation to decrease the nonlinearity. Moreover, by this approach, voltage swing is also not disturbed. These characteristics aid in possessing zero input impedance, infinite output impedance, and a constant current gain. Also, the current amplification will result in a high level of static power consumption; therefore, the current gain of ideal current mode circuits is set to unity. This constant current gain up to large frequency range enables current and voltage mode circuits to be used for high-frequency applications.

The circuit proposed here is implemented using 0.45 nm CMOS technology. Replacing the input bias current increases the gain–bandwidth product and reduces the power consumption which is the basic need of today's analog circuit. The following figures illustrate the current mode and voltage mode current conveyor implementation in CMOS, respectively (Figs. 4 and 5).

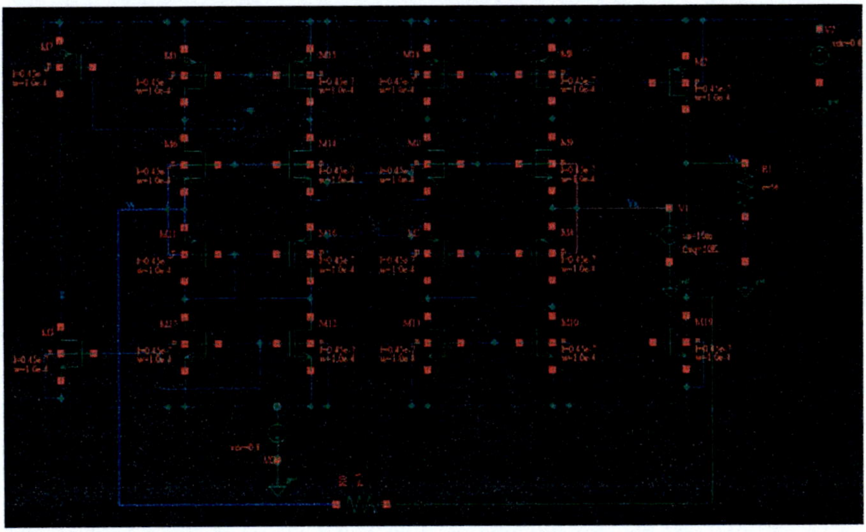

Fig. 4 Fully differential current mode current conveyor implementation in CMOS

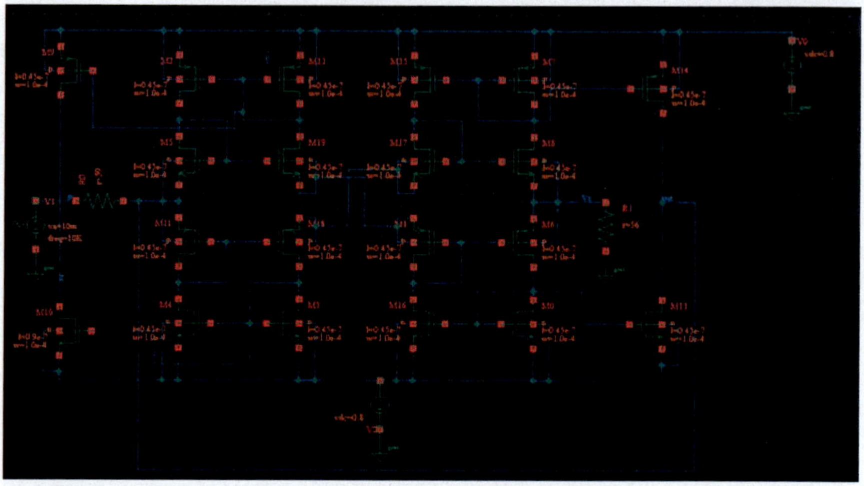

Fig. 5 Fully differential voltage mode current conveyor implementation in CMOS

3.4 Current and Voltage Mode Design

In this study, the design of fully differential amplifier in third-generation conveyor is done, and the proposed design will increase gain–bandwidth product and reduce power consumption in an analog circuit. In addition, this paper deals with current conveyor-based design criteria for the low power operational transconductance amplifier (current and voltage mode). Its transconductance has been controlled by using third-generation current conveyor, replacing input bias current (I_{bias}). The performance is obtained through Cadence simulations (circuit layout-DSCH2 and Microwind3) which show the usability of the proposed circuit for high frequencies. The proposed FDACV-CCIII circuit is implemented using 0.45 nm CMOS technology. Replacing input bias current increases the gain–bandwidth product and reduces power consumption which is the basic need of today's analog circuit (Figs. 6 and 7).

The layout designs for proposed FDACV-CCIII are shown in Figs. 8 and 9. The layout depicts the series connection of CMOS devices for input power supply applied. Further, this layout design contains 19 CMOS transistors for cascade signal transmission. The layout design is developed in the DSCH2 software, and further, it is converted into Verilog file and implemented in the Microwind3 software. Performance evaluation of the proposed FDACV-CCIII is carried out using the application of Microwind3, and the determination of impedance levels, transfer characteristics of current and voltage mode, evaluation of magnitude and bandwidth of the proposed FDACV-CCIII are carried out.

The CCIII has a variable voltage gain by changing the resistors R and R_L. The voltage gain can be theoretically computed as:

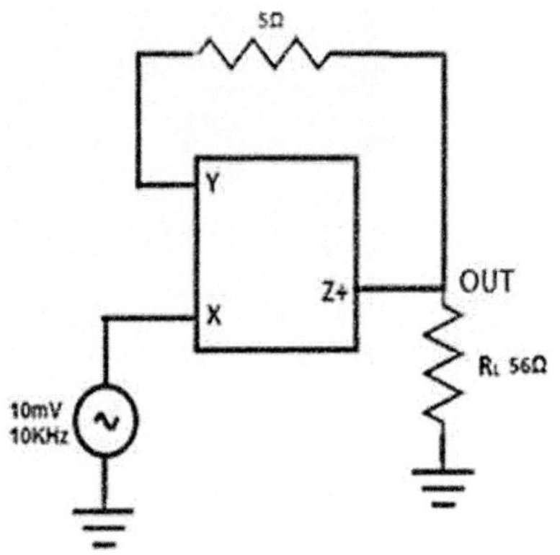

Fig. 6 Circuit for current mode current conveyor

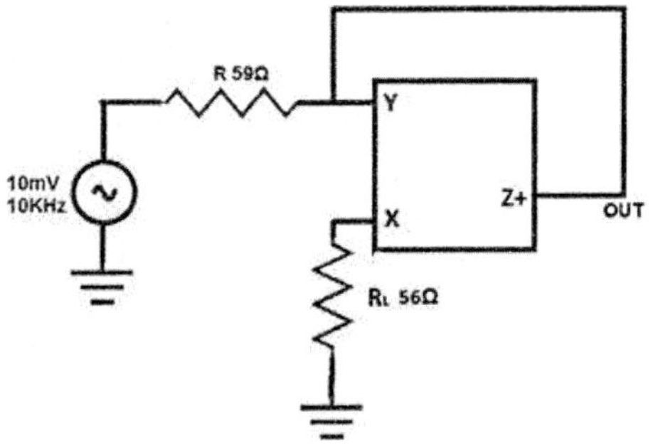

Fig. 7 Circuit for voltage mode current conveyor

$$\beta = \frac{R_L}{R}$$

Assuming that the output resistances of the transistors M15–M18 are equal to R_o, the resistances seen at input and output terminals of the CCIII at mid-frequency region can be found as follows:

$$\begin{aligned} R_X &= \frac{1}{\frac{r_{o13}+r_{o19}}{r_{o13}r_{19}} + g_{m15}\frac{r_o}{2}(g_{m13}+g_{m19})|} \\ &\cong \frac{2}{g_{m15}r_o(g_{m13}+g_{m19})}, \end{aligned} \tag{1}$$

$$R_Y = \frac{r_{o13}r_{o19}}{r_{o13}+r_{o19}}, \tag{2}$$

$$R_Z = \frac{r_{o10}r_{o16}}{r_{o10}+r_{o16}}, \tag{3}$$

where r_{oi} and g_{mi} are the output resistance and transconductance of the ith CMOS transistor, respectively.

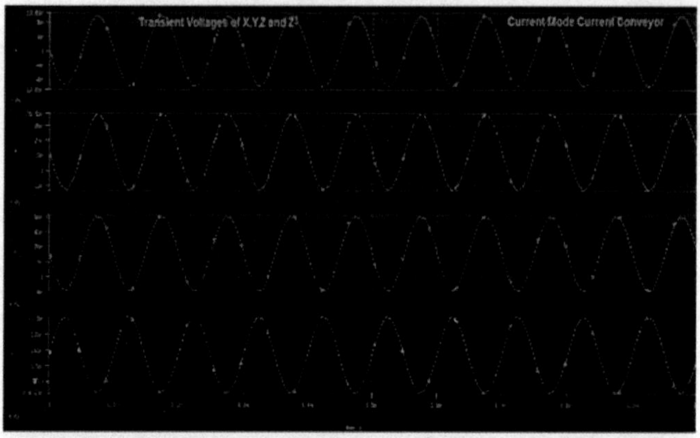

(a)

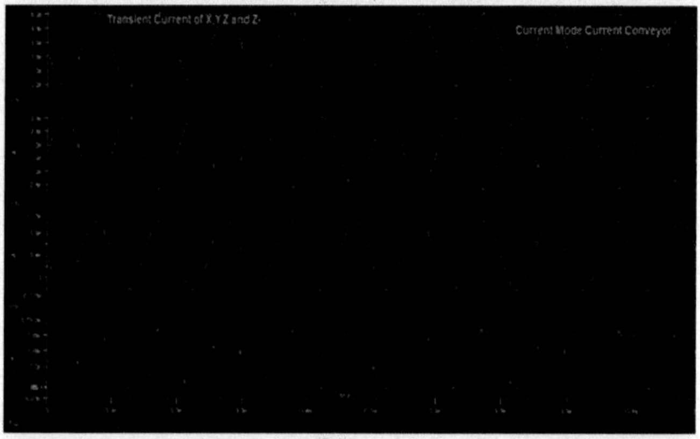

(b)

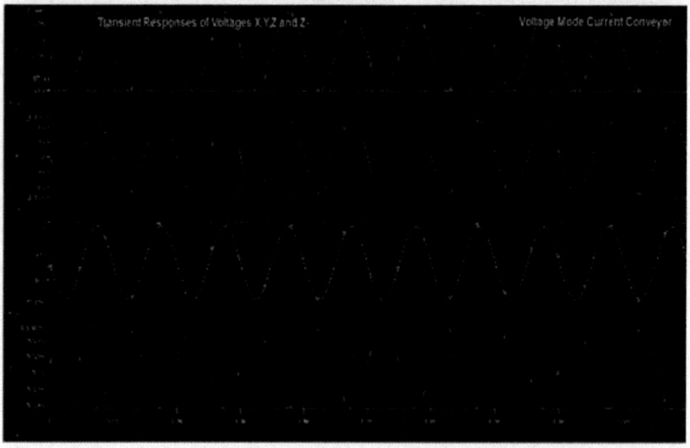

(c)

Fig. 8 Transient response

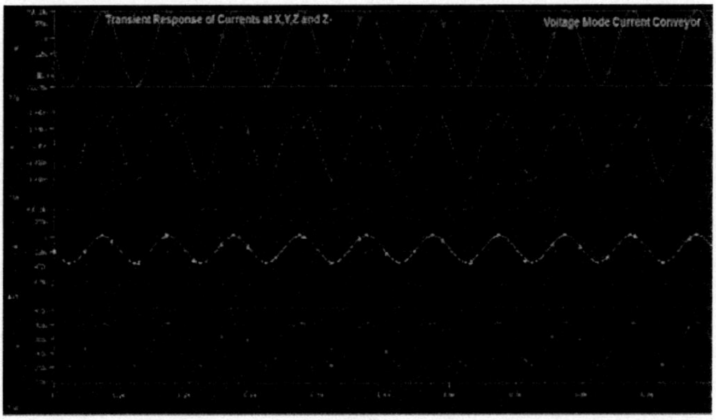

(d)

Fig. 8 (continued)

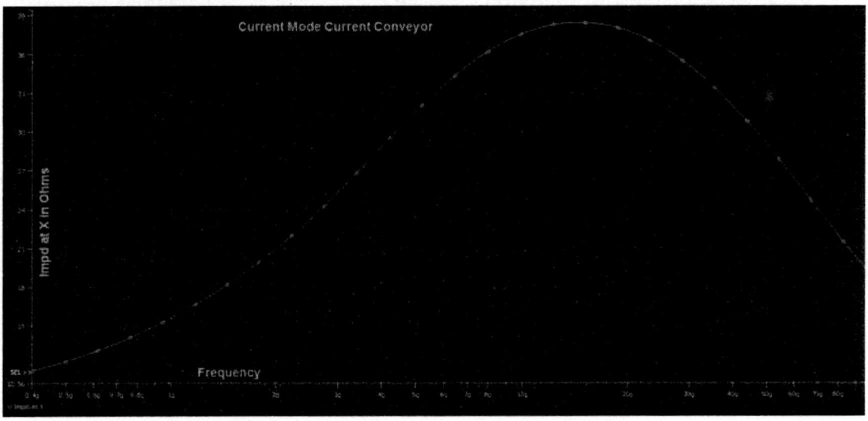

Fig. 9 Impedance at *X* over frequency

The current transfer characteristics are given by

$$\beta = \frac{gm3gm6gm7 + gm4gm5gm8}{gm5gm6(gm3 + gm4)} \cong 1 \tag{4}$$

if $gm5 = gm7$ and $gm6 = gm8$

The impedance level at *Y* node can be ensured by employing good biasing sources showing high resistances as:

$$Z_y = \left(\frac{ro1}{1 + gm1ro1} + R_i bias1\right) || \left(\frac{ro2}{1 + gm2ro2} + R_i bias2\right) \cong \frac{R_i bias1}{R_i bias2} \tag{5}$$

Impedance level at port X,

$$Z_x = \frac{1}{gm3 + gm4 + \frac{ro3+ro4}{ro3ro4}} \cong \frac{1}{gm3 + gm4} \tag{6}$$

Impedance level at port Z is typically high, and it is given as,

$$Z_z = \frac{ro7ro8}{ro7 + ro8} \tag{7}$$

The transconductance of the ith CMOS transistor can be obtained by the below equation as follows:

$$g_m = \sqrt{\beta I_D} \tag{8}$$

where β denotes the current gain of the transistor and I_D represents drain current.

The transconductance from drain to source for the transistor is represented by the equation as follows,

$$g_{ds} = \lambda I_D \tag{9}$$

$$g_{mb} = \frac{g_m \gamma}{K} \tag{10}$$

In which λ denotes channel length modulation parameter and the drain current, I_D and K is given by

$$K = \frac{V_t - V_{t0}}{\gamma} + \sqrt{\beta I_D},$$

in which V_t and V_{t0} represent the voltage at time t and t0, respectively. Hence, the impedance levels at the ports X, Y, and Z can be determined by Eqs. (5), (6), and (7). Further, the transconductance of each terminal of each transistor can be determined from Eqs. (8), (9), and (10). Additionally, the current transfer characteristics can be evaluated using Eq. (4).

4 Experimental Results

This section covers the description of the parameters that are employed in the proposed FDACV-CCIII. Further, the experimental results are conducted in Cadence software by the implementation of 0.45 nm CMOS technology. The proposed model is designed using DSCH2 software, and further, the design is converted into Verilog file, and the performance evaluation is carried out using Microwind3 simulation software. Subsequently, the simulated results are depicted along with the interpretations and analysis of the FDACV-CCIII's performance.

4.1 Parameter Description

The following Table 1 lists down the description of the parameters utilized in the design and implementation of the proposed FDACV-CCIII. This study made use of suitable values for simulating the circuits in order to attain the optimal performance and efficiency of the proposed FDACV-CCIII.

The results of the performance of proposed FDACV-CCIII are shown in Table 2, which was simulated using Cadence. Aspect ratios of transistors are given in the above table, and 0.45 nm CMOS technology process is employed. Considered supply voltages are $V_{\mathrm{DD}} = V_{\mathrm{SS}} = 1.25$ V and I_B is set to 200 mA. Current following action of the CCIII at node X when the input voltage Y is probed from −1 V to 1 V is obtained. It can be seen that the offset voltage at node X is less than 1 mV when Y is grounded. Moreover, the proposed FDACV-CCIII operates in a saturation region, and the transistors are operated at different threshold values. The obtained average threshold voltage is found to be having negative value; hence, the insensitivity of the circuit toward the varying threshold values can be interpreted. Since the drain current (I_D) values are higher, the minimum usable frequency can be obtained.

Table 1 Description of parameters

Description	Variable	Value
Technology used	CMOS	0.45 nm
Channel width	W	1.0 e^{-6}
Channel length	L	0.45e^{-7}
Supply voltage	V_{DD}	1.25 V
Bias current	I_B	50μ
Oxide capacitance	Cox	3.9
Voltage gain	A_v	0.55
Current gain	A_I	0.55
NMOS process transconductance parameter	λ_n	0.02
PMOS process transconductance parameter	λ_p	0.12
Transconductance parameter	$K1$	0.4

Table 2 Transistor aspect ratio specifications of proposed FDACV-CCIII

Trans	Type	Region	Vgs (mV)	Vds (mV)	Vth (V)	gm (μS)	gds (μS)	gmb (μS)	Id (uA)
M0	n	Sat	246.74	246.714	0.265	245.12	3.12	46.74	168
M1	p	Sat	−182.96	−1.069(V)	−0.215	437.67	59.62	83.4	447.2
M2	n	Sat	246.714	1.598(V)	0.2386	3500	643.47	671.6	32
M3	p	Sat	−471.842	−1.069(V)	−0.4632	111.37	3.86	21.23	32.55
M4	p	Sat	−471.842	−471.842	−0.375	1300	516.48	245.64	4.4
M5	p	Sat	−182.968	−800.65	−0.201	243.21	18.41	46.37	146.9
M6	n	Sat	446.454	129.073	0.54	1300	86.81	246.67	4.4
M7	n	Sat	527.242	374.31	0.466	1300	86.81	780	4.4
M8	n	Sat	446.454	446.454	0.31	1900	187.74	362	9.4
M9	p	Sat	−471.842	−471.842	−0.375	1300	516.48	245.64	4.4
M10	p	Sat	−182.968	−182.968	−0.1788	56.44	0.99	10.76	8.29
M11	p	Sat	−624.033	−306.652	−0.485	1900	1100	362.76	9.3
M12	p	Sat	−471.842	−624.773	−0.585	1500	714.05	288.82	5.9
M13	n	Sat	415.777	415.777	0.308	1500	119.01	288.83	5.9
M14	n	Sat	46.23672	663.924	0.0503	56.43	0.165	10.76	8.36
M15	p	Sat	−295.875	−295.875	−0.2775	253.04	19.93	48.25	162.8
M16	p	Sat	−289.66	−306.52	−0.271	243.24	18.42	46.38	167.69
M17	n	Sat	446.454	530.527	0.4416	2100	221.86	394.35	11.9
M18	n	Sat	246.109	246.109	0.2635	243.24	3.07	46.38	151.97
M19	n	Sat	246.714	263.574	0.2649	253.04	3.32	48.25	166

4.2 *Simulation Results*

The performances of the proposed current conveyor circuits shown in the table are simulated using PSpice. Transistors' aspect ratios are given in the above and 0.45 nm CMOS technology process is employed. Supply voltages are $V_{DD} = V_{SS} = 1.25$ V and I_B is set to 100 mA. Current following action of the CCIII at node X when the input voltage Y is scanned from −1 to 1 V is obtained. It can be seen that the offset voltage at node X when Y is grounded is less than 1 mV.

In order to prove the practical utility of all the proposed circuits, the PSpice\Cadence simulations using the 0.45 nm CMOS process parameters are performed. The supply voltages used are $V_{DD} = V_{SS} = 1.25$ V and $V_B = -0.5$ V.

Figure (a) depicts the transient responses of current at terminals, *X*, *Y*, and *Z*, and Fig. (b) depicts the transient responses of currents at *X*, *Y*, *Z*, and *Z*- in current mode current conveyor. Similarly, Fig. (c) represents the transient responses of currents at terminals, *X*, *Y*, *Z*, and *Z*− , and Fig. (d) represents transient responses of voltages at *X*, *Y*, *Z*, and *Z*− in voltage mode current conveyor. From the above graph, it can

Table 3 Transient response of current and voltage mode

S. no.	Configuration	Variable	Amplitude	Frequency	Voltage offset	Phase shift
1	Current mode	V_x	10 mV	10 kHz	0	0
	Voltage mode					
2	Current mode	V_y	65 mV	10 kHz	13 mV	0
	Voltage mode					
3	Current mode	V_{z+}	70 mV	10 kHz	21 mV	0
	Voltage mode					
4	Current mode	V_{z-}	−94 mV	10 kHz	−157 mV	180
	Voltage mode					
5	Current mode	I_x	2 mA	10 kHz	3.2 mA	0
	Voltage mode					
6	Current mode	I_y	2 mA	10 kHz	−1.4 mA	180
	Voltage mode					
7	Current mode	I_{z+}	1 mA	10 kHz	0.4 mA	0
	Voltage mode					
8	Current mode	I_{z-}	0.72 mA	10 kHz	7.29 mA	0
	Voltage mode					

be interpreted that the transient responses show steady-state behavior expect for the phase shift in the voltages at *X*, *Y*, *Z*, and *Z*– (Table 3).

The transient analysis for both current and voltage mode circuits has been evaluated by considering the sinusoidal input of 10 kHz frequency, and undistorted outputs were obtained and validated in the below sections. While simulating at frequencies higher than 10 kHz, the phase shift is found in the output. Hence, the actual range of frequency is about 10 kHz. From the table, it can be interpreted that phase shift of 180° is found in current mode at terminals V_z- and I_y. Moreover, the obtained average voltage offset value is found to be zero (i.e., $V_{IN} = V_{OUT}$), in between the input terminal and output terminals. From the transient response, the effectiveness of the designed circuits can be interpreted.

4.3 Performance Evaluation of Impedance for Current Mode Current Conveyor

The figures below depict the impedance levels of the current mode current conveyor at ports *X*, *Y*, and *Z*, respectively. The obtained impedance levels at all the ports aided in attaining higher gain for the transistors employed. Moreover, the varied impedance levels facilitated to obtain the desired current following action. This feature helps the

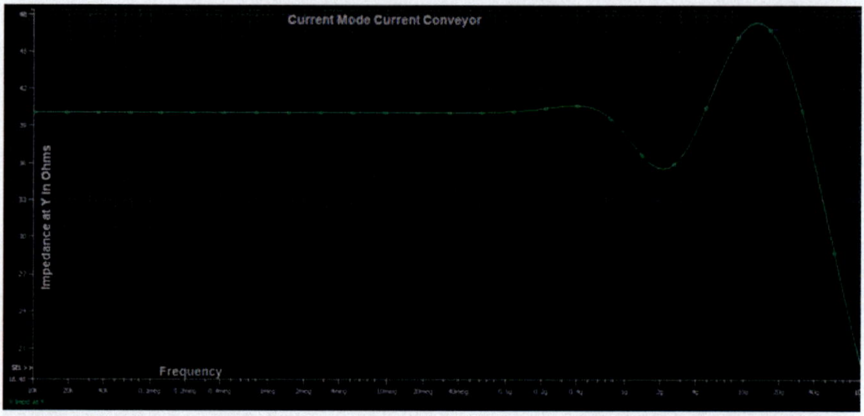

Fig. 10 Impedance at *Y* over frequency

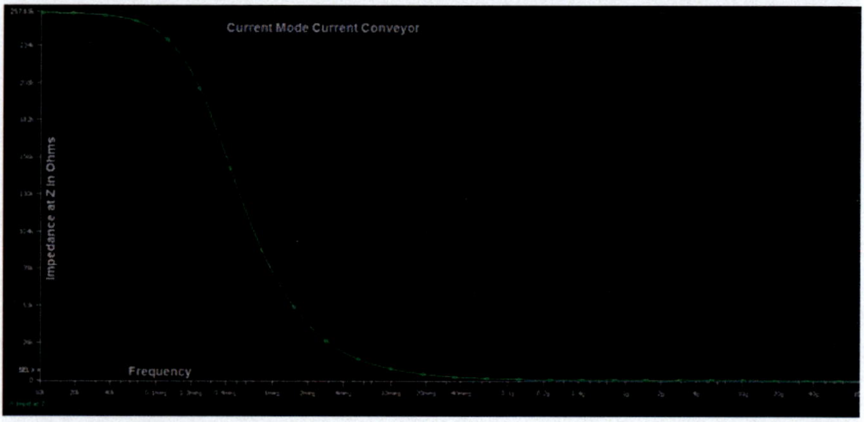

Fig. 11 Impedance at *Z* over frequency

proposed FDACV-CCIII to be used in the development of applications like impedance converter (Figs. 10 and 11).

4.4 *Performance Evaluation of Impedance for Current Mode Current Conveyor*

The graphical plots of the impedance levels at the ports *X*, *Y*, and *Z* over frequency for voltage mode current conveyor are presented in Figs. 12, 13, and 14, respectively. The responses are found to be varying for the different input impedances. The responses are found to either exponentially increase or linearly decrease and rarely does it tend

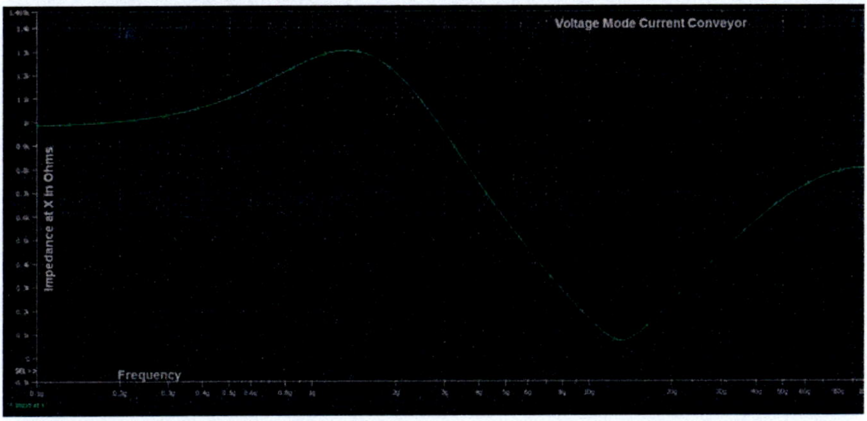

Fig. 12 Impedance at *X* over frequency

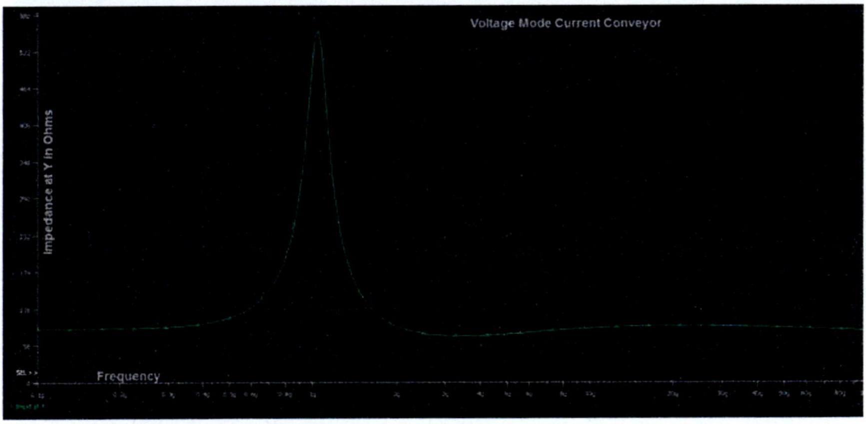

Fig. 13 Impedance at *Y* over frequency

to touch zero. Table 4 briefly depicts the response of the impedance over frequency response.

The impedance level at ports *X*, *Y*, and *Z* for the proposed FDACV-CCIII is shown in Table 4. Impedance levels at all the ports of current conveyors play a significant role in ascertaining its characteristics. The current mode experiences an exponential increase or rapid decrease and voltage mode mostly experiences a flat response and further tends to reach zero. The obtained low impedance levels facilitate in reducing the total power consumption followed by low voltage swing. Moreover, in current mode, higher impedance levels are obtained, which aid in implementing it within many applications (Figs. 15, 16, 17, and 18).

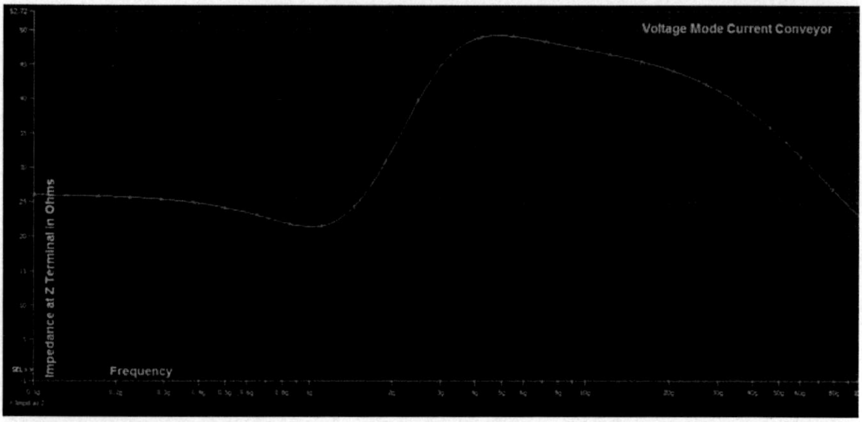

Fig. 14 Impedance at Z over frequency

Table 4 Impedance versus frequency response

S. no.	Configuration	Variable	Description of performance	
1	Current mode	Impedance at X	Exponentially increase from 10 Ω at 10 kHz to 36 Ω at 12 GHz	Linearly decrease from 36 Ω at 12 GHz to 19.5 Ω at 80 GHz
2	Voltage mode		Exponentially increase from 980Ω to 1.3 KΩ at 1.5 GHz and linearly decrease to 75 Ω at 14 GHz and exponentially increase to 800 Ω at 100 GHz	
3	Current mode	Impedance at Z	Decreasing sigmoid from 257.65 KΩ at 10 kHz and tending zero from 0.4 GHz	
4	Voltage mode		Flat 27 Ω till 0.4 GHz and from 1.5 GHz linearly increase to 49 Ω at 5 Hz and nonlinearly decrease	
5	Current mode	Impedance at Y	Flat 40 Ω from 10 kHz till 0.15 GHz with ripple from 0.15 GHz to 15 GHz and rapidly decreases linearly to zero at 100 GHz	
6	Voltage mode		Exponentially increase from 87 Ω and peak to 550 Ω at 1.1 GHz and exponentially decrease back to 87 Ω	

It can be interpreted from the plotted graphs that the proposed FDACV-CCIII with the implementation of CMOS has the optimal current transfer accuracy when operated at a supply voltage of 2 V. The obtained voltage transfer accuracy is better than conventional CCIII, which can be interpreted from the voltage and current transfer characteristics. Moreover, the graph depicts that after a particular input voltage (V_x) the output voltage becomes constant in current mode current conveyor and the same characteristics are obtained in the voltage mode current conveyor.

The above Table 5 depicts the transfer characteristics of both current and voltage mode current conveyors. In voltage mode, voltage exponentially increases from V_x

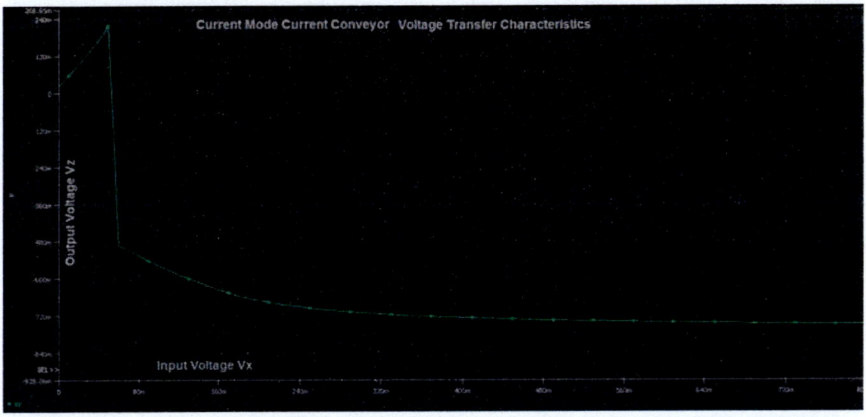

Fig. 15 Voltage transfer characteristics between V_z and V_X (current mode current conveyor)

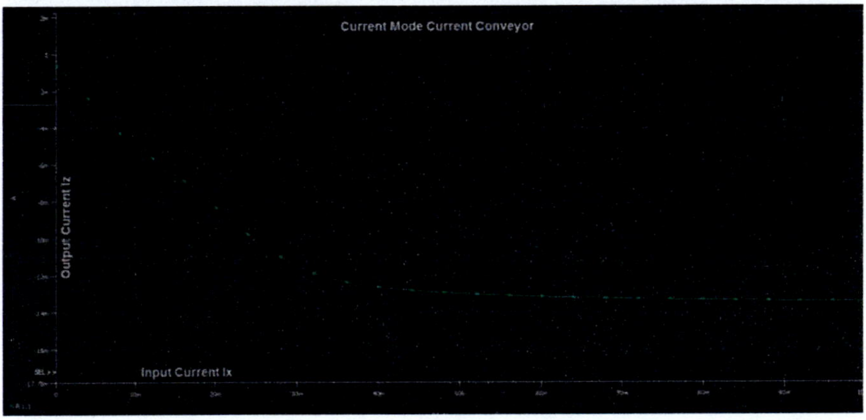

Fig. 16 Current transfer characteristics between I_z and I_X (current mode current conveyor)

$= 0.01$ with $V_y = 0$ and in current mode, current exponentially decreases with $I_x = -0.5$ mA at $I_x = 0$. Hence, this approach obtains better and effective performance.

4.5 *Voltages and Current Measurements*

Above Table 6 depicts the magnitude and bandwidth obtained for both current mode and voltage mode current conveyor and further frequency attributes are obtained. The frequency response of the voltage gains V_z/V_x when the node X is terminated by R_x is shown in Fig. 15. The voltage gains V_z/V_y when the node Z is terminated by R_z is also presented. The attained frequency characteristics are shown for the cases when

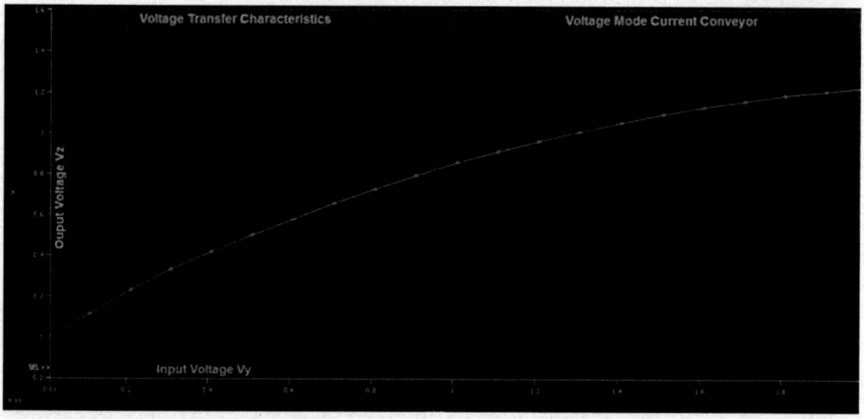

Fig. 17 Voltage transfer characteristics (voltage mode current conveyor)

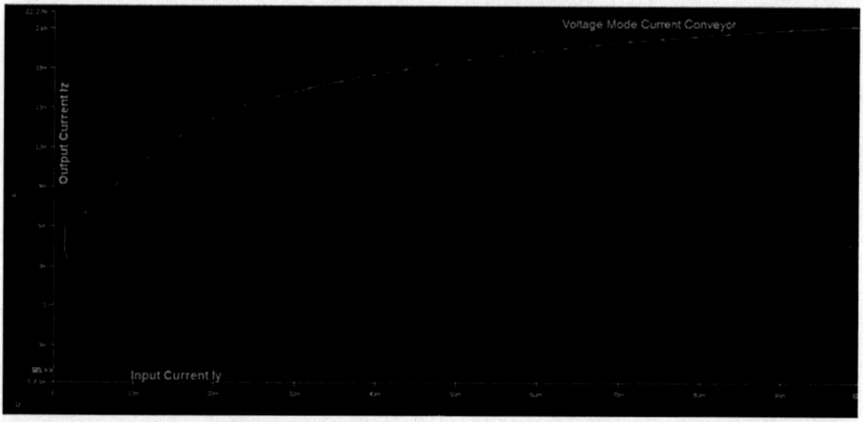

Fig. 18 Current transfer characteristics (voltage mode current convey)

Table 5 Transfer characteristics

S. no.	Configuration	Variable	Description of performance			
1	Current mode	V_z Vs V_x	$V_x < 0.466$ (Vt)	Linearly increasing	$V_x > 0.466$ (Vt)	Exponentially decreasing
	Voltage mode		Exponentially increasing from $V_x = 0.01$ with $V_y = 0$			
2	Current mode	I_z Vs I_x	Exponentially decreasing with $I_x = -0.5$ mA at $I_x = 0$			
	Voltage mode		Exponentially increasing from $I_x = 1.2$ mA with $I_y =$ 5.8 mA $I_x < 1.2$ mA, $I_y = 0$ approximately			

Table 6 Magnitude and bandwidth characteristics of the proposed model

S. no.	Variable	Parameter	Current mode current conveyor	Variable	Voltage mode current conveyor
			Input is at *X*		Input is at *Y*
1	V_z/V_x	Magnitude	4.2 dB	V_z/V_y	0.877 dB
		bandwidth	900 MHz		1.35 GHz
2	V_{z-}/V_x	Magnitude	6.67 dB	V_{z-}/V_y	−2.81 dB
		bandwidth	900 MHz		1.65 GHz
3	V_y/V_x	Magnitude	4.2 dB	V_y/V_y	−5.04 dB
		bandwidth	900 MHz		1.51 GHz
4	I_z/I_x	Magnitude	−5.25 dB	I_z/I_y	5.73 dB
		bandwidth	9 GHz		122 MHz

the CCIII of Fig. 3b, c is used. The bandwidth extends beyond 800 MHz for the case when the CCIII of Fig. 3c is used. It is clear that the bandwidth is not affected by an increase in the gain, which is a main drawback in the op-amp-based architectures. The improvement in the CCIII current bandwidth due to the absence of the PMOS current mirror is clearly observed. Figure 9 shows the current I_y vs. I_x for the CCIII circuit. It is seen the current following action between the nodes *X* and *Y* is excellent over the wide current range.

The bandwidth of the input impedance at *X* terminal, *Y* terminal, and the *Z* output terminal impedances of CCCIII were found to be 9 GHz for the current mode and 1.35 GHz, 1.65 GHz, 1.51 GHz, 122 MHz for voltage mode. The experimental results show that the *X* and *Y* terminal input impedance is suitable for current following action.

The above Table 7 depicts the comparison of the parameters of the proposed FDACV-CIII with the suitable existing literature models. Through the comparison results, it can be interpreted that the proposed FDACV-CCIII in both current and

Table 7 Comparison results of proposed model with existing models

Parameters	Units	Mita et al. [33]	Madian et al. [34]	Kumngern and Khateb [35]	Proposed FDACV-CCIII
CMOS technology	(μm)	0.35	0.35	0.18	0.45
Power supply (V_{DD}, V_{SS})	(V)	(1.5, −1.5)	(0.75 V, −0.75 V)	(0.5, −0.5)	(1.25, −1.25)
No. of transistors	(count)	23	22	11	19
Offset voltage variation	(mV)	−0.125 to 0.125	−0.3 to 0.3	−0.007 to 0.007	−157 to 13
Offset current variation	(mA)	−0.03 to 0.03	−0.17 to 0.2	−0.9 to 0.4	−1.4 to 7.29

voltage mode performed effectively in terms of variations in offset voltage and offset current, for the considered 0.45 nm CMOS technology.

5 Voltages and Current Measurements

The present study designed a circuit based on current and voltage mode fully differential amplifier using third-generation current conveyor (CCIII) to increase gain–bandwidth and reduce the power. In this proposed design, transconductance was regulated using CCIII with input bias current (I_{bias}) replacement that caused an increase in gain–bandwidth.

The layout design of this proposed circuit has 19 transistors (CMOS) for the cascade. Signal transmissions are designed using the DSCH2 software. Moreover, the constructed layout designs are then converted into Verilog file, and the implementation is performed using 0.45 nm CMOS technology in Microwind3 software. The performance evaluation of this proposed design including the magnitude and bandwidth evaluation, current and voltage mode transfer characteristics, impedance level determinations is done in this study. The experimental results of the proposed design have the average magnitude 4.2 DB in current mode and 1.65 dB in voltage mode with the higher frequency range of about 900 MHz. Moreover, proposed design achieves advantages such as high gain, lesser voltage exemption, high linearity, and reduced power consumption. Hence, with these beneficial characteristics, it could be used in biomedical applications as an instrumentational amplifier. Additionally, the current conveyor realizations are more practical than the op-amp realizations which made this developed design to be used in the IC configuration applications.

Hence, due to its capability of analyzing the complicated analog functions, current mode circuits have the remarkable influences in the analog field including nonlinear dynamics. However, there are only a few studies concentrating on this nonlinear dynamics with the current conveyor. Therefore, it is necessary to focus on this area in future to enhance the performance of the proposed design and to increase its applications in the analog field.

References

1. Toumazou C, Lidjey FJ, Haigh D (1990) Analog IC design: the current-mode approach. Peter Peregrinus, UK
2. Palmisano G, Palumbo G, Pennisi S (1999) CMOS current amplifiers. Kluwer Academic Publishers
3. Mucha I (1995) Current operational amplifiers: Basic architectures, properties, exploitation and future. Analog Integr Circ Sig Process 24:243–255
4. Bruun E (1997) Analysis of the noise characteristics of CMOS current conveyors. Analog Integr Circuits Signal Process 12(1):71–78
5. Maheshwari S, Singh SV, Chauhan DS (2011) Electronically tunable low-voltage mixed-mode universal biquad filter. IET Circuits Devices Syst 5(3):149

6. Myderrizi I, Minaei S, Yuce E (2011) An electronically fine-tunable multi-input–single-output universal filter. IEEE Trans Circuits Syst II Express Briefs 58(6):356–360
7. Hu J, Xu T, Zhang W, Xia Y (2005) A cmos rail-to-rail differential voltage current conveyor and its applications. In: Proceedings of 2005 international conference on communications, circuits and systems, vol 2, p 1079
8. Jaikla W, Sornklin P, Siripruchyanun M (2008) An electronically controllable dual-mode universal biquad filter using only single CCCCTA. In: APCCAS 2008—2008 IEEE Asia Pacific conference on circuits and systems,pp 1144–1147
9. Singh AK, Bhaskar DR, Senani R (2015) Current conveyors: variants, applications and hardware implementations. Springer, Switzerland
10. Biolek D, Senani R, Biolkova V, Kolka Z (2008) Active elements for analog signal processing: classification, review, and new proposals. Radioengineering 17(4):15–32
11. Chen HP (2007) Universal voltage-mode filter using only plus type DDCCs. Analog Integr. Circuits Signal Process 50(2):137–139
12. Khateb F, Lahiri A, Psychalinos C, Kumngern M, Kulej T (2015) Digitally programmable low-voltage highly linear transconductor based on promising CMOS structure of differential difference current conveyor. AEU - Int. J. Electron. Commun. 69(7):1010–1017
13. Fabre A (1995) Third-generation current conveyor: a new helpful active element. Electron Lett 31(5):338–339
14. Sedra AS, Roberts GW, Gohh F (1990) The current conveyor: history, progress and new results. IEE Proc G Circuits Devices Syst 137(2):78
15. Imran A, Hasan M, Akram MW (2010) Design of high frequency low power CMOS dual-output current conveyor at 32 nm technology node. In: 2010 International conference on advances in recent technologies in communication and computing, pp 200–203
16. Wang C, Zhang Q (2008) Current controlled fully balanced current conveyor. In: IEEE Asia-Pacific conferences circuits system proceedings, APCCAS, vol 30, no 7, pp 1062–1065
17. Jaikla W and M. Siripruchyanan (2006) Dual-outputs current controlled differential voltage current conveyor and its applications. In: 2006 International symposium on communications and information technologies, pp 340–343
18. Chanapromma C, Jaikla W, Siripruchyanun M (2009) A digitally programmable voltage-mode multifunction filter using CCDVCCs. In: 2009 9th International symposium on communications and information technology, pp 897–900
19. Khateb F, Dabbous BAS, Vlassis S (2013) A survey of non-conventional techniques for low-voltage low-power analog circuit design. Radioengineering 22(2):415–427
20. Carvajal RG et al (2005) The flipped voltage follower: a useful cell for low-voltage low-power circuit design. IEEE Trans Circuits Syst I Regul Pap 52(7):1276–1291
21. El Feki NB, Masmoudi DS (2009) High performance dual-output second and third generation current conveyors and current-mode multifunction filter application. In 2009 6th International multi-conference on systems, signals and devices, pp 1–6
22. Khan IA, Masud MI, Moiz SA (2015) Reconfigurable fully differential first order all pass filter using digitally controlled CMOS DVCC. In: 2015 IEEE 8th GCC conference and exhibition, 2015, pp 1–5
23. Shahsavari S, Biabanifard S, Largani SMH, Hashemipour O (2014) A new frequency compensation method based on differential current conveyor. In: 22nd Iranian conference on electrical engineering. ICEE 2014, vol 1, no Icee, pp 361–365
24. Chen C-C, Lin K-Y, Lu N-K (2008) Low voltage, high performance first and third generation current conveyor in 0.18μm CMOS. In: APCCAS 2008–2008 IEEE Asia Pacific conferences circuits system vol 1, no 3, pp 1498–1501
25. Herencsar N, Koton J, Vrba K, Cicekoglu O (2013) New current-mode all-pass filter with grounded capacitor based on gain-variable CCIII. In IEEE africon conference, pp 0–3
26. Alaybeyoglu E, Guney A, Kuntman H (2013) A new CMOS ZC-CDBA realization and its new filter application. In: Eurocon 2013, pp 1814–1820
27. Mahmoud SA, Elwan HO, Soliman AM (2000) Low voltage rail to rail CMOS current feedback operational ampli ® er and its applications for analog VLSI. Analog Integr. Circ. Signal Process. 25(1):47–57

28. Pandey R (1956) Design of CMOS current conveyors for analog VLSI' master of technology in VLSI. Thapar University
29. Vijay V (2017) Second generation differential current conveyor (Dccii) and its applications. Vignans University
30. Senani R, Bhaskar DR, Singh AK, Singh VK (2013) Current feedback operational amplifiers and their applications. Springer, New York, New York, NY
31. Senani R, Bhaskar DR, Singh AK (2015) Current conveyors. Springer International Publishing, Cham
32. Kumar M (2012) Realization of some novel active circuits. Jaypee Institute of Information Technology
33. Arora TS, Sharma RK (2016) Realization of current mode KHN-equivalent biquad employing third generation current conveyor. In: 2016 International conference on computational techniques in information and communication technologies (ICCTICT), pp 558–562
34. Madian A, Mahmoud, S, Soliman, AM (2006) New 1.5-V CMOS second generation current conveyor based on wide range transconductor. Analog Integr. Circ. Signal Process.
35. Kumngern M, Khateb F (2015) 0.5-V bulk-driven fully differential current conveyor. ISCAIE 2014 – 2014 IEEE symposium on computer applications and industrial electronics

A New Adiabatic Multiplier for Low-Power Application-Specific Signal Processors and Estimation of Performance Parameters

Samik Samanta, Rajat Mahapatra, and Ashis Kumar Mal

Abstract Complementary metal–oxide–semiconductor-based circuits are dominating the device industry for more than decades. The designers have estimated that the power performance of future generation systems will become very important and crucial. Many approaches are already available which can be utilized to minimize the dissipation of power. The approaches are reduction of supply voltage, minimization of load capacitance and decreasing the switching activities. All the methods have their drawbacks. A new classical method of decreasing power comes into play. This is known as adiabatic or energy recovery logic. The principle of operation of this logic depends on reversible computing. In arithmetic, multiplications always play a vital role. Multipliers are used as the main component of various useful blocks. The main arithmetic operations like discrete cosine transform, Fourier transform and design of digital filters depend on multipliers. In the present scope, we have presented one adiabatic Vedic multiplier which has good power and delay efficiency.

Keywords Adiabatic · ECRL · PFAL · Vedic mathematics · CMOS · EDA

1 Introduction

Theoretical studies already proved that charge recovery in electronic systems can be applied to low-power VLSI designs. The systems operate at substantially low power. Circuit complexity and speed of operation are increasing day by day. This results rise of silicon area, heat generation [1]. Designers are trying to minimize lifetime of battery and supply voltage. But they are trying to improve packaging density. Very

S. Samanta (✉)
ECE Department, Neotia Institute of Technology Management and Science, Kolkata, West Bengal, India
e-mail: samik.nitmas@gmail.com

R. Mahapatra · A. K. Mal
ECE Department, National Institute of Technology, Durgapur, West Bengal, India

V. Nath and J. K. Mandal (eds.), *Proceedings of the Fourth International Conference on Microelectronics, Computing and Communication Systems*, Lecture Notes in Electrical Engineering 673, https://doi.org/10.1007/978-981-15-5546-6_63

high temperature generation results gate dielectric breakdown, failure of silicon and breakdown of gate dielectric.

If we reduce power supply voltage, we can achieve low power at the cost of device breakdown. Therefore, some methods are needed to reduce power and improve functionality.

In adiabatic logic, half of the power submitted by power source is utilized and transferred back and forth to the logic and power source required.

Multiplications are very useful operation in analog and digital signal processing. The performance of multiplier is crucial. They are very important for 3D graphics design, multimedia applications and algorithms of signal processing [2].

In the current work, we have used adiabatic logic or energy recovery logic to design the Vedic multiplier, and the use of adiabatic Vedic multiplier shows more than 80–90% of performance improvement.

2 Adiabatic Logic

Adiabatic is a thermodynamic term. In ideal adiabatic process, no energy transfers and receives.

In VLSI, adiabatic technology is very important. Transfer of electric charge happens between various nodes. Node switching activity is the basic parameter of adiabatic process [3].

Adiabatic process is a low-power method. Charge recycling takes place here. Recycling is done between source and logic. Recycling means here reuse of power for node transitions. In this process, conventional sine wave-based power supply can be used. But it is better to use trapezoidal power source for real and good adiabatic operation [4]. In CMOS technology, clock supply and power supply are sent from different buses [5, 6]. In adiabatic process-based circuit, a common bus is used for power and clock supply.

Generally, CMOS-based circuits have major three losses of power. They are dynamic, standby and DC power loss. Dynamic power dissipation contributes to 80% of total power dissipation [7]. By reducing the switching activities, we can reduce the switching power. If we reduce the energy of signal to reduce system power, it will modify the background noise. So reducing power by recycling method and adopting adiabatic or energy recovery process is essential.

In adiabatic process, a current source which is constant is used. It is just opposite of CMOS process where a constant voltage source is applied [8].

In Fig. 1, the basic adiabatic process is demonstrated.

Basically, there are two types of adiabatic logic circuits. Fully adiabatic logic and semi adiabatic logic. In fully adiabatic logic, 100% charge has been recovered, whereas in semi-adiabatic logic, less than 50% of charge has been recovered.

Pass-transistor adiabatic logic (PAL) is fully adiabatic logic. Positive feedback adiabatic logic (PFAL) is the example of semi-adiabatic logic. All the systems have

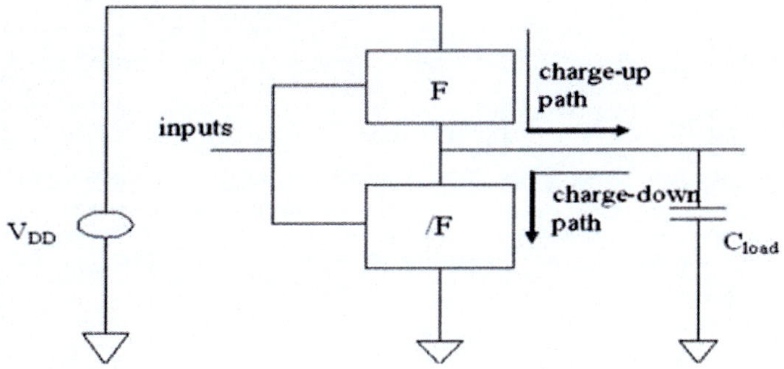

Fig. 1 CMOS logic

their effects on environment. Systems can never be isolated from environment. Therefore, 100% charge is impossible to recover. So, fully adiabatic logic circuits are very hard to implement (Fig. 2).

2N2N2P is partial energy recovery logic. NMOS transistors are arranged in cross-coupled mode. There is 2N2P latch. NMOS and PMOS transistors are used to avoid the floating gate nodes [9].

In ECRL adiabatic logic, precharge and evaluation are performed simultaneously. It replicates CVSL structure. Figure 3 shows the basic ECRL inverter.

PFAL is partial energy recovery logic. It has transmission gate (TG) like structure [10]. It has cross-coupled structure [11]. It has lowest power dissipation among cross-coupled adiabatic inverters. It builds the dual-rail network. It is driven by four-phase

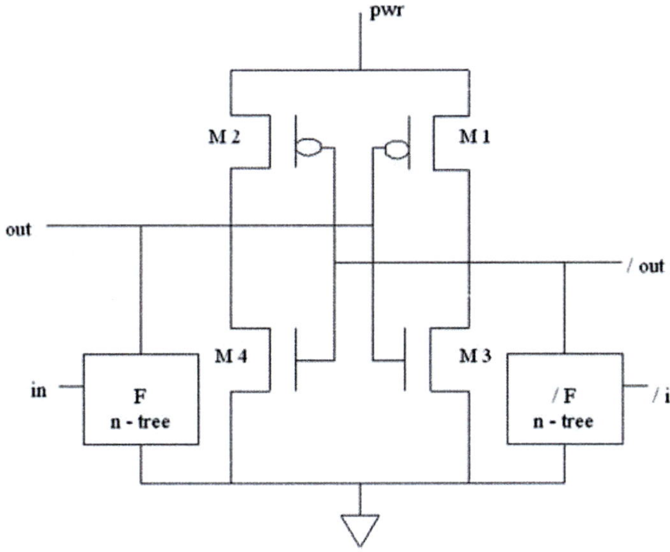

Fig. 2 2N2N2P inverter

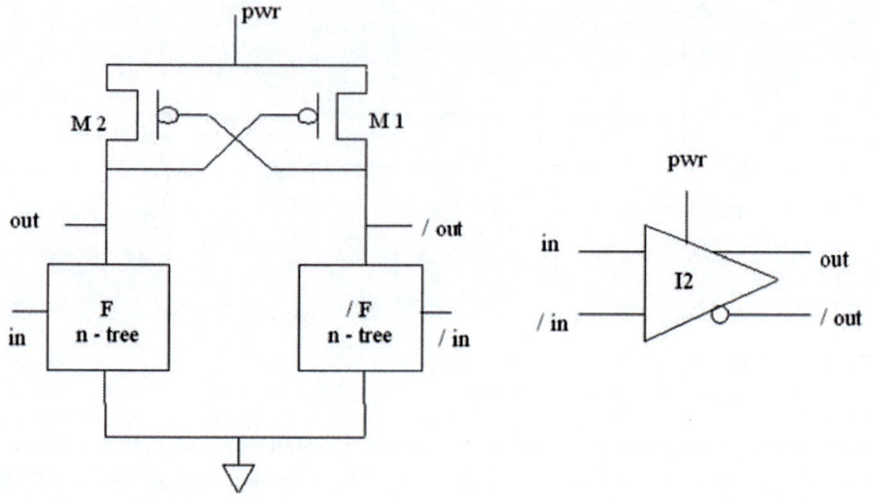

Fig. 3 ECRL inverter

clock [12]. In Fig. 4, we have shown the basic PFAL inverter. Figure 4 shows basic PFAL inverter.

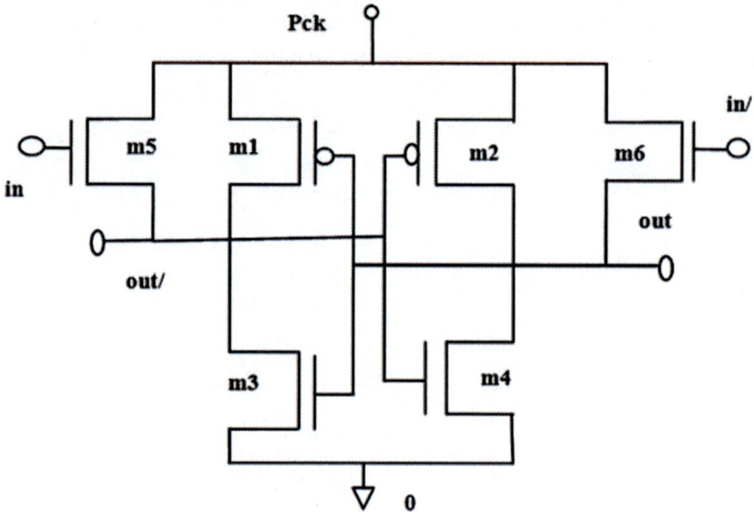

Fig. 4 PFAL inverter

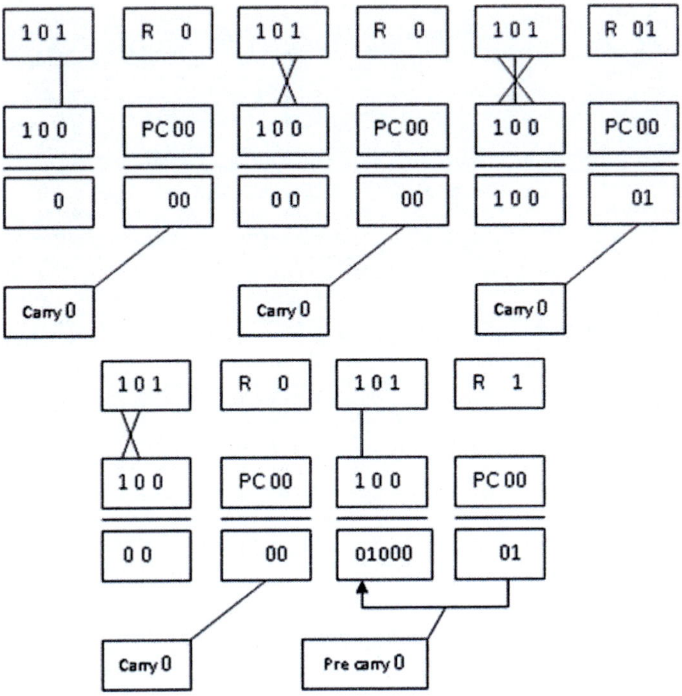

Fig. 5 Vedic multiplier

3 Adiabatic Vedic Multiplier

"Vedic" word comes from "Veda." Veda means the storehouse of knowledge. Vedic mathematics is an ancient methodology. Vedic mathematics is literally vertical and crosswise mathematics [13]. Here, concurrent additions are made of various partial products. Right-hand digits (R.H.D) are taken and multiply this to get LSB. Now, LSB of top number is multiplied with second bit of LSB of bottom number. Figure 5 shows the basic Vedic multiplier.

4 Simulation Results

Power dissipation of CMOS multiplier and adiabatic multiplier has been calculated using VLSI EDA Tool. After power dissipation, we have calculated the delay and energy saving factor. The major advantage of adiabatic logic-based multiplier is their best behavior for lower generation of switching noise. In Vedic multiplier, switching occurs with the minimum voltage drop across devices and node voltage changes very slowly. To calculate the performance of adiabatic logic circuits, there is a parameter

Table 1 Power dissipation comparison of CMOS and adiabatic multiplier

Frequency (MHz)	CMOS multiplier (μW)	Adiabatic vedic multiplier (μW)
20	3.32	2.58
50	5.01	3.22
100	6.58	4.02
300	7.38	5.23
400	10.21	6.03

Table 2 Delay comparison of CMOS and adiabatic multiplier

Frequency (MHz)	CMOS multiplier	Adiabatic vedic multiplier
20	2.72 μS	0.93 μS
50	2.81 μS	1.33 μS
100	3.18 μS	1.51 μS
300	4.18 μW	2.14 μW
400	6.03 μW	3.12 μW

called energy saving factor (ESF). Comparing the power of static CMOS and energy recovery CMOS. ESF is the measure for how much energy is used in a static system compared to an adiabatic system. ESF can be considered as efficiency of an adiabatic system (Tables 1, 2 and 3; Fig. 6).

Table 3 ESF of multiplier

Frequency (MHz)	ESF (%)
20	34
50	47
100	48
300	51
400	52

Fig. 6 ESF estimation

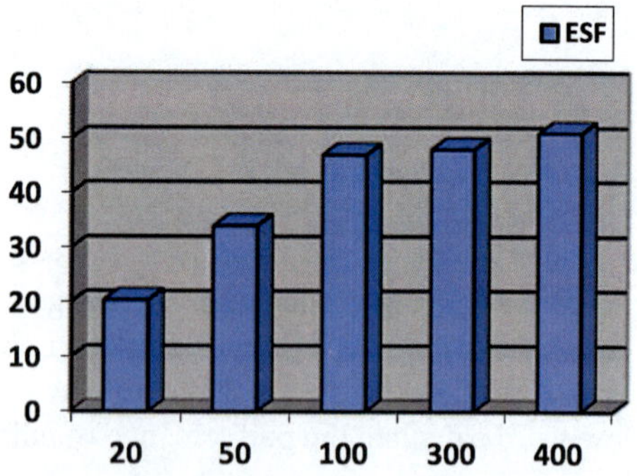

5 Conclusion

Generally, multipliers have large area, long latency and consume considerable power. Fast multipliers are essential parts of digital signal processing systems. The speed of multiplication operation is of great importance in digital signal processing as well as in the general-purpose processors.

From the simulation results, it is found that adiabatic Vedic multiplier has less power delay product than conventional CMOS multiplier. Vedic multipliers also have good ESF characteristics in high frequencies. So, this type of multiplier is more suitable for DSP, embedded and SOC architectures. Power delay product is the key parameter to estimate the device performance. So, adiabatic Vedic multiplier-based application devices are more suitable for low-power wearable and IOT applications.

References

1. Athas WC, Svensson L, Koller JG et al (1994) Low-power digital systems based on adiabatic-switching principles. IEEE Trans VLSI Syst 2:398–407
2. Moon Y, Jeong D-K (1996) An efficient charge recovery logic circuit. IEEE J Solid-State Circ 31:514–522
3. Weste N, Eshraghian K (1993) Principles of CMOS VLSI design: a systems perspective, 2nd edn. Addison—Wesley, New York
4. Chandrakasan A, Brodersen RW (1995) Minimizing power consumption in digital CMOS circuits. In: Proceedings of the IEEE vol 83, pp 498–523
5. Dickinson AG, Denker JS (1995) Adiabatic dynamic logic. IEEE J Solid-State Circ 30(3):311–315
6. Starosel VI (2002) Adiabatic logic circuits: a review. Russ Microlectron 31(1):37–58
7. Kramer A, Denker JS et al (1995) 2nd Order adiabatic computation with2N-2P and 2N-2N2P logic circuits. In: Proceedings of the 1995 international symposium on low power design, 191–196
8. Takahashi Y, Fukuta Y, Sekine T, Yokoyama M (2006) 2PADCL: two phase drive adiabatic dynamic CMOS logic. In: Proceedings of IEEE APCCAS, pp 1486–1489
9. Khandekar PD, Subbaraman S, Patil M (1994) Low power digital design using energy-recovery adiabatic logic. Inter J Eng Res Ind Appl 11(III):199-208, 94–97
10. Thapliyal, Srinivas MB (2005) An efficient method of elliptic curve encryption using ancient Indian vedic mathematics. In Proceedings of IEEE midwest symposium on circuits and systems, Cincinnati, pp 826–829
11. Samanta S (2009) "Adiabatic Computing" a contemporary review. In: 4th International conference on computer and devices for communication: codec09, Kolkata
12. Navi RK, Reza Saatchi Md, Daei O (2009) A high-speed hybrid full adder. Eur J Sci Res 26(1):29–33
13. Svensson L (1995) "Adiabatic switching" in low power Digital CMOS Design. Kluwer Academic Publisher

Smart Public Transport Using BT5

Kashyap Anup, Abhijith Nethi Satya Sai, Gambhir Shivam, and R. K. Mugelan

Abstract In today's world, time plays a major role in everyone's life. The wastage of time by waiting for a bus at a particular bus stop can be reduced by discovering methods which help people to acquire information about public transport. People using inner city public transport would want to know what bus stop they are in, what buses come to the particular bus stop, when the bus arrives, the route it takes, when it reaches the destination, etc. They would like to know the travel timings of the bus both while waiting at the bus stop and also while traveling. So, it will be very useful if a product is made which can be used to know all the information related to the bus. In this work, a smart public transport information system using Bluetooth 5 is developed which helps users to connect to the app very easily as it does not require an active Internet connection. This is developed as a very reliable and a power-efficient system using the latest technology Bluetooth 5. An IoT sensor kit named, Nordic Thingy 52 is the main component in this project. We implemented it in a way to connect the city buses to users' phones, which can process information in an interface they are comfortable with. It is believed that the use of public transport will increase once this product is released, because it solves many of the present problems which stop people from using public transport.

Keywords IoT · Nordic Thingy 52 · Bluetooth 5 · Public transport information

---

K. Anup (✉) · A. Nethi Satya Sai · G. Shivam · R. K. Mugelan
School of Electronics Engineering, VIT University, Vellore, India
e-mail: anupkashyap7@gmail.com

A. Nethi Satya Sai
e-mail: abhijithsatyasai@gmail.com

G. Shivam
e-mail: shivamgambhir97@gmail.com

R. K. Mugelan
e-mail: mugelan.rk@vit.ac.in

V. Nath and J. K. Mandal (eds.), *Proceedings of the Fourth International Conference on Microelectronics, Computing and Communication Systems*, Lecture Notes in Electrical Engineering 673, https://doi.org/10.1007/978-981-15-5546-6_64

1 Introduction

The rapid development of cities and increase in their population has led to a serious difficulty in commuting from one place to another within the city itself due to traffic. Lately, government has brought a lot of awareness among people to use public transport as it partly solves a very serious issue of traffic. Identifying this problem and the need to attract more people toward the use of public transport, a product has been developed which makes it much easier for users as they can get the information regarding the bus and bus stops like where it is, the route it takes, etc. Instead of waiting in a bus stop for several minutes not knowing whereabouts of the bus, this app makes it much simpler for people, especially aged people to use public transport. The main aim is to make it offline because it can be used in areas where there is no Internet connection and to save battery in phones. For this purpose, we have used Nordic Thingy 52 IoT sensor kit which operates on Bluetooth 5, one of the latest wireless technologies present. This kit has a battery life of about 10 years when kept idle which makes it very power efficient and easy to use.

The primary objective of this project is introduced as easy, convenient, user-friendly and power-efficient solution in public transportation domain. Our approach makes it easier for any public transport user to get the information he needs regarding the bus, bus stops and bus routes in order for him to reach from source to destination in a hassle-free manner. The traditional approach was to put up sign boards and information posters in the bus stops. But in the current age of rapid digitization, the information has moved from physical displays to Web sites and apps that the user can access from wherever he/she is.

The novelty in our work is that, instead of waiting in a bus stop for several minutes not knowing whereabouts of the bus, this app makes it much simpler for people, especially aged people to use public transport. The main aim is to make it offline because it can be used in areas where there is no Internet connection and to save battery in phones. For this purpose, we have used Nordic Thingy 52 IoT sensor kit which operates on Bluetooth 5, one of the latest wireless technologies present. This kit has a battery life of up to 10 years (idle mode) which makes it very power efficient and easy to use.

2 Literature Review

After a detailed study of related materials and papers, we have come across four of such approaches which have been published in IEEE journals (Refer Table 1) that aim to solve the problem that we set out to solve. The previously provided solutions were as follows:

Table 1 Comparison with previous works

Features	Product 1 [1]	Product 2 [2]	Product 3 [3]	Product 4 [4]	Our Product
Information Display	✓	✓	✓	✓	✓
Bluetooth					✓
GPS / Internet / GSM	✓	✓	✓	✓	
Smart-phone app	✓	✓			✓
Live tracking	✓	✓	✓	✓	✓
Low energy					✓
Automatic detection		✓			✓

1. *A Smart Bus Tracking System Based on Location-Aware Services and QR Codes* [1].

This system uses a QR code that is placed at each bus stop. The user who arrives at the bus stop scans the QR code and based on that they get the information of the buses arriving there using Internet and the GPS data from the bus. This approach needs cloud to gather location data from the buses and pass it on to the phone. The disadvantage is the usage of Internet and the possibility of a situation where the QR code is tampered at the bus stop.

2. *Development of Bus Location System Using Smartphones* [2].

This is the most frequently used approach where a network of buses is created over the Internet with location data of all the buses. An algorithm is also included that helps in estimating the possible delay of the buses.

3. *Smart Bus Station-Passenger Information System* [3].

This is an approach which is currently implemented in some cities. The location data, instead of being sent to the phone will be sent to the mini-computers at the bus stations, which will process the information and display it on a screen. This approach adds more cost to the state for installation and may not be a viable solution in places where there is no continuous electricity throughout the day.

4. *Smart Public Transport System* [4].

The approach used is very similar to the previous method, but this research article focuses more on passenger and driver safety by monitoring the speed of the bus and reporting if the bus is found to be over speeding.

After analyzing all these alternatives, we have concluded our findings in the form of this table:

We observed that our work is around 40% similar to theirs in terms of information display, smartphone app, live tracking, etc., but the unique selling point of our work would be that we used BT5 and made it very user-friendly as there is no need to have an Internet connection to use this app. We understood one of the problems of users through our survey that they find Internet connection not feasible in remote areas where other apps fail to work. So, we addressed this vital issue and used Nordic Thingy 52 IoT sensor kit which uses the latest technology BT5 which is very power efficient.

3 Methodology

3.1 Project Description

In our approach for smart public transport system, beacons are mounted both on the bus and also on the bus station. An app is created using the official Android development IDE, Android Studio 3.3.1 which uses Bluetooth of the mobile phone [5] to connect with these kits. For an instance, when a person waits at a bus stop, the Thingy (proprietary name of an individual unit of the kit) mounted on the bus stop connects to his mobile application via Bluetooth, and he will immediately be able to know what all buses arrive at that particular bus stop and where they go. He will also be notified as soon as he reaches his destination. We have implemented Google Maps SDK for live tracking of the bus and also to know what all bus stops are on the way from the users' source to destination.

The novelty in our work is that, instead of waiting in a bus stop for several minutes not knowing whereabouts of the bus, this app makes it much simpler for people, especially aged people to use public transport. The main aim is to make it offline because it can be used in areas where there is no Internet connection and to save battery in phones. For this purpose, we have used Nordic Thingy 52 IoT sensor kit which operates on Bluetooth 5, one of the latest wireless technologies present. This kit has a battery life of up to 10 years (idle mode) which makes it very power efficient and easy to use (Refer Fig. 1).

This digital approach has entered the market and is rapidly growing. Although the currently available apps are very helpful, the main disadvantage is that, they need an

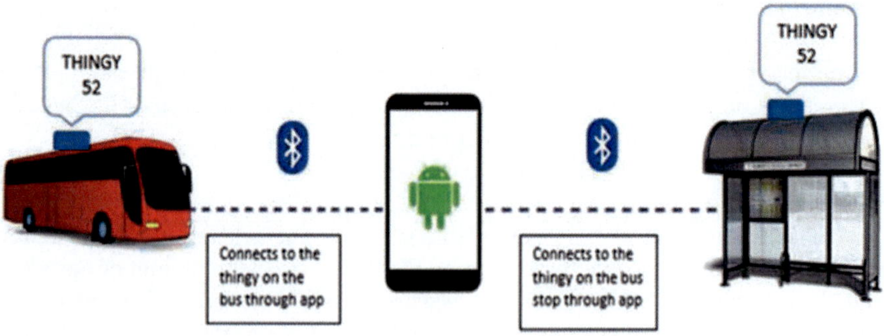

Fig. 1 System overview

active Internet connection and GPS access in order to get the required information. Both these features—mobile data and GPS are high power-consuming features on all devices. A reliable Internet connection might also not be available everywhere and hence a solution that works without these services is a unique and intuitive approach. As a substitute for GPS and mobile data, our approach makes use of Bluetooth which is relatively less power consuming. Our approach gives a more localized approach rather than providing wide range of information. Hence, ensuring that, the user gets the information that he needs.

3.2 Technical Specifications

The hardware and software aspects of our project are as follows:

3.2.1 Hardware

- **Nordic Thingy 52 (IoT sensor kit)**

The Nordic Thingy: 52 (nRF6936) is a compact multi-sensor, power-optimized device designed to collect various types of environmental data. It is also a user-friendly development platform designed to help you build IoT prototypes and demos without the need to build hardware from scratch or write firmware. Thingy is built around Nordic Semiconductor nRF52832 Bluetooth 5 SoC.

The Thingy comes with a 32-bit ARM Cortex-m4F CPU. It uses 512 kilobytes of flash memory and 64 kilobytes of RAM and incorporates a Bluetooth low energy radio. The kit also supports floating point instructions and digital signal processing instructions. It is powered with a rechargeable lithium–ion polymer battery with a capacity of 1440 mAh. The battery can be charged with USB.

Thingy can sense the quality of movement, orientation, temperature, moisture, air pressure, light, color, and air. It can also play sound from its microphone through its speaker and stream sound to the host. Thingy's functionality can be configured over-the-air via a Bluetooth API, making it possible to create demos and prototypes without the Nordic Thingy: 52 programming itself.

- **User's Smartphone Device**

We have specifically targeted the Android ecosystem, since it is popular, and remains stable across all make and models of mobile phone hardware. The app that is built for our project works across all hardware specifications of the phones. The app also works on phones that do not have Bluetooth 5, as the app detects and works with a lower Bluetooth standard seamlessly.

3.2.2 Software

- **Android Studio**

Android studio is a popular IDE used for making applications for Android OS. Ever since Android became popular, Google has worked to make this IDE better and more user-friendly, to maintain their proprietary ecosystem (Google Play Store) and its profitability. Since it is quite a student-friendly and requires Java, we also have used it to improve the novelty of our project, and since we were more focused and algorithm and UI, Android Studio helped us reach the final outcome of the project.

- **Google Maps API**

Google Maps is the most popular platform for using and developing apps with geographical maps. In order to make use of Google Maps in the app, an API key needs to be requested from Google [6]. The API key needs to be mentioned in the manifest of the app in order to be able to make requests to the map's platform. Google Maps offers up to 25,000 requests per day for free and charges for requests exceeding that. This loads the maps and enables us to use multiple functionalities such as zooming into a particular location, adding a marker on the map, moving the marker to particular app, etc. Additionally, other APIs like places API, directions API [7] can be integrated to get more features like requesting direction between two places, getting to know the distance, time required, route, etc. We had to get the location of the user by requesting the location from the Android system. An OnLocationChanged () receiver was used in order to keep track of the user location which is constantly changing when he is traveling in the bus.

- **User End Android App**

The app that we have used (Refer Fig. 2) is specifically designed to make commuting by bus an easy task. Extremely friendly UI, coupled with a simplified algorithm, makes even a complex process of boarding the bus easy, for a daily commuter.

For example, when we open our app, this is the splash screen:

The app is designed to work on phones as old as the ones running Android 5.0 Lollipop and above (API level 21+) which covers almost 95% of the phone currently used around the world. Since the system requirements of the app are very less, it runs smoothly of low-cost phones without any flaw.

3.3 *Design Approach*

The design approach that we have taken (Refer Fig. 3) to design our system has been done all the way from the algorithmic level, and each step and process has been carefully designed, keeping the UI and the overall user experience very friendly to

Fig. 2 Screenshot from the app

the user. Our algorithm process seamlessly ensures that the user gets what he wants, with the simplest UI design, while a more complex algorithm runs in the background of the user's phone.

The key elements in our design are:

1. User's phone
2. IoT kit on bus
3. IoT kit on the bus stop.

While designing, we decided to have these set of features (based on our literature survey that we have conducted):

1. The user should be as close as possible to our system, so we need an app.
2. App can use phone's GPS easily.
3. We used the IoT kits as proximity beacons. This conserves energy.

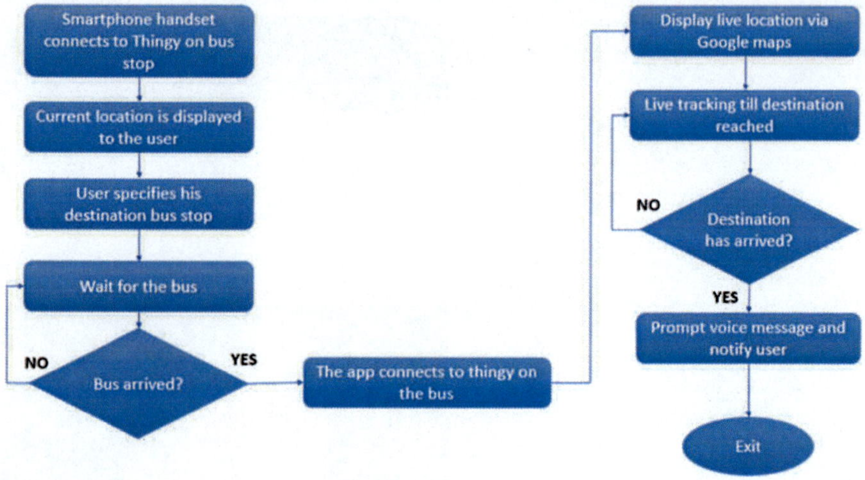

Fig. 3 Design flow

Our overall algorithm flowchart is:

The beacons are installed in buses and bus stops and are set to be discoverable all the time. Each beacon is provided a unique name as an identifier to the bus stop or bus. Considering only 10 characters are used, it gives the freedom of using up to 3 quadrillion buses and bus stops. The identifiers are assigned such that it is easy to differentiate between buses and bus stops. For example, all the bus stop names have the first character as a number, whereas the buses have a letter as the first character. When the app is launched by the user the app will first enable Bluetooth [8], then enable discovery mode and then will start looking of devices in range. If a device is found with the identifier matching a bus stop, it stops looking for devices and redirects the app to a new activity.

This new activity shows the user what bus stop he is at and other details pertaining to that particular bus stop. It shows the available destinations in a selectable list. Once he clicks on a particular destination, the app sets up a broadcast receiver again to get the Bluetooth devices in range. It scans for every new Bluetooth devices that enters the range and compares to the name to the identifier given to the buses that have the route from the current bus stop location to the destination the user has chosen. Once the device is found to be in range, the app notifies the user that the bus he has to get onto has arrived and launches the Google Maps activity. The maps are initiated and a marker is first set to the destination showing the user the place he has to reach. It then starts listening to OnLocationChanged () listener and keeps updating the second marker on the map that shows the current location of the phone.

The user now continues to get the live location and can see how long before he reaches the destination. The app does not necessarily have to be open at this point of time as the app is made to work in background as well. At the same time, another broadcast receiver is set up in order to listen for Bluetooth device found action.

Similar to the previous scans, the app matches the names of the Bluetooth devices in range and looks for the identifier that is set to the bus stop. As soon as the bus stop appears in the range of the bus, the app sends out a notification both on the screen and an audio notification telling the user that the destination has arrived. Hence, to summarize, the app first looks for the bus stop beacon, then looks for the bus beacon based on the user choice and thereafter looks for the destination bus stop beacon.

3.4 Codes and Standards

The main set of standards that apply to our project is the "Bluetooth Wireless Communication Standard (IEEE 802.15.1)". We are using the latest Bluetooth 5.0 wireless communication standard. Although it commonly used for peripherals like wireless keyboards, mice, game controllers, etc., it has found its application in smart home and IoT applications. Bluetooth 5 has noticeable improvements to the Bluetooth low energy specification which was introduced in the previous standard (Bluetooth 4.0) in order to increase the energy efficiency of Bluetooth peripherals. It was used mainly for wearable devices, beacons and other low energy device. The other benefits of Bluetooth are its improved speed and great range enabling it to communicate over greater distance when compared to the previous standards. Bluetooth 5 is claimed to be having four times the range, twice the speed and eight times the broadcasting capacity when compared to Bluetooth 4.0 all of which are applicable to the Bluetooth low energy standard.

Bluetooth 5.0 support data transfer speeds of up to 2 Mbps and has a range of around 800 feet (0.24 km). However, if there is no line of sight available, the range reduces due to attenuation. With Bluetooth 5.0 devices have the freedom and can prefer to have longer range or more speed. Since Bluetooth is a full protocol stack, the bottom layer is the physical layer (PHY). Bluetooth 5 offers three variants to the physical layer of the stack, namely LE 1 M, Le2M and LE Coded (Refer Fig. 4).

LE 1 M which is actually low energy at 1 Ms/s, uses Gaussian frequency shift keying is the old standard. LE 2 M is a new variant introduced which supports double

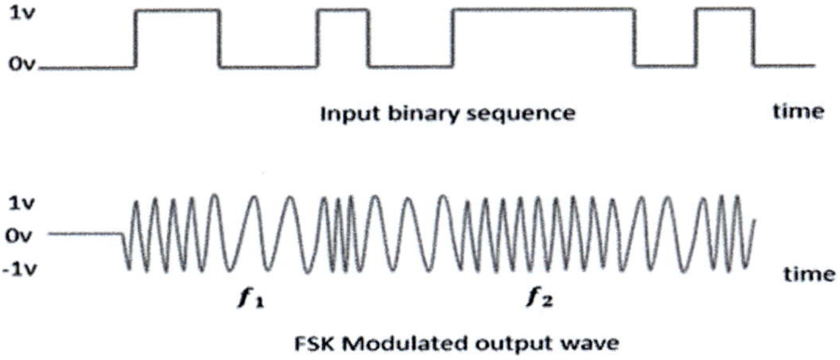

Fig. 4 FSK modulation

the data rate of LE 1 M. It is characterized by using double the symbol rate and therefore uses two-level Gaussian frequency shift keying (GFSK) with a decrease in the carrier frequency by a giver frequency deviation representing a binary zero and an increase in the carrier frequency by a particular frequency deviation representing a binary one (as shown in Fig. 4). It uses a frequency deviation of at least 370 kHz to mitigate the inter-symbol interference that will be caused if higher symbol rates are used at 185 kHz (the frequency deviation used in LE 1 M).

4 Results

As of now, we seem to get a linear desired output from our system. After running multiple tests, the Nordic kits seem to show perfect required amount of signal strength (which is configurable). Also, the pictures presented (Refer Fig. 5) are few screenshots from the app (step by step).

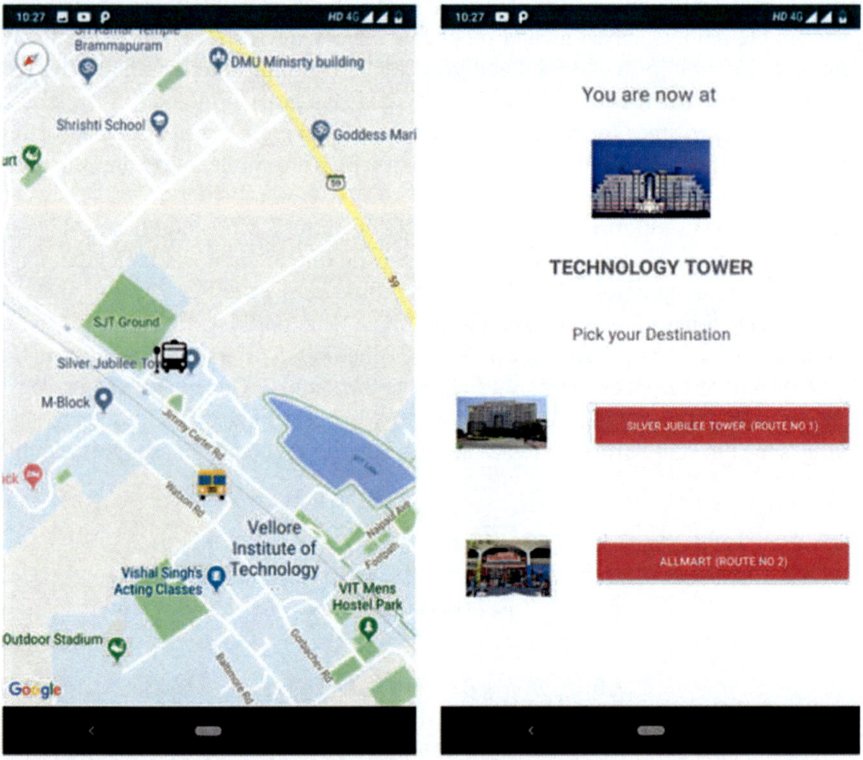

Fig. 5 Screenshot from the app with Google Maps integration

5 Conclusion

The approach has a very wide scope for development and can be scaled to work in many other areas [9, 10]. This paper keeps road transport in mind, although the same methodology can be extended to trains and metros. And also, it is possible to cater all public transportation services through the same app making it very convenient for the user as he will be having to install only one app in the place of two or three apps. This also opens up wider range of features that can be included. For example, if a person has to go from point A to point C, then the app could suggest him to take metro to point B and then take a bus from point B to point C. These applications when developed to work without Internet will be extremely effective.

One of the issues that we might come across is the possibility of someone tampering the system. A Bluetooth device can be brought in range to imitate the bus or bus stop beacon. This problem can be solved by using dynamic identifiers that change over time. It can be set to change frequently and the new identifier can be set through an algorithm based on the time at that moment again eliminating the use of Internet. This adds a security layer to the product making sure that no one hacks the system. Another further scope for this product is to integrate a ticketing option which automatically deducts money or credits from a wallet. The payment method must be encrypted with the latest standards for the system to be foolproof.

References

1. Smart public transport system—IEEE conference publication. Ieeexplore.ieee.org, 2019. [Online]. Available https://ieeexplore.ieee.org/document/6953110. Accessed 06 Mar 2019
2. Smart bus station-passenger information system—IEEE conference publication. Ieeexplore.ieee.org, 2019. [Online]. Available https://ieeexplore.ieee.org/document/7120749/. Accessed 06 Mar 2019
3. Development of bus location system using smart phones—IEEE conference publication. Ieeexplore.ieee.org, 2019. [Online]. Available http://ieeexplore.ieee.org/document/5602380/. Accessed 06 Mar 2019
4. A smart bus tracking system based on location-aware services and QR codes—IEEE conference publication. Ieeexplore.ieee.org, 2019. [Online]. Available https://ieeexplore.ieee.org/document/6873634/. Accessed 06 Mar 2019]
5. Bluetooth overview | Android developers. Android Developers, 2019. [Online]. Available https://developer.android.com/guide/topics/connectivity/bluetooth. Accessed 06 Mar 2019
6. Get Started | Maps SDK for Android | Google developers. Google Developers, 2019. [Online]. Available https://developers.google.com/maps/documentation/android-sdk/start. Accessed 06 Mar 2019
7. Developer guide | Directions API | Google developers. Google Developers, 2019. [Online]. Available https://developers.google.com/maps/documentation/directions/intro. Accessed 06 Mar 2019
8. Bluetooth low energy overview | Android Developers. Android Developers, 2019. [Online]. Available https://developer.android.com/guide/topics/connectivity/bluetooth-le. Accessed 06 Mar 2019

9. Smart public transport for smart cities—IEEE conference publication. Ieeexplore.ieee.org, 2019. [Online]. Available: https://ieeexplore.ieee.org/document/8389288/. Accessed 28 Mar 2019
10. Development of bus location system using smart phones—IEEE conference publication. Ieeexplore.ieee.org, 2019. [Online]. Available https://ieeexplore.ieee.org/document/5602380/. Accessed 28 Mar 2019

GLAD Assisted In_2O_3 NW-TiO_2 NW Heterostructure for Enhanced UV-Vis Absorption

Vinod Kumar Yadav, Vinit Kumar Yadav, Amitabha Nath, Rahul Raman, Priyanka Choudhury, and Mitra Barun Sarkar

Abstract The experimental fabrication of highly ordered perpendicular TiO_2 and In_2O_3 nanowires (NW) on n-type silicon (Si) substrate has been reported in this paper. The glancing angle deposition technique (GLAD) has been employed to synthesis In_2O_3 NW-TiO_2 NW heterostructure and ordinary TiO_2 NW on n-Si substrate inside e-beam evaporation chamber. An average of ~1.4 fold enhancement in absorption at UV and visible spectra have been observed in case of hetero-structured In_2O_3 NW-TiO_2 NW/TiO_2 TF/n-Si compared to simple TiO_2 NW/TiO_2 TF/n-Si samples. The main band gap transition in case of hetero-structure In_2O_3 NW-TiO_2 NW has been observed at ~3.3 eV and that in case of simple TiO_2 NW has been observed at ~3.1 eV under open-air room temperature.

Keywords GLAD · Optical absorption · Tauc plot

V. K. Yadav (✉) · V. K. Yadav · A. Nath · R. Raman · P. Choudhury · M. B. Sarkar
Department of ECE, National Institute of Technology Agartala, Agartala 799046, India
e-mail: sanikumar12054@gmail.com

V. K. Yadav
e-mail: vinityadav94@gmail.com

A. Nath
e-mail: nathamitabha851@gmail.com

R. Raman
e-mail: ramanrahul355@gmail.com

P. Choudhury
e-mail: priyanka.choudhury22@gmail.com

M. B. Sarkar
e-mail: barun06@gmail.com

V. Nath and J. K. Mandal (eds.), *Proceedings of the Fourth International Conference on Microelectronics, Computing and Communication Systems*, Lecture Notes in Electrical Engineering 673, https://doi.org/10.1007/978-981-15-5546-6_65

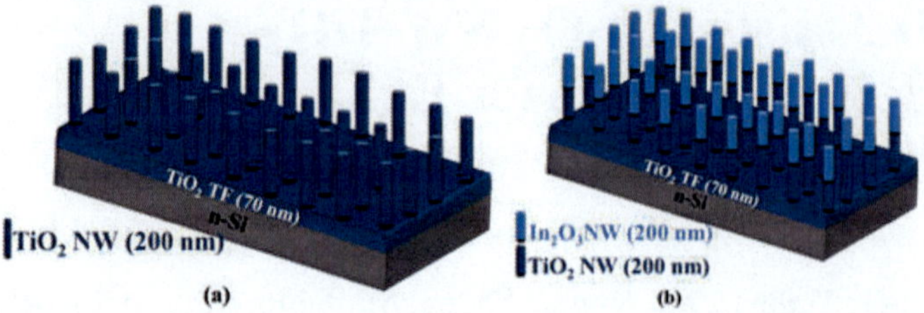

Fig. 1 **a** Proposed structure of simple TiO_2 NW. **b** Proposed hetero-structured In_2O_3 NW-TiO_2 NW sample

1 Introduction

One dimensional (1-D) metal oxide nanostructures are the new emerging trend in nanotechnology since they are the most frequent material on the earth due to their specific shapes, chemical, physical properties and their composite structure [1]. One of such prominent semiconductor material is indium oxide (In_2O_3) with wide band gap (indirect band gap of 3.5–3.75 eV and direct band gap of 2.6 eV) have been widely used due to their optical and electrical properties [2–6]. Further, nanocrystalline In_2O_3 material is very well known due to its potential for various applications such as UV detector, gas sensor, photo-catalysts, deflectors, and electroluminescent [7]. Whereas titanium dioxide (TiO_2) is a transition metal oxide material which demonstrates characteristics such as abundance, relatively inexpensive, nontoxic, and impede to chemical reaction [8]. It is being accredited to be the appropriate technique as a photocatalytic owing its strong oxidant property which can decompose recalcitrant substrates and for being economic environment-friendly compared to other conventional techniques [9]. Nanowire of oxide material is more useful due to the fact that nanowire enhances good electrical, optical properties, and magnetic properties. TiO_2 also exhibits photocatalytic activity under UV radiation, which depends on its crystal structure, surface area, and size charge distribution.

In this report, a comparative study has been done for two distinct structures of In_2O_3 NW-TiO_2 NW/TiO_2 TF/Si (n-type) and TiO_2 NW/TiO_2 TF/Si (n-type) which has fabricated by GLAD assisted electron beam (e-beam) evaporation method and subsequent improvement in absorption properties have been discussed in relation to numerous sub-bandgap transitions due to their structural discrepancies. Bandgap tailoring for such distinct structures has also been discussed elaborately. The schematic diagram of the fabricated structure is depicted in Fig. 1a, b.

2 Experimental

Perpendicular deposition technique has been followed to fabricate two distinct structures of hetero-structured In_2O_3 NW-TiO_2 NW/TiO_2 TF/n-Si and simple TiO_2

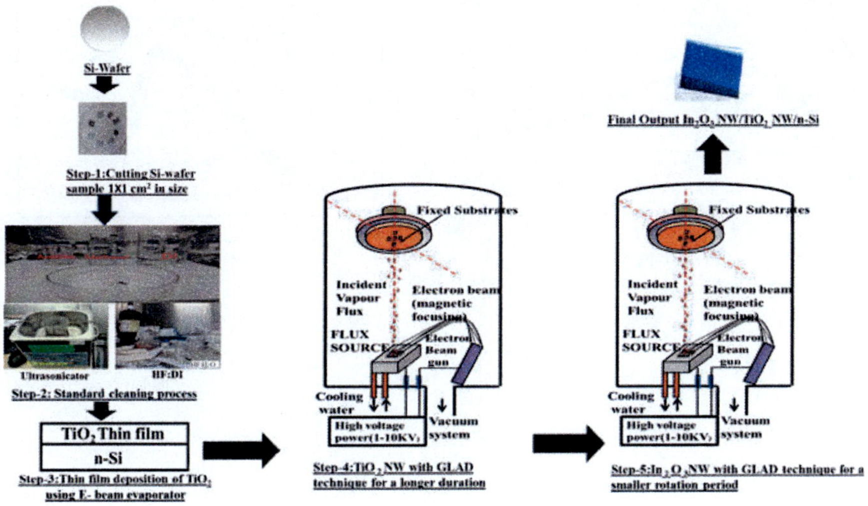

Fig. 2 Steps involved in the fabrication of sample

NW/TiO_2 TF/n-Si as shown in Fig. 1a, b. Firstly, TiO_2 thin film of 70 nm was deposited over n-Si wafer with the help of e-beam evaporator at room temperature and base pressure maintained at 1×10^{-5} mbar. After that TiO_2 NWs of 200 nm were synthesized using the GLAD technique with an appropriate GLAD angle and in similar fashion In_2O_3 NWs of 200 nm was grown on top of 200 nm TiO_2 NWs using e-beam evaporation technique to fabricate In_2O_3 NW-TiO_2 NW hetero-structure. The deposition rate for both thin film (TF) and nanowire was kept 1.2 $Ås^{-1}$.

Optical absorption experiments were observed separately in UV-Vis spectrophotometer (Lamda950, Perkin Elmer) for the wavelength lying between (300 and 700 nm) at room temperature. Subsequently, bandgap calculations for both NWs structures have been done using the Tauc plot. The various steps involved in the fabrication of the proposed sample structure are shown in Fig. 2.

3 Result and Discussion

The optical absorption measurement of hetero-structure In_2O_3 NW-TiO_2 NW and simple TiO_2 NW was observed separately in UV-Vis region whose wavelength lies between (300 and 700 nm) at room temperature and the result is shown in Fig. 3a. Absorption enhancement of ~1.4 times below 370 nm is due to absorption of light caused by excitation of an electron from the valence band to conduction band and Fig. 3b shows $(\alpha h\nu)^2$ versus energy graph of hetero-structure In_2O_3 NW-TiO_2 NW and TiO_2 NW where $h\nu$ is photon energy and α shows absorption coefficient. The main band transition for hetero-structure In_2O_3 NW-TiO_2 NW was obtained close to the band gap of 3.3 eV. Similarly, the main direct band transition of TiO_2 NW was

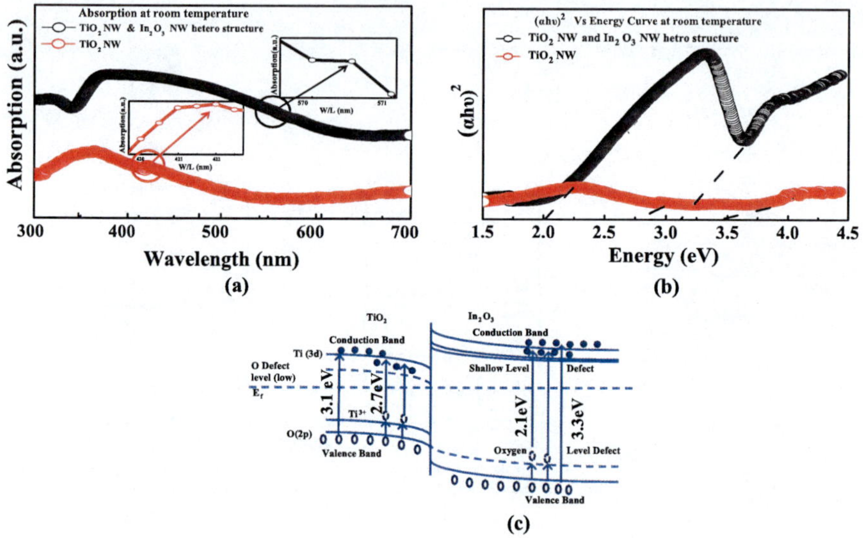

Fig. 3 **a** Absorption of hetero-structured In_2O_3 NW-TiO_2 NW and TiO_2 NW sample, **b** combined $(\alpha h\nu)^2$ energy graph, **c** energy band diagrams of In_2O_3 NW-TiO_2 NW hetero-structure

obtained close to 3.1 eV [10–15]. A sub-band transition is also obtained for TiO_2 NW at 2.7 eV due to the presence of Ti^{3+} and other dangling bonds and that for hetero-structure In_2O_3 NW-TiO_2 NW at 2.1 eV due to the presence of oxygen level defects as well as shallow level defects which have been depicted in Fig. 3c.

[16–23] The uniqueness of this paper can be confirmed with the help of comparison of the present work and the already reported work on some standard platform (Tables 1, 2).

4 Conclusion

In conclusion, hetero-structured In_2O_3 NW-TiO_2 NW has been successfully fabricated by using GLAD technique and proliferation in absorption method has been compared with the ordinary TiO_2 NW structure. The absorption intensity has increased in In_2O_3 NW-TiO_2 NW hetero-structure compared to TiO_2 NW at 365 nm. An average of ~1.4 fold enhancement in absorption at UV and visible spectra have been observed in case of hetero-structured In_2O_3 NW-TiO_2 NW/TiO_2 TF/n-Si compared to simple TiO_2 NW/TiO_2 TF/n-Si samples. The main band gap transition has been observed at ~3.3 eV in case of hetero-structured In_2O_3 NW-TiO_2 NW and at ~3.1 eV in case of simple TiO_2 NW under open-air room temperature. Visible spectra absorption has been increased at 590 nm due to the inclusion of shallow level defect states. Hence, the proposed NW-based hetero-structure can be used for efficient UV-Vis detection purposes.

Table 1 Data collected from other papers reported by standard publications

Sl. No.	Structure	Oxide used	Phase	UV absorption (nm)	Visible absorption (nm)	Band gap (eV)	Defect states	Optical	Fabrication process
1	n-Si/TiO_2 TF [13]	TiO_2	Rutile	380	–	3.2	Ti^{3+}	UV detection	Sol-Gel
2	p-Si/Al_2O_3 NP [14]	Al_2O_3	–	320	450 nm	3.2 (UV), 2.8 (VIS)	O^-, Ti^{3+}, Ti^{4+}	UV-Vis detection	PVD
3	TiO_2 NW/GO TF [22]	GO, TiO_2	–	366	440	3.3	Ti^{3+}	UV-Vis detection	Spin coating, GLAD
4	ITO/In_2O_3 NW [23]	ITO, In_2O_3	Poly crystalline	378	515	3.75	O-defect	UV-Vis detection	GLAD

Table 2 Data obtained in the present work

Sl. No.	Structure	Oxide used	Phase	UV absorption (nm)	Visible absorption (nm)	Band gap (eV)	Defect states	Optical	Fabrication process
1	TiO_2 NW/n-Si	TiO_2	Anatase	–	400 and 460	3.1 and 2.7 (UV)	O-defect	Visible detection	GLAD PVD
2	In_2O_3NW TiO_2 NW/TiO_2 TF/n-Si	In_2O_3, TiO_2	Bixbyite	365	590	3.3 (UV) 2.1(VIS)	O-defect, shallow defect	Visible detection	GLAD PVD

Acknowledgements The authors would like to thank ECE department and Dr. Biswajit Saha, NIT Agartala (Department of Physics) for providing the facility to sample testing for the absorption measurement.

References

1. Shougaijam B, Swain R, Ngangbam C, Lenka TR (2016) Enhanced photodetection by glancing angle deposited vertically aligned TiO_2 nanowires. IEEE. https://doi.org/10.1109/tnano.2016.2536162
2. Sarkar MB, Choudhuri B, Bhattacharya P, Barrman RN, Ghosh A, Diwedi SMMD, Chakrbartty S, Mondal A (2018) Improved UV photodetection by indium doped TiO_2 thin film based photodetector. J Nanosci Nanotechnol 18:4898–4903
3. Choudhuri B, Mondal A, Dhar JC, Singh NK, Goswami T, Chattopadhyay KK (2013) Enhanced photocurrent from generated photothermal heat in indium nanoparticles embedded TiO_2 film. Appl Phys Lett 102:233108
4. Yi S, Tian S, Zeng D, Xu K, Zhang S, Xie C (2013) An In_2O_3 nanowire-like network fabricated on coplanar sensor surface by sacrificial CNTs for enhanced gas sensing performance. Sens Actuators B 185:345–353
5. Shchukin DG, Caruso RA (2004) Template synthesis and photocatalytic properties of porous metal oxide spheres formed by nanoparticle infiltration. Chem Mater 16:2287–2292
6. Reyes-Gil KR, Reyes-Garcia EA, Raftery D (2007) Nitrogen-doped In_2O_3 thin film electrodes for photocatalytic water splitting. J Phys Chem C 111:14579–14588
7. Kityk IV, Liu Q, Sun Z, Fang J (2006) Low-temperature anomalies of two-photon absorption in In_2O_3 nano crystals incorporated into PMMA matrixes. J Phys Chem B 110:8219–8222
8. Girtan M, Cachet H, Rusua I (2003) On the physical properties of indium oxide thin films deposited by pyrosol in comparsion with films deposited by pneumatic spray pyrolysis. Thin Solid Films 427:406–410
9. Lee S-Y, Park S-J (2013) TiO_2 photocatalyst for water treatment applications. J Ind Eng Chem 19:1761–1769
10. Jimenez C, Mendez A, Paez S, Ramierz E, Rodriquoz H (2006) Production and characterization of indium oxide and indium nitride. Braz J Phys 36(3b):1017–1020
11. Ellmer K (2012) Past achievements and future challenges in the development of optically transparent electrodes. Nat Photonics 6:808–816. http://dx.doi.org/10.1038/nphoton.2012.282
12. Gan J, Lu X, Zhai T, Zhao Y, Xie S, Mao Y, Zhang Y, Yanga Y, Tonga Y (2011) Vertically aligned In_2O_3 nanorods on FTO substrates for photo electrochemical applications. J Mater Chem 21:14685
13. Tang H, Levy F, Berger H, Schmid PE (1995) Urbach tail of anatase TiO_2. Phys Rev B 52:7771
14. Song X, Qu P, Yang H, He X, Qiu G (2005) Synthesis of γ-Al_2O_3 nanoparticles by chemical precipitation method. J Central South Univ Technol 12(5):536–541
15. Konstantatos G, Sargent EH (2010) Nat Nanotechnol 5:391
16. Sarkar MB, Mondal A, Choudhuri B (2016) Presence of capacitive memory in indium doped TiO_2 alloy thin film. J Alloys Compd 654:529
17. Sarkar MB, Mondal A, Choudhuri B, Mahajan B, Chakrabartty S, Ngangbam C (2014) Enlarged broad band photodetection using indium doped TiO_2 alloy thin film. J Alloys Compd 615:440
18. Ju S, Ishikawa F, Chen P, Chang HK, Zhou C, Ha Y, Liu J, Facchetti A, Marks TJ, Janes DB (2008) High performance In_2O_3 nanowire transistors using organic gate nanodielectrics. Appl Phys Lett 92:222105
19. Sabino FP, Oliveira LN, Wei S-H, Da Silva JLF (2017) Optical and fundamental band gaps disparity in transparent conducting oxides: new findings for the In_2O_3 and SnO systems. J Phys Condens Matt 29:085501 (7)

20. Sekiya T, Kurita S (2008) Defects in anatase titanium dioxide. Adv Mater Res 121–141
21. Chinnamuthu P, Mondal A, Singh NK, Dhar JC, Chattopadhyay KK et al (2012) Band gap enhancement of glancing angle deposited TiO_2 nanowire array. J Appl Phys 112:054315
22. Deb P, Dhar JC (2018) Enhanced absorption and photoemission from TiO_2 nanowire/graphene oxide thin-film hetero-structure. J Electron Mater 47(10)
23. Mondal A, Shougaijam B, Ngangbam C (2014) Dispersed Ag nanoparticles on TiO_2 nanowire clusters for photodetection. In: TENCON 2014—2014 IEEE region 10 conference

Diabetes Diagnosis Prediction Using Ensemble Approach

Kavita Agrawal, G. Bhargav, and E. Spandana

Abstract Diabetes is considered as one of the most dangerous diseases in the world. It may also aid in causing heart attacks, blindness, etc. So, instead of taking medication for a long time after it has occurred, it is better if we can predict its occurrence at an early stage so as to prevent it. In this paper, we have used an ensemble approach using multiple classifiers to predict the result. We have trained the dataset using a number of classifiers. The dataset used in this paper is based on the parameters that are likely to cause diabetes in India. We got accuracies of each model separately and tried to achieve a better accuracy using the ensemble approach.

Keywords Support vector machine · Random forest · Confusion matrix · Naive Bayes · Cross validation

1 Introduction

The number of people having diabetes is increasing day by day and has now become a very commonly occurring disease. As diabetes affects various parts of the body including heart, eyes, nerves, etc., it is important that it has to be predicted before it can occur or at an early stage. Using decision support system, mining techniques, big data analytics, and association rules helps the medical physicians to identify and predict whether the patient will be affected by this acute disease or not at an early stage itself. If diabetes is predicted correctly, then we can prevent the effects caused by it by making required changes to our lifestyle.

K. Agrawal (✉) · G. Bhargav · E. Spandana
Department of Computer Science and Engineering, Chaitanya Bharathi Institute of Technology, Hyderabad, India
e-mail: akavita@cbit.ac.in

G. Bhargav
e-mail: chinnu.srk40@gmail.com

E. Spandana
e-mail: spandanasweety78@gmail.com

V. Nath and J. K. Mandal (eds.), *Proceedings of the Fourth International Conference on Microelectronics, Computing and Communication Systems*, Lecture Notes in Electrical Engineering 673, https://doi.org/10.1007/978-981-15-5546-6_66

A lot of models have been proposed using various classifiers taking either only one or more than one together. Using ensemble models [1], one can increase the overall accuracy by combining predictions of various individual classifiers. Decision tree, KNN, Naïve Bayes, support vector machine, random forest [1] are examples of few of the classifiers. Random forest is itself an ensemble of decision trees. Accuracies can be calculated using the confusion matrix obtained for each classifier predictions.

Rest of paper is organized as follows: Section 2 describes about the related work done in this area. Section 3 describes proposed method in detail. Section 4 deals with implementation part. Experimental results and analysis are given in Sect. 5. Finally, a conclusion of the work is given in Sect. 6.

2 Related Works

Several models have been proposed so far [1–4] for prediction of diabetes mellitus. In the work of Panwar et al. [5], the classifier used is KNN classifier and the main focus was on data preprocessing and supplying a clean data to the classifier for training and testing. The dataset used in this was the Pima Indian diabetes dataset (PIDD) obtained from UCI repository. It has nine attributes including the class label and 768 samples. The attributes available in this dataset are number of times pregnant, plasma glucose concentration a 2 h in an oral glucose tolerance test, diastolic blood pressure (mm Hg), triceps skin fold thickness (mm), 2-h serum insulin (mu U/ml), body mass index, diabetes pedigree function, age (years), class. Preprocessing was done and then the classifier was used for prediction after training it. The dataset consists of records of woman.

Kr et al. [6] used decision tree, Naïve Bayes and support vector machine and found individual accuracies. The decision tree was implemented using the WEKA tool, and the Naïve Bayes and SVM were applied using R-tool. SVM gave higher accuracy than the others and the ensemble of these three gave an even better accuracy. Preprocessing and fuzzification were done before deploying the classifiers. The dataset used in this model is also the Pima Indian diabetes dataset with nine attributes and 768 records.

Joshi et al. [7] used an ensemble of random forest, KNN, J48 (decision tree) and Naïve Bayes classifiers. A hybrid model was made using these classifiers which gave a boost to the accuracy. The dataset used here is also the PIDD from UCI repository. Missing values and duplicate data were first removed.

All the models discussed focussed their model, and based it on PIDD but in a country like India, there are a number of other factors like lifestyle of a person and his/her food habits that can also affect the occurrence of diabetes so our work is basically focussed in this area and to collect a dataset that consists of these attribute and then perform classification. Based on our study [8–12], there are a number of other factors that can cause diabetes which are:

- Different cholesterol levels
- Low density lipo-protein

- High density lipo-protein
- Systolic and diastolic pressure
- Previously occurred accidents or other injuries
- Alcohol consumption
- Stress/tension levels
- Poor or healthy diets
- Tobacco consumption.

We were able to obtain a dataset consisting of few of the above-mentioned factors and then performed the classification and ensemble task on it. We used Python for the classification and ensemble part and used a library present in Python called scikit learn which consists of different classifier functions. These can be applied to the dataset and the confusion matrix can be obtained and accuracy can be calculated. Some of the algorithms are described below which we are going to use in our dataset.

(i) **Decision Tree**: Dataset was applied in WEKA tool. Out of "N" classifiers in the existing market, here three main classifiers are employed for the prediction. Qualities of decision tree include easy interpretation and explanation. Outliers are handled by the decision tree without having any trouble providing the possible outcomes. Feature selection can be done using this classifier [5]. Dataset preparation is simple in this classifier. Non-relationship attributes will affect the result.

(ii) **Naive Bayes**: Bayesian classification has got brilliant qualities. (i) Training the dataset is enough to predict the result, (ii) it is in a semi-supervised learning too, (iii) in the embarrassing situation, it can predict an optimal result and (iv) time consuming compared to other new classifiers. Expert knowledge and intuition are the inbuilt quality for the Bayes classifier.

(iii) **Support Vector Machine**: SVM has a machine learning classifier provides always high accuracy results because of the linear and nonlinear segmentation. Over-fitting can be handled very gently by this classification. Normally, the classifier used in text mining, opinion mining and sentiment analysis. But in the class of supervised learning, it always works better. Hence, N-dimensional spaces were deployed to predict the accurate result up to 93%. A support vector machine (SVM) is a supervised machine learning algorithm that can be employed for both classification and regression purposes. SVMs are more commonly used in classification problems and as such, this is what we will focus on in this post. SVMs are based on the idea of finding a hyper-plane that best divides a dataset into classes, as shown in the image below (Fig. 1).

Support vectors are the data points nearest to the hyper-plane, the points of a data set that, if removed, would alter the position of the dividing hyper-plane. Because of this, they can be considered the critical elements of a data set. The purpose behind applying three classifiers for the same trained dataset is to check the real accuracy of the recommendation. The type of applied methodologies called ensemble model. The variation between the three classifiers is very low. Hence, the recommendation

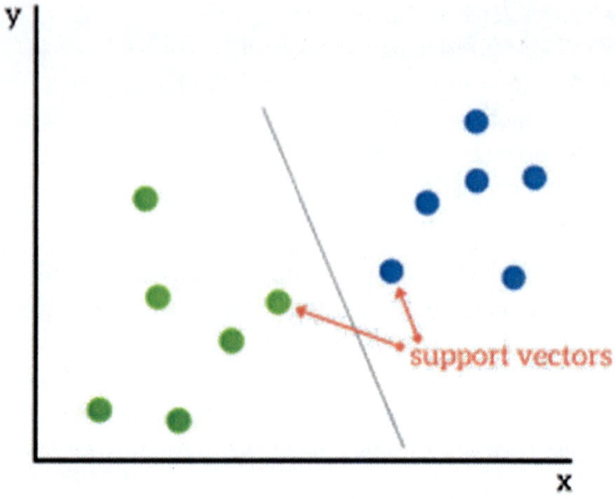

Fig. 1 Support vector machine

has got a high result. SVM though was seen to give better accuracy when compared to rest though the difference being less.

To check the real accuracy, three classifiers for the same trained dataset have been considered. The type of applied methodologies called ensemble model. The variation between the three classifiers is very low. Hence, the recommendation has got a high result. We recommended medicine and their dosage. The disadvantage is many new classifiers are there which can be used in the ensemble model to give even better accuracy.

(iv) **Random forest (RF)**: It is a popular and adaptable algorithm used in ensemble technique. It is the best and popular machine learning algorithm in the concept of hybrid model for the improving performance and prediction accuracy. RF is easy to handle large data and high dimensionality. The samples are selected arbitrarily.

(v) **KNN**: The idea behind the KNN algorithm is to dynamically identify K samples in the training data set that are similar/near to a new sample (in terms of the independent variables) and classify this new sample by a majority vote of its neighbors where an object is assigned to the most common class of K nearest neighbors (Fig. 2).

2.1 Confusion Matrix

Terms in use [12]

1. True Positive: whose actual and predicted class, both are true.
2. True Negative: whose actual and predicted class both are false.
3. False Positive: whose actual class is false but predicted value is true.
4. False Negative: whose actual class if true but predicted value is false.

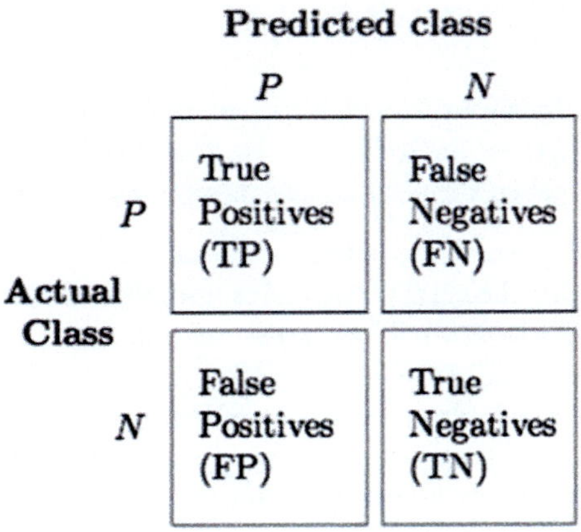

Fig. 2 Confusion matrix

All the above parameters are part of the confusion matrix which is thus used to calculate the various performance metrics of the classifiers which are as follows along with their formulae:

Accuracy:

$$\mathrm{ACC} = \frac{\mathrm{TP} + \mathrm{TN}}{\mathrm{TP} + \mathrm{TN} + \mathrm{FN} + \mathrm{FP}} = \frac{\mathrm{TP} + \mathrm{TN}}{\mathrm{P} + \mathrm{N}}$$

Sensitivity:

$$\mathrm{SN} = \frac{\mathrm{TP}}{\mathrm{TP} + \mathrm{FN}} = \frac{\mathrm{TP}}{\mathrm{P}}$$

Specificity:

$$\mathrm{SP} = \frac{\mathrm{TN}}{\mathrm{TN} + \mathrm{FP}} = \frac{\mathrm{TN}}{\mathrm{N}}$$

2.2 *Cross Validation*

In cross validation [13] process, the dataset is split into k parts called folds. If supposed $k = 10$, then the dataset is split into ten equal parts. Out of these, nine are used to train the classifier and the remaining one is used to test the classifier. This whole process is repeated until each part is tested against the remaining nine parts used to train it in the particular combination. The number of folds can be changed as well which also affects the accuracy.

2.3 Dataset

Under our survey, we have tried to analyze the parameters that are already available in the Pima Indian diabetes dataset. The dataset that is available is the PIMA Indian diabetes dataset containing following attributes [3]:

1. Number of times pregnant
2. Plasma glucose concentration 2 h in an oral glucose tolerance test
3. Diastolic blood pressure (mm Hg)
4. Triceps skin fold thickness (mm)
5. 2 h serum insulin (mu U/ml)
6. Body mass index (weight in kg/(height in m)^2)
7. Diabetes pedigree function
8. Age (years).

So far, a large number of prediction models were built based on only this dataset but in this paper the goal is to extend the dataset by considering other parameters that are predominantly diabetes aiding in India. There was one dataset that was available in data.gov.in that had at least few of the parameters we were planned to take into account. Clinical, Anthropometric and Bio-chemical (CAB) 2014 Survey conducted in various districts of various states in India and a total of 53 attributes were considered in this survey. Out of these, 53 attributes there were 14 attributes that could be considered as part of the dataset we are looking to retrieve.

The following are the 14 parameters:

- Rural/urban
- Sex of the person
- Age
- Weight of the person in kg
- Length of the person in cm
- Hemoglobin_level: outcome of hemoglobin test (in percentage grams)
- BP systolic reading 1
- BP systolic reading 2
- BP diastolic reading 1
- BP diastolic reading 2
- Pulse rate reading 1
- Pulse rate reading 2
- Diabetes_test: whether diabetes test was conducted or not
- Fasting glucose: status of fasting blood glucose level measurement.

3 Proposed System

The first part involves filling up of the missing values. We used WEKA tool for this process and filled the missing values. Next, we applied different classifiers on the

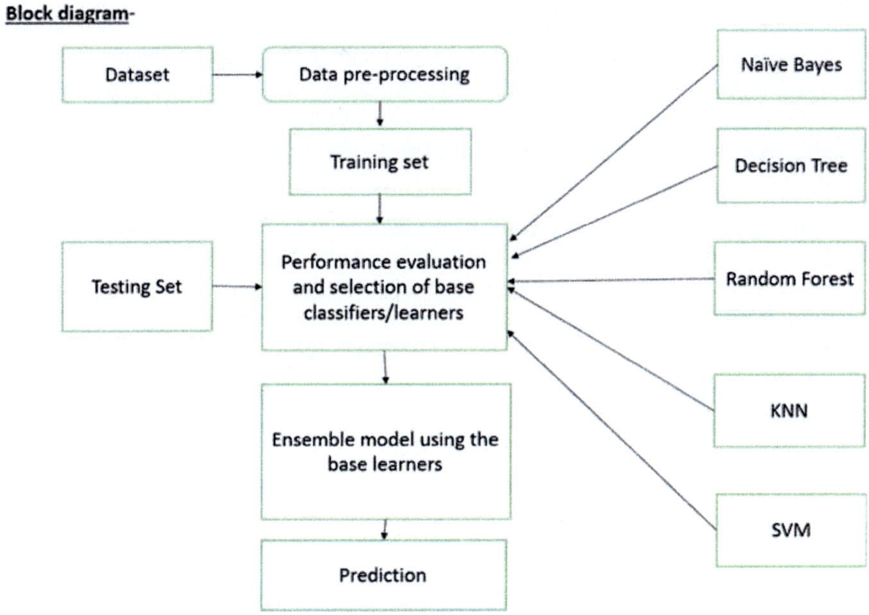

Fig. 3 Diabetes diagnosis prediction using ensemble approach

dataset and obtained the confusion matrix and also the accuracy of the classifier. After this, we took a variety of combinations for the voting classifier giving all the classifiers a weightage of 1. We found the combination that gave the best accuracies from it. The combination consisted of 2, 3 and 4 classifiers in the voting process. Once these results were obtained, weightages were given to each classifier based on individual performance of the classifiers, i.e., more weight to classifier that gave more accuracy. Using this weight new accuracy for different models was calculated (Fig. 3).

3.1 Generating Dataset

Dataset is obtained by studying various factors that may result in diabetes and their normal and harmful levels in humans in different cases. Combine all these values for each record and then we get the dataset for diabetes based on conditions or factors causing it in our country. This dataset must contain values for both people with and without diabetes and hence respective variations in different attributes based on it.

3.2 Data Preprocessing

(i) *Data Cleaning*

Solving missing value problem using pattern matching

1. Consider a *p*th record.
2. For this particular record, check from the top as to which row has maximum common values for different columns.
3. Once the closest pattern is found, then replace the missing value for that particular column value of the *p*th record with the value of the column value in the closest pattern match row.
4. Repeat for all the missing values in the dataset.

(ii) *Dimensionality Reduction*

Feature selection approach

1. Either greedy approach or best first approach is used to select the attributes that affect the prediction.
2. The ones which do not cause much change or do not affect much are ignored.
3. Then the subset of attributes are obtained which can be used for the training process.

3.3 Training and Classification

Once we get the enhanced dataset, the following steps are to be performed;

1. Train the dataset for each of the classifiers, i.e., decision tree, Naïve Bayes, AdaBoost with decision Stump, SVM, KNN, artificial neural networks, random forest.
2. Obtain the confusion matrix for each of the classifiers by supplying a test dataset.
3. Use the values to calculate the accuracy, specificity and sensitivity of the classifiers from the values of the confusion matrix using the respective formulae given below.

4 Implementation

Using www.dhsprogram.com website, we can get survey records of various countries. The survey done in India had few attributes which could affect diabetes occurrence based on our research. The data needs to be requested and then permission has to be given by the administrators of the website.

The following are the attributes of use found based on our study and their particular value description in the dataset

- Age: numeric (people of different ages).
- State: each state is given its own code. Union territories are also included.
- Rural/urban: where urban assigned value 0 and rural with 1.
- S_cig: if he smokes cigarette value assigned is 1 if no then 0.
- S_bidi: if he smokes bidis value assigned is 1 if no then 0.
- S_pipe: if he smokes pipe value assigned is 1 if no then 0.
- Ch_tob: if he chews tobacco value assigned is 1 if no then 0.
- Uses_snuff: if he uses snuff value assigned is 1 if no then 0.
- S_cigar: if he smokes cigar value assigned is 1 if no then 0.
- Gutka: if he eats gutka value assigned is 1 if no then 0.
- Paan_w_tob: if he consumes pan with tobacco value assigned is 1 if no then 0.
- S_other: if he smokes something other than above value assigned is 1 if no then 0.
- No_smoke: if he does not smoke value assigned is 1 if he does then 0.
- District: there are number of districts in this data and each as its own numeric code.
- D_alcohol: if he drinks alcohol value assigned is 1 if no then 0.
- Fre_alco: frequency of drinking if daily 1, once a week 2, less than once a week 3, not drink 0.
- Tadi_madi: if he drinks tadi_madi value assigned is 1 if no then 0.
- Coun_liq: if he drinks country liquor value assigned is 1 if no then 0.
- Beer: if he drinks beer value assigned is 1 if no then 0.
- Wine: if he drinks wine value assigned is 1 if no then 0.
- Hard_liq: if he drinks hard liquor value assigned is 1 if no then 0.
- Other: if he drinks anything other than above ones value assigned is 1 if no then 0.
- f_mik_curd: frequency of consuming milk and curd, 0 if does not consume at all, 1 if regularly, 2 if weekly, 3 if occasionally.
- f_pul/bean: frequency of consuming pulses and beans, 0 if does not consume at all, 1 if regularly, 2 if weekly, 3 if occasionally.
- gr_lf_veg: frequency of consuming green leafy vegetables, 0 if does not consume at all, 1 if regularly, 2 if weekly, 3 if occasionally.
- Fruits: frequency of consuming fruits, 0 if does not consume at all, 1 if regularly, 2 if weekly, 3 if occasionally.
- Eggs: frequency of consuming eggs, 0 if does not consume at all, 1 if regularly, 2 if weekly, 3 if occasionally.
- Fish: frequency of consuming fish, 0 if does not consume at all, 1 if regularly, 2 if weekly, 3 if occasionally.
- chik_meat: frequency of consuming chicken and meat, 0 if does not consume at all, 1 if regularly, 2 if weekly, 3 if occasionally.
- fried_food: frequency of consuming fried food, 0 if does not consume at all, 1 if regularly, 2 if weekly, 3 if occasionally.
- aerated_drinks: frequency of consuming aerated drinks, 0 if does not consume at all, 1 if regularly, 2 if weekly, 3 if occasionally.
- sys_pre1: systolic pressure reading 1.

- dias_pre1: diastolic pressure reading 1.
- sys_pre2: systolic pressure reading 2.
- dias_pre2: diastolic pressure reading 2.
- sys_pre3 systolic pressure reading 2.
- dias_pre3: diastolic pressure reading 2.
- glucose_level.
- class: 0 non-diabetic and 1 if diabetic.

Using the dataset consisting of the above attributes, we computed accuracies using all the classifiers mentioned in the previous sections. Since we used numeric encoding to attributes like state, district, frequency of various attributes regarding diet, the decision tree might have considered them to be numeric attributes and not categorical. Hence, we decided to use hot line encoding where each value was taken as a separate attribute like daily, weekly and occasionally were taken as three attributes and values 0 and 1 were assigned to them based on the record values for them. Then we got a total of 66 attributes including the class labels. Also, since there are lot of states and districts resulting in increase of attributes if we do hot line encoding for them, we decided to not consider them and removed them from the dataset. Also, we applied a k-fold cross validation with 100 splits (Fig. 4).

Naïve Bayes Classifier:

CM	Predicted:NO	Predicted:YES
actual:NO	397	353
actual:YES	87	226

Accuracy: 0.5868181818181817

Decision Tree:

CM	Predicted:NO	Predicted:YES
actual:NO	585	165
actual:YES	160	153

Accuracy: 0.6948181818181819

The above decision tree is just a part of the original tree that was obtained by keeping the max number of leaf nodes limited to 16. Since there are 65 attributes, the decision tree is actually very big and cannot be visualized in a single screen. Graph viz is used to visualize this tree.

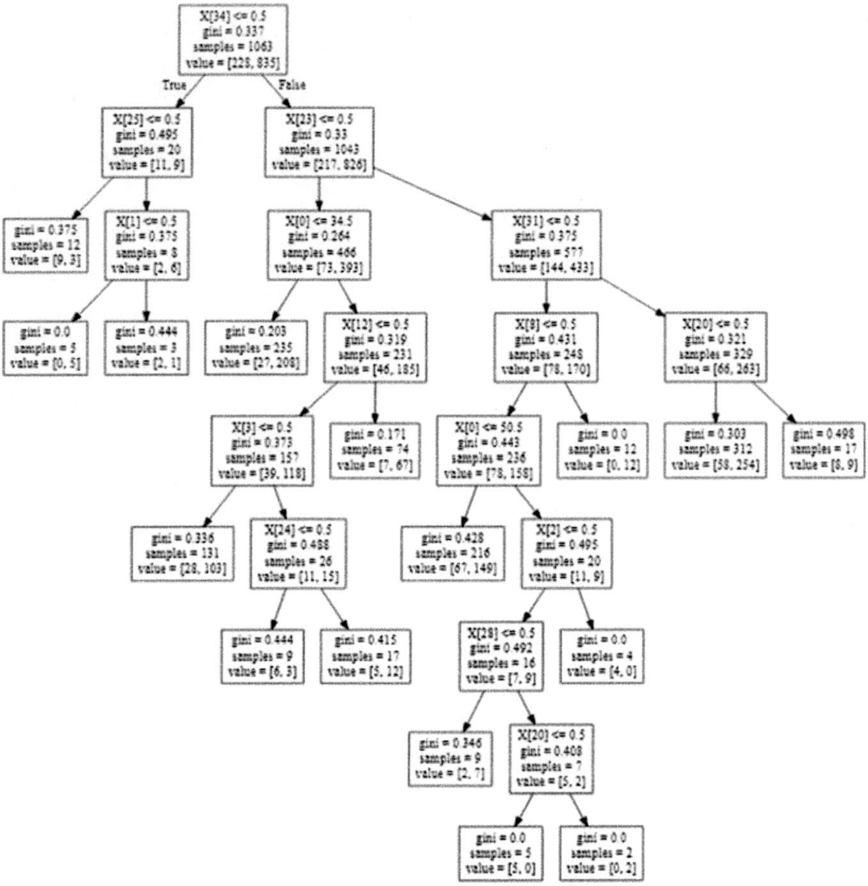

Fig. 4 Decision tree

KNN Classifier:

CM	Predicted:NO	Predicted:YES
actual:NO	709	41
actual:YES	235	78

Accuracy: 0.7401818181818183

Table 1 Output for one classifier

Classifier	Accuracy	TP	TN	FN	FP
Random forest	77.33	98	724	215	26
KNN classifier	74.01	78	709	235	41
AdaBoost classifier	73.47	119	665	194	85
Decision tree	69.48	153	585	160	165
Naïve Bayes	58.68	226	397	87	353

Random Forest:

```
+------------+-------------+--------------+
|     CM     | Predicted:NO | Predicted:YES |
+------------+-------------+--------------+
| actual:NO  |     724     |      26      |
| actual:YES |     215     |      98      |
+------------+-------------+--------------+
Accuracy: 0.7733204020454946
```

AdaBoost Classifier:

```
+------------+-------------+--------------+
|     CM     | Predicted:NO | Predicted:YES |
+------------+-------------+--------------+
| actual:NO  |     659     |      91      |
| actual:YES |     191     |     122      |
+------------+-------------+--------------+
Accuracy: 0.7347998589314053
```

Our next step after getting the accuracies was to check which combination of ensembles gave better accuracy compared to each of the classifier used in the ensemble. Hence, we tried all the possible combinations for the above classifiers and noted them down in different tables. We have chosen the Voting classifier which is one of the ensemble models and provided it with various combinations of classifiers as arguments. First, we started off with combination of two classifiers (Tables 1, 2, 3, 4 and 5).

5 Results

Since random forest was found to be the best classifier out of the remaining, we decided to give a weight parameter to the ensemble giving more weight to random forest that is twice of the other classifiers. When a weight of 2 was given to random forest in various combinations, the following were the results:

Out of all the accuracies, we found maximum accuracy was for the ensemble with random forest, AdaBoost classifier and decision tree. Random forest was considered

Table 2 Output for two classifiers

Classfiers accuracy		TP		TN	FN	FP
Random forest	KNN	74.60	58	735	255	15
Random forest	AdaBoost	75.30	78	723	235	27
Random forest	Decision tree	75.69	79	726	234	24
Random forest	Naïve Bayes	75.79	85	721	228	29
KNN	AdaBoost	74.84	63	733	250	17
KNN	Decision tree	74.89	65	731	248	19
KNN	Naïve Bayes	74.59	68	725	245	25
AdaBoost	Decision tree	75.14	88	711	225	39
AdaBoost	Naïve Bayes	73.67	103	680	210	70
Decision tree	Naïve Bayes	72.7	124	648	189	102

Table 3 Output for three classifiers

Classifiers: Accuracy				TP	TN	FN	FP
Random forest	KNN	AdaBoost	75.97	93	715	220	35
Random forest	KNN	Decision tree	76.54	96	718	217	32
Random forest	KNN	Naïve Bayes	76.50	109	70	204	45
Random forest	AdaBoost	Decision tree	76.29	113	698	200	52
Random forest	AdaBoost	Naïve Bayes	75.08	128	670	185	80
Random forest	Decision tree	Naïve Bayes	73.70	148	635	165	115
KNN	AdaBoost	Decision tree	76.30	110	701	203	40
KNN	AdaBoost	Naïve Bayes	74.16	124	664	189	86
KNN	Decision tree	Naïve Bayes	72.96	145	630	168	120
AdaBoost	Decision tree	Naïve Bayes	71.59	163	597	150	153

Table 4 Output for four classifiers

Classifiers				Accuracy	TP	TN	FN	FP
Random forest	KNN	AdaBoost	Decision tree	75.60	81	723	232	27
Random forest	KNN	AdaBoost	Naïve Bayes	75.59	87	717	226	33
Random forest	KNN	Decision tree	Naïve Bayes	76.26	89	722	224	28
Random forest	AdaBoost	Decision tree	Naïve Bayes	75.76	101	704	212	46
KNN	AdaBoost	Decision tree	Naïve Bayes	76.30	102	709	211	41

for the ensemble as it had the highest accuracy when compared to the rest of the classifiers. AdaBoost was chosen because it had almost a good number of all the four prediction classes and decision tree because it gave lesser False Negatives also better True Negatives when compared to Naïve Bayes classifier which gave fewer

Table 5 Output for classifiers with weights

Classifiers				Accuracy	TP	TN	FN	FP
1 Weight = 2	2 Weight = 1	3 Weight = 1	4 Weight = 1	–	–	–	–	–
Random forest	KNN	–	–	76.53	99	715	214	35
Random forest	AdaBoost	–	–	76.53	99	715	214	35
Random forest	Decision tree	–	–	76.53	99	715	214	35
Random forest	Naïve Bayes	–	–	76.53	99	715	214	35
Random forest	KNN	AdaBoost	–	75.5	83	720	230	30
Random forest	KNN	Decision tree	–	75.9	84	724	229	26
Random forest	KNN	Naïve Bayes	–	76.1	92	718	221	32
Random forest	**AdaBoost**	**Decision tree**	**–**	**77.68**	**118**	**708**	**195**	**42**
Random forest	AdaBoost	Naïve Bayes	–	75.08	128	670	185	80
Random forest	Decision tree	Naïve Bayes	–	76.24	94	717	219	33
Random forest	KNN	AdaBoost	Decision tree	76.25	96	715	217	35
Random forest	KNN	AdaBoost	Naïve Bayes	76.7	103	713	210	37
Random forest	KNN	Decision tree	Naïve Bayes	76.9	104	714	209	36
Random forest	AdaBoost	Decision tree	Naïve Bayes	76.74	113	703	200	47

True Negatives. The accuracy improved by 0.3 only but the number of False Negatives were decreased which is very important in medical related predictions. Also, there was in increase in the True Positives that was found.

Many other combinations can also be seen which have higher accuracies in ensemble when compared to the individual contributing classifiers, and hence, a successful ensemble model can be created using them also but the one mentioned above has a higher accuracy.

6 Conclusion

We have successfully created an ensemble model using the voting classifier and random forest, AdaBoost classifier and decision tree as the contributors to it. We achieved better accuracy compared to individual models and were also able to reduce the number of False Negatives.

For the future work, we can work on class imbalance problem which will the help to reduce the False Negatives even better and also include diet predictor for people who may have a chance to get diabetes by looking up their attribute values and suggesting the diet and medication.

References

1. Seni G, Elde J (2010) Synthesis lectures on data mining and knowledge discovery. In: Ensemble methods in data mining: improving accuracy through combining predictions
2. Lee C-S, Wang M-H (2011) A fuzzy expert system for diabetes decision support application. IEEE Trans Syst Man Cybern Part B (Cybernetics) 41(1):139–153
3. Jayalakshmi T, Santhakumaran A (2010) A novel classification method for diagnosis of diabetes mellitus using artificial neural networks. In: 2010 International conference on data storage and data engineering (DSDE). IEEE
4. Temurtas H, Yumusak N, Temurtas F (2009) A comparative study on diabetes disease diagnosis using neural networks. Expert Syst Appl 36(4):8610–8615
5. Ribeiro UC et al (2015) Diabetes classification using a redundancy reduction preprocessor. Res Biomed Eng 31(2)):97–106
6. Kr S, Kamaladevi M, Sethuraman J, Ravichandran Ks (2017) A novel approach for finding diabetic mellitus using ensemble model for an optimized classification. Asian J Pharma Clin Res
7. Joshi R, Alehegn M (2017) Analysis and prediction of diabetes diseases using machine learning algorithm: ensemble approach. Int Res J Eng Technol (IRJET)
8. Joshi SR (2015) Diabetes care in India. Annals of Global Health Icahn School of Medicine at Mount Sinai, Mumbai, India
9. Kapoor D, Bhardwaj AK, Kumar D, Raina SK (2014) Prevalence of diabetes mellitus and its risk factors among permanently settled tribal individuals in tribal and urban areas in northern state of Sub-Himalayan region of India. Int J Chronic Dis
10. Khan N, Gaurav D, Kandl T (2013) Performance evaluation of levenberg-marquardt technique in error reduction for diabetes condition classification. Procedia Comput Sci
11. Patel M, Patel IM, Patel YM, Rathi SK (2013) A hospital-based observational study of type 2 diabetic subjects from India. Indian J Clin Prac 24(2)
12. Patil TR, Sherekar SS (2013) Performance analysis of Naive Bayes and J48 classification algorithm for data classification. Int J Comput Sci Appl 6(2)
13. Lingqiao L, Huihua Y, Qian H, Jianbin Z, Tuo G (2012) Design and realization of the parallel computing framework of cross-validation. In: 2012 International conference on industrial control and electronics engineering

An 86 DB Gain 18.06 mV_{rms} Input-Referred Noise LNA for Bio-Medical Applications

G. Revanth Kumar, K. Naga Sunanda, and M. Durga Prakash

Abstract This paper admits a low-noise amplifier (LNA) designed by taking bio-medical applications into considerations. The amplifier is designed based on two gain stages, supply insensitive gain stage and inverter gain stage. Input-referred noise of the proposed amplifier is 18.02 mV_{rms} and it consumes a power of 0.012 mW. The amplifier produces a gain of 86.5 dB. Bandwidth of the proposed amplifier is 227 Hz with cut-off frequencies as 227 Hz (higher) and 1 MHz (lower). The entire system is built in 45 nm technology with supply voltage of 0.6 V.

Keywords CMOS (Complementary Metal Oxide Semiconductor) · LNA (Low Noise Amplifier) · BMR (Beta Multiplier Reference) · OTA (Operational Transconductance)

1 Introduction

Electronic devices have very huge demand in biomedical applications like neural recording systems, ECG, EKG etc. LNA is critical for interfacing the machine with the extracellular signals. As these signals are very weak, they exhibit the amplitudes between 50 and 500 μV, and noise levels between 5 and 10 μV [1]. Designing extracellular interfaces for observing brain and heart activity became very tough task for designers. There is a great demand for LNA's in biomedical engineering that allows technicians to continuous monitoring for extracellular signals. The purpose of

G. R. Kumar (✉) · K. N. Sunanda · M. D. Prakash
Department of ECE, VR Siddhartha Engineering College, Vijayawada, Andhra Pradesh, India
e-mail: revanth7766@gmail.com

K. N. Sunanda
e-mail: naga.sunanda5@gmail.com

M. D. Prakash
e-mail: mdprakash@vrsiddhartha.ac.in

V. Nath and J. K. Mandal (eds.), *Proceedings of the Fourth International Conference on Microelectronics, Computing and Communication Systems*, Lecture Notes in Electrical Engineering 673, https://doi.org/10.1007/978-981-15-5546-6_67

LNA is that they must be capable of observing and amplifying the very low amplitude and low frequency signals.

Low-noise amplifiers are also called as neural amplifiers and bio-potential amplifiers that are only preferred for amplifying the very low frequency and amplitude signals. Misreading or mismatching of neural signals from the neural systems to the neural interfaces results failure in monitoring patients health activity; for the prevention of these situations, we need a neural interfaces like low-noise amplifiers.

From the last years, there are several bio-amplifiers are investigated by the research fellows to achieve high bandwidth, and small chip area. Quadir et al. designed an ultra-low-power low-noise bio-amplifier in 0.13 μm standard RF CMOS process. It has a bandwidth of 9.51 kHz with input-referred noise of 0.91 μV_{rms} and a power dissipation of 198.6 nW [1]. Harrison and Charles designed low-power, low-noise amplifier in 1.5 μm technology. It has a bandwidth of 7.2 kHz with an input-referred noise of 2.2 μV_{rms} and dissipates a power of 80 μW [2]. Salhi et al. designed low-voltage, low-noise bio-signal amplifier using 130 μm CMOS technology. It produces a gain of 75 dB and absorbs 1.24 μW with input-referred noise of 0.91 nV^2/Hz [3]. Kim et al. designed low-power neural amplifier in 0.35 μm CMOS process. It has a bandwidth of 11.1 kHz with an input-referred noise of 14.5 μV_{rms} and a power dissipation of 220 nW [4]. Gosselin et al. proposed a low-power bio-amplifier with active frequency suppression. The bio-amplifier is designed in 0.18 μm CMOS technology. It provides a mid-band gain of 50 dB and achieves an input-referred noise of 5.6 μV_{rms}. It dissipates a power of 8.6 μW and requires 1.8 V to operate. It occupies nearly 0.050 mm^2 chip area [5]. Holleman and Otis presented a sub-microwatt low-noise amplifier for neural recording application. The amplifier is designed in 0.5 μm SOS Bi-CMOS technology. It exhibits 3.5 μV_{rms} of input-referred noise and provides the gain between 36 and 44 dB. It requires 1 V to operate and consumes only 805 nA [6]. Ruiz-Amaya et al. presented a low-noise amplifier for neural spike recording interfaces. The system is designed based on capacitive feedback network using two stages OTA. The amplifier is designed using 0.13 μm CMOS technology. It exhibits 3.8 μV_{rms} of input-referred noise and provides the gain between 46 dB. It consumes 1.92 μW and operates at 1.2 V. It provides a bandwidth of 192–7.4 kHz [7]. Figure 1 represents the whole block diagram of bio-signal processing system. The LNA is theoretically simulated using the small signal analysis [8]. An inverter-based gain stage is used in designing the LNA, which helps in producing the high gain to the amplifier [9]. Proposed LNA has been simulated practically in transient

Fig. 1 Block diagram of bio-signal processing system

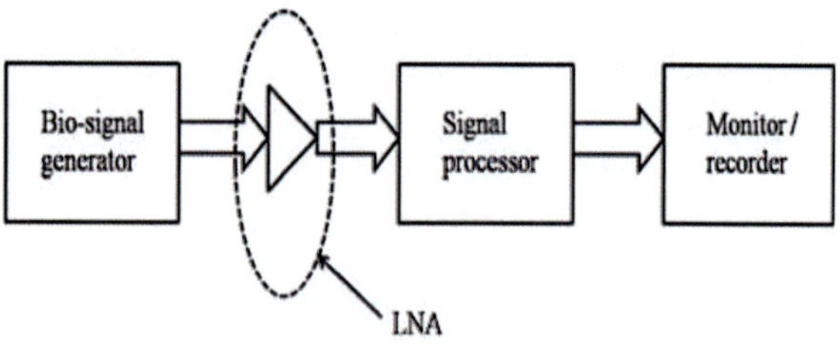

analysis, AC analysis, DC analysis and noise analysis for calculating gain, cut-off frequencies, power consumption, band width, etc. [10]. Baishnab et al. designed a low power, LNA for neural signal processing. It uses an OTA which is built using folded cascode architecture. The amplifier presents a good noise performance with input-referred noise 3.22 $\mu V/\sqrt{Hz}$ and a mid-band gain of 42 dB. The LNA dissipates very low power nearly 630 nW [11]. Fang et al. designed an ultra-wideband LNA which produces a gain of 13 dB only with noise figure of 3.4 dB. The designed LNA consumes a DC power of 12.9 mW [12]. Yang et al. designed a low-power LNA with chopping technique. The measured input-referred noise for the LNA is 39 nV/Hz, with a power consumption of 117 μW [13]. Limei et al. presented a LNA for ultra-wideband applications. The LNA produces a 23.2 dB gain and the power consumption is 12.2 mW with a supply voltage of 1.2 V. The LNA is designed in 0.13 μm technology [14]. Li et al. presented a low-power LNA for neural signal acquisition. The system is designed in 0.18 μm technology. The LNA has an input-referred noise of 2.19 μV_{rms} with a power consumption of 55.8 μW and has 2.53 NEF (noise efficiency factor) [15].

This paper presents a low-noise amplifier which is designed based on beta multiplier reference (BMR)-based differential first gain stage and a CMOS inverter-based second gain stage. The proposed amplifier is designed in 45 nm standard CMOS process, capable of recording the biomedical signals in between the range of 0.001–227 Hz. The amplifier has a gain of 86.7 dB. It consumes a power of 0.012 mW with 0.6 V supply voltage and the input-referred noise is 18.02 mV_{rms}.

2 LNA Design

The amplifier is designed using nMOS differential input stage ($M_x - M_y$), a beta multiplier reference (BMR) circuit ($M_1 - M_4$), and a CMOS inverter ($M_5 - M_6$) [1]. The schematic structure of the proposed LNA is shown in Fig. 2.

2.1 Beta Multiplier Reference Circuit Gain Stage

The differential amplifier circuit produces the first amplification stage. The beta multiplier reference circuit should be used for supply independent biasing. From the BMR circuit, we can write [10],

$$V_{\mathrm{GS},M_4} = V_{\mathrm{GS},M_3} + V_M R, \tag{1}$$

where I_M represents reference current of the BMR circuit, R represents bias resistance, V_{GS} represents gate-to-source voltage of the particular MOSFET. The gate-to-source voltages of M_3 and M_4 in (1) must be related as [1], $V_{\mathrm{GS},M_4} > V_{\mathrm{GS},M_3}$.

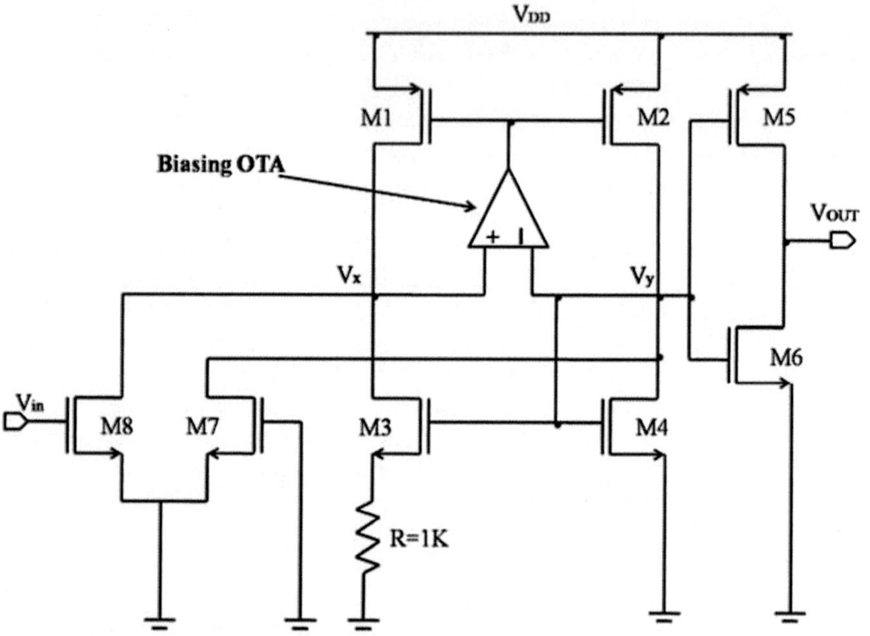

Fig. 2 Schematic diagram of LNA

The amplifier is designed to operate in sub-threshold region. The sub-threshold drain current can be calculated as [9],

$$I_M = I_{D_0}\left(\frac{W}{L}\right)\exp\left(\frac{qV_{GS}}{nkT}\right) \tag{2}$$

where I_{D_0} represents technology current, q represents electrical charge, n represents sub-threshold slope parameter ($3 > n > 1$), k represents Boltzmann constant, T represents absolute temperature and the aspect ratio of the MOSFET is represented as (W/L).

To make biasing point highly insensitive to temperature and supply voltage, the BMR circuit uses an operational transconductance amplifier (OTA). To calculate the first stage gain, we have to perform small signal analysis.

The output resistance of the half-folded cascode circuit is [8],

$$R_{\text{cascode}} = r_{01} \| g_{m3} r_{03} R, \tag{3}$$

where R_{cascode} represents equivalent output resistance of the cascode amplifier, g_{m3} represents transconductance and r_{03} represents output resistance of the corresponding MOSFET, i.e., M_3.

Now, the first stage gain can be calculated as [4],

$$A_{V_1} = -g_{mx}(r_{01} \| g_{m3} r_{03} R), \tag{4}$$

Here, g_{mx} represents transconductance of the corresponding MOSFET, i.e., M_x, A_{V_1} is the gain of first stage of amplifier. From (4), it is clear that gain is constant for any variation input.

2.2 *CMOS Inverter-Based Second Gain Stage*

To obtain high gain for the amplifier, we have to use a CMOS inverter in the middle of transistor region. Hence, for the second amplification stage, we can use a CMOS inverter [9]. The small signal gain of CMOS inverter is [9],

$$A_{V_2} = -(g_{m5} + g_{m6})(r_{05} + r_{06}), \tag{5}$$

Here, A_{V_2} represents gain of CMOS inverter, g_{m5}, g_{m6} and r_{05}, r_{06} represents transconductance and output resistance of the corresponding MOSFETs M_5 and M_6.

Hence, LNA overall gain can be calculated by multiplying the gain of both stages,

$$A_V = A_{V_1} A_{V_2} \tag{6}$$

By using (4) and (5), we will get,

$$A_V = -g_{mx}(g_{m5} + g_{m6})(r_{01} \| g_{m3} r_{03} R)(r_{05} \| r_{06}) \tag{7}$$

where A_V represents the total gain of LNA.

3 OTA Design

The schematic diagram of the LNA is shown in Fig. 3. It utilizes a transconductance amplifier (OTA). Figure shows the schematic diagram of the proposed OTA. There are several topologies available in the design of OTA. This OTA is designed using two stage architecture for high open-loop gain and large swing. For better flicker noise performance and minimized gm/ID ratio, it uses a telescopic structure with PMOS input transistors biased in sub-threshold region in its first stage. In order to reject the common mode and power supply noise, a conventional pair is used. This design employs a low-power low-noise OTA configuration in order to achieve efficient power, noise trade off making use of supply current. The amplifier is designed in a way that it should be capable of recording neural spikes or local field potentials.

Using the resisters in OTA schematic causes to produce noise in source degeneration current sources and by choosing resistance properly, we can reduce the noise levels to much smaller the noise levels produced from MOS transistors operating

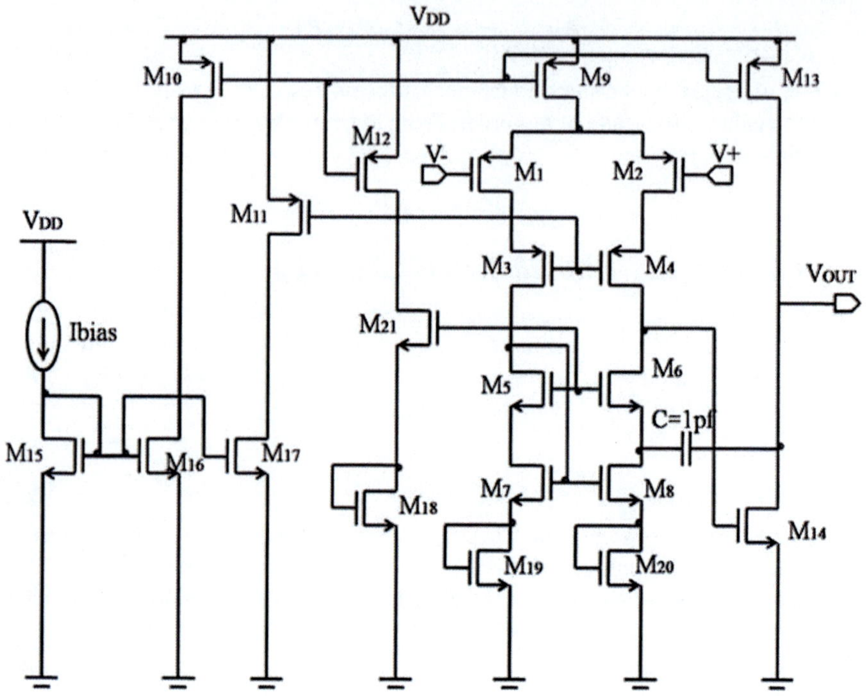

Fig. 3 Schematic diagram of OTA

with same current. To achieve low input-referred noise, transconductance of the OTA must be made maximum for a given total current. If the transistors M_1 and M_2 are operating in sub-threshold region, the transconductance will be maximized for a given current. By cascoding M_3 and M_4 with the input differential pair transistors results increase in their output impedances. The source degenerated current sources are designed to have large output impedances by M_{16}, M_{17}.

In addition, for improved noise performance, source degeneration resistors are used in current mirror loads. Frequency compensation is done employing miller capacitor CC with the common gate current buffer. This strategy is used to reduce the required value of CC and improves the bandwidth of OTA.

4 Results and Discussions

The LNA is simulated using Cadence Virtuoso Tool in 45 nm technology. Proposed LNA produces a gain of 86.7 dB. Gain response plot of the proposed LNA is shown in Fig. 4. It allows the signal ranges from 0.001 to 227 Hz.

Input-referred noise response of the proposed LNA is shown in Fig. 5. The LNA achieves a gain of 86.7 dB which is very good and better than previous works. From

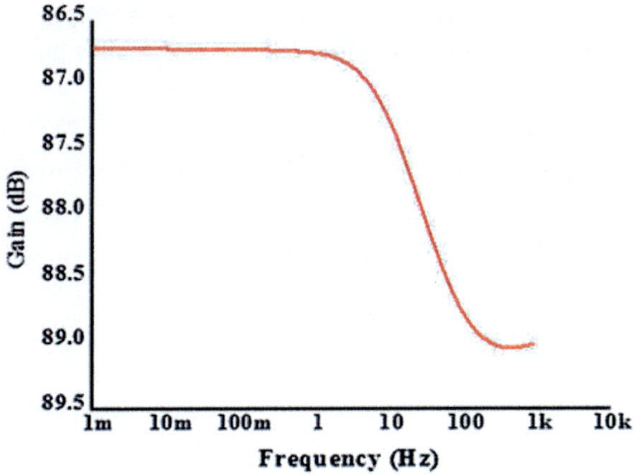

Fig. 4 Gain response of proposed LNA

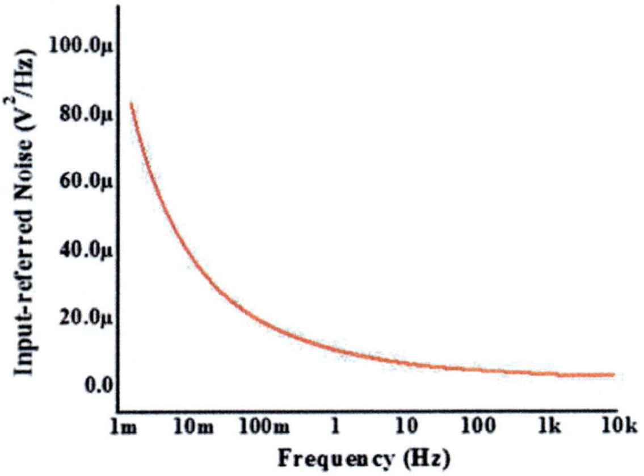

Fig. 5 Input-referred noise response of the LNA

Fig. 4 it is clear that it has a constant gain up to 80 Hz and then it decreases slowly. Gain is constant from 1 MHz, which is very low frequency. To calculate bandwidth, lower cut-off frequency and higher cut-off frequency we have to add −3 dB to the maximum output or dB. By adding −3 dB, we obtained the bandwidth 227 Hz, lower cut-off frequency and higher cut-off frequency are noted as 1 and 227 Hz.

Table 1 represents the overall performance of low-noise amplifier and compares with the previous work. The parameters like gain, bandwidth, lower cut-off frequency, higher cut-off frequency, power consumption and input-referred noise of the LNA are well defined. Gain of the LNA is very good when compared to the other works bandwidth of the LNA is also good enough as its job is to amplify very weak signals. The input-referred noise (output noise divided by amplifier gain) of the LNA is also better than previous works. Additionally, table describes about the different CMOS technologies used in previous works and their operating voltages.

Table 1 Comparing LNA parameters with previous work

Parameters	This work	[1]	[2]	[3]	[4]
CMOS technology	45 nm	0.13 μm	1.5 μm	0.13 μm	0.35 μm
Supply voltage	0.6 V	0.6 V	±2.5 V	±0.6 V	±0.9 V
Power consumption	0.012 mW	198.6 nW	80 μW	1.24 μW	220 nW
Gain	86.7 dB	24.94 dB	40 dB	75 dB	28.9 dB
Bandwidth	227 Hz	9.51 kHz	7.2 kHz	18.8 kHz	11.1 kHz
Higher cut-off frequency	227 Hz	9.51 kHz	–	19 kHz	11.2 kHz
Lower cut-off frequency	0.001 Hz	0.01 Hz	0.13 Hz	–	0.1 Hz
Input-referred noise	18.02 mV_{rms}	0.91 μV_{rms}	2.2 μV_{rms}	0.91 nV^2/Hz	14.5 μV_{rms}

μm = micro meter, V = volt, nm = nano meter, mW = mille watts, μW = micro watts, nW = nano watts, Hz = hertz, K = kilo, V_{rms} = rms value of voltage

Input-referred noise of proposed LNA is varies for different frequency levels. Figure 5 is obtained from different input-referred noise levels at particular frequencies. Here, frequency is taken on horizontal axis and input-referred noise is taken on vertical axis. Input-referred noise values are represented in V^2/Hz and frequency values are represented in Hz.

Figures 6 and 7 represent the input noise response and output noise response of the low-noise amplifier. These graphs are obtained by performing noise analysis ranging from 1 to 10 kHz frequencies in cadence virtuoso tool. Above graphs describe different noise values at particular frequency levels. Noise values are represented in V^2/Hz on vertical axis and frequency is representing in Hz on horizontal axis. The decrease in noise levels from input noise response to output noise response is very clear from Figs. 6 and 7.

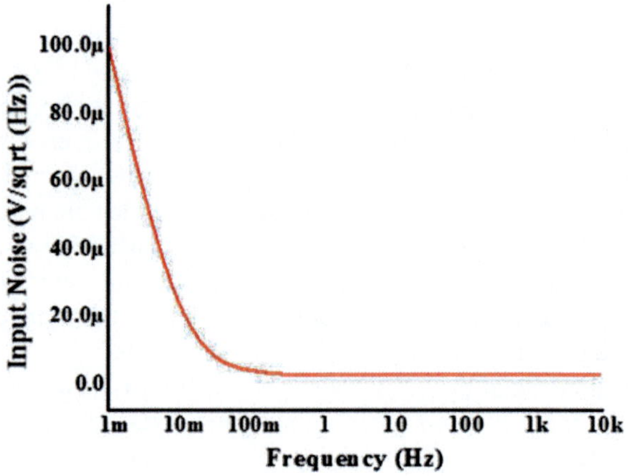

Fig. 6 Input noise response of the LNA

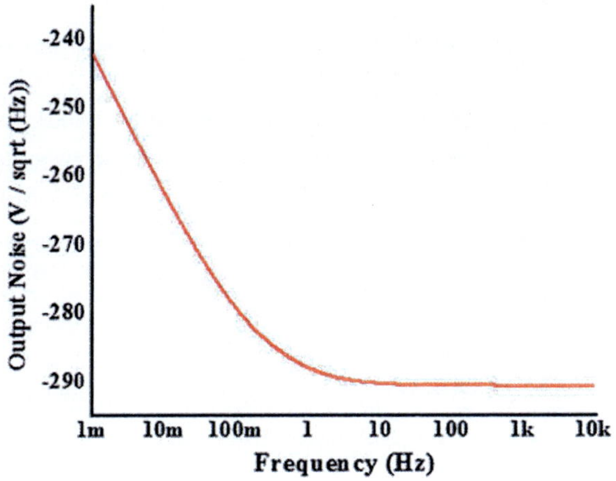

Fig. 7 Output noise response of the LNA

5 Conclusion

A new LNA has implemented for biomedical applications. The LNA is designed based on two stages, first gain stage is beta multiplier reference (BMR)-based differential stage and the second gain stage is based on CMOS inverter. The total gain will be multiple of two stages. The LNA is implemented in 45 nm CMOS technology. The gain of the amplifier is 86.7 dB operates at 0.6 V and consumes only 0.012 mW. The input-referred noise of the proposed LNA is 18.02 mV_{rms} and the bandwidth of the proposed LNA is 227 Hz which is helpful for bio-medical applications.

References

1. Quadir MdSE, Haider MR, Massoud Y (2012) A low-power low-noise bioamplifier for multi-electrode neural recording systems. In: IEEE international symposium on circuits and systems. IEEE, pp 2557–2560
2. Harrison RR, Charles C (2003) A low-power low-noise cmos amplifier for neural recording applications. IEEE J Solid State Circuits 38(6):958–965
3. Salhi D, Godara B (2010) A 75 db-gain low-power, low-noise amplifier for low-frequency bio-signal recording. In: Fifth IEEE international symposium on electronic design, test and application, pp 51–53
4. Kim J, Chae MS, Liu W (2009) A 220 nw neural amplifier for multichannel neural recording systems. In: IEEE international symposium on circuits and systems, pp 1257–1260
5. Gosselin B, Sawan M, Chapman C (2007) A low-power integrated bioamplifier with active low-frequency suppression. IEEE Trans Biomed Circuits Syst 1(3):184–192
6. Holleman J, Otis B (2007) A sub-microwatt low-noise amplifier for neural recording. In: 29th Annual international conference of the IEEE engineering in medicine and biology society, pp 3930–3933
7. Ruiz-Amaya J, Rodriguez-Perez A, Delgado-Restituto M (2015) A low noise amplifier for neural spike recording interfaces. Sensors 15(10):25313–25335
8. Razavi B (2000) Design of analog CMOS integrated circuits, 1st edn. McGraw-Hill

9. Allen PE, Holberg DR (2002) CMOS analog circuit design, 2nd edn. Oxford University Press
10. Baker RJ (2007) CMOS circuit design, layout, and simulation, 2nd edn. Wiley-IEEE Press
11. Baishnab KL, Guha K, Chanda S, Laskar NM, Biswas D (2017) A low power, low noise amplifier for neural signal amplification in SCL 180 nm. In: International conference on electron devices and solid-state circuits (EDSSC). IEEE, pp 1–2
12. Fang C, Law CL, Hwang J (2007) A 3.1–10.6 GHz ultra-wideband low noise amplifier with 13 dB gain, 3.4 dB noise figure, and consumes only 12.9 mW of DC power. IEEE Microw Wirel Compon Lett 17(4):295–297
13. Yang X, Yang J, Lin L, Ling C (2010) Low-power low-noise CMOS chopper amplifier. In: International conference on anti-counterfeiting, security and identification
14. Limei S (2009) A 1.2 V 3.1–10.6 GHz CMOS low noise amplifier with 24 dB gain and 2.4 dB noise figure. In: International conference on wireless communications and signal processing. IEEE, pp 1–4
15. Li M-Z, Tang K-T (2009) A low-noise low-power amplifier for implantable device for neural signal acquisition. In: Annual international conference of the IEEE engineering in medicine and biology society. IEEE, pp 3806–3809

Electrical Characterization of Sb_2Se_3 for Memory Applications

N. Shylashree, Adithya Thonse, Aditya Madhavan, and B. G. Sangeetha

Abstract Sb_2Se_3 thin films of 160 nm thickness were prepared and subjected for various characterizations to prove its usability. Glass was taken as the substrate, over which aluminum was deposited to serve as bottom electrode. A sandwich of Sb_2Se_3 is made between two layers of aluminum. The thin film was then subjected to various potentials using both the layers of aluminum as the contacts. The paper draws inferences from the electrical characterization of the thin films through current–voltage characteristics, dynamic response, temperature characteristics, semiconductor analysis, and resistance–voltage characteristics.

Keywords Antimony triselenide · Chalcogenide · Sandwich configuration · Phase change random access memory

1 Introduction

The demand for higher endurance and number of write cycles have been a primary attribute for mitigating the setbacks of the traditional memory devices and led to the discovery of phase change memory devices (PCM). The capabilities of PCM materials have been explored and exploited to create new technologies that could replace the old memory designs [1]. One such material that best exploits these properties is found in chalcogenide glasses that undergo phase changes to store Logic 0 or 1.

Conventionally, a flash storage is used for solving memory requirements, which uses a Floating Gate MOSFET. FG-MOS has been proven to have high read/write speeds. However, the process involves usage of high positive potential at the gate. This increases the overall power consumption. Moreover, the endurance of these

N. Shylashree (✉) · A. Thonse · A. Madhavan
Department of Electronics and Communication Engineering, RV College of Engineering, Bangalore, India
e-mail: shylashreen@rvce.edu.in

B. G. Sangeetha
Department of Electronics and Communication Engineering, RNSIT, Bangalore, India

V. Nath and J. K. Mandal (eds.), *Proceedings of the Fourth International Conference on Microelectronics, Computing and Communication Systems*, Lecture Notes in Electrical Engineering 673, https://doi.org/10.1007/978-981-15-5546-6_68

types of memory devices is limited to 100,000 write cycles. Hence, there is a need to introduce a new type of a storage device. Chalcogenide materials are the primary considerations next to MOSFET devices [2]. Hence, this project is aimed at verifying the usability of these materials. Consequently, the device is fabricated with 160 nm thickness and various characterizations are performed on the material.

2 The Experiment

With memory being the integral part of all the electronic devices that are in the market, attributes like speed and endurance are of pivotal importance and due to which, the need for ultra-fast switching and higher data retention technologies is focused. It is necessary for us to look for an alternate memory other than the ones from traditional semiconductors. One such device is the phase change material (chalcogenide). The change in the state could be used to store two different logic levels. A cleaned glass slide is taken as a substrate and a thin film of aluminum is deposited on it [3]. This serves as the bottom electrode. It is masked and a layer of antimony triselenide is deposited. Finally, a thin layer of aluminum is deposited after making the material. The third layer acts as the top electrode. All the depositions are carried out using thermal vapor deposition technique whose setup is shown in Fig. 1. After completing the fabrication process, the material is subjected to physical characterization tests to check its structural properties and to confirm the thickness of the device fabricated. Electrical characterizations are then carried out which include I-V characteristics, R-V characteristics, dynamic response, and temperature stability to check the viability of its use in memory industry.

The fabrication process begins with careful measurement of the molecular ratio of antimony and selenium (2:3), mixing the ratio in a sealed ampoule and evacuating a vacuum chamber and then heated to 800 °C and rotated continuously. The sealed ampoule is then cooled and opened to get antimony triselenide [4]. Next, glass slides are cleaned using ultrasonic cleansers and vacuum coater is evacuated to a vacuum of around 3.7×10^{-5} torr. The device fabrication process begins by initially attaching around 8 slides to a circular holder. After the required vacuum is achieved, the deposition of the aluminum bottom electrode is done. Aluminum is in the form of a

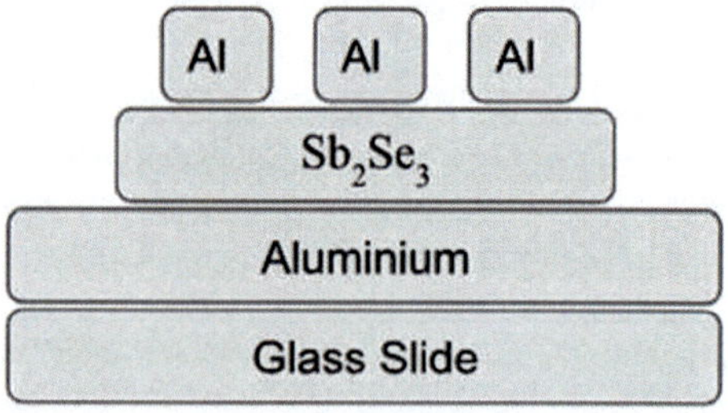

Fig. 1 Lateral view of the 160 nm thin film Sb_2Se_3 device which is sandwiched between two aluminum electrodes on a glass substrate

Fig. 2 Glass slides coated with the sandwich configuration of Al-Sb_2Se_3-Al, where the two aluminum layers are 200 nm thick and the Sb_2Se_3 layer is 160 nm thick

bead, which weighs 0.476 g, is kept in a tungsten basket and fixed inside the coating chamber. Temperature is 660 °C during coating. The thickness of the aluminum layer is 200 nm [5]. Sufficient settling time is provided. Physical characterization of the material requires thin film of Sb_2Se_3 alone without electrodes. Therefore, few glass slides are added to the substrate holder and direct deposition of Sb_2Se_3 is carried out.

Mask is prepared using aluminum foil or OSP sheet and is placed onto the base material with the help of Kapton tape. This is done to provide electrical isolation of the bottom electrode [6]. After this, for the deposition of the second layer, i.e., Sb_2Se_3, the material in the required ratio is placed in a molybdenum boat. The coated aluminum slides along with glass slides are placed inside the vacuum coater. The boat with the Sb_2Se_3 is fixed inside the coater and the vacuum is brought to 10^6 torr. During the deposition of Sb_2Se_3, the temperature inside the coater is 630 °C. The thickness of the Sb_2Se_3 layer is set to be 160 nm. After this, the final layer of aluminum is coated as the top electrode with the help of an appropriate mask. The mask is aluminum foil or OSP sheets with circular holes in it cut with the help of a scalpel. The procedure for coating aluminum is the same as before. The final glass slides are shown in Fig. 2. Settling time of 8 h was required in vacuum environment so that device settles down properly and adheres to the substrate. After the fabrication, the sandwich configuration slides were ready for characterization.

3 Results and Discussions

A part of the sample which are only coated with Sb_2Se_3 are then annealed to check the structural morphology. Annealed samples are to be crystalline and the un-annealed ones are to be in the amorphous form. Unit cell dimensions were calculated using

Table 1 Strongest three peaks of unannealed sample

No.	Peak no.	2 Theta (°)	d(A)	I/I1	FWHM (°)	Intensity (counts)	Integrated intensity (counts)
1.	22	17.3350	5.11148	100	0.21000	7	118
2.	6	11.7700	7.51277	86	0.22000	6	71
3.	8	12.7100	6.95918	86	0.22000	6	90

Table 2 Strongest three peaks of annealed sample

No.	Peak no.	2 Theta (deg)	d(A)	I/I1	FWMH (°)	Intensity (counts)	Integrated intensity (counts)
1.	17	27.2366	3.27157	100	0.52670	68	1950
2.	5	16.6620	5.31639	84	0.48400	57	1513
3.	24	34.0100	2.63391	53	0.66000	36	1195

XRD analyzer [7]. Electrical characteristics of the fabricated Sb_2Se_3 device such as I-V, V-R, temperature stability, and endurance using semiconductor device analyzer are interpreted and analyzed.

3.1 X-Ray Diffraction

X-ray power diffraction (XRD) studies were carried out on the samples coated with Sb_2Se_3 of 160 nm approximately [8]. The run range provided for the XRD evaluation is 20–60°. The peaks were plotted, and the strongest three peaks were used to determine the cell dimension "a." Annealed and unannealed samples were illuminated with X-rays inside a shielded box, and the strongest three peaks were tabulated in Tables 1 and 2.

The results of annealed and unannealed samples were compared. As in amorphous or the unannealed sample, there is no long-range order [9]. Therefore, intensity of the peak is less when compared to that of the crystalline material or annealed sample. Unit cell dimensions of the crystalline Sb_2Se_3 can be determined.

3.2 I-V Characteristics

The I-V characteristics for Sb_2Se_3 device in voltage sweep mode are plotted as shown in Fig. 3. The voltage was sweeped from DC 0 to DC 5 V. Switching from amorphous to FCC and conversion to crystalline was observed at 2 V. Therefore, 2 V was the threshold voltage for Sb_2Se_3 device. The I-V characteristics of the device resemble

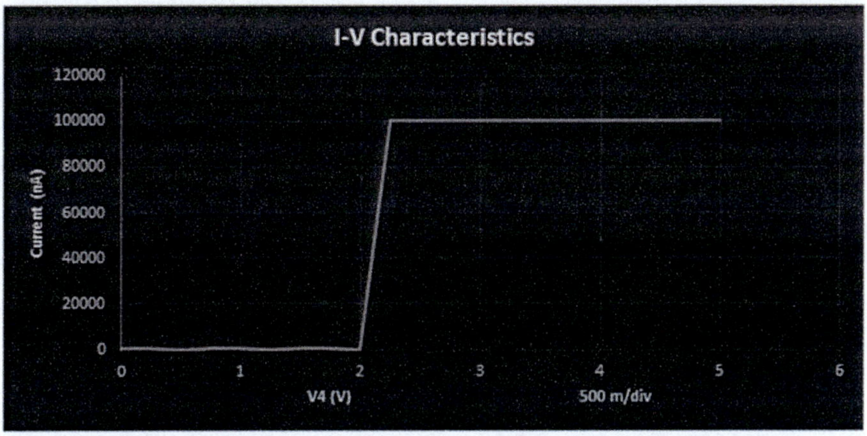

Fig. 3 I-V characteristics observed for the sandwiched 160 nm Sb_2Se_3 thin-film device

the I-V characteristics of memory elements and a very sharp transition from RESET (amorphous) to SET (crystalline) state is observed, which states that it can be used for memory application.

3.3 Dynamic Response

The dynamic response of the device confirms memory switching operation [10]. We can confirm this behavior from Fig. 4. SET (crystalline state with Logic 1) requires

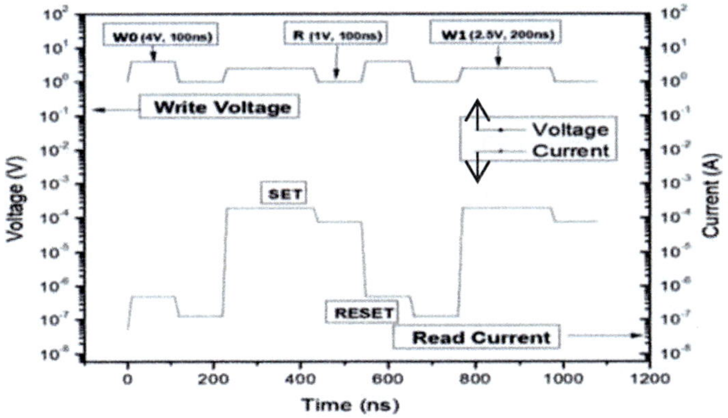

Fig. 4 Voltage–time relationship (dynamic response) of the Sb_2Se_3 device. An initial pulse "*W*" to the material converts it to amorphous state, due to the higher potential and a lower retention period, irrespective of its previous state

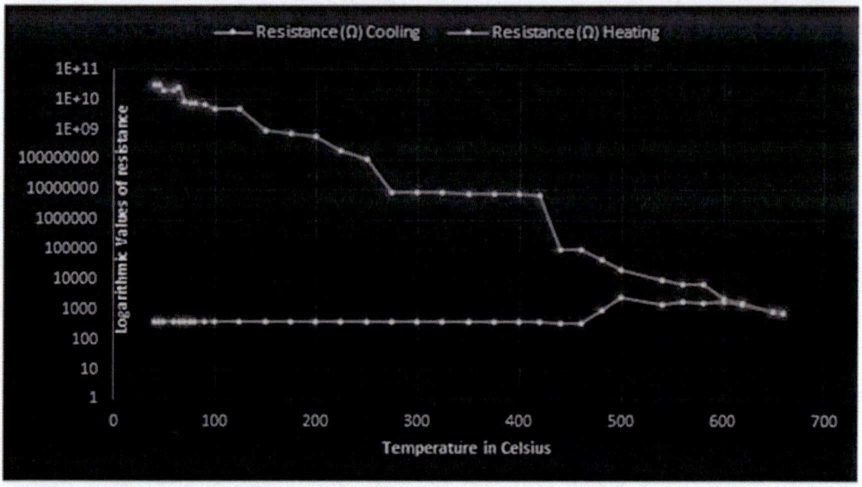

Fig. 5 R-T characteristics of the Sb_2Se_3 160 nm thin-film device

a supply temperature of 220 Â°C which can be practically supplied using a low electrical potential, and then cooling down gradually. The resistance at crystalline state is in the order of Ohms. RESET state (amorphous state with Logic 0) requires a much higher supply at melting point so that the material can be melted and then instantaneously cooled down to obtain a porous-like nature, which can be concluded as in amorphous state [11]. The resistance at amorphous state is in the order of mega Ohms.

3.4 *R-T Characteristics*

The resistance–temperature characteristics are gained by supplying voltage to the device and plotting the corresponding resistance. The threshold voltage from Fig. 5 represents the shift from RESET to SET state [12]. Owing to crystallization, one can perceive resistance change from orders of 10E6 to hundreds of Ohms. Consequently, the current increases and the resistance in the cooling curve saturates at approximately 100–1000 Ohms, thereby allowing the device to store Logic 1. This confirms that it can function as a memory device.

3.5 *Temperature Stability (Resistance versus Time)*

Figure 6 illustrates a relationship between resistance of the material and time for different operating temperatures. It can be deciphered that the graph implicitly represents the temperature stability of the material [13]. The graph accounts for resistance

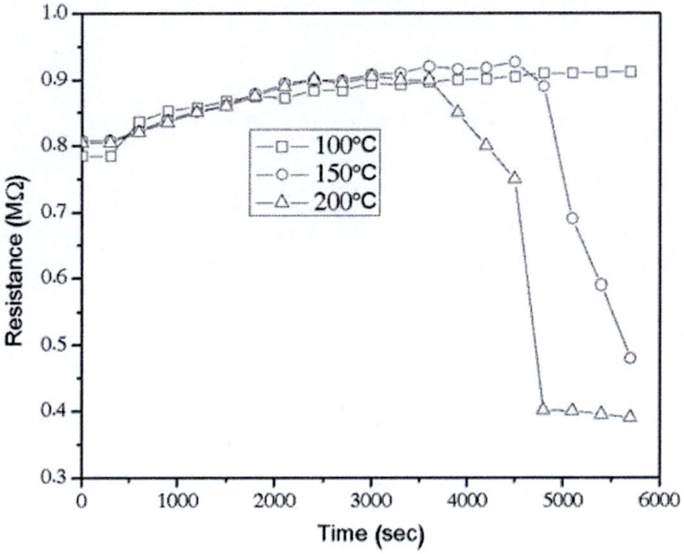

Fig. 6 Resistance–time characteristics of Sb_2Se_3 160 nm thin-film device

of the material only in its amorphous state because the resistance in crystalline state is insignificant. The decrement in resistance is inversely proportional to the operating temperature. At 100 °C, the device is stable for over 6000 s. As the temperature increases, the resistance starts to decrease at earlier instants. However, the least stability can be seen only after 3600 s of usage at elevated temperatures of around 200 °C.

3.6 *Endurance (Resistance versus No. of Cycles)*

Figure 7 represents the nature of non-volatility of the material with response to write cycles. The experiment is conducted for 1200 write cycles beyond which the curve can be extrapolated to measure the exact wear-out point of the device. It can be inferred from the graph that the resistance of the material at RESET state is gradually decreasing and that of the material during SET state is gradually increasing as a number of write cycles increase. The point when both the curves meet refers to the endurance of the device which is in the order of 10^{13}.

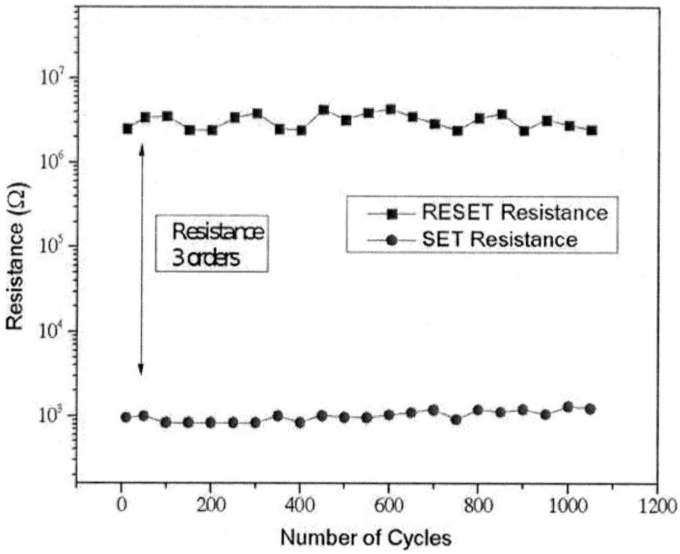

Fig. 7 Endurance characteristics of Sb_2Se_3 160 nm thin-film device

4 Conclusion

Increased demand for memory devices has led to the investigation of various materials that are triggered by physical stimuli. Phase change material (PCM) has been a novel choice for it. In particular, an alloy of antimony and selenium in the ratio of 2:3 (Sb_2Se_3) is proven to work at its best. Thin films and devices of antimony triselenide (Sb_2Se_3) for phase change random access memory (PCRAM) applications have become an imperative talk of the day. Current–voltage (I-V), V-R, temperature stability, and endurance measurements on Sb_2Se_3 devices also verify their great advantage. It was confirmed that antimony triselenide at 160 nm thickness can prove to be a long-lasting, highly reliable memory unit. The capabilities of chalcogenide glass thin-film devices are being continually explored and can be constructed to a product stage for using it as a mainstream memory device. Moreover, the coalesced material differentiation into cells can be exploited.

Acknowledgements We would like to acknowledge the facilities provided by Rashtreeya Vidyalaya college of Engineering. Interdisciplinary research center (IDRC) and IISC.

References

1. Sangeetha BG, Shylashree N (2018) Electrical studies of $Ge_4Sb_1Te_5$ devices for memory applications. AIP Conf Proc 1966(1):020031
2. Kostadinova O, Yannopoulos SN (2009) Raman spectroscopic study of SbxSe100-x phase-separated bulk glasses. J Non Cryst Solids 355(37–42):2040–2044
3. Shylashree N, Uma BV, Dhanush S, Sagar A, Nisarga A, Aashith K, Sangeetha BG (2018) Preparation and characterization of Sb_2Se_3 devices for memory applications. AIP Conf Proc 1966:020032
4. Kalkan N, Yildirim S, Ulutas K, Deger D (2008) Electrical switching in $TlSbSe_2$ chalcogenide semiconductors. J Electron Mater 37(2):157–160
5. El-Sayad EA, Moustafa AM, Marzouk SY (2009) Effect of heat treatment on the structural and optical properties of amorphous Sb_2Se_3 and Sb_2Se_2S thin films. Phys B 404(8–11):1119–1127
6. Pirovano A, Redaelli A, Pellizzer F, Ottogalli F, Tosi M, Ielmini D, Lacaita AL, Bez R (2004) Reliability study of phase-change nonvolatile memories. IEEE Trans Dev Mater Reliab 4(3):422–427
7. Raoux S, Xiong F, Wuttig M, Pop E (2014) Phase change materials and phase change memory. MRS Bull 39:703–710
8. Sangeetha BG, Joseph CM, Suresh K (2014) Preparation and characterization of $Ge_1Sb_2Te_4$ thin films for phase change memory applications. Microelectron Eng 127:77–80
9. Yu D, Wu J, Gu Q, Park H (2006) Germanium telluride nanowires and nanohelices with memory-switching behavior. J Am Chem Soc 128(25):8148–8149
10. Bang KS, Lee S-Y (2014) Formation of threshold switching chalcogenide for phase change switch applications. Appl Sci Convr Tech 23:34–39
11. Hyota B, Poupineta L, Gehannoa V, Desreb PJ (2002) Analysis of writing and erasing behaviours in phase change materials. J Magn Magn Mater 249:504–508
12. Lacaita AL (2006) Phase change memories: state of the art, challenges and perspective. Solid State Electron 50:24–31
13. Ju C, Cheng XM, Miao XS (2012) Amorphization and amorphous stability of Bi_2Te_3 chalcogenide films. Appl Phys Lett 100:1–4

Novel Single Balancing Circuitry for Modular Cell for Electric Vehicle Applications

Ginu Ann George, M. V. Jayan, Fossy Mary Chacko, and A. Prince

Abstract Battery state monitoring and cell balancing are crucial requirements to achieve full cycle life of lithium-ion battery systems especially in EV/HEV applications to improve the overall efficiency of the system. This paper proposes a novel modular balancing system with a single balancing circuitry, and the performance of the proposed method is validated under MATLAB Simulink environment. A four-module battery system with three cells in each module is considered for simulation-based performance analysis. Through this new topology, balancing can be achieved in an efficient manner.

Keywords Electric vehicle · Hybrid electric vehicle · Battery management system · Modular cell balancing

1 Introduction

Electric vehicle (EV) is more superior to conventional IC engine from comparison based on pollution and efficiency. The main elements of the EV are an electric motor,

G. A. George (✉) · F. M. Chacko · A. Prince
Department of Electrical Engineering, Rajiv Gandhi Institute of Technology, Government Engineering College Kottayam, APJ Abdul Kalam Technological University, Kottayam, Kerala, India
e-mail: ginuanngeorge@rit.ac.in

F. M. Chacko
e-mail: fossy.mary@rit.ac.in

A. Prince
e-mail: prince@rit.ac.in

M. V. Jayan
Department of Electrical Engineering, Government Engineering College Trissur, APJ Abdul Kalam Technological University, Trissur, Kerala, India
e-mail: jayan@gectcr.ac.in

V. Nath and J. K. Mandal (eds.), *Proceedings of the Fourth International Conference on Microelectronics, Computing and Communication Systems*, Lecture Notes in Electrical Engineering 673, https://doi.org/10.1007/978-981-15-5546-6_69

an electronic control unit, an energy storage system (battery), a battery management system, etc. [1]. Hybrid electric vehicle, plug in hybrid electric vehicle, and all electric vehicle are the main classification of electric vehicle based on energy source as shown in Fig. 1.

The more prevalent choice of batteries used in EV application is Li–ion batteries due to its numerous advantages compared to other battery types [2, 3]. Li–ion cell is a metal lithium having the highest negative potential and lowest atomic weight, which provides the EVs highest features in terms of performance efficiency. Due to the intercalation and removal of smaller ions, there are a few structural changes of the anode and cathode of Li–ion battery during normal operation. Batteries like lead-acid (Pb-acid) and nickel metal hydride (NiMH) can withstand overcharge condition to some extent. But overcharging of Li–ion batteries leads to battery damage either in the form of electrode decomposition or electrolyte decomposition. Battery management system (BMS) [4] plays a vital role to ensure the safe operation of Li–ion battery. Battery state monitoring and balancing are the other two main functions of BMS apart from protection both in normal operation as well as in critical condition like accidents. More likely, Li–ion charging system is integrated with BMS to ensure the protection of battery system from overcharging.

This paper has been structured as follows; the cell balancing strategy is explained in Sect. 2. Proposed single cell balancing circuitry is described in Sect. 3. Simulation results are shown in Sect. 4. Finally, Sect. 5 contains conclusion.

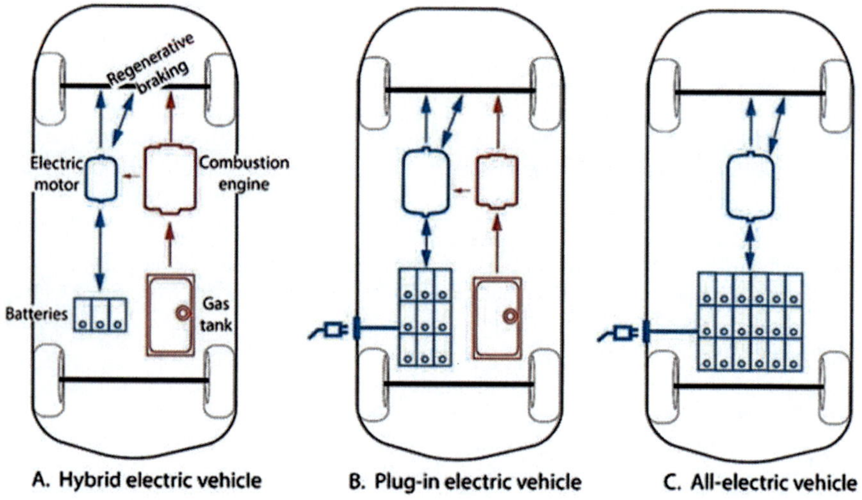

Fig. 1 Different types of electric vehicle

2 Cell Balancing Strategy

Several hundreds of batteries are connected in series and parallel to get the required power source for most of the applications. In this type of battery strings, small voltage variations among the cells may happen due to manufacturing variations environmental conditions or state of charge (SOC) variations. SOC is the remaining capacity of the cell as defined in Eq. (1).

$$\%\mathrm{SOC} = \frac{Q_{\mathrm{remain}}}{Q_{\mathrm{rated}}} \times 100 \tag{1}$$

For battery balancing, accurate estimation of the state of charge is crucial [5, 6]. Coulomb counting is most common method of estimation as shown in Eq. (2)

$$\mathrm{SOC}(t) = \mathrm{SOC}(t_0) + \int \frac{i(\tau)\mathrm{d}\tau}{C_n} \tag{2}$$

where $\mathrm{SOC}(t)$ is real-time SOC, $\mathrm{SOC}(t_0)$ is initial SOC, i is discharge current, and C_n is rated battery capacity.

This unbalance condition will accumulate over time in every charge or discharge cycle which leads to overcharge or under discharge condition of weaker cells. Figure 2 shows a battery pack with four cells having unequal capacities. During charging, some of the cell charge fast than the others which lead to overcharge condition at the end of charging cycle.

Similarly during discharge condition, some of the cells in a pack may drain fast before the others lead to under charge condition while completing the cycle. Therefore, there is always an unused capacity of the string causing the reduction in the battery cycle life. Finally, un-balancing causes a premature failure of the whole battery. In order to balance the voltages of series string, many cell balancing strategies are adopted thus to extend cycle life of battery.

Balancing methods are broadly classified into two passive and active [7–12]. The main advantage of passive balancing method is its simple control. In passive balancing method, excess energy is dissipated through the resistors connected parallel

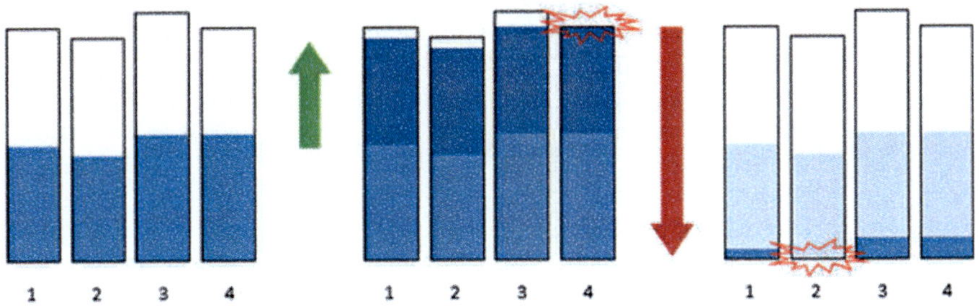

Fig. 2 Unbalanced state of battery pack

to each cell. This method of balancing is seen in lead-acid and nickel-based batteries. Passive method is effective only for small power applications.

For the effective utilization of energy storage system used in high-power applications such as EV requires active balancing method. The principle behind active balancing method is the transfer of energy from the highest cell to the lowest cell. Many types of active balancing methods are available based on the type of charge transfer within the cell and components used for charge transfer such as transformers, capacitors, inductors, and converters.

3 Proposed Single Balancing Circuitry

The major part of proposed novel single balancing circuitry is an improved switched capacitor type balancing method and its control algorithm. Switched capacitor type is superior in terms of its performance, simple control circuitry, reduced component size, and associated losses [13–15]. Conventional method consists of $n - 1$ cell balancing circuit for n series connected cells. Figure 3 shows the single balancing circuitry connected to four battery modules. Each module consists of three series connected Li–ion cells. S1 is a module selection switch which selects the weakest module and then connects the module to the switched capacitor active balancing method.

After balancing that module, the circuitry again check the next weakest module. Module selection switch then connects that module to the balancing circuit. Likewise, entire system is balanced through this module-wise balancing using single balancing circuitry. The proposed method is applicable to n number of modules with any number of series connected cells as seen in EV applications.

3.1 Switched Capacitor Balancing Method

Figure 4 shows switched capacitor circuitry. In order to balance n number of cells, this topology requires $2n$ switches and $n - 1$ capacitors. It consists of two stages. In one state, C_1 will be charged or discharged through the parallel connection with B_1 to obtain the same voltage as B_1. After that in the next state, C_1 will be paralleled with B_2 to obtain the same voltage as B_2. After cycles of this process, B_1 and B_2 will be balanced. The same thing will happen to next capacitor and so on. Finally, the entire battery pack will be balanced.

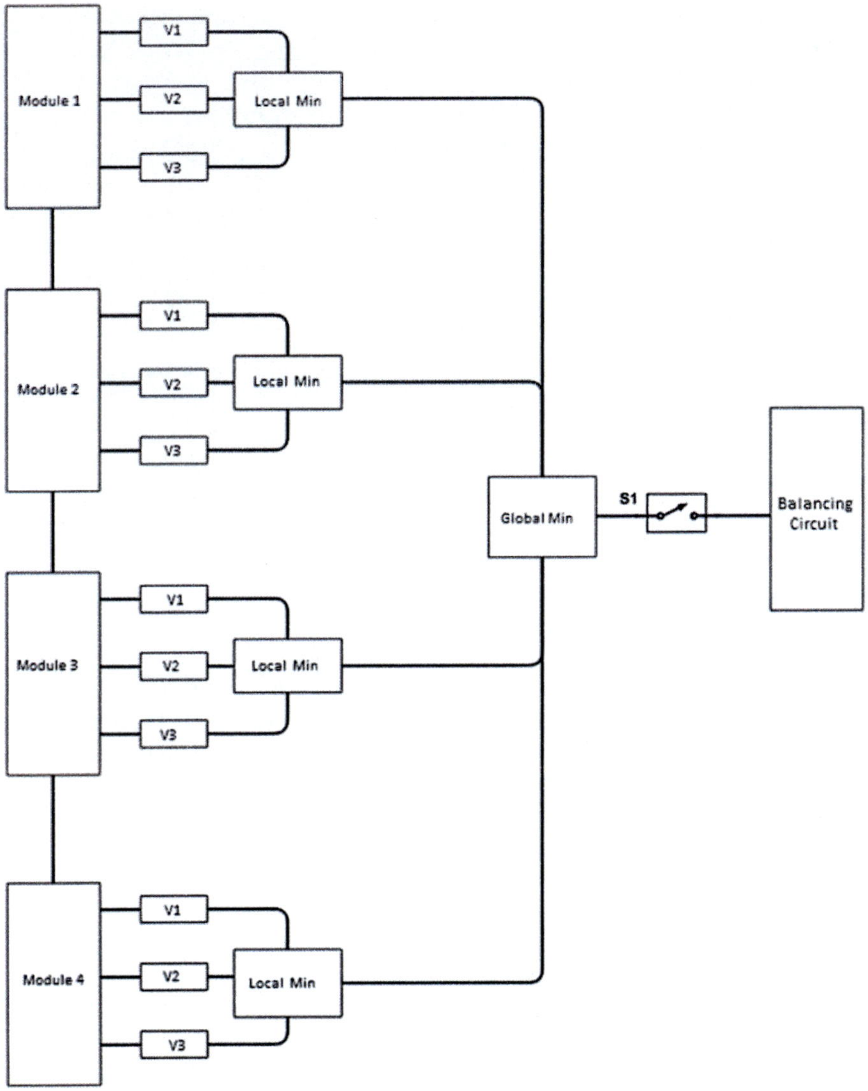

Fig. 3 Block diagram of novel single cell balancing circuitry

3.2 Flowchart of Balancing Algorithm

The control algorithm of improved switched capacitor is explained with the help of flowchart shown in Fig. 5. Various parameters of battery are continuously measured and monitored. Individual cell voltage and SOC are main parameters considered for module and the balancing circuitry.

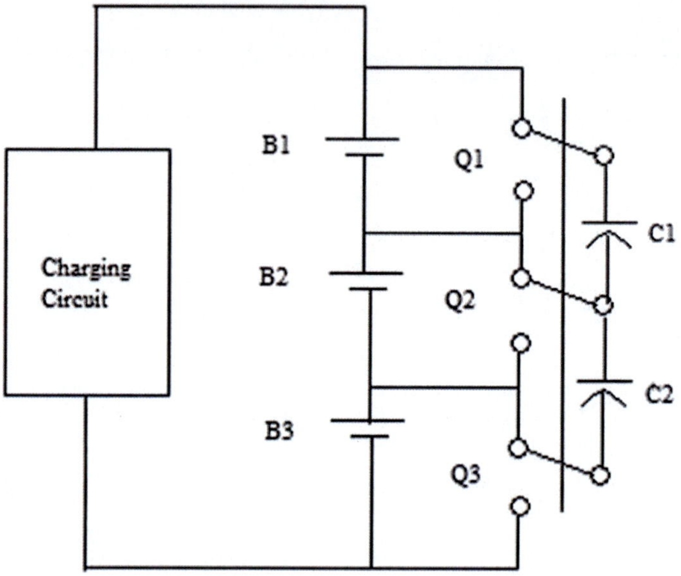

Fig. 4 Overview of switched capacitor balancing circuit

Control strategy for module selection switch is in such a way that it will connect the module having lowest voltage and lowest SOC to the switched capacitor control algorithm. Firstly, control algorithm identifies the module having minimum voltage and minimum SOC. The module selection switch is connected between the cell or balancing circuit. In switched capacitor topology, capacitor stores the excess energy from the highest cell and delivers it to the lowest cell. After balancing the module having lowest cell voltage and SOC control algorithm search the next module with lowest cell voltage and SOC. This process continues until all the modules are achieved a balanced state.

4 Simulation Results and Analysis

The simulation works are carried out under MATLAB

Simulink environment. For simulation analysis, 12 Li–ion batteries are connected in four modules in series arrangement. Table 1 explains the parameters of Li–ion battery and the value of elements in the switched capacitor balancing circuitry initial voltage, and SOC of all the 12 Li–ion batteries is shown in Table 2. Detailed simulation results are shown in Table 3. Results shows that all the cells in the four modules are balanced. SOC variations and balanced voltages are shown in Figs. 6 and 7.

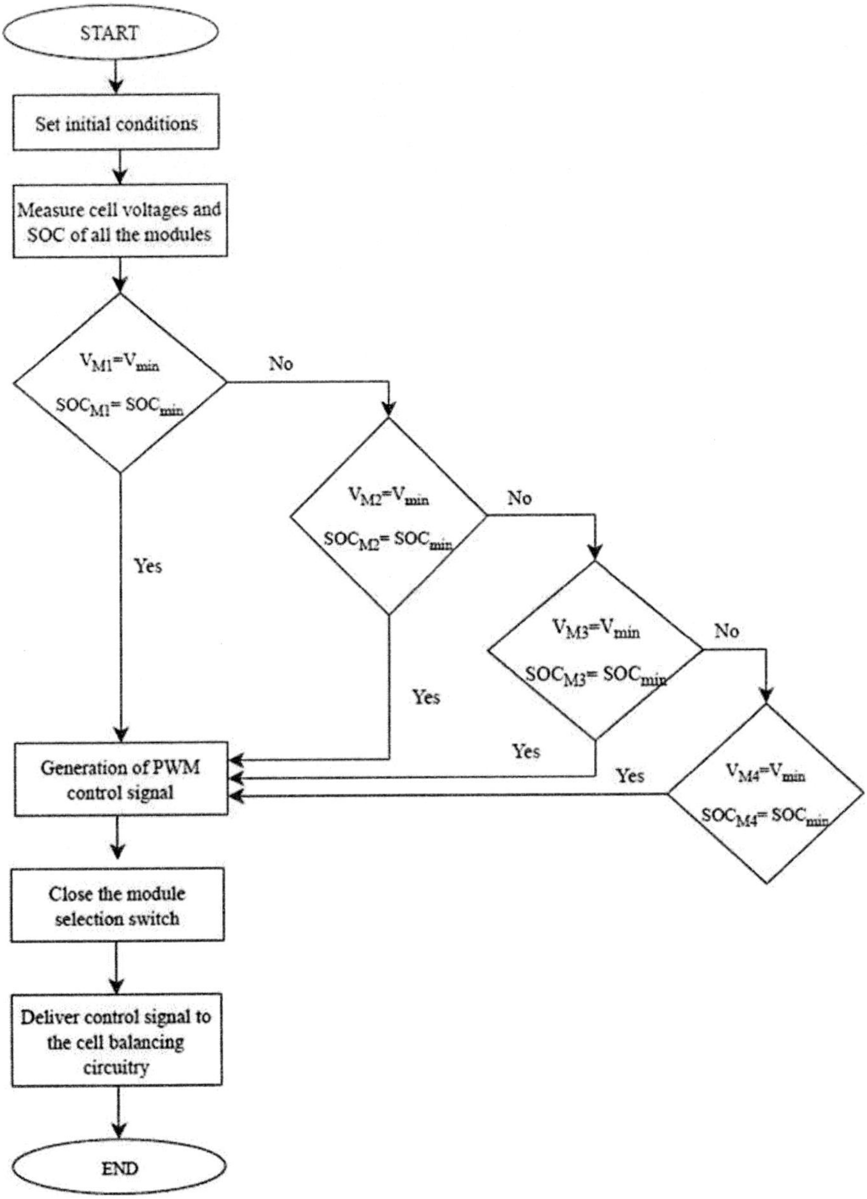

Fig. 5 Flowchart of single cell balancing control algorithm

Table 1 Specifications of Li–ion

Parameters	Rated values
Nominal voltage (V)	4
Maximum charge voltage (V)	4.2
Nominal capacity (mAh)	2600
Initial SOC (%)	100
Nominal discharge current (A)	0.52
Internal impedance (mΩ)	70
Capacitors (μF)	470
Switches	MOSFET
No of switches	4

Table 2 Initial voltage and SOC of cells in each module

Parameters	Battery	Voltage (V)	SOC (%)
M1	B1	3.7	88
	B2	3.5	83
	B3	3.5	83
M2	B1	4	95.2
	B2	3.9	93
	B3	4.1	97.5
M3	B1	3.3	78.5
	B2	3.5	83
	B3	3.2	76
M4	B1	3.8	90.4
	B2	3.7	88
	B3	3.6	85.7

5 Conclusion and Future Scope

A novel single balancing circuitry is proposed for modular type series connected batteries used for high power applications such as EV. Improved witched capacitor active balancing method is applied in proposed circuitry. For simulation analysis, four-module structure is selected, all four modules are balanced, and this can be extended to any number of cells. This paper also eliminates the module-wise balancing circuit with a module selection switch and single balancing circuitry, thereby reduction in size and switching loss are attained. This will improve the cycle life and efficiency of the battery system. Practical implementation is much easier due to the simple control of switched capacitor circuit. As the number of module

Table 3 Simulation results

Module	Battery	Initial voltage	Initial SOC	Balanced voltage at 2000 rpm	Balanced voltage at 1000 rpm	Balanced voltage at 500 rpm
M1	B1	3.7	88	4.098	3.795	3.649
	B2	3.5	83	4.038	3.823	3.72
	B3	3.5	83	4.038	3.824	3.72
M2	B1	4	95.2	4.022	3.837	3.747
	B2	3.9	93	4.038	3.823	3.72
	B3	4.1	97.5	4.038	3.824	3.72
M3	B1	3.3	78.5	4.023	3.837	3.747
	B2	3.5	83	4.038	3.823	3.72
	B3	3.2	76	4.038	3.824	3.72
M4	B1	3.8	90.4	4.022	3.837	3.748
	B2	3.7	88	4.038	3.823	3.72
	B3	3.6	85.7	4.038	3.824	3.654

increases, the balancing speed may reduce, but this can be overcome by using an intelligent control-based switched capacitor circuit.

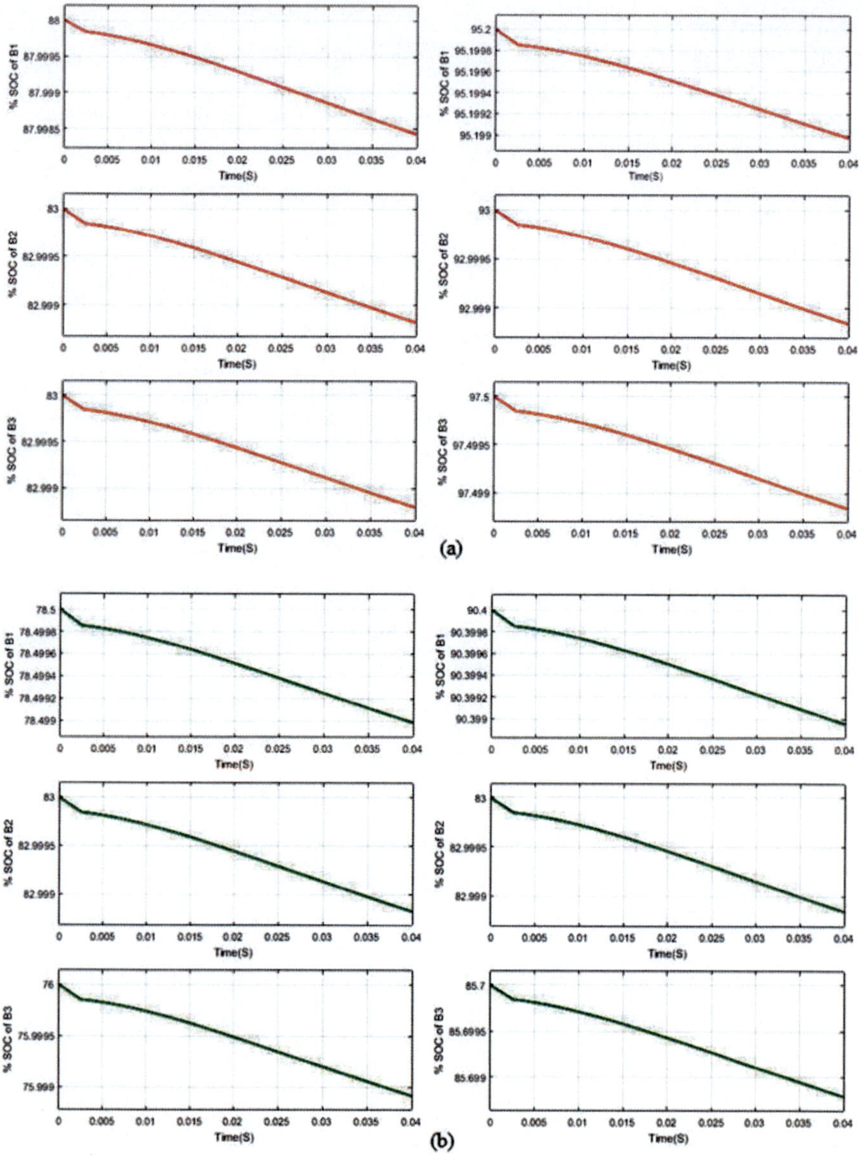

Fig. 6 **a** SOC variations of batteries in Module 1 and Module 2, **b** SOC various of batteries in Module 3 and Module under the speed condition of 1000 rpm

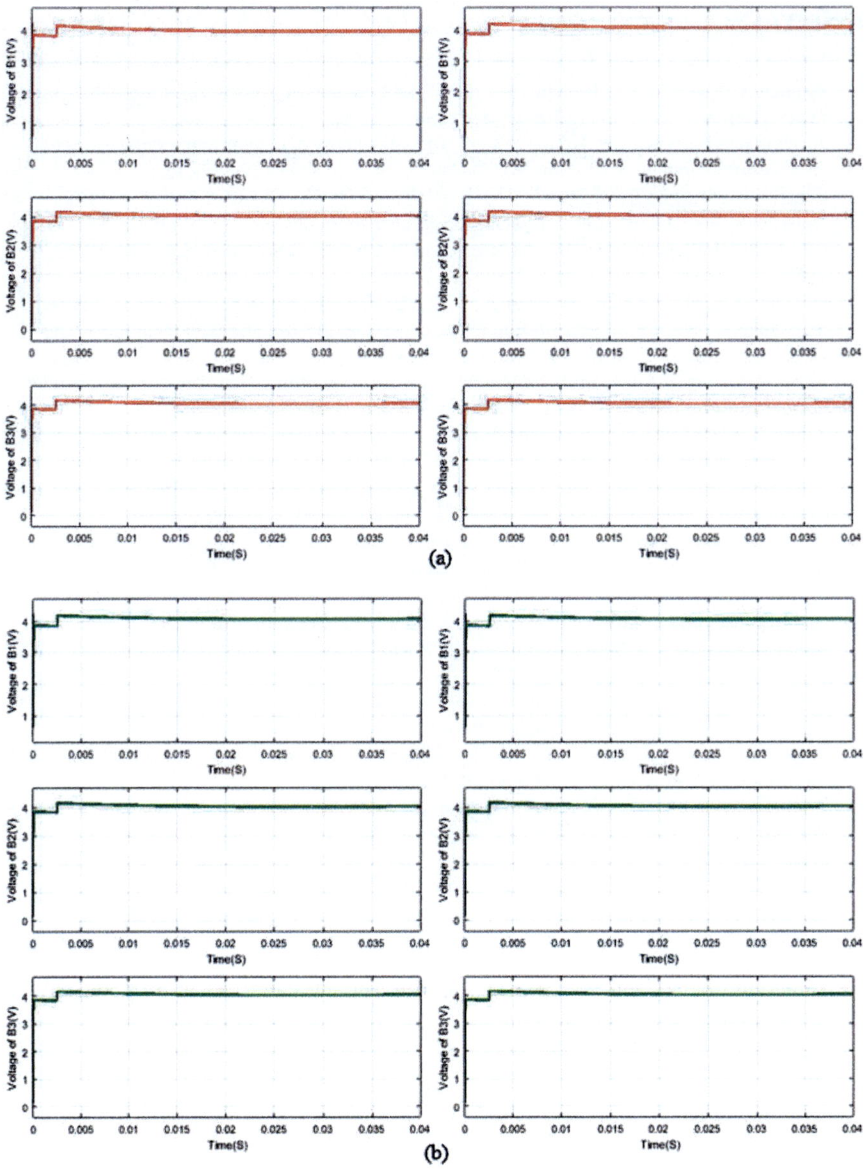

Fig. 7 **a** Balanced voltages of all batteries in Module 1 and Module 2, **b** balanced voltages of all batteries in Module 3 and Module under the speed condition of 1000 rpm

References

1. Hannan MA, Azidin FA, Mohamed A (2014) Hybrid electric vehicles and their challenges: a review. Renew Sustain Energ Rev 29:135–150
2. Aifantis KE, Hackney SA, Kumar RV (2010) High energy density lithium batteries, materials, engineering, applications. Wiley-Vch Verlag GmbH Co. KGaA, Weinheim
3. Omar N, Verbrugge B, Mulder G, Van den Bossche P, Van Mierlo J, Daowd M, Dhaens M, Pauwels S (2010) Evaluation of performance characteristics of various lithium-ion batteries for use in BEV application. In: Vehicle power and propulsion conference (VPPC). IEEE, pp 1–6
4. Hannan MA, Hoque MM, Hussain A, Yusof Y, Ker PJ (2018) State-of-the-art and energy management system of lithium-ion batteries in electric vehicle applications: issues and recommendations. IEEE Access 6:19362–19378
5. Zhang Z, Cheng X, Lu Z, Gu D (2018) SOC estimation of lithium-ion battery pack considering balancing current. IEEE Trans Power Electron 33(3):2216–2226
6. Xiong R, Cao J, Yu Q, He H, Sun F (2018) Critical review on the battery state of charge estimation methods for electric vehicles. IEEE Access 6:1832–1843
7. Einhorn M, Roessler W, Fleig J (2011) Improved performance of serially connected Li–ion batteries with active cell balancing in electric vehicles. IEEE Trans Veh Technol
8. Moore SW, Schneider PJ (2001) A review of cell equalization methods for lithium ion and lithium polymer battery systems. In: Proceedings of SAE 2001 world congress, Detroit, MI
9. Kutkut NH, Wiegman HLN, Divan DM, Novotny DW (1999) Design considerations for charge equalization of an electric vehicle battery system. IEEE Trans Ind Appl 35(1):28–35
10. Moo CS, Hsieh YC, Tsai IS (2003) Charge equalization for series-connected batteries. IEEE Trans Aerosp Electron Syst 39(2):704–710
11. Gallardo-lozano J, Romero-cadaval E, Milanes-montero MI, Guerrero-martinez MA (2014) Battery equalization active methods. J Power Sour 246:934–949
12. Baronti F, Roncella R, Saletti R (2014) Performance comparison of active balancing techniques for lithium-ion batteries. J Power Sour 267:603–609
13. Kimball JW, Krein PT (2005) Analysis and design of switched capacitor converters. In: Proceedings of 20th annual IEEE applied power electronics conference and exposition, vol 3, pp 1473–1477
14. Pascual C, Krein PT (1997) Switched capacitor system for automatic series battery equalization. In: Proceedings of 12th annual applied power electronics conference and exposition, vol 2, Atlanta, GA, pp 848–854
15. Ye Y, Cheng KWE, Fong YC, Xue X, Lin J (2017) Topology, modeling, and design of switched-capacitor-based cell balancing systems and their balancing exploration. IEEE Trans Power Electron 32(6):4444–4454

Classification of Big Data Using Spark Framework

Ritesh Jha, Vandana Bhattacharjee, and Abhijit Mustafi

Abstract The enormous amount of data being generated through social networking sites, satellite and medical images, bank transactions, and many such applications needs to be handled in ways much different from those required for regular size data. Big data is typically characterized by its volume and variety, among others. The use of machine learning techniques to handle this complexity is becoming a very sought after research area. Classification is another challenging task in big data and this is the problem we are aiming to solve using multinomial logistic and multilayer perceptron approaches. It is observed that the accuracy of MLP is much higher than multinomial approach.

Keywords Big data · Multilayer perceptron · Multinomial logistic BFS · Spark

1 Introduction

Big data is characterized by 5 V's dimensions. The first is volume (amount of data), second is velocity (rate at which data is generated), third is variety (format, nature of data), fourth is veracity (quality of data), and last one is value (insights of data). This is big data [1], and it is very complex, and therefore, it is important that to analyze and store the data we require a framework like Apache Hadoop [2] and Spark [3].

The biggest challenge in big data is parallelization, storage, and scalability, and the machine learning algorithms apply a distributed approach so that big data can be analyzed [5].

R. Jha (✉) · V. Bhattacharjee · A. Mustafi
Department of Computer Science and Engineering, Birla Institute of Technology Mersa, Ranchi, Jharkhand, India
e-mail: riteshjha@bitmesra.ac.in

V. Bhattacharjee
e-mail: vbhattacharya@bitmesra.ac.in

A. Mustafi
e-mail: abhijit@bitmesra.ac.in

V. Nath and J. K. Mandal (eds.), *Proceedings of the Fourth International Conference on Microelectronics, Computing and Communication Systems*, Lecture Notes in Electrical Engineering 673, https://doi.org/10.1007/978-981-15-5546-6_70

The objective of machine learning approach in big data is to identify the hidden patterns. This can be done either by classification or clustering techniques. Problem arises when the data is imbalanced and multiclass. In this paper, multinomial logistic regression algorithm has been applied for classification. Further, the multilayer feedforward peceptron algorithm also has been applied for the same data set and comparative results have been presented.

The organization of the paper is as follows: Sect. 2 presents related work, Sect. 3 explains the methodology including the Spark Frameworks, and Sect. 4 describes the experimental setup and algorithm. Section 5 discusses the results and finally Sect. 6 presents the concluding notes for the paper.

2 Related Work

The multiclass logistic regression approach has been applied for medical areas in cardiac arrhythmia which predicts serious vascular diseases [7]. In this paper, authors aim to predict the cardiac disease. Another work is done in the field of classification of images using multinomial logistic regression approach [8], and in this paper, the authors have proposed a new semi-supervised learning algorithm which assigns both hard and soft labels. In the case of social media, the research work has been done using sentiment analysis to identify the opinion and comments in social media. The authors collect Twitter data and transform features into feature vector. Multinomial logistic regression is then applied [9].

To scale multinomial logistic regression to big data which does not fit in single machine, hybrid parallelism approach has been adopted [10]. To classify the real-time problem, logistic regression via variable splitting and augmented Lagrangian (LORSAL) algorithm for sparse multinomial logistic regression is presented on commodity graphics processing units (GPUs) using Nvidia's compute unified device architecture [11].

For multilayer perceptron, the work has been done in this area to avoid high computational cost and time involved in using MLP for classification of big data having large number of features. A parameterized multilayer perceptron (PMLP) has been proposed where the weight matrix has been parameterized using periodic function [12].

In [13], authors review error functions and analyze their mathematical properties. For data classification purposes, a new error function inspired by Z-EDM algorithm is proposed. Experimental results show performance improvement. CNN approach is very impressive in large-scale machine learning domain. Nowadays, in big data context, the size of data increases everyday in terms of increasing number of features, and this problem is known as high-dimensional data. Apart from increasing features, a number of classes are also increasing. So the task of classification or predictive analysis remains very challenging. This approach has been applied for large image classification [14].

3 Methodology

This section describes the methodology and the framework used in this work.

3.1 Spark Framework

In Fig. 1, the diver is the main program or main function. It is the main controlling body. To execute any job, the driver passes the logic to the workers with the help of Spark Context. Each worker has the executor which does the job of executing the RDD jobs. The driver program assigns the job across all the worker. Each worker has task and executor to execute the task or job. Each worker works simultaneously to execute the job, and final output from different workers is collected by driver program.

Spark Core

In Fig. 2, Spark Core is the one special type of main engine for parallel and distributed data processing. Its sole responsibility includes memory management for the jobs, fault tolerance or recovery, interacting with various storage system devices, task-scheduling and monitoring the jobs which executed on a cluster. Spark Core uses special kind of data structure known as RDDs. RDDs are known as resilient distributed data sets (RDDs). RDDs have a large collection of items which are distributed across many workers nodes. These items can be manipulated and accessed parallelly. The Spark Core provides a special kind of APIs, which can manipulate these collections.

Fig. 1 Apache framework [15]

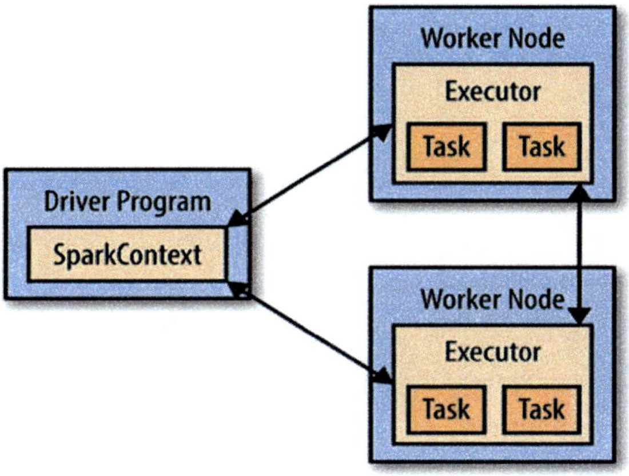

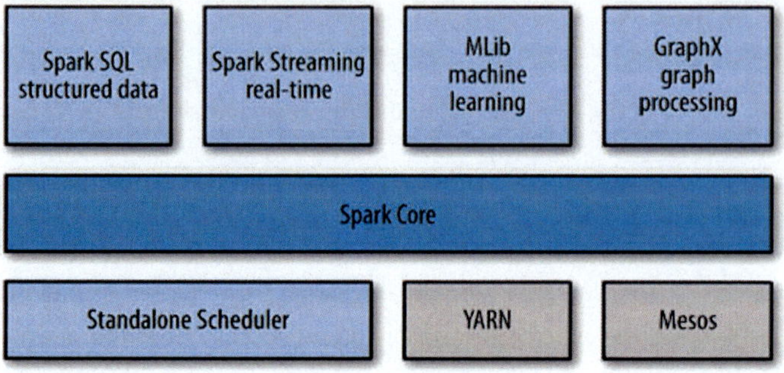

Fig. 2 Spark stack [4]

Spark SQL
To deal with structured data, a special package that Spark provides is called the Spark SQL. It provides the query-based operation to access the data.

Spark Streaming
To process the data which is being generated from CCTV camera, network devices, IOT-based sensors and satellite data which is coming continuously in the form of stream, Spark provides streaming approach like task of classifying the streaming data, regression analysis, and other machine learning techniques that can be applied on streaming data.

3.2 *Multinomial Logistic Algorithm*

Multinomial takes the function as $f(k, i)$ to predict the probability where i is the number of samples that have labels or outcomes k:

The function has following form:

$$f(x) = \ominus_{0,k} + \ominus_{1,k}x_{1,i} + \ominus_{2,k}x_{2,i} + \ldots + \ominus_{M,k}x_{M,i}$$

where $\ominus_{m,k}$ is a regression coefficient. It represents the mth feature and the kth outcome or label.

The regression coefficients and feature variable are grouped into vectors of size $M + 1$, and the predictor function is defined as:

$$f(k, i) = \ominus_k.x_i$$

where $\ominus_k$ is the set of regression coefficients with labels k, and x_i (a vector) is the set of features with i number of samples (Fig. 3).

Fig. 3 Algorithm for logistic regression

Algorithm : 1
Input : Data set
1. Load the data in RDD
2. Split the data into training, testing = randomsplit(0.8,.02)
// data sample has been spllited into 80% training and testing 20 % in ratio.
3. Apply LogisticRegisticBFS (training) function to train the model
4. Collect the training time
5. Predict the class label using train model for testing data

3.3 Multilayer Perceptron Model

Multilayer perceptron classifier (MLPC) is same as feedforward neural network. In this network, three special layers exist, which are input layer, hidden layer, and output layer. Each layer has a different number of neurons, and each layer is connected with next layer through the neurons. Neurons in the each layer are known as information passing elements. The input layer represents the features of the data set, and output layer represents the class label of the data sets.

For the data set having n features, the input layer will have n number of neurons.

The information flows input to output layer by means of a linear combination of the inputs which having neurons' weights w and bias b. The function which transforms the input to output is called the activation function.

For the $k + 1$ layers in MLPC, the matrix model is given as:

$$y(x) = f_k\left(\ldots f_2\left(w_2^T f_1(w_1^T x + b_1) + b_2\right) \ldots + b_k\right)$$

where $y(x)$ represents the function output from all preceding layers.

Information passes in hidden layers through the neurons, follows the sigmoid function defined as:

$$f(z_i) = \frac{1}{1 + \mathrm{e}^{-z_i}}$$

Information passes in output layers through the neurons, follows softmax function, and is defined as:

$$f(z_i) = \frac{\mathrm{e}^{z_i}}{\sum_{k=1}^{N} \mathrm{e}^{z_k}}$$

Fig. 4 Algorithm for MLP

Algorithm : 2
Input : Data set
1. Load the data in RDD
2. Split the data into training, testing = randomsplit(0.8,.02)
// data sample has been spllited into 80% training and testing 20 % in ratio.
3. Define input layer of size 10 (features),
two intermediate of size 5 and 4 and output of size 10(classes)
layers = [10, 5, 4,10]
4. Apply the multilayerPeceptronClassifier with given parameters
5. collect training time.
6. predict the class label using train model for testing data.

4 Experimental Setup

The experimental setup was designed for conducting the experiment consists of 5 computer systems: 1-master node and 4-worker nodes. Each system has the same following configuration: 8 GB DDR3 RAM, Intel Core i7 5th Generation Processors and 1 TB Hard disk drive. The input parameters for the MLP were as follows: maxIter = 100, blockSize = 128, and seed = 1234 (Fig. 4).

The experiment was conducted on Linux Ubuntu-18 operating system with Apache Spark version −2.4.3 and language supported Python.

5 Results

Poker data set [6] was used for the purpose of this experiment. The data set contained 1,000,000 rows in the testing data set (size: 23 MB) and 25,000 (size: 400 KB) rows in the training data set.

The multinomial logistic and MPL has been applied on the following varying sample of Poker data set.

The results are shown in Table 1.

Where S-DB1, S-DB2, S-DB3, S-DB-4, S-DB5, and S-DB6 were extracted from the original data set. The training and testing data were in the ratio of 80% and 20%.

From Fig. 5, the training time for MLP is higher than multinomial. This is due to the hidden layer computation of neurons.

From Fig. 6, we can observe that the accuracy for MLP is much higher than multinomial approach. But there is drop in accuracy for data sample 1,000,000. This is an indicator that at this size there are scalability issues and need further resolving techniques like advanced hardware, etc.

Table 1 Experimental results

Poker data set	Size	Logistic multinomial BGS		Multilayer peceptron	
		Training time (s)	Accuracy	Training time (s)	Accuracy
S-DB1	2000	0.9098	49.51	1.5144	49.09
S-DB2	10,000	1.9966	50.34	2.411	52.06
S-DB3	20,000	1.2470	49.86	4.1922	51.5
S-DB4	50,000	1.3994	50.14	8.0400	51.66
S-DB5	100,000	2.3470	50.44	15.66	53.51
S-DB6	1,000,000	14.4382	50.19	36.66	51.75

Fig. 5 Training time plot

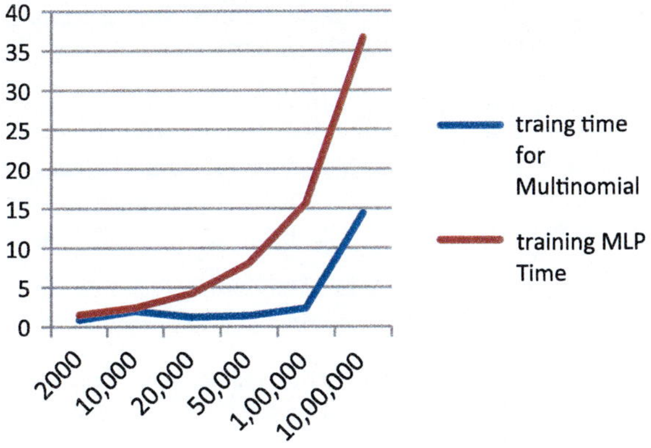

Fig. 6 Accuracy plot

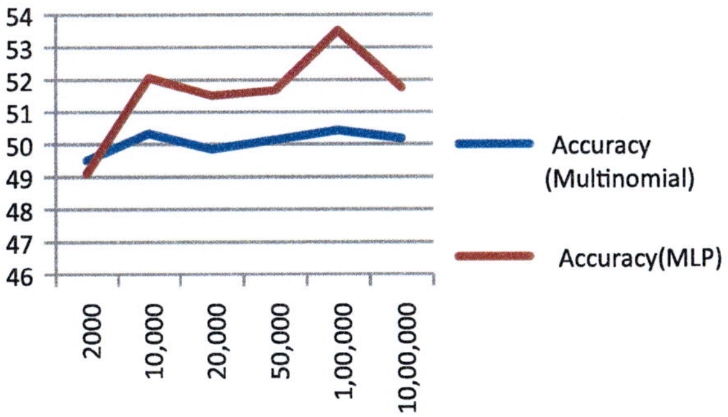

6 Concluding Remarks

The two algorithms have been tested for the Poker data set. We found that MLP dominates in terms of accuracy than the multinomial algorithm. But the training time is higher than that required for the multinomial algorithm. As part of our ongoing work, we propose to apply deep learning techniques to overcome this limitation.

References

1. Sagiroglu S, Sinanc D (2013) Big data: a review. In: International conference on collaboration technologies and systems (CTS), pp 42–47
2. Apache Apache hadoop. https://hadoop.apache.org
3. Salloum S, Dautov R, Chen X, Peng PX, Huang JZ (2016) Big data analytics on Apache spark. Int J Data Sci Anal 1:145–164
4. Karau H, Konwinski A, Wendell P, Zaharia M (2015) Learning spark: lightning-fast big data analysis, 1st ed., O'Reilly Media
5. Zhou L, Pan S, Wang J, Vasilako AV (2017) Machine learning on big data: opportunities and challenges. Neurocomputing 237:330–361
6. https://archive.ics.uci.edu/ml/datasets/Poker+Hand
7. Behandada O, Trovati M, Georgis K, Korkontzelos Y (2017) A multinomial logistic regression approach for arrhythmia detection 8:17–33
8. Yang L, Yang S, Jin P, Zhang R (2013) Semi-supervised hyperspectral image classification using spatio-spectral laplacian support vector machine 11:651–655
9. Ramadhan WP, Novianty STMTA, Setianingsih STMTC (2017) Sentiment analysis using multinomial logistic regression. In: 2017 International conference on control, electronics, renewable energy and communications (ICCREC). IEEE
10. Raman P, Srinivasan S, Matsushima S, Zhang X, Yun H, Vishwanathan SVN (2019) Scaling multinomial logistic regression via hybrid parallelism. In: KDD'19 proceedings of the 25th ACM SIGKDD international conference on knowledge discovery & data mining, pp 1460–1470. ISBN 978-1-4503-6201-6
11. Wu Z, Wang Q, Plazo A, Li J, Wei Z (2015) Real-time implementation of the sparse multinomial logistic regression for hyperspectral image classification on GPUs. IEEE Geosci Remote Sens Lett 12:1456–1460
12. Chandra B, Sharma RK (2015) Fast learning for big data applications using parameterized multilayer perceptron. In: IEEE international conference on big data (Big Data). ISBN 978-1-4799-5666-1
13. Silva LM, de Sá JM, Alexandrecd LA (2008) Data classification with multilayer perceptrons using a generalized error function. Neural Netw 21:1302–1310
14. Botalb A, Moinuddin M, Al-Saggaf UM, Ali SSA (2018) Contrasting convolutional neural network (CNN) with multi-layer perceptron (MLP) for big data analysis. In: 2018 International conference on intelligent and advanced system (ICIAS). ISBN 978-1-5386-7269-3
15. https://spark.apache.org/

Encryption and Authentication of Data Using the IPSEC Protocol

Jayendra Kumar, Mohit Kumar, Deepak Kumar Pandey, and Rishikesh Raj

Abstract This paper presents an overview of the Internet Protocol Security (IPSec) protocol [13, 14] implemented in a Dynamic Multipoint Virtual Private Network (DMVPN) to implement security features at the network layer to provide authenticated and confidential communication using various authentication and encryption protocols operating between devices located at a far away distance from one another on the Internet. In order to provide secure communication between various branches of a company located at distant locations across the Internet and to protect the data from unauthorized access, we create a virtual tunnel connecting the branches and send data across the branches in encrypted packets encapsulated within Internet Protocol (IP) datagram which are hidden from the Internet. The packets pass across to the other branch as if they are in one private network which overlays the Internet. We use the IPSec protocol for providing authentication and encryption to the IP packets. We create a Generic Routing Encapsulation (GRE) tunnel to create a DMVPN connecting the branch offices to one another. We have used GNS3 for the simulation of the scenario which is an emulator software for routers, computers, switches and various other network devices. With the help of GNS3, we were able to access the configuration terminal of various network devices and configure appropriate commands for implementing the DMVPN solution. We also used Wireshark which is a packet analyzer software that is used for tracking the IP packets which pass through the network and analyze their header, protocol information and the data that is encapsulated within it.

J. Kumar · M. Kumar · D. K. Pandey · R. Raj (✉)
Department of Electronics and Communication Engineering, National Institute of Technology Jamshedpur, Jamshedpur, Jharkhand 831014, India
e-mail: rajhrishi12345@gmail.com

J. Kumar
e-mail: jkumar.ece@nitjsr.ac.in

M. Kumar
e-mail: mohitkumar.jsr15@gmail.com

D. K. Pandey
e-mail: dpaknitjsr@gmail.com

V. Nath and J. K. Mandal (eds.), *Proceedings of the Fourth International Conference on Microelectronics, Computing and Communication Systems*, Lecture Notes in Electrical Engineering 673, https://doi.org/10.1007/978-981-15-5546-6_71

Keywords PSec · DMVPN · GRE · AH · ESP

1 Introduction

Internet Protocol Security (IPSec) [1, 2] protocol contains a collection of open standards which enable the services of data confidentiality, integrity and authentication between communicating network devices at the network layer. [3] IPSec enables the following features on communicating devices such as:

Data privacy—it means the IPSec sending device can perform encryption of the IP packets before sending them over the Internet; data integrity—it means the IPSec receiving device can perform authentication of the packets sent by the IPSec sending device to make sure that the information has not been corrupted during transmission by an unauthorized person; data origin authentication, which means that the IPSec receiving device can verify the sender of the IPSec packets. This feature is closely related to the data integrity service and anti-replay, which means that the IPSec receiving device is able to recognize and drop duplicate packets.

IPSec has two major components: Authentication Header (AH) and Encapsulating Security Payload (ESP). In addition, IPSec utilizes many encryption algorithms to build a protocol suite. Some of which are discussed later in this paper.

This paper presents a DMVPN solution to connect the branch offices of a multinational company and implementation of the IPSec protocol which provides the services of authentication and encryption to protect the data that is sent through the Internet.

2 Related Works

In this paper [3], an analytical study of IPSec encryption and authentication algorithms is provided which also includes various limitations of these protocols. HMAC-MD5 is shown to be more suitable for providing authentication services as compared to the more complex HMAC-SHA1 algorithm. Encryption is also provided along with authentication. Encryption is done only after the message passes the authentication test.

In this paper [4], implementation of the IPSec protocol is presented to ensure that data is transmitted securely. Transmission of packets encapsulated in IPSec requires more time as compared to packets without IPSec encapsulation. The research also proposes simulation result and theoretically that ESP encapsulated data packets require more time due to implementing encryption services than that of AH encapsulated packets.

3 Description of the IPSEC Protocol Standard

Authentication Header (AH) is used to provide the services of authentication and integrity to the IP packets exchanged between two devices. [3, 5]. The sending device applies a one-way hashing algorithm to the IP packet to generate a hash. The hash so generated is attached to the IP packet and sent along with it. The receiving device passes the received IP packet through the same hashing algorithm to generate another hash. The hash generated is compared with the hash received. If both the hashes match, the sender is authenticated. Figure 1 shows the AH header format. AH can also implement the anti-replay feature by necessitating that a receiving device sets the replay bit in the header to indicate the sender that the packet has been received. The hash function is applied only to that part of the IP packet that does not change during transmission through the network. For example, the hash function cannot be applied to fields such as time to live (TTL) which is modified by each router along the network path.

The features of confidentiality (encryption), information origin authentication, integrity and anti-replay are provided by Encapsulating Security Payload (ESP). [5] ESP gives privacy by performing encryption at the Internet layer. It employs various symmetric encryption algorithms like 3DES.

In computing, the crucial part in the implementation of IPSec is the Internet Key Exchange (IKE) [6]. Two parties, namely initiator and responder, exchange secret key through negotiation in Internet Key Exchange protocol. The resulted secret key

Fig. 1 AH header format

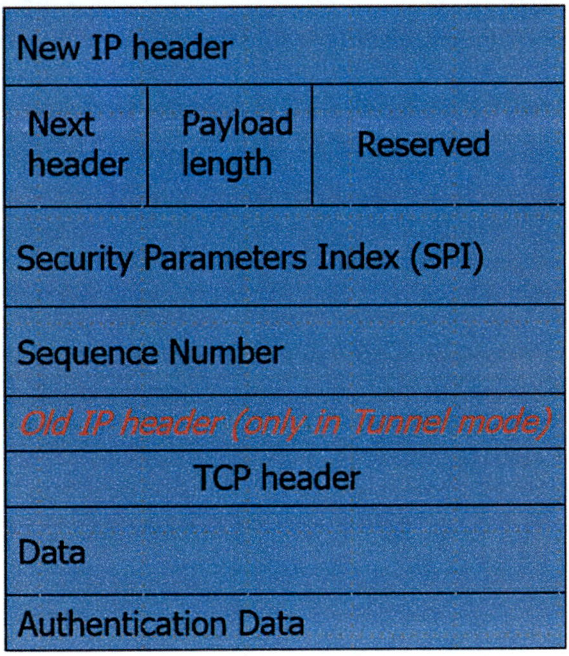

generated using IKE protocol is used to define certain security associations that contain information about how the outgoing packets and incoming packets between two parties are to be safely delivered.

Through IKE protocol, mutual authentication of both the parties is done by verifying provided digital signatures [7]. If in any certain case, certificates (public keys) are missing, then the device can be checked with a pre-shared key. In one sentence, we can say that IKE protocol is used when an outgoing packet or traffic requires IPSec but does not yet contain a security association. IPSec can be implemented without IKE but IKE adds extra features and increases flexibility of IPSec with added ease of configuration for the IPSec standards.

The actual procedure followed for key exchange is defined by several modes of IKE. The keys are generated in this step using the Diffie–Hellman key generation algorithm [8, 9]. The main mode and aggressive mode are the two frequently used modes.

The main mode and aggressive mode differ from each other in terms of the length of the implemented algorithms. Out of these two, the mode chosen for execution of security-related elements is called phase-1 mode of IKE. When the phase-1 is completely executed, the negotiation activity of phase-2 can start. The main aim behind phase-2 exchange is to provide a fresh key material to generate security associations. When IKE phase 2 is completed with all its constituent element and IPSec SAs have been decided by quick mode, then sensitive information can be exchanged by an IPSec tunnel. Encryption and decryption of IP packets are done by utilizing negotiated encryption algorithms defined in the IPSec SAs.

When the security features are negotiated, the two IPSec peers must decide upon which procedures to use for every service that IPSec provides (e.g., DES [10] or 3DES for encryption, MD5 or SHA for integrity). The security association is the strategy that IPSec uses to provide each one of the services of interest concerning a given IPSec session.

In the IPSec process, first, an IPSec tunnel is set up between two devices that wish to exchange data. To define the data that needs the services of IPSec, access lists are set up appropriately on the interfaces. IKE is used to negotiate security parameters and establish keys for authentication between two devices. In IKE phase, one negotiation that follows an IKE security association is set up between the two peers using either main or aggressive mode. Now, the tunnel can be used to transfer data using either AH or ESP encapsulation or both.

Encapsulation is encryption of data and adding appropriate header to it so that the receiving device can decrypt it and generate hashes to be used by other device for verifying that the data was not tempered during transit.

ESP uses DES or 3DES for encryption. MD5 or SHA [11] is used for generating hashes to provide data integrity. For providing anti-replay feature to the IP packets, sequence members are assigned to them and sliding window protocol [12] is used along with it.

4 Proposed DMVPN Solution

Figure 2 shows the DMVPN scenario design. The aim of the proposed solution is to show how by implementing the IPSec protocol in a Dynamic Multipoint Virtual Private Network we can exchange sensitive data securely using authentication and encryption services provided by the IPSec protocol suite which includes various sub-protocols such as AH, ESP, IKE and 3DES between various branches of a multinational company. The branches dynamically create a Generic Routing Encapsulation (GRE) tunnel between themselves and the headquarter site. The GRE tunnel overlays on the Internet and connects the head office with the branches logically in one virtual private network. The traffic that passes through the GRE tunnel is encapsulated within an IPSec packet which adds an IPSec header which has various fields which implement the authentication, encryption, data integrity and anti-replay services.

Security solutions based on VPN are very effective and secure for protection of sensitive data that passes through an insecure channel like the Internet. Companies with big enterprise network requirements need IPSec solutions that are scalable and dynamic for connecting sites through the Internet with minimized latency, optimal bandwidth utilization and network performance.

Dynamic Multipoint VPN solution has the capability to build dynamic IPSec tunnels between sites and provide optimal network performance with reduced latency. It triggers automatic IPSec encryption through the GRE tunnel assuring zero loss of packets. The GRE tunnel can handle multicast traffic and is equipped with VPN routing and forwarding capabilities. It can use protocols which provide load balancing capability to increase network performance and reliability.

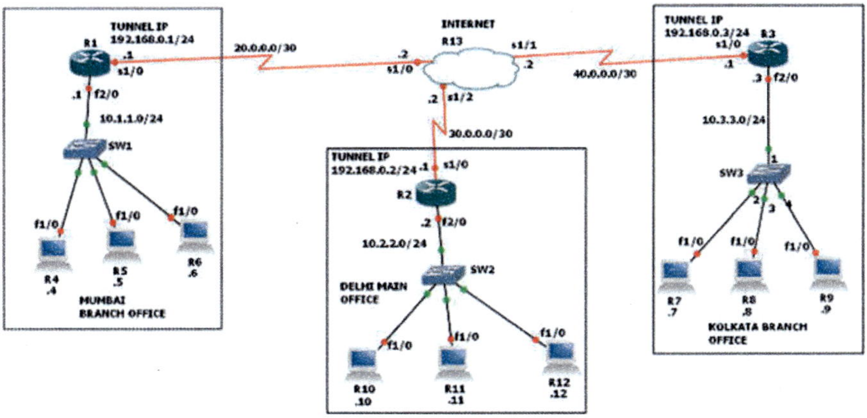

Fig. 2 DMVPN implementation scenario

```
0000  0f 00 08 00 45 00 00 c8 03 10 00 00 fe 32 7c f2   ....E........2|.
0010  28 00 00 01 14 00 00 01 fe 94 bf ec 00 00 00 c6   (...............
0020  cc 84 89 81 2f ff 8a 5d 0f 50 6d c7 8d 2d 78 75   ..../..].Pm..-xu
0030  17 64 9f 72 3f 3b 7c 2a 20 f7 03 93 e3 d5 4f 85   .d.r?;|* .....O.
0040  df 9e cd 89 8a 68 8a da 75 96 19 68 02 f0 6b 3d   .....h..u..h..k=
0050  a2 83 b1 54 e5 c7 11 8e 7f 64 b9 7d 8e 34 29 85   ...T.....d.}.4).
0060  34 67 a9 53 bc 4c 45 7d 7f c4 3f a6 0c ab 52 a2   4g.S.LE}..?...R.
0070  dd 96 ec 25 bc cb 0b 92 eb 02 59 3d e8 93 c8 b1   ...%......Y=....
0080  24 36 c2 6a 2c 1f b3 92 79 db 9a 20 14 7e 89 52   $6.j,...y.. .~.R
0090  e4 1a 7d 16 dd 5d 01 2b 83 0d 1b 76 1a 7e 60 63   ..}..].+...v.~`c
00a0  42 bc 21 61 2b 50 98 c1 1c 90 4c b8 d5 aa 44 f8   B.!a+P....L...D.
00b0  b0 2d c1 5e fb 5c af c2 8a 56 cf f1 56 27 17 e9   .-.^.\...V..V'..
00c0  4e bd 04 1f 93 9c 15 79 f3 ed 56 58               N......y..VX
```

Fig. 3 Data in encrypted form sent through the IPSec tunnel

Major components of the DMVPN solution are as follows:

- Generic Routing Encapsulation (GRE) Protocol: GRE is a protocol which does the job of encapsulating IP unicast, multicast and broadcast packets. GRE IP protocol number is 47.
- Next Hop Resolution Protocol (NHRP): NHRP is a client–server resolution protocol that works on layer 2. It is used to translate a tunnel IP address to a global IP address. It can have static or dynamic bindings.
- Dynamic Routing Protocol: such as EIGRP and OSPF. Its primary purpose is to communicate information about private networks within the DMVPN.
- Standards-based IPSec Encryption Protocols: Protocols used for the protection of the data that passes through the tunnels in the DMVPN.

Figure 3 shows the data in the encrypted form which is sent through the IPSec tunnel. Each branch office first establishes an IPSec tunnel that is permanent to the headquarter site. The branch offices register their global IP address as a client to the NHRP server at the headquarter location. The headquarter site has a NHRP database which stores the public IP address of each branch office. When a branch office device has to send data to another branch office device, it queries the headquarter site for the public IP address of the other branch office device so that it can build direct tunnels to connect to it. The NHRP server replies by querying into its NHRP database. After locating the public IP address of the destination device, the source device initiates to create a dynamic IPSec tunnel to the destination device. With the help of the combination of the multipoint GRE (m GRE) interface, NHRP and IPSec protocols, a direct and dynamic tunnel is established for having secure communication over the DMVPN network.

5 Conclusion

In this paper, we proposed a DMVPN-based solution for enabling the protection of sensitive data that travels through an insecure communication channel like the Internet. It is a very effective, dynamic and scalable solution with fast convergence and easy deployment of the IPSec protocol in which IPSec tunnels can be established on demand and minimize latency and provide optimized network performance and utilization of the bandwidth of the channel. In this paper, we also described the IPSec protocol standard in detail along with its component protocols and standards. We also described the DMVPN solution architecture, its components and working in detail.

References

1. Deshmukh D, Iyer B (2017) Design of IPSec virtual private network for remote access. In: 2017 international conference on computing, communication and automation, Greater Noida, 5–6 May 2017
2. Zhang Y, Singh B (2000, August 14–17) A multi-layer IPSec protocol. In: USENIX security symposium, California, USA, pp 213–228
3. Elkeelany O, Matalgah MM, Sheikh KP (2002) Performance analysis of IPSec protocol encryption and authentication. In: IEEE international conference on communications, New York, USA, 1–2 May, pp 1164–1168
4. Dhall H, Dhall D, Batra S, Rani P (2012) Implementation of IPSec protocol. In: 2nd international conference on advanced computing and communication technologies, Rohtak, Haryana, India, 7–8 January 2012
5. Zhang Y (2004) A multilayer IP security protocol for TCP performance enhancement in wireless networks. IEEE J Sel Areas Commun 22(4):767–776
6. Pearlman R, Kaufman C (2001) Analysis of the IPSec key exchange standard. In: Tenth IEEE international workshop on enabling technologies, Cambridge, USA, 20–22 June 2001
7. Nozaki H, Motoyama M, Shimbo A (2001) Implementation of RSA algorithm based on RNS Montgomery multiplication. In: International workshop on cryptographic hardware and embedded systems, Paris, France, 14–16 May 2001, pp 364–376
8. Kocher PC (2001) Timing attacks on implementations of Diffie-Hellman, RSA, DSS and other systems. In: Annual international cryptography conference, San Francisco, USA, 13 July 2001, pp 104–113
9. Joux A (2000) A one round protocol for tripartite Diffie-Hellman. In: International algorithmic number theory symposium, Netherlands, 2–7 July 2000, , pp 385–393
10. Yang B, Wu K, Karri R (2004) Scan based side channel attack on dedicated hardware implementations of data encryption standard. ITC, Washington DC, USA, 26–28 October 2004, pp 339–344
11. Khan E, El-Kharashi MW, Gebali F (2007, December 12) Design and performance analysis of a unified, reconfigurable HMAC hash unit. IEEE Trans Circ Syst 2683–2695
12. Chkliaev D, Hooman J, de Vink E (2003, February 28) verification and improvement of the sliding window protocol. In: International conference on tools and algorithms for the construction and analysis of systems, pp 113–127
13. Al-Khatib AA (2018, September) Impact of IPSec protocol on the performance of network real-time applications. Int J Netw Secur 811–819

14. Alshamsi A, Saito T (2005) A technical comparision of IPSec and SSL. In: 19th international conference on advanced information networking and applications, Taiwan, 28–30 March 2005

Location Privacy Protection in Mobile Wireless Networks Through Hidden Forest and Fog Computing

Deepanjali Kumari, Kishan Kumar Singh, and Vijay Kumar Jha

Abstract As we know that the key to personal freedom is mobility and due to which many providers begin to offer location-based services and these services enrich our mobility experiences and with this the purpose of security increases as the location-based services can easily track the location of the user. Many of the current limitations would inherit several upcoming features as security and privacy issues have not fully been addressed by standardization bodies. Existing methodologies propose some method of protecting location privacy in mobile wireless networks but it has some drawbacks. So in this paper, drawbacks have been resolved as shown in comparative study, and in proposed methodology, two technologies have been used, i.e. hidden forest and fog computing where hidden forest method is used for protecting location privacy in mobile wireless networks and to increase more security, i.e. for 2-layer security, fog computing technology has been used to create illusion and mislead the attacker. To find hidden forest, Chinese remainder theorem (CRT) method has been used and able to protect the location privacy of the mobile wireless networks. And fog computing technology has been used to create false data and illusion so that our actual data will be safe. Fog computing based on its networking edge computes feature where the location-based service and applications are the most popular. For simulation tools, cloud analyst and CloudSim are used.

Keywords Hidden forest · CRT · Fog computing · Location-based services · Prime matrix

---

D. Kumari (✉) · K. K. Singh · V. K. Jha
Department of Computer Science and Engineering, Birla Institute of Technology Mesra, Ranchi, Jharkhand 835215, India
e-mail: deepanjalikumari94@gmail.com

K. K. Singh
e-mail: kishansingh1392@gmail.com

V. K. Jha
e-mail: vkjha@bitmesra.ac.in

V. Nath and J. K. Mandal (eds.), *Proceedings of the Fourth International Conference on Microelectronics, Computing and Communication Systems*, Lecture Notes in Electrical Engineering 673, https://doi.org/10.1007/978-981-15-5546-6_72

1 Introduction

As in the recent years, the location-based application and services have been used rapidly in day-to-day life and have brought a lot of convenience in today's busy life. The location-based services can track the location and can solve several queries of the common man. For the use of location-based applications and services, the mobile users have to send their correct information of their location to service provider. The location service provider's server is not credible, so the user's location information will be vulnerable to theft. After hacking the location information of the user, the hacker, via location tracking to other public information (such as geographic database, encoding the phone book), may be able to confirm the user's identity and gain additional private information. This paper also focuses on 2-layer security to increase more privacy and protection, so use of fog computing technology is also discussed. It is assumed that attacker will break the hidden forest and then apply fog computing into hidden forest. As fog computing will create illusion and mislead the attacker, the actual data will be protected. Location protection is very much important nowadays; the most average people do not understand how location services use their personal information.

This paper focuses on $n \times$ square patches which is called hidden forest. An example of a 2×2 hidden forest is given. Data sets are also generated based on hidden forest. The specific four visible trees obscure this hidden forest. It also resolves the drawbacks of hidden tree as discussed in the comparative study, and the hidden tree is not able to satisfy latent tree conditions. And if latent tree conditions are not satisfied, then attacker can easily trace the nearest neighbour points as observed from hidden tree. So, the drawbacks of hidden tree are minimized through the construction of hidden forest.

2 Literature Review

Gao, Zhu, Gong, Hengsong and Tan proposed in [1] HRHF method to protect location privacy. It uses BFS method to meet certain requirements of the ring in the graph. HRHF method provides mobile users with average quality location-based services but it also faces some drawbacks as in simple network; it will be able to provide protection but in complex it is quite difficult and also latent tree conditions are not satisfied in the hidden tree. It is suitable for smallest neighbour point, but for large, it is quite difficult. Yu, Xu, He, Zhang, Li, Xu proposed in [2] L2P2 problem where different users can define dynamic and diverse privacy protection. The main aim of this problem is to find small area cloaking for each user-requested location service so that diverse dimensions are fulfilled over temporal dimensions for each and every user. Some test methods are proposed in this, but accurate data sets for real-life tracking system are not proposed. Liao, Li, Sun, Zhang, Chang proposed in [9] vehicular social networks for 5G which fulfilled location services in an advanced way and also maintained trajectory privacy protection for vehicles. It used M Femtocell technique and design a novel for 5G VSNs. And it also proposes DGD algorithm but providing privacy is not sufficient in framework of mix group. Austingoodrich Dongfengfang, Yiqian and Rose qingyanghu proposed in [5] traditional cellular networks compared with the 5G wireless networks security. It also gives information about different technologies applied to 5G, and it also shows the existing schemes and latest development. It also faces some challenges in privacy protection. Farhang, Hayel and Zhu proposed in [4] and investigate about preserving privacy through access point selection method. In this, decentralized algorithm is used for location protection but location privacy is a critical issue for some mobile users. It is quite difficult to provide privacy in complex mobile networks.

3 Proposed Methodology (1) [Hidden Forest]

In this section, hidden forest work is presented. Hidden forest method describes about invisible lattice points. It uses Chinese remainder theorem to find hidden forest which uses $n \times n$ square patches. The application of $n = 2$ cases is also discussed further. To increase more privacy in hidden forest, fog computing is used as a 2-layer security.

3.1 Hidden Forest

As we know that approximately 60% of origin is visible from integer lattice points and hence integer lattice points are hidden in the origin which is approximately 40%. 40% of the invisible lattice points compromise a variety of problems involving lattice point visibility searching for patterns. Square patch is one of such patterns which is known as hidden forest.

Hidden forest is known to exist arbitrarily in integer lattice.

3.2 Density of Visible Lattice Points in Z^2

A criterion for the visibility of an integer lattice point can be recast in the number-theoretic setting as the following theorem

Theorem 1 *Let (a, b)* $\in$ $Z^2\backslash\{(0,0)\}$*. Then (a, b) is visible if and only if gcd (a, b)* = *1.*

Proof Let (a, b) be a non-origin point in Z^2. Suppose $d = gcd$ (a, b). If $d > 1$ then $(\frac{a}{d}, \frac{b}{d})$ lies strictly between the points (0, 0) and (a, b), and hence (a, b) is not visible from the origin. Then, there is a point $(a_0, b_0) \in Z^2$ such that $(a, b) = (c\, a_0, c\, b_0)$

for some integer $c > 1$. That is, c divides both a and b. But $d = 1$ is the greatest common divisor of a and b, contradicting that $c > 1$ Thus if gcd $(a, b) = 1$, then (a, b) is visible.

3.3 The Traditional Method to Find Hidden Forests

This section finds arbitrarily large patches of hidden square regions in Z^2 using the known technique, which is called the CRT algorithm, since the main tool in this technique is the Chinese remainder theorem (CRT). The strategy is to find two sets of n consecutive integers.

$A = \{ a_1, a \ldots, a_n\}$ and $B = \{ b_1, b_2, \ldots, b_n\}$ such that $A \cap B = \emptyset$ and $gcd\ (a_i, b_j) > 1$ for all $1 \le i, j \le n$. Then, it is clear that the n^2 points in the set $\{(a_i, b_j) | 1 \le i, j \le n\}$ yield the desired hidden square region. To this end, first, some necessary preliminary definitions have been discussed.

Definition (*Hidden forest*). An $n \times n$ hidden forest in Z^2 is a square patch of n^2 invisible integer lattice points that have width and height of n consecutive lattice points. Denote this hidden forest by the symbol $H^n_{(a,b)}$ where (a, b) is the closest corner lattice point of the square to the origin.

Definition (*Prime matrix*). Let $\{ k_1, k_2 \ldots., k_{n^2}\}$ be the set of the first n^2 primes. Construct an $n \times n$ matrix with these primes by filling row i with the n primes k through $k_{(i-1)n+n}$ for each $1 \le i \le n$ to yield the following:

$$\begin{pmatrix} k_1 k_2 \ \cdots & k_j \cdots & k_n \\ k_{n+1} k_{n+2} \cdots & k_{n+j} \cdots & k_{2n} \\ & & \\ k_{(i-1)n+1} k_{(i-1)n+2} \cdots & k_{(i-1)n+j} \cdots & k_{(i-1)n+n} \\ & & \\ k_{(n-1)n+1}\ k_{(n-1)n+2} \cdots & k_{(n-1)n+j} \cdots & k_{n^2} \end{pmatrix}$$

Note that the prime $k_{(i-1)n+j}$ is located in row i and column j of the matrix. Call this $n \times n$ matrix a prime matrix and denote it k_n.

3.4 The CRT Algorithm

The following theorem is the primary tool used in the CRT algorithm to find hidden forests of arbitrary size.

Theorem 3.4 *There exist two sets of n consecutive natural numbers* **A**= { $a_1, a_2, \ldots, a_n$} *and* **B**= { $b_1, b_2, \ldots, b_n$} *for each n* ∈ *N such that* **A** ∩ **B**= ∅ *and gcd* $(a_i, b_j) > 1$ *for all* $1 \le i; j \le n$.

Proof Fix n ∈ *N*. Consider the prime matrix k_n. Let R_i and C_j be the product of the entries in row i and column j, respectively, so we have

$R_i = \prod_{m=1}^{n} k(i-1)n + m$ and
$C_j = \prod_{m=0}^{n-1} kmn + j$

Since they share no primes in common, the row products $R_1, R_2, \ldots, R_n$ are pairwise relatively prime. Similarly, the column products $C_1, C_2, \ldots, C_n$ are pairwise relatively prime. Consider the following pair of systems of linear congruences:

$$\begin{cases} a+1 \equiv 0 \pmod{\mathrm{R}_1} \\ a+2 \equiv 0 \pmod{\mathrm{R}_2} \\ \vdots \\ a+n \equiv 0 \pmod{\mathrm{R}_n} \end{cases}$$

$$\begin{cases} b+1 \equiv 0 \pmod{\mathrm{C}_1} \\ b+2 \equiv 0 \pmod{\mathrm{C}_2} \\ \vdots \\ b+n \equiv 0 \pmod{\mathrm{C}_n} \end{cases}$$

Observe that $R_1 \cdot R_2 \ldots. R_n = C_1 \cdot C_2 \ldots. C_n = \prod_{i=1}^{n^2} k_i$, which denote M. By the CRT, there exist solutions a_0 and b_0 to the left and right systems, respectively, such that a_0 and b_0 are unique modulo M. Let $\mathbf{A} = \{ a_0 + 1, a_0 + 2, \ldots, a_0 + n\}$ and $\mathbf{B} = \{ b_0 + 1, b_0 + 2, \ldots, b_0 + n\}$. Then, claim that none of the integers in **A** are pairwise relatively prime to any of the integers in **B**. For an arbitrary $a_0 + i \in$ **A** and $b_0 + j \in$ **B**, these two elements by construction are multiples of R_i and C_j, respectively, and hence, the prime that lies in the intersection of row *i* and column *j* in the matrix, namely $k_{(i-1)n+j}$, divides gcd (a_0 + i, b_0 + j). Thus, *gcd* (a_0 + i, b_0 + *j*) > 1 as desired.

Observe that for $n \ge 2$ the sets **A** and **B** are necessarily disjoint. Otherwise, if **A** ∩ **B** ≠ ∅, then some element a ∈ **A** is relatively prime to some element l ± 1 ∈ **B** since gcd (l, l ± 1) = 1, contradicting gcd($A_0 + i$, $B_0 + j$) > 1 for all $1 \le i, j \le n$. For the trivial case when $N = 1$, the algorithm above yields **A** = **B** = {2}. So set **B** = {4} and hence **A** ∩ **B** = ∅ (Figs. 1 and 2).

By Theorem 3.4, the prime matrix k_n yields the hidden forest $H^n_{(a,b)}$ comprised of the n^2 points (a_i, b_j) where $a_i = a_0 + i \in A$ and $b_j = b_0 + j \in B$ for $1 \le i, j \le n$.

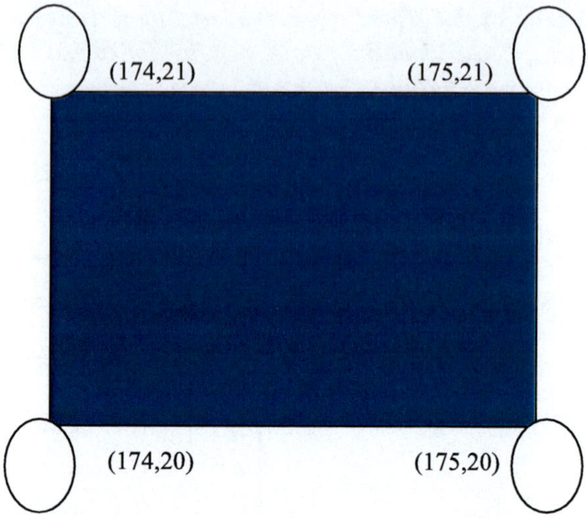

Fig. 1 Hidden forest coordinates

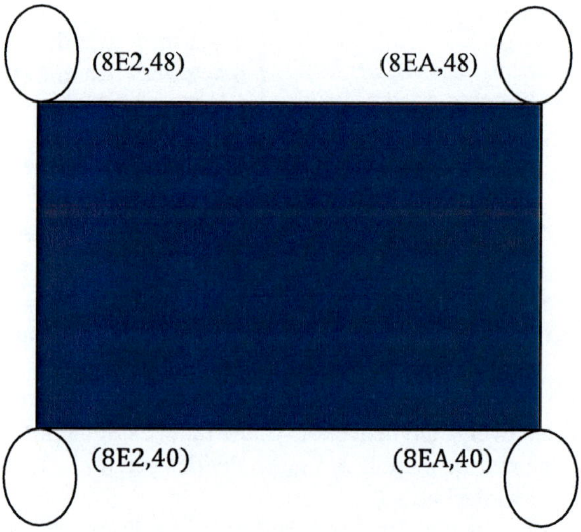

Fig. 2 Fog computing coordinates

3.5 An Algorithm to Construct a Hidden Forest Using CRT Algorithm

1. Fix a value $n \in N$
2. Construct the prime matrix k_n.
3. Apply the Theorem 3.4 to k_n to yield sets A and B.
4. Construct the hidden forest $H^n_{(a,b)}$ from A and B.

3.6 An Application: The N = 2 Cases

In the 2×2 case, using **Theorem 3.4**, set $n = 2$ and the prime matrix is

$$k_2 = \begin{pmatrix} 2 & 3 \\ 5 & 7 \end{pmatrix}$$

The row products are $R_1 = 6$ and $R_2 = 35$, while the column products are $C_1 = 10$ and $C_2 = 21$. Hence, the corresponding linear congruences need to solve are:

$a + 1 \equiv 0 \pmod{6}$
$a + 2 \equiv 0 \pmod{35}$
$b + 1 \equiv 0 \pmod{10}$
$b + 2 \equiv 0 \pmod{21}$

By the CRT algorithm, the left and right systems have the unique solutions $a_0 = 173 \pmod{210}$ and $b_0 = 19 \pmod{210}$, respectively. Set $\mathbf{A} = \{174, 175\}$ and $\mathbf{B} = \{20, 21\}$. Then $\mathbf{A} \cap \mathbf{B} = \emptyset$ and gcd $(a_i, b_j) > 1$ for all $1 \le i, j \le 2$. Thus there is a hidden forest $H^2_{(174,20)}$ of four trees at (174, 20), (174, 21), (175, 20), and (175, 21).

4 Proposed Methodology (2) [Fog Computing]

As we can define fog computing as an architecture that carries out a computation, locally communication storage and Internet backbone routed over with edges devices. The another name of fog computing is fog networking or also known with the name of fogging.

The use of fog computing in location privacy is indicated to the fog clients of the location privacy. And the use of the fog clients or who using fog computing is that whatever tasks fog computing do it mainly offloads to the fog nodes nearest one and the tasks to whom fog node offloaded can tell that it is close or far away from other fog nodes.

Fog computing technology is mainly used in this paper to maintain the privacy of the exact or true data and create false data and to confuse the attacker or hacker. So that attacker or hacker will be mislead and true data will be protected.

4.1 An Algorithm to Implement Fog Computing

1. Take the coordinate points, i.e. in the form of decimal integer
2. Reverse the coordinate points
3. Apply the 4-bit binary-coded decimal (BCD)
4. Reverse the 4-bit BCD
5. Convert the reverse BCD to hexadecimal code

6. New coordinates generated.

4.2 Fog Computing Technology Applied into Hidden Forest

To apply fog computing technique, take four coordinates of hidden forest $H^{n}_{(a,b)}$. $H^{2}_{(174,20)}$ = (174, 20), (174, 21), (175, 20) and (175, 21). Then, apply algorithm of fog computing.

4.3 Final Coordinates Generated

This is how fog computing is applied into hidden forest.

5 Comparative Study

Here, this section shows the comparative study. Comparative study is based on theoretical analysis and study and also the drawbacks of all these papers are resolved through hidden forest method and fog computing (Table 1).

All these limitations are resolved in this paper.

Table 1 Comparative study

S. No.	Journal	Title	Drawbacks
1.	IEEE Access	Security for 5G mobile wireless networks (Dongfeng Fang, Yi Qian and Rose Qingyang Hu)	It faces many challenges. Privacy protection is very challenging not able to provide security
2.	Springer	Location privacy protection algorithm for mobile networks (Kun Gao, Yiwei Zhu, Songjie Gong and Hengsong Tan)	Hidden tree unable to hide user hidden position and also through hidden tree low value is generated
3.	IEEE	PHY-layer location privacy-preserving access point selection mechanism in next-generation wireless networks (SadeghFarhang, YezekaelHayel and Quanyan Zhu)	Location privacy is critical issue for some mobile users. It is not providing privacy protection in complex networks

Table 2 Data sets of hidden forest

k_2	Coordinates
2	(174, 21)
3	(175, 21)
5	(174, 20)
7	(175,20)
11	(1287, 748)
13	(1292, 748)
17	(1287, 741)
19	(1292, 741)
23	(4669, 2139)
29	(4588, 2139)
31	(4669, 2146)
37	(4588, 2146)

6 Results and Data sets

It is observed that two coordinates are successfully generated. It is also observed that hidden forest coordinates provide with higher value points than hidden tree [1] and drawbacks are successfully resolved. Data sets of hidden forest are shown and it is analysed that as k_2 increases coordinates points also increases. Through this, high-value coordinate points are generated and privacy is better. And also hidden forest is better in protecting privacy as shown in comparative study and if in the case hidden forest might break by attacker then 2-layer security has been used, i.e. fog computing. Fog computing coordinates are used to confuse the attacker and attacker will think this is only the actual coordinates. And false coordinates are generated in place of actual coordinates to mislead the attacker (Table 2).

7 Conclusion and Future Scope

As we know that in today's world mobile phones or devices are increasing its popularity day by day, location-based applications or services are also increasing. And we also know that how location-based services are dealing with public facilities. And through advancing technologies, mobile users' location-related information can be accessed from anywhere at anytime. But some location-based applications or services can get attacked by hackers or crackers or by new threat and steal information of users and misuse it. So it is very important to protect our location in mobile wireless networks from stealing information. So, this paper has given the concept of use of traditional method to find hidden forests, through CRT method and also focuses on finding closer hidden forests. $n \in N$ and a prime matrix k_n, there exists an $n \times n$

hidden forest $H^n_{(a,b)}$. And also to increase more privacy and protection, fog computing concept, i.e. 2-layer security has also been discussed. And through $n = 2$ cases, the concept of hidden forest is successfully tested and also fog computing technique is successfully applied into hidden forest.

For the future directions, we would like to do practical implementation of hidden forest and fog computing concept and apply practically into LBS and also we would like to test for more higher value of data sets and cases.

Acknowledgements I give thanks to all Almighty God for giving an opportunity for doing research. And I wish to acknowledge Dr. Vijay Kumar Jha of Birla Institute of Technology for explaining the details of the experiments, Suggestions and valuable comments which motivated to me to use the concept of hidden forest and fog computing technique in location privacy protection.

References

1. Gao K, Zhu Y, Gong S, Tan H (2016) Location privacy protection algorithm for mobile networks. EURASIP J Wirel Commun Netw 2016:205
2. Wang Y et al (2012) L2P2: location-aware location privacy protection for location-based services. In: 2012 Proceedings IEEE, INFOCOM. IEEE
3. Shin KG et al (2012) Privacy protection for users of location-based services. IEEE Wirel Commun 19(1):30–39
4. Farhang S, Hayel Y, Zhu Q (2015) PHY-layer location privacy-preserving access point selection mechanism in next-generation wireless networks. In: Proceedings of IEEE conference on communication network security (CNS), pp 263–271
5. Fang D, Qian Y, Hu RQ (2018) Security for 5G mobile wireless networks. IEEE Access 6
6. Yang X, Yin F, Tang X (20147) A fine-grained and privacy-preserving query scheme for fog computing-enhanced location-based service. J Sens 17:1611
7. Yi S, Hao Z, Qin Z, Li Q (2015) Fog computing: platform and applications. In: 2015 third IEEE workshop on hot topics in web systems and technologies
8. Liao D, Li H, Sun G, Zhang M, Chang V (2018) Location and trajectory privacy preservation in 5G-Enabled Vehicle social network services. J Netw Comput Appl 110:108–118
9. Gruteser M, Liu X (2004) Protecting privacy in continuous location-tracking applications. IEEE Secur Privacy 2(02):28–34
10. Huang L, Matsuura K, Yamane H, Sezaki K (2005) Enhancing wireless location privacy using silent period. In: IEEE wireless communications and networking conference, 2005, vol 2. New Orleans, LA, pp 1187–1192
11. Hoh B, Gruteser M (2005) Protecting location privacy through path confusion. In: First international conference on security and privacy for emerging areas in communications networks (SECURECOMM'05), Athens, pp 194–205
12. Kido H, Yanagisawa Y, Satoh T (2005) Protection of location privacy using dummies for location-based services. In: 21st international conference on data engineering workshops (ICDEW'05), Tokyo, Japan, pp 1248–1248
13. Hwang R, Hsueh Y, Chung H (2014) A novel time-obfuscated algorithm for trajectory privacy protection. IEEE Trans Serv Comput 7(2):126–139
14. Li P, Peng W, Wang T, Ku W, Xu J, Hamilton JA Jr (2008) A cloaking algorithm based on spatial networks for location privacy. In: 2008 IEEE International Conference on Sensor Networks, Ubiquitous, and Trustworthy Computing (sutc 2008), Taichung, pp 90–97

15. Shen Q, Liang X, Shen X, Lin X, Luo H (2014) Exploiting geo distributed clouds fore-health monitoring system with minimum service delay and privacy preservation. IEEE J Biomed Health Inf 18(2):430–439
16. Jung T, Mao X, Li X, Tang S-J, Gong W, Zhang L (2013) Privacy preserving data aggregation without secure channel: multivariate polynomial evaluation. In: Proceedings of INFOCOM. IEEE, pp 2634–2642
17. Yang Y, Li H, Liu W, Yang H, Wen M (2014) Secure dynamic search symmetric encryption with constant document update cost. In: Proceedings of GLOBCOM. IEEE
18. Cao N, Wang C, Li M, Ren K, Lou W (2014) Privacy-preserving multi-keyword ranked search over encrypted cloud data. IEEE Trans Paral Distrib Syst 25(1):222–233
19. https://support.google.com/websearch/answer/173733?hl=en
20. Song DX, Wagner D, Perrig A (2000) Practical techniques for searches onencrypted data. In: Proceedings of S&P. IEEE, pp 44–55
21. Li R, Xu Z, Kang W, Yow KC, Xu C-Z (2014) Efficient multi keyword ranked query over encrypted data in cloucomputing. Future Gener Comput Syst 30:179–190
22. Li H, Liu D, Dai Y, Luan TH, Shen X (2014) Enabling efficient multi-keyword ranked search over encrypted cloud data\through blind storage. IEEE Trans Emerg Topics Comput
23. Wang C, Cao N, Li J, Ren K, Lou W (2010) Secure ranked keyword search overencrypted cloud data. In: Proceedings of ICDCS. IEEE, pp 253–262

Design and Characterization of DC-to-DC Converters Using Active Inductor

Om Prakash, Kumar Ankit, Rohan Kumar, and Vijay Nath

Abstract Digital to digital converters are methods for power transformation. They move a DC voltage level to another level dependent on the prerequisite. The fundamental circuit utilized for a DC/DC converter is one which is used to regulate the switching actions. This type of circuit is mainly responsible for relocating the energy from input terminal to the output terminal using various components like diodes, inductors, electronic switches etc. They are capable of stepping-up, stepping-down and inverting a given voltage parameter. The components which constitutes this regulating circuit can be arranged in different configurations to achieve desired outputs like that of step-down converters, step-up converters etc. The operating voltage requirement of various electronic devices, for example, Integrated Circuits (ICs) can vary over a broad range, making it mandatory to supply a particular voltage for each device. A DC/DC converter makes modification to the input DC voltage from a source into another DC voltage which is given as output. The Buck Converter yields a voltage lower than provided, while converse is the case with a Boost converter. This research article sees the comparison of DC/DC Converter using passive inductor with the one using active inductor. Switched mode DC–DC converter makes modification to the voltage level value by hoarding the energy either in form of magnetic field or electric field from input source for some temporary amount of time and then emancipating that stored energy at another voltage level. Whatever output voltage is received at the output terminal, it is exploited to match the power requirement of the load. These converters are mainly used in the vehicles, chargers and Video players.

O. Prakash · K. Ankit (✉) · R. Kumar · V. Nath
VLSI Design Group, Department of Electronics and Communication Engineering, Birla Institute of Technology Mesra, Ranchi, Jharkhand 835215, India
e-mail: ankitsfsj@gmail.com

O. Prakash
e-mail: opabhi913@gmail.com

R. Kumar
e-mail: rohan28kumar@gmail.com

V. Nath
e-mail: vijaynath@bitmesra.ac.in

V. Nath and J. K. Mandal (eds.), *Proceedings of the Fourth International Conference on Microelectronics, Computing and Communication Systems*, Lecture Notes in Electrical Engineering 673, https://doi.org/10.1007/978-981-15-5546-6_73

Keywords DC/DC converter · Switching regulator · Buck converter · Boost converter · Buck-Boost converter · SEPIC converter · CUK converter

1 Introduction

Firstly we shall discuss the various topologies that has been discussed in this article. Those are:

(a) Buck Converter
(b) Boost Converter
(c) Buck-Boost Converter
(d) CUK Converter
(e) SEPIC Converter

Above mentioned circuit design are discussed independently as:

(a) **Buck Converter**

A buck converter commonly referred to as a step-down converter owing to the fact that this converter is able to lower down the voltage level to a lower level than the input. The conventional voltage source inverter (VSI) likewise alluded to as a buck inverter, is presumably the most imperative power converter topology [1]. It is utilized in numerous particular mechanical and commercial applications. Among these applications, uninterruptible power supply (UPS) and air conditioning engine drives are the most essential. One of the qualities of the buck inverter is that the instantaneous voltage is dependably lower that the input dc voltage.

The Buck Converter finds its primary usage in the switched mode power supply (SMPS) network designs whereby the DC output voltage should be less as compared to the input DC voltage level. The transistor which connects the input terminal to the output terminal switches between on and off position at a very high and fair amount of frequency value. When the transistor is in on position, energy gets stored in the inductor. This stored energy is used to provide a continuous output even when the transistor is in off position [2]. The circuit task basically relies upon what often addressed to a Flywheel Circuit. This is on the grounds that the circuit demonstrations rather like a mechanical flywheel such that the wheel keeps spinning very smoothly when provided with energy pulses separated at a regular space and it spins at a fair amount of steady rate.

The figures below show the schematic of a Buck converter and also the output plot of it (Figs. 1 and 2).

(b) **Boost Converter**

This subsection deals with a boost type DC–DC converter circuit which is responsible for the increment of output DC voltage as compared to that of the input one. It nosedive under the class of switched mode control supply (the SMPS) which may

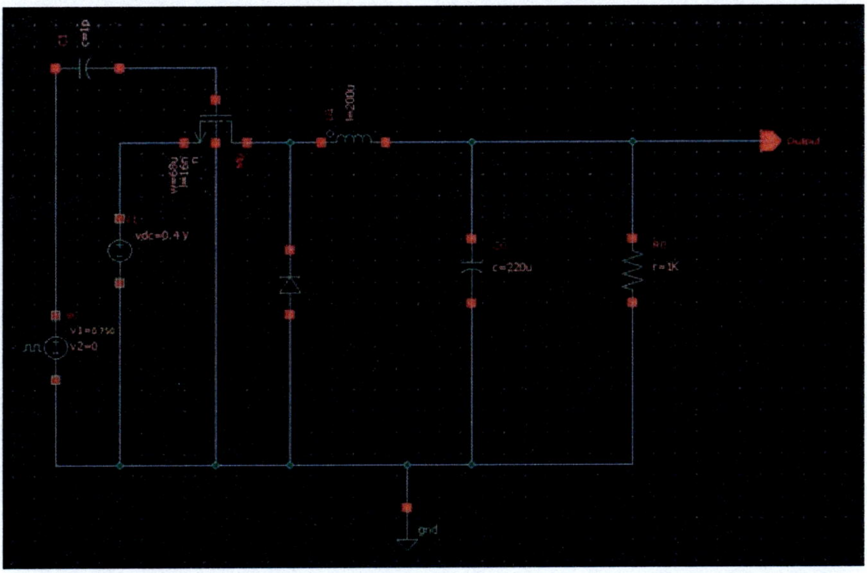

Fig. 1 Schematic figure of Buck converter

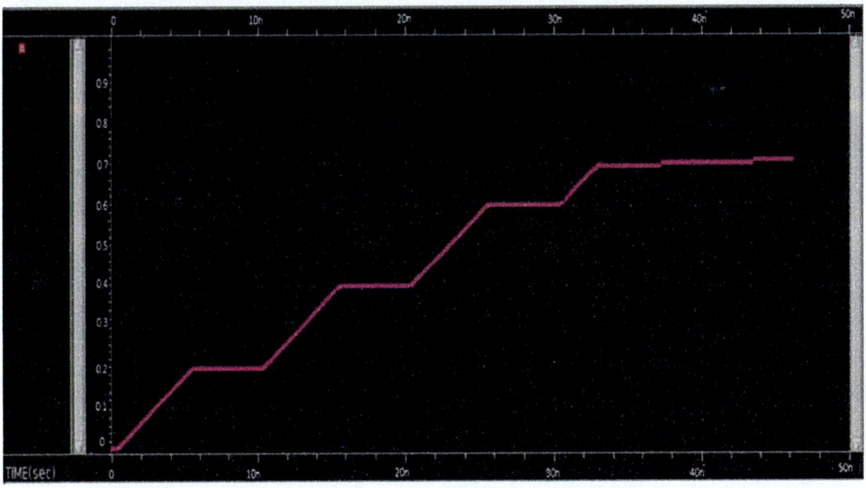

Fig. 2 Output plot of Buck converter

house around two semiconductor based elements (a diode & a transistor) and additionally an energy storing element namely an inductor, a capacitor or both of them together. In order to lower down the voltage ripple, capacitive filters (here and there in combination with the inductors) are often added at the output of such a converter (load side filter) and input (supply side filter). Since it is a conservation law that the power ($P = V * I$) must remain safeguarded so that's the reason for the output current value being less in magnitude than the input counterpart [3].

DC/DC Converters are one of the reason for the implementation of the switched mode power supplies. Although, a DC power supply source is capable for providing power but it might not be sufficient for the converters being considered in this article. The space requirement of the battery banks, their large weights etc. makes them not very convenient to be utilized. This is a major problem in today's world where space and weight are one of the important constraints. Thus, to overcome these problems, the best possible solution is the use of boost DC/DC transfigure circuit. Another issue with usage of batteries (either expansive or small) is that their output voltage differs when the available charge is utilized and ultimately the battery voltage supply turns out to be too feeble to be even considered for powering the circuit that has been provided. By employing certain boost converters if this considerably low level of the output can be supported a back up to the valuable dimension once more, we can definitely enhance the battery life [4, 5]. The DC input source for this converter can be a numerous sources such as batteries: let's say rectified AC from the mains supply or for that matter DC from sunlight based boards, energy components, dynamos and DC generators. This converter differs from its buck counterpart in terms of the output voltage it produces, because it produces a voltage which is the incremented version of the input voltage. It is very much important to keep in mind that the overall power remains constant as the voltage increases but on the other hand the value of current decreases.

The figures below show the schematic of a Boost converter and also the output plot of it (Figs. 3 and 4).

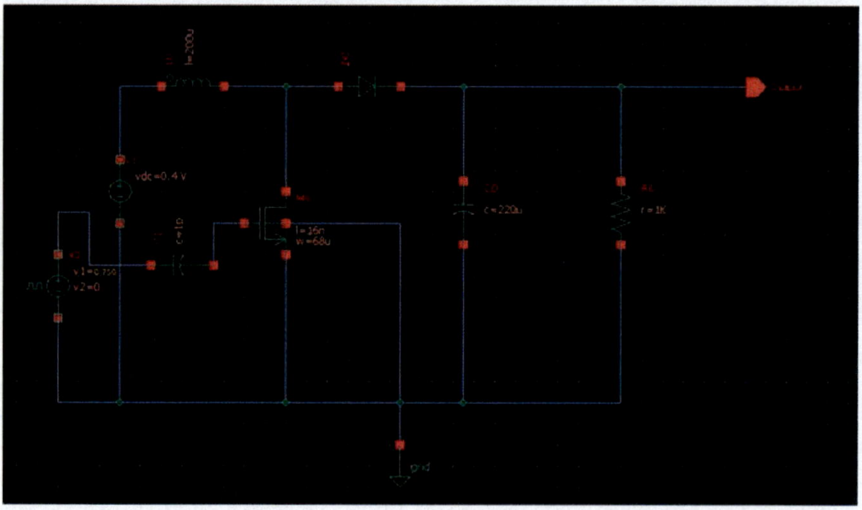

Fig. 3 Schematic figure of Boost converter

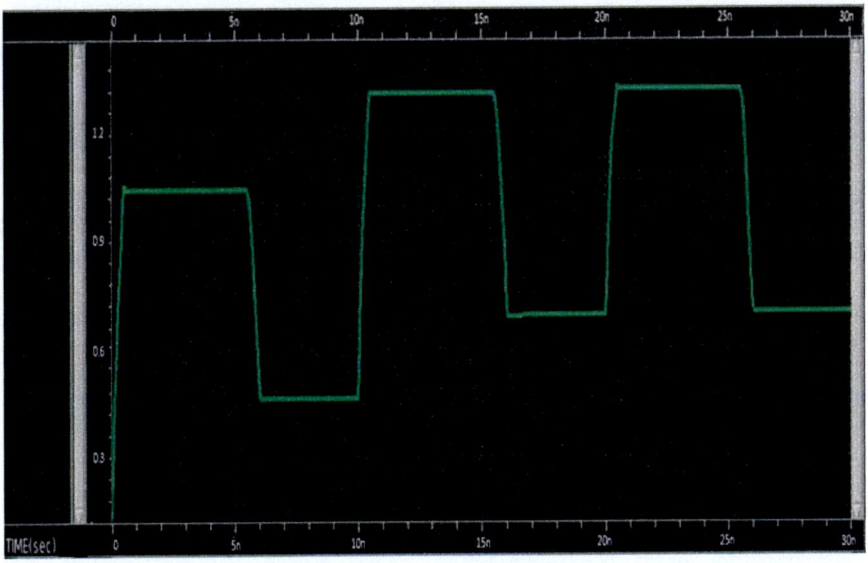

Fig. 4 Output plot of Boost converter

(c) Buck-boost Converter

Another most important converter is discussed in this article, which is the buck-boost converter which is responsible to modify the output voltage value greater than or less than the input voltage value. Thus, this can be considered as a combination of the previously discussed two converters. This translator is a genus of Switched Mode Power Supply (SMPS) converter which is identical in topology to buck and boost converter. The output is adjusted according to the duty cycle of the switching transistor.

By cascading the two topologies, it can be adjusted to a wide range of input voltage that give the voltage output of magnitude higher or lesser. Both buck and boost converter have basically the same components, just simply rearranging these components, the output voltage can be obtained. The input voltage source is connected to a FINFET directly which acts a switch and the second switch is a diode. The diode connection is reverse to the current flow from the input source and the capacitor [3, 5]. The diode which is used has a low forward voltage drop and very high switching speed to reduce the spikes in output voltage. For the switch either a FINFET or BJT can be used depending on the application, such as FINFET for a high-frequency application and BJT in case of low-frequency application.

The figures below show the schematic of a Buck-Boost converter and also the output plot of the converter (Figs. 5 and 6).

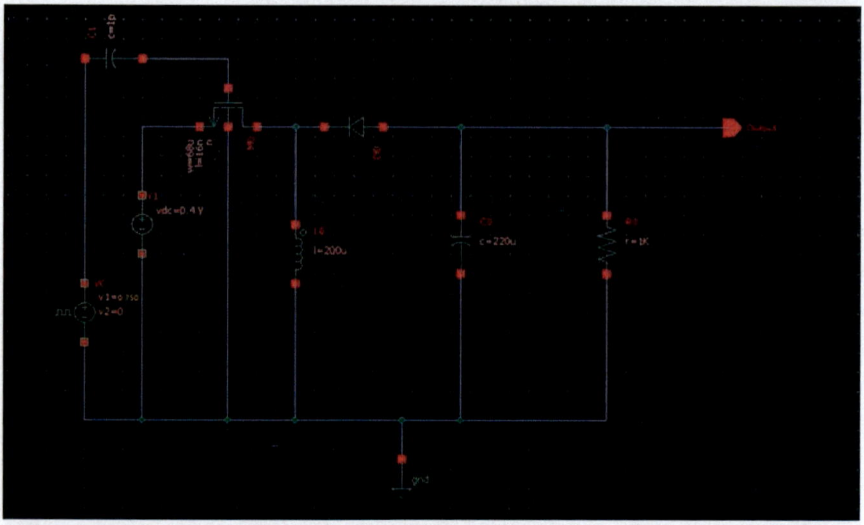

Fig. 5 Schematic figure of buck-boost converter

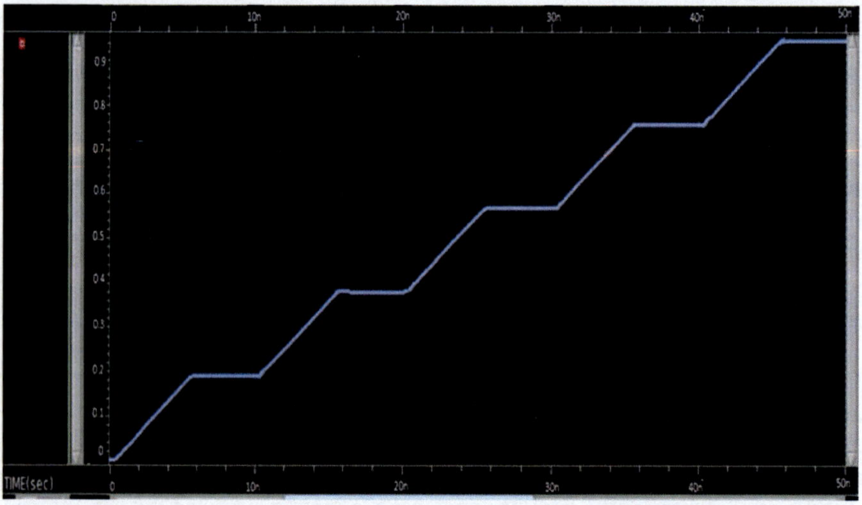

Fig. 6 Output plot of buck-boost converter

(d) **CUK Converter**

There is another topology of DC–DC converter called the Cuk converter which gives an output voltage that is the inverted version of the input voltage source. The basic principal of this type of converter is that it comprises of the earlier discussed two types of converter, i.e., boost which is cascaded with buck converter with a coupling capacitor to couple the energy. In this type of converter, the main coupling part is the capacitor contrary to previous converters in which inductor was the main coupling part.

The major benefit of this converter is that it has a low-switching loss making it all the more very effective. The drawback of this converter is that it incorporates a high number of reactive segments (*L* and *C*). Cuk converter is a capacitor based topology so energy put away in capacitor and furthermore cuk converter is overshoot of the output voltage. In any case, buck-boost converter is inductor based topology so energy is stored in inductor [4, 6]. This converter is used for getting an output voltage has a wide range of output. This implies that the output is having an inverted polarity with voltage more or less than the input. The work of filter is done by the input inductor for the dc supply to avert extensive harmonics in the current. The energy exchange is primarily dependent upon a capacitor rather than an inductor. As can be seen from the figure of the Cuk converter, in the OFF state of the switch, the coupling capacitor is being charged through inductor L1 by the given input supply. And in the ON state of the switch, the coupling capacitor transfers energy to the output capacitor with the help of inductor L2. The capacitor C is utilized for the transfer of energy from input terminal to the output terminal with the reciprocation of transistors and diodes. The main purpose of the two inductors is to convert the input and output voltages into respective currents. In fact, the inductor is treated as a current source for a short span of time as it can withstand a steady current. Had this arrangement not been there and the capacitors were connected directly to the voltage sources, there would have a greater energy loss [7, 8]. When charging the capacitor with current source, i.e., the inductor in this context, the energy loss is less as it prevents the resistive current loss associated with the capacitor.

The figures below show the schematic of a CUK and the output plot of the converter (Figs. 7 and 8).

(e) **SEPIC Converter**

Some of the applications of converters need to either boost or buck the input voltage and in this case the respective converters are utilized. But, often some applications require that the output voltage is varies around the input voltage. In this case, the most suitable practice is to use a converter that can either enhance or lower the input voltage. Some converter such as Buck-Boost is cheap because of the lower component requirements such as one inductor and capacitor. But, the drawback of this type is that their current has a high amount of ripple in them. Because of this ripple, there are harmonics created, restricting their use in many applications and requires the use of an LC filter. Because of this, the buck-boost converter are not effective. In many applications, there is a requirement to either increase or decrease the voltage [7, 9]. In these cases, the SEPIC converter is worth the price for the extra components required in this converter.

This topology of the converter proffers a positive output voltage as compared to that of the input voltage. It can regulate the input voltage to be either larger or smaller than the input. But the drawback of this converter is that it is difficult to comprehend

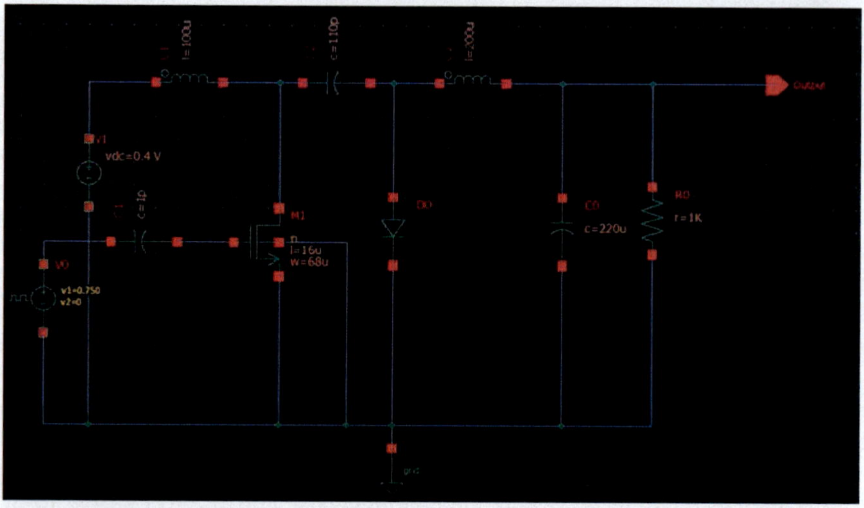

Fig. 7 Schematic diagram of Cuk converter

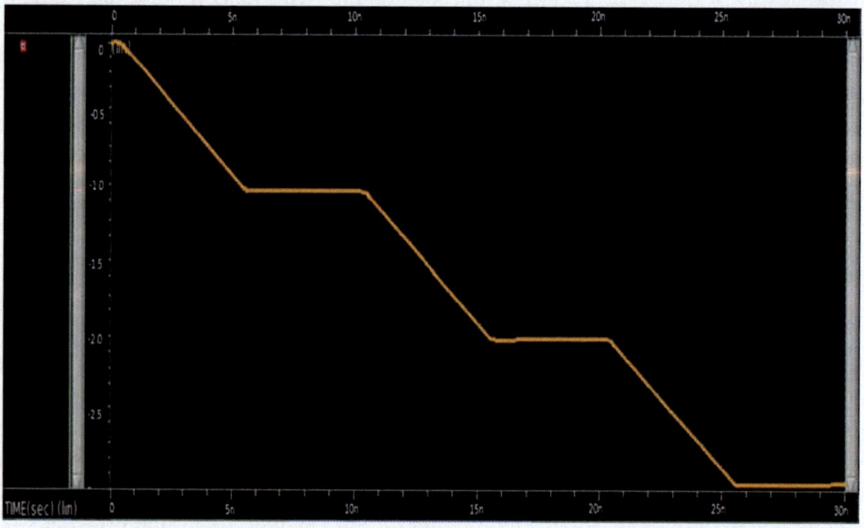

Fig. 8 Output plot of Cuk converter

the schematic since it has two inductors in it resulting in a larger amount power requirement. Some manufacturer of inductor have stated to manufacture coupled inductor in place of single inductors at a slightly higher cost than individual inductor which results in a reasonable cost for this converter. These coupled inductors also have the advantage of low power consumption and lower inductor ripple current.

As shown in figure below is a schematic of the SEPIC converter. Similar to any other converter, the SEPIC also exchanges energy between inductor and capacitor to get the desired value of the output. The switch which is used is mainly a FINFET which has the advantage of providing higher input impedance and drops a low amount of voltage as compared to the Bipolar Junction Transistor (BJT). The FINFET don't need a biasing resistor because the FINFET switches based on the voltage as opposed to current which is used in BJT.

The performance of SEPIC converter is highly dependent on the switching time and IR drop of diode D1 specially its reliability and efficiency. To reduce the high voltage spikes in the inductor, the switching time of diode need to be very fast. For this purpose, conventional fast diode or Schottky diode can be used.

The ripple in output and efficiency of the converter is also dependent on the resistance in capacitor and inductor. Inductors which have small series resistance dissipates less energy as heat which results in a greater efficiency. This should also be followed in the capacitors where the series resistance should be low so that ripple is minimized and also helps in preventing heat build-up, particularly for C1 in which current changes direction frequently [10]. A capacitor having high value of capacitance and greater capability to handle current is required since this converter transfer energy with the help of series capacitor.

The figures below show the schematic of a SEPIC and the output plot of the converter (Figs. 9 and 10).

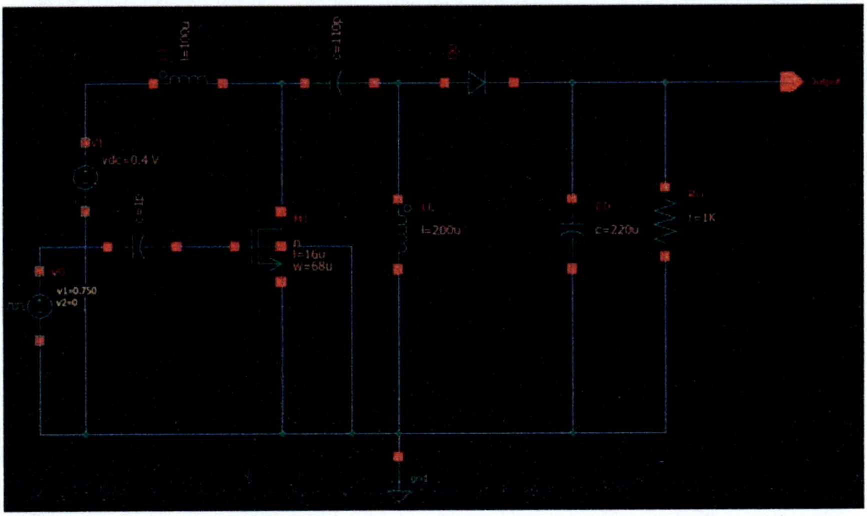

Fig. 9 Schematic figure of SEPIC converter

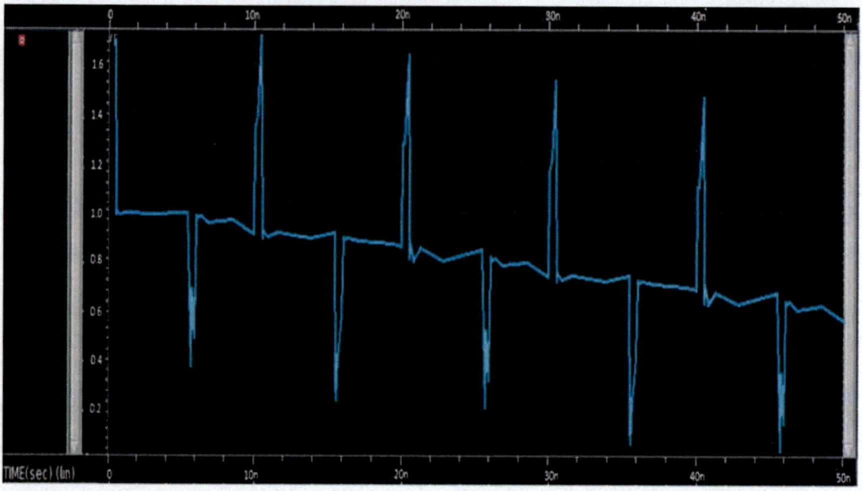

Fig. 10 Output plot of SEPIC converter

2 Active Inductors

The main inspirational fact behind the proposal of Active Inductors is to avoid the usage of spiral or metal wire bonding inductors in integrated circuits because of these main reasons. Large Area requirement, Large Power Consumption and Limited *Q*-factor. These are the most important drawbacks of the monolithic inductors. To overcome these limitations, a new type of inductors have been proposed, the Active Inductors which are based on the transistors based technology. The basic principle of modeling an active inductor is the Gyrator-C network [11]. A gyrator is a circuit which has two back to back connected OTA (Operational Transconductance Amplifier). When this network has a capacitance attached to its one end, the entire circuit is known as the Gyrator-C network and is shown in Fig. 11.

From Fig. 12, the values of the various components can be calculated by just calculating the admittance as seen from node 'b' [11].

Fig. 11 Schematic of a gyrator-C network

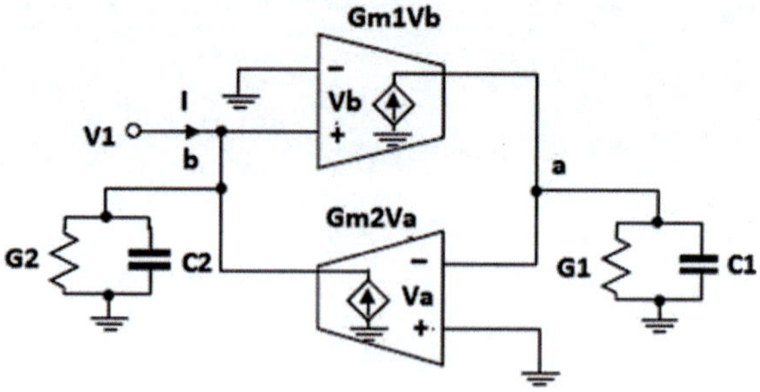

Fig. 12 Component level equivalent of gyrator-C topology

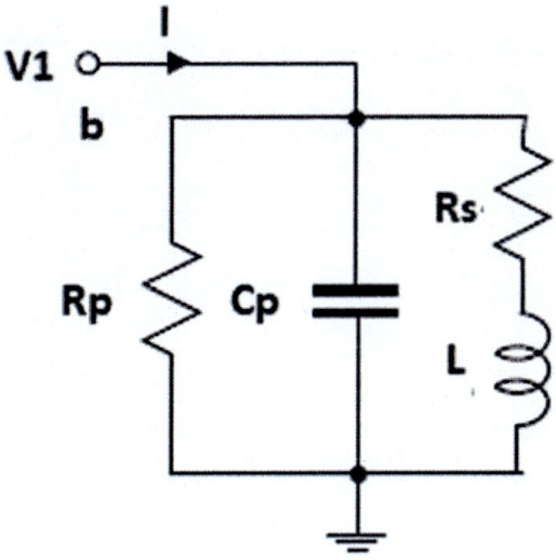

$$Y = sC2 + G2 + \frac{1}{s\left(\frac{C1}{Gm1Gm2}\right) + \frac{G1}{Gm1Gm2}} \tag{1}$$

From Eq. 1, it can be seen that the values of components are as depicted by Eqs. 2–5.

$$Rp = \frac{1}{G2} \tag{2}$$

$$Cp = C2 \tag{3}$$

$$Rs = \frac{G1}{Gm1Gm2} \tag{4}$$

$$L = \frac{C1}{Gm1Gm2} \tag{5}$$

The various topologies of active inductors discussed in this paper are as follows:

3 Basic Active Inductor

Figure 13 shows the circuit schematic of Basic Active Inductors based on Gyrator-C topology. It is known fact that the model of field effect transistor is based on the Trans-conductors. In the Gyrator-C topology, two back to back amplifiers were connected to form a network. The straightforward amplifiers can be the transistors itself [4].

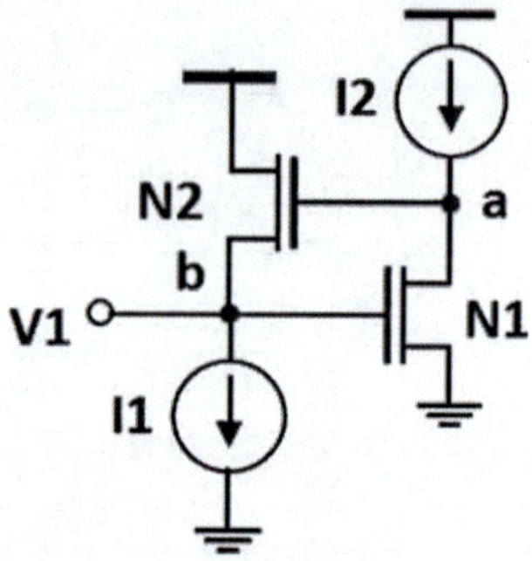

Fig. 13 Schematic of basic active inductor

Table 1 Simulation results of basic active inductors

$I1$ (μA)	100	200	500
L (nH)	16.38	13.67	9.46
Q-factor	2.58	2.79	3.10

The simulations have been performed by fixing the value of current source $I2 =$ 500 μA and varying the value of $I1$ from 100, 200 and 500 μA and at 2.4 GHz input source frequency. The simulation results are shown in Table 1.

Figure 14 shows the plot of Q-Factor of the basic active inductor versus the frequency.

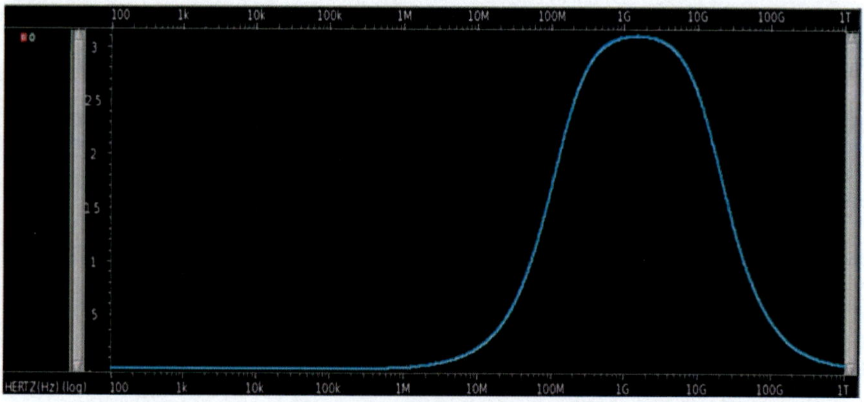

Fig. 14 Plot of Q-factor versus frequency for basic active inductor

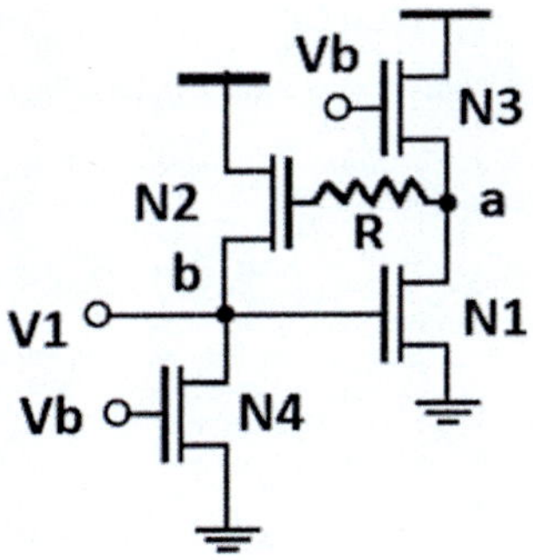

Fig. 15 Schematic of resistive feedback active inductor

Table 2 Simulation results of resistive feedback active inductor

R (Ω)	500	1000	1500	2000
L (nH)	4.79	5.23	6.71	7.89
Q-factor	6.45	13.69	18.72	21.95

4 Resistive Feedback Active Inductor

Figure 15 shows the circuit schematic of Resistive feedback Active Inductors based on Gyrator-C topology. In the previous topology it has been observed that the quality factor was not very large, so in order to increase the quality factor along with keeping in mind the low power consumption criteria as well, another topology has been developed which has a resistor as a feedback element [4, 12].

The simulation of this active inductor topology is performed at varying *R* values of 500 Ω, 1 kΩ, 1.5 kΩ and 2 kΩ. All simulations have been performed at an input frequency of 2.4 GHz. Table 2 shows the simulation results for this topology.

Figure 16 represents the graph of *Q*-Factor of the resistive feedback active inductor vs the frequency.

5 Regulated Cascade Active Inductor

Figure 17 shows the schematic for the regulated cascade active inductor. This topology is too based upon the principle of Gyrator-C network, but along with that it has got some modifications in order to improve our main parameter of interest, i.e., the quality factor and the power consumption. The transistors N4 and N5 acts as the constant current source. Transistors N1 and N3 are normal transistors which are present in a Gyrator-C network. Transistor N2 is placed at the drain terminal of the transistor N3 because to reduce the transconductance at the node 'b' because this high transconductance node was earlier responsible in restraining the quality factor

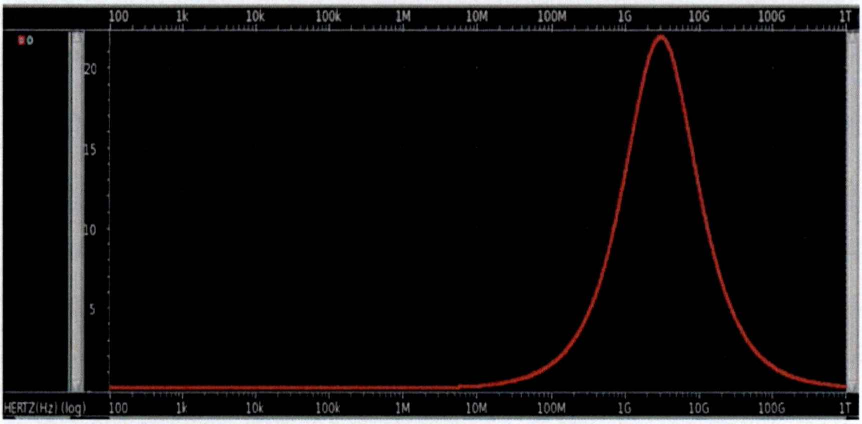

Fig. 16 Plot of *Q*-factor of resistive feedback topology versus frequency

Fig. 17 Schematic of a regulated cascode active inductor

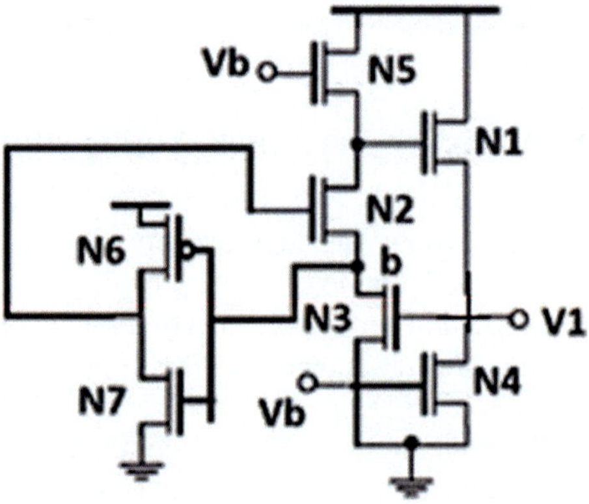

[13]. Now this cascade structure of N3 and N2 lowers the transconductance at node 'b'. Beside all these the main component which is responsible for the enhancement of the quality factor is the presence of the negative amplifier consisting of pFET (N6) and nFET (N7), which acts as a feedback component in this schematic.

From the facsimile conclusion as shown in Table 3, it is concluded that at $W6 =$ 68 nm and $W7 = 68$ nm, maximum quality factor of 22.09 is obtained at a inductance

Table 3 Simulation results of regulated cascode active inductors

*W*6 (nm)	34	34	34	68
*W*7 (nm)	34	52	68	68
L (nH)	6.36	7.41	7.96	7.53
Q-factor	11.46	14.31	18.64	22.09

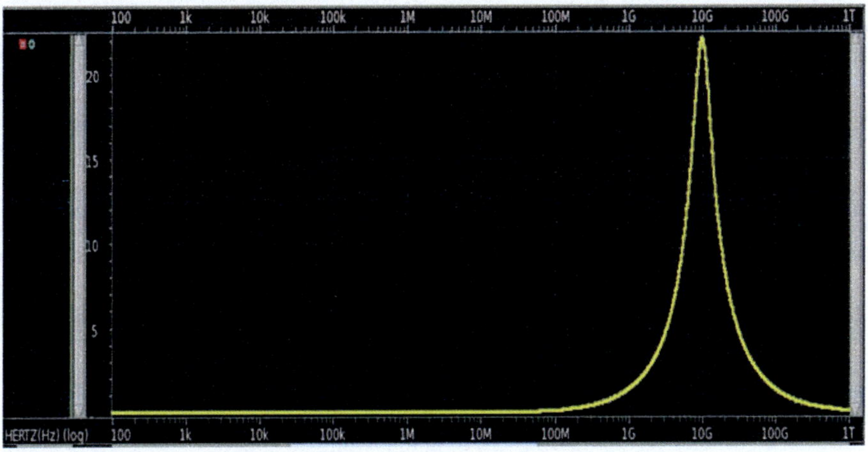

Fig. 18 Plot of *Q*-factor versus frequency for regulated cascode topology

value of 7.536 nH. The resonant frequency is about 9.8 GHz. This simulation was done at an input frequency of 2.4 GHz.

Figure 18 shows the plot of *Q*-Factor of the basic active inductor vs the frequency.

6 Weng-Kuo Active Inductor

Another most important and effective active inductor is the Weng-Kuo active inductor. The schematic of Weng-Kuo active inductor is shown in Fig. 19. This topology finds many similarity with the regulated cascade topology but the only difference between the regulated cascade topology and the Weng-Kuo topology is the presence of extra current source which is represented by *I*3 in the schematic x.18. The main role of this extra current source is to reduce the transconductance at

Fig. 19 Schematic of a Weng-Kuo active inductor

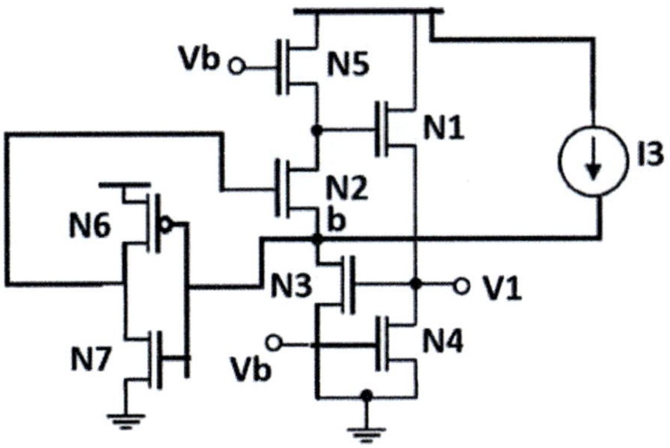

Table 4 Simulation results of Weng-Kuo active inductor

Current I3 (μA)	100	200	500
Q-factor	14.97	19.39	26.43
L (nH)	7.32	8.08	8.64

node 'b' which will further increase the value of the main parameter of interest, i.e., *Q*-Factor [4, 13].

For calculating the results for this topology, the width of *N*6 and *N*7, i.e., *W*6 and *W*7, respectively, are both fixed at value of 68 nm as they were giving very high value of quality factor. Now upon this improved value, one more current source has been added in this topology and in this simulation the value of that current source *I*3 is varied. From the simulation results shown in Table 4, it can be observed that the maximum quality factor is 26.43 at a current value of 500 μA with an inductance value of 8.64. The resonant frequency of the Weng-Kuo active inductor is about 16.32 GHz.

Figure 20 shows the plot of *Q*-Factor of the basic active inductor vs the frequency.

Thus, the conclusion is that despite of having slightly less quality factor value, **Regulated Active Inductor** is most suitable to be used in a circuit despite of the fact that its *Q*-factor plot is less narrow than the Weng-Kuo topology but the most important reason is its less power consumption [11].

Table 5 shows the comparison of *Q*-factor results obtained in this work which uses FinFET technology with the reference [11] which uses FINFET technology.

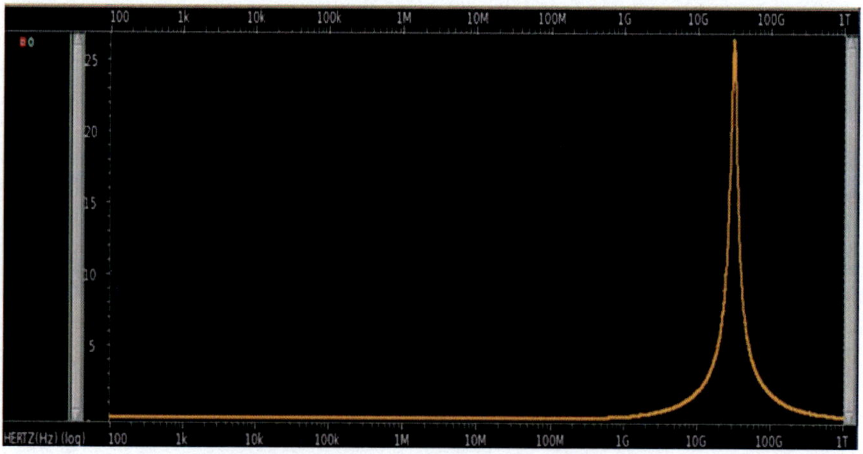

Fig. 20 Plot of *Q*-factor versus frequency for the Weng-Kuo active inductor

Table 5 Comparison between Q-factor obtained in this work and Ref. [11]

Active inductor topology	Q-factor in our work (FinFET technology)	Q-factor in Ref. [11] (FINFET technology)
Basic	3.10	2.50
Resistive feedback	21.95	18.30
Regulated cascode	22.09	17.0
Weng-Kuo	26.43	22.0

7 DC–DC Converters with Active Inductors

All the topologies of the five DC–DC converters have been implemented by replacing the spiral inductors with active inductor. Since the different architectures of active inductor has already been simulated and observed that regulated Cascade implementation has the highest advantage in terms of performance like the high quality factor values and less power.

Consumption, we have replaced the passive inductors in the DC–DC converter with this topology of the active inductor. There is a reasonable improvement in the performance of the DC–DC converters by substituting the passive inductors with this active inductor. Since the quality factor is high, the current delivering capability of the inductors have increased by a significant amount which results in a higher output voltage and moreover the quality of the output has improved by reducing the ripple so that a more robust device can be obtained and will be more reliable. There are some other benefits also with this substitution of the passive inductors with active inductor like the area and power consumption of the overall device which will be fabricated since the conventional passive inductors consume a huge area and require a large amount of power for their operation. The above said benefit of output is supported with graph for all the five topologies of the DC–DC converter below:

The Plot for Output Voltage as a function of time for Boost DC–DC Converter is shown in Fig. 21.

The Plot for Output Voltage as a function of time for Buck DC–DC Converter is shown in Fig. 22.

The Plot for Output Voltage as a function of time for Buck-Boost DC–DC Converter is shown in Fig. 23.

The Plot for Output Voltage as a function of time for CUK DC–DC Converter is shown in Fig. 24.

The Plot for Output Voltage as a function of time for SEPIC DC–DC Converter is shown in Fig. 25 (Table 6).

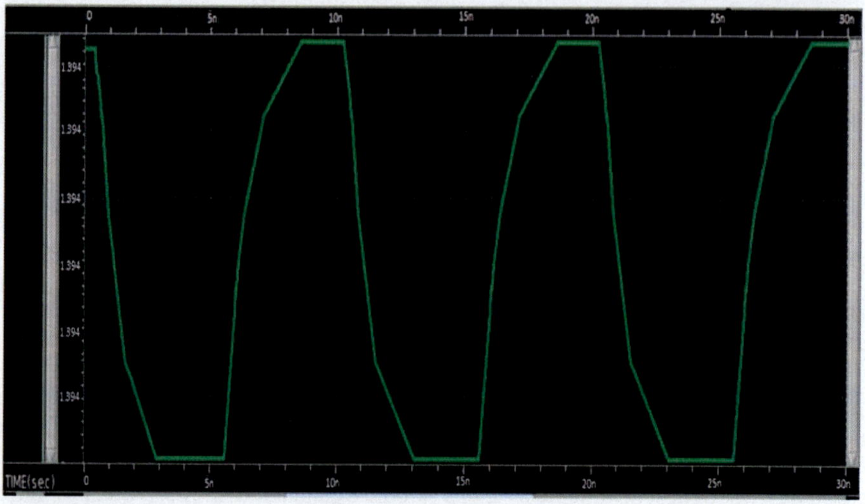

Fig. 21 Plot for Boost converter using AI

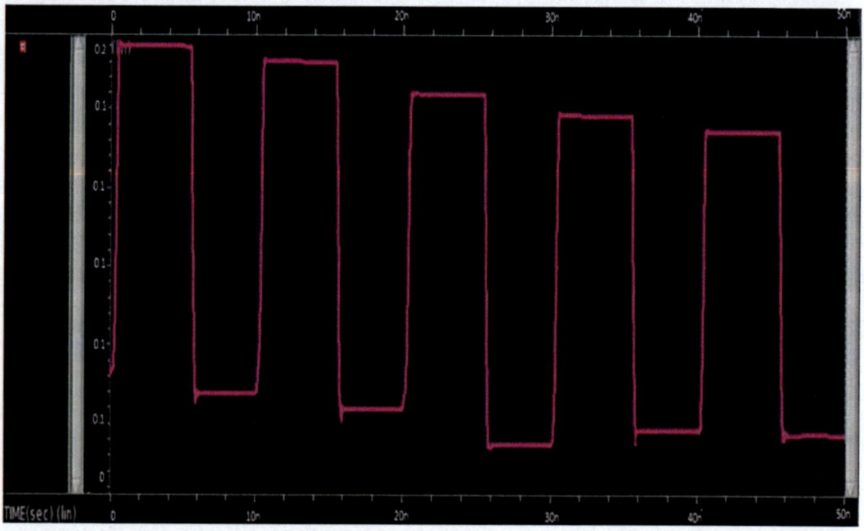

Fig. 22 Plot for Buck converter using AI

8 Result and Conclusion

In this work various topologies of DC converters have been implemented; a comparative study between their implementation with spiral inductor versus active inductor is done. Among the Active inductors implemented the regulated cascade based AI proved to be most efficient. The plots and tables above clearly are in line with the

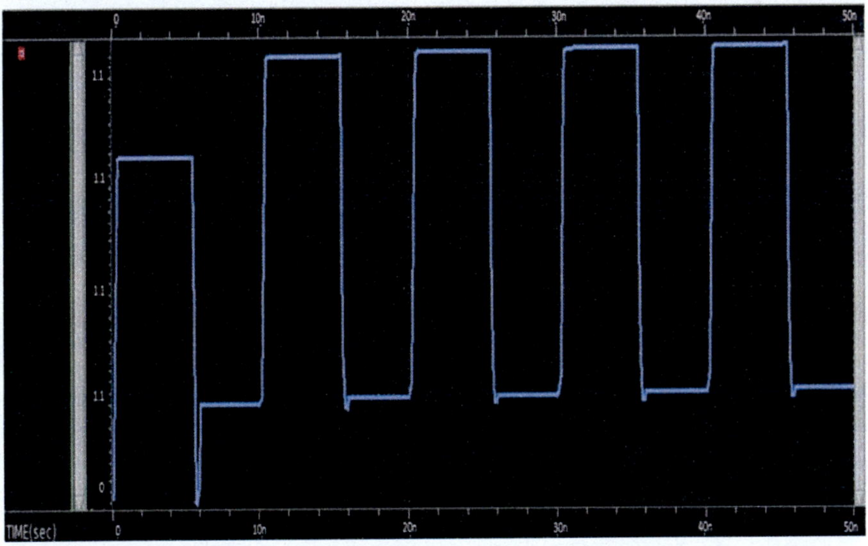

Fig. 23 Plot for Buck-Boost converter using AI

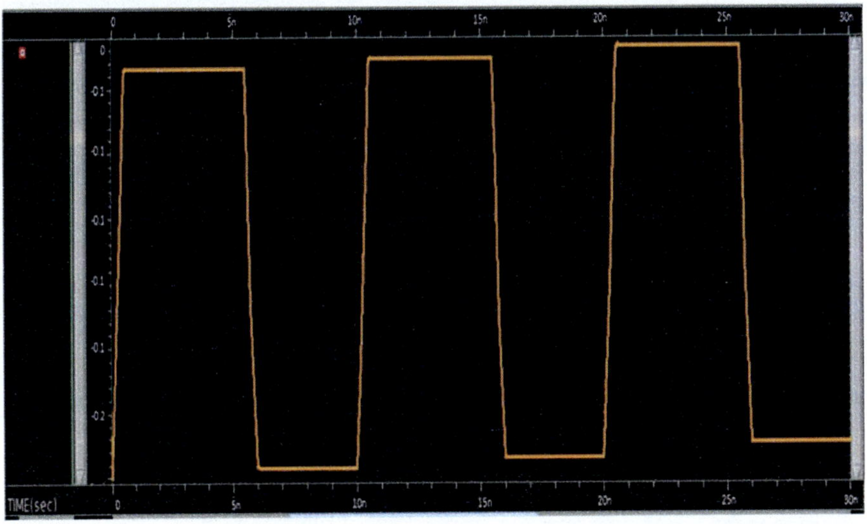

Fig. 24 Plot for Cuk converter using AI

theoretical advantages stated. The incorporation of AI in DC–DC converter brings in several advantages which could be listed as:

1. Reduction in the area required for the implementation of the design since MOS based inductor requires less area as compared to spiral inductors

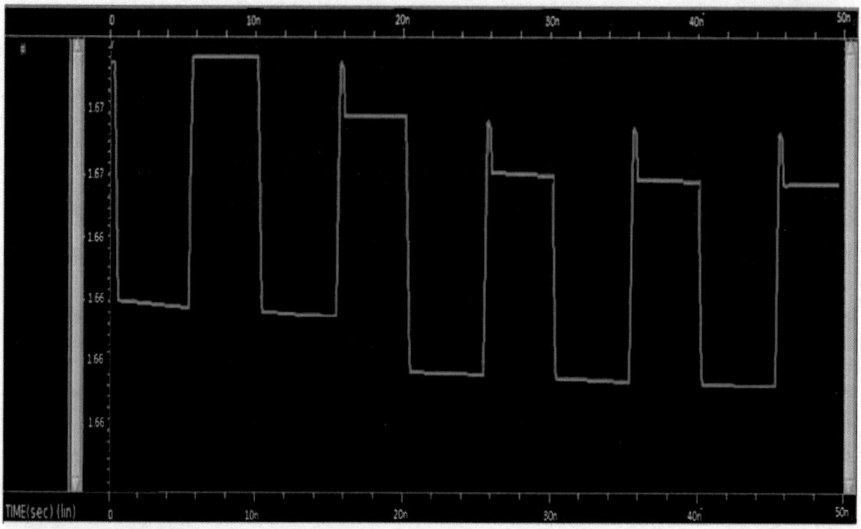

Fig. 25 Plot for SEPIC converter using AI

Table 6 Comparison of output voltage for different topologies with and without AI

DC–DC converter topology	Without AI (voltage fluctuation) [V]	With AI (voltage fluctuation) [V]
Boost	0.7–1.3	1.3941–1.3948
Buck	0–0.7	0.12–0.21
Buck-Boost	0–0.95	1.09–1.17
CUK	0–(−3.0)	(−0.05)–(−0.25)
SEPIC	0.1–1.65	1.662–1.674

2. The quality factor of an AI is tunable which provides the design a unique adaptability
3. The power loss in the circuit also reduces since the silicon based inductor is less leaky as compared to a spiral bound inductive coil.

Acknowledgements The authors are thankful to Department of Electronics and Communication Engineering BIT Mesra, Ranchi for providing the laboratory facilities. They are also thankful to their Vice-Chancellor, Dr. M. K. Mishra and their Head of the Department, Dr. S Pal for their constant inspiration and encouragement.

References

1. Yodprasit U, Ngarmnil J (2000) Q-enhancing technique for RF CMOS active inductor. In: 2000 IEEE international symposium on circuits and systems. Emerging technologies for the 21st century. Proceedings (IEEE Cat No. 00CH36353), vol 5, Geneva, 2000, pp 589–592
2. Shen ZJ, Okada DN, Lin F, Anderson S, Xu C (2006) Lateral power FINFET for megahertz-frequency, high-density DC/DC converters. IEEE Trans Power Electron 21(1):11–17. https://doi.org/10.1109/TPEL.2005.861111
3. Wu T, Lai Y, Hung J, Chen Y (2008) Boost converter with coupled inductors and buck–boost type of active clamp. IEEE Trans Ind Electron 55(1):154–162. https://doi.org/10.1109/TIE.2007.903925
4. Thanachayanont A, Payne A (1996) VHF CMOS integrated active inductor. Electron Lett 32(11):999–1000. https://doi.org/10.1049/el:19960669
5. Costinett D, Nguyen H, Zane R, Maksimovic D (2011) GaN-FET based dual active bridge DC-DC converter. In: 2011 twenty-sixth annual ieee applied power electronics conference and exposition (APEC), Fort Worth, TX, pp 1425–1432. https://doi.org/10.1109/apec.2011.5744779
6. Hirakawa M, Nagano M, Watanabe Y, Andoh K, Nakatomi S, Hashino S (2009) High power density DC/DC converter using the close-coupled inductors. In: 2009 IEEE energy conversion congress and exposition, San Jose, CA, pp 1760–1767. 10.1109/ECCE.2009.5316389
7. Tombak A, Baeten RJ, Jorgenson JD, Dening DC (2008) Integration of a cellular handset power amplifier and a DC/DC converter in a Silicon-On-Insulator (SOI) technology. In: 2008 IEEE radio frequency integrated circuits symposium, Atlanta, GA, pp 413–416. 10.1109/RFIC.2008.4561466
8. Wens M, Steyaert M (2008) A fully-integrated 0.18 μm CMOS DC-DC step-down converter, using a bondwire spiral inductor. In: 2008 IEEE custom integrated circuits conference, San Jose, CA, pp 17–20. 10.1109/CICC.2008.4672009
9. Harada K, Sakamoto H, Shoyaraa M (1987) Phase controlled DC-AC converter with high frequency switching, In: 1987 IEEE power electronics specialists conference, Blacksburg, VA, USA, pp 13–19. 10.1109/PESC.1987.7077158
10. Musunuri S, Chapman PL (2005) Optimization of CMOS transistors for low power DC-DC converters. In: 2005 IEEE 36th power electronics specialists conference, Recife, pp 2151–2157. 10.1109/PESC.2005.1581930
11. Suresh DR (2014) Design of CMOS active inductors and their use in tuned narrowband and wideband extension low noise amplifier. In: 2014 Escola Technica Superior d'Enginyeria Industrial de Barcelona
12. Hara S, Tokumitsu T, Aikawa M (1989) Lossless broad-band monolithic microwave active inductors. IEEE Trans Microwave Theor Techniques 37(12):1979–1984. https://doi.org/10.1109/22.44111
13. Maksimovic D (1991) A MOS gate drive with resonant transitions. In: PESC '91 record 22nd annual ieee power electronics specialists conference, Cambridge, MA, USA, pp 527–532. https://doi.org/10.1109/pesc.1991.162725

Analysis of Dispersion Compensation Techniques Used in Optical Fiber Communication

Nidhi Jha and Soumya Sidhishwari

Abstract In optical fiber, communication signal can be transmitted through one place to another by optical fiber in the form of pulses. When signal or light pulse is transmitted by the optical fiber, it suffers from some loss due to the attenuation and dispersion. Due to this effect, the system bandwidth decreases. Recent advancement in telecommunication services demands higher bandwidth which can be achieved by lowering the dispersion effect, i.e., pulse spreading. By using OptiSystem software and MATLAB, variant types of dispersion compensation techniques used in optical fiber communication have been analyzed. In our work, we are using OptiSystem, version 13.0 (Optiwave) for performance analysis of different compensation techniques. We have analyzed dispersion compensation using dispersion compensating fiber (DCF) and fiber Bragg grating (FBG) techniques for different fiber lengths. The parameters such as bit error rate (BER), quality factor (QF), and eye diagram have been investigated at the receiver end.

Keywords Optical fiber · Dispersion · Dispersion compensation techniques · Bit rate · Quality factor · Eye diagram

1 Introduction

Due to several advantages like enormous bandwidth, small size and weight, electrical isolation, immunity to interference and crosstalk, signal preservation, low transmission loss, elasticity, etc., optical fibers have its great application in communication model, sensor, etc. In optical fiber, communication signal can be transmitted through one place to another by optical fiber in the form of pulses. When signal or information is transmitted through the optical fiber, it suffers from some loss due to the

N. Jha · S. Sidhishwari (✉)
Department of ECE, Birla Institute of Technology Mesra, Ranchi, Jharkhand 835215, India
e-mail: soumya@bitmesra.ac.in

N. Jha
e-mail: nidhi.rs11@gmail.com

V. Nath and J. K. Mandal (eds.), *Proceedings of the Fourth International Conference on Microelectronics, Computing and Communication Systems*, Lecture Notes in Electrical Engineering 673, https://doi.org/10.1007/978-981-15-5546-6_74

attenuation and dispersion. Dispersion is defined as pulse widening of the optical signal as it accesses along a fiber. Pulse widening causes the neighboring pulses to overlap and results in inter-symbol interference. Due to this result, the system bandwidth decreases. Recent advancement in telecommunication services demands higher bandwidth which can be achieved by lowering the dispersion effect, i.e., pulse spreading. There are different techniques to compensate the dispersion effects along the length of the optical fiber. Different methods of dispersion compensation techniques in optical fiber have been analyzed by using OptiSystem software and MATLAB.

Several works have been done for analyzing different compensation techniques. B. Prasad has discussed fiber Bragg grating as a dispersion compensator used in an optical transmission model using OptiSystem software. From the simulation result, he concluded that the fiber length and the input power are directly proportional to the signal power [1]. Md. Jahidul Islam has proposed MATLAB design work to compensate dispersion up to 300 km fiber length. They used FBG technique for dispersion compensation [2]. Md. Yusoff has investigated the signal performances of pre-, post-, and symmetrical compensation using chirped fiber Bragg grating for L-band dispersion compensator. The author concluded, the signal performance at 10 Gbps data rate after spreading through a 90 km transmission fiber has been highlighted [3]. R. S. Asha has compared the performance of dispersion compensation with a linear chirped FBG, except for dispersion at higher input powers with FBG and DCF [4]. B. J. Eggleton has used fiber Bragg grating for dispersion compensation in transmission. He has demonstrated for the first time that an FBG in transmission can be used to compensate dispersion [5]. Ashwani Sharma has proposed a method to compensate dispersion using different techniques which happen during transmission of signals through a distance of 120 km at 100 Gbps [6]. Kishore Kumar Panda has proposed simulation and performance analysis of dispersion compensation using DCF in different schemes for 40 Gbps DWDM network. The author concluded that the symmetrical compensation gives a better performance than other schemes [7]. There are many review papers on dispersion compensation methods [8, 9]. They discussed DCF, FBG, EDC, and digital filter. Simulation of transmission model to compensate dispersion in an optical fiber by chirp gratings has been proposed by several authors [10, 11]. M. Sumetsky has proposed an overview of FBGs fabrication principles and applications with emphasis on the chirped FBG used for dispersion compensation in high-speed optical communication systems [12]. Md. Talha Hassan has proposed a comparative study of different dispersion compensation techniques in long-haul communication [13].

In our work, we are using OptiSystem, version 13.0 (Optiwave) for analyzing the performance of different dispersion compensation techniques. OptiSystem software can be used as a platform for designing and simulation of different optical fiber systems. We have analyzed the performance of dispersion compensation process using dispersion compensating fiber (DCF) and fiber Bragg grating (FBG) for different fiber lengths. The value of parameters such as bit error rate (BER), quality factor (QF), and eye height has been investigated at the receiver end. All the results have been analyzed using OptiSystem, version 13.0 (Optiwave) at bit rate (BR) 10 and 20 gigabits per second (Gbps) transmission systems using Mach–Zehnder modulator.

2 Theory

Dispersion

Dispersion effects broaden the pulse as it propagates along the fiber. After large distance, the pulses overlap (intersymbol interference-ISI) and become indistinguishable limiting the bandwidth. Generally, dispersion can be classified as [14].

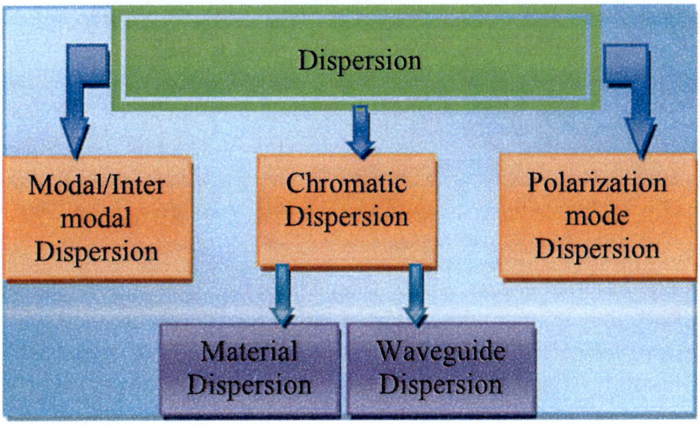

Intermodal dispersion

Intermodal dispersion is defined as the change of the group velocity for every mode at a single frequency. Hence, each wave will reach at the exit of the fiber at different times causing dispersion. It appears only in multimode fibers.

Chromatic dispersion

Chromatic or intra-modal dispersion happens in single-mode optical fiber, and it results from the finite spectral emission width of an optical source. The spectral emission width is the band of wavelength over which the source emits light. As it depends on wavelength, the signal distortion due to this increases. It is further divided into two types:

Material dispersion

It is caused by the variation of the refractive index of the core material as a function of wavelength or frequency of light. It is directly proportional to the frequency bandwidth of the transmitted pulse.

Waveguide dispersion

This arises when the modal propagation constant β is a function of the ratio between the core radius and the wavelength.

Polarization mode dispersion

It occurs because the light signal energy at a given wavelength in a single-mode fiber occupies orthogonal polarization states or modes. Each mode will travel at a slightly different velocity due to lack of perfectly symmetric and an isotropicity of the fiber giving rise to pulse spreading.

In optical fiber communication model, dispersion compensation before detecting the signal is an essential requirement to overcome the inter-symbol interference. It can be achieved by using different dispersion compensation methods such as dispersion compensating fiber (DCF), fiber Bragg grating (FBG), electronic dispersion compensation (EDC), digital filters, and optical phase conjugation technique (OPCT).

In our work, we are using DCF and FBG to analyze dispersion compensation with the help of OptiSystem software.

Dispersion compensating fiber: The component of DCF is more stable and does not easily get affected by temperature and bandwidth. So, it is a suitable way to compensate dispersion. Dispersion compensating fibers have a high negative dispersion. Therefore, it can be tuned to the transmission fiber having the positive dispersion coefficient such that the overall dispersion of the link becomes zero. This technique is used in long-haul WDM communication model. Dispersion compensating fiber has low insertion loss and higher performance. It has three compensation schemes as shown in Fig. 1a–c [6].

Pre-compensation scheme: When DCF is located before the single-mode fiber (SMF) to compensate the positive dispersion of the optical fiber, it is termed as pre-compensation scheme.

Post-compensation scheme: Post-compensation scheme is the method in which DCF is located after single-mode fiber (SMF) to compensate the positive dispersion of the optical fiber.

Symmetrical compensation scheme: If DCF is located before as well as after single-mode fiber (SMF) to compensate the positive dispersion of the optical fiber, then we have symmetrical compensation scheme.

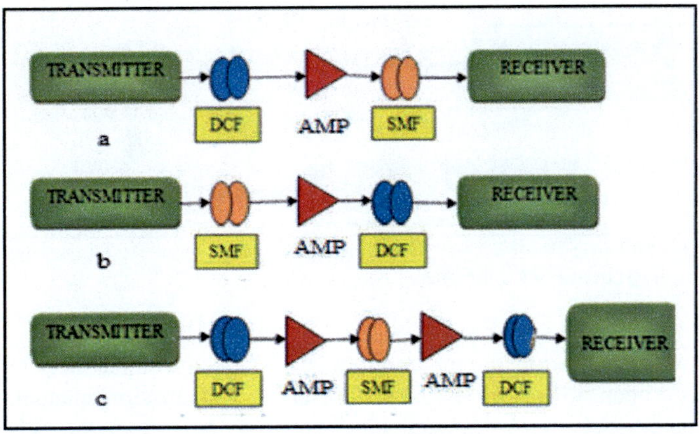

Fig. 1 a–c—Block diagram of dispersion compensation scheme of DCF technique **a** pre-compensation, **b** post-compensation, **c** symmetrical compensation

Dispersion compensation using fiber Bragg grating: FBG is the most important technology being used for dispersion compensation. It is most useful, reliable, versatile, practical, and attractive passive device in the field of optical fiber communications. It is also a Bragg reflector which creates reflection for a particular wavelength. The pulse which enters into the grating, depending upon the wavelength of pulses got reflected by grating. More distance will be traveled by the signal having larger wavelength in grating before getting reflected; on the other hand, less distance will be traveled by the signal having smaller wavelength in the grating before getting reflected. Hence, the pulse spread due to chromatic dispersion in a single-mode fiber can be reduced using FBG. The Bragg wavelength is expressed by the relationship shown in Eq. (1)

$$\lambda_{\text{Bragg}} = 2n_{\text{eff}}\Lambda \tag{1}$$

where n_{eff} represents refractive index of the grating in the fiber core and Λ is grating period.

Block diagram to compensate the dispersion using fiber Bragg grating is shown in Fig. 2.

2.1 Simulation of Dispersion Compensation Techniques

The simulation layout mainly consists of three sections like transmitter, transmission channel, and receiver. Transmitter consists of sequence generator, pulse generator, modulator, etc., receiver part consists of BER generator, eye diagram generator,

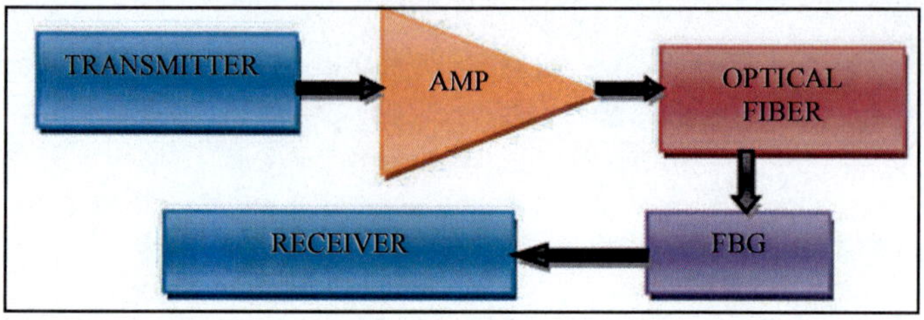

Fig. 2 Block diagram of FBG techniques

filter, etc., and erbium-doped fiber amplifier (EDFA), single-mode fiber (SMF), and dispersion compensating fiber (DCF) are present in transmission channel.

The transmitter section as shown in Fig. 3 consists of data source which generates pseudo-random bit sequence at the rate of 20 Gbps. This has amplitude 1 a.u. and width 0.5 bit. The bit sequence is fed to Gaussian pulse generator that produces electrical signals. This signal is modulated by using Mach–Zehnder modulator. Mach–Zehnder modulator is used to control the modulation of data pulses. The modulator is driven by a CW laser that has work frequency of 193.1 THz and 0 dBm power. The bit sequence is fed to Gaussian pulse generator that produces electrical signals.

This signal is modulated by using Mach–Zehnder modulator. Mach–Zehnder modulator is used to control the modulation of data pulses. The modulator is driven by a CW laser that has work frequency of 193.1 THz and 0 dBm power.

Transmission channel is shown in Fig. 4. Here, we can change length, wavelength, dispersion parameters, attenuation, etc.

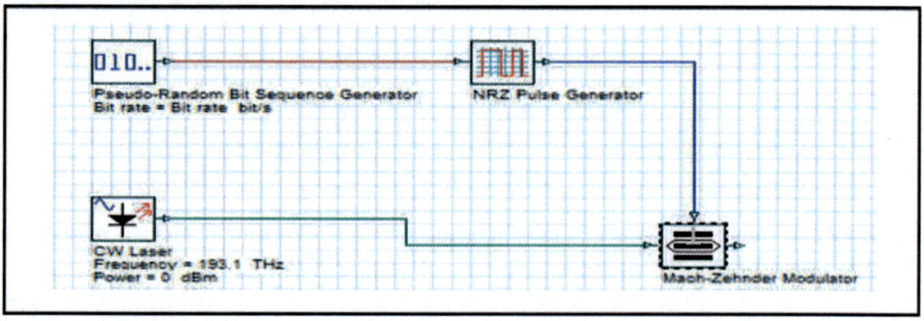

Fig. 3 Transmitter section

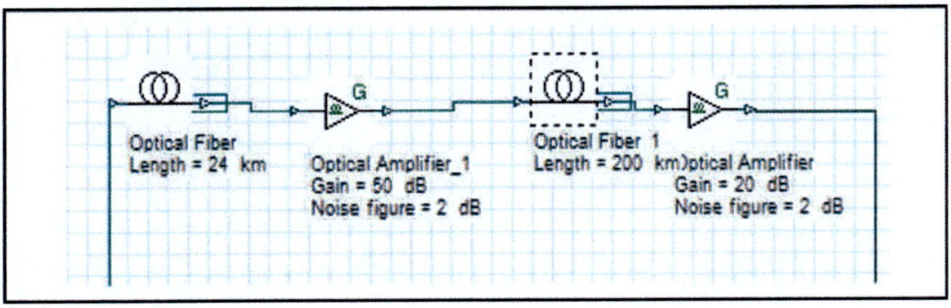

Fig. 4 Transmission channel

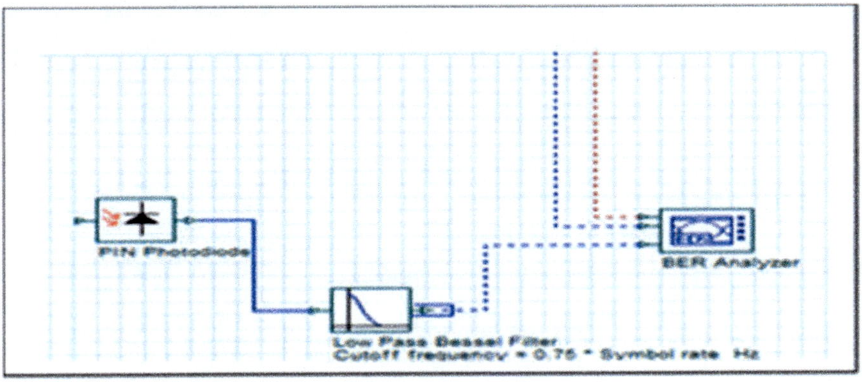

Fig. 5 Receiver section

For analyzing the result, receiver section is used as shown in Fig. 5. In this section, result of optical signal can be measured by visualizer library component like OTD visualizer, eye diagram analyzer, etc.

3 Simulation Setup

Figure 6a–c shows the simulation setup for different schemes of dispersion compensating fiber (DCF) technique.

Figure 7 is representing the simulation layout to compensate dispersion using fiber Bragg grating.

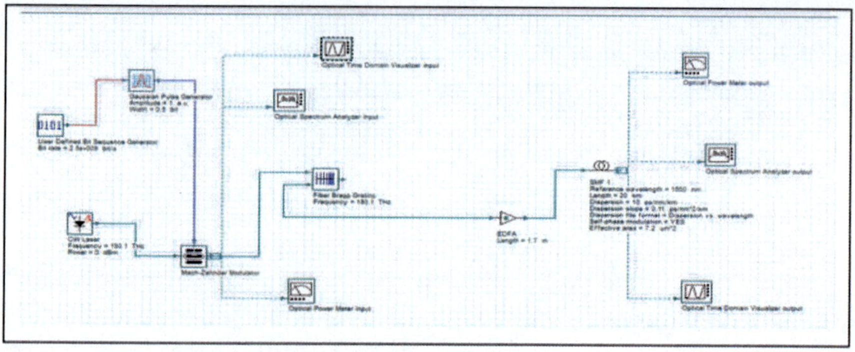

(a) Pre-Compensation Technique Layout

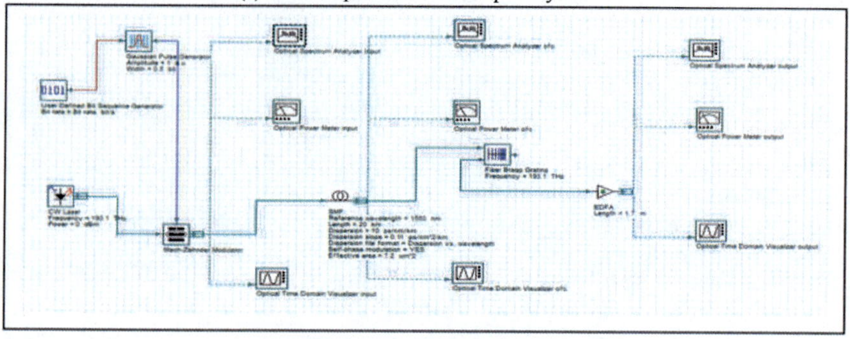

(b) Post-Compensation Technique Layout

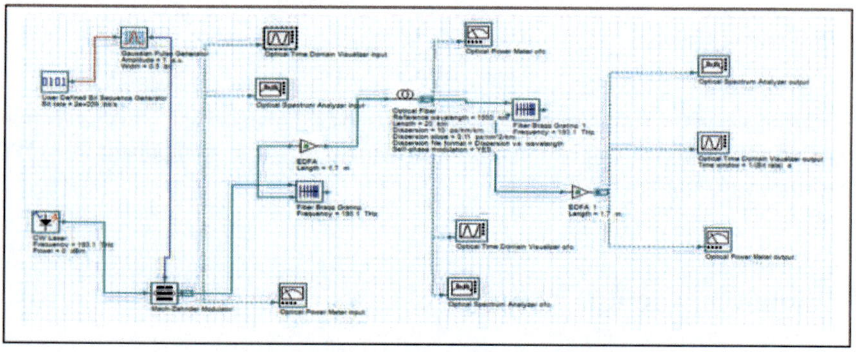

(c) Symmetrical- Compensation Technique Layout

Fig. 6 Design layout of DCF-**a** pre-compensation, **b** post-compensation, **c** symmetrical compensation

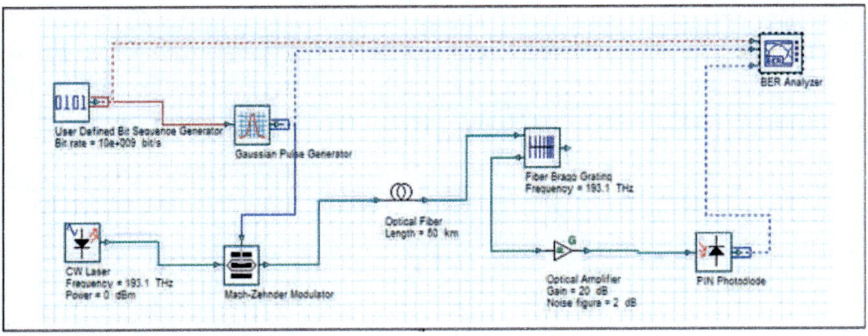

Fig. 7 Simulation layout of FBG

4 Result and Discussion

4.1 Simulation Result of DCF and Their Different Schemes

In our work, we have analyzed dispersion compensation methods (DCF and FBG method) of optical fiber communication system. We have used Gaussian pulse generator as input signal which creates a sequence of optical pulses modulated by an input digital signal. The frequency and power of input signal are set as 193.1 THz and 0 dBm, respectively. After transmission of input signal through transmission channel, some changes will occur which is received at BER analyzer. In receiver section, photodetector, filter, and BER analyzer are used. Different parameters like bit error rate, quality factor, and eye height from BER analyzer have been investigated for different fiber length. Figure 8a, b show the eye diagram result of DCF technique used for fiber length of 50 Km. Figures 9a–b, 10a–c, and 11a–c show the disper-

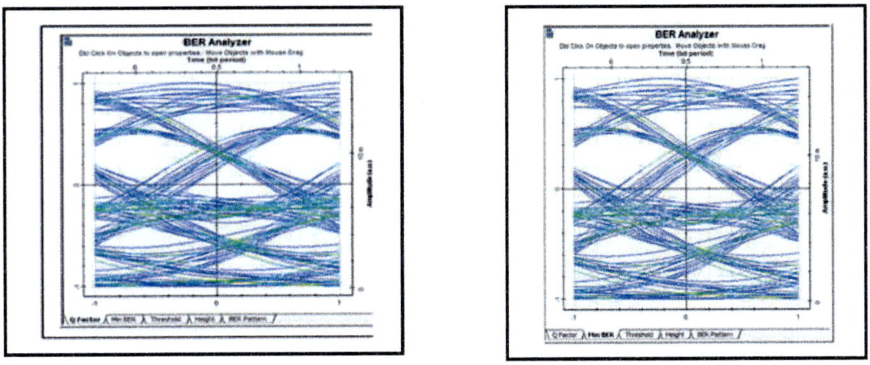

(a) (b)

Fig. 8 **a**, **b** Eye diagram result of DCF technique showing BER and quality factor for 50 km

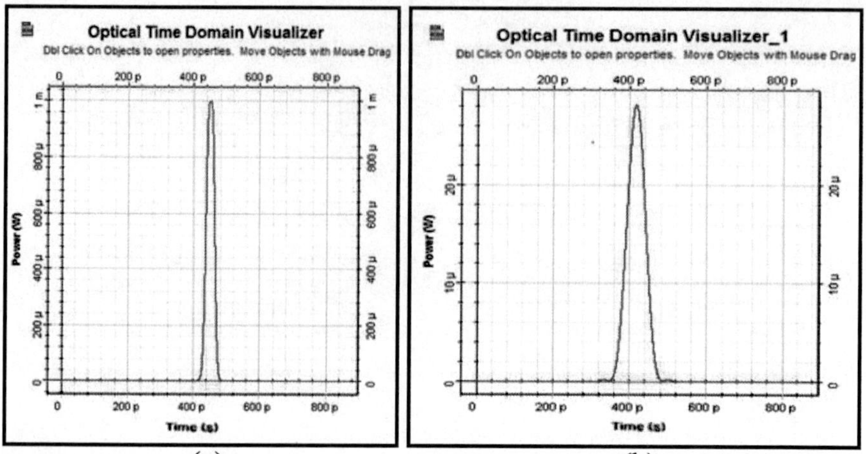

(a) (b)

Fig. 9 **a**, **b** Dispersion compensation of signal in DCF techniques at **a** input, **b** output level of pre-compensation

sion compensation of signals in pre, post and symmetrical dispersion compensation schemes.

4.2 Simulation Result of Dispersion Compensation Using Techniques FBG

Bit error rate (BER) is the number of bit errors per unit time and is defined as ratio of the numbers of bit errors divided by the total number of transmitted bits during a performance time interval. It is expressed by Eq. (2)

$$\text{BER} = \text{Errors/Total number of transmitted bits} \tag{2}$$

The relationship between quality factor and BER is shown in Eq. (3)

$$\text{BER} = \frac{1}{2}\text{erfc}\left(\frac{Q}{\sqrt{2}}\right) \tag{3}$$

where $\text{erfc}(x)$ is error function, which is expressed by Eq. (4)

$$\text{erfc}(x) = \frac{2}{\pi}\int_{x}^{\infty} \text{e}^{-y} \tag{4}$$

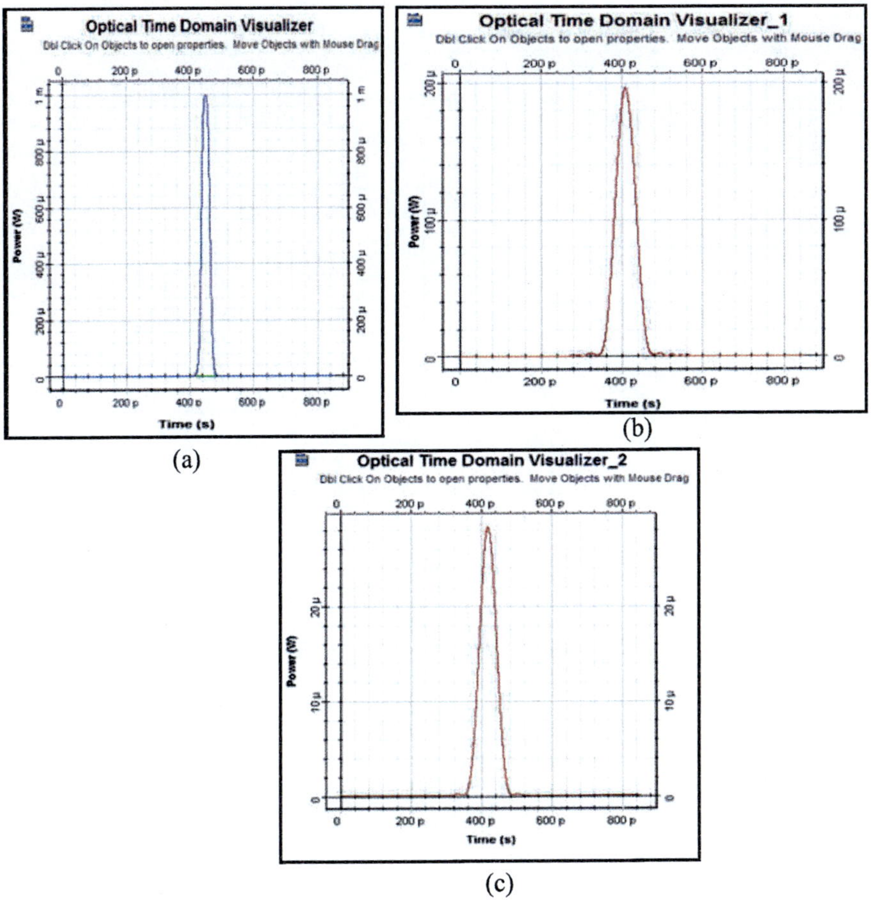

Fig. 10 **a–c** Dispersion compensation of signal in DCF techniques at **a** input, **b** mid, **c** output level of post-compensation

Therefore, from equation, we find that when the quality factor increases the BER decreases [14]. Eye diagram provides the overall communication efficiency and information about the noise. Eye height measurement determines eye closure due to noise. Figure 12a–b show the eye diagram result of FBG technique used for fiber length of 50 km.

Table 1 shows the comparison of pulse width for input signal, signal after OFC, and output signal from which we can determine how much dispersion compensation has been done. Table 2 shows the comparison of two different dispersion compensation techniques (DCF and FBG) for different fiber lengths (50, 100, 200 km).

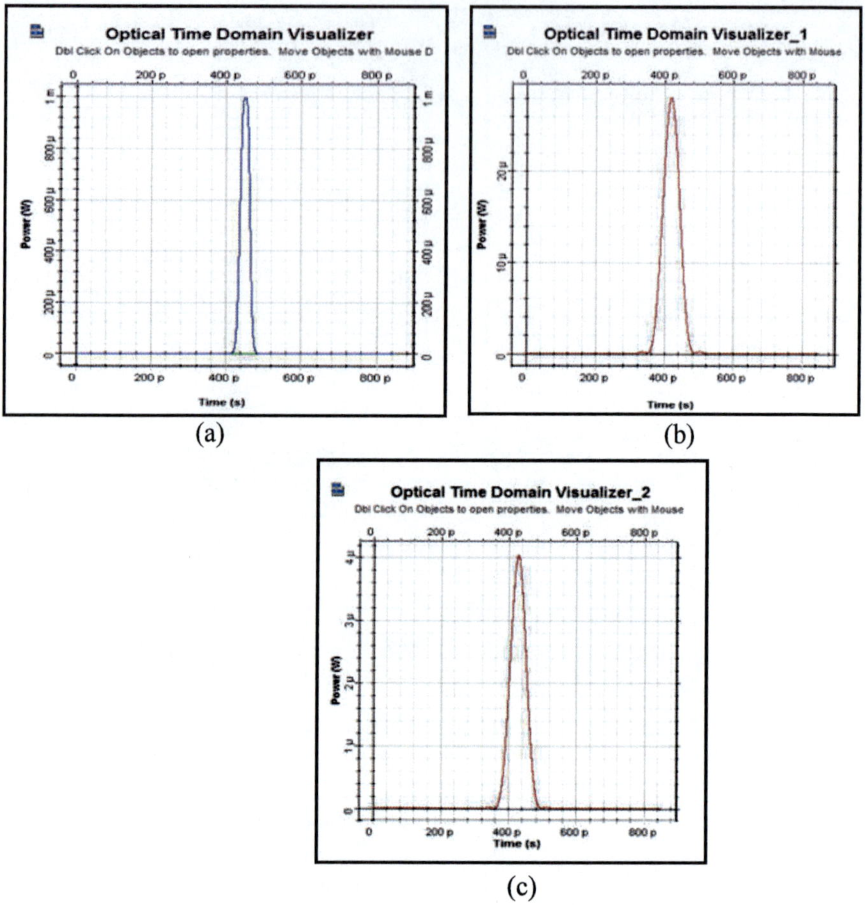

Fig. 11 **a–c** Dispersion compensation of signal in DCF techniques at **a** input, **b** mid, **c** output level of symmetrical compensation

5 Conclusion

Transmission of information through optical fiber is associated with dispersion, attenuation, etc. To minimize the dispersion DCF and FBG dispersion compensation techniques have been studied. This paper compares different techniques on the basis of different parameters like fiber length, quality factor, bit error rate, bit rate, etc. We have also analyzed the width of the input signal at different positions, i.e., before and after dispersion compensation. From comparison table, we have concluded that the symmetrical compensation of DCF techniques is more advantageous than other.

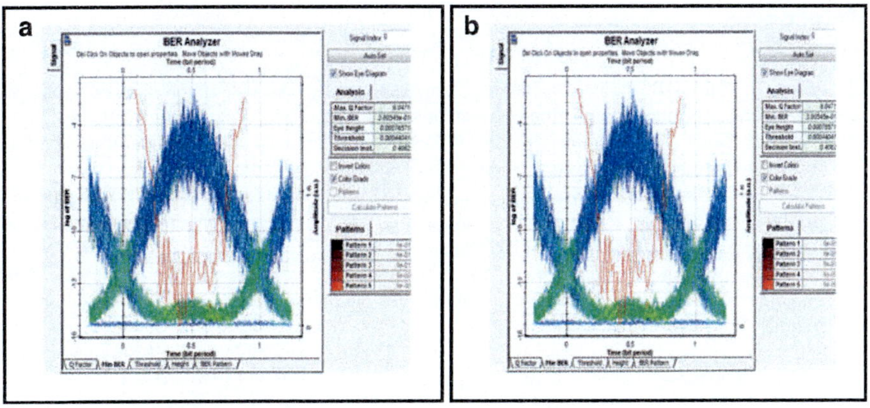

Fig. 12 **a**, **b** Eye diagram result of FBG technique showing BER and quality factor for 50 km

Table 1 Comparison of pulse width for different compensation

Techniques	Type of input signal	Width of input signal (ps)	Width of signal after OFC (ps)	Width of output signal with DCF (ps)
Pre	Gaussian pulse	1.55	–	2.07
Post	Gaussian pulse	1.55	2.07	2.05
Symmetrical	Gaussian pulse	1.55	1.7	1.65

Table 2 Comparison of different dispersion compensation techniques using Mach–Zehnder modulator

Techniques	Length (km)	Bit rate (Gbps)	BER	Quality factor	Eye height
FBG	50	10	3.80545e−016	8.04715	0.000765718
		20	1	0	0
	100	10	3.26972e−005	3.89825	−0.00412555
		20	1	0	0
	200	10	1	0	0
		20	1	0	0
DCF	50	10	0.00787682	2.4133	−0.00802285
		20	0.000731183	3.18071	0.393657
	100	10	0.00907506	2.31835	−0.0194126
		20	1	0	0
	200	10	0.000954338	3.10409	3.70926e−006
		20	0.00176702	2.91679	−3.06974e−006

References

1. Prasad B, Mallick B, Parida AK (2014) Fiber Bragg grating as a dispersion compensator in an optical transmission system using optisystem software. Int Res J Eng Technol 2:9–14
2. Islam MJ, Islam MS, Rahman MM (2012) Dispersion compensation in optical fiber communication using fiber Bragg grating. Glob J Res Eng 12:25–29
3. Yusoff NM, Mohd Sharif N, Sulaiman AH, Mahdi MA (2015) Investigation on the signal performances of pre-, post- and inline compensation of L-band dispersion compensator using chirped fiber Bragg grating. IEEE 1:107–110 (Digests 1st International Conference on TAFGEN)
4. Asha RS, Jayasree VK, Mhatli S (2018) Fiber dispersion compensation in single side band optical communication system using ideal fiber Bragg grating and chirped fiber Bragg grating without DCF. Opt Soc India 1–6
5. Eggleton BJ, Stephens T, Krug PA, Dhosi G, Brodzeli Z, Ouellette F (1996) Dispersion compensation using a fibre grating in transmission. Electron Lett 32:1610–1611
6. Sharma A, Singh I, Bhattacharya S, Sharma S (2019) Performance comparison of DCF and FBG as dispersion compensation techniques at 100 Gbps over 120 km using SMF. Springer, Berlin, pp 435–448
7. Panda KK, Mishra P, Shukla A, Nayak SP (2017) Simulation and performance analysis of dispersion compensation using DCF in pre, post and symmetrical arrangement for 40GBPS DWDM network. Int Res J Eng Technol (IRJET) 4:225–229
8. Singh M (2015) Different dispersion compensation techniques in fiber optic communication system: a survey. Int J Advanced Res Electron Commun Eng (IJARECE) 4:2236–2240
9. Kaur K, Kaur B (2016) Dispersion compensation techniques: a review. Int J Eng Sci 20:58–61 (International Conference on Innovative Trends in Electronics Engineering)
10. Mohammadi SO, Mozaffari S, Shahidi MM (2011) Simulation of a transmission system to compensate dispersion in an optical fiber by chirp gratings. Int J Phys Sci 6(32):7354–7360
11. Khairi K, Lambak Z, Samsuri NM, Hamzah Z, Hann FK (2011, December) Investigation on the performance of pre- and post-compensation using multi-channel CFBG dispersion compensators. In: International RF and microwave conference (IEEE), pp 254–255
12. Sumetsky M, Eggleton BJ (2005) Fiber Bragg gratings for dispersion compensation in optical communication systems. Optical Fiber Commu Rep 2:256–278
13. Talha M, Zaman HU, Salam KMA (2015, November) A comparative study of different dispersion compensation techniques in long haul communication. In: International conference on computer & information engineering, IEEE, pp 58–61
14. Agrawal GP (2002) Fiber Optic Communication System, 3rd ed

MBFF for Power Reduction of an SoC Partition

V. Anandi, G. Shanmathi, and Muthuswamy Ramesh

Abstract Physical design is process of transforming RTL netlist into a layout which is manufacturable [GDS or GDSII]. An efficient physical design flow is typically divided into five major steps: floorplanning, placement, clock tree synthesis, routing, and timing closure. Power optimization is always one of the most important design objectives in modern nanometer IC design. System clock signal in electronics product consumes the important part of dynamic power, among them 70% is spent by clock buffers. This critical problem can be optimized using a technique namely clock gating. Recent studies have shown that applying MBFF is an effective means in reducing clock network power. This paper uses MBFF technique to optimize dynamic power and later clock gating on the same design to achieve better results. Experimental results on the technology used indicate that MBFF is very effective and efficient method in deep sub-micro-design to reduce power, save area, and improve routing of clock tree. Amount of power being reduced is ~5 to 15% when the power is between ~25 and 75 mW.

Keywords MBFF · Dynamic power · Power reduction · Low power

V. Anandi (✉)
Department of Electronics and Communication Engineering, Ramaiah Institute of Technology, Bangalore, India
e-mail: anandi.v@msrit.edu

G. Shanmathi
Management School, Presidency University, Bangalore, India
e-mail: shanmathig@gmail.com

M. Ramesh
ECE Department, MSRIT, Bangalore, India
e-mail: ramandsur@gmail.com

V. Nath and J. K. Mandal (eds.), *Proceedings of the Fourth International Conference on Microelectronics, Computing and Communication Systems*, Lecture Notes in Electrical Engineering 673, https://doi.org/10.1007/978-981-15-5546-6_75

1 Introduction

Modern system on chips (SoCs) integrate huge number of transistors in the design and thus power reduction or thermal minimization plays a significant role during the design process.

Chips that require high performance are more prone to power-related issues. Dynamic power, leakage power, and short-circuit power are the three main types of power that are measured during the design process. Dynamic power is the largest among all and is dependent on square of the voltage, clock frequency, capacitance, and switching activity factor. Clock circuitry contributes to a large amount of dynamic power due to the switching activity of the clock [1].

Power consumption has been increasing due to the following factors:

- Gate density increasing rapidly
- Supply voltage and interconnect trace length is reduced at a slower rate.
- Leakage power of the transistors [2].

There are many techniques that are commonly used to reduce power consumption of a SoC. Buffer sizing, clock gating, register sizing or banking, multivoltage design, and voltage scaling are the most commonly used techniques to reduce dynamic power. Power gating has been used recently to reduce leakage power of the design. Recently, multibit flip-flops (MBFF) have been used to reduce power and area [3].

Multibit flip-flop consists of two or more 1-bit flip-flops merged together sharing a common clock driver. Figure 1 shows how two 1-bit register cells can be replaced with a single 2-bit register cell. Initially, MBFF was implemented in the design phase of IC, than during physical synthesis of the design. MBFF can also be implemented in the placement stage of the physical design [4]. Apart from power reduction at cell level, MBFF has many more advantages:

- The number of clock sinks in gate-level netlist with MBFF is much lower which leads to a simpler clock network. The number of clock buffers in the clock tree is reduced thus saving the clock dynamic power.
- MBFF provides efficient clock synthesis due to reduced number of levels thereby reducing the clock tree synthesis (CTS) runtime, skew, and latency.
- Reducing the clock and scan nets helps in improving the routing congestion [5].

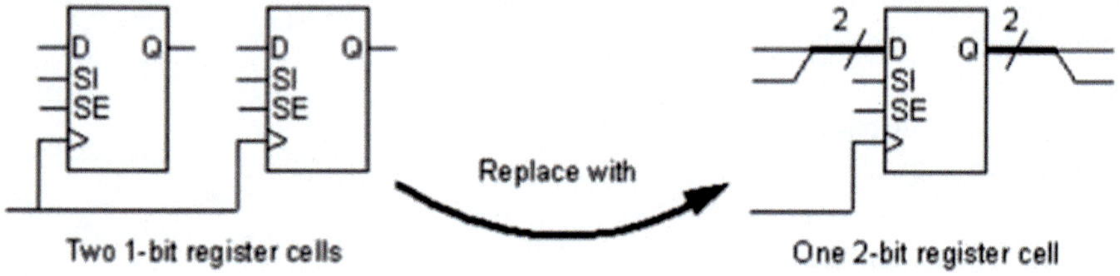

Fig. 1 Replacing two 1-bit registers with 2-bit MBFF

In this paper, we will experiment MBFF implementation in placement stage of physical design and in both synthesis and compare the results. Section 2 will contain the details of MBFF implementation. Section 3 will contain implementation of clock gating post MBFF implementation in order to achieve better results. Section 4 will contain the results and analysis of the experiments, and Sect. 5 will conclude this paper.

2 MBFF Implementation

Synthesis and physical implementation tools can organize multiple register bits into groups called "multibit components" in the RTL bus inference flow or "banks" in the placement-aware flow, and this process is also known as vectoring [4]. To support multibit register flows, the logic library and physical library must contain both single-bit and multibit library cells, and the multibit cells must meet certain requirements so that the tools can recognize them as functionally equivalent to a group of single-bit cells [6]. Design Compiler and IC Compiler of Synopsys were the tools used for implementation of MBFF [7].

(A) *Library requirements for implementing MBF*F:

(1) To perform mapping from single-bit to multibit registers, the tool checks for matching pin functions and naming conventions in the multibit register pins as shown in Fig. 2. For example, if a single-bit register has Q and QN

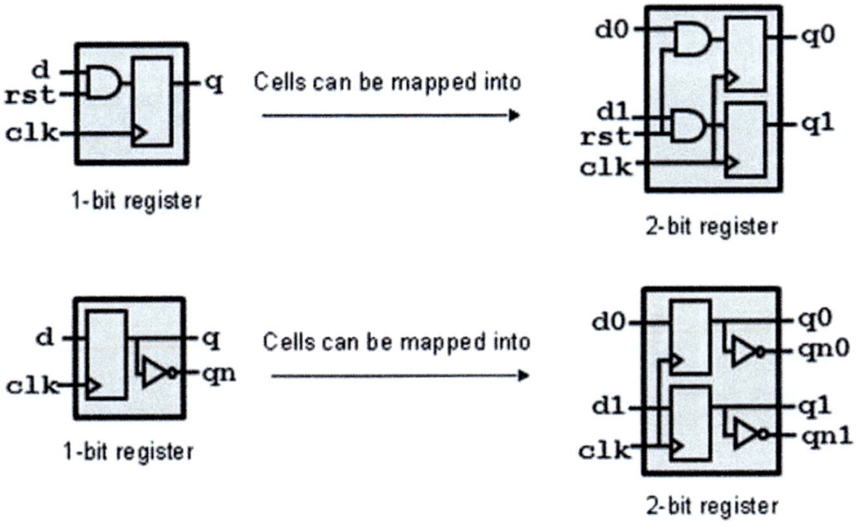

Fig. 2 Mapping of single-bit to multibit cells with the same I/O pins

outputs, the tool can replace this register only with a multibit register that also has Q and QN outputs for each output register bit.

(2) The Design Compiler Graphical and IC Compiler tools do not support multibit registers that have a control pin for an individual bit, as shown in Fig. 3.

(3) For scan cells, you can use a multibit register cell with single scan input and single scan output for the whole cell, with the scan bits daisy-chained inside the cell; or with one scan input and one scan output for each register bit [4]. Both types of multibit scan register configurations are supported. Dedicated scan output signals are also supported. Figure 4 shows how two single-bit scan cells can be mapped into either of two different compatible multibit scan cells, which have different scan configurations [8].

(B) *Implementation of MBFF*:
The following procedure was used to implement MBFF:

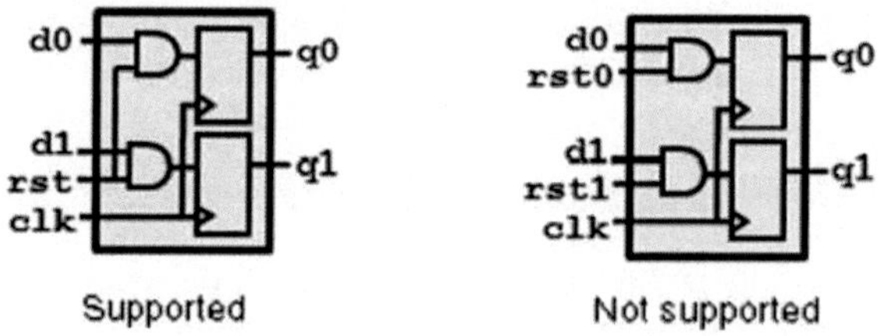

Fig. 3 Multibit registers with a control pin for an individual bit are not supported

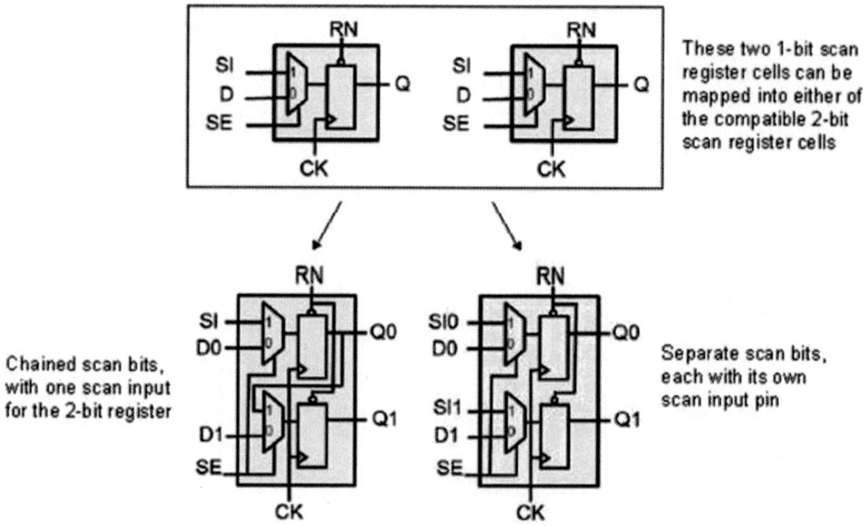

Fig. 4 Single-bit scan cell and compatible multibit scan cells

(1) Exclude the cells from getting swapped with its corresponding multibit version. The cell types include RTL specified cells, scan cells, macros, enable registers, debug cells, and design specific cells.
(2) Set the bounding bow for the pair search.
(3) Set the maximum capacitance difference allowed for clustering the cells.
(4) The size of the merged cell should be controlled based on timing or power requirements. For timing, the smallest cell with C_{max} bigger than the original cell is taken whereas for power the smallest cell with largest C_{max} smaller than the original cell is taken.
(5) The single-bit registers groups should be identified that can be replaced by multibit registers and then modify the netlist accordingly or generate a banking script file which can then be sourced back into the tool to replace single-bit cells with multibit cells. This is achieved using an input mbit-map file.
(6) New multibit cell from a list of registers or latches are created in the current design. All the single-bit cells in list are replaced by one multibit cell.
(7) It should be verified that the cells in the list exist in the design and have valid locations, and do not have a "dont_touch" or "fixed" attribute.
(8) The order of the specified cells determines the pin connection order of the new multibit cell. For example, a net connected to the third specified cell in the list will be connected to the third bit of the multibit cell inserted by the command.
(9) If the multibit library cell has a larger bit width than the total bit width of the specified cells, the pins of the unused bits of the cell are left dangling. If a pin of a specified cell does not have a corresponding pin in the multibit library cell, the pin is disconnected.
(10) For the final merging of cells, the multibit register is created, and the single-bit cells are disconnected from the nets and then those nets are reconnected to the multibit registers.
(11) Legalize placement of the multibit flip-flops.

In experiment 1, MBFF was implemented only in the placement stage and results were tabulated. In experiment 2, we tried implementing MBFF in both synthesis and placement stage and results were tabulated as shown in Table 1.

(C) *Commands used for MBFF*:
(1) Identify_register_banks

Table 1 MBFF results

(Power mW)	No vectoring	Vectoring in place	Vectoring in syn + place
Fub A	31	27.5	26.86

The identify_register_banks command assigns the single-bit registers in the design into groups according to the input map file and register group file. It does not actually replace the single-bit registers. Instead, it writes out a banking script file containing create_register_bank commands. To perform register banking and change the design netlist, you need to execute the script file [9].

(2) Create_register_bank

To prepare for replacing single-bit registers with multibit registers, the tool writes out a banking script containing create_register_bank commands. You need to execute the script to carry out the actual replacement of single-bit cells with multibit cells. You can insert, delete, and edit the create_register_bank commands in the script file to modify the banking behavior of the script. You can also execute the create_register_bank command by itself to perform a specific banking task [7].

3 Results and Discussion

Figure 5 shows the layout of the design on which experiments were performed. Figure 6 shows how for 1-bit registers are swapped into a single 4-bit register from the same technology library. Table 1 shows the results of multibit flip-flop implementation. Initially, without any power optimization, the total power of the design was 31 mW. When vectoring (use of MBFF) was implemented only in placement stage of physical design, the power was 27.5 mW. We had recovered 11.3% of total power. Later in another experiment on the same design, we implemented vectoring in both synthesis and placement stages of physical design and the result obtained was 26.86 mW of power, i.e., 13.35% of power was recovered. Hence, we conclude that implementation of MBFF in both synthesis and placement stages will yield better results than implementing only in placement stage. Power and area can be recovered at the cost of increased runtime in the latter case.

Figure 7 shows the report of multibit implementation in both synthesis and placement generated using the "report_multibit" command in IC Compiler. The vectoring percentage of sequential cells, also known as sequential cells banking ratio is 64.93% and that of flops, known as flops banking ratio is 67.03%.

4 Conclusion and Future Scope

Experimental results on the technology used indicate that MBFF is very effective and efficient method in deep sub micro design to reduce power, save area and improve routing of clock tree. Amount of power being reduced is ~ 5 to 15% when the power is between ~25 and 75 mW. (It is also design specific). In this paper, we have implemented MBFF and clock gating on the design. In our design, the power

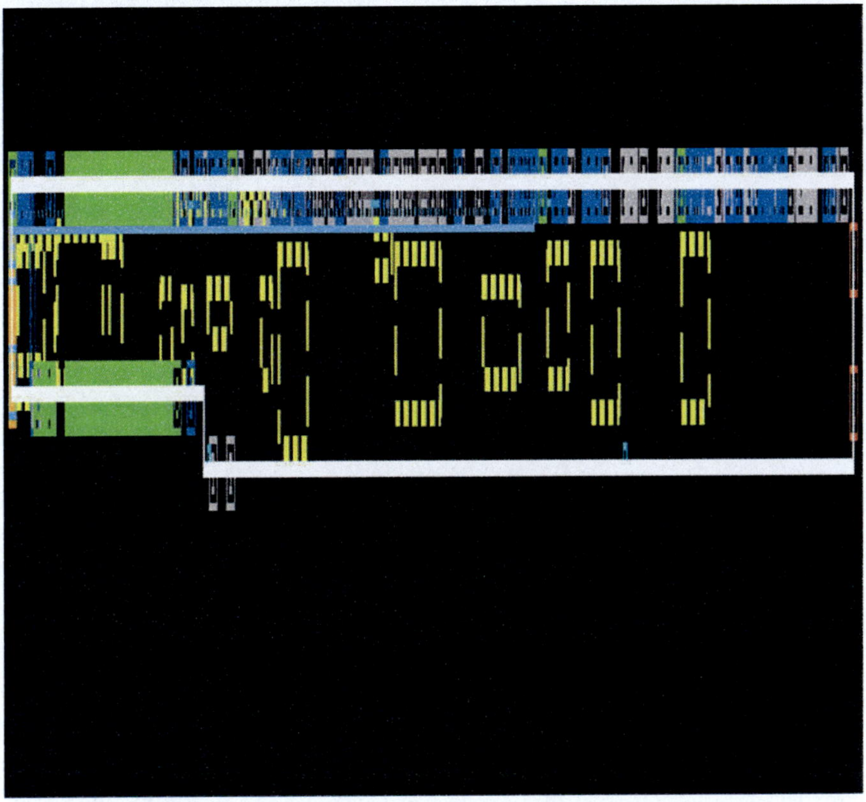

Fig. 5 Layout of design

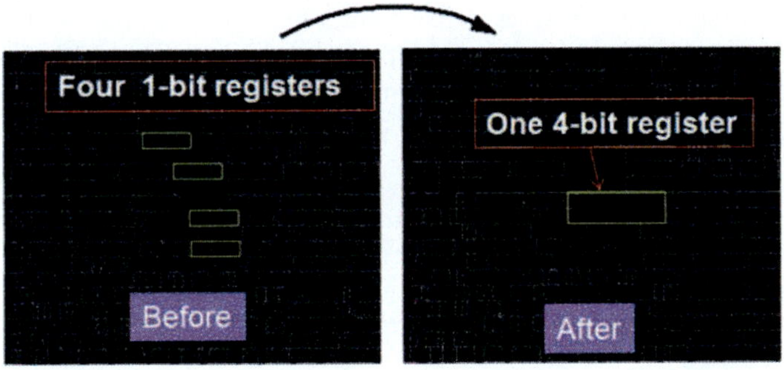

Fig. 6 Four 1-bit registers swapped into a single 4-bit register

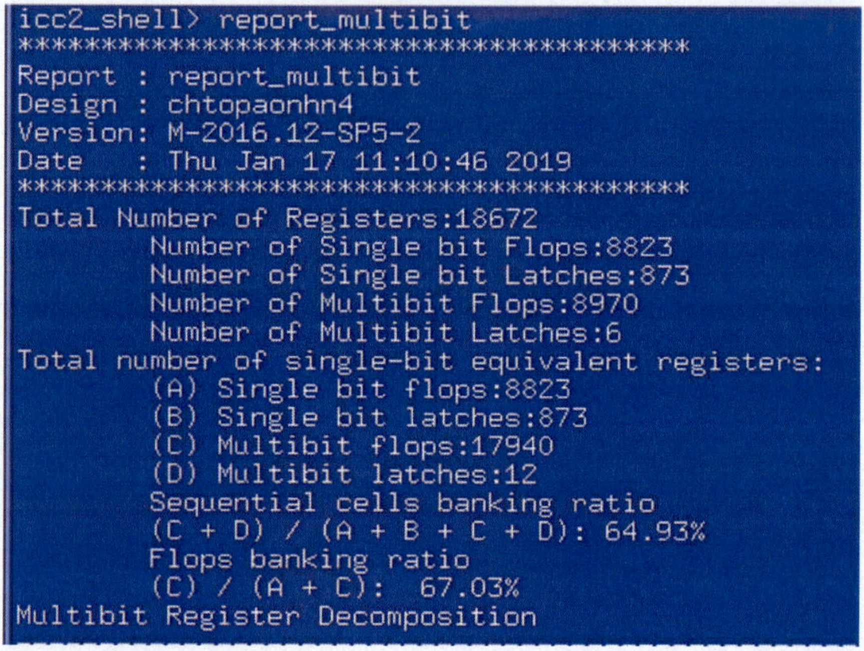

Fig. 7 Vectoring percentage

before applying MBFF was 31 mW and after applying MBFF in both synthesis and placement, the power is 26.86 mW. We have recovered 13.35% of power using MBFF. Thus, implementing MBFF in both synthesis and placement followed by clock gating will result in more power saving. These techniques can be further implemented in complex cores and IPs as well as in partitions where the area constraints are more. MBFF with 4-bit and 8-bit can be used if the cells are available in the technology library. Also, power gating technique can be used in combination with MBFF in order to achieve better results.

References

1. Shinde J, Salankar SS (2011) Clock gating—a power optimizing technique for VLSI circuits. In: Annual IEEE India conference, https://doi.org/10.1109/indcon.2011.6139440, 16–18 December 2011
2. Madhushree, Rajan N (2017) Dynamic power optimization using look-ahead clock gating technique. In: 2nd IEEE international conference on recent trends in electronics information & communication technology (RTEICT), India, 19–20 May 2017
3. Chakraverty M, Harisankar PS, Ruparelia V (2016) Low power design practices for power optimization at the logic and architecture levels for VLSI system design. In: 2016 International conference on energy efficient technologies for sustainability (ICEETS), IEEE, 6th October 2016, pp 727–733
4. Hsu C-C, Chen Y-C, Lin MP-H (2015) In-placement clock-tree aware multi-bit flip-flop generation for power optimization. IEEE Trans Comput-Aided Des Integr Circ Syst 34(2)
5. Cherif L, Chentouf M, Benallal J, Darmi M, Elgouri R, Hmina N (2018) Usage and impact of multi-bit flip-flops low power methodology on physical implementation. In: 4th International conference on optimization and applications (ICOA). https://doi.org/10.1109/icoa.2018.8370498, 26–27 April 2018
6. Mark P-HL, Hsu C-C, Chen Y-C (2015) Clock-tree aware multibit flip-flop generation during placement for power optimization. IEEE Trans Comput-Aided Des Integr Circ Syst 34(2)
7. IC Compiler User Guide: Implementation, M-2016.12-sp4 ed., Synopsys Inc., March 2016
8. Design Compiler User Guide, B-2008.09 ed., Synopsys Inc., September 2008
9. Santos C, Reis R, Godoi G, Barros M, Duarte F (2012) Multi-bit flip-flop usage impact on physical synthesis. In: 25th symposium on integrated circuits and systems design (SBCCI)
10. Garg L, Sahula V (2016) Accurate and efficient estimation of dynamic virtual ground voltage in power gated circuits. In: 2016 29th international conference on vlsi design and 2016 15th international conference on embedded systems (VLSID), IEEE, 17th March 2016, pp 252–257
11. Hirofumi Kawauchi, Ittetsu Taniguchi, and Masahiro Fukui A new approach for accurate RTL power macro-modeling. J Semicond Technol Sci 10(1)
12. Mei LY, Rosdi BAB, Kok LC (2011) A methodology for automation structured datapath placement in VLSI design. In: 2011 IEEE symposium on industrial electronics and applications (ISIEA), IEEE, 22nd December 2011, pp 273–278
13. Zhao P, Wang Z, Hang G (2010) Power optimization for VLSI circuits and systems. In: 2010 10th IEEE international conference on solid-state and integrated circuit technology (ICSICT), IEEE, 13th December 2010, pp 639–642
14. Liu X, Liu Y-F (2009) Intel Corporation, "The extraction and measurement of on-die impedance for power delivery analysis". In: 18th conference on electrical performance of electronic packaging and systems, EPEPS '09, IEEE, 17th November 2009, pp 195–198
15. Muhtaroglu A, Aizik Y, Taylor G, Arabi T, Yagil Y (2006) Extraction of power components from measurement in digital CMOS VLSI circuits. Intel Corporation

Analysis and Enhancement of Biometric-Based Multi-Server Authentication Scheme Using Chebyshev Chaotic Map

Ashish Kumar and Hari Om

Abstract Earlier authentication protocols used to ask a user to register with all the servers with whom he/she wanted to access the services. Nowadays, an efficient way is provided by multi-server authentication techniques which remove the burden of individual registration and ask a user to register only with one registration center and access services provided by different servers. So far, many multi-server authentication protocols have been given in the literature. But most of them are having drawbacks. Recently, Chatterjee et al. gave a multi-server authentication protocol that used Chebyshev chaotic map approach for providing authentication. This paper analyzes the abovementioned protocol and found that it suffers from eavesdropping and user impersonation attacks. Besides that, there are flaws in registration, login and authentication phases of the protocol. Thereafter, we propose the modifications in the protocol without increase in the computation cost in order to make it more robust and secure.

Keywords Security · Authentication · Multi-server · Cryptography · Attacks

1 Introduction

With the exponential increase in the size of computer networks, focus to information security has also been increased. Today, a user can remotely access various services of his/her choice from the servers deployed on the network. To provide security and check the legitimacy of users and servers, various authentication techniques have been introduced. In traditional password-based authentication techniques, a user needs to register with all the servers with which he/she wants to obtain the

A. Kumar (✉) · H. Om
Department of Computer Science and Engineering, Indian Institute of Technology (ISM), Dhanbad, Jharkhand 826004, India
e-mail: ashish.iitdhn@gmail.com

H. Om
e-mail: hariom4india@gmail.com

V. Nath and J. K. Mandal (eds.), *Proceedings of the Fourth International Conference on Microelectronics, Computing and Communication Systems*, Lecture Notes in Electrical Engineering 673, https://doi.org/10.1007/978-981-15-5546-6_76

services. As a result, users have to keep separate identity and password pairs for different servers, which is fairly complicated. In order to avoid this complication, multi-server authentication techniques have come into picture, in which a user is required to register with only one central server to avail the services from multiple servers. The multi-server authentication schemes, presented in the literatures, use different approaches such as cryptographic hash functions, symmetric cryptosystem, public cryptosystem, elliptic curve, bilinear pairings and chaotic maps. Chaotic map is a cryptographic tool which is lightweight, cost efficient and unpredictable. In this paper, security of Chatterjee et al.'s [1] multi-server authentication protocol, which uses Chebyshev chaotic map approach, has been analyzed.

This paper is sketched as follows: Sect. 2 and Sect. 3 present the related works and preliminaries, respectively. Section 4 describes the protocol [1], and Sect. 5 analyzes its security. In Sect. 6, we propose the modifications in protocol [1]. Section 7 gives the performance comparisons. Lastly, the paper is concluded in Sect. 7.

2 Related Works

First, we give a brief description of different authentication techniques proposed in the literature. Lamport [2] presented first protocol to provide password-based authentication. This protocol used password table to verify the passwords entered by users. However, it was vulnerable to the attack on password table, i.e., stolen verifier attack. Thereafter, many two-factor protocols [3–6] had been proposed that included smartcard to remove the need of password table. These two-factor protocols were susceptible to user's impersonation attack in case of theft of the smartcard, because passwords are guessable. Thus, various authentication protocols [7–10] were later proposed, in which user's biometrics was included as third factor due to its unguessable nature. After that many multi-server protocols were proposed in the literature, because the traditional authentication protocols failed in fulfilling the requirement of a user. Li et al. [11] presented first protocol for multi-server authentication that uses the concept neural networks. However, the time consumption in the training of neural networks made it inefficient. Later, ElGamal digital signature-based protocol was proposed by Lin et al. [12]. But it was also not very efficient due to the use of many system parameters. After this, Juang [13] proposed another protocol in which the concept of symmetric cryptosystems and lightweight cryptographic hash functions was used. However, this protocol was found to be unsafe against various attacks [14]. Thereafter, many authentication protocols for multi-server architecture were proposed based on symmetric cryptosystems [15–17] and cryptographic hash functions [18–20].

In the last few years, several authentication protocols [22–24] based on chaotic map have been proposed. Reference [21] presented an authentication protocol that used self-relying timestamp. After that, Reference [22] gave an improved protocol after finding out that the protocol proposed in [21] is susceptible to known-plaintext and insider attacks. Unfortunately, Reference [23] found some flaws in protocol [22], such as it was suffering from spoofing attack, and also wrong password modification and unauthorized login was possible in it. Again, Reference [24] recognized that protocol provided in Reference [23] are susceptible to session-specific temporary information, malicious server and stolen smartcard attacks. Recently, protocol [1] has been proposed by Chatterjee et al. for multi-server architecture which claims to be efficient and secure. In this paper, we have analyzed the protocol [1] and have found that it suffers from eavesdropping and user impersonation attacks. In addition, registration, login and authentication phases of the protocol have some flaws.

3 Preliminaries

This section elucidates different cryptographic details needed to understand and analyze protocol [1] as well as the modifications proposed in this paper.

(A) *One-way hash function*

A one-way hash function h: X → Y with collision resistance is an algorithm that takes random length input $x \in \{0, 1\}^*$ and gives fixed length output $h(x) \in \{0, 1\}^n$. It is said to be one way because of hardness in getting the input x from the known hash value $y = h(x)$. The collision resistant property states that for any pair of inputs $(x_1, x_2) \in (X, X)$, if $x_1 \neq x_2$ then $h(x_1) \neq h(x_2)$.

(B) *Chebyshev chaotic map*

For a given integer p and a variable $x \in [-1, 1]$, $T_p(x)$: $[-1, 1] \rightarrow [-1, 1]$ is defined as $T_p(x) = \cos(p.\cos^{-1}(x))$. The recurrent relation of Chebyshev polynomial map is defined as $T_p(x) = 2x.T_{p-1}(x) - T_{p-2}(x)$. The enhanced Chebyshev polynomials [18] is defined as $T_p(x) = 2x.T_{p-1}(x) - T_{p-2}(x) \pmod N$, for $p \geq 2$, $x \in (-\infty, +\infty)$ and large prime N. Its semi-group property says that for given variable x and integers p and q,

$$T_{pq}(x) = T_p(T_q(x)) = T_q(T_p(x)).$$

(C) *Biohashing*

Biohashing takes user's biometric impression as input and gives a unique output. In general, two biometric impressions of a same person may have some variations. Biohashing always give the same output if two biometric impressions as inputs vary little bit. In it, false rejection rate is low.

(D) *Attacker model*

In authentication schemes, all communicating messages are transmitted over an insecure channel. Thus, it becomes advantageous for an attacker to perform various attacks. Below, we mention a few capabilities of an attacker.

- Attacker can intercept all the messages communicated over a public channel between the involved entities of an authentication scheme.
- Attacker may delete, modify and retransmit the intercepted messages.
- Attacker can be a genuine entity of the scheme.
- Attacker may have the capability to take out all the stored parameters of a smartcard using some techniques discussed in [26, 27].
- Attacker can guess a secret identity or password of low entropy. However, it is computationally infeasible to guess both of them at the same time.
- An attacker has complete knowledge about the description of the scheme.

4 Review of Protocol [1]

In protocol [1], smartcards, user passwords and biometrics are used to provide multi-server three-factor authentication. This protocol employs the cryptographic hash function and the Chebyshev chaotic map and includes user, common registration center and service offering server as participating entities.

(E) *Server's registration phase*

In this phase, a newly deployed service offering server S_j executes the following steps to register itself with the registration center *RC*:

Step 1: S_j selects an identity ID_{Sj} and transmits it securely to *RC*.

Step 2: *RC* firstly chooses Sec_S and Sec_U as secret values for all registered servers and users, respectively. It also selects a secret value *Xj* for each individual server S_j. Then, *RC* calculates T_{Xj} (Sec_S) and T_{Xj} (Sec_U) as Chebyshev polynomials.

Step 3: *RC* secretly transmits $\{\mathrm{ID}_{Sj}, T_{Xj}(\mathrm{Sec}_S), T_{Xj}(\mathrm{Sec}_U), Xj\}$ to S_j.

In this protocol, it has been assumed that there are *m* number of servers initially deployed, and *n* more servers can be deployed in future. Thus, *RC* allows first *m* servers to choose its identity, while for remaining *n* servers, *RC* will choose their identities and secret values for them, i.e. $\{\mathrm{ID}_{Sj}, Xj \mid m + 1 \le j \le m + n\}$. It also calculates the Chebyshev polynomial T_{Xj} (Sec_S) for the remaining *n* servers.

(F) *User's registration phase*

A newly joined user U_i executes the following steps to register with *RC*:

Step 1: U_i selects an identity ID_i, password P_i, and imprints his/her biometrics information BI_i on the sensor. Also, U_i selects a random value RAN_i and performs the following calculations: $\mathrm{PID}_i = H(\mathrm{ID}_i \mid \mathrm{RAN}_i \| T_i)$, $\mathrm{BIO}_i = \mathrm{BH}(\mathrm{BI}_i)$, $\mathrm{RPW}_i = \mathrm{H}(\mathrm{ID}_i \| P_i \mid \mathrm{BIO}_i \mid \mathrm{RAN}_i)$, $K_i = H(\mathrm{BIO}_i \mid \mathrm{RAN}_i \mid \mathrm{PID}_i)$, and $C_i = \mathrm{RAN}_i \oplus H(\mathrm{BIO}_i \mid \mathrm{ID}_i \| P_i)$. Here, T_i is the timestamp generated at the time of registration. After that U_i securely transmits $\{\mathrm{PID}_i, T_i, K_i, C_i, \mathrm{RPW}_i\}$ to *RC*.

Step 2: *RC* selects secret values *Xi* and Sec_{Ui}. After that it performs the following calculations: T_{Xi} (Sec_{Ui}), T_{Xi} (Sec_U), $SK_i = K_i \oplus Xi$, $P = \mathrm{Sec}_S \oplus H(Xi \| K_i)$, $A_i = H(\mathrm{ID}_i \mid \mathrm{RPW}_i \| T_i \| T_{Xi}(\mathrm{Sec}_{Ui}) \| X_i \| P)$. *RC* stores $\{\mathrm{ID}_i, T_i, A_i, T_{Xi}(\mathrm{Sec}_{Ui}), C_i, SK_i,$

P, T_{Xi} (Sec$_U$), {(ID$_{Sj}$, T_{Xj} (Sec$_S$))| $1 \le j \le m + n$}} in a smartcard, and transmits it securely to U_i.

Step 3: RC stores pseudo-identity PID$_i$, unique registration number Ur_i, and a calculated value $Uh_i = H$ (T_{Xi} (Sec$_{Ui}$)||Ur_i) of all users U_i to each server S_j.

(G) *Login phase*

This phase executes the following steps:

Step 1: U_i inserts his smartcard to a local system and inputs ID$_i$, P_i and BI_i. Using biohashing function BH(.), the value BIO$_i$ = BH (BI$_i$) is calculated. Thereafter, the following values are calculated: RAN$_i'$ = $C_i \oplus H$ (ID$_i$ ||P_i| BIO$_i$), PID$_i'$ = H (ID$_i$|RAN$_i'$||T_i), RPW$_i'$ = H (PID$_i'$||P_i|BIO$_i$|RAN$_i'$), X_i' = H (BIO$_i$|RAN$_i'$|PID$_i$) $\oplus$ SK_i, A_i' = H (ID$_i$|RPW$_i'$||T_i||T_{Xi} (Sec$_{Ui}$)||X_i'||P). After that, it checks whether A_i' and the stored value A_i of smartcard are equal or not. If true, then it will proceed to Step 2.

Step 2: After having successful verification, the following values are further calculated: T_1 = T_{Xi} (T_{Xj} (Sec$_S$)), K_1 = H (T_{Xj} (Sec$_S$)|PID$_i$|ID$_{Sj}$|TS$_i$), Sec$_S$ = $P \oplus H$ (X_i||H (BIO$_i$|RAN$_i$|PID$_i$)), M$_1$ = {PID$_i$, ID$_{Sj}$, E_{K1} (PID$_i$|ID$_{Sj}$||T_1||T_{Xi} (Sec$_S$)||T_{Xi} (Sec$_U$)||T_{Xi} (Sec$_{Ui}$)||R_i||H (BIO$_i$|RAN$_i$|PID$_i$)), TS$_i$, H (H (BIO$_i$|RAN$_i$|PID$_i$) | TS$_i$|PID$_i$| ID$_{Sj}$| R_i| T_{Xi} (Sec$_U$)||T_1)}. Here, TS$_i$ and R_i are the current timestamp and random nonce of U_i. U_i sends the value M_1 over insecure channel to S_j.

(H) *Authentication phase*

This phase executes the following steps:

Step 1: S_j firstly checks if |TS$_i^*$ − TS$_i$| $\le \Delta T$, where ΔT is the tolerable time delay and TS$_i^*$ is the current timestamp of S_j. If condition is true, then it computes K_1' = H (T_{Xj} (Sec$_S$|PID$_i$|ID$_{Sj}$|TS$_i$) and performs decryption to get T_1, T_{Xi} (Sec$_S$), T_{Xi} (Sec$_U$), T_{Xi} (Sec$_{Ui}$), R_i and H (BIO$_i$|RAN$_i$|PID$_i$).

Step 2: S_j extracts Ur_i for the respective PID$_i$ and calculates Uh_i = H (T_{Xi} (Sec$_{Ui}$||Ur_i) to check whether the computed value is equal to any of the stored values. If true, then it proceeds further. S_j computes H (H (BIO$_i$|RAN$_i$|PID$_i$)|TS$_i$|PID$_i$|ID$_{Sj}$||R_i||T_{Xi} (Sec$_U$)||T_1) and T_1' = T_{Xj} (T_{Xi} (Sec$_S$)) and then compares with the hash value of M_1 and extracted value of T_1. If the compared values are equal, then S_j authenticates U_i successfully.

Step 3: S_j thereafter computes the following: T_2 = T_{Xj} (T_{Xi} (Sec$_{Ui}$)), Y = H (BIO$_i$|RAN$_i$|PID$_i$) $\oplus$ T_2, K_2 = H (T_{Xi} (Sec$_{Ui}$)|ID$_{Sj}$|PID$_i$|TS$_i$|TS$_j$||R_i||T_1), T_3 = T_{Xj} (T_{Xi} (Sec$_U$)), M$_2$ = {PID$_i$, ID$_{Sj}$, E_{K2} (PID$_i$|ID$_{Sj}$||Y|| T_{Xj} (Sec$_U$)||R_j||T_3), TS$_j$, H (R_j|TS$_j$||Y||T_3||T_{Xj} (Sec$_U$))}. Here, TS$_j$ and R_j are the current timestamp and random nonce of S_j. S_j computes the session key SK_{ij} = H (PID$_i$|ID$_{Sj}$|TS$_i$|TS$_j$||R_i||R_j||T_1||T_2||T_3) and sends the value of M_2 over insecure channel to U_i.

Step 4: U_i checks if |TS$_j^*$ − TS$_j$| $\le \Delta T$, where TS$_j^*$ is the current timestamp of U_i. If condition is true then U_i computes K_2' = H (T_{Xi} (Sec$_{Ui}$)|ID$_{Sj}$|PID$_i$|TS$_i$|TS$_j$||R_i) and performs decryption to get Y, T_{Xj} (Sec$_U$), R_j and T_3. U_i then computes T_2' = H (BIO$_i$|RAN$_i$|PID$_i$) $\oplus$ Y and T_3' = T_{Xi} (T_{Xj} (Sec$_U$)) and authenticates S_j successfully, if computed value T_3' and the extracted value T_3 are equal. The session key SK_{ij} is calculated as $SK_{ij} = H$ (PID$_i$|ID$_{Sj}$|TS$_i$|TS$_j$||R_i||R_j||T_1||T_2'||T_3').

5 Security Analysis of Protocol [1]

(I) *Flaws in registration, login and authentication phases*

In user's registration phase, RC computes $A_i = H$ (ID$_i$|RPW$_i$||T_i||T_{Xi} (Sec$_{Ui}$)||X_i||P) after receiving {PID$_i$, T_i, K_i, C_i, RPW$_i$} from U_i. However, the real identity ID$_i$ of U_i is unknown to RC, because it has only received U_i's pseudo-identity PID$_i$. Thus, it is not possible for U_i to calculate the value of A_i.

In login phase, U_i has wrongly calculated the value of RAN$_i'$. In Step 1, RAN$_i'$ is calculated as $C_i \oplus H$ (ID$_i$||P_i|BIO$_i$). However, the value of C_i stored in the smartcard is equal to RAN$_i$ $\oplus$ H (BIO$_i$|ID$_i$||P_i). Thus, the actual value of RAN$_i'$ will be $C_i \oplus$ H (BIO$_i$|ID$_i$||P_i) instead of $C_i \oplus H$ (ID$_i$||P_i|BIO$_i$). It can be noticed that the wrong value of RAN$_i'$ will lead to the incorrect calculation of A_i', which will in turn stop all the legal users to perform successful login.

In Step 3 of Authentication phase, S_j performs the symmetric encryption with the key $K_2 = H$ (T_{Xi} (Sec$_{Ui}$)|ID$_{Sj}$|PID$_i$|TS$_i$|TS$_j$||R_i||T_1). However, in Step 4, U_i performs the symmetric decryption with the key $K_2' = H$ (T_{Xi} (Sec$_{Ui}$)|ID$_{Sj}$|PID$_i$|TS$_i$|TS$_j$||R_i). Both key values are not same; hence, user U_i cannot find the correct plain values which are encrypted by S_j.

(J) *Eavesdropping attack*

The smartcard of a legitimate user stores the pair (ID$_{Sj}$, T_{Xj} (Sec$_S$)) for all the servers, i.e., $1 \le j \le m + n$. If an adversary eavesdrops the message M_1 communicated over insecure channel by U_i, then he/she will extract the values PID$_i$, ID$_{Sj}$, TS$_i$. After that, the key K_1 can be calculated as H (T_{Xj} (Sec$_S$)|ID$_i$|ID$_{Sj}$|TS$_i$). From the calculated value of K_1, adversary will decrypt E_{K1} (PID$_i$|ID$_{Sj}$||T_1||T_{Xi} (Sec$_S$)||T_{Xi} (Sec$_U$)||T_{Xi} (Sec$_{Ui}$)||R_i||H (BIO$_i$|RAN$_i$|PID$_i$)) and extract the values T_1, T_{Xi} (Sec$_S$), T_{Xi} (Sec$_U$), T_{Xi} (Sec$_{Ui}$), R_i, and H (BIO$_i$|RAN$_i$|PID$_i$). Further, if the same adversary eavesdrops the message M_2 sent by S_j to U_i, then K_2 will be calculated as H (T_{Xi} (Sec$_{Ui}$)|ID$_{Sj}$|PID$_i$|TS$_i$|TS$_j$||R_i||T_1). From the calculated value of K_2, adversary will decrypt E_{K2} (PID$_i$|ID$_{Sj}$||Y||T_{Xj} (Sec$_U$)||R_j||T_3) to get Y, T_{Xj} (Sec$_U$), R_j and T_3. Then he/she will calculate T_2 as $Y \oplus H$ (BIO$_i$|RAN$_i$|PID$_i$). Now, the session key can be easily calculated as $SK_{ij} = H$ (PID$_i$|ID$_{Sj}$|TS$_i$|TS$_j$||R_i||R_j||T_1||T_2||T_3).

(K) *User impersonation attack*

If an adversary is a legal user and steals U_i's smartcard, then it is possible for him/her to impersonate U_i and perform all the calculations that U_i is capable of doing. For this, we assume that adversary has already calculated the $K_i = H$ (BIO$_i$|RAN$_i$|PID$_i$) from the eavesdropped message M_1 communicated in any previous session. Now, the adversary can easily calculate X_i and Sec_S as $Xi = K_i \oplus SK_i$, $\text{Sec}_S = P \oplus H$ $(Xi||K_i)$. Here, SK_i and P are the values stored in the U_i's smartcard. Now, adversary is capable to do all the computations given in Step 2 of Login phase and Step 4 of Authentication phase on behalf of U_i.

6 Modifications proposed in Protocol [1]

In this section, we have proposed some modifications in the protocol [1] in order to remove its consistency and improve security. To do so, it should be kept in mind that any adversary, who may or may not be a legal user, should not be able to calculate the value of K_1, K_2 and K_i. For this, the following modifications are suggested:

(1) In the registration phase, the value of A_i should be equal to H $(\text{PID}_i|\text{RPW}_i||T_i||T_{Xi}\ (\text{Sec}_{Ui})||X_i||P)$ instead of H $(\text{ID}_i|\text{RPW}_i||T_i||T_{Xi}\ (\text{Sec}_{Ui})||X_i||P)$.

(2) In login phase, the modified value of RAN_i', A_i', K_1 and M_1 should be calculated as $\text{RAN}_i' = C_i \oplus H\ (\text{BIO}_i|\text{ID}_i||P_i)$, $A_i' = H$ $(\text{PID}_i|\text{RPW}_i||T_i||T_{Xi}\ (\text{Sec}_{Ui})||X_i||P)$, $K_1 = H\ (T_{Xi}\ (T_{Xj}\ (\text{Sec}_S))|\text{PID}_i|\text{ID}_{Sj}|\text{TS}_i)$ and $M_1 = \{\text{PID}_i,\ \text{ID}_{Sj},\ E_{K1}\ (\text{PID}_i|\text{ID}_{Sj}||T_1||T_{Xi}\ (\text{Sec}_S)||T_{Xi}\ (\text{Sec}_U)|$ $T_{Xi}\ (\text{Sec}_{Ui})||R_i||H\ (\text{BIO}_i|\text{RAN}_i|\text{PID}_i|\text{TS}_i)),\ T_{Xi}\ (\text{Sec}_S),\ \text{TS}_i,\ H\ (H$ $(\text{BIO}_i|\text{RAN}_i|\text{PID}_i|\text{TS}_i)|\text{TS}_i|\text{PID}_i|\text{ID}_{Sj}|\ R_i|\ T_{Xi}\ (\text{Sec}_U)||T_1)\}$.

(3) In authentication phase, the modified value of K_1' should be calculated as H $(T_{Xj}\ (T_{Xi}\ (\text{Sec}_S))|\text{PID}_i|\text{ID}_{Sj}|\text{TS}_i)$. After that, S_j performs decryption to get T_1, $T_{Xi}\ (\text{Sec}_S)$, $T_{Xi}\ (\text{Sec}_U)$, $T_{Xi}\ (\text{Sec}_{Ui})$, R_i and $H\ (\text{BIO}_i|\text{RAN}_i|\text{PID}_i|\text{TS}_i)$. Then, S_j computes $H\ (H\ (\text{BIO}_i|\text{RAN}_i|\text{PID}_i|\text{TS}_i)|\text{TS}_i|\ \text{PID}_i|\text{ID}_{Sj}||R_i||T_{Xi}\ (\text{Sec}_U)||T_1)$ and $T_1' = T_{Xj}\ (T_{Xi}\ (\text{Sec}_S))$, and compares with the hash value of M_1 and extracted value of T_1. If the compared values are equal, then S_j authenticates U_i successfully. After verifying the authenticity of U_i, S_j computes the following: $T_2 = T_{Xj}\ (T_{Xi}\ (\text{Sec}_{Ui}))$, $Y = H\ (\text{BIO}_i|\text{RAN}_i|\text{PID}_i|\text{TS}_i) \oplus T_2$, $K_2 = H\ (T_{Xj}\ (T_{Xi}$ $(\text{Sec}_U))|\text{ID}_{Sj}|\text{PID}_i|\text{TS}_i|\text{TS}_j||R_i||T_1)$, $T_3 = T_{Xj}\ (T_{Xi}\ (\text{Sec}_U))$, $M_2 = \{\text{PID}_i,\ \text{ID}_{Sj},$ $E_{K2}\ (\text{PID}_i|\text{ID}_{Sj}||Y||T_{Xj}\ (\text{Sec}_U)||R_j||T_3),\ \text{TS}_j,\ T_{Xj}\ (\text{Sec}_U),\ H\ (R_j|\text{TS}_j||Y||T_3||T_{Xj}$ $(\text{Sec}_U))\}$.

(4) The modified value of K_2' computed by U_i should be equal to $H(T_{Xi}\ (T_{Xj}$ $(\text{Sec}_U))|\text{ID}_{Sj}|\text{PID}_i|\text{TS}_i|\text{TS}_j|\ R_i||T_1)$. After performing decryption with K_2', the values of Y, $T_{Xj}\ (\text{Sec}_U)$, R_j and T_3 are extracted. U_i then computes $T_2' = H\ (\text{BIO}_i|\text{RAN}_i|\text{PID}_i|\text{TS}_i) \oplus Y$ and $T_3' = T_{Xi}\ (T_{Xj}\ (\text{Sec}_U))$ and authenticates S_j successfully, if computed value T_3' and the extracted value T_3 are equal. The session key SK_{ij} is calculated as $SK_{ij} = H\ (\text{PID}_i|\text{ID}_{Sj}|\text{TS}_i|\text{TS}_j||R_i||R_j|$ $T_1||T_2'||T_3')$.

Table 1 Computation cost comparisons

Protocol	Computation cost	Computation time (ma)
[25]	$6C_E + 24C_H + 6C_M$	442.65
[1]	$8C_E + C_B + 16C_H + 6C_C$	224.74
Our	$8C_E + C_B + 16C_H + 6C_C$	224.74

7 Performance comparisons

The modifications proposed in this paper is not affecting the computation cost of the protocol [1]. The computation cost of the modified protocol and the protocol [1] are same. Table 1 clearly shows that the computation time of the modified protocol and protocol [1] are better than protocol [25]. The notations C_E, C_H, C_C, C_M and C_B used to compare the protocols are the computation cost of performing symmetric encryption and decryption, one-way hash functions, Chebyshev polynomial, elliptic curve point multiplication and biohashing, respectively. The values for C_E, C_H, C_C, C_M and C_B (in second) are 0.0087, 0.0005, 0.02102, 0.063075 and 0.02102, respectively [1].

8 Conclusions

In this paper, Chatterjee et al.'s multi-server authentication protocol, which uses Chebyshev chaotic map to provide authentication, has been analyzed. Even though the protocol is protected against various attacks, but this paper pointed out that it is suffering from eavesdropping and user impersonation attacks. In addition, we found that there are flaws in protocol's registration, login and authentication phases. Thus, we propose the modifications in the protocol in order to make it more robust and secure. The modified protocol does not increase the computation cost.

References

1. Chatterjee S, Roy S, Das AK, Chattopadhyay S, Kumar N, Vasilakos AV (2018) Secure biometric-based authentication scheme using Chebyshev chaotic map for multi-server environment. IEEE Trans Dependable Secure Comput 15(5):824–839
2. Lamport L (1981) Password authentication with insecure communication. Commun ACM 24(11):770–772
3. Hwang MS, Li LH (2000) A new remote user authentication scheme using smart cards. IEEE Trans Consum Electron 46(1):28–30
4. Das ML, Saxena A, Gulati VP (2004) A dynamic ID-based remote user authentication scheme. IEEE Trans Consum Electron 50(2):629–631
5. Sun HM (2000) An efficient remote use authentication scheme using smart cards. IEEE Trans Consum Electron 46(4):958–961

6. Shen JJ, Lin CW, Hwang MS (2003) A modified remote user authentication scheme using smart cards. IEEE Trans Consum Electron 49(2):414–416
7. Fan CI, Lin YH (2009) Provably secure remote truly three-factor authentication scheme with privacy protection on biometrics. IEEE Trans Inf Forensics Secur 4(4):933–945
8. Li X, Niu JW, Ma J, Wang WD, Liu CL (2011) Cryptanalysis and improvement of a biometrics-based remote user authentication scheme using smart cards. J Netw Comput Appl 34(1):73–79
9. Chuang MC, Chen MC (2014) An anonymous multi-server authenticated key agreement scheme based on trust computing using smart cards and biometrics. Expert Syst Appl 41(4):1411–1418
10. Jiang Q, Khan MK, Lu X, Ma J, He D (2016) A privacy preserving three-factor authentication protocol for e-Health clouds. J Supercomputing 72(10):3826–3849
11. Li L, Lin I, Hwang M (2001) A remote password authentication scheme for multi-server architecture using neural networks. IEEE Trans Neural Netw 12(6):1498–1504
12. Lin IC, Hwang MS, Li LH (2003) A new remote user authentication scheme for multi-server architecture. Future Generation Computer Systems 19(1):13–22
13. Juang WS (2004) Efficient multi-server password authenticated key agreement using smart cards. IEEE Trans Consum Electron 50(1):251–255
14. Chang CC, Lee JS (2004) An efficient and secure multi-server password authentication scheme using smart cards. In: Proceedings of the third international conference on cyberworlds, pp 417–422
15. Tsaur WJ, Wu CC, Lee WB (2004) A smart card-based remote scheme for password authentication in multiserver internet services. Comput Stand Interfaces 27:39–51
16. Irshad A, Sher M, Rehman E, Ch SA, Hassan MU, Ghani A (2015) A single round-trip SIP authentication scheme for voice over internet protocol using smart card. Multimedia Tools Appl 74(11):3967–3984
17. Irshad A, Sher M, Faisal MS, Ghani A, Ul Hassan M, Ashraf Ch S (2014) A secure authentication scheme for session initiation protocol by using ECC on the basis of the tang and Liu scheme. Secur Commun Netw 7(8):1210–1218
18. Zhang L (2008) Cryptanalysis of the public key encryption based on multiple chaotic systems. Chaos Solitons Fractals 37(3):669–674
19. Kumar A, Om H (2018) An improved and secure multiserver authentication scheme based on biometrics and smartcard. Digital Commun Netw 4(1):27–38
20. Mishra D, Das AK, Mukhopadhyay S (2014) A secure user anonymity-preserving biometric-based multiserver authenticated key agreement scheme using smart cards. Expert Syst Appl 41:8129–8143
21. Tsaur WJ, Li JH, Lee WB (2012) An efficient and secure multi-server authentication scheme with key agreement. J Syst Softw 85(4):876–882
22. Lee CC, Lou DC, Li CT, Hsu CW (2014) An extended chaotic-maps-based protocol with key agreement for multiserver environments. Nonlinear Dyn 76(1):853–866
23. Li X, Niu J, Kumari S, Islam SH, Wu F, Khan MK, Das AK (2016) A novel chaotic maps-based user authentication and key agreement protocol for multi-server environments with provable security. Wirel Pers Commun 89(2):569–597
24. Irshad A, Chaudhry SA, Xie Q, Li X, Farash MS, Kumari S, Wu F (2018) An enhanced and provably secure chaotic map-based authenticated key agreement in multi-server architecture. Arab J Sci Eng 43(2):811–828
25. Odelu V, Das AK, Goswami A (2015) A secure biometrics-based multi-server authentication protocol using smart cards. IEEE Trans Inf Forensics Secur 10(9):1953–1966
26. Messerges TS, Dabbish EA, Sloan RH (2002) Examining smart-card security under the threat of power analysis attacks. IEEE Trans Comput 51(5):541–552
27. Kocher P, Jaffe J, Jun B (1999) Differential power analysis. In: Advances in cryptology CRYPTO 99, LNCS 1666, pp 388–397

Characteristic Analysis and Pattern Recognition of Arc Sound Under Typical Penetration Status in Mig Welding

Punam Kumari and Kartik Mahto

Abstract Targeting proposing a web based checking strategy for infiltration status in MIG welding, discernible bend sound sign under halfway entrance, precarious infiltration, full infiltration and extreme entrance over the span of level butt Soldering with shower move was gathered, prepared and investigated. And afterward 11 trademark constraints, which can describe weld entrance status from the points of view on schedule, recurrence, geometry-areas and cepstrum, were separated by utilizing wavelet de-noising and brief timeframe windowing. MIG butt welds of IS 2062 E250 mellow steel plates to be welded utilizing CO_2 as protecting gas. Soldering speed is select as procedure variable while circular segment voltage, Soldering current, wire feed rate separation between the spout and the plates are fixed right now.

Keywords IS 2062 · CO_2 gas · Arc sound · MIG soldering · Characteristic analysis · Pattern recognition · Arc sound · Mild steel · Soldering parameters · Audible sound · Penetration status · A mild steel

1 Introduction

GAS METAL ARC Soldering (GMAS) is a procedure that melts and joins metals by warming them with an arc built up between a persistently nourished filler wire terminal and the metals. The procedure is utilized with protecting from a remotely provided gas and without the use of weight. In the 1920s, the fundamental idea of GMAW was presented. Be that as it may, it was not industrially accessible until 1948. From the outset it was viewed as, in a general sense, a high-current thickness, little width, exposed metal cathode process utilizing an inactive gas for arc protecting.

P. Kumari · K. Mahto (✉)
Department of Electronics and Communication Engineering, Birla Institute of Technology Mesra, Ranchi, Jharkhand 835215, India
e-mail: Kartik_mahto@rediffmail.com

P. Kumari
e-mail: ajitbecs@gmail.com

V. Nath and J. K. Mandal (eds.), *Proceedings of the Fourth International Conference on Microelectronics, Computing and Communication Systems*, Lecture Notes in Electrical Engineering 673, https://doi.org/10.1007/978-981-15-5546-6_77

The essential utilization of this procedure was for Soldering aluminum. Therefore, the term MIG (Metal Inert Gas) was utilized is as yet a typical reference for the procedure. Resulting process improvements included activity at low-current densities and beat direct current, application to a more extensive scope of materials, and the utilization of receptive gases (especially CO_2) and gas blends. This last improvement has prompted formal acknowledgment of the term GMAS for the procedure on the grounds that both latent and receptive gases are utilized. There are two activity methods of GMAS. All financially significant metals, for example, high-quality low composite steel, hardened steel, aluminum, copper and titanium compounds can be welded in all situations with this procedure by picking the proper protecting gas, terminal, joint plan and Soldering factors.

2 Literature Review

Numerous researches' have been accomplished research chip away at various materials for getting greatest yield quality and elasticity. A reaction surface model was created by Faseeulla et al. to contemplate thein fluence of procedure parameters of weld holding on elastic shear quality of the weld-obligation of 2 mm thick aluminum amalgam 6061 T651 sheets. Relieving time, Soldering current, soldering time and anode pressure are enhanced for greatest tractable shear quality of the weld bond. Padmanaban and Balasubramanian built up an exact relationship to adequately foresee the elasticity of beat current gas tungsten bend welded AZ31B magnesium compound joints at 95% certainty level.

The huge procedure parameters, for example, top current, base current, beat recurrence and heartbeat on time were considered. Yahya sees that the weld quality of thermoplastics, for example, high thickness polyethylene and polypropylene sheets is affected by rubbing mix-Soldering parameters. The assurance of the Soldering parameters assumes a significant job for the weld quality. The outcome additionally shows that for the persuasive utilization of the thermoplastics joints, the weld ought to have satisfactory quality.

A successful system of Response Surface Methodology (RSM) has been used for finding the ideal estimations of procedure parameters while enlistment solidifying of AISI 1040 under two distinct states of the material, i.e., rolled and standardized by Amit and Hari.

The examination plan depended on rotatable, Central Composite Design (CCD). Benyounis and Olabi examined that Soldering input parameters assume a critical job in deciding the nature of a weld joint.

Beal et al. researches the laser combination of a blend of H13 and Cu powders. The Response Surface Methodology (RSM) was 2006 Int. J. Mech. Eng. and Rob. Wang et al. [1] demonstrate the identification of different laser welding penetration status based on multi-sensor fusion. Shuang et al. [2] define the study on the method of vision detection on penetration status in Laser-MIG hybrid welding. Liao et al. [3] define a brief view on the development of the penetration detection and control for bead. Wu et al. [4] shows the determination of characteristic parameters of weld penetration in electron beam welding of aluminum alloy. Inoue et al. [5] show the measurement of arc sound with burn-through in MAW welding. Cudina et al. [6] demonstrate the evaluation of the sound signal based on the welding current in the gas-metal arc metal welding process. Ma et al. [7] shows the Characteristics analyzing and parametric modeling of the arc sound in CO_2 GMAW for on-line quality monitoring. Schiebeck et al. [8] explain the audible range acoustic diagnosis of the MAG welding arc. Manz et al. [9] show the welding arc sounds. Saini et al. [10] define the investigation of gas metal arc welding sound signature for on-line quality control. Ladislav et al. [11] demonstrate the feasibility study of acoustic for on-line monitory in short circuit gas metal arc welding. Wang et al. [12] sound sensing of the keyhole behaviors in plasma arc welding. Liu et al. [13] shows the correlation of acoustic signals and weld depth in laser welding. Brown et al. [14] demonstrate the application of artificial intelligence techniques to resistance spot welds.

3 Signal Collection

The graphic of circular segment sound sign observing framework for MIG Soldering entrance tests, exhibited right now, appeared in Fig. 1. The entire framework is made out of receiver, molding module, assortment unit, mechanical actualizing organizations, industrial PC (IPC), soldering supplies, etc. What's more, it fills in as follows:

(1) Collect circular segment sound sign in soldering method by utilizing AWA14423 receiver.
(2) Communicate signal gathered to IPC over preamplifier molding unit.
(3) Excerpt infiltration attributes of curve sound signal by embracing current advanced sign preparing system and build artificial neural network.

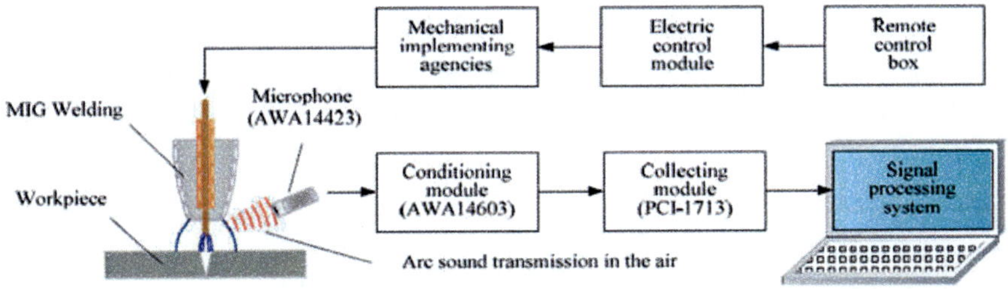

Fig. 1 Schematic of arc sound signal monitoring system for MIG soldering penetration tests

4 Wavelet de-Noising

Bend sound sign in MIG Soldering receipts on high unpredictability and nonlinearity, since it is affected by various cooperation. It is hard to naturally and essentially depict the altering principles between bend sound sign and infiltration status, and exploit it. So as to viably remove trademark data of entrance rank, simultaneously, to take out the negative effect of commotion on consequent sign preparing and investigation, wavelet change which has incredible favorable circumstances of managing non-stationary irregular sign was embraced. Figure 2 illustrations difference of time-area waveform of circular segment sound sign when de-nosing in MIG Soldering with shower move.

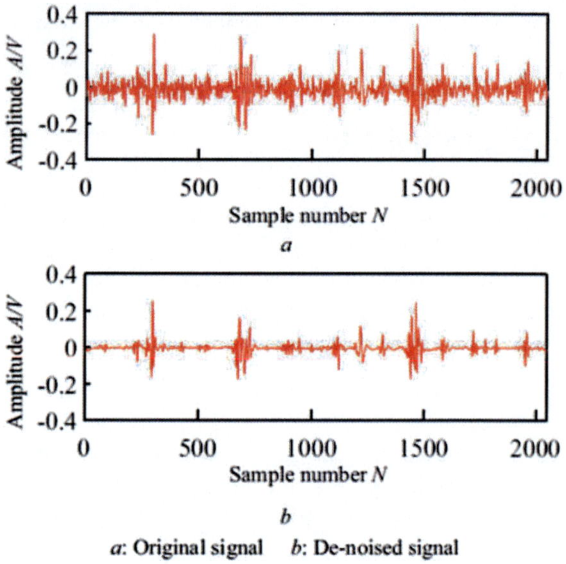

Fig. 2 Time-domain waveform of arc sound signal de-noised by wavelet

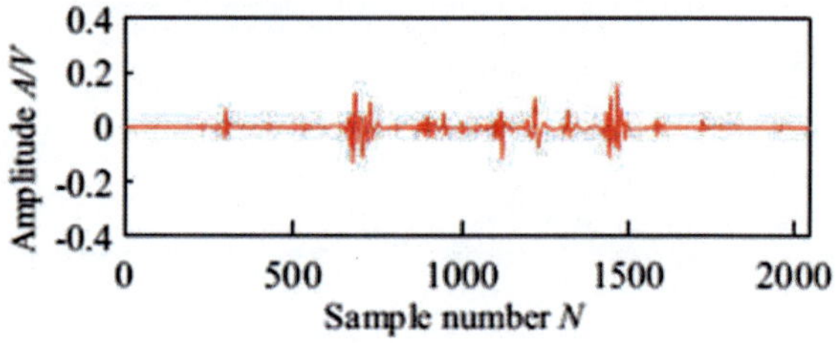

Fig. 3 Waveform of windowed arc sound signal using hamming at frame 1

Fig. 4 DAQ system of arc sound signal in MIG welding

5 Short-Time Windowing

Circular segment sound, originating from a quickly altering of plasma stream among cathode and anode for the duration of soldering, is a sort of time-differing non-stationary irregular signal. Nevertheless, current investigation strategies and speculations are just for stationary signal. On a fundamental level, curve sound signal may be seen as brief period stationary or semi-stationary signal, that is, brief timeframe examination may be adjusted to it. The over-all exercise is to block bend sound signal to shape investigation outline by utilizing a window work with restricted length. For the denoised circular segment sound signal in Fig. 4, the after effect of picking Hamming window is appeared in Fig. 3.

6 Feature Extraction

To understanding on-line acknowledgment of entrance status in MIG Soldering by utilize circular segment sound sign, the reason is that curve sound sign with high repetition ought to be communicated as substantial parameters which mirror idea of infiltration. As it were, it is first basic to manage extraction and evaluation of trademark constraints of curve sound sign.

(A) Time-domain

The most nature and direct technique to investigate curve sound signal is to require some serious energy t as free factor.

(B) Frequency-domain

Attributes of circular segment sound sign, somewhat, have portrayed the progressions of weld entrance status, yet are insufficient to be vital state of infiltration arrangement.

7 Experimental Procedure

Mild steel plates (IS2062 E 250A) with the Soldering on of size 1 inch × 2 Inch, 5 mm thick 50 metal sheets. arranged The direction of the Soldering anode concerning the weld joint was Arc sound Soldering Pressure of protecting gas chamber is taken as 50 kg/cm^2 and 4 kg/cm^2 understand the interrelationship between process variable weld speed m/min and weld dot profile different parameters are kept consistent.

Chemical composition % of the Product analysis of grade IS 2062E250 (Fe 410WA)

C	Si	Mn	N	P	S	Cr	Mo
0.23max	0.40max	1.50	0.012 max	0.045 max	0.045 max		

Mechanical properties of grade IS2062 E 250A

Grade	Thickness (mm)	Min yield (Mpa)	Tensile (Mpa)	Elongation (%)	Min impact energy
IS2062E250A	<20 mm	Min 250	410a	23	0
IS2062E250A	20–40 mm	Min 250	410	23	0
IS2062E250A	>40 mm	Min 230	410	23	0

Equivalent Grades of Is2062 E250A

EN10025(93)	EN10025(90)	DIN17100	NFA35-501	BS4360	JIS3101
S275JO	Fe430C	St44-3U	E28-3	43C	–

8 Results

S. No	Soldering voltage (V)	Soldering current (AMP)	Soldering speed (MM/MIN)	Wire speed (MM)	Gas flow rate (min)
1	20	120	16.7	2.6	14
2	20	130	17.2	3.0	14
3	20	150	18.5	3.8	14
4	20	160	19.5	4.2	14
5	20	170	19.9	4.7	14
6	20	180	20.0	5.2	14
7	20	190	19.3	5.6	14
8	20	200	19.5	5.8	14
9	20	210	22.0	3.2	14
10	20	220	22.5	3.6	14
11	20	230	28.9	4.0	14

9 Conclusions

Connection between circular segment sound sign in MIG butt Soldering with splash move and weld entrance was talked about. Significant discoveries were abridged as follows: perceptible curve sound sign picked progressively and trademark parameters removed may be viewed as quantization element during time spent MIG welding. PCA-based combination innovation of highlight level constraints, it tends to be realized that low-dimensional eigenvector re-orchestrated, conveying the most extreme entrance data, can all the more likely delineate the coordinating connection between circular segment sound sign and infiltration status. Both RBF and BP neural system models, in view of decreased dimensional eigenvector by PCA innovation, understood on-line assessment of infiltration position.

References

1. Wang CM, Wu SP, Hu LJ et al (2007) Identification of different laser welding penetration status based on multi-sensor fusion. Chin J Lasers 34(4):538–542
2. Shuang YQ, Shan JG, Chen WZ et al (2008) Study on the method of vision detection on penetration status in Laser-MIG hybrid welding. Electric Weld Mach 38(12):45–49
3. Liao P, Yang WJ, Wang W (1997) A brief view on the development of the penetration detection and control for bead. J Jiamusi Inst Technol 3:245–248
4. Wu QS, Feng JC, He JS (2006) Determination of characteristic parameters of weld penetration in electron beam welding of aluminum alloy. Trans China Weld Inst 27(2):15–18

5. Inoue K, Takahashi Y, Zhang J (1993) Measurement of arc sound with burn-through in MAW welding. Weld Int 7(10):770–775
6. Cudina M, Prezelj J (2003) Evaluation of the sound signal based on the welding current in the gas-metal arc metal welding process. Proc Inst Mech Eng 217(5):483–494
7. Ma YZ, Ma WB, Qu M et al (2006) Characteristics analyzing and parametric modeling of the arc sound in CO_2 GMAW for on-line quality monitoring. China Weld 15(2):6–13
8. Schiebeck E (1991) Audible range acoustic diagnosis of the MAG welding arc. Weld Int 5(7):572–576
9. Manz A (1981) Welding arc sounds. Weld J 60(5):23–27
10. Saini D, Floyd S (1998) An investigation of gas metal arc welding sound signature for on-line quality control. Weld J **77**(4):172–179
11. Ladislav G, Janez G, Ivan P (2004) Feasibility study of acoustic for on-line monitory in short circuit gas metal arc welding. Int J Machine Toll Manuf 44:551–561
12. Wang YW, Chen Q, Sun ZG et al (2001) Sound sensing of the keyhole behaviors in plasma arc welding. Chin J Mech Eng 37(1):53–56
13. Liu JL, Chen YB, Xu QH (2006) Correlation of acoustic signals and weld depth in laser welding. Trans China Weld Inst 27(1):72–80
14. Brown JD, Rodd MG, Willms NT et al (1998) Application of artificial intelligence techniques to resistance spot welds. Ironmaking Steelmaking 25(3):199–204

A 0.48 mW High Performance 4-Bit Flash ADC for System-on-Chip Applications in 90 nm CMOS Technology

S. Sanjay Kumar, Vidushi Goel, Deepak Prasad, and Vijay Nath

Abstract In the current research article, a complete new design of 4-bit flash ADC is proposed and discussed. The proposed flash ADC can be integrated with CMOS sensors where obtained outputs are analog in nature. This paper presents the design of 4-bit flash ADC using Transistor Inverter Quantization (TIQ) comparator, which is far better comparator than power hungry conventional resistive ladder network. Three different types of encoders are used for designing and on showing best result is being proposed. The differential non-linearity (DNL) and integral non-linearity (INL) have been tested and found to be 0.42 and 1.76 which are well within the acceptable limits. The FFT analysis has also been done using Cadence tools. To check the robustness of proposed design in real environment, process corner analysis has been performed. In the above analysis, dynamic parameters being used are ENOB, SNDR and THD. The presented 4-bit flash ADC utilizes an active area of 0.0107 mm^2 with 0.48 mW power consumption. The proposed flash ADC is implemented in Cadence virtuoso analog and digital design environment using 90 nm CMOS technology. For the proper operation of the circuit, a power supply of +1 V is used.

Keywords MOSFETs · ADC · DNL · INL · FFT · ENOB · SNDR · THD · TIQ

1 Introduction

The analog signals found in nature are quite difficult to process as compared to digital signals. Rapid advancement in the integrated circuit technology has led to the

S. Sanjay Kumar (✉) · V. Goel · D. Prasad · V. Nath
Department of ECE, Birla Institute of Technology, Mesra, Ranchi, JH 835215, India
e-mail: sanjaysurishetty@gmail.com

V. Goel
e-mail: vidushigoggle@gmail.com

D. Prasad
e-mail: prasaddeepak007@gmail.com

V. Nath and J. K. Mandal (eds.), *Proceedings of the Fourth International Conference on Microelectronics, Computing and Communication Systems*, Lecture Notes in Electrical Engineering 673, https://doi.org/10.1007/978-981-15-5546-6_78

development in the field of digital signal processing. Therefore, a conversion from analog to digital form is required to make the processing easier. An analog to digital converter is an interfacing system for analog signal and digital signal processing systems. ADCs are chosen due to its conversion process of analog signal into digital signal. There are various ADCs among which the flash ADC is preferred for its low power and high speed design such as aerospace and satellite communication applications. Flash ADC due to its parallel operation is most suitable for these applications. Conventionally, $2^n - 1$ comparators are required by N-bit flash ADC. The architecture of a conventional resistive ladder flash ADC is being shown in Fig. 1. The reference voltages for comparators are given externally by using resistive ladder. They are spaced at equal intervals from largest to smallest reference voltages. With common analog input to all the comparators placed in parallel manner so that it produces output in one-clock cycle. This output is in some specific manner known as "thermometer code". Encoder follows next and converts thermometer code to binary code [1]. This paper presents the design of 4-bit flash ADC using Transistor Inverter Quantization (TIQ) comparator. TIQ comparator generates an internal reference voltage. The main challenges that come while designing ADC for SoC application are low power, low voltage and high speed. Power reduction is major necessity in today's world in conventional flash ADC as number of bits increases proportionally,

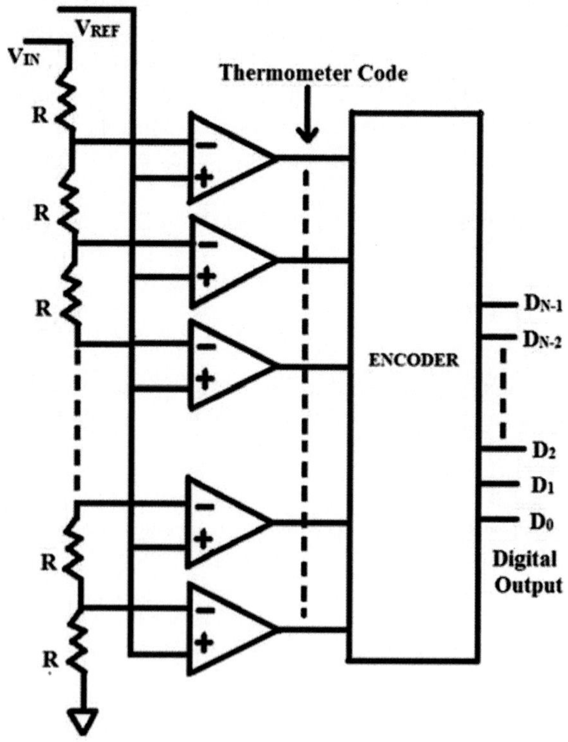

Fig. 1 Architecture of N-bit flash ADC

the number of comparators required increases which results in large die size and large power consumption due to usage of more resistors. So, for reduction in power consumption, we are using TIQ comparator in the proposed design in which reference voltage is generated internally. The comparator output (thermometric code) is fed into encoder and converted to respective binary value. Various encoders can be used for this conversion like multiplexer-based encoder, Wallace tree encoder, fat tree encoder, ROM type encoder and logic style encoder. While designing such encoder, we must take care of parameters like error handling capability and power dissipation. Bubble error in thermometer code is the result of offset voltage in comparator. The presence of 1 s in series of zeroes or 0 s in series of ones in thermometer code is termed as bubble error. Different thermometer code to binary code convertors have different bubble error correction approach and different power consumption, speed and transistor delays. First encoder used is Wallace tree encoder being designed using full adders. To reduce glitches in output of Wallace tree encoder-based design, multiplexer-based design has been proposed. Furthermore, to diminish the effect of bubble error, two-step thermometer code to binary code converter has been introduced. The differential and integral non-linearity have been tested and found well within the acceptable limits. The FFT analysis is done using Cadence tools and then INL and DNL tests are performed using MATLAB. The remainder of the research article is organized as follows. Section 2 portrays the circuit configuration of the conventional flash ADC and proposed 4-bit flash ADC. Section 3 is explaining the post-layout analysis and chip layout. The further post-layout analyses along with robustness and reliability verification have been done in Sect. 4.

2 Circuit Configuration

2.1 Threshold Inverter Quantizer (TIQ) Comparator

Here, in our approach towards 4-bit flash ADC, TIQ comparator has been adopted over traditional comparator because of features like low power, reduced size, high noise margin and ease of fabrication using digital CMOS technology [2]. TIQ comparator is the voltage comparator in which the input voltage is compared with built-in reference voltage. For TIQ comparator, inverter's output becomes VDD/2 at the static voltage known as reference (threshold) voltage. It is a cascade of two CMOS inverters as shown in Fig. 2. The first inverter stage is used to set quantization level of input analog signal by changing the "Voltage Transfer Curve" (VTC) by means of transistor sizing [3, 4]. With the use of second inverter stage, we obtain an increase in gain, preventing an unbalanced propagation delay and logic level inversion to make circuit behave as a comparator. Generally, in aspect ratio, length is kept constant while width is changed during the design process. To maintain the same DC threshold levels as the first inverter stage and to keep the linearity in balance for the voltage rising and falling intervals of high frequency input signals, the second

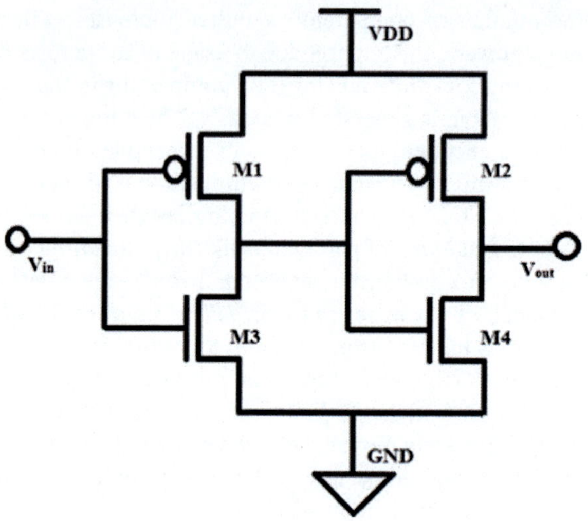

Fig. 2 TIQ comparator

stage must be exactly the same as the first one [5]. When we design inverter, if we increase the NMOS transistor's width, the inverter's threshold voltage will decrease which works oppositely for PMOS. A careful selection of the transistor widths is required to minimize non-linearity errors. The mathematical threshold equation for giving exact values of TIQ comparator's width is given by

$$V\text{th} = \frac{V_{DD} - V_{tp} - V_{tn} * \sqrt{k_n/k_p}}{1 + \sqrt{k_n/k_p)}} \tag{1}$$

where v_{tp} and v_{tn} denote the threshold voltage of PMOS and NMOS, respectively. Similarly, $k_n = (w_p/l_p)\mu_n\text{Co}_x$ and $k_p = (w_p/l_p)\mu_p\text{Co}_x$. Not all the values are used while designing due to limitations on device dimensions.

2.2 *Wallace Tree Encoder*

Following the TIQ comparator, Wallace tree encoder helps in direct conversion of TC to BC as shown in Table 1. Wallace tree encoder serves as one of the method in removing bubble error, presence of 1 s in series of 0 s or vice versa, in thermometer code which results due to offset voltage in comparator. This design requires adder cells [6]. Number of logical 1 s entering each adder cell is being counted to give a 2-bit binary-coded output. We add up these 2-bit words of adjacent cells as shown in Fig. 3 to produce a 4-bit binary output [7]. The number of adder cells required to this encoder design for an N-bit ADC is given by:

Table 1 4-bit truth table TC to BC

$T15$	$T14$	$T13$	$T12$	$T11$	$T10$	$T9$	$T8$	$T7$	$T6$	$T5$	$T4$	$T3$	$T2$	$T1$	$B4$	$B3$	$B2$	$B1$
0	0	0	0	0	0	0	0	0	0	0	0	0	0	0	0	0	0	0
0	0	0	0	0	0	0	0	0	0	0	0	0	0	1	0	0	0	1
0	0	0	0	0	0	0	0	0	0	0	0	0	1	1	0	0	1	0
0	0	0	0	0	0	0	0	0	0	0	0	1	1	1	0	0	1	1
0	0	0	0	0	0	0	0	0	0	0	1	1	1	1	0	1	0	0
0	0	0	0	0	0	0	0	0	0	1	1	1	1	1	0	1	0	1
0	0	0	0	0	0	0	0	0	1	1	1	1	1	1	0	1	1	1
0	0	0	0	0	0	0	0	1	1	1	1	1	1	1	0	1	1	1
0	0	0	0	0	0	0	1	1	1	1	1	1	1	1	1	0	0	0
0	0	0	0	0	0	1	1	1	1	1	1	1	1	1	1	0	0	1
0	0	0	0	0	1	1	1	1	1	1	1	1	1	1	1	0	1	0
0	0	0	0	1	1	1	1	1	1	1	1	1	1	1	1	0	1	1
0	0	0	1	1	1	1	1	1	1	1	1	1	1	1	1	1	0	0
0	0	1	1	1	1	1	1	1	1	1	1	1	1	1	1	1	0	1
0	1	1	1	1	1	1	1	1	1	1	1	1	1	1	1	1	1	0
1	1	1	1	1	1	1	1	1	1	1	1	1	1	1	1	1	1	1

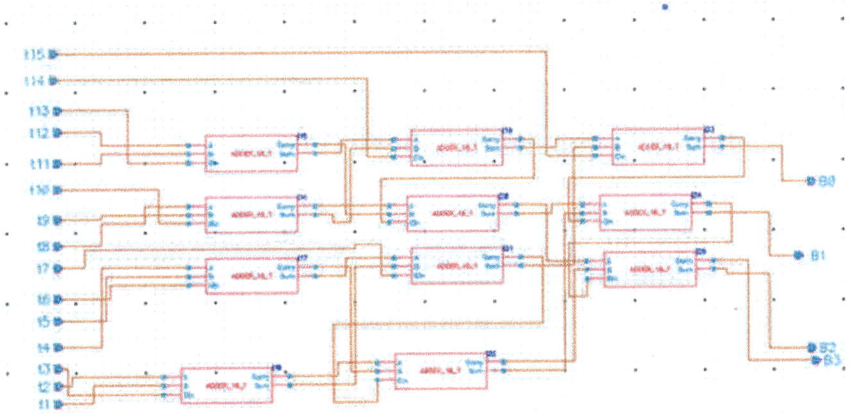

Fig. 3 4-Bit Wallace tree encoder

$$X_N = \sum_{i=1}^{N} (i-1)2^{N-i} \tag{2}$$

So, for 4-bit, we require 11 adder cells which are shown in Fig. 3. The TG logic is used in an adder cell. When we apply high voltage to n-channel transistor, the output voltage becomes $V_{dd} - V_{tn}$. Similarly, when we apply low VSS to p-channel transistor, the V_{tp} drop is present in output which is undesired [8]. Eighteen transistor-based full adders have been utilized as it have optimum threshold voltage drop.

2.3 Multiplexer-Based Encoder

While using Wallace tree encoder for flash ADC design, with an increase in resolution, the required number of full adders increases leading to an increase in area and overall power dissipation. The hardware used here is less and delay is also less which makes it better compared to Wallace tree encoder [9, 10]. Furthermore, this design reduces glitches produced in the output of Wallace tree encoder-based design. The conventional 2:1 multiplexer requires 12 transistors which occupy larger area, more power and more delay. So to minimize these, we used the multiplexer which uses TG logic consisting of six transistors. The TG logic works as when selector line S0 is zero, then PMOS1 and NMOS1 are on and output is D0. Similarly, when S1 is one, then PMOS2 and NMOS2 are on and output is D1. In this way, six transistor-based multiplexers with TG logic are used here [11, 12]. Now, this TG logic-based multiplexer is used in converting TC to BC. Firstly, TC is converted to GC using the following truth Table 2 and respective equations. From these equations, circuit is designed as shown in Fig. 4. Eleven multiplexers are used in this encoder [13]. BC is obtained from GC using exclusive-OR logic.

$$G4 = T8 \tag{3}$$

$$G3 = \overline{T8} \cdot T4 + T8 \cdot \overline{T12} \tag{4}$$

$$G2 = \overline{T8} \cdot \left(\overline{T4} \cdot T2 + T4 \cdot \overline{\overline{T6}}\right) + T8 \cdot \left(\overline{T12} \cdot T10 + T12 \cdot \overline{T14}\right) \tag{5}$$

$$\begin{aligned} G1 = {} & \overline{T8} \cdot \left[\overline{T4} \cdot \left(\overline{T2} \cdot T1 + T2 \cdot \overline{T3}\right) + T4\left(\overline{T6} \cdot T5 + T6 \cdot \overline{T7}\right)\right] \\ & + T8\left[\overline{T12} \cdot \left(\overline{T10} \cdot T9 + T10 \cdot \overline{T11}\right) + T12\left(\overline{T14} \cdot T13 + T14 \cdot \overline{T15}\right)\right] \end{aligned} \tag{6}$$

2.4 Proposed Design

Our design is based on directly converting thermometer code to binary code with intermediate grey code which is useful in bubble error correction which is shown in Fig. 5. Use of grey code provides less probability of error than binary code as it has only one bit difference in every successive input. TG logic used in multiplexer-based encoder has slower speed of operation and also reduces output voltage swing. Compared to multiplexer-based encoder, here, we use simple logic gates which

Table 2 4-bit truth table for TC to GC

T15	*T14*	*T13*	*T12*	*T11*	*T10*	*T9*	*T8*	*T7*	*T6*	*T5*	*T4*	*T3*	*T2*	*T1*	*T0*	*G4*	*G3*	*G2*	*G1*
0	0	0	0	0	0	0	0	0	0	0	0	0	0	0	1	0	0	0	0
0	0	0	0	0	0	0	0	0	0	0	0	0	0	1	1	0	0	0	1
0	0	0	0	0	0	0	0	0	0	0	0	0	1	1	1	0	0	1	1
0	0	0	0	0	0	0	0	0	0	0	0	1	1	1	1	0	0	1	0
0	0	0	0	0	0	0	0	0	0	0	1	1	1	1	1	0	1	1	0
0	0	0	0	0	0	0	0	0	0	1	1	1	1	1	1	0	1	1	1
0	0	0	0	0	0	0	0	0	1	1	1	1	1	1	1	0	1	0	1
0	0	0	0	0	0	0	0	1	1	1	1	1	1	1	1	0	1	0	0
0	0	0	0	0	0	0	1	1	1	1	1	1	1	1	1	1	1	0	0
0	0	0	0	0	0	1	1	1	1	1	1	1	1	1	1	1	1	0	1
0	0	0	0	0	1	1	1	1	1	1	1	1	1	1	1	1	1	1	1
0	0	0	0	1	1	1	1	1	1	1	1	1	1	1	1	1	1	1	0
0	0	0	1	1	1	1	1	1	1	1	1	1	1	1	1	1	0	1	0
0	0	1	1	1	1	1	1	1	1	1	1	1	1	1	1	1	0	1	1
0	1	1	1	1	1	1	1	1	1	1	1	1	1	1	1	1	0	0	1
1	1	1	1	1	1	1	1	1	1	1	1	1	1	1	1	1	0	0	0

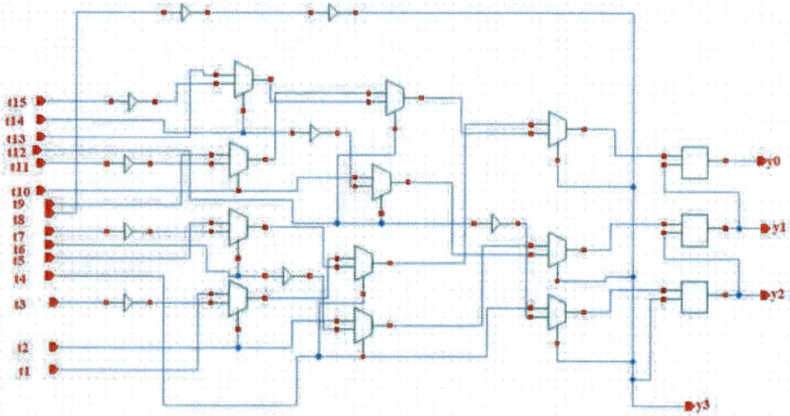

Fig. 4 Multiplexer-based encoder

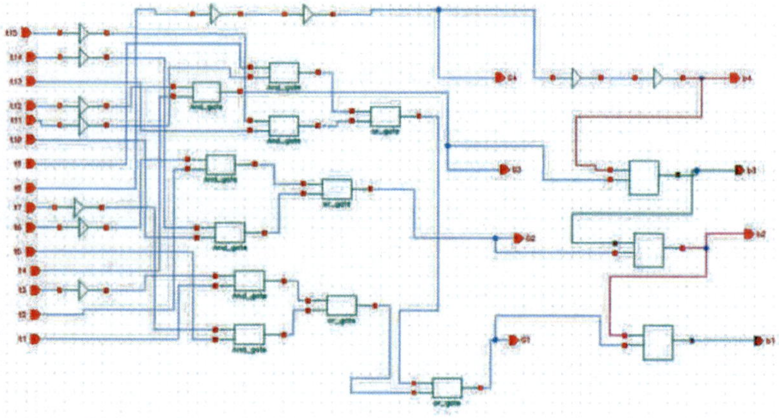

Fig. 5 Proposed encoder

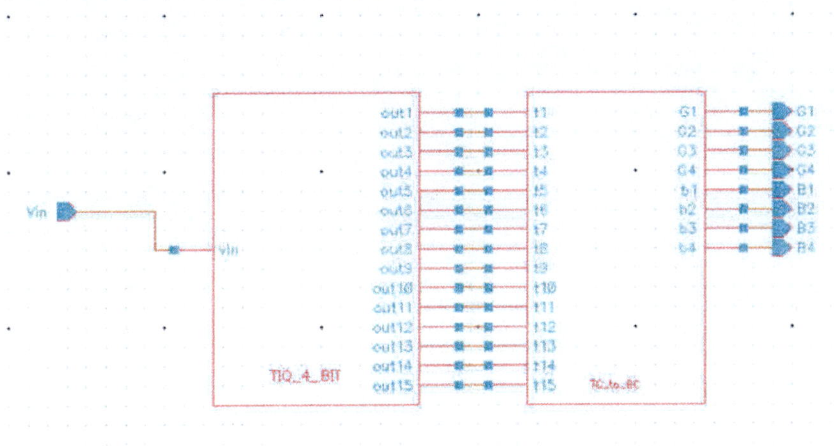

Fig. 6 Flash ADC using proposed encoder

reduce die area and power consumption. TC to GC and then GC to BC method are preferred for designing flash ADC because of its capability of bubble error tolerance, less power consumption, acceptable power delay product and higher speed of operation. The complete structure of proposed flash ADC is shown in Fig. 6 and its output is shown in Fig. 7. On the other hand, Table 3 compares the different encoder in terms of transistor utilized and power consumed.

Ta. Simple logic gates used here are according to the equations fro and above truth tables as follows:

$$G4 = T4 \tag{7}$$

$$G3 = T4 \cdot \overline{T12} \tag{8}$$

$$G2 = T2 \cdot \overline{T6} + T10 \cdot \overline{T14} \tag{9}$$

$$G1 = T1 \cdot \overline{T3} + T5 \cdot \overline{T7} + T9 \cdot \overline{T11} + T13 \cdot \overline{T15} \tag{10}$$

The grey code is further converted to binary code using exclusive XOR logic.

$$B4 = G4 \tag{11}$$

$$B3 = B4 \oplus G3 \tag{12}$$

$$B2 = B3 \oplus G2 \tag{13}$$

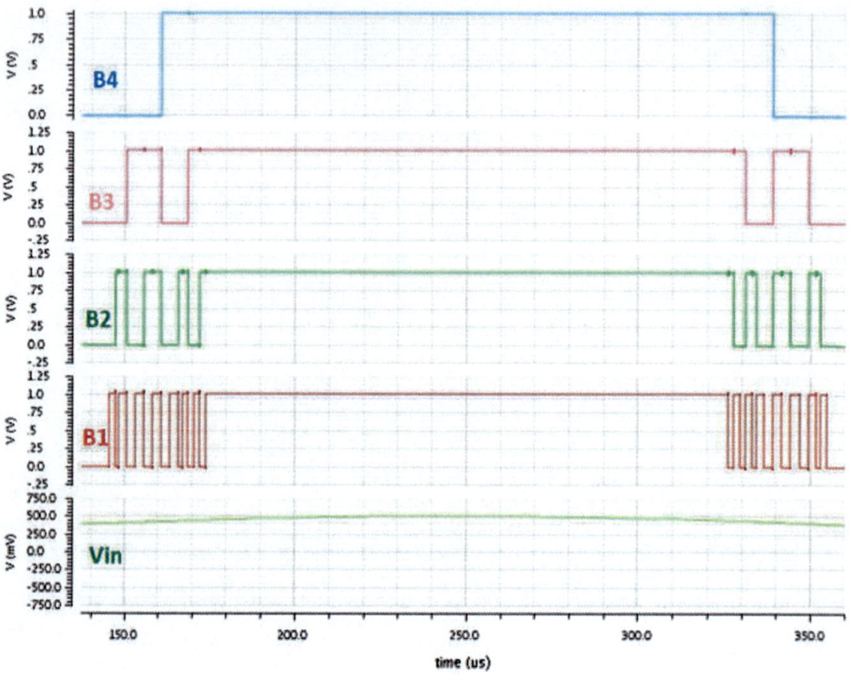

Fig. 7 Output waveform of flash ADC with proposed encoder

Table 3 Comparison between three types of encoders

	Wallace tree encoder	Multiplexer-based encoder	Proposed encoder
Number of transistors	198	132	136
Power consumed by ADC (mW)	2.62	0.878	0.48

$$B1 = B2 \oplus G1 \tag{14}$$

3 Realization Measurement

A full layout is designed using 90 nm CMOS technology to demonstrate the design's suitability. The proposed circuit occupies an active area of 0.0107 mm^2. The layout is shown in Fig. 8 and the micrograph of the chip is shown in Fig. 9.

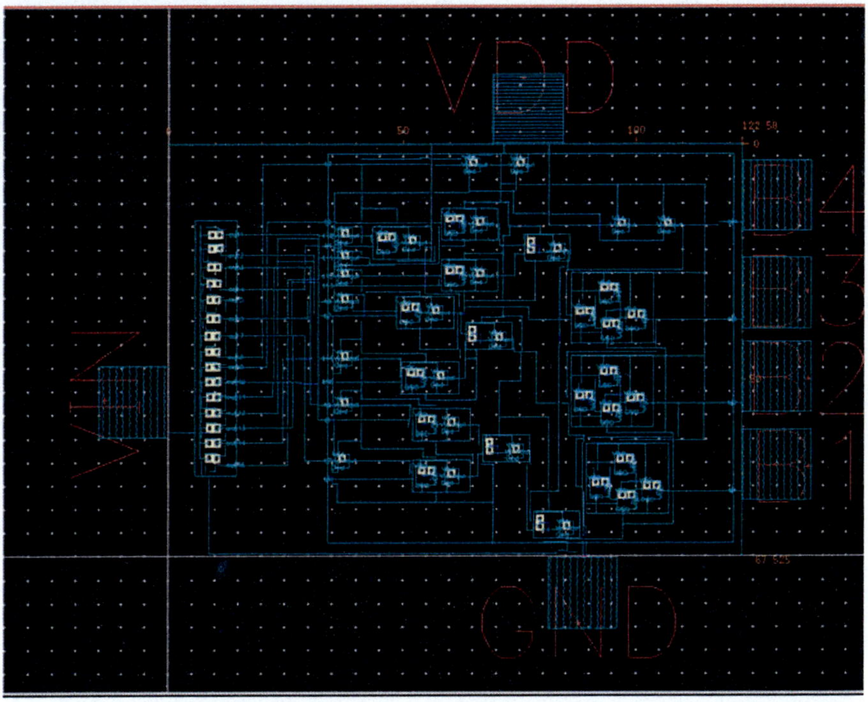

Fig. 8 Layout of proposed flash ADC

4 Measurement Results

4.1 DNL and INL Analysis

4.1.1 Differential Non-Linearity (DNL)

DNL error is the deviation of width of an actual quantization step from the ideal value of 1 LSB. Specifying DNL defines the entire characteristics of ADC. An ideal ADC DNL is equal to 0 LSB. Absolute value of DNL should be less than 0.5 is sufficient condition to ensure no missing code [14, 15]. Large values of DNL can limit the ADC performance in terms of signal-to-noise ratio (SNR). The analog voltage versus output digital code is shown in Fig. 10. The DNL plot is shown in Fig. 11. The DNL for an N-bit ADC is calculated as:

$$\text{DNL}[K] = \frac{(W_K - 1\text{LSB})}{1\text{LSB}} \tag{15}$$

where 1 LSB = 4.2 units from graph, and W_K is the width of code k for $0 < k < 2^N - 1$.

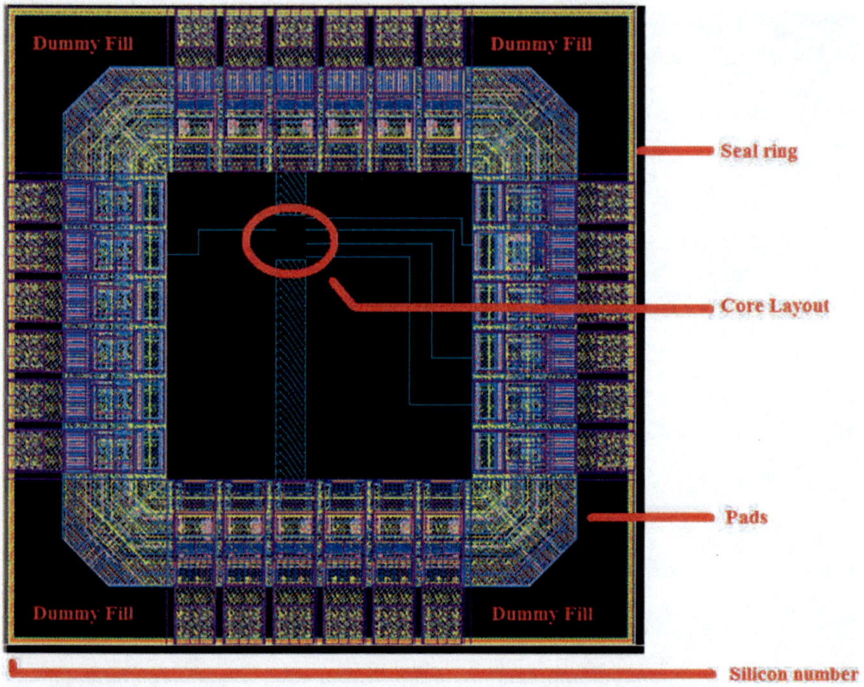

Fig. 9 Micrograph of chip

4.1.2 Integral Non-Linearity (INL)

INL error is the deviation of the code transition w.r.t the ideal transition. It can also be stated that INL is the running sum of DNL [16, 17]. The INL plot is shown in Fig. 12. The INL for an N-bit ADC is calculated as:

$$\mathrm{INL}[K] = \frac{\left(\sum_{i=1}^{k-1} W_i - k * \mathrm{LSB}\right)}{1\mathrm{LSB}} \tag{16}$$

where 1 LSB = 4.2 units from graph, W_i is the width of ith code for $0 < k < 2^N - 1$.

Power comparison plot for Flash ADC using three types of encoders is shown in Fig. 13.

4.2 Process Corner Analysis

Process corner refers to a variation of fabrication parameters. It represents the extremities of these parameter variations within which circuit should function correctly even

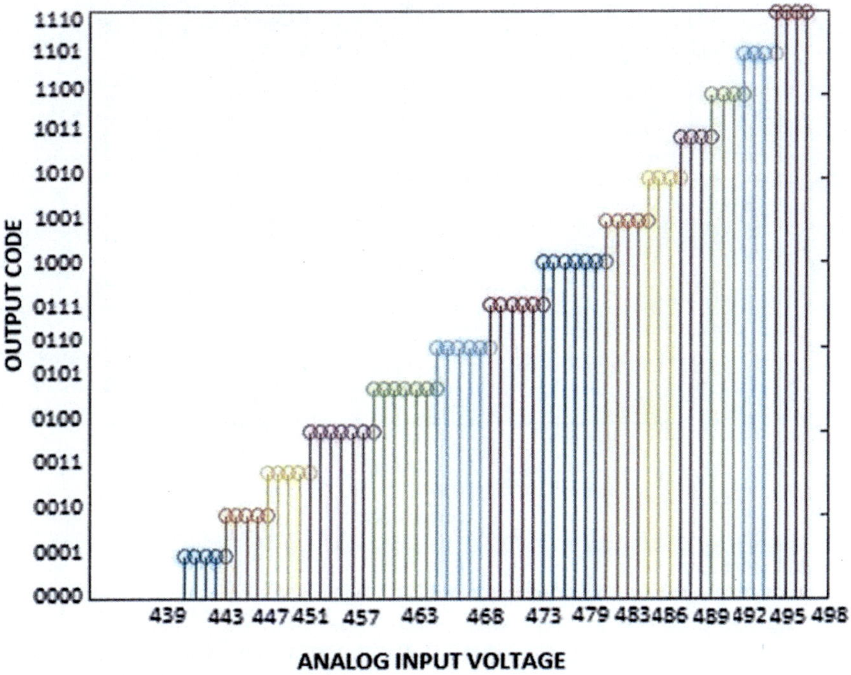

Fig. 10 Digital output versus analog input

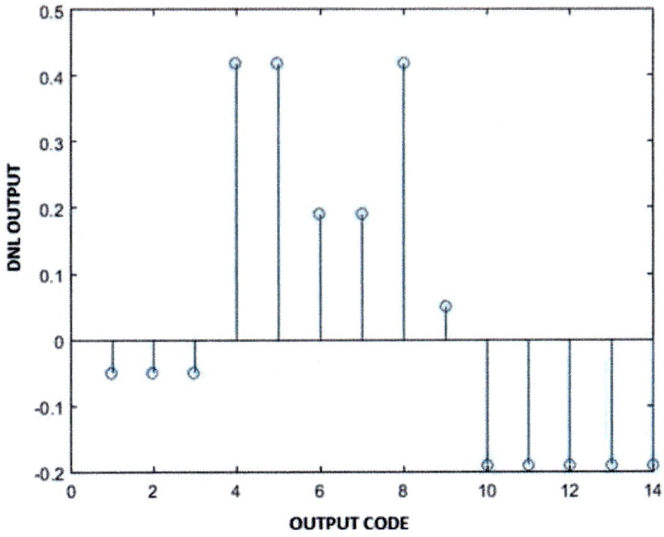

Fig. 11 DNL output

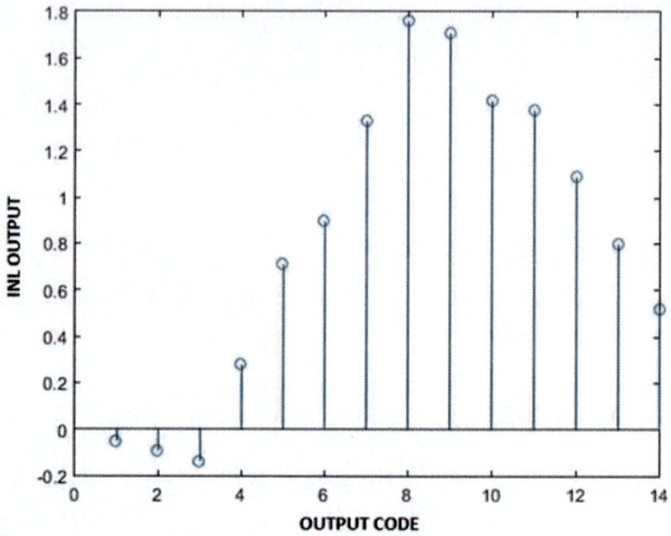

Fig. 12 INL output

if transistors work slower or faster due to temperature voltage variations [18]. Analysis of parameters like effective number of bits (ENOB), signal-to-noise distortion ratio (SNDR) and power consumption has been made. The complete process corner analysis is tabulated in Table 4. On the other hand, FFT analysis and total power drawn graphs are shown in Figs. 14 and 15.

4.2.1 Effective Number of Bits

ENOB measures dynamic range of an ADC. The ENOB is calculated by weighted sum of ENOBs obtained through FFT analysis of binary outputs using cadence tools.

4.2.2 Signal-to-Noise Distortion Ratio

SNDR or SINAD is a ratio of root mean square of signal amplitude to mean value. SNDR is calculated using the following equation [19].

$$\mathrm{SNDR} = 6.02 * \mathrm{ENOB} + 1.76 \tag{17}$$

Also, $\mathrm{SNDR} = 20\log\left(\frac{S}{N+D}\right)$.
Signal-to-noise ratio, $\mathrm{SNR} = 20\log\left(\frac{S}{N}\right)$

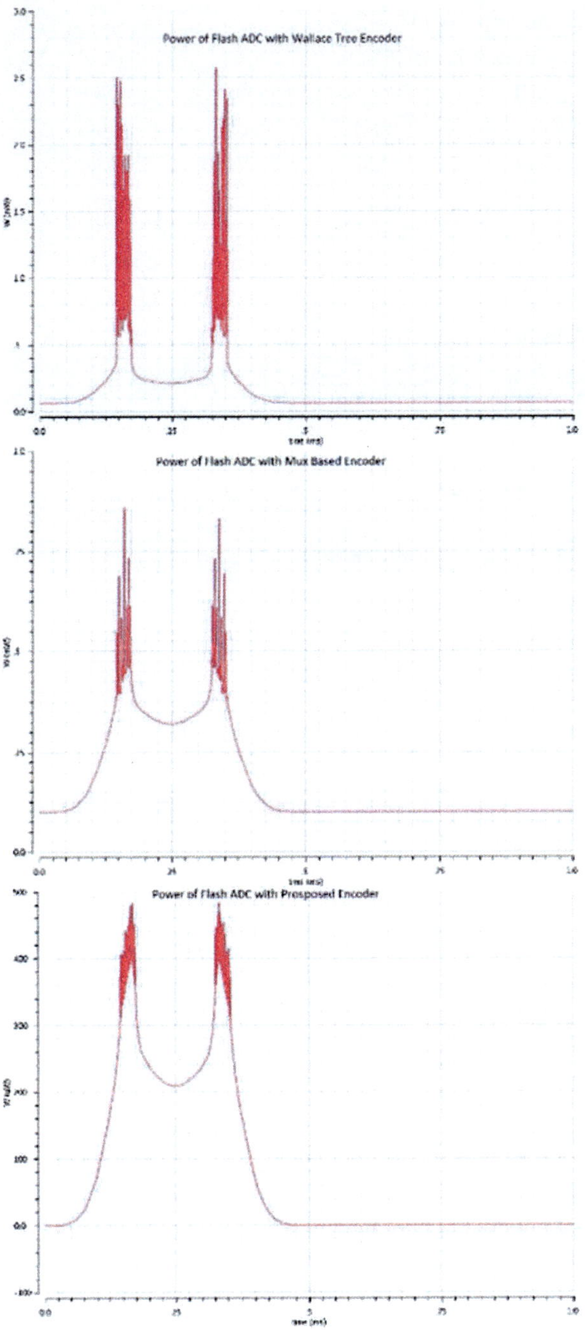

Fig. 13 Power comparison plot

Table 4 Process corner analysis

Parameter	Process corner				
	TT	SS	FF	SNFP	FNSP
ENOB	2	1.967	2.01	2.34	2
SNDR (dB)	13.8	13.6	13.8	15.8	13.8
THD (dB)	14.08	13.86	14.14	16.29	14.08

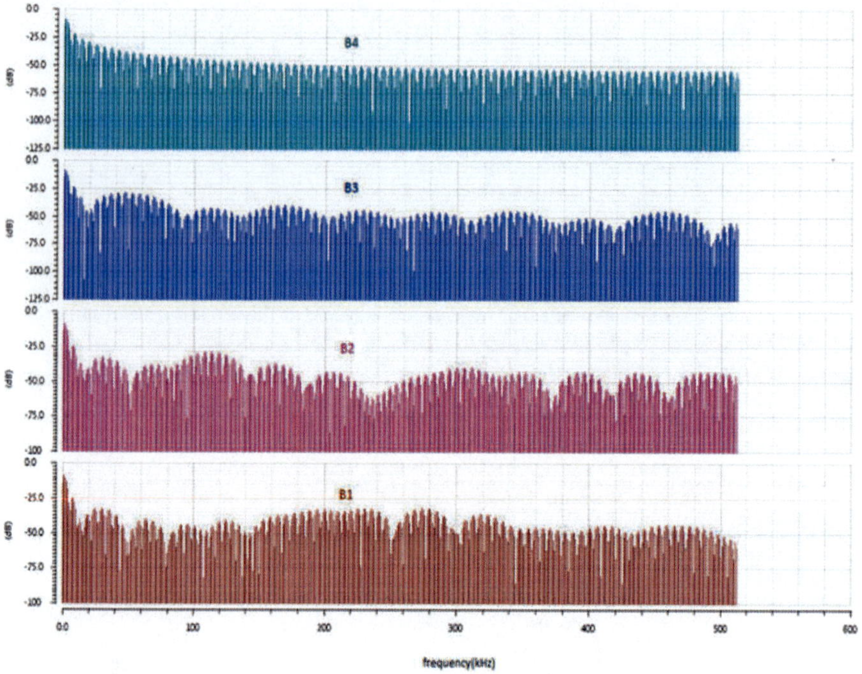

Fig. 14 FFT analysis

4.2.3 Total Harmonic Distortion

THD specifies the ratio of the rms value of the fundamental signal to the mean value of the rms of its harmonics. THD of an ADC is also usually specified with the input signal nearly full-scale.

$$\mathrm{THD} = 20\log\left(\frac{S}{D}\right) = -10\log\left[10^{-\mathrm{SINAD}/10} + 10^{-\mathrm{SNR}/10}\right] \tag{18}$$

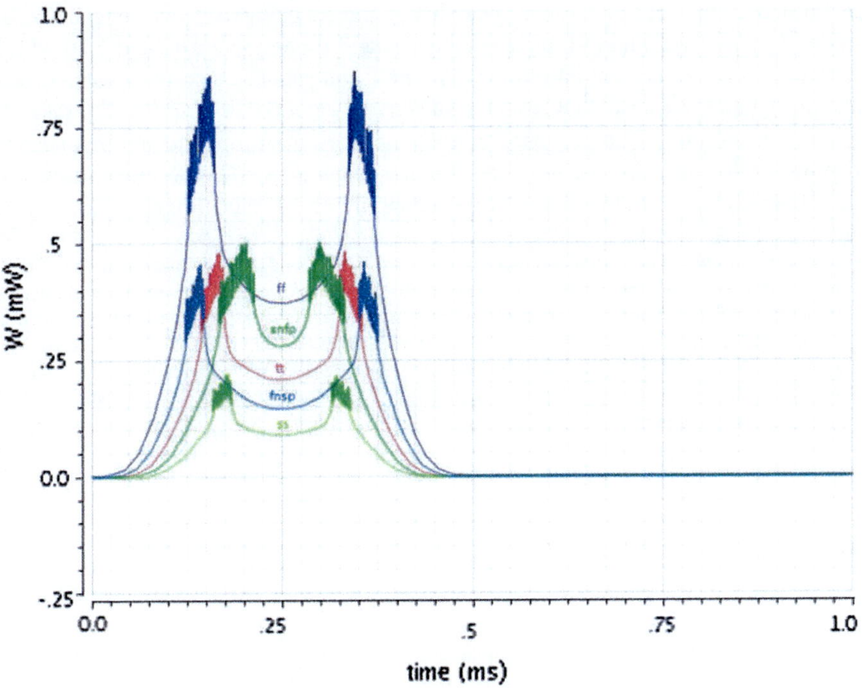

Fig. 15 Total power drawn graph

5 Conclusion

This research paper concludes with a 4-bit flash ADC design consisting of TIQ comparator with proposed encoder. Table 5 shows the comparison of proposed design with some recent research works. The flash ADC works fine with low power consumption and shows high accuracy in the system-on-chip (SoC) temperature sensor where analog output is being fed from the sensor unit to this ADC. To check the robustness of proposed design in real environment, process corner analysis has been

Table 5 Comparison table

Characteristic	This work	[20]	[21]	[22]	[23]	[24]
Technology	90 nm	180 nm	90 nm	350 nm	180 nm	130 nm
Resolution	4 bit	4 bit	4 bit	4 bit	4 bit	4 bit
INL	1.76	1.1	–	–	–	–
Supply voltage	1 V	–	1.2 V	3.3 V	1.8 V	−1/+ 1 V
Power consumption	0.48 mW	20 mW	23 mW	12.4 mW	24.6 mW	30 mW
DNL	0.42	0.4	–	–		–

performed. In the above analysis, dynamic parameters being used are ENOB, SNDR and THD. The DNL and INL have been tested and found to be 0.42 and 1.76 which are well within the acceptable limits. The FFT analysis has also been carried out. The presented 4-bit flash ADC utilizes an active area of 0.0107 mm^2 with 0.48 mW power consumption. The proposed temperature sensor is implemented in Cadence virtuoso analog and digital design environment using 90 nm CMOS technology. For the proper operation of the circuit, a power supply of +1 V is utilized.

Acknowledgements The authors are heartily thankful to Respond, Indian Space Research Organization (ISRO) Bangalore, India, for funding this research work. The sanction number is ISRO/RES/3/679/16-17. A piece of thanks also goes to Council of Scientific and Industrial Research (CSIR), New Delhi, India, for providing the senior research fellowship to the research scholar. The sanction number is 09/554(0042)/2018-EMR-I, Ack. No. 143183/2k17. The authors also want to thanks Vice-Chancellor, Dr. M. K. Mishra and Head of Department, Dr. S. Pal for their constant support and encouragement.

References

1. Njinowa M, Siadjine M, Bui H, Boye F (2013) Design of low power 4 Bit flash ADC based on standard cells. In: IEEE 11th international new circuits and systems conference. France
2. Prasad D, Nath V (2018) Design of CMOS integrator circuit for sigma delta ADC for aerospace application. In: Industry interactive innovations in science, engineering and technology. India, pp 377–383
3. Mashhadi SB, Lotfi R (2014) Analysis and design of a low-voltage low-power double-tail comparator. In: IEEE transactions on very large scale integration (VLSI) systems, pp 343–352
4. Chandra KS, Kumar VGR (2015) Low power and high speed 4-bit flash analog to digital converter using dynamic latch comparator technique. Int J Adv Res Electr Electron 4:6881–6886
5. Velagaleti S (2009) A novel high speed dynamic comparator with low power dissipation and low offset. M-tech thesis, NIT Rourkela, India
6. Saravanan SV, Ganesh RS, Sasipriya S (2017) Design and performance analysis of low power 4 bit flash analog to digital converter. J Adv Res Dyn Control Syst 01:198–208
7. Varghese GT, Mahapatra K (2016) A low power reconfigurable encoder for flash ADCs. Procedia Technol 25:574–581
8. Pardhu T, Manusha S, Sirisha K (2014) A low power flash ADC with wallace tree encoder. In: 11th international conference on wireless and optical communications networks (WOCN). India
9. Kiranmai M, Patnaikuni VYSSS, Mouli K, Sai BM, Nancharaiah V (2017) CMOS full adder and multiplexer based encoder for low resolution flash Adc. IOSR J Electron Commun Eng 12:20–27
10. Veeramachanen S, Kumar AM, Tummala V, Srinivas MB (2009) Design of a low power, variable-resolution flash ADC. In: 22nd IEEE international conference on VLSI design, India
11. Mayur SM, Siddharth RK, Kumar YBN, Vasantha MH (2016) Design of low power 5-bit hybrid flash ADC. In: IEEE computer society annual symposium on VLSI, USA
12. MAX108 data sheet, Rev. 1, 5/99, Maxim integrated products
13. MAX108EVKIT data sheet, Rev. 0, 6/99, Maxim integrated
14. Johns D, Martin K (1997) Analog integrated circuit design. Wiley, Hoboken
15. Sanchez-Sinencio E, Andreou AG (1999) Low-voltage/low-power integrated circuits and systems: low-voltage mixed-signal circuits. IEEE Press Mark

16. van de Plasche R (1994) Integrated analog-to-digital and digital-to-analog converters. Kluwer Academic Publishers, Dordrecht
17. Inamdar A, Sahu A, Ren J, Setoodeh S, Mansour R, Gupta D Design and evaluation of flash ADC. In: IEEE
18. Mano MM, Ciletti MD (2006) Digital design, 4th edn. Prentice Hall, Upper Saddle River
19. Yoo J (2003) A TIQ based CMOS flash A/D converter for system-on-chip applications. Ph.D. dissertation, Pennsylvania State University
20. Banik S, Gangopadhyay D, Bhattacharyya TK (2006) A low power 1.8 V 4-bit 400 MHz flash ADC in 0.18 μm digital CMOS. In: 19th IEEE international conference on embedded systems and design, USA
21. Megha R, kumar KAP (2014) Implementation of low power flash ADC by reducing comparators. In: IEEE international conference on communication and signal processing, India
22. Chauhan SS, Manabala S, Bose SC, Chandel R (2011) A new approach to design low power CMOS flash A/D converter. Int J VLSI Des Commun Syst (VLSICS) 2
23. Halim ISA, Hassan SLM, Akbar NDBM, Rahim AAA (2012) Comparative study of comparator and encoder in a 4 bit flash ADC using 0.18 μm CMOS technology. In: International symposium on computer applications and industrial electronics (ISCAIE), Malaysia
24. Fan S, Tang H, Zhao H, Wang X, Wang A, Zhao B, Zhao GG (2011) Enhanced Offset Averaging Technique for Flash ADC Design. Tsinghua Sci Technol 16:285–289

Design of 4-Bit Multiplexer-Based Encoder for Analog to Digital Converter

Madhu Kumari Ray, Sanjay Kumar Surshetty, Vidhusi Goel, Deepak Prasad, and Vijay Nath

Abstract A multiplexer-centered thermometer to binary code encoder for 4-bit flash ADC is suggested. The proposed encoder configuration utilizes 2:1 multiplexes (MUX) and two-input XOR gates are used for convert the thermometer code to binary code. Primarily, 2:1 MUX is utilized to change over thermometer code to comparable dark code and resulting dark code is then changed to binary code via XOR gates. The proposed designed encoders delay is less than ordinarily encoders. The proposed encoders circuit are designed using UMC 90 nm CMOS technology at Cadence Virtuoso platform. This circuit is simulated at 1V power supply and resulting power consumptions is 1.55 mW.

Keywords Multiplexer · Encoder · Thermometer code · Gray code

1 Introduction

The analog signals found in nature are quite difficult to process as compared to digital signals. Rapid advancement in technology has led to development in the field of digital signal processing. Therefore, a conversion from analog to digital form is required to make the processing easier. An ADC is an interfacing system for analog signal and digital signal processing systems. ADCs are one of the fastest methods to change analog information into digital information. There are various ADCs among which flash ADC is one of the fastest approaches to adapt analog

M. K. Ray (✉) · S. K. Surshetty · V. Goel · D. Prasad · V. Nath
VLSI Design Group, Department of Electronics and Communication Engineering, Birla Institute of Technology, Mesra, Ranchi, JH 835215, India
e-mail: madhuray0601@gmail.com

V. Goel
e-mail: vidushigoggle@gmail.com

V. Nath and J. K. Mandal (eds.), *Proceedings of the Fourth International Conference on Microelectronics, Computing and Communication Systems*, Lecture Notes in Electrical Engineering 673, https://doi.org/10.1007/978-981-15-5546-6_79

data into digital data. Flash ADC due to its parallel operation is most suitable for various applications. Conventionally, N-bit flash ADC requires comparators. The architecture of a conventional resistive ladder flash ADC is being shown in Fig. 1. While designing such encoder, we must take care of parameters like error handling capability and power dissipation. Bubble error in thermometer code is the result of offset voltage in comparator. The presence of 1s in series of zeroes or 0s in series of 1s in thermometer code is termed as bubble error. Different thermometer code-to-binary

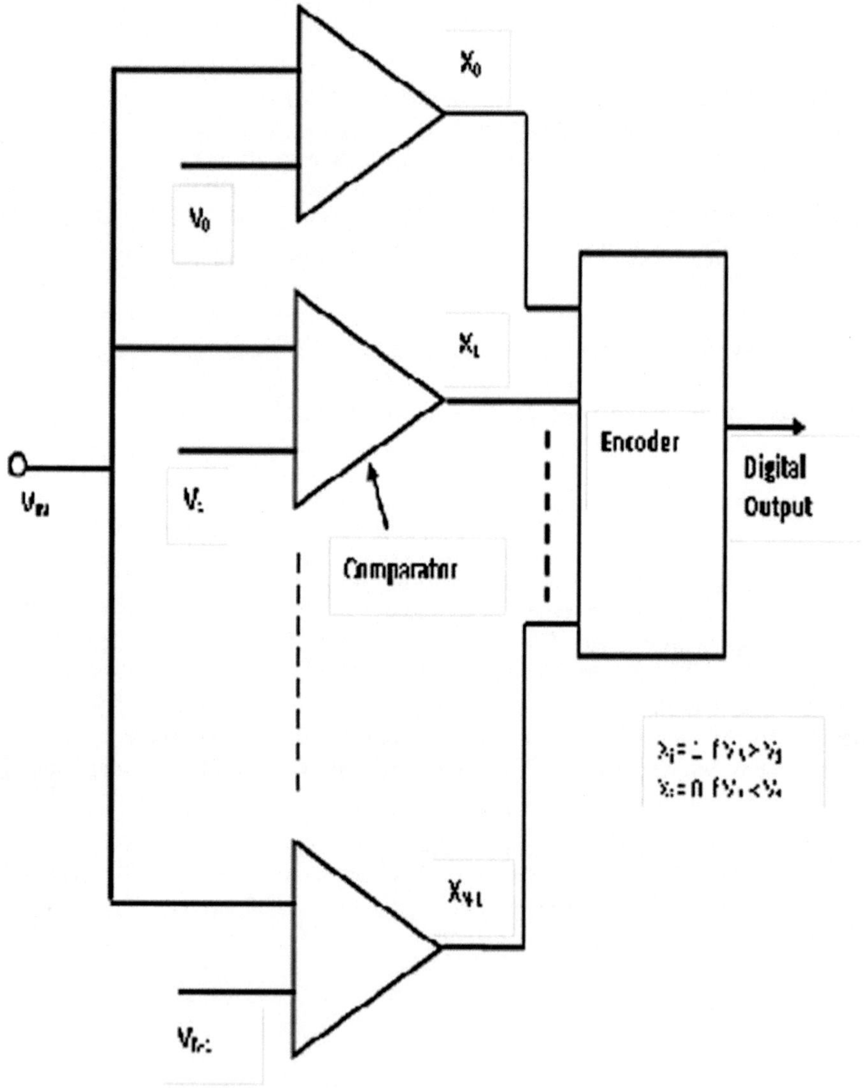

Fig. 1 Block diagram of general analog to digital converter

code converters have different bubble error correction approaches and different power consumptions, speeds, and transistor delays.

In this paper, we have described about MUX-based encoder.

In this encoder to reduce, the bubble error is reduced.

2 Multiplexer-Based Encoder

While using Wallace tree encoder for flash ADC design, with an increase in resolution, the required number of full adders increases leading to an increase in area and overall power dissipation.

The hardware used here is less and delay is also less which makes it better compared to other common encoders [1, 2]. The conventional 2:1 multiplexer requires 12 transistors which occupy larger area, more power, and more delay. So to minimize these, we used the multiplexer which uses TG logic consisting of six transistors (Figs. 2, 3 and Table 1).

The TG logic works as when selector line S0 is zero then PMOS1 and NMOS1 are on and output is D0. Similarly, when S1 is one, then PMOS2 and NMOS2 are on and output is D1. In this way, six transistor-based multiplexers with TG logic are used here [3]. Now, this TG logic-based multiplexer is used in converting TC to BC. Firstly, TC is converted to GC using the following truth Table 2 and respective equations. From these equations, circuit is designed as shown in Fig. 4. Four multiplexers are used in this encoder. BC is obtained from GC using exclusive-OR logic [7–17].

$$G3 = \overline{T8} \cdot T4 + T8 \cdot \overline{T12} \tag{1}$$

$$G2 = \overline{T8} \cdot (\overline{T4} \cdot T2 + T4 \cdot \overline{\overline{T6}}) + T8 \cdot (\overline{T12} \cdot T10 + T12 \cdot \overline{T14}) \tag{2}$$

$$\begin{aligned} G1 = {} & \overline{T8} \cdot [\overline{T4} \cdot (\overline{T2} \cdot T1 + T2 \cdot \overline{T3}) + T4(\overline{T6} \cdot T5 + T6 \cdot \overline{T7})] \\ & + T8[\overline{T12} \cdot (\overline{T10} \cdot T9 + T10 \cdot \overline{T11}) + T12(\overline{T14} \cdot T13 + T14 \cdot \overline{T15}) \end{aligned} \tag{3}$$

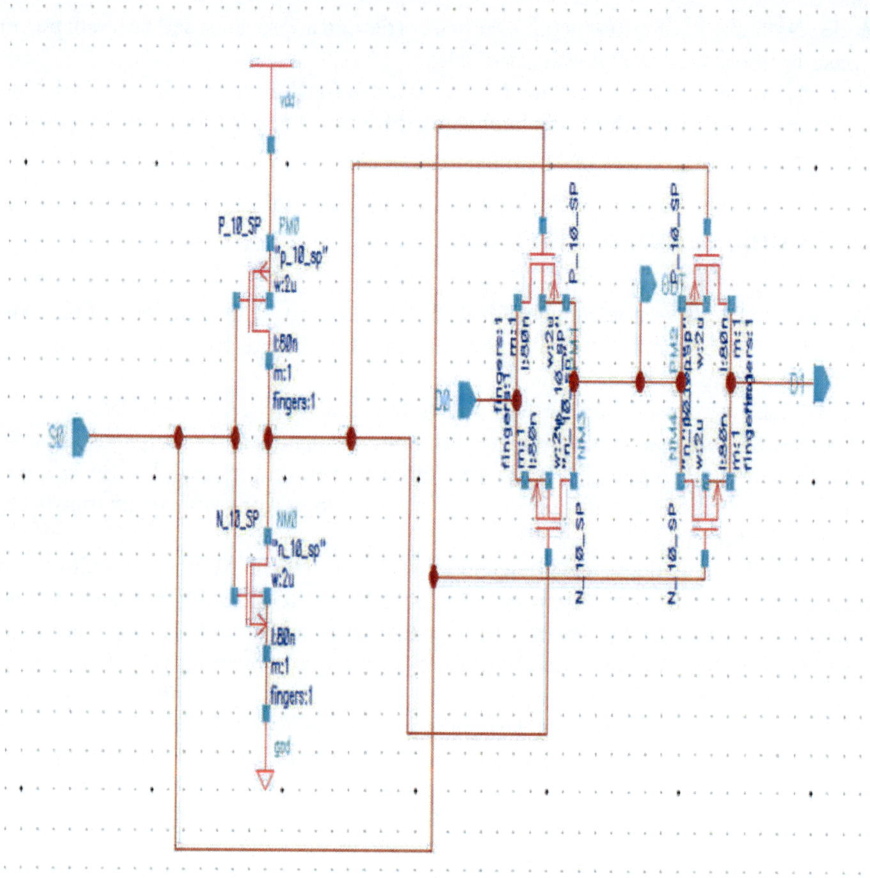

Fig. 2 2:1 Multiplexer

3 Simulation and Result

See Fig. 5 and Table 3.

4 Conclusion

This paper concludes with a 4-bit multiplexer-based encoder design. The design of circuit and simulation results are done using UMC 90 nm technology on cadence virtuoso environment. The proposed design is consuming 1.55 mW power and there are no glitches in the encoder output.

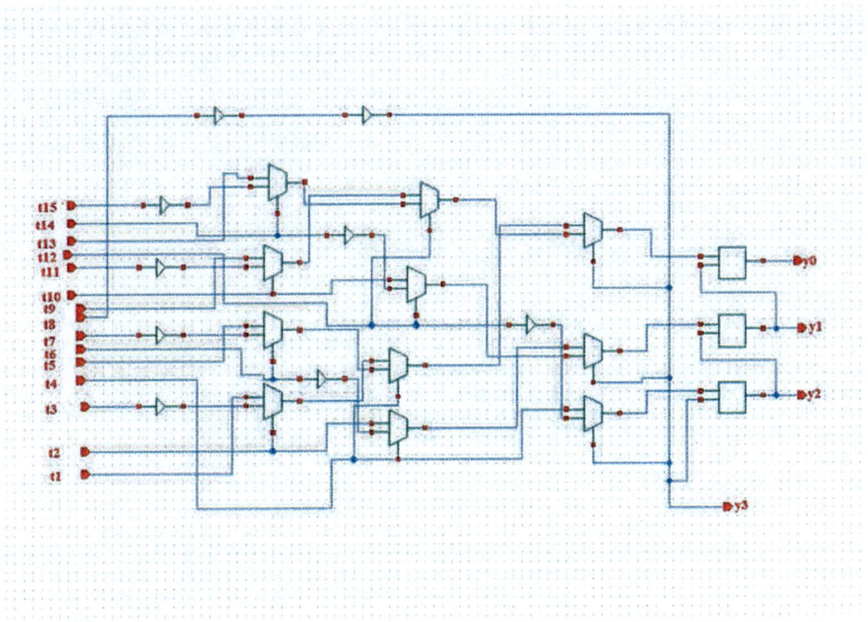

Fig. 3 Multiplexer-based encoder

Table 1 Thermometer code

*G*4	*G*3	*G*2	*G*1
0	0	0	0
0	0	0	1
0	0	1	1
0	0	1	0
0	1	1	0
0	1	1	1
0	1	0	1
0	1	0	0
1	1	0	0
1	1	0	1
1	1	1	1
1	1	1	0
1	0	1	0
1	0	1	1
1	0	0	1
1	0	0	0

Table 2 Gray code

*T*15	*T*14	*T*13	*T*12	*T*11	*T*10	*T*9	*T*8	*T*7	*T*6	*T*5	*T*4	*T*3	*T*2	*T*1	*T*0
0	0	0	0	0	0	0	0	0	0	0	0	0	0	0	1
0	0	0	0	0	0	0	0	0	0	0	0	0	0	1	1
0	0	0	0	0	0	0	0	0	0	0	0	0	1	1	1
0	0	0	0	0	0	0	0	0	0	0	0	1	1	1	1
0	0	0	0	0	0	0	0	0	0	0	1	1	1	1	1
0	0	0	0	0	0	0	0	0	0	1	1	1	1	1	1
0	0	0	0	0	0	0	0	0	1	1	1	1	1	1	1
0	0	0	0	0	0	0	0	1	1	1	1	1	1	1	1
0	0	0	0	0	0	0	1	1	1	1	1	1	1	1	1
0	0	0	0	0	0	1	1	1	1	1	1	1	1	1	1
0	0	0	0	0	1	1	1	1	1	1	1	1	1	1	1
0	0	0	0	1	1	1	1	1	1	1	1	1	1	1	1
0	0	0	1	1	1	1	1	1	1	1	1	1	1	1	1
0	0	1	1	1	1	1	1	1	1	1	1	1	1	1	1
0	1	1	1	1	1	1	1	1	1	1	1	1	1	1	1
1	1	1	1	1	1	1	1	1	1	1	1	1	1	1	1

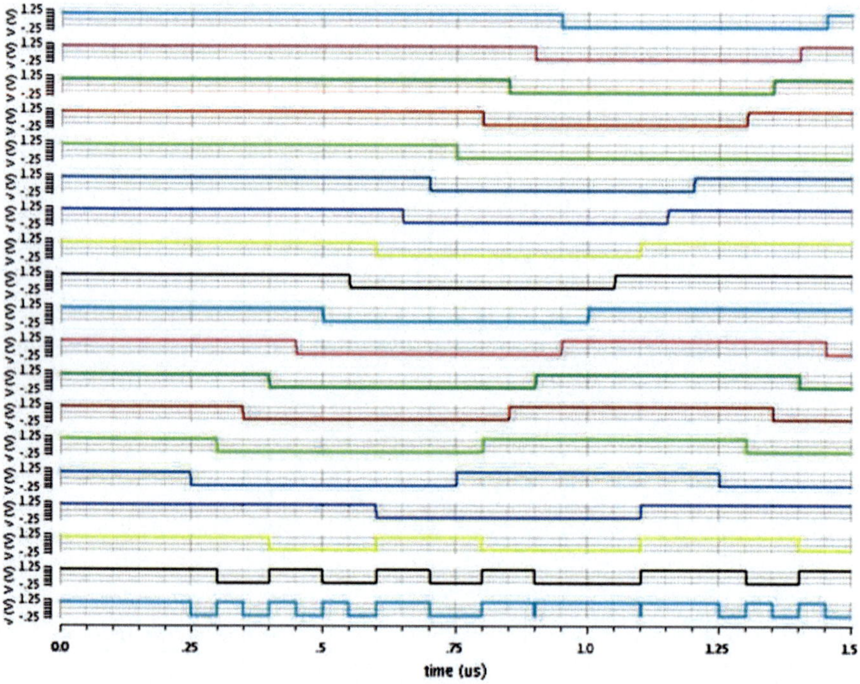

Fig. 4 Output waveform

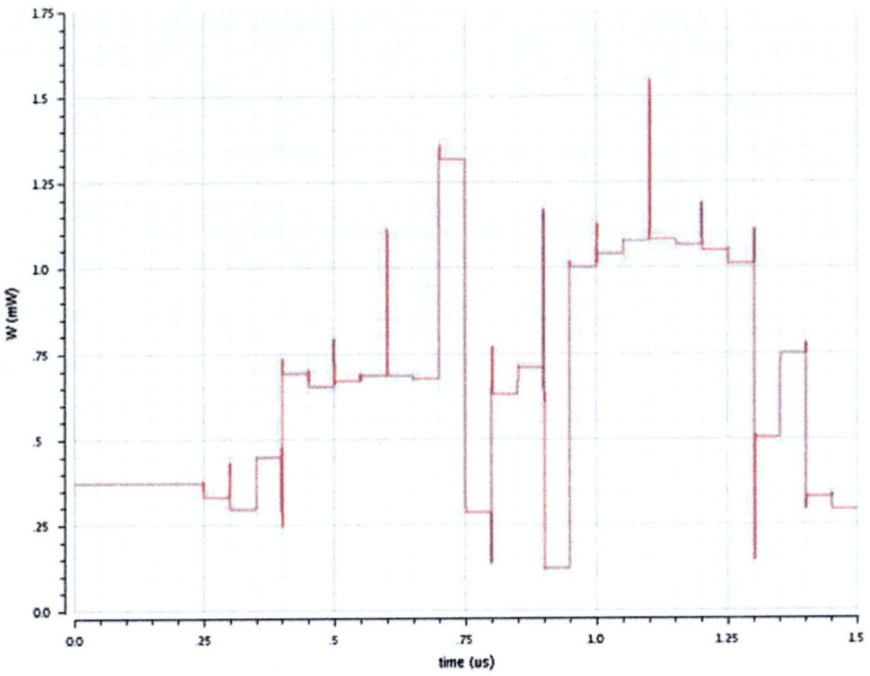

Fig. 5 Power consumption

Table 3 Comparison with other work

Parameter	This work	[4]	[5]	[6]
Architecture	Flash	Flash	Flash	Flash
Resolution	4 bit	4 bit	4 bit	4 bit
Power supply	1 V	1.8 V	1.8 V	1.8 V
Power consumption	1.55 mW	4.19 mW	4.43 mW	36.98 mW

References

1. Kiranmai1 M, Sudir Patnaikuni VYSS, Mouli K, Manikanta sai B, Nancharaiah V CMOS full adder and multiplexer based encoder for low resolution flash Adc
2. Veeramachanen S, Kumar AM, Tummala V, Srinivas MB Design of a low power, variable-resolution flash ADC
3. Mayur SM, Siddharth RK, Kumar YBN, Vasantha MH (2016) Design of low power 5-bit hybrid flash ADC. In: 2016 IEEE computer society annual symposium on VLSI
4. Kalita T, Das B (2016) A 4-bit quantum voltage comparator based flash ADC for low noise applications. In: IEEE
5. Hosur KN, Dariyappa, Shivanand, Vijay, Nagesha, Attimarad GV, Kittur HM (2015) Design of 4 bit flash ADC using TMCC & NOR ROM encoder in 90 nm CMOS technology. ISBN: 978-1-4673-6667-0/15

6. Kulkarni M, Sridhar V, Kulkarni GH (2010) The quantized differential comparator in flash analog to digital converter design. Int J Comput Netw Commun (IJCNC) 2(4). https://doi.org/10.5121/ijcnc.2010.2404
7. Al A, Reaz M, Jalil J, Ali A (2014) An improved a low power CMOS TIQ comparator flash ADC
8. Pardhu T, Manusha S, Sirisha K A low power flash ADC with wallace tree encoder
9. Varghese GT, Mahapatra K A low power reconfigurable encoder for flash ADCs
10. Design and performance analysis of low power 4 bit flash analog to digital converter. Design and performance analysis of low power 4 bit flash analog to digital converter
11. Velagaleti S (2009) A novel high speed dynamic comparator with low power dissipation and low offset. M-Tech thesis, NIT Rourkela, India
12. Morris Mano M, Ciletti MD (2006) Digital design, 4th edn. Prentice Hall, Dordrecht. ISBN: 978-0-13-198924-5
13. Yoo J (2003) A TIQ based CMOS flash A/D converter for system-on-chip applications. Ph.D. thesis, Pennsylvania State University
14. Maxim integrated products, INL/DNL measurements for high-speed analog to-digital converters (ADCs)
15. Njinowa M, Siadjine M, Bui H, Boyer F (2013) Design of low power 4 bit flash ADC based on standard cells. In: IEEE 11th international new circuits and systems conference (NEWCAS)
16. Babayan-Mashhadi S, Lotfi R (2014) Analysis and design of a low-voltage low-power double-tail comparator. IEEE Trans Very Large Scale Integr (VLSI) Syst 22(2):343–352
17. Patil H, Raghavendra M (2016) Low power dynamic comparator for 4-bit flash ADC. In: 2016 IEEE international conference on computational intelligence and computing research (ICCIC)

Design of a 4-Bit Wallace Tree Encoder for Flash ADC in 90 Nm CMOS Technology

Krishna Datta, Deepak Prasad, Sanjay Kumar Surshetty, Vidhusi Goel, and Vijay Nath

Abstract This paper proposes a Wallace tree encoder that converts thermometer code (TC) to binary code (BC) for 4-bit flash ADC. The design of encoder we proposed here utilizes full adders for doing the conversion. The simplicity in the design makes it usable in flash ADC design. The CMOS Technology used here is UMC 90 nm and the implementation of the proposed encoder is done using this technology. The design and simulation are done using Cadence Virtuoso ADE. A supply voltage of 1 V is used. The results of simulation show power consumption of 2.1 mW.

Keywords Wallace · Encoder · Thermometer code · Binary code

1 Introduction

In nature, usually, signals are analog whereas modern-day devices like computers work with digital signals. So, there is a necessity to convert signals from their analog form to digital form. Therefore, ADCs are basic building block in many applications of communication and instrumentation. There are various ADCs among which the flash ADC or parallel-type ADC is preferred because it shows higher speed and lower power than the other types. They are useful in various large bandwidth requiring applications. $2^n - 1$ [1, 2]comparators with common analog input to all the comparators are placed in parallel manner so that they produce output in one clock

K. Datta (✉) · D. Prasad · S. K. Surshetty · V. Goel · V. Nath
VLSI Design Group, Department of Electronics and Communication Engineering, Birla Institute of Technology, Mesra, Ranchi, JH 835215, India
e-mail: krishnadatta596@gmail.com

V. Goel
e-mail: vidushigoggle@gmail.com

V. Nath
e-mail: vijaynath@bitmesra.ac.in

V. Nath and J. K. Mandal (eds.), *Proceedings of the Fourth International Conference on Microelectronics, Computing and Communication Systems*, Lecture Notes in Electrical Engineering 673, https://doi.org/10.1007/978-981-15-5546-6_80

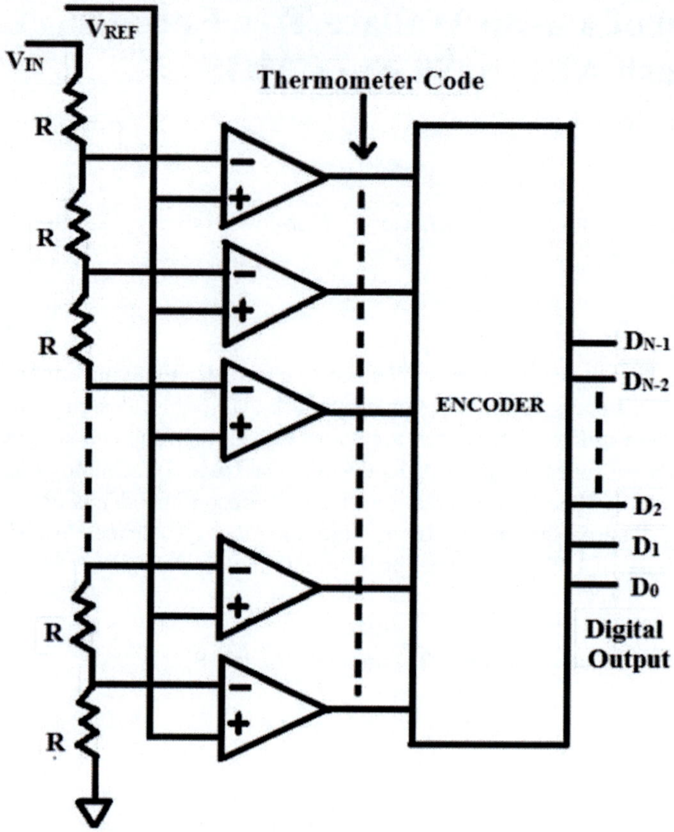

Fig. 1 Flash ADC

cycle. This output is in some specific manner known as "thermometer code [3–6]." Encoder follows next and converts thermometer code to binary code.

Various encoders can be used for this conversion like Wallace tree encoder, ROM-type encoder, fat tree encoder, logic-style encoder, and multiplexer-based encoder. While designing such encoders, we must take care of parameters like error handling capability and power dissipation. In this paper, we have described about Wallace tree encoder (Fig. 1).

2 Wallace Tree Encoder

Following TIQ comparator, Wallace tree encoder helps in direct conversion of TC to BC [7] as shown in Table 1. Wallace tree encoder serves as one of the methods in removing bubble error, presence of 1 s in series of 0 s or vice versa, in thermometer

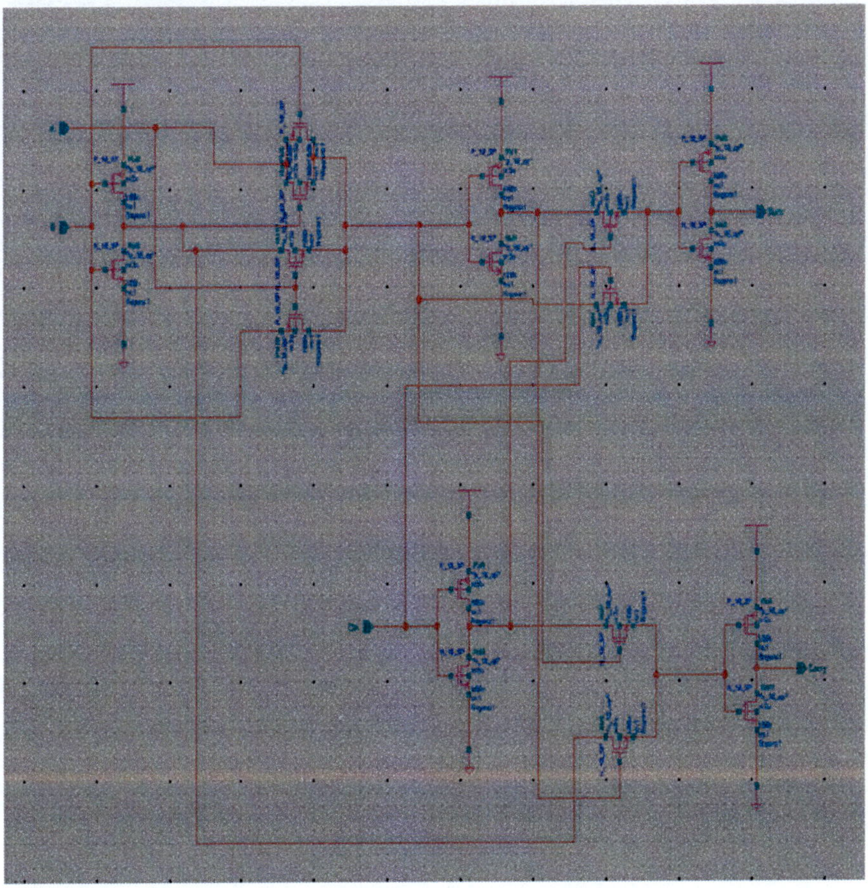

Fig. 2 Full adder

code which results due to offset voltage in comparator [8]. This design requires adder cells [9]. Number of logical 1 s entering each adder cell is being counted to give a 2-bit binary-coded output. We add up these 2-bit words of adjacent cells as shown in Fig. 3 to produce a 4-bit binary output [10] (Fig. 2).

The number of full adders required to this encoder design for an M-bit ADC is [11, 12]:

$$X_M = \sum_{j=1}^{M} (j-1)2^{N-j}$$

So, for 4-bit, we require 11 adder cells. The TG-logic is used in an adder cell. When we apply high voltage to n-channel transistor, the output voltage becomes $Vdd - Vtn$. Similarly, when we apply low VSS to p-channel transistor, the Vtp [13, 14]

Table 1 Binary code

*B*4	*B*3	*B*2	*B*1
0	0	0	0
0	0	0	1
0	0	1	0
0	0	1	1
0	1	0	0
0	1	0	1
0	1	1	1
0	1	1	1
1	0	0	0
1	0	0	1
1	0	1	0
1	0	1	1
1	1	0	0
1	1	0	1
1	1	1	0
1	1	1	1

drop is present in output which is undesired [15]. We used 18 transistor-based full adders as they have optimum threshold voltage drop.

3 Simulation Results

See Figs. 4, 5 and Table 2.

4 Comparative Study

Wallace tree encoder	Power dissipation (10% of power consumption)	Power consumption
Using 28T full adders [16]	1.506 nW	15.06 nW
Using 14T full adders [16]	939.43 pW	9394.3 pW
Using 28T full adders [15]	–	–
Using 18T full adders	0.21 mW	2.1 mW

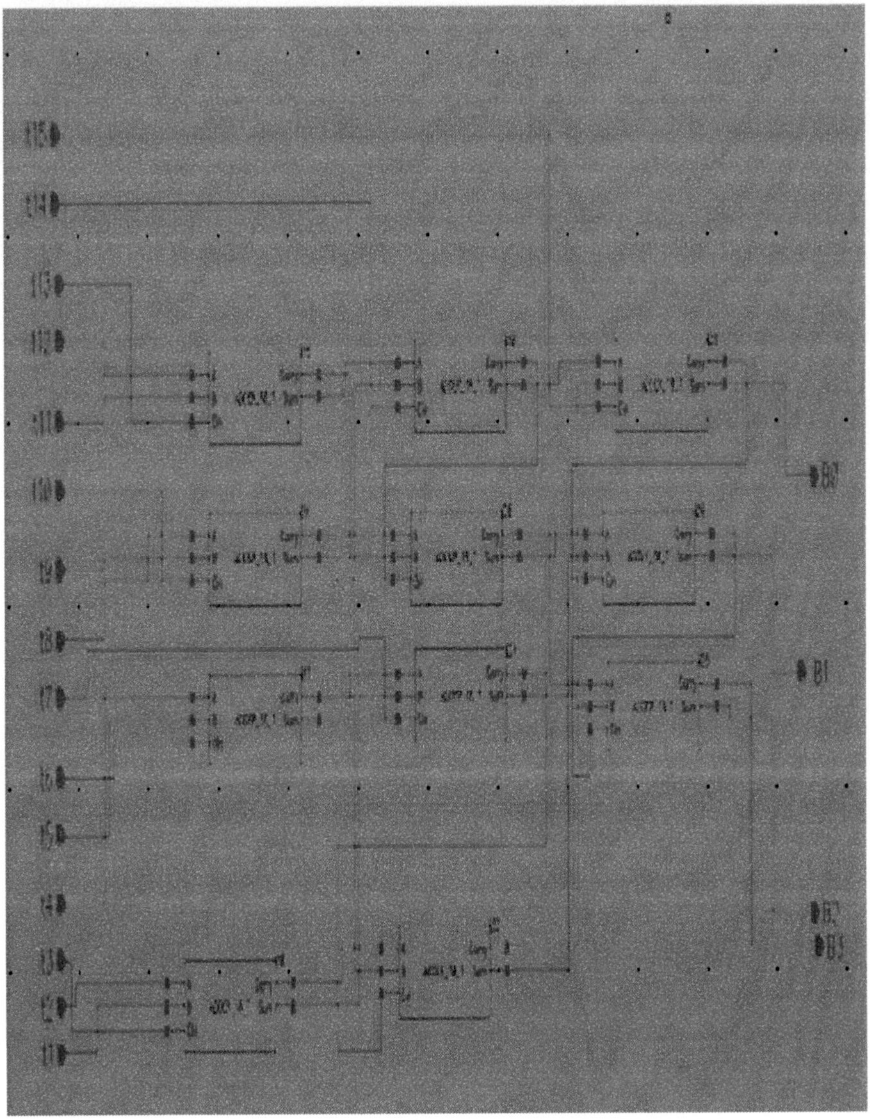

Fig. 3 Wallace tree encoder

5 Conclusion

This paper concludes with a Wallace tree 4-bit encoder design. The design of circuit and simulation results are done using UMC 90 nm technology on cadence virtuoso environment. The proposed design is consuming 2.1 mW power.

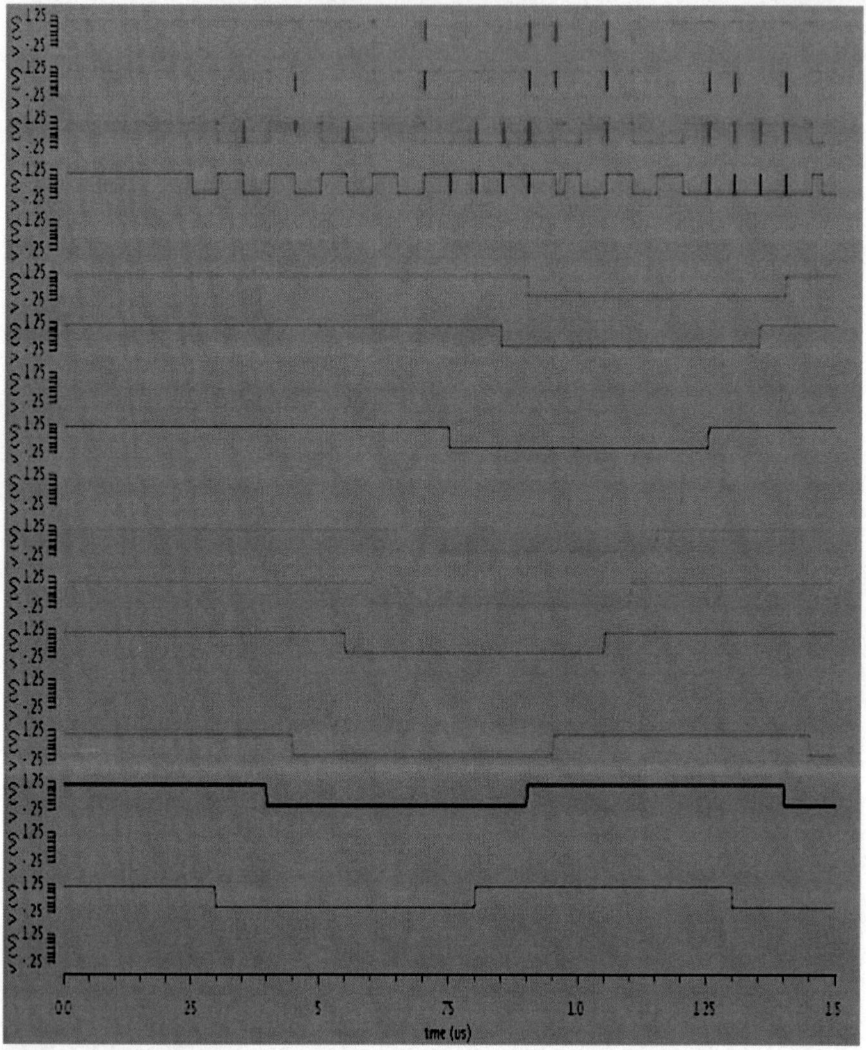

Fig. 4 Output waveform

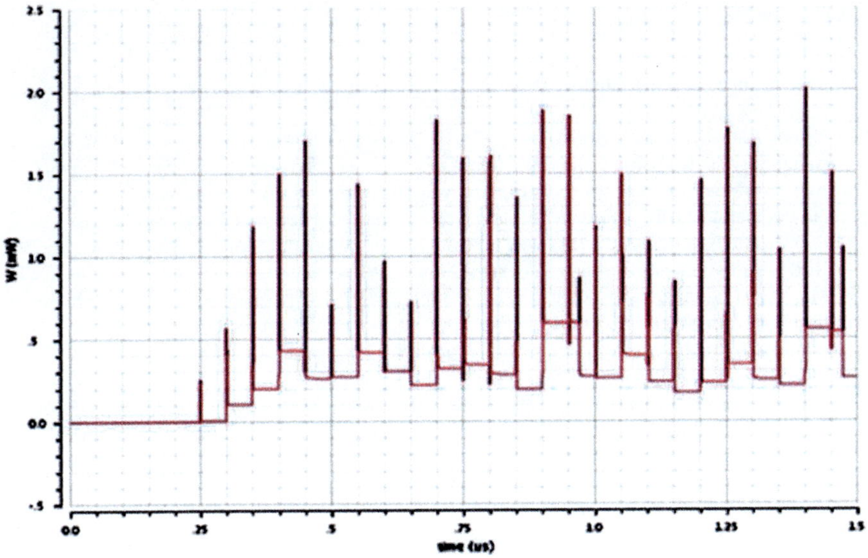

Fig. 5 Output power

Table 2 Thermometer code

$T15$	$T14$	$T13$	$T12$	$T11$	$T10$	$T9$	$T8$	$T7$	$T6$	$T5$	$T4$	$T3$	$T2$	$T1$	$T0$
0	0	0	0	0	0	0	0	0	0	0	0	0	0	0	1
0	0	0	0	0	0	0	0	0	0	0	0	0	0	1	1
0	0	0	0	0	0	0	0	0	0	0	0	0	1	1	1
0	0	0	0	0	0	0	0	0	0	0	0	1	1	1	1
0	0	0	0	0	0	0	0	0	0	0	1	1	1	1	1
0	0	0	0	0	0	0	0	0	0	1	1	1	1	1	1
0	0	0	0	0	0	0	0	0	1	1	1	1	1	1	1
0	0	0	0	0	0	0	0	1	1	1	1	1	1	1	1
0	0	0	0	0	0	0	1	1	1	1	1	1	1	1	1
0	0	0	0	0	0	1	1	1	1	1	1	1	1	1	1
0	0	0	0	0	1	1	1	1	1	1	1	1	1	1	1
0	0	0	0	1	1	1	1	1	1	1	1	1	1	1	1
0	0	0	1	1	1	1	1	1	1	1	1	1	1	1	1

Acknowledgements A deep and sincere gratitude goes to Prof. S. Pal, Head of Department, Electronics and Telecommunication for his invaluable support. An expression of gratitude also go to Prof. M. K. Mishra, vice-chancellor, Birla Institute of Technology, Mesra, Ranchi, for giving us the opportunity, infrastructure, and facility to do our research.

References

1. Maxim integrated products, INL/DNL measurements for high-speed analog to-digital converters (ADCs)
2. Njinowa M, Siadjine M, Bui H, Boyer F (2013) Design of low power 4 bit flash ADC based on standard cells. In: IEEE 11th international new circuits and systems conference (NEWCAS)
3. Babayan-Mashhadi S, Lotfi R (2014) Analysis and design of a low-voltage low-power double-tail comparator. IEEE Trans Very Large Scale Integr (VLSI) Syst 22(2):343–352
4. Patil H, Raghavendra M (2016) Low power dynamic comparator for 4-bit flash ADC. In: 2016 IEEE international conference on computational intelligence and computing research (ICCIC)
5. Mayur SM, Siddharth RK, Kumar YBN, Vasantha MH (2016) Design of low power 5-bit hybrid flash ADC. In: 2016 IEEE computer society annual symposium on VLSI
6. Inamdar A, Sahu A, Ren J, Setoodeh S, Mansour R, Gupta D (2015) Design and evaluation of flash ADC. IEEE Trans Appl Supercond
7. Veeramachanen S, Kumar AM, Tummala V, Srinivas MB Design of a low power, variable-resolution flash ADC
8. Yoo J (2003) A TIQ based CMOS flash A/D converter for system-on-chip applications. Ph.D. thesis, Pennsylvania State University
9. Design and performance analysis of low power 4 bit flash analog to digital converter
10. Varghese GT, Mahapatra K A low power reconfigurable encoder for flash ADCs
11. Al A, Reaz MBI, Jalil J, Ali MA (2014) An improved a low power CMOS TIQ comparator flash ADC
12. Kiranmai1 M, Sudir Patnaikuni VYSS, Mouli K, Manikanta Sai B, Nancharaiah V CMOS full adder and multiplexer based encoder for low resolution flash ADC
13. Velagaleti S (2009) A novel high speed dynamic comparator with low power dissipation and low offset. M-tech thesis, NIT Rourkela, India
14. Morris Mano M, Ciletti MD (2006) Digital design. 4th edn. Prentice Hall, Upper Saddle River. ISBN 978-0-13-198924-5
15. Pardhu T, Manusha S, Sirisha K A low power flash ADC with wallace tree encoder
16. Ajanya MP, Varghese GT Low power wallace tree encoder for flash ADC

Opinion Mining of Restaurant Reviews and Comparison of Different Classifiers

Ananya Sinha, Manila Oraon, Sneha Anand, and Vandana Bhattacharjee

Abstract The growth of Internet world wide, social media and networking sites, blogs, discussion forums, e-commerce websites has gained a huge importance these days and has provided platform for users to express and share their views on entities and their features and aspects. These days, many users give their reviews, good or bad about food items, services and ambience in the restaurants to express their actual opinion. The true reviews are very important in order to take decisions to other users. Retrieving and analyzing these many opinions manually are difficult and tiresome. So, the need of time is an automated tool or technique to overcome this problem. Opinion mining is a method to analyse these reviews and classifies them as positive, negative, and neutral. We presented lexicon-based sentiment analysis, that is, classical and VADER method for opinion mining of restaurant reviews and sentence relevance score-based method for opinion summarisation. We also implemented different machine learning classifier algorithms for the classified data. This detailed analysis hopefully provides insights into choosing the most appropriate restaurant.

Keywords Opinion mining · Text mining · Sentiment analysis · Machine learning classifier algorithms

---

A. Sinha (✉) · M. Oraon · S. Anand · V. Bhattacharjee
Department of Computer Science and Engineering, Birla Institute of Technology, Mesra, Ranchi, JH 835215, India
e-mail: ananya.sinha093@gmail.com

M. Oraon
e-mail: manilaoraon129@gmail.com

S. Anand
e-mail: anandsneha1010@gmail.com

V. Bhattacharjee
e-mail: vbhattarcharya@bitmesra.ac.in

V. Nath and J. K. Mandal (eds.), *Proceedings of the Fourth International Conference on Microelectronics, Computing and Communication Systems*, Lecture Notes in Electrical Engineering 673, https://doi.org/10.1007/978-981-15-5546-6_81

1 Introduction

Internet acts as an important source of information for users due to growth in science and technology. These days, every business like shopping, booking for a movie or a restaurant is done online. Many of these websites like www.zomato.com allow users to rate and review the restaurants accordingly. As large number of users expresses their opinions, there is huge amount of data along with impurities floating on the Internet. Therefore, the extraction and analysis may take a lot of time and may not produce desired results. So, it is necessary to present correct and concise information to the users which can help them to make correct decisions.

Opinion mining is a type of text classification that is an approach to natural language processing which is used to detect the hidden sentiments of people about a particular product from the reviews. The reviews are classified as positive, negative, and neutral. Also, a cumulative rating for every entity is found out to present the opinion in short and summarised form. Opinion mining has wide range of applications ranging from recommendation systems to decision-making systems, etc.

In this paper, we carry out the sentiment analysis of restaurant reviews using classical method (rule based) and Valence Aware Dictionary and sEntiment Reasoner (VADER) [1] which classifies the reviews as positive, negative, and neutral on the basis of sentiment score calculated. It then combines the score of the reviews for a particular restaurant to give a final rating which is used to find best-rated, average-rated, and worst-rated restaurants. We also present the comparison of different classifier algorithms like Naïve Bayes, multinomial Naive Bayes, Bernoulli Naïve Bayes, k-nearest neighbour, decision tree, support vector machine, and logistic regression [2] on the basis of accuracy, training time, and prediction time. Comparison between classical and VADER method is also presented.

We have used Python and Java languages for this project. The tools and libraries used are ANACONDA NAVIGATOR-Jupyter, Numpy, Pandas, Scikit [3], VADER, etc. The most important library is NLTK which contains many other text processing libraries.

2 Related Work

Many researchers have carried out sentiment analysis on different domains data like Twitter [4–6], Movies [7], Business [8], etc.

Extensive work on VADER sentiment analysis is performed for quick sentiment analysis of streaming data and its better performance than other sentiment analysis tools like LIWC [9], SWN [10], etc.

Machine learning classification algorithms such as Naive Bayes [11–13], SVM [12, 14], K-NN [13], and decision tree [15] are used to predict the classes of test data set.

3 System Architecture and Classification

The proposed architecture shows classification of reviews as positive, negative, and neutral using two methods, first classical method which consists of review retrieval, preprocessing, and opinion summarisation and second method passes the reviews to a function to calculate the score of the sentiment. Both the systems rate restaurants on the basis of cumulative scores generated. Evaluation of classifier algorithms by machine learning is carried out on the classified data (Fig. 1).

3.1 Review Collection

The reviews are collected by Web crawling techniques applied on www.zomato.com from the location Mumbai. Here, we have used import.io for this purpose. The collected reviews are stored in csv file. The data set contains 12,800 reviews of 1347 restaurants.

3.2 Review Classification

3.2.1 Classical Sentiment Analysis

In this method, we have used splitter for sentence segmentation, tokenizer for tokenization, stop word removal, POS tagging, and dictionary tagging. Here, dictionaries for positive, negative, inverter, increment, and decrement words are created by domain knowledge. The above steps are a part of preprocessing.

The next module is opinion summarisation which calculates the score of each review and finally divides reviews into positive, negative and neutral (Table 1).

3.2.2 VADER Sentiment Analysis

VADER is a library of NLTK toolkit. It handles emoticons, slangs, idioms, phrases, capitalisation, punctuation, etc. It also measures the degree of positivity, negativity, and neutrality of the reviews. It is very fast sentiment analysis method as compared to classical method and does not suffer from speed-performance trade-off. The positive, negative, and neutral reviews are classified on the threshold value of compound score which is equal to 0.05 here. The value 0.05 is taken because it gives highest accuracy in maximum number of samples in minimum time. The reviews having compound score greater than 0.05 is considered as positive review, less than −0.05 is negative review, and between −0.05 and 0.05 is considered to be neutral. It is a valence-based method. The number of neutral reviews significantly reduces in VADER (Table 2).

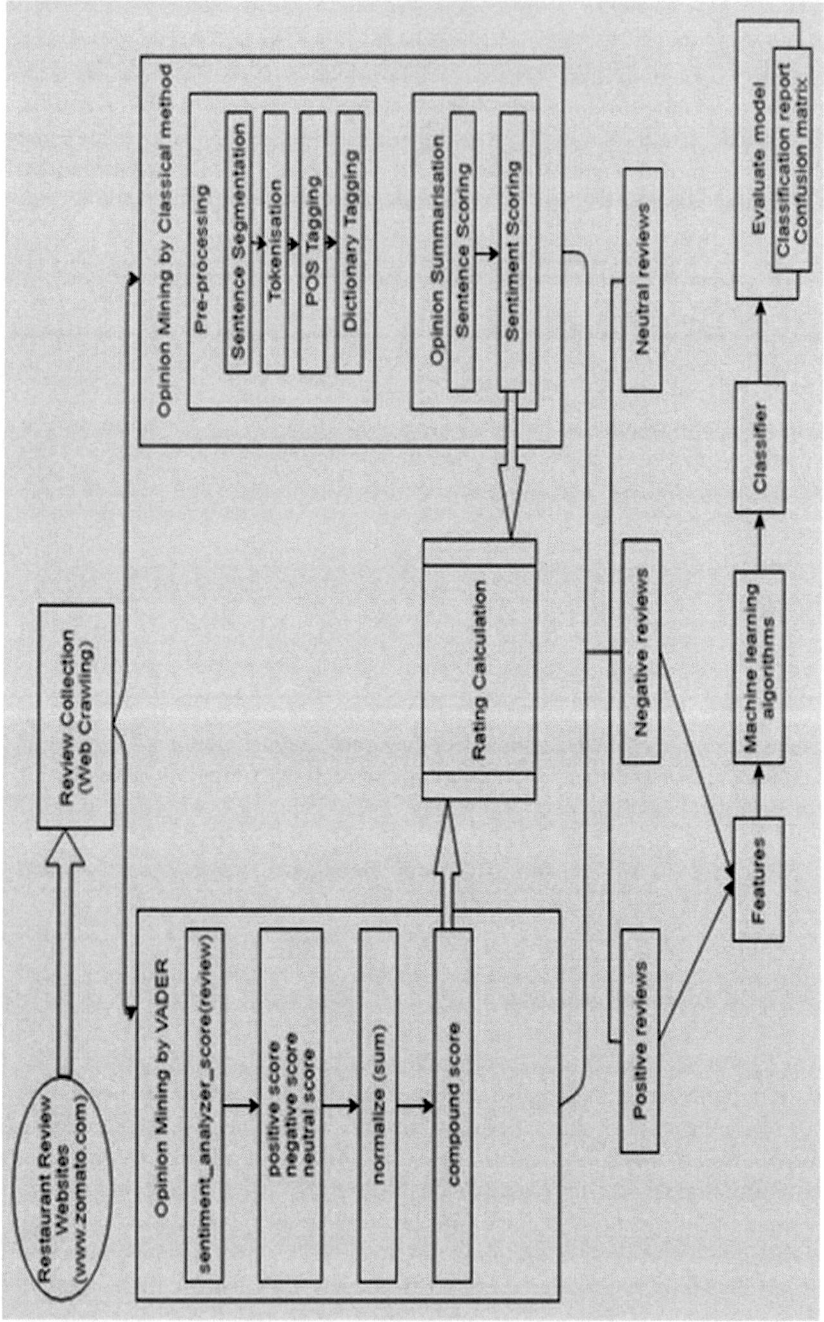

Fig. 1 System architecture

Table 1 Classified reviews by classical method

	Without using INV.YML, INC.YML, and YML	Using INV.YML, INC.YML, and DEC.YML
Negative reviews	869	1059
Positive reviews	11,216	11,095
Neutral reviews	716	647
Time taken	11.62806247141626	10.307204059494

Table 2 Classified reviews by VADER method

Negative reviews	1067
Positive reviews	11,558
Neutral reviews	175
Total time	3.25 min

3.3 Rating Calculation

The rating of a restaurant is calculated by following formula:

$$\text{Rating} = \text{average} + \text{count} * \text{bias}$$

where, average is sum of sentiment score of all the reviews for that restaurant divided by total number of reviews for that restaurant.

count is equal to number of reviews for that restaurant.

bias is equal to total number of reviews for a particular restaurant divided by total number of reviews in the dataset.

Based on this, we find best-rated, average-rated, and worst-rated restaurant.

4 Analysis and Comparison of Different Classifier Algorithms

Here, we have shown the comparison of different machine learning algorithms on the basis of prediction time, training time, and accuracy (Figs. 2, 3, 4 and 5).

Variation in accuracy with respect to number of neighbours in K-NN algorithm is also represented.

The prediction model is evaluated on the basis of confusion matrix and classification report which is depicted by the following Fig. 6.

The comparison between classical method and VADER method is shown by the following table.

Graph 1 Training time versus different classifier algorithms

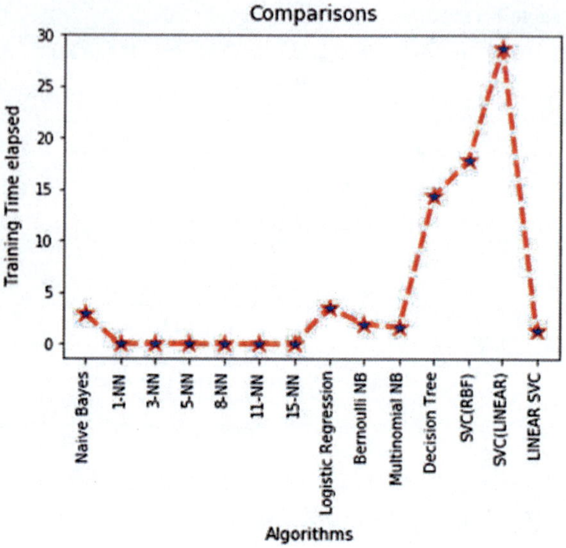

Graph 2 Prediction time versus different classifier algorithms

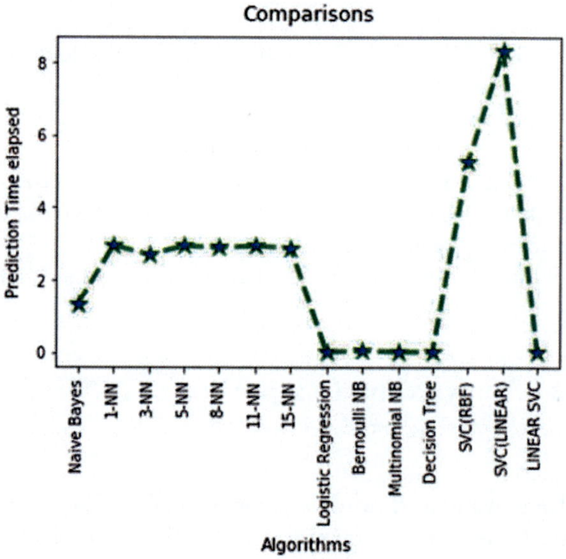

Algorithms implemented	Classical (rule based) method	VADER
Naive Bayes	Accuracy: 92.89% Training time: 2.861015 s Prediction time: 1.327940 s	Accuracy: 91.45% Training time: 3.735648 s Prediction time: 2.007427 s

(continued)

(continued)

Algorithms implemented	Classical (rule based) method	VADER
8-NN	Training time: 0.014029 s Prediction time: 2.908042 s Accuracy = 92.00%	Training time: 0.016029 s Prediction time: 3.280307 s Accuracy = 92.46%
Logistic regression	Training time: 3.572531 s Prediction time: 0.019014 s Accuracy = 93.12%	Training time: 3.605556 s Prediction time: 0.016012 s Accuracy = 92.62%
Decision tree	Training time: 14.339162 s Prediction time: 0.032040 s Accuracy = 89.31%	Training time: 20.063349 s Prediction time: 0.036007 s Accuracy = 90.28%
Linear SVC	Training time: 1.404992 s Prediction time: 0.019014 s Accuracy = 94.74%	Training time: 1.598114 s Prediction time: 0.026019 s Accuracy = 93.57%

5 Conclusion

Classical method (rule based) is a polarity-based approach because it treats the words 'good' and 'excellent' same whereas VADER method is a valence-based approach where whereas 'excellent' would be treated as more positive than 'good.' Classical method does not give the degree of positivity, negativity, and neutrality whereas VADER does. In case of classical method, when inverse, incremental, and decrement dictionaries are not used, the number of neutral reviews comes out to be more than when these dictionaries are used because in the former case many positive and negative words simply cancel out to give neutral review. This number of neutral reviews further decreases on using VADER as it is more accurate in classifying positive and negative words along with intensifiers. Time taken for opinion mining in classical (rule-based method) is very large as compared to the VADER method. VADER handles emoticons, slangs, punctuation, and idioms whereas classical method does not. Accuracy of classical method is slightly greater than VADER method.

Training time of Naive Bayes classifier, logistic regression, Bernoulli NB, multinomial NB, SVM (rbf), SVM (linear), and linear SVC is greater than their prediction time because these algorithms are eager learners. Prediction time of k-NN classifier is greater than its training time because it is a lazy learner. Decision tree classifier gives less accuracy because it is not practically feasible to divide real-world data into many classes and data also contains impurities.

The best classifier algorithm for this sentiment analysis problem is Linear SVC followed by Naive Bayes classifier. The algorithm K-NN for $K = 1$ gives lowest accuracy. In case of K-NN, accuracy first increases on increasing the number of neighbours, reaches to a maximum value, and then starts decreasing but after that remains nearly constant. SVC with linear kernel takes maximum time to train and predict among all models but linear SVC takes a very small time because former uses

Graph 3 Accuracy versus different classifier algorithms

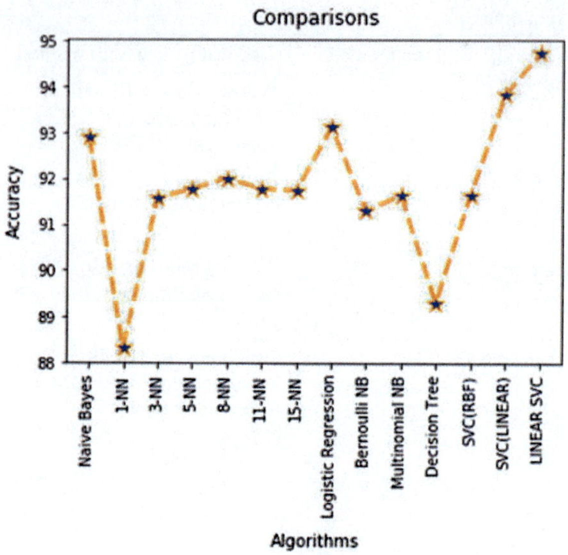

Graph 4 Accuracy versus number of neighbours

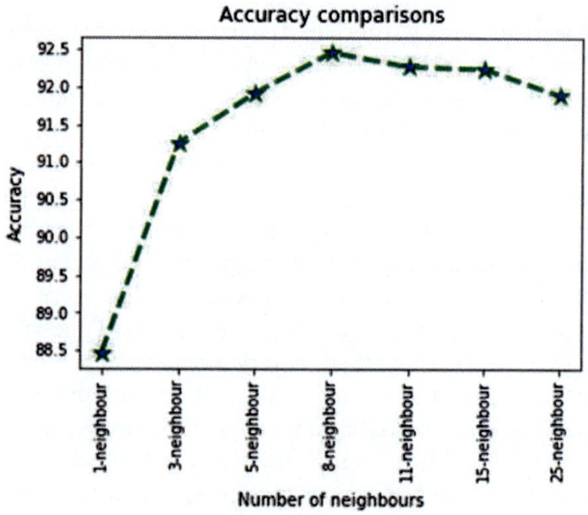

the estimator libsvm (time complexity $O(n^2)$ or $O(n^3)$) and later on liblinear (time complexity $O(n)$). Our ongoing work will focus on improvising the algorithms for getting better efficiency.

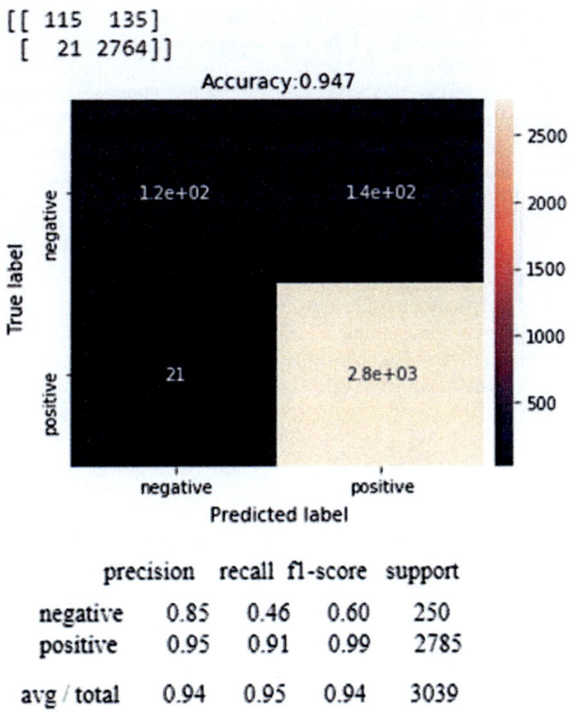

	precision	recall	f1-score	support
negative	0.85	0.46	0.60	250
positive	0.95	0.91	0.99	2785
avg / total	0.94	0.95	0.94	3039

Fig. 2 Confusion matrix and classification report

References

1. Hutto CJ, Gilbert E (2014) VADER: a parsimonious rule-based model for sentiment analysis of social media text. In: Proceedings of ICWSM
2. Tyagi A, Sharma N (2018) Sentiment analysis using logistic regression and effective word score heuristic. Int J Eng Technol (UAE) 7:20–23. https://doi.org/10.14419/ijet.v7i2.24.11991
3. Scikit-Learn Available: http://scikit-learn.org/
4. Saif H, He Y, Alani H (2012) Semantic sentiment analysis of twitter. In: Cudré-Mauroux P et al (ed) The semantic web—ISWC 2012. ISWC 2012. Lecture Notes in Computer Science, vol 7649. Springer, Berlin
5. Barbosa L, Feng J (2010) Robust sentiment detection on twitter from biased and noisy data. In: Proceedings of the 23rd international conference on computational linguistics: posters, pp 36–44
6. Sahayak V, Shete V, Pathan A (2015) Sentiment analysis on twitter data. (IJIRAE) ISSN: 2349-2163
7. Liu CL, Hsaio WH, Lee CH, Lu GC, Jou E (2012) Movie rating and review summarization in mobile environment. IEEE Trans Syst Man Cybern Part C: Appl Rev 42(3):397–406
8. Salinca A (2015) Business reviews classification using sentiment analysis, pp 247–250. https://doi.org/10.1109/synasc.2015.46
9. López-López E, del Pilar Salas-Zárate M, Almela Á, Rodríguez-García MÁ, Valencia-García R, Alor-Hernández G (2014) LIWC-based sentiment analysis in spanish product reviews. In: Omatu S, Bersini H, Corchado J, Rodríguez S, Pawlewski P, Bucciarelli E (eds) Distributed computing and artificial intelligence, 11th international conference. Advances in intelligent systems and computing, vol 290. Springer, Cham

10. Esuli A, Sebastiani F (2006) Sentiwordnet: A publicly available resource for opinion mining. In: Proceedings of the 6th international conference on language resources and evaluation (LREC'06), pp. 417–422
11. Pang B, Lee L, Vaithyanathan S (2002) Thumbs up? Sentiment classification using machine learning techniques. In: Proceedings of the conference on empirical methods in natural language processing (EMNLP)
12. Preety SD, Dahiya S (2015) Sentiment analysis using svm and Naïve Bayes algorithm. IJCSMC 4(9):212–219
13. Dey L, Chakraborty S, Biswas A, Bose B, Tiwari S (2016) Sentiment analysis of review datasets using naïve bayes and K-NN classifier. Int J Inf Eng Electron Bus (IJIEEB) 8(4):54–62. https://doi.org/10.5815/ijieeb.2016.04.07
14. AnalyticsVidhaya: https://www.analyticsvidhya.com/blog/2015/10/understaing-support-vector-machine-example-code/. Accessed 12 Nov 2016
15. Liu J, Seneff S, Zue V (2012) Harvesting and summarizing user-generated content for advanced speech-based HCI. IEEE J Sel Top Signal Process 6(8):982–992

A Novel Deployment Scheme to Enhance the Coverage in Wireless Sensor Network

Vishnu Anugrahith Sateesh, Aniket Kumar, Rahul Priyadarshi, and Vijay Nath

Abstract The efficiency of any sensor network is heavily depending on the coverage delivered by the deployed sensors. Wireless sensor network (WSN) represents the gathering of certain number of sensors that are closely deployed in a recognizable area. This paper suggested the development of "deployment approach" to increase the WSN coverage, connectivity, and reliability. This method is based on the "distance between" each sensor node and its neighboring sensors. It aims to improve the nodes coverage in steps after a primary arbitrary deployment. In each step, a sensor node is appealed in the direction of its neighbors that have lower distance. This method capitalizes on the coverage of the detected area by forcing the sensor to change its position toward the area with a lower sensors density. The simulation results were compared with the GSO results. Our results showed that this deployment approach could deliver high coverage, full connectivity, and good dependability. Such results could be achieved with less number of iterations.

Keywords Coverage · Connectivity · Deployment · Region · Placement · Area

V. A. Sateesh (✉) · V. Nath
Department of Electronics and Communication Engineering, Birla Institute of Technology, Mesra, Ranchi, JH 835215, India
e-mail: vishnuanugrahith@gmail.com

V. Nath
e-mail: vijaynath@bitmesra.ac.in

A. Kumar
Department of Electrical and Electronics Engineering, Manipal Institute of Technology, Manipal, Karnataka 576104, India
e-mail: aniketkmr1999@gmail.com

R. Priyadarshi
Department of Electronics and Communication Engineering, National Institute of Technology, Patna, Bihar 800005, India
e-mail: rahul.glorious91@gmail.com

V. Nath and J. K. Mandal (eds.), *Proceedings of the Fourth International Conference on Microelectronics, Computing and Communication Systems*, Lecture Notes in Electrical Engineering 673, https://doi.org/10.1007/978-981-15-5546-6_82

1 Introduction

Wireless sensor networks (WSNs) are constituted with tiny electro mechanical devices, i.e., sensors nodes. Sensor nodes communicate via RF signals with one or sinks that are more powerful called base stations (BSs). In the most recent years, numerous scientists center on the "WSNs" developments and applications. WSNs have potential applications in various essential territories. WSNs speak to frameworks that include certain quantities of remotely connected sensors that are spatially sent in a specific area [1]. Sensors can be sent in enormous numbers because of their acknowledged cost and little size. One of the key difficulties confronting the WSNs applications is the means by which to decide the physical areas of the conveyed sensor hubs.

This issue speaks to a functioning examination point as of late. The incomparable answer for such an issue is to give the sensor hubs by "Worldwide Positioning System (GPS)". GPS is exorbitant, power expending, and confined for open-air applications. To discover the area of the ideal sensor, many nonlinear conditions must be settled to arrive at the best possibility reason in some cases [2]. The adequacy of this methodology is the manner by which to explore the arrangement of the sensors via looking for ideal or close ideal arrangements. Once placement of nodes completed, then communication of data from node to BS should be completed [3]. It can be called single hope if they can contact directly to base station or can be multi-hop communication in which intermediate nodes takes part for communication from sensor node to base station [4–10]. Formally, sensor networks on basis of their functionality are classified into two types one is proactive and the other is reactive networks. Proactive networks are passive in nature and well suited for data-aggregation-type applications.

In this type of networks, nodes sense and send data on periodical basis, i.e., on regular intervals. In reactive networks, opposite to passive networks sensor nodes respond immediately and only to changes in the significant considerations of interest. To prolong lifetime of sensor networks, it will be more proficient if placement of sensor nodes should be in a proper manner. That is the reason why sensor networks need management planes. As per requirement, three management planes will work which are named as mobility, power, and task management plane. These planes solve objectives like first to make sensor nodes work collectively in a power efficient way, second to perform routing of data in wireless sensor network, and to stake resources among them. Without these three planes, each sensor node will act and work as an individual. Power management plane ensures to manage how should a sensor node uses its power like when to turn off receiver; when to broadcast low power message; and to quit from routing so that sensor nodes can save energy for sensing.

The mobility management plane is in use when sensor nodes are not stationary in networks. Mobility plane helps to maintain a path back to the user, and sensor nodes could keep track of their neighbor sensors. The task management plane divides and schedules the tasks among sensors, which are distributed in monitoring region. In sensor networks, nodes are deployed with high density. Therefore, nodes deployed in nearby locations measure approximately same values about the environment [11,

12]. Removal of redundant information helps in increasing energy efficiency and network lifetime. Further, changes in environment occur very slowly so successive readings contain related values about the parameters. Self-organization: Sensor nodes must operate autonomously so that sensors can self-organize and self-heal under different circumstances. Communication range: Sensor nodes have small communication range only a few meters whereas range of ad hoc networks is in hundreds of meters [13, 14].

2 Related Work

Authors in [15] proposed a method with objective inclusion dependent on cross-section examine. They isolated the region into grids. At that point, the best cross-section was chosen as a situation to the following sensor. This strategy utilized the most number of hubs to accomplish the inclusion objective and to improve the situations for the sent sensor hubs. The arrangement approaches in WSN was isolated into two gatherings as "inadequate" and "thick organizations." In the given field of intrigue, the "scanty organization" can be utilized with minimal number of hubs conveyed, while the "thick sending" can be utilized in high number of sent sensors. The "thick sending" procedure was utilized because of each occasion needing to be seen or the organization of various sensors to cover a similar zone. The scanty arrangement is reasonable to accomplish most extreme detecting territory with least number of sensors. The sensor hubs are either static or portable; in this manner, they can change their situations with time. Sensors are either conveyed in certain field by finding them in "foreordained spots" or haphazardly sent. "Arbitrary sensors arrangements" approach is one kind of the "thick organizations." WSNs with versatile sensors are typically started with an "irregular sending." They change their areas to the best reasonable area because of their portability.

In "static sending," sensors are static and their areas must be chosen because of certain "improvement technique." Sensors areas will remain fixed and do not change during the activity of the system. Deterministic advances and arbitrary organizations can accomplish "Static arrangement" procedure. This methodology (the "deterministic sending") begins by reviewing the zone of interests and proceeding with the arrangement procedure [16].

A significant utilization of a "powerful arrangement" is its use in robots. Versatility helps in going to the detecting objective by letting the sensor moves toward a greatest detecting execution. In "irregular sending," the greater part of the hubs is hurled in the initial step, before an uncommon usage and reorganization estimations are led to choose the following development. Many created calculations have been constructed, for example, the "virtual force oriented particles algorithm," "simulated annealing algorithm," "particle swarm optimization algorithm," and "simulated annealing genetic algorithm." The coverage represents the "quality of service (QoS)" for the sensing function of the WSN. Its main function is to answer the vital question:

"How well do the sensors observe a physical space"? Therefore, the position of the sensor is the essential factor related to the "coverage problem."

The sensor nodes must be located in an optimal position to ensure complete utilization and increase its sensing ability. Such process helps in maximizing the network sensing covered area. Coverage value can be estimated as the fraction of dividing the covered area to the area of the "region of interest (ROI)." In many cases, the coverage problem is considered as a minimization problem. The stated objective function in this case is to minimize the area not covered by the sensors. These two approaches need hard calculations and computations in a longer time. The "length measurement" is easier and needs smaller time than the "area computation" [17]. Thus, the "length measurement" is widely used in calculating the WSN coverage problem. Sampling techniques can be applied in the case of existing fixed number of points, which can be used in evaluating the coverage. Grid is one of the frequently used sampling techniques.

3 Proposed Scheme

Numerous methodologies have been proposed and executed to accomplish certain inclusion availability and unwavering quality. Creators in [18] for instance recommended a sensor arrangement framework dependent on "glowworm swarm enhancement" to improve the inclusion after an "underlying arbitrary sending" of the sensors. They considered every sensor hub as "singular glowworms producing a luminant" body called "luciferin" and the quality of the "luciferin" relies upon the space secluding the sensor hub with its neighboring. A sensor hub can be spoken to its adjoining with lower quality of "luciferin." The sensor that will in general move toward the region having lesser sensor focus then the detecting inclusion territory will be boosted.

Their recreation results indicated that "GSO-based sensor arrangement approach" can convey incredible inclusion with confined number of the sensor developments. They closed the accompanying GSO results: When 50 sensors were conveyed in a focal point of 100 × 100 m region with detecting scope of 5 m and correspondence scope of 10 m, they found an inclusion level of about 35.4%. At the point when 100 sensors were conveyed at a similar territory and a similar detecting and correspondence ranges, they found the inclusion of 64.6%. At the point when they utilized 200 sensors with a similar zone, a similar detecting and correspondence ranges fulfilled the inclusion of 92.2%. This strategy neglected to accomplish availability and/or unwavering quality with less quantities of sensors sent arbitrarily in such area [19].

The optimum number of sensors required to cover all the areas of interest according to the grid distribution to achieve maximum coverage area, minimum overlapping coverage, full connectivity and good reliability can be estimated. In this paper, a suggested approach was created to estimate the optimum required number of sensor nodes to achieve maximum coverage, minimum overlapping, 2-connectivity, and certain limit of reliability with minimum number of sensors (minimum cost).

The suggested approach in this paper aimed to find the required number of sensors in a simulation manner; 100 sensors were found to be suitable for the area of 100 × 100 m with sensing range of 5 m and communication range of 10 m. This algorithm estimates the resulted coverage area, which is 77.63% and the full connectivity with minimum of 2 connections (2-connectivity).

In this paper, a new developed deployment approach [20] has been suggested, implemented, and evaluated using the same GSO approach parameters.

The suggested approach to "sensor deployment technique" was based on the space between the sensor node and its "neighboring sensors" to improve the coverage area after a random deployment. A sensor node was appealed toward its neighbors that have a large distance and transferred it toward one of them. In this way, the coverage area of the sensing area can be maximized as the sensor nodes move toward the region having "lower sensor density" [21, 22].

4 Simulation and Results

The proposed scheme is simulated in the MATLAB simulator. We tested the algorithm with three cases as follows:

Case 1: Less than the optimum required number of sensors: In this case, 40 sensors were used to be deployed in 100 × 100 m area with 10 m for coverage area and 15 m for communication range. These sensors were deployed randomly in the suggested area. The developed algorithm in this paper was applied to reach the best deployment. The wanted deployment aimed to ensure good connectivity, good coverage, less overlapping, and good reliability. The results were reached after 135 iterations. Figure 1 shows the resulted deployment positions.

Case 2: Equal optimum required number of sensors: In this case, 80 sensors were used to be deployed in 100 × 100 m area with 10 m for coverage area and 15 m for connectivity. These sensors were deployed randomly in the suggested area. The final results were reached after 182 iterations. Figure 2 shows the resulted deployment positions.

Case 3: Greater than the optimum required number of sensors: In this case, 120 sensors were used to be deployed in 100 × 100 m area with 10 m for coverage area and 15 m for connectivity. These sensors were deployed randomly in the suggested area. Figure 3 shows the resulted deployment positions.

The final results were reached after 220 iterations. Table 1 shows the outcomes of all the three case. Figure 4 shows comparison between the proposed algorithm and GSO approach. From the results, it is clear that the developed approach provides a good result compared with the GSO. The number of iterations used in our approach is too less than the number of iterations used to reach the GSO.

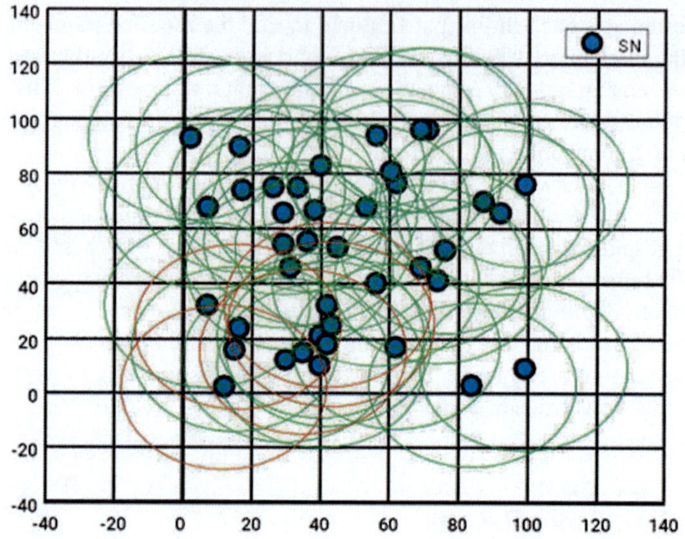

Fig. 1 Deployment position for case 1

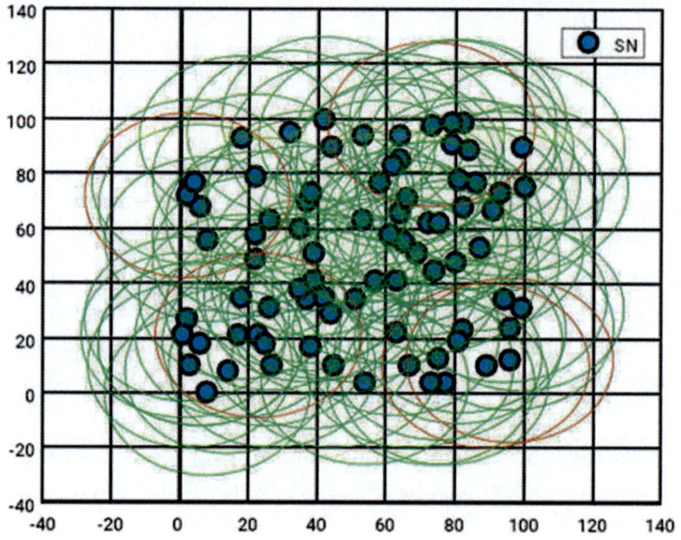

Fig. 2 Deployment position for case 2

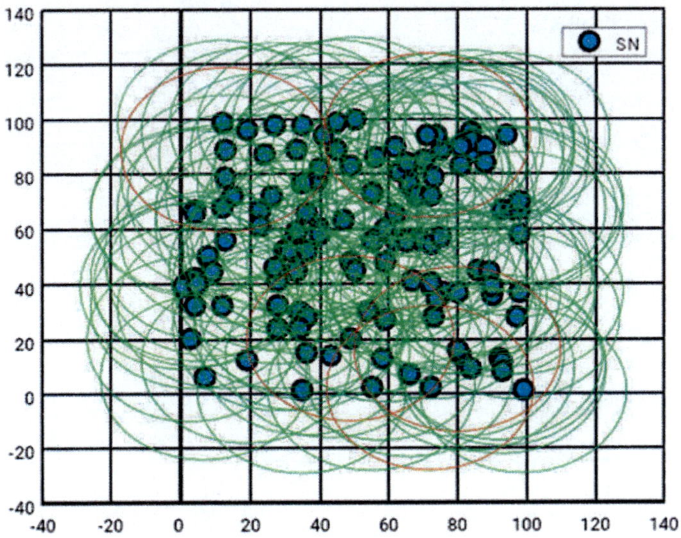

Fig. 3 Deployment position for case 3

Table 1 Comparison with different parameters value

Cases	Number of sensors	Field area	Sensing range (m)	Communication range (m)	Total coverage (%)
Case 1	40	100 × 100	10	15	32
Case 2	80	100 × 100	10	15	65
Case 3	120	100 × 100	10	15	94

5 Conclusion

In this paper, the established sensor deployment method based on spaces (distances) can make the most of the coverage, diminish the overlapping, ensure full connectivity and achieve good reliability level. These issues may be accomplished in slight movement numbers after an "initial random deployment." The developed method in this paper benefits in growing the network capability to achieve its sensing. The established method has a benefit that it does not want "centralized control" of the "deployed nodes" consequently; it is simply scalable for large "sensor networks." The developed approach results dominate the GSO results in all the used metrics in addition to the reduction in the number of the sensors positions movements.

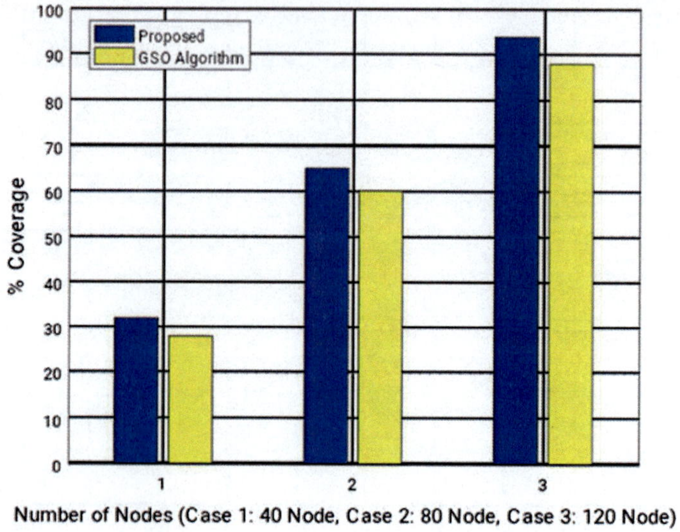

Fig. 4 Proposed and developed approach

References

1. Zou Y, Chakrabarty K (2003) Sensor deployment and target localization based on virtual forces. In: Proceedings of the IEEE INFOCOM. San Francisco, USA, pp 1293–1303
2. Li SJ, Xu CF, Pan WK, Pan YH (2005) Sensor deployment optimization for detecting maneuvering targets. In: Proceedings of the 7th international conference on information fusion. Stockholm, Sweden, pp 1629–1635
3. Zhang H, Hou JC (2005) Maintaining sensing coverage and connectivity in large sensor networks. Wirel Ad Hoc Sens Netw 1(1):89–124
4. Priyadarshi R, Soni SK, Sharma P (2019) An enhanced GEAR protocol for wireless sensor networks. In: Lecture Notes in Electrical Engineering. https://doi.org/10.1007/978-981-13-0776-8_27
5. Priyadarshi R, Singh L, Kumar S, Sharma I (2018) A hexagonal network division approach for reducing energy hole issue in WSN. Eur J Pure Appl Math 118
6. Priyadarshi R, Rawat P, Nath V (2018) Energy dependent cluster formation in heterogeneous wireless sensor network. Microsyst Technol. https://doi.org/10.1007/s00542-018-4116-7
7. Priyadarshi R, Bhardwaj A (2017) Node non-uniformity for energy effectual coordination in WSN. Int J Inf Technol Secur 9(4):3–12
8. Priyadarshi R, Soni SK, Nath V (2018) Energy efficient cluster head formation in wireless sensor network. Microsyst Technol 24(12):4775–4784. https://doi.org/10.1007/s00542-018-3873-7
9. Priyadarshi R, Singh L, Sharma I, Kumar S (2018) Energy efficient leach routing in wireless sensor network. Int J Pure Appl Math 118(20):135–142
10. Priyadarshi R, Tripathi H, Bhardwaj A, Thakur A (2018) Performance metric analysis of modified LEACH routing protocol in wireless sensor network. Int J Eng Technol. https://doi.org/10.14419/ijet.v7i1.5.9146
11. Priyadarshi R, Yadav S, Bilyan D (2019) Performance analysis of adapted selection based protocol over LEACH protocol. In: Luhach AK, Hawari KBG, Mihai IC, Hsiung P-A, Mishra RB (eds) Smart computational strategies: theoretical and practical aspects. Springer, Singapore, pp 247–256. https://doi.org/10.1007/978-981-13-6295-8_21

12. Priyadarshi R, Yadav S, Bilyan D (2019) Performance and comparison analysis of MIEEP routing protocol over adapted LEACH protocol. In: Luhach AK, Hawari KBG, Mihai IC, Hsiung P-A, Mishra RB (eds) Smart computational strategies: theoretical and practical aspects. Springer, Singapore, pp 237–245. https://doi.org/10.1007/978-981-13-6295-8_20
13. Priyadarshi R, Singh L, Singh A, Thakur A (2018) SEEN: stable energy efficient network for wireless sensor network. In: 2018 5th international conference on signal processing and integrated networks, SPIN 2018. https://doi.org/10.1109/spin.2018.8474228
14. Priyadarshi R, Singh L, Randheer, Singh A (2018) A novel HEED protocol for wireless sensor networks. In: 2018 5th international conference on signal processing and integrated networks, SPIN 2018. https://doi.org/10.1109/spin.2018.8474286
15. Wang X, Han S, Wu Y, Wang X (2014) Coverage and energy consumption control in mobile heterogeneous wireless sensor networks. IEEE Trans Autom Control 58:975–988
16. Wan P, Yi C (2006) Coverage by randomly deployed wireless sensor networks. IEEE Trans Inf Theory 52(6):2658–2669
17. Zou Y, Chakrabarty K (2003) Sensor deployment and target localization based on virtual forces. In: Proceedings of the IEEE INFOCOM
18. Krishnanand KN, Ghose D (2009) Glowworm swarm optimization: a new method for optimizing multi-modal functions. Int J Comput Intell Stud 1(1):93–119
19. Li G, Wang R, Zhang Y (2005) Integrated coverage and connectivity configuration for energy conservation in sensor networks. ACM Trans Sens Netw 1:36–72
20. Purohit GN, Verma S, Sharma M (2010) Hexagonal coverage by mobile sensor nodes. Int J Comput Netw Secur (IJCNS) 2(4):41–44
21. Priyadarshi R, Gupta B (2019) Coverage area enhancement in wireless sensor network. Microsyst Technol 1–10. https://doi.org/10.1007/s00542-019-04674-y
22. Priyadarshi R, Gupta B (2020) Deployment techniques in wireless sensor networks: a survey, classification, challenges, and future research issues. J Supercomput. https://doi.org/10.1007/s11227-020-03166-5

Fractional Frequency Reuse Scheme for Noise-Limited Cellular Networks

Vishnu Anugrahith Sateesh, Iti Dutta, Rahul Priyadarshi, and Vijay Nath

Abstract The fractional frequency reuse (FFR) method generally used to investigate the noise-limited cellular network in the system. In this paper, we drive the two different strategies, which are related to secrecy outage probability with the best base station (BS) serve and the nearest BS serve, by analysing the data of channel gains. For the noise, limited network in secrecy outage probability without FFR, two transmission schemes are analysed, i.e., nearest BS and best BS. Simulation results show that both the plans of transmitting with the best BS and the utilization of the FFR are gainful for physical layer security in the downlink cell systems.

Keywords Cellular networks · Fractional frequency reuse (FFR) · Nearest BS · Best BS · Eavesdroppers

1 Introduction

To boost the capacity and performance of cellular system, FFR was proposed as an inter-cell interference coordination strategy with the absence of an eavesdropper in [1] and [2]. The basic impression of FFR is to split bandwidth of cell to decrease the

V. A. Sateesh (✉) · V. Nath
Department of Electronics and Communication Engineering, Birla Institute of Technology, Mesra, Ranchi, JH 835215, India
e-mail: vishnuanugrahith@gmail.com

V. Nath
e-mail: vijaynath@bitmesra.ac.in

I. Dutta
Department of ECE, RTCIT, Ranchi, JH 835215, India
e-mail: itidutta3@gmail.com

R. Priyadarshi
Department of Electronics and Communication Engineering, National Institute of Technology, Patna, Bihar 800005, India
e-mail: rahul.glorious91@gmail.com

V. Nath and J. K. Mandal (eds.), *Proceedings of the Fourth International Conference on Microelectronics, Computing and Communication Systems*, Lecture Notes in Electrical Engineering 673, https://doi.org/10.1007/978-981-15-5546-6_83

interference at the cell users. Cell-edge users can make free from the interference of the adjacent cells by using FFR technique. Another concept to improve the security is the sectorized transmission which was studied in [3, 4] and [5]. By using directional antennas, the sectorized transmission scheme was performed by transmitting independent signals in several sectors of the cell [6]. Based on the strategy of sectorized transmission, the effect of adopting multiple antennas is to improve the security in the interference-free scenario. The combination of sectoring and artificial noise was further analysed in.

The transmitter transmits the source information in the sector where the target user is situated, and sends artificial noise to other sectors. Although the approach of FFR is similar to that of sectorized transmission as they both divide the whole cell into different sectors. There is need of FFR to the Base Stations (BSs) to transmit with different sub-bands of the total bandwidth to different sectors. Further, from the view of security, FFR will also contribute towards increasing the secrecy rate by decreasing interference at the cell users, due to the partition of frequency bands. The study in [7] concluded physical layer security in cellular systems by comparing the approaches of communicating with nearest BS and optimum BS, by ignoring interference and the channel fading. To improve physical layer security in the network, combined effect of artificial noise and sectoring was investigated in [8], by using the same frequency for all the sectors. Study of secrecy outage probability in the downlink of cellular systems was considered in [9] by approximating the PDF of interference. In sensor network to cover entire region [10] various coverage algorithms [11] as well as various routing protocols have been proposed [12–22] recently but that is not suitable for noise-limited cellular networks.

1.1 System Model

We reflect downlink of a cellular system where the BSs are organized according to the repulsive property, which is thinned from a parent homogeneous parent point process (PPP) ϕ_P with intensity λ_P. The parent point process $\phi_P = \{(x_i, m_i)$ where $i = 1, 2, 3 \ldots\}$ is a marked PPP, where $x_{i,}$ denotes the ith point of the process and m_i stands for its marr.

The repulsive point process assurances the minimum distance d between any two BSs in the network. ϕ_M is obtained by removing all the points in ϕ_P that have a neighbour within distance d that has a smaller mark. The cellular network distribution is shown in Fig. 1, where r_l and r_e denote the distance from the transmitting BS to the real receiver and the eavesdropper, respectively. The eavesdroppers are distributed as an independent PPP, i.e., ϕ_e, with their intensity being represented as λ_e. The BSs are supposed to transmit with the same power P. We consider the security at a typical real user. We are assuming that all the BSs are transmitting with the same power P.

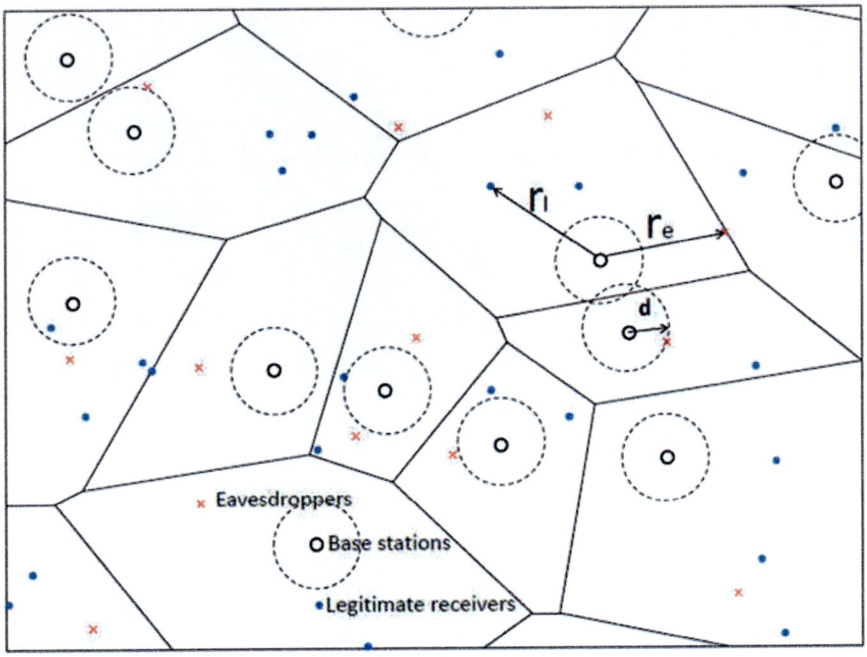

Fig. 1 Deployment of cellular networks

1.2 Fractional Frequency Reuse(FFR)

In the circumstances with FFR, we adopt that each of the Voronoi cell in the cellular network is separated into L sectors and each of the sectors is aided by its BS with a sub-band of the total bandwidth [23]. The set of interfering base stations are base stations in other real receivers and eavesdroppers. To differentiate from scheme without FFR, the interference in the cellular network with FFR at the real user and the worst-case eavesdropper are denoted as $\hat{I}_l$ and $\hat{I}_e$.

Figure 2 shows that the transmission structure of using FFR in a Voronoi cell from Fig. 1. The same transmission structure is also organized in other Voronoi cells of the cellular network. For a typical real receiver, the disadvantage of its secrecy rate is from eavesdroppers that communicate using the same sub-band.

2 Problem Formulation

We consider a state where the serving BS sends messages to a typical real user in existence of randomly distributed eavesdroppers. By considering that there is no agreement among the eavesdroppers, secrecy capacity of transmission from serving BS to real user is given by [24].

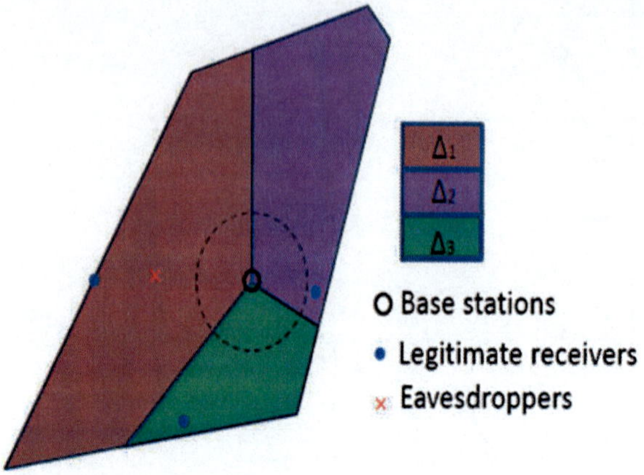

Fig. 2 Transmission scheme of a voronoi cell with FFR

$$R_s = \left[(1+\gamma_l) - (1+\gamma_e)\right] \tag{1}$$

$$P_{\text{out}}(\tau) = 1 - P[R_s > \tau] = 1 - P\left[\frac{1+\gamma_l}{1+\gamma_e} > 2^{\tau}\right] \tag{2}$$

where τ is the target secrecy rate. The PDF of γ_l and γ_e are denoted by $f_{\gamma l}$ and $f_{\gamma e}$, respectively. At the real user,

$$P_{\text{out}}(\tau) = P\left[\frac{1+\gamma_l}{1+\gamma_e} < 2^{\tau}\right] \tag{3}$$

$$= P\left[\gamma_e > 2^{-\tau}(1+\gamma_l) - 1\right]$$

$$= 1 - \int_0^{\infty} f_{\gamma l}(y) \int_0^{2^{-\tau(1+y)-1}} f_{\gamma e}(x)\mathrm{d}x\mathrm{d}y$$

$$= 1 - \int_0^{\infty} f_{\gamma l}(y) F_{\gamma e}\left(2^{-\tau}(1+y) - 1\right)\mathrm{d}y$$

where $F_{\gamma e}$ is the CDF of γ_e.

In the noise-limited cellular networks, the interference from other BSs can be disregarded or has the same level of power compared to the noise [25]. The analysis in this section is planned by the secrecy outage probabilities of the downlink cellular networks which is derived by inspecting two different schemes, i.e., with and without FFR.

To avoid misunderstanding, we denote SNR at typical real user and eavesdropper as $\hat{\gamma}_l$ and $\hat{\gamma}_e$ which can be stated as

$$\hat{\gamma}_l = \frac{P|h_l|^2 r_l^{-\alpha}}{\sigma_l^2} \tag{4}$$

And

$$\hat{\gamma}_e = \frac{P|h_e|^2 r_e^{-\alpha}}{\sigma_e^2} \tag{5}$$

2.1 *Secrecy Outage Probability without FFR*

In this portion, we inspect secrecy outage probability of downlink transmission in two situations, i.e., served by the nearest and the best BSs. Selection of the serving BS according to different standards is executed through the support of the BSs by replacing the position and distinctiveness info of the real users [26].

2.1.1 Nearest BS

Representing the path loss from the nearest BS to the real user as $r_l^{-\alpha} = x_l$, the PDF of x_l is given by

$$f_{xl}(x) = \pi\lambda_M \delta s^{\delta-1} \exp\left(-\lambda_M \delta s^{\delta}\right) \tag{6}$$

where $\delta = \frac{d}{\alpha}$.

According to the description of secrecy capacity, the SNR at the worst-case eavesdropper is taken into justification to examine the secrecy outage probability. Note that, in this state, the eavesdroppers are trying to understand the source information from the serving BS.

The secrecy outage probability of the real user communicating with the nearest BS can be explained as

$$P_{n\_\mathrm{out}}(\tau) = 1 - P\left[\frac{1+\gamma_l}{1+\frac{P}{\sigma_e^2}\frac{1}{\xi_e}} > \tau\right] \tag{7}$$

For the case of high SNR, secrecy outage probability of real user communicating with the nearest BS can be simplified as

$$P_{n\_\mathrm{out}}(\tau) = 1 - P\left[\xi_e > \frac{2^{\tau}}{\zeta_l}\right] \tag{8}$$

2.1.2 Best BS

In this scheme, the real user will be served by the BS from which the user can get the maximum channel gain. Similar to the examination of the channel gain at worst-case eavesdropper, PDF for the path-loss procedure with fading from the best BS to the real user can be given as

$$f_{\xi l}(s) = \delta A_l s^{\delta-1} \exp(-A_l s^{\delta}) \tag{9}$$

where $A_l = \pi \lambda_l \delta \frac{\Gamma(\delta+m)}{m^{\delta}\Gamma(m)}$. The secrecy outage probability of real user served by best BS can be explained by

$$P_{n\_\mathrm{out}}(\tau) = 1 - P\left[\frac{1 + \frac{P}{\sigma_l^2}\frac{1}{\xi_l}}{1 + \frac{P}{\sigma_e^2}\frac{1}{\xi_e}} > \tau\right] \tag{10}$$

Similarly, for the case of high SNR, the approximation of secrecy outage probability of the real user communicating with the best BS is given by

$$P_{bst\_\mathrm{out}}(\tau) = 1 - P\left[\frac{\frac{1}{\xi_l}}{\frac{1}{\xi_e}} > 2^{\tau}\right] \tag{12}$$

2.2 *Secrecy Outage Probability with FFR*

In this scheme of FFR, the intensity of the path-loss process with fading is reduced by frequency reuse, which works as autonomous withdrawing of the eavesdroppers. The number of eavesdroppers in each sector is $\frac{1}{L}$ of the total number in the cell, i.e., the intensity quantity of each sector is reduced by L. Accordingly, the parameter A_e will also be affected by the frequency reuse. To avoid confusion, we denote the value of A_e in the scheme of FFR as $\hat{A}_e$, then $\hat{A}_e = \frac{A_e}{L}$. The real users in each sector of the Voronoi cells are assigned a common sub-band, Δ_x, of the total bandwidth transmitted from the BS. To analyse the conclusion of executing FFR, the real user is supposed to be located at a distance of r_0 from its nearest BS. The secrecy outage probability at the real user served with the plan of FFR can be derived as

$$P_{\mathrm{sct\_out}}(\tau) = 1 - P\left[\frac{1 + \frac{P}{\sigma_l^2}|h_l|^2 r_o^{-\alpha}}{1 + \frac{P}{\sigma_e^2 \xi_e}} > \tau\right] \tag{13}$$

$$= \int_0^{\infty} \exp(-y)\left(1 - \exp\left(-\hat{A}_e \frac{\frac{P}{\sigma^2}}{2^{-\tau}(1 + \frac{P}{\sigma^2} r_o^{-\alpha} y) - 1}\right)\right) dy$$

For the case of high SNR, the secrecy outage probability at the real user served by the strategy of FFR can be simplified as

$$P_{\text{sct\_out}}(\tau) = 1 - P\left[\frac{|h_l|^2 r_o^{-\alpha}}{\frac{1}{\xi_e}} > \tau\right] \tag{14}$$

$$= \int_0^\infty \exp(-y)\left(1 - \exp\left(-\hat{A}_e\left(\frac{2^\tau}{y r_o^{-\alpha}}\right)^\delta\right)\right)\mathrm{d}y$$

The expression in (14) is similar to that in (8), and only the parameter $\hat{A}_e$ is affected by FFR due to the change of intensity.

3 Simulation and Results

For the scenario of the noise-limited cellular network, the secrecy outage probabilities of the downlink transmissions from the best and nearest BSs are plotted in Fig. 3. Secrecy outage probability is worse when intensity of eavesdroppers λ_e is higher.

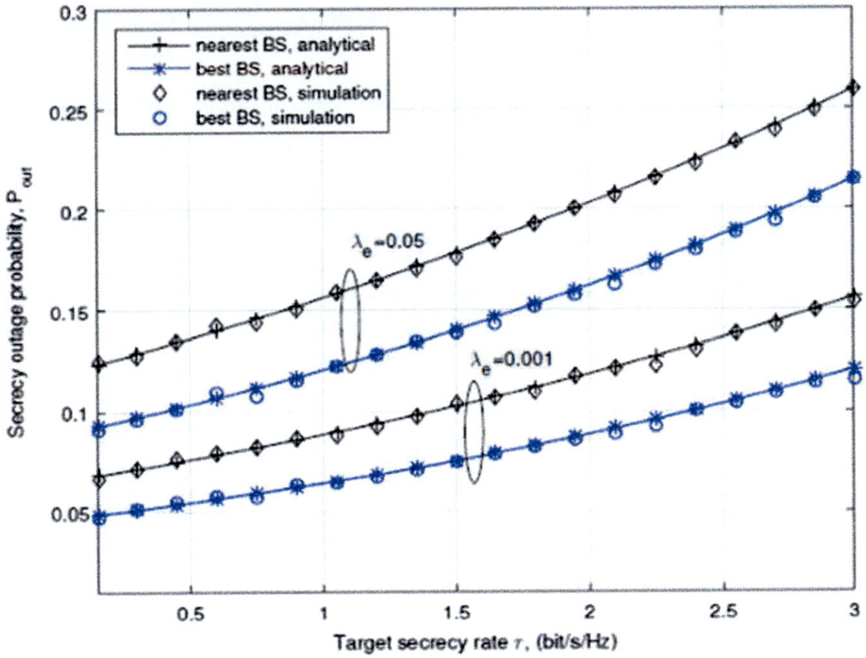

Fig. 3 Secrecy outage probability with the best BS vs the nearest BS with $\lambda_p = 0.2$ and $d = 2$

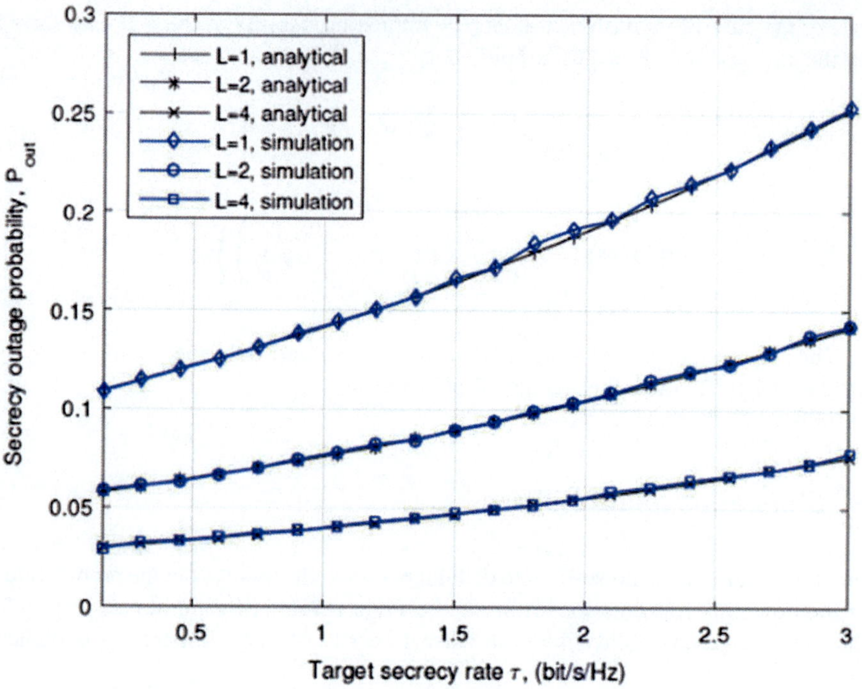

Fig. 4 Secrecy outage probability with and without FFR, with $\lambda_p = 0.2$ and $d = 2$

Increase of λ_e will reduce the probable distance from the transmitting BS to the worst-case eavesdropper, which will accordingly reduce the SNR of the eavesdropper, i.e., γ_e. Figure 3 shows that the strategy of transmitting with the best BS is always helpful to improve the secrecy outage probability. However, the improvement is less beneficial when λ_e is higher. It is because that the increase of λ_e reduces the expectation of distance from BS to worst-case eavesdropper.

Figure 4 describes secrecy outage probability of the downlink cellular networks using FFR based on different frequency reuse factors. It shows that in the noise-limited cellular networks, the use of FFR can improve the secrecy outage probability suggestively, especially when number of frequency reuse factor is large. This is due to the application of frequency reuse which helps to reduce λ_e with an equivalent effect of independent withdrawing.

4 Conclusion

In this work, we studied physical layer security in the downlink transmission of cellular networks by comparing the schemes with and without FFR. The effect of frequency reuse on secrecy outage probability in the noise-limited networks were investigated. Although increasing the frequency reuse factor contributes to improve the security, the improvements will be less important when the number of frequency reuse factor is large. Besides, the strategies of transmitting with best BS, increasing intensity of BSs, and reducing repulsive distance of BSs are helpful to develop secrecy performance in downlink cellular network.

5 Future Work

For more practical usage of security in wireless networks, the application will lead to the attention of more conservative challengers, i.e., eavesdroppers might be active rather than passive and conspiring to maximize their received signal powers. The remainder of this section contains some interesting future research directions of physical layer security that can make our work more general and practically attractive. As we know that eavesdroppers are usually supposed to be passive and act independently. According to these models, still need more allowance on the assumption to solidify the adversary model, such as considering active eavesdroppers and their mutual collaboration. For simplicity, eavesdroppers are usually assumed to be separated from each other and cannot transfer information.

References

1. Novlan TD, Ganti RK, Ghosh A, Andrews JG (2011) Analytical evaluation of fractional frequency reuse for OFDMA cellular networks. IEEE Trans Wireless Commun 10(12):4294–4305
2. Ullah R, Fisal N, Safdar H, Maqbool W, Khalid Z, Khan AS (2013) Voronoi cell geometry based dynamic fractional frequency reuse for OFDMA cellular networks. In: IEEE international conference on signal and image processing applications (ICSIPA), pp 435–440
3. Pinto PC, Barros J, Win MZ (2012) Secure communication in stochastic wireless networks-part I: Connectivity. IEEE Trans Inf Forensics Secur 7(1):125–138
4. Jeong Y, Quek TQS, Shin H (2013) Stochastic wireless secure multicasting. In: IEEE international conference on communications (ICC), pp 4718–4723
5. Zhang X, Zhou X, McKay MR (2013) Enhancing secrecy with multi-antenna transmission in wireless ad hoc networks. IEEE Trans Inf Forensics Secur 8(11):1802–1814
6. Wang H, Zhou X, Reed MC (2013) Physical layer security in cellular networks: a stochastic geometry approach. IEEE Trans Wireless Commun 12(6):2776–2787
7. Geraci G, Dhillon HS, Andrews JG, Yuan J, Collings IB (2014) Physical layer security in downlink multi-antenna cellular networks. IEEE Trans Commun 62(6):2006–2021
8. Csiszar I, Korner J (1978) Broadcast channels with confidential messages. IEEE Trans Inf Theory 24(3):339–348
9. Bloch M, Barros J, Rodrigues MRD, McLaughlin SW (2008) Wirelessinformation-theoretic security. IEEE Trans Inf Theory 54(6):2515–2534
10. Priyadarshi R, Gupta B (2020) Deployment techniques in wireless sensor networks: a survey, classification, challenges, and future research issues. J Supercomput. https://doi.org/10.1007/s11227-020-03166-5

11. Priyadarshi R, Gupta B (2019) Coverage area enhancement in wireless sensor network. Microsyst Technol 1–10. https://doi.org/10.1007/s00542-019-04674-y
12. Priyadarshi R, Soni SK, Nath V (2018) Energy efficient cluster head formation in wireless sensor network. Microsyst Technol 24(12):4775–4784
13. Priyadarshi R, Rawat P, Nath V (2019) Energy dependent cluster formation in heterogeneous wireless sensor network. Microsyst Technol 25(6):2313–2321
14. Priyadarshi R, Soni SK, Sharma P (2019) An enhanced GEAR protocol for wireless sensor networks. In: Nanoelectronics, circuits and communication systems. Springer, Singapore, pp 289–297
15. Priyadarshi R, Singh L, Singh A, Thakur A (2018) SEEN: stable energy efficient network for wireless sensor network. In: 2018 5th international conference on signal processing and integrated networks (SPIN). IEEE, pp 338–342
16. Priyadarshi R, Singh L, Singh A (2018) A novel HEED protocol for wireless sensor networks. In: 2018 5th international conference on signal processing and integrated networks (SPIN). IEEE, pp 296–300
17. Priyadarshi R, Singh L, Sharma I, Kumar S (2018) Energy efficient leach routing in wireless sensor network. Int J Pure Appl Math 118(20):135–142
18. Priyadarshi R, Tripathi H, Bhardwaj A, Thakur A, Thakur A (2017) Performance metric analysis of modified LEACH routing protocol in wireless sensor network. Int J Eng Technol 7(1–5):196
19. Priyadarshi R, Bhardwaj A (2017) Node non-uniformity for energy effectual coordination in WSN. Int J Inf Technol Secur 9(4):3–12
20. Priyadarshi R, Singh L, Kumar S, Sharma I (2018) A hexagonal network division approach for reducing energy hole issue in WSN. Eur J Pure Appl, Math, p 118
21. Priyadarshi R, Yadav S, Bilyan D (2019) Performance analysis of adapted selection based protocol over LEACH protocol. In: Smart computational strategies: theoretical and practical aspects. Springer, Singapore, pp 247–256
22. Priyadarshi R, Yadav S, Bilyan D (2019) Performance and comparison analysis of MIEEP routing protocol over adapted LEACH protocol. In: Smart computational strategies: theoretical and practical aspects. Springer, Singapore, pp 237–245
23. Mukherjee A, Swindlehurst AL (2013) Jamming games in the MIMO wiretap channel with an active eavesdropper. IEEE Trans Signal Process 61(1):82–91
24. Zhou X, Ganti RK, Andrews JG (2011) Secure wireless network connectivity with multi antenna transmission. IEEE Trans Wireless Commun 10(2):425–430
25. Trappe W (2015) The challenges facing physical layer security. IEEE Commun Mag 53(6):16–20
26. Priyadarshi R, Singh MP, Bhardwaj A, Sharma P (2017) Amount of fading analysis for composite fading channel using holtzman approximation. In: 2017 fourth international conference on image information processing (ICIIP). Shimla, pp 1–5. https://doi.org/10.1109/iciip.2017.8313759

Real-Time Monitoring of PM2.5 with IoT

Vishal Kumar and Kamal Kant

Abstract The main aim of researchers is to make a standalone device that measures the real-time concentration of particulate matter and provided with an Android application for monitoring the concentration in surrounding environment. Particulate matter is the complex mixture of tiny dust particles and liquid droplets suspended in the atmosphere. Their main compounds are nitrogen oxides, mineral dust, carbon monoxide, organic matter, soot or black carbon. The dust particles vary in size that is relevant to the potential causing health problems. Particulate matter enters the human respiratory system that causes serious illness. The device is microcontroller-based system that is connected to the sensor that monitors the concentration of PM2.5. The device is connected through Internet facilitates to send data over cloud. The researchers conduct test in location given by CPCB.

Keywords PM2.5 · Microcontroller · IoT · GSM module · CPCB device

1 Introduction

In recent years, the study-related health effects due to the air pollution have been increased. Due to exposure to airborne particulate matter, there is an increase in mortality and hospital admissions for disease related to respiratory system and cardiovascular problems. Gradually, over the years airborne particles have changed size and composition which in turn increases their toxicity. The airborne particulate matters are PM10 which are smaller than 10 μm and PM2.5 which are smaller than 2.5 μm in size. These particles can penetrate the respiratory system as well as the lungs of human being [1]. The air pollution effects on environment are of great interest

V. Kumar (✉) · K. Kant
Department of ECE, B.I.T. Mesra, Ranchi, JH 835215, India
e-mail: vishal2400034@gmail.com

K. Kant
e-mail: kamal.kant2291@gmail.com

V. Nath and J. K. Mandal (eds.), *Proceedings of the Fourth International Conference on Microelectronics, Computing and Communication Systems*, Lecture Notes in Electrical Engineering 673, https://doi.org/10.1007/978-981-15-5546-6_84

on global discussion. Today, it is becoming a necessity to monitor the air pollution on regular basis [2]. As dust particles vary widely from region to region, a real-time measurement of the particulate matter should be considered along with the multi-measurement approach to access the impact on the climate. Specifically, we use detailed measurement from satellite network of ground-based instruments [3]. Particulate matter consists of a heterogeneous mixture of solid and liquid particles suspended in air which are hazardous to health. It depends on meteorological conditions and emission sources of a particular region [4]. The important atmospheric pollutants are PM2.5 and PM10. The particulate matter is a mixture of solids and liquids suspended in the air. Basically, it is a complex mixture of organic or inorganic components such as dust, soot, pollen and liquid droplets [5]. The main constituents of PM2.5 are nitrates, ammonium salts, sulfates, carbonaceous particles, metal particles, minerals and organic compounds. The main constituents that made up the PM10 are carbonaceous substance and crystal material. The sources for their emission include exhaust, organic compounds they adsorb [6, 7]. The dust sensor can measure the concentration of particulate matter in the air. The basic principle of the sensor is based on light scattering. Light scattering can be used when the dust particles go through the detecting chamber. The scattered light is transformed in electrical signals which is then processed and give out the digital output as the raw data of the sensor. The dust sensor is capable of measuring PM2.5 [8, 9].

This research is aimed to develop a model of cloud-based air quality monitoring system, which can be accessed on mobile application and personal computer. Specifically, the project model will take the form of standalone device which monitors the air quality nearby and sends the data to the cloud and can easily be accessed by public. The device can be provided power supply from any source of 12 volts supply. The collected data will be stored on the cloud platform and will be used for future references. This data will also be beneficial for the officials working for controlling air pollution. The collected data is plotted on the graph according to the time and date on the cloud platform. By monitoring and identifying the level of PM2.5 suspended in the air, they can access and determine the actions needed to reduce or prevent the identified pollutant. The main intention of this study is to monitor the level of PM2.5 pollutants and also consists of creating a GUI for the mobile phone that will be used to view the monitored data.

The research is limited to measurement of PM2.5 concentration and not giving the average air quality of the environment and does not cover the maintenance of the air quality of environment.

2 Methodology

The device is based on microcontroller that connects to the required sensor modules according to the need. The device is connected to the dust sensor which provides the data of particulate matter in the atmosphere. The sensor processes the collected raw data and sends it to the cloud platform. The cloud platform used it accessed using an API. This also provides the graphical format of the collected data over time. The data can be accessed on cloud platform as well as Android phone. The user can also access the past data and compare it. The sensor sends the real-time data to the microcontroller used which processes the data for further use. The main objective of this prototype device is to provide real-time data of particulate matter suspended in the atmosphere and is easily accessible to the user.

2.1 Flowchart

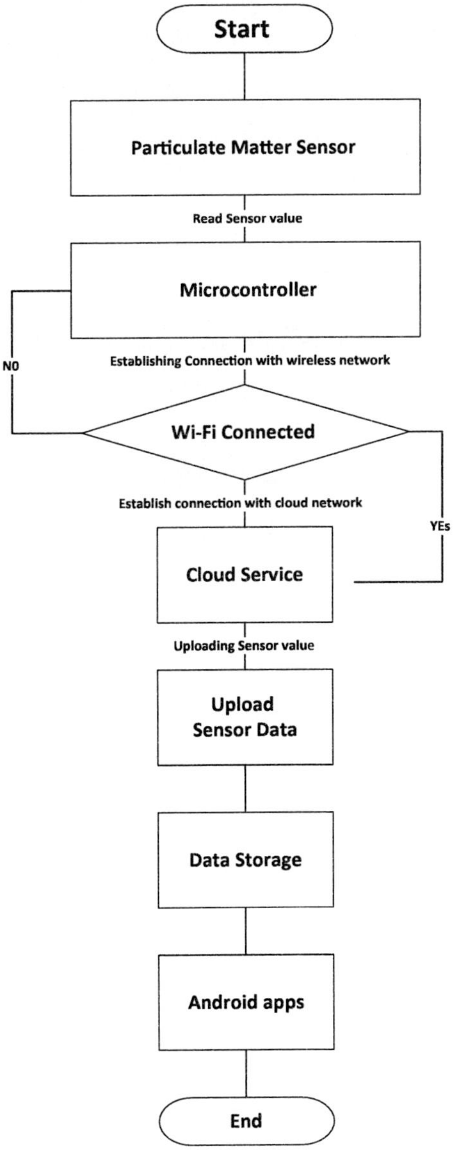

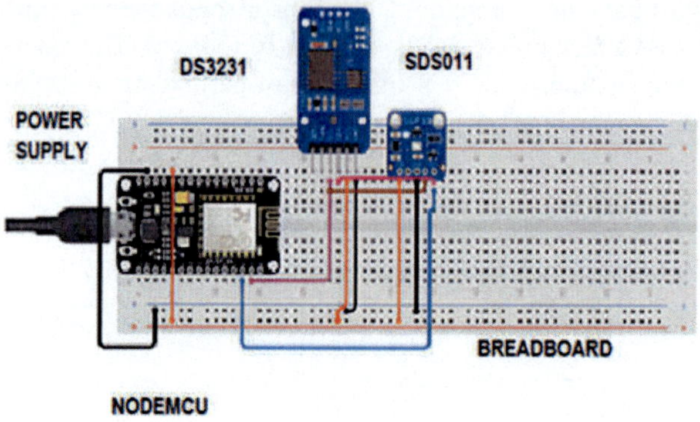

Fig. 1 Circuit configuration of the microcontroller interfacing with the sensors

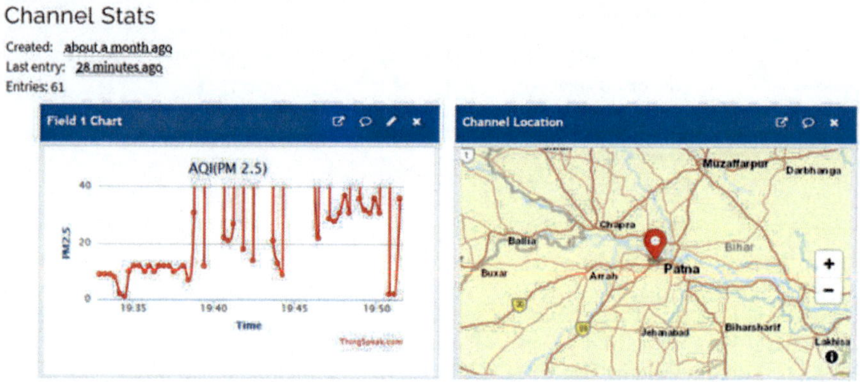

Fig. 2 Data plotted on graph through the cloud platform from the AQI meant for measurement of concentration of PM2.5. The picture also contains the live location during the monitoring of PM2.5 concentration

2.2 *Diagram*

See Figs. 1 and 2.

2.3 *Working*

For monitoring of the level of particulate matter concentration suspended in the air, the prototype has a sensor and a Wi-Fi module connected to the microcontroller. The dust sensor is connected to the microcontroller which gives out the electrical signals to the microcontroller which further transformed in the concentration value of the particulate matter. The microcontroller has input–output pins to configure the circuit and gives out the required data; similarly, Wi-Fi module connected to the microcontroller is used for uploading sensor data over an open-source cloud platform which analyzes and visualizes the sensor data and let others to retrieve the data stored over Android apps. The open-source Internet of Things (IoT) platform is used with API to store the data from things using the HTTP protocol over the Internet for which we are using GSM module. The GSM module is used to provide the Internet connectivity to the microcontroller to send the data over cloud. The monitored data will be sent over the open-source cloud platform from where the user can access the data easily. Using API link, we can access the real-time data on mobile application describing the concentration of the particulate matter in the surrounding environment along with some preventive measures in the situation of increased concentration of particulate matter. The formed prototype device is provided power supply with 5 V adapter.

2.4 Calculation of Particulate Matter Concentration

The measurement of particulate matter (PM2.5) is done with the help of a dust particle sensor that is connected to the microcontroller. The principle of working for dust sensor is scattering of light. Light scattered can be interfered at the area where particles pass through the detecting area. The scattered light inside the sensor is converted in the form of electrical signals, and further, the signals are amplified and processed. The counted number and the diameter of dust particles can be obtained by analyzing according to the relation of the signal waveform with the particles diameter.

The scattered light is transformed into electrical signals for further processing. It interprets the data in terms of and high byte for PM2.5. The sensor gives out the data in form of digital output.

PM2.5 value ($\mu g/m^3$): {(PM2.5 High Byte * 256) + PM2.5 low Byte)} [10].

The measured data by the sensor is send to the microcontroller. The microcontroller then sends the data over cloud using the Wi-Fi module that will be used for monitoring air quality index. The Wi-Fi module is provided Internet connection using the GSM module (Fig. 3).

2.5 Validation of Results

Table 1 shows the data of the results obtained from the cloud platform monitored concentration of PM2.5. The concentration of PM2.5 is arranged according to time mentioned in the real-time monitored data.

Table 2 shows the compared data of prototype and CPCB device.

Fig. 3 Hardware model of the prototype along with the cloud data available on the laptop display

The prototype device output data is calibrated and compared with the data of CPCB device [11]. The graph is plotted on the MS-EXCEL. The collected data is plotted on the graph using MS-EXCEL. The data must be arranged sequentially before importing the data to plot the graph. Select insert chart then label the axis on the graph obtained on MS-EXCEL (Fig. 4).

3 Comparative Study

The study of this paper is compared to the research related on this topic.

The study related to Ref. [2] uses complex circuitry and the prototype device formed is high cost and the power supply is provided by solar panel, whereas this prototype device uses NodeMcu microcontroller which has inbuilt Wi-Fi module and also the network connectivity can be provided by GSM module as well as the public Wi-Fi access or mobile hotspot. This device is low cost and the device formed is portable.

The study of Ref. [12] has used the Thingspeak platform for sending data to cloud and ESP8266 Wi-Fi module for sending data for specific period of time, whereas this study prototype uses Thingspeak platform for sending data to cloud in real time and the device is portable and can used by mobile users. The data flow of monitored particulate matter is sent to the cloud server in real- and stored for future reference. The device formed is compact and can be used by individual as it is portable and the

Table 1 Monitored data of PM2.5

S. No.	PM2.5	Time
1	9	2019-04-15 14:02:38 IST
2	9	2019-04-15 14:02:56 IST
3	9	2019-04-15 14:03:14 IST
4	9	2019-04-15 14:03:33 IST
5	8	2019-04-15 14:03:51 IST
6	2	2019-04-15 14:04:10 IST
7	1	2019-04-15 14:04:28 IST
8	10	2019-04-15 14:04:46 IST
9	12	2019-04-15 14:05:04 IST
10	12	2019-04-15 14:05:22 IST
11	10	2019-04-15 14:05:44 IST
12	12	2019-04-15 14:06:02 IST
13	10	2019-04-15 14:06:20 IST
14	12	2019-04-15 14:06:38 IST
15	12	2019-04-15 14:06:57 IST
16	12	2019-04-15 14:07:15 IST
17	10	2019-04-15 14:07:34 IST
18	12	2019-04-15 14:08:10 IST
19	7	2019-04-15 14:08:29 IST
20	31	2019-04-15 14:08:48 IST

Table 1.2 Comparison data of prototype and CPCB device

Date time	Prototype data	CPCB device
2019-04-17 13:15:12 IST	22	24
2019-04-17 13:30:30 IST	24	24
2019-04-17 14:30:49 IST	22	24
2019-04-17 14:45:40 IST	24	24
2019-04-17 15:00:59 IST	22	24
2019-04-17 15:30:34 IST	18	28
2019-04-17 15:45:07 IST	24	28
2019-04-17 16:00:18 IST	22	28
2019-04-17 16:15:37 IST	24	29

power supply can be given by adapter, power bank which makes it easy access for mobile users.

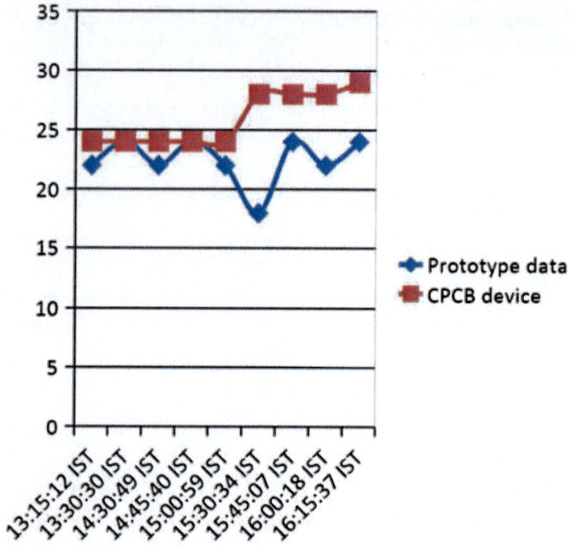

Fig. 4 Comparison graph for results of prototype data and CPCB device

4 Conclusion

The device formed gives the monitored data of the particulate matter concentration suspended in the air. The results are arranged in the tabular form. Based on the results of the research, the researchers monitored the particulate matter concentration suspended in the air using the specified sensor. The sensor used is laser-based SDS011 sensor. This dust sensor can measure the concentration level of particulate matter using laser on the principle of light scattering. The microcontroller processes and computes the concentration level of particulate matter in the air and it is connected to the Wi-Fi module that sends the monitored data of particulate matter to the server with the help of GSM module that provides the Internet connectivity. The microcontroller used was Node MCU. The dust sensor data was compared in order to match data with the device of Government CPCB device. The CPCB device measures the air quality index of IGSC Planetarium complex, Patna, India, with 24 h clock. The users can easily access the real-time measurement of the monitored data via Think Viewer Android application, along with the data of past. The cloud platform also provides the graph of the monitored data.

Acknowledgements I am thankful to International Conference on Microelectronics, Computing and Communication Systems (MCCS) for encouraging me for this project. I would like to pay my gratitude toward Dr. Shiv Kumar Choubey, Department of Electronics and Communication Engineering, Birla Institute of Technology-Mesra, Patna campus for guiding me on this project. I am thankful to my parent and friends who supported me throughout the working on this project.

References

1. Brunekreef B, Holgate ST (2002) Air pollution and health. Lancet 360(9341)
2. Avendano GO, Dela Cruz JC, Ballado AH, Ulyzes LGR, Atienza ACP, Regala BJG, Uy RC (2017) Microcontroller and app-based air quality monitoring system for particulate matter 2.5 (PM2.5) and particulate matter 1 (PM1). In: IEEE 9th international conference on humanoid, nanotechnology, information technology, communication and control, environment and management (HNICEM)
3. Kaufman YJ, Tanré D, Boucher O (2002) A satellite view of aerosols in the climate system. Nature 419(6903)
4. Janssen NA, Hoek G, Simic-Lawson M, Fischer P, van Bree L, ten Brink H, Keuken M, Atkinson RW, Anderson HR, Brunekreef B, Cassee FR (2011) Black carbon as an additional indicator of the adverse health effects of airborne particles compared with PM_{10} and $PM_{2.5}$. Environ Health Perspect 119(12):1691–1699
5. Gieré R, Querol X (2010) Solid particulate matter in the atmosphere. Elements 6(4):215–222
6. Harrison RM, Jones AM, Lawrence RG (2004) Major component composition of PM_{10} and PM_{25} from roadside and urban background sites. Atmos Environ 38:4531–4538
7. Celis JE, Morales JR, Zaror CA, Inzunza JC (2004) A study of the particulate matter PM_{10} composition in the atmosphere of Chillán, Chile. Chemosphere 54(4):541–550
8. Suruthimadevini T, Chitra S (2015) Int J Innov Res Sci Eng Technol 4(2)
9. Holstius DM, Pillarisetti A, Smith KR, Seto E (2014) Field calibrations of a low-cost aerosol sensor at a regulatory monitoring site in California. Atmos Meas Tech 7(4):1121–1131
10. Wang Q, Du R (2018) Study on indoor simulation method of atmospheric turbulence and aerosol environment. Lanzhou University
11. www.cpcb.nic.in
12. Pasha Sharmad (2016) Thingspeak based sensing and monitoring system for IoT with Matlab analysis. Int J New Technol Res (IJNTR) 2(6):19–23

FPGA Implementation of Digital Watermarking System for Robust Detection Using Discrete Wavelet Transform

Muthumanickam Shanmugam and Arun Chokkalingam

Abstract Watermarking is the process of providing authentication rights by ingraining ownership information in the form of text, audio, video or images called watermarks on the original image. The paper aims to implement an efficient watermarking algorithm using second-level discrete wavelet transform (DWT) for increased robustness. A hardware description language (HDL) has been developed for Haar wavelet due to less computation complexity. The proposed watermarking embedding algorithm has been implemented in two FPGA boards. First one is a low-cost and low-power ARTIX FPGA-based Zynq 7000 SoC-based ZedBoard. Next one is Kintex FPGA-based FPGA Genesys 2 board which is suitable for DSP applications. An input image of size 16×16 and a watermark image of size 8×8 are used, and results of device utilization, power and time delay are compared. The results obtained suggest that the correlation and PSNR between the input image and the watermarked image are high, and the minimum device utilization indicates an effective watermarking algorithm.

Keywords FPGA · DWT · HDL · Watermarking · Authentication

1 Introduction

The brisk pace with which the technology is growing has made communication more accessible and faster. But the enormous amount of data being created every second due to this swift growth has made the digital media more vulnerable to attacks that include unauthorized duplication. This problem can be overcome by the digital watermarking technique which is the process of embedding digital content like

This project has been funded by Space Application Centre—Indian Space Research Organisation (ISRO) No. B.19012/76/2014-II Dated 26/11/2014.

M. Shanmugam (✉) · A. Chokkalingam
R.M.K. College of Engineering and Technology, Chennai, India
e-mail: muthusushant@gmail.com

V. Nath and J. K. Mandal (eds.), *Proceedings of the Fourth International Conference on Microelectronics, Computing and Communication Systems*, Lecture Notes in Electrical Engineering 673, https://doi.org/10.1007/978-981-15-5546-6_85

text, audio, video or images onto the input data to prevent copyright infringement. The watermarks of varying degrees of clarity are embedded to mark the data for authentication purposes. The embedding process can be made in both spatial and frequency domains. Watermark is added onto randomly selected pixels of the image in case of spatial domain whereas, in the frequency domain, transformations like discrete cosine transform (DCT) and discrete wavelet transform (DWT) are used to embed the watermark. Based on the usage, the type of watermarking can be chosen from blind, semi-blind, visible, invisible, symmetric, asymmetric and so on.

A real-time implementation of the watermarking algorithms requires the use of field-programmable gate arrays (FPGAs). FPGAs are continuing to gain strength since there invention in 1984 where they were viewed as simple logic chips. Now they are widely used in signal processing and control applications. These hardware-based implementations are more advantageous than the software-based implementations. Because in software implementations, there is a delay between capturing of image and embedding of the watermark, thus making it susceptible to attacks. The other assets of using FPGAs are that they are flexible, reusable and more importantly capable of performing highly parallelized tasks. The parallel processing allows multi-channel processing, thus making it competent for performing intensive calculations with high speed and minimum power requirement. These make the FPGA a well-suited option for realizing the watermarking algorithms. The HDL language is commonly used for programming the design based on the required algorithm. Once the implementation is optimized and the desired results are achieved, it can be downloaded into the FPGA. The rest of the paper is organized as follows. Summary of literature survey is explained in Sect. 2. The DWT mathematical analysis is presented in Sect. 3. The architecture of FPGA implementation is given in Sect. 4. The details of the finite state machine is given in Sect. 5. Experimental results are analyzed in Sect. 6. Concluding remarks are given in Sect. 7.

2 Literature Survey

Joshi et al. [1] have developed an algorithm for digital watermarking using non-blind detection scheme. The algorithm has combined the advantages of both the frequency and spatial domain watermarking techniques. The robustness provided by the wavelet-based frequency domain watermarking against Internet threats and the reduced computational complexity provided by the Bitplane slicing scheme of spatial domain watermarking have been incorporated in the algorithm. The implementation of the algorithm in FPGA on 64 $\times$ 64 sized grayscale images showed efficient hardware utilization. An application-specific integrated circuit (ASIC)-based low-power robust invisible binary image watermarking 8-bit processor using HDL was developed by Karthigai Kumar et al. [2]. The watermarking algorithm is implemented in spatial domain in both the 0.35 and 0.13 μm technologies with a power consumption of 404.15 and 3.65 μW. The watermarking algorithm is prototyped using both Xilinx FPGA and ASIC. Ghosh et al. [3] have proposed a VLSI architecture of reversible watermarking encoder and decoder chips that are capable of performing reversible watermarking. The advantages of rhombus interpolation-based watermarking technique by difference expansion have been investigated and implemented on Xilinx Virtex-7 FPGA, Zynq SoC and ultra-scale FPGA platforms. The implementation gives very compatible and suitable result of the reversible watermarking system on chip. Jamal Ahmed et al. [4] have proposed an architecture for image watermarking algorithm using wavelet coefficient quantization. The architecture is pipelined in nature, and the watermarking algorithm is implemented in FPGA using XSG. The design optimization in latency and hardware resources was brought about by utilizing the system generator in algorithm design. The design produced a PSNR value of 42 dB against attacks and a design frequency of 141.9 MHz. A Haar wavelet-based JND model is used by Shenghui Liu et al. [5] which reduces the complexity in embedding watermark into integrated circuits. The fast pipeline architecture is modeled with HDL and evaluated on Altera EP2C35 FPGA device. When compared with full-band JND-based architecture, the hardware cost of 1460 logic cell combinations and 330 registers is smaller. The system utilized 120 MHz clock and a 75% time saving compared to the full-band JND algorithm. Williams Antonio Pantoja Laces et al. [6] proposed an efficient FPGA-based hardware implementation of low complexity steganography where the transparency and embedding rates are considered. Digital images are used as host signal for the steganographic system using additive interpolation-error expansion. The said hardware implementation is 64 times faster than a software implementation. Comparatively, the 40 nm technology utilized less hardware resources and has higher frequency of operation than the 90 and 60 nm technologies but has greater power consumption. Elamaran et al. [7] have implemented image enhancement algorithms utilizing spatial filtering techniques. The FPGA using Xilinx System Generator (XSG) is used for convolution filtering operations because of its high performance in implementing DSP algorithms. The results by Simulink and Xilinx DSP tools show that the resource utilization by the algorithm is 436 slices, 1620 flip flops, 1575 LUT and 37 IOBs indicating effective implementation. A steganographic technique for transmitting captured images using DWT and least significant bit (LSB) algorithm is proposed by Jadhav et al. [8]. The algorithm uses XPS and VB for the hardware implementation. The MATLAB simulation result shows that the system has a high PSNR value of 55 dB and MSE of 0.20 and thus is robust against malicious attacks on the image. Prabhishek Singh et al. [9] have presented an overview about the various frameworks and techniques for digital watermarking. The classification of digital watermarking based on aspects like robustness, domain and host signal is discussed. The advantages and disadvantages of various algorithms like LSB, Correlation. Patchwork, texture mapping coding, DCT, DWT and DFT were analyzed along with the wide applications of watermarking in the field of medicine, communication and security. A DCT-based blind digital image watermarking technique is proposed by Waikhom Mona Chanu et al. [10]. Sonjoy Deb Roy et al. [11] proposed video watermarking algorithm using DCT coefficients. In this method, the original image is not required for watermark recovery; instead a watermark sample is inserted to a pair of DCT coefficients. Thus, it is resistant to attacks and compression. The watermarking system is implemented in Xilinx Spartan FPGA board. The MATLAB simulation reports the minimum hardware utilization of 32,492 slices, 249 flip flops, 62,482 LUTs and 2117 IOBs indicating minimum number of devices.

3 DWT Mathematical Analysis

Wavelet the term defines small waves with limited duration for performing time–frequency analysis. Time domain states about the variations in the pixels at a particular time, whereas the scaled-version wavelets allow us to analyze the signal on a different scale. It gives variable resolution (higher frequencies are better resolved in time, and lower frequencies are better resolved in frequency). The wavelet transform analysis is computed separately for different segments of the time-domain signal at different frequencies. Multiresolution analysis (MRA) analyzes the signal at different frequencies giving different resolutions. It is good for the signal having high-frequency components for short durations and low-frequency components for a long duration, for example, images and video frames.

The DWT has been decomposed into different levels. The mother wavelet $\psi(x)$ is

$$\psi_{(a,b)} = [a]^{-1/2}\psi\frac{(t-b)}{a} \tag{1}$$

with $a, b \in R$ and $\psi \in L^2(R)$. It is necessary to express the continuous dilation and translation parameters a and b in terms of discrete values:

$$a = a_o^j, \quad \text{and} \quad b = kb_o a_o^j \tag{2}$$

where j affects the scaling of wavelet transform and k is the translation of wavelet function. After substituting the values of a and b into Eq. (1), then

$$\psi_{a,b}(x) = a_o^{-\frac{j}{2}}\psi(a_o^{-j}x - kb_o) \tag{3}$$

by $a = 2^j$ and $b = k2^j$ and then the wavelet in orthonormal basis function is

$$\psi_{j,k}(x) = 2^{-\frac{j}{2}}\psi(2^{-j}x - k) \tag{4}$$

With the above expression, the discrete wavelet transform coefficients are defined as

$$W_{\phi(jo,k)} = \frac{1}{\sqrt{M}}\sum_{x=o}^{M-1} f(x)\phi_{jo,k}(x) \tag{5}$$

$$W_{\psi(j,k)} = \frac{1}{\sqrt{M}} \sum_{x=o}^{M-1} f(x)\psi_{j,k}(x) \quad \text{for } j > jo \tag{6}$$

where $W_{\phi(jo,k)}$ is the approximation coefficient and $W_{\psi(jo,k)}$ detailed coefficient

$$\phi_{jo,k}(x) = \Sigma \alpha_n \phi_{jo+1,n}(x) \tag{7}$$

$$\psi_{j,k}(x) = a_o^{-\frac{j}{2}} \phi\left(a_o^{-j}x - kbo\right) \tag{8}$$

The below equation is the reconstruction form of DWT from the coefficients.

$$f(x) = \frac{1}{\sqrt{M}} \sum_k W_{\phi(jo,k)} \phi_{jo,k}(x) + \frac{1}{\sqrt{M}} \sum_{j=jo}^{\infty} W_{\psi(j,k)} \psi_{j,k}(x) \tag{9}$$

The two-dimensional DWT can be implemented in $f(x, y)$ using digital filters and down samplers.

3.1 Block Diagram of Watermark Embedding Process

The watermark embedding process is shown in Fig. 1. The entire process is developed using VHDL. The grayscale host image with the size of 16 × 16 and watermark with

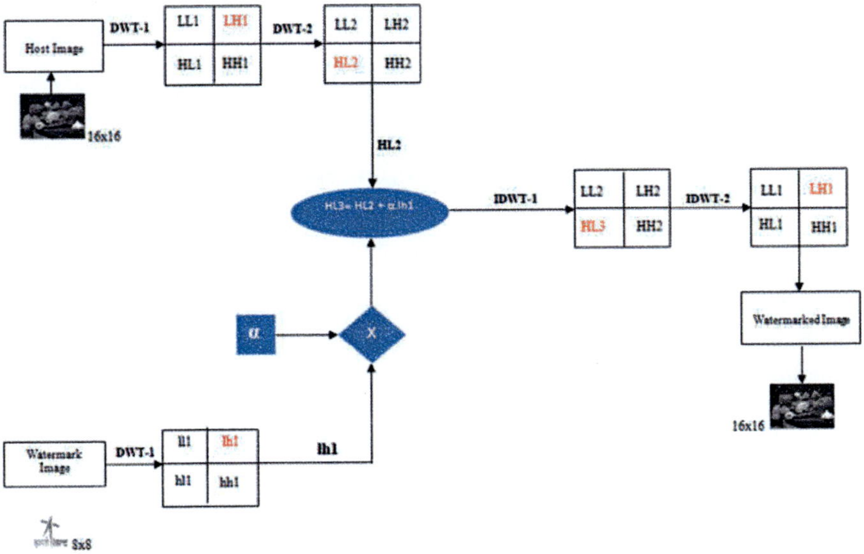

Fig. 1 Block diagram of watermark embedding process

the size of 8 × 8 are taken for watermarking. The process is explained in the section followed by the block diagram.

Embedding Process

Step 1: Grayscale host image (peppers 16 × 16) is taken as input image.
Step 2: First-level DWT is applied to host image by applying horizontal and vertical filters. This results into four sub-bands given by **LL1, LH1, HL1** and **HH1**. The size of each sub-band is 8 × 8.
Step 3: Second-level DWT is applied to **LH1,** and this results into four sub-bands given by **LL2, LH2, HL2** and **HH2** with the size of each band 4 × 4.
Step 4: Grayscale watermark image (ISRO logo 8 × 8) is taken.
Step 5: First-level DWT is applied to watermark image, and this results into four sub-bands given by **ll1, lh1, hl1** and **hh1** with the size of each band 4 × 4.
Step 5: Now the matrix dimension of second-level DWT of host is equivalent to first-level DWT of watermark. This eases the further mathematical calculation in FPGA.
Step 6: The intensity factor 'α' is selected as 0.5 and multiplied with **lh1** band of watermark image. This is the modified component of **lh1**, whose intensity value is reduced to 50%.
Step 7: The modified **lh1** band is added to **HL2** of host image which is given in Eq. (10). This is the proposed embedding process of watermarking algorithm.

$$\mathrm{HL3} = \mathrm{HL2} + (\alpha * \mathrm{lh1}) \tag{10}$$

Step 8: First-level inverse DWT is applied to **HL3** to reconstruct **LH1** sub-band by taking **LL2, LH2** and **HH2** bands.
Step 9: Second-level inverse DWT is applied on **LH1** to reconstruct the watermarked image by taking **LL1, HL1** and **HH1**. The dimension of watermarked image is equivalent to host image.

4 Architecture Flow of FPGA Implementation

The complete proposed algorithm has been implemented in FPGA, and the functional architecture is shown in Fig. 2. The five modules in this architecture are data storage ROM, FSM, watermark embedder, result BRAM and VGA. The watermark embedder is the main module which instantiates other four modules. The pixel values of the images are converted to hexadecimal values using MATLAB and stored as coefficient (COE) file. The COE files are moved to the corresponding address in the specified ROM for host and watermark images. The result will be stored in the specific address of block random access memory (BRAM). The stored pixel values are displayed in the monitor via VGA using vertical synchronization and horizontal synchronization control signals. The complete operation is established using internal

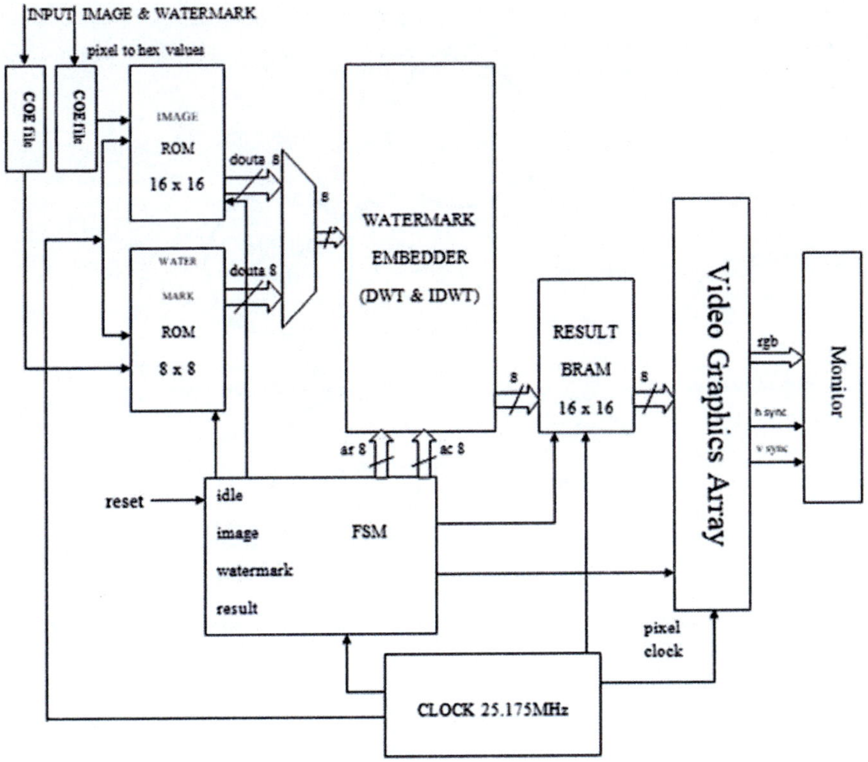

Fig. 2 FPGA architectural flow of watermarking

clock 25.175 MHz which is suitable for VGA. The entire architecture is explained in the following sessions.

4.1 Finite State Machine

FSM is developed to control the entire flow with four states such as idle, image, watermark and the result is given in Fig. 3. The detail flow diagram for each state is enlightened using timing diagram shown in Fig. 4. When reset signal is enabled, the state machine will be in idle state and memory units are cleared. The processing is enabled when reset signal is '0' with operating clock frequency 25.175 MHz. The signal coutcnt is a counter to count clock cycles for each state used in FSM. The entire processing units are completed with 588 clock cycles. Once it attains 588 clock cycles, then the FSM will be in idle state. During idle state, the FSM reads pixels from BRAM and displays output to VGA. Wait state requires 12 clock cycles to fetch the pixels to result state. Coutrun is the signal, which calculates the running

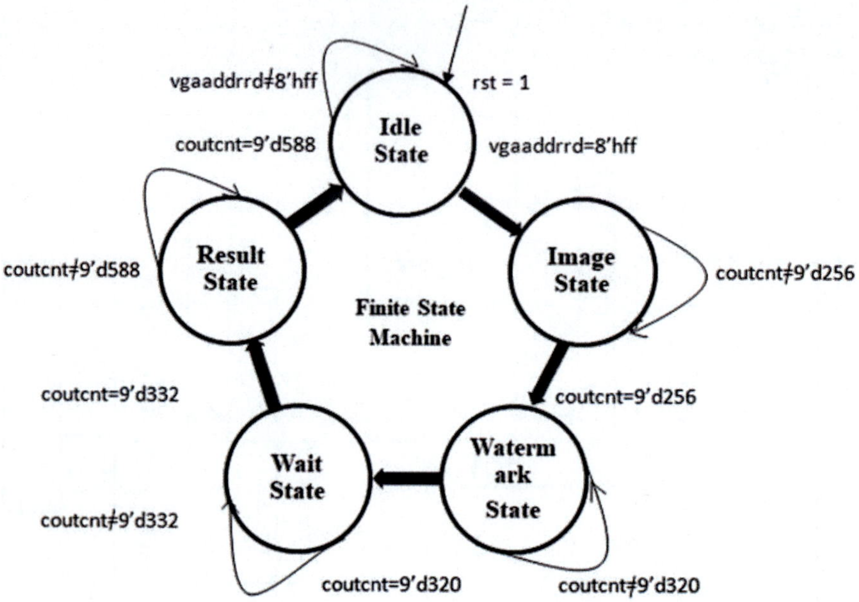

Fig. 3 State diagram of finite state machine

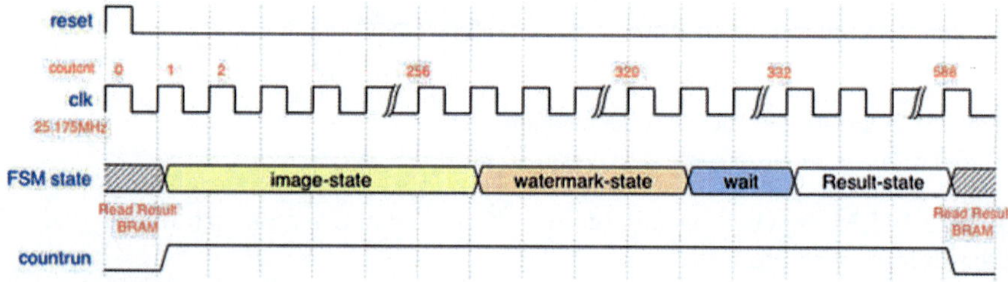

Fig. 4 Timing diagram of FSM

clock cycles required for the complete process. The processing time is calculated from Eq. (11).

$$\text{Processing Time} = 588 \times \frac{1}{25.175 \times 10^6} = 0.023\,\text{ns} \tag{11}$$

4.1.1 Image State

Image state is the next immediate process of idle state. This state reads hexadecimal values of host image from COE files and also stores in ROM. The signals xcounts

Table 1 (a) Arrangement of pixels in host image and (b) address generation for pixels

(a)

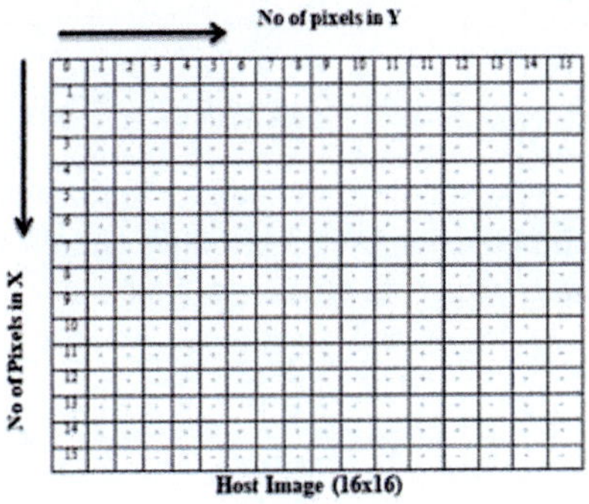

(b)

Clock Count (Coutent)	Row	xcounts	ycounts
1		0	0
2		0	1
3	0	0	2
.		0	.
.		0	.
16		0	15
17		1	0
18		1	1
19	1	1	2
.		1	.
.		1	.
32		1	15
33		2	0
34		2	1
35	2	2	2
.		2	.
.		2	.
48		2	15
.	.	.	.
.	.	.	.
255	15	15	14
256	15	15	15

and ycounts latches are used to generate the row and column addresses, respectively, for the host image. The pixels with the address generation are given in Table 1a and b. The host image size is 16 * 16; i.e., the number of pixels in row is 16; the number of pixels in column is 16. The total number of pixels is 256, and it requires 256 clock cycles to store input host image in ROM.

4.1.2 Watermark State

The watermark state is used to read hexadecimal values of watermark image from COE files and also stores in ROM like image state. The signals xcounts and ycounts latches are used to generate the row and column addresses, respectively, for the watermark image in this state. The pixels with address generation are given in Table 2. The watermark image size is 8 * 8; i.e., the number of pixels in row is 8; the number

Table 2 (a) Arrangement of pixels in watermark image and (b) address generation for pixels

(a)

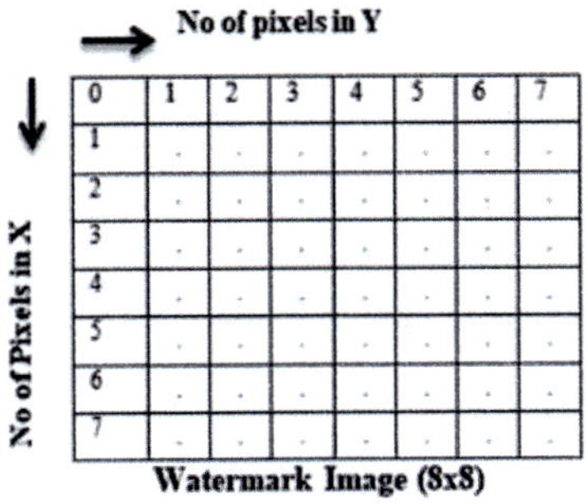

(b)

Clock Count (Coutent)	Row	xcounts	ycounts
1		0	0
2		0	1
3	0	0	2
.		0	.
.		0	.
8		0	7
9		1	0
10		1	1
11	1	1	2
.		1	.
.		1	.
16		1	7
17		2	0
18		2	1
19	2	2	2
.		2	.
.		2	.
24		2	7
.	.	.	.
.	.	.	.
65	7	7	6
64	7	7	7

of pixels in column is 8. The total number of pixels is 64, and it requires 64 clock cycles to store watermark image in ROM.

5 Implementation of Haar Wavelet

In this section, the two images are decomposed into various bands and these require different memory locations to store the coefficients of Haar wavelet.

5.1 Memory Allocation for Host and Watermark Images

The process flow of initialization of clock, reset, memory unit of input images for watermarking algorithm is shown in Fig. 5. If reset is enabled, then cout is '0'; when reset is released and coutrun is '1', the coutcnt starts running. If cout is less than 256 clock cycles, then input host image is stored from ROM to RAM1 in watermark embedder module. At this clock, the FSM state will be in image state. If cout is between 256 and 320 clock cycles, then input watermark image is stored from ROM to RAM8 in watermark embedder module. Now the FSM is in the watermark state.

5.2 First-Level DWT Decomposition of Host Image

If the signal cout is equal to 141 clock cycles, then host image is decomposed by horizontal filtering and the resultant bands H and L are stored in RAM2 by looping from 15 to 0. From Fig. 6, if cout is equal to 142 clock cycles then it is decomposed by vertical filtering. The resultant sub-bands HH1, HL1, LH1 and LL1 of the host image are stored in RAM3 by looping from 7 to 0. This process completes the first-level DWT of host image using Haar wavelet (Fig. 7).

5.3 First-Level DWT Decomposition of Watermark Image

Figure 8 shows the process flow of first-level DWT for watermark image. If cout is equal to 145 clock cycles, then horizontal filtering is applied over watermark image. This results into two bands given by H and L, which are stored in RAM9. If cout is equal to 146 clock cycles, then vertical filtering is applied and results into LL, LH, HL and HH sub-bands and stored in RAM10, and it is shown in Fig. 9.

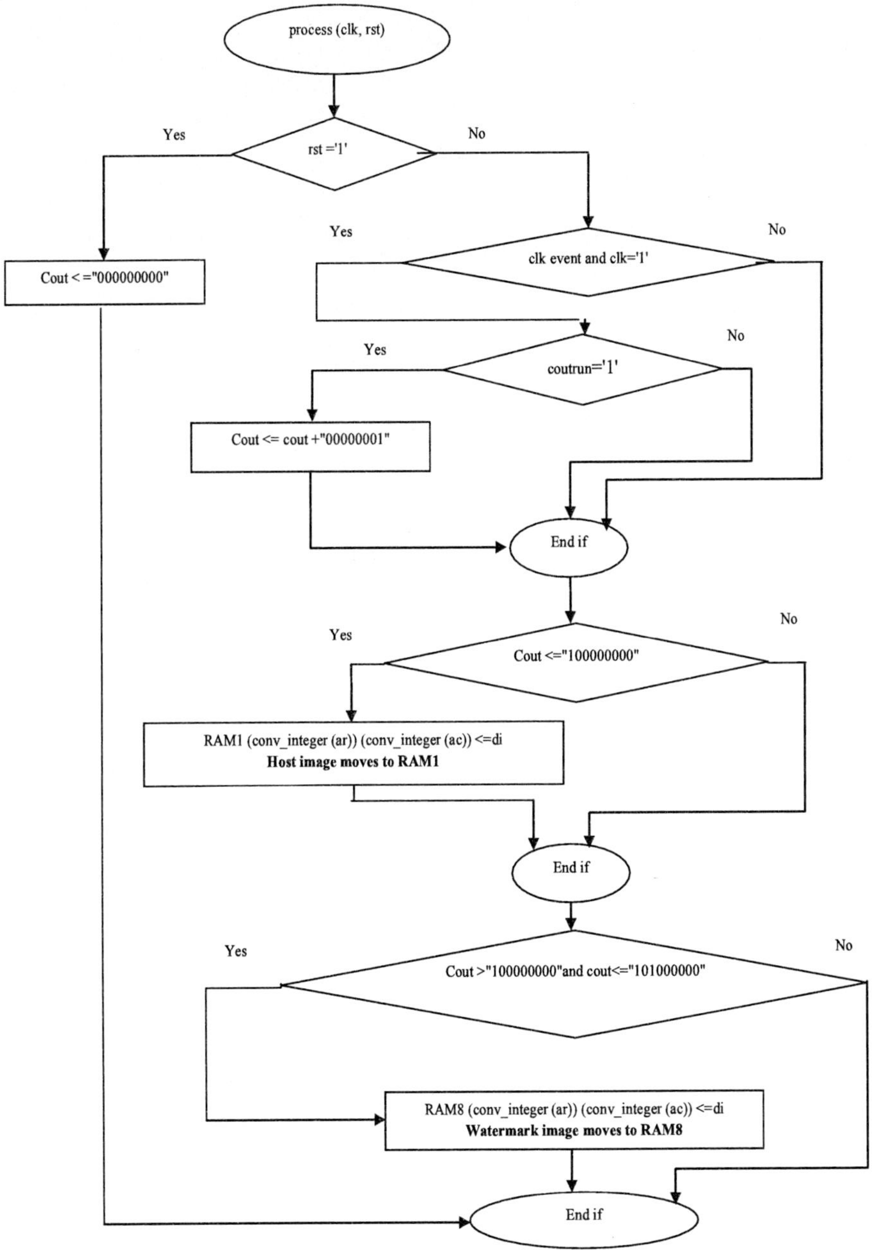

Fig. 5 Flowchart for memory allocation—host and watermark images

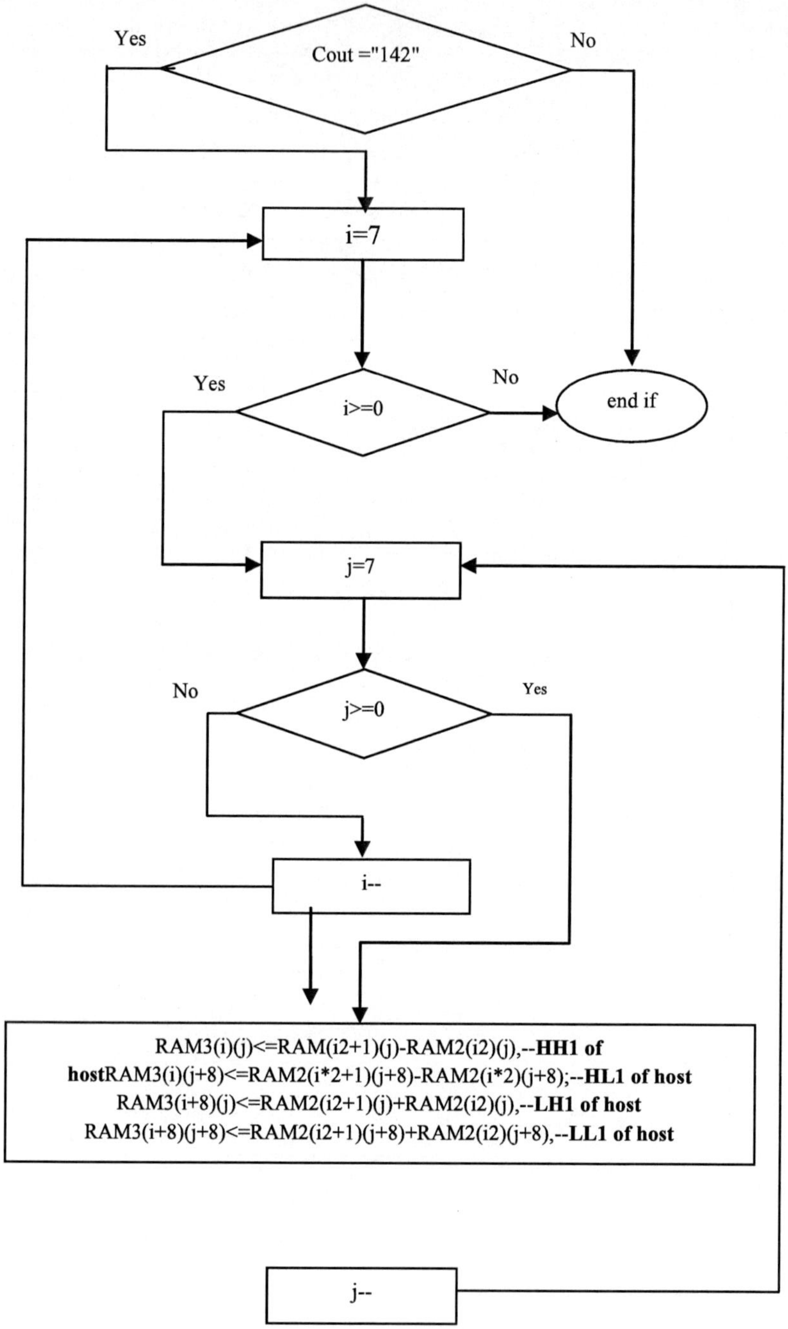

Fig. 6 Flowchart for calculation of LL, LH, HL and HH components of host image

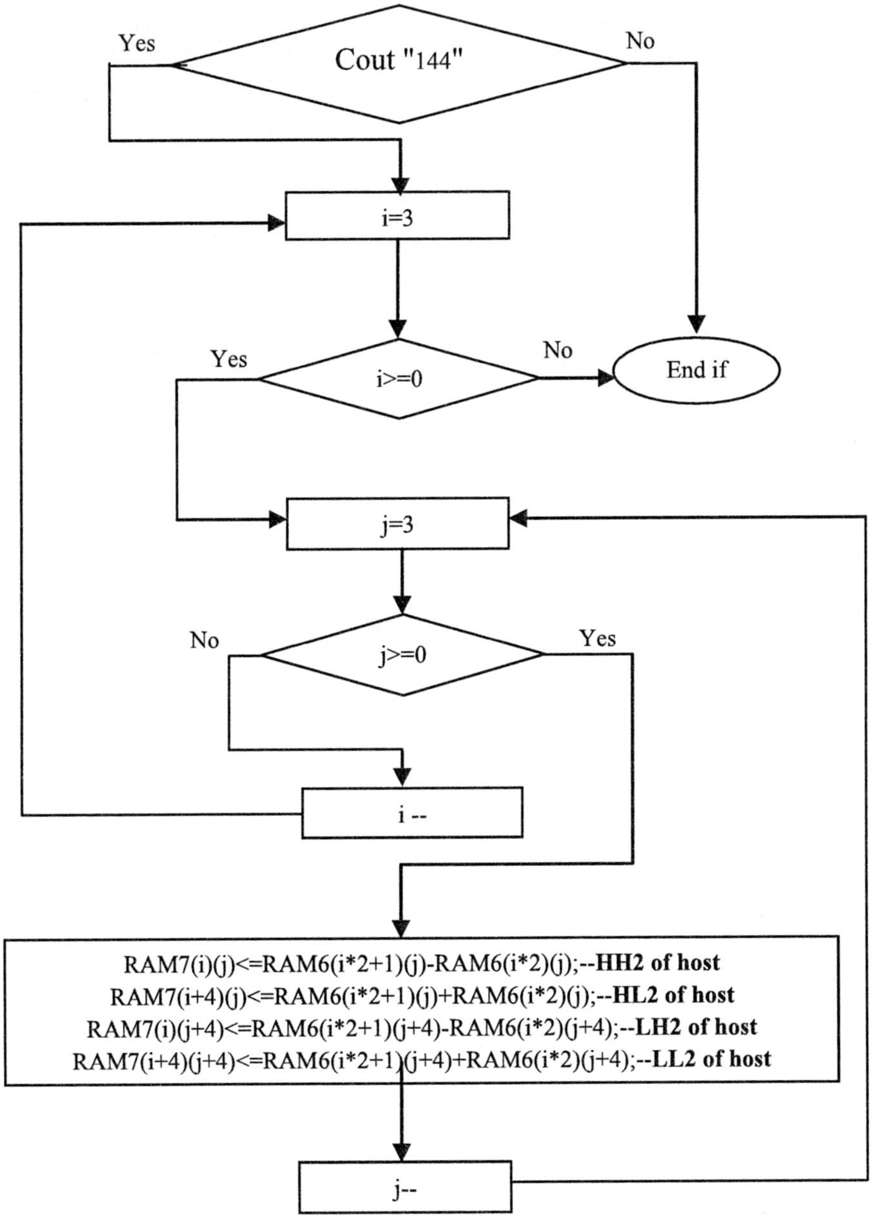

Fig. 7 Flowchart for calculation of LL, LH, HL and HH components of LH band of host image

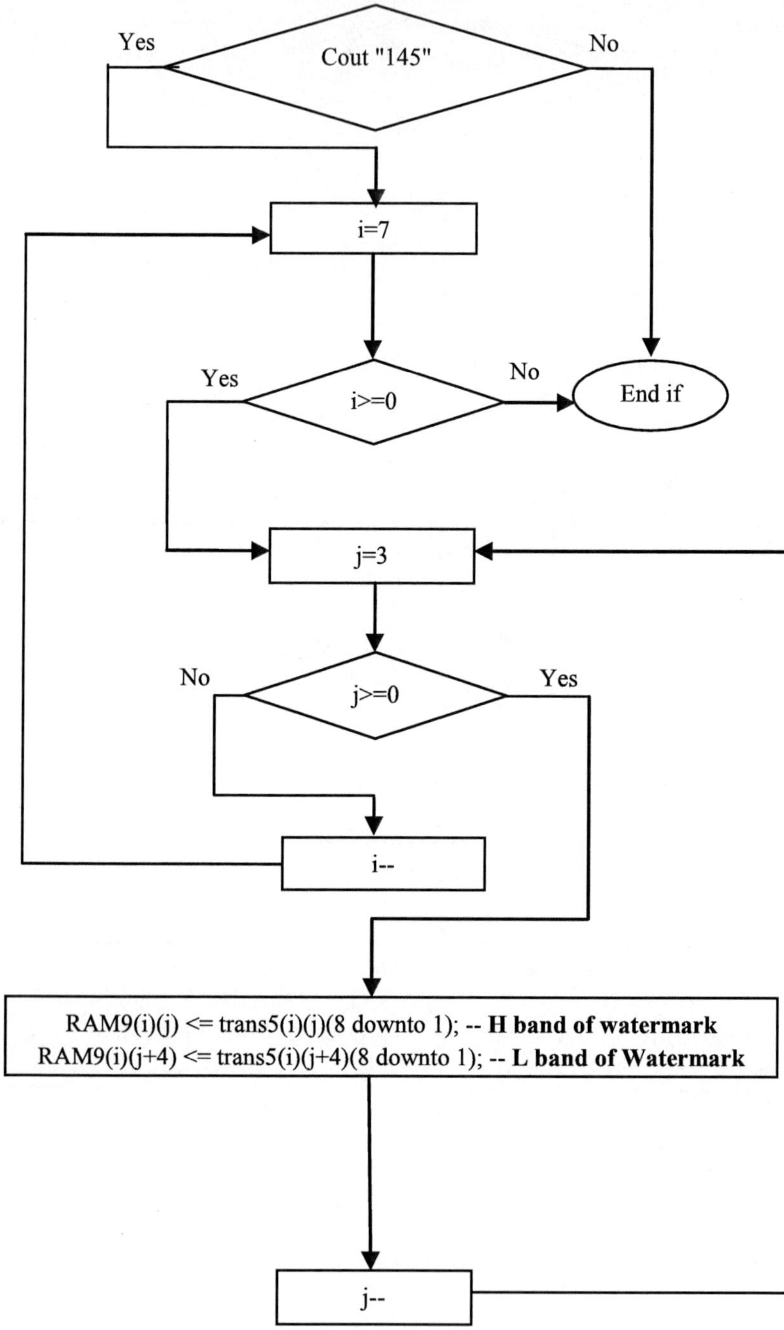

Fig. 8 Flowchart for calculation of L and H bands of watermark

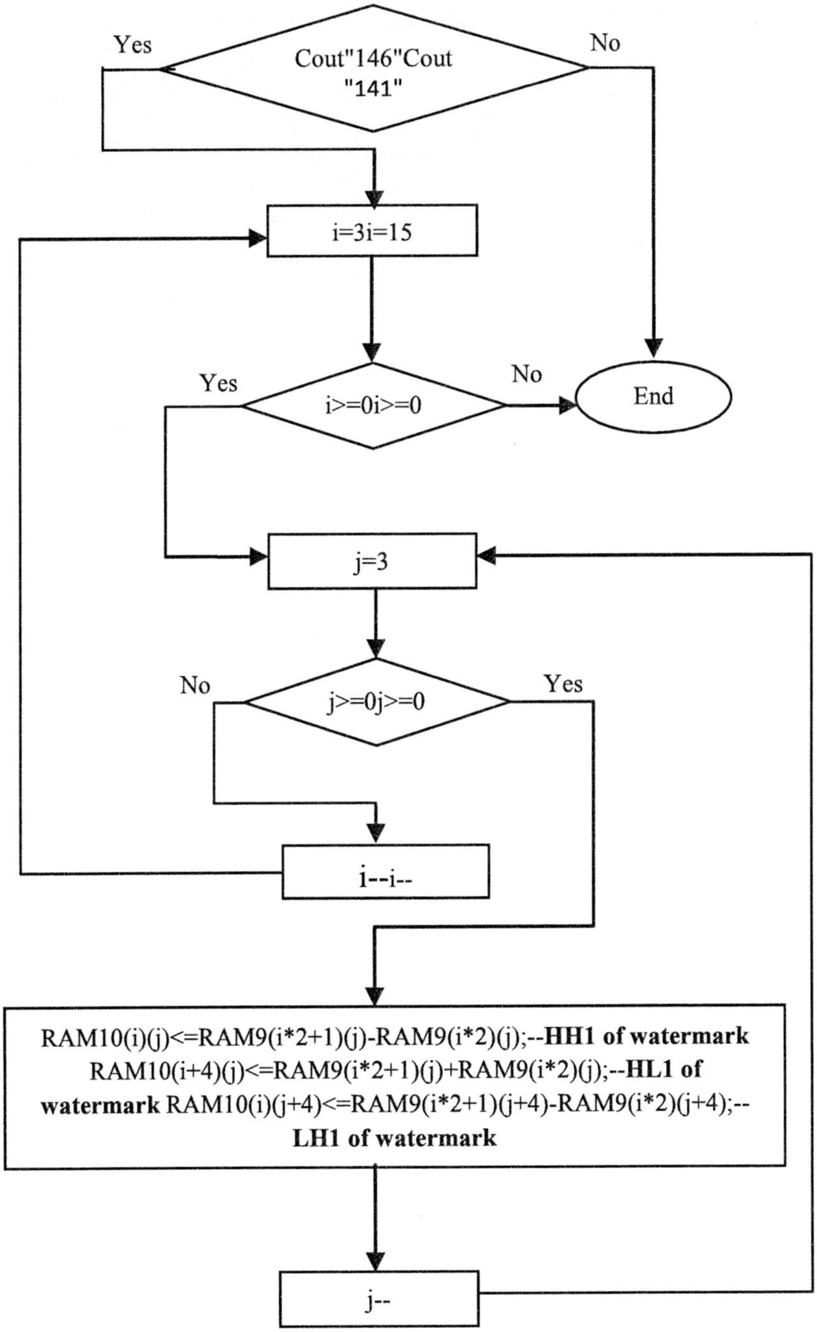

Fig. 9 Flowchart for calculation of LL, LH, HL and HH bands of watermark

5.4 Watermark Embedding Process

If cout is equal to 147 clock cycles, then alpha value (0.5) is multiplied with lh band of watermark. This is the modified lh band and stored in RAM12. Here, the intensity of watermark is reduced to 50% to minimize visibility of watermark. Figure 10a shows the details of embedding process. If cout is equal to 148 clock cycles, then the modified lh band is added with HL2 of host image; this results into modified HL2 named as HL3. This is the processing of watermark embedding process given in Eq. (12). The component HL3 is stored in RAM7, and it is shown in Fig. 10b.

$$\text{HL3} = \text{Host(HL2)} + \alpha \times W(\text{lh}) \tag{12}$$

where HL3 is the modified HL2 band.

$\alpha = 0.5$ is the intensity value to make invisible watermark.

W (lh) is diagonal matrix of watermark image.

Host (HL2) is the diagonal matrix of host image.

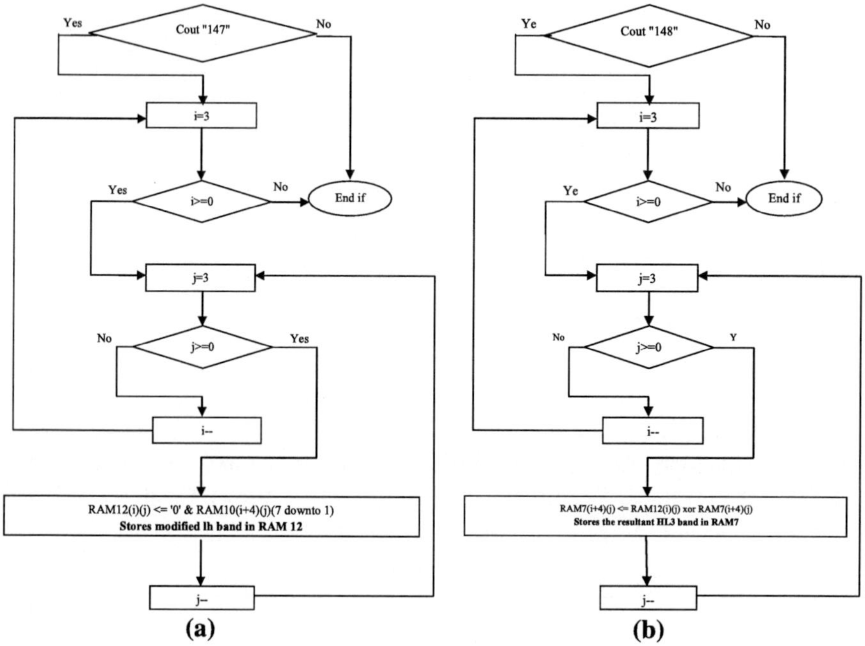

Fig. 10 **a** Flowchart for multiplying alpha value with LH band of watermark. **b** Addition of modified lh (watermark) with host HL band

5.5 *Inverse DWT for Host Image*

Inverse wavelet transform is the exact reverse process of wavelet transform. If cout is equal to 149 clock cycles, then first-level inverse DWT of host is performed by combining HL3 along with the sub-bands LL2, LH2 and HH2. This is the reconstructed LH2 band of host image. These sub-bands are stored in RAM3, and it is shown in Fig. 11a. Similarly, when cout is 151 cycles, then second-level inverse discrete wavelet transform of host is calculated by combining LH2 with sub-bands LL1, HL1 and HH1. This provides the watermarked image; then, it is moved to RAM1 as shown in Fig. 11b.

6 Results and Discussion

This section elaborates the simulation and implementation process. The result of clocking wizard, interfacing VGA, transition of state machine, RTL schematic and bit stream generation is discussed for implementation of entire watermarking algorithm in FPGA boards.

6.1 *Initialization of Clocking Wizard*

The input board clock frequency is 200 MHz. The clocking wizard of Vivado generates 25.175 MHz. Since VGA supports 25.175 MHz, this is used as output clock for the design. The wizard generates 25.175 MHz from 200 MHz in Genesys 2 board. The clock setting is shown in Fig. 12.

6.2 *Simulation of VGA Timing*

The Simulated waveform obtained from Vivado system Edition with VGA timing, hout and vout signals are shown in Fig. 13. When vout signal is high, then redout, greenout and blueout signals carry data to VGA. Similarly for every high state of vout, this timing sequence repeats.

(a)

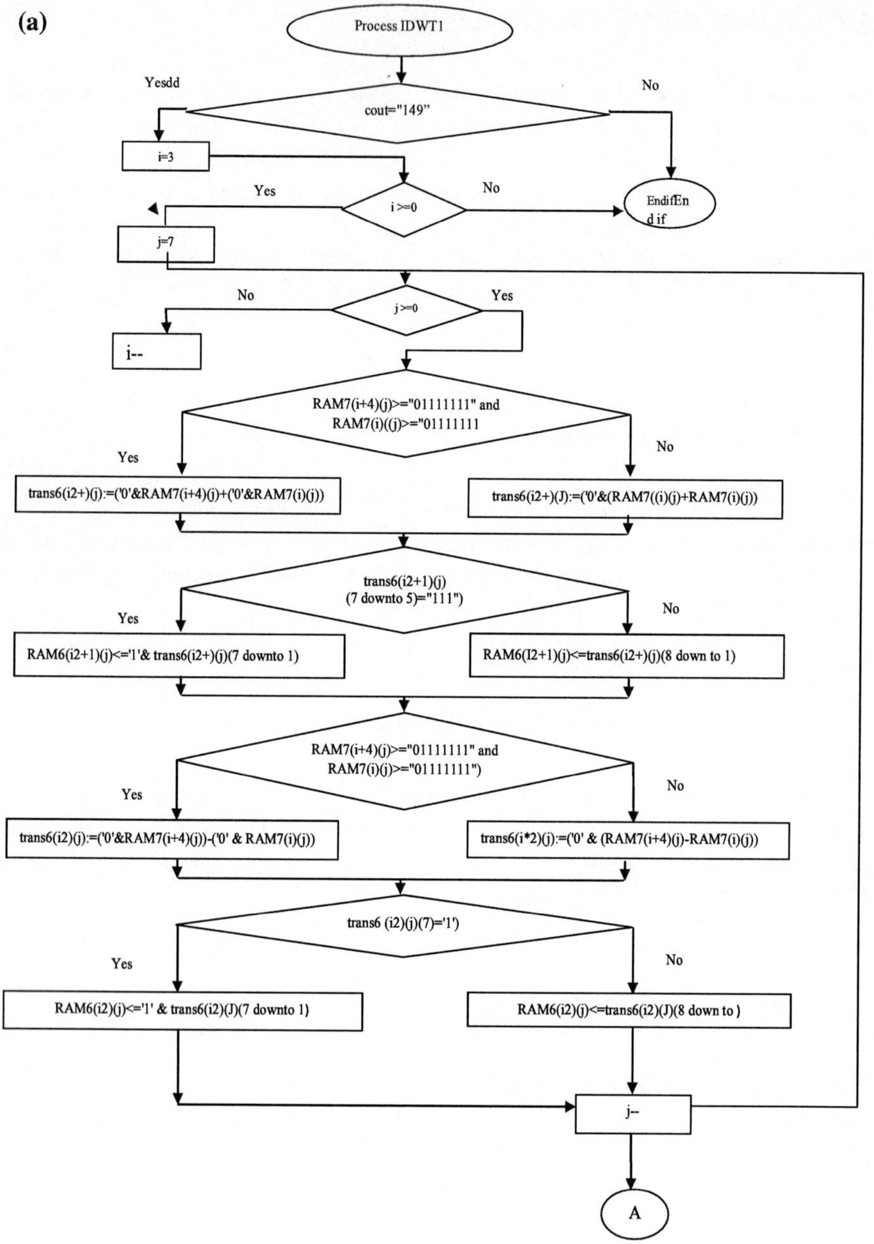

Fig. 11 **a** Flowchart for first level IDWT of host image. **b** Flowchart for second-level IDWT of host image

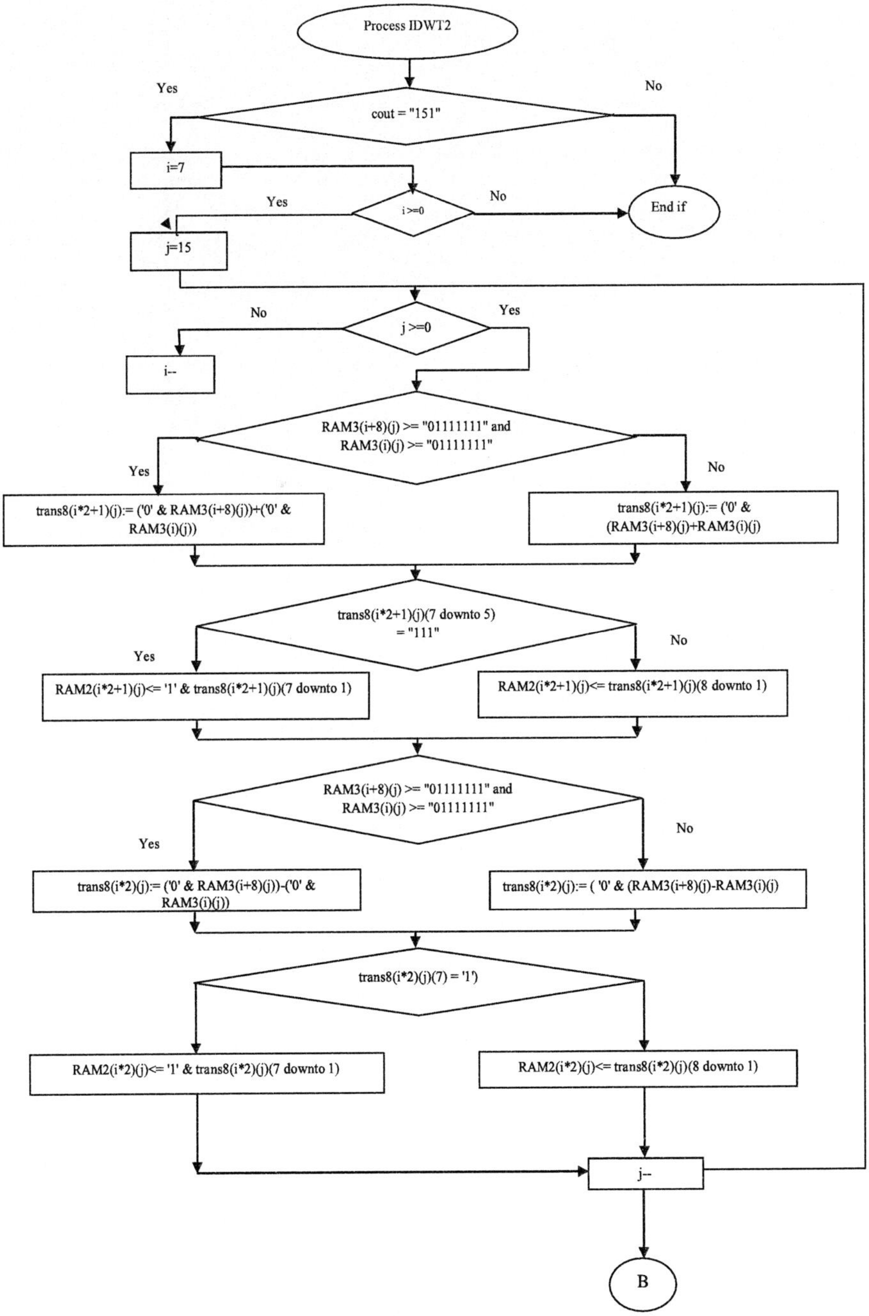

Fig. 11 (continued)

(b)

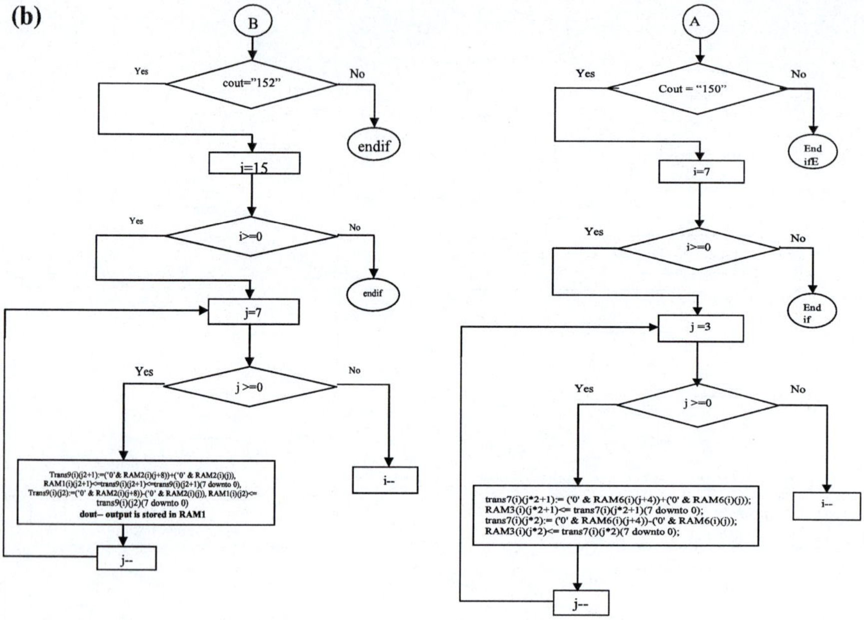

Fig. 11 (continued)

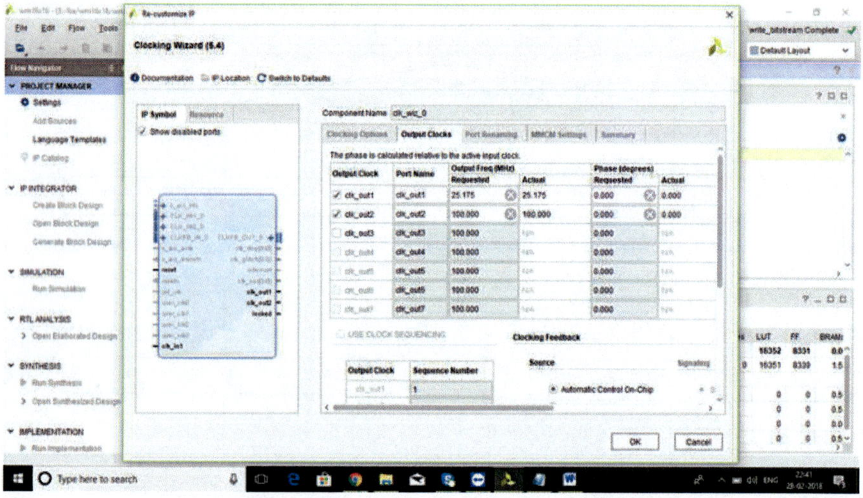

Fig. 12 Initialization of clocking wizard

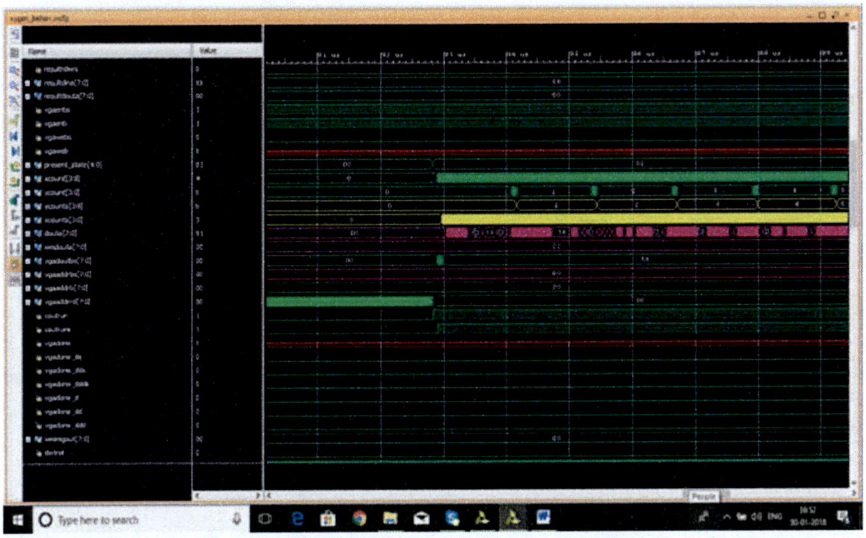

Fig. 13 VGA synchronization signal

6.3 *Simulation of State Machine Timing and State Machine Transition*

The state machine shows the transition of each state by initiating clock cycles. The FSM takes an image from image ROM in image state and gives input to watermark embedder module. Then, state machine feeds watermark data from watermark ROM and stores in watermark module. In wait state, 12 clock cycles are taken for watermarking. In result state, output values are shown from dout signal. This is explained in state machine flow timing diagram as shown in Fig. 14.

Then, next state signal (pink color waveform) shows the transition of states—(0) idle, (1) image state, (2) water mark state, (3) wait state and (4) result state. RTL schematic of result and VGA module is shown in Figs. 15 and 16, respectively. The bit stream generation with hardware implementation of entire algorithm is shown in Fig. 17.

The experimental setup of real-time hardware implementation of watermarking is shown in Fig. 18.

The availability of device utilization for various components like LUTs, memory register, flip flops and multiplexers is compared for the two processors and given in Table 3. For large memory usage, Genesys 2 board is comfortable since the board has 2.03 lakh available LUTs. This consumes 8.09% of available LUTs. The SoC-based ZedBoard has low power consumption and low cost of implementation. Also, the dynamic and static powers have been compared and listed in Table 4. From this analysis, the ZedBoard consumes low on-chip power when compared with Genesys 2 board.

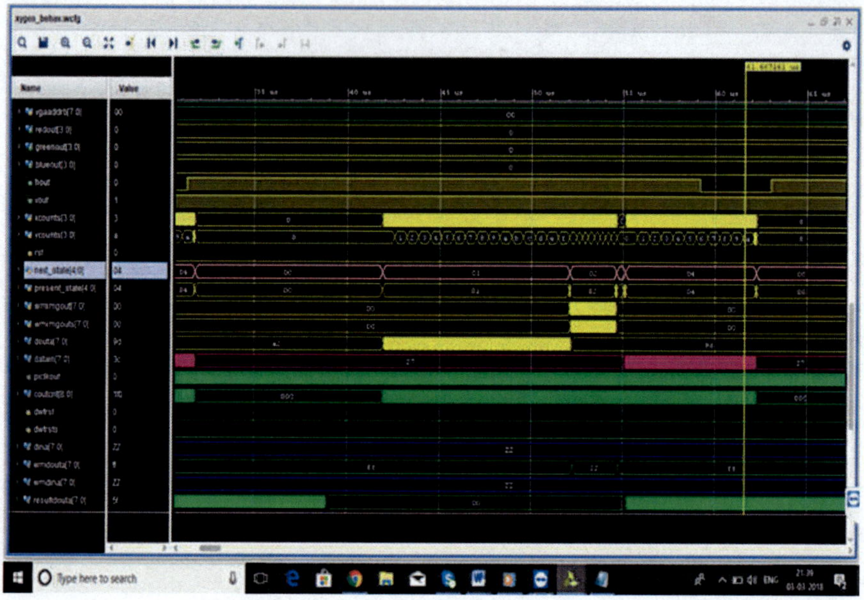

Fig. 14 State machine timing and transition

7 Conclusion

Design and implementation of the hardware architecture of a novel digital image watermarking system to authenticate image in real time were presented in this paper. DWT and IDWT for conventional method have been written in VHDL and implemented in Kintex 7 Digilent FPGA board and low-power ARTIX FPGA-based Zynq 7000 SoC-based ZedBoard using Vivado 2017.4, and the results have been viewed via VGA. Device utilization and power analysis have been compared. It is observed that on-chip power consumption in both the processors is less (0.25–0.3 W). Robustness of the proposed method has been verified for non-blind and blind watermark extraction process using MATLAB [11]. The proposed scheme is highly robust to salt and pepper noise, and the PSNR of about 47.3 dB is achieved. The algorithm successfully extracts the watermark with good PSNR of 46 dB and correlation closer to 0.99 against various attacks. This work will be extended to future work for a real-time video signal, where continuous video frames will be taken by watermark image kept in another block RAM.

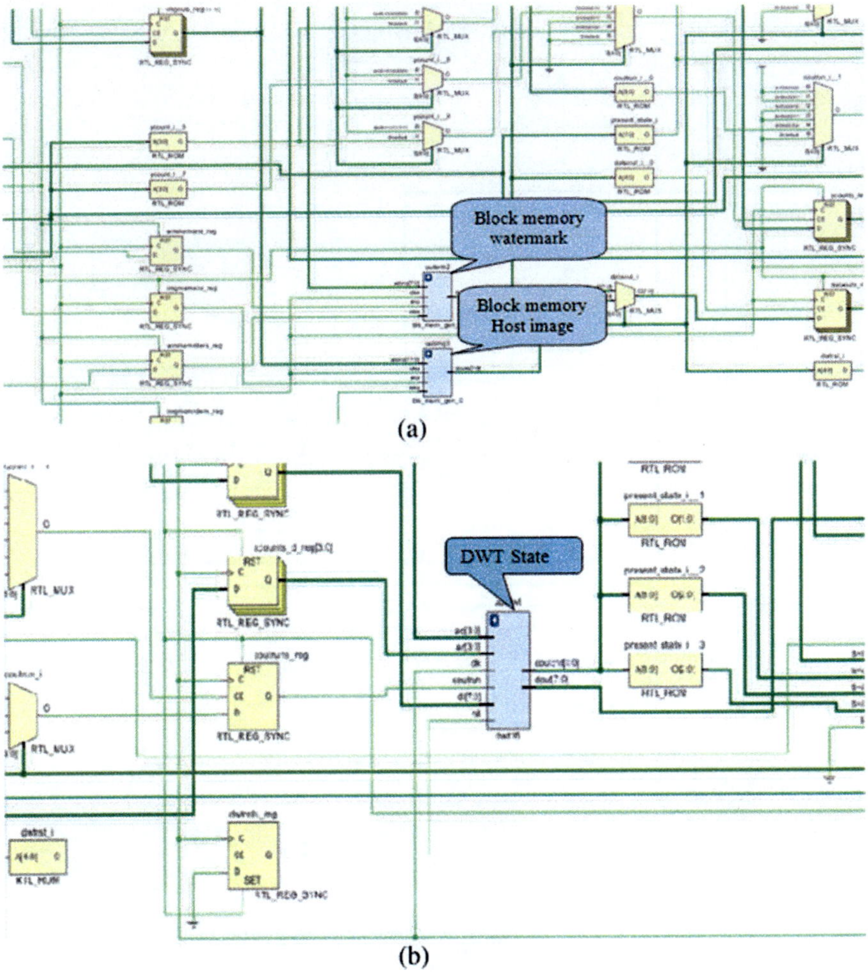

Fig. 15 **a** RTL schematic of memory module and **b** RTL schematic of DWT module

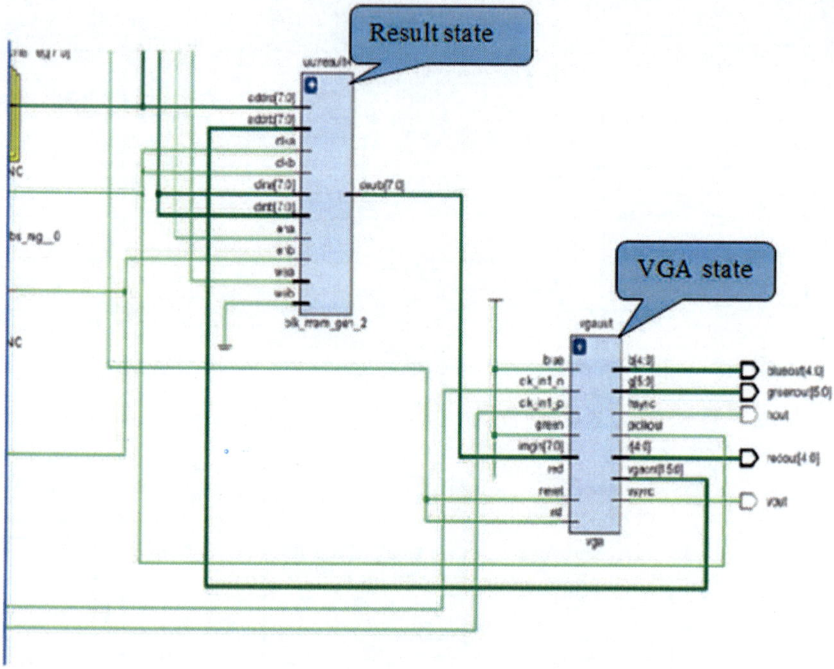

Fig. 16 RTL schematic shows result and VGA module

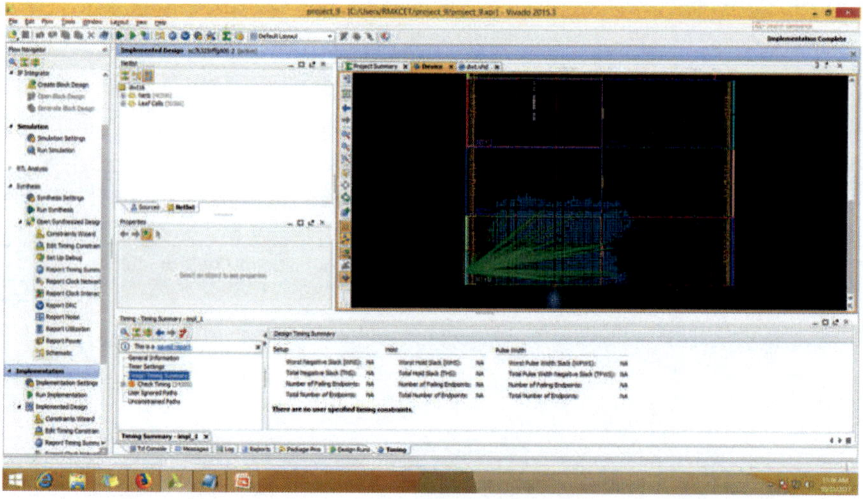

Fig. 17 Hardware implementation of entire DWT algorithm in Kintex 7 Genesys 2

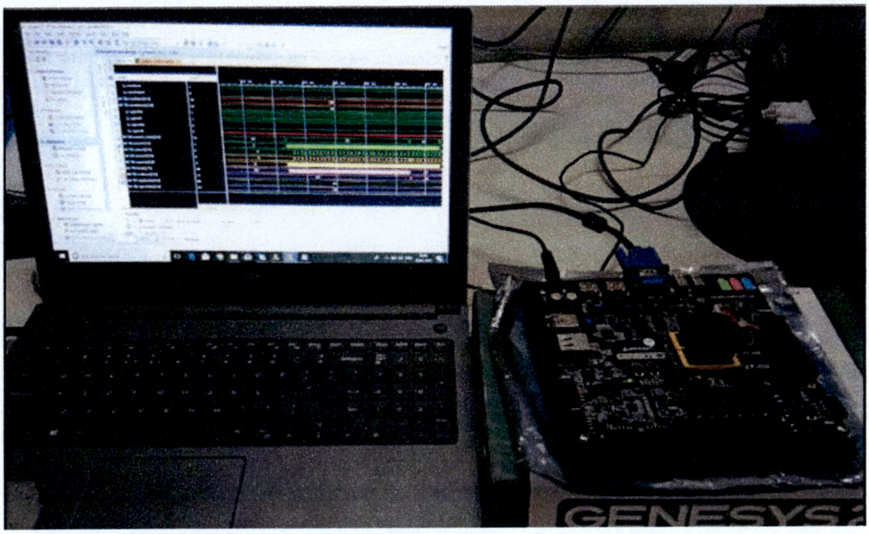

Fig. 18 Experimental setup for real-time implementation

Table 3 Comparison of device utilization of ARTIX and Genesys 2 board

S. No.	Site type	ARTIX FPGA-based Zynq 7000 SoC-based ZedBoard			Kintex 7 Genesys 2 FPGA board		
		Avail.	Used	Util%	Avail.	Used	Util%
1	Slice LUTs	53,200	16,362	31	203,800	16,494	8.09
2	Slice registers	106,400	8359	8	407,600	8378	2.06
3	Register as flip flop	106,400	8333	8	407,600	8360	2.05
4	Register as latch	106,400	26	1	407,600	18	<0.01
5	F7 muxes	26,600	248	1	101,900	323	0.32
6	F8 muxes	13,300	124	1	50,950	136	0.27
7	BRAM	140	2	1.5	445	40.5	9.10
8	IO	200	16	8	500	21	4.20
9	MMCM	4	1	25	10	1	10

Table 4 Comparison of power analysis of ARTIX and Genesys 2 board

S. No.	Type	Kintex 7 Genesys 2 FPGA board	ARTIX FPGA-based Zynq 7000 SoC-based ZedBoard
1	Total on-chip power (W)	0.293	0.25
2	Dynamic power (W)	0.130	0.127
3	Device static power (W)	0.163	0.123
4	Effective TJA (C/W)	1.8	11.5
5	Max ambient temperature (C)	84.5	84.5
6	Junction temperature (C)	25.5	25.6

Acknowledgements A special thanks to Mr. Riteshkumar Sharma and Mr. Rahul Sharma, Scientists of Space Application Center, Indian Space Research Organisation (ISRO), Ahmedabad, India, for giving the opportunity and constant support for the funded research project.

References

1. Joshi AM, Darji A, Mishra V (2011) Design and implementation of real-time image watermarking. In: 2011 IEEE international conference on signal processing communications and computing (ICSPCC), pp 1–5
2. Karthigai Kumar P, Baskaran K (2011) An ASIC implementation of a low power robust invisible watermarking processor. J Syst Archit 57(4):404–411
3. Ghosh S, Das N, Das S, Maity SP, Rahaman H (2014) FPGA and SoC based VLSI architecture of reversible watermarking using rhombus interpolation by difference expansion. In: Proceedings of the annual IEEE India conference, pp 1–6
4. Ahmed J, Aziz A, Akhtar P (2014) FPGA based efficient architecture for image watermarking using wavelet coefficients quantization. In: 2014 international conference on open source systems & technologies, pp 105–112
5. Liu S, Liu Z, Huang H (2012) FPGA implementation of a fast pipeline architecture for JND computation. In: 2012 5th international congress on image and signal processing, pp 577–581
6. Pantoja Laces WA, Garcia-Hernandez JJ (2015) FPGA implementation of a low complexity steganographic system for digital images. In: 2015 IEEE/ACIS 14th international conference on computer and information science (ICIS), pp 319–324
7. Elamaran V, Praveen A, Reddy MS, Aditya LV, Suman K (2012) FPGA implementation of spatial image filters using Xilinx system generator. Procedia Eng 38:2244–2249
8. Jadhav AA, Babar RV, Gaikwad MS (2015) Hardware implementation of digital watermarking system for real time captured image transmitting. In: 2015 international conference on pervasive computing (ICPC), pp 1–4
9. Singh P, Chadha RS (2013) A survey of digital watermarking techniques, applications and attacks. Int J Eng Innov Technol (IJEIT) 2(9)
10. Chanu WM, Chakraborty B, Marur DR (2014) Implementation of blind image watermark encoder in FPGA. Int J Inf Comput Technol 4(5):545–552. ISSN 0974-2239
11. Roy SD, Li X, Shoshan Y, Fish A, Yadid-Pecht O (2013) Hardware implementation of a digital watermarking system for video authentication. IEEE Trans Circuits Syst Video Technol 23(2)

Smart Communication in Coal Mines

Jaya Anand, Sneha Chowdhury, Deepak Prasad, and Vijay Nath

Abstract This paper demonstrates personnel safety and maximizing the mining process communication system. Environment monitoring in underground coal mines has been a mandatory requirement to ensure safe working conditions in the mines. Hence, active communication and information network need to be developed, which can accurately provide locational references and evacuate workers from the danger zone. Many of the technologies are common on the surface of the Earth but have limitation issues underground. This paper outlines the major implications for wireless communication systems in underground coal mines and also addresses the recent technological developments in the area of communication and environment monitoring in underground coal mines.

1 Introduction

Mining is one of the oldest endeavors of mankind along with agriculture. Underground coal mining is a dangerous job, as the miners work thousands of feet below the ground underneath tones of rocks. They are surrounded by very high voltage electric lines, darkness, and dust. Mine disasters lead to trapping of survivors underground. Accurate knowledge of underground communications, environmental conditions and actual location, and conditions of victims must be known [1]. Some method of knowing the location of trapped miners and means of communicating with them is essential. Although the technology involved in removing materials from below the

J. Anand · S. Chowdhury · D. Prasad (✉) · V. Nath
VLSI Design Group, Department of Electronics and Communication Engineering, Birla Institute of Technology, Mesra, Ranchi, Jharkhand 835215, India
e-mail: prasaddeepak007@gmail.com

J. Anand
e-mail: jaya.anand2k10@gmail.com

V. Nath
e-mail: vijaynath@bitmesra.ac.in

V. Nath and J. K. Mandal (eds.), *Proceedings of the Fourth International Conference on Microelectronics, Computing and Communication Systems*, Lecture Notes in Electrical Engineering 673, https://doi.org/10.1007/978-981-15-5546-6_86

Earth's surface has a long history, communication systems in underground workings are relatively new to the industry. Communication equipment did not begin appearing in underground mines till early 1990s. During the early days, phones were used in underground mines. The only difference between these phones and the ones that were used on the surface of the earth was that these phones were enclosed in cast iron boxes as protection against moisture, acid fumes, and gases. In the recent history, the most common method of underground mine communication consisted of loud-speaking or paging-type telephones. Early shaft communications consisted of bells or whistles signaling systems. Today, these systems have been replaced by radio and carrier current systems that allow two-way voice communication to and from personnel in the mines. The most important reason for developing the systems is the safety of coal miners. They work under such harsh conditions so certain device has been invented to ensure their safety. The equipment is so developed in order that it can work in the environmental conditions of high pressure and high temperature in the underground mines. These devices may not work on the surface as the specifications of the devices may vary.

2 Historical View of Communication on Coal Mines

The communication system has evolved a lot in the past few years. The Bureau of Mines along with the Clinch field Coal Co-operation in the year 1976 signed a consensus for the growth of the systems used for the purpose of communication which can be used in the field of mines [2]. For the purpose of expulsion of the coal, which is produced to an out by portage conveyor, a segmented mobile conveyor was incorporated in a mine of that kind. At the Clinch field's Splash dam, three ambulatory aqueduct conveyors are tethered as one segment by two ambulatory aqueduct bearers. This segment is responsible for the unburdening of the coal onto the extendible prolonged John Belt. Each aqueduct bearer requires one machine minder which is similar to the one used by the persistent miner [3]. The region of operation of the persistent miner is performed several feet away from the machine and this operation is conducted by the help of remote dominant box which consists of wired system. The depth of the coal deposition varies from 30 to 48 in. The equipments are so developed in order that it can work in the environmental conditions of high pressure and high temperature in the underground mines. These devices may not work on the surface as the specifications of the devices may vary. There have been several instances which have been surveyed where the rescue teams faced a number of problems as a result of mine disasters [4]. One of the tedious and dangerous jobs is to search a mine after it has been exploded or caught in a fire. One of the famous examples of the rescue team killed was the incident from the Scotia mine where the entire team of 13 members was killed because of the mine disaster. In spite of the danger of losing their lives, the rescue team must decide where to begin its search operation from, because generally the area to be searched is quite large. The miners who manages to escape the explosion can inform the rescue team to those parts where the other trapped miners are present. The entire techniques of locating the survivors are based on the heat and trial method. Thus, there is need for the accurate knowledge about the position of the trapped miners so that the rescue team can rescue them with minimal hazards. Without any sort of accurate knowledge, it would be very difficult for the team to search in such dangerous areas. An electromagnetic system has been developed, the basic structure of which has been completed and the equivalent hardware is in the phase of testing [5]. The system consists of a transmitter system which is carried on by the miner and a receiver system which is present on the surface and receives the signals from the earth or from the mines. Two long wall sections are present in the working sections which consist of room and pillar in the mine which is run by Robinson near Shinnston. A world record was made in January 1973 when 23 workers who were working on the long wall sections produced 12,387 tons of coal in only 24 h time. There are a number of long wall systems in Europe but on the other hand, there are only around 75 longwall systems in USA and about less than 5% of the country's coal production is produced through the longwall systems. Majority of the equipment are shipped from other counties and about $1–2 million per machinery are invested by these countries. Overall, the importance of the longwall systems is very small in USA, but they represent the country's expertise role in underground coal mining [6]. Despite of the not so impressing stats, the Federal Bureau of Mines and the US mining industry are undertaking various steps for the promotion of the longwall mining in the country. There are mainly three reasons behind this impressive step undertaken by these two bodies of national importance. The first reason being the very high potential of this longwall mining to produce large amount of coals and because of this, there is a belief that this technique will see a very rapid growth in the upcoming years. As discussed in the above sections that the machinery for the longwall mining requires the equipment to be shipped from other countries as a result of which there has been a decrease in the use of this technique as it was generating more economical burden on the state, but now, the US companies are starting to produce the equipment in their own country [3]. Thus, this is one of the major reasons why everyone is encouraging longwall mining. And this is a very important reason as it can reduce human interference and can reduce the hazards.

Communication is one of the indispensable parts of the mining process. Without any sort of communication process, it is very hard to imagine any mining process. The main focus of this research article is to discuss the communication systems which have been incorporated in the mining process.

3 Need for Voice Communication

There is immense need for the voice communication in mining areas. The main aim for the communication in these regions is to transmit information between the trapped miners and the rescuers and this information could be well enough to save somebody's life [7]. The information that can be transmitted is any unusual situations being faced by the trapped minors or any sort of advice that could be provided by the rescuers to the trapped miners until their arrival, would be very useful. Thus, the communication plays an important role in this field as it can be used to track the location of the trapped miners so that they can be located and rescued. The most important reason for developing the systems is the safety of coal miners [8]. They work under such harsh conditions so certain device has been invented to ensure their safety. The equipment is so developed in order that it can work in the environmental conditions of high pressure and high temperature in the underground mines. These devices may not work on the surface as the specifications of the devices may vary. The equipment is so developed in order that it can work in the environmental conditions of high pressure and high temperature in the underground mines. These devices may not work on the surface as the specifications of the devices may vary.

4 Modern Means of Communication in Coal Mines

The process of coal extraction from the coal deposition depends on the depth at which the coals are present, also upon the quality of the deposition and various other factors like geographical factors along with the environmental factors. The various process of extraction of coal has been distinguished from one another by whether they are performed on the surface or underground [9]. The coals which are extracted from the either surface or underground mines need washing in a plant where the preparation of coals takes place. The two methods which are the most common ones for the extraction of coals are the surface mining and deep underground mining methods for surface mines and underground mines, respectively. The decision for the mining methods is done on the basis of the depth at which the coals are present, their density, and their extent.

4.1 Extraction's Method

4.1.1 Modern Surface Mining

An economical and convenient method for the extraction of the coals deposited near the surface is the method of open-cut which is also known as methods of open cast or open strip dynamos and explosives are used in this mining method in order to break the surface and achieve an access to the inside of the surface. Once the coal deposition gets exposed, then the process of drilling is performed which is followed by fracturing and which is further followed by thorough mining of the coals in strips. These extracted coals are then loaded onto large vehicles for the purpose of transportation to either the plant where the preparation of coals takes place or to the place where the direct consumption of coal takes place.

4.1.2 Underground Mining

Majority coals are penetrated deep inside the ground and as a result of which opencast mining seems to be a difficult process and thus their extraction requires the process of underground mining which is at present responsible for 60% of world's coal production. The process of deep mining incorporates the method of room and pillar or pillar or board along with it and the support to the mine roof is provided by the pillars and timbers. Today's section of pillars makes use of equipment which is controlled using remote and these also include huge hydraulic supports for the roof. These hydraulic supports are responsible for preventing cave-ins as long as the area designated for work has been left by the miners along with their machineries [10].

4.2 Communication Techniques

There are various systems which find application in the mining industries. Some of the equipments have been discussed in the following section.

4.2.1 Automatic Telephone Exchange

It is evident from the name itself that it is an automatic dialing-type telephone which is indistinguishable from the one used on the surface. The telephone line can be protracted to the surface as well as to the standpoints which are situated underground. The extension to the underground points can be done by the help of interfaces [11]. These interfaces are similar to the one used in C.D.S. systems.

4.2.2 Telephone System Powered by Sound

Short distances communication between point to point such as loading bounder and loading faces in mines which are relatively small in size can be performed with the help of telephone system which are powered by sound. These telephone systems operate without the use of battery. Electronic impulses are generated by the sound which strikes the microphone. These generated impulses are responsible for facilitating the transmission of signals over very short distances. Figure 1 shows the telephone system powered by sound.

4.2.3 Telephone System Powered by Battery

This type of telephone system shows resemblance to its sound-powered counterpart and they are also used for the point-to-point communications. The difference between

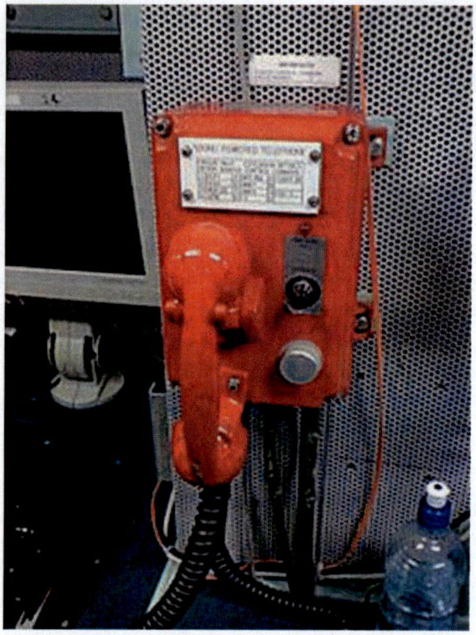

Fig. 1 Sound-powered telephone system

the sound-powered and battery-powered telephone system lies in their range of operation, the sound powered can be used for small distance because signal attenuation takes place for large distance which makes the communication inefficient whereas the battery-powered telephone system can be used for large distance without much attenuation [12]. The reason for the transmission of the signals without much attenuation is the use of battery in these systems which provide enough energy for the transmission of the signals. In order to achieve more versatility, the several point-to-point communication systems can be connected through the means of exchange. Figure 2 shows the battery-powered telephone system.

4.2.4 Radio Paging System

The radio paging system is mainly used at the surface. These systems are used for communications between key personnel situated at different points on the surface itself. There are two versions of the paging system. The first one being a simpler version which helps in the transmission of only coded message which notifies the users to talk from the nearest telephone exchange. The other one being a complex version, which helps in the transmission of messages which consists of data which is present in the form which can easily be understood. It also facilitates the movement to the nearest telephone exchange through the means of message that can be transmitted from the user itself. The component diagram for the radio paging system has been shown in Fig. 3.

Fig. 2 Battery-powered telephone system

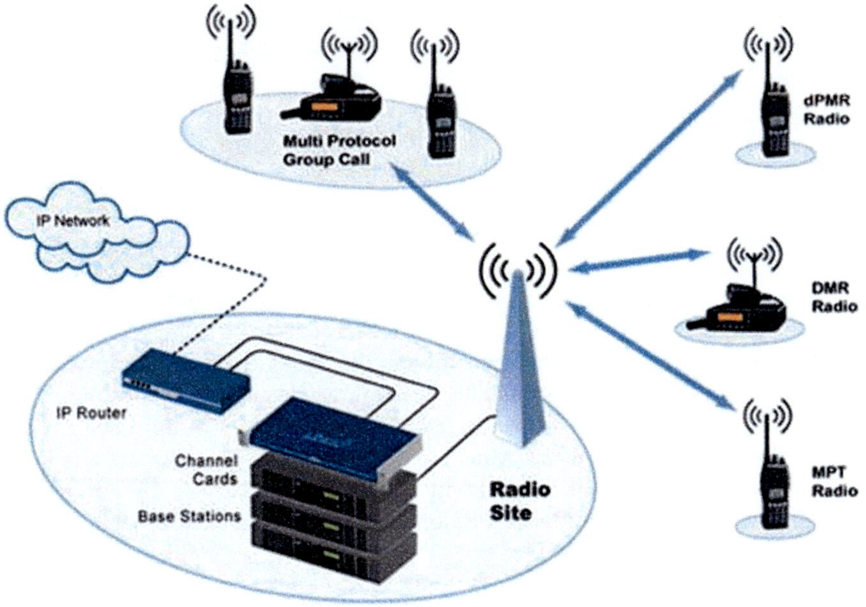

Fig. 3 Radio paging system

4.2.5 VHF Communication System

Very low frequency systems, as the name signifies can be used in very low frequency operations and these can be utilized to locate the position of the trapped miners, as these systems can generate radiations which can pass through rigid bodies.

Communication is one of the vital components of mining industries. It is essential from protection point of view as well as for maximizing the productivity in the field of mining. Thus, in order to connect to the demands of the mines like the infrastructure of the communication systems, an impact suite has been developed in the twenty-first century which is known as the Mine Site Technologies. Most of the precious and important resources can be obtained in the underground regions. Thus, locating these underground resources is very essential and keeping this in mind, a tracking system has been developed. This tracking system makes use of Wi-Fi, for the purpose of tracking the resources and miners present in the underground regions. The sprightly dockets, which are carried by the miners or are associated with some equipment, can be tracked with the help of these systems. In underground mines, low frequency signals are used because they can easily pass through the grime and rocks [13]. These low frequency signals can be transmitted with the help of antennas, repeaters, and reticulation alignment. The communication through these elements is the line of sight communication. Such radio signaling having low frequency signals are known as the Through-The-Earth (TTE) signaling. In order to increase the penetration of the signals through the hard rocks, ultra-low frequency signals can be used which can penetrate to a greater depth than the low frequency signals.

5 Personal Emergency Device

Among various mines disasters, the incident from the mines of Australia at Moura number 4 coal mine, 1986, was a famous one because it lead to the initial stage development of a device which came to known as the Personal Emergency Device (PED). This device finds further development after the Moura number 2 coal mine. PED is a device which finds its application in the one-way communication which supports text message forwarding. This device is widely used in Australia, USA, Argentina, Norway, Bermuda, Sweden, etc. The development of this device began in 1987 by an Australian firm, "Mine site technologies." This device was made available to the personals associated with the mining industry in 1991 by Health Administration (MSHA). All these developments in the field of PED came after the ignition of coal mines in 1994.

6 Impact Tracking System and Its Core Components

6.1 Wi-Fi Access Points as a Readers

In order to meet the draconian necessity of the industries carrying out underground mining, a system has been designed which is known as the impact tracking system. This system has proven its reliability on every ground which is necessary for a system to be used successfully in these industries. This system is very manageable, pitted, and meet most of the safety caliber of the underground mining [5]. In case of any exigency, such systems play a very vital role in this field. Besides providing augmentation to the safety, this system also improves the coherence in the working of any underground drudgery.

6.2 Zone Display Units

The tagging particulars, if necessary, are exhibited at definite sites with the help of colossal LED screens which are situated in the stockade made up of stainless steel. These colossal LED screens are known as the Zone Display Units (ZDU's). Some of the definite sites could be cardinal elocutionist locations, entrances, exits of mine, etc.

6.3 Collision Avoidance Systems

Any mine whether it is an underground mine or surface mine, it requires tracking system for monitoring the location of the trapped miners. Thus, one such tracking system is Collision Avoidance System and this is also known as the Proximity Detection System or the Collision Avoidance Technology. These systems can also be utilized for the purpose of tagging. Such systems can be utilized to minimize the threat of the people advancing into association with automobiles in an untrammeled approach. The investment on the cognitive chattels and the consummate received with the existing Impact tracking and tagging technology paved the road for the invention of the Collision Avoidance System. The crucial unit of the Collision Avoidance System consists of the Vehicle perspicacious system which also allows the real-time verdict observation of the vehicles. The system discussed above finds reinforcement from the Australian Coal Association Research Program (ACARP). This system is also responsible for receiving feed in from large amount of coal and mines [14, 15] situated on hard rocks in Australia. The communications are necessary from the individual's safety point of view and their efficiency. The safety can be ensured by the messages which are sent by the trapped miners. These messages may be in the form of short text messages, voice over Internet protocol phone, and radio frequency identification tags. These various forms of messages are received by devices which are capable of receiving it. Personal emergency device and personal pager are the example of such devices [16].

7 Conclusion

The conclusion which can be drawn from this research article is that how the communication systems evolved over the period of time. In earlier days, they were not efficient enough, they were not advanced, and there were many limitations as a result of which they were unable to rescue the lives of the miners. But with the advancement in the field of communication systems, several advanced techniques have been invented, several advanced equipment have been generated which are capable enough to rescue a large number of trapped miners and can minimize the amount of hazards.

Acknowledgements We are thankful to Department of Electronics and Communication Engineering BIT Mesra, Ranchi, for providing the laboratory facilities. We are also thankful to our Vice-Chancellor, Dr. M. K. Mishra and our Head of the Department, Dr. S Pal for their constant inspiration and encouragement. Without their support, this research article would not have been possible.

References

1. Kumar A, Kingson TMG, Verma RP, Kumar A, Mandal R, Dutta S, Chaulya SK, Prasad GM (2017) Application of gas monitoring sensors in underground coal mines and hazardous area. Int J Comput Technol Electron Eng 23–34
2. Forooshani AE, Bashir S, Michelson DG (2013) A survey of wireless communications and propagation modeling in underground mines. IEEE Commun Surv Tutor 1–22
3. Lagace RL (1980) Detection of trapped miner electromagnetic signals above coal mines
4. Bandyopadhyay LK, Chaulya SK, Mishra PK (2010) Wireless communication in underground mines. RFID based sensor networking. Springer, New York
5. Misra P, Kanhere S, Ostry D, Jha S (2010) Safety assurance and rescue communication systems in high-stress environments: a mining case study. IEEE Commun Mag 48:66–73
6. Durkin J (1984) Apparent earth conductivity over coal mines as estimated from through-the-earth electromagnetic transmission tests
7. Schiffbauer WH, Brune JF (2006) Coal mine communications. American Longwall Mag
8. Guo Hongzhi, Sun Zhi, Zhou Chi (2017) Practical design and implementation of metamaterial-enhanced magnetic induction communication. Access IEEE 5:17213–17229
9. Gogoi AK, Raghuram R (1996) Analysis of VLF loop antennas on the Earth surface for underground mine communication. In: Antennas and propagation society international symposium 1996. AP-S. digest, vol 2, pp 962–965
10. Wait JR (1992) Antennas in the geophysical environment—some examples. Proc IEEE 80(1):200–203
11. Stolarczyk LG (1991) Emergency and operational low and medium frequency band radio communications system for underground mines. IEEE Trans Ind Appl 27(4):780–790

12. Machado VM, Brandão Faria JA (2015) Accurate electromagnetic analysis of MF TL backup communications in cylindrical tunnels. IEEE Trans Antennas Propag 63(5):2032–2040
13. Cypriani M, Delisle G, Hakem N (2015) Wi-Fi-based positioning in a complex underground environment. J Netw 10
14. Maurya DK, Kumar A, Kaunoujiya S, Prasad D, Nath V (2019) Study and design of smart industry: a review. In: Nath V, Mandal J (eds) Nanoelectronics, circuits and communication systems. Lecture notes in electrical engineering, vol 511. Springer, Singapore, pp 591–598
15. Sanjay Kumar S, Khalkho A, Agarwal S, Prakash S, Prasad D, Nath V (2019) Design of smart security systems for home automation. In: Nath V, Mandal J (eds) Nanoelectronics, circuits and communication systems. Lecture notes in electrical engineering, vol 511. Springer, Singapore, pp 599–604
16. Mishra PK, Prasad D, Roy LB (2015) Study of roof monitoring instruments for underground coal mines. MINETECH J 36(3). ISSN: 0970-7204

FPGA-Based Smart Irrigation System

Madu Kumari Ray, Alisha Oraon, Renuka Kumari, Shradha Shreya, Deepak Prasad, and Vijay Nath

Abstract In this paper, the main focus is on smart irrigation system which is a vital need in today's world. The twenty-first century deals with technology and automation, making our lives easier, adaptable, and cost effective. In our country, India is divided into two groups of people, the agro-based farmers whose livelihood depends on food production and on the other side, the consumers who depend solely on the consumption of food produced by the farmers. Both have an interdependent link. In an agriculture-based country, it is very important that the methods used for agriculture are efficient to satisfy the increasing demand. This led to the origination of smart irrigation system which is one of the smartest method used in agriculture. It provides automatic irrigation of crop field and does not require human involvement. It keeps the track of the water level as well as soil moisture content. This ensures proper and healthy growth of plants. The major factor involving agriculture is the depletion of water which keeps on increasing day by day. The smart irrigation system also resolves this issue as the water is conserved by the automated water system consisting of sensors which sense the climatic data such as humidity, soil moisture, and water level. The agriculture field is irrigated based on this data and water flow.

Keywords FPGA (Field Programmable Gate Array) · Automatic Irrigation System · Humidity · Moisture · Climatic data

M. K. Ray (✉) · A. Oraon · R. Kumari · S. Shreya · D. Prasad · V. Nath
VLSI Design Group, Department of Electronics and Communication Engineering, Birla Institute of Technology, Mesra, Ranchi, Jharkhand 835215, India
e-mail: madhuray0601@gmail.com

A. Oraon
e-mail: alishaoraon97@gmail.com

V. Nath
e-mail: vijaynath@bitmesra.ac.in

V. Nath and J. K. Mandal (eds.), *Proceedings of the Fourth International Conference on Microelectronics, Computing and Communication Systems*, Lecture Notes in Electrical Engineering 673, https://doi.org/10.1007/978-981-15-5546-6_87

1 Introduction

The traditional irrigation system has still been used by farmers for watering crops in the farm. These processes are inefficient and consume a lot of time and effort. Smart irrigation technique has proved to be a boon for modern irrigation system. Smart irrigation system includes smart watering system which waters those areas of farm which rally required water and it also provides correct amount of water at suitable time to get best productivity. Water scarcity is one of the biggest problems that is faced in today's world. The increase in population has increased the demand for food production to feed the billions of people especially for the country like India which is still a developing one. Therefore, water should be used judiciously which can only be achieved by a smarter system like smart irrigation system. Smart irrigation is an intelligent and feasible system for operating an irrigation system for proper measurement and estimation of moisture content. In conventional system, the farmers have to handle the process of irrigation. Farmer has to keep a check of soil moisture level and based on this estimation, they turn on the pump to irrigate the respective agriculture field. Farmers keep waiting for the water to flow sufficiently to the agriculture field. After a certain period of time farmers switch off the motor. The conventional system of irrigation consumes time as well as effort from the farmer's side especially when it is required to irrigate multiple crop fields in different areas.

By doing this, we can increase the yield and quality of the crop. For this, a new system was proposed in this paper to improve the irrigation system. By using this system, the farmer can monitor soil moisture, water level, and temperature, humidity, and dew point sensors for the better cultivation of the crop. The required control system is designed in XILINX 8.1-series FPGA using Verilog HDL. The sensors are integrated to the FPGA through in-built A/D converter (ADC). The communication between various devices takes place through RS232. Simulation and synthesis are performed to the remote monitoring system by using suitable tools and analyzing the performance of the system by comparing speed constraints. This system will monitor all these parameters through different sensors. Soil moisture sensor will measure the water content in the soil, i.e., and it will check whether the soil is dry or wet. Water level sensor senses the water in the water source. Temperature sensor and humidity sensor are used for forecasting the weather conditions. Dew point sensor will convert the water vapor into liquid state. The major disadvantage of this type of system is that the plants are prone to many types of diseases. There should be a proper watch over the water level and requirement needed for plants depending on different terrain which may be sloppy or hilly. Hence, there is an automated system which can fulfill the purpose. As the system also employs memory in which all the sensor values are stored and accordingly switches pumps on or off by using relays [1]. A sensor is used to sense the changes in the parameters like temperature, humidity, and water level. The sensor is followed by FPGA which reads the data based on climatic parameters [2]. The block diagram of proposed system is shown in Fig. 1. On the other hand, block diagram of smart irrigation system is shown in Fig. 2.

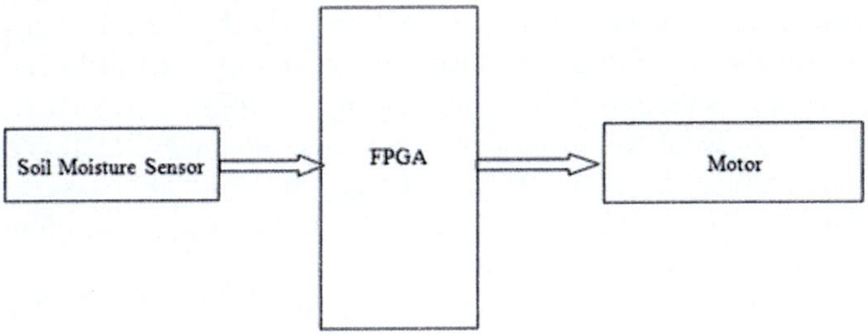

Fig. 1 Block diagram of proposed system

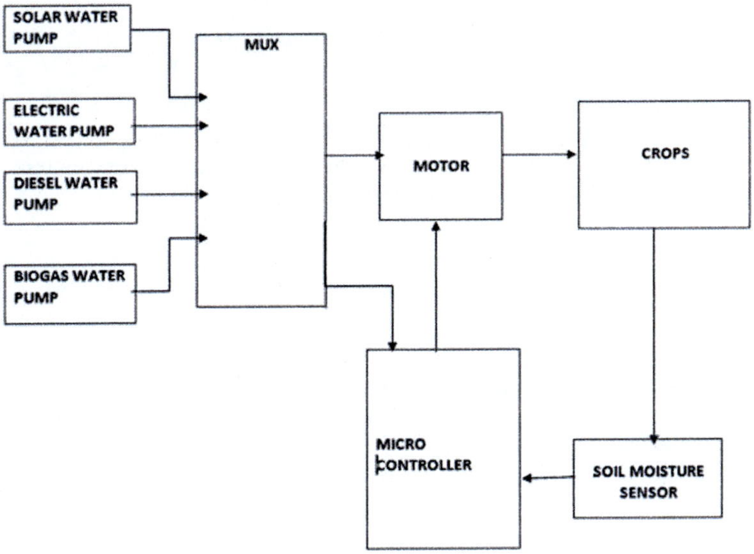

Fig. 2 Block diagram for smart irrigation system [12]

The automated irrigation system is very advantageous for the farmers in terms of time consumption and it minimizes their need of interference and tracking in the agricultural fields [3]. It ensures proper irrigation and uniform distribution of water for the crops. Hence, the work is done with higher efficiency once an automatic irrigation system is installed in the agriculture field irrespective of types of crop planted in that particular field.

2 Significance of Irrigation

In country like India, the farmers entirely depend on monsoon for agriculture. Rainfall is an important factor which controls agriculture, as monsoon rainfalls are not very predictable and are uncertain and irregular and thus the lives of the farmers (i.e., producers) and the consumers are co-dependent on monsoon. Irrigation is the essential factor that affects the productivity and quality of crops, so uneven and unequal rainfall affects the livelihood of farmers. Normally, the total rainfall occurs in the month of mid of June to mid of October. Thus, it becomes very essential to keep checking the water level of the agriculture field in rest of the months of the year and irrigating the field as per the requirement [4].

3 Methods of Irrigation

Various different types of method have been adopted for irrigating the agriculture field for different types of crop field. The basic methods used by farmers are channel system, sprinkler system, and drip system. Among the above-mentioned methods, channel system of irrigation is being considered as the traditional method. The smart irrigation system evolved as new technology for irrigation of farm field which automatically irrigates the field according the requirement [5].

3.1 Channel System

The channel system is the traditional and largely used method irrigation which is being adopted from the very beginning. One of the advantages of this method is that the total capital investment required is not very high and can be easily implemented. The channel system consists of pipes which is connected with a water pump and while pumping starts, it allows water to flow from the lake, river, or bore well to farming field as shown in Fig. 3. The farmers on the other hand are fully engaged for irrigating the crop field with help of number of workers [6]. This system is a fail as it requires large amount of workers for watering and also huge amount of water gets wasted.

3.2 Sprinkler System

The sprinkler system method of irrigation proves to be more useful as it provides better irrigation even when the quantity of water available is very less. The pump consists of main pipes and perpendicular pipes. When pump starts, the water starts flowing through the main pipe and then it starts flowing through the perpendicular pipes connected with the main pipe as shown in Fig. 4. A nozzle is present on the top of perpendicular pipe which rotates automatically at regular intervals and thus helps in irrigating the field. Sprinkler system is also useful on the sandy soil. This system

Fig. 3 Figure of channel system method

Fig. 4 Sprinkler system method

requires a smaller number of workers and wastage of water is less as compared to traditional channel irrigation.

Fig. 5 Drip system

3.3 Drip System

In drip system method, waterfalls drop by drop at the position of the roots. It is one of the best methods of watering fruit plants, garden, and trees. Pump consists of main pipes and sub-pipes. Firstly, water flows through a main pipe and then flow gets divided into sub-pipe1. Nozzles are attached to the top surface of these sub-pipes. This drip system method is highly efficient than channel and drip irrigation method which h is shown in Fig. 5. Wastage of water is very less in this method of irrigation and no workers are needed for irrigating. Farmers keep track of the status of the farm field then the motor is started accordingly, and the directions of nozzles are chosen as per the requirement. Then automatically the agriculture fields are watered and at particular intervals, the farmer keeps checking the status of the field, while the whole crop is irrigating to turn off the motor when there is no need.

3.4 Smart Irrigation System

The three methods of irrigation system mentioned above generally operate by a user, but a smart irrigation is a system which is controlled by autonomous means. Smart irrigation represents a system automatically controlling the total irrigation system irrespective of the presence of farmers in the field and send messages to the farmer about the status of the farm field and also about the changes in operation of the

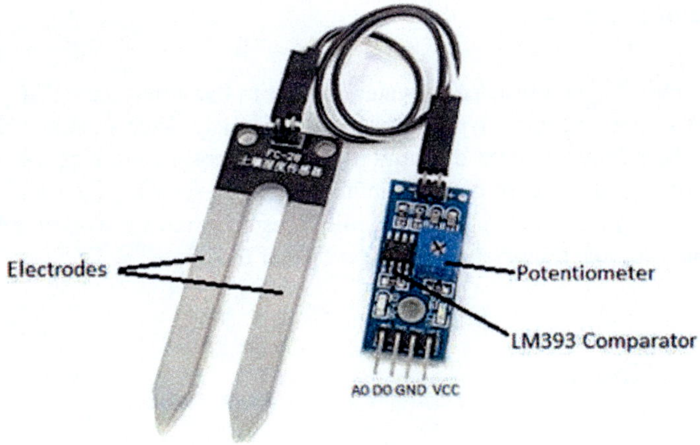

Fig. 6 Soil moisture sensor

farm field as the system is fully automated no manual involvement is required. This system proves to be the most efficient technique in a modern-day era for farmers amongst all the outdated methodology as there is a minimal chance of wastage of water. The operation fully depends on the condition that is being analyzed by the sensors attached in the full system. One of the biggest advantages is also that since there is no involvement of human, the precious time of farmers is saved.

4 System Components

4.1 Soil Moisture Sensor

In any irrigation system, one of the most important steps is to first measure the amount of water present in the soil which further helps in determining how much irrigation is needed in the field. For increasing the productivity, yields, and quality of the crops, management of soil moisture content is of great necessity [7]. To know the amount water level in the soil, soil moisture sensor is used to get the exact numerical value of the water content as shown in Fig. 6. Thus, accordingly at the final stage, decision can be made whether to run the motor or not, or for how much motor should be run.

4.2 Temperature Sensor

A temperature sensor is used to measure the temperature of the fields. This is most effective field parameter which has greater influence on other parameters [8].

4.3 Humidity Sensor

Humidity refers to the presence of water content in the atmosphere. The humidity sensor is also referred to as hygrometer. It continuously detects the humidity level and gives the relative content of humidity in the atmosphere which then helps in finally predicting the occurrence of rainfall. Humidity sensor gives back the result of measurement in terms of relative humidity present in the air. Also, it helps in the measurement of both temperature and humidity of the air.

4.4 Field Module

Field module is a sensor-based system [9–21] used in irrigation system which finally performs the operation on the data obtained from the sensors in previous stages. To get the input data, moisture sensor, humidity sensor, and temperature sensor are being interfaced with the FPGA board to get all the required information of irrigation field. Operation in FPGA board is performed in two formats, i.e., digital and analog. While measuring in digital mode, it is being checked whether the output value is higher or lower with respect to the reference voltage being considered. For the case of absence of moisture, the value of output obtained is higher than the reference value. While measuring in analog mode, the numerical value of moisture content obtained is more accurate and appropriate one. So, in many cases, analog mode is being considered for the measurement for more accurate output.

4.5 Control Module

Control module plays the most vital and final stage of this irrigation system. Control module is the one which has the control on the working of motor. It basically determines when to run the motor and for how much time motor should be run to provide the proper irrigation to the agriculture field as per the requirement. In this, the FPGA board is interfaced with the motor and voltage supply is provided to the motor according to the analog output obtained from the FPGA board. The board has inbuilt digital to analog converter which gives an 8-bit analog voltage output which is then fed to the motor. The flow diagram representing the working of motor is as shown in Fig. 7.

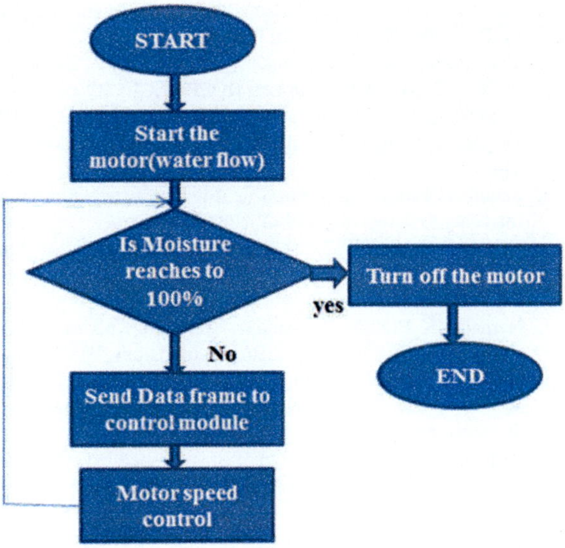

Fig. 7 Block diagram representing control module

5 Conclusion

The main motive of doing this project is to introduce an enhanced system of irrigation which can provide an autonomous control on irrigation of different crops. FPGA implementation of irrigation control system leads to the improvement in yield and overall productivity of the crops. The system will monitor all the field parameters thus providing the necessary information to the farmers. It becomes very efficient to use in rural areas. The use of FPGA board in field module and control module helps in processing and interfacing the real-time values of physical parameters which involve the environmental conditions and thus enhances the irrigation system by increasing the efficiency of motor. The usage of this FPGA enhances the application area of sensor networks in all agricultural environments by improving the speed of operation. This irrigation system has high processing speed and has lower complexity as compared to the previous ones developed in other papers. Thus provides efficient irrigation system for the better production of crops.

Acknowledgements We would like to thank the professors of our department, Dr. S. Pal, Head of Department of Electronics and Communication Engineering and Dr. M. K. Mishra, Vice Chancellor. Without their constant support, it would not have been possible to complete this research article.

References

1. Sindhu K, Sri chakrapani Y, Kamaraju M (2014) FPGA implementation of irrigation control system. Int J Sci Eng Res 5(12)

2. Ranade P, Takale SB (2016) Smart irrigation system using FPGA based wireless sensor network. Int Res J Eng Technol (IRJET) 03(5)
3. Darshna S, Sangavi T, Mohan S, Soundharya A, Desikan S (2015) Smart irrigation system. IOSR J Electron Commun Eng (IOSR-JECE) 10(3)
4. Husni MI, Hussein MK, Zainal MSB, Hamzah SAB, Nor DB, Poad HB (2018) Soil moisture monitoring using field programmable gate array. Indones J Electr Eng Comput Sci 11(1):169–174
5. Salam Al-Ammri A, Ridah S (2014) Smart irrigation system using wireless sensor network. Int J Eng Res Technol (IJERT) 3(1)
6. Behera P, Sahu CK (2015) A low cost smart irrigation control system. In: IEEE 2nd international conference on electronics and communication system (ICECS 2015), 25–27 Feb 2015, Coimbatore, India
7. Namala KK, Prabhu KK, Math A, Kumari A, Kulkarni S (2016) Smart irrigation with embedded system. In: IEEE Bombay section symposium (IBSS), 21–22 Dec 2016, Baramati, India
8. Nidhi N, Prasad D, Nath V (2019) Different aspects of smart grid: an overview. In: Nath V, Mandal J (eds) Nanoelectronics, circuits and communication systems. Lecture notes in electrical engineering, vol 511, pp 451–456
9. Razi Q, Nath V (2019) Design of smart embedded system for agricultural update using internet of things. In: Nath V, Mandal J (eds) Nanoelectronics, circuits and communication systems. Lecture notes in electrical engineering, vol 511. Springer, Singapore, pp 372–382
10. Fatma S, Nath V (2019) Study and design of smart embedded system for train track monitoring using IoTs. In: Nath V, Mandal J (eds) Nanoelectronics, circuits and communication systems. Lecture notes in electrical engineering, vol 511. Springer, Singapore, pp 385–395
11. Kumar A, Nath V (2019) Study and design of smart embedded system for smart city using internet of things. In: Nath V, Mandal J (eds) Nanoelectronics, circuits and communication systems. Lecture notes in electrical engineering, vol 511. Springer, Singapore, pp 397–408
12. Anand S, Nath V (2019) Study and design of smart embedded system for remote health monitoring using internet of things. In: Nath V, Mandal J (eds) Nanoelectronics, circuits and communication systems. Lecture notes in electrical engineering, vol 511. Springer, Singapore, pp 409–414
13. Kumar C, Nath V (2019) Design of smart embedded system for auto toll billing system using IoTs. In: Nath V, Mandal J (eds) Nanoelectronics, circuits and communication systems. Lecture notes in electrical engineering, vol 511. Springer, Singapore, pp 415–424
14. Nidhi N, Prasad D, Nath V (2019) Different aspects of smart grid: an overview. In: Nath V, Mandal J (eds) Nanoelectronics, circuits and communication systems. Lecture notes in electrical engineering, vol 511. Springer, Singapore, pp 451–456
15. Nidhi N, Prasad D, Nath V (2019) A high-performance energy-efficient 75.17 dB two-stage operational amplifier. In: Nath V, Mandal J (eds) Nanoelectronics, circuits and communication systems. Lecture notes in electrical engineering, vol 511. Springer, Singapore, pp 469–474
16. Raj U, Nidhi N, Nath V (2019) Automated toll plaza using barcode-laser scanning technology. In: Nath V, Mandal J (eds) Nanoelectronics, circuits and communication systems. Lecture notes in electrical engineering, vol 511. Springer, Singapore, pp 475–481
17. Chaudhary S, Prava A, Nidhi N, Nath V (2019) Design of all-terrain rover quadcopter for military engineering services. In: Nath V, Mandal J (eds) Nanoelectronics, circuits and communication systems. Lecture notes in electrical engineering, vol 511. Springer, Singapore, pp 507–513
18. Mohan R, Suraj AK, Agarawal S, Majumdar S, Nath V (2019) Design of robot monitoring system for aviation. In: Nath V, Mandal J (eds) Nanoelectronics, circuits and communication systems. Lecture notes in electrical engineering, vol 511. Springer, Singapore, pp 535–547
19. Hari Charan EVV, Pal I, Sinha A, Baro RKR, Nath V (2019) Electronic toll collection system using barcode technology. In: Nath V, Mandal J (eds) Nanoelectronics, circuits and communication systems. Lecture notes in electrical engineering, vol 511. Springer, Singapore, pp 549–556

20. Goel V, Riya, Kumari P, Shikha P, Tanushree, Prasad D, Nath V (2019) Design of smartphone controlled robot using bluetooth. In: Nath V, Mandal J (eds) Nanoelectronics, circuits and communication systems. Lecture notes in electrical engineering, vol 511. Springer, Singapore, pp 557–563
21. Sinha PK, Saraiyan S, Ghosh M, Nath V (2019) Design of earthquake indicator system using ATmega328p and ADXL335 for disaster management. In: Nath V, Mandal J (eds) Nanoelectronics, circuits and communication systems. Lecture notes in electrical engineering, vol 511. Springer, Singapore, pp 565–572

Smart Healthcare System Using IoT

Rashi Patel, Nishu Sinha, Kuhu Raj, Deepak Prasad, Abhishek Pandey, and Vijay Nath

Abstract Healthcare is an important aspect of any developing economy. In a country like India, poverty and lack of resources in rural areas prevent the inhabitants from getting appropriate medical facilities. In this paper, we have proposed a system that helps find the necessary medical help in rural areas whenever required using IoT. This is accomplished by creating a database that contains information of all the doctors and can be used to find which doctor is free. The medical records of patients can be sent to that doctor and he/she can be given immediate attention. This will make healthcare system more time efficient, safe and cost efficient. We can also add alarms to make sure that the different medical conditions of the patient are above par. An alarm should be generated if his condition deteriorates. This will make response time faster.

Keywords Database · Immediate response · IoT · Smart healthcare · Automated alarms

---

R. Patel · N. Sinha · K. Raj · D. Prasad · V. Nath
VLSI Design Group, Department of Electronics and Communication Engineering, Birla Institute of Technology, Mesra, Ranchi, Jharkhand 835215, India
e-mail: rashipatel831@gmail.com

D. Prasad
e-mail: prasaddeepak007@gmail.com

V. Nath
e-mail: vijaynath@bitmesra.ac.in

A. Pandey (✉)
Faculty of Engineering and Applied Science, Usha Martin University, Ranchi, Jharkhand, India
e-mail: a.p.bitmesra@gmail.com

V. Nath and J. K. Mandal (eds.), *Proceedings of the Fourth International Conference on Microelectronics, Computing and Communication Systems*, Lecture Notes in Electrical Engineering 673, https://doi.org/10.1007/978-981-15-5546-6_88

1 Introduction

Visualise a world where several objects can sense, communicate and share information over private or public networks. These interconnected objects collect the data at regular intervals, analyse and initiate required action—providing an intelligent network for analysing, planning and decision making. This is known as the Internet of things (IoT). The aim of a smart healthcare system is to add an IoT framework which gives secure health awareness checking. The main aim of our system is to provide immediate care to patients. This paper proposes to achieve this objective by sensing the health parameters of a patient by a wireless sensor network (interacting with smart phone through Bluetooth module in the microcontroller nRF52832). Wireless sensor network is used to refer to a group of dedicated sensors dispersed spatially for monitoring and recording the physical conditions of the environment and organising the collected data at a central location. In this paper, the central location will send the data to the database of hospitals via Internet to provide hospitals with immediate awareness. The sensors then are used for transmitting data to a smart phone, which will be stored in the cloud-server via internet using CoAP (Constrained Application Protocol). Doctor and nursing staff will use CoAP as well as SOAP (Simple Object Access Protocol) to view the details of the patients as well as to feed the information regarding availability of ambulances, doctors and empty beds on the database to ensure the patients are provided with immediate care, by comparing the data in the database.

2 Related Work

The Internet of things [1–6] has been driving the innovation in technology since its name was coined in 1985. Many IoT-based systems have been proposed in the past for a smart healthcare system or smart hospital management. There was use of RFID-based system for mobile object positioning in hospitals; this system was limited only to monitoring of medical devices in hospitals and nursing homes [7]. A survey was undertaken to understand the evolution of WSN towards IoT in which devices where connected to each other through WSN. The evolution of WSN networks and their roles in today's smart healthcare system is shown in this survey [8]. Various wearable tags were combined together to form a night-care system for monitoring disabled and elderly people [9]. This system relied on use of UHF RFID which estimated sleep parameters, human activity and identify emergencies during night-time. A smart healthcare system was proposed in [10]. The authors created a RFID and WSN integrated SHS which used 6LoWPAN devices. The system, here, makes use of RFID tags to tag patients and devices which need to be carried over patient's body, that can cause discomfort in some situations, as well as RFID tags that cannot be programmed, thus limiting their use in large numbers if cost efficiency of system is considered. An augmented module was proposed for smart environmental sensing

[11]. Ambient light and temperature sensing were the main motivation behind this system. In our proposed system, we have assured immediate care by employing sensors in hospitals and a GPS tracking device in the ambulance, as well as creating a database and monitoring it to make sure that the current information reaches the needful in an a financially economical manner, thus providing a cost efficient and time efficient solution [12].

3 System Overview

The architecture, proposes an alert, time efficient and cost-efficient structure with the help of the components mentioned below connected in a network reflected in the form of block diagram shown in Fig. 1. Wireless sensor networks are employed at both patient's end as well as doctor's end and are updated on the database on the main server. The vacancy of hospitals and ambulances is known by the wireless sensors connected to the remote server, updating it on the main server (in the block diagram the PC server). The sensors giving analog data, it will then be signal conditioned and fed to the appropriate microcontroller [13]. At the patients end, Nordic microcontroller is employed ensuring low-cost and significant Bluetooth range. While at the hospitals end, any microcontroller which suits the Bluetooth range required by the hospitals and cost supported by them can be employed. The microcontroller will be connected to an alarm system which will start buzzing in case the patients' data through sensors go beyond the alarming condition state, set previously as references to compare with, in microcontrollers. Need of immediate care will be indicated and resources can then be employed. The data from microcontrollers are then sent to the

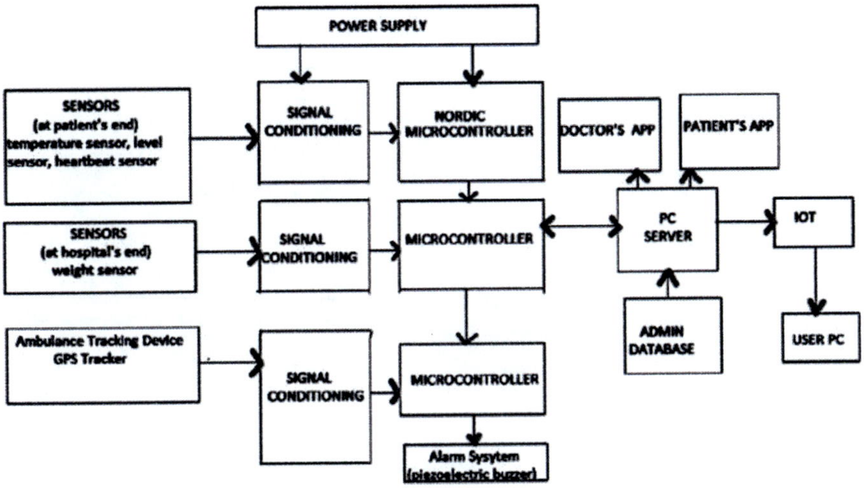

Fig. 1 Block diagram representing the system architecture

PC server from where it goes to the individual user servers at every hospitals end [14]. The patient and doctors can both be aware of the necessary data at the same time connecting the two ends using an app having patients as well as doctors interface. The doctors can send their availability at hospital through PC user or put a constant check on patients while at Home through the doctors app. Thus, the patients are informed about the nearby hospitals where they can receive appropriate and immediate care. The next section discusses the architecture and explains how the objective of this project is achieved.

4 Architectural Details

4.1 Nordic nRF52832 Microcontroller

Nordic nRF52832 microcontroller is the powerful low-cost microcontroller available with built-in Bluetooth module having hardware support for Bluetooth supported by a 2.4 GHz transceiver [15, 16]. It consists of 512 kB flash and 64 KB RAM, two data rates of 1 Mbps and 2 Mbps, 12-bit ADC and 3 Master/Slave SPI. Its low-cost and built-in Bluetooth module help to cut the down the cost supported by the patients.

4.2 Sensor Network

4.2.1 At Patient End

Sensing network shown in Fig. 2 consists of three sensors namely; temperature sensor, heartbeat sensor and level sensor. These sensing points of these sensors are

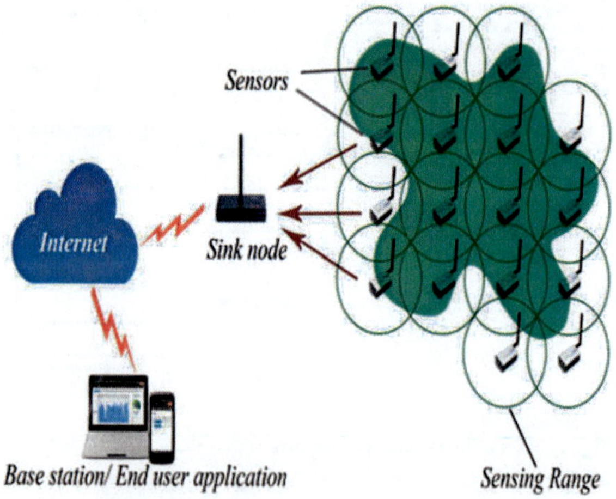

Fig. 2 Wireless sensor network

attached to the patient's body and the end is connected with the microcontroller [17]. The various sensors, i.e. temperature sensor, heartbeat sensor and level sensor, sense the temperature, heartbeat and saline level of the patient, respectively. All these sensors work in the real-time system and send real-time parameters [18]. Then the microcontroller sends these values to the patient's app via Bluetooth. Heartbeat values are displayed in the form of graph. Saline level is detected by level sensor. Threshold values are already assigned to the microcontroller.

4.2.2 At Hospital End

The hospitals will also consist of wireless sensors in the operation theatre as well as patient's beds and seats inside doctor's cabin. The WIFI weight sensor as shown in Fig. 3 is a connected weighing scale which can measure weights of various load capacities and upload the measured weight periodically to the remote server whenever the weight changes by set threshold [19]. That will indicate whether the doctor or operation theatre is occupied with patients or not. It can be operated from mains or with batteries.

4.3 Piezoelectric Buzzer

For temperature sensor, heartbeat sensor and level sensor, the microcontroller compares all these real-time parameters with the threshold values, and if some serious condition occurs, then the alarm is generated through buzzer which is also connected to the microcontroller, an "alert" message is send to the doctor through WiFi. Otherwise, SMS is sent through Android SMS chat.

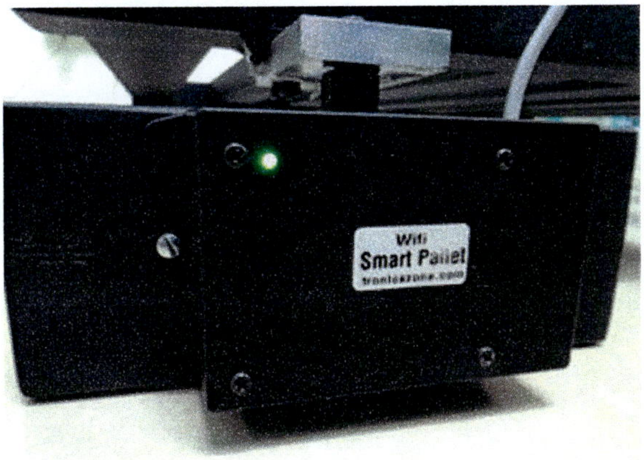

Fig. 3 WiFi smart pallet—WiFi weight sensor

Fig. 4 Flash track HW (hardwired GPS vehicle tracker)

4.4 *User Interface*

4.4.1 Doctor's App

Doctor will login to the Android app. He can check patient's health parameters through this app. Also, he can upload his timings and days during which he will be available in the hospital.

4.4.2 Patient's App

Patient's app will contain all of the patient's parameters. Login Id will be created for each patient. The values sensed by the sensors will be sent to the patient's app and would alert the urgency when present in the server.

4.4.3 Admin Application

The admin application consists of a major operation: monitoring and updating the database. The desktop admin application manages the server and the database [20]. It registers the patients, updates information and manipulates the data.

4.5 *Ambulance Tracking Device*

The ambulance's availability position can also be tracked by a GPS tracker as example of one given in Fig. 4 and uploaded by the remote server (hospitals server) to the main server for the patient's information who can also use the ambulance in case his location is close to that of the ambulances location. Input range is from 8 to 30 V DC [21–24]. Thus, it can be supported using normal batteries as well as since cost is low it can be supported by the belonging to the hospital of rural areas too.

5 Conclusion

With the help of above architecture, an alert, time efficient and cost-efficient structure can be developed available to each and every patient in need. Any lag between the patients to be operated will be thus eliminated except the travel time. Travel time lag can also be eliminated in case where lag cannot be tolerated, i.e. in the case of urgent requirements, in which a direct communication can occur between the required doctor and patient through the patient id suggesting the small relaxations or relief possible at that time.

Acknowledgements We would like to thank the professors of BIT Mesra, the Vice Chancellor Dr. M. K. Mishra, Head of Department of Electronics and Communication Engineering Dr. Srikanta Pal for their constant support and inspiration. Without their motivation, this work would not be possible.

References

1. Paradiso R, Loriga G, Taccini N (2005) A wearable health care system based on knitted integrated sensors. IEEE Trans Inf Technol Biomed 9(3):337344
2. Nidhi N, Prasad D, Nath V (2019) Different aspects of smart grid: an overview. In: Nath V, Mandal J (eds) Nanoelectronics, circuits and communication systems. Lecture notes in electrical engineering, vol 511. Springer, Singapore, pp 451–456
3. Nidhi N, Prasad D, Nath V (2019) A high-performance energy-efficient 75.17 dB two-stage operational amplifier. In: Nath V, Mandal J (eds) Nanoelectronics, circuits and communication systems. Lecture notes in electrical engineering, vol 511. Springer, Singapore, pp 469–474
4. Raj U, Nidhi N, Nath V (2019) Automated toll plaza using barcode-laser scanning technology. In: Nath V, Mandal J (eds) Nanoelectronics, circuits and communication systems. Lecture notes in electrical engineering, vol 511. Springer, Singapore, pp 475–481
5. Chaudhary S, Prava A, Nidhi N, Nath V (2019) Design of all-terrain rover quadcopter for military engineering services. In: Nath V, Mandal J (eds) Nanoelectronics, circuits and communication systems. Lecture notes in electrical engineering, vol 511. Springer, Singapore, pp 507–513

6. Mohan R, Suraj AK, Agarawal S, Majumdar S, Nath V (2019) Design of robot monitoring system for aviation. In: Nath V, Mandal J (eds) Nanoelectronics, circuits and communication systems. Lecture notes in electrical engineering, vol 511. Springer, Singapore, pp 535–547
7. Shirehjini AAN, Yassine A, Shirmohammadi S (2012) Equipment location in hospitals using RFID-based positioning system. IEEE Trans Inf Technol Biomed 16(6):1058–1069
8. Occhiuzzi C, Vallese C, Amendola S, Manzari S, Marrocco G (2014) NIGHT-Care: a passive RFID system for remote monitoring and control of overnight living environment. Procedia Comput Sci 32:190–197
9. De Donno D, Catarinucci L, Tarricone L (2014) RAMSES: RFID augmented module for smart environmental sensing. IEEE Trans Instrum Meas 63(7):1701–1708
10. Tyagi S, Agrawal A, Maheswari P (2016) A conceptual framework for IoT-based health care system using cloud computing. In: Sixth international conference cloud system and big data engineering, Jul 2016, pp 503–507
11. Archip A, Botezatu N, Serban E, Herghelegiu PC, Zal A (2016) An IoT based system for remote patient monitoring. In: International Carpathian control conference, June 2016, pp 1–6
12. Nitha KP, Sreekanth KU (2016) A study on health care in internet of things. Int J Rec Innov Trends Comput Commun 4:44–47
13. Hassanalieragh M, Page A, Soyata T, Sharma G, Aktas M, Mateos G, Kantarci B, Andreescu S (2015) Health monitoring and management using internet-of-things (IoT) sensing with cloud-based processing: opportunities and challenges. In: IEEE international conference on services computing, Aug 2015, pp 285–292
14. Raizul Islam SM, Kwak D, Humaun Kabir MD, Hossain M, Kwag K-S (2015) The internet of things for health care: a comprehensive survey. IEEE Access 3:678–708
15. de donno D, Palano L (2015) An IoT-aware architecture for smart health care system. IEEE Internet Things J 2(6):515–526
16. Occhiuzzi C, Valleseb C, Amendolab S, Manzarib S, Marroccob G (2014) Night-Care: a passive RFID system for remote monitoring and control of overnight living environment. In: Fifth international conference on ambient systems, networks and technologies, Jun 2014, vol 32, pp 190–197
17. Benharref A, Serhani MA (2014) Novel cloud and SOA-based framework for e-health monitoring using wireless biosensors. IEEE J Biomed Health Inform 18(1):46–55
18. Ming L, Shucheng Yu, KuiRen YZ, Lou W (2013) Scalable and secure sharing of personal health records in on cloud computing using attribute-based encryption. IEEE Trans Parallel Distrib Syst 24(1):131–143
19. Mao Y, Chen YN, Hackmann G, Chen M, Lu C, Kollef M, Bailey TC (2011) Medical data mining for early deterioration warning in general hospital wards. In: IEEE 11th international conference on data mining workshops (ICDMW), Dec 2011, pp 1042–1049
20. Matthews R, Park C, Chou PH, Bai Y, Hibbs A (2006) An ultra-wearable, wireless, low power ECG monitoring system. In: Biomedical circuits and systems conference, BioCAS 2006, Nov 2006. IEEE, pp 241–244
21. Chen Y, Zhao Q (2005) On the lifetime of wireless sensor networks. IEEE Commun Lett 9(11):976–978
22. Hari Charan EVV, Pal I, Sinha A, Baro RKR, Nath V (2019) Electronic toll collection system using barcode technology. In: Nath V, Mandal J (eds) Nanoelectronics, circuits and communication systems. Lecture notes in electrical engineering, vol 511. Springer, Singapore, pp 549–556
23. Goel V, Riya, Kumari P, Shikha P, Tanushree, Prasad D, Nath V (2019) Design of smartphone controlled robot using bluetooth. In: Nath V, Mandal J (eds) Nanoelectronics, circuits and communication systems. Lecture notes in electrical engineering, vol 511. Springer, Singapore, pp 557–563

24. Sinha PK, Saraiyan S, Ghosh M, Nath V (2019) Design of earthquake indicator system using ATmega328p and ADXL335 for disaster management. In: Nath V, Mandal J (eds) Nanoelectronics, circuits and communication systems. Lecture notes in electrical engineering, vol 511. Springer, Singapore, pp 565–572

Author Index

V. Nath and J. K. Mandal (eds.), *Proceedings of the Fourth International Conference on Microelectronics, Computing and Communication Systems*, Lecture Notes in Electrical Engineering 673, https://doi.org/10.1007/978-981-15-5546-6